An open letter about "Gravity's Prescription"

ISBN-13: 978-1535085915

ISBN-10: 1535085916

WRITTEN BY PEET (P.S.J.) Schutte

By P.S.J. (Peet) Schutte

© KOSMOLOGIESE EN ASTRONOMIESE TEGNIKA

THIS LETTER IS THE ANNOUNCING OF THE BOOK IN SEVEN VOLUMES CALLED

MATTER'S TIME IN SPACE: THE THESIS ISBN 0-9584410-8-1 **VOLUMES 1-7**

This letter was the letter that was sent to near enough eighty Universities through out the world in regard of announcing a new cosmic theory. It now is turned into a separate and individual commercial book.

TO WHOM IT MAY CONCERN,

An open letter **TO SELECTED ACADEMICS** ISBN 0-9584410-9-X Is THE

ACADEMIC NOTIFYING OF

MATTER'S TIME IN SPACE: THE THESIS ISBN   0-9584410-8-1 Written by PEET SCHUTTE

Dear Professor,

I am Petrus Stephanus Jacobus Schutte going by the name of Peet and who is the author of the above-mentioned book(s). I hope you find your reading of this book presented as an open letter a most fruitful experience. I feel I need to warn you, the person reading this letter, that the work contained herein strays widely from mainstream science and for that there is a very good reason.

My aim in writing this letter is to confess my doubt about the manner in which science regard gravity forming. I have serious reservations about the matter in as much as how science assumes that mass is the reason behind whatever is forming gravity. I equally hold strong doubts about gravity being a force of pulling or gathering. This institutionalised concept has formed the basis of science in physics during the past three hundred and fifty years but in the light of evidence that came to the attention of science in the last hundred years and more so during the past fifty years there is reasons that puts those assumptions previously formed in the past about gravity in doubt. The new evidence points away from the previous direction and assuming. Modern evidence calls for re-examining in regards to this matter. The facts that are now available to science to form any prognoses makes the old principles that was used for such assuming much more questionable. One only has to study aviation principles to form the doubts. Aircraft flight disproves the assumption that mass is instituting gravity and in that context the notion forming the concept of mass and gravity becomes extremely doubtful. I am about to prove in my writing of this letter that mass is motion discrepancy between two bodies and gravity is motion pure and simple not a force. Any object holding any mass of whatever description can become airborne with the correct conditions applying. It only depends on motion coming about achieving a speed over and above the earth speed that must be sustained. Mass of whatever magnitude can fly. There is no limit on the mass becoming airborne. What is required to achieve such a fete is a speed. Only that limits flying. A Jumbo jet can get into flight as easy as a micro light aircraft. The only requirements are a differentiation in velocity of air movement above and below the aircraft wings.

Mass aside, laden or not, aircraft of all description will fly. The wings determine the required velocity while supporting all mass. The aircraft support the same mass it has after becoming airborne as it had before it was airborne and the mass remain in the air because of the speed which enable the wings to have air pushed past the surface of the wing and that is what allows gravity to be countered by flight. Science thinks of movement as momentum but in fact it is antigravity since it counteracts the containing of Earth gravity. Momentum in fact is only an increase of gravity on the object that is taxed by mass, which extends the gravity of the burdened smaller factor beyond the limitation of Earth restraining by mass. An object has mass because it moves slower than the Earth and is therefore dragged along by the Earth. Once the object moves faster than the Earth even within the Earth atmosphere the influence of mass abates. Mass is only the friction caused by speed differentiation whereas the sound barrier is the limit on the other end of the speed differentiation scale. At the moment science differentiate in terminology used to describe what I deem to be the same issue but that puts a huge question mark over the use of terminology and the concept of mass playing any part in forming gravity. That what is considered as momentum is gravity in addition. Notwithstanding the pulling and the tugging brought about by the mass, the ultra light plane finds as much flying capabilities as does the enormous cargo jet aircraft. The pulling of the mass is many times more in the case of the heavy cargo plane than what it is in the micro airplane but still at speed the plane overcomes the pulling of the mass by only establishing motion differences between the top of the wing and the bottom of the wing. When flying having more mass does not implicate or discriminate in any way because with the required motion both craft keeps in the air equally effortlessly. Once the craft is airborne it is a matter of maintaining speed that would secure the flying ability and not the loss in mass that would increase such flying possibilities. In the case of a satellite such flying demonstrates my argument even much better. The space station Mir was in space for many years but it fell when the space station could no longer maintain a specific orbiting speed. While it was in space in an orbit there was no changes in mass required to maintain an orbit. The changing of the mass could not secure Mir a longer orbit life. At the point where it started to plummet to the Earth, the mass did not increase but the orbit motion decreased and that allowed the plummeting. The fuel eventually ran out and by not further being able to accelerating in order to maintain a specific required velocity. Without the speed requirements needed to match the motion equilibrium that was required to maintain a steady orbit between the Earth and Mir, differentiation set in and that caused mass, which came as a fall towards the Earth whereby Mir met its end in a fireball. It had nothing to do with mass. It had everything to do with speed being at an equal pace between the Earth in orbit and Mir's orbit. While the satellite is able to hold a speed in a ratio with the speed the Earth manages, an orbit is secured and such securing of an orbit has no bearing on the mass that is up there flying through the sky. It has everything to do with matching a speed in relation to the distance from the centre of the Earth and in relation to the motion of the Earth that requires a specific velocity at that distance. It has everything to do with the time it takes the space to go through the motion from one point to another point. It is the required heat in the space that the space needs to move at a certain distance in a specific time of motion. Gravity is a relation between space and the heat required to provide motion which is $a^3 = T^2 k$. It is space (a^3)-time ($T^2 k$) and that brings me directly to Kepler. The only way mass becomes an issue is where a bigger object such as the Earth frustrates or hinders the natural motion of any individual particle in its capture (atmosphere). Then the gravity or movement results in a tendency to move but still the tendency is manifesting the motion where mass is the restraining thereof. There always has to be motion but there is not always mass. The tendency of motion, which becomes the mass when the motion is restricted, remains a part of free movement the frustrated object still possesses as motion. There always remains a tendency to move. The Earth set a speed of motion and in order to beat mass the object has to excel by providing amplified motion in excess of the Earth motion granted to the object and gravity extending beats the restricting effect that mass has on any body connected to the centre of the Earth. That is motion and I am about to show that gravity is all about motion.

The value of space that Kepler's indicated as a third dimension a^3 does not depend on indicating a structure forming a^3 that is in rotation T^2 but only needs one position having a constant of some sorts. There has to be a space to point to a space in defining. Any point where **k** as a line is ending at that point **k** indicates a position between the start of the line k^0 and the end k^0 of the line **k** where one will find a spot with the value matching a^3 and the matching location will fit T^2 at that point. The line **k** is coming from the centre and ends at a point a^3 but such a point establishes another space in which a^3 rotates. The line **k** that forms by a^3 rotating T^2 is the relation there is in the solar system between all planets and the Sun . The Sun always indicates the centre k^0 and the planets always indicate the rotation. But $a^3 = T^2 k$ is only producing a relevancy of three dimensions that is equal to two plus one dimension. The fact of a^3 being at the end k^0 of the line **k** is how a^3 secures **k** and where **k** then secures a^3. The line **k** cannot have validity without proving the validity of a^3 just as much as the rotation T^2 of the space a^3 defines **k**. There is no possible removing of any of the factors without removing all three factors equally. The space a^3 serves to prove T^2 that defines the line **k** that presents a^3 an existence.

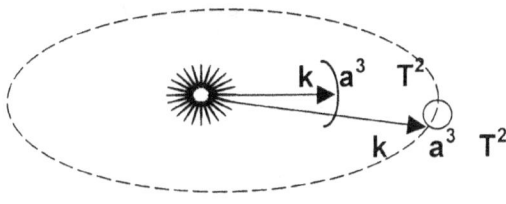

From the Sun there are three points moving between two points from one point to two other points giving the six dimensions we find in space. It is space in time or space converting space through the movement of time. It is a location of a point in the third dimension a^3 that will move according to the second dimension T^2 that will implicate **k** as a reference in the first dimension. It is about dimensions in reference to one another.

Let us take it from a point where the Sun provides a centre as one starting edge of **k** then that centre **k** will provide a line from the centre and the line **k** will provide three spots in a formation that produces a structure by the square T^2 of the dimension. Not once did Kepler indicate size as a contributing factor to a^3. That means every single point that **k** indicates there are three positions a^3 implicating sides of a double dimension. In the same manner is **k** not limited to distance or is T^2 lesser by size $k = a^3 / T^2$. That is what Kepler said. There are three dimensions a^3 between any two points T^2 flowing as time from the centre of the Sun , which is indicated by the line **k.** The implication of the relevancy produced by the use of the formula $k = a^3 / T^2$ brings about that when dividing T^2 into a^3 there is k left. The fact is that a^3 is a three dimension (^{3}) of single k (^{1}) showing one or T^2 is two dimensions of k being the one dimension it means that k is a part of space a^3 or T^2 which is time. It is the same thing in a double dimension or space being a triple of k then k is one factor and k cannot show a position of zero. If $k = 0$ then there is no possibility of $k = a^3 / T^2$ because $k = 0$ then $0^3 / 0^2 = 0$. That does not make sense. Mathematically space cannot be zero because those being of the opinion of space being zero or nothing must first prove mathematically that space is zero. Moreover they then must prove mathematically how does zero grow through the Hubble constant. By translating Newton's vision of the circle in completing a cycle would become zero through rotation…well that does not count the use of the formula a^3. If k cannot be zero then k could not start from zero. With $k = a^3 / T^2$ no point can be zero because k shows space $a^3 = k\,T^2$ is no reference to the volumetric mathematical formula used to calculate $a^3 = 4/3\;\Pi\;r^3$. Nor does it show the use of the circle ion the second dimension being $a^2 = \Pi\,r^2$. In the case of the Newton formula the circle factor becomes the square as indicated by the duration of the time T^2. The factor standing in for the line which normally would be r and then be the square value is in the case of Kepler not the value indicating the square. That means Kepler never indicated a circle of mathematical procedure but said mathematically the distance of the planet from the Sun k holds space a^3 in relation to time T^2 **Lines mathematically cannot start at zero because there is no evidence of zero as a factor in mathematics. Should you disagree with my statement** the question in need of answering is this: **What will the length of the shortest hypothetical line imaginable be and moreover, what would the total overall length be in that case?**

Locating zero

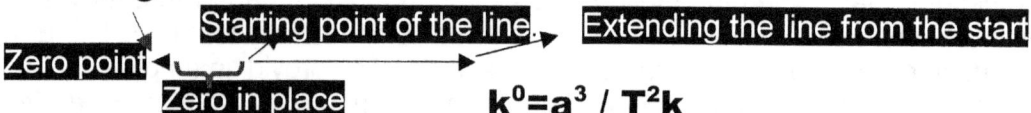

$$k^0 = a^3 / T^2 k$$

The fact of form proves that the sphere captured all sides that can possibly influence the sphere. The sphere therefore holds $k^0 = a^3 / T^2 k$ within the boundaries designated to the sphere. When a body is placed in a location on the outside of such spherical borders that object seem to float in any direction. There is no control one can establish which will secure movement in any specific direction of preference except by releasing heat to counter act the required motion in a specific direction of choice. We all have seen what happens to any object that comes into the border area of a sphere. The object suddenly is motivated by motion to follow a specific designated direction and the motion leads the object to move towards the centre of the sphere. It is as if the support of the six opposing sides has lost one side where the sphere took over the control and movement starts in the direction of the Earth centre. The support of one side is literally removed by the centre of the earth where Einstein claimed the strongest gravity is and the motion of the object starts in that direction. There is no pulling on the object but there is removing of space by the centre of that specific point leading the object and the space it is in as well as the space it carries to move to the centre spot. In the sphere the borders the sphere holds are deliberate and very distinctly placed edges forming a specific distance from the centre. The centre is also proven beyond any debating. The centre of any sphere has to be at the very point where space completely falls away. That will put that space at that point in the single dimension and centre is the single dimension.

The fact in the matter suggests that all three factors hold the same identifiable measure and the differentiation between the factors is in name alone. In what I explain at a later event the true value should stand connected to singularity and in singularity the value is allocated to singularity is Π. The formula should read that $\Pi^0 = \Pi^3 / \Pi^2 \Pi$ and when used with the correct connotation the formula becomes more sensible. But be as it may the factors indicate the same measure carried in different dimensions and should all read the same value. The cube is a loosely connected structure with any form possible but the only precondition is that there must be at least six sides connecting. The six sides hold a relevancy or a responsibility to one another and provide a Universal accepted form maintaining the universe. From the structure one can see gravity is not strongly present. All six sides support what ever are inside evenly form all sides. The sphere is the form securing gravity. In the centre of the sphere there is a point where space vanishes. At that point where space vanishes gravity is the strongest. From the centre point where gravity is the strongest gravity hold the sphere true to form. At the edges of the sphere there are also point lining in 90^0 and 180^0 holding relevancy and responsibility to one another but the centre spot being the gravity point positions all the points in a location that the centre point allocate. In the centre where all lines cross one will locate singularity but I am explaining that fact a little tater on.

Newton changed the symbol of **k** by using the mathematical equated symbols G (m + m$_p$). This is just a longer and probably a more detailed manner of indicating **k** and better defining of **k** but it symbolises precisely to the point what **k** stands for nonetheless. I wish to draw your attention to the matter of Johannes Kepler's findings that Mainstream science considers as resolved and closed for many a century while it is not. My investigating Kepler helped me to resolve other unresolved matters but it was only possible by using Kepler's work. This changed the aspect of gravity in cosmology fundamentally and as I am about to show most and totally incorrectly.

Let us for one minute leave Newton's surmising about Kepler's failure out of the picture and concern us with what Kepler found long before Newton thought about what Kepler found. Kepler said that the space a^3 is equal to the motion T^2 of the space a^3 distant from a specific centre **k**. That then is $a^3 = T^2$ **k**. Reading this mathematically encrypted coded formula of the cosmos given to Kepler and keeping it removed from Newton it reads as the space a^3 is equal to = the motion T^2 of the space a^3 in ratio to a centre **k**.

What this proves is that gravity is the motion of space provided by time being the liquid.

Please allow me to explain. In the formula $a^3 = T^2$ **k** the space forms as the space is in motion.

Newton suggested that $\dfrac{dJ}{dt} = 0$ where he stopped time to have the motion of the circle demolish the

work that the circle does. That means he got time standing still or being T^1 and the motion T= 0. Let us ponder on that thought for a while, while remaining with the formula Kepler suggested it will seem that according to Newton $a^3 = T^2k$ and in that T^2 then becomes **1**. Should that be the case then we have space going flat because $a^3 = T^2k$ where $a^3 = T \times k =$ forming a square instead of a cube, and the Universe we have is a three dimensional cube in every aspect there is. If time stands still then time becomes $T^0 = 1$ and **k** being in one dimension all the time it too has to become zero.

$a^3 = T^0k^0$
$a^3 = T^0k^0 =$ **being** $(T^0 \times k^0) = 1 \times 1 = (1)^2$.

That is mathematically incoherent and a complete fallacy. To proclaim that a cube can be a square at the same time it is a cube is ridiculous, even coming from a man that has the stature of Newton! By taking the thought further we find the same blunder in removing **k** in Kepler's formula as a factor.

$a^3 = T^2k$, **then** $a^3 \div k = T^2k \div k$
being $a^2 = T^2$.

That is totally going against all mathematical principles. It would be more correct when saying that $a^3 = T^2k$. Looking at the formula in this way we find that $a^3 = T^2k$ bringing about that $a^3 = T^2k$. That proves the following:

$T^2 = a^3 \div k$ **or** $T^{-2} = k \div a^3$. $(T^2 = a^3 / k$ **or** $T^{-2} = k / a^3)$

From the implementing of $a^3 = T^2k$ we can see that:

(circle)			
$k = k^{3-2} = k^1$ $a^3 = a^{2+1} = a^3$ $T^2 = T^{3-1} = 2$	$k = a^3 / T^2$	$a^3 = T^2 k$	$T^2 = a^3 / k$
	$k = a^{3-2} (T^2)$	$a^3 = T^2 k^1$	$T^2 = a^3 / k^1$
	$k = a^{3-2} = k^1$	$a^3 = T^{2+1} (k^1)$	$T^2 = a^{3-1} = T^2$
	$k = k^{3-2} = k^1$	$a^3 = a^{2+1} = a^3$	$T^2 = T^{3-1} = 2$
	is the same as	**is the same as**	**It is all the same**

$k = k^{3-2} = k^1$ is in direct relation to $a^3 = a^{2+1}$ and that is, is in direct relation to the formula $a^3 = T^2 = T^{3-1=2}$.

With this information staring mainstream science in the face and scream pleading at them to recognise the information they turn around and ask why can man not fly off to other galactica at the speed of light

$k = k^{3-2} = k^1$

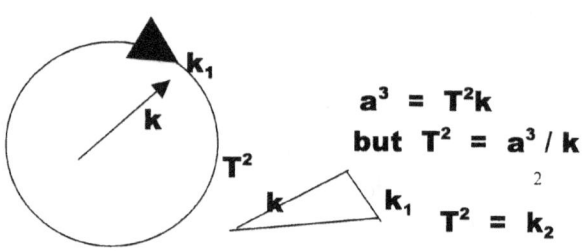

It takes time for space to fill **k** in the distance. In fact, it takes the distance that **k** developed since the Big Bang $k = k^{3-2} = k^1$ to fill the distance.

It also takes time $T^2 = T^{3-1=2}$ to produce the distance forming k^2

It takes space $a^3 = a^{2+1} = a^3$ to form k^3 since coming from the Big Bang

$a^3 = T^2 k$

but $T^2 = a^3 / k$

$T^2 = k_2$

That would give gravity meaning and that would explain not only gravity but also the Coanda affect which to my mind is gravity by principle.

The fact is that although the symbols Kepler used were not the same, the value they measured were very much the same. The values of the symbols were interchangeable.

From the implementing of $a^3 = T^2 k$ we can see that:

$$
\begin{aligned}
&k = k^{3-2} = k^1 \\
&a^3 = a^{2+1} = a^3 \\
&T^2 = T^{3-1=2}
\end{aligned}
$$

$k = a^3 / T^2$	$a^3 = T^2 k$	$T^2 = a^3 / k$
$k = a^{3-2} \, (T^2)$	$a^3 = T^2 k^1$	$T^2 = a^3 / k^1$
$k = a^{3-2} = k^1$	$a^3 = T^{2+1} (k^1)$	$T^2 = a^{3-1} = T^2$
$k = k^{3-2} = k^1$	$a^3 = a^{2+1} = a^3$	$T^2 = T^{3-1=2}$
is the same as	**is the same as**	**It is all the same**

$k = k^{3-2} = k^1$ is in direct relation to $a^3 = a^{2+1}$ and that is, is in direct relation to the formula $a^3 = T^2 = T^{3-1=2}$.

With this information staring mainstream science in the face and scream pleading at them to recognise the information they turn around and ask why can man not fly off to other galactica at the speed of light

$k = k^{3-2} = k^1$

It takes time for space to fill **k** in the distance. In fact, it takes the distance that **k** developed since the Big Bang $k = k^{3-2} = k^1$ to fill the distance.

It also takes time $T^2 = T^{3-1=2}$ to produce the distance forming k^2

It takes space $a^3 = a^{2+1} = a^3$ to form k^3 since coming from the Big Bang

Man could create motion but at first, such motion was far less than that motion which the Earth

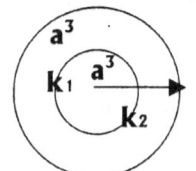

provides. The motion of man's ability was vested in what his muscle power could provide. But a very short while ago man grew wise to machines and the fact that machines can provide more motion much faster than could animal muscle bring

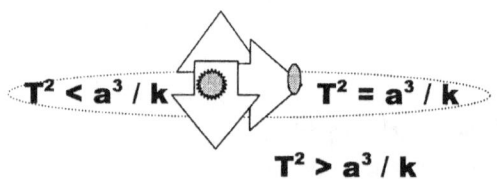

$T^2 < a^3 / k$ $T^2 = a^3 / k$

$T^2 > a^3 / k$

about motion. By supplying machine motion it gave man extra ability whereby man extended the relation between what the object has when in normal contact with space and when extended by extra motion allowing more space to apply to the surface of the object, thus enlarging the object surface in the relevancy brought on by motion.

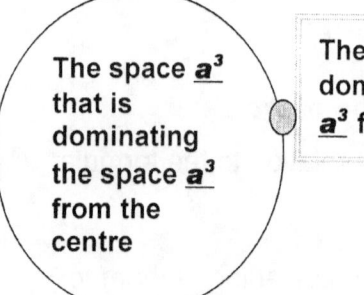

The space a^3 that is dominating the space a^3 from the centre

The space a^3 that is dominated by the space a^3 from the centre

The smaller space a^3 is distinctly distinguishing the larger space a^3 as the larger space a^3 is housing the smaller space a^3 where the factor **k** is as much indicating by length the larger space a^3 as much as it is indicating the end of the length of the larger space a^3 at the location of the smaller space a^3 by directly pointing at the position the smaller space a^3 holds. Where the larger space a^3 ends the smaller space a^3 is. The two remain as an inseparable single unit in double motion where the motion identifies the unit as much as distinguishing the separateness in the unity and always remain in absolute relevancy.

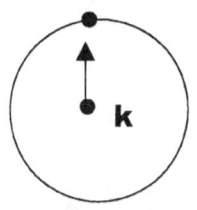 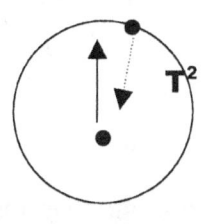 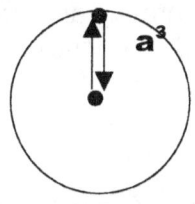 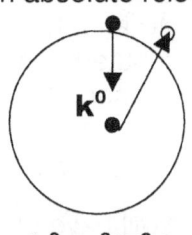

$$k = a^3/T^2$$
$$T^2 = a^3/k$$
$$a^3 = T^2 k$$
$$k^0 = a^3/T^2 k$$

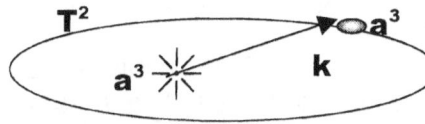

By duplicating the space of any particle sharing space within a larger cosmos structure such as an atom inside a star or a human inside the Earth there are two relations applying.

The independent object serves as the outer relevancy while the centre forms the contact with the independent object starts moving however this flexibility is only increase the limits on the space starts to stretch and such

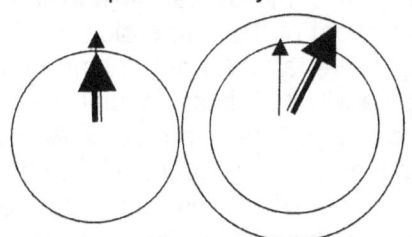

contact the Earth has with its containing singularity. When the the atmosphere allows flexibility, limited to a point. As the speed that retains the object in motion stretching has definite limits.

Kepler's formula is the exclusive Newton's version of Kepler's misleading and a deception of the truth.

display of _space-time_ and formula is plainly and commonly

There is another correct connotation to Kepler's work that divert completely from Newton's vision. In the picture to the left we see the all-familiar Moon that every one blessed with the ability to see has seen at some point in his or her life. Any person not agreeing totally with what I have to say in the next sentences knows less about mathematical principle than a newborn baby. The Moon is according to Kepler's studies, (and discounting Newton's inconsistency if not totally flagrant rubbish) from our perspective in the space a^3 that is equal to the time T^2 it is in placed at the distance between us by the factor

k. The Moon consists of the **space a^3** the **time T^2** ad the distance **k in time not space.**

The observing "size" of what we see depends on the time factor **k** relating to the space a^3 we see in relation to the time T^2 we

see it in. The bigger the space in relation to what we see gets the smaller the time factor is that allows the time frame or the motion that frames the space. From serving my cosmic position allocated to the space I claim in the time I hold Kepler stated space-time is $a^3 = kT^2$ and therefore mathematically it is

correct to say $k = a^3/T^2$.

When I observe the Moon the Moon is at a distance **k** from me where I see the space a^3 in relation to the time or the moving or the spinning T^2

The figure to the left I covered the area of the Moon that I use to indicate the space a^3 is served by the sketch to the left. The symbol to the left of this point is what I use to indicate the area time T^2 is in which the space moves or the picture to the left depicts the motion aspect of the space.

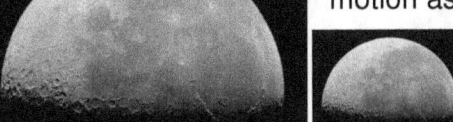

The bigger a^3 will seem the less **k** has to be in order to reduce the motion or time T^2. A large space a^3 will bring about a small time T^2 factor because of a large distance **k**. That means that $10a^3 = T^2 \times 10k$, and $a^3 / 10 k = 10T^2$ or then the distance **k** is **10** times less. $k = a^3 / T^2$ then $k / 10 = 10a^3 / T^2$ or from another perspective it is $k /10 = 10a^3 / T^2$ that would bring about space that is 10 times bigger in the time that is relating to the space. That is space-time and that forms gravity and that is what I am about to prove in this letter. The qualities Newton attributed to mss and the principle what Newton tried to pin to mass is just the opposite of what can be pinned to mass. That, what Newton regarded to be mass, serves the very opposite to mass because gravity is motion. Gravity is the moving of objects through a container every one considers being space but the truth is that the container is time.

An object coming closer will have the relative factor **k** indicating a negative indicator as a k^{-1}. In such an event the space will increase by decreasing the time aspect and this Kepler's formula most accurately show to be $k^{-1} = T^2 / a^3$. In such an event the time putting distance between the object and the space would reduce allowing the space to relate much stronger in terms of the time as part of the overall picture.

In the overall picture $a^3 = T^2k$, the time factor, which we incorrectly see as a distance **k**, has gone very small and this is putting the space a^3 as an enormous part of the overall view and gravity reduce the time T^2 as a small fragment of the entire picture. The very same picture is used but by reducing a^3 we take it that the time factor **k,** which we mistakenly put as a distance and which it is not, as being huge and that large factor increase change our relevant factors in the entire picture completely. According to our use of mathematics by increasing **k**, that reduced a^3 because also T^2 reduces a^3 and by reducing a^3 T^2 gets a bigger share of the entire picture. That means the space a^3 stands related to the overall time being **k** as well as T^2.

That is unmistakably exactly, precisely and to the iota correctly what Kepler said when Kepler said one is observing space (a^3) –time (T^2k). However from that we can see time increases and when time increases it is the **k** factor in the relation of space-time $(a^3)= (T^2k)$, that reduce the space (a^3) in relation to the allocated position the onlooker has. Reducing the value of **k** allows the object (a^3) to get closer by the reduction of the time **k** between the object and the

onlooker reduces making the relevancy falling to the space a^3 larger and by the same measure does the time factor in the overall picture take a smaller part. By the same token does the time factor T^2 reduce to allow the space more prominence in the entire picture. The second picture shows where the time factor **k** reduces giving the space factor a^3 more prominence. This will indicate a much reduced time factor **k,** which

reduced the time factor T^2 and this gives the space factor a^3 much more prominence. The time factor is connected to both aspects of time being (T^2k) and that affects the space factor.

In our minds and us being those blessed with life this picture shows the Moon closer or further away. That is the image life holds. To the cosmos there is no misplacing of displacing and every aspect comes with the development of time. Travel is not a concept in the Universe and only Newtonians see that as an option but as all Newtonians do that also is a pipe dreams of fools. It takes time to get the Moon that much further away from the Earth. The Moon was much closer when the cosmos was younger because it started when everything was much closer. It time that is pushing the Moon and the Earth apart. In that there is proof of what time is, what space is and what the concept is dividing the two cosmic differences. Space is defined while time is eternal. That black stuff there is filling the night sky is time and not space. Since that is time and not space, that time not being space is eternal without ending. By Newton throwing the time aspect relevancy away by declaring it as a value of zero, Newton threw the baby away and kept the bathwater instead.

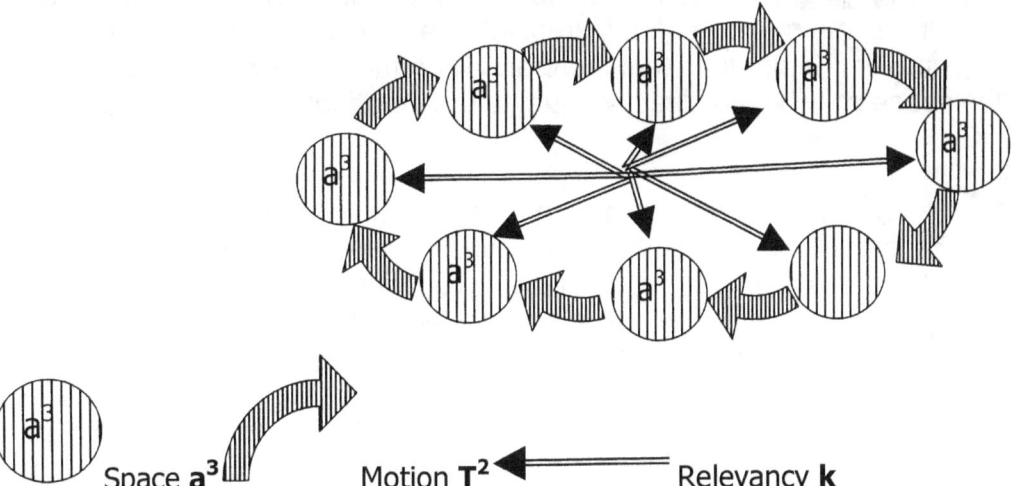

Space a^3 Motion T^2 Relevancy k

That Sir, Madam, is mathematics on a much higher level than the mathematics that Newton could understand. That is the mathematical interpretation of words put in mathematical terms that the cosmos used to speak to us. That is a language that goes beyond the language of us mortals. That mathematical interpretation is a language that makes sense when one who is schooled in mathematics has a mind to read what is said. That is the language that the cosmos used to teach us what the cosmos wished us to know. That is how the cosmos told Kepler what Newton never came to understand. That is words Newtonians still cannot read because Newtonians wish to interpret into the cosmos what Newtonians want to tell the cosmos what the cosmos has to be and then humans would accept the cosmos the way they want it to be.

If science assert them and truly display a need to know instead of a need to inform the cosmos in the effort they display when studying the cosmos, science would find progress. What is gravity? After three hundred and fifty years of research there is no one that vaguely knows anything about gravity. Someone came up with an idea of having a graviton but that is even more elusive than finding the gravity that the graviton supposedly is producing. The cosmos told us what gravity is but Newton knew more than the cosmos so Newton told the cosmos that the cosmos has mass. Newton told the cosmos that gravity is mass inflicted. The result is that after three hundred and more years we are as close to determining what gravity is than Newton was and Newton admitted he had no idea what so ever what gravity is! Today with all the

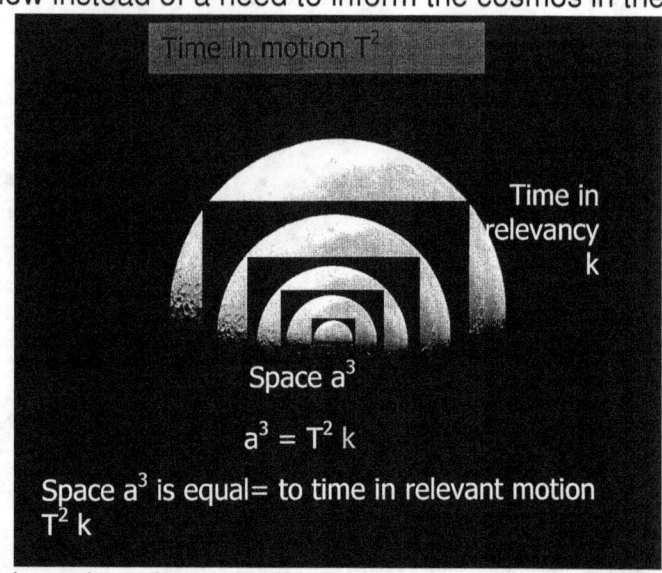

Time in motion T^2

Time in relevancy k

Space a^3

$$a^3 = T^2 k$$

Space a^3 is equal= to time in relevant motion $T^2 k$

wisdom there is going around there also still is not one Newtonian that knows more about gravity than the nothing Newton knew. That alone should guide Newtonians out of the dark ages but it does not. It only confirms the Newtonian stubborn nature to tell the cosmos what they want the cosmos to be without listening for one instant what the cosmos said. This letter is about listening to what the cosmos said on the principle Newton named gravity.

The Big Bang today is the cosmic development profile that is accepted by most and correctly so too, indeed. But in the cosmos development the Hubble constant is represented by the factor Kepler indicated as **k**, the linear time between objects. More important is that once any one accepts that this factor in time **k** grows, and then the factor cannot respond alone but has to influence the

Time T^2 k

Space a^3

other factors it is related to by the same measure. If **k** increased, then so too had space a^3 had too but also included the rotating time T^2. Kepler said that there are three factors in very close relation and the three factors are the space a^3, which is directly equal to the time factor T^2 that is placed relevant to the time factor **k.**

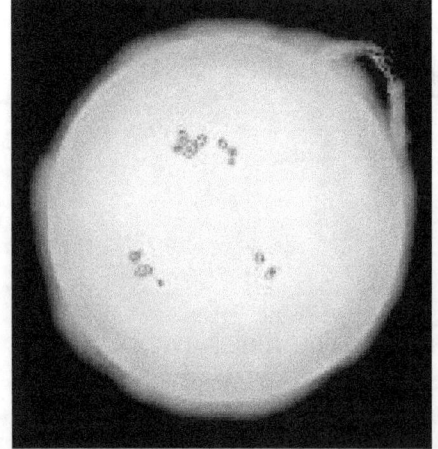

The logic behind giving the reason that substantiates this claim this argument seems almost an insult to the intellect of others. How does a Sun the size we now have fit into a Universe the size the Universe was at the event of the Big Gang when the Big Bang was the size of a Neutron?

Every claim any person ever made when suggesting the Big Bang theory was that the "distance" between the objects in the Universe grew and it is still growing to this day. It is accepted as the Hubble constant. If **k** in $a^3 = T^2k$ grows, then the material a^3 must grow but also to the same measure must the circle forming time also increase to fit into the total relevancy of space-time or then $a^3 = T^2k$. It is clear that the time factor **k,** which we connect to the distance between objects are as important in the formula Kepler handed down as any other part of space-time. That completely goes against the grain of Newtonian interpretation of this significant formula. The more **k** would find significant the more k will increase T^2 and the more **k** will reduce a^3. This conclusion is part of Kepler's formula $a^3 = T^2 k$ which is declaring that $k = a^3 / T^2$. Time decreases space when time increases relative motion.

All spinning matter has the point where the spin is still there but the radius is to small to measure by any means. That point is standing still in relation to the rest of the spin. In relation to that logic I do not except Newtonian science holding the radius of s spinning object unaccountable in the spin, whether the spin is applying or not.

Applying Newton's second law F=ma

One arrive at the formula

$GMm / r^2 = m (\omega^2 r)$

By replacing $(\omega^2 r)$ with $2\Pi / T$ we obtain Kepler's third law

This law predicts that $T^2 = a^3$

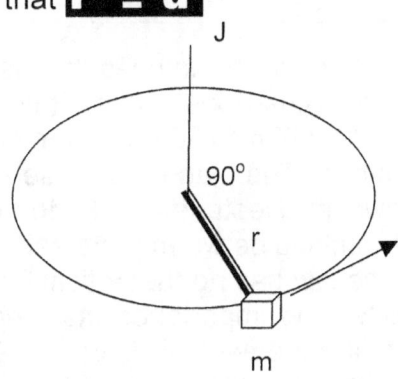

$p = m . v$

The mass (m) multiplying the speed (v) forms a new value J AND THEREFORE j CONTINUOUS TO IMPLY $J = I \omega$

$J = r \times p$ where $p = (v = r \times \omega)$

$J = r.m.v = m.r^2 .\omega = I. \omega$ and becomes interpreted as $J = I \omega$

This establishes that $r = dJ / dt$

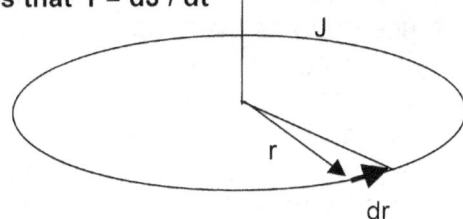

$r = dJ / dt$ In the case of planets in orbit around the sun r forms a value of zero because $dJ / dt = 0$.

What this statement implies is that r does not exist. When anything has a value of zero it is for all purposes non-existent. Only when an object is following s straight line can the radius be non-existent because the radius alters value through time development.

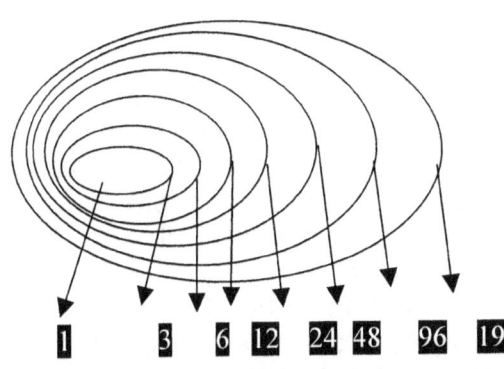

If we have a close look at Kepler's formula we find the space is in motion centred by a specific distance, which forms a conclusive part of the formula. We can see that the space as well as the time effecting, the space varies in accordance to the distance **k** and it is the distance **k** that produces the time, which produces the space. How is that possible you might ask when Newton has k down to zero?

We find that the distribution of planets relating to the Sun is all in precise sequence and the name of such a sequence is the Titius Bode law, and a law it surely is! The gravity extending from the Titius Bode law forms the entirety of the building of the Universe by constructing the Universe in the using of the atoms to form the Universe in the entirety thereof. Bode's Law:

Outer space cannot be a vacuum because if it were a vacuum it would implode unless it was secured with very stringent outer walls. There are no such retaining walls and therefore there is no vacuum. It is again another Human myth transported into outer space that has no validity. The absence of material proves the abundance of structural composition and if matter is reduced then something else is absolutely overbearing to fill the absence of material. That what fills outer space is what keeps outer space rigid.

Planet	Mercury	Venus	Earth	Mars	Ceres	Jupiter	Saturn	Uranus
Bode's law distance	4	7	10	16	28	52	100	196
Actual distance	3.9	7.2	10	15.2	28	52	95	192

A numerical sequence announced by J.E. Bode in 1772, which matches the distances from the Sun of the six planets then known. It is also known as the Titus-Bode law, as it was first pointed out by the German mathematician Johann Daniel Titius (1729-96) in 1766. It is formed from the sequence 0,3,6,12,24,48,96, and 192 by adding 4 to each number. The planets were seen to fit this sequence quite well – as did Uranus, discovered in 1781. However, Neptune and Pluto do not conform to the 'law'. Bode's Law stimulated the search for a planet orbiting between Mars and Jupiter that led to the discovery of the first asteroids. It is often said that the law has no theoretical basis, but it does show how orbital resonance can lead to commensurability. The importance that becomes known is the sequence the Titius Bode law saw in the number arrangement of 3; 6; 12; 24; 48; 96 etc. The incorrect application of the Titus Bode law lies in subtracting the figure of 3 from 10 leaving 7. The other way of reasoning is to add four each time to the first value of three starting with 3 and so on.

This puts the entire assortment of planets in neatly allocated and precise arrangements according the composition and outlay of outer space. The location of the planet does not depend on size neither does it depend on mass. The planets are arranged irrespective of any particular factors in accordance with anomalies or other uniqueness on the side of the planet. The arrangement of the located position runs by way of number and the distance doubles every time the number is one more in position to the Sun . I know I am on record of saying distance has no value in the cosmos and it has no merit other than to be a man made device formulated by man to be used in some insignificant calculation concerning some needs man may have. This arrangement in the layout I do explain later on but at this point I wish to show that there is absolute prominence in the composition of outer space and the planets has no particular preference accept to be where they are in the cue they take from the centre of the Sun . Let's for one minute forget the childish Newtonian argument of outer space being zero because I cannot afford to waste time in this book in disputing the validity of such an argument. I concentrate in another book where on this issue of explaining. In order to have a precise arrangement or pattern of positioning the fact of outer space must present some merit in as much as its composition. The planets as such can even be fragments because a structure forming a unit indicates that it does not participate in the issue. There is an arrangement of positioning in outer space, which shows that outer space holds the measure and gives the relevance.

Outer space is not just another vacuum of no importance but has absolute say when put into context

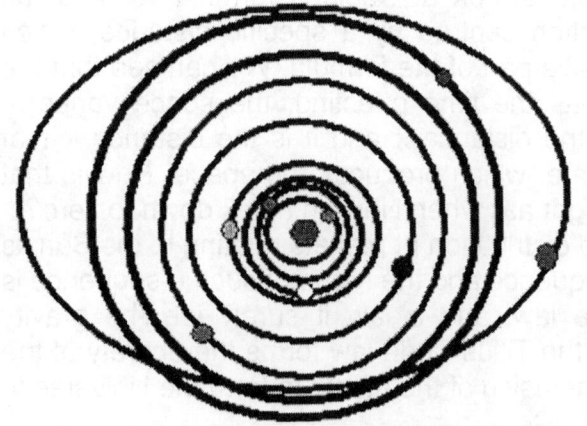

with the layout is has related to the position of the Sun . The proof is there in much more undeniable prominence than the unproven mass mesmerising nonsense Newton presents. The Titius Bode layout shows clearly no mass has any validity in the position allocated to any specific planet and therefore mass as attraction or force can be ruled out totally. It is not there while the Titius Bode law is there.

Furthermore we can find that the Kepler research proves beyond denying that **k** as a factor goes beyond dispute in contradiction to Newton's claims.

T^2 k a^3

PLANET	PERIOD (Years) (T)	MOVEMENT (T^2)	DISTANCE	SPACE (a^3)	RATIO k
Mercury	0.241	0.058	0.39	0.059	0.983
Venus	0.615	0.378	0.728	0.381	0.992
Earth	1.000	1.000	1.000	1.000	1.000
Mars	1.881	3.54	1.524	3.54	1.000
Jupiter	11.86	140.66	5.20	140.6	1.000
Saturn	29.46	867.9	9.54	868.25	0.999
Uranus	84.008	7069	19.19	7067	1.000
Neptune	164.8	27159	30.07	27189	0.999
Pluto	248.4	61703	39.46	61443	1.004

The factor **k** has a value of a^3 / T^2 and not even Newton can take such an argument away. Studying the evidence that was in hand even before Newton's birth where we find that the arrangement is such that it clearly defines the tempo in relation to the space of orbit at the location where such orbit can take place. The space has the dynamics of the other two time factors and the time factors indicate that the three factors are inter reliant on one another. The one gives the other value and the one takes value from the other. The space $a^3=T^2$ **k** to the time in motion. This means the space receives its value in relation to the value of the motion, which is in relation to the centre of the Sun . Again mass is a pipe dream.

	Distance (AU)	Radius (Earth's)	Mass (Earth's)	Rotation (Earth's)	# Moons	Orbital Inclination	Orbital Eccentricity	Obliquity	Density (g/cm^3)
Sun	0	109	332,800	25-36*	9	---	---	---	1.410
Mercury	0.39	0.38	0.05	58.8	0	7	0.2056	0.1°	5.43
Venus	0.72	0.95	0.89	244	0	3.394	0.0068	177.4°	5.25
Earth	1.0	1.00	1.00	1.00	1	0.000	0.0167	23.45°	5.52
Mars	1.5	0.53	0.11	1.029	2	1.850	0.0934	25.19°	3.95
Jupiter	5.2	11	318	0.411	16	1.308	0.0483	3.12°	1.33
Saturn	9.5	9	95	0.428	18	2.488	0.0560	26.73°	0.69
Uranus	19.2	4	17	0.748	15	0.774	0.0461	97.86°	1.29
Neptune	30.1	4	17	0.802	8	1.774	0.0097	29.56°	1.64
Pluto	39.5	0.18	0.002	0.267	1	17.15	0.2482	119.6°	2.03

The **space** $a^3=T^2$ **k** to the time in **motion,** which would then have **k** $= a^3/T^2$ and in another ratio it is $T^2= a^3/$ **k**. That makes mathematical sense. Compare this mathematically to what Newton suggested and compare the mathematical sanity that Newton portrays when Newton declare that $a^3=T^2$ **k** = $a^3=T^2$ **X 0 (because according to Newton k=0)** $a^3=T^2$ **X 0** = $a^3=T^2$. It is blatant corruption of mathematical law and any child that would put this down, as a valid mathematical argument in any maths class will find that child has failed the examination. Yet Newton is allowed to go insane. $a^3=T^2$ **k** can only be happening if there were two relating factors in the entire issue. It can only have meaning if outer space stands in relation to the Sun and no planet has any meaning in this partnership.

Again there is no trace of mass of individual objects having any significance in the outlay or in influencing the time of rotation to go faster or slower on the value of the mass or that the orbit is bigger or slower in relation to the size of the planet. If one scrutinizes the whole issue as a unit, it seems as if there is no significance in the fact that there are planets participating in the orbit process. The allocated position shows order but the order concerns the planet and its location in relation to the Sun and one inner marker. I explain that later on. The Sun it seems could care less about any planets being in the formation of outer space. If they were all in a specific allocated position that proves a given formula as to where they find such a position and the allocated position is in no way served by the influence of mass or size, then that proves buoyancy. Only when buoyancy enters the equation will size and mass not reflect any influence. If they rotate in precise equal fashion and the orbits in which they orbit shows identical relation to the space and the time we can surely claim that it is more the soup that matters that the size of the crumbs in the soup. In relation to the Sun every planet is not only at the same timing but also at an equal relevant timing of $k = a^3/T^2$. That means in ratio to the Sun all planets are at an even distance and that could only be if the Sun regards the outer space arena as liquid where the lot is floating in a bowl of liquid with a specific density applying. That puts the arguments about gravity in a totally different principal. Where liquid stand in affect to a solid we find the Coanda effect in place.

Newton stated that $a^3 = T^2$. That too is corrupting Kepler's fact as far as possible only to give veracity to the falsified claims Newton presented as trueness.
If that were the case then the space divided by the time would leave on $T^2/a^3 = 1$.
That is not the case because we have the tables to prove that it is a fallacy. Neither is $k = 0$ or is $T^2/a^3 = 1$ because $k^{-1} = T^2/a^3 \neq 1$.

PLANET	SEMIMAJOR AXIS $A(10^{10}m)$	PERIOD T (y)	T^2/a^3 $(10^{-34} y^2/m^3)$
Mercury	5.79	0.241	2.99
Venus	10.8	0.615	3.00
Earth	15.0	1.00	2.96
Mars	22.8	1.88	2.98
Jupiter	77.8	11.9	3.01
Saturn	143	29.5	2.98
Uranus	287	84.0	2.98
Neptune	450	165	2.99
Pluto	590	248	2.99

From Kepler's space-time $a^3 = T^2k$ formula we find that the relevancy of all planets $k^{-1} = T^2/a^3$ in relation to the Sun is alike. This is only possible when the planets are floating in buoyancy because when in buoyancy all objects are equal in relation to the water holding them. There is no big or small but just those having specific density in relation. If there was no buoyancy mass or size would form some sort of resistance that would allow more and less restriction in some or other form to be present. In the sea and to the sea in that case there is no big fish or small fish but there is only fish. To us humans we think in perception of distance but in the cosmos space in the form of distance is the measure of time developed. The same goes for temperature and mass. Mass is good and mass is a product of the Human mind to put perception when it is needed but mass is dysfunctional in relation to the cosmos. All the planets are more or less 299 in ratio from the Sun, which makes the time affecting all the planets, which is then $T^2 = a^3 \times 299$ and in relation to space $a^3 = T^2 / 299$. What this does is it puts all the space in ratio at an equal distance to the centre of the Sun and it puts the time in motion rotating at an even period around the Sun.

From where we stand the planets are assorted but from the Sun it is liquid of outer space spinning in relation to the Sun that is serving as the solid. Being liquid and solid is having motion or not. The Sun is at a level with all the planets parading past the Sun in the given ratio of 299. All the planets are precisely the same "distance " $299 = T^2/a^3$ from the Sun and has precisely the same "mass" $a^3 = T^2 / 299$ in relation to the Sun and the rest also floating about the Sun while they all travel at the same velocity $T^2 = a^3 / 299$ around the Sun.

The lot is in a bowl of liquid and it is the liquid that keeps a regard to the Sun where the Sun forms the solid as the regard to the liquid. In that regard gravity loses all it's mystery as the Coanda effect starts to explain gravity and the purpose of gravity. In accordance with and depending on what Kepler's law indicate, we find space-time is $a^3 = k T^2$. I wish to remind you where Newton came up with the idea that $a^3 = T^2$ when he disclaimed Kepler's findings to validate his position on mass and the way he interpreted gravity. Let us again go back to investigate….

Applying Newton's second law F=ma
One arrive at the formula
$GMm / r^2 = m (\omega^2 r)$
By replacing $(\omega^2 r)$ with $2\Pi / T$ we obtain Kepler's third law
This law predicts that $T^2 = a^3$

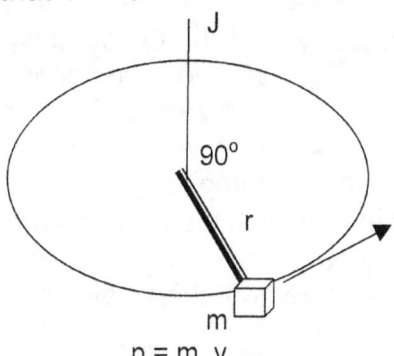

p = m .v

The mass (m) multiplying the speed (v) forms a new value J AND
THEREFORE j CONTINUOUS TO IMPLY J = I ω
$J = r \times p$ where $p = (v = r \times \omega)$
$J = r.m.v = m.r^2 .\omega = I. \omega$ and becomes interpreted as $J = I \omega$

This establishes that r = dJ / dt

r = dJ / dt **In the case of planets in orbit around the sun r forms a value of zero because dJ / dt = 0. The big incorrectness is that this presumption works on stopping time. That is impossible to achieve.**
This is totally going against the grain of what Kepler said because Kepler said the motion of the space is equal to the space. On this rests the entire Universe because as soon as time stops, space collapse. The motion provides time a space to move and that is the Coanda effect.

r = dJ /

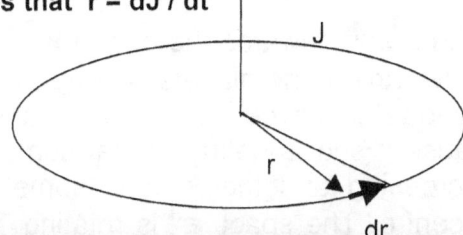

$$r = \frac{a^3}{T^2 k}$$

In the case of planets in orbit   **around the sun r forms a value of**
zero because
dJ / dt = 0.

Then Mainstream science still has the audacity to say Newton was never proven to be incorrect...it should rather be said that Newton was never investigated and Kepler was left undiscovered because Newton raped everything about Kepler. It is as if Newton stole a vision Kepler got by studying but never saw. Let us for once seriously reflect on the matter while forgetting about Newton and Kepler.
If it is true that $a^3 = T^2 k$ and we dismiss Newton's obscenity while going back to basic mathematical principles we find that:
$a^3 = T^2 k$

Taking the argument back to Kepler's law,

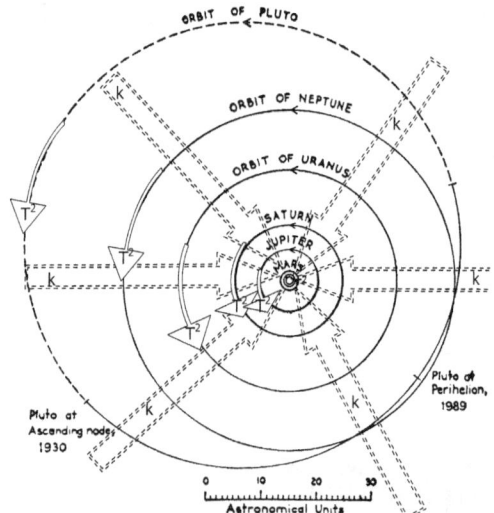

$a^3 = k T^2$
$a^3 / k T^2 = 1$

T^2
k

R^3

$T^2 = (4\Pi^2 / GM) r^3$
Π^2
Π

That means the space will move in a straight line while it circles. That can only indicate a controlled expanding because $a^3 = T^2 k$ (also is); that the line between the centre k^0 and a^3 / T^2 is increasing at a rate of $k = a^3 / T^2$ which indicates that expanding is happening at this very moment. The line **k** is not zero as Newtonian madness would suggest or the line is not 1 ($a^3 = T^2$) but the line is gaining by one dimension every rotation that is completed.

$T^2 = a^3 / k$ which means the time it takes to move is in ratio with the distance the space will move.

$k = a^3 / T^2$ which means the distance the space would move depends on the time allowed for moving. However if that is true and it is a mathematical statement then it also must be true that:

$k^0 = a^3 / T^2 k$ there is a appropriated centre

Time can increase by manipulation as well as time that can reduce by manipulation. One can travel by increasing or decreasing the valid time.

$k^{-1} = T^2 / a^3$ The distance is moving between the objects. If the objects are not coming closer it must be the substance holding the objects that is coming closer. When we have a table indicating a shift towards the centre $k^{-1} = T^2 / a^3$ that does not involve any of the planets coming towards the Sun it would then show the space holding the planets are moving inwards because it shows definitely that something about the radius is decreasing and that means something is shifting towards the centre. The space a^3 is rotating T^2 and the only other counter action could be when the space holding the rotating space is reducing $T^2 / a^3 = 299$ by the margin space is expanding $a^3 = T^2 k = k = a^3 / T^2$ that proves the Big Bang is in progress. **Kepler told so much about so many by using so few syllabifications that a mathematician such as Newton was unable to comprehend the full implication.** Allow mw to explain with the aid of a manmade machine.

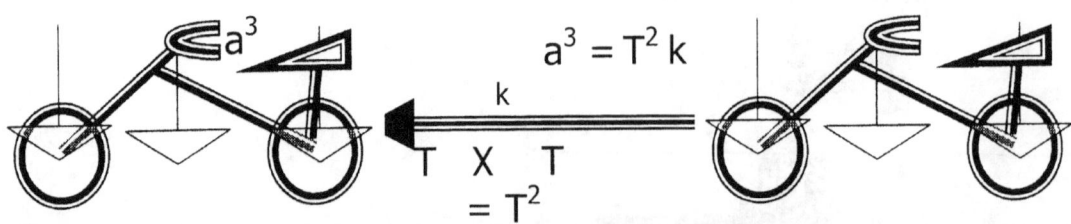

a^3
$a^3 = T^2 k$
k
$T \times T = T^2$

We know time connects to the moving bicycle where the position changes from where the position allocated is putting the bicycle from the one instance to the next instance. The distance the bicycle moves is associated with time because the distance than **k** has is the duration in time that it took the bicycle to replace a^3 during T^2 all across the length of **k**. We know the bicycle cannot jump from the one position to the next position notwithstanding whatever drive we connect to the bicycle. The bicycle constitutes of a massive number of atoms forming the structure that forms the bicycle.

The get the bicycle from one point to another point we have to take the one lot of atoms forming the bicycle structure that is the bicycle unit to the next lot that forms the same bicycle at another location.

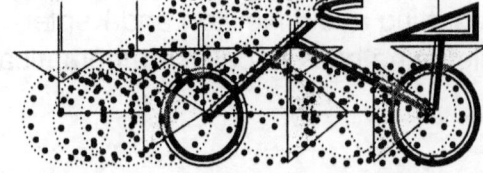

That means to every atom there is a replacing of the one atom to the next location of the same atom at a distance of **k**, which is time during a period, which is time. Time therefore has two components

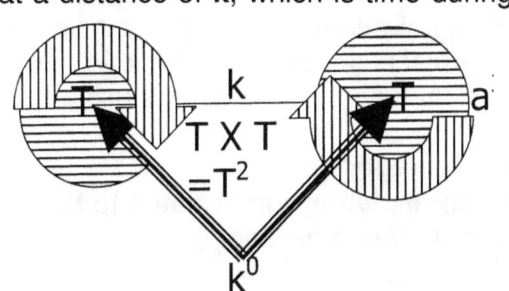

being **k** and being T^2 and all the while **k** multiplied by T^2 forms space a^3. If Newton had no regard for Kepler then at least he should have had some regard for mathematics.

It takes the time that **k** allows to move a^3 from **T** to (X) **T** during T^2 and that is what Kepler said about space-time. Kepler mathematically said that $a^3 = T^2k$. Time is the motion that time allows space to move from and towards a^3 = (T to (X) **T**= T^2) **k** It proves that the factor **k** shares all the dynamics which the factor T^2 presents in allocating the position in time of space a^3

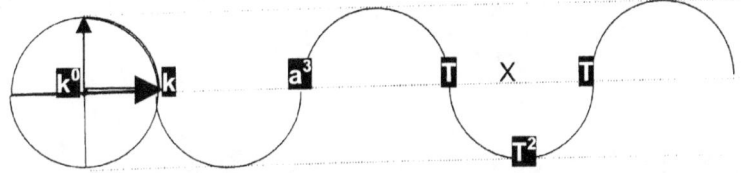

The motion of the atom follows the exact curves of the graph. Because there is motion the motion is

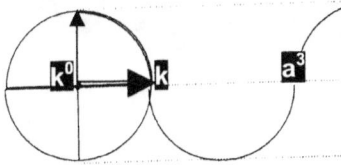

in the square and since the square is present we can refer the result to the triangle. The motion is always 90^0 to line of connecting the centre to the motion and where 90^0 become present the law of Pythagoras becomes the dominant mathematical factor.

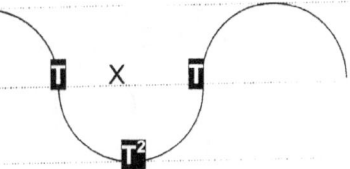

The factor **k** is always an extending line from where **k** as a straight line started to where **k** then will end. The line that forms the basis for the triangle is k^0 going on to **k**. If **k** is implicated as zero as Newton said then the entire Universe loses all its validity in being. That throws the entire cosmos into not being and that is the only thing that cannot ever be a factor in the cosmos since the cosmos is everything that is between infinity and eternity. The motion of a^3 is from where k^0 starts the line to where **k** ends the line and in between the start and the end there is a beginning and an end **T** to (X) **T**. That is also forming the other end of what totals as time T^2, but it holds time in duration T^2.

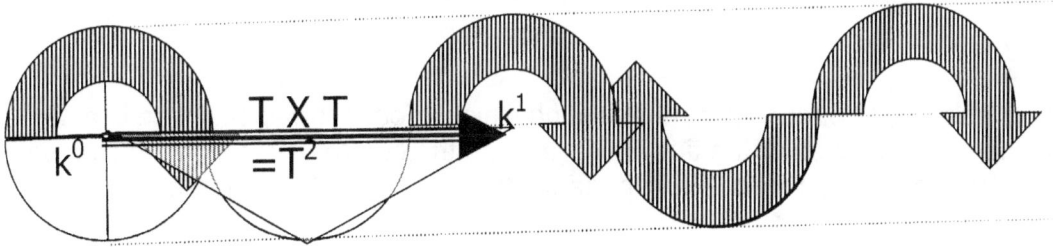

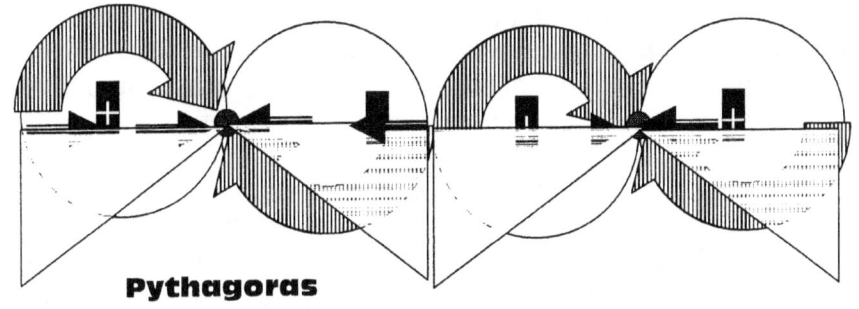

Pythagoras

The movement of the bicycle concerns the unit as a structure that we recognise as the bicycle but the bicycle comprises of all the atoms forming the bicycle. The atom is in motion $k = a^3 / T^2$ but also within the atom forming the atom there is motion of the atom $T^2 = a^3 / k$ and the totality in reference form space-time a^3 =(T^2) **k.** The formula implicates that the space a^3 = is equal to the position that the space in the next movement will have (T^2) **k.** The space becomes the motion of the space and in the space there has to be motion producing the liquid as well as rotary motion that gives the space a definition, which presents the space with motion in relation to the liquid, and then there is the space defining the purpose of the liquid.

$$a^3 = (T \text{ to } (X) T = T^2) k$$

Where there is a square present there Pythagoras is also present. I am sure Newton knew that…what is on the one side of the line forms a relation to what is on the other side of the square. The referring of motion in relation to a centre proves an establishing of a line in relation to a point. It serves a line being in relation to a specific centre and that centre projects the start of the line in a 90^0 angle with the point where the line must end. This is a mathematical reflex of Pythagoras using the line in forming a square that comes in relation to a centre.

That is the essence of Pythagoras.
The relevancy of k forms two correlating points where one relates to the centre and another relates to distance moved

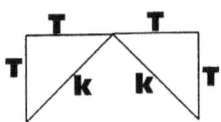

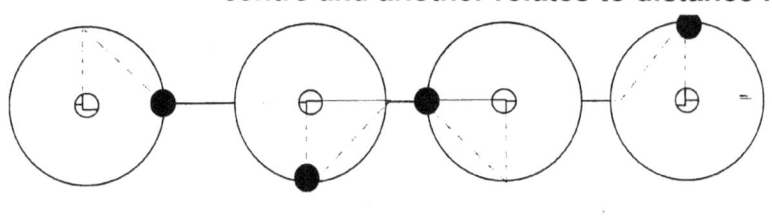

The motion of the outside to a the atom and honouring such of the centre has according to the Pythagoras

the atom reposition point in the centre of since the electron is a centre any moving to effect the atom in dynamics

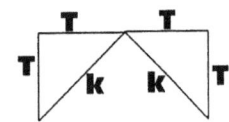

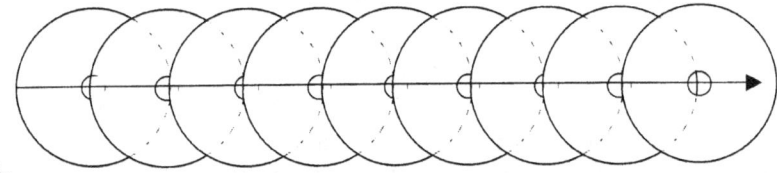

demands.
The moving of the atom is formed by a specific relocation as well as a specific reproduction of the atom in structure as well as in allocated positioning in accordance to every aspect of its surroundings. The atom must remain precisely associated with all aspects outside in order to find the precise duplicating of time in every aspect. That is gravity.

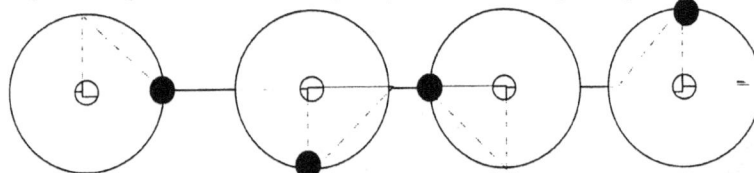

From this point the atom orbits as it honour the centre around which it focus the spin. There is a relation by four points where the distance or the time will align with the centre as rotation focuses on such a centre. Every time the realigning forms 90^0 with the centre we will find Pythagoras applying the law aspect of the aligning.

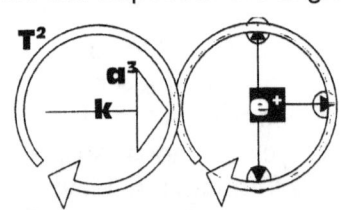

In relation to the outside there is more motion that has an effect on the atom where the centre point that the electron focus on shifts in accordance with the atom motion as a unit. It takes the atom time in duration to remove all the material forming the atom to shift from one point to another point in order to allow motion to take place.

At the same time it takes the electron time to move distance the electron is from the proton serves as the apply when on cycle of rotation is completed. The rotation since it takes light 3×10^5 km / sec to reposition the light is not time but is in relation to time. Einstein proved that

around the centre and the measure of time that will connects to time and photon the speed of light

the speed of light is time but that cannot be since it then must take light zero time to cross the entire Universe. Since light can only travel at 3×10^5 km / sec and the Universe is slightly wider than that, Einstein's argument goes up in smoke. To light, as it is the photon, the light finds time standing still but that explaining is long and is off the point as far as this discussion goes. Since light is not time but is connected to time like the rest of all material it takes the electron time to circle the atom. One may argue it is insignificant but to the atom the cognisance is one year or one day or one period of time achievement. It is not only beauty that is in the eye of the beholder but everything about anything comes down to the measure of the beholder. To produce the space a^3 of the atom the atom must have the electron circle T^2 once around the atom border at the relevance of **k**. The electron is focussing the relevance of the atom a^3 where it secures the border T^2 at the relevant time position of **k**. However **k** as a factor is in relation with the start and the start of any line is not zero but it is infinite. Therefore **k** comes all the way from k^0 to where **k** ends to position T^2, which then secures the border of the atom in a^3.

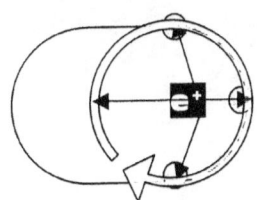

When the atom centre moves while the electron rotates such movement of the centre has a direction and this direction must influence the rotation because it will change the position of the rotation in relation to the centre of the atom. In the direction the centre moves the space between the electron and the centre will

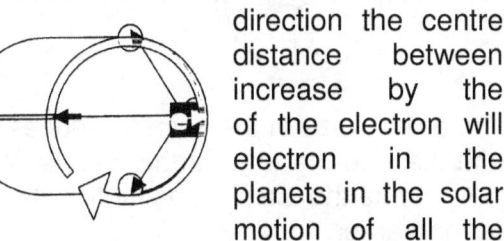

reduce in accordance with the direction the centre moves. At 180^0 from that point the distance between the centre and the electron has to increase by the margin the distance decreased 180^0 from there. The shift of the electron will be influenced by the intensity of the motion of the electron in the direction the electron moves. Since all the value of all the planets in the solar system mounts to $a^3/ T^2 = \pm\ 300$ that means all the motion of all the planets related to the Sun is moving at an equal pace in time with the directional flow of time. The rotations around their individual axis might be different but the rotations they have in relation with the Sun are all equal. That proves that from the Sun we have all planets orbiting equal at $k= a^3/ T^2 = \pm$ **300**.

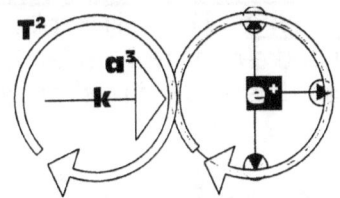

The fact of the relevancy is not merely argumentative but is a mathematical reality. Changes in the motion of the atoms will affect the atom in the rotation time it takes the electron to circle bout the centre. By shifting the centre the time will reduce the ration period in one direction as much as it will increase the rotation in the other direction. This will have a sever timing implication on the atom as much as the time in motion increase the space it holds per time unit in rotation will decrease.

As the motion of the atom increase the electron allocate position while rotating

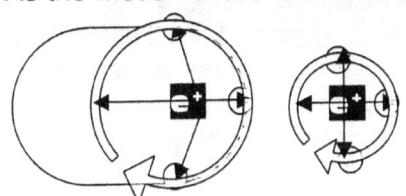

in alignment has to readjust with the shifting centre. If **k** shifts then **k** will insist on T^2 readjusting and with T^2 readjusting we will find the borders of a^3 reposition the circle of rotation. The more motion there is in relation to another fixed centre that represents the starting point that serves as k^0 the smaller will the allocated distance **k** be that we find between the rotating electron T^2 and the shifting proton in the centre.

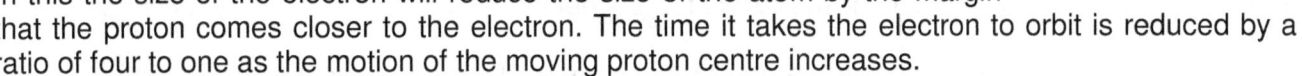

In this the size of the electron will reduce the size of the atom by the margin that the proton comes closer to the electron. The time it takes the electron to orbit is reduced by a ratio of four to one as the motion of the moving proton centre increases.

That is why gravity or motion reduces the atom as it accelerates the atom in relation to a more compact space. As the space in which the atom moves through time by duplicating increase in density that projects the ratio to reduce the rotating of the electron and with that the electron reduced

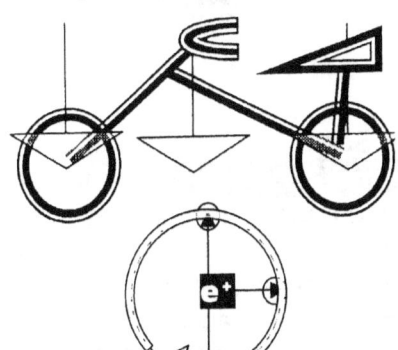

the claimed space the atom holds. But it is not where the significance ends. In that way gravity increase time as gravity reduces space where gravity is motion accelerated or decelerated by change of the density of the heat through which the space moves.

By reducing the atom it relieves the atom of controlled heat within and such heat is ejected to the outside of the atom. The atom heat reduces by claiming less space because it holds less heat. Since the movement aligns with another centre k^0 that represents the link between k^0 and **k** and therefore institutes the triangle the space held by the electron remains an allocated value in relation to the Earth. Bo body shrinks because it removes heat from the atom but the atom still hold the heat in relation to what the Earth measures. That is the essence of the sound barrier. That is the essence of the heat blanket that covers the aircraft.

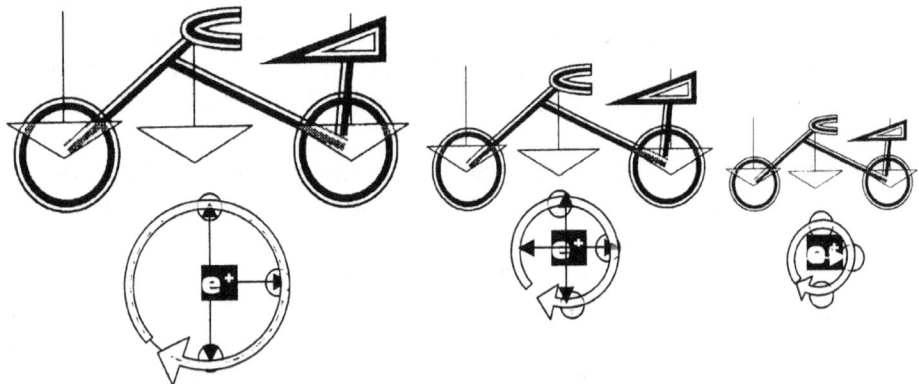

The atomic size of the aircraft reduces while the structure space of the aircraft remains in accordance with the size that the Earth permit and heat fill the vacancy that motion produce. The aircraft has no mass since the aircraft has motion. Newtonian misconception can never establish any reason for the sound barrier.

That is what time implicates but all that is lost to Newton trying to fixate every aspect in cosmology on mass and by making the value of **k** redundant he removed the concept of time from the cosmos altogether. The value of **k** is space-time and that is $k = a^3/T^2 = \pm 300$. **What this suggestion implicates would put the cosmos in a bowl of liquid where all swim alike**. That will explain gravity. That proves outer space has a specific density relation with the Sun and a density In as far as material within outer space goes. Every object floats just like the rest with no differences in as far as size or mass goes and outer space is liquid/gas.

By reading what Kepler said correctly so many centuries ago the effort brings all the answers to so many unrealised questions…but it does not involve looking at what Newton said about what Kepler said… it is all about looking at what Kepler said. Not only did he introduce space-time a^3/T^2 but he also placed space a^3 and time T^2 in a relevancy long before Einstein did and placed gravity in space-time a^3/T^2 even before Newton named gravity. Kepler was the person who placed gravity as the ingredient in the universe that determines space a^3 and time T^2 and much more. Kepler was the first one that saw that gravity comprises of two factors being **k** or linear gravity and circular gravity or

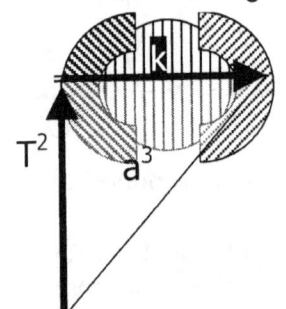

Kepler was the very first person to mathematically introduce space a^3 in a time relevancy by **k** and related directly at all time in the universe that determines space a^3 and time T^2 and much more. Kepler was the first one that saw that gravity comprises of two factors being k or linear gravity and circular gravity or T^2.

Kepler was the one that discovered space / time as
$$k = a^3/T^2$$
Kepler was the one that discovered gravity holding space-time relative
$$k = a^3/T^2$$
Kepler was the one that discovered there in the centre of a sphere is singularity as $k^0 = a^3/T^2\ k$

Gravity then is $a^3 = T^2k$ where space is equal to the time it moves in. Space has to move or space will collapse. That we see in the case of a Black Hole. Gravity then is also $T^2 = a^3/k$ Time forming is valid since the object is unable to depart any further distance and is captured at that distance by the restricting of the space in motion. Gravity is keeping space relevant to determining the relevancy of motion in relation to the centre. Gravity then is also $k = a^3/T^2$ because gravity is space in division of time which comes down to space having a velocity that time holds space in motion too. Gravity is motion of space at a specific speed or velocity, which is moving across a distance in relation the time it takes. Again that is space-time. Gravity then is also $k^0 = a^3/T^2\ k$ where the Universe is being in the centre of space from where the Universe can and is claiming and controlling space- time. Translating Kepler's mathematical expression $a^3 = T^2 k$ correctly to the verbal statement in English Kepler said that there is a space a^3 which is equal $=$ to the motion in the time duration T^2 thereof where the motion of space takes up the time it uses to go between two specific points which holds a relation to a centre where from where there forms a straight line k and is located on the spot where space begins the circle therefore that centre spot has the least space. Forget for one minute what Newton said. Then take my challenge and prove what I said is not mathematically sane while you also know that $\dfrac{dJ}{dt} = 0$. I challenge you to prove how, where and why this can mathematically be sustained anywhere and under whatever mathematical law.

Place whatever you wish in a relevancy and have to outcome mathematically reach zero and see how mathematically appropriate that will be.

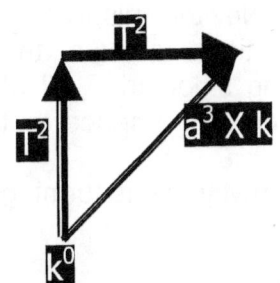

Kepler stated the very opposite of what Newton saw. Kepler had direct opposing ideas about the circle and what factor should be using the square because in Kepler's method of expression the circle indicator T^2 goes square ^{2} and the diameter indicator **k**, which replaces r remains single in the face of the volume being a cube a^3.

The two masters was using different dialects of the same mathematical language spoken by all

It is about translating mathematical equations and being correct in interpretations of mathematical expressions. Other factors are about certain mathematical deductions that were made in the past but were incorrectly presumed. A line cannot and therefore does not start with zero. Should you think such a statement is trivial then this book is even more especially for you because that changes where one presumes the Universe came from at the very beginning of the cosmic conception? Kepler answered all the questions we have...and much more!

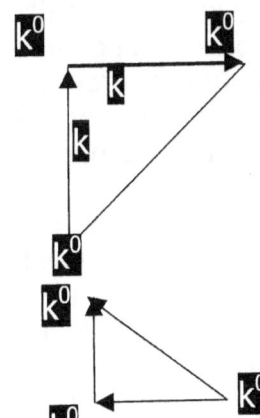

In order to have any line running there is a vertical response to the horizontal implication as much as there is a horizontal response to any vertical implication. The one response implicates the other. What ever there is there has to be another to implicate, which is. If there is **k** there has to be k^0 to begin **k** as much as there has to be k^0 to end **k**

It is a mathematical fact that where there is a line of any description such a line will stand between two points holding infinity because a line can only start at infinity. That fact of a line puts a triangle in place because a triangle has the equal value of a straight line. If there is a straight line there has to be the presence of a triangle because both are equal to 180^0 and therefore are equal in all respects. Also in response there will have to be a responding line of 180^0 crossing by the value of 90^0 in relation to 180^0

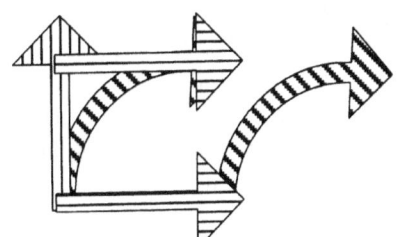

There can be no line between two zeros because only zero is between zero and another zero. All lines have to start at infinity and end at infinity and with infinity connecting lines all lines stand related to each other at the least of the value infinity. If that was not the case then the Universe had no validity. The first line assembled a Universe of possibilities.

The universe is possibilities waiting for a hold. Where you now are centre of a raging star at history. For all we know planet teeming with life a star life was The Universe is not

chance to be whatever it can could and will become the a time when life is not even the Sun could have been a and as the Sun developed into desecrated by development.

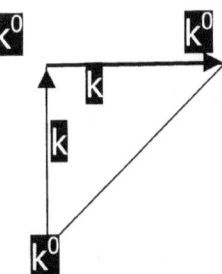

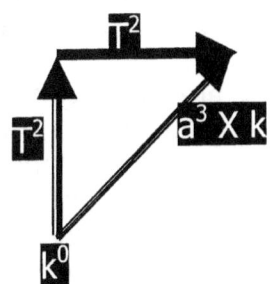

nothing but is every possibility waiting to happen.

The moving of one line establishes another line because by departing from singularity such motion establishes two dimensions. Motion goes by the square, which means that no matter how long the line is the line also has to be wide to be a line. Being wide in the face of being long puts the line in a square and with the line in the square the line immediately represents the other factor of time where time is in the square. Therefore drawing a line has the same implication than enticing Pythagoras because every aspect of the cosmos is the cosmos because of the law of Pythagoras. The cosmos is Pythagoras but that is not what the cosmos introduced when the cosmos introduced it to Kepler. The cosmos presented gravity, the glue that keeps the cosmos together.

Pythagoras and the cosmos is $a^3 \times k = T^2 \times T^2$ and that is putting the three dimensions of space in motion that is in relation to time towards where it will be (T^2) while k). However in this book we do as a letter is dedicated to Yet that is the Coanda where the space by three connected by line this is meaningless if it is not is defined by time

That is the atom and that is time material in time.

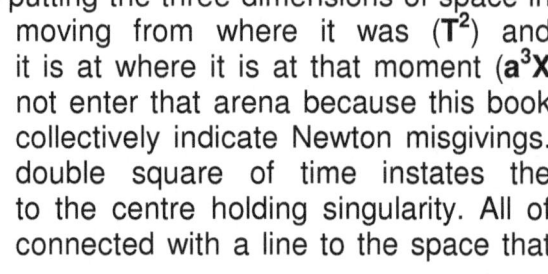

moving from where it was (T^2) and it is at where it is at that moment ($a^3 X$ not enter that arena because this book collectively indicate Newton misgivings. double square of time instates the to the centre holding singularity. All of connected with a line to the space that

and that is gravity that is forming

Planets orbiting a given centre move by a line that measures by a triangle, which forms a circle. That is the conclusion of Pythagoras but that is far better explained in another book.

The gravity controlling stars work different from gravity working planets but gravity is the motion of space in time in relation to time. That puts all motion in context to the triangle and that puts all motion in

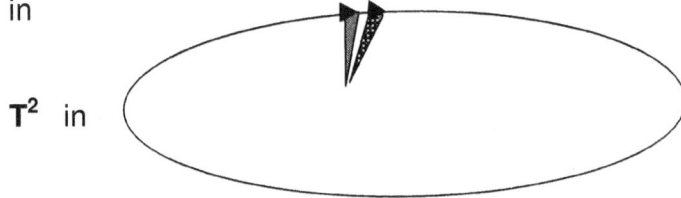

T^2 in

relation to Pythagoras. The gravity, which was referred to when Kepler's studies revealed gravity was the motion of space a^3 through time relation the a centre. The line running from the centre confirms the square of time by the distance. This is what Newton did not see.

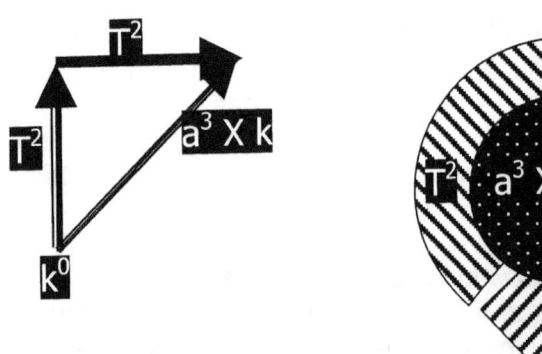

If we investigate the Coanda effect it is clear from what effect the Coanda principle institute that the space of the circle depends in size on the distance that **k** forms but that distance reverse with the action of the motion, which the liquid produces.

I challenge you Academics in physics too show how much the Moon , is coming closer as the product of the mass of the two objects namely the Earth and the Moon having a mass that is pulling the Moon towards the Earth, and the Earth towards the Moon bilaterally. I challenge you to show how mass influences orbiting planets by contributing or increasing / decreasing or to what degree the mass is influencing the orbit velocity of any planet in any way or manner. The orbits are in ratio. Notwithstanding whatever mass, the lot are still very alike and in fact they are the very same. Notwithstanding the enormous mass discrepancies between the planets where Jupiter is 318 times the mass of the Earth and Pluto has only 0.0025 the mass of the Earth they all orbit at a shared equality in ratio of about 300. I challenge you to mathematically prove that there is a gravitational constant. I challenge you to prove that the gravitational constant is responsible for evening out the mass differences between orbiting planets and that this is the cause why planets all orbit in a compatible fashion using the same velocity notwithstanding mass differentiations. Remember this evidence I mention opposes Newton in no uncertain way by contradicting Newton at the very heart of physics and yet with all the contradicting evidence coming from the cosmos as such Mainstream science remain of the opinion that Newton was never yet contradicted!

The force is the measure whereby the radius in the square will vanish and this argument becomes the cornerstone on which all physics hold its foundation. The gravitational constant is used as the complement of the mass in conjunction with the gravitational constant to produce such a force value, which then supposedly produces a force able to destroy the radius between the objects. Then Newton went even further at the time and gave the concept a name: he called it gravity. In the way the formula is presented the gravitational constant increases the effect of the mass of all the planets, which is forming the force of contraction between the various objects and the Sun.

The only way this can be present is when outer space is equal in relation to the Sun while to the Sun planets are just floating in outer space and is the same as outer space is. The Sun holds no planet in any different way that it holds outer space in regard to the Sun. All planets are at an equal distance where the planets are point in an area where no point has special value. There is no mass in any consideration that the Sun gives credence to the planets.

Considering that neither distance nor size proves to be any different in relation to the Sun the only conclusion that can come from this is that the part outer space must perform as a liquid or a gas of sorts giving the planets a buoyancy regard put them all in equalisation. That can only be true if there is buoyancy in outer space, which makes the one part perform as a liquid T^2k while the other part serves as space a^3. This brings the Coanda effect to mind. It clearly shows a density factors as all the evidence of mass or size disappear. The only way that can happen is when the Sun becomes the space a^3 part and outer space becomes the liquid T^2k.

By expressing the statement in a mathematical equation it reads as follows $F = G\dfrac{M \times m}{r^2}$ **F=G (M X m) / r²** The evidence that this is a fable and that mass is not pulling the objects closer was already established with Kepler introducing his work. Newton knew that the planets orbit in equilibrium by the measure if T^2/a^3 **= 300** and all of that is in a ratio of about 300. That disputes Newton claim on mass forming influences because it clearly shows equanimity.

Mercury   ° From what we can see the Sun holds the space a^3 equal to the

Venus   O time T^2 at any given distance from the Sun . The ratio of space in

Earth   O relation to the motion in time is at an equal footing. That means to

Mars   ° the Sun the only thing that is out there is outer space, which is

Jupiter

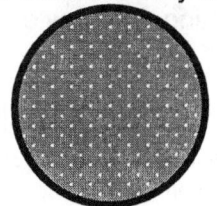

representing a unit in relation to the Sun forming

Saturn

another unit. That is exactly what Kepler inferred with

Neptune

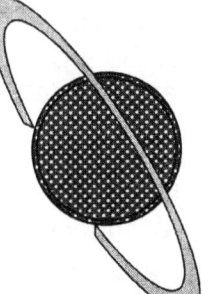

his formula $a^3 = T^2k$. From the space the Sun holds the Sun is

Uranus

equal to the motion of the space and the Sun forms the one

Pluto   ○ factor of space **a³** where the other equal factor is then that, which is moving. Every planet is in ratio as far as any other planet, which puts all planets on equal footing. From the time of relation to the size in equality to the distance they are in relation to the Sun all proves a marked pattern of evenness.

Then when all the planets orbit at an even pace and all being the same ratio, where not one is found to come any closer to the Sun or to each other, your officially accepted reason for explaining that, is that the gravitational constant is evening out the mass differences and yet, according to the official formula you use, the gravitational constant is not evening out the mass but is complimenting the mass by improving the so called force. If the gravitational constant is evening out the mass the constant firstly has to divide into the mass to wither the effect that the mass produce and by nullifying the mass to the extent that Kepler's work proves, there is no mass factor present left to use. The

formula then should read $F = \dfrac{M \times m}{G \times r^2} = 1$ F=(M X m) / (GX r²)=1 which in normal verbal English states

that the product of both masses is in equilibrium when divided by the conjunction of the gravitational constant and of the square of the distance parting the objects where the product of the gravitational constant and the square of the radius between the planets is in division with the product of the mass which then will produce the force that cancels any influence of the mass factor that both of the planet's will have on the orbit of every planet orbiting evenly around the centre structure.

That is very much not what your formula you use $F = G\dfrac{M \times m}{r^2}$ suggests. Your formula shows that the

gravitational constant is helping the mass product to destroy the square of the radius. In the cosmos however and in reality that destroying of the radius is the last thing that is happening. To explain this ridiculous deception Mainstream physics go on to explain that the gravitational constant which is supposedly the gravity that is keeping outer space in a unit, where that gravitational constant nullifies the mass influence on the force. Newton said mass is all-important but this explanation now suggests that the mass is nullified by the intervention of the gravitational constant.

This is a statement that is not in line with the mathematical equations Mainstream science present to serve as truth. Making such claims that is so far away from the actual mathematical statements, which came into place serving as yet another attempt to cover up Newton fraud with more deceit and becomes compatible with mafia like behaviour. This criminality is the behaviour of all physics paternity in control of all physics worldwide! This is the group of men that is in control of the thoughts in the minds of the brightest developing brains in the world. They harvest the best intellectual group of all students this Earth can offer! To the students questioning or rejecting the Newton claims is not an option.

The students accept what is taught or face examination failure and total rejection of the institute where they are educated. If that which is presented that the statement would suggest about their explaining how there is such planet mass apathy that will not influence orbit details of planets making them go faster or slower it is based on yet another Newtonian lie. If the gravitational constant were evening out the mass then it would have the gravitational constant in multiplication with the square of the radius in dividing the product of the mass. This will then be evening out the product that the multiplying of the mass put in place. If that was said and translated in mathematical terms, then it would mathematically equate to showing where the force will be in equilibrium because the radius is supported by the addition of the gravitational constant which increases the radius in the square and the total of that is in equilibrium with the product of both object's mass. If that was true then the gravitational constant must be placed where it increases the radius in every event and it would be stating that $(G \times r^2) = (M \times m)$ (G X r²) = (M X m) is applying in all the cases so that the mass has no longer an influence on the outcome of the equation because out there in the real cosmos Kepler proved that the mass has no implication on the orbit of any planet **T² / a³ = 300**.

Kepler never even thought of a mass of any description and I can assure you that his not mentioning any mass was not the result of his retarded mind or not because he was mentally inferior to one Isaac Newton. With the corrected implementation of Kepler I can and I do support such statement of equality but then in that event the equation then should read that the gravitational constant is complimenting and supporting the square of the radius and mathematically such a statement is $(G \times r^2) = (M \times m)$ (G X r²) = (M X m). That removes mass as a cosmic factor. There is nothing wrong with this supposition but the way it is presented. The way it is presented dies Newton's claim on mass, which then should run through to the abolishing of his formula and the discarding of his theory. It shows that firstly k is not insignificant as zero but presents much significant proof that the Sun places all planets on an equal footing of 10 / 7. I shall explain that statement in a short while. Every distance of every planet in relation to the Sun is equal and every orbit of every planet is in harmony. Mass does not implicate or profess to have any influence which is strongly suggesting the characteristics of a density factor as particles would have when they display buoyancy characteristics in water.

You know very well that in the planet orbit ratio the space factor (space **a³**) divided by time **T²** leaves a space **(a³)** – time **(T²)** displacement factor ratio between space and time of round about three hundred in all cases, $\left(\dfrac{T^2}{a^3} = \pm 300\right)$ (T²/a³ = ± 300 , which disqualifies mass discrepancies altogether.

With that the cosmos put Newton in dispute! So why then bring in the use of mass in any event if it does not change the result but has only one purpose and that is to cover Newtonian's misconception and to hide Newton's fraud? However, the formula used by mainstream science clearly shows that the gravitational constant is in effect increasing the mass by multiplication of the mass. The formula undeniably put the gravitational constant in a multiplying position with that which the mass of both structures hold.

$F = G\dfrac{M \times m}{r^2}$ F= G (M X m) / r² which then in ratio is decreasing the square of the radius. That is pure rubbish and I challenge you to prove me wrong by proving your mathematical interpretation. Then all the while, while promoting such rubbish you still maintain that Newton has never been proven incorrect. When Newton is taken into the cosmos Newton physics goes bizarre. There is no evidence of mass influencing the motion of planets in any manner. On the contrary all evidence show a clear equilibrium present in the relations of the Sun and planets.

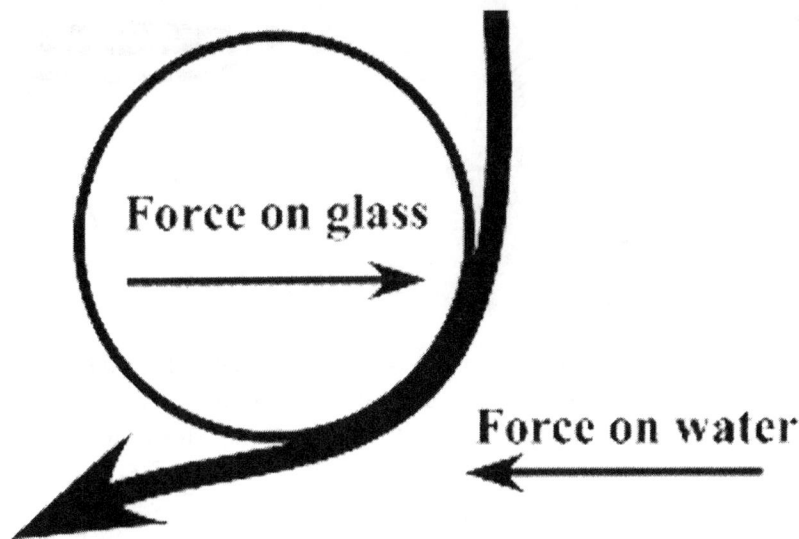

Force on glass

Force on water

Newton is not incorrect as it is used in normal applied physics but in cosmology there are no grounds in supporting of mass or gravity in the manner Newton saw or in the way Newton suggested mass to apply. On Earth and on the ground Newton's work is impeccably correct and that I admit without any reservations on any aspect of physics in whatever manner. However taking Newton's ideas into space becomes fraud. Mass only serves as a human concept to allocate differentiation when required when gravity is obstructed and in the cosmos gravity is not obstructed.

In the Coanda effect we find that when a liquid such as water runs down the side of a round shaped surface, which is a glass amongst others the water follow the contour of the glass. Where water normally is used, the water would fall directly to the ground as gravity effect the water and only allow the water to run in a straight line towards the Earth. When the water is poured over the round surface that the glass provides, the water follows the contour of the roundness of the glass.

This part Newtonian physics admit and they put two forces to explain the action but fail to give any further explaining as to why the forces come about or what would initiate the forces to counteract and disturb the gravity pull that should have the water running around the surface of the glass at the top end and then as soon as the water would find the opportunity, it will get a gravity pull and run straight down ignoring the interrupting of the glass contour since it then no longer blocks the path of the running water.

 If we return to Kepler's formula we find the answer, but the answer again totally contradicts what Newton claimed about a circle having no effect on motion. According to Newton the rotation of an object is not time related and has no influence on the surroundings of the object.

When a circle of an object rotating produces the motion there are four points on each side that plays a part. Each point produces a direct relation as to the coming towards / the going away, the passing the front and the crossing the back marker.

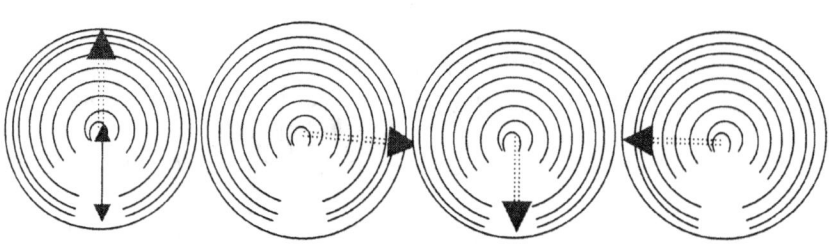

The Coanda effect proves otherwise when studied closer.

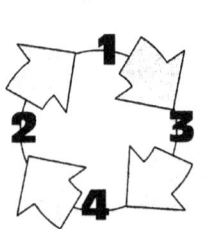 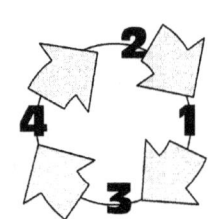 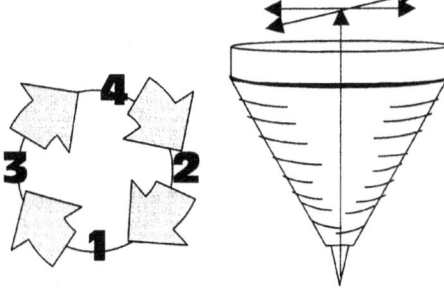

We can see that there are some coherency between the time and the rotation of objects.

That coherency we think of as time. The rotation of the object finds new alliances with the surrounding and if that was 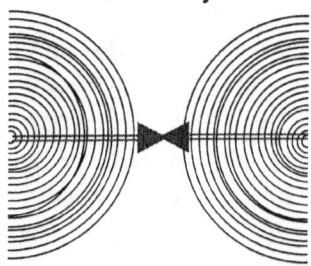 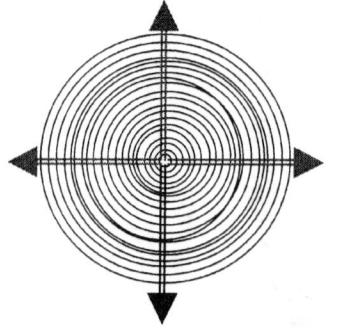 not the case then no fluid drive of objects was possible. You may discard this remark as not applying to the cosmos, but later in the book I am about to prove that outer space is lesser-concentrated form of heat and the atmosphere of the Earth is just a higher concentration of heat.

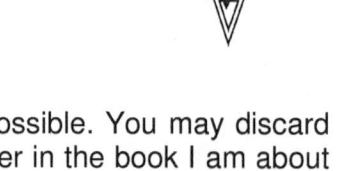

Please allow me to explain the relation between time and rotation by a simple experiment. Let's take a top and think of the same top as having two ends in opposition to each other.

 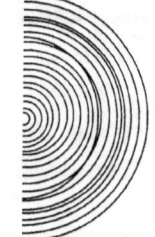

When we take an object in rotation as having one side running left which it has, and another side running right, which it also naturally has. The side heading left is totally opposing the side running right because of conflicting directions in the spin.

If we put the same objects two sides in position, as they are when they rotate we find the tow sides is contradicting each other by not being the same as one another. The one is a clone of the other but the clone is an opposing duplication of the other. This is part of the characteristics of all rotation and in this rotation it entirely depends on a liquid state that limits the space where the space includes the liquid $a^3 = T^2k$. **Gravity is merely the motion of particles in space.**

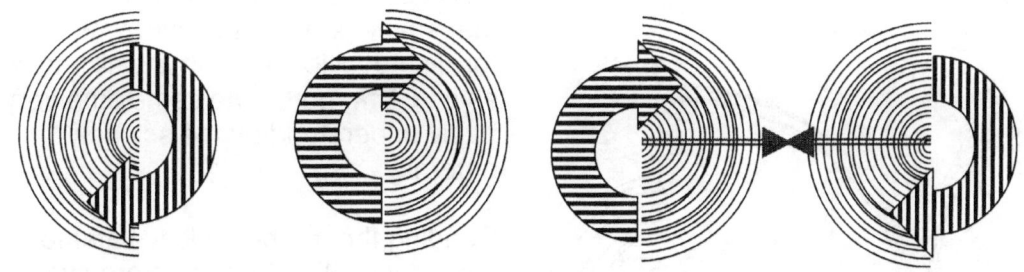

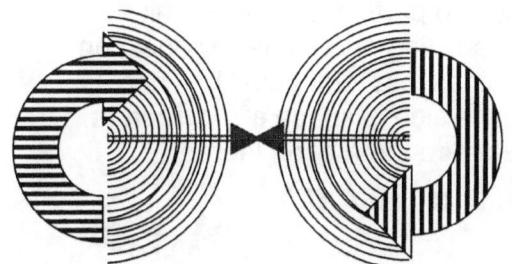

When any and all objects spin the spin is in opposing relevancy notwithstanding being part of the same object or being the in the location where the next allocated position is precisely where the opposing direction at that point forms the opposite of what it will be as the next part of the same but ongoing

rotation produces the rotation graph. That is because no object can be twice in the same Universe and every object is in two parts of the same Universe in different locations. No object share Universes but forms duplication by motion of the same Universe on two sides of the same and one unifying Universe.

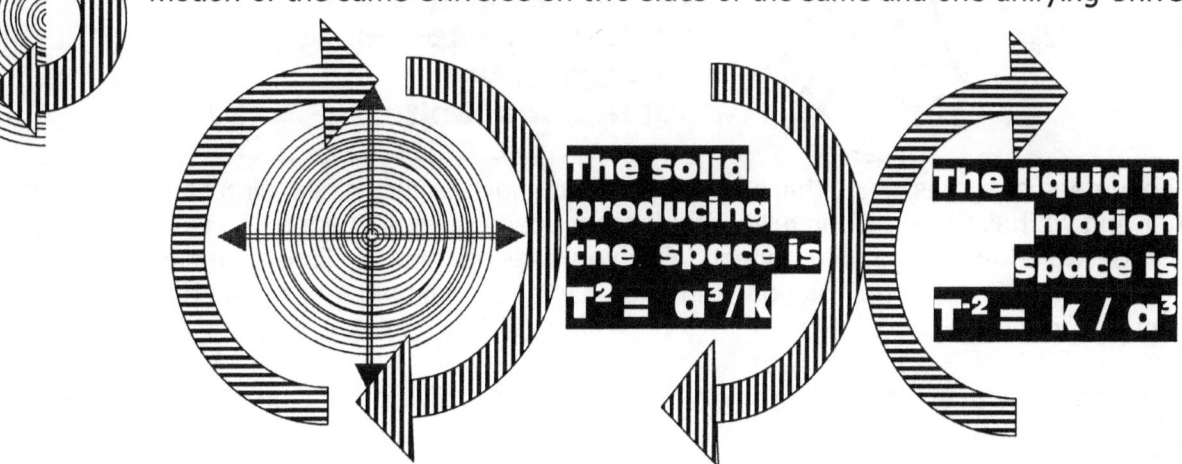

The solid producing the space is $T^2 = a^3/k$

The liquid in motion space is $T^{-2} = k / a^3$

Taking this one step further in order to explain gravity there are two factors found in sharing the same unit. The one factor is the liquid to motion in relation to the other substance forming in the relation the part of the solid or the stable not moving part.

Taking a considered look at the spin procedure we find not forces but a rotation ratio at work which

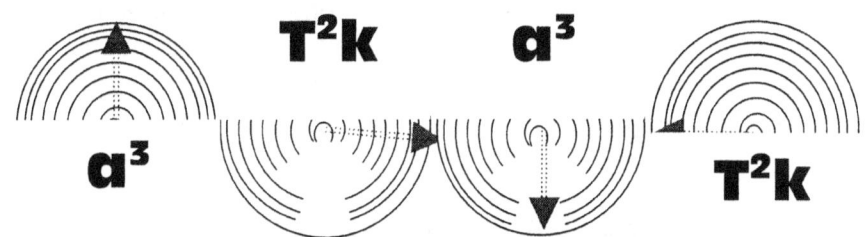

T^2k a^3

a^3 T^2k

produces the opposing of two sides in exact opposition of each other where the sides in opposition will cyclic interchange and become the other party as the direction of the spinning objects change. In the spin we find a divide running

as a line in the centre of the spin and is the point on which the ration is grounded. From that line of division two sides oppose each other all the time. On the one side we find **k** being the distance having a relation with time in division of space and on the other side we find space being in division with time.

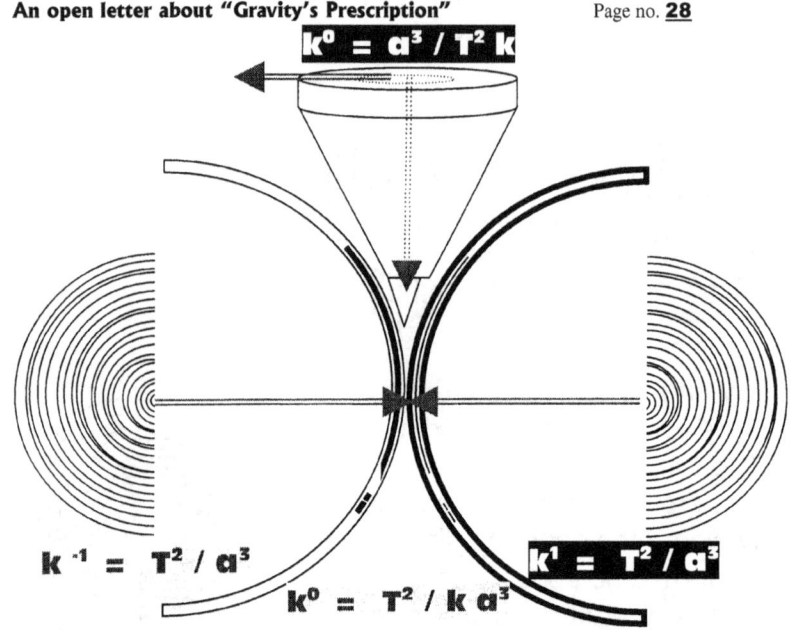

$$k^0 = a^3 / T^2 k$$

$$k^{-1} = T^2 / a^3$$

$$k^1 = T^2 / a^3$$

$$k^0 = T^2 / k\, a^3$$

If we take Kepler's formula on individual merit the cosmos told Kepler that space is equal to time by motion form a specific centre. It is well known as $a^3 = T^2k$, but at the same time very badly understood by the science paternity.

Reading the mathematical statement correctly we find that there are two opposing parts of the same unit contradicting each other by being on two sides in a total equal balance. On the one side is space a^3, **which is equal** to the motion thereof .

$$T^{-2} = k / a^3$$

$$T^2 = a^3/k$$

The comet is in a position of expanding by setting the duplication limits.

It is documented at this point that any object spinning has two sides directionally opposing one another and we have the cosmos telling Kepler that space divides in opposing sides as space a^3 and T^2k motion which forms the time aspect. We even use that side as time by indicating a cyclic year from the completion of one rotation and monthly and seasonal positional points indicating the different periods.

In this we have the one side holding **k** related to space a^3 in division of time T^2 being the relevancy of $k = a^3/ T^2$ There then is an opposing relevancy to this because of the opposing we find in the direction of the same object and being in opposition it then must be $k^{-1} = T^2/ a^3$. However with that being the case we then find objects are in opposing relations to time as well with time going positive and negative or going bigger in relevancy or smaller in relevancy. $T^{-2} = k / a^3$ and $T^2 = a^3/ k$. This proves that the object in rotation around a specific centre will reduce the relevancy or distance by reducing or increasing the time and space relation. That is the reason why comets do not slam into the Sun by force instigated on grounds of mass as Newton would suggest but rather cyclically circling the Sun in rotation by going away further and afterwards coming back to come much closer. Guarding this very jealously is the balance of the line being $k^0 = a^3/T^2 k$ and every time the rotating object crosses this dividing line the relevancy changes to the opposite because the direction changes to the opposite.

This we find as a principle in the orbit patter of the Sun and all planets where the relevancy shifts from the one onto the other. But mass has no role to play in this because the structures do not have mass, they have gravity.

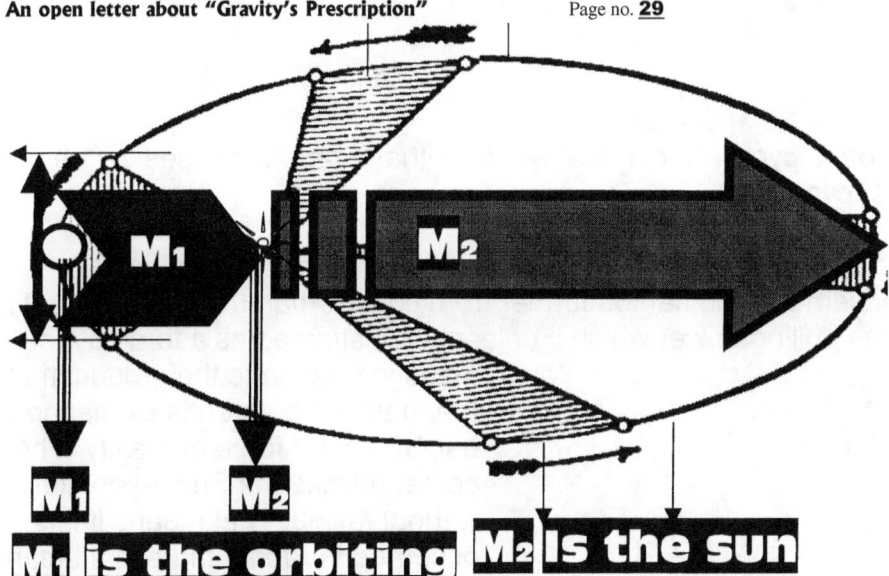

M₁ is the orbiting M₂ Is the sun

What Kepler said in hid formula $\mathbf{a^3 = T^2\ k}$ is directly opposite to what Newton suggested with $\dfrac{dJ}{dt} = 0$. This means that the spin is uninfluenced by that which is forming the surrounding of that which is spinning.

Newton stated that $\dfrac{M_1 \times M_2}{r^2} G$ = Force and the mass in each case remains the same as well as the gravitational constant, then why would the radius change every time of each year and not draw closer (or further).

$\mathbf{F = G(M\ .\ m)/r^2}$ where:

G = the gravitational constant,

M = the mass of the body,

M = the mass of the lesser body

$\mathbf{r^2}$ = the radius between the two bodies.

Newton stated that $\dfrac{M_1 \times M_2}{r^2} G$ = Force, because of the fact that:

1.   The value of M_1 in both cases is equal because that is the mass of the earth.

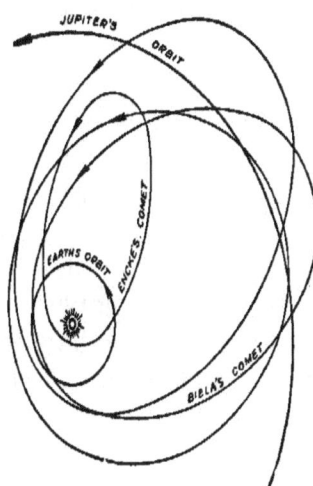

2.   The value of r^2 is equal because the two objects have been dropped at an equal distance.

3.   The value of M_2 is still a mass filled with gravitons and therefore has to effect the relation in the same way as both objects consist of different compositions of materials that are used to manufacture the different objects.

4.   That means Force one has to have a different value to that of force two ($F_1 \neq F_2$). In contrast time duration $P_1 = P_2$ and this cannot apply in the case of gravity because the greater the force are through more mass, the bigger the impact has to be on the time duration.

I think in all fairness it now is rather obvious even to children that orbiting comets do not destruct as they collide with the Sun by the pulling force of gravity. Except in extra ordinary cases where some interference changed the orbit cycle of comets we find that comets passes by the Sun unscathed to go into another cycle.

In such manner Halley was able to predict the cyclic return of the comet named after him.
As far as cosmology physics is concerned, little can be further from correct than the idea that just it is the mass bringing destruction by the pulling power which it unleashes instigated as a force by the mass in relation to the product that the mass on both ends exchange for becoming the force of gravity. The comet misses the Sun every time without failing. That means if it is occurring in repeat there is a balance in place.

On the one side we find that the structure controlling the contracting aspect holds dominance in providing the pivotal circle relevance $k^0 = a^3/T^2 k$, then this switches around where the other plays the key factor providing the gravity as $k^0 = a^3/T^2 k$. Seen from the Sun the one side of the rotation would be time in progress of becoming more by applying $T^2 = k/a^3$ and in this the other orbiting structure would provide the relevance. The factor then would be in relation $k^{-1} = T^2/a^3$. As the orbiting object crosses the centre divide the complete opposite comes into action where the time factor once more grows and $T^2 = a^3/k$ applies.

However at that point from the other side we find that the factor also grows in $k = a^3/T^2$ Since the factors inter act by taking on different roles through the orbit we do not have collisions as Newton suggested but controlled orbit interacting as Kepler suggested.

By applying the Coanda principle there are two relevancies at work and the motion as well as the direction time takes the motion will place either one of the two relevancies in prominence at any particular given time. There is absolutely no chance that both the factors may apply simultaneously and either of the relevancies has an opportunity to dominate but with motion applying stronger, there is a situation that presents the contracting to overshadow the expanding but that I explain in a later stage of this letter.

$$k^0 = a^3 / T^2 k$$

$$T^2 = a^3 / k$$

$$T^{-2} = k / a^3$$

Time grows to accommodate space (comet)

Time shrinks to accommodate space (Sun)

$$k^{-1} = T^2 / a^3$$

$$k^1 = a^3 / T^2$$

$$k^0 = a^3 / T^2 k$$

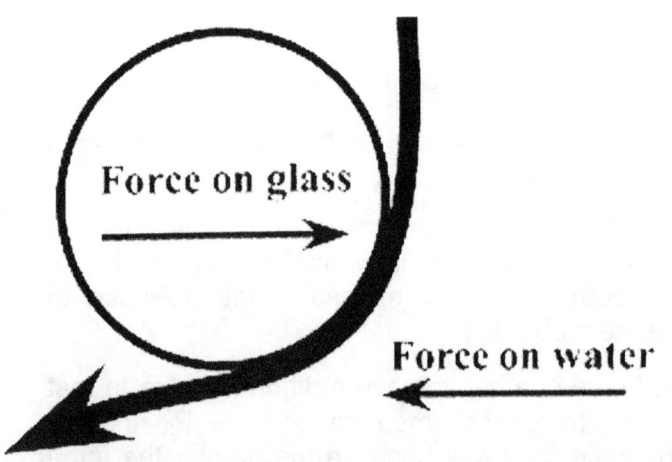

Force on glass

Force on water

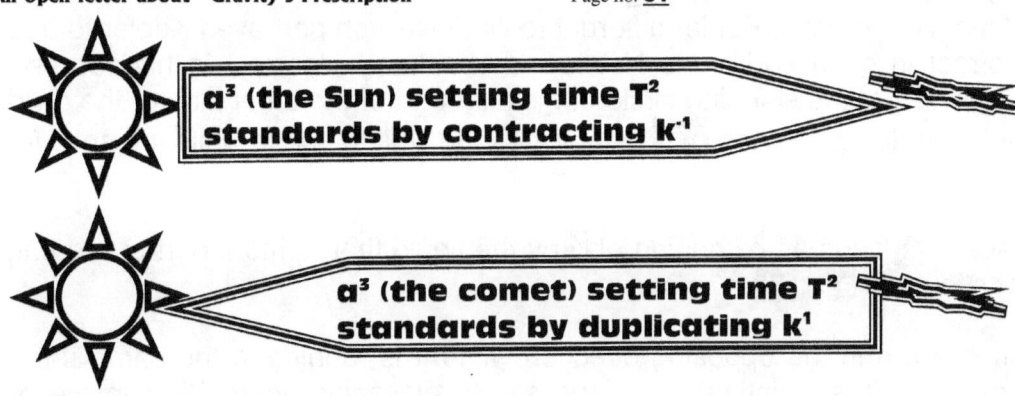

a^3 (the Sun) setting time T^2
standards by contracting k^{-1}

a^3 (the comet) setting time T^2
standards by duplicating k^1

From this comes the working principle we at present gave the name to as the Coanda effect. In the Coanda effect we have full gravity working as a compliment when the liquid provide motion to implement the gravity action.

It would be far more prudent if we left Newtonian forces (and this I say in spite of notwithstanding all the rejection I thus far encountered from the celebrated mainstream physics paternity by my criticizing Newton on cosmology) and witchcraft to the Middle Ages from where it comes and see what gravity is by standards applying in the modern age.

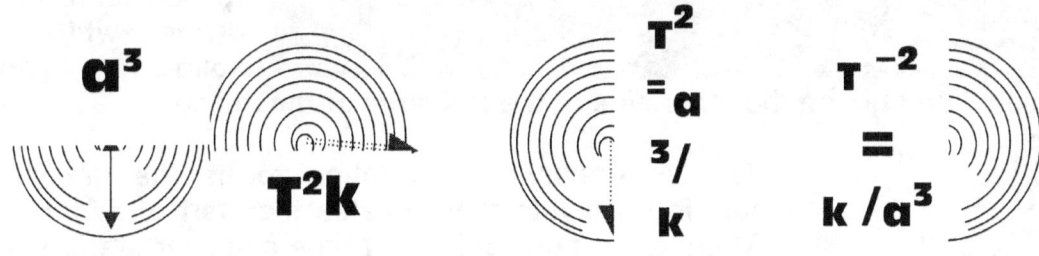

a^3

$T^2 k$

$T^2 = a^3 / k$

$T^{-2} = k / a^3$

In the Coanda effect we have the same opposition forming the same unit but two different substances taking on roles in the opposing sides and from this springs gravity.

Going back to the spin action we saw that one side is opposing the other side while in rotation and while forming a single unit. This we take back

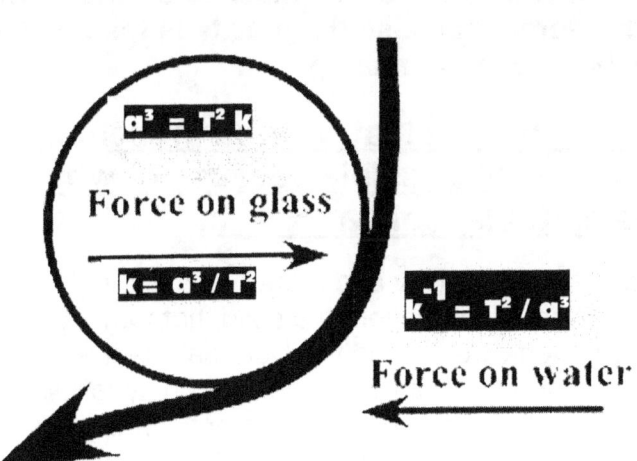

$a^3 = T^2 k$

Force on glass

$k = a^3 / T^2$

$k^{-1} = T^2 / a^3$

Force on water

to gravity and in that we find what forms the motion we find in gravity.

As I explained previously there ate two action forming part of four factors bringing about a rotation action.

While on the one side of the divide $k^0 = a^3/T^2 k$ we have $k = a^3/T$ where the time factor provide space with a limit there are the other side where the time factor being on the other side of the divide $k^0 = a^3/T^2 k$ extends the limit of the space by the liquid to provide the limit to space at $k^{-1} = T^2/a^3$.

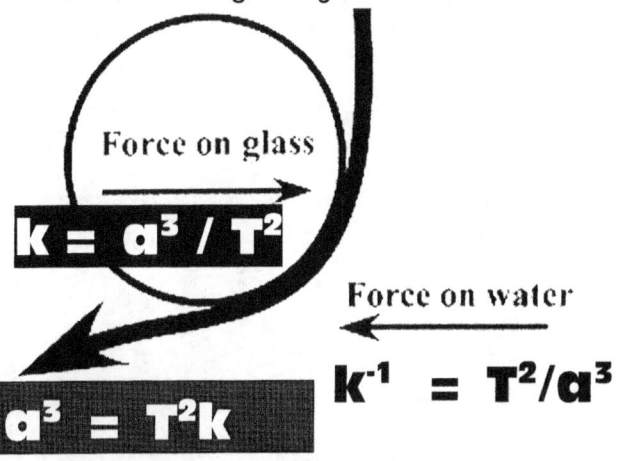

Force on glass

$k = a^3 / T^2$

Force on water

$k^{-1} = T^2/a^3$

$a^3 = T^2 k$

The **space a^3** of the unit is **defined k** by the **flow T^2** of the liquid

In all of this the lot that I mentioned prove Kepler different to what Newton portrayed Kepler to be and also it proves Kepler correct in gravity whereas Newton is absolutely incorrect in his surmise on cosmic gravity. The liquid is part of the spinning circle but is also part of the other side of the divide $k^0 = a^3/T^2 k$ and on that side the flow or motion $T^2 k$ establishes the limit to space while the solid forms the space.

On the other hand the space a^3 finds an extending of k by the liquid that forms a part of the unit as the motion part in Kepler's formula $a^3 = T^2 k$

Because the liquid seems to run in the opposing direction while it is running in the same unit and therefore in the same direction it is in influence by the same relevancy factor k but in the other direction k^{-1}.

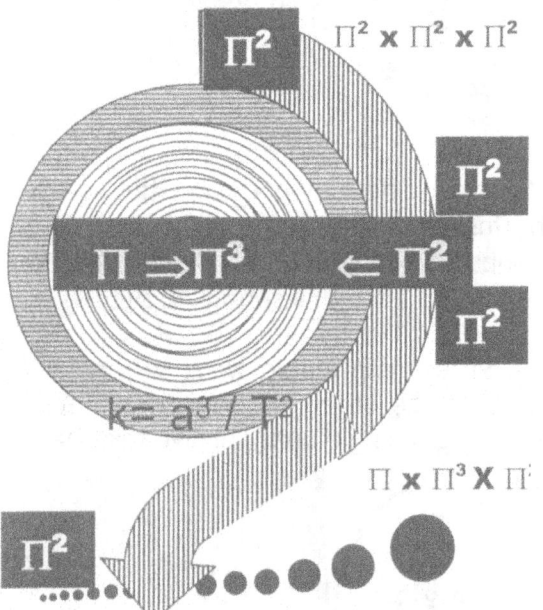

The Coanda effect #1
JL Naudin - 09-26-99

The Coanda effect #2
JL Naudin - 09-26-99

Although both the solid and the liquid forms part of the same circle and is in one unit the liquid as one part serves one side of the Universe or the divide while the solid serves another part of the same unit but on the other side of the Universe or the divide.

Π^2

$\Pi^2 \times \Pi^2 \times \Pi^2$

Π^2

$\Pi \Rightarrow \Pi^3 \quad \Leftarrow \Pi^2$

Π^2

$k = a^3/T^2$

Π^2

$\Pi \times \Pi^3 \times \Pi^2$

That is gravity. The rotation forms the gravity by forming a divide that initiates the start of a Universe. There is no possibility that time can ever stand still or that motion of rotation has no influence on the gravity bonding that forms because the gravity is the result of the divide that the rotation brings on.

Sir / Madam, the fact that Newton present k as zero is, is quite impossible because of what Kepler said. Kepler said a³ = T² k.

The Coanda effect is gravity and moreover it is the movement of a liquid in relation to a solid that forms gravity. In the phenomena we named the Coanda effect we find that water runs down in a straight line seemingly following a direct route to the centre of the Earth or to where it will contact the soil.

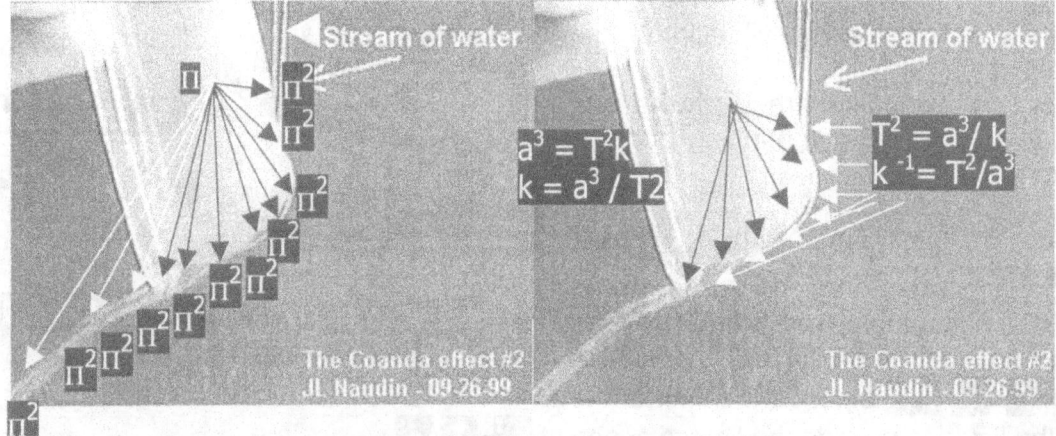

Stream of water

Stream of water

$a^3 = T^2 k$
$k = a^3 / T2$

$T^2 = a^3 / k$
$k^{-1} = T^2/a^3$

The Coanda effect #2
JL Naudin - 09-26-99

The Coanda effect #2
JL Naudin - 09-26-99

When motion produced, space and motion came because of singularity that is unable to produce any sort of motion of any nature whatsoever motion appointed three positions holding singularity that formed a relevancy with other positions that produced the space in relevance to singularity as a factor.

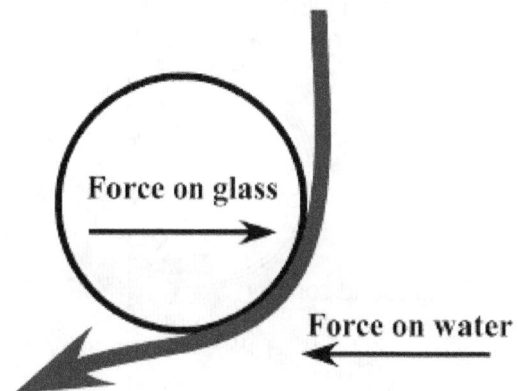

Force on glass

Force on water

The reduction of space will bring about heat. Injecting fuel in a place where such reducing of space already exist it increases the heat level further and the fuel will "spontaneously" ignite. The igniting creates a rise in the heat level to a point where such a level that one can only find that level in the stars. Fuel will establish conditions, that is in accordance with the laws of cosmology such heat will only apply to stars where such heat levels will indicate the enormity of the gravity present that is generating the massive gravity accumulated in the absence of space. Gravity is the concentration of heat in the utmost reducing and concentration of space thus a star is born in the gravity on Earth. In this example, **k** increases as thrust pushing space **a³**, which the **k** creates to a new **T²**. That started the Coanda effect but that also started the Universe before the Universe and space – time. The Universe then was spinning at speeds faster that the speed of light and everything in the Universe only had form. Later on with the event of the Big Bang the Universe, received dimensions but form remains the template of dimensions and even today that is how we interpret timesaving three positions to a centre. We still have the evidence, which is even more obvious than most other certainties we uncover in the Universe.

The Coanda effect is creating gravity. It is not replacing gravity. It is not recreating gravity. It is not substituting gravity. What the Coanda effect is is what gravity is. The Coanda effect is gravity.

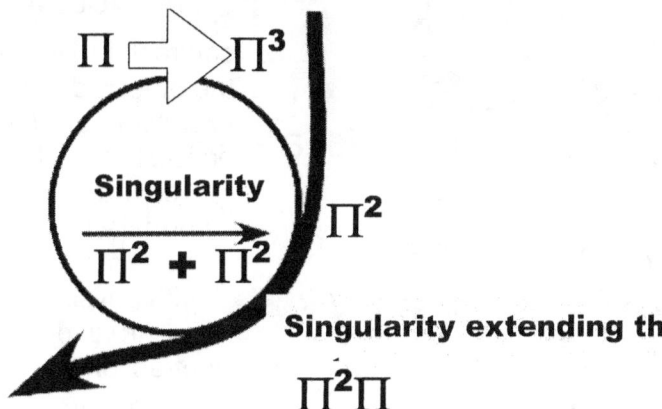

$$\Pi \Rightarrow \Pi^3$$

Singularity

$$\Pi^2 + \Pi^2$$

$$\Pi^2$$

Singularity extending the influence on flowing

$$\Pi^2\Pi$$

The Coanda effect is also the perfect example of the curvature of space-time brought about by the extending of singularity influencing due to the shape that imitates or duplicates the value of singularity and again conforms Π. By establishing a new value of singularity as Π, singularity can once again take control and establish a new Π^2 as gravity in the new Π^3 forming space.

Looking at the Coanda principle we find two distinctly separate parts forming one Unit. Looking at Kepler's formula there too we find two distinct sides forming the same unit. In the Coanda principal there is a space **a³** forming a basis to which a moving **T²k** attach. The **T²k** extends the space limit while the liquid provides the motion, which the space uses to become larger. The space **a³** is one side with the motion of the liquid **T²k** being the other side. As I shall indicate later on is that Kepler did not introduce a mathematical calculating measuring formula as Newton would have suggested, but it is a cosmic principle on which the entirety of the Universe is built. It is gravity by all measures there can be. It is does not mean that the one side is in precise measure the other side but it states a principle on which all aspects of the cosmos rests.

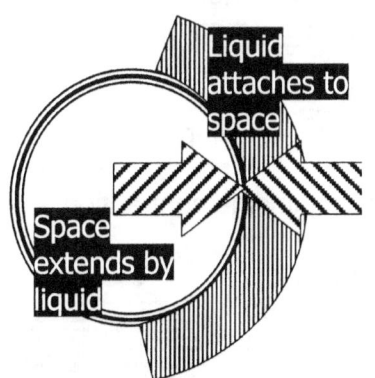

Liquid attaches to space

Space extends by liquid

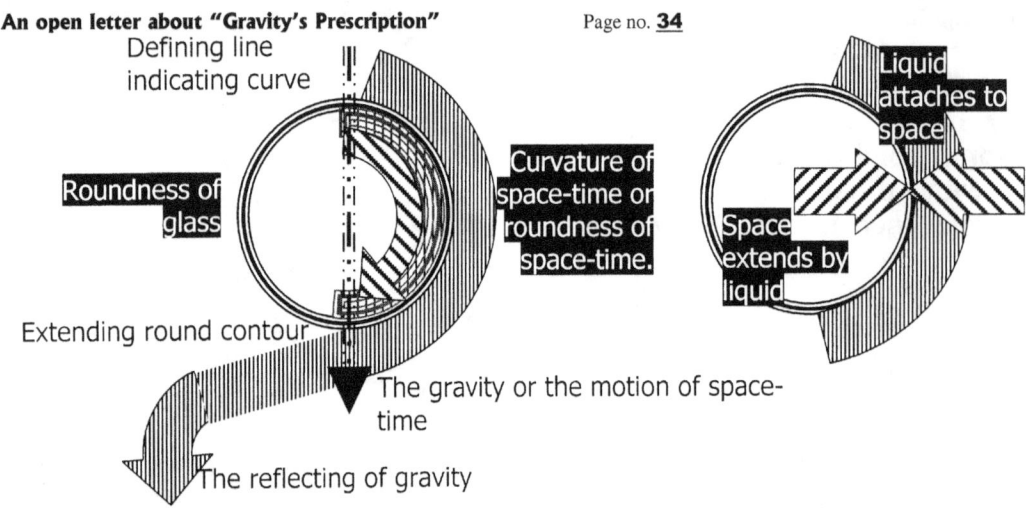

Defining line indicating curve

Roundness of glass

Curvature of space-time or roundness of space-time.

Liquid attaches to space

Space extends by liquid

Extending round contour

The gravity or the motion of space-time

The reflecting of gravity

$$k = a^3 / T^2$$

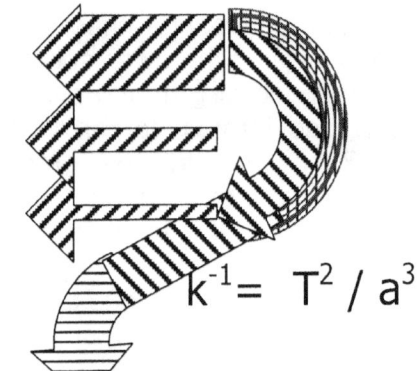

$$k^{-1} = T^2 / a^3$$

We find a line running through the middle of the round glass, which is dissecting the two parts in two directions. The line is $k^0 = a^3 / T^2 k$ where there is space a^3 on one side of the divide k^0 and a substance providing motion $T^2 k$ on the other side of the divide. On the one side there is an attraction $k^{-1} = T^2 / a^3$ where the liquid is controlling the space border and on the other side there is $k = a^3 / T^2$ where the space is identifying the position the liquid has as the liquid then becomes a part of the solid space.

Looking closely at the principle there are two principles where the one involve a contracting to the centre and the other form an expanding or a rejecting of the centre. There is one side that openly and exclusively

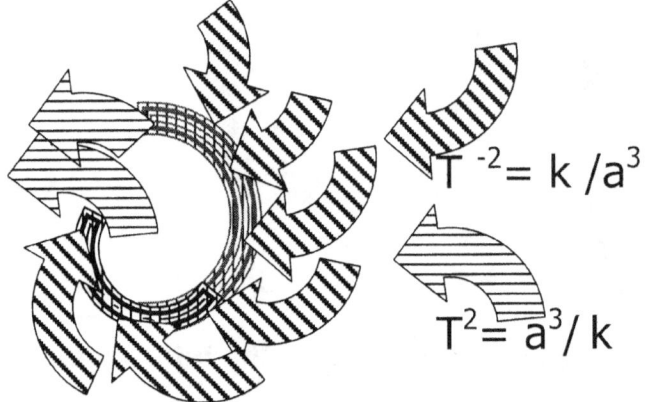

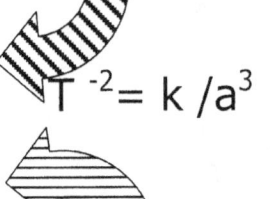

favours there is provides

In the two there balance interfere find a shift

$$T^{-2} = k / a^3$$

$$T^2 = a^3 / k$$

contracting while the other side that expansion the possibilities to be. contradictions normally is a fair but as life with physics we in favour of contraction as the motion part.

motion life provides throws the balance towards the

The contracting and expanding balance is totally in line the liquid that provides the balance. In normal gravity between contracting and expanding is divided in the line in place. As motion excels it favours the contracting to expanding.

However when saying this we have to realize the is in relation to the Earth providing the gravity conditions

with the motion of motion the balance that puts division the decline in the

symptoms we find and by excelling

the normal motion the conditions then would amplify the contraction.

The one part is forming a motion and the motion is attributed to the fact that the part is a liquid. By improving the motion the contraction benefits but that is because the motion that favours contraction turns the balance in the favour of motion. The higher the motion the more the motion will overburden the expanding and reduce the solid factor. With us

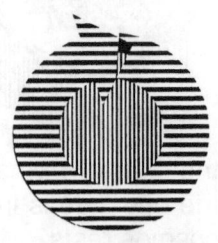

completely engulfed and in the control of the atmosphere we experience a complete contracting with no possible expanding.

The slightest inclination of the solid object also spinning that will favour the contraction and reduce the expanding part.
This is a fundamental part of the nature of motion as Kepler's formula introduces such a principle.
 It is the motion of the spinning that produces and substantiates the attracting or the expanding.

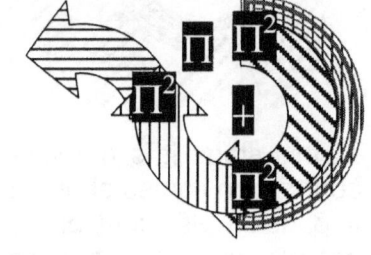

With only gravity working in the motion sector will reduce the contraction to only one sector while moving the solid as well as the liquid will bring altogether favour to the contracting part.

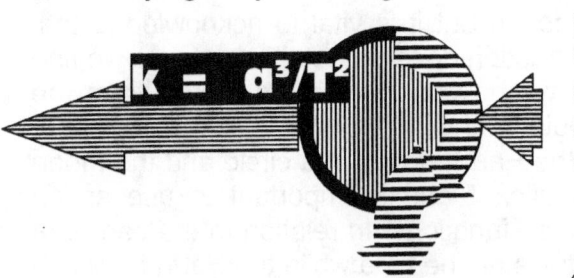

no force or else whatever there is It is directional motion of a circular govern the relevancy in strength as of inclining. There is an expanding to and the expanding is in direct relation to the motion of the liquid, which brings about the contraction

That is gravity and that is motion. There is then is a force. nature that well as direction the contracting

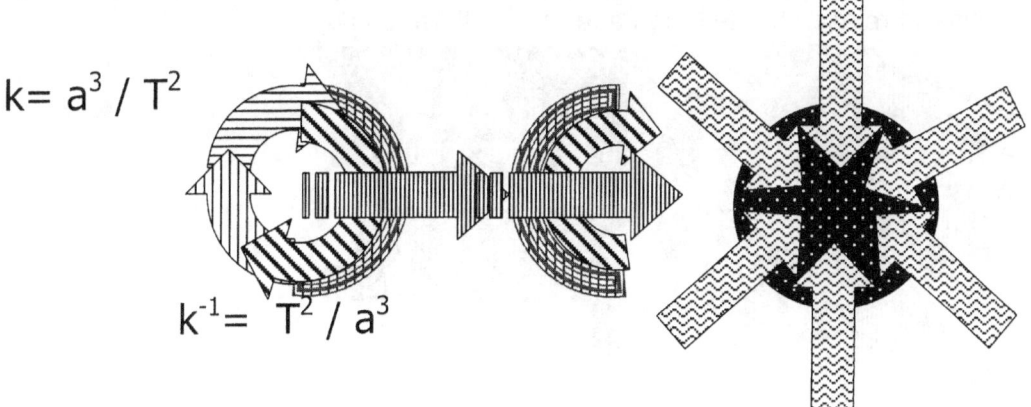

$$k = a^3 / T^2$$

$$k^{-1} = T^2 / a^3$$

Even the direction of motion needs not to change because it is the relevancy when the motion crosses the divide that sets the motion apart from what it previously favoured.

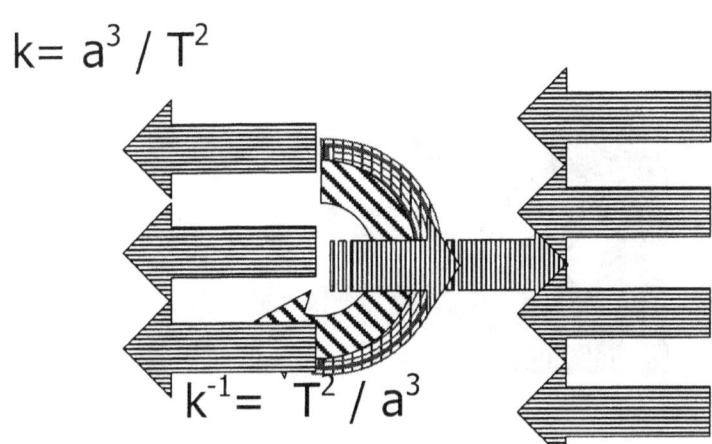

$$k = a^3 / T^2$$

$$k^{-1} = T^2 / a^3$$

It is always a liquid that brings on the motion when the liquid in motion is connected to a solid that gives the liquid stability and the liquid provides the space in question with a limited definition of securing a precise border to end the space. It is a liquid in defining a solid that produces the influence of the contraction in relation to the expanding and in that we find gravity. It has no bearing on mass what so ever. It is motion of a liquid relating to the solidness securing the position of a solid.

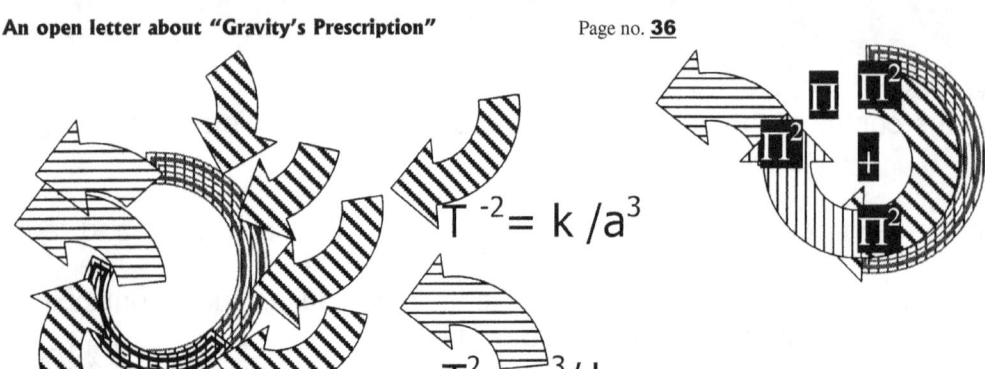

$$T^{-2} = k/a^3$$

$$T^2 = a^3/k$$

At this stage it is not important to study the direction of motion but it is vital to acknowledge that contraction comes about from one side where as where contraction does not totally dominate we find expansion also part of the equation. From Newton's time we have observed the gravity from the position we hold as trapped belongings within the Earth's liquid motion.

In the relevancy there is between the Moon and the Earth the Earth moves in a circle and the moon  moves in an orbit. This is very significant and is important to use as an example to understand how the cosmos functions. In relation to the centre of the Earth the Moon is changing positions all the time while the Earth is locked in one position that does not change. The Earth stands steady while the moon is orbiting around the Earth. The Earth is the solid while the Moon is a part of the liquid. Then from the vantage that the Moon centre singularity holds the Moon is standing still because the Moon is not rotating around a personal axis. The Earth however is rearranging its position constantly by providing motion, which is realigning with the Moon centre by every rotation motion in the minutes of moving. In this instance again the Moon is the solid while the Earth forms the liquid by securing a motion free centre in relation to the Moon and the Earth motion.

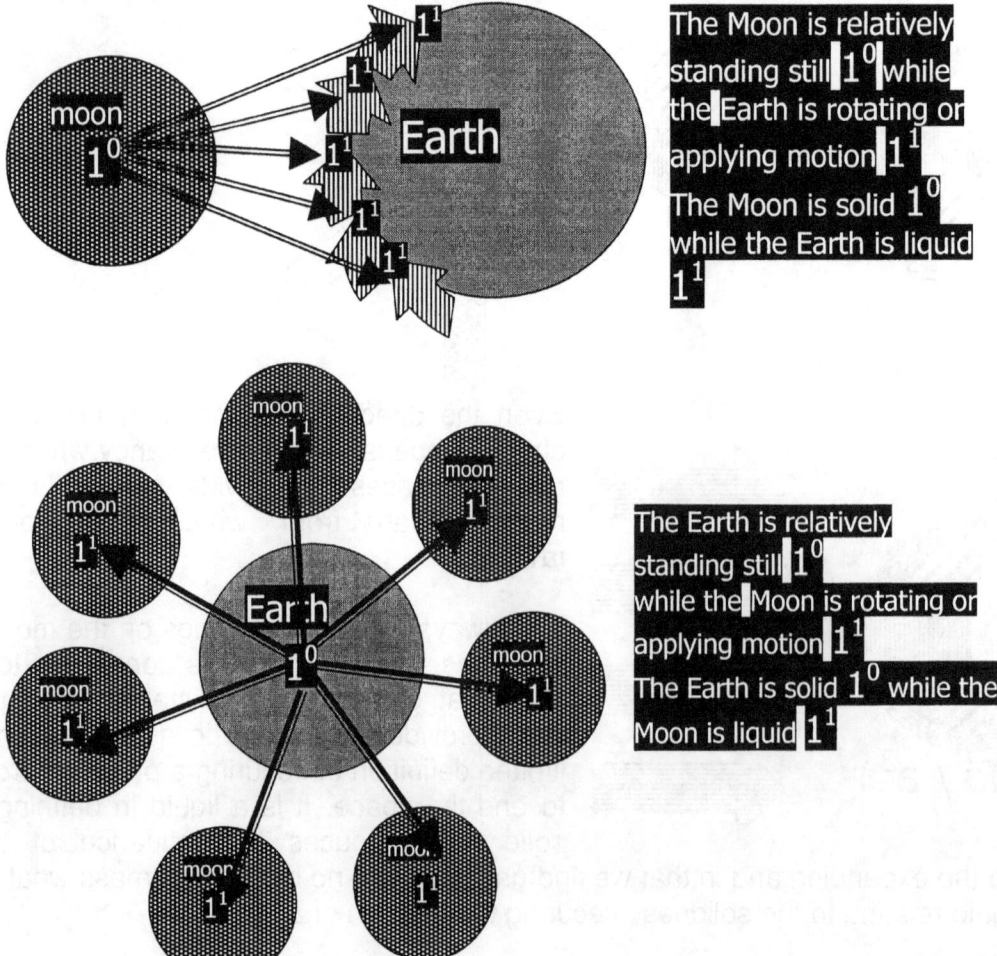

The Moon is relatively standing still 1^0 while the Earth is rotating or applying motion 1^1
The Moon is solid 1^0 while the Earth is liquid 1^1

The Earth is relatively standing still 1^0 while the Moon is rotating or applying motion 1^1
The Earth is solid 1^0 while the Moon is liquid 1^1

To outer space the Earth is liquid as the Earth is rotating about its axis is contracting outer space

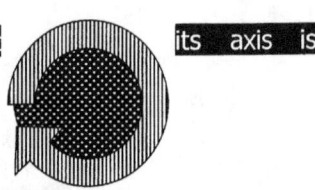

Being a liquid or a solid depends on what provide the anchor or pivotal role and what provide the motion within the relation. The planet moves in orbit as well as around its axis. This duel capacity is all motion while it is also all solidity. The cyclic rotation forms a liquid in relation to the point where the Earth meets outer space but since the Earth at that point is connecting to singularity by seven it is outer space that then carries the motion at the point. The contact point at the earth's end remains the same although it moves because it is solid, but the outer space are show a new point in relation every time although it is steady. To the Earth it is outer space that shifts while in fact it is the Earth that rotates but that rotation does not affect the centre since the centre remains directly aligned with the edge of the Earth solidity. However the Earth renew a contact point with outer space which is the motionless part in the relation at that point but serves as the changing factor in the relation.

To the Earth outer space is liquid as the Sun is contracting outer space

With the Earth rotating while the Earth considered its position as stable and solid the motion is reflected to outer space, which at that point is solid. Outer space, which is stable, is facing a new position in relation to the Earth but since it is not part of the solid structure of the Earth, outer space shows changes and diverts its position in ratio to the centre of the Earth. This ratio gives outer space the liquid partnership.

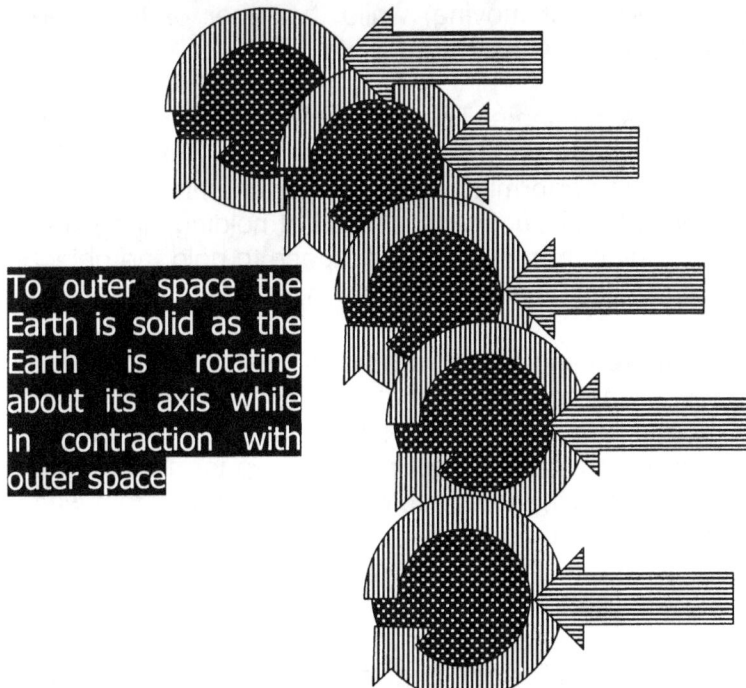

To outer space the Earth is solid as the Earth is rotating about its axis while in contraction with outer space

But then the Earth shows another side in the affair where the Earth in orbit tears through outer space being without motion. This brings about that outer space will allow the Earth to move and while the Earth is moving the Earth is aligning with the centre of the Sun . To outer space the Sun represents all that can be stable and in that regard it takes the Earth as another solid principle. In that way the Earth again proceed as the solid or stable factor while outer space holds the motion.

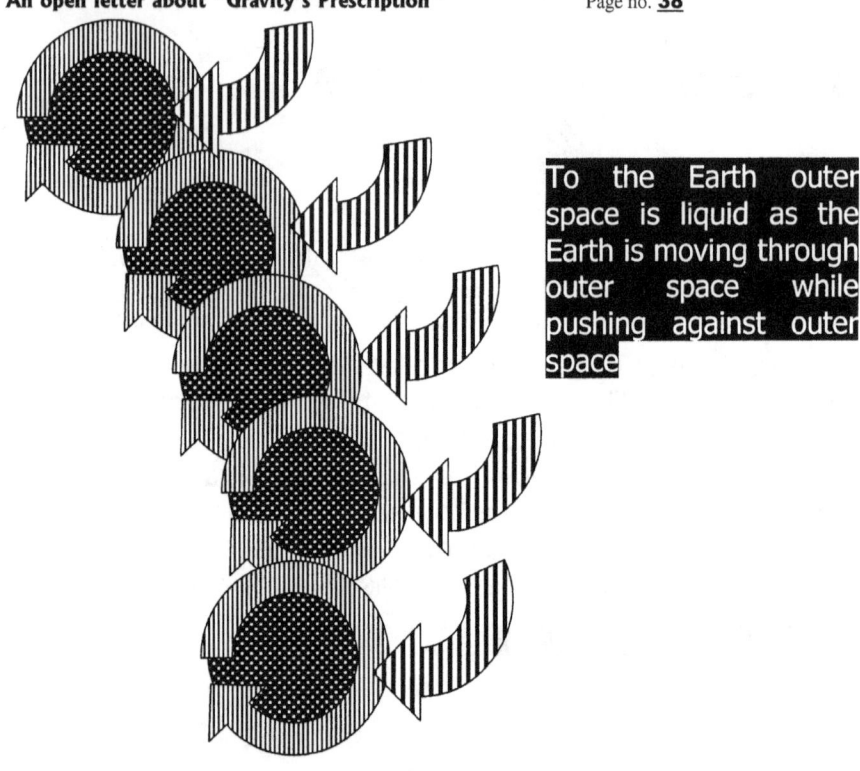

To the Earth outer space is liquid as the Earth is moving through outer space while pushing against outer space

That is not yet the end of the Coanda partnership. The Earth is pushing against the outer space and while outer space is inherently a liquid outer space give road to the moving Earth. Therefore while it is the Earth that pushes against outer space it is outer space that is moving away to allow the Earth the motion it insist to have. Therefore again outer space is moving in relation to the Earth which is pretending to be solid and the total result from all the activity is that outer space seems to move at a rate of 10 X 10 X gravity which is Π^2 and that gives space-time a value of space (a^3) / time (T^2) is (=) 298 but I shall come back to this when much more information is exchanged. The whole debate now in this part resonates around the fact that mass never comes into the argument and mass is no factor in the Universe. It is all about motion and being solid (not moving) while the other party in the equation is moving (being the liquid). It is as Kepler said $a^3 = T^2 k$ or then $k = a^3 / T^2$.

That is why in all cases the ratio of orbit in space-time is the same. The planet rotating the movement of outer space is in relation to the liquid position the Sun give outer space and from the vantage point that outer space holds the orbiting object is in space ratio performing as a solid partnership in relation to the motion it allows outer space to have as the liquid. The motion is a result of holding singularity steady as a solid while moving through outer space which then allows outer space to hold the object singularity as a solid reference point while taking on the liquid part of the relation.

My first nut I cracked in cosmology as an individual standing apart from what I was reading about cosmology through the avenue of mainstream science was concerning motion in relation to the speed of light. Today the fact that it took me a full six months to solve is a joke. That it took me so long to get to such a simple answer is in hindsight not very complimentary but please keep in mind that at the time I had a blank paper in front of me, which was blank in more than one way. However that was what set me on the way to be able to crack the first code. I must admit if I did not break the seal I would be totally lost in cosmology but still such a simple solution took me six months of head breaking arguments with myself.

Einstein said that if light were travelling at the speed of light for one year it would be away from the source of origin by a distance of one year. That is acceptable even to a person with my mental capacity. Then came the jawbreaker. The two photons travelling in opposing directions will also at that instant be at a distance of one year apart. It takes one light year to go in one direction while it takes the other photon one year to move in the other and opposite direction and yet the two is one year from the light source it left while also being one year apart from each other. That baffled me into almost madness. It was just way above what I could mentally cope with. The two photons opposed each other while travelling but at the same time moved apart only by one light year of total motion.

The total that should add to a double was the same as the single, which was the same as three totally different points. Something told me in this was the key to understanding cosmology.

Then one day the simplicity about the whole argument hit me between the eyes like a ton of bricks. I was staring at outer space while viewing outer space as a distance. Outer space is time and not distance in as much as forming space. It takes space a^3 time T^2 to bridge time k^1 and reach space a^3. The time it takes a^3 is in relation to move at time T^2 to cross time k^1. We all confuse space and time. Time is what is between the Moon and the Earth and not the distance. The time lapse was one year and therefore the time that parted the objects and the source of origin was one year but also the time that parted the two photons travelling merrily was one year.

Light travelled in as straight line as well as a half circle and where light is connecting to time the half circles 180^0 is equal to the straight lines 180^0 which is equal to the triangles 180^0 that means outer space has nothing to do with space but is all about time while time is all about motion. That is the key to solving the riddle we call cosmology and by using that key I found a way to unlock so many answers. The light flowed in a straight line, which is 180^0 while they move apart by a half circle also to the value of 180^0 and while being in a triangle position in relation to the point wherefrom they came which was also a value of 180^0.

Time was moving and if time was moving then time has to be liquid. Since time and space is not the same space then space has to be a solid holding time as a liquid in relevance and knowing that it is time that is between the Earth and the Moon it made that which is between the Earth and the Moon the liquid and that made the Earth and the Moon solids. How simple can everything be if one takes the correct line of arguing? Even I being who I am could start to understand what everyone should understand. I feel obligated to explain my referring to myself in the position I have. When saying this about myself I must request that you should never forget while reading that I am only a motor mechanic and that is all I can ever be.

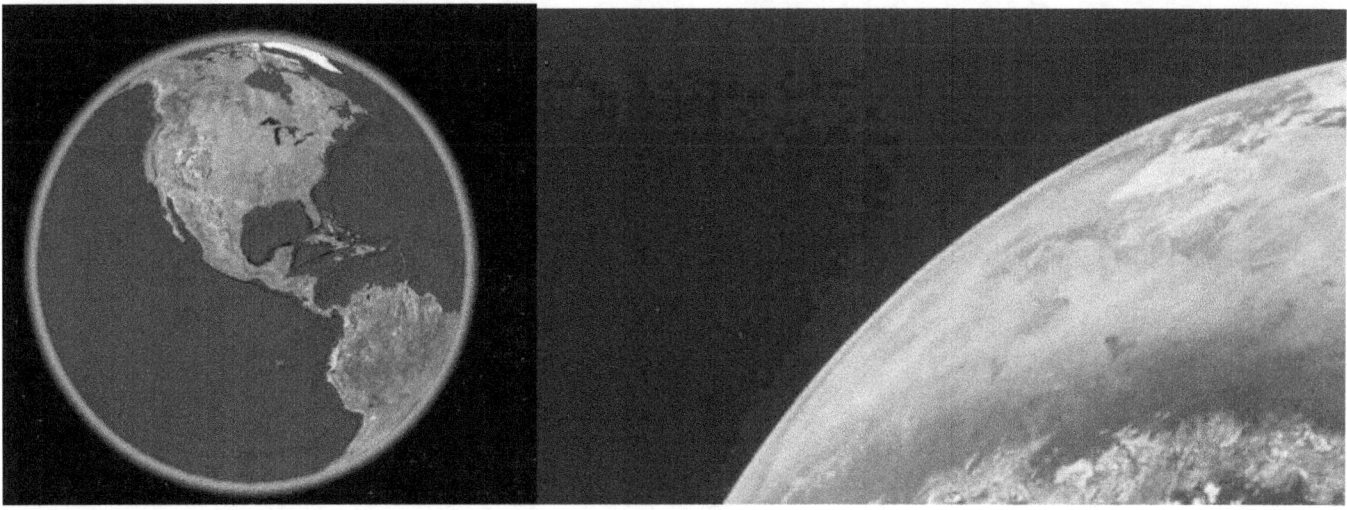

We are part of a thin top layer, the part that is contracted without really being part but only part as an extended part of the Earth. That is part of Newton's gravity where the liquid we are in and the liquid that we are becomes part of the Earth as it is confined to the Earth by the atmosphere of the Earth that forms the liquid restricting us to the Earth. That is why Newton and his apple were on the ground very much secured.

We are in the motion, which forms a liquid. We are part of the liquid because we are part of what moves. Every aspect surrounding us brings proof that we are contained in liquid and we are preserves as being part of the liquid. Where the wind blows, it indicate a wave pattern and a liquid leaves a wave pattern When looking at a mirage we distinctly see that where the atmosphere becomes more dense as the heat at that point becomes more intense, that what we see and which we call a mirage is water, is a liquid substance floating in waves where the concentration in density varies.

The liquid engulfs us and in being part of the liquid we are experiencing only the contracting aspect since we are totally secured by and in the motion the Earth provides. The earth atmosphere puts us on the ground as the atmosphere secure our positions on Earth. We are part of the $k^{-1} = T^2 / a^3$ while the earth being in motion and providing the motion forms $k = a^3 / T^2$. That is the time aspect, which the Earth provides and that time aspect enable us to read the time by the measure of Galileo's pendulum.

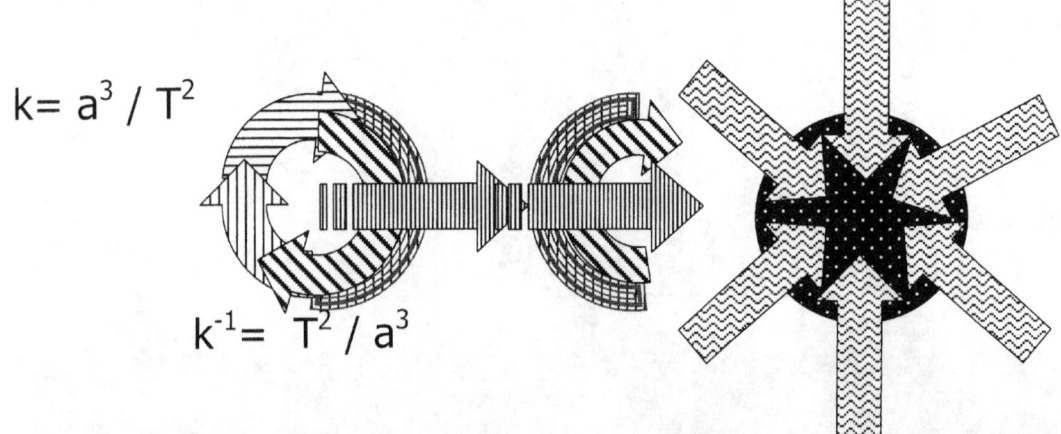

$$k = a^3 / T^2$$

$$k^{-1} = T^2 / a^3$$

By the motion in rotation as well as the motion in the lateral we find that the liquid being air confines us. It is not the particles in the air that is the liquid but the substance separating the particles in the sir that form the liquid. By moving around the axis while the axis is moving around the Sun forms motion, which confines us to a position that only Newtonian gravity, apply. There is only one direction flowing towards the centre of the Earth. That flow is the space a^3 that the liquid T^2 secures k to the space $k = a^3 / T^2$. In this however there are no grounds that support the suggestion that mass is producing the motion or that mass is indicating the flow.

By observing a fire we can see where and how the flames bring intensity to the heat in the air. The flames are the densest form that heat can have while being in a liquid form. The flames are so dense a liquid it provides light and light is pure heat in minute quantities of liquid space. In the picture we see three forms of liquid heat where in each case the element responsible for producing the liquid heat contains the heat in a different form. We see the flames souring and that is the responsibility of the nitrogen where the nitrogen being $_7$ expands heat into space. Then there is the oxygen that contains the heat in a dense material we think of as smoke. The next we can see is the wood or carbon $_6$ that keeps the heat contained in the particles. Every layer in a star has this duty that the substance hold in providing and managing the heat within the stars structure. Taking the idea of air (not particles in the air but atmospheric substance containing the particles in the air) to further proof of finding a liquid we again have to go to the Coanda principal.

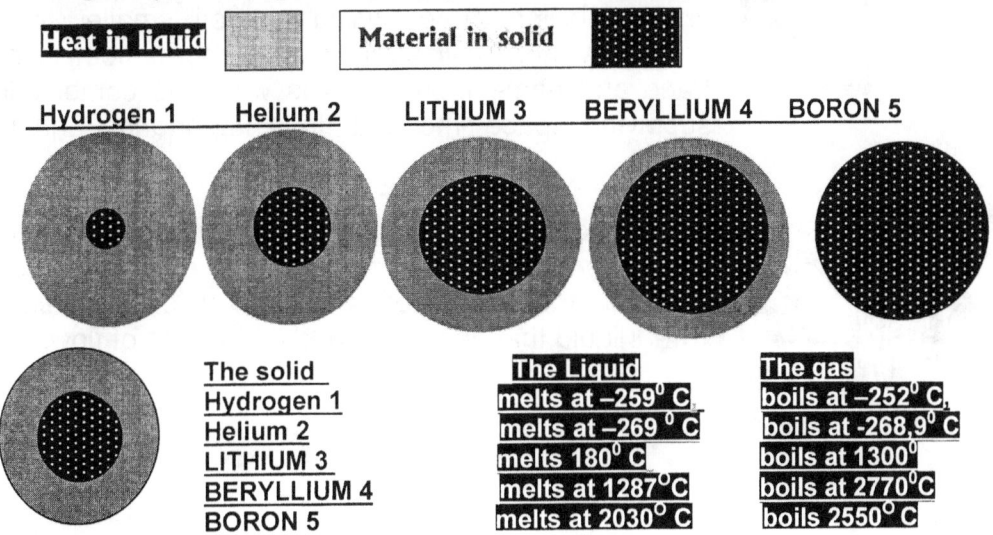

Heat in liquid		Material in solid	

Hydrogen 1	Helium 2	LITHIUM 3	BERYLLIUM 4	BORON 5

The solid	The Liquid	The gas
Hydrogen 1	melts at –259° C	boils at –252° C
Helium 2	melts at –269 ° C	boils at -268,9° C
LITHIUM 3	melts 180° C	boils at 1300°
BERYLLIUM 4	melts at 1287°C	boils at 2770°C
BORON 5	melts at 2030° C	boils 2550° C

The particle (atom) secures the solid basis that provides the motion which enable us to categoriser elements according to our perception. It is not the truth or cosmic reality but it is our perception through culture that teach us about gasses, liquids and solids, about noble gasses and heavy metals and non ferromagnetic or good conducting and immeasurable other characteristics we attach to elements except what truly is important as far as cosmology goes. All material are solids as much as they are gasses or liquids. It depend on the concentration of the heat surrounding the element in that particular element that turns the element at that point and time (temperature) to be either a solid, a liquid or a gas. The state that the element is in is a response to the conditions, which the heat levels bring on, and that is a response to the Coanda gravity motion that serves the atom as a liquid at the moment of response.

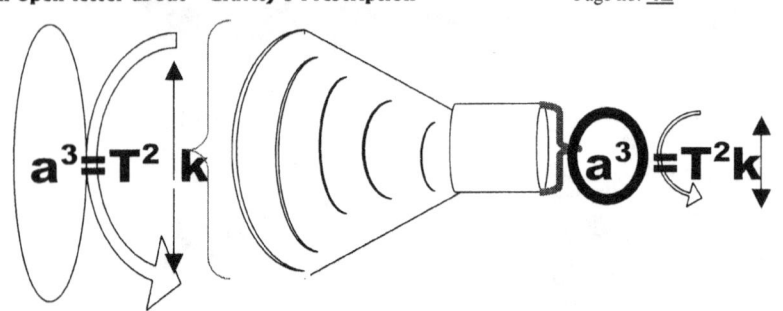

However when saying that it is a liquid in motion around a solid we also find that the principle drives turbine engines and the air compresses within the turbine to cause heat. It is the air that flows when driven by the turbine rotor, which provide the sold. Then from that we must deduct that the air producing the flames forms the liquid, which provide the contraction when the solid spins the contracting turbine and the spinning turbine serves as the solid part.

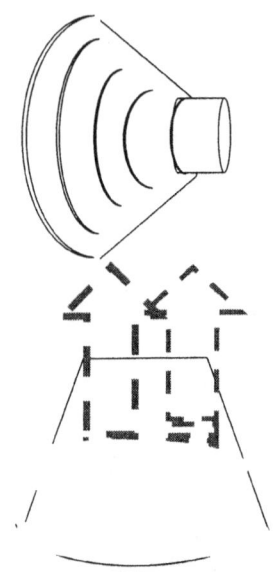

The Coanda principle works in two parts where there has to be a solid securing space and there has to be a liquid performing the motion that result in contraction. The ingredient is about a liquid in motion T^2 capping a limit or an end k or k^1 where the space $k = a^3/T^2$ holding the solid extends to appreciate space $k^{-1} = T^2/a^3$ that holds the liquid. It reduces space by motion providing contraction to a point where the space goes liquid in finding flames. That is what Henri Coanda first saw when he tested his new enclosed propeller. It is all so exciting but what the Newtonians of the day and those today completely misses is that for the Coanda effect to be functional one needs a solid and a liquid. The solid part the rotary propeller provides and with such motion it contracts the atmosphere into a compressed flaming liquid. The atmosphere (not the particles in the atmosphere) compresses to become a liquid.

That is the conditions applying to the atom, which provides the conditions applying in the atom. We differentiate the particles by giving every part a name and try to find meaning in the particle combination. "discover" another now there waits discovered. We will go and still we would not provides each other in naming the lot individually.

In a hundred years from now we are about to million smaller particles and in a thousand years from another billion more, which are all smaller to be on discovering until eternity once more meet infinity trace and name them all. It is what the combination supplying space-time that finds importance and not

The major function that the combination of the electron provides is the expanding and contracting of solids in relation to time being a liquid that has a flow and a direction of flow.

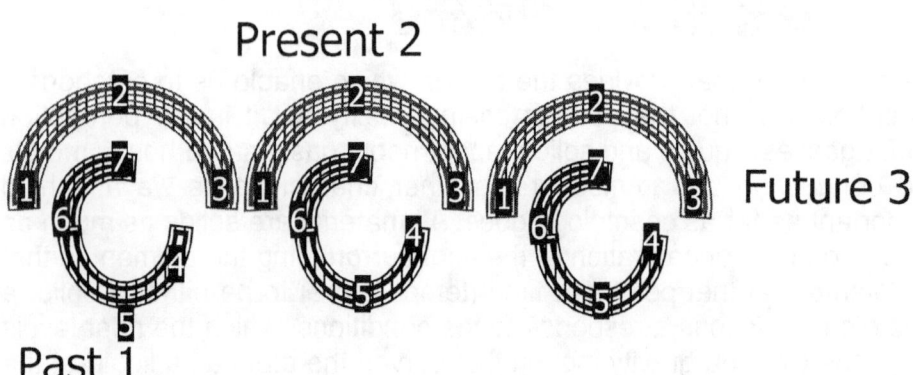

Present 2

Future 3

Past 1

It then becomes pertinent to find what is liquid and what is a solid in cosmology. The first aspect we have to abandon is our preconceived notion about what liquids are and what solids are. Everything that moves or forms part of that which moves or may move while another in relation is standing still becomes that which is a liquid while the part forming the motionless factor then becomes the solid. From that gravity becomes reality and gravity is motion that has no bearing on mass in any way possible. It is the restriction of mass that prevents gravity from becoming a reality and because the Earth restricts our motion as we find motion that the Earth prescribes we have mass and not gravity. An object may have some part vested in mass while having another part acting in gravity but mass does not bring about gravity. In gravity there has to be motion of space in time and when time ends space collapses. In this letter I am about to introduce you to gravity being the motion around a solid that provide the contraction.

Gravity has two parts where the one part is expanding by motion and the other part is controlling the expanding by contracting the motion. It is $k^{-1} = T^2 / a^3$ and it is $k = a^3 / T^2$ but above all to find gravity we must find the centre of the Universe being singularity at $k^0 = a^3 / T^2 k$

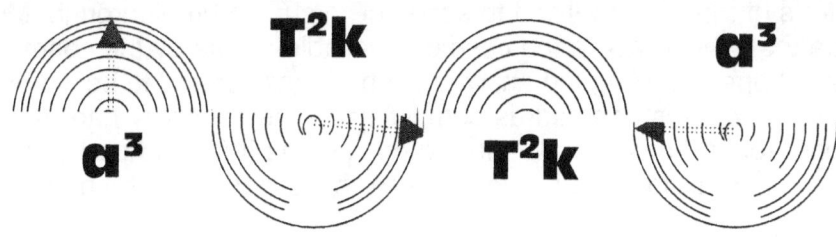

The liquid in motion provides the gravity and by initiating the motion the gravity contracts the solid as much as the solid is extended by the motion, which the liquid provides and the liquid is the gravity or the motion of the solid where the solid provide

restriction or mass to the liquid.

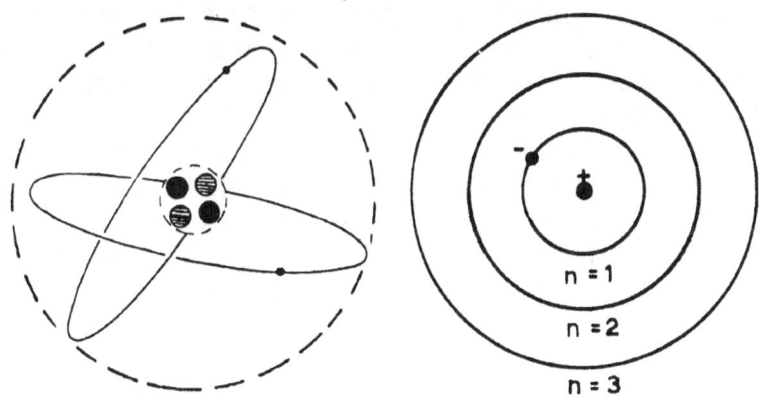

It is in this manner that the entire Universe operates because from the spinning atom to the most prolific galactica and to the other end of the spectrum where Back Holes destroy time the entirety out there works in this manner.

I first started my studies in the field of Cosmology as a spontaneous development of my natural curiosity spawned from childhood interests in the field of cosmology, which I developed even before I went to school. The studies were a reaction (I would imagine) that was part of my personal childhood development in how I was forming a personal concept of a lifelong interest that followed me into my future. At first I conducted all my earlier studying mostly on the basis that inspired me to find out more about what made the Universe tick, with no intention ever on my part to reach a point where I would be writing books on the subject. At first I was investigating cosmology on a part time basis. This went on, on and off, or the best part of twenty odd years (*as* time and *when* time would permit).

Then in later life with my health deteriorating I committed myself to more intense investigation and my effort developed onto involving a study using time that is only permitted by a person when that person is involved in such a quest on a full time basis. That quest has now been going on for the last seven years in full devotion and if one includes all the years invested on my part including the twenty odd years before, part time, then the time I have spent in completing my theory when adding all in comes down to almost twenty eight years. This is to say that I did not come to realise what I am about to introduce on a light-hearted conclusion. I mention this because I wish to ensure the reader that he should have no doubt about my most sincere commitment in producing a cosmic theory on matters concerning the start and the working of the Universe during and before the Planck era. At first I began by arguing that there is something that is blocking our progress.

There is some barrier preventing humans passing a threshold whereby our understanding will pass such an obstacle. If there were any way that any one may break through that barrier there is that is preventing normal research to go pre-Big Band, it would be accomplished by finding the barrier whereby then our vision we use to focus would pass such a limit. If we wished on progress in our pursuit of the very first cosmic moment then we have to find and cross the barrier that blocks our view. We have to look deeper and in another direction should the desire driving us be strong enough to commit us to reach into the very birth of the cosmos. We have to rethink the strategy that we use. Max Planck was one of the most brilliant men of all times and even he, notwithstanding all his personal brilliance, accomplished little.

There are parts missing in what we have and that which we have at our disposal to use, because if there was no such an obvious barrier then the Wise-Men involved in science would by now have found the way to break through the seal that is locking us out of the critical past which will uncover the origin of the Universe's infancy stage. I went about trying to find what everyone since Adam, (meaning all of the rest of mankind and myself) were missing throughout the ages of speculating and interpreting while philosophising about whatever we find inspirational. The obvious we saw; that was clear. Therefore I had to find a route that would lead into the not so obvious that all of us were missing, notwithstanding the best efforts of the best qualified to accomplish such a breakthrough. My effort involved trying to accommodate that which was in the cosmos available to use by the cosmos in all phases of developing. If I had any hope of finding the answer, such an answer had to be simple because I am not very inclined to unravel what is deemed as complicated. The simplicity had to be locked in what was not yet understood about that which was in the cosmos as it formed part of the process used in forming the cosmos. My realising this brought me to focus not on that which we understand.

There is not a lot we actually understand because even gravity is very poorly understood. In fact gravity is so poorly understood that there is not one person alive that can claim the prestige of understanding gravity and among the dead there is even less that can make such a claim. There are several phenomena that are presented in nature and acknowledged by science but also discounted by science and therefore not presented as accepted science. By admitting that that what we have available to us to use concerning our research of cosmology in an attempt to better our understanding of cosmology, is useless to use, then one realises that not having what there might be makes what we already have useless. It then is useless to use what there is as part of the big picture we are trying to paint because what we use is not really part of the picture. This leads one to believe that the picture of the cosmos Mainstream science is painting, is being painted without painting a full picture.

In my first attempt to understand the full picture of what science was painting I found so many colours missing there was no picture painted that anyone could appreciate. This is what made me decide to go on researching the 'unknown' in the hope it might clarify the 'known' and as the book unfolds. You as the reader may agree that I was correct in pursuing the misunderstood and rejected phenomena. Finding the missing phenomena helped me to place the phenomena mentioned above in a theory where the principles also mentioned above form a part of the overall gravity used in binding the Universe. I believe what is in the Universe is not able to be coincidental because of too many influences contributing to what there is - notwithstanding the fact that that is the manner which science uses when they refer to the Bode law. What is in the Universe has a role as it had a role, which is the same role that phenomena has had and in future will have. This is establishing a very new idea about the working relationship between particles and in explaining it by using Kepler's studies. Redefining the work of Kepler's views brings a new Universe to light involving new concepts that are based on old principles but principles in updating man's view about cosmology are very new in that capacity. Through that new vision I was able to come to realise what the reasons might be why Kepler never saw it fitting to include the measure of Π in his formula. I do not suggest his neglect thereof was intentional, nevertheless the formula he devised without using Π proved that there was no need for the inclusion of Π since his figures brought about a correct answer in the final end result leaving a well concluded fitting answer. The numbers he produced brought about a specific space a^3 contained in a circle T^2 at the distance of **k** from a defining centre thus the calculations did not require the use of Π to find a meaning. In that Kepler did not see a need to include Π. I would not go as far as declaring with absolute certainty on his behalf that he did it deliberately, however there never arrived such a necessity. It is prudent to agree on whether or not such a need is necessary, because if one is agreeing about such changing not being required a new Universe emerges. The circle that Kepler discovered came about without ever forcing Π into the frame because it is clear that the circle formation came about as a natural consequence and came spontaneously delivering an equation while he was working. In this book I prove that the reason for adding Π to the rest of Kepler's formula is unnecessary. This unnecessary addition is because when going one step further in the investigation one will find that **k** and **a** and **T** are symbolising the same value with the only difference being that each one represents a different dimension to our six dimensional or six sided Universe we enjoy.

In fact I shall show that Π replaces **"a"** and **"k"** and **"T"** and that Π is the true value that should be replacing each factor as to indicate the correct value to the sides nominating Π. We humans work on a numerical base using ten as a basis where we count to nine and re-establish a new decimal numbering line by adding a nought behind the number in value. This is using the numerical basis of ten, which I suspect we took from ancient knowledge about cosmology and not from using our fingers and toes as the earliest calculating processors. In this letter there is unfortunately no room to explain my suspicion but another fact I do prove is that the cosmos uses Π in the cosmic numerical basis as a means to measure and quantify. Therefore in fact the Kepler formula should read instead of $a^3 = T^2 k$ as it does it must be $\Pi^3 = \Pi^2\,\Pi$ where I shall show that Π represents singularity wherefrom the entire Universe sprang from Π and by forming as $\Pi^3 = \Pi^2\,\Pi$ it is confirming that space is equal to the motion thereof. Kepler's greatest achievement was showing that the cosmos is space –time $a^3 = T^2\,k$ while time is the motion of space in space. The value of Π is the primeval and most basic of measures applying as an accepted cosmic legal value that the cosmos used exclusively in the very beginning and as it does today. The measure of Π in the Universe, values particle development that brought about all development ever conducted in the Universe. Only after this stage did the rest come including mathematics and went on to freeze spilled singularity into frozen material. Reading this statement may sound suspiciously senseless but as the book unfolds the sensibility will become apparent.

The full implication of such a statement will become clear when one dissects different facts coming from studying Kepler. My discovery of this fundamental basis of legal valuing ensured me again that there was no need for someone the likes of Newton to add Π in any form to the work of Kepler because Kepler discovered the ultimate Π in the Universe, the Π giving the Universe form and gravity. The concept of Π that is the only single form of all other forms available that can by duplication of Πs assemble the value of gravity. When replacing the symbols with Π the facts of the Universe become self-explanatory because the most basic form that forms the cosmos has a definitive and uncompromising value.

But getting this far took me down roads overgrown by ignorance and which I had to uncover myself as if hacking away miles of overgrowth with a machete chopper. All of the disbelief science showed to my work in the past and their refusal to see past Newton made any and all attempts on my part as bad as they could be, strangling and smothering my attempts to announce my uncovering of the newly found insight on my part.

For decades I tried to come to terms with the inability there is in science to explain the cosmos in real terms, when using the science of official reputation. That which there is makes a mockery of science because the undisputable clues left in the cosmos makes what little correct explaining there is available, seem like a comedy of errors, when it is mixed in with all the other near Dark Age errors we still use after so many centuries that provided countless opportunities to revise the old muck. By applying current accepted Astronomy as such the phenomenon found all over the cosmos is still beyond the explaining ability of Mainstream science. This is true and it is a shame because it also is an undeniable fact in spite of the vast knowledge and progress in other forms of science taken in the manner science uses when it approaches cosmology. Cosmology truly lagged behind while the understanding and advancing of physics, mathematics and chemistry as subjects were flourishing. By comparison I saw how little there was available in explaining cosmic phenomenon and how much improvement in understanding the other departments such as chemistry, electronics, medicine etc. could offer as results were coming about from research. Even where there is a little explaining available in cosmology it turns out that such explaining is confusing to say the least and at best it highlights the manner in which science is applying double standards. For decades photographs were the only progress forthcoming as an addition to improve the meagre field in cosmology and that improvement was artificially stimulating cosmology. By providing a false impression of advancement, everyone missed what and how much was missing…To the connoisseur desperately looking for more than the obvious stirred in with some out-dated misinformation dating back to the Middle Ages, it all seemed as if it was a picture portraying the ridiculous to make the sublime look good. The pictures only proved the opposite of what progress in cosmology will represent. In truth and as such in cosmology the cover up that was hiding the lack of progress about the science of true cosmology was only forthcoming in the improving of electronic optical telescopic advances and spectroscopic progress. There were only photographs carrying beautiful pictures which pleased the less informed except the photographs did not bring progress to cosmology at any intellectual level by promoting insight.   The explaining that the photos demanded about the subject had the opposite effect of installing hope because what it did do was underline what lack in any notable progress there truly is in our understanding of cosmology and laws in the cosmos.

While such Hubble telescopic images might seem to be as clear as daylight it was more than clear there was little academic value to them. To the person in need of more stimulation than being impressed with pictures of God's marvellous Creation and the sightseeing that always accompanies such pictures, such persons always felt very disappointed. The pictures did give satisfaction to those more easily impressed, but the rest of us seeking knowledge accompanied by understanding the images left us despondent. Although they leave the vast majority in total amazement there are those less impressed about not knowing the 'why' and the 'how' in such amazing pictures. I know the group I fall into may be the greater minority and the majority may only demand the portraying of the images, which is what that easily satisfied group demand. The rest of us rouse with anguish at the lack of information about what is known and what lies behind what those pretty pictures are conveying. Nevertheless there can be no real progress in scientific understanding about the images portrayed by the Hubble telescope, and others, if no one is able to show the slightest clue of a deeper understanding of what is going on in the Universe. Everyone is almost breathless waiting the commentating by the most informed which accompanies the magnificent cosmic portraying of God's Creation. When we are portraying the new images, we should also be investigating that what we see that the cosmos is at the moment portraying. The lack of actual believable explanation coming from investigating by means of telescopic imaging should impress one and all, but the impressing must not be based on the colours in the images but the sensible information attached to the image investigated. It is **that** that we wish to see. What we wish to see must at least be accompanied by scientifically backed information, which provides the proven understanding coming from science. When science is employing new explanations with such photos it should also be discarding senseless baggage carried over from the past. Most images contradicted Newton and for saying that, every Academic I ever came across in the past ostracized me. That bothers me little! I know I cannot possibly be the only person absolutely discontented with what Mainstream science accepts as science. Here I refer to the out of date theorising Mainstream science still accepts amongst many others as how they suggest stars and planets are forming. One cannot promote cosmology in honesty and advocate scientific fact whilst dishing up such fairy-tale nonsense to students. Moreover I hold the opinion that amongst Academics in particular there must be many if not most that share my personal serious doubts or have an inclination to share some of them. This I say when considering the overall doubtful picture painted about what there is and what one believes there should be. I just cannot believe those forming the most intellectual group of mankind are unaware of the mismatching facts seen over the broader picture because the contradiction and lack of a plan, makes what there is so very doubt provoking. Newton dismissed the formula Kepler presented as all factors forming motion. That is where the apple cart derailed.

In honesty we have to realise that we cannot dismiss the whole formula that Kepler produced as being motion. It is so much more than just motion. It is $a^3 = T^2 k$:

That is what Kepler brought into civilization for all time to come. He saw space a^3 being in isolation due to the time it uses to move T^2 claiming such space forming independence according to the lines k indicate.

Let us look at the factors in more detail before we proceed with the rest of the book.

a^3 symbolises a mathematical interpretation of implicating the three-dimensional space.

T^2 is representing the period or time that Kepler suggested we should use to calculate time that holds the orbiting planet in direct contact with the space in relation to a very specific centre.

k is the space taken from the centre to the end of the line from which the planets must have grown if one accepts the Big Bang growth of particles and the affect of the Hubble constant on all cosmos material. The specific value about the centre is most important because from the specific centre gravity always applies the strongest influence.

One cannot justify Newton's dismissing of Kepler's formula as that all factors only contribute to the motion indicated because that is misleading. We all accept that the true cosmic form *would be* and most probably *is* a sphere. Everyone accepts the Universe as a whole as a sphere…but why would the sphere form? What would be the reason why the original form that we devote to the Universe would take on a sphere as a natural form? Apparently our imagination grabs the sphere as form. In all natural events the gravity in that space which stands apart and independent from all other space takes on by cosmic pre-casting the sphere as form of shape … **it is because gravity chooses the smallest space to hold the strongest force**.

I am of the opinion that gravity is about dismissing space to the advance of heat increasing in such a specific and concentrated space using the concentration as measure for the heat as well as the space holding the heat in space. According to Kepler that is what he found to be true. Space a^3 will always be circling space around as T^2 in any position from the centre k. That is what Kepler said when he said $a^3 = T^2 k$. Kepler indicated space a^3 will forever fight for independence and show separate individuality in remaining apart as identifiable cosmic components by means of motion. Every space will cling to independence indicated by k through fighting off the integrating of another coverall unifying unit by applying the motion of T^2! The problem we have to solve in this letter is what will the cosmos use to secure such independence between all particles? What sets space apart from the rest of space? First we have to admit that Kepler was the one that introduced the following.

Kepler gave us the answer to the following but no one ever took notice!

Kepler was the one that discovered **space / time** as $k = a^3/T^2$

Kepler was the one that discovered **singularity** as $k^0 = a^3/T^2 k$

Kepler was the one that discovered **gravity** is holding **space-time** relative by the measure of distancing k as $k = a^3/T^2$ and $k^{-1} = T^2/a^3$

Everyone able to read mathematics has to realise that Newton suggested collisions between cosmic structures must eventually come about as gravity erodes the distance separating the cosmic structures multiplied by the product of the mass of both structures from both ends. Newton said the multiplying mass of both structures destroys the distance between the structures by using the eroding force of gravity in the square. The cosmos then must end in a Big Crunch with all material joining together but that joining is not forthcoming at all…and that only indicates how much insufficient understanding there is on offer in cosmology by the educated–to-be-wise-about-these-matters. There is precious little available to explain about their field of cosmology amongst the ranks of Astronomers. So…let's us return to the beginning of cosmology before every one became oh so wise and see what there is to see. Let us see how the humble bicycle can teach the so wise about what gravity is because it is easy to demote the prominence of riding a bicycle when it can be so easily explained as just being a balance.

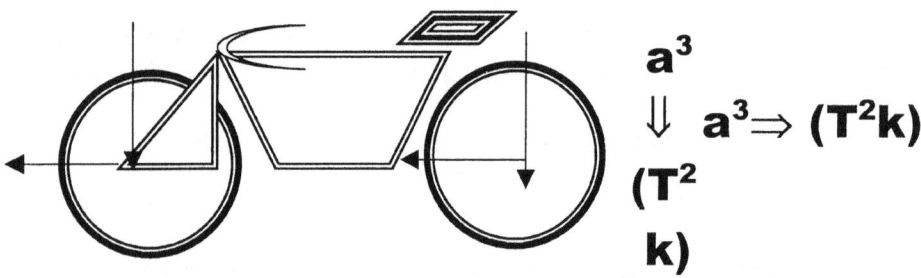

$$a^3$$
$$\Downarrow \quad a^3 \Rightarrow (T^2k)$$
$$(T^2$$
$$k)$$

A person that acquired the skills of peddling while staying upright on the bicycle has achieved the method of rearranging gravity within singularity. Without motion the bicycle falls on the spot it holds. When the bicycle is put in motion the bicycle can maintain the upright stance as long as the motion applies. When the motion stops the bicycle drops. To introduce motion to the bicycle the motion brings about a stable unsupported upright stance where balance can result from the motion the Earth enforces to the balance coming about by the bicycle using independence gained from motion of the space holding the bicycle. The space that the stationary bicycle holds is the direct result of the Earth providing the motion.

That is why it then will adhere to the gravity or motion that the Earth will enforce. The motion restricts the static bicycle to one allocated position that the Earth supplies to the bicycle. When the bicycle starts to move the bicycle gains a cosmic independence. The gravity effecting the redirecting of the Earth gravity response comes about as the result of additional motion that is introduced to the bicycle, This is the very same process that the aircraft need to get air born because it replaces or repositions the singularity the Earth holds to the singularity the bicycle develop in motion. The aircraft only takes the change in direction of what the gravity is insisting on through changing direction in motion through phase one and into phase two. It all is still part of the Coanda effect. With more motion contributing to acceleration the bicycle will become airborne on condition that it is also given the advantage of a set of wings to increase the effect of creating space-time to the advantage of the motion requiring the change in singularity direction.

I specifically chose to use a bicycle in my explaining because the bicycle is the object that relies the most on singularity achieving the required balance in which to operate. It is singularity, which puts space in balance of the time the space uses to duplicate. The singularity create space-time and such space-time results in a balance of space and time $a^3 = T^2k$. It is through the Coanda effect that marries the motion to the space that gives the balance that keeps the bicycle up right while it is singularity that allows the bicycle to move or duplicate the material by relocation through time. It is an act of balancing singularity that gets the bicycle as a machine working properly. It is also the next best thing to illustrate how singularity by motion provides gravity in addition to that which the Earth already produces. In the Coanda principle there are two factors where one is motion and the other is space and the two provide both duplication as well as contraction of space-time.

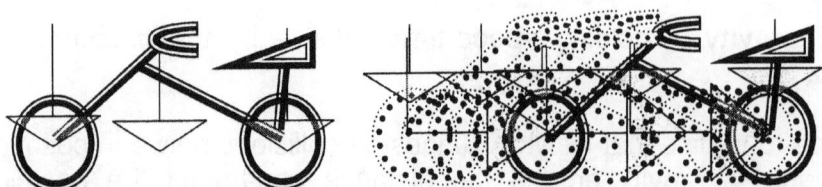

In the normally applying of gravity, we find contracting lines running vertically as the lines connect with the Earth centre. There are numerable lines running from the outer regions to the centre in diverting by 7^0. Motion provides extending of the 7^0 establishing the centre connecting points to the Earth to which it connects. When it is only the Earth providing the motion there is only one spot of space allocated to the bicycle in the time frame applying at the time. The Earth takes on a specific size by which it duplicates the space it holds in relation to what it renders the bicycle also sharing the space, which the Earth has. By that standard the bicycle also holds an exact relation of space and volumetric size related to its position within the Earth. **This is where motion through duplicating**

changes the dimension in equation. Motion is the duplicating of existing space from time in the past through time in the present towards in the future to time.

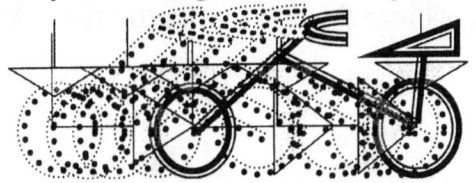

The instant the bicycle starts to move independently it increases its share of space it holds within the Earth. By duplicating the space allocated to the bicycle at a higher premium than the Earth does with the motion the Earth provides, the space the bicycle charge increases in ratio to that which the Earth charge because the bicycle maintain the duplication that the Earth grant but then still add to that space by enforcing more space provided by the motion addition. The motion of the bicycle not only extends the vertical connecting lines and not only changes the direction of the vertical connecting lines, but does both. The value added and the change in direction contributed is what brings about flying and moreover is the cause of the sound barrier.

The Earth takes the position that was before the motion of the independent object came about previously and held by the Sun . By establishing the directional motion in accordance with the k^0, which the Earth then provides instead of as previously provided by the Sun , relevancies replace previous ones.   In normally applying gravity, we find contracting lines running vertically as the lines connect with the Earth centre, which is the gravity we confuse with mass. It is a state of contraction and is the result of space being confirmed in relation to the motion of liquid time. Motion provides extending of the 7^0 establishing the centre connecting points to the Earth to which it connects.

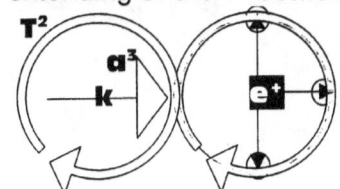

With The bicycle moving the bicycle change associations and where **k** was the factor securing the bicycle as part of the Earth that factor **k** then goes 90^0 in relation as the bicycle in motion then is in association with motion putting it then in terms of what forms a part of the liquid.

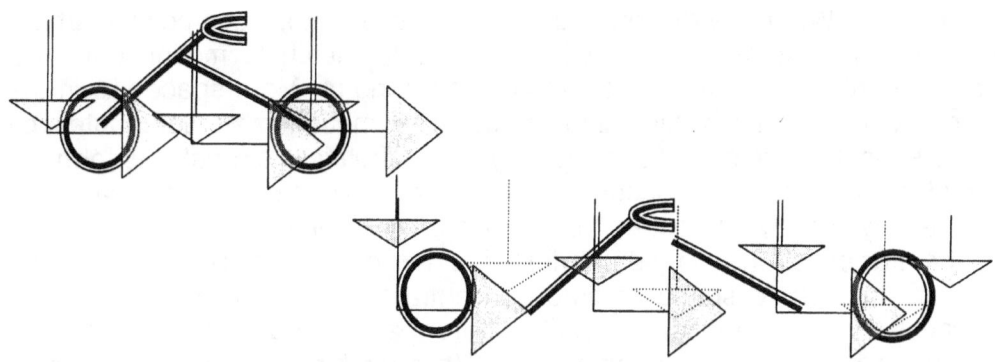

The motion provides the bicycle that is already confined to share space with the Earth its due in motion. The bicycle can only have independent motion if and when the bicycle has the correct number of atoms filling the space the bicycle holds that will grant the bicycle independent motion. Without the unit being able to concentrate the correct amount of heat that will enable the space to generate motion, such motion under cosmos standards does not exist. In fact under cosmos standards the entirety of the bicycle unit does not exist. Only when the unit forming the bicycle can hold the number of atoms, which will produce the amount of motion through which the required heat will be subtracted from time in order to generate the gravity or motion needed, will the bicycle under cosmic standards gain motion. Before that the cosmos does regard the bicycle as the Earth and only the atoms standing independent from the Earth seems to regard the bicycle not being pert of the Earth. As the bicycle confirm the required and the displacement to launch an independent duplication wills the cosmos regard the bicycle in such a manner. When the atoms forming the bicycle unit are able to condense from time enough heat to give drive in order to provoke independent motion does the cosmos put the emphasis on the unit as a star. But in order to be the carrier of such independence the bicycle will have to grow. On its own as it is there is no way in hell that the bicycle can get fired up and start going as a star cosmos style. If not for life it will go nowhere but be consumed by the Earth to become a part of the Earth in time. In the cosmos the atoms forming the unit provide the motion and only when the total effort of the combined unit atoms manage to move abruptly and with the required confidence can the gravity be generated where the gravity generated is the motion of the entire unit. That means by having motion the fact of having motion grants the bicycle unduly respect from the Earth. The bicycle having motion of any status enlarges the bicycle in atomic space in ratio to what it has when without motion in relation to what the Earth has when only the Earth provide motion and when the earth stands in size in relation to a static bicycle.

When the bicycle is motionless, the bicycle is part of the Earth by gravity applied. As soon as life steps in and brings about separate and artificial motion but still uses the support of the motion that the Earth provides it will inevitably do better than the Earth as long as the motion that life provides is not in conflict with the motion the Earth provides. The bicycle becomes an object with the ability to transform the direction of the Earths domineering motion by redirecting gravity there in find the ability to change the direction of gravity. **When gauging what happens we must also admit that it is highly unlikely to find running rocks on Venus or moving craters on Mars. The motion that applies to the bicycle is an extension of the second force in the Universe, which is not part of the Universe and only affects a very small part of the Universe within the Universe. Life giving motion is an alien product and gives an unrealistic adding to the Universe. The motion however has nothing to do with mass but is only extending what was to what will be through what is. It is refurnishing what will be with what was before it now is present.**

When the motion of the bicycle accelerates such points that are forming by the increase in motion the forming connections extend to match to motion putting a standard of duplication per time unit extra to what previously was the norm. More space fills in the same period or the period reduce to match the filled space. The motion then contributes by increasing the space factor to keep the commitment with gravity valid. The bicycle breaks its form but because it is structurally bonded. Other aspects concerning gravity have to commit to the breaking of space. When expressed extremely crudely it is put as follows but is very bluntly stated. Yet, it still is the best way to explain the basics of the sound barrier. The bicycle is the compiling of the independent space within the atmosphere space or time concentrated to be more exact, that is holding the motion in duplication where the motion is continuing from a facet going to the next facet by duplication what was through what is to what is going to be. While the bicycle is filling the space, in motion that is part of the space holding all aspects within by the atmosphere and the atmosphere is holding all that is in it together in the atmosphere of the Earth. That is time performing in ratio to space filled by material. Because the conflict the gravity experiences by having motion within motion, gravity first tries to break the object that is in independent motion. As the motion continuous, as motion extends the reflex to the situation is then to contain by the breaking to the connecting devices such as the sound waves in the adjoining space. The atmosphere does the breaking of the relevant **k** on behalf of the object in motion since the moving space holding the object in motion as a unit shows much stronger bonding in structure unifying. We experience such breaking of space as the breaking of sound, which is showing motion or gravity differentiation.

The wing holding singularity while maintaining singularity puts the object in a Universe apart from other Universes. The aircraft becomes a separate identity maintaining a singularity and as all singularity does, such a singularity insist on two factors. The one factor is duplicating by motion the production of space-time while the other is the dismissing if space-time by controlling of space-time. The contact with space-time allows the motion to present the wing (and the aircraft) as being much bigger than in reality because it is not only the size of the space-time that maintains the singularity but it is also the contact or relevance which holds the dimensional area of control which stands as a controlling factor. These factors combined make the aircraft become bigger as the capability of motion suddenly allows the relevant size of the aircraft to grow in stature. The contact that the air or space has on one side which is more than the contact that the wing has on the other side where airflow is restricted, makes the side having more airflow larger by motion that the other side has in size by restricting motion. The motion enlarges or restriction reduces the contact and therefore the size per time unit. This alters the size as much as it redefines the balance and that makes the craft fly.

From the allocated position we hold the bicycle seems to be stationary when it is not moving. When we stand still it seems to us that we are standing still when we are not moving. However that is a human conception like mass is and is far adrift from a cosmic reality. We move as fast as the Earth spins. We move as fast as the Earth rotates around the Sun . We move as fast as the Sun rotates around the Milky Way. We move as fast as the Milky Way is rotating around another common centre because there shall be such a common centre that is allocated to order another common centre and this role diversification goes on running up to eternity.

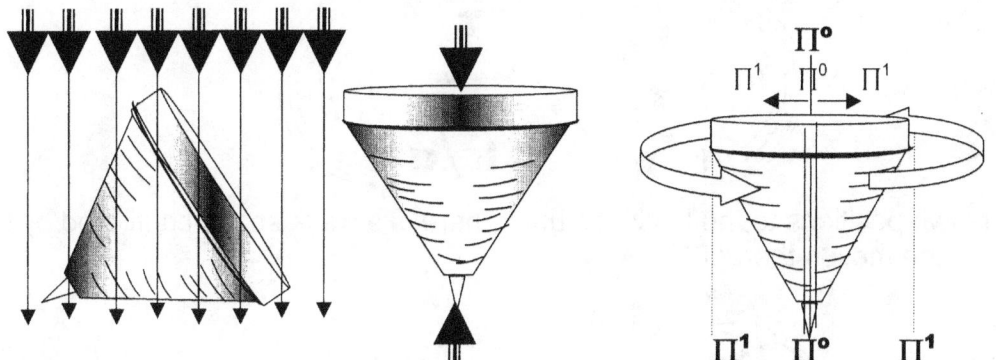

When any object is not moving the object form a part of the object, which holds the first object, captured. The cosmos disregard the existing of the first object in the event of it being stationary. However as soon as motion apply the cosmos grant the object having motion a position of existing by recognising it as an independent Universe that is entitled to all the privileges granted to a Universe.

As soon as the motion enters the equation the bicycle gets cosmic status because the bicycle generate time in the manner of parting singularity with time. The bicycle achieves cosmic independence and Universal recognition as an independent cosmic Universe. The bicycle is keeping upright because it forms a relation between time and singularity. The position is changing because time is changing singularity by the movement through time that gives the bicycle the opportunity to remain upright. The position the bicycle had in the past taking the bicycle through the present into the future is what is keeping the bicycle in balance because the one position in time is supporting the next position, which is supporting the next position. It is the fact that time has one position, which supports the next position by supporting the next position, and this support forms a line, which brings about the balance by which the bicycle stays in line. Once the motion is no longer present, the bicycle – Universe collapses.

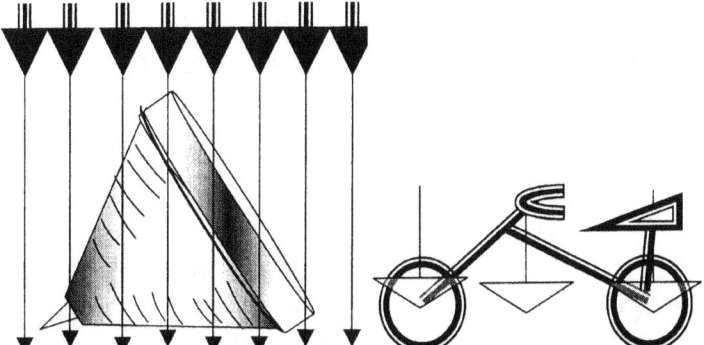

As the bicycle is standing still it is still in motion but holds a position to time where it fills a certain volume of space in that given time in motion.

It is Kepler's $a^3 = T^2k$. Let us now forget the fact that life is responsible for the motion and pretend it is all a cosmic affair.

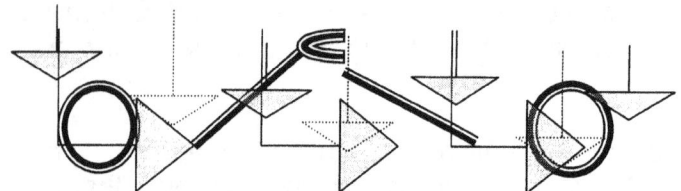

Since the bicycle moved faster than what it did when it was within the Earth motion it now fill more space than what it did before the individual motion commenced. In having individual motion on top of the motion the Earth supply, the bicycle is filling more space than it did before when it was stationary.

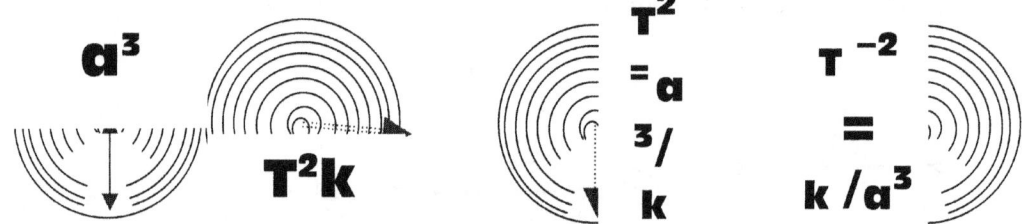

As Kepler stated there are two positions to the Universe unit. There is always space confirmed by the motion thereof in relation to the motion thereof.

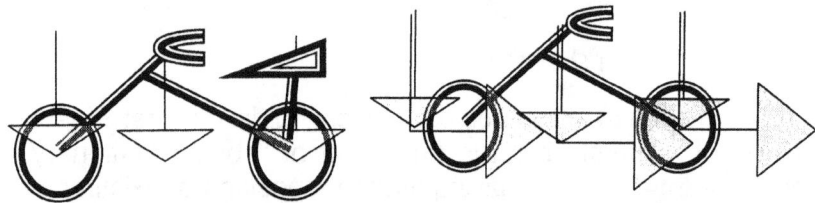

The motion fills space the Bicycle fills but also the bicycle fills space the Earth provides. It is filling more space within the Earth, which is space of the Earth than it did when only the Earth provided such space. Their stand bigger in it's filling of the earth space than it did before it started to move independently. It has more of the gravity going around than it had before when it was only dependant on the Earth to allow gravity going around. The bicycle grew bigger by the same margin that the Earth grew smaller in ratio to each other.

The bicycle without motion forms part of the space, which the liquid space confirms as part of the space. The bicycle then forms part of the other side of the Universe that is part of the solid.

In the motion we have two contributing factors that plat crucial part in the dynamics as to how the material in forming the atom construction reacts on such motion. The atom comprises of spin around the proton by the electron.

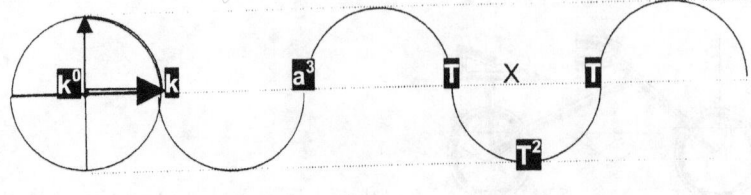

Any motion exerts an influence by all factors on one another and the is irrefutably connected.

dynamics Yet at the so small it and yet it is same time the factors are can never be detected so huge it spans across an entire Universe. Remember every atom is a Universe in its individual making.

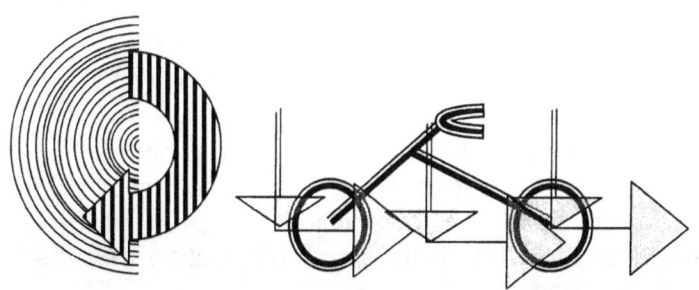

The moment the bicycle move the bicycle switches sides and switches allegiances. The bicycle then becomes part of the liquid that the space confirms not as space but as the liquid that extends the space. It then forms the gravity adding space instead of the gravity confirming space. It becomes $k^{-1} = T^2 / a^3$ instead of $k = a^3 / T^2$. What applied before then does not apply any longer because the bicycle is then on the other side of the Universe. Yet it is much more complicated than that because when stationary the bicycle was part of the Earth extending solid space which is gravity in relation to contracting being $k = a^3 / T^2$. As soon as it moved it became $T^2 = a^3 / k$ which is relative to the old position $T^2 = a^3/k$. This too has no principles that it can share in the mass applying gravity idea. It is about being stationary in relation to moving.

Taken from the Earth perspective there are lines (**k**) running at 90^0 to the rotation of the Earth T^2. The earth gives the bicycle one line in singularity which to confirm its position as far as the space the Earth grant the bicycle to manage. That gives the bicycle a specific space to hold in relation to what the Earth has and in relation to what the Earth offers the bicycle. That is gravity. That however have no principles it shares with mass pulling mass. It is being on the one side of the Universe which space holds motion in relevance $a^3 = T^2k$. In moving the alliances switches where the bicycle forms an legions with motion by duplicating the space it holds in relation to that which the Earth grants $k = a^3 / T^2$ and therefore becomes part of motion where motion forms part of time $T^2 = a^3 / k$.

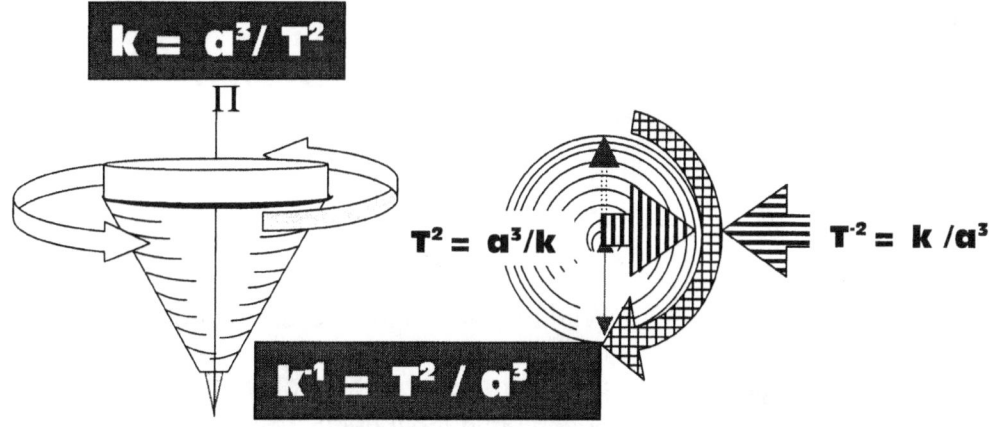

There are always two sides of the same Universe forming one Universe. There is the space extending to confirm the liquid by producing a solid and there is the liquid attaching to the space to extend the space by motion that secures the space. The one stands related to the other by opposing the other.

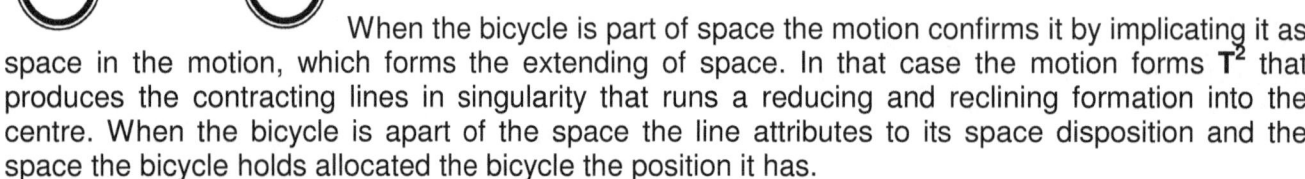

When the bicycle is part of space the motion confirms it by implicating it as space in the motion, which forms the extending of space. In that case the motion forms T^2 that produces the contracting lines in singularity that runs a reducing and reclining formation into the centre. When the bicycle is apart of the space the line attributes to its space disposition and the space the bicycle holds allocated the bicycle the position it has.

There are always two opposing time lines forming one united space. The one is the line **k** and the other is the half circle T^2 where from those perspectives there then form the triangle a^3. By moving the bicycle then forms the 90^0 cross reference to the allocated position it had before. It is **$k = a^3 / T^2$ or it is $T^2 = a^3 / k$.**

The moving of the bicycle involves duplicating the space and the position the bicycle has in relation to the space the Earth allocates and the position the Earth allocates to the bicycle. The faster the bicycle goes is actually the number of times such repositioning of the space the bicycle holds are in response to what the Earth allocate and what the Earth takes in a specific period of motion of the Earth. When faced with the question of how the bicycle manages to stay upright it always comes down to charging the achievement to a balance…but a balance of what? What goes into balance to achieve the upright position? The bicycle repeats its position in relation to the position the Earth grants and when the repositioning of the bicycle is faster than the re-allocating of the earth, the duplicating of the bicycle from one position to the next will sustain that the bicycle can cross the vertical lines faster than the vertical motion will effect the stance of the bicycle. The bicycle firstly crossed its allegiance by no longer forming a partnership with space but becomes a factor of liquid presenting motion.

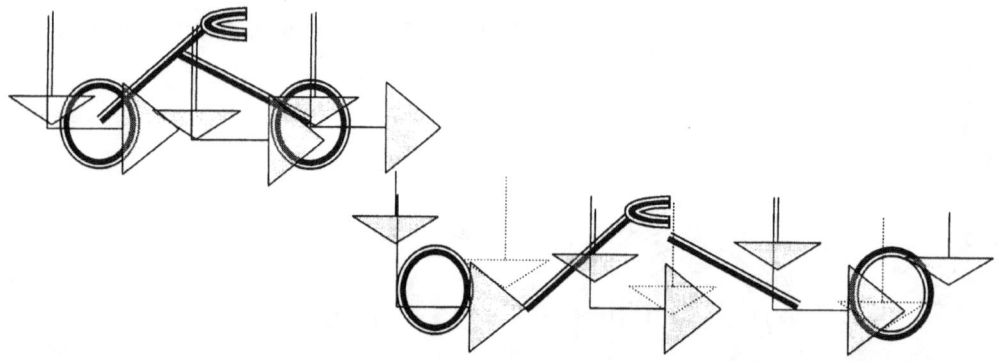

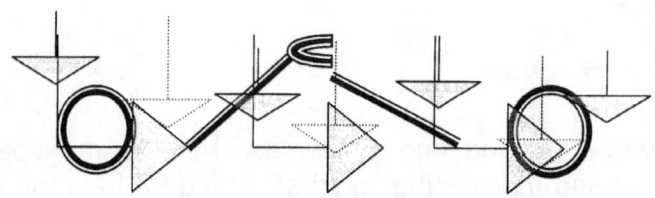

With the motion the bicycle is duplicating in ratio more space than what it had while being part of the Earth. The motion holds more space because it holds less space per time unit and there are more units of space per time unit. Since the bicycle is propelled at a faster pace than what it was when the earth alone supplied the space and forced the time by establishing the duplication tempo in the time contracted. By supplying more motion the bicycle grew larger in ratio to that which it had when the Earth was the sole space-time provider. After accelerating the bicycle then has more space in relation to what the true status is. The bicycle holds a larger part of the Earth by which the Earth has to reduce the space it offers the bicycle. The earth had to shrink in order to provide the bicycle with more space.

In the normal relation that the bicycle has with the earth when the bicycle is motionless (or in the manner we think about the status of the bicycle in terms of only its position within the scope the Earth provides) being motionless and then started moving the space increased rapidly as the Earth space decreased rapidly. With the new motion the bicycle finds much more duplicated space and that disturbs the ratio of space shared by the bicycle within the confinement of the Earth. The earth now presumes the bicycle to be much bigger than what it was. By moving the bicycle physically got larger and this is a fact not only by relevance but also by actual annexing and capturing of space in any given period of equal ness. The earth provides a certain value but as the bicycle moves faster the bicycle annexes more of the Earth space and that improves the size of the bicycle in a volumetric and physical measurement. It is a^3 grows bigger therefore the earth a^3 has to compensate by reducing that much actual space. In this comes a problem.

The bicycle does not even contribute a morsel of space when compared to what the Earth delivers. The earth that actually became smaller resents it becoming smaller by the demand of such an outrageous exploiter. To the Earth the bicycle is motion and the motion is liquid and therefore taking space contravenes the being liquid part. Being liquid is also being heat but being heat means becoming hotter when becoming more. To the earth the moving bicycle is liquid motion and being more makes it being hotter and being hotter is therefore is more volatile. Therefore the Earth refuse to become smaller and the bicycle being space claimed cannot become hotter without destroying its independent molecule unit.

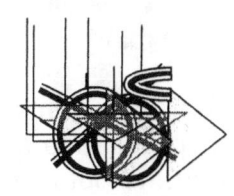

By the size the earth holds the Earth will not allow such a renegade to grow bigger and the Earth sanctions the time in space in accordance with the Earth generated governing singularity. Since the generated singularity the bicycle has still adheres to the dominance that the Earth generate, the bicycle reduces it proportional space in the face of the Earth showing such a strong reluctance to abide by the will of the smaller bicycle. The earth crushes the bicycle in response to the bicycle growing and when the response is more than what the bicycle can withstand, the bicycle crushes buy reducing space. That is what happened to Challenger when it entered the Earth and was liquefies by turning into gas in time.

From the past   In the present   Onto the future

$k^{-1} = T^2/a^3$	$k^0 = T^2 k / a^3$	$k^1 = a^3/T^2$

The ratio between space and heat goes into an imbalance where the liquid that is standing as ($k^{-1} = T^2/a^3$) totally dominates the space factor $k = a^3 / T^2$. The cosmos informed Kepler of another gravity, which the cosmos applies much more widely and is used by nature all over the Universe. Being with life and being part of life we humans take motion in content. Life is the manipulation of space-time by motion and since we can move because life is about motion we generalize motion with contempt. Motion is the most complicated process in the Universe because the Universe is motion. But why would motion occur because the fact of

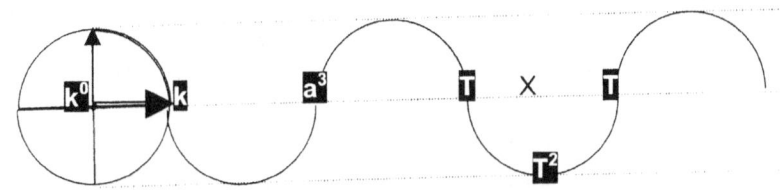

motion proves the presence of a Universe.

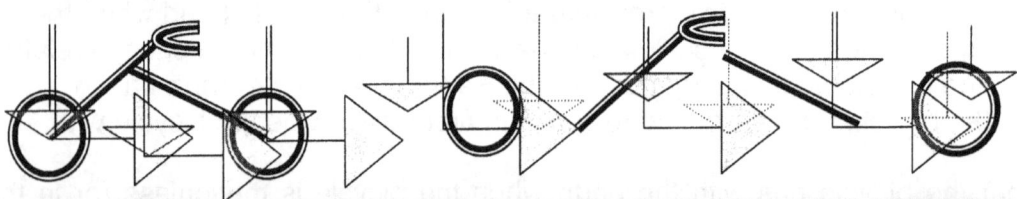

Motion distributes space and therefore decreases heat. By spreading the space over a larger area the heat in the space is reduced because the density of the heat allocated to the space reduces. Motion decreases the heat and therefore the Sun is the coldest place in the solar system while outer space is the hottest part of the Solar system

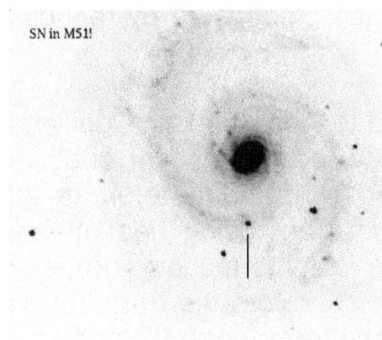

SN in M51!

Supernova stars are stars that overheated. The overheating came as a result of the motion in duplication that was not in ratio with the control in contraction. The expanding of the space the supernova held before had nothing to do with mass. The expanding had nothing to do with Newtonian gravity. It is all a ratio that puts a relation between the space that is taken by the material in concentrated heat (or time) and the time in ratio to the duplicating material. If the material forming the motion is not very highly dense the motion is poor. By the same token is the material density high when the motion is volatile. It all depends on the motion that the atoms forming the Unit generate and that motion is forming the gravity. There is no mention imagination of the Newtonian mind.

of mass except in the

There is no pulling between

particles compressing them into plums of pressure. That is nonsense because a star has no containing wall on the outside to capture the pressure on the inside. It all depend on the duplication of material confirming the containing of the unit in relation to the distributing of the material in ratio with the liquid time with which it is in partnership. It is all about containing heat

and distributing heat at the same time. The more motion that is present the more heat is condensed and is contained.

By duplicating at a higher ratio the heat contained is spread over a greater part of time, which reduces the heat contained in the material as it is distributed in a larger ratio to the liquid heat that is maintaining the balance.

There is a defined ratio between liquid heat that is the basis for time and solid heat, which forms the norm for space. The higher the ratio favours the liquid the colder the solid will be and the lesser the solid is in contact with the liquid the hotter will the solid be. Stars are ice-cold ice cones floating in outer space because of the motion they generate. If we pump air into a compressor the air gets more inside the compressor. The compressor gets hot while the air gets more. The size or the compressor remains the same while the air gets more inside the container and that means the compressor is shrinking while the air is remaining the same because the air cannot get more while the compressor is unaffected.

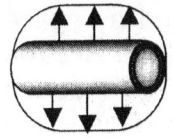 In relation to the heat the heat gets more because the surface of the compressor remains the same while the size of the compressor within the relation shrinks. Because the size of the compressor that shrinks the space outside the compressor has to accommodate the flow of heat because equilibrium has to be re installed. The size will remain reducing up to a point where the compressor is just to small to accommodate the air. The air then will expand. However the air was always expanding from the pumping started because the compressor inside became smaller as the heat ride to expand the size of the inside of the compressor to match the outside of the air.

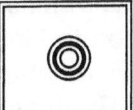 The same goes for material blown by wind to reduce heat. The object has an indicial size to start with. Then we put heat to the object and the heat makes the object increase in size. That is hardly the increase worth noting because the relevancy of heat in the air to the heat in the heating object goes array. The heat has to increase the size of the object in relation to the match it has to find in the space it is within.

With the heat coming into the object the relation the object has with the heat or air outside makes the object that many times bigger because the ratio in the heat balance is disturbed. If we blow air over the object we increase the size of the object by allowing the surface  of the object to make contact with much more air in the same period of time, which will bring the size of the object back to normal because in relation and considering the contact with air the object expanded by the motion of the air in contact with the object.

In the normal flow of time the object has a heat to space relation set by the time the Earth dictates. Then we go and increase the heat on the object and in that event we actually increase the size the body has in relation to the heat in the air. By blowing air over the body we increase the air and therefore we increase the size of the body in the same period of time.
There is now a dispensation of many times the body   carrying more heat and contacting many

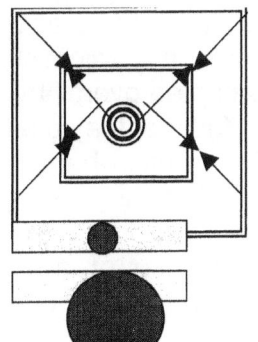 times the heat or air which bring the equilibrium back to normal. There was a body size and by applying heat the balance shifted to the reducing of the body size in relation to the heat. The body then had to expand in heat because the body was too small to incorporate the large heat. Blowing the air over the body increase the size of the body and heating the body decreases the size of the body in comparison with the air it comes in contact with. The body is either expanding or the body is redefining and the balance in heat places the body in relation to either gravity cooling by contraction or expanding by overheating. The very same principle applies in the sound barrier.

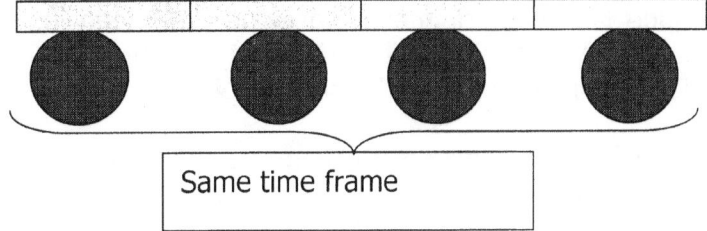

Same time frame

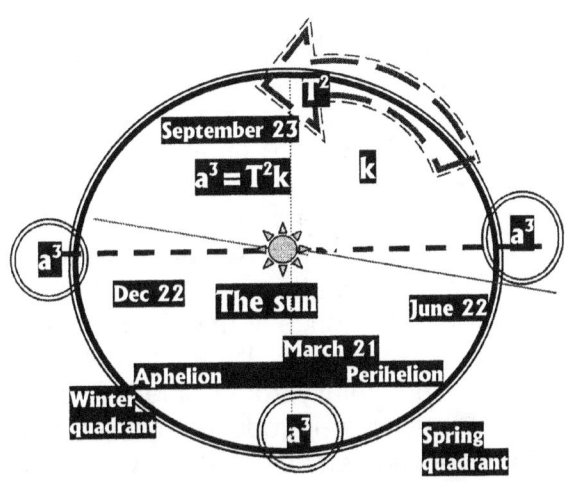

In spite of outer space being as expanded as anything can get we still regard outer space as being incredibly cold. Anything expanded to its limits and which can heat no more is as hot as anything will ever get. Outer space is the very edge of expanding of space where heat cannot expand into space any more. On the other hand we fin that concentrating heat is producing cold making anything in the atom in stars as cold a there can be. That heat filling the outer space lacks motion and is therefore space in another form of material that could conduce by diverting from space to constrain further expanding through motion and therefore was unable to marry the union of space by becoming more space. One must look at outer space and judge outer space from the findings only considering outer space. By motion space duplicates and by space halving it removes heat in space as well as by dismissing space. In the case of material the electron is spinning at the speed of light to contain the heat inside the atom at a higher rate than the speed of light. It is containing heat at a greater motion than what light can travel The atom by motion is the condensing of heat by contraction in relation to expanding or duplicating of heart. The concentration or release of space with heat or space from heat is a direct contribution of the singularity in control of the space-time. The regard of the singularity stipulates the conducing of heat in space or the release of heat to form space by means of bisecting the occupied space. By applying motion the space duplicates what it is from the past through the present and into the future. This is no hypothetical suggestion but is the actual flow of time coming from outer space as a liquid and is incorporated into the spinning atom on the way to confirm singularity. While we are in gravity the manner in which gravity applies in our use of gravity makes us part of the Earth by mass forcing us onto the Earth as a semi unit with all other Earth belongings. Is that which we have truly gravity? By using mathematics, the cosmos spoke to Kepler personally and by the use of mathematics as the medium, it provided Kepler with information about the cosmos coming directly from the cosmos.

The picture we see coming from the Hubble telescope shows why, in the perfect Universe...but can the Universe be perfect when... we see a radius between the Sun and individual planets is not using a regular distance as one would expect of gravity in being a force driven by the mass and in that sense the mass is producing the gravity that always remains even because the mass doesn't alternate. As the mass is never changing on either side, that steady mass has to keep the gravity steady. But in our imperfect understanding of the Universe we find that the radius that should be constant varies considerably proving either that mass somehow adds by measure unnoticed while the structure is in orbit and later allows the same amount of mass to escape undetected; or it's the seasons adding and removing mass at will. This is an absolute contradiction to reality if mass was the factor determining the radius we find between the Sun and the planets. This suggests strongly that we'd better be getting very suspicious about the idea of mass contributing to gravity. But in contrast to this, science is unshaken about their confidence in the perfection about facts they use in terms of correctness. It is well known amongst all persons that science only uses dependable and ultra reliable facts coming from sources beyond doubt. Referring to any work done by any scientist will find a remark about science only accepting facts they use to work with. It is accepted overall by all communities that in science those in science use one hundred percent accurate facts or they use no facts at all. If our view was as perfect as science would lead us to believe it then must be the Universe that is imperfect as it otherwise would not behave so mystifyingly. The unshaken confidence science uses has us believing at first consideration that the drawing of gravity should produce an even diameter positioned between the Sun and the planets because of ever dependable evenly distributed gravity... but I believe there is a perfect Universe and our understanding carries the doubtful suspicions.

Delving deeper uncovers even more contradictions and the level of accuracy contained by our scientific understanding then arouses more suspicion about the correctness of science. Remember Newton changed what the cosmos told Kepler leaving much suspicion as to how far the misdirection takes science. We have to correct the facts we doubt because when correcting the facts they use in science concerning our view about science, such correcting brings along a better understanding and then the Universe has to become ever more perfect as one learns to understand the perfect Universe even better. But it does require an open and clear mind and it needs no culture driven preconception that should confirm interpretations about facts surmised even before they are carefully studied. It becomes obvious that Newton never gave careful attention to Kepler's findings because if he did he would have seen what gravity was. Kepler described gravity without using the name that later was given as 'gravity'. Kepler did not give the name gravity, but Kepler's studies gave Kepler the insight to coin the concept of gravity. Nevertheless it was a name and not the concept that was later named by Newton. The naming was the contribution of the Englishman. The concept that Newton later introduced is totally incompatible to the concept that Kepler introduced. What he (Newton) introduced as the force of gravity, he connected to mass, which diverts totally from Kepler's findings. With giving a name, the Englishman also changed the concept that Kepler introduced. Kepler made no mention of size or mass as part of the phenomenon that later was named as gravity, yet it must be gravity that holds the Universe together. The concept the Englishman changed when he introduced what he introduced with the name he introduced. That what he introduced, he corrupted beyond recognition. The concept that accompanied his new name strayed completely from what Kepler introduced. Newton brought in something that was mismatching what Kepler saw in Kepler's view of the phenomenon that holds the Universe true to form.   The name was dominant but even more dominant and totally inaccurate was the other concept Newton introduced. In truth Newton only gave the world a name of an idea, which he then corrupted as far as cosmic physics are concerned. It is important to admit that as far as cosmology is concerned Newton gave the concept the name but *only* the name and not the concept of gravity. Newton's persuasion on matters of gravity as gravity functions between cosmic structures orbiting one another as we find in outer space is inaccurate. What Kepler saw, Newton saw differently and used the opportunity that Kepler left by not giving any name to the process he (Kepler) and Tycho Brahe worked on for two life spans. Newton did seize the opportunity to name what he, Newton, saw but that what Newton saw did not include that which Kepler uncovered. In Kepler's era the name or title was lacking but Kepler established the concept of gravity and the formulation thereof. The concept came from Kepler even before the name gravity was used by Newton to describe in the concept of whatever we today (after Newton) became accustomed to believing what the concept of gravity is about. With the help of Newton everyone since Newton confused Kepler and Newton on the issue of gravity and this confusion even begins with Newton. Gravity might not have been named but became a proven concept and factor after Kepler formulised it, which is before Newton named it. The concept of gravity that Kepler saw is about the manner in which the structures orbit because there is a space that circles around a centre and this process has kept planets secured, connected and rotating around the Sun which is the same concept that is keeping the Universe secure and comes about with a process Newton later named as 'gravity'. What Kepler saw is not the same as what Newton saw when he saw two objects drawing closer by pulling on each others mass. Then later on Newton named, what he thought he saw as the force that Kepler saw but introduced another completely different concept. Kepler saw cyclic formations keeping the Universe together and never approaching each other. Newton ignored what he wished not to see but he changed as he saw fit and what he thought that should be. His experience as a young man drove him to establish a process he formulated as the process that is keeping the Universe together. In that act he corrupted as much as ignored the work of Kepler, which he also named as the same gravity that he saw as a young man. Why he chose to ignore Kepler's findings on gravity we shall never know but why the world still chooses to ignore Kepler's findings about gravity almost four hundred years after the fact I shall never know. My saying this has literally made Academics ignore me as they would avoid the plague. I am not pretending nor do I exaggerate when I say there were those in Academic institutions that questioned my mental development. Some went as far as seeing me as a joker of sorts and I have correspondence to show evidence to that fact. I know by now while Newtonians are reading this letter I have aroused the tempers of every Academic reading this far, therefore let's see what is being ignored by the Academics which I blame to do just that. .

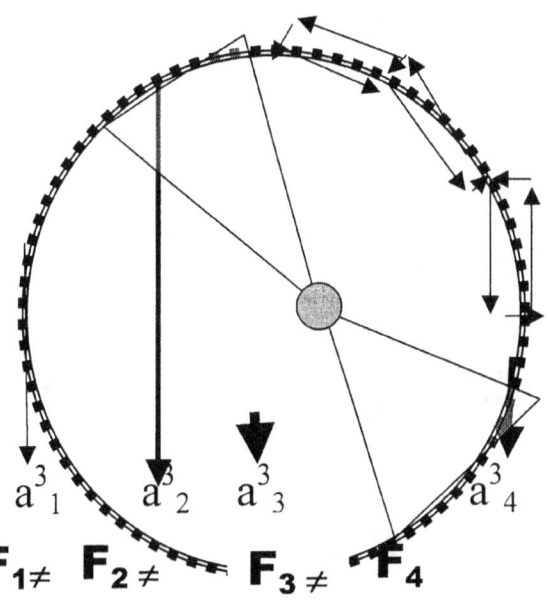

$$F_1 \neq F_2 \neq F_3 \neq F_4$$

We live through seasons which comes from being that at one point, (a^3_1) the distance between the **Sun** and the earth **is less than** at another point we call a^3_3

Let us put a value **of a^3_1 =** one and a^3_3 = three. This means that each year, for the past 4 500 000 000 years the effect of the common gravity between the earth and the Sun has a greater effect than at another point six months later.

That means at one point the earth should be drawn or pulled closer to the Sun   and after another six months interval the earth should stand less effected by the Sun 's gravity, therefore it should move away from the Sun .

Kepler said gravity in space is about the area a^3 that would always keep equilibrium with the time T^2 it takes to travel the distance of the full circle position placed by the indicator **k**, therefore adjusting **k** as the need arrives. With **k** shifting in length **a**³ will have to readjust and therefore **T**² will find a new relating value each time. This was the finding of Kepler and came after his intense study of orbiting planets.

Before I attempt any investigation into this matter there must be coherence in our agreeing about what gravity is. If you the reader insist that the falling of objects is the only gravity found, your further reading will convince you little. Anything we do decide upon must support the fact that it is gravity that prevents planets from dislodging from the grip the Sun has on them. Gravity is not about the Sun trying to catch the Earth by attracting the Earth…no, there is so much more to gravity. We must be under no illusions about what gravity is and that being the focus of our discussion and where that gravity is because we have to identify and not confuse the gravity we are looking at. We are now discussing the gravity, which is keeping planets circling around the Sun , and stars around specific galactica centre. In that we do not find one example to use as proof in connection to stars coming tumbling down on galactica centres and crushing into galactica centres. If that is gravity keeping structures in orbit around specific centres we must look at the behaviour of the structures in gravity. We have to find a reason why the planets do not reduce the radius between them as Newton suggested but we must trace the reason why it is gravity, which is keeping them apart because if anything, they are departing as they extend the radius connecting them to the Sun . That is gravity because it applies throughout the Universe. The gravity Kepler found is the general gravity that is keeping structures from colliding and in that the principles are avoiding collision or on the other hand avoiding abandoning each other. It is about confirming respect for one another's independence and clearly staying at a predetermined distance while at the same time both are sharing a common space unit. That then must be the defining of gravity we have to study to find the Universal enticing gravity holding the Universe together. By close investigation one will find three factors in urgent need of investigation. There is firstly a centre that draws the object closer. This gravity is clearly a synonym to what Newton saw as gravity. If it were not drawing the object closer the object would not be orbiting around the centre and applying motion. It will draw and absorb all rotating things in its field of gravity.

The fact it does not draw the object into its ranks is because there is another gravity standing alongside this first mentioned gravity. Our recognising the first gravity forces us to accept the presence of another part of gravity. This forces us to recognise the second gravity. When saying this we are not using Newton's cosmic formula concept $F = G\,(M.m)/r^2$ because that can barely be what is out there happening. What Newton saw was falling. If that what Newton saw is the only gravity then whatever Kepler saw including all other parts of everything out there that are spinning around some centre must come closer to one another and connect in collisions. While that is not happening we must start to look past Newton to new grounds we can investigate. We have to go beyond Newton and admit there is more than that what Newton led us to believe because it is clear that what Newton had us believe…is not happening. That confirms the presence of the second gravity. The fact proves that everything is departing and not arriving. Even the moon is drifting away from the Earth and this information comes about from the most advanced investigation up to date, including a moon visit and the placing of measuring devices there.

Looking at the gravity intensely we find the roving structure travels in a straight line, which repeats another circle around another centre but because of the influence of a centre keeping the roving structure attached to such a centre the motion allows a circle to form by reforming motion from the original straight line to that of a partial circle. There is centre; a connecting line travelling between what the two points establishes the specifics of a centre within circle and the end of the circle. According to Newtonians the centre supposedly draws the rotating object closer. That is half the story.

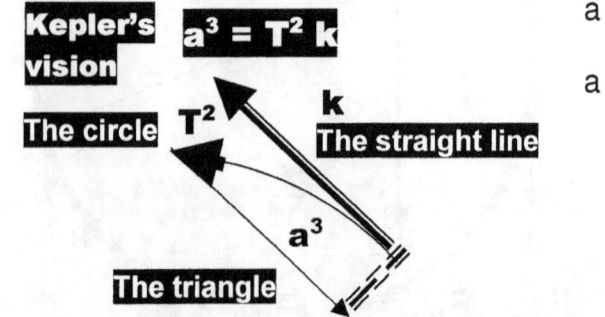

I suggest we do some deliberation and in deliberating may I remind you THAT NEWTON'S OWN LAWS ARE IMPLIED, and again the planets disobey these laws completely!! In the modern age all evidence points away from contracting and favours eternal expanding.

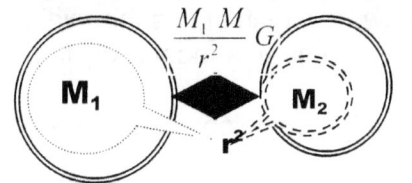

Newton's vision $F = \dfrac{M_1 M}{r^2} G$
─────────

Newton saw his apple fall and then went on to blame everything on mass… and you think it is all that plain and simple? Kepler on the other hand said (and let's forget what Newton said about what Kepler said for a bit) that a space of cubic proportions a^3 that will always keep equilibrium with the time T^2 it takes to travel the distance of the full circle at the distance (or relevancy) in ratio or relevance **k** with an indicator pointing the distance the circle is from a specific centre. That means **k** is as crucial as T^2 in positioning a^3. In placing the allocated position a^3 requires in determining the sectors (we think of it as seasons) a need arrives to predetermine **k** in order to measure T^2. Every spot a^3 fills is located at T^2 and is allocated where **k** indicates. When **k** shifts relevancy (from Earth to mars or even from season to season) the space in a cube a^3 will have to relocate as well as readjust and with that T^2 will be redefined. This was the findings of Kepler after annualising the work dome by Tycho Brahe and then later himself. The line forming looks straight from the onset but the line

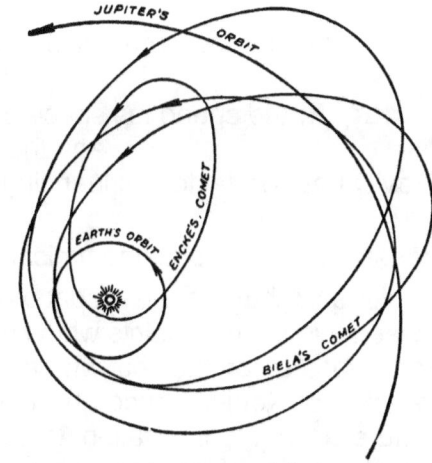

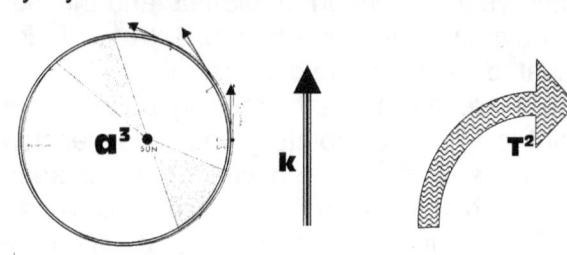

never moves straight but goes bended in relation to the relevancy that **k** indicates. In what **k** contributes as a factor it introduce T^2 and from the alliance of the product that T^2 **k** delivers we find that a^3 forms an eternal circle about an eternal centre.

That is gravity if you wish to call it gravity by name. That is the force that is no force that prevents planets escaping from the grip the Sun has on every planet. Gravity is what puts order to what would be the most chaotic arena there can ever be. That gravity has no bearing on mass because with mass pulling that order which gravity then would bring will result in complete destruction. That is not happening. On the other hand if we don't do what Newton did by putting words in Kepler's mouth we find that Kepler gave gravity a completely other meaning. Kepler said (when ignoring Newton's uncalled for interfering and meddling) that the motion T^2 puts the space a^3 at a distance from a centre and the relevance factor **k** prevents a^3 to come close or drift further away but stay in the allocated position where T^2 locates a^3. The factor **k** has it at task to prevent a^3 coming closer and orders T^2 complying with a measured value. By denouncing **k** everything Kepler said goes array.

Every object in outer space holds as much turning as it commences its run in a straight line. It turns as much as it goes straight. The comet coming towards are no different from the comet going away and by passing a dividing line they change direction from coming towards too going away but in that they remain equal.

A space remains between the comet and the Sun that define the line it crosses when the comet changes from coming to going. That gives the comet a cyclic approach and departure and does not put the comet on any collision course with the mass of the Sun . There is no death defying all destructive route calling for a disastrous end with no chance of avoiding the immanent disaster colliding that is unavoidable predestined to happen as Newton's formula would suggest. As the comet approach it is the relevancy brought on by the changing of **k** that puts a^3 in a ratio with T^2. It is **k** that sets the approaching limit as it sets the departing limit and it is **k** that prevents a collision as it prevents an escaping departure. That is what Kepler said…and that is not what Newton said Kepler said. What Kepler said is quite a different story from what Newton said Kepler said…

> $a^3 = T^2$ **k** then $k^3 = k^2$ **k** and this is showing that the space k^3 is equal = to the motion k^2 **k** of the space k^3 seen form one specific point.

In Kepler's formula $k = a^3/T^2$ the smaller **k** becomes the smaller a^3 becomes and the bigger T^2 as far as movement goes. T^2 represents the gravity that positions the space a^3 at distance **k** from the centre capturing the structure through gravity applying T^2. We all are very aware which star is the mighty gravity producer. So has mass the least say when gravity is generated? It seems most likely to be true.

$k=a^3/T^2$; the distance depends on the position that the orbiting object develop space-time

$a^3= k\ T^2$;the space depends on the distance the space develop from the centre and the speed the space moves around the centre.

$T^2= a^3/\ k$; the speed the space orbits around the centre depends on the distance of development and the size into which the space developed.

Gravity has two factors influencing space, which are a straight-line **k** and a circle going around the centre T^2. $a^3 = k\ T^2$

Translating Kepler's mathematical expression $a^3 = T^2k$ correctly to the verbal statement in English

Kepler said that there is a space a^3 which is equal = to the motion in the time duration T^2 thereof between two specific points which holds a relation to a centre wherefrom there forms a straight line **k** and is located on the spot where space begins the circle. Therefore that spot has the least space. The value of Kepler's space he indicated as a third dimension a^3 does not depend on indicating a structure a^3 that is in rotation T^2 but only needs one position having a constant of some sorts. Any point where **k** may indicate a position one will find a value matching a^3 and the matching location will fit T^2 at that point. That is the relation there is in the solar system between all planets and the Sun. The Sun always indicates the centre and the planets always indicate the rotation. But $a^3 = T^2\ k$ is only producing a relevancy of three dimensions that is equal to two plus one dimension.

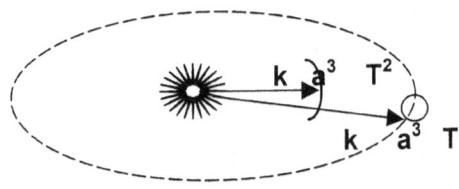

From the Sun there are three points moving between two points from one point to two other points giving six dimensions we find that is forming space. It is space in time or space converting space through the moving of space in time. It is locating a point in the third dimension a^3 that will move according to the second dimension T^2 that will implicate **k** as a reference in the first dimension. It is the duplicating of space by time providing the dimensions to do so.

Let us take it from a point where the Sun provides a centre as one starting edge of **k** then that centre **k** will provide a line from the centre and the line **k** will provide three spots in a formation that produces a structure by the square T^2 of the dimension. Not once did Kepler indicate size as a contributing factor to a^3. That means every single point that **k** indicates there are three positions a^3 implicating sides of a double dimension. In the same manner is **k** not limited to distance or is T^2 lesser by size. **k** $= a^3/T^2$

There are infinitely more implications in the statement Kepler delivered than what is merely a contribution to motion and only motion as Newton was of the opinion. What is there mathematically not correct in my interpretation of Kepler's manner of translating mathematics to English and why is any changing thereof by Newton or any other person necessary in any way?

We can test any of the following symbolic values in the mathematical expression and also test the principals behind the expression in which Kepler stated them. By such testing we will find that time after time there were never any corrections in the translations required since the translation thereof was never incorrectly presented and in that a case asked for no alterations to secure the correct reporting of the cosmic information being translated. By taking the formula on face value it can change as follows: $a^3 = T^2\ k$ can become $k = a^3/T^2$

When translating Kepler's mathematical expression into English we can see what Kepler said also read as $k = a^3/T^2$ where **k** is one point from a centre point that is space a^3 relating to time T^2. From a centre comes space-time. The centre **k** brings space a^3 in ratio to time T^2, which is space / time a^3/T^2. Reading this correctly cannot bring any dispute…yet it does…and it's been doing it for centuries on end!

The cosmos spoke to Kepler about space-time coming from singularity. Kepler gave us his findings. Any discomfort that may come when we read what is revealed must be set aside, because we must remember it is not me, or Kepler, but the Cosmos that is doing the revealing and lending us the tools we can use to decipher what the cosmos is trying to make us understand. Kepler translated what the cosmos told him (Kepler) as $a^3 = T^2k$. Translating Kepler's mathematical expression $a^3 = T^2k$ correctly to the verbal statement in English Kepler said that there is a space a^3 which is equal $=$ to the motion in the time duration T^2 thereof between two specific points which holds a relation to a centre where from there forms a straight line **k.** What is there mathematically not correct in Kepler's expression and why is any changing thereof necessary in any way? It says where there is space such space has to move. Test the following symbolic values in the mathematical expression and test the principal behind the expression in which Kepler stated them. Convince yourself about the evidence that Newton saw what Kepler saw where the translation thereof that was done by Kepler is mathematically incorrectly translated by Kepler's interpretation from mathematics to English:

a^3 The fact that any symbol uses a value to the third power indicates space or a volumetric established and separate unit which is serving an under dividable dynamically separate space being within a space. Although being apart the two in space sharing a unit can never be apart but serves as a unit by division of motion. It is space because it is volume using the third dimension. But since the space is smaller than the Universe it must be space being within space, which is within space. There a relevancy is forever present.

T^2 Is an indication of space apart from the surrounding space by granting the independent space by establishing borders through motion, an ability of moving from one point to another point or following a flat distance between two points. It is motion that is taking time in the second dimension.

k^1 Is the symbol used to indicate a straight line between two points with a definite beginning and a specific end position. The two points is valid only by re-aligning an eternal straight line to the figuration of a circle through alternating as well as recognising the control coming from such a centre. It is Pythagoras by the triangle, half the square and the straight line sharing value in the 180^0 they represent. Kepler introduced this absolute basic mathematical principle.

That is what Kepler said. There are three dimensions a^3 between any two points T^2 flowing as time from the centre of the Sun , which is indicated by the line **k.**
The implication of the relevancy produced by the use of the formula $k = a^3 / T^2$ brings about that when dividing T^2 into a^3 there is **k** left.
The fact is that a^3 is a three dimension (^{3}) of single **k** (^{1}) showing one or T^2 is two dimensions of **k** being the one dimension. It means that **k** is a part of space a^3 or T^2, which is time. It is the same thing in a double dimension or space being a triple of **k** then **k** is one factor and **k** cannot show a position of zero.

If $k = 0$ then there is no possibility of $k = a^3 / T^2$ because $k = 0$ then $0^3/ 0^2 = 0$. That does not make sense. Mathematically space cannot be zero because those being of the opinion of space being zero or nothing must first prove mathematically that space is zero.

Moreover they then must prove mathematically how does zero grow through the Hubble constant. By translating Newton's vision of the circle in completing a cycle would become zero through rotation…well that does not count the use of the formula a^3. If **k** cannot be zero then **k** could not start from zero.

With $k = a^3 / T^2$ no point can be zero because **k** shows space $a^3 = k T^2$ is no reference to the volumetric mathematical formula used to calculate $a^3 = 4/3 \Pi r^3$.

Nor does it show the use of the circle in the second dimension being $a^2 = \Pi r^2$.

Newton's mathematical vision was the way to calculate the space by using a mathematical formula used as

$$a^3 = 4\pi \text{ X } (r^3/3)$$

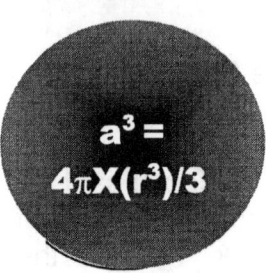

$$a^3 = 4\pi X(r^3)/3$$

Kepler's cosmic vision was that in the formula $a^3 = k\ T^2$ the space a^3 is equal to the movement $T^2 k$ of the space, which comes about as time T^2 in relation to a distance k

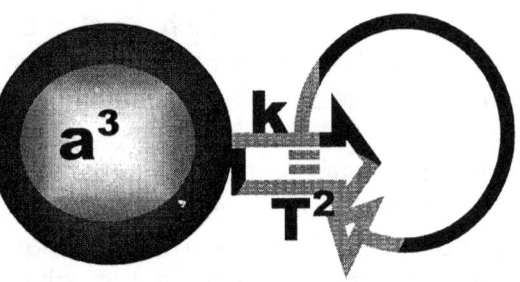

In the case of the Newton formula the circle factor becomes the square as indicated by the duration of the time T^2. The factor standing in for the line which normally would be r and then be the square value is in the case of Kepler not the value indicating the square. That means Kepler never indicated a circle of mathematical procedure but said mathematically the distance of the planet from the Sun k holds space a^3 in relation to time T^2

May I remind you THAT NEWTON'S OWN LAWS ARE IMPLIED, and again the planets disobey these laws completely!! **In the modern age all evidence point away from Newton's vision**
The lot is more likely moving away from the Sun . The lot is not coming closer!

The lot is more evidently moving further apart

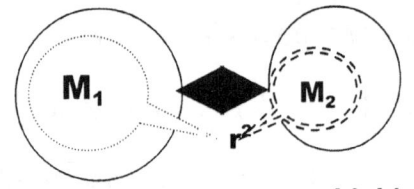

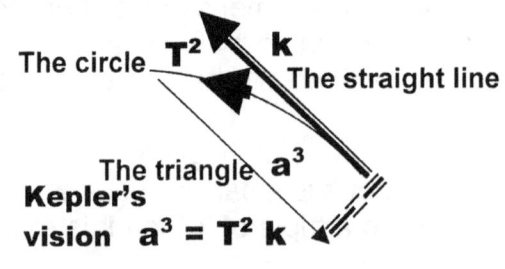

Newton's vision F $= \dfrac{M_1\,M}{r^2}\,G$

The circle T^2 k The straight line
The triangle a^3
Kepler's vision $a^3 = T^2\ k$

$$\dfrac{M_1\,M}{r^2}\,G$$

F $= \dfrac{M_1\,M}{r^2}\,G$ This is the suggested formula confirming the behaviour of planets used by Newtonian scholars underlining the argument that contraction is coming about between all cosmic objects. What Newton witnessed, if my memory serves me correctly was an apple falling from a tree where both the apple and the tree were part of the Earth and this did not constitute - or lead to - or come as a result of a catastrophic cosmic event happening. In the mathematical sense it does not make sense when Newton's argument is taken out and used in outer space. What Newton saw with his falling apple was a mass influencing another mass to reduce the distance as the influencing involved motion that came about. In outer space there is another gravity where in the case of those cosmic structures in outer space there is no mass pulling each other about or pulling one another onto each other. In the case where there is particles falling from space onto the Earth, that falling also results from gravity, as much as it varies from the cosmic gravity. There is another type or form of gravity different to the concept Newton introduced. The concept Newton introduced is not the cosmic gravity Kepler formulised. What Kepler introduced is a duel where both objects are clearly in an eternal compromise therefore neither party relents its position. Newton saw just the opposite...Newton saw both compromising their individual as well as each other's position. But since the mass in both cases is unchanged and the mass is the factor that is establishing the force that is used by the circle to hold the radius steady and in place, these facts point to a balance that formed bringing about the above-mentioned steadiness. In the view of science however it is the mass that either draws the orbiting objects closer or is keeping them apart. The mass does not change and since that mass of both produces the radius between both, the logic is that there has to be an even and steady radius that develops. The radius has to be equal all the time since the mass never changed throughout the rotation. The radius must be the same from any and all given points that form the rotating circle which must keep the radius equal from every angle...yet we know that Kepler proved this not to be the case even before Newton's naming and changing of Kepler's work came about. What we see is that there is one factor that is trying to run away being a lesser space within the pulling powers of a larger space (the second factor) trying to capture and control and a referee (the third factor) is seeing to it that the even-handedness is at all times applying in the fight. That gravity which I am familiar with and know is there). In some part but not in all out representing all the gravity there might be because I cannot see the jerking, as much as I do not feel it. That is then most probably another gravity I can see and which is Kepler's gravity which $a^3 = T^2 k$ represents. We have a motion of pulling...yes and that is what Newton saw...but then there is another motion of establishing a motion trying to depart, leaving the centre by tearing away from the centre and thirdly there is a motion that sees to it that the balance evolves as rotation. That is what Kepler said when he saw all three factors whereas Newton saw but one of the three. The one space is filling the next space as the space duplicates the position it had in the next moving moment that brings about the next position through motion. This eventually will have confined the next point by using a circle motion, which at first was intended to be a straight line, which is stopped by another straight line. The quest in this book is to find out why the other two factors apply in outer space as only one of the factors comes about on Earth under normal applying conditions.

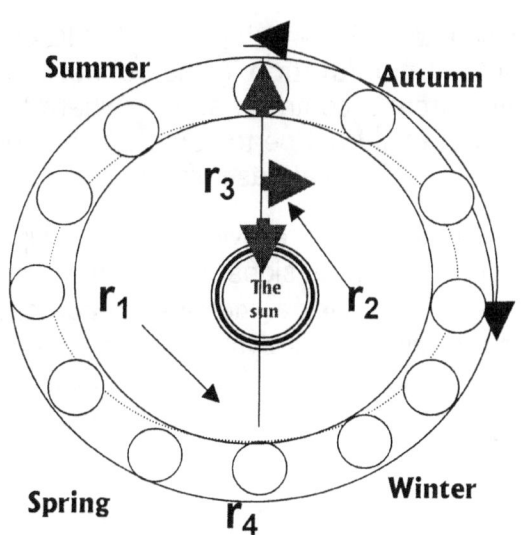

As the two factors are in a motion directional dispute there is obviously one of the two factors or strengths fighting to cut loose from the other one's grip and run off. If there were not such a force trying to escape, the first force would have a quick and decisive victory by reeling in the loser just as Newton predicted. The fleeing object and its matching fighting partner has a third party referee that allows the fight to go in a specific direction as long as there is no decisive victor.

This book, which I produced in the form of an open letter, is on a quest to find the missing two factors and I can declare with some delight and with even more certainty that I found the missing factors. By Newton's introducing gravity as a force with the formula $F = G \frac{(M_1 . M_2)}{r^2}$ a precedent was set of gravity being a contracting force forcing distances supposedly to grow smaller. Apply Newton's view to comet behaviour. Newton insists that the Sun has gravity reducing distance between the objects and while lecturers are teaching this during the day, at night they all witness how the comet follows this principle in detail showing Newton as a prophet. No sooner does the final conclusion draw near by orchestrating the final demise of the distance separating the two cosmic components when the opposite changes all concepts taught by institutions of science, the next minute out of the blue with no pre warning of the comet changing its mind, the comet defies all logic in scientific circles that apparently even included defying Newton and his logic. Because at the very point you'd think there is no chance of any return where gravity supposedly should peak because the comet is so close to the Sun and due to that fact makes the collision unavoidable...then the comet chooses that very point to dart away into the blackness of outer space, missing the definite collision by miles. By the time the collision is truly unavoidable with the radius between the Sun and the comet being as small as it realistically can be the comet starts gaining on the radius distance in spite of Newtonian denial of any possibility that such an event can in fact take place. The radius that should be shrinking further is instead enlarging. The radius that now begins to stretch proves Newton incorrect and it even depicts Newton as possibly being a fraud. The gravity applied that focussed on the comet reducing the radius between it and the Sun was not acting predictably by maintaining the reducing of the distance until collisions come about as Newton insisted on. In our reading the Newton formula in English it says that $F = G (M_1.M_2)/r^2$ which when one translates that which is said in mathematics to a verbally spoken linguistic dialect, the translation then suggests that a force is committing the material that forms the factors involved, and forcing the material into a path that is leading to a collision. It says that the two will eventually collide because of the non-retractable mass inside each one that enforces the pulling which by the mass in each case is creating the force. The unchangeable ability of the mass and the unavoidable pulling each mass creates would bring about such a collision. The mass contributes a force making a collision imminently unavoidable. The collision is beyond any attempts of diverting any oncoming objects away from the inevitable possibility of contact. The force that mass contributes is ruling out all possible evading each other or avoiding the destruction. (By enforcing a mass it created force that removes all chances from diverting away from the collision that is about to occur). Such a force then removes all possibilities of avoiding the oncoming collision. The force will not allow any attempt to try and bring into the equation other possibilities in as much as rerouting the approaching object and changing the course in the imminent collision that is due and in due course will come about between the comet and the Sun . That which I explained is what Newton mathematically suggested with the formula. That is not what Kepler said notwithstanding so many arguments with Academics that I had in the past who tried to prove to me that the two visionaries views were equal and the same. Well…it's not the same because when we go onto translate Kepler to the verbal English the letters that come out do not even spell the same words.

It is conducive to remember that there is another part of the two relevancies applying where one is a^3 that is relevant to k but also there is the point where k has a duty to place a relation to a^3

The correct Translation of Kepler's mathematical expression will be $a^3 = T^2k$ which proves that Kepler said that there is a space a^3 which is equal $=$ to the motion in the **time duration T^2** thereof between two specific points which is a straight line k that holds a relation from a centre to an end where the two ends run from the beginning of k to connect at the end of k. I might not be the smartest boy on the block but I'm not that stupid either. I know how to translate… and I translate as follows:

a^3 must have a volumetric interpretation because the third dimension is sure evidence of multiple conjunctions of dimensions put together in three sides opposing three sides having the third dimension in place. The fact that any symbol uses a value to the **third power a^3** indicates **space** or a volumetric established and separate unit. Using a cube by three dimensions symbolises a cube, a room, a space to be filled, a unit able to hold other ingredients on the inside when empty or partly filled. It is space because it is volume using the third dimension.

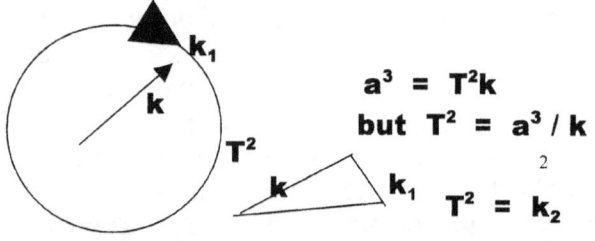

$$a^3 = T^2k$$
$$\text{but } T^2 = a^3 / k$$
$$T^2 = k_2$$

T^2 is an indication of something having a cubic nature other than the square forming motion that is provided by the motion the square indicates, which is where the moving object is representing a third dimensional object that is moving from point to point and it is this point to point that multiplies into the square. The space is moving as a unit from one point to another point and the moving between the points are represented by a flat square or following a flat distance between two points. The cubic space was in one instant in one place and then the second instant in the other and because time can never stand still or become single dimensional (this I am about to prove as the letter unfolds) insisting that time must always support the motion it consists of or time cannot be. It is motion that is taking time, which is motion in the second dimension moving the space in the cube.

k^1 is the symbol used to indicate a straight line between two points with a definite beginning and a specific end position. It is the location where the cube is holding space and where the space was and where the cube in space is going to be in very the next split instant that follows. That will then in multiplying form the square that indicates the time the journey took to move the cube of space from one point where **k** is indicating the location of the space to where the next indicating of **k** will shift the space being the cube pointing at the end of **k**. Since time represents the square and with **k** being the distance that proves that the **k** represents the distance the space representing the cube went to take the time represented by the square through the motion. It is the distance moving space in the cube to complete time in duration in the square of motion; therefore **k** is permitted to be in the single dimension.

There are infinitely more implications in the statement Kepler delivered than what is merely a contribution to motion and only motion as Newton was of the opinion. What is there mathematically not correct in my interpretation of Kepler's manner of translating mathematics to English and why is any changing thereof by Newton or any other person necessary in any way?

We can test any of the following symbolic values in the mathematical expression and also test the principals behind the expression in which Kepler stated them. By such testing we will find that time after time there were never any corrections in the translations required since the translation thereof was never incorrectly presented and in that a case asked for no alterations to secure the correct reporting of the cosmic information being translated. By taking the formula on face value it can change as follows: $a^3 = T^2 k$ can become $k = a^3 / T^2$

That proves that the establishing of distance **k** will produce space a^3 and set space a^3 in motion T^2 where such motion is in opposition to singularity, which means gravity or contraction is the deliberate opposite of expanding $a^3 /k = T^2$. In the beginning the expanding then also involved three more points all just outside the border of singularity but within the atom exclusivity. It extends **k** while it introduce a returning relevancy back to singularity k^0 by creating motion in spin and duplicating space by reducing space.

With this mathematical reality what then later formed the grounds for any individual to develop any need to change Kepler's translations from the cosmic given to mathematics and then from mathematics to English while the guilty party is renowned for his superior skills in mathematics?

Kepler translated what he found to be the cosmic given to mathematics which we humans are able to interpret from the mathematical expressed to the verbally pronounced and written but Newton still saw a need to change what the cosmos said about how the cosmos is presented and by no one less than by its own interpretation of its self structured composition.

When viewing my interpreting of what Kepler said I might have asked myself countless times what did I not translate correctly from the mathematical expressed to English after encountering a battery of Academic onslaught and resentment on my Newtonian views because after all it is directly diverting strongly from the teachings presented by Mainstream science and the diverting is not coming in a small way.

In truth from my diverting I came across very new ideas I am able to prove. By my translating Kepler's work correctly I came upon answers not yet uncovered by Mainstream Science

Kepler gave the World mathematically translated cosmic answers he received from the cosmos that Kepler uncovered long before Newton, Einstein and others got wise about cosmology...and later the wise came up with old news (old views as far as Kepler expressed their views before they, the wise were born with the purpose of coming to the conclusion that those wise men eventually did) and where the conclusions that the wise concluded brought much surprise to the world with the originality of the later Masters' initiative while Kepler said the same thing ages before...!)

Such is the advantage of recollecting Kepler facts that it does answer many questions, which went unnoticed and therefore not spoken about up to now and some were previously never even thought about.

Newton said a sphere is $a^3 = 4/3 \, \Pi \, r^3$, which is mathematically correct, however

Kepler said the cosmos told him a cosmic sphere is $a^3 = k\,T^2$ There are the two distinct possibilities which Newton saw and which Kepler saw and both are most valid. Between the two concepts there is literally one Universal difference and the two can never be mistaken as promoting the same principles. 'Ever try to answer facts about the Universe in as much as…what brings about the expanding? Kepler said the Universe plus its entire content is expanding centuries before Edwin Hubble realised what he was seeing through his telescope.

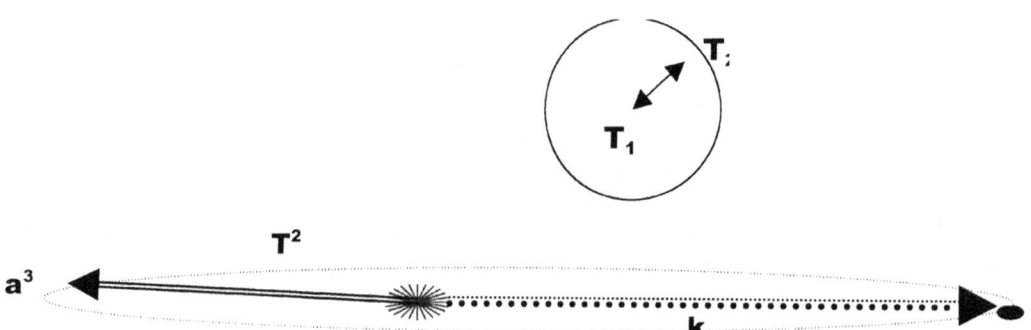

Kepler was the very first person to mathematically introduce **space a^3 centre k** and **time T^2**. Not only did he introduce **space-time a^3 / T^2** but he also placed **space a^3** and **time T^2** in a relevancy long before Einstein did and placed **gravity in space-time a^3 / T^2** even before Newton named gravity. He showed that space **k** is growing in the measure of what means the Universe attend to by promoting space-time as $a^3 / T^2 = k^1$. Kepler was the person who placed gravity as the ingredient in the Universe that determines **space a^3** and **time T^2** and much more. Kepler was the first one that said that gravity comprises of two factors being **k** or linear gravity and **circular gravity or T^2** as gravity keeps space in form while all is staying together.

Although not one Academic has ever openly admitted to me that they as members and part of Mainstream science are more aware than I am of all the facts and doubts I point out to them, such evidence then becomes clear whenever I mention the matter to them I get more than the impression it does not come as a surprise to them and hit them like a brick between the eyes. The lack of surprise and initial doubt they should show at first when they discover the incorrectness of evidence in their theory is a telltale sign confirming my suspicions about their evidently knowing all this information all along. They clearly seem very agitated about every detail I show when I bring the mistakes and double talk to their attention in the hope that they may confirm my doubts.

 Never is there a whisper of a surprise or a hint of a suggestion that would initiate an argument carried on by the bewilderment or the astonishing surprise they should feel confirming my arguments because there is a mild complacency in their voices. My jumping them total unexpectedly about matters they never contemplated in the least leaves them unturned. The rush in blood pressure that should be a factor on their part and part of the instant where total surprise will bring about some confusing thoughts that will inspire the unleashing of an argument in defending their holy grail should at least carry a surprise in an attempt to save what they believe as being the Gospel in science and with that defending their honour. They lack embarrassment, which they should have in their disputing of my claim as they fight off my allegations with a countering of denial claiming foul on my part as they are in shock when finding out about any doubts. A lack of true emotion on their part is a telling sign that they also may have some serious thoughts on the quiet about any inclination presenting a flawed view about what they always thought they knew to be true.

There is only that eerie dismissing of the seriousness and the lack they show in excitement that would deny or support my credibility as I present my findings. If they know about the inconsequential facts in science why is it not generally acknowledged and pronounced as a matter of fact? Why is there the covering up and hiding facts that we associate with some professional criminals such as politicians. The fact that Academics are aware of this evidence in general terms about the misinformation and doubting evidence about Newton's cosmic vision but moreover underlying this is their total denial of knowing about it and that is what is so seriously unforgivable. The fact that all Academics are aware of my evidence even before my presenting them with such evidence is beyond doubt. If that is the case then why are they forever trying to kill my viewpoint and forever try to silence me where I am only the messenger because I bring the solution and the answer? Please note that the answer and the solution are unbelievably simple and unsophisticated. It lacks all the splendour and grandeur expected by all Academics concerned. It is because it is so simple that it went amiss for four hundred years. It is because it is so simple that it misses the grandeur that will entice them. Instead every academic accuses me of not understanding Newton while they can't show me what part it is that I can't understand and I on the other hand can't see what there is not to understand..

Newton said that it is the reducing of the distance between the objects that would bring about the un-reversible reducing that will end in a total demolishing of the radius that is between the cosmos structures, but instead we find the gravity applying in outer space is one of the instances where gravity provides an orbit circle that gravity seems never to completed as the orbiting objects follow from closing any circle that is leading into a following circle up to where the circle is completed in cyclic precision. That is not the gravity that Newton identified although Newton admitted that here is a presence of a centre forming a point in the middle between the two objects. He was unable o know what caused or even the presence of the Coanda principle, which forms so critical a part of my theory. The formula concerning cosmic balanced gravity however leaves no room for the admitting of such a point and by not leaving a possible inclusion of such a point in his formula Newton did by such gesture in principle repeal his admission of such a centre. This had me cast doubt on what is taught at institutions of learning. It motivated me to venture back to an era before Newton came to influence science. I came to acknowledge Kepler as I came to understand Kepler. The accepting of that what I understand in Kepler involves much more reading into what Kepler said by finding what Kepler did not say in the way that he did say what he said than the reading about what Kepler said as it is written in the precise detail and to the letter used in his statements. He never directly stated what he said. Again I must stress this point: when I refer to what Kepler said it most likely means reading into the part that is being a part of the part that he did not say when he was saying what he said but I accept that he meant to say what I am reading and translating from Kepler as part of what he did not say but meant to say. I have to read more with my mind than with my eyes. This comes as a result of interpreting Mathematics to the verbally expressed. I had to learn to read with my mind and not my eyes and I found that that is the manner in which one has to approach cosmology. From the first time I discovered what manner one should use if one wished to read into Kepler's findings I saw Kepler was all about uncovering the unknown. Realising that, the conclusions I drew by reading in such a way cemented my better understanding of Kepler's work, which then helped me improve my insight into Kepler's work as it increased my understanding about cosmology several fold. This helped me to realise what implications were to be found underneath Kepler's discoveries. From my realising what approach I should use, it helped me to improve my cosmic realising by using the method of reading Kepler and from that I could come to appreciate what Kepler introduced.

Only then did it bring insight and proof to me as a student of Kepler and this proof I found by dissecting what Kepler **did not** say instead of what he **did** say, which I now present to you with this letter, you being a superior intellectual person. Kepler said $a^3 = T^2 k$ and that correctly translates to a mathematical expression $k^0 = a^3 / T^2 k$ which in the verbal statement in English translates that Kepler said that there is a **space a^3** which is **equal =** to the motion in **the time duration T^2** thereof between two specific points which holds a relation onto a centre k^0 where from there forms **a straight line k** that is centred on the spot where space begins from k^0 **that produces k** as well as producing the circle therefore that spot $k^0 = a^3 / T^2 k$ has hold k^0 at a value of having the least space. The line **k** is centred onto a spot where space begins specifically at k^0. This point not only produces the line k^0 but represents also the space that forms the eventual circle T^2. Therefore from the centre holding k^0 , k^0 leads to **k** that forms the roving space a^3, which is rotating at a distance **k** where T^2 forms the outer limit of k^0. Mathematically $a^3 = T^2 k$ will be $k^0 = a^3 / (T^2 k)$ because $k^0 = 1$. But $k^0 = 1$ also present the single dimension where all factors are a product of one. If one can locate k^0 one will find singularity. That is where gravity is because gravity is strongest where space is least. Then that suggests that gravity is strongest at k^0 because space is least. That is gravity because that is what keeps the orbiting object in orbit but also that is what Newton completely missed when he changed Kepler's work. Newton failed to recognise gravity as the only ingredient in Kepler's formula. He admitted he missed this because he admitted he did not know what gravity is while Kepler explicitly showed what gravity is. Gravity is what keeps the orbiting object orbiting. **k = a^3 / T^2** is **distance^1 = space ^{3}/ time^2**

forming from a pivoting centre k^0. That is a cycle and moreover it is a cycle formed **by space/time**. What Kepler said is that space is a^3 **in motion** T^2 **k.**

That says **space**^{3} **(a^3)** relates directly to **time**^{2} that uses the symbol T^2. This is also what I refer to when I say one has to read what Kepler did *not* say when one wishes to see what he *meant* to say. Kepler introduced space^3 –time^2 long before Einstein's date of birth appeared on any calendar although Einstein is credited with the formulating of the concept of space-time and giving it a name. Going even further Kepler stated that the space a^3 is on the move T^2 around in a circle at a distance **k.** That is what that comet we are discussing is doing. The space^3 (Comet) is circling the Sun using a radius **k** to establish the cyclic time^2 as a period of continuous motion and continuous motion is gravity. That reads much more correctly and closer to the truth than what Newton predicted what according to him (Newton) was happening in space. Remember in this statement I am separating cosmic principles applying from the way that gravitational principles apply on Earth. I distinguish that which is the rule in the cosmos from what we find ourselves trapped in on Earth. The two just don't mix. I am removing cosmic physics from normally accepted physics because the gravity concerned is not the same.

The proof I bring is real however simple it may seem. It has none of the mind-blowing complexities normally associated in the presenting of investigative analyses of Astronomy. I realise the information in this book carries the arguments in a childlike manner which are very simple to follow, and for that in the past I have been blamed over and over again as being unprofessional. In my answer to that I can only reply by using another question: Are only professionals adequately equipped with minds that make them (the professionals) the only ones able to think? We being part of the human race are all thinkers. Everyone as a human being can think. Every person on Earth is a thinking thinker that uses his brainpower by exploring thoughts mainly and normally to his or her personal benefit. It is what we think about that produces the results of our efforts by which we accomplish what ever we are thinking about. I have met professional Academics that I found foolish as much as there are other cases where the so-called amateurs can credit themselves with much more wisdom and insight. Albert Einstein as a patent clerk was that much but to name one. Please understand that I do not compare my achievements or myself in any way, shape or form with the likes of a Master such as Einstein although I speak my mind when not being totally in agreement with some of his or other views. My unsophisticated retracing of Mainstream physics concerning the Big Bang in detail helps to reinvestigate established principles and moreover investigate proof in the light of modern evidence. In principle I distinguish between Kepler and Newton in that Newton is one hundred percent correct concerning gravity on Earth but as far as outer space forms gravity the conclusions of Kepler and Newton do not match and they had totally different ideas about what they saw in gravity.

I am in disagreement with some basic principles that science acknowledges and I divert strongly from all accepted roads Mainstream physics follow. My doing that prompted those who are considered and accepted as self-proclaimed members of Mainstream Physics have categorised my views in the past as incoherent. That I do not accept. I admit that my line of thought is extraordinary and controversial but only to Mainstream science and not to the standards lay down by nature. Since the concepts I follow start at the beginning, and I take Kepler at the point where modern cosmology began and in that mindset I re-evaluate Kepler's work. I start by tracing a new approach as to what I see Kepler found. The main condition of my investigation is to establish a divorce between what Kepler said and what Newton thought to add to what Kepler said. It is this divorce I create that Mainstream science finds repugnant or even in some persons' opinion repulsive. I believe the repugnancy does not come from or is not manifested in any part of my work to the letter as such, but rather what my work suggests and who is doing the suggesting. To my view in cosmology such adding to Kepler by Newton was unnecessary and it diverts Kepler's work away from cosmology. But as the generations moved on Newton became religiosity in the mind of science wherever science was taught. To students there is little or no choice in the matter since the only choice left to them is one of understanding by forcefully accepting or die an academic death since Newton is academically accepted without asking questions or raising an opinion. For the second choice, the less accepting students are greeted with a Dear John good-bye letter sending them off into the unknown Sun set that such a future outside physics will bring them. That is brain washing.

From studying Kepler I saw that we have to gauge what we find in the Universe. What we find is not that what we realise with our eyes but that what we observe by using our minds to translate from visions coming from our eyes to our minds. We have to test the part that we are seeing much more than merely accept what there is to see on face value. We have to not only see what other life beings blessed with much less insight most probably also should see. We must stop using our eyes in the same manner as animals do and start seeing with our mind, as humans should do. By being the superior evolved species that we are, it gives us the ability to read into that which only we can see and that we only can see by using our intellectual mindset. By seeing with an intellectual understanding what there is to see when we see what we can observe, we should therefore have the ability to be in understanding by looking at what we can see but moreover understand that which we cannot see. It is the same as playing chess. See what there should be moved instead of noticing an object not having an ability to move by own initiative.

This I first found to be true about Kepler's work and when I started projecting this method of observing what the Universe is, as it scattered most previous perceptions I found that using the new method brought along answers so fast I could sometimes hardly keep up with the interpreting thereof. But as is the case with Kepler so is the case with the entire study of cosmology: One should see what there is about the cosmos, which is unseen to us and then we may find so much more in the cosmos unseen to us representing that which we cannot see and that which we cannot read because we have to learn to read what is not written in light. Armed with this realising I then proceeded from that point by further arguing and debating the full implication of Kepler's contribution. Kepler placed cosmic structures in relevance to one another and so does the Big Bang Theory. The backbone of the Big Bang is that relevancies apply in dynamics and such dynamics are placing all structures without any reservations independent from each other. As the Big Bang progresses all filling the Universe to the inside is in the same Universe that was then at the time of the Big bang just as much as it will always be and the lot remain the same, however the relations that the elements comply to bring across new relevancies with new positions to fill. The father of the Big Bang concept is a person by the name of Father LE MAÎTRE, GEORGE ÉDOUARD (1894-1966) who was a Belgian priest and cosmologist. He was the first person to embrace the fact that the Universe expanded from an infant stage.

His model of an expanding Universe (1927) was superior to that of W. de Sitter in that it took into account mass, gravitation and the curvature of space. Similar models were proposed in the early 1920s by the Russian mathematician Alexander Alexandrovich Friedmann (1888-1925) but Friedman compiled various such possibilities. Lemaître argued further (1931) that the quantum theory supported an origin in the explosion of a 'primeval atom' or 'cosmic egg' into which was originally concentrated all mass and energy. As modified by A.S. Eddington, Lemaître's model provided the springboard for G. Gamow's Big Bang theory. In the wider picture of science in general a lot changed to just allow such turnabout in thought since the day of Isaac Newton. From Newton's attraction and contraction many things came into place that allowed change in the most hardened minds. Accepting facts about the Big Bang concept is quite radical. By promoting expansion the Big Bang theory contradicts gravity and our accepting of the Big Bang has to change all other concepts. By accepting the Big Bang other changes are also involved.

KEPLER, JOHANNES (1571-1630)
The German mathematician and astronomer KEPLER, JOHANNES (1571-1630) became Tycho Brahe's assistant in Prague in 1600 A. D. where he undertook to complete the tables of planetary motion Tycho had begun. Kepler first calculated the orbit of Mars. He spent much time trying to reconcile Tycho' s accurate observations of the planet with a circular orbit, but concluded (in Astronomia nova, published in 1609) that Mars moved instead in an elliptical orbit. Thus, he established the first of his laws of planetary motion. A theory that the Sun controlled the planets by a magnetic force led him to the second and third of his laws, which were published as part of his treatise on theoretical astronomy, Epitome astronomiae Coernicanae (1618-21). The Rudolphine Tables (named after Tycho's patron, the Holy Roman Emperor Rudolph II) of planetary motion appeared in 1627 and were still in use in the 18th century. Kepler also wrote De Stella nova, on the supernova of 1604 and Diptirce on optics and the theory of the telescope. The overall view followed in this book **an open letter To Selected Academics ISBN 0-9584410-9-X** places the true significance of his work in true contents. In KEPLER'S EQUATION is the equation that relates the

eccentric anomaly of a body in an elliptical orbit to its mean anomaly. The equation is $E - e \sin E = M$., where E is the eccentric anomaly, M the mean anomaly, and e the eccentricity of the orbit. It is important as one of the mathematical relations enabling the position of a planet about the Sun , or a satellite about is planet, to be calculated from the orbital elements for any time. However this only relates to the solar system, and KEPLER'S LAWS only apply in the contents of the solar system. The three laws governing the orbital motions of the planets, discovered by J. Kepler is as follows: The first law states that the orbit of a planet is an ellipse with the Sun at one focus of the ellipse. The second law states that the radius vector joining planets to the Sun sweeps out equal areas in equal times. The third law states that the square of the orbital period of each planet in years is proportional to the cube of the semi major axis of the planet's orbit. The first law gives the shape of the planet's orbit; the second describes how the planet must continuously vary its speed as it follows its orbit, moving fastest at perihelion and slowest at aphelion. The third law gives the relationship between the planets' average distances from the Sun and their periods of revolution.

Instead of studying the true value and contribution of to Kepler's laws an Englishman going by the name of I. Newton placed his own interpretation to Kepler's laws, and in doing this, he wilfully destroyed the principle working of the Creation. Saying this I hear the alarming hooters announce Newtonian dismay. In the past my experience was that all the revered Academics lost their appetite for any further investigation of my work. That is sad as much as it is regrettable. Through Newton's tunnel vision, he applied his own misinterpretations to the correct presumptions of Kepler and through the Newtonian tunnel vision Academics did not move an inch away from repeating the same procedure. In the past it was this that had Academics shying away from me because at the point where I raise criticism of the Newtonian viewpoint I am rejected. The point where I declare my suspicions concerning they're accuracy and the correctness about their theories, which is where I should then be raising their doubts about their way of thinking is the point where in stead I raise their suspicions about my way of thinking. That is what caused the rejection of my criticism about Academic Newtonian science and evoked their criticism in the past about my views instead of them following the logic by investigating what I said.

Their rejection of self-investigating got me and my work rejected to a point where the applecart lost its wheels on every occasion. It is where Academics read my remarks and what brings (seemingly in an instant) wrath to Academics. I say this because I realise that reading my remarks or hearing me remarking about this notion brought much resentment on their part and if the reader at the present moment is a Newtonian, boiling his/her blood. It is blood boiling because I believe they see my remarks as belittling that which they feel they have accomplished. This is not the case but still my remarks have the same effect on the Academic as pouring icy cold water down the back of his shirt. I mention this because I know it has happened many times before and if possible I wish to avoid this response. Therefore I ask you kindly to please be warned about the negativity you must feel towards me where you are the Newtonian and I am not. Before you lose interest in reading this letter any further please allow me to finish. In the past Academics thought me to be presumptuous and that normally became the point where all the Academics find their interest vanishes. That should not be because if Newton's work is as utterly accurate as those with faith in his work believe it is, then every aspect about Newton should stand above any and all reprimanding or any form of doubt causing a notion to reprimand. The testing of Newton's work should withstand all testing notwithstanding the person or the prominence of such a person's social or academic standing in the Academic society or even the prominence that such testing will deliver.

From what I see about Kepler's work it is a flow of circumstances that lead to Academics neglecting Kepler's work and the realising of the theory I suggest is not forthcoming due to my personal brilliance. I do not consider myself to be the brilliant in any way as to be the one that can remove the verbal splinter from the eye of the Academic. Yet…if there is a splinter what else should I then do…Newton reduced the implication that Kepler findings hold by introducing to the law of gravitation. He then went about and changed it to three laws of motion. It is clear that while he formulated the laws on motion he missed the way Kepler introduced gravity as space a^3 coming about through motion T^2 and that gravity is space a^3 within space k within motion T^2. Newton also missed the fact that gravity is at its strongest where motion and space cease to be. This is most important to recognise about gravity in one of the two forms it has. I. Newton generalized Kepler's first law, verified the second law, and showed that the third law should be amended to the form; $4\pi^2 a^3/T^2 = G$

$(m + m_p)$. In this, the value of "**T**" and "**a**" are the period of revolution and semi major axis of the orbit of a planet of mass m_p about the Sun of mass m, and G is the gravitational constant.

It should be clear to any person investigating Johannes Kepler and his work that Isaac Newton hijacked Kepler's work and any time there is the slightest referring to Kepler about the research Tycho Brahe and Johannes Kepler did such referring to Kepler always lead to and always include the mentioning of Isaac Newton changing the work of Johannes Kepler. It is as if the World never could acknowledge Johannes Kepler because the work of Johannes Kepler would be completely wrong and misleading if it were not for the intervention of Isaac Newton saving the skin of the less admirable Johannes Kepler. This comes in the midst of every one realising that Kepler used the information he received directly from the cosmos.

I do stress this on many occasions throughout the letter because the embarrassing part is that Newton changed the work of The Universe and not of the man called Kepler. Should you reading the letter entertain the opinion of Newton and feel any urge to defend Newton you should ask the question as to who is standing corrected, is it Kepler or is it the cosmos that gave Kepler the information he concluded? The cosmos supplied all the information by using mathematics, which Kepler then had to translate. But Newton destroyed the accuracy by altering what the cosmos said and directly by adding to that what he (Newton by name) thought that the cosmos left out. This set a precedent by Newton in cosmology and also set a trend, which was retained in all future cosmological development and it lasted in cosmology for three hundred and fifty years. In this book you are reading, I am about to show that such practise should no longer be accepted in cosmology. In the process the world of Mathematics developed by the world of cosmology stood still for almost four hundred years. Faculties contributing to cosmology and feeding off cosmology improved as much as they developed, but when cosmologists see the Roche limit in action in the lens of the Hubble telescope and refer to the event as "stars blowing bubbles" being the ultimate response coming from those persons who are supposedly the Masters of cosmology affairs, then the truth of what I just said comes down on you like a ton of bricks.

Everyone having any remote interest in cosmology will find they are being very disillusioned by such "official" testimony about the evidence the Ultra Wise report about. This book is about showing how great Johannes Kepler was and how enormous his work was. It will show he preceded all ideas of everyone that came later and officially introduced the novelty of such ideas. Back during the time Kepler was introducing his work the stature and the magnitude of his work was beyond any person's understanding (including Isaac Newton) and this prevailed for most of half a millennium. I do not say I am the brilliant one to uncover Kepler in the face of everyone failing that came before me, but as I am not a Newtonian such bias was not part of my repertoire and denying me the fortune of being a Newtonian added to my fortune of realising Kepler.

Yet as you will notice, the work I contribute is much below the sophisticated norm of modern investigative research and the levels that modern research accomplishment demands to better the effort of the understanding ability in the splendour that investigative research work should deliver in view of our modern times. It is only pure neglect in science circles that moved science past Kepler. Not seeing and therefore not investigating through almost half a millennium has paved a road past the inferior levels that the researching of Kepler's work holds because it was rocket science four centuries ago but the brilliance of it has faded since then. My contribution holds no astonishing flair that may add to science in general. Only failure to notice what I see on the part of those truly brilliant can explain my being able to present my contribution about my work in investigating Kepler. Only by their passing such degrading levels of the Academic establishment in the past and the present can bring the blame for such an obvious discrepancy because any involvement in the work at such an inferior level as that which I bring cannot interest and excite a salted Academic and when thinking about it, the idea is totally unthinkable.

This letter, although it is on this inferior level is about correcting this tendency and has in mind the effort to put in writing what would place Kepler in the greatness and glory he deserves. As I already said, if Kepler was wrong then the cosmos was wrong about facts and applying relevancies and tendencies in the cosmos. I yet again wish to reiterate we should never for one moment forget that Kepler received his information directly from studying the cosmos so how could the cosmos stand corrected? In spite of all the brilliance attributed to Newton nonetheless if Newton had the mind to change Kepler's work and my saying this includes all persons agreeing with such changing by Newton of the work of Kepler those persons admit that he or she or Newton never took any time to really and truly investigate what the cosmos told Kepler. Through understanding the work of Kepler I prove gravity, the Titius Bode law, singularity, space-time, space-time relevancy, the Lagrangian system, the Coanda effect and the Roche principle, the sound barrier, the principle behind the Black hole. The precondition for my ability in doing so is that I have to remove Newton's opinion about Kepler's work from Kepler's work.   Whenever cosmology comes into question and all the phenomena, which I mentioned just now remains unexplained and by that token alone it shows to what degree did cosmology remain undeveloped. Whenever there is any mention of Newton, Kepler is never mentioned. But the reverse is always applying. Mainstream physics holds the opinion that Kepler may only have an opinion if Newton can change the opinion. Kepler gave space-time, gave gravity, gave singularity, gave the Plank theory, gave the theory on relativity but no one ever found Kepler's work deserving enough to launch any investigation such as I did. I belabour this because of what revulsion my rejection of Newton unleashed. That is one barrier much unnecessary but it has been an insurmountable barrier this far.

NEWTON, ISAAC (1642-1727) and NEWTON'S LAWS OF MOTION
An English physicist and mathematician who developed his principal theories about gravitation, optics and mathematics between 1665 and 1666. In 1668, he made the first working reflecting telescope. Most of his work remained unpublished for long periods, partly because of criticisms by c. Huygens and the English scientist Robert Hooke (1635-1703) of his early work on the corpuscular theory of light. However, in 1684 E. Halley persuaded him to organize his work on the celestial mechanics of the Solar System, which was published as the Principia. Newton's other major work, Opticks, was not published until 1704. It contains his corpuscular theory of light, and the theory of the telescope. His greatest mathematical achievement was his invention of calculus, independently of the German mathematician Gottfried Wilhelm Leibniz (1646-1716). His profound influence on physics and astronomy is reflected in the phrase 'Newtonian revolution'. Three laws published in 1687 by I. Newton concerning the motion of bodies.

1. A body continues in a state of uniform rest of motion unless acted upon by an external force.
2. The acceleration produced when a force acts is directly proportional to the force and takes place in the direction in which the force acts.
3. To every action there is an equal and opposite reaction.
4. However there is one more law on motion that went undetected by   Newton…This book is not about trying to disprove Newton…it is about adding too science more than there now is available without removing any that science already accumulated.

In this book I use Kepler's formula to either prove or to disprove the following accepted principals in cosmology and if any person in the past gave only the slightest attention to Kepler's work, many statements would have come much sooner delivered by someone else or may never have come at all. By applying Kepler's formula correctly in this letter I can either agree with or in other cases deny the following principles.

It began with NICOLAUS COPERNICUS who changed the status quo. COPERNICUS, NICOLAUS (1473-1543) was, according to the Anglo Americans, a Polish churchman and astronomer although this is just more politically inspired propaganda because his parents were both German (in Polish, Mikolaj Kopernigk). While he was completing his studies, he had realized that the Earth revolves around the Sun and not vice versa. Such a view was in that time, held to be heretical. As I pointed out in the first few articles, the Church regarded the geocentric world-view of Ptolemy as consistent with its doctrines. Copernicus set down his basic ideas around 1510 in the Commentariolus, which he circulated anonymously, because of the Islam link. In 1512-- 29 he conducted his study and concluded the observations that he needed to support his theory, while carrying out ecclesiastic and local administrative duties. In this time, he had to defend his mother in court on charges of witchcraft. In 1539, the Austrian astronomer and mathematician Georg Joachim von Lauchen (1514-74), known as Rheticus, became a pupil of Copernicus and began to spread his ideas. The published work was openly spread as the Copernican system, in spite of the life-threatening dangers connected with such a "crime", in 1543 in the book De revolutionibus orbium coelestium. However, the reality of a heliocentric Solar System was only commonly accepted, after the work of Galileo and J. Kepler. The ideas introduced developed along and proved to be correct until such a time it met a solid wall with the investigation of Max Planck.

PLANCK CONSTANT

(Symbol h) A constant that relates the energy of a photon to its frequency. It has the value 6.62076 $\times 10^{-34}$ Js. It is named after the German physicist Max Karl Ernst Ludwig Planck (1858 – 1947). PLANCK ERA. In the Big Bang theory, the fleeting period between the Big Bang itself and the so-called Planck time when the Universe was 10^{-43} s old and the temperature were 10^{34}K. In this period, quantum gravitational effects are thought to have dominated. Theoretical understanding of this phase is virtually non-existent. It is named after Max Planck (1858-1947).

PLANCK'S LAW

A mathematical description of the energy radiated at different wavelengths by a black body: $E = hf$, where E is the energy of a photon and f its frequency. It was formulated in 1900 by Max Planck (1858-1947), who realized that energy is radiated in discrete packets, which he called quanta, and it formed the basis of quantum theory. The quantum of light is a photon, the energy of which depends on its wavelength.

There is one rule which is well established and which Mainstream science agrees about. It is one aspect, which forms the very principle that holds the theory about the cosmic starting together under the covering of a verbal blanket. All in science agree that it all started with singularity but I manage to go one step further where I prove that it is also where it ends, as singularity reunites space-time, which is from where Creation split in the very beginning.

Singularity is as follows: Singularity: a mathematical point at which certain physical quantities reach infinite values, for example, according to the general relativity, the curvature of space-time becomes infinite in a black hole. In the big bang theory the Universe was born from singularity in which the density and temperature of matter were infinite. From singularity flows space-time.

Space-time is as follows: Space-time is a four dimensional position of the Universe where the position of an object is specified by three coordinates in space and one position in time. According to the theory of special relativity there is no absolute time, which can be measured independently of the observer, so events that are simultaneous as seen from one observer occur at different times when seen from a different place. Time must therefore be measured in a relative manner as are positions in three-dimensional Euclidean space, and this is achieved through the concept of space-time. The trajectory of an object in space-time is called world line. General relativity relates to curvature of space-time to the positions and motions of particles of matter.

SPECIAL THEORY ON RELATIVITY

A theory proposed by A. Einstein in 1905, based on the proposition that the speed of light in a vacuum is constant throughout the Universe, and is independent of the motion of the observer and the emitting body. A consequence of this proposition is that three things happen as an object's velocity approaches the speed of light: its mass goes up, its length shortens in the direction of motion, and time slows down. Hence, according to special relativity, no object can ever reach the speed of light because its mass would then become infinite, its length would become zero, and time would stand still. In addition, Einstein concluded that the mass of a body is a measure of its energy content, according to the famous equation $E = MC^2$, where c is the speed of light. This equation describes the conversion of mass into energy in nuclear reactions within stars.

GRAVITATIONAL COLLAPSE

The collapse of a body that is unable to support itself against its own gravity. Gaseous bodies undergo such collapse if they are not hot enough for their gas pressure to balance gravity. This can happen in the early stages of star formation, or when nuclear burning ceases in a star's core. The time taken for such collapse decreases rapidly with increasing density, varying from about 100 000 years for the birth of a new star to less than a second for the formation of a neutron star. Star clusters may undergo a similar collapse if the random motions of their constituent stars are insufficient to offset gravitational effects, either during their formation or at an advanced stage of their evolution.

GRAVITON

A hypothetical particle or quantum of gravitational energy, predicted by the general theory of relativity.   gravity - motions have not been observed but are predicted to travel at the speed of light and to have zero rest mass and charge. A graviton is the gravitational equivalent of a photon. It is this anti-photon-being-a- gravity - motion by just merely swapping direction and all is proved that I find not very indigestible in modern science. One of the main issues that I wish to protest by my writing of this is my argument that if the Universe can be compressed back to the size it had at the point of 10^{-38} seconds after the Big Bang the daily outdoor temperatures of 10^{27} K will also come about once more. The expansion was the result of compressed space, which then formed into heat and in turn resulted in finding a Universe with all the insufficiency of space less ness prevailing throughout and wherever space was needed. By that it forced space-time to come into being. Space-time came about at the time of endless time duration without space availability, which brought about the period of the Big Bang wherein space growth was the converting of such heat to space. If the Universe was in a vacuum as big as being available now then what was the temperature of the vacuum while it was empty before material filled it later.

Then I presume the vacuum was there present as it is now in this present day. If the Universe then employed the space of say one atom, the impression comes through that from edge to edge and from Universal border to border the space occupied was the same as one atom will claim in our present day and age. Normal gravity started at 10^{-43} seconds. The Universe was the size of a neutron or somewhere in that vicinity. The Big Bang began and GUT, or the grand unified theory, produced the attempt to describe the strong and weak nuclear forces and electromagnetism in one single mathematical theory. Somewhere before 10^{-12} seconds of counting the Universe cooled to about 10^{15} K the electromagnetic and the weak interactions acted as one single physical force. Science reckons that unification may come about at temperatures of 10^{27} K, which was the temperature of the day at 10^{-38} seconds after the Big Bang. This statement echoes my viewpoint but one has to look carefully for that to surface.

In the suggestion the presumption claims that all the space that the Universe made available at that time was the total space one atom might take up today. If that might be the case then where was the rest of the space that now fills the Universe? Or was the rest of the space we now find in the Universe and what is now explained away as the vacuum, also available back then. Did the Universe only have that one tiny hot spot it filled with huge volumes of heat? Was the rest of the space vacant being out there all along during all the time running to the present date but filled with emptiness standing around as a big vacuum with no better to do than sucking on the Universe while the Universe was exploding at the speed of light. Then that statement suggests that in this hot Universe there were light-years upon light-years of vacuum waiting to be filled by the intense heat soaring in the smallest spot. If that is the case then why did the vacuum not fill in the blink of an eye by all the exploding expanding material growing at the speed of light?

Was the Universe overall bitterly cold where the vacant space was locked in with one spot of the vacuum filled with temperatures so hot we can only produce it in numbers suggesting a value but never claim to be able to digest the reality thereof in the human mind? If so what happened to the natural consequence that heat flows in the direction of cold and equalise between hot and cold. Was the space being available at present available then or was the hot space the only space available at the time. If so what prevented the heat from instantaneously filling the eternally cold vacuum because with the rules controlling vacuum in affect, it should have filled in such a manner in less than a heartbeat? I believe that singularity formed space-time and space-time developed from the overflowing thereof at the time it was extending. With time marching onwards and outwards to this day space-time developed. Space-time developed another product that everything in the cosmos has to have. It must be in such large quantities everything imaginable in the Universe has to have it and that is space using time to move about. I suggest that it is space that is holding heat in a quantity providing density and ratio to space available and in relevance to the space being available to quantify the presence of the heat and which then proves to form the time factor. The container and contained all together mixed by motion. From that very first separating of heat and space, which is what formed from singularity to produce space-time. The Universe was full... It was overflowing by the speed of light in the beginning...so where and when did vacuum or nothing enter the Universe as a factor if and when the Universe was so full.

The answer to that is absolutely crucial because how did the Universe decide to fill some parts with a variety of something and decide to fill some parts within the in-between with nothing? If that is true why did gravity not prevent the vacuum filling because no gravity that came about since can beat the force that gravity had back then? This leads to another question following the previous one in asking why did gravity at the time when it was so strong with r^2 so much compromised not fill the nothing immediately as it entered with something that could absorb the nothing. At the very beginning the mass that was pulling on the mass by force was immeasurable and none quantifiable. Even more to the point is the question to be asked in how big was the radius between the materials with the immeasurable mass placed in such a little space. This is all the more important in the light that the smaller the radius is the bigger the force will become from the immeasurable mass pulling.

With the immeasurable mass that was producing the first gravity between the particles divided by an almost non-existing radius the gravity produced had to be in gigantic proportional quantities and with the separation of the radii being in the infinite measure that it was at that point then how did the Universe establish the chance to expand. It did expand, as we all are witnesses to in spite of this contraction of gravity that had to have been compromising the expanding factors. Still the expanding filled the unknown part of the unoccupied Universe, which at the time was there or was not there and where it was not there it was then filled with nothing. If the nothing was not "nothing" then the nothing that was not being nothing was also filling the rest of the vacant Universe that was or that was not because if it was it was filled with nothing and if it was not then it was nothing.

This is then taking into account that then all the reducing that is resulting from Newtonian contraction and that was going about in the space available at that time was something filled with nothing and surrounded by more nothing. With everything in the Universe being that much crowded and crammed where and how did nothing enter the Universe and fill the rest that was unfilled? What factors introduced nothing into the picture since the entire Newtonian concept finds its base on the principle that matter reduces using gravity by force which then brings about reducing or the removing of the many nothing between particles, which will then lead to nothing that has to vanish even before nothing can enter the space.

This question may seem small-minded belonging to the mentality of a child or to that of the mentally impaired with not much factual appreciation developed yet. Please do not see it that way. If you think in those lines it will be because you do not have an answer to challenge these silly questions. Beware, silly as they are they represent official backing by the Wise-and-Informed. If the space is nothing and if the space was as large as it is at present then there was no need for such a small area to fill with something leaving only the rest filled with nothing at first since all the space we know about was there present and by being present it was there then for the taking. What ever filled the Universe had to start at the centre of the Universe and fill the entire Universe all over from a centre as it moved outwards filling from the inside outwards. This is a natural human instinctive realisation but is beyond proving by using Accepted Scientific policy. But that leaves Newtonian science with a massive unsolved problem: where is such a centre at the present time and where does the centre produce the limits or border it apparently has to form as it expands.

By expanding there is an additional contribution too that which was when that was, it was receiving more than there was before the addition increased that which was and by then becoming more than there previously was it had to be improving the border from where it must have been before the adding took place to where it was after what that was added was added. When that was less than it became when it was added, it was at the limit that was there before it was added too and that limit there was, was a limit that is the limit that I am referring to as a border being there. The cosmos is filled with unrecognised borders. The expanding has to be an ongoing filling that is at the same time expanding from the inside towards the outer limits of the Universe. Since nothing can enter from the outside where nothing is, the filling of nothing as a substance that would take up vast quantities of room had to fill from the very centre spot where all other filling came from. This filling of nothing with material has to be well mixed. The truth about cosmology is that space forms no borders but by using any Newtonian centre from where mass is attracting we must find a point where there has to be the ultimate Universal centre which is the cardinal point in the entire Universe and it is the first, the prime position to locate coming before any other concept one wish to put forward because all concepts has to start with locating that cardinal centre. There has to be the ultimate r^2 radii located precisely between the ultimate mass drawing the other ultimate mass closer. If there was a Big Bang then there has to be the spot where from the Big Bang developed therefore there has to be such a centre connecting the past to that ultimate centre with the line of development flowing onwards to this day. The fact that science is Newtonian proves that in the meantime Mainstream science is still of the opinion that there was the specific centre in the Universe that is nowhere to be found as it was filling the unknown with nothing coming from nowhere, but which somehow is still somewhere in the centre of all of that which is something. On the opposite side of nowhere there is an outer border in space producing a limit to nothing and serves nothing with a specific point to stop being nothing because that point is precisely where nothing ends and forms a beginning of a Universal border or a Universal end. How one will stop vacuum being no longer nothing was a question everyone comfortably missed to ask therefore no one ever seemed to deliver any form of answer.

One night some years ago very close friend of mine had a meal at his restaurant and as the conversation progresses he asked me about space and where it must end. I tried to explain to him what I believed in comparing to what Mainstream physics believed, but soon saw I was not gaining in his understanding. Then I decided to jot it down on paper and he could read it at his leisure as he saw fit. That led to the first book written by me (in Afrikaans my native language). What I tried to explain to Johan Boonzaier that night, is that if the Universe was the size of say even a tennis ball with only the size of a tennis ball being the very all of space there is available, then yes, it must take time to expand from that having the excessive heat there was back then in all the space we have at present.

It then is converting heat into space bringing about the expansion. But one will find most expanding within the atoms, as the atom must grow since the Universe in all was the size of what one atom is today. The space in the atom pushed the space outside the atom but there must be plenty more to the growth. Something outside the atom contributed in it own right because there is more expanding than there can be blamed on coming from the atom. But the space then also developed as the Universe developed and if space developed then it cannot be total vacuum filled with nothing because "nothing" cannot develop. You the reader must judge whom is correct between my view that space developed with the Universe as part of the Universe and reject the official view about space being nothing or otherwise you the reader must then decide that I am wrong, but should you do that, then find a reason why the Big Bang started out small and filled all the available vacuum or what is contemplated as vacuum that we have with the motion of time. When Mainstream science accepted the Big Bang as the principle that will take science into the future the view about such a Big Bang concept unlocks a different door to another view on the cosmos from birth to end. It calls for the revision of all aspects of the entire history on cosmology and change that what is dead wood and that, which needs to be chucked out. Most of all it was my following the lead I got from Kepler that unlocked the doors I now present to you. I claim there is no gravity - motion as there is no gravity forming weight or forming mass. I hope the sketch contributes to my explaining effort:

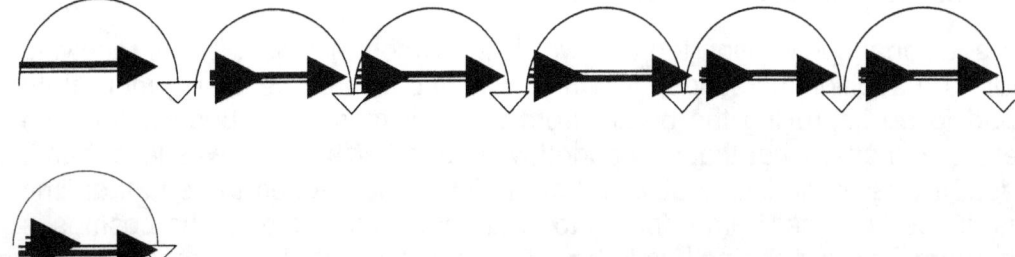

The duplicating frequency the Earth shows as k$_1$
The frequency of motion duplicating my body maintains as k$_2$.

k$_1$ minus k$_2$
The frequency of motion difference my body has minus to what the Earth has where that difference in motion becomes my mass. It is the sum total of the reducing of the motion that my body has in comparison with the motion capability of the Earth that is the mass value.

In k_1 as well as k_1 the symbol represents motion, however in the case where k_1 minus k_2 that shows an incapability of motion, which is motion, frustrated or a more commonly used name would be <u>mass</u> is created

The new **k** that is applying the relevance, must link the space a^3 being equal to the motion T^2 to singularity k^0 in order satisfy k^0. The flying of the aircraft is then unequal to the motion in the previous relation that was in place where it was part of the Sun and the Earth motion relation and the new motion will bring a correcting in the relevant distance **k** to put the motion in balance with space. Space will always demand a correct establishing of the miss-interpretation of the equilibrium that is needed to sustain the effected singularity because of the space-time factor.

There is a point where the two points forming the relevancy unite in shared singularity. It comes because of shared motion. In all space-time, one finds at least two relevancies where one is at the centre. That part Newton saw and formulised. He missed another part. Crossing a limit of inclusion is the limit of division and such limits are in distinction by motion producing the gravity, which is parting the two objects. Motion brings about a relevancy where two positions no longer share a common point in singularity. That is what Newton missed. That is the gravity aspect Newton and all other Newtonians miss.

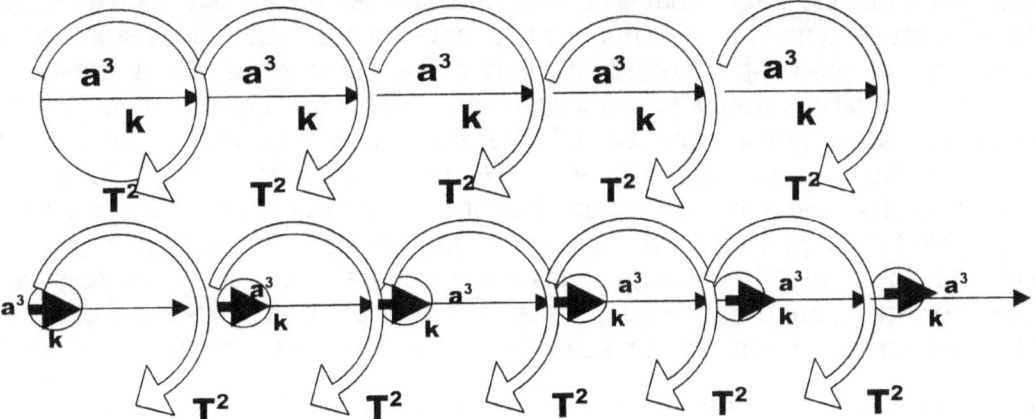

Two objects of substantial size differences are travelling at the same time but one has a space, which it has to move when it travels that is considerably different from the larger space. The larger space will produce an extending line equal to the space it moves while the smaller space will also produce a line in ratio to fit the space it holds relevant and which it has to move.

Mass has precious little to do with the whole affair except to be an obstacle intended to restrain the motion of the hosting space. The difference in size between the one in circular motion and the space in contracting motion must bring about that the smaller object has to move about a circle much closer to the centre because the larger space form the centre hosts it. However there is no large or small in the cosmos but only those better developed or those poorer developed. By duplicating there is more to duplicate in the better developed than in the lesser developed. When the lesser-developed space is duplicating the less developed space would hold a lesser extending from point to point forming a shortfall by distance in comparison.

The motion being extended needs less extending and should therefore be closer to the centre in relation to what the better developed space would need in extending by a duplicating effort. This is the principle we find behind the sound barrier. The motion the aircraft produces forms an increase in the duplication of the aircraft, which extends the duplication of the aircraft splitting the Earth and the duplication that is producing an extension of the aircraft. The splitting does not align gravity lines with the Earth as it did before. The aircraft is reproducing more in a shorter time duration by duplicating and extending space filled by material that goes beyond the attempt of the Earth's extended of duplication by such motion.

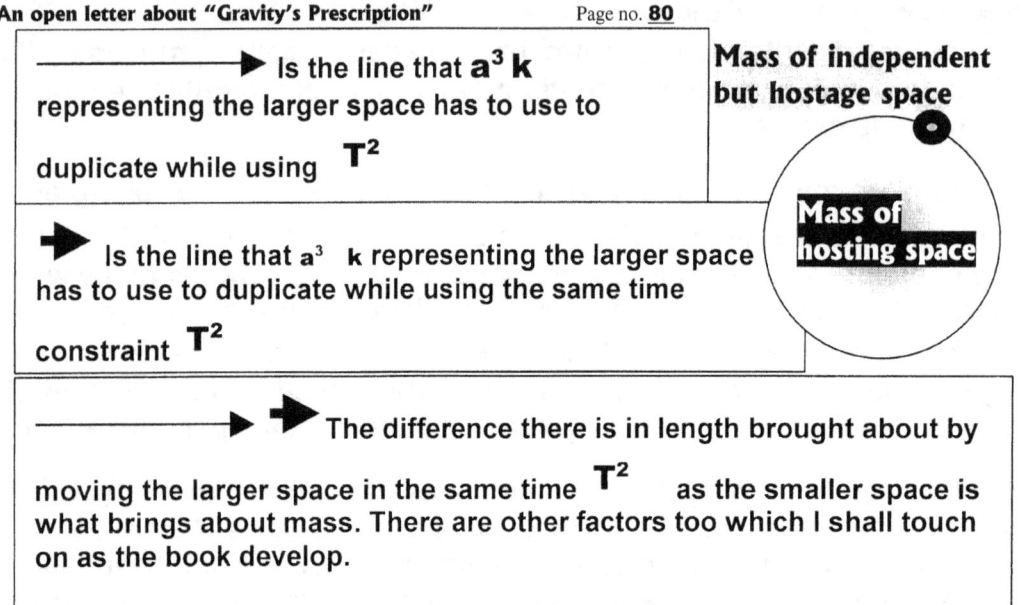

I know this may sound barely believable but please hear me out. While we use gravity, the use of gravity as such makes us part of the Earth. We see gravity as some influence or force producing mass and that mass is forcing us down on the solid ness and onto the Earth. By having the mass we become a semi unit with the Earth. That is how we on Earth see gravity but when investigating gravity in outer space we must come to a basic question: Is that what we experience as gravity on Earth truly gravity? Much of the proof about gravity is part of our perception about gravity because we experience certain conditions with gravity while we find ourselves bogged down on Mother Earth. But are our perceptions about gravity truly correct? We experience mass but are the mass the result of gravity or are the mass the product of gravity. We only experience gravity, as a factor from the position we have on Earth and the conclusions we form is a product of a perception we formed while we are being forced to be part of Earth. It's as if we are upside down and have to decide on which route we should follow. I want to make a suggestion, which I aim to prove in the following pages. My personal being on the ground and having mass that is keeping me on the ground comes about because of the speed that I travel through space being the very same as that which the Earth has.

By me not applying a speed difference I then inherit the speed the Earth places on me. But my space which I use $a^3 = T^2k$ to travel and the space which I use tot travel through is much smaller than that which the Earth burdened with to move and to move through. By me having a smaller space to move $a^3 = T^2k$ the space a^3 being moved k in the time it would take to move T^2 will produce less space a^3 to shift k and therefore a smaller distance k to replace all the space a^3 that is moved in the time T^2 the space a^3 needs to enable it to move k.

To duplicate by motion the smaller space requires a smaller distance to shift the space but the motion will take up, as much time to complete than would the larger space take to complete though the space the larger space has to duplicate will require a longer distance to complete the total duplication of the larger space. A large space a^3 will produce a large extending k when using a^3 the same time duration T^2 when using the same time factor as that which the smaller space is required to use when under obligation to use same time constrain. Behind this is the most basic principle hiding which allow us the fortune to be able to fly using a flying machine. It is all about motion supplying relevance and forcing on time constraints.

Because my body that I have is travelling so much slower than the Earth is travelling due to my size in relation to the size the Earth has and although I am using the same time as the Earth does to move, such a speed difference is not in the time differences it takes to complete but in the space differences that has to be completed in the same time but is unable to fill and the space is trying to crush me into the Earth where I am forced toward the centre. If I were able to penetrate the soil solidness I would reach a point where my speed as zero would equal my space I occupy.

The space I duplicate by moving from one position and placing the space I hold in the next position while keeping my space I move as it is identical in the next spot but located in the next position. Such moving by duplicating takes a certain time to move from one spot to the following spot and it will use a certain frequency that will have the same ratio in bridging the gap from one point to the next point as that which the Earth has. My speed of duplicating by motion has to be even in frequency because I am within the duplicating space, which the Earth is duplicating and as part of the space that the Earth is duplicating but the duplicating of my space I do myself. But in size there is a massive difference between the space I hold and the space the Earth holds but to duplicate will take me as long as it takes the Earth. Notwithstanding this common factor the Earth has to use equal time in duplicating its massive space, as I have to duplicate my small space when we both have to share a frequency that will keep us duplicating evenly. Therefore the frequency of duplicating using the same time period will be a lot different to my much shorter frequency of duplicating space.

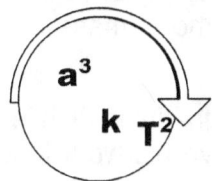 The difference is between me being in mass and me being in the correct position in the space-line the Earth has will place me in the correct position but the heat that then will surround me will fry me into non-existence. Fortunately, for life, the soil forms a barrier through which I cannot fall any further as to correct my location. Being where my position would have no mass would allow me to float there in that location in the same manner as I would float in water. I would be buoyant. It is because I do not harmonise the displacing frequency that I should that I have mass.

My having weight is what Mainstream Physics use to give me my gravity. Science purposely switch my having mass and confusing my mass with my having weight to explain what is beyond explaining. It is said that while I float in outer space in state of suspending hanging above the Earth in the weightlessness I still have all the mass that I had on Earth. But in order to prove that those in science will give me a mass even in outer space whether I deserve it or not. By that token science first has to cheat all logic by reasoning in some bizarre way that I take my mass up there to where there is only micro gravity. They firstly claim that all of a sudden I take my mass to outer space and in their next argument they say I have micro gravity in outer space since my body is floating as if it is in the sea. But if I stop floating and start falling to the Earth my body and I did not gain any mass. My falling then comes as the result of my motion being much smaller in relation to the space I claim and my motion then is being less than what is required to keep me in the position I have which in I maintain my orbit up there. By moving to slow I fall. I do not fall because my mass grew. But science has been proven wrong by their work without any of them aver admitting to such a defeat.

All the satellites fall if the satellite motions are not reset. The satellites do not gain or lose mass. They gain or lose motion. By amplifying either my space (using a Hot Air balloon) or by accelerating my motion that I have in relation to that which the Earth forces me to have, I will break free from my weight or mass. I shall become airborne and float as if I am in outer space. By pretending my mass can be multiplied many times over in using a process, which then is called not gravity but momentum. But motion and gravity is all the same because motion is gravity that is redirected, which then forms another part of gravity where gravity again is also only motion applying. Science maintain the argument that when I am in outer space and am no longer part of the Earth I then will only have mass. But since there is only micro gravity I will be in a state of weightlessness. My mass is what gives me gravity and while being up there I take my mass long with me. But with my mass up there I will only have micro gravity. I am floating with my mass and it is my mass that is responsible for my gravity and I am floating above the mass of the Earth, which is right down below me, but still I have micro gravity. That is true if I wish to incorporate the dubious use of double standards by separating mass from weight.

The mass my body will have in a Black hole will be a billion times more than a trillion times (at least) more than what it is on Earth. With that the Black hole destroys the fact propagated in science that my mass will be the same everywhere. That is more than permitting double standards. Because our motion is much slower than the Earth is spinning, we place a breaking effort on the velocity the Earth has and that breaking effort we accept as the mass we have. The truth is that my mass comes about from the lack of motion I have in relation to the space I occupy and has nothing to do with any gravity - motions pulling me down. If I increase the motion I have there shall come a point where my motion will be sufficient to pull me into the air, as I then will have the required velocity to lift from the ground. That motion being in excess of what I have and is complimenting the motion that I receive from the Earth counteracts the motion of gravity that is containing me. The motion I adopt then release me from the motion containing me and if motion can release me by only becoming more, then gravity is my motion not being enough in the first place to keep me onto the Earth. Nowhere and at no time does my mass ever gain by having more protons that will get me back to the ground as if I am bigger or carrying more material or does my mass reduce to get me into the air as if I am smaller or carrying less material. Please note that this is my way of explaining to you about the fact of bodies having weight or mass. It is not mass or the lack thereof or any means to measure occupied space within the atmosphere of a larger body that pins me onto the ground. My body is claiming space by motion in space. Gravity is the result of motion because it is in the motion that bodies have that gravity affects them. This is proved because by adding motion the mass does get more but the body never gets bigger or hold more material, and in defiance of that statement by increasing the motion my body lifts and flies. The reality is that my body in motion has more mass being momentum but still my body lifts when motion allows my body to lift. This statement confirms Kepler that a^3 becomes more (massive) when motion T^2k becomes more (moving).

Mainstream physics admits all along that nobody, human or otherwise knows what gravity is. While investigating Kepler's work with employing much motivation and detail in order to give his work the much duly credit it deserves it will also serve a valiant purpose when by the same token we try to establish what gravity is, because I believe Kepler possibly answered that mystery. We have to start with the person that introduced gravity or so does everybody acknowledge. Newton saw an apple fall from a tree and he subsequently realised there is some force pulling the apple to the Earth.

Although he still was a student he announced his findings and became a genius on the spot. The concept he introduced as gravity gave him instant admiration from which he became the legend he is today and that reputation he gained there at that moment would last him from that day he instantaneously unveiled his mastermind, and that same genius still serves him in his honour to this day long after his death. He found that this force has to have some thing to do with the weight and the mass of that particular object and the mass of the Earth. There is some force pulling that apple as much as the force is pushing the apple and the same goes for the Earth because the mass the Earth has is doing the same to the apple. Between the two objects facing gravity there is a force that develops where such a force is pulling the apple on a constant basis towards the Earth even after the apple is already in a steady state on the Earth. That forms the mass and the mass forms gravity. He concluded that the mass is responsible for the pulling. Remember this observation came three point five centuries ago when knowledge and brilliance carried a much different defining than what such defining of brilliance is worth today. He realised the pulling on that apple brings about weight that brings about mass because the apple departs from its location and arrives at its end location when the falling is completed. Then he went out convinced all that was in line of finding the needed convincing because no body before Newton thought of what Newton thought quite in the way that Newton thought about gravity.

Newton succeeded because he found a way in presenting science with the fact that objects move closer because of some force. He went one step further and named the force he fathered as gravity. But there it stopped! Any and all other further defining the matter or going into any possible observations of whatever magnitude concerning the topic never realised any motive to go further. Inspiration to further commitment just flew out the window as the essence to do so immediately expired as far as the rest of science is concerned. What might he have missed if he missed anything? We all fall down when we are unstable and out of balance. He never realised that balance is more crucial than brutal gravity because that part is the defining part about gravity. No one ever gave a thought about the balance part even centuries later even as we grew into all the sophistication we now enjoy. What brought about the balance that secured objects in an upright stance and supplied some form of control over the managing of a position? Any other position than being flat on the floor would have a better defining than being just at the mercy of the force gravity. Standing tall is a stance that defies gravity so there is another force other than the pulling of gravity. Admit tingly the force would first and foremost have to aspire to the rules of gravity and then comply with other demands. True enough is the fact that that position would ultimately and firstly by all accounts have to satisfy gravity before any further motion could commence. Yes but then by balance motion defies gravity by changing gravity's force of pulling everything straight down towards a visionary centre between the objects. In affect this means somehow there is control over gravity and gravity does not leave objects beyond outside control. Gravity is manageable and can be controlled; we just have to find a way...

Years later some one came up with the novelty of hot air ballooning. Ballooning proved that there is antigravity but that part was missed by all even to this day. Some people speak of antigravity as if that is some mystifying mysterious concept that is so well hidden in the secret annals of the hidden Universe that only Ali Baba and his magic words can reach it. Please consider the following statement. If gravity was bringing the object down, because of the affect of gravity which is that what we experience as the gravitational sensation and that is what we interpreted as gravity by our sensation and observation, then that is only coming about by our bodies that is in a state of being dragged down. The dragging down of the body is in the direction of the Earth centre. That sensation of being firmly locked onto the ground constitutes to what we believe we experience as gravity.

When some influence brings about the very opposite effect, which then results in establishing the opposite result, it deserves to be anti. In example we feel dragged down but anti will be the lifting of the body into the air. Anti will be going in an opposing direction of the motion that gravity inflicts. It will counter the influence that gravity applies. Such motion has to indicate antigravity. The counter acting of the mass dragging us down must be anti gravity pulling us up into the air above the ground. Antigravity must come from such an opposing influence that will bring about the lifting of my body. If hot air ballooning gave the object an opportunity lift, then ballooning must be antigravity. The balloonist and the entire balloon found a manner to counteract the pulling of gravity enforcing weight. The balloon can lift what gravity depresses and if Newton said gravity is the falling then later Newtonians must agree that the opposite of falling is flying or lifting. A balloon is lifting-and- flying. If gravity is pulling down objects in the direction of the centre of the Earth then flying is antigravity. Moving away from the Earth by means of motion and in particular flying is using whatever means to defy gravity where the lifting can also be the hoisting of a body by a crane. Lifting by ballooning in a hot air such balloons escape from gravity where the balloon constitutes to bring about the effect of establishing antigravity. Climbing up mountains must fall into the antigravity department because parachuting down the mountain definitely falls in the gravity department. Nevertheless it still does not answer the question of what gravity is.

Let us look at antigravity because the antigravity is releasing the object from the gravity that controls the object by an Earth fed force. The balloon starts flying when the confined space of the balloon is veraciously and violently heated in access. The balloonist shows us that in order to overcome gravity we have to introduce heat. That is the only manner in which we can defeat gravity. Even by an engine driving an aeroplane such flying can only result if an engine combust solid fuel by creating motion as the fuel mixture is turned into heat. It is heat that makes the difference. That is the very thing that Kepler said. Expand the space a^3 and the motion T^2 will move further increasing k. Blowing hot air into the balloon is increasing space within the balloon a^3 which then results in providing the balloon with a larger distance k from the Earth centre k^0 that still holds time with in the Earth atmosphere with the Earth T^2 within the space of the Earth k. Using Kepler provides us with insight and the ability to see what gravity is by showing us what antigravity is (a^3 gets bigger and that will bring in a larger k). But moreover the larger space in enough compensation to bring about extra motion that will defeat gravity by the extending of k. If that is not antigravity then we can forget about Ali Baba and his magic rhymes too.

The balloon assists us to escape the Earth's hold on our body, because there has to be the force producing motion countering the motion of the Earth gravity. The balloon shows that releasing enormous quantities of heat into an inclusive area excluding space such as that which the balloon canvas provides, which is establishing the release from the gravitated containing force on the body giving the body a means to escape by floating about above the ground. The motion is at that point breaking free from the containing gravity by moving in a specific direction, other than the direction the Earth gravity inclines the body to travel. By concentrating the releasing of heat into the balloon, the direction of motion starts to contradict the enlisting of the Earth gravity and the heat breaks the balloons confining properties while the balloon is released from the Earth as the balloon and us lift up into the air and away from our confining to the Earth.

At the point of explaining we arrive at the point where we can say what we think the difference is between the balloon floating in the air above the Earth and a body suspended in outer space floating above the Earth's atmosphere. The difference is the heat that is in the confined air per volumetric ratio favouring the heat being more in the space than what the heat is outside the confined space. If we had any method to put the required heat we need to escape from the limits of the Earth to outer space into the canvass of the balloon there was no canvass left to contain the heat. The heat is available to do the job but the means to do the job with the tools in hand is unavailable as far awe can use the balloon. By having more heat in the one area than there is in the other area beats of the pulling of gravity. Obviously it is antigravity that keeps the balloon in the air and what keeps the balloon in the air is having a larger volume of heat per space unit than what is in the atmosphere. The balloonist shows us that by applying more heat we can defeat gravity more. Someone took the advice, because the next minute the Germans had rockets. The launching of rockets brought about the ultimate defeat of gravity but it involves almost the ultimate releasing of heat.

In antigravity we find heat more concentrated in one definitive area than the heat concentration is elsewhere. The more the heat is that we release into space the more the antigravity is that we achieve and the more release such antigravity can produce. But what connection can gravity have with heat and if there were any connection between heat concentrated and gravity, what would such connection be? The history behind Carl Benz should bring the answer but more so would be the story behind James Watt and steam although the James Watt story may not be that thought provoking because it is much less filled with the ever popular cheap thrill only sensational gossip can provide…Still both stories cover the same principles. In the Carl Benz story a housewife leaves a pot of benzene fuel on a coal stove. The pot with benzene heats up where the pot with benzene becomes hot and under pressure. This performing of heat increase, such increasing expands as space and releases the heat as newly creates space, which then removes the housewife with her house from the neighbourhood she used to regularly frequent as her residential address. Afterwards almost the entire neighbourhood is not there to tell the tale or ask why...

It was a stupid tragedy that brought about the end of steam and the rise of the internal combustion engine and on Earth billions on billions of human souls are in torment not to please or suffer for the advantage of coal Barons any longer but now they are dying and suffering in agony to please the wishes and desires of oil Barons. How much did the world not change…While it is no longer the coal Barons shackling us in chains and telling us democracy broke our burden of slavery, we have now the pleasure of the oil Barons enslaving us with democracy and telling to be happy because we are the fortunate slaves, there are others circumstances in which they can enslave us that will leave us worst off. All this came just because the pot of fuel created a houseful of space that was enough to remove the house from the address the house previously enjoyed. But Mainstream science neglects to appreciate this. They see the heat, they see the antigravity but they fail to add the heat, the anti gravity and the space that no longer housed the house of the naive and rather impractical thoughtless housewife. They call the tragedy an explosion but then again everything that expands while using a noise during the expanding is an explosion. Adding of new space to the space holding the house at first altered everything that was previously proportional positioned in the space where the house was. Such exchanging of heat to accumulate and introduce more space in the process referred to as an explosion was bringing in more space that came directly as a consequence from the explosion which was producing more space where the increase in space brought disorder because the well organised material distribution and placing was before the event filling just enough of the required space arrangement that was holding every object in a prearranged order of tidiness.

Then suddenly out of the blue the space which held the house in a tidy arrangement had to accommodate more space therefore the ratio of material per space volume increased dramatically many times over in the favour of the space in the balance. That part no one ever acknowledges. However the losing of the house was not much surprising to Mainstream science back then and even today because who cares about old news. All of Mainstream science was at the time, as they are today, very familiar with all explosions because of wars and bombing that leads to maiming and killing and all the unspeakable monstrosities we associate with war so that the dirt poor can suffer and die to leave the disgustingly rich even richer. The poor has not the means to pay science to be clever and devise methods to save their lives, so the rich does the poor the favour of paying science to find methods whereby more poor could be killed as long as the rich saw it as a good investment with great capital gain on the part of the rich. Therefore science is well established in the method of creating more elaborate and destructive explosions that the rich pay them to invent. In the explosion caused by our housewife no one put up money to investigate what happed during the explosion but money went to why the explosion happened.

That inspired an investigation in connection with the fact of the finding more about what takes place during the carnage as more money goes to finding means to create more carnage per money unit spent. At least that is why the poor were invented and that is why wars are invented. It is invented so that no money goes wasted on saving the poor people except if the poor has the money ready and available to pay the rich for medicine to enable the poor to stay alive. So science goes out and develops more fuel for carnage but fails to find out why the housewife and her house are no longer part of the neighbourhood she used to frequent. With the loss of the presence of the ignorant housewife with her house her neighbourhood and all were a normal way of leaving us with a new way of tapping and harvesting energy and untold riches which was born with the death of the absent minded housewife. But according to the mindset of science they saw not what the incident presented in space producing for to their view nothing new came about since it was just another exploding of fuel...so no body bothered as to finding out how. What they missed was the part the coal stove played in the whole tragedy. Without the intervention of the coal stove producing the heat that turned the liquid fuel to liquid heat liquefying the space that turned the liquid space into a gaseous space where the liquid space revealed its true incentive in nature by turning out as space and the newly created space that was in fact liquid space that went on to become more space, well that space was providing the one main factor in space-time relevancy.

The stove's heat was producing space by transferring the heat from the stove to the pot filled with volatile fuel and the transfer of the heat to the fuel brought about the expanding of the space that the fuel claim to need in the pot filled with the heating and volatile fuel. The fuel space requirements became more as the heat filled the space that was filled with fuel and that took up more space, which the pot could not cope with since the pot had no room to allow such an increase on the demand for more space. The fuel expanding as such was claiming new space to sustain and accommodate the growing requirement for more space to be created in order so that the volumetric increase of heat added to the fuel could be accommodated. At one point the asking for space became a claim on new space, which we see as an explosion. The heat transformed to new space by an exhilaration of breathtaking increase in heat forming space and this increase of newly formed space was transforming all other surrounding space within the room, the house and the neighbourhood in general.

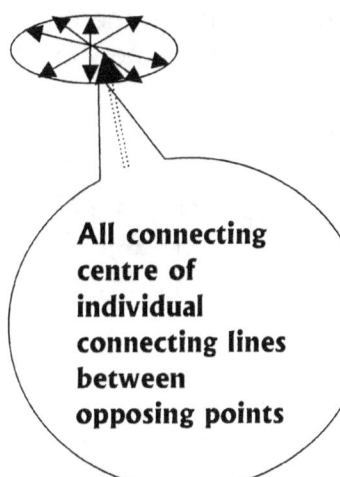

All connecting centre of individual connecting lines between opposing points

By increasing the volumetric quantity of the space it rearranged all space, which included some of the space held by material that was in a solid form and scattered the rearrangements as fragments in all other designated places far away from each other. This meant there was excessive and all around rearranging and relocating as well as re-allocating of space in general. This to science sees as shock waves resulting from an explosion but is merely heat expanding space to set new required standards. It is rearranging every aspect that contains space or that space contains. It will bring a much different looking end. Everything about this concept is missing from Newtonian science because Newtonian science failed to investigate Kepler. Kepler said space a^3 is equal to the motion T^2k thereof and then that says without Kepler directly saying it, it says that if space a^3 goes bigger as a result of the explosion then such increasing in space will constitute to more space a^3 which has to produce an increase in motion T^2k where more motion T^2k will bring about faster displacing space. This is one small fact that Newton robbed the world of realising with his ignoring of Kepler's work.

From every point there may form on the outer circle line of every part of the circle structure and all structural positions of the circle in all circles, all circles refer to the centre in perfect aligning. Every point wherever located on the sphere has a matching and equal but an opposing point on the other side of the circle but in equal position on the other side of the circle. Between the two controlling points runs a precise straight line connecting the two opposing points in counter balancing. When drawing the connecting line between the two controlling points and connecting such points on further edges of the circle by lines formed the lines will all cross the centre? From wherever a line may cross and from every point forming a line to the other side of the circle rim holding the connecting points there has to be a counter point located on the very opposing side that when connected by a line, such a line crosses in the centre. In the middle the centre spot bonds all sides coming from any and every direction there can possibly be. The line will run to an equal point on the other side across the same distance from such a centre and that then has to be where the strongest gravity can be located.

We are now serving time in the twenty first century. One Professor once told me I must realise that Newtonian science took man to the moon and back several times and in such a view I am rather annoying presumptuous to criticize Newton. The Professor missed the point. I criticize Newton on what he did not give us, which he gave us as incorrect by his own admitting that it is mostly guesswork on his (Newton's) part and his guessing about the facts where later that guesswork became institutionalised facts believed by all concerned to be correct and to be proven to a degree of correctness that is far beyond doubt. Newton gave us gravity but Newton never gave us the explanation about gravity. At the time Newton met strict opposition from his colleges and piers because others felt his introducing of an unexplained force was taking Science back in time, which of course it did. Many scientists at the time accused Newton by name of dragging science back in the wrong direction of progress by introducing unexplained forces acting in a superstitious and mediaeval manner. I went one step further by asking myself the question: If space becomes more when heat becomes uncontrolled why can space not become heat when space is under control? If space becomes more as we see with every explosion of every kind and such heat forming space releases energy, then why would space being managed not form heat being under control and produce energy. We only have to see what Kepler said gravity is. Motion gives us energy.

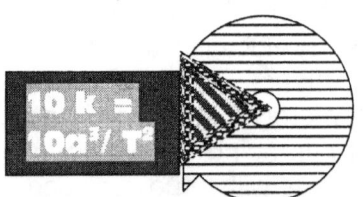

In the normal flow of events there is a liquid time that stands related to a specific generated by the motion of the liquid in solid that binds the liquid to the solid. That representing the factor **k** in the Kepler

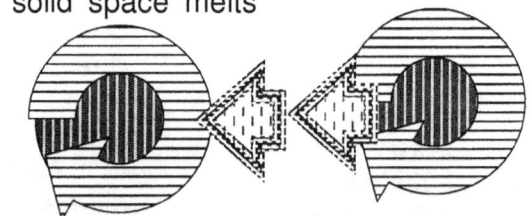

certain ratio of gravity, which is relation to the uses symbol formula $a^3 = T^2 k$.

This symbol is standard on both sides but due to the inherent nature of rotation qualities, it opposes that what it was by crossing the divide there is. The liquid attach to the solid by diminishing the relative facto bring about that **k** goes negative $k^{-1} = T^2/a^3$.

The solid on the other hand confirms the liquid attaching to the space by the motion thereof and withthat the solid recognises the shift in limits where there forms a new boundery and a new location where space that include the attaching liquid then ends.

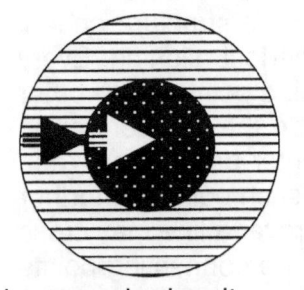

When an explosion occur the solid space melts and the lot in the space becomes liquid. That is a direct result from the expanding of the limiting or relative factor **k** that reposition the boundaries of the liquid turning space into an altogether new dynamic.

As the solid melts it amplify the space held by the heat many fold by tarnishing the spin T^2 that increase the space a^3 and reduce the density of the heat in the space a^3 where the velocity of the motion T^2 is reduces as the factor **k** is increased in distance. By increasing the claim on space as the spin velocity reduces and the density intensity decreases the liquid part becomes more since the overheating destroyed the solid part and that which we know as gravity reduces because that we know as concentrated space reduces in intensity. The expanding is due to more overheating than what the governing singularity can control.

Since there is an excessive demand on space through the overheating which reduces the density of the solid and turns the solid into liquid the reduction in heat density by lost of singularity control. The increase in liquids demands more space in the light of the decrease in liquid density and the motion reduces as the density diminishes.

It will take years of increasing the controlling ability of the singularity governing space-time and a rebuilding of the solid under the control of the motion that increases while the space reduces, which then confirms the increase in density and the improvement of the control of space-time.

Where space is the least, which is in the centre of the circle, gravity is the strongest. The gravity located in the circle's space less centre holds not only the sphere together but all that is in the surrounding of the sphere outside the sphere as well. It is from there in a giro action that gravity bonds all atoms forming the structure of the sphere as one unit together in a unit as well as distributing a specific alliance in shape and form. How the atoms manage that we will get to in a while, but there is a law allowing for that to take place.

Gravity is the strongest in all cosmic structures holding the form of the sphere and gravity controls all around from that very centre where space is the least, therefore the more material there is to generate motion within a star the more secure would the generated centre be in any star where such centre produces gravity. It is not the material but the motion the material accomplishes that becomes the factor of gravity. The smaller the star is as far as volumetric occupation goes, the stronger the gravity is that is coming from such a centre. The less the space there is the less the motion is and therefore the stronger and more deliberate the motion is evoking gravity. From the centre in the middle where space is absolutely at a premium the gravity grows stronger as it draws all material.

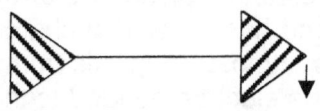 If outer space was nothing then the crossing of light in outer space was impossible because of the total destruction nothing leaves when contacting anything (1X0=0). Light crosses space at so many points holding infinity that it becomes the speed of light or the ability of light to convey space by motion thereof.

The motion is one of confining the space to a centre by the moving or trying to move the flow of space and whatever is in the space into the centre where the space is least. Take the Neutron star and the Black Hole as an example and compare that with the Sun and the answer is simple. I claim that gravity is all about reducing space and not attracting matter but that I explain a little later on. Therefore the matrix of gravity must be permanently located in the location where space is the least. Looking at a sphere we find that what holds the sphere true to form is placed in the centre of the sphere, which then has to be the most intense point of gravity. Gravity is confirming the round shape without favouring any specific point. Such evenness of gravity comes from what is applying at such a centre and is in control of the surroundings. The centre that secures all of the space and material in the space holding the specific form has to be round if it is anything. That shows that in the sphere one can see that the sphere as a form is dominated or controlled from one specific location in the centre. The explanation about the reason there is control coming from the centre, has a very childlike simple answer.

The Big Bang was where gravity held the Universe in the least space there ever was. To find the original gravity we therefore have to reduce the sphere to the circle and reduce the circle from there narrowing the circle down to as far as one can go. The Universe is a magnitude of spheres constructed by a complexity of circles. This is because everything sprouted from singularity. To narrow any circle down will be the same as narrowing down the Universe. In our reducing of the Universe we must first acknowledge that the Universe constitutes many spheres, which is giving the Universe gravity as a combining unifying part which is the part of the sphere giving the sphere form (or gravity) and that confirms that the sphere is a circle in many times over multiplying the positions from where gravity secures form. If we wish to go back in time by taking the Universe back down the same route and at the same time maintain some coherency we must concentrate on a single circle because a sphere is a circle by millions of possibilities linked together by just a name that changes the concept. When one takes this accepted route in thinking that by reducing the connecting line to the connecting circle point in the centre of the lot, it must take us back in time at the same time as the circle reduces to the time during the Big Bang.

During the Big Bang where all circles were as small as they can get, we run into an unknown substance we came to know as antimatter. This theory is propagated according to Mainstream science but what is most surprisingly is that I do agree with this part of the statement. All material produces gravity. I go one step further and say all material applies motion where some motion may be to contain by using gravity attributing to the contracting that leads to the reducing of their space. Then as everything in Creation has an opposing to restore and maintain balance, there had to form another or other material that did not by our lamentable standards produce gravity because those material produce antigravity, a concept beyond human discernment. Antigravity must be the expanding in counteracting contracting. A counter action to contracting is where expanding provides growth of space to that which has then reduces the gravity effects. Forming pappy provides more space by losing density to the advancing of their space. Material either have gravity by solidifying or concentrating the space they hold in ratio to the material within the space they hold whereas others lose their solidness by entertaining more space within the ratio of material to space where such material becomes liquid and in more extreme cases they become gas. Being a gas they float which gives that material a high degree of antigravity being airborne. It is however not clear if antimatter produced gravity as it did when it went to lunch on and ate up all material in the immediate surrounding. It was cannibalistic but the unanswered question is this: was it a gravity producing predator or a non gravity-producing carnivore. Did material find a comrade in their gravity forming of form or did the gravity it produced bring on the demise that subsequently followed the event as is reported by the highly informed.

The Accepted statement on antimatter reads that matter composed of anti particles where such subatomic particles that have identical rest mass to corresponding particles of ordinary matter but opposing charge and are opposing in other fundamental properties. One example given is that an electron would have a positron, which then functions as the anti particle and has a positive charge compared to the electron's negative charge. That is put bluntly in its utmost simplistic form. Unanswered and tough questions arise from such a statement. What kept the electron bonded to the atom since the protons must by implication produce expanding or by definition be repelling the atom and surroundings instead of the normal contracting or confirming of form. What is a positive compared to a negative charge, because it is human concepts that put the directional qualities of material into a positive or a negative contexts as we did with hot and cold. It is human standards that humans brought about to make all human inadequacy by lamented human understanding better but it is not applied cosmos principle.

If there is extracting electrons performing in the capacity as antimatter, then there better be protons by other name in service to the anti electrons, which then of course serves the anti electron in the capacity of an anti proton with an equal but negative charge to that of the proton. When matter and anti matter meet, the two opposing particles annihilate each other until one vanishes from the Universe. I have to add that at the time this theory was devised the first computer games became a crazy fashion played by young and old, those wise and those foolish all alike. This game was called the packman and the packman ate up all the skulls and after eating left nothing as evidence. The theory about antimatter has some very striking similarities to that packman game. It still does not answer the most ardent questions: What makes a positive electron different from a electron in the working place each has and can any person show such an object found in nature.

Can people take a positive electron to an investigative bureau and are acclaimed for bringing about such evidence? It is unwise to substitute nature with human concepts just to further mathematical equations. This was apparently presented as normal as nature was when nature developed with the Big Bang and nature then did behave this oddly just after the Big Bang came about. But one huge misgiving in this argument is declaring that everything the antimatter had as a meal vanished and even moreover then antimatter went and vanished too.

Where could the combination that was produced when the matter and antimatter collided go after it disappeared and did it form the by-product of antimatter science is talking about, which since then apparently vanished too. What a bloody none-intellectual fairytale that is on the in addition as well as one of those made–up-as-they-go-along stories, which is told by persons that supposedly should know of better. Since there is no place other to find a location to be within than being in a place inside the Universe it is hardly possible to vanish from the Universe except in fairy tales because for one simple fact: there is no other home to have but the home we have which we call by the name the Universe and we have no where to escape to but within the walls that the Universe provide for such a purpose.

There is one Universe containing all and preserving the lot. Mainstream physics is accepting this fact. But then by the same margin they accept a principle that allows property that once was part of the Universe to leave the Universe and go somewhere outside the only Universe. They create a loophole whenever it suits them to misplace what they cannot explain readily and logically. In Creation to their and my thinking there can be no hiding of anything but in the Created Universe. This they admit and confirm although with the same breath those very same intellectuals also admit that there is another place outside of what we are able to find in the Universe. When someone comes up with the marvel where such a person can declare in all honesty that the product of antimatter or singularity escaped from the Universe to God knows where that person should leave the field of science and go for fantasy writing such as fairy tales or reporting about politicians inner deepest chastity and integrity. That is what we can find outside the spectrum of what the Universe can deliver. With such a statement on anti matter or the loss of any Universal product disappearing from the Universe then alarm bells should go off in the mind of the trained and professional Scientist working with such matters.

Yet those in charge do not once belabour the question on the validity of a statement that involves stating losses occurring of substance being in the Universe going lost or being removes from the Universe. Their surprise of pressruns stating the loss of factors and declaring the possibility that there was a possibility of such losses where those factors now are outside of what once was part of the only place there ever can be. They can read mathematical calculations and agree on an outside the Universe without stating it in an explanation what happened to the lost and found or they're ability of introducing the concept as a reality, which they claim it is. That such factor can go outside the Universe and leave the Universe by causing a Houdini vanishing act of never –to-be-repeated-again status. Science would have us to believe this antimatter went into hiding in a manner that is out of the Universe.

They applaud this thoughtless presumption while fully knowing that at the time they do this acknowledging that there is no other place for anything wishing for a place to be within then having to be in another place other than inside the Universe. If it was ever anywhere it still is within the Universe merely because there is no other place to go than to be inside and part of the known Universe! There cannot be some factor and then misplace it as if a valid factor calculate the value can prove the disappearance and by disappearing it no longer is. If it was in the Universe it must still be in the Universe somewhere. Then we better start looking for it.

Another big issue is that what ever the Big Bang produced must be in equal terms everywhere. The Big Bang was a process that had the Universe act as a high-speed cocktail mixer of no repeating ever again. Whatever the Big Bang was of all that it was, the most it was in the beginning was that it was one massive mixer mixing everything in it at the speed of light. The relevancies might change slightly and balances may change favouring opposing ends...yes and known appearances did change...yes. But in the end all the factors must always be present everywhere through out the Universe. By this lacking of a fundamental explanation about what antimatter will look like when found Mainstream is incredibly poorly judged by scientific standards. Those mathematicians calculating physics suggest that science should take antimatter as a cosmic fact and then in disregard of other realities they dispose the truth by discarding its properties onto the unknown.

That hardly suggests plausible science by any one's admitting. In the cosmos is, was and will be all the material there can ever possibly be. Our concepts we put forwards can be faulty but nature cannot ever be at fault. Our arrangement of our ideas can be at fault, but we cannot pull a vanishing act on certain cosmic products and in doing that then dismiss the existing of such a factor or factors, which we then claim, have vanished in the further developed Universe. Our concepts of what they became may be at fault and by changing some basic principals such changing may produce a better understanding about what we think we read into mathematics. Mathematics is purely a language and mathematicians are purely translators. Mathematicians translate from the language they read to the verbal equivalent they speak. As in all translations made, certain concepts may become misinterpreted.

The terminology used to explain this is "lost in translation" Mathematicians must see what there is in the translation and try to incorporate what there is available in the cosmos to what the Mathematician sees in his mathematical calculations. The Universe was full of heat and it was full of material but it was not full of free space. If that is the case then where did the heat come from and where did the heat go? Hiroshima and Nagasaki taught us many things about the horror of human nature but most of all it taught us that material is heat secured in atoms and atoms are heat tightly wrapped in a cocoon, which we named the atom. Heat in any form cannot have anti in another form. The package holding heat wrapped can unwrap as it does with nuclear atomic demise. But the anti to heat is cold and cold is space.

The undeniable fact about the Big Bang theory is the accepting of a growing state in which the entire cosmos seems to be in. With all the expansion that went on we came to the point where we now are at and in such growth all aspects in the Universe must grow in relation to quantifiable progress in all different aspects, which takes us to that which is seen and that is unseen and which came along as products in the Universe where everything took everything on a growing spruce by unveiling space. That is where we now are. Such expansion include all there is including everything and not just with outer space growing. The dynamics of outer space alone cannot grow by leaving the growth of material behind. Should we wish to see where we came from we have to reduce that which we now see in our surroundings to apply to the measures that once applied in all aspects of the cosmos. Mainstream physics is over pronouncing the growth of space and with that suppresses the part matter must play in such growth by simply ignoring the issue. That is the reason why they prefer to ignore the evidence that material is growing notwithstanding that material is growing or that their disbelief about the matter of material growing do not change that material is growing in any case. Because they cannot find any reason why material should grow they refuse to admit that material does grow.

This is hiding from the truth by hiding the truth. If space grows and the Universe is getting bigger then all space grows to allow the Universe to get bigger. That includes matter and space not in matter. Space can only grow if materials that also hold space also grow within the space that is growing with the growing space. It means that stars get bigger by the cosmos growing from the Big Bang onwards and outwards to the moment in which we are at the present. But if stars grow then the atoms forming the stars are doing all the growing as they secure more space within the space they claim.

If Hubble saw space grow, the growth of space must include the growth of space holding material as well. In studying the Hubble's expanding theory we come across evidence that makes it clear that all material expand in a manner as if the expanding comes from the centre of each and all particles within the expanding space and the expanding grows outwards from every particle centre. It is using every star centre to grow from in all directions proportionally in all directions evenly. This leads one to believe that gravity is this securing of space in the material just as Kepler showed it to the world. It proves a connection with deliberate implications coming from every as well as in every specific centre. It proves that the centre $k^0 = a^3 / T^2 k$. It becomes apparent if and when separating Kepler from what Newton thought about the work of Kepler which Newton accepted as being inferior and all incorrect.

That is making the Universe small and as man grows man allows the Universe also to grow in relation and corresponding to man's ability to comprehend. We see the cosmos as a circle and we accept the circle because the circle is what gravity implement when the choice of form is coming from material that has all options to freely choose from. By taking the circle back one will follow will trace the rout of the cosmos to where it then started.

All stars are many circles in many dimensions, which form when all circles join into what we call a sphere, but that leaves us only with the circles in the plural. Taking the cosmos back can only lead to one point and that Kepler told us we will find singularity $a^3 = T^2 k$ which is $k^0 = a^3 / T^2 k$. We can only reach $k^0 = a^3 / T^2 k$ if we repeat $1/k = T^2 / a^3$ in a continuing manner indefinitely. When one does the effort of reading this correctly, it says that when distance k brakes from singularity $1 = k^0$ that is then $(k^0 = 1) / k = T^2 / a^3$ where the space a^3 produced a time T^2 equal to singularity k^0 and singularity k^0 is equal to eternity which was where all was equal to a never changing cosmos that was holding the single form into one dimensional space that included all the filled and vacant material filling in from all sides.

This is one way of looking at the issue and by doing that I am about to prove that singularity is Π. I am about to prove that not only is the planets adhering to the Titius Bode rule of seven over ten and ten over seven in relation to the Roche limit but that the Roche limit explains the very, very first instant the Universe experienced outside eternity. The atoms relates to space in the very same manner of seven singularity positions to ten points and from this motion of material interacting with space is securing material on the inside as well as on the outside. By that motion gravity comes about finding the value of Π^2. Gravity uses the relation of the Titius Bode seven on ten and ten on seven as well as the Roche factor to form gravity and gravity is always Π^2.

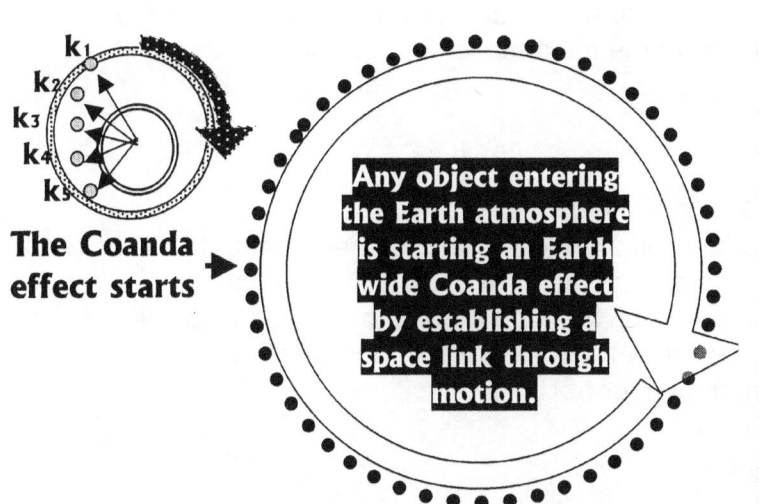

The Coanda effect starts →

Any object entering the Earth atmosphere is starting an Earth wide Coanda effect by establishing a space link through motion.

This I see by reading Kepler's work as Kepler produced the work and introduced the work as $a^3 = T^2 k$. With this formula $k^0 = a^3 / (T^2 k)$ must also be true because $a^3 = T^2 k$ is a relevancy that has to be in relation to singularity and therefore singularity must be $k^0 = 1$. Where will we find $k^0 = 1$? When an object is within the singularity alignment of the Earth, meaning it is either stationary, or free falling on pure "gravitational" momentum it holds the $4 \times \Pi^2$ in relevancy to Π. What I try to say by that is that the Earth holds all objects within the atmosphere to a displacement value of $3\Pi^2$ within the seven are where it is using 10 as the liquid in space. Because the aircraft does not leave the Earth's atmosphere the 6^2 and the 10^0 does not come into effect.

$$a^3 = (T^2 k)$$

Mass is the refusing of any object to dismiss the form it has and to join the Earth solid structure. Mass cannot and does not contribute to the establishing of gravity except by depleting space through motion and such numbers of the protons in a space forming an exclusive unit.

$$k^0 = a^3 = (T^2 k)$$
$$k^0 = a^3 = (T^2 k)$$

It is when the motion exceeds the mass the aircraft has the ability to break the sound barrier. Galileo proved that no mass is present in falling, which is also matter in the process of flight and because of that can the sound barrier become some form of constant.

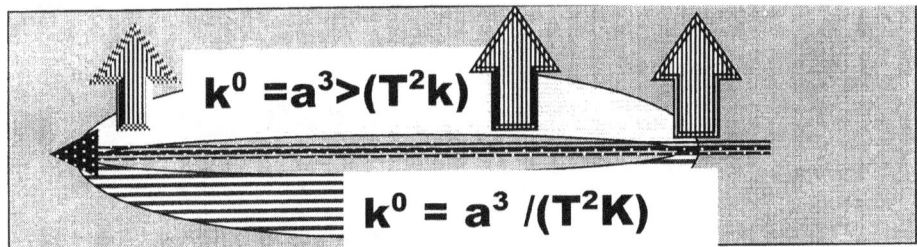

$$k^0 = a^3 > (T^2 k)$$
$$k^0 = a^3 / (T^2 K)$$

The establishing of independent motion of the craft secures an individual gravity and such individuality leads to the breaking of the sound barrier because the one gravity can no longer subdue the smaller motion, which is producing gravity.

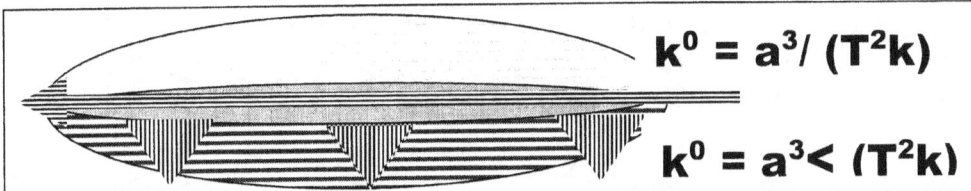

$$k^0 = a^3 / (T^2 k)$$
$$k^0 = a^3 < (T^2 k)$$

At a height of 31000 km above the Earth the mass of the wing becomes compensated only by a motion of a relevancy that comes about at 2500 km per hour. In that case the craft has to apply motion at a rate of 2500 km / hour just to create the required velocity to keep the aircraft in motion in the sky. Motion creates gravity just as Kepler said when he said gravity is about $a^3 = T^2 k$, which translates to the dismissing of space and the motion, duplication establishes a centre that controls the balance that the newly secured singularity will provide. When the aircraft stands still the Sun provides such a pivoting centre but when independent motion comes about the point shifts from the Sun to the Earth centre where there is a line contact between the singularity that the Earth holds which then forms a new relation in respect to the singularity activated by the independent motion of the moving body which the aircraft takes on a trip in motion and with it a position that the relevant singularity is claiming which are released as part of the minor space.

$7 (3\Pi^2)\Pi^0 = 207.2$ km/h. $\mathbf{R^2 /T=} 7 (3\Pi^2)\Pi$ and $\mathbf{R/T = } \Pi^0$

The Roche limit also changes to accommodate this change and becomes either Π^0, Π, or $(\Pi^2/2)$. Falling "free" will then mean that the object holds a position of $7 (3\Pi^2)\Pi^0$ to singularity. In the half circle applying two of the quarters in time, the position is in the triangle of singularity placing the half circle in the first quarter. Any further linear movement will follow the triangle second line by multiplying the line by the value translated as a number of Π^0.

$7 (3\Pi^2)\Pi^0 = 207.2$ km/h or $(3\Pi^2) 2\Pi^0 = 414.5$ km/h.

$\mathbf{R^2 /T=} 7 (3\Pi^2)$ and $\mathbf{R/T = 2\Pi^0}$ This is the first barrier but we only see the linear part as a value and peaks at $5\Pi^0$

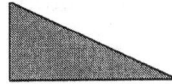 

$7 (3\Pi^2)(\Pi)= 651.13$. In this $\mathbf{R^2 /T=} 7 (3\Pi^2)$ and $\mathbf{R/T = } \Pi$ where then the linear component Π becomes the second line in the triangle of singularity. The linear component or negative displacement can go as high as 2Π but then another barrier would come about and in this we star to locate the principle behind the breaking of the sound barrier. There are definitely TWO barriers to comply with when breaking the sound barrier

$(\Pi^2/2)$
$7 (3\Pi^2)(\Pi^2/2)= 1022.79$. In this $\mathbf{R^2 /T=} 7 (3\Pi^2)$ and $\mathbf{R/T = 2\Pi}$ where then the linear component Π becomes the second line in the triangle of singularity.

$2(\Pi^2/2)$
$7 (3\Pi^2)2(\Pi^2/2)= 2045.59$.

From this point on and above there are a boundary no one can cross because singularity will not allow the crossing of the third quarter by any object.

Everything that is in the Universe is heat. There is liquid heat that can flow and there is solid heat that cannot flow. The solid state or liquid state has no bearing on human perception but is in motion. When the body moved in

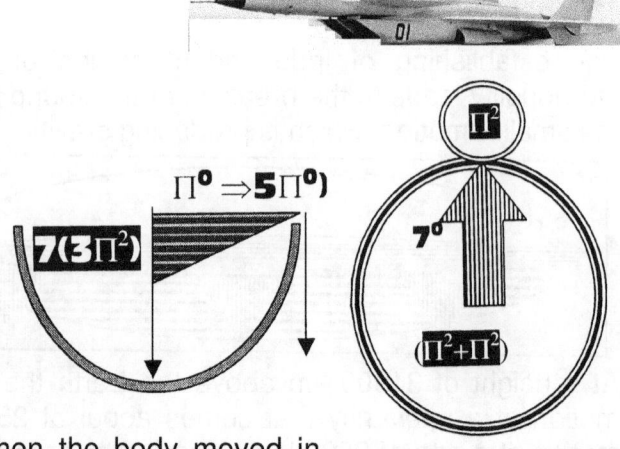

relation to the body that does not move the body that moves are liquid and the body that is relevantly stationary is a solid. Movement is liquid and immobility is solid. '

The sound barrier is in principle Galileo's pendulum arm

The sound barrier is directly related to the swing of Galileo's pendulum arm. By standing still the arm points directly down to a centre point within the Earth. As the motion becomes a factor in the aircraft or any moving object for that matter, the motion will push a diverting of the stationary line as the moving object changes relation to the lines, which the Earth dictates.

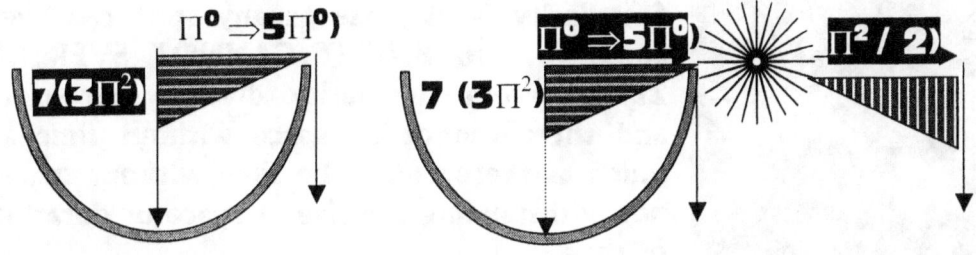

The line holds a velocity of $7\ (3\Pi^2)$ and that is the motion the space has in relation to the time the Earth dictate by contracting motion of time to the centre of the Earth. Any more motion that will underline the independence of the object in motion will show a diverting from the Earth singularity line by measure of Π^0.

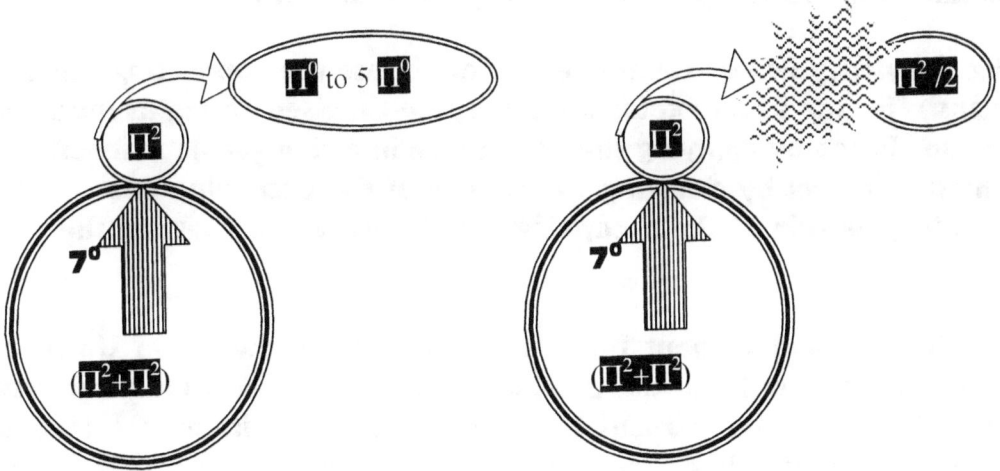

Motion is connected to heat in every principle applying. It takes heat to drive motion and it takes drive to establish independence of singularity. By additional motion brought on by adding heat the motion will show how far the independence divert from the Earth prescribed singularity. The sound barrier comes about when the second form of the Roche limit is crossed at $\Pi^2/2$

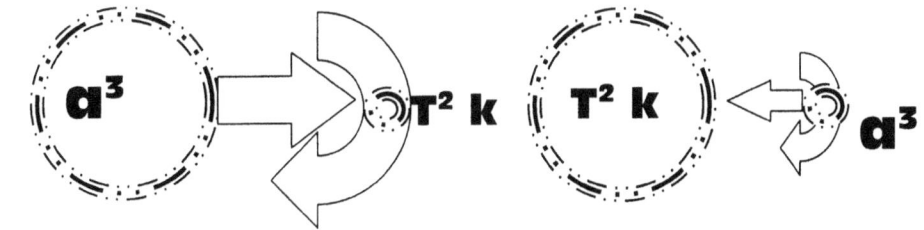

Every aspect however small or however large is serving as an atom that is maintaining singularity. Every atom is a galactica and every galactica is a star. Everything moving is containing heat and everything containing heat is preserving singularity where singularity is the Universe.

THE SPOT THAT'S HOLDING THE LOT.

Every position in the universe either holds singularity in a form, or relates to singularity. There can be no position unrelated to singularity therefore every aspect of the cosmos is space-time in various forms under the provision of singularity connecting. Matter cannot be if not surrounding singularity

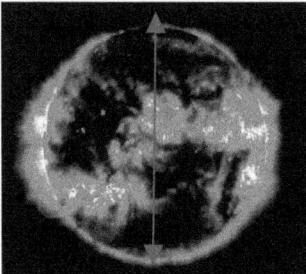

Singularity is as close as any spot can ever come to zero BUT IT CANNOT EVER BE ZERO. From singularity diverts space-time and there cannot be space without time as much as there cannot be time without space, not withstanding the size of space or duration of time.

Through space-time singularity connects as much as relates linking the universe into a network of influences beyond what ever we can ever conduct. There can be no spot that does not participate in the curvature of space-time. From the point of singularity runs space holding time to the prescription singularity dictates.

With singularity connecting singularity will or cannot relieve or release the connecting other than by a method we humans refer too as an explosion, but space-time separating as much as joining singularity dividing can change by applying time too space in a changeable manner, stretching and shrinking the time aspect by changing the density of the occupying heat, which creates the space allowing the spin of the occupying heat creating space setting the time.

There is a Universe that is limitless and without boundaries as much as there is a Universe that is within limits, neither one can ever be in the position we think they are. Our Universe is limited between the Titius Bode law of 7/10 and 10 / 7/. Once the limits of 7 / 10 and 10/7 are crossed and that region in time is passed either way our Universe stops. It is not the Universe that stops but it is only our Universe that stops.

While singularity is the ultimate endless space parted from the infinitive spinelessness singularity presents that which is between the two forms the Universe we have.

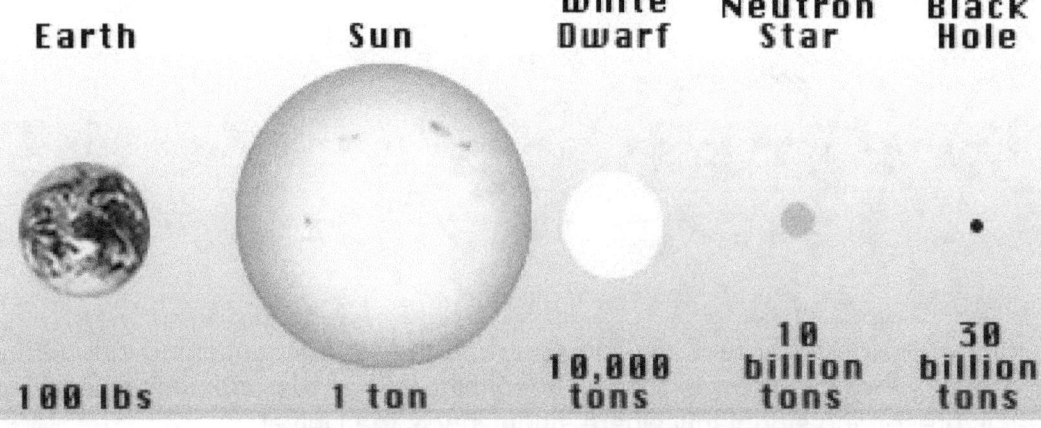

Earth	Sun	White Dwarf	Neutron Star	Black Hole
100 lbs	1 ton	10,000 tons	10 billion tons	30 billion tons

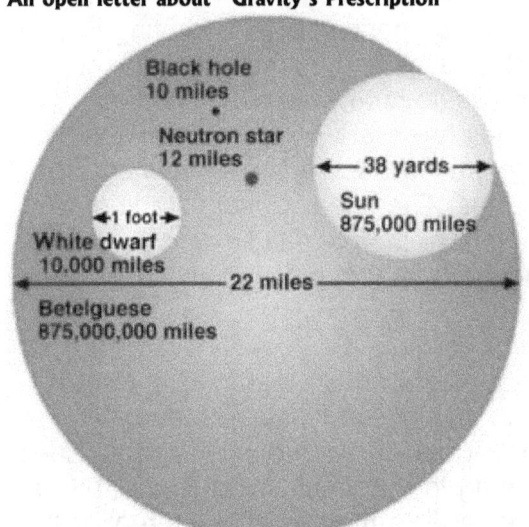

The statements taken from the pictures to the top and to the left is indicating that science does not attach size with mass any longer. The smaller the diameter of the structure is, the larger science think the mass is.

The spinning top is all the evidence any one needs to come to such a conclusion. In the past when I have acknowledged the fact that I have no academic background in the field of cosmology, this admitting brought about discontent, or rather dismay and I might even add some scorn from academics. In my opinion this trend of behaviour is uncalled for as every person is entitled to be opinionated. Every person has an opinion and it is the opinion that has validity, not the persons social standing. It is the way one absorb and reflect a personal view about matters that are important and not the manner in which the person arrived at such conclusions but it is the validity of the conclusions that should carry importance. The cosmos contains matter, space and time in heavens or dimensions standing in relevancy to one another. Whether you refer to dimensions as dimensions or heavens it is no different because it is the same thing. Singularity brings about heavens as it brings about dimensions and singularity is a mathematical fact. A straight line cannot start at zero and still be a straight line because zero extending to wherever brings about a full zero. A straight line starts at the point where the pen point meets paper. That point may be any distance from infinity to a measurable dot, but it cannot be zero.

Any straight line is also half a square because the line forming the square cannot start at zero for the reasons I just mentioned. That is singularity pointing an eternal direction from a point of infinity and that is the basis of the cosmos as much as that is the basis of mathematics. To escape from nothing one has to become something and by doing that one could not have been in nothing in the first place. If one holds a point in nothing one cannot become something because of the nothing value.

$$180^0 \times 2 = 360^0$$

To back this argument that no line can ever start at zero is to ask the simple question: what will the length of the shortest possible line be. It must be a line where the starting point is so close to the ending point the distance parting the two is incalculable yet there is the line therefore the end and the start is apart still sharing the same spot.

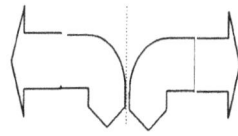

To this end any shortest line will start run and end in infinity Should a straight line start from infinity and never be able to reach a point of zero (because that will bring in the factor of no line ($0 \times 1 = 0$)) that means the line dips into infinity and it has to come out on the other side leading in the opposite direction of the first line in an attempt of avoiding zero.

The length cannot be zero because zero means no line. The starting point and the ending point may be inseparably the same point with virtually no space between the two points but neither of the three points can be zero simply because there is a line (be it infinitely small it is there). If the point is zero, then the line will be shorter than the shortest possible line. If there is a line and the two points starting the shortest possible line and being the continuing of the line it may still be the same point and even by sharing the spot as the point ending the shortest line possible the line must be there and the line holding the start and finish is next to zero but can never be zero otherwise there is no line. If that is the case then the one side of the square also presents the point of singularity to the other line holding the other dimension. That then makes the straight line not 180^0 because it is half that of the complete straight line and indicating the point of singularity of the other line and that results in any form holding only one aspect of the full form. But because the two lines are in a relevancy to a common starting point, both will enjoy the remaining distance of flow still holding the relevancy in comparison to each other and that brings about a triangle that also has 180^0 in number. That means any straight line is also a half square which is a triangle bringing about 180^0.

The co-ordinates of referral will only hold three references as surface contact areas pointing in one direction of the possible six surface areas available. That is mathematics and mathematics cannot be bias or lie. Every object in the Universe holds three points of face value to any direction possible. You can only see three sides of any six-sided object. Because the lines has a starting point in infinity that starting point has to represent singularity outside the sphere because at some point the lines cross the border of being the shortest lines possible to being normal lines. That means every starting point represents singularity by measure of r instead of Π. Every point is also the point indicating singularity at another point pointing in the direction of a point holding singularity as the pointing line as well as the supporting line.

I explain r later on but r does not in any way refer to radius, as does the r that I use for that purpose. By the same measurement, I also provide the third line with a singularity as reference and a singularity of direction. The prominence of this will become most apparent when explaining the Titius Bode law in relation to matter relating to space and matter relating to matter claiming space, but as usual, I am getting ahead of myself. What I have just indicated is pure mathematics moreover the most basic mathematics there is. There are always two lines running in supporting but opposing directions claiming space from that one point of singularity and a third line confirming (controlling) space in a third direction.

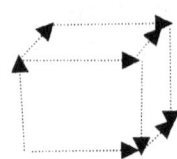 Every line is as much part of a square than it is pointing in another direction relating to a position holding singularity. That brings about my conclusion in changing Kepler's formula of $a^3 = T^2 k$ to my own which is $R^2 / T \times R / T = 1$ where any of the three components can form a square in relation to a mutual point of singularity where each one is the starting point on that lines individual singularity

Squares self-action is where Π forms r. 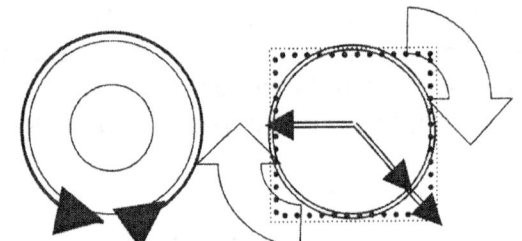 rotating will find a point where it goes into destruction through faster motion. This imbalance but in principle it has a point forms a meeting with r and Π at that point

That is space-time holding one square of space in time and directing time in space in a specific direction of flow. You may not agree with me on this issue at this point but my reasons for such a claim comes in other parts of the book and there I substantiate the claim with much more explicate mathematical detail. Space-time is matter-claiming space in time while directing space in that same duration of time from any point forming singularity. Now one arrives at the question of positioning singularity. Singularity cannot be in space because space is claimed from the point of singularity and space is confirmed from that point of singularity. If space is claimed and controlled Einstein must be at fault looking for singularity in outer space.

To find such a point one should once again look at the possible dimensions available. A square is six sides holding three to any direction. A circle is a square without edges. That is what we are looking for. A square without edges can only be a circle with a definite un-varying point of singularity indicating a specific location. A square can place singularity at any position because space holds no specifics where as a circle holds specifics at a defiant point. Space provides no specifics because an astronaut can and does drift in any direction when not being attached to a sizable object. That throws outer space out as a possibility.

Take the top with some astronaut to outer space and ask him to spin the top in outer space. We do not even have to bother such a busy person because we know the answer. The spinning top will not be able to turn because there is no stability factor in outer space. The top needs support from the needle running upwards, therefore the top will then have thrust running downwards, and the accumulative effort will bring about a point where the opposing points will allow a spin to occur. In outer space there is no such a point therefore the point we seek is within matter. A top cannot spin in outer space. We already established the fact about singularity being the position where two points originate in a square and the third will show direction. The top is a circle.

The difference between the circle and the square is in the direction the indicator follows and a square cannot spin, but through contact with a sphere as a circle cannot be motionless The factor of Π indicates eternal motion and NOT zero motion. There is a massive difference in that concept. If no line can have a zero point to start with where will the circle get the zero to indicate motion! This principle is the most basic mathematic rule and even I the ILL EDUCATED can see this. When the end of the rotation arrives the end rotation also announces the beginning of another rotation and not nullifying of the previous rotation because the rotation will have a line showing the effort it made and as it forms a wave, the wave will be there forever. The pitch may decline to a straight line, but the line remains. When calculating the motion a triangle does the honours.

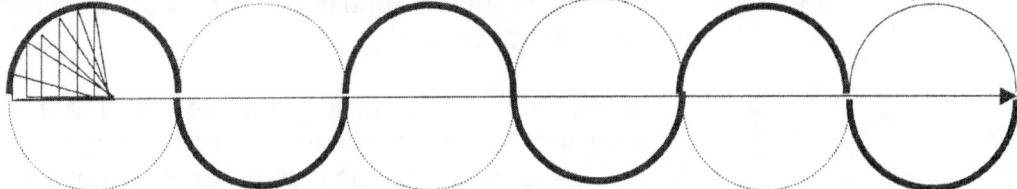

The wave confirms rotating directions followed by the circle as it spins. By stating that a wheel has a relevancy of zero by completion of a rotation such a claim denies the wave its entitlement of existing. The wave going flat, as it becomes a straight line also has an indication to singularity. All spinning matter has the point where the spin is still there but the radius is to small to measure by any means. That point is standing still in relation to the rest of the spin. In relation to that logic I do not accept that science holding the radius of a spinning object unaccountable in the spin, whether the spin is applying or not. It is in the very fact that the spin places the reference factor in the graph that makes the graph that astonishing accurate. It is mainly mathematics through the graph, which affect day and night and not the Earths standing in relation to the Sun that brings about climatically and weather changes and Earth development conditions.

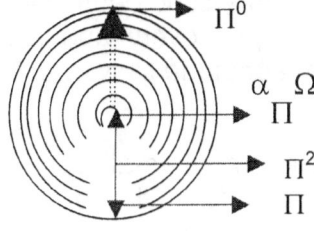

In the centre there is the singularity from which all stems. That is the centre of the Universe carrying the value of Π^0. By rotation Π^2 a line forms Π to Π with the value of 2Π. The forming of the line by initiating Π^2 is the realisation of gravity and the constituting of the Universe, which releases the Universe from Π^0 to Π by Π^2. The motion spawns the gravity because the motion is the gravity. To block the motion is to resist the gravity and such resisting forms the mass aspect that counteracts gravity.

Nothing said so far is high tech or mind bending complicated. All the above arguments from the first page to this point reached are simple and there's only ordinary primary school mathematics involved that every scholar should know. One does not need a brain fitting Einstein to come to these conclusions but just thinking about everyday issues.

All circles have something in common with squares. It has a surface but where the square holds a surface pointy the circle holds no points. To calculate a square there has to be a point where the line starts and that line will hold a value of at least infinity running from that point in two opposing directions. Arithmetic presents the possibility of zero and mathematics excludes such possibility. That is the difference there is between arithmetic and mathematical science. It is where mathematics departs from arithmetic and is as basic as counting is in arithmetic. There is nothing outrages about that which I have mentioned. Neither does one need the brainpower of a person like Einstein to come to such basic conclusions nor yet it completely destroys the claims.

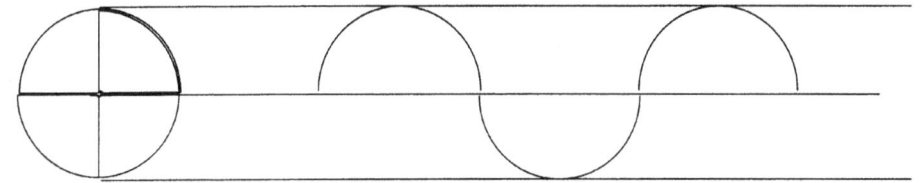

From the graph one can establish the link in the circle's rotation around a conforming unit being singularity.

Being a circle means the thing must be round and spinning. In that case, let us take an example well known to all, the spinning top. The top spins on the thinnest of points, and still maintains a balance.

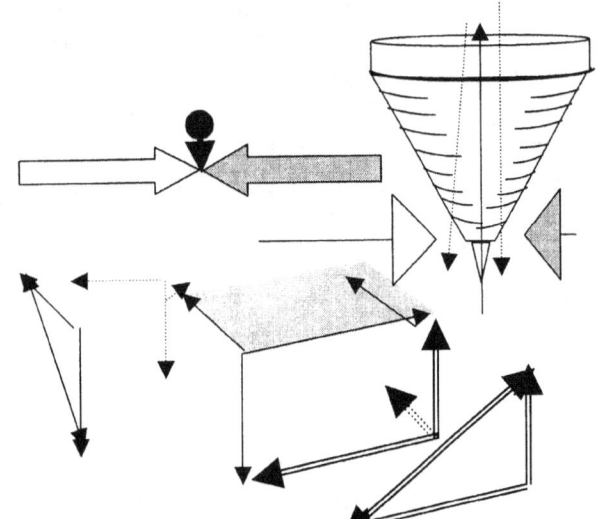

All circles have something in common with squares. It has a surface but where the square holds a surface pointy the circle holds no points. To calculate a square there has to be a point where the line starts and that line will hold a value of at least infinity running from that point in two opposing directions.

In the scenario depicting the square there is always some three pointing triangle everywhere and the common denominating figure is therefore three.

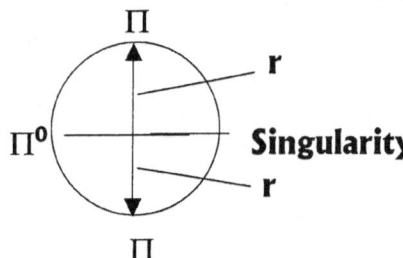

In the circle, there are also two lines where each line holds one point to singularity. Splitting the line in the two opposing directions, so in that way it is the same as a square, but the third line indicating direction brings about a difference that distinguish the circle from the square. The circle direction indicator is always Π placing the pointers at $r \times r = r^2$

Applying the second law $F = m\,a$ one arrive at the formula $G(M \times m) / r^2 = m\,(\omega^2 r)$ claiming a zero influence between the radius and the orbit of planets. I shall later again return to this issue but firstly I would like to take your mind to one other thought that seems to have escaped every body. It concerns the medium science rely most on for the gain of facts and information about the Universe. It is the influence of light.

When realising the error of science in accepting a value as zero to be legitimate in mathematics, one can establish from that that the circle does not employ zero as a value after the completion of one rotation therefore $F = G\,(M_1 \times M_2)/r^2$ is invalid, one has to return to Kepler's $a^3 = T^2 k$ and establish a value from that.
From the graph one can establish the link in the circle's rotation around a conforming unit being singularity.

Saying that one therefore has to admit that the smallest spot has to hold space because the most insignificant dot can transmit light and being able to accomplish that, one must accept it to carry a value of something. If that spot had the value of nothing, it means that spot was not there to begin with. Holding space-time one should return to the original formula indicating space-time in as much as $a^3 = T^2 k$ where $a = R$ and $T = T$. Being time it has to alternate positions and that can therefore only apply to k where k will indicate a relation to the space-time in question or the relevancy to singularity being $k^0 = 1$. This reality has serious implications on the speed of light we take as a constant.

Have you as you sit reading this part at this minute sat back and gave a thought about the light enabling you to read? Such a thought brings to mind the most simplistic answer one can imagine. The light hits the page bounces from the page and contact the lens of my eye where the lens conveys the photons becoming electricity to a part of the brain that translate the electricity to an understandable message and that makes one read. It is as simple as that! Ever gave a broader thought about light streaming across the night sky, coming from ends of the Universe we do not even realise is there? How does the photons manage to convey one complete picture coming from as far apart and as wide an area as it does? With a few photons connecting the eye or lens no one ever noticed the wonder of light. The photons reflect a view that seems as if coming from all the billions upon billions of stars. But most is coming from darkness covering an area no man can measure. Yet how many photons can actually connect to the lens of the camera or to the eye? Still a few photons coming from a single direction directly ahead eventually tell the entire storey. It is very simple to take the process of seeing by means of photon conducting very lightly and I have never heard one of the Brainy Bunch really in sincerity dissect the process to its potential. It is impossible that light from such an array of assorted sources can simply come together at the eye lens and show a picture of objects spanning across a Universe as wide as our mind can receive where the objects they reflect is beyond human measurement and the quantity is inconceivable many.

Light is much more than the medium science takes it to be. Light connects the Universe in a way we cannot contemplate. Light being far apart originating from regions not in the same time or universal space connects in a way that present us with a picture holding the Universe in an understandable content. From the point we stand and we watch the Universe the significance of what we see surpasses the sense of understanding of what we are experiencing. How can the few photons that our lenses catch coming from such a vast area covered by the night sky cover transmit the complete picture of what we see. Take a few seconds and inspect the picture of the night sky then rethink the picture applying the full content in the picture to what the size of your eyes. Think how big the picture is that your eyes take in and translate that area to the size of your eyeball in an effort to determine a ratio. One will be forgiven if one thinks of the ratio as eternal to nothing. Yet a few pages back I showed that according to mathematics there couldn't be anything as nothing. Consider the path the light followed from the source connecting to light from all other sources where all particles of the other light may come from and bringing a full picture to the lens one use to look through. In your mind connect a line from every atom producing light and connect the lines to your eyeball and see how you can manage to fit all the lines, as small as the lines may be.

Scientists think of outer space as geodesic zero, with nothing in outer space but space. Geodesic zero means the light travels in a straight line from where it originates unhindered all across space to where the light connects the eye. Such an idea by itself is outrageous because the stream of photons reduce in space to such a minute quantity, that taken the area the photons travel and the space in vastness it covers, the chances of one photon coming across many hundreds of light years through billions upon trillions of cubic kilometres of space and selecting my eye to convey the electricity is less than infinite. Yet such conveying takes place every second of every minute. The position of the location of the second singularity, which is the precise duplication of the first singularity but in a diminished capacity, is obvious to miss when one is not applying a detective mentality, as one should in scrutinizing the cosmos. Culture will have us believe that when one sees a colour shining from an object the colour is associated with the object. Logic tells a different tale. A yellow dot is all the colours in the spectrum but yellow because it is disassociating with the yellow. That goes for red blue and all other colours we may visualise. I think the norm accepts this as scientific fact with very little argument or substantiating proof.

If light came as individual streams of photon flurries our visage would translate that as such shown in the fragmented picture above. It would be a picture unconnected bringing across some photons in the manner where every object stands apart not being related in any way and that will be what we see, if it is anything that we see. That we know is not the case but that means geodesic zero is as much rubbish as anything Newtonians regard with simplicity and with careless thought. Geodesic zero means nothing and how can I see nothing as darkness because "nothing" is not darkness, nothing is "nothing" and the darkness I see is darkness showing the darkness as something.

What then about colours that are technically not colours as is the case with black and white? White is simple. By spinning all the colours in the spectrum the colour white shines through. Black is quite another matter. A friend of mine whom is one of the best painters I have ever come across told me that one couldn't paint black but have to make black a dark blue to show shade on the canvass. That apparently is his success in achieving the realism.

He also went on to explain how many variations of dark blue form the shadows in one simple tree. This remark set my mind in motion. One cannot see black because black has no colour to show, but black is the colour most prevalent in the universe. One can see only by colour and since black is not a colour we should not see black, but we do.

If the darkness was the representation of "nothing", then that should be exactly what we must see, nothing but the stars. Taken from the top picture some stars and leaving the rest to nothing is what we see in the picture below. A blind person sees nothing but when we look at space, we see something that we think nothing of as we see as space. One cannot have the ability of sight and see nothing except by closing your eyelids and then you see nothing. But in that case you do not see "nothing" in contrast of "something" you see "nothing" without it contrasting to "something".

Nothing is all about not being and not "not seeing".

By the ability to see the darkness renders the darkness something other than nothing and that changes the acquired value of the darkness from nothing to something. There is an eternal difference between something in infinity and nothing.

The arguments introduced up to this part of the introduction prologue only touches the most basic aspects of my work and by no means can such an introduction secure an opinion. Yet, not once through all my long investigation in the past thirty or more years have I found any other person claiming such views that I have brought about even in this skimpy way as I do in the prologue.

The arguments introduced up to this point of the introduction prologue only touches the most basic aspects of my work and by no means can such an introduction secure an opinion. Yet, not once through all my long investigations in the past thirty or more years have I found any other person claiming such views that I have brought about even in this skimpy way as I do in the prologue. As it applies with all things, so it does in this case as well that when delving deeper into any issue. The complexity of the issues truly comes to the fore ground when analysed in more detail. I wish to advise the reader to treat the seven books as seven different works and in that light I have separated each work in volumes of seven separate books with individual I.S.B.N. numbers with adding one part, the one you are reading, with one sole purpose and that is to bring about an academic introduction to clarify a quick perspective. Then the next three parts being of a general introductory nature there are overlapping in some sense but each highlighting issues in a different manner as to clarify facts used in the last three parts bringing conclusion to different cosmic perspectives. Yet the work is seven parts of one thesis and as such it serves.

I WISH TO DEFINE THE CATAGORISING I USE AS PART OF THE BOOK.

I have the utmost admiration for Scientists and I shall never dream of placing me in the same category as academics mainly because of their intellect and achievements. To substantiate this segregation I use some referring to place distinction between the highly schooled super trained academics that spent most if not all of their lives in preparing to further their minds. Because I tip the opposite of the scale and spent as little time in an official capacity on learning and education I have to be on the "other end". From where I stand and admire you, I can only see intellect: being the academic's common denominator. If that is the common denominator on the one side, the joining factor on the other side *"my side"* must then be the class of stupidity. To you and your class such a remark would be an insult but to me and therefore my class it rings truth and that makes it not an insult but a norm we should accept and learn to live with.

It is rather a pity that while the SUPER CLASS will never say it to our faces; the SUPER CLASS is strongly of such opinion that we on the one side of the Universe have no minds to think in any way, and it is our duty as much as privilege to accept what you the ones occupying the other side of the Universe inform us to accept and you live by that idea. As I said I have to live with it too and if I am the illiterate, then the SUPER CLASS must be the SUPER–EDUCATED; where I am the class amounting to stupidity the SUPER CLASS must be the Brainy Bunch. It all comes from the fact that there is such a huge differentiation between us. You consider one with merely one Baccalaureate degree a stable boy; think how low your opinion must then be of my type including me that never even came that far. To distinctly point to grouping or class or whatever you wish to consider the division there is between you and me I refer to your side of the Universe by the names I use above. Further more when I refer to mistakes that I do prove to be mistakes in the book as we go along I refer to it as Xepted mistakes to clear another distinction of necessity.

Introducing the book *Matter's Time In Space* written with facts about the creation in mind produces a problem because the complete picture that I introduce has nothing in common with current accepted science. The issue remains comprehensive even by using it in a very simple form. In spite of this, I shall explain three of the four unrecognised phenomena I use in proving my statements in this very book aiming at a theme of simplistic introduction being "an Academic Letter". Then in the following books I go into extensive detail proving all of the Cosmic Pillars. With my introduction of the phenomena, which I named the four cosmic pillars you will find it obvious why science do not accept them even if it is documented throughout the Universe and is quite commonly found. Applying them totally annihilates Academic's formula of the basis on which science rests in the formula being $F = G (M_1 M_2)/r^2$. The four cosmic pillars are the following:

1.   Roche-Lobe
2.   Titius Bode principal
3.   The Lagrangian principle
4.   The Academics Gravitational Concept forming the atomic relevancy.

The Universe is a combination of many material formations holding positions in space. Some of such material was covered in the blanket of heat, distributing into more spacious surroundings as the material expanded from the centre flowing outwards. Hubble's constant is proof that the space between cosmic structures are departing from many centred positions between such objects and this is a trend being located between all the objects throughout space but also indicating a definite growth in the radius and such radius growth follows a pattern where the growth seems to flow from any such a centre point away from the centre. With out the absolute and undeniable proof coming from the Hubble constant bringing proof beyond any possible doubt in any one's mind that expanding is very much and a very big part of all Cosmic activity. The accepting of the Big Bang would not be in place. $F = G (M_1 \times m_2)/r^2$ is in essence a big issue about contraction while Hubble showed the space was not dividing. The space was multiplying. The stars are growing apart and so is the galactica. This then brings in the question of space available.

With me not whishing to go into the formation of structures at this point in the book I would like to point to the fact that my following referring to the solar system is actually referring to a similar solar system that is somewhere and is now a part of a galactica we do not know about. I bring this in to disqualify any academic loophole that may come about from an argument about the solar system coming into place at a later stage of the cosmic development and therefore the argument I am about to present that such an argument does not apply. To avoid such a loophole we now use a hypothetical but real solar system in space, which formed as the Big Bang took place. There are those who avoid admitting to inconsistencies by arguing that my argument about growth is invalid because the solar system was not in place at the Big Bang. To them we now present a solar system that is identical to the one we know in precise duplication thereof. But it represents a precise duplication of our solar system and was in place ever since the Big Bang. That means with the solar systems being apart in millions of kilometres there was a time the planets and the Sun were apart by the measure of kilometres. The Big Bang shows a growth in space. Then there must have been a time when the planets were between fifty-nine and fife hundred and ninety kilometres away from the Sun . How did the planets being the size they are at present fit into such a space and still be apart? Material too must be part of the growing. This line of arguing I suppose is much below the Academics pursuit of matters but since I am much lesser in mental standards of development than they are, such reasoning prompted me to go on some investigative journey. Light journeys through out the cosmos and it will be sensible to follow the travelling of lights.

With objects being apart at some distances and light flowing in straight lines between them it must take light a straight line to travel between cosmic objects. The distance the light has to cover depends on the radius there is between the objects and as such the Universe is then about structures claiming space and space setting objects apart being the radius standing between those objects. The objects are circles by dimensions and the space is also dimensions that are crossed by lines travelling through the dimension. With light being a line and the Big Bang coming from a situation that was a lot more cramped for space than at present, the correct path to follow if I wish to trace the steps of the Cosmos back to the Big Bang is to reduce the straight line between the structures and find where such a line will no longer be a line. The same procedure will apply to the material structures all being in a sphere form. A sphere is a lot of circles forming a unit but not repeating any occupation of the space, which any one particular one claimed. Such a circle also applies a straight line only known by another name but still serves the same purpose. Reducing the line will lead us to the beginning of time.

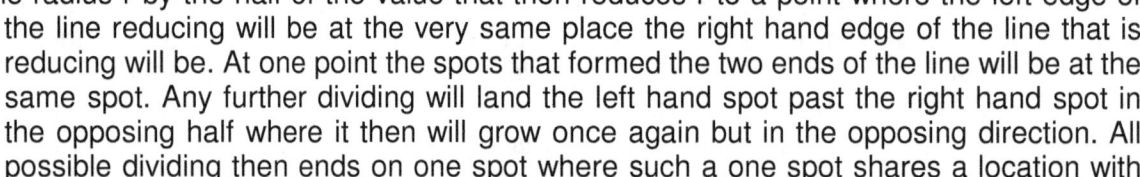

By dividing the radius r by the half of the value that then reduces r to a point where the left edge of the line reducing will be at the very same place the right hand edge of the line that is reducing will be. At one point the spots that formed the two ends of the line will be at the same spot. Any further dividing will land the left hand spot past the right hand spot in the opposing half where it then will grow once again but in the opposing direction. All possible dividing then ends on one spot where such a one spot shares a location with all other possible sides. The centre then physically is in the single dimension applying as one spot to share a location for all sides. At such a point there is no further dividing possible. On several occasions in the past I have been accused of manipulating the argument to produce none-existing or overrate facts. That is not the case. I am not manipulating facts to create an argument as so many accuse me of. What I am talking about is a mathematical fact that any one can prove by calculating following a very simple procedure. A child is capable of using the two times table and dividing by two every time is the most simple form that mathematics may be used. It is a mathematical fact that a line will reach a point where all sides are at one spot and as such cannot divide any more. At that point all sides share but all sides prevent zero becoming as factor since the sides share on spot. While the different sides are in one place the factor and value is one to all.

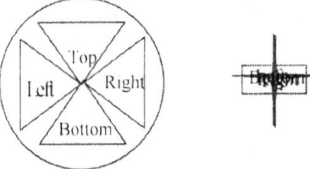

Reducing the radius r from all angles possible throughout the circle will bring about that all possible direction will eventually land on the very same spot with no more dividing possible. Yet zero cannot be a factor since the sides still hold value. A point arrive where more reducing will land the one side on the opposite side of the line but it will not bring about zero in the equation

What this argument further proves is that the circle reducing must then come from all points because the radius might be a line but that line represents a circle through 360^0. Taking that into account it is important to recognise that notwithstanding the size of a line, which any radius of any size is there is another line (or dot) eternally bigger as well as eternally smaller than the line in question. While we are in the third dimension being part of the third dimension then allows that all parts of the third dimension forever can be divided once more until the line in the third dimension is no longer part of the third dimension. When such a line leaves the third dimension it is still dividable because it might not be part of our dimension any more but it can still reduce further as part of the second dimension. By that time it has left our scope by miles. It does not mean that it end there because from our perspective that is where it ends. Yet it can still reduce infinitely more until it has left the second dimension and then at last forms part of the first dimension. Only then when the line reaches the first dimension, no further dividing of that line is longer possible.

We can never grasp the size of a line that forms the utmost or the least of possibilities and therefore size belongs to the human mind forming conceptions of big and small, but it has no place in the cosmos at large. This concept not only applies to size, but to all limits and divides we wish to create forming borders that we can appreciate. When looking at the circle in the conventional manner, we persist with errors brought about in culture and not by applying some significant modern logic. Take a circle and reduce such a circle constantly to where it no longer can reduce. Reduce it to a point where only form remains part of the circle because the radius has gone beyond human measure and becomes so small it is not noticeable with what ever tools man may use, then what remains is pi since pi does not indicate size but indicates form, and form is all that then will remain. In any circle or sphere the size only depends on the fluctuation of r, as a component to the circle or sphere but that does not affect the form by indication of Π in any way there may be. The conclusion I drew from following this process is that from this line no start can be at zero because that will be a mathematical impossibility since no line can ever reduce to zero. A line will forever be able to reduce further becoming smaller but it can never reach zero because zero is not on the scale of lines. If a line cannot reduce to zero it then cannot start at zero. A line or spot starting at zero would therefore be shorter than the shortest line possible. For obvious reasons can no line, or any line grow or extend from zero because such a line must then quit zero and become something, thus abandon its original value. That would mean the start of the line has a different value to the end and a line holds conformity through out. When any line is starting from point zero it can never leave zero because of the influence of being zero disqualifies any possibility of growth. If the line then had to grow in all directions at the same pace the line must then become a circle or being three-dimensional, then forms a multi circle we name a sphere. Since the Universe is about circles and lines connecting circles, I came to conclude that flowing from this fact is that in the Universe there can be no zero improvising as a filling ingredient for the space of a point or be unfilled space. In the case of the growing sphere the value of the circle is Π, and that is where creation must have started. That gave me the clue where to start looking for singularity. One would find singularity in the value Π and the value Π will be in all things rotating in a circle. As usual I am again shooting the gun before the hunt started. Lines in mathematics do not start from zero and that is no discovery on my part that was a realisation I came to.

UNIVERSE
Everything that exists, including space, time, and matter. The study of the Universe is known as cosmology. Cosmologists distinguish between the Universe with a capital 'U', meaning the cosmos and all its contents, and Universe with a small 'u' which is usually a mathematical model derived from some physical theory. The real Universe consists mostly of apparently empty space, with matter concentrated into galaxies consisting of stars and gas. The Universe is expanding, so the space between galaxies is gradually stretching, causing a cosmological redshift in the light from distant objects. There is growing evidence that space may be filled with unseen dark matter that may have many times the total mass of the visible galaxies. The most favoured concept of the origin of the Universe is the Big Bang theory, according to which the Universe came into being in a hot, dense fireball about 10-20 billion years ago.

UNIVERSAL TIME (UT)
A worldwide standard time-scale, the same as Greenwich Mean Time. Universal Time is the mean solar time on the meridian of Greenwich. It is defined as the Greenwich hour angle of the mean Sun plus 12 hours, so that the day begins at midnight rather than noon. It is closely linked to Greenwich Mean Sidereal Time (GMST), since the mean sidereal day is a precisely known fraction of the mean solar day. In practice, UT is determined by a formula from GMST, which in turn is derived directly from such observations of the meridian transits of stars. The version of UT derived directly form such observations is designated UTO, which is slightly dependent on the observing site. When UTO is corrected for the variation in longitude due to the Chandler wobble, a version of Universal Time, UT1, is derived which has genuine worldwide application. When UT1 is compared with International Atomic Time (TAI), it is found to be losing approximately a second a year against TAI. Broadcast time signals use the time-scale known as Coordinated Universal time (UTC). This is TAI with an offset of a whole number of seconds. The offset is adjusted when necessary by the introduction of a leap second, and UTC is always kept within 0.9 s of UT1. On this issue there is much more to explore than the meagrely mentioned. Time stands related to the position an object holds to a centre such an object refers too while in rotation. Kepler found for instance that T^2, which holds the orbit to a rotation specific, is directly dependent on **k** to value the space a^3.

Einstein proved that in the presence of a strong gravity time slows down. Surprisingly, with that evidence being around this long, nobody since then in science took those statements and made any further progress from there. It was left in some drawer to dry. Science still sticks to its change that time did not change slightly since the beginning of the time and holds the same pace ever since. With the entire Universe including all the gravity now present and not excluding one Black hole or dust speck pressed in an area possibly the size of a lepton the gravity extending from that must have been beyond what words can ever describe. If the gravity was that high and Einstein already proved gravity slows time down, then there is one logical conclusion and that is that time was n fact standing still. Mathematically it is incorrect to allow gravity to compress the Universe into a spot smaller that an atom and exclude any other factors and relevancies to change. But before coming to the mathematics I would first like to bring your attention to the practical side. I am promoting a theory in which I am able to prove there is as much contraction (moving in the direction of the Big Crunch) taking the cosmic Universe back to the size it had during the Big Bang as there is expansion (moving apart by Hubble's Constant) and the contraction is as much part of the expansion. By contracting the Universe is expanding and everything is based on gravity providing both actions. The Universe rides on a balance and we have to locate such a balance. To prove my theory I firstly had to locate the centre of the Universe. Even admitting to such a notion sounds like madness or in the least a tasteless joke, but please give me a chance to explain in more detail. I realised that my effort to locate the point holding singularity only stood any chance of success if the reducing of the line enabled me to backtrack the exploding Universe to its origins. By applying some basic effort I have located the position from where all movement came and the direction it took moving forward in time…and yes, during my search on locating the centre of the Universe I also stumbled on time as such.

Let us find the smallest possible line first. Reducing the line will eventually leave all sides on the same spot. Such a spot must be round in form. The line being the smallest line will start off as a dot. A line so small it has reached a point not dividable any more will have all sides literally on the preside same spot, and I have located singularity in just such a spot. I came to the conclusion that the spot I found had to be singularity purely on the grounds that that spot holds only one side to serve as a start to the starting point of all directions possible. There in that side is only one spot is only one side applicable and one dimension present. With all the factors given one can only come to one conclusion and that is that there can be only singularity. In such a case more dividing by two will land further positions on the other side of the divide. That point serving as a position for all point and cannot allow further dividing is the smallest line or spot there may ever be. This spot is the result of a most basic process of reduction as the Hubble constant is a most basic process of doubling up during a matter of time. By reducing the line constantly the only value that will eventually remain without dispute from any party arguing about the facts is Π. By only having Π and a radius as one square (the radius effectively becomes one holding any and all sides on one point) of any significant measure as the radius it will be an evenly spaced dot. From the smallest ever possible dot will grow a line in every imaginable direction relating to a prospect of Π not favouring one direction that puts all directions at equilibrium meaning that any form of what ever might develop from such a spot will have the end and the start being in the same position, which will also have to be a sphere as the flow outward will be equal in all directions.

Please think clearly, is that not precisely the commitment we find in gravity, where gravity is flowing from singularity outwards but never favouring any side? This reasoning prompted me to look for singularity in such a spot because if the prime spot from which all came was a spot holding all, then the spot must hold the shortest line but more prominent it will hold the smallest form including the smallest circle or for that matter the smallest sphere. With gravity always being in the centre of a sphere where the space is least available in the entire structure (there is not even space left to fill) one finds a flow of gravity from that centre spot outwards in all possible direction even-handedly. The fact that the original gravity will begin as a circle or will be a circle is the direction it will take when being the first spot created. All progress will be evenly in all direction because no direction will stand out or be in favour above any other direction at first.

The spot forms a full circle, but the line running through the circle is forever present because that is the future radius of the circle that will one day develop the circle, which is equal to the present diameter. The fact of the presence of such a possible line in such a possible circle dividing the possible circle into two parts makes the centre line equal to the half circle. The line forms the half circle but not only that the line presents the half circle as much as the line is the half circle. The line then is 180^0 and the half circle is 180^0 because in singularity the two factors are the same. The same value is of course $\Pi^0 = 1$.

In this half circle of the future, which is no half circle as yet because of a lack of space there are three future points indicating the space less ness that will go on to become space filled with something. On top of such a circle to form must be a marker indicating an awaiting boundary or future border and at the bottom of the future circle there also must be a similar marker that is no marker as yet. Between the two possible points that are not there yet is a future line running that is not there yet.

Then indicating the possibility of a position to come that will bring about the half circle being a future distance apart from the future line indicating a diameter that will one day be there a third such a marker must be established for the future. That forms a triangle with two more sides being connected by either a line being one or half pi being one. From singularity comes about that the line is the same as the half circle is the same as the triangle and all has one value being 180^0. From this come the most basic principles in as much as forming the ground rules of the law of Pythagoras.

When drawing a line such a line then starts off with a dot serving the spot that holds all sides equal. That means the line serving as the future radius will be equal to the half circle which is then Π. The only aspect of the point that stands in for the end of the single line forming the radius of the circle is that we then mathematically reach the single dimension. We decreased the line to where a circle being Π formed on the single dimension. This dimension also hold the circle dividing line because from there the radius must once again generate a value and by such a gesture that the extending would form the circle that forms the sphere that eventually leads to the formation of particles. This leaves a problem to investigate.

 With no line possible there had to be another dot that formed since the Universe has many dots that formed lines. But let us not to get confused and lost in the range of possible diversions but let us stick to two dots. One dot was next to the dot next to the dot, but as I said we stick to one dot next to the second dot. $M \times M / r^2$ is the first step gravity began with. That leaves us with a huge problem in as much as when $r = 0$ then $r^0 = 0$ and 0 dividing any value will leave 0 as the answer. If the particles were inseparable at the start it must bring about that gravity would not be forming since the distance will not permit any dividing. By allowing the distance separating the particles to be zero, the particles melt into a unit.

Again this is Mathematics and not my incoherency as some Academics dismissed my work. Let me run through the argument one more time because I have been insulted by Academics in the past telling me I am bending mathematic rules with my applying double values to try and produce some argument. The two particles formed by an inseparable unit separated by a sharing of a spot. We know that at least two spots formed because there are many more than just two that remained to become part of the visual Universe. Let us name the spots because that is what humans do best if they do not know what to do with what they have to do.

Let us call the one em and the other one spot next to em we then call emtoo. Between em and emtoo there were nothing because em and emtoo were inseparable. By they're being inseparable we would naturally be inclined to think that the separation value should be nothing or at least zero. But putting zero in that place is a mathematical excluding procedure leaving future mathematics excluded. With m multiplying m_2 and then dividing $\div$ r with zero (r=0) such a procedure will leave the lot at zero and with that nothing is going nowhere. That means although we think the space between the two parts are nothing the non-existing space has to be at least one to be a future factor.

Every part of the argument is sound but was never yet used. I repeat once more if my argument reflects on inconsistencies those inconsistencies are not about my work. In order to disprove my argument replace Mass one and Mass two with any number possible, then divide such a number with the square being zero. If there was no space then the value of the particles had to be one. If there was no space between the particles the particles then had to form a unit. But if there is a mathematical possibility of reducing a line to the single dimension then there had to be a factor representing r as a factor of one. Take $(M_1 \times m_2) / r^2$ and substitute any of the factors with zero and the result coming about has to be zero. The factors in the equation have to have any and all the elements at a value of at least one. Only if r was a factor of one can gravity bring about any mathematical equation developing from this argument.

That means the mass on both sides must have a factor of one being a limit, which does not allow such further reduction of r and any further reducing of r beyond the limit will not be tolerated. Only if r = 1 then r^2 can be 1 and mass can be apart. Like it or not but believing in the Big Bang must also bring about the accepting that the cosmos moved apart somewhat. The fact that r brought increase in the space separating the mass produces a problem that was solved already. About a century and a half ago Roche found just such a limit.

Once again I were confronted by zero becoming growth. There is a huge hole that needs filling when bringing into a relation any forming of an alliance between a cosmos coming from nothing and filling with nothing and a cosmos growing spontaneously through balance shifting prominence. Mathematically the fact of applying nothing serving as a factor applying in the cosmos is not a strong and convincing argument. The minute one brings in zero as a multiplying factor forming a definite value working into the calculations of the cosmos, growth disappears. If growth was not a factor, the zero factors could be involved with some form of maintaining stability and where then further growth will accept the responsibility of zero

The region surrounding each star in a binary system, within which any material is gravitationally bound to that particular star. The boundary of the Roche lobes is an equipotential surface, and the lobes touch at the inner Lagrangian point, L_1, through which mass transfer may occur if one of the components expands to fill its lobe. It names after the French mathematician Edouard Albert Roche (1820-83).

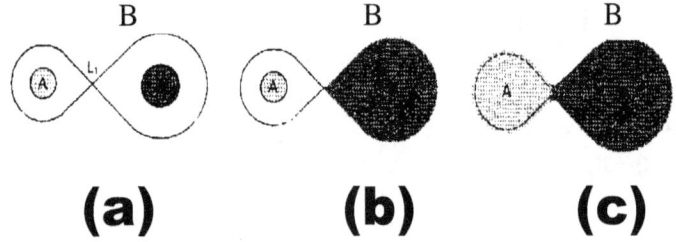

(a)   (b)   (c)

THE ROCHE LOBE: In a binary system, the Roche lobes of components A and B meet at the L_1 Lagrangian point. (a) In a detached system, neither star fills its Roche lobe. (b) In a semidetached system, one massive component, B, fills its Roche lobe. (c) In a contact binary, both components overfill their Roche lobes and share a common envelope. As with the graph I can see the two sides forming a connection therefore relevancy has to apply, all contradicting Newtonian claims of no connection but through mass attractions. The mass does not attract but one interferes with the other total influencing the space surroundings.

The closest encounter worth noting we ever had with this law in the modern age of news and Television was the Shoemaker-Levy 9 incident during the previous decade. At the time and even in the present no one drew any similarities but after completing this book the reader should find why I could draw such a similarity, which there is between this incident and the Roche limit. Even the phenomenon called the Sound Barrier became clear when applying the Roche factor with the laws governing the influence of singularity.

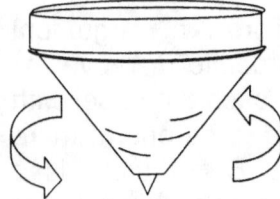

The gravity Newton suggested and the gravity Mainstream science is in search of is not there. In fact there is no gravity if we consider what science would wish to locate as gravity. Mass does no instigate, initiate or create gravity by any means or measure. Gravity is the independence of matter in relation to its depending on matter in order to realise its independence. The motion that produces gravity is what inspires a new Universe into being

Every time a top or anything else starts to spin the spin create a newly establishe4d Universe which fills that Universe with the surrounding time in that Universe with a new centre and the centre is the gravity that rule and dominate that Universe.

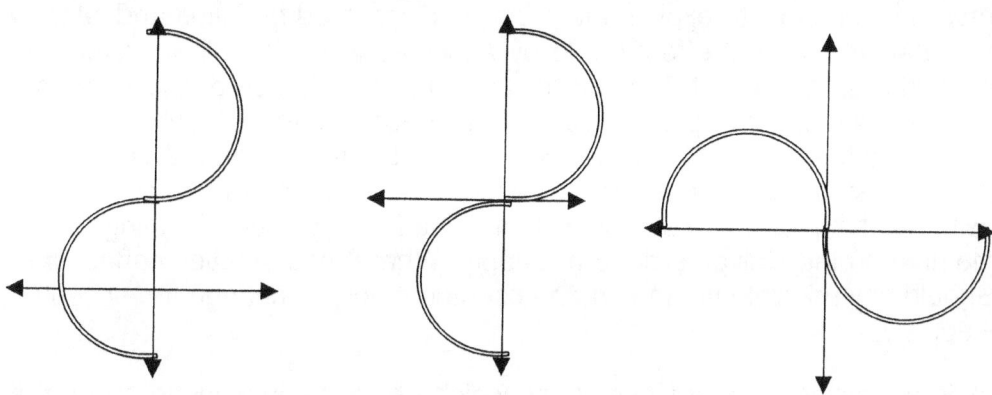

The graph is the result of motion where singularity form a centre but the centre formed is not zero. The centre formed is Π^0 and from such a centre space-time forms by expanding **space a^3** to the value of **Time T^2k. In the centre of the graph is Π^0.** This point, which I now am referring to, is the point where Π a fully appreciated value while the diameter D still remains a dimensional factor of one. His is the dawn of the second dimension where space was there but space was sparsely shared in some cases. It is when Π^0 shifted to become Π for the very fist time. The point without movement, the point holding singularity must have a value of Π being the eternal dot but since the dot has no dimension in having form the Π that indicates the dot must be Π^0. From such a point there has to be to the side of the centre point be a point where space do start. That point will then receive a diameter but that point will have form only in being a circle. In that point there is a shift from in relevance from Π to the centre Π^0 and for the first time it brought about two separate values for Π.

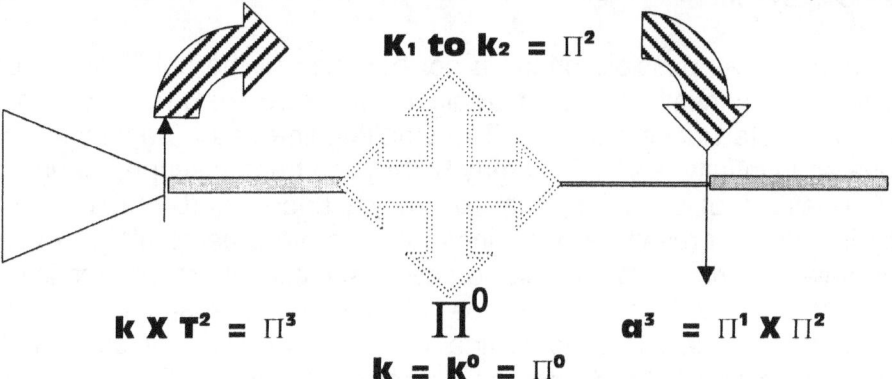

$$K_1 \text{ to } k_2 = \Pi^2$$

$$k \times T^2 = \Pi^3 \qquad \Pi^0 \qquad a^3 = \Pi^1 \times \Pi^2$$

$$k = k^0 = \Pi^0$$

Because the three points existed on equal terms in singularity sharing a same spot the coming out of singularity will enforce that equal value comes to all. That means the circle gets to become Π, the diameter becomes Π and the distance setting the structures apart will also become Π. This is what the coming from one point brings along. Only when being part of the second dimension can there start being separate values.

While the form was still being in the single dimension from the one side of the form the dots had to establish identities apart but not separated yet. The one circle had a factor of $\Pi^0 = 1$ and the centre had to have a value of ($\Pi^0 / 2$) extending past the very next object but also cutting such an object into a square double half value that was going to come about as soon as the other dimensions came into form.

The only definite place one will locate zero is in between the starting point of the lines going in opposing directions in the position the lines hold before there was the least of directions applied, but that is only because there is no such a position, not because any line is coming from there. The two lines are still one holding the opportunity of parting as an option but have not yet parted and therefore are on the very precise same spot. The line coming from there is already there because it already has the choice of going in any and all opposing directions and when it starts running it will place filled space in that location because the space was already filled with a line starting and not with a line not there at all. When reversing a line we might find a better idea of what is in place and where it is in place. Gravity is officially a force without limits going past and through borders and has an unlimited reach. It seems to remain even and this is conflicting with the flow of perceptions about mathematics.

The formula $F = G \, (M_1 . m_2) / r^2$ is unable to explain the principle discovered by Titius and later by Bode and in contrary to all statements to that effect made by Accepted Science policy makers the Titius Bode principle is not coincidental. In fact it is one of the four most adhered and important cosmic pillars holding the cosmos structural in place. From the two above comes gravity. In a few pages I prove how one can arrive at the facts that prove how the Titius Bode Principle leads us in the direction the origins of the solar system. But before we can accept the influence of the Titius Bode Principle we have to deal with "Nothing" and as such dismiss nothing from science. "Nothing" in the Universe is coincidental; "nothing" in the Universe does not apply. Where mathematics connects to lines nothing disappears. Should any principle not match an accepted theory or change the accepted theory, the theory does not apply.

The content of my work holds a new view about Cosmology, which I have been working on for the past twenty-seven years and exclusively for the past six years. I always had a problem with the idea that space constituted of nothing, while I came to realise that lines mathematically couldn't start at zero because there is no evidence of zero as a factor in mathematics. Should you disagree with my statement the question in need of answering is this: What will the length of the shortest hypothetical line imaginable be and moreover, what would the total overall length be in that case? The shortest possible line (hypothetically) must be so short it must have an initial and ultimate point sharing the same spot. The two points must be one and only then can further reducing of any line not occur. If it used zero as a start, the zero part would not count, because the line will only start at a point past zero where the line then will start forming an infinitely small dot. I press this point in urging the understanding because there is such a point, but in an attempt to recognise the point, I have to convince the reader to abolish four or five thousand years of accepted and practised mathematical culture and that is no easy feat.

Taking the line back as far as possible brings a dot because of the equilibrium that will stem from such a position. The dot is in infinity, however small, it is not zero. Zero ultimately means not existing and then that point, as a start does not exist. The smallest line has a beginning and an end at the very same spot located in infinity, and infinity may be beyond human scope, though infinity is still not zero. Infinity may constitute of something we do not yet understand, but we may not define our human misunderstanding not present in our minds and therefore as nothing. In this aspect lies the difference there is between arithmetic and mathematical science where arithmetic can have position such as zero since arithmetic excludes the cosmos calculating numbers only. Cosmology is not about numbers because no one can calculate the number of stars. Cosmology is all about lines and angles positioning objects, and in those there features no zero. No line can be zero long and forming a position of zero degrees in relation to another object.

A man may have that many oxen or so many sheep and even this amount of wives, (in Africa) or not have any therefore having then a total of nothing, but there cannot be nothing between the Sun and its orbiting structures. The having and have-nots are part of arithmetic. Light will indicate a line flowing between the Sun and whatever planet, following dot after dot thereby proving the existing of the possibility of something going about by a straight line, and any straight line in relation to other straight lines will be under the law of Pythagoras in as much as obeying the rules of trigonometry. There is no possibility of a straight line not forming in space. If there is space, there can be a straight line. The mere fact of two spots having different positions in space gives the two dots different values. If the line has the length of zero it is not present. If the triangle has one angle at a value of zero it is no triangle because the zero would dismiss all the other angles. Mathematics converts the values of integrating lines according to Pythagoras and arithmetic is about numbers to be added or subtracted. By mathematically excluding zero from cosmology a new Universe opens to the human mind. With the distance between the Sun and Pluto being roughly one hundred times more than the distance between Mercury and the Sun, the distance must hold something more than pure vacuum filled with nothing except one atom here and there occupying the vacuum between them and the Sun. If space supposedly comprises of nothing how can nothing then become plural forming more or be multiplied by a number as to indicate a growth in something not even existing. As the one becomes one hundred the one cannot substitute a value of nothing but then must be part of something. If the one substituted the nothing, all laws of mathematics will go in disarray because when one multiplies any number by zero it becomes zero placing both planets in the Sun. If Pluto was one hundred times closer than it is at present was it one hundred times nothing closer? $100 \times 0 + 0 = 0$. That is mathematics!

By allowing the three hundred a value the nothing must form one making that which is between Pluto and the Sun not to be nothing but there has to be something. This argument follows mathematics to the letter and in precise detail. With Pluto and the Sun being apart that being apart has to have one of something in place forming the being apart from each other's cosmic position one time multiplied by the many ones we find in that space standing relative to other space regarding whatever the space becomes what is between the Sun and Pluto. That factor cannot stand in for not one, which is the same as nothing as that is because one cannot take the place in the position that zero secures. By excluding nothing from the equation space becomes something bringing in a value lying inside the realms of the infinite that must form singularity. As the zero becomes a dot, something else becomes clear about the dot. Looking at the night sky we find darkness overwhelming the space in relation to the stars bringing across light.

My approach to cosmology shall prove to be somewhat unconventional but through the abandoning of the accepted, it enabled me in locating the precise location of singularity that forms the connecting basis of the Universe (and this I say with some degree of confidence). There **are two locations** but I shall **first concentrate** my explaining effort on **the prime singularity**. Singularity did not vanish into the unknown after the completion of the Big Bang development but is in a place science incorrectly valued and classified incorrectly and in that, there is something hiding the truth. If singularity was or is where the beginning is we have to go back and see just where such a beginning was. I cannot accept that the Universe started at zero and neither does anything else in the Universe start at zero. My excluding the possibility of zero includes that the Universe is not filled to the top with nothing and neither is nothing part of outer space.

The Universe is about lines allowing light to flow from one point to another point and in following that line it has to continue in the line as the line has to represent something. The Universe is all in relation about lines indicating distances between cosmic structures. The cosmos is in short about lines connecting points in space being apart. It is about a line starting and continuing from such a start. But science advocates their opinion that such a start of a line flowing between any and all objects can hold zero because according to them the Universe are full of nothing. If the Universe in as much as outer space is a container filled with nothing at the present moment, and there is no place anything that was part of outer space previously could release to and there was no emptying of what ever filled it before, then it could not get rid of what was in the outer space when it started with what it started off with. We must then accept from what is not in the Universe was not in the Universe at the time during the start that at is present at present according to science because it then still must contain the same nothing and must have that same filling from the start to the present. If it was nothing it still must be nothing and that same substance being nothing is what it also used to grow using it as it grew because it filled outer space with nothing growing from and growing to nothing. Is that true? The filling of the Universe could not go anywhere so one has to presume it started off from nothing and from there it kept filling with nothing since what ever was in the Universe at the start had no place to escape to or no place through which to escape.

That is only applying if it is nothing filling the Universe at large. Can nothing grow as much as a line is growing from a start of nothing? The answer is that such lines not only indicate a distance but since the Universe came from such a small space as science propagates with the theory of the Big Bang then all particles in the Big Bang Universe were rather cramped for space when the Universe started from that small line between particles and is now the same line but is now so big. In the past it seemed being so small and showing the space between particles to be awfully short at one time. It was short but how short was it? Did it start off as nothing? Is the line starting at nothing as science wishes us to believe? If it does then all lines must start from nothing so we better investigate this trend with the start of a line. In this following I show my argument with which I hope to prove the counter part of what science believes. I am about to prove that which science sees as nothing in space and in material is the very location of singularity.

Lines mathematically cannot start at zero because there is no evidence of zero as a factor in mathematics. Should you disagree with my statement the question in need of answering is this:

What will the length of the shortest hypothetical line imaginable be and moreover, what would the total overall length be in that case?

Locating zero

Zero point   Starting point of the line.   Extending the line from the start.

Zero in place

Let us duly test my statement by taking the line back as far as possible. The shortest possible line (hypothetically) must be so short it must have **an initial and ultimate point sharing the same spot.** The line that **cannot reduce** any **further** must be **so short** that **directions flowing away** from each other **are located** in the **same position**. Any theoretical line being the shortest possible line cannot have the line holding the initial starting point at point zero and advance from there. Mathematics simply will not allow it. If the point had zero as all it had to offer, such a point is not present. The zero means there is no such a position. If it used zero as a start, the zero part would not count, because the line will only start at a point past zero where the line then will start. Zero ultimately means not existing and then that point, as a start does not exist and where the line then stars is a point in existing. When the line **has a beginning and an end at the very same spot** and it wishes to extend the position as to further the possibility it has, which direction should it favour. Extending the line in any one direction will favour one direction without any clear reason not extending in other directions. The fact of direction being present only proves and is proved by another point established, which is placed in relevance to such a second pointing a position already established by the relevancy of two point located in a direction to one another. But if one point starts one line there is no favour of direction since there is no established direction yet. The only mathematically sensible option about extending any line starting at a pre-designated point without any other point to establish a pre determined direction will be non-bias progress in all directions equally in order to give a meaningful flow of mathematical equilibrium. Not one direction stands superior to other directions and all directions are equal with no bias anywhere. Of this statement the Pythagoras mathematical principle is proof of and that I explain later.

Let us dissect nothing, as we find nothing in the presence of the cosmos. The distance between the Sun and Pluto is roughly one hundred times more and if the distance between Mercury and the Sun , but both has nothing between them and the Sun . The space filling the distance from the Sun to Mercury has nothing more than the space between Pluto and the Sun . That means the distance between the Sun and Pluto is as equal in relevancy than the distance from the Sun and Pluto since both is the measure of nothing. If the one substituted the nothing, all laws of mathematics will go in disarray because when one multiplies any number by zero it becomes zero placing both planets in the Sun . The distance between the Sun and Pluto **is Pluto is 5900 X 10^6** kilometres of space, but in that statement we take it that the one of a kilometre is present in such a multiplication. The one constitutes the presence of fact being a statement of a value. By saying the distance constitutes of nothing we have to substitute the one factor with a factor of zero. Then the calculation must read **Pluto is 5900 X 10^6 X 0 = 0.** Including nothing as to state the presence of that part contained by the calculation delivers the total of zero. By excluding nothing from the equation space becomes something bringing in a value lying inside the realms of the infinite that must form singularity. Applying this logic to the Lagrangian system and interpreting that information to the law of Pythagoras a clear pattern comes about.

The reaction responding from my argument is that it is silly, but should that be your personal opinion too then test where the silly part applies. Bring the zero into the calculation, the zero that science so eagerly places in outer space and see the mathematical result. By applying the distance one accepts automatically that the figure become calculated with one as it represents one in being a calculating part of the cosmos. The calculation as all calculations normally are is in order to calculate something and the something will at least stand in as one in relation to the rest being part of the calculation. But saying that the factor of one in fact represents nothing since nothing is so much the part in the calculation being calculated, then the zero has to replace the one as the fact of being calculated.

The claim becomes obvious when observing the connection between the half circle, the straight line and the triangle, which could also promote all the qualities lurking behind the pyramid. Consider the connection between 180^0 sharing three different forms all part of mathematics where each is different in form, but equal in value and then one may realise in considering the very basic in mathematics being the Law of Pythagoras on which all mathematics are focused. The triangle stands in for one factor represented by one at a value of 180^0. So does the straight line become a factor of one and the half circle also becomes one where the factor of one equals all 180^0. All three are most seriously part of shapes in the cosmos. Revalue any one form to zero and the rest too must follow and share the same value. The Law of Pythagoras is about angles in relation to lines and not one angle can represent zero because that will reduce all the lines also to zero. The measure of angles between stars at a distance uses parsec as the indicator, but the parsec between the stars indicating an angle has to represent an angle whereby one may measure distance and such a distance cannot be zero because then the parsec will be equal to zero. Again it is multiplying the factor with the measure but if the measure is about a factor of zero, then the factor too becomes zero. That is as basic mathematics as I can present.

If the argument seems ridiculous it is not my mentioning such a fact that is ridiculous but the mere fact of the reasoning also becoming a recognising of an argument accepted by science making it as such ridiculous. If space is nothing then it has a number to use indicating just that value being zero or the capitol O indicating zero. Try and indicate what is measured and calculated in space, but not by simply not thinking about the fact and therefore simply ignoring that what is measured forming the sole value of space, but put the value of nothing as part of the distance in calculation because that is what is measured. When stating the distance between the Earth and the Sun place on paper what will allow the kilometres measured to represent the factor that is being measured. If represented by one being the total of one by hundred and forty nine million kilometres of nothing put that language in the International language of mathematics that spans all dialects spoken on Earth. Put it in mathematical terminology by saying there are 149 000 000 X 1 (multiplied by the kilometres) multiplied by what it is being measured which is 0 and what will the total come too... a full zero.

149 000 000 X 1 (km) x 0 (indicating what the km are made of) = 0

Mathematics says it. If there is something to be measured then the least value the measurement can have in relation to what is used in the measuring has to be one. It cannot be zero and be measured...and we do measure outer space! It sounds as if something here is at fault. It is not with my mentioning the inconsistency one should find fault but the fault is with the fact that it is there and no one noticed! I am not to blame just because I am mentioning it, but the blame must go where it belongs.

I think it is by now little understood although I imagine not nearly accepted that by adding a million of nothing to one nothing there will remain one nothing and that is still nothing. Nothing cannot accumulate therefore I cannot accept anything holding the vastness of space being able to constitute nothing as the major component.

When reducing the circle in size one have to reduce the radius or the diameter because the pi is the indicator of the form as being a circle. Divide the r until there can be no dividing any further and that cannot in the end indicate zero because no matter how small, in that will forever be a value in place.

There will forever be subatomic particles building atomic particles because the reducing of space goes down as far and to infinity. There will always be some material forming a part of more material that builds into something which ultimately becomes the atom as a unit. In every centre of every subatomic particle running down into infinity there will be a centre composing singularity and the group will establish a centre governing singularity.

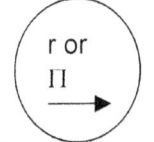 ●r / 2 ● r / 2 • r / 2 dividing r reduces r to infinity but not ∏ as ∏ remains

stable, protected by the rotation of matter forming a circle around singularity

Taking that into account it is important to recognise that notwithstanding the size of a line, there is another line (or dot) eternally bigger as well as eternally smaller than the line in question. We can never grasp the size of a line that forms the utmost or the least of possibilities and therefore size belongs to the human mind forming conceptions of big and small, but it has no place in the cosmos at large. This concept not only applies to size, but to all limits and divides we wish to create forming borders we can appreciate. When looking at the circle in the conventional manner, we persist with errors brought about in culture and not by applying some significant modern logic.

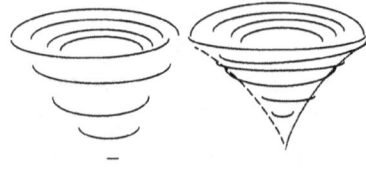

By reducing r indefinitely to the tune of half each time, r would become infinitely small, beyond human calculating means, however as mentioned in the case of the smallest dot holding one spot, r would become insignificant beyond human comprehension even, but never reaching zero and still Π would remain intact and dictating form. I believe one can begin to see where my suspicions are heading because the flaw comes about in the manner mathematics are practised for thousands of years. But before coming to the mathematics I would first like to bring your attention to the practical side. I am promoting a theory in which I am able to prove there is as much contraction going on in the cosmic Universe as there is expansion and the contraction is as much part of the expansion. The Universe rides on a balance and we have to locate such a balance. To prove my theory I firstly had to locate the centre of the Universe. Even admitting to such a notion sounds like madness, but please give me a chance to explain in more detail. If I wish to achieve success that would depend on my ability to convince all that outer space comprises of material and as such we can locate such material even if we are unable to see such material.

To find the invisible I had to locate singularity. I realised that my effort to locate the point holding singularity enabled me to backtrack the exploding Universe to its origins.

By applying some basic effort I have located the position from where all movement came and the direction it took moving forward in time

The reversing of the circle radius is not alien to nature at all. An observation coming instinctively to mind one may recognise is that the form reminds rather explicitly of natural phenomena such as hurricanes, water whirls and even the shape most commonly favoured to express the cosmic object referred too as a Black Hole. The similarity may be more than coincidental. Let us consider the statement in the reverse. In our calculating of a circle we apply two formula methods. The one use an r to indicate the radius and the other use a D to indicate the diameter, which is double the radius and therefore needs to be divided by a four to eliminate the Newtonian inverse square law amounting to the difference there will be between the two. The one using the radius is Πr^2 and the other formula is using the diameter is $\Pi D^2 / 4$.

In any circle or sphere the size only depends on the fluctuation of r in the square as a component to the circle or sphere but that does not affect the form by indication of Π in any way there may be. The conclusion from this is that no line can start at zero because that will be a mathematical impossibility. A line or spot starting at zero would therefore be shorter than the shortest line possible. For obvious reasons can no line, or any line grow or extend from zero because such a line must then quit zero and become something, thus abandon its original value. That would mean the start of the line has a different value to the end and a line holds conformity through out. When any line is starting from point zero it can never leave zero because of the influence of being zero disqualifies any possibility of growth. If the line then had to grow in all directions at the same pace the line must therefore be a circle or being three-dimensional, a sphere. Flowing from this fact is that in the Universe there can be no zero point or unfilled space. In the case of the growing sphere the value of the circle is Π, and that is where creation started. That gave me the clue where to start looking for singularity. One would find singularity in the value Π and the value Π will be in all things rotating in a circle. You might wonder how does that apply to the cosmos and moreover to gravity? In my search I stumbled on two accepted but not intergraded laws and when I found and located singularity the two laws became very much plausible and factual. Take a circle and reduce such a circle constantly to where it no longer can reduce. Reduce it to a point where only form remains part of the circle because the radius has gone beyond human measure and becomes so small it is not noticeable with what ever tools man may use, then what remains is pi since pi does not indicate size but indicates form, and form is all that then will remain.

I believe one can begin to see where my suspicions are heading because the flaw comes about in the manner mathematics are practised for thousands of years. But before coming to the mathematics I would first like to bring your attention to the practical side. I am promoting a theory in which I am able to prove there is as much contraction going on in the cosmic Universe as there is expansion and the contraction is as much part of the expansion. The Universe rides on a balance and we have to locate such a balance. To prove my theory I firstly had to locate the centre of the Universe. Even admitting to such a notion sounds like madness, but please give me a chance to explain in more detail. I realised that my effort to locate the point holding singularity enabled me to backtrack the exploding Universe to its origins. By applying some basic effort I have located the position from where all movement came and the direction it took moving forward in time…and yes, even time as such.

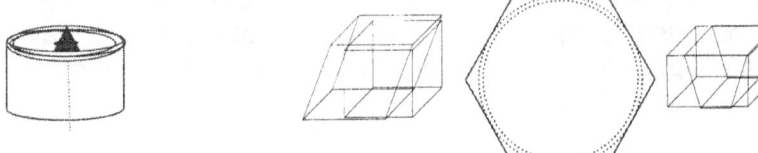

Anything occupying space in the cube will apply r and by r I mean just a distance not using Π because Π serves as a form indication while the collective product of r will determine form as well as accumulative dimension total. Notwithstanding the name used confirming the shape or r named as length width or height, it is all just a straight line bringing about the cube with all its other names that may find attachment to specific form but nevertheless still remains only a six-sided cube with connecting lines applying different angles changing in some cases. The normal perception is that any circle growing spontaneous would grow by the radius, which is r. In mathematics that may be true but it is not true in nature. In nature that cannot be the case because r is an indication of a straight line. By growing with the aid of a straight line from the centre to circle the influence that that would have on the circle would result in many circles following one another and not a continuous growth.

Gravity is the dimensional changing of space holding r as reference in the cube as to the sphere holding Π as the reference. In order to generate spin producing time in matter occupying space, therefore creating dimensional change, Π has to be a factor indicating the possibility of spin because implementing Π the circle sides will follow one another without establishing separation. The answer must be in finding Π, and thereby locating singularity. If singularity is in affect the original point of the cosmos birth, the reducing path we should follow will indicate the whereabouts such a point must be.

There are two standard formulas used to calculate a circle. The one uses an r to indicate the radius and the other uses a D to indicate the diameter, which is double the radius and therefore needs to be divided by a four to eliminate the Newtonian inverse square law amounting to the difference there will be between the two. The one using the radius is Πr^2 and the other formula using the diameter is $\Pi D^2 / 4$. However one looks at the mathematical expressions and Kepler's formulating of space-time, there is an exceptional difference between the two scientific uses. When investigating Kepler's formula one do find it appreciably differs from the normal Mathematical equation like $a^2 = r^2 \Pi$ and $\mathbf{a^3} = 4/3\, \Pi r^3$. In the normally used mathematical expressions such equations tend to concentrate on the volumetric aspect. In the case of Kepler's expression it is something else that wants to surface. It is another idea that is coming to mind. In Kepler's formula $\mathbf{a^3}$ stands to symbolise the third dimension and such a third dimension becomes equal to two other dimensions grouping and sharing value to equal $\mathbf{a^3}$ efforts.

It is not the circle of the rotation because with such a normal circle the radius is in the square and Π evaluates form. Here there is no mention of a factor Π, which one would suspect to be somewhere applying since the circle is Π and Π is the circle and the two are inseparable, but not in Kepler's $\mathbf{a^3}$, where there is no mention of Π at all. The fact that there is a radius of some sorts used to indicate a position cannot hold the square as it normally does in the case of the normal equations. In the mathematical equation the factor indicating the position of the circle edge has the square value being called the radius or in some cases the radius doubles and which then is the diameter, and the circle indicator is Π. But in this event the formula value will bring about a square value to the answer one receives. It will bring a value to the surface of the circle. In Kepler's formula it specifically does not.

I realised before starting my quest that one possibility that the shortest line or smallest spot can never have is having a starting point on the zero mark. If the mark of zero holds the start it must also hold the end because the end and the beginning has the same position. If the position of zero then is the beginning, the end will also be zero leaving the line or spot without an end as well as without a beginning. Such a spot will constitute all of nothing. Any line starting from zero would inevitably start from a point where it ignores the zero mark because the fact of zero does not implicate a start or a size of value, but only the not being there of that position. All lines would form a duplication of another line sharing value since there will always be a possibility of yet another line in the realms of singularity lying between the two lines in question reducing the size infinitely to either side of the divide we humans create. Boundaries therefore are human and as man made substances it does not belong to the cosmos outside the influence of man and must be discarded. No mathematics will ever measure the thickness, because as the line that is standing still it cannot have a width at all. The moment a width appears which one can measure or calculate, the line will become part of the factor forming the divided and not the divide.

The instant when space connects, the spin direction will produce the partisanship of space and spin. Any form of space (even in the most minute) will expand as it favours a direction but changing the direction is by rotary motion. The moment there is an area there is a measurable rotating brought about and no longer a non-interfering divide. Such a line holds space in a position that runs far beyond the boundaries and limits of the three-dimensional. Another factor of such a line would be that the radius (let us substitute the radius r with the using of Kepler's **k**), **k** would be immeasurably small. The factor **k** cannot be zero because infinitely close to that first **k** is the start of the third dimension where time plays the part as the fourth quarter. The presence of **k** is undeniable and recognisable yet it is not visible. The fact that **k** is there albeit stripped of any influence, disqualifies it from being zero and therefore not being there. With **k** already beyond any measurable space, leaving a^3 as a factor of one and not being able to pin any volume measure to that one **k** will have to be to the power of 0 being k^0. In Kepler's formula $a^3 = T^2 k$ the area a^3 would be one because of the dimensional non-existing of measured sides in any direction. If $k^0 = 1$ and $a^3 = 1$ the only alternative T^2 could possibly have is also one. The factor of T^2 identifies the time in the formula and when the formula indicates time as one, the time component must therefore be eternal. Only time in eternity does not change

The real formula applying when the calculation of the sphere volume is calculated is $a^3 = 4/3\, \Pi r^3$ where it places one third of the dimensional (but lesser) factor in direct relation to another third dimensional relation held by the radius and all aspects about the factors being in relation is to acknowledge the form that is applying and serving as a sphere. However there is no criss-cross matching of dimensional accumulating. It places time in the square directly in relation to space in the cube in association where time shows two distinct qualities. The one factor is time in the circle rotating while the other is in the linear or the straight line implicating the position that the other would have. In all instances of measuring the distance the orbit travels around the Sun as the space displaces or space covered by travelling in the time it is covered and dividing such a ratio one find the distance of the orbiting object from the Sun the in relation to the other factors form one or very close to one. It is relevancies carried from the Sun and the Sun is the governing singularity representative for the entire solar system. This is about relevancies applying throughout the Universe. This balance is much, much more than what the figures say. It underlines and it explains gravity as a life form in the cosmos other than what we consider our life to be.

The German mathematician and astronomer KEPLER, JOHANNES (1571-1630)
German mathematical and astronomer became Tycho Brahe's assistant in Prague in 1600 A. D. where he undertook to complete the tables of planetary motion Tycho had begun. Kepler first calculated the orbit of Mars. He spent much time trying to reconcile Tycho's accurate observations of the planet with a circular orbit, but concluded (in Astronomia nova, published in 1609) that Mars moved instead in an elliptical orbit. Thus, he established the first of his laws of planetary motion. A theory that the Sun controlled the planets by a magnetic force led him to the second and third of his laws, which were published as part of his treatise on theoretical astronomy, Epitome astronomiae Coernicanae (1618-21). The Rudolphine Tables (named after Tycho's patron, the Holy Roman Emperor Rudolph II) of planetary motion appeared in 1627 and were still in use in the 18th century.

Kepler also wrote De Stella nova, on the supernova of 1604 and Diptirce on optics and the theory of the telescope. The overall view followed in this book **Matter's Time in Space** places the true significance of his work in true contents. In KEPLER'S EQUATION is the equation that relates the eccentric anomaly of a body in an elliptical orbit to its mean anomaly. The equation is $E - e \sin E = M$., where E is the eccentric anomaly, M the mean anomaly, and e the eccentricity of the orbit. It is important as one of the mathematical relations enabling the position of a planet about the Sun , or a satellite about is planet, to be calculated from the orbital elements for any time. However this only relates to the solar system, and KEPLER'S LAWS only apply in the contents of the solar system. The three laws governing the orbital motions of the planets, discovered by J. Kepler is as follows: The first law states that the orbit of a planet is an ellipse with the Sun at one focus of the ellipse. The second law states that the radius vector joining planet to Sun sweeps out equal areas in equal times which as it says refers to time and not the circle.

The third law states that the square of the orbital period of each planet in years is proportional to the cube of the semi major axis of the planet's orbit. The first law gives the shape of the planet's orbit; the second describes how the planet must continuously vary its speed as it follows its orbit, moving fastest at perihelion and slowest at aphelion. The third law gives the relationship between the planets' average distances from the Sun and their periods of revolution. Instead of placing the true value to Kepler's laws, I. Newton placed his own interpretation to Kepler's laws, and in doing this he wilfully destroyed the principle working of the Creation. Through Newton's tunnel vision, he applied his own miss interpretations to the correct presumptions of Kepler. Newton reduced the implication that Kepler's findings hold, by using Newton's variation of what Newton wished Kepler's work would provide Newton when Newton was introducing his interpretation the law of gravitation. He then went about and changed it to three laws of motion. I. Newton generalized Kepler's first law, verified the second law, and showed that the third law should be amended to the form; $4 \pi^2 a^3 / T^2 = G (m + m_p)$. In this, the value of T and a are the period of revolution and semi major axis of the orbit of a planet of mass m_p about the Sun of mass m, and G is the gravitational constant. The major aim of this book is to correct these misgivings of Newton. I shall return to the statement about $4 \pi^2 a^3 / T^2 = G (m + m_p)$

What Kepler observed and formulated was more of a dimensional coming together by the cosmos in nature. He saw more of the principles guiding the cosmos than what one will find in the cosmos. It was practical by form Newton first and third dimension duplicate mathematical 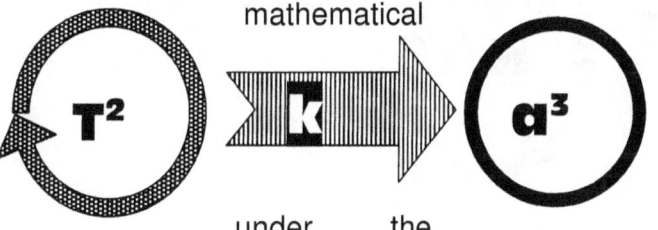 under the forms of symbols finding value instead of true figures, which wished to see. Kepler saw the second dimension forming the dimension by allowing the third space to flow and room to control of singularity.

In the argument Kepler made he had hidden so much more facts into one formula than what I think even he realised. Well, it is much more than what the Accepted Policy Protectors Of Science ever came to realise. He officially formulated space-time, he officially coined not the name but the origins of the Universe being the Big Bang and he was the first to put the speed of light in relation to cosmic development…and all of that with his rather simple formula. He said the space a^3 not the circle (a) or the circumference a^2 but in the circle a^3… where such a circle represents a factor in the third dimension.

The formula he compiled was not rather but very specific about the area being a third dimension area and to prove it beyond doubt he placed it in the relevancy of the formula in a ratio of presenting the third dimension in space. He said a^3 is equal to $T^2 k$. Newton and Newtonians came afterwards and played with mathematical toys as to challenge their mental capabilities. Newton introduced a $4\Pi^2$ to indicate the presumed circle on the one hand and on the other hand he brought this lot equal to {G $(m + m_p)$} which he then presumed to be the general Universal gravity constant (G) and the sum total of the two structure mass. Newton saw a ring circling around a centre having $4\Pi^2$ to indicate such a ring outside a centre and he positioned {G $(m + m_p)$} where the two mass factors combine the gravity effort in the general grand gravity constant in space. I have had so much resistance in the past from all Academics but that is not what I see what Kepler saw. I shall trace this back to the centre of creation.

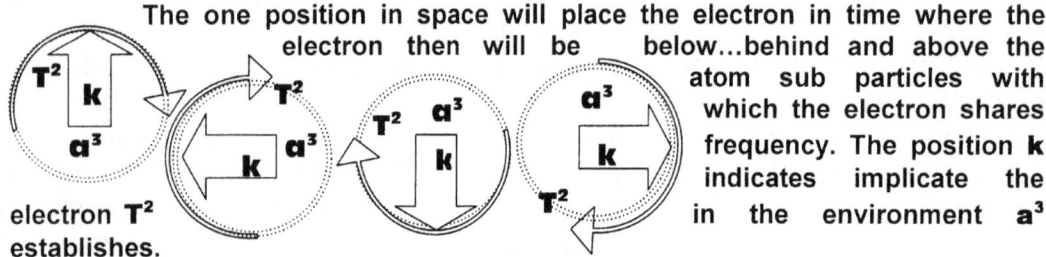

The one position in space will place the electron in time where the electron then will be below...behind and above the atom sub particles with which the electron shares frequency. The position **k** indicates implicate the in the environment **a³**

electron **T²** establishes.

In their eagerness to calculate they calculated a formula to measure the circumference **a²** of a circle being $\Pi\, r^2$. I have seen an Astro physics examination where they use **4Πr³ / 3** as the formula to calculate the Sun and other stars volumetric space! They formulated the measuring procedure of the circle being in the third dimension that will show how big the volumetric space is of a sphere at **a³** being measured with the procedure being **4Πr³ / 3**. This too was a fanciful devise allowing mathematicians to be much superior to the rest of the commoners and to dictate to the lowlife how and what they should think when they think and if they indeed can think of anything to think of. Then some Mathematician and an Englishman of Substance came onto the idea of gravity. Being a mathematician the Englishman placed the Universe at the feet of mathematicians. He saw circles where Kepler saw three dimensions. He saw three dimensions where Kepler saw nothing. He knew time had to be somewhere as something and then covered it by denouncing the circle as nothing.

What then is it that Kepler saw as he formulated **a³ = T² k.** At the normal flow of time it takes the electron a certain time to spin around the atom. The atom uses space **a³** and the atom is a certain length **k** that forces the distance the electron has to travel in one cycle period **T²**. The atom **a³** connects the electrons travel **k** to gravity **T²**. The relevance **k** produces to support **a³** is to point **T²** to two positions the electron will be in the duration of one specific time. The electron travel will be cyclic and periodic in relation to the space the atom holds. The space stands related to the gravity with which the Earth confines the space of the atom to the space and speed with which the liquid heat confines the atom space.

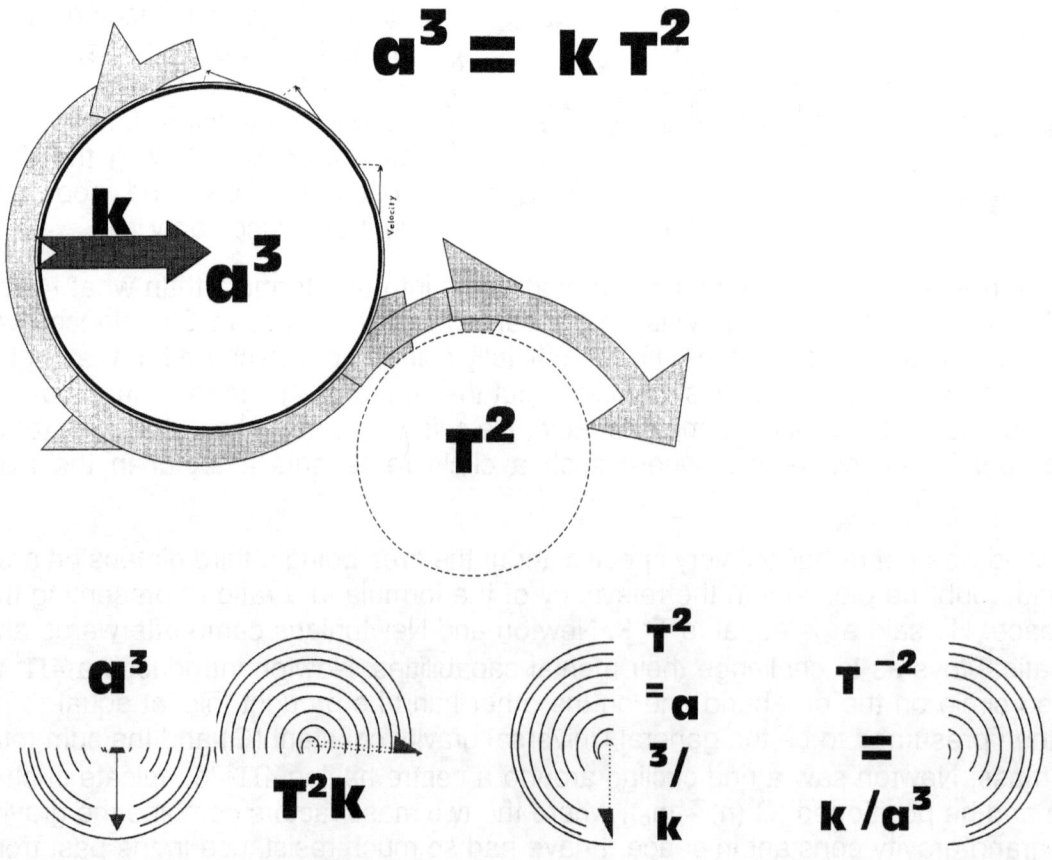

$$a^3 = k\,T^2$$

The Universe divides between space that was and will be where time is and time where it just released space is where it is accepting space. The one side is while the flanks are in motion of releasing or accepting the position space has.

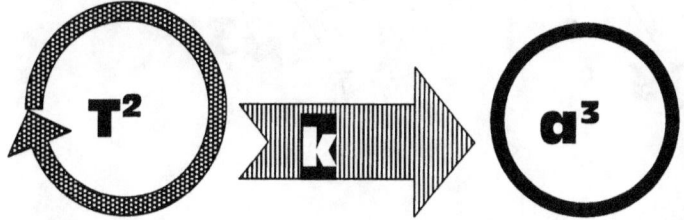

Planet	Period T years	T^2	Distance k	Space a^3	Ratio
Mercury	0.241	0.058	0.39	0.059	0.983
Venus	0.615	0.378	0.728	0.381	0.992
Earth	1.000	1.000	1.000	1.000	1.000
Mars	1.881	3.54	1.524	3.54	1.000
Jupiter	11.86	140.66	5.20	140.6	1.000
Saturn	29.46	867.9	9.54	868.25	0.999
Uranus	84.008	7069	19.19	7067	1.000
Neptune	164.8	27159	30.07	27189	0.999
Pluto	248.4	61703	39.46	61443	1.004

At the first glance Kepler's formula seems to be numbers and positions applying between the sun and specific but different planets in the solar system.

The time frozen on paper in a single t is effective in remembering the viewer of an event but that is not the event in the present any longer. That was how the event occurred during the time from where the camera shutter opened T_1 to where the camera shutter closed T_2 and the time frame T^2 was then during the open period of the camera shutter. But afterward it represented **t** when looking at the picture and the looking of the picture became an event during a specific T^2 that went from where one is taking the first look to where one is looking away from the paper carrying the first dimensional image of an event gone by and that is at that stage a representation of **t** in another milieu of $a^3 = T^2 k$. The **t** in the single is when mathematically presented as only t indicating a mathematical single flat dimensional view of time and is then correctly applied because it represents a reminder of a four dimensional event $a^3 = T^2 k$ that went single dimensional because the moment in the fourth dimension was then frozen in a single dimension on paper while the fourth dimension $a^3 = T^2 k$ soldiered on and time will always be representing T^2 as Kepler stated in the square allocated to space having a cube $a^3 = T^2 k$ at a time even before gravity got a name.

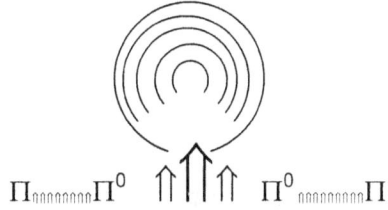

With singularity placed in infinity within the centre of every rotating object every atom and its relation to its surroundings including other atoms form space-time diverting from the point holding singularity as far as rotation goes because every object holds three relative positions in as far as where it was, where it is and where it will be in relation to singularity providing time. I elaborate on this else where.

Newton said a sphere is $a^3 = 4/3 \; \Pi \; r^3$

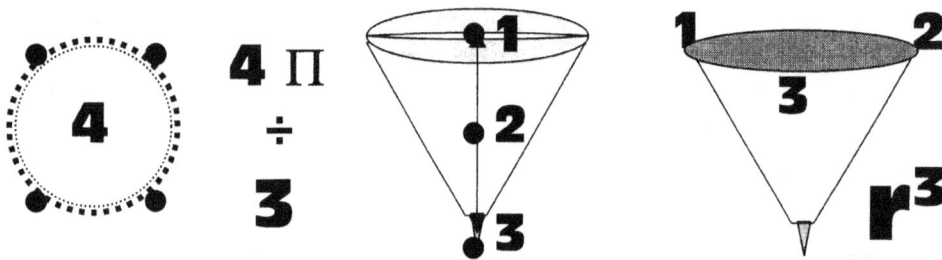

In $4 \pi^2 a^3 / T^2 = G (m + m_p)$

$a^3 = T^2 k$

$a^3 / k = T^2$ but at the same margin is

$k / a^3 = 1 / T^2$

$k = a^3 / T^2$ singularity

$a^3 / T^2 = G (m + m_p) / 4 \pi^2$

and $a^3 / T^2 = k$

then $k = G (m + m_p) / 4 \pi^2$

But I showed that $k = a^3 / T^2$ and Newton's claim is that $a^3 / T^2 = G (m + m_p) / 4 \pi^2$

The only definite place one will locate zero is in between the starting point of the lines going in opposing directions in the position the lines hold before there was the least of directions applied, but that is only because there is no such a position, not because any line is coming from there. The two lines are still one holding the opportunity of parting as an option but have not yet parted and therefore are on the very precise same spot. The line coming from there is already there because it already has the choice of going in any and all opposing directions and when it starts running it will place filled space in that location because the space was already filled with a line starting and not with a line not there at all. When reversing a line we might find a better idea of what is in place and where it is in place.

In the action of the inseparable drawing closer and moving closer gravity finds the dual value of linear and circular gravity. There is no separation of the two factors acting as one but both have different applications and values in the unit. This is the result of singularity having three parts acting as one but giving three distinctions in application. Gravity is as much part of dismissing space as it is about making contact with space in time.

But since the connection comes about as a circle, the connecting points will relate to Π as the value.

Due to the spinning nature of such a point with all surrounding the point will be alternating direction favouring change every second and in that the value of such a point can only be Π because of its constant changing. Using r would specifically oppose another r from every angle because the use of r will bring about a static relation to the previous and following instant and therefore it will cancel the constant spin flow.

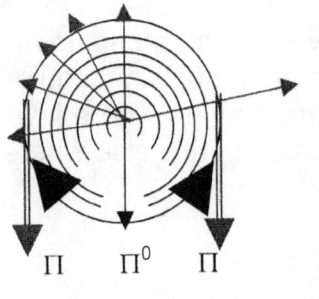

The new direction pointing to a new location in relation to the previous point will oppose the previous point it had in relation to direction considering the centre point.

Pinpoint positioning of singularity Π^0 with Π positioning space to either side forming the border set by singularity

The new direction pointing to a new location in relation to the previous point will oppose the previous point it had in relation to direction considering the centre point.

The motion of a liquid confirms a centre and the confirming of the centre provides a flow of space-time in either direction, which produces the gravity. Without establishing and activating such a centre the centre is not active. It is there but it is inactive in being present.

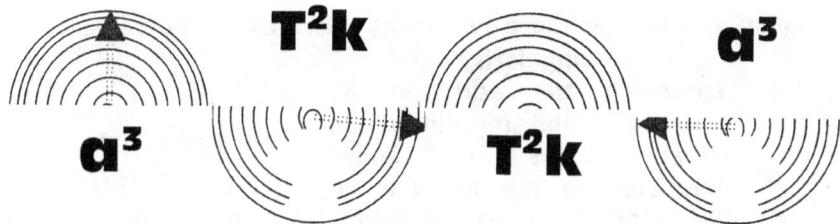

If the one side is a^3 then the other side is T^2k because the one side opposes the other side by providing control a^3 to the motion T^2k on the other side

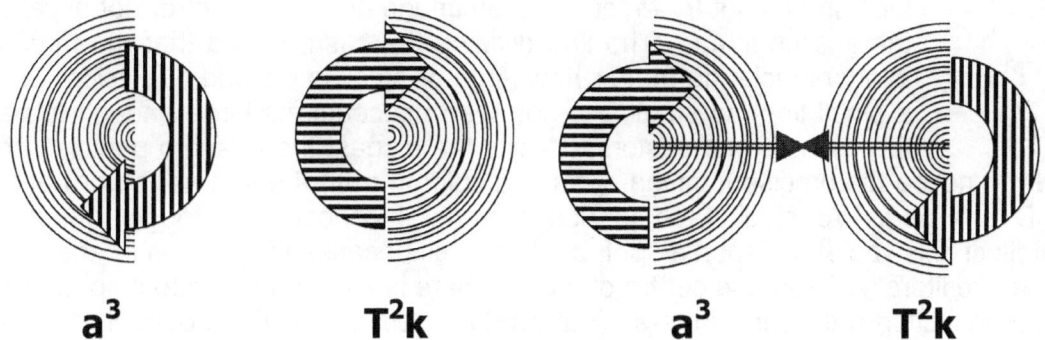

$$a^3 \qquad T^2k \qquad a^3 \qquad T^2k$$

The same motion is contradicting the motion it was or will be and because of that the contracting part is going to be expanding while the expanding part is going to contract. The one will accept while the other will release but the cycle can never be broken. Still in the end it is the same motion and the motion never interrupts the flow of time.

By taking k into a negative the space will reduce the time because the space cannot sustain the demand of space growth.

$k^0 = a^3 / T^2k$

$1/ k^0 = T^2k / a^3$

$a^3 / k = T^2$

$\Pi^0 = \Pi^3 / \Pi^2 \Pi$

$1/ \Pi^0 = \Pi^2 \Pi / \Pi^3$

$\Pi^3 / \Pi = \Pi^2$

In all my other work, I make exclusively use of the value of singularity Π since it makes a lot more sense, but when I use the value of singularity, which is Π then no one seems to have a remote idea to which I am referring.

$k^0 = a^3 / T^2k$ forms

$k^0 / k = T^2 / a^3$ that becomes

$k^{0-1} / a^3 = a^3 / T^2 / a^3$

$k / a^3 = 1 / T^2$

The replacing of the symbols Kepler used with the value of singularity the mathematic equation comes into practise.

$\Pi^0 = \Pi^3 / \Pi^2 \Pi$

$\Pi^0 / \Pi = \Pi^2 / \Pi^3$

$\Pi / \Pi^3 = 1 / \Pi^2$

However, keeping Π as one ($\Pi^0 = 1$) we keep the Universe in the first dimension.

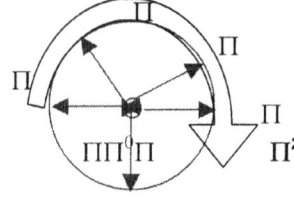

This point, which I now am referring to, is the point where Π is a fully appreciated value while the diameter D still remains a dimensional factor of one. This is the dawn of the second dimension Π^2 where space was there but space was sparsely shared in some cases. It was when Π^0 shifted to become Π for the very fist time.

In the sketch below the circle to the right would come about from a straight line r growing influencing the appreciation of Π, but to influence Π would lead to a breakdown in r as Π and r are different entities. The circles to the left shows a continuous growth by extending Π every time and since Π is the same part as the previous Π, only extending that billionth of a millimetre or many times smaller each time, the circle will be truly continuous without any signs of a break. In the context of dimensions one finds coming from the centre Π^0 an established eternal flanking of Π to six positions since Π^0 forms the centre to the six sides and all six sides not having a diameter yet must apply Π to indicate specific value. In the very centre, which I am referring to, rotation must end or start depending from what vantage point the relevance is placed.

However, the equation looks far more sensibility when using the value of singularity

$k^0 = a^3 / T^2 k$ forms

$1/ k^0 = T^2 k / a^3$

$1/ (k^0 k) = T^2 k / (a^3 k)$

$1/ k = T^2 / a^3$

Expressing the equation by using the value singularity has instead of the symbols Kepler designated to the formula he introduced it makes far better sense expressed mathematically

$\Pi^0 = \Pi^3 / \Pi^2 \Pi$

$1/ \Pi^0 = \Pi^2 \Pi/ \Pi^3$

$1/(\Pi^0 \Pi)= \Pi^2 \Pi/(\Pi^3 \Pi)$

$1/ \Pi = \Pi^2 / \Pi^3$

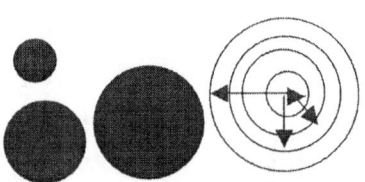

One should not try to focus on an image of such a spot or dot because there is no image. The line dividing the cosmos and that run through every particle, no matter how large or small is beyond our vision. Such a small line, so small it is not even noticeable is large enough to part the cosmos into sectors. It splits the biggest there is into particles and we are not even able to notice the precise location of such a split. In truth there is no top or bottom that we living in 3D can see. We shall have to use a general conception brought about by intelligence. Your intellect tells you about such a spot, but that is all because that spot is on the other side of the Universe (quite literally). From the centre of the dot there is a top and a bottom spot. From those points there is connection with four quarters. That produces six connecting points that are all aligning to the centre.

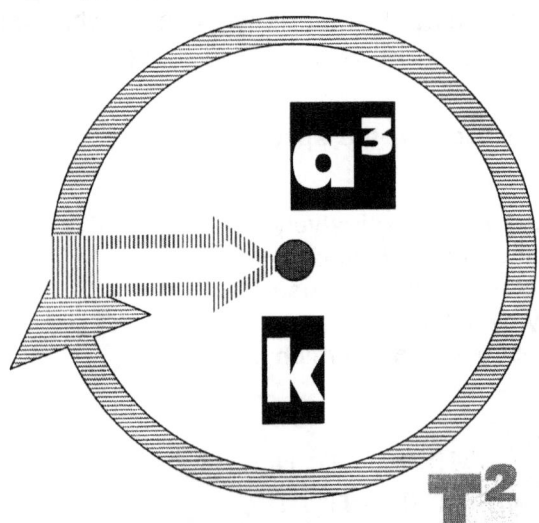

a^3 forms the space the atom claims while travelling in the Earth spinning all the way and travelling with the Earth around the Sun

k positions the electrons travel in the space relating to the space the atom holds while travelling with the Earth in the Earth around the Sun in relation with a specific position k will indicate in relation to the Sun .

T^2 is the time it takes the electron to be relevant to the position the Earth places on T^2 while the Earth captures the space of the atom by providing the space for the atom to be within while the Earth travels from one point T_1 to another point holding T_2 in frequency to the atoms T^2 relating to the Earths T^2 in perfect harmony with the Sun having another T^2 relevant to all the other factors we call cosmic particles. Big or small it is only about cosmic particles holding space in time in relevancy.

Looking at the effect of gravity it shows the precise quality of no distinctive point, as gravity never seems to end at a point but flows all over affecting all that holds a position in its sphere of influence. The gravity coming from China meets the gravity coming from America at no particular spot but intermingles without distinction.

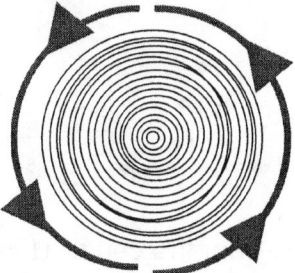

The very centre form an eternal divide that will not allow what is on the one side to present an influence on the other side. It divides spin. It divides direction of spin. It divides all rotation from the outside that one may detect and such divide is there because at one point spin will run to the left coming from the right and just immediately next to that point must run a direction from left to right. It cuts without contributing or participating in movement. It divides without any favour.

By expressing a wish to accomplish time travel such a person wishes to accomplish that material must collide with itself a mean feat if ever there was one. Such a person wishes to have one side collide with the other side as he stops and reverse time while the rest of time is motoring on.

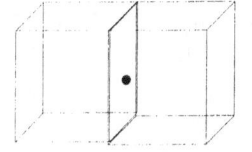
Taking the outlook from the point the sphere is holding from that centre out into space there are ten points connecting to the centre. In that are the dimensions of singularity connecting to space where five connects to space in the second dimension of singularity, and five connect in the third dimension of singularity

That is singularity not having a dimension of space and not having a dimension of time, or a radius connecting the rotating distance to Π. Every rotating object holds a centre from where the rest of the rotating direction will differ at any and all given points. Not one point is exactly the same, but in the very middle, the centre no one can draw, measure or see is a point not in motion.

In the centre runs an axis line that forms the division of rotation. No one human will
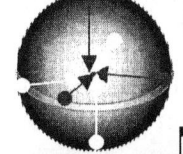
ever be able to indicate the precise line, but such a line must exist because of our logic telling us about such a line. In the centre one will always find one more line smaller than the outside but forever also always bigger as it is towards the inside.

The sphere holds six sides in relation to form as unit and as does all other shapes and forms. All forms have to have at least six sides indicating different exposures to the Universe. But with gravity having a free choice, gravity always chooses the sphere. As I shall prove later on gravity is the strongest where the form produces the least evenly distributed space.

The first condition for gravity is even-handedness through out the sphere holding the applying gravity and the second is to have most or the strongest gravity located where the space is least. That gravity then has a position in the very centre of the sphere and from that centre the gravity produces all the edges or borders that the sphere consist of. In the case of the sphere this factor makes the sphere much more dominating than any other form does. From the centre point controlling all sides is gravity and with gravity applying control the sphere has seven sides to the square in any other possible form having at least six sides.

The cube can come in whatever form there may be but the sphere adheres to precise measure and behind this principle is all that forms the Universe. The cube has six sides connected loosely and can change form just by changing the relevancy between one side (or more) in relation to the distance brought about by the other sides. The sphere being a complex circle stands related where the sides has to apply precise measure in equality. This becomes a law because in the precise middle one will find the strongest gravity as that gravity holds the object in form and true to form. If there is even gravity spread in all directions the form must be a sphere and the sphere insist on seven points relating to sides or borders.

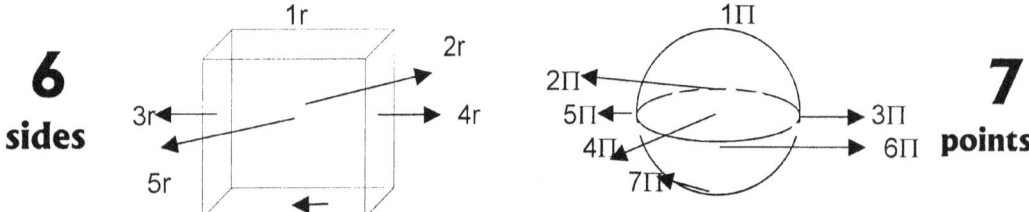

6 sides   **7** points

In the sphere there are no radius but only the extending of Π from the centre Π in six opposing directions relating to one another by the square but remaining Π because of the unity the matter holds in relating to space. It is not possible to draw a precise line that would form a precise ring and not cut some atoms in parts. Because there will always be an atom disallowing the precise positioning of the circle the circle continues on a solid basis holding Π as a positional reference and not r. In every sphere there then are the seven Π relating in precise dimensional and positional equality forming equilibrium to the centre Π as well as to one another by 90^0 and 180^0 implicating the dimensional positioning. Therefore the sphere holds 7^{Π} and the cube holds $6 \times r^2$. Where space comes into contact with the sphere the cube loses one of the six dimensions it has to the more dominating seven dimension of the sphere whereby the seven dimension in equilibrium will dominate the six dimensions loosely connected by r bringing about that the cube then has 5 sides to the seven of the cube. Because the space surrounding the sphere takes on the shape of the sphere and not the other way round where the sphere resolves in accepting the form of the cube, one may presume the form of the sphere is the most dominant of the two choices.

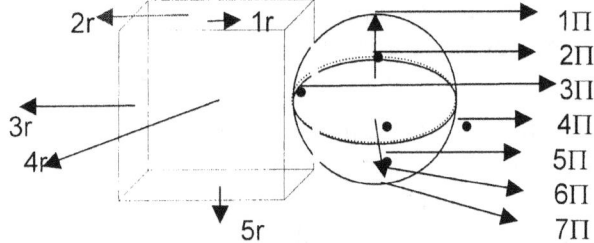

5 sides in the cube vs. **7** sides in the sphere

The sphere is a multitude of innumerable circles that forms one unit with all the innumerable circles that compile a sphere all put together. The circle is a constellation of Π where every Π flow from one into another and such flowing varies the number arrangement of r. In order to measure the surface of a sphere the radius carries the torch by going square. It is the radius, which is just another line that come from the outside and run all the way inside to bring the value of the line into the circle.

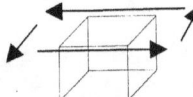

In a circle, there is a radius that initiates the circle. The calculation of such a circle is Π X r^2.

precise distance
equal distance
that initiates the

Only the circle has a point on the outer edge of the form that is at a from the centre and every point in the form measures a precise and from the outer rim. In a circle, there is a radius circle. The calculation of such a circle is Π X r^2.

The radius r runs from the circle outwards, from a circle centre point towards Π, the value of the circle. In the centre of the circle, there is a point where the radius starts. It runs outwards from that point in all directions towards the circle Π. Technically, there then has to be a point where r is infinite and not zero, an absolute infinite. However, the circle therefore remains Π. The circle does not disappear; it remains there for all to see. It is only the radius that almost disappears into the infinite, but it does never become zero! $\frac{\Pi r^2}{r^2} = \Pi$. If one removes the radius from the circle, the circle remains, only holding the value of Π. By removing the value of r, Π becomes singularity with no place to be. Singularity is the place where there is no space to be in place. However, Π remains because once r receives the slightest of space Π will find space. Then the circle will grow to Πr^2 and r would determine the space. Without space, there is no r but there is a circle with the value of Π. Singularity is in every single rotating object, be it the proton or the Universe. This situation is part of any and all circles and is therefore part of any and all spheres. The line will end at a point where the line starts going in the opposing direction and that value is indicated by the use of $r^0 = 1^0 = 1^1$

To that end the shortest possible line (hypothetically) must be so short it must have **an initial and ultimate point sharing the same spot.** Any theoretical line being the shortest possible line cannot have the line holding the initial starting point at point zero and advance from there. If it used zero as a start, the zero part would not count, because the line will only start at a point past zero where the line then will start. Zero ultimately means not existing and then that point, as a start does not exist. At one point the reducing attempt of the line would start making the use of mathematics seem silly. The reducing would seem tedious and leading nowhere. But as sturdy as mathematicians can be they would carry on (or so I am made to believe…).

Then when the man doing the calculations gets carried off in a straight jacket, while the man is making funny noises, when he is totally cracked mentally, the calculations can still go on and on and on and…and that is where sanity prevail and someone says "drop the affair". It is at that point I would have loved to see Einstein carry on counting stars in so many galactica in his attempt to determine the critical density joke. That is not where we get into infinity. That is where man's brain gets blistered but infinity is still far off. As any one can see the Universe is far beyond some insignificant and senseless formulae invented to impress Academics while others are kept busy and free from boredom, but in the real Universe the attempt is not worth the thought it takes to disregard the attempt.

When the line **has a beginning and an end at the very same spot** and it wishes to extend the position as to further the possibility it has, which direction should it favour. Extending the line in any one direction will favour one direction without any clear reason not extending in other directions. The only mathematically sensible option about extending will be in all directions equally in order to give a meaningful non-bias flow of mathematical equilibrium. That is where one would have to go look for the beginning of the Universe. The Universe is a about lines connecting but where does that which does the connecting end. The Universe used form to this point in development, but then at some point the line came and established the presence it still has. The first form was moving from $\Pi^0 \Rightarrow \Pi$. Again I wish to press the issue, at that stage form was in use and not mathematics. The Universe was just simply too big to measure. If radius did apply, one could use r and r2 but since only Π was in use there was no radius to be used. There is forever one circle leading to the next circle, which is followed by the following circles. Where the light does not reflect the image that is there, we will still find a concentration of circles leading another and another and another. The end is eventually endless.

When working with concrete and heavy metalled would show as a crack distinctly parting solid Π indicate a continuous flow of solidness giving overall and continuous structural strength, yet never recognised this difference. By confirming Π employs singularity in all components and to be a much stronger support as building choice

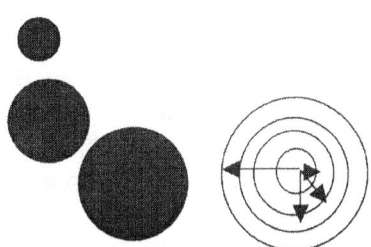

solid objects r structures, while the material an engineering the circle therefore proves than other

shapes.

In order to understand the difference there are between Π forming g gravity and a radius, think of the difference there is in working performance between air pressure and oil pressure. Pneumatics use r as a pressure indicator

and hydraulics use Π therefore air can compress and liquids cannot but act as the toughest solid found specifically because of it uses a relevance in the applying of Π and not r bringing conformity evenly.

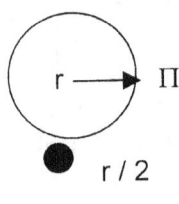

r / 2

r / 2

r / 2

The circle to the left would come about from a straight line r growing influencing the appreciation of Π, but to influence Π would lead to a breakdown in r as Π and r are different entities. The circles to the right shows a continuous growth by extending Π every time and since Π is the same part as the previous Π, only extending that billionth of a millimetre each time, the circle will be truly continuous without any signs of a break. When **the circle reduces**, the **value** located to **r** will become implicated because **r determines specific size**. **Not so** in the **case of Π, because Π** in the true sense only **indicate that the circle is a square without corners** and therefore **Π dictates form and not size**. By **reducing size** only **r comes into contest** and will point to such reduction. By **reducing** the circle **radius r by half continuously** will lead to an **infinite small circle** but **Π will remain because the circle as a form remains** even being infinitely small

The spot forms a full circle, but the line running through the circle is forever present because that is the future radius of the circle that will one day develop the circle, which is equal to the present diameter. The fact of the presence of such a possible line in such a possible circle dividing the possible circle into two parts makes the centre line equal to the half circle. The line forms the half circle but not only that the line presents the half circle as much as the line is the half circle. The line then is 180^0 and the half circle is 180^0 because in singularity the two factors are the same. The same value is of course $\Pi^0 = 1$.

In this half circle of the future, which is no half circle as yet because of a lack of space there are three future points indicating the space less ness that will go on to become space filled with something. On top of such a circle to form must be a marker indicating an awaiting boundary or future border and at the bottom of the future circle there also must be a similar marker that is no marker as yet. Between the two possible points that are not there yet is a future line running that is not there yet. Then indicating the possibility of a position to come that will bring about the half circle being a future distance apart from the future line indicating a diameter that will one day be there. A third such a marker must be established for the future. That forms a triangle with two more sides being connected by either a line being one or half pi being one. From singularity comes about that the line is the same as the half circle is the same as the triangle and all has one value being 180^0. From this come the most basic principles in as much as forming the ground rules of the law of Pythagoras.

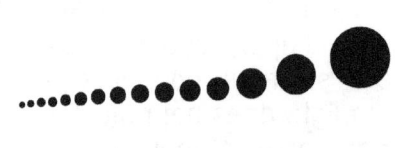

In the circle using $r^2\Pi$ the r has to have distinctive qualities placing it as a factor apart from Π. Where the growth shows no separate distinction but a continuous flow from the precise centre to the precise edge the flow would become in relation with Π depicting the circle and Π replacing r as reference to any point on the circle. By using r distinction in the circle is possible but by using Π there is no distinction possible.

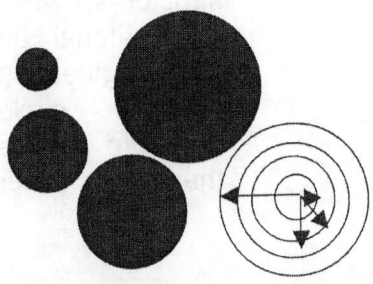

When working with concrete and heavy metalled solid objects r would show as a crack distinctly parting solid structures, while Π indicate a continuous flow of solidness giving the material an overall and continuous structural strength, yet engineering never recognised this difference. By confirming Π the circle employs singularity in all components and therefore proves to be a much stronger support as building choice than other shapes.

Everything at the time that was outside singularity and was in form at that time was equal. Think how big they were. They filled larger parts of the Universe than our brains can cover by thought. They formed the holding tanks that still hold us in the massive Universe. They were at the time too small to have size, but since then they grew into structures that are too big to have size. Those dots still are bigger than mathematics can apply because there still is no measure quantifiable mathematics can reach. Just because they compare with what we seem to preserve as small in cosmic relation they are too big and too large for us to comprehend. Even if they were immeasurably many they filled an immeasurable Universe in the same way they still fill the immeasurable Universe and we are so small we and our surroundings are measurable and quantifiable. They were so enormous there were no relevancies applying to compensate for distinguishing. Distinguishing only followed later when size started to matter. That meant the Coanda effect was in place without the Roche or the Titius Bode law. It was the start of the relevancy principle from which the atom later came and which is the result of the Kepler expression.

It eventually gets so small we humans can fit into it... remember we are seeing the reverse of the truth. It was never so big that it contained nothing because all that and we came afterwards being smaller than the dot that filled the dot.

With no line possible there had to be another dot that formed since the Universe has many dots that formed lines. But let us not to get confused and lost in the range of possible diversions but let us stick to two dots. One dot was next to the dot next to the dot, but as I said, we stick to one dot next to the second dot Π is the first step where gravity began with. That leaves us with a huge problem in as much as when $r = 0$ then $r^0 = 0$ and 0 dividing any value will leave 0 as the answer. If the particles were inseparable at the start it must bring about that gravity would not be forming since the distance will not permit any dividing. By allowing the distance separating the particles to be zero, the particles melt into a unit. Again this is Mathematics and not my incoherency as some Academics dismissed my work. Let me run through the argument one more time because I have been insulted by Academics in the past telling me I am bending mathematic rules with my applying double values to try and produce some argument. The two particles formed by an inseparable unit separated by a sharing of a spot. We know that at least two spots formed because there are many more than just two that remained to become part of the visual Universe. Let us name the spots because that is what humans do best if they do not know what to do with what they have to do. Let us call the on dot and the other one spot next to dot we then call dot two. Between dot and dot two there were nothing because dot and dot two were inseparable. By they're being inseparable we would naturally be inclined to think that the separation value should be nothing or at least zero. But putting zero in that place is a mathematical excluding procedure leaving future mathematics excluded. With m multiplying m_2 and then dividing $\div$ r with zero (r=0) such a procedure will leave the lot at zero and with that nothing is going nowhere. That means although we think the space between the two parts are nothing the non-existing space has to be at least one to be a future factor.

At fist $\Pi^0 = \Pi$. Then after a while $\Pi^0 = \Pi^3 / \Pi^2\Pi$ and gravity comes about forming space-time by motion of form. Being the sphere that formed the 7 holding relevance in the form of the sphere took shape. But the Universe is layer in dimension forming the next layer in dimension forming the following layer in dimension. The Universe was $\Pi = \Pi^3 / \Pi^2$, which is taken from Kepler's formula he received from the cosmos as $k = a^3 / T^2$. Where there is a sphere involved there is a natural tendency to grow by developing the sphere.

Every part of the argument is sound but was never yet used. I repeat once more if my argument reflects on inconsistencies those inconsistencies are not about my work. In order to disprove my argument replace Mass one and Mass two with any number possible, then divide such a number with to the square being zero. If there was no space then the value of the particles had to be one. If there was no space between the particles the particles then had to form a unit. But if there is a mathematical possibility of reducing a line to the single dimension then there had to be a factor representing r as a factor of one. Take $(M_1 \times m_2) / r^2$ and substitute any of the factors with zero and the result coming about has to be zero. The factors in the equation have to have any and all the elements at a value of at least one. Only if r was a factor of one can gravity bring about any mathematical equation developing from this argument. That means the mass on both sides must have a factor of one being a limit, which does not allow such further reduction of r and any further reducing of r beyond the limit will not be tolerated. Only if $r = 1$ then r^2 can be 1 and mass can be apart. Like it or not but believing in the Big Bang must also bring about the accepting that the cosmos moved apart somewhat. The fact that r brought increase in the space separating the mass produces a problem that was solved already. About a century and a half ago Roche found just such a limit.

Once again I were confronted by zero becoming growth. There is a huge hole that needs filling when bringing into a relation any forming of an alliance between a cosmos coming from nothing and filling with nothing and a cosmos growing spontaneously through balance shifting prominence. Mathematically the fact of applying nothing as a vale applying in the cosmos is not a strong and convincing argument. The minute one brings in zero as a multiplying factor forming a definite value working into the calculations of the cosmos, growth disappear. If growth was not a factor, the zero factors could be involved with some form of maintaining stability and where then further growth will accept the responsibility of zero.

Any point will be opposing itself within the **rotating of 180°** where it **then change every aspect** of its **previous flowing** characteristics it had or **will once again have** in 360^0 from there. While in rotation from the view point of a bystander it all may seem static and never changing but to the object in spin every next instant in time will be diverting from every aspect it had every second passing, and the direction it held in relation to the direction it held the previous mille, mille second will totally be incompatible with the direction it holds the very next mille, mille second of rotation. This is why we can use degrees measuring the circle by (6^2) (forming the square relating to matter through singularity) X 10 (square if space) = 360^0 however it is always in motion. That proves no point can be static or constant, though it may seem that way to outsiders. Although matter is matter, matter can also be anti-matter and moreover form its own anti-matter at the same time. This degeneration of structure is very likely to occur with overheating. Revaluing Π to Π^2 will bring about a new contact point where Π meets **r** forming another relation in Π^2 **Time is** the **changes in relation** where Π **contacts a different r** not withstanding the many r points there may form because **every r constitutes a different value** to the universe through other ratios and relevancies brought about **by heat and light. Time is the duration it takes Π to rotate between any two given points of r** and therefore must always amount to **a square (T^2)** moving from point to point through the **cube of space (a^3)** in that **duration of time (k)**. With that it proves **Kepler's a^3 (space) =T^2 k (time in the instant of motion)** but motion must continue through a specific value in space where the space-time is maintaining relevant equilibriums throughout singularity connecting.

In the circle using $r^2\Pi$ the r has to have distinctive qualities placing it as a factor apart from Π. Where the growth shows no separate distinction but a continuous flow from the precise centre to the precise edge the flow would become in relation with Π depicting the circle and Π replacing r as reference to any point on the circle. By using r distinction in the circle is possible but by using Π there is no distinction possible.

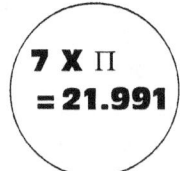

The sphere is 7 X Π = but from the other is singularity relating to ten. That starts the principal which in relation to the Roche limit use Π^2 / 4 to form gravity Π^2. However this is involved, as it seems because the one because of the consequence of the other

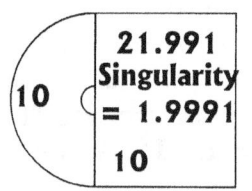

view the sphere Titius Bode at the limit of Π a little more sprouted bringing the one into the Universe as a relative that will sprout to bring about gravity.

However if the attachment was 7 + 10 + Π^2 / 2 = 21.93 / 7 = Π is the circle that serves as an attachment meant being a sphere and holding as well as sharing singularity. This is what Newton saw, but this is not gravity in the cosmic sense.

7 A sphere being formed by the six ends crossing as it incorporates the centre singularity
+ 10 anther sphere with an identifiable motion keeping the independent singularity apart but within the relevancy of the unit formed.
+ Π^2 / 2 Singularity by gravity shared by two in one unit.
= 21.93 / 7 still holding the unit to form where the overall containing form will be a cosmic sphere.

The sphere holds many dimensions relating to seven but also by the square of space which is 5+5 = 10. This brings about gravity generated by means of the Titius Bode law and is principle proof of this statement because it indicates an infinite number of numerical positions influenced by quarterly divided sectors around a point holding singularity.

Boys playing games will never realize scientific breakthrough explaining and grown ups do not play with toys. In this little toy played everywhere everyday by almost every one is the answer most brilliant of human Brainpower seek answers about all the cosmic riddles no one seem to understand.

Newton said the rotation delivers no work and therefore the effort of the r0tation results in a zero. Firstly it bring us back to the zero idea where with all the reasoning in the world and all the leniency I allow I cannot find zero as a value being part of mathematics. Let's move back to the circle to try and find the zero Newton saw in the rotation.

$\Pi \times r^2 = \text{CIRCLE}$

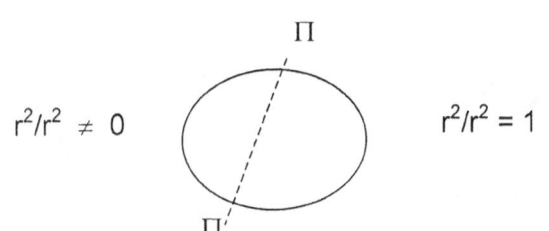

If you remove r it then is $\Pi \times r^2 / r^2 = \text{CIRCLE}$.

You cannot then say $r^2/r^2 = 0$ and therefore $\Pi \times 0 = 0$. That is nonsense. $\Pi r^2/r^2$ will always be $\Pi \times 1$, and that is the eternal circle. When looking at any rotating object, there has to be a point of no rotation and no rotation means "no rotation", not no existence. No rotation means a factor of 1, not zero.

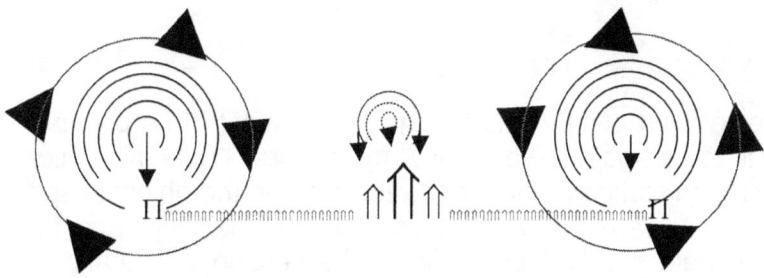

Not only does atomic individual singularity maintain self preservation, but in doing that it also sustain a governing singularity holding structural composition and form within a cluster of matter for example a star. As there is between stars so there are in the same manner a mutual or bonding singularity between atoms in stars, which we see as fusion.

Any object in rotation will have a middle point, a very specific centre point that does not spin. That point once again hypothetical but none the less must be standing still because every line running from that pint in opposing directions are also in opposing directional spin to each other

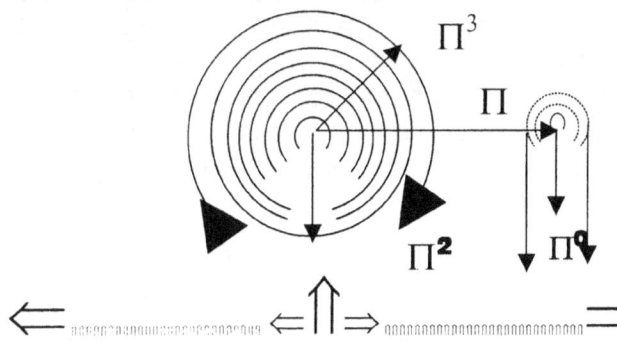

As the stop starts to spin the motion establish the centre line, which activates singularity, which activates space-time that activates gravity at a specific relevancy. Where we locate singularity there is not nothing because gravity cannot come from nothing because only nothing comes from nothing.

After all it is gravity that keeps the top as it is spinning in an upright position while it is spinning because it is gravity that stabilises the cosmos.

Moreover, what is actually in progress from the top spinning is the Coanda principle activating gravity and that happens in accordance with Kepler's formula

This means that in the cube at the point of contact between the cube and the sphere the cube experience such a contact point as if the "bottom falls out" of the cube and without a "bottom" to support objects they fall to the sphere as objects does fall to the Earth. Remember that a body "floats" in space, but at one specific point it starts to "fall" to the Earth. That is gravity and it is a dimension change much more than any force. I shall explain this last remark later on. That too is the Lagrangian system with five cosmic structures holding relevancy to the centre structure where the centre structure stands in for seven positions diverting from singularity and the orbiting structures standing in for five positions in space.

In the centre runs an axis line that forms the division of rotation. No one human will ever be able to indicate the precise line, but such a line must exist because of our logic telling us about such a line. In the centre one will always find one more line smaller than the outside but forever also always bigger as it is towards the inside.

From such a point every other point will be opposing any other point not pointing in the direction to which the first point is pointing, whereby it extends the direction it holds. No matter what the point is or where the point leads, such a point holding a specific direction will be unique in the direction it is rotating because at that or any other specific point wherever, it will be directing not in the direction it spins but in the direction flowing from the centre point outwards.

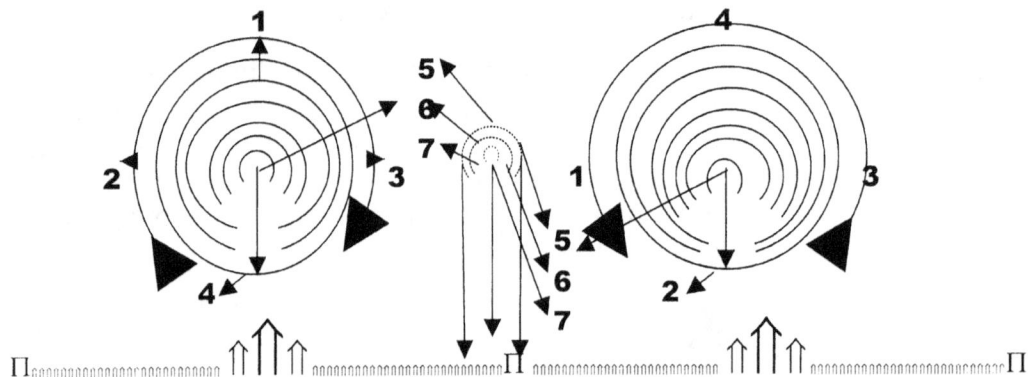

Any point will be it opposing itself within the rotating of $180°$ changing every aspect of its previous flowing characteristics it previously had or will once again have in $180°$ from there. While in rotation from the point of an outside observer all may seem static and never changing but to the object in spin every next second will be a diverting from every aspect it was in very second passing, and the direction it held in relation to the direction it held the previous mille, mille second will totally be incompatible with the direction it holds the very next mille, mille second of rotation. That proves no point can be static or constant, all though it may seem that way to outsiders.

In the very centre of the sphere the form of the sphere dictates that the shape will relinquish space as the line run from the outside towards the very centre. With this natural state of affairs the sphere are naturally inclined to dismiss all space that it can form in the form as the sphere holds space inside and the form will finally be without dimension. All that I attribute to the line shrinking by reducing actually takes pace in every sphere as the diameter reduces to the centre. In the centre where the radius line goes single the form relinquish the three dimensional form it has inside. Being without dimension in the very centre means that at a point in the extreme centre of all spheres there are a point that holds singularity because this point with no space has a mathematical position although it is invisible since there is no sides to such a point to give that point any dimensions. The shape of the sphere is calculated by using the formula $4\Pi (r^3) / 3$. By reducing r to a point where r is r^0 singularity steps in because only the form remains as Π.

Going even further we find that there then comes a point where Π goes singular Π^0. At that point absolute singularity is present but so is absolute gravity present at that point. When holding the strength of the shape of the sphere in mind as well as taking into account that all cosmos objects of importance is in the form of planets or stars and they are all in the form of a sphere, we therefore may contemplate that it is where gravity originate. We now only have to find the reason why gravity will hold a base in a space less ness as Einstein predicted. It is clear to be seen that gravity is in the centre of the sphere controlling from the centre everything that is outside the space less centre. We can reason with confidence that gravity is the strongest where space is the least. We can further reason that it is gravity that is holding the sphere in true form and since the sphere allow gravity the best working opportunity, gravity can form the sphere in as strong a shape and form as the sphere seems to have. From every point on the surface of the sphere is where that point connects with the other side of the surface of the sphere by a line that runs through the space less ness of such a centre of the sphere. Such a line also connect by an angle of $180°$ as well as $90°$ to six other lines running from top to bottom, right to left, and back to front, where all join and cross in the centre of the sphere. There are therefore six lines crossing and connecting by a centre from any given point on the surface of the sphere. Such points connects in total six surface points on each side of the sphere while they all support one another through the space less centre. In that absolute space less ness in the centre holding singularity we find gravity supporting and controlling all space within the sphere as well as space connected to the sphere. That is where gravity control and guide the space, which falls in the parameters as well as under the influence of the form of the sphere. In the gravity centre space goes singular meaning space becomes space less or flat.

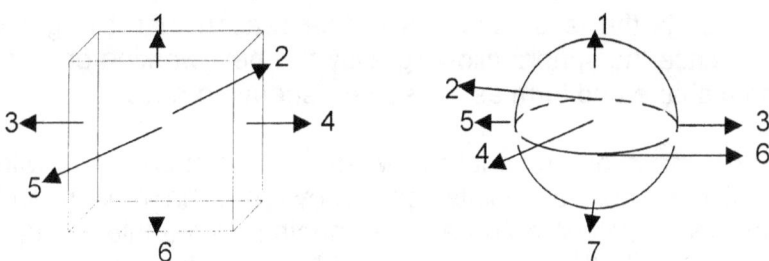

Also it is true that the entire form that is the sphere is controlled from a centre within the sphere. That centre holds the sphere in form and shape. Therefore the strong form is dictated from that space fewer centres where there is no space and no form left. The natural inclining is in the form of the sphere. It is part of the roundness that the overall shape of the sphere represents and this structural strength is carrying down to the very centre. Because the circle is forever reducing that reducing which is inherently part of the form of the sphere becomes a tool in distorting of space in the sphere and is eventually removing all forms of space from within the centre of the sphere. The very centre ends up as having no space because of the reducing that continuous down to become the space less inner centre. The all roundness is the ingredient that forms the backbone of the absolute strength that the sphere has and that is the component that the sphere is so famous for. The form the sphere has allows the sphere to have a control that is coming from the centre deep inside the sphere where the space vanishes and being without space seems to keep the entire structure rigged. The strength of the sphere comes from the centre of the sphere, which is inherent of the shape. That is why the sphere has such and the fact that all connecting sides refer to a centre brings the strength that the shape has. How does it work in its most basic analyses?

It is from the layout that the sphere uses as a natural form that we are able to locate singularity. In the case of the sphere the material naturally reduces by measure of the radius becoming smaller to a point where the radius is r^0. At that point the line that will form the radius has gone single dimensional r^0 and that is equal to 1^0, which is singularity.

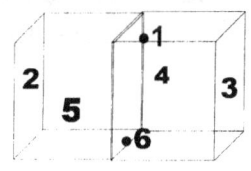

There is one more point in the sphere in the centre forming an addition in the sphere. That point holds gravity secure.

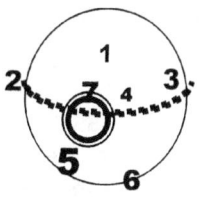

The cube has sides and the sides form a rather weak and flat surface that connects four corners. The flat surface produces a rather indifferent contact point with no special features on the surface. The corners connect to other sets of corners and those corners form a weak structure without any direct support coming from the other five sides. Without material to fill the body of the cube the cube has no direct connecting between any of the sides other than corners connecting at the edges of the sides.

Taking the vantage from the point the sphere is holding from the centre out into space there are ten points connecting to the centre. In that are the dimensions of singularity connecting to space where five connects to space in the second dimension of singularity, and five connects in the third dimension of singularity. On the other hand, the cube does show a very different characteristic, which involves only six sides (at least) connected.

In the very centre of the sphere the form dictates that the shape will relinquish all grounds in space that it can hold and the form will finally be without dimension. Being without dimension means that at a point in the extreme centre of all spheres there is a point that holds singularity because this point with no space has a mathematical position although it is invisible since there are no sides to such a point to give that point any dimensions. When holding the strength of the shape of the sphere in mind as well as taking into account that all cosmos objects of importance are in the form of planets or stars and they are all using the form of a sphere, we therefore may contemplate that it is where gravity originates. We now only have to find the reason why gravity will hold a base in a space less ness as Einstein predicted. It is clear to be seen that gravity is in the centre of the sphere controlling from the centre everything that is outside the space less centre. We can reason with confidence that gravity is the strongest where space is the least. We can further reason that it is gravity that is holding the sphere in true form and since the sphere allows gravity the best working opportunity, gravity can form the sphere in as strong a shape and form as the sphere seems to have.

From every point on the surface of the sphere is where that point connects with the other side of the surface of the sphere. All other possible points connect by a line that runs through the space less ness of such a centre of the sphere. Such a line also connects by an angle of 180^0 as well as 90^0 to six other lines running from top to bottom, right to left, and back to front, where all join and cross in the centre of the sphere. There are therefore always no less than six lines crossing and connecting by a centre from any given point on the surface of the sphere. Such points connects in total six surface points on each side of the sphere while they all support one another through the space less centre. In that absolute space less ness in the centre holding singularity we find gravity supporting and controlling all space within the sphere as well as space connected to the sphere. That is where gravity control and guide the space, which falls in the parameters as well as under the influence of the form of the sphere. In the gravity centre space goes singular meaning space becomes space less or flat. That is where Einstein's Universe goes flat because that is where gravity is at its strongest. However my bringing up this statement brings me directly to the point where I get very confrontational about how the brilliant mathematicians treat those they suspect are less inclined to think.

By examining the form of the sphere, we find that there are 6 points on the surface of the sphere that is holding the form at a specific and equal distance from the centre. Lines run from the centre into space at 90° and 180° angles of each other from six opposing sides. There then are six lines at 90° and 180° connecting to the centre from six points on the outside edge of the sphere. As a result of the basic shape that a sphere has, there is a spot in the extreme inner centre of the sphere where the lines in 90° relevance cross each other and others connect by 180°. There is also at that point a spot where all space relinquishes a position and only singularity 1^0 as form remains. At such a point we find the measure of the sphere being Πr^0 with $r^0 = 1^0$. That is where the line that represents the radius as a line disappears, as it becomes singularity r^0. After more reducing continue we get to such a point where we find only Π^0 left. At that extreme point is where space in all form disappears, as the circle providing the sphere the form the sphere has, removes all possible form by going into singularity $\Pi^0 = 1^0$.

Then in that area all form of any possible space disappeared leaving only the dimensions of singularity 1^0. I cannot delve deeper into the argument. However, from such a point there runs lines that connect to space on the outside where six points on the outside points connect to the space less point in the inside. In this book I take this argument much further but for now I leave the argument at that. Those lines carry the structural strength the sphere has. Contact with one point has support of six other points across the whole structure where the other six support every one of the six by singularity and the support runs through the entire sphere including the middle. Where there is no space, there must be singularity 1^0 just because the space filled with material removes zero and only material filled space is present. That means material fills the lot although in singularity 1^0. If zero was a factor where all space finally halted in zero as the value, then zero would be able to remove the space from the centre and such removing would continue to remove the space until all space was removed. It will finally abolish all space in the sphere and it would remove the sphere. Zero removes all possibilities of anything coming about. Since the sphere is there, a zero factor in the centre cannot be present. Only infinity can be a factor from where space may grow because infinity can extend and grow into and up to eternity.

The implication of this is that following the line down to the centre of the sphere we located the centre of the Universe. That is where gravity is. There is a lot more to that but be patient, we are getting there. In every centre we find a point, which is in truth not there but is the mainstay of all that is within the sphere. The mathematical value of such a point is $\Pi^0 r^0 = 1^0$ and 1^0 is singularity. That is the point where the Universe started and that is where the Universe will finally end. That is the Universe without space-time. That is $k^0 = a^3 / T^2 k$ which proves the Universe is without doubt a sphere…and we just located the centre of the Universe!

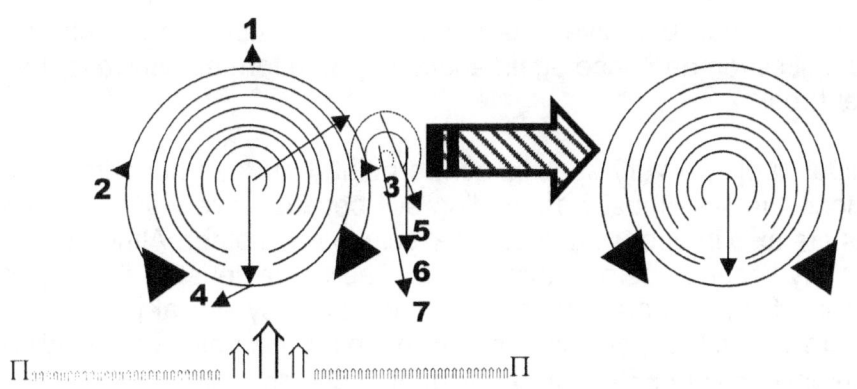

As one can see with the spinning top delivering the Coanda principle, every point overheating can spawn space-time by centralising singularity. One can see from the top that singularity is established wherever spin occur. The motion generates a position of seven in relation to ten and singularity manifests as 1.9991 as is explained elsewhere. That means any point formed by the sphere spinning can and does start a centre in which no motion holds no space and of which motion surrounds such a point by forming space. Although everything at the time was in the form as a multiple circle, which results in a sphere, the sphere was not the only form present. This too has to do with singularity interpretations. We see a cube, as we know the cube but at first when form came about the cube were not yet a form.

While the one sphere forms on this spot where the dominating sphere secures an edge the dot may be reserved as an edge marker to the dominating sphere. To the forming sphere in progress of emerging heat gathers at that point because the rotation is a result of duplicating and duplicating is the tendency of naturally growing in space-time $k = a^3 / T^2$. In order to find duplicating coming about there has to be heat in order to duplicate what will form heat. The duplicating process is a process of one factor going softer or less solid and therefore more dynamic than the other. To have singularity is to have gravity but to have gravity there has to be a point of motion and a point of sturdiness. The point of sturdy may be in the centre of singularity, but then the solid must be motion. However even today it still apply: what moves forms liquid in the presence of a solid and at that point singularity presented the solid therefore what we might think of as solid was the liquid because it moved around the solid. Where the one factor is duplicating the other factor is compressing $k^{-1} = T^2 / a^3$

The points duplicating is four moving around a centre by the square of gravity. The motion is the

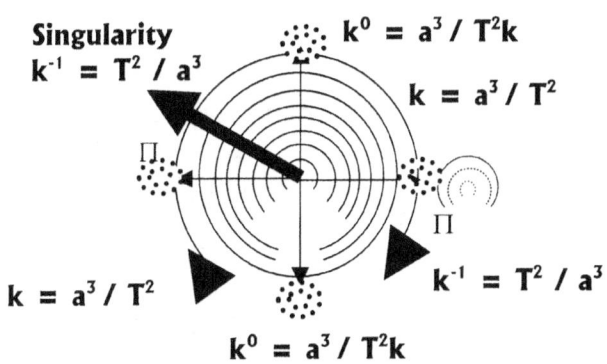

sources of heating because the heat is bringing about the movement. The heat growth therefore provides the action because the action is what energises the points to provide the motion. The motion is purely is space-time duplicating and the duplicating is feeding heat to the centre from the four points overheating thus the points that shows expanding.

But also the duplication leads to the spawning of one point of singularity that provides the installing of the next centre for the next sphere.

Because of the principal in which the Coanda works the motion will centralise a new sphere and by appointing six position around the centre three points will not move while four will move about the three points forming the centre line. The result is that the four points by duplication will reserve the point moving as the next point in singularity because of $k = a^3 / T^2$ singularity will be a natural result of the motion. Then that point will secure a position $k^{-1} = T^2 / a^3$ which will secure six points about such a centre. The centre will bring about four points spinning around three points holding a line singularity. The line in singularity will stand in relevance the contacting factor $k^{-1} = T^2 / a^3$ and the duplicating by expanding points will be four and serve the relevancy by contributing $k = a^3 / T^2$ as space-time only in form. From this the rest of the Universe burst into the next phase of Creation.

The gravity is in relation to the spin, which is in relation to the four points spinning which are $\Pi^2 / 2$ and that is the Roche limit. It is the dividing of singularity sharing space-time just as we on Earth share singularity by division between the Earth and us others that is not part of the Earth. The total that forms from the point that spawns is seven plus five plus pi square in division of four totalling twenty one that stands related to the first seven and once again another sphere formed. However this is an eternal relevancy that can never break.

Any object in rotation will have a middle point, a very specific centre point that does not spin. That point once again hypothetical but none the less must be standing still because every line running from that point in opposing directions are also in opposing directional spin to each other. Although the points had the same characteristics only seconds before, they oppose the characteristics it had just before and just after the very second in which they are and to which they relate by similar points also in rotation. Due to the spinning nature of such a point with all surrounding the point very varying second, the value of such a point can only be Π because of its constant changing.

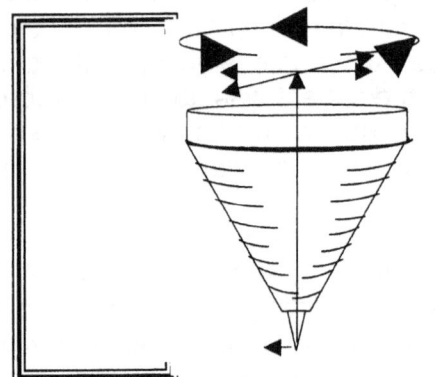

The sphere has seven points. The cube without truly being a cube but is just in consideration of having a cube in form holds five points to singularity. In the centre runs singularity to the value of Π^0, which means that which surround Π^0, holds a position of Π^2. The spinning sphere activates the seven points, which places gravity in relation to a centre. Outside the centre there are five sides by dimension. The sphere has seven points of which four is spinning. The four spinning stands related to the gravity of spin,

an ending point where the limit attach or dissolve the material motion. Such a point liquefies the location. That is the principle Roche limit but also not only that; principle behind the Coanda is closer than what the Roche at $\Pi^2/4$ will allow becomes liquid and removed into the unit sphere.

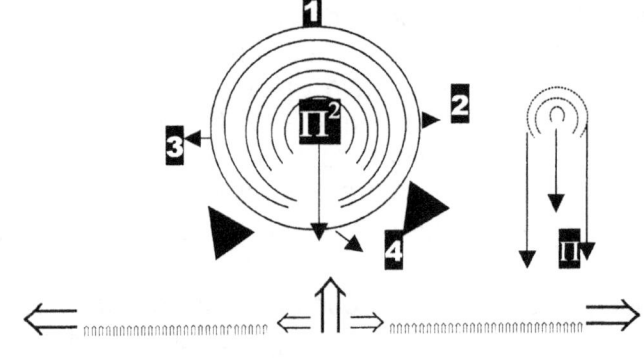

The extending of the motion also produces will either next to the what ever is in behind the it is also the effect. All that limiting point dissolved as forming the

Using r would specifically oppose another r from every angle. From such a point every other point will be opposing any other point not pointing in the direction to which the first point is pointing, whereby it extends the direction it holds. No matter what the point is or where the point leads, such a point holding a specific direction will be unique in the direction it is rotating because at that or any other specific point wherever, it will be directing not in the direction it spins but in the direction flowing from the centre point outwards. Any point will be it opposing itself within the rotating of $180°$ changing every aspect of its previous flowing characteristics it previously had or will once again have in $180°$ from there. While in rotation from the point of an outside observer all may seem static and never changing but to the object in spin every next second will be a diverting from every aspect it was in very second passing, and the direction it held in relation to the direction it held the previous mille, mille second will totally be incompatible with the direction it holds the very next mille, mille second of rotation. That proves no point can be static or constant, all though it may seem that way to outsiders. Although matter is matter, matter can also be anti-matter at the same time.

At this stage time was still eternity being interrupted by infinity. To say the Universe is or was 13.5 X 10^9 years old is shear Newtonian thinking. Was it 13.5 X 10^9 years and how many days in the year of our Lord and what about all the years that passed since this date was revised? Time was flowing according to interruptions in eternity changing from what was to what is to what will be. Time is a norm that comes as things in the Universe change about things that are places around and scattered throughout the Universe. We may presume time at this point somewhere became a factor since sphere sprouted from points on sphere edges and differentiation in development came in place. Considering the role that the Roche limit played one can see how points in singularity grew from contraction and secured ever stronger centres by divulging hear points within the realm of singularity control. When a point in form developed at a position that was close than the original Π^0 to Π, the singularity in control took control.

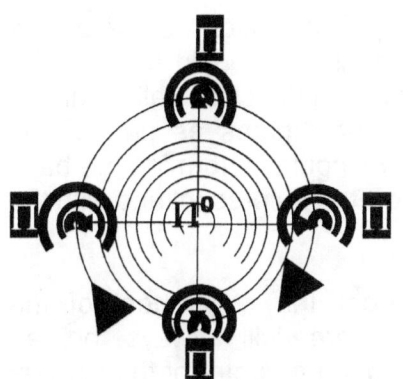

With every one of the four points taking position on the side of Π^0 running to the allocated position of singularity extending that forms the value of Π at a measure of Π /2 each brought about the Roche value of Π^2 /4 in relation to the developing centre. One has to remember that the star of today takes on the characteristics of the form of that era.

Consider what happens to a star that developed closed than the Roche limit of Π to Π^2 / 4 would allow, it is easy to see how the singularity centred grew by concentrating the heat the points in singularity brought about

$(\Pi^2 + \Pi^2)$
$(\Pi^2\Pi) = 7$
**positions
holding
singularity**

$(\Pi^2 + \Pi^2)$ $(\Pi^2\Pi)$ 3. =1836, which after wards the atom was about and the Big Bang proceeded

I again wish to repeat the centre in the sphere and the centre of the sphere because in this is where the realisation comes from how the Cosmos started. If there were gravity at the very first instant then there was a sphere at the very first instant because gravity can only be in the sphere because of what the sphere represents. If that which extends from singularity does not meet the form of singularity by measure of $\Pi \Rightarrow \Pi^2$ in precise duplication it will tend to destruct and that we call imbalanced spin.

The truth about gravity in the cosmos and that gravity will contract the cosmos one day is the fact that the sphere has singularity as a natural substance. The entire form that is the sphere is controlled from a centre within the sphere. That centre holds the sphere in form and shape. Therefore the strong form is dictated from that space fewer centres where there is no space and no form left. The natural inclining is in the form of the sphere. It is part of the roundness that the overall shape of the sphere represents and this structural strength is carrying down to the very centre. Because the circle is forever reducing that reducing which is inherently part of the form of the sphere becomes a tool in distorting of space in the sphere and is eventually removing all forms of space from within the centre of the sphere. The very centre ends up as having no space because of the reducing that continuous down to become the space less inner centre. The all roundness is the ingredient that forms the backbone of the absolute strength that the sphere has and that is the component that the sphere is so famous for. The form the sphere has allows the sphere to have a control that is coming from the centre deep inside the sphere where the space vanishes and being without space seems to keep the entire structure rigged. From the centre the sphere shape shows strength that the shape as tough as it is. How does it work in its most basic analyses?

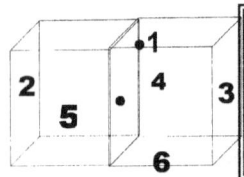

There is one more point in the sphere in the centre forming an addition in the sphere. That point holds gravity secure.

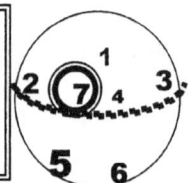

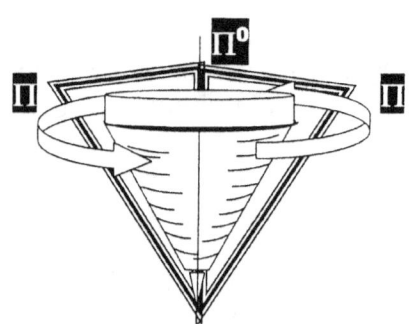

If the cosmos had any other shape the might be available contraction in the end as the final conclusion would then not be possible. Only in the sphere and more so in the circles all forming a sphere is singularity present. Singularity comes as part of the construction we find in the sphere. By singularity forming the base of the sphere the Coanda effect will forever be present and with the Coanda effect gravity applies.

attraction by singularity controlling Such control has the name of liquid (10) and solid (7) in the such control by motion. But why is moreover by motion?

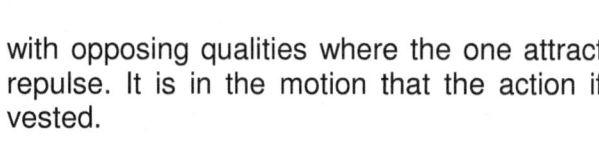

As is evident from the experiment of the Coanda effect there will always be an the space-time from the centre of the sphere. gravity, but it is no force. It is a combining of presence of three in time that will produce the Coanda effect arose by rotation and

It is because one Universe arises as the other gravity is

with opposing qualities where the one attract repulse. It is in the motion that the action if vested.

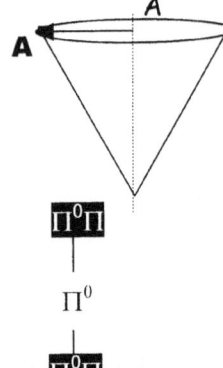

The spinning of Π^0 around the centre Π^0 establishes Π and Π is what produces the singularity. Singularity is always present in the form the sphere presents however; in the sphere singularity presents no influence except in the case where the sphere starts to spin. In the spinning a line comes about which caries the term if an axis. The axis has never been acknowledged as the most vital piece of any particle in the Universe manly because the role that the axis plays has been neglected and subdued by the incorrect attributing Newton placed on the motion and the axis.

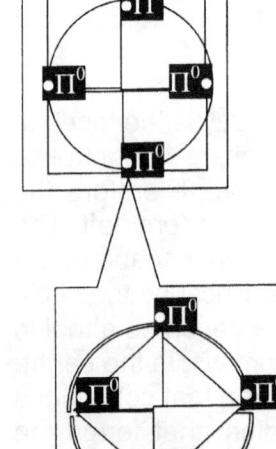

The moving of Π^0 to Π involved relegation and not motion as we consider motion. It was Π^0 getting a side and that is all. There was no true side but only a form that came into place. Singularity (A) received singularity (A) and no more of anything but the shift to comply with having a relevancy forming in relation to singularity. The dots had no sides, had no length or diameter. There was not measurable space or measurable time involved. The time could have been a micro, micro second as much a trillion millennium because time had no relevance. It was eternity interrupted by infinity, as it still is the case, however the line that eternity followed was no line because there was no space to hold the line.

The line was momentarily interrupted by infinity, however with no one there, there was no one to notice. The lines were not lines but relations to sides being formed.

Inherent to the form the sphere offers, there is a specific location of singularity where the radius first goes single $r^0 = 1$ and then form goes into the realms of singularity $\Pi^0 r^0$. The cube also may have such a pint bur having such a point does not connect directly to six points located on the edges of the cube or any other form the is.

In relation to such a centre where $\Pi^0 r^0$ forms singularity there are always four cubes related to such a centre where the centre is part of seven points in total representing the sphere.

Every cube gas lost one side to a point of the sphere where the sphere takes control of form and removes one side of the cube. In relation to the time factor that is inherently part of singularity by the extending of singularity there are five sides connecting to four points standing related to singularity by the Π^0 factor and that gives $5 \times 4 = 20$. That is always directly in relation to seven points singularity offers.

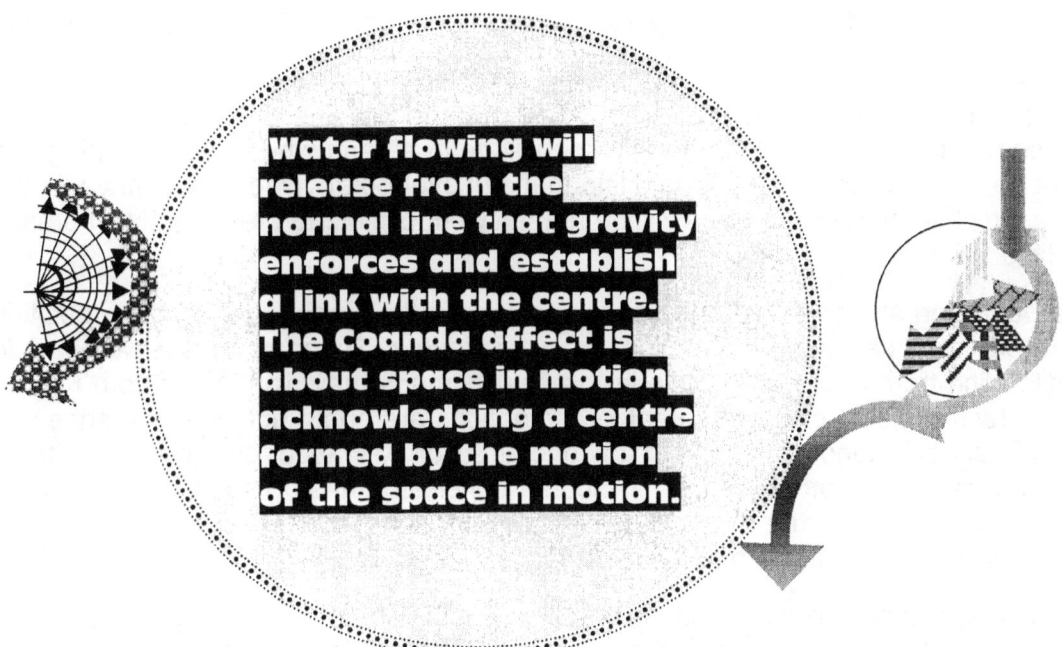

Water flowing will release from the normal line that gravity enforces and establish a link with the centre. The Coanda affect is about space in motion acknowledging a centre formed by the motion of the space in motion.

This should lead any person to investigate a centre that forms because evidently, there is a centre but that centre comes about by motion of rotating around a fixed point serving all points in motion. That leads us to the centre of everything in rotation because everything in the Universe is in rotation. As Kepler said the Universe is centred by $k^0 = a^3 / k T^2$ and we have to find $k^0 = 1$

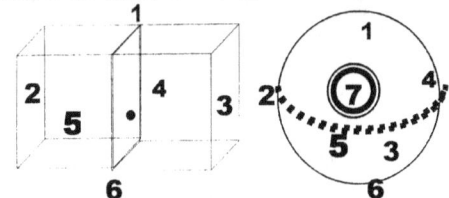

Kepler's formula also indicates that a sphere is within a cube that is holding a sphere at singularity $k^0 = a^3 / k T^2$ with all the centres being all the same $k^0 = 1$

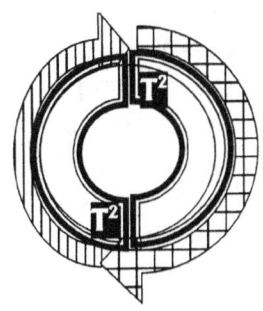

At this point Newton's second law come into affect. Motion by means of the Coanda effect introduced space as motion introduced time. For the first time ever time was interrupted when motion provided time the space to interrupt. From motion by the way of the Coanda principle gravity came about as a centre formed a point where motion surrounded space, By motion space-time was established in relation to singularity

If the universe did start from one single point and time matter and space flowed from that point, then that point must have a relative connecting base because such a point holding singularity must be eternal as space matter and time link eternal. There therefore must be one point linking the entire universe when regarding the fact of singularity. Then according to the theory off relativity there has to be one exact point holding time in a relevance notwithstanding the fact that time depart from that position and relate differently to all space-time away from such a point.

Every person I have discussed facts about creation recollects images in the trend depicted in a presentation as one may find to the above. That would be the most unlikely way Creation came in place. The recalling of pictures representing images about creation must have form, but to mathematics it had no form. From this thought the very opposite arise where Creation came from nothing but such an idea is mathematically simply not possible.

The thought of nothing is just what it is, a thought of nothing and although it is in the human mind common nature to present nothing as a value in the recalling of something, nothing is a presentation of the figment in the human mind. There can be no number such as nothing and that was (possibly) Newton's biggest error. Nothing represent non-existing and that is just what nothing is, it is non-existing.

In order to prove my point I wish to ask the reader to define the shortest line there can theoretically be. If he should answer anything but that the shortest line will be at a point where the beginning and is the very same spot he will be wrong. The shortest line that can ever be anywhere must have a start and finish holding the exact same spot. The line will be humanly impossible to create but we humans are capable of very little.

When the line has a beginning and an end at the very same spot and it wishes to extend the position as to further the possibility it has, which direction should it favour. Humans in the west would naturedly think of extending from left to right while in the east humans may want to go from right to left. Some persons will tend to go up or down, but all of the options are about human preference and not mathematical conclusions. Extending the line in any one direction will favour one direction without a conclusion about not extending in other directions. Such a conclusion has no sound mathematical foundation. The only option about extending will be in all directions equally in order to give a meaningful non-bias flow of mathematical equilibrium

The shortest line in the realm of possibilities must have a start and finish holding one spot and such a line will also be a dot or a circle. Not favouring one direction puts all directions at equilibrium meaning that any form what ever may be can develop from such a spot with the end and the start being the same. This reasoning prompted me to look for singularity in such a spot because if the prime spot from which all came was a spot, then the spot must hold the shortest line but more prominent it will hold the smallest form including the smallest circle.

One possibility that the shortest spot can never have is having a starting point on the zero mark. If the mark of zero holds the start it must also hold the end because the end and the beginning has the same position. If the position of zero then is the beginning, the end will also be zero leaving the line without an end as well as without a beginning.

While very line is circling bringing about time in space to the value of Π repeating Π to form Π^2 at the same time Π is extending in one specific centre to the value of Π^0 and only the spin value keeps Π not becoming r The spin keeps the immovability from becoming Π and maintaining Π^2 by performing duplication, but with any slightest reduction in spin reducing Π^2 to $\Pi^{2/4}$, Π will start extending and as one can see from the behaviour shown in the Roche limit, the heat will be concentrated at the centre and the singularity in the centre will grow four time in concentration. Only at points exceeding Π in diameter was time as Π^2 able to retain form and also grow. From that space slowly developed because at Π could Π^2 bring about a form which provided motion. In the centre there developed Π^2 and Π^2 kept all form at a safe distance of Π to bring about the needed solid immovable centre with the form $\Pi^2\Pi$ about the double Π could $\Pi^2 + \Pi^2$. That secured the makings of the atom by applying the Coanda principle of enticing gravity in the centre of motion, which then provides space-time by measure of $(\Pi^2 + \Pi^2)(\Pi^2\Pi)$. This totalled seven in dots and with three of those seven circling singularity at 1.9991 the atom came about. **I again wish to repeat the centre in the sphere and the centre of the sphere because in this is where the realisation comes from how the Cosmos started**. If there were gravity at the very first instant then there was a sphere at the very first instant because gravity can only be in the sphere because of what the sphere represents.

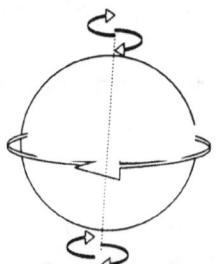

On the inside, there are the seven markers of which singularity is the focus point in the centre of the centre. The markers are representing one aspect of space, which for argument's sake let us call it cold. Then there are three more markers on either side being part of the space but not captured in the space. It is space in motion by the influence of the motion of the Earth.

By not having motion the lines also have no space as the space extends to form

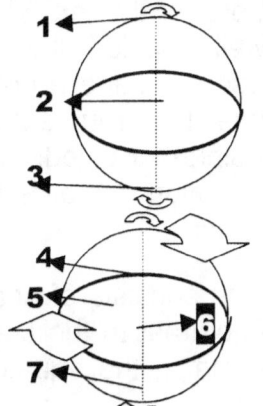

space forms space and the line includes serving the three points to the outside. Where there is no motion, there is no space and where there is little motion, there is little space. The only space the line may relate to can be a point that is on the border of the sphere that is crossing singularity and connecting the two edges on either side of the sphere that is forming the sphere. That means the line from one point holding singularity to another point holding singularity that line will cross the centre line which gives the line in singularity valid space-time to control. Singularity does not have the ability of motion therefore singularity does not hold space. Singularity is also eternally indifferent to motion and motion can excite singularity but singularity cannot be shifted by motion.

Three points form the line.

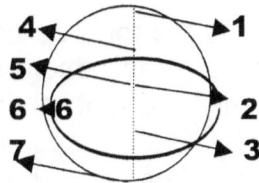

Every time motion takes place, the centre line holding positions 1, 2 and 3 stands still. There is little generating going on and reducing to a point where there is no generating going on. On the outer edge of the rim however there are four points that do shift. The points shift from one location to the next location by generating space

All this is happening while the crossing is all concerning singularity moving from one sector of singularity to the other sector of singularity which is (Π/2) X (Π/2).

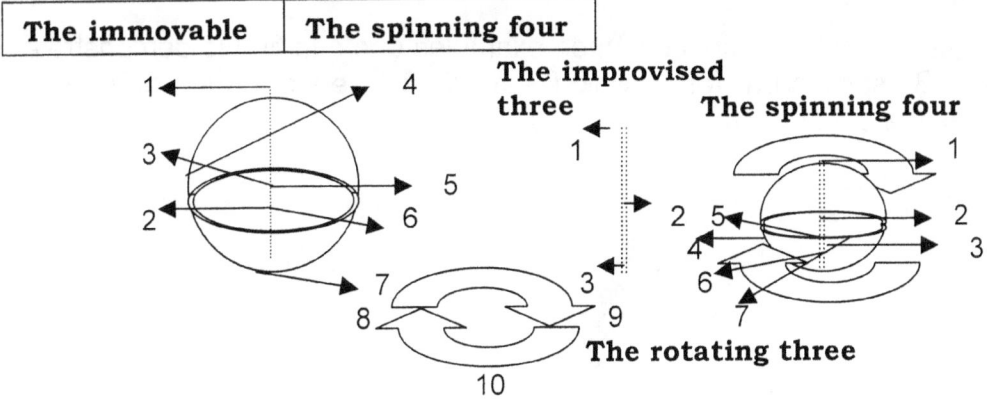

The relevancy forms part of the duplicating and dismissing displacement of space-time we call gravity. In that, we are looking at relevancies and no precise specifics. However, the Universe was built block by block in this manner. As it was but is no longer only form that applies in the Universe but concrete measurements also come into play therefore even the relevancies may apply in different relations as they switch over to compensate for other factors alternating as they are coming into prominence. The lesser developing sphere orbits the dominating sphere and between them, there are definitive relevancies. The centre circle singularity line of three is unaffected by spin which I shall call the immovable three. However, the immovable three holds such a stout position as far as the centre sphere is concerned. In relation to the orbiting circle the centre line is part of the building and destructing process that manifests as duplication as the centre singularity maintain domination and control over the orbiting structure. In addition, it has a major part in the motion building of the sphere in orbit and moreover building by generating the singularity line that generates the lesser and the orbiting sphere. In that relation the centre sphere reflects the centre line to serve the orbiting sphere by supplying the reference needed to establish motion in the orbiting sphere as one Unit. In that there is an undisputable reference of seven orbiting the centre and the centre providing three as a reflection of the seven which in all accounts for ten relating to the four which also is spinning as time and in total forms the seven taken in relation from the orbiting ten.

Points seen from the side form a line that never moves and is a line in singularity holding three points

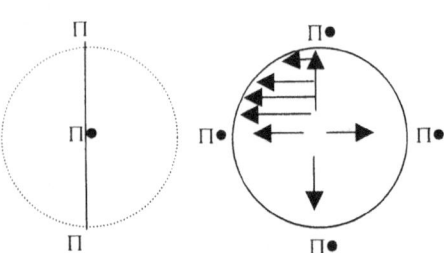

The seven can never totally separate from the ten, but by singularity being the same but being on the other side it is withdrawing space-time altogether. See it as seven (let us think of that as the cold basis of space) spinning or turning in the ten (which then will represent the hot part in the cold basis) and the ten is part of the seven but the seven is not part of the ten. The third factor is the axis around which hot as well as cold will turn. Therefore when reading the next page, please envisage a cold base turning in a hot and cold space. The purpose of this is not to define whether the argument is correct or not but it is to help the reader gain understanding of the process principles involved. But motion also converts space to relate to space by changing relevancies through motion matter is in relation (part of) to the total dimension of space but is not the total dimension of space.

Space-time is allocated in progress from the line singularity offer. That space-time consists of heat

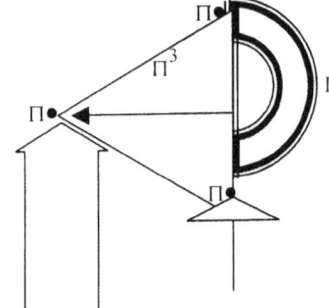

and can therefore store more heat, which strands in contrast to the cold than singularity presents. By providing, more space-time at the equator there is more space-time to allocate to heat being there

In this one can clearly see that it is the motion that sets the top free and independent from the gravity of the Earth. By motion the top generates individual gravity that allows the top an individual gravity and that motion frees the top from the gravity by which the Earth restrains the top. The motion gives the top independence that the top immediately loses when the motion subsides. This fact is the utmost important issue of all physics in the Universe.

The top is one of (perhaps) the easiest and most common examples one can find to demonstrate the cosmic generating of gravity. By spinning a "force" which is no "force" keeps the top erect and it is only motion that accomplishes the act of gravity.

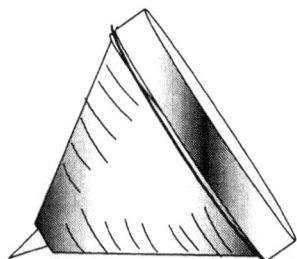

The top lying still holds the same singularity principle that the sphere holds because if the shape the top has. The roundness protects singularity at a seventh position deep inside. However the top is a dot Π or even going down to a spot Π^0 and is only by the form the material has which puts singularity in place.

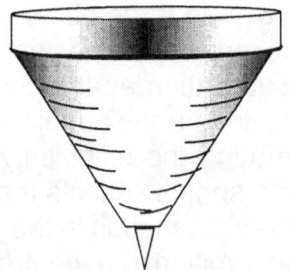

According to Newton it takes no effort $\dfrac{dJ}{dt} = 0$ to get the top from where the top was motionless to where the top is spinning. I say this on the work that Newton suggested comes about from the effort it takes the top to circle.

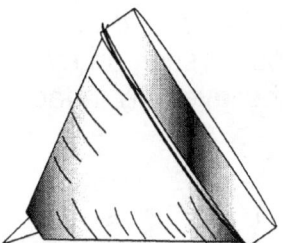 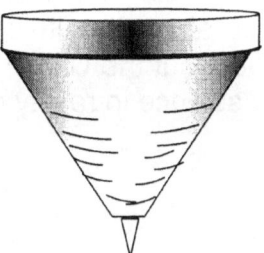

From a position where the top is lying down being collapsed the top generates a position by spinning and the motion puts the top erect. It takes motion and not nothing to establish such an independent and secure position. It puts the top in the centre a newly established and independent Universe as the top then finds courage to fight the gravity of the Earth up to the last "breath" is fought.

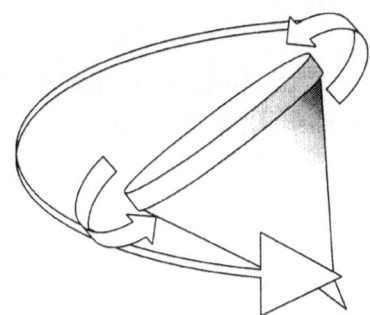 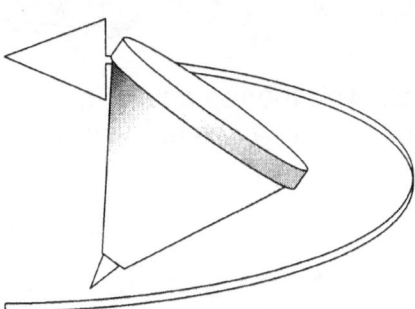

In the motion a line comes to life running in the centre of the top. This line is not just another line but can focus the top to spin upright and erect. The line was not there when the top was on its side. By motion the line can concentrate an effort that will unleash such dependence to the top that the top will come into a position where the top has the tenacity to take on the Earth gravity in a struggle for life and supremacy.

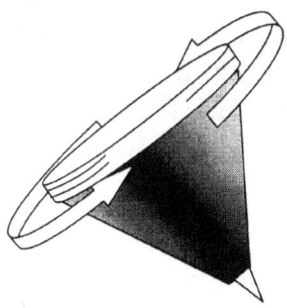 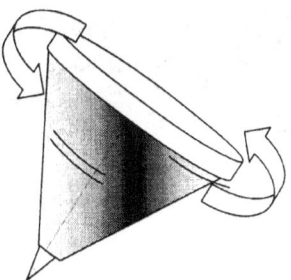

From whichever angle one looks at the top, the top seems possessed and I can even be slightly forgiving towards Newton for calling it a force because although not a force the stance the top takes when spinning upright leaves on with an impression of forcefulness being part of the situation. One must not see a force but one should see the manipulating qualities of life extending to the top and by life's ability to manipulate space-time and control motion in space-time with space-time, the throwing of the top is as little a cosmic event as the apple Newton saw falling from the tree. In every event, the top as well as the apple the drive was life controlling events and as far as there is proof there is no possibility of such an event taking place anywhere in the Universe by something as small as the top or the apple. In the case of stars such development does start the star on a course of independence and it sets the star development on the rode of becoming independent from the galactica. The drive generates gravity but the driving that allow such rotation is inspired by the accumulating spin of the entirety of all the atoms in motion within the young developing star. The heat blanket in which the star cradles has a lot of influence but it is the atoms becoming a driving factor that inspires the motion and such inspiration allows the top the initiative to form independent gravity as a star. In the case of the top the spin is completely cosmic unnatural. If you start to imagine about life in the Universe you may just as well start believing in ghosts fairies and all other fantasy creatures. Science must decide whether they wish to speculate about the life's abundance and being a dime a case, found all over and everywhere you look throughout the Universe, but in such an event distance their fantasies from science and reality, or stick to science in reality and believe only in facts as science presents facts.

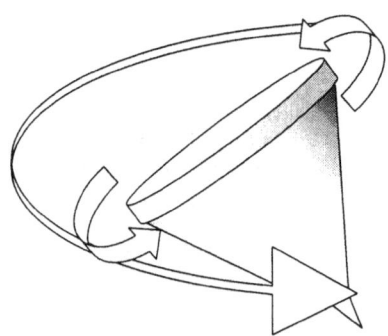 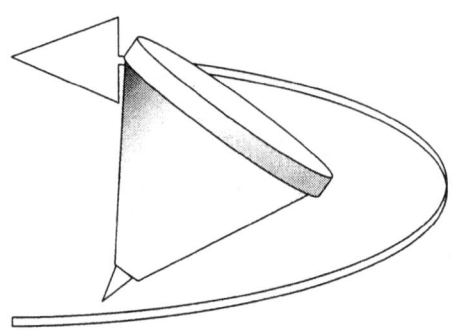

Let's consider what are facts with the top spinning as the top does. This is no fantasy or life coming from same imaginary source but it is a cosmic reality which life found a way to manipulate.

6 sides 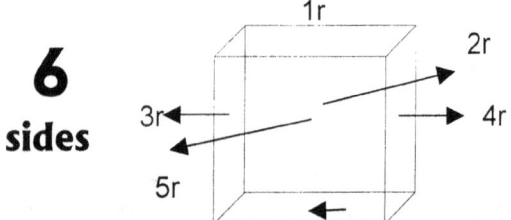   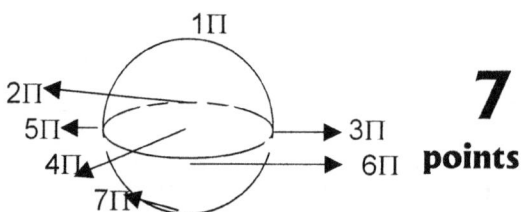 **7** points

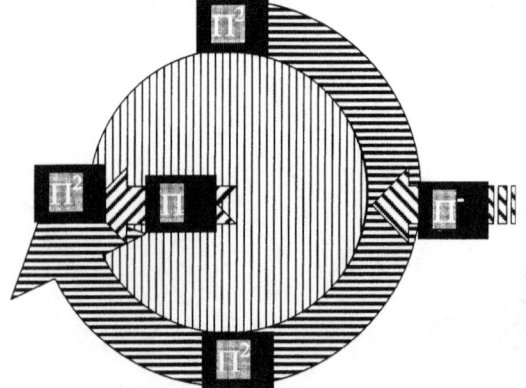

The effort it takes the top to spin gives the top a distinction of extreme significance. The top is promoted by the motion initiative to that of a star in motion because it charges singularity into existence where singularity then controls space-time. There is no difference between the top spinning and the Coanda effect and in both cases singularity is generated by motion that is charging a centre to activate a Universe.

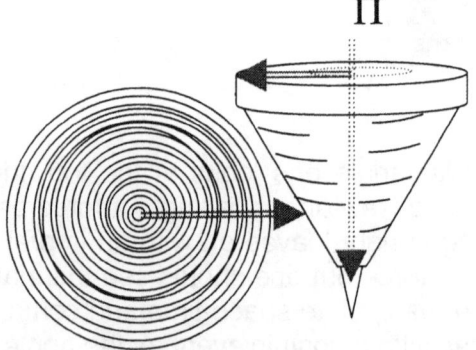

Singularity is a mathematical point hidden in every sphere. Singularity is very much inactive in every sphere but to keep a centre of structural bonding in every sphere. Something happens to the top in having a singularity just like any other sphere has to a point that takes charge with all the cosmic dynamics the sphere may show.

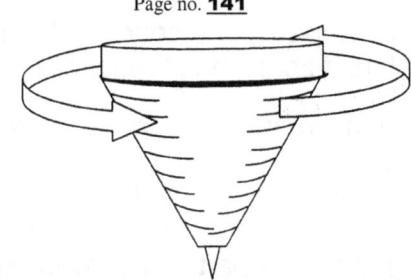

The top is charged with an energy, which not only takes charge of the top as well as the body of the top, but also the immediate space surrounding the body of the top. When a person with skill manages to put a high degree of spin into the motion of the top the top then spins in the surrounding space so vigorously the top stars to whistle vigorously

The linear remains linear because the linear redirects its intentional direction because of the rotational change that the linear motion always ends up doing. The line forms an eventual circle because the linear line must constantly entertain the centre.

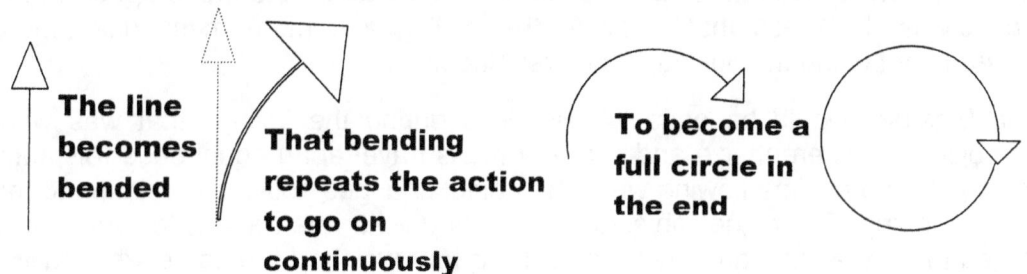

The line becomes bended

That bending repeats the action to go on continuously

To become a full circle in the end

Our gravitational falling to the Earth is a result of a circle going straight and forcing us straight down to an everlasting directional alternating circle, we have as we spin with the Earth as we spin around the Sun. As we fall straight down we, change direction while we are falling straight down because that point we are heading to what we are falling to is changing too. From the centre of the axis, everything seems neutral. The axis does not spin at all, because the axis brings about spinning motion changing eternally. That is in nature and not man-made motion.

As the top comes to motion the top finds the characteristics that the top shows very indicative of the characteristics that all moving objects in the cosmos show. The top spins in a straight line that bends by a 7^0 inclination.

Apparently an idea concerning the subject of gravity Einstein came to was about him falling off a multi story building and the gravity mass that he then would experience. Remember, this was long before flying and parachuting with free fall acrobats showing on TV how a man and a car with a man in the car can descend five or six kilometres while falling side by side. This happened while Einstein was still being a patent clerk in his younger days. Apparently Einstein was looking out a window of the multi story patent office, when Einstein suddenly realised that had he, Einstein fall out of the window from the roof to the ground of the patent office where he was working at the time, then he (Einstein) would feel as if he was weightless during the time of his fall. By falling with him, those articles would feel equally weightless should they accompany his fall down as being part of the falling process in his imagination. As the objects were travelling alongside Einstein down the building to the ground the lot would travel at the same speed from the top to the bottom of the building. Then I went one step further by supposing the Einstein group's falling was real and no imaginary thoughts were set in the fall, then what was the imaginary factor then? Let's pretend Einstein did fall with his pen, his chair and his desk and Einstein was not imagining his fall.

Einstein as a human being can imagine but his falling companions can't. If Einstein was imagining his weightlessness, it might be psychological, but in the case of the other travelling companions it was not possible to imagine anything. There is an immense difference in size between the falling companions and that notwithstanding they travelled the same speed while descending. If they travelled the same speed as Galileo proved and they all hit the Earth the same time, which then indicated that their weight and mass, that which gravity used to drive and what propelled them downwards and that which was causing the drawing of what the mass was instigating to allow the motion of fall to commence, was equal. Kepler found space a^3 being equal to the motion thereof T^2 in relevancy to a centre point **k**. Kepler found space had to move.

When reading this that evening so many years ago, I came to realise that Einstein could only feel weightless if it was true that he (Einstein) was weightless. He could not feel as if as if was part of his imagination because he was truly falling, and in truly falling the falling was then without his imagination doing the pretending. Einstein had to feel his weightlessness as a cosmic fact in the true sense because if he was truly falling, then the part, which was the falling experience, was what he was experiencing in reality by three dimensions with one dimension in time. If Einstein was experiencing weightless ness, it would be because he was weightless while falling, then Einstein would not imagine the weightless ness because Einstein was truly falling, thus carrying out his cosmic state he was in. His body being in motion ($a^3 = T^2k$) was at that moment truly weightless while experiencing unrestricted gravitational motion. Einstein, the pen, and the chair had the same weight since they were all weighing the same in falling. If there were any mass differences there had to be speed differentiation for the force of the one would generate more motion than the force of the other onto the different mass components but since there is not mass discrepancy amongst the falling while falling, the lot is having the same state of weightless ness, they adopt the same speed in the fall. After all it supposedly is the mass that is doing the pulling and more mass does more pulling...except if the mass is not doing the pulling in the first place.

All four items including Einstein, would be equally weightless during the falling...that was what Galileo found because objects of different size and different mass travel at an equal pace (distance over time or space moving divided by time flowing while the object changes position in relation to the Earth ($a^3 = T^2k$)) while descending. The bigger objects do not fall quicker than a smaller object and that can only be attributed to one fact; it can only be true if the four weighed the same while falling and no one weighed anything while falling. That means the gravity applied while time flow in relation to the space that was applying the motion, which was what gravity is $k = a^3/T^2$ according to Kepler. The single line falling is represented by the factor **k** being the relevance of space a^3 that was relocating its cosmic position while all that was happening in relation to the motion of the Earth T^2, which was in relation to the Earth spinning around the Sun and that rotation gives us our time T^2. While in motion the four different objects weighed the same since they travelled at equal speed downwards. By standing still the objects had mass differences and when they were in motion they weighed the same.

When the motion became frustrated by being blocked by another space that was also filled with material and that was holding the spot too where the motion was directed, they then had different weight. The pushing resulted from the bodies striving to remain independent. The two objects were in a fight to claim the position each desired, and that was to fill the centre of the Universe. Being ($a^3 = T^2k$) was being in the centre of the Universe because the centre of the Universe was $k^0 = a^3/T^2k$. Then one may conclude that gravity is motion of space and mass is the restricting of the motion of space. Having mass does not bring about gravity but it does restrict gravity's motion, which is what brings about the mass and weight. Gravity produces mass but mass does not produce gravity or in fact mass produce weight but mass is not responsible for the intended motion. The intent on moving while being blocked by another object is frustrating the motion of gravity in both cases and the higher the frustration on motion is, the more mass there is coming the way of the bigger object who then has the greater desire to move. The reason why it has the desire to move and why space is equal to the moving in time of the space in relevance to the centre of the Universe (which at that point might be the Earth or be the Sun) is what the have the effort to explains. Mass is the restraining of motion and gravity is material moving about by committing gravity. Mass only comes into the application thereof when two objects filled with space moves into a position where both want to claim the very position in space the other occupies.

It is the motion and the independence they show to hold onto their individuality as independent cosmic structures that prevent them the sharing of space which in turn prevent further motion that causes mass. Gravity is in essence where mass is present, still in a tendency to commit motion but is then in the frustration of motion and gravity at such a point is the commitment to move once the blocking of space is relinquished. Because the one object that has more "mass" would put in a more assertive effort to move in relation to a smaller object and the effort to move will constitute to a greater resisting effort by the blocking objects in a fight not to relinquish its position on the space both object claim that the tendency to move and the tendency to block the movement will bring the effect of greater or smaller mass being present during the effort and in line of resisting the effort. However while any space is in motion, the gravity of motion is equal to all and puts everything on an equal basis. Therefore there are no big and small and the big Sun does not pull the small Earth closer. Mass is when the motion is prevented that a differentiation in motion effort becomes part of the picture.

Do not be fooled by the seemingly innocent explanation that space is the motion thereof which is what gravity produces because of all things the cosmos creates, motion of space through time is the utmost complex manoeuvre and without bringing a restraining of mathematics into science, it is so complex there is no viable explaining in physics about how the cosmos produces the act of motion of space in time. In order to get every atom to spin as every atom follows the lead of the atom in front. This gives direction to follow to the atom just behind while giving coherency to the structure. By following the one in front and being followed by the one behind the lot of atoms are holding as an individual unit times the units there are going around in the entire Universe. The measure of this complexity is beyond what the human mind can absorb. While the atom in front is vacating space to fill the space of the atom in front is vacating at that instant, the atom behind is filling the space that the atom in front has vacated in order to vacate and relinquish the previous position in favour of the following position to honour the direction gravity is insisting upon. Times that with every atom there is in the Universe and one may grasp the significance of the calculation. Removing material from space by filling material into a position of new space sounds simple because the complexity has never been realised. I am in the hope that in this matter I will be able to will reveal what the factors are in understanding the commitment of material to move through time. This was all a result of understanding the dynamics of Einstein's arguing about gravity and mass. Then I kept this information in mind and it helped me to further realise gravity is motion differentiation between objects.

It is the independent motion providing a different speed while sharing a common centre of attracting that allows a discrepancy to establish mass under specific conditions applying between the two in relevancy. While falling the gravity applies as moving of space that is putting time in relation to the distance travelled. That means there is a speed relevancy between particles in motion and synchronised motion, which would bring about equal orbit around a shared centre. That is the result of gravity functioning. While the object falls, the motion confirms gravity. When motion ends mass sets in and becomes the constraining of the object preventing further motion. The motion is still there but now it is reduced to a tendency to move thus establishing the object mass as the limiting of further motion. Preventing the motion by implementing mass is the resting of objects against each other by resisting the motion to continue, which then is where the mass takes the place of the motion. Where a confronting of objects restricts gravity, the action then implements an introducing of the mass as a substituting factor to motion that then replaces motion as substitute to the motion that would be and the mass is providing the tendency of gravity being the motion of space.

However mass then restricts motion and becomes motion in a tendency to apply motion. While falling, gravity applies and motion neutralizes size, mass or weight. Mass counters motion being when the Earth restrains further motion of the falling object and the moving object is stopped from further movement where mass is then preventing or hindering gravity. This is the result of objects claiming an individual and personal claim to space occupied in a dual or in fighting for their individuality and independence of each other while wanting to be in the centre of the Universe. While falling or moving there is no opposition to the body being independent. When the motion seizes, the falling object remains individual and still tends to move while Earth individuality resists further movement of the falling body's movement. Further movement is disallowed as other material fills space that falling body wants to lay claim to. The only manner to remain independent by the falling object will be to relinquish to motion in the securing of mass as a substitute to motion where it then finally comes to rest. Mass then sets in not causing the motion but substituting the motion and from that motion restriction becomes resistance that becomes mass. While falling the object is experiencing gravity because the object is in gravity but when on the soil the object experience mass which is the restricting of gravity or motion by other space filled with material.

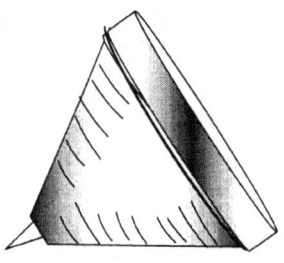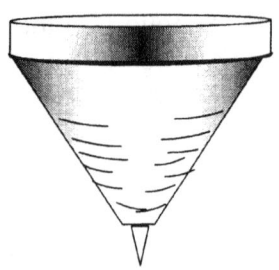

Looking at the top spinning and not spinning, brings a question to mind: why would the motion of the top beat the gravity of the top and that of the Earth hands down when the top is spinning. Surely the mass is in effect while the top is spinning just as much as the mass is in effect when it is not spinning and yet when it is spinning the pulling subsides to give way to free the top from the mass restriction and charge top with much excitement. The excitement is so much it seems to relieve the top of the pulling there is between the top and the Earth. That even strengthened my suspicions more about the fact that gravity is motion. By spinning the top finds additional motion on the side of the top that brings the side of the top in a more favourite position than was the case before the spinning commenced. The top finds additional gravity in the spinning and the gravity in addition has to bring reconsideration to the position the balance in gravity sets in margins. The top secures a better margin or stretches the parameters of location because the top inherits the Motion of the Earth and in addition to the gravity motion that the Earth provide, the two secures more gravity motion. With more gravity motion in addition to the Earth's gravity motion the top has to have more gravity motion than what the Earth has. That will allow the top the spin while facing such enormous disadvantage there is on the side of the top in mass difference when considering the size disproportions. Let's face it, if it was about mass pulling the top in relation to the top pulling the Earth, then the top had no chance ever to move by the very slimmest of chances there may ever be.

However if it is the gravity motion that the top inherits and with the aid of what motion life can add in addition to the motion already the which is the motion the Earth already contributes and considering the mass the Earth provides then it will not require that much a bigger effort to get the top going. Singularity charges motion by instigating motion without ever moving. Coming from a spot to a dot and then producing a line running from dot to dot though a spot signify the birth of the cosmos. A line holding time by the square and by the square of ninety degrees announce a birth in the Universe of a birth of the Universe. That which was not there suddenly is there by not being there. That which was undetectable suddenly is detectable by being undetectable. It is not my forte to write riddles but in this case there is a Universe within a Universe, which is not in the Universe and does control the Universe from a point no one may ever locate inside the Universe. The Universe is built up by innumerable dots and each dot is charged with being the Universe while being in a representative position since it is not in the Universe.

Space-time is a four dimensional position of the Universe where the position of an object is specified by three coordinates in space and one position in time. According to the theory of special relativity there is no absolute time, which can be measured independently from the observer, so events that are simultaneous as seen from one observer occur at different times when seen from a different place. Time must therefore be measured in a relative manner as are positions in three-dimensional Euclidean space, and this is achieved through the concept of space-time. The trajectory of an object in space-time is called world line. General relativity relates to curvature of space-time to the positions and motions of particles of matter.

In view of the definition of space-time I wish to elaborate on my view of singularity and my deriving of space-time from the likeliness that singularity may produce space-time. In the past singularity was mentioned in the manner one would speak of a ghost hiding in a haunted Black Hole. Let's put singularity in the clear. Singularity is within every sphere due to the natural shape or form the sphere is committed to.

While the toy top is, spinning one will find singularity by moving the rotating line or radius progressively to the middle by reducing the length the line has from the edge to the middle. At

one point all further reducing must end but the ending cannot include zero or nothing because the rest of the line still attach the rest of the top.

Locating and finding the presence of singularity

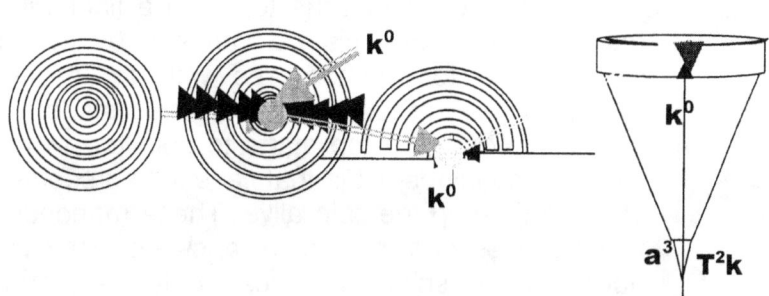

What is in the Universe is spinning. In the **precise middle** of all **objects in rotation** is a precise centre dividing the object in sectors that will **start the spinning initiation** from that centre point.

$k^0 = a^3 / T^2 k$ states that whatever is, is also spinning in order to be present.

Thus, the spinning object **will have a middle point,** a very specific **centre point that does not spin** and only holds Π as a specific value because no radius can apply. But also the one value such a line **cannot have is zero** because the line **is there and holds contact** to the rest of the material bringing about that **zero does not start any** line and therefore the **value of the line must be infinite,** just as described in accordance and by **the definition of singularity**

As I am introducing a very new idea, I wish to explain in better detail what I try to convey. **That point** albeit hypothetical, is also as much a reality none the less and is placed where that point **must be standing still** because every line **running from that point in opposing directions** is also **in opposing directional spin the other or opposing side.**

As the rotating direction moves inwards, the rings holding Π will become smaller and smaller. The reducing of the radius r will eventually end where the spin direction ends at Π^0. However that point where the directional spin ends is the point where the actual spin takes place. The spinning is on the precise location the point is not spinning.

The definition of space-time is as follows:

According to Einstein singularity is a mathematical reality within the Black Hole but much more so in every sphere. Einstein may be the first to name it and Galileo (unwittingly) may have been the first to define it as Kepler was the first to formulate singularity, but in mathematical terms singularity is the most basic principle. At this point I wish to establish a fact that seems lost in all other grandeurs of cosmology. When tracing the radius down into the sphere the radius stars where all lines start and a straight line cannot begin at zero or nil it can only start at infinity. Such a statement will hardly seem appropriate but the relevancy of this fact has no limits. If gravity is motion then motion starts with a line. Let us follow the line as motion abides by the rules of the line.

POINT OF INFINITY

If the line started at zero there was no line to start because whatever results in zero as the answer. That must also be starting point. Einstein introduced such a point and named singularity. When looking at the cosmos from whichever indications lead to the fact that the whole cosmos is in entirety. It is forever spinning and it is going too as much as Everything is on the move and always encircling something

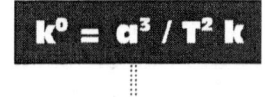

$$k^0 = a^3 / T^2 k$$

zero multiplied by the cosmic that point angle all motion in its it is coming from. of greater importance. A top can spin but the parameters of its spin are limiting the motion it can apply. By not spinning the top is still spinning as the Earth is doing the spinning on its behalf.

When the top starts by spinning to fast it is clear that the top is in a fight with something that is restricting its spin. This we can see by the re-aligning and the swaying the top manoeuvres with to try and circumvent the restriction. As the top spins something starts to tarnish and erode the spin. When the top starts to spin too slowly the top tries the same manoeuvres but in that case it then seems as if the top is in a struggle to keep the spin alive. These manoeuvres the top display triggers questions in need of answers. Why would the top stand upright when spinning? It can only be that the spin activated singularity into manifesting the gravity of motion.

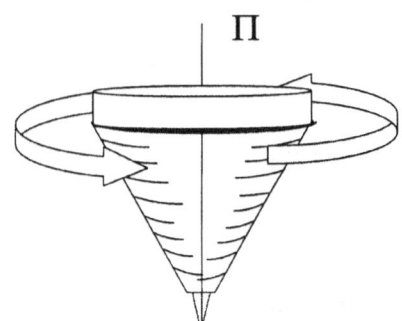

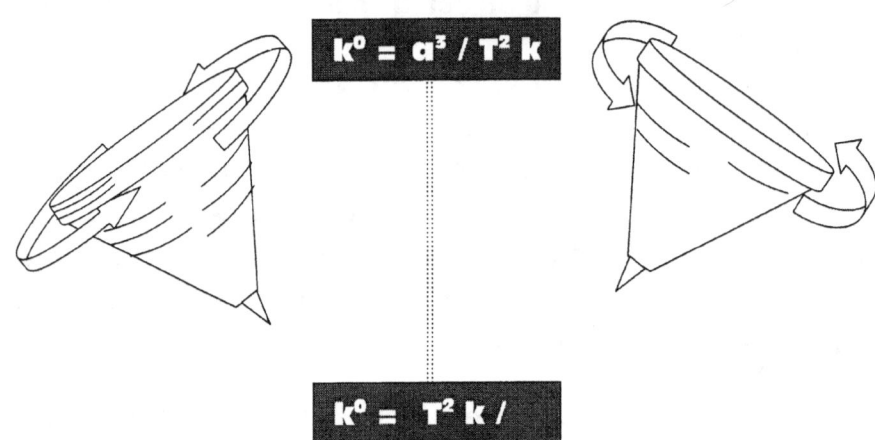

$$k^0 = a^3 / T^2 k$$

$$k^0 = T^2 k /$$

The spinning of the top is all the evidence one needs to come to a conclusion that the motion establishes a drive and the drive establishes independence of the surrounding space the top finds a control over. The top divides the Universe by establishing a generated Universe independent of the Universe we think of as the Universe.

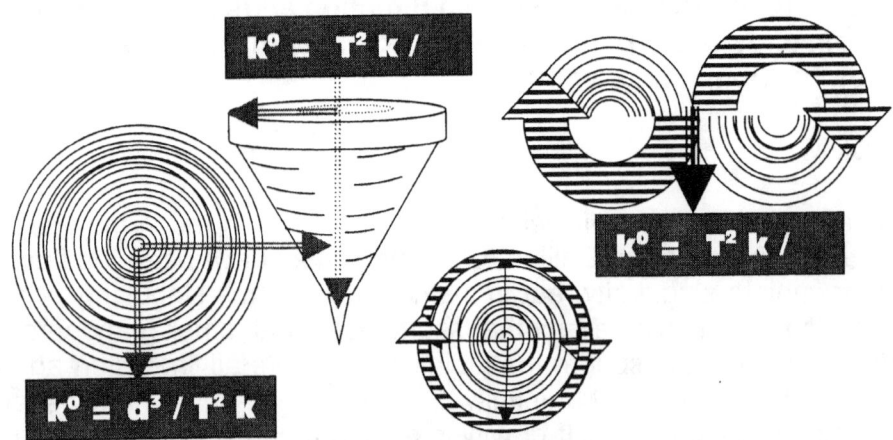

$$k^0 = T^2 k /$$

$$k^0 = T^2 k /$$

$$k^0 = a^3 / T^2 k$$

The centre is never there because the centre is eternally there.

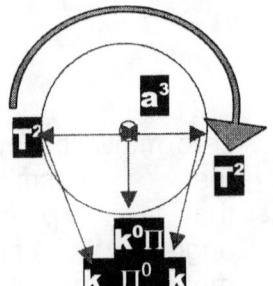

The centre is drawn into action by motion but the centre does not change in principle. That which activates the centre changes by directional contradicting of its nature.

The motion establishes singularity, which implicates the Coanda effect as much as the motion establishes the Coanda effect. The spin realises a space limit while the space limit attaches the motion in the form of time onto the space of the material, which then allows time inside connecting to time outside the space filled with material.

$$\Pi^1 \quad \Pi^0 \qquad \Pi^1$$
$$\Pi^1 \quad \Pi^2 \quad \Pi^0 \quad \Pi^2 \quad \Pi^1$$

$$a^3 < T^2 k$$

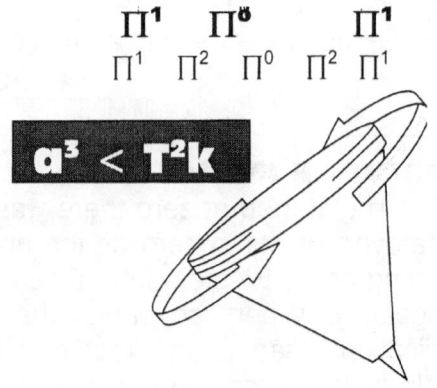

Judging from the behaviour of the top while the top is spinning it seems very obvious that not only is there a singularity running inside the centre of the top but another point holding singularity forms next to the top. It is this singularity point that seems to carry the restraining the top wishes to fine release from.

In dimensional terms, which I explain later on the value of $2k$ relates to T^2. That relation extends to the next value where T^2 relates to k, which relates to T^2. The first space in the circle will then be $T^2 k$. From the centre being in infinity, one can realise by applying mental power the single dimension factor not seen but present all the same. Extending that into the 3D comes six k and any one of the six will further extend to form a seventh point as T^2 All this is a multiplying of k^0 $= a^3 / (T^2 k) = 7$

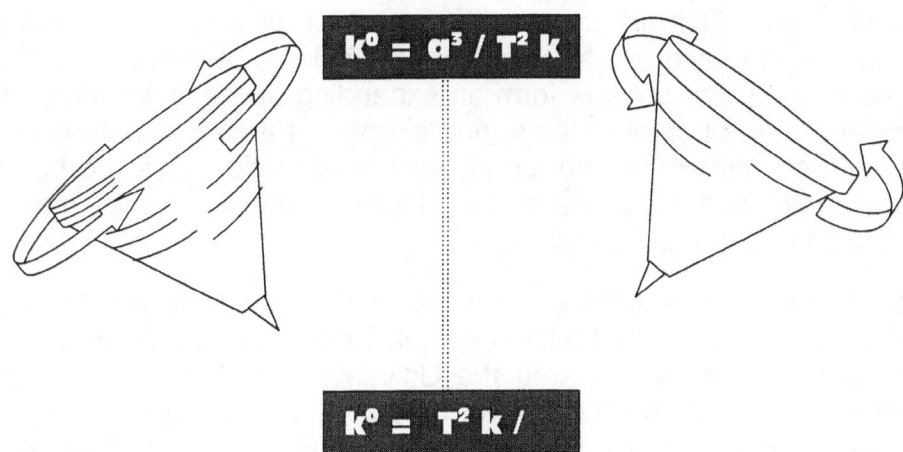

$$k^0 = a^3 / T^2 k$$

$$k^0 = T^2 k /$$

The Coanda principle indicate that the gravity described in the previous page is generated by motion of liquid in relation to a solid anywhere motion can produce gravity. There is no mention of mass because mass is a derogative of the gravity which the motion creates.

A centre is formed where the surrounding space-time forming the one group is relating a position from the "centre point". That forms one inclusive relevancy between points within the gravity field. The gravity field is holding "back" and "front" running through "the centre" where the other line is relating from "side" to " side" running through the "centre point". The fact of the line in the centre is that "it is there", but we cannot see it. Try as you may, no one will be able to calculate the very position that forms the lines, but as they change all particle characteristics, the lines are a reality as the spin of the matter is real.

Being too small to hold atoms, the space holding such a centre line is no space at all and with that knowledge we may presume then therefore what ever the line constitutes of must become part of singularity, where singularity is a spot in the centre with two lines crossing the spot at an angle of 90^0. That is the basis of singularity, and since all the positions still relate too a centre of a circle, forming a part of a spinning circle, Π must form the basic value.

The second major reality that one has to recognise is that the only way singularity was broken was by motion. The only way motion can come about and break space less ness is by establishing heat which establishes expansion and the Universe became a possibility and later a reality by expansion. The heat swell into space and the space swelling is the motion that produces the gravity we find visible in the Coanda principle. The space at first was presumably filled with material because the expanding could only be material. The Coanda principle alters time and establishes with such alterations to space-time a new Universe with borders and all. By introducing motion it sets a new time standard by which the space created will apply a newly generated gravity.

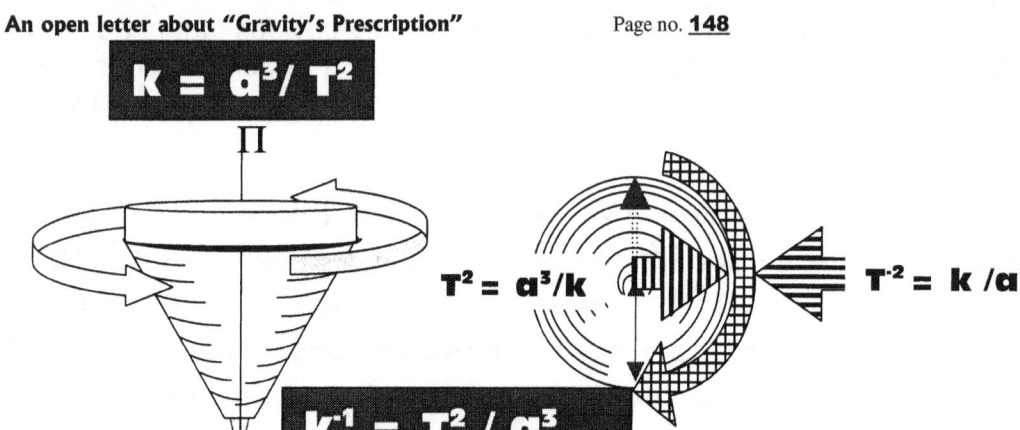

$$k = a^3 / T^2$$

$$\Pi$$

$$T^2 = a^3/k$$

$$T^{-2} = k /a^3$$

$$k^{-1} = T^2 / a^3$$

The motion activates singularity but also establish singularity by creating limits and borders. Those limits and borders serve as the gravity applying a differentiation the spin direction brings about by normal flow of rotational spin. This gravity establishing factor we call the Coanda principle and the division also divide the liquid relation as a factor from the solid as a factor. Wee find that the centre secures a point of control and the borders form an expanding of the limits. When the spin exceeds the limits, the expanding of the borders tries to find a way of release or relieve from the controlling centre. When the limits of rotation can no longer sustain a motion, it is the borders that become unable to balance the control and the centre control diminish the spin. The liquid reduces as it wishes to contract while the space claims as it expands.

We have to be clear about what we think of when we think of the Universe. Most people think of a picture recalling the black night sky when thinking of the Universe and that thought is most incorrect. Einstein was most correct when he declared the Universe was going flat where gravity is at its utmost, but the concern we should have is not with the mathematics being valid or not but with the vision about the Universe being what we think of and where we place the Universe. The Universe is in the centre of what is spinning and the biggest single particle that is spinning in total independence of the rest of what forms a total Universe is the atom.

The atom spins and by the motion the atom evokes the Universe forming what must be the group effort of all the atoms then spin by the motion the atom renders the rest of the larger Universe. The Universe is the part that allows the rest of what the Universe establishes to spin. What spin you may ask. Kepler said it without saying it: $k^0 = a^3 / T^2k$ and not even Einstein with his super human mathematical skills could say it better or more accurately.

The motion established by singularity results in the implicating of the Coanda effect as much as the motion establishes the Coanda effect. The spin realises the space limit while the space limit attaches the motion to the space in the time within the time.

With the top spinning the Coanda effect steps in and do justice to Kepler's formula.

Time is always a displacement of space in relation to the implication of singularity, and comes about between two points in space relating to the centre of singularity as positioned by **k**, either to the value of **k** or to k^0.

With the top spinning as it establishes the Coanda principle it brings justice to Kepler's formula.

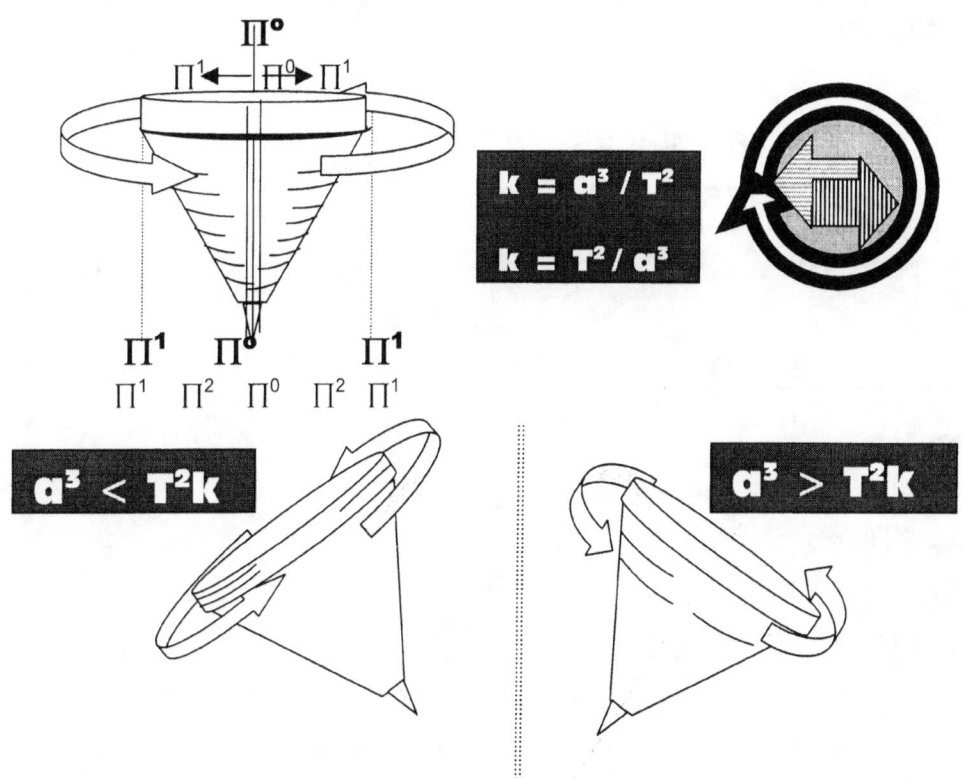

$$k = a^3 / T^2$$

$$k = T^2 / a^3$$

$$a^3 < T^2k$$

$$a^3 > T^2k$$

Time is always a displacement of space in relation to the implication of singularity, and comes about between two points where time forms time by being divided by space. Time in the centre is generated by parting from time in motion where that establishes space as **k** departs from k^0

We not only have to start thinking exclusively space-time but we have to stop not thinking space as a substance standing apart from time because it is the same thing. The one is not merely complimenting the other it is providing the other with substance of being something. Without space there is no time and without time there is no space. It is having a left side when there can only be a left side when there is a right side to match the left side and without the right side the left side disappear. We cannot accept space without accepting dimensions and by accepting dimensions we accept space because time is the duplicating of sides to form the space that time is matching. It is space doubling or it is time halving but it is the same thing.

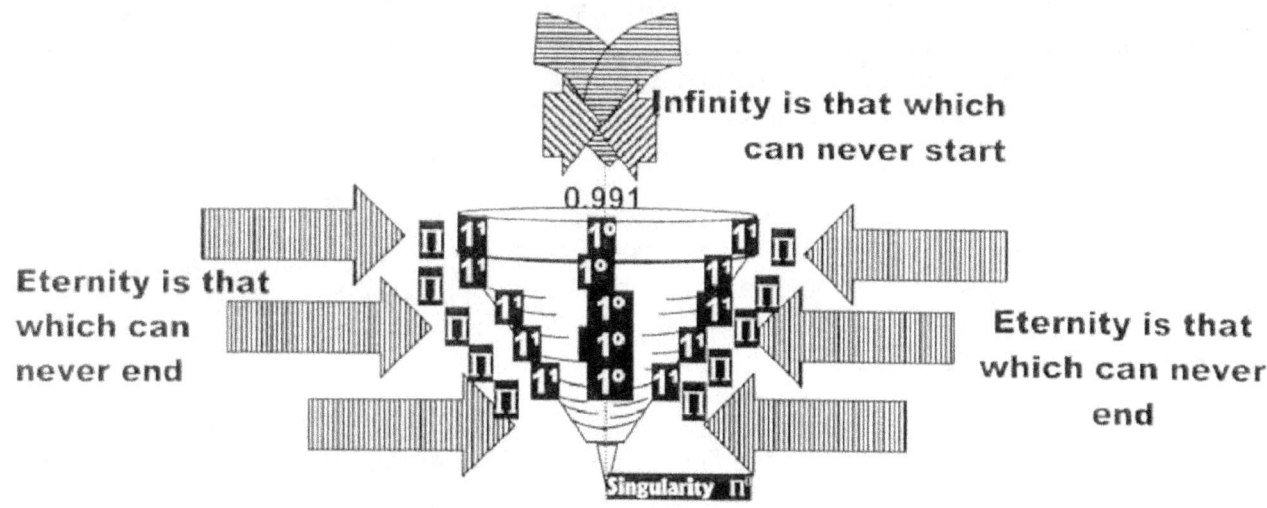

When space falls away inside a Black hole this comes about as time goes eternal inside the Black hole. The Neutrons leaving the Neutron star provides the Neutron star with time to dismiss the neutrons as it finds the time to provide the neutron to space to leave. It is not only modern and cool to think space-time (as I one day herd a lector tell his class) but it is of utmost understanding the concepts to think space-time because there is no other way of thinking than to think of space as time and time as space.

Singularity by Time

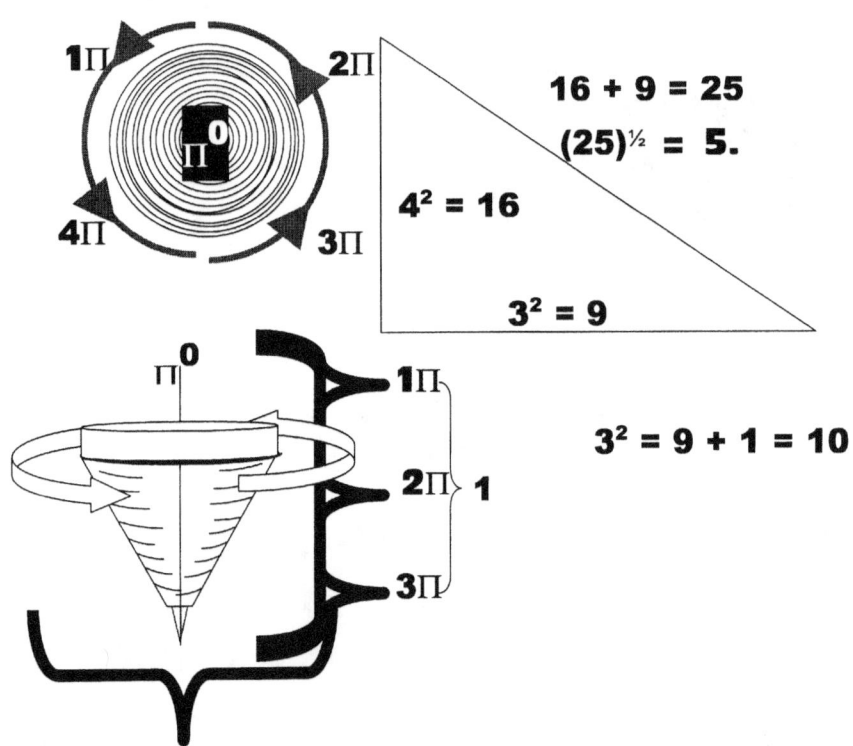

At the point where space began time began because time and gravity is the same thing. It is motion that is creating space. That is why the universe is so today is what it is. It is antigravity applying heat and will always expand. It is securing heat to cool   the rest through applying gravity or space conservation. But material grows   as space grows and as **k** extends   so   will T^2 and a^3 extend. When the dimension walls reduce space reduces   but   so does time increase because the gravity providing the density claiming the space increases. Space compressed is heat denser and in that is time because time is

motion of heat in space. By accepting that there is some conductor (not the ether of old) between two cosmic structures can one accept that there are certain invisible undetectable influences on the edges outside the surrounding of material. One then can see how space conforms as it converts to liquid heat. In the same effort one can see how material confirms heat from space to

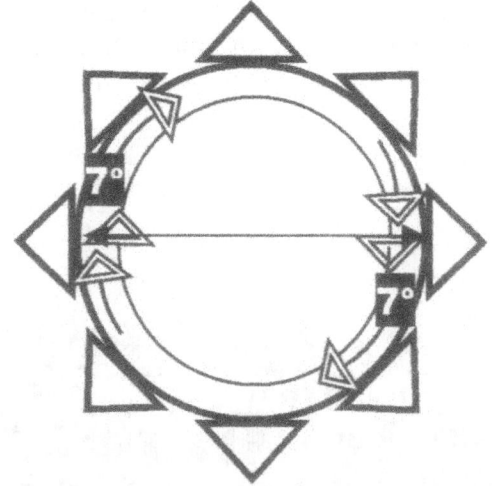

material. By reducing and confining the heat drawn to the centre the space becomes more concentrated as the heat levels begins to rise. Take any bicycle pump and compress the plunger

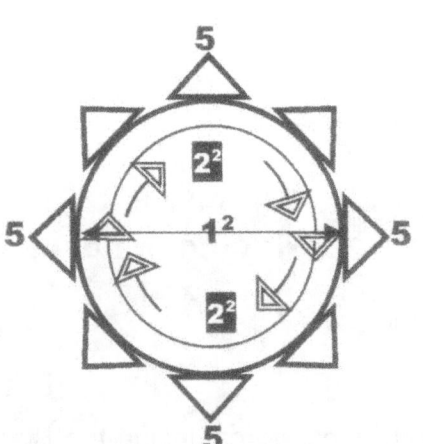

and the result will be that the heat created by such action will burn the finger you use to cover the valve hole. The heat comes about from concentrating the space, which holds the air. But the air does not concentrate because one does not bring in more air than there was before. The relevancy changes as the space reduce to change the space back to heat. This action is the very opposite of an explosion. But reducing pace the action brings about that the space turns to heat. This is most crucial in accepting because this is the precondition about the understanding as much as accepting a new concept, which I try to introduce, and at the same time I try to produce a concerned effort in dismissing myths from cosmology.

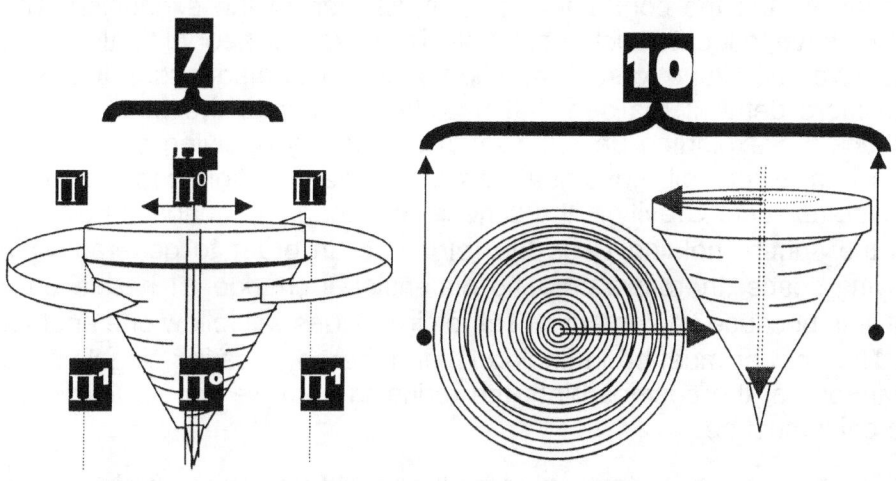

Gravity is about turning space into hotter denser space as it reduces space and that is the reason why there is no visible or measurable stronger gravity in the centre of objects. The centre does not indicate more gravity because the gravity that is the measure of the accumulation of heat that the gravity produce in that space $a^3 = kT^2$. The heat increase as the gravity becomes more intense because the more intense heat is the gravity increase. By the reducing of the space coming down towards the smaller area such coming down leads to space reducing, which brings about heat increases. The stronger gravity personifies in the denser heat produced by the reduced space. As space increases heat dissipates and that we find is what happens in an explosion. The bigger the heat release the more space becomes available as winds (shock waves to use the name hiding the truth) blowing across fields. The winds come about as space multiplies through the release of heat creating new space that was not there before the time. In the explosion will the heat decrease be the decrease of gravity that was before the explosion bounding the heat into condensed space and the explosion is the release or antigravity reducing the heat as it increases the space. In my search I stumbled on two accepted but not intergraded laws and when I found and located singularity the two laws became very much plausible and factual.

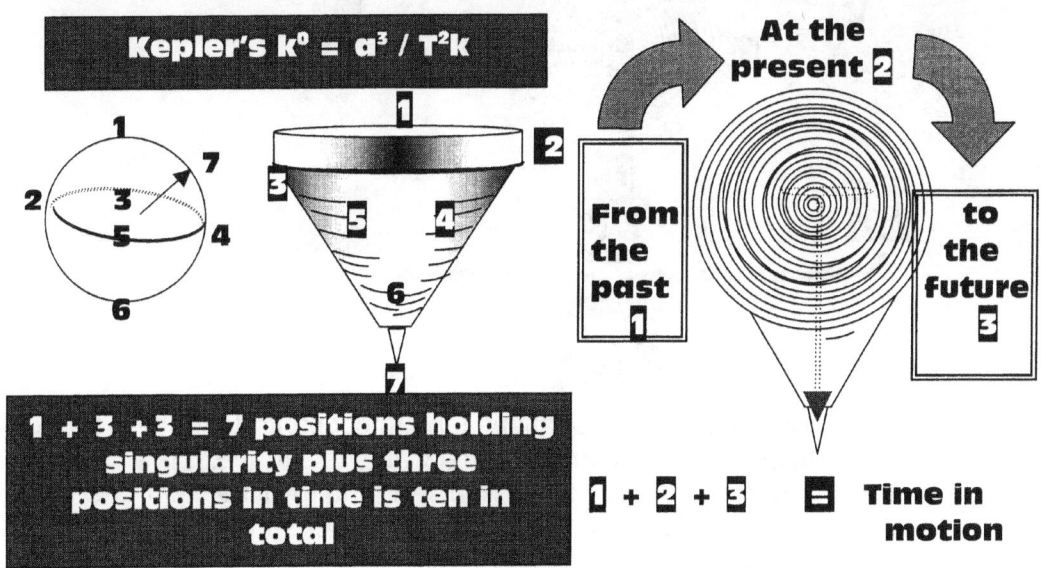

Kepler's $k^0 = a^3 / T^2k$

1 + 3 + 3 = 7 positions holding singularity plus three positions in time is ten in total

At the present 2

From the past 1

to the future 3

1 + 2 + 3 = Time in motion

Take a circle and reduce such a circle constantly to where it no longer can reduce. Reduce it to a point where only form remains part of the circle because the radius has gone beyond human measure and becomes so small it is not noticeable with what ever tools man may use, then what remains is pi since pi does not indicate size but indicate form, and form is all that then will remain. I believe one can begin too see where my suspicions are heading because the flaw comes about in the manner mathematics are practised for thousands of years. Space is nothing because that means space is a standard fit all issued out before the time. Space cannot increase and winds are ghost blowing their breath. Winds are as much antigravity returning reduced space back to increased space. Before coming to the mathematics I would first like to bring your attention to the practical side. I am promoting a theory in which I am able to prove there is as much contraction going on in the cosmic universe as there is expansion and the contraction is as much part of the expansion. The universe rides on a balance and we have to locate such a balance. To prove my theory I firstly had to locate the centre of the universe. Even admitting to such a notion sounds like madness, but please give me a chance to explain in more detail. I realised that my effort to locate the point holding singularity enabled me to backtrack the exploding universe to its origins. By applying some basic effort I have located the position from where all movement came and the direction it took moving forward in time...and yes, even time as such. Gravity is the dimensional changing of space holding r as reference in the cube as to the sphere holding Π as the reference. In order to generate spin producing time in matter occupying space, therefore creating dimensional change, Π has to be a factor indicating the possibility of spin because implementing Π the circle sides will follow one another without establishing separation. The answer must be in finding Π, and thereby locating singularity. If singularity is in affect the original point of the cosmos birth, the reducing path we should follow will indicate the whereabouts such a point must be.

In the normal applied mathematics there are two standard formulas used to calculate a circle. The one use an r to indicate the radius and the other use a D to indicate the diameter, which is double the radius and therefore needs to be divided by a four to eliminate the Newtonian inverse square law amounting to the difference there will be between the two. The one using the radius is Πr^2 and the other formula using the diameter is $\Pi D^2 / 4$. However one looks at the mathematical expressions and Kepler's formulating of space-time there is an exceptional difference between the two scientific uses. When investigating Kepler's formula one do find it appreciably differs from the normal Mathematical equation like $a^2 = r^2\Pi$ and $a^3 = 4/3\ \Pi r^3$.

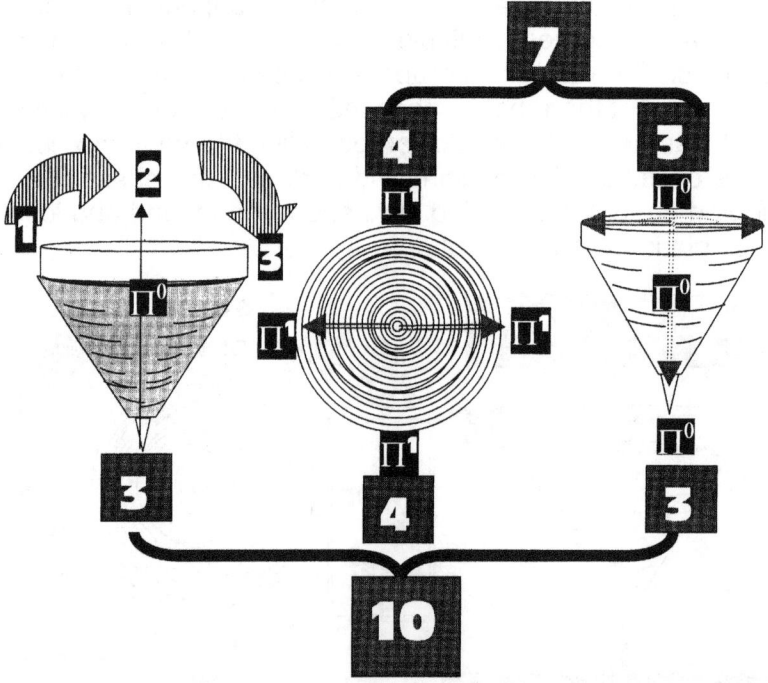

In the normally used mathematical expressions such equations tend to concentrate on the volumetric aspect. In the case pf Kepler's expression it is something else that wants to surface. It is another idea that is coming to mind. In Kepler's formula a^3 stands to symbolise the third dimension and such a third dimension becomes equal to two other dimensions grouping and sharing value to equal a^3 efforts. It is not the circle of the rotation because with such a normal circle the radius is in the square and Π evaluates form. Here there is no mention must of a factor Π, which one would suspect to be somewhere applying since the circle is Π and Π is the circle and the two are inseparable. But not in Kepler's a^3, where there is no mention of Π at all. The fact that there is a radius of some sorts used to indicate a position cannot hold the square as it normally does in the case of the normal equations. In the mathematical equation the factor indicating the position of the circle edge has the square value being called the radius or in some cases the radius doubles and which then is the diameter, and the circle indicator is Π. But in this event the formula value will bring about a square value to the answer one receives. It will bring a value to the surface of the circle. In Kepler's formula it specifically does not. I am not the one that brought Newton into disrepute. Before me the cosmos did. The comets with they're not colliding did, and so did Roche and Lagrangian principles. Hubble was another one and it becomes apparent that every one that made a study about matters in the cosmos was in some disagreement about Newton. By Newton's effort to improvise on behalf of Kepler Newton made a statement that Kepler never made. In all honesty nature reacted strongly against the claims Newton made on behalf of Kepler and not about Kepler's work but about Newton's modifying of Kepler's work. In short: how can a comet sail past the Sun time after time without colliding and still apply a contraction in the manner which Newton suggested by the one claiming a freezing grip on the other? This strongly contradicts F = G (M.m) / r² How can five structures as the LAGRANGIAN POINT form around a centre structure while the centre structure keeps the five in position at equilibrium? By rejecting Newton's improvising this strongly contradicts F = G (M.m) / r²

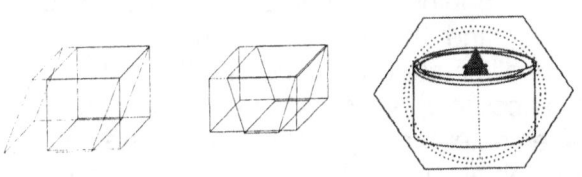

Anything occupying space in the cube will apply r and by r I mean just a distance not using ∏ because ∏ serves as a form indication while the collective product of r will determine form as well as accumulative dimension total. Notwithstanding the name used confirming the shape or r named as length width or height, it is all just a straight line bringing about the cube with all its other names that may find attachment to specific form but nevertheless still remains only a six-sided cube with connecting lines applying different angles changing in some cases.

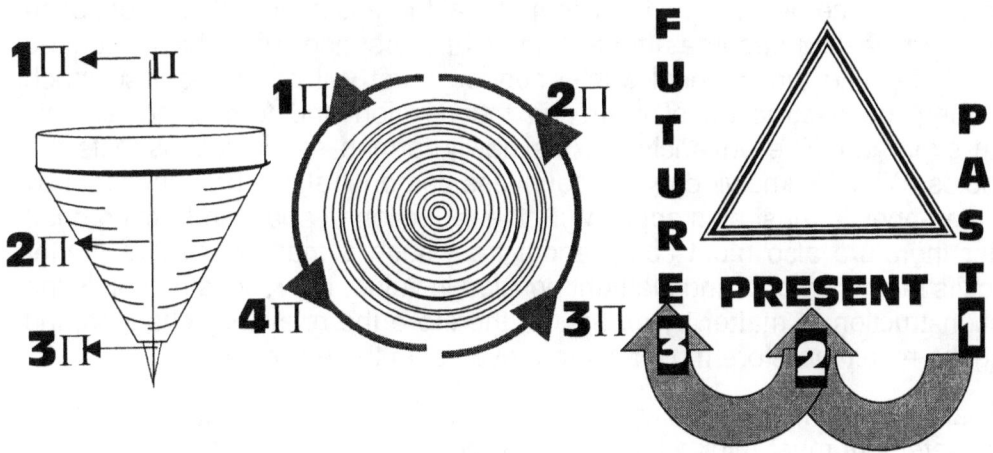

The normal perception is that any circle growing spontaneous would grow by the radius, which is r. In mathematics that may be true but it is not true in nature. In nature that cannot be the case because r is an indication of a straight line. By growing with the aid of a straight line from the centre to circle the influence that that would have on the circle would result in many circles following one another and not a continuous growth.

If we wish to believe in the Big Bang and we wish to accept the factor of singularity then we have to accept that there was a period where there was no space. We can backtrack the space to a point where the space is no longer space on the precondition that the space is coming about from motion. If more heat comes to such a centre the centre will produce more motion. The motion will produce more duplication of space and the duplication of space is the gravity we experience as a contracting direction of motion where as heating is the expanding of space through motion. But it had to have started with a space less motionless dimension-less Universe wrapped in singularity. The differentiation coming from motion is a dimensional barrier that changes many aspects in cosmology. The dimensions came about as the Universe came about and each had its individual introduction period. Space and time parted at **(∏³)²=961**, material formed identities at **∏x∏²x∏³/5=192** and **∏²x∏²x∏²/5=192** where space either had material or had heat without material and space separated from heat and matter at space holding **10/7π²/2(π²+π²)=139** material **7(π²+π²)=138** and space having liquid within **7/10 π²/2(π²+π²)=136** This is suggesting that these are meaning this was the first time liquid became part of the cosmos while all were still part of the same unit as the Roche principle would suggest **(π²/2)** as well as the **(7/10)** and the **(10/7)** .

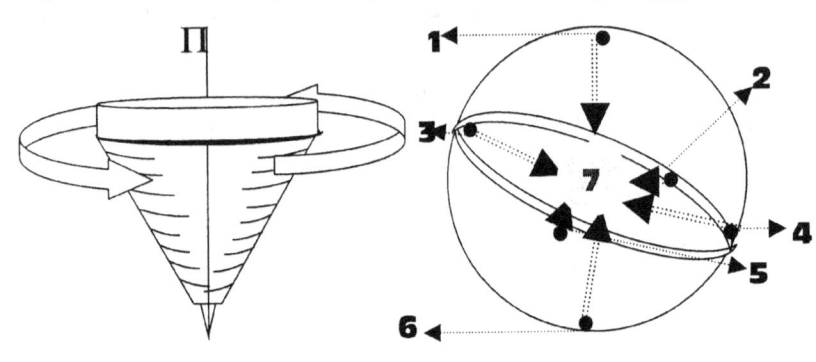

Material seems to get glued to the earth by some force, which holds the name of gravity. Moving such a gravitated particle needs some drag by motion. The secret of lessening the effort in applying motion to the object in need of shifting is reducing the drag that is not drag at all but motion not in motion and hiding behind the name of mass. Let us find this drag in nature and work from there to find a better natural understanding of being in a solid state on ground or a liquid flowing down into the ground or a gas floating above the ground. It is accepted by science that water can rub together and form static electricity. Can you believe respectable men say this shameless! The scientist that thought this one up had some or other big problem with his hair and thought that rubbing his hair with a plastic comb will be the same as water rubbing against water. I cannot believe that science can indorse such shit and shit it is! How can water form static electricity because lightning is the product of gas going into liquid by motion by turbulence of heat. Let's remove gravity and find mass.

It is well documented that where heat turns to space and space becomes more than the space which the matter is occupying before when the heating. By heating material there is an introducing of space because the space needed after heating becomes more than what the space would have required before the heating started. By heating the material with a sudden burst will bring about so much space available it brings along the destruction of matter in the position and form it holds. This destruction we know as an explosion. The advancing of the newly formed space reconstructs the position layout the matter holds. With the knowledge and countless demonstrations brought about by war and other destruction, the opposite must also apply. With the reduction of space in forming heat, the reconstruction of matter therefore also must come about from such a manner. Where matter removes heat to reconstruct its element worth and element position in value, the reconstruction is the direct opposition to the deconstruction of matter by explosion, therefore the relevancy changes and with the relevancy changing the result therefore must become reversal to the explosion.

In outer space an object floats. On the moon nothing solid will float but there is no liquid air either. Those not familiar with this statement must think what the difference is of space in outer space and space in the atmosphere and why objects entering the atmosphere suddenly acquire the ability to heat up and burn out. The reason why outer space is a gas is because the density applying between space and material is very little in comparison to what we are use to in the atmosphere. Outer space is not colder but hotter, much hotter. In space all objects are very loosely connected and move quite freely about. When this occur it reminds one of a gas because in a liquid there is much more density in the matter relation and when solid the matter is as close as can be found. Therefore the conditions in outer space form a gas and a gas is the hottest of the three conditions there are available to substance material. Comparing the likeness with anything we can compare too with our vision of what is on earth, we must move to something we all consider to be a natural in all three forms, one being solid ice (very cold), two being liquid water (less cold) and three being gaseous steam (very hot). Conditions in outer space come down to steam because there is much space between the particles bringing about more space in the density than there is material. By introducing heat to water, water changes from being a solid we call ice where there is much more material in the ratio between space not filled and material filling space whereas with through liquid such as we call water is more space unfilled than there is in solids but much less there is than in gas and gas we call steam.

By introducing heat to water we change water from a solid (cold) to a liquid (less cold and more hot) and with the introducing of much more heat we get the heat to become a gas such as it is in outer space. By introducing heat to water we get clouds forming. By introducing heat to air we get clouds moving, moving excessively, where the movement in fact displays a density increase. Because there is a density increase the wind can uproot large trees. To suggest that something as light as say oxygen and nitrogen can blow down a tree with the quantities present in such a density as one find in space proves how little science are able to think! With more increase of density in the wind we find spiral motion in lateral movement. With wind circling it has terrific density because in such a form it not only uproots trees but also takes on houses and much of what man can build. The wind in access blowing extensively produces the same qualities by producing destruction damage than water flooding in a river can match. When the density increase by adding motion much of the increase goes along with vapour, a form of air that is thick with water, (I distinguish the terminology because why not only use one word, either steam or vapour. After all it is the same thing!) In clouds we find lots of vapour but we find little water. The difference between water and vapour is that vapour has more unoccupied space and less space filled with water material in ratio. The thick density is there, but the air is so thick the vapour and the air combines to form a gaseous liquid we can see as a cloud. Remove the heat in the cloud (which the cloud needs to be being vapour) the water returns as a liquid and produce a form of heat we named lightning, The vapour liquefies too electricity as liquid heat and fall as a solid in the form of water we named rain. The liquid which was the water did not vanish but became liquid air which we call lightning. The question about density increase always comes from material being more prominent and more abundant in such a space. With the increase of density it always accompanies the increase of heat and the discharging of heat.

If one would think that it is vapour in the wind that increased to such extend that the wind can uproot trees, then why does hail with such a lot of solid water not uproot the trees. There is a world of difference between windstorms and hailstorms because of the abundance of electricity or more bluntly phrased heat in spinning motion. Windstorms having the ability too uproot the trees have a very sticky substance between the molecules and the more sticky evidence there are, the bigger the ability to cause damage. The air substance shows a bigger resistance to part or create space than the tree shows to remain secured by its roots. The substance can be sticky to the point where it breaks braches that will require an effort of many hundreds of Newton meter to break.

Even spraying the tree with water which man created artificially by pressurising the water would hardly break the branches and less hardly uproot the tree. If it is done, the water flow will be enormous but I have seen many times trees uprooted without one drop of water visible. The only logic remaining supplier of such a density increase must be the heat. In this there are changing relevancy dynamics, which I then introduced as equal to the substance found in atoms.

When one takes Kepler's equation into consideration the whole process of motion starts to make sense. It is motion that keeps the top erect and that was accepted from the time of Newton. Now however, by close scrutiny as well as considering Kepler's equation where the statement emphatically reads that the space a^3 is equal= to the motion of the space in relation to a very specific centre. But this relation works both ways and not only from one side. It has nothing to do with pulling because if the two were pulling the top had no chance of any motion. The top uses the motion of the Earth to the advantage of the top and then on top of the Earth's motion, the top applies individual initiative and claims even more independence than the structural independence it had before. If it were merely gravity of mass and nothing more, then the gravity disappearance the Earth produces would be such an imbalance that the top would never stand a chance of committing individual motion. However should my view be correct about motion and mass being the frustration of motion hindering the motion of gravity, then yes, the top by motion free the distorting of the Earth mass, which is a frustration to the top gravity motion and that would enable the top to spin even with such slight energy

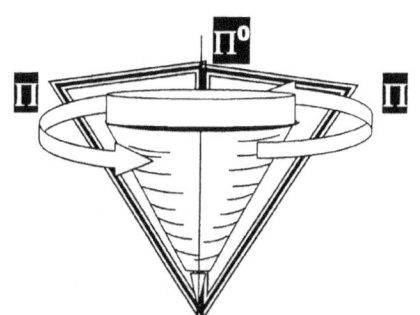

applied. Merely taking into account that the top has to overcome the considerable mass f the Earth brings the Mass theory into dispute because how can the top with such slender energy find freedom from the enormity of the Earth mass.

Every round object has a point establishing a very centre, a middle dividing one side from the other. That division determines the space from one side away from the other side. At one point there must be a point that does not fall on either side of the divide. Such a point will still be a circle, because from that side the circle divides into two sectors.

In every object there is a centre but the centre we find in the sphere is the only centre that can taker complete charge of gravity because it is the only centre that is controlling every point on the edge of the circle at any point the circle can offer.

In all units there is a singularity seeking independence in relation to singularity elected seeking dominance. From one side and any singularity there will be present in the unit one factor of singularity carrying the value of 1 but also there will be one divided by space square singularity absent because only one singularity can apply to the unit in dimension. Therefore, one tenth of the space is absent where singularity is one and holding .9991 valid as part of the other side of the Universe it is attached to but not connected to. With the unit being connected to motion the motion will stand related to seven from the one side as well as the full ten from the other side on both sides of the Universe. That relates as ten in space-time on both sides of the Universe (10 + 10 = 20) plus one factor of singularity present and one factor in the tenth not present (1+.9+.09+.009+.0009+.00009−.0001=1.99991)

Many dots spawned from the spot and such spawning was in flurries but equal.

Then the four cosmic pillars set laws and progress started happening as the cosmos applied and stuck to conditions set under these principles.

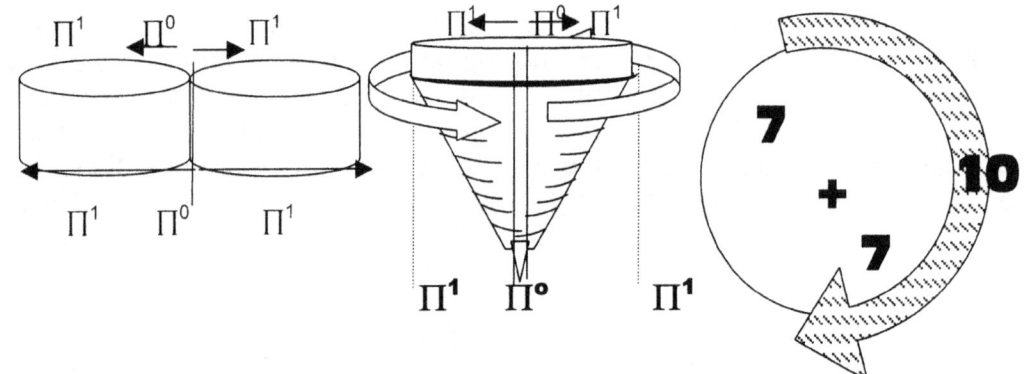

Some singularity formed a dominating role as they were in dominating some particles formed a subordinate role but space-time formed since there had to be motion and motion provide form which can be space. But I wish to place a distinction between form and space because by calling what was in progress space, the mediate human connection would be to adapt space by dimension to what was in place. If there were space it was Π, which is form. There were no mathematical equations because for mathematics to be in place there has to be space. There was no space of that particular type yet invented.

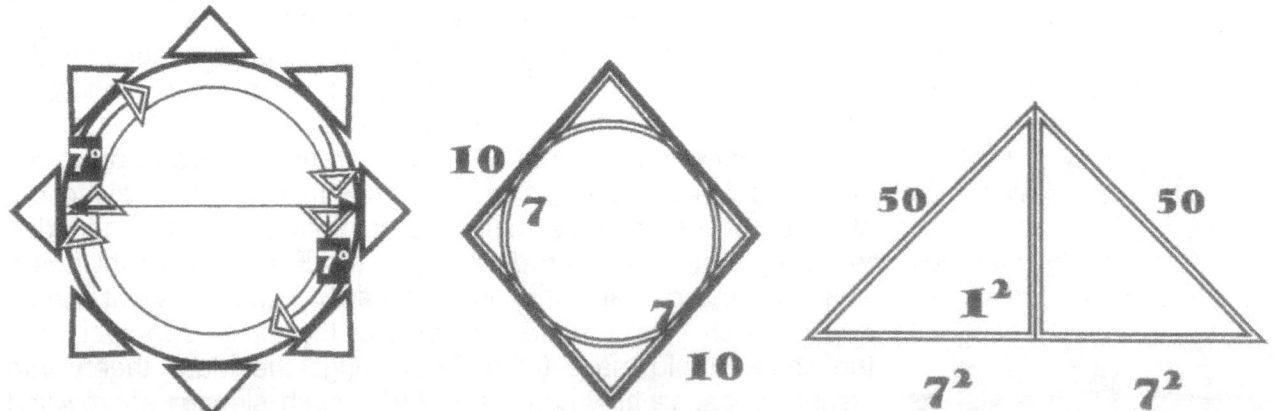

Please think clearly as this is very important, is that not precisely the commitment we find in gravity, where gravity is flowing from singularity outwards but never favouring any side? This reasoning prompted me to look for singularity in such a spot because if the prime spot from which all came was a spot holding all, then the spot must hold the shortest line but more prominent it will hold the smallest form including the smallest circle or for that matter the smallest sphere. With gravity always being in the centre of a sphere where the space is least available in the entire structure (there is not even space left to fill) one finds a flow of gravity from that centre spot outwards in all possible direction even-handedly. The fact that the original gravity will begin as a circle or will be a circle is the direction it will take when being the first spot created. All progress will be evenly in all direction because no direction will stand out or be in favour above any other direction at first.

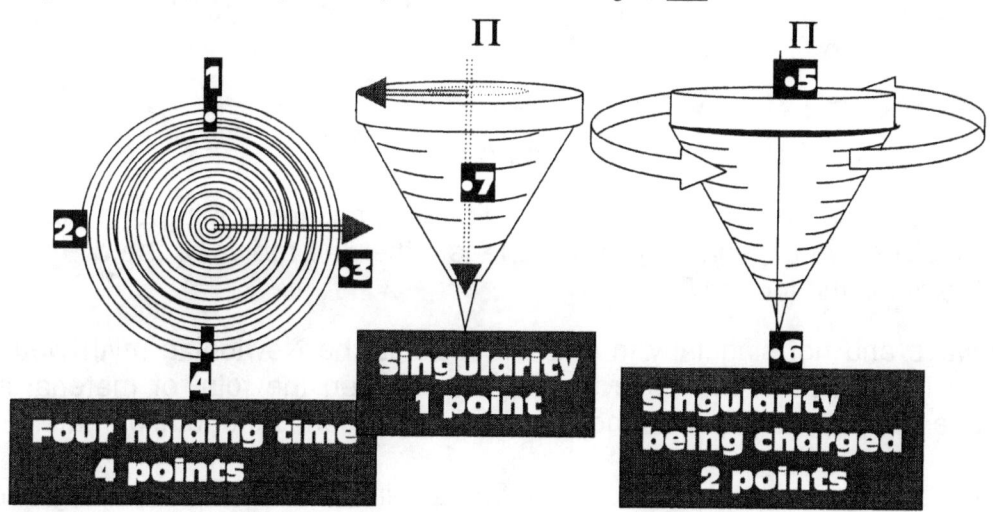

Four holding time 4 points

Singularity 1 point

Singularity being charged 2 points

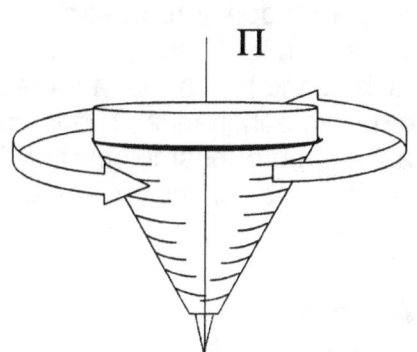

When the line **has a beginning and an end at the very same spot** and it wishes to extend the position as to further the possibility it has, which direction should it favour. Extending the line in any one direction will favour one direction without any clear reason not extending in other directions. The only mathematically sensible option about extending will be in all directions equally in order to give a meaningful non-bias flow of mathematical equilibrium.

That is where one would have to go look for the beginning of the Universe. The Universe is a about lines connecting but where does than connecting end. The Universe used form to this point in development, but then at some point the line came and established the presence it still has. The first form was moving from $\Pi^0 \Rightarrow \Pi$. Again I wish to press the issue, at that stage form was in use and not mathematics. The Universe was just simply too big to measure. If radius did apply, one could use r and r^2 but since only Π was in use there was no radius used.

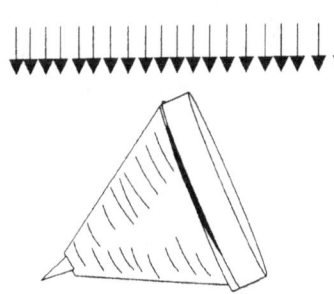

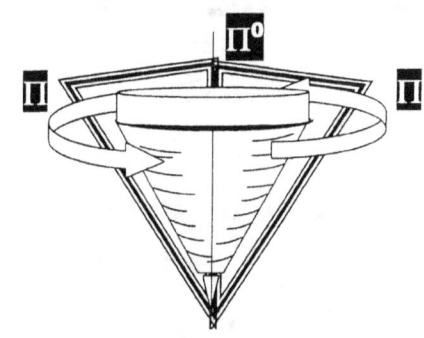

There is one Universe of difference between the top lying down showing no independent motion and the top spinning erect. A top on its side that is not spinning exists as part of another Universe. The top's entire Universe collapsed when the motion seizes to activate singularity in infinity. While being in a motionless state the top submits to the singularity lines running towards the centre of the Earth. There are gravity lines invisible as they are still they are there running through the top at 180^0 to the Earth placing the lines at 900 angle in relation to the top. These lines suppress the structure of the top to confirm in mass the gravity the Earth applies. The top succumb to the flow of the lines running to the centre of the Earth and it is the responsibility of these lines to eventually burry the top as part of the structure of the Earth. That is the purpose of the mass because by applying mass to the top the Earth will eventually have the top relinquishing its structural independence to the Earth and totally submit all individuality to become part of the Earth. But when motion is added to the structure the top holds another dimension of independence are accomplished and the top receive almost the same cosmic independence as the Earth has. It is the task of the top to uphold the motion and maintain independent singularity while the Earth singularity will fight to submit the chances of independence by restraining the top even to a point where the top is liquid.

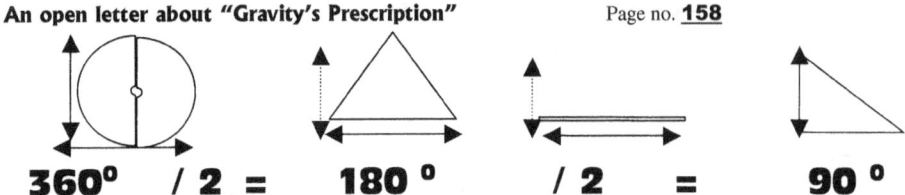

360°   / 2 =   180 °   / 2   =   90 °

The circle is a square holding a round shape, as the straight line is a square holding one side to infinity. Calculating a circle involves two aspects where the one is either the radius or the diameter that is double the radius. The other is the factor Π

Because gravity work both ways and not singularly in one direction as the Newtonian myth would have us believe, there is the interaction in the neutron position between the total of material in relation to time formed in space as space and time formed in space in relation to the total of material.

In contrast to Newtonian view about a spontaneous effort there are in the cosmos of joining and sharing, quite the contrary is true. There is a natural tendency to remain independent t and away from each other and where the tendency of staying apart is bridged, there is a tendency to destroy and conquer, to control and delete the lesser by an onslaught of the more superior. There is no mass fighting to join mass and to become one in all. That part is fiction as much as the part about a force is fiction. There is a struggle for superiority and there is a fight for freedom from dominance. The whole idea about masses joining and uniting runs very much against the basic fabric of cosmology and in particular the Big Bang theory, The sound barrier principle, the Coanda affect of motion bringing about space-time control and so many more.

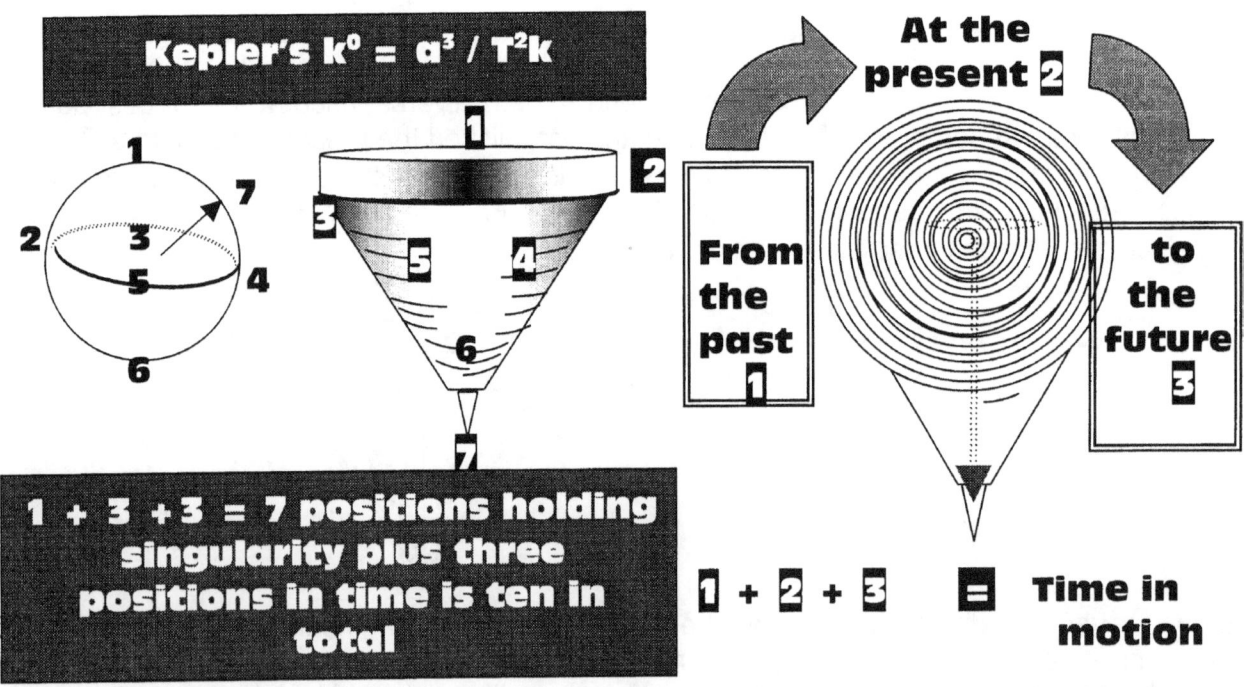

If the joining were with merit, we would by now not have known a moon orbiting apart and on its own coarse around the Earth around the Sun around the Milky Way. While there are those attachments they are only attachments and not obsessions of joining and uniting. The moon holds a separate identity, which it refuses to relinquish.

This refusal we call a lunar cycle.

The moon is on a running spree ever since it's Independence Day. The moon is taking a route that would progressively carry the moon further way from the Earth as the Earth is rerouting its orbit further way from the Sun. The question to ask is what makes the Sun, the Sun and the Earth, the Earth and the moon the moon. It is $a^3 = T^2 k$ or better put it is $\Pi^3 = \Pi^2 \Pi$.

It is the space-time as collected by singularity using the Coanda affect. If the joining were with merit, we would by now not have known a moon orbiting apart and on its own coarse around the Earth around the Sun around the Milky Way. While there are those attachments they are only attachments and not obsessions of joining and uniting. The moon holds a separate identity, which it refuses to relinquish. This refusal we call a lunar cycle. The moon is on a running spree ever since it's Independence Day. The moon is taking a route that would progressively carry the moon further way from the Earth as the Earth is rerouting its orbit further way from the Sun. The question to ask is what makes the Sun, the Sun and the Earth, the Earth and the moon the moon. It is $a^3 = T^2 k$ or better put it is $\Pi^3 = \Pi^2 \Pi$. It is the space-time as collected by singularity using the Coanda affect

The spin was going on for eternity because the spin does not apply, it has a value of infinity and infinity at the time was combined with eternity.

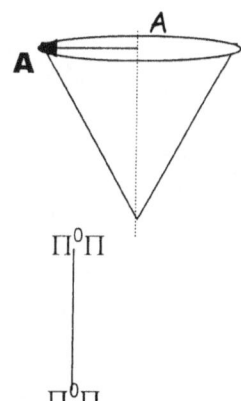

The moving of Π^0 to Π activate a line that was not there before. It was Π^0 moving to Π that evoked a line but the fact of the matter is that the line is still not there. The line shows a presence in *Singularity (A)* establishing (A) and that shift announced the rise of another Universe. The movement did not do nothing as Newton would indicate, but the motion evoked the birth of an entire Universe surrounding singularity.

The motion brought the top into space just as Kepler announced with his formula $a^3 = T^2k$ where it says space is produced in equal measure of the motion of the space...and the top is the undeniable proof of Kepler's statement. The top by motion brings space into the Universe and without motion the space is denounced as a Universe by the Earth motion. Should the motion of the Earth end all space accountable to the Earth will seize. It is once more proof that time cannot stand still as Newton, Einstein and Mainstream science would declare because if time stands still, all fall back into and to singularity. Singularity is one being $k^0 = a^3 / T^2k$ and if T^0 then all the other factors would follow the same path.

The top has space but there is a space in which the top spins that covers the time part of space. That is the time part Einstein identified (1) as coming from (2) being at and (3) going to and the position holding singularity is represented by the entire body that holds all the space of the spinning top. As soon as the spinning of the top commences, the time aspect releases space in which the top spins from the space holding the time of the Earth and the rest of the Earth within that time. It is this space in time that becomes so hot when the aircraft is speeding because the motion takes the time back to what the time was when time was nearer to the Big Bang. By receiving space, singularity received a value from where it was in eternity Π^0 to just one point outside eternity as Π. But the motion of seven relating to ten brought about gravity as Π^2

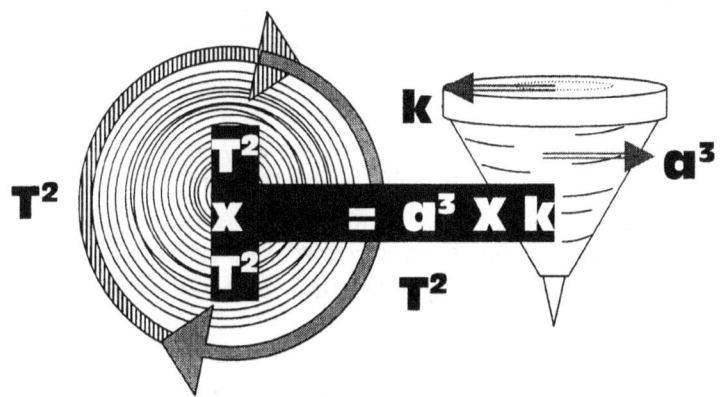

With everything in a cube or a circle or a potential of the two, brings about the implication of eternity in a form of singularity or the point of creation. Removing the radius of a circle does not remove the circle, because the circle is there, securing the ring. If the line (or imaginary line if you wish) holding the value of $\Pi^0 = 1$there has to be a point where the circle is no longer in infinity but claims existing outside the imaginary. At that point the radius may be lightly more than infinity, but to all calculating purposes it still remain as infinity. The spin was going on for eternity because the spin does not apply, it has a value of zero and zero is another expression for eternity. The full square of the motion $T^2 X\ T^2 =$ is equal to the full space $a^3\ X\ k$ created by the motion and that relevance became the atom.

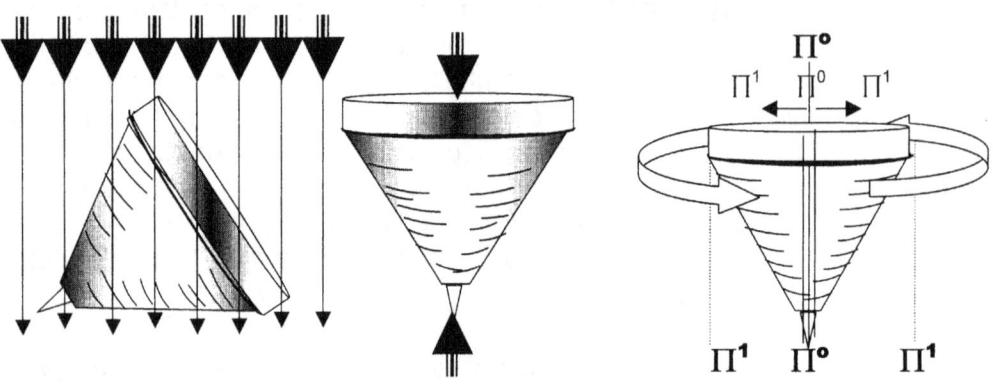

Singularity form lines running towards the centre of the Earth.

Singularity by Time

Singularity in eternity

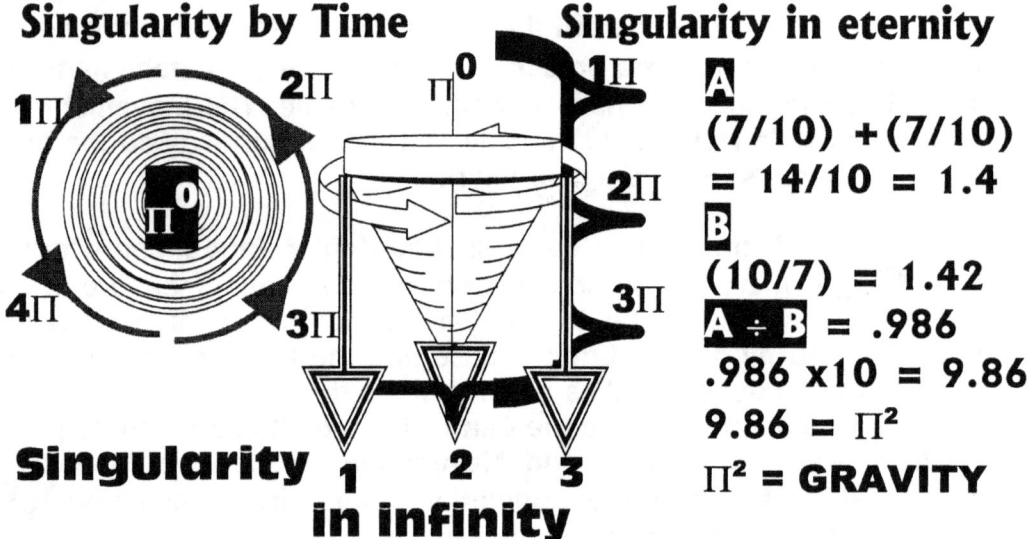

A

(7/10) + (7/10)
= 14/10 = 1.4

B

(10/7) = 1.42

A ÷ B = .986

.986 x10 = 9.86

9.86 = Π^2

Π^2 = GRAVITY

Singularity in infinity

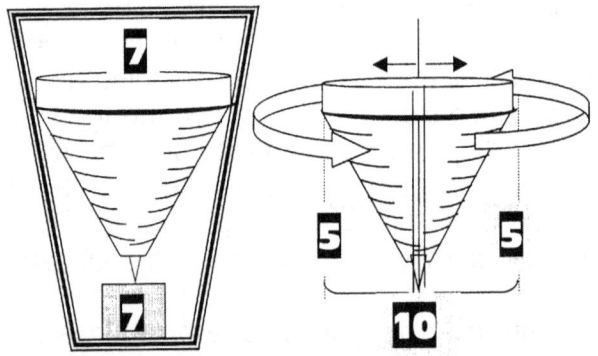

Having edges where Π^0 duplicate to present the edges singularity lost the value of Π^0 to the value of Π^1 with the same value singularity had being Π^1 to the one side and Π^1 to the other side, Π^0 must be the point splitting singularity into two parts of eternity, the eternal value of the first dimension outside eternity. It was the square of Π^1 being Π^{1+1}. That was the first dimension outside singularity Π^0 where singularity has a value of Π^1 in the form of $\Pi^{1+1=2}$. The first claim to space had a value of Π^2. This applied to both sides of the claim to space outside singularity, and the double proton became the dominant factor on matter.

The seven is part of infinity parting eternity. Infinity is a point that one find inside any and all solid spheres and the point is where all lines cross at a 180^0 as well as 90^0. The points form as rotating motion establish a charged line that take control of space –time. Then where singularity at seven points ends and the eighth point holding singularity begin another sector comes into action. This holds points in time eternity at eight nine and ten. The points are physical but needs to be generated while it never actually is there.

In they're using of such logic makes science appear foolish. Since the time of Newton, the arguments made by those in the time of Newton tarnished from being brilliant to clever to fair too poor and a hundred years ago it reached the point of being stupid. That is what Kepler's formula is all about? That is what Kepler indicated with his formula $a^3 = T^2 k$. The space of an object (a^3) is equal to the time (T^2), which it is in, in every given instant (**k**). If the space becomes smaller, the time duration becomes longer every instant of time's progress. The motion follows the graph in relation to motion and time.

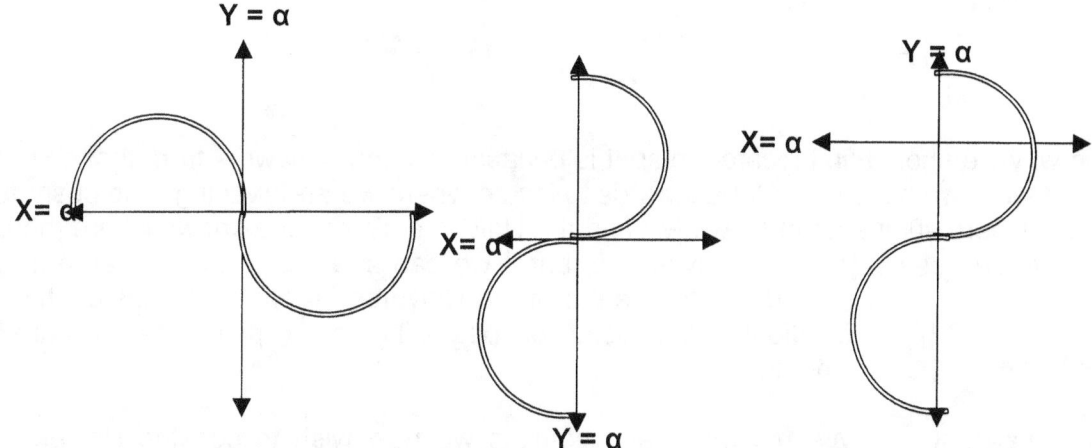

From the graph one can establish the link in the circle's rotation around a conforming unit being singularity. Saying that one therefore has to admit that the smallest spot has to hold space because the most insignificant dot can transmit light and being able to accomplish that, one must accept it to carry a value of something. If that spot had the value of nothing, it means that spot was not there to begin with. If the graph connected by zero that would mean there is no connection at all. With no connection the graph would be a mathematical tool with no value or use in any way.

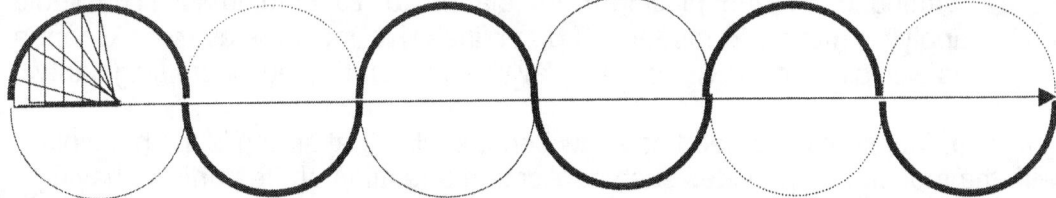

Holding space-time one should return to the original formula indicating space-time in as much as $a^3 = T^2 k$ where a = R and T = T. Being time it has to alternate positions and that can therefore only apply to **k** where **k** will indicate a relation to the space-time in question or the relevancy to singularity being $k^0 = 1$. By receiving k on top of the already $k^0 = 1$ that is in place the top becomes an atom by erecting the line of singularity from $k^0 = 1$ to $k^0 = a^3 / T^2 k$

It started with a dot, because that is the only form, size and dimension mathematical logic will allow our brain to accept. From the one dot had to come a second dot and a third dot. The dynamics of such a dot is smaller than we can understand because such a dot is in negative relation to what we see Π to be, and the deeper we delve in finding the smallest fragment where space started, in the spot where time is still eternal as much as we can accept eternity to be.

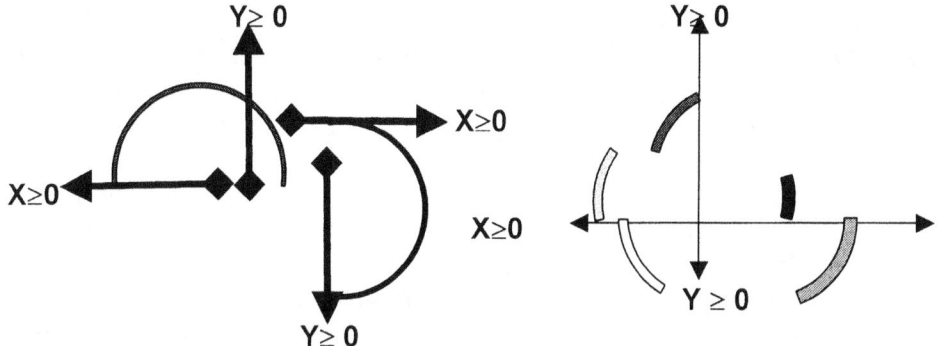

The graph with no connation between points because zero or nothing connects the points will render the use there of in mathematical terms quite obsolete.

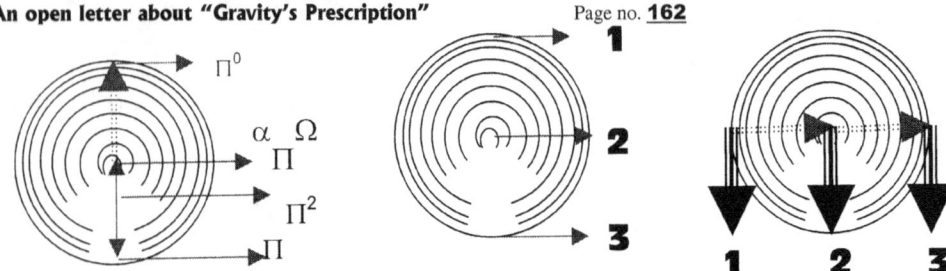

The reason why we should first locate the spot is because we can only work from that point forward. By working forward we have to work backwards to locate where we are heading. The cosmos started at a point and where such a point is, we will find the Universe. Every one knows where the Universe is, because we can see where the Universe is, but if we can see where the Universe is, then we should find the centre of the Universe in that spot. Einstein theoretically positioned the point of beginning at a place he indicated where singularity should be.

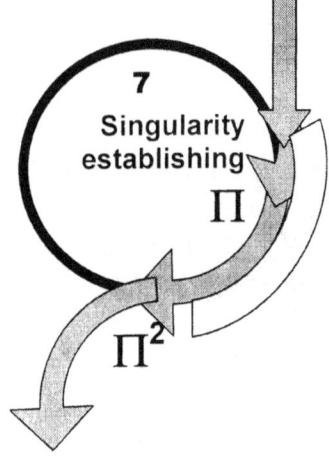

We not only have both as we now wish to see the Universe. By the duplication it therefore insists on relevancy because without relevancy there can be no motion and no motion means no space. The strongest proof there is about this is the manner in which the Coanda principle applies the reproducing of space taking shape from a round object and involving motion to produce such duplication. The relation forming the duplication of singularity is a duplication but applies as a dimensional forming of Π and placing 7 in relation to 10 forming Π^2 The liquid applying motion forms the 10 disciplines. No motion leaves no Coanda as well as no gravity because gravity is motion that duplicate singularity

The Coanda principle which in fact should be seen as a law because it is that strong is the principle of gravity duplicating with the motion that provides such duplication a relation of the particles having the seven factor and such a factor of seven produces through motion another three dimension. This total that material fill while in motion is ten and when ten crosses the line of singularity too duplicate the seven in the other side of the Universe the crossing cuts singularity in two as much as it puts singularity in the square. But it involves the motion of concentrating space to be or hold fluids around solids that may or may not move. In this must be a solid, a round basis Π, fluids concentrated in space and motion applying to one or all of the factors.

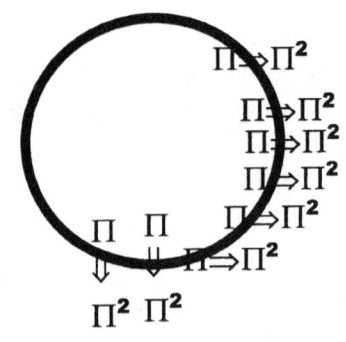

Conditions that prescribes the enactment of the Coanda effect is that the one surface has to duplicate singularity by establishing Π as a form. The round surface Π will bring about the shape of singularity Π that becomes enticed by the action of the motion of the liquid or of the solid or the motion of both around Π, which then establish and confirm singularity by form. The next factor is the presence of liquid. Air or atmosphere is liquid and water is liquid. Heat is liquid. The third factor being just as important is the motion establishing Π^2 by duplicating singularity as singularity becomes relevant through the applied motion that produces gravity from the singularity spot that provides the form.

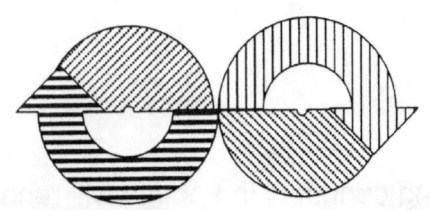

That too forms the answer about the question concerning the Titius Bode gravity implicating of cosmology. The seven sides are linked by rotation nothing changes because there is a steady linking to the inside centre of the sphere. But it is to the outside that this rotation brings about dimensional complications. There are five T_1 points moving to five T_2 making contact with five moving points. The moving non fixed points is the point before reducing by five to the point after reducing by five that bring along the ten points in stead of the five to one point as it is the case with the Lagrangian system.

The heat brings about expanding singularity from a one sided affair to filling a volumetric Universe. But all of it is a relevancy where ten positions will sacrifice individuality and compromise singularity in order to secure two positions in singularity. The spin that comes about from such expanding and the duplication has the end as the Coanda principle where in the same motion of the same unit opposition spin forms both factors in gravity. It is this quality that forms the Coanda effect where two forms (solid and liquid) bong as a unit. The solid in relation to the liquid substantiate the contradicting nature there is in circular motion. The liquid will ale\ways substitute the solid because the liquid always contradicts the solid.

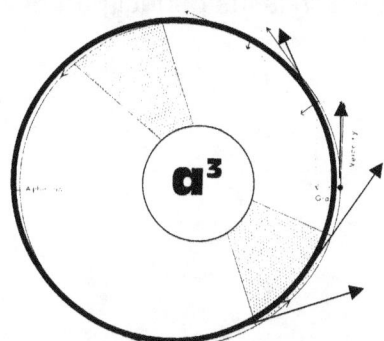

The orbit of any cosmic object around a controlling centre object is a fight between two relevancies born from the conflict in rotation. In every rotation there is one part that moves in the opposing direction of the other part notwithstanding that it is the same object or that it is changing relevancies or that the one will very shortly be in the role the other side has at the present motion. It is the one going to the future while the other is in the present holding singularity and the third I coming from the past. One can not literally give value or connect a measure to either but giving an non-relating example just for the sake of making conversation it is similar to the one coming from the past being $k^{-1} = T^2/ a^3$ and the other is at the same time on the other side of the Universe at that time $k = a^3 / T^2$ and the present confirm singularity at $k^0 = kT^2/ a^3$. There is always an opposing to the other side of the present form. While the one is striving to advance by confirming new space, the other is conforming space by retracting and the third holds the balance of circling around. In that we find the Coanda principle.

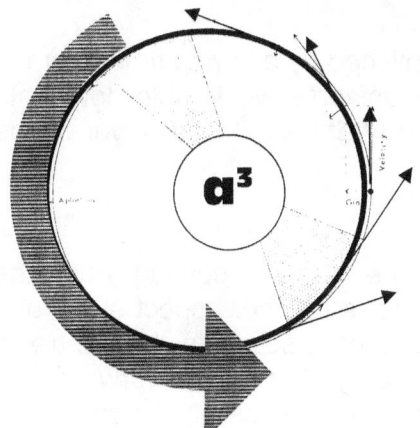

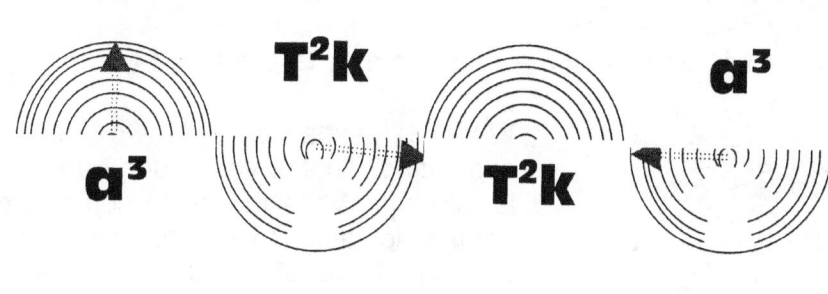

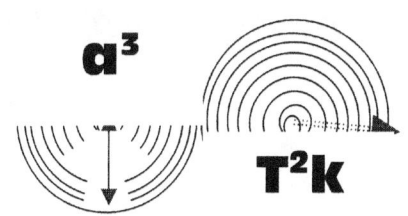

There are always the conflicting sides, which is a built in characteristic of rotation. The one side forms the space while the next is in motion of going away or coming towards at the same time. In that we find the Coanda effect producing gravity by applying opposing directions in the same unit as a result of the spin contracting as well as expanding simultaneously but on different sides of the divide.

The lot is more evidently moving further apart

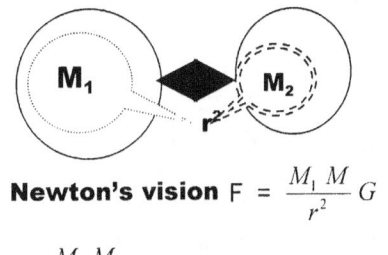

Newton's vision $F = \dfrac{M_1 M}{r^2} G$

$\dfrac{M_1 M}{r^2} G$

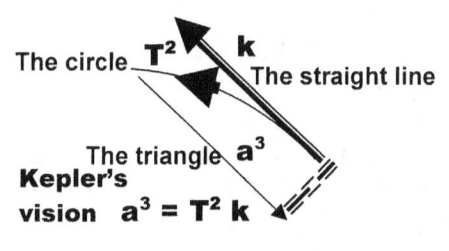

The circle T^2 k The straight line

The triangle a^3

Kepler's vision $a^3 = T^2 k$

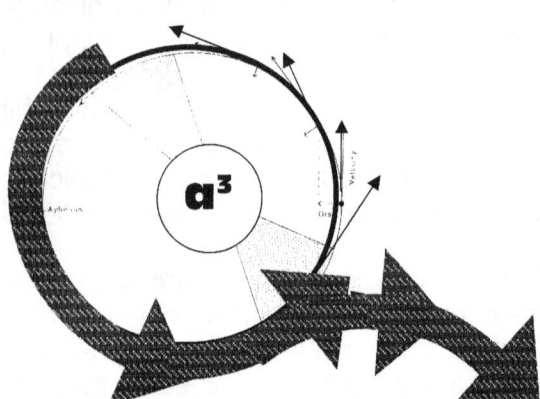

We have this tendency of a dual in rotating action that is the principle that is bringing about parts of the same unit orbiting other part of the same unit and still being in conflict with the other part of the same purpose and that is to rotate a centre.

If we look at the rotation we find from a human aspect that we put much claim to the circle. The circle holds importance but however important the circle may be, the circle forms part of eternity, which is the perfect part of creation.

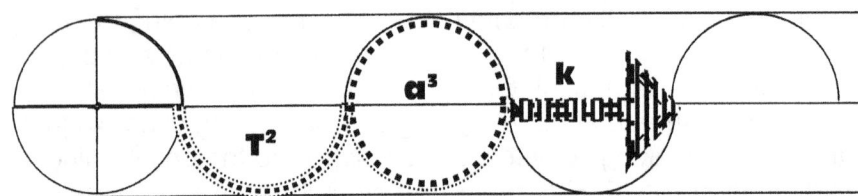

From the graph one can establish the link in the circle's rotation around a conforming unit being singularity.

Newton said that the rotation brings no influence and that he accomplished by allowing time to stand still. What he said was quite true because if time stood still all the Universe will tumble down into singularity or into the centre where from our perspective we find that **dJ / dt = 0.** This however will be a totally destructed cosmos

When $k = k^0 = 1$ then at the time also $a^3 = T^2$ and $a^3 = 1$ leaving $T^2 = 1$

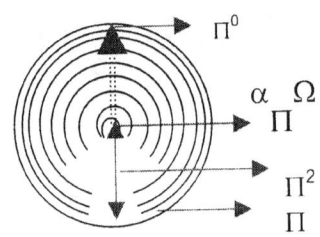

When Π^0 expands to Π we think of the rotation coming from the expanding. That is very much true but there is another aspect every one is missing. While the circle comes about time also shifts the centre to another location. In the duplication time brings the motion forward and although "forward" would not be a direction that is possible to point at, so is singularity not a point visible. Time in infinity does bring about the expanding of the sphere in the infinitive number of circles but the motion in time in eternity repositions such location to a new relevancy where the entirety of all the Universe rematch to find altogether new relevancy all over.

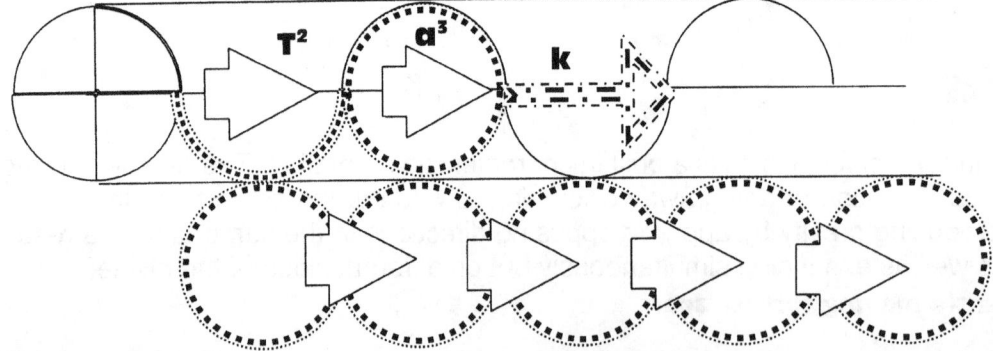

Matter or material is the concentration of heat that has gone dense because of time delay. I do not wish to elaborate that at this moment but there is a book out in which I explain this statement in much detail. It is the time delay of the shift in time from the past running through the present into the future that produces the material density we call the Universe in matter space and time. The duplication of the material is so extensive that the entirety of the Universe has to be demolished and again re-established to allow motion to take place. The one atom pulls the other atom while the one atom pushes the other atom to take its place the very next instant. Singularity cannot move but is totally rigid. Look at the centre of the top activated and one can see how immovable singularity truly is.

The entirety of time that we think of as the Universe is singularity holding every possibility that it can become when generated by heat. The circle remains the circle but the circle as a unit relocates while the circle complete the circle motion. By the way, that black stuff we see at night is not black stuff but it is light so bright our eyes kill of the brightness to allow us to see by daylight. That black stuff we see at night is heat and heat is time in motion.

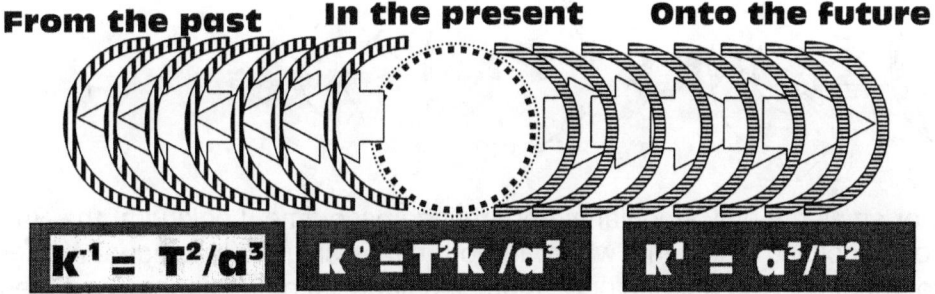

From the past   **In the present**   **Onto the future**

$$k^{-1} = T^2/a^3 \qquad k^0 = T^2k/a^3 \qquad k^1 = a^3/T^2$$

The material complete a circle by rotation and thereby confirm the heat that is sealed in the unit as the unit and that forms the unit we call material. That however is half the story where the other half is the directional redirecting of all that has spun into new allocated locations where each fin a relevancy that only apply at that specific moment. The motion redirecting movement plays as much a critical role as does the circle confirming the rotation and sealing the heat into the confined space that confirms material.

As the material duplicate and the duplication forms motion the material drags what it left behind as much as it pushes what it is catching in the future. The atom is following the other atom but to fill the time slot in the future the following atom has to relinquish the position it holds in the next one's future. In order to move into the next position in the future pretender the atom has to relinquish the claim it has on the present location. The only easy way to do that is to drag the next tenant into the vacant to

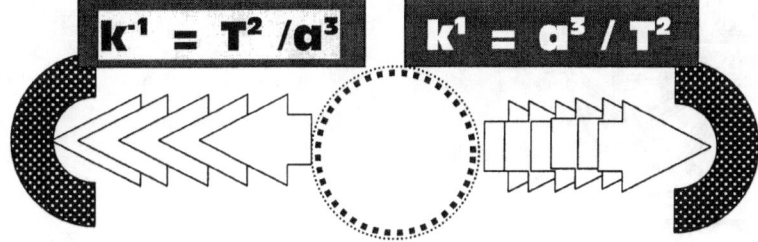

$$k^{-1} = T^2/a^3 \qquad k^1 = a^3/T^2$$

be slot the atom at that instant fills. To move it has to push the atom in front in to a new location while dragging the one behind to fill what it wishes to vacate. With that action we find t hat the relevancy k goes minus and goes plus which allow space to grow as time decline or to allow space to shrink as time becomes more prevalent.

Apparently according to informed sources we will find the Sun viewed from each planet indicated by name, as the photo would suggest. From Mercury the Sun seems large therefore the Sun is close and from Pluto the Sun seems small which we think of as further.

Mercury

◄ Earth

◄ Pluto

That way of thinking is very indicative of Newtonian thinking with a very explicit Earthly connotation. It is far from cosmology.

When an object such as one of the photos suggest shift further away by getting smaller, it is not the distance we should consider because the distance has no meaning in cosmology. It is the time it would take to reach the object that has to be considered.

The "further away" the object seems the longer it would take to reach the object and the "closer" the object seems the less time it would take to reach the object. There is more time between the Sun and Pluto than the time being between the Sun and Mercury.

In the case of Mercury time is more and therefore it divides space into a smaller factor $100k = a^3 / 100\,T^2$ where as the space the Sun has seems to be more space at Mercury because $k = a^3 / T^2$ or in relevance to Pluto the Sun must be a **100** times bigger with the time being a hundred times less of a factor
$k = 100a^3 / T^2$. That mathematically with the aid of Kepler proves that the Black stuff being the Biblical light that was mentioned at the

explaining of creation is time and not space. Material fills space but the filled space of material move through time we incorrectly think of as space.

When the one is pulling the other is in opposing mode not only by spin but also by having a singularity to protect.

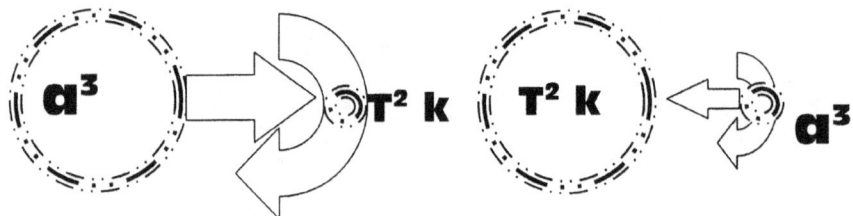

In the cosmos there is no big or small but only singularity generating space-time, Looking at the Sun and any planet we see two points holding singularity where in both cases singularity is equal at $k^0 = a^3 / T^2 k = 1$. On the one side of the divide the one factor holding singularity take prominence and then crossing the divide the other point holding singularity holds dominance. Crossing the divide of singularity puts one of the two in dominance as far as controlling time and controlling space. The one holding lesser space would affectively control lesser time and the one controlling more space holds more time in control.

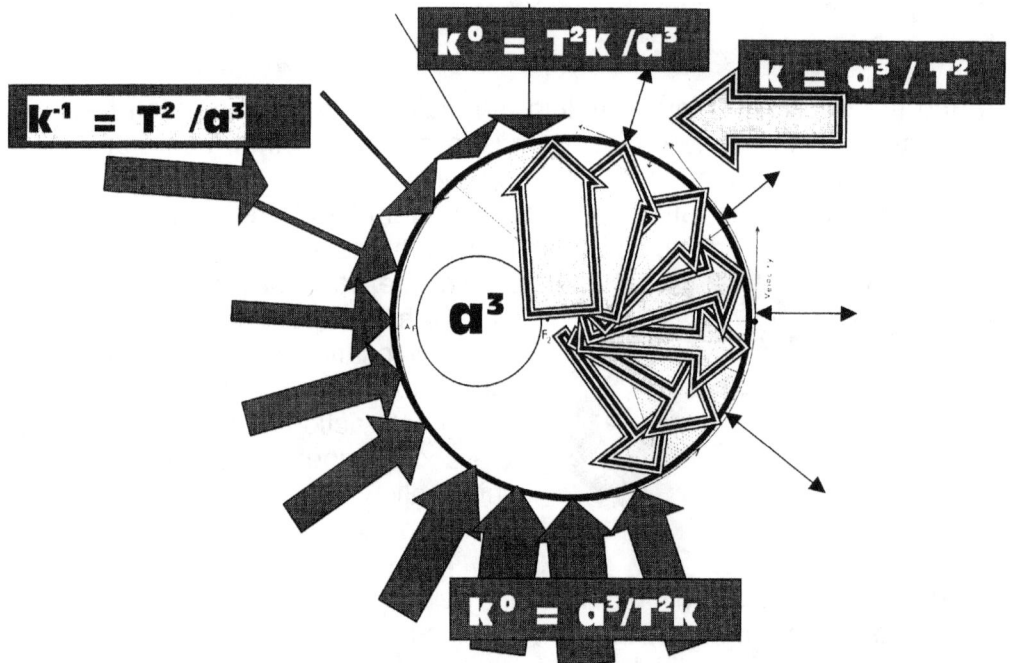

As I mentioned a while ago, material is the delay of time bringing about heat where the motion contain the heat in confirming the location of the preserved heat. The material fill time with space by relocating singularity while singularity generate material filling time in a specific allocated location. Through the motion there will be some area where the expanding $k = a^3 / T^2$ is more frivolous and where the material is pushing the next out of position to fill the next position. The there is the other where the delay becomes more because the delay is caused by matter not generating space filling fast enough so that the follow on filling can come about becoming motion or heat relocation of time.

In this changing of the relevancies we find that at one point the Sun gives the planet a nudge onwards and the next part where the planet go past the centre singularity the Sun drags the planet onwards. The one part of the circle that the planet form in orbit is smaller than the other part that is bigger but small and big is no issue. It has more time concentrated by motion effort or less time concentrated by motion effort. In that way the Coanda effect forms gravity and locates the time intervals as structures orbit a centre of concentration.

• Before time moved there was a spot that repeated by being ⌢ perfect.
Then the imperfect entered creation and the spot moved to a dot.
►By returning to the spot the dot advanced to form a new dot.

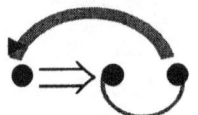

The motion of parting from the perfect to move into a position held by being imperfect gave time in infinity the opportunity to part from time in eternity.

The returning to the previous position was always there and is part of the eternal perfect. Moving in one direction brought about a future that institutionalised time by implementing heat. It was the departing from the position the perfect held that is time but that time is a delay caused by time going imperfect. As time becomes more imperfect so would the relevancy of repeating the perfect in ratio of the imperfect grow and space filled with material will compact from that to become more, denser and dominating the entirety of the Universe.

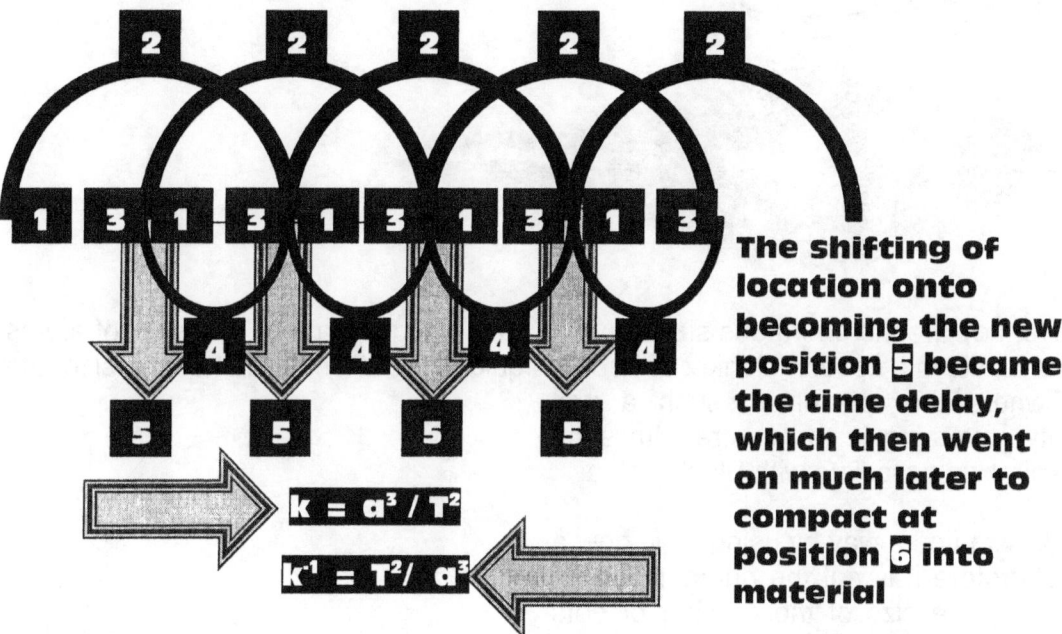

$$k = a^3 / T^2$$

$$k^{-1} = T^2 / a^3$$

The shifting of location onto becoming the new position 5 became the time delay, which then went on much later to compact at position 6 into material

Even high and low tides has nothing to do with "the pull of gravity". If it did have anything to do with the pull of gravity then there was no reason for cyclic change since the mass of both the Earth and the moon remains at a constant. The changing of tides is the rotational cyclic contradicting nature that motion has and the Moon crosses the divide twice daily where the cycle then begins to oppose what it was before. It remains just the crossing of singularity where singularity indicates the divide the motion establishes.

T^{-2} T^2

Planet orbit

$$k^{-1} = T^2 / a^3 \qquad \text{Sun} = k^1 = a^3 / T^2$$

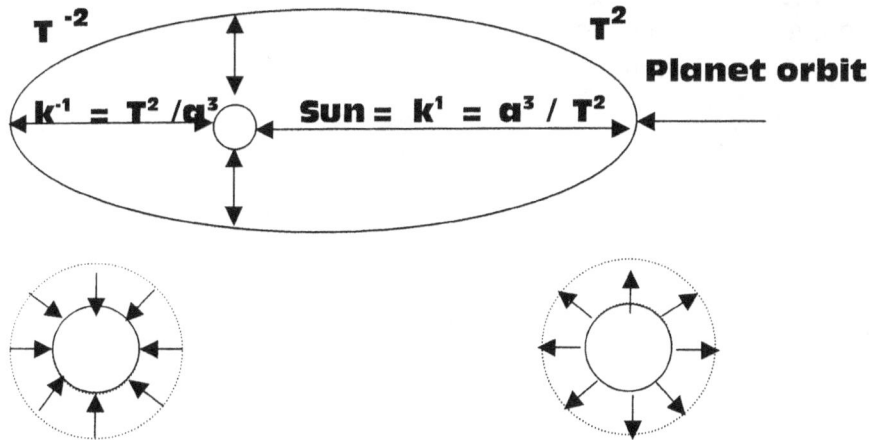

Gravity is about reducing space

Expanding is all about heating. Heating takes up more space and gravity reduces space.

The same pattern is still very much visible in the way structures follow the centre of contraction. We still have the relevancy shifting from one to the other with not one point holding singularity absolutely domineering, We still have the one trying to expand while the other is trying to preserve and the relevancies does not go all out the way of the controlling structure. We find in this manner that the straight line goes bended by 7^0, and the curve follow the guidelines of singularity by measure of Π.

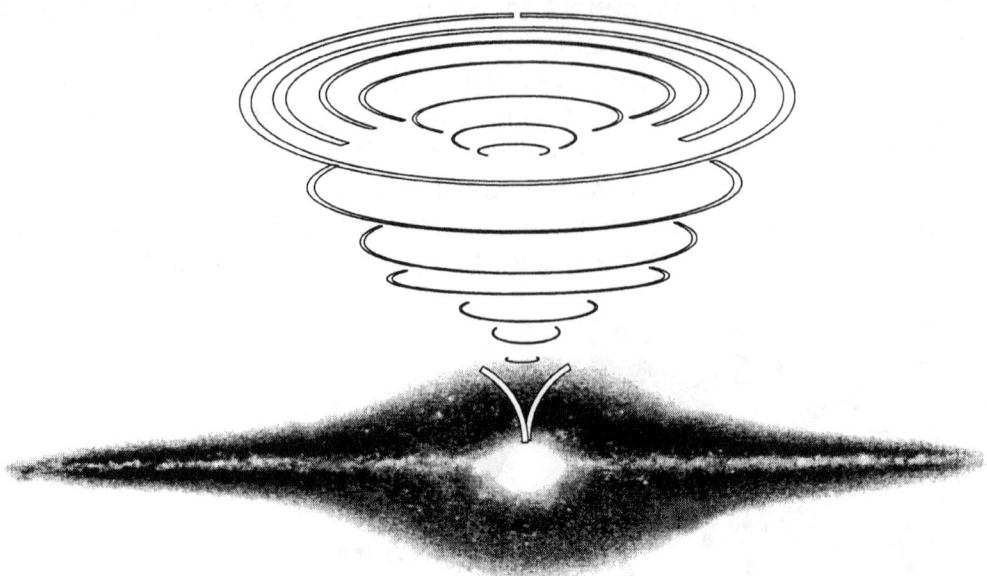

Even in galactica where they are the enormous size they are, they still generate by measure of atoms spinning in future stars that is in a cocoon blanket of heat or liquid time. Still notwithstanding size, the atoms form a unit where the rotation of such a unit generates a singularity governing in the centre with such intensity it forms the spiral we associate with Black Holes.

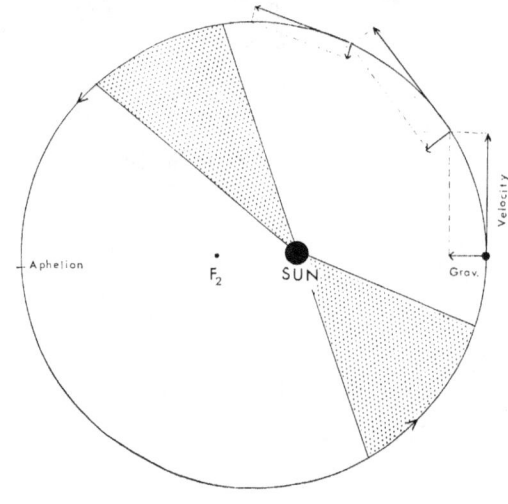

Let us investigate and try to find a way by using logic how a star applies gravity. Therefore it is not the number of dots that is important. It is not the size of the number of dots occupying the position or the size of the space the dots occupy that is prominent. It is the relation in the dismissing of space and the duplicating of space that becomes important. The less space there is the more the favour will be to reduce the space because of the advantage the dots have in securing space-time that will prevent overheating. On the other hand the more space secured will also prevent overheating and therefore those will opt to duplicate space in order to find space to secure and prevent overheating.

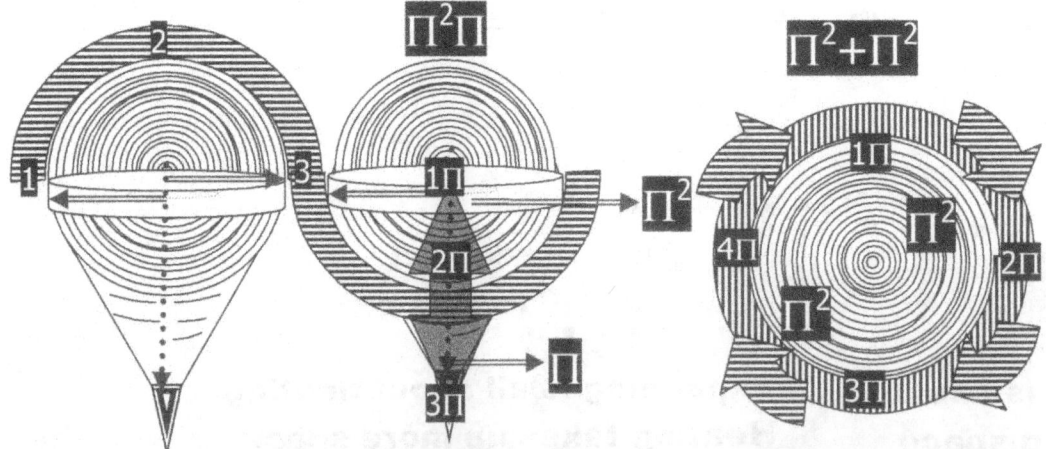

Since the Earth has no singularity demand that is much better developed than the universe sustains, we find on Earth a relevancy of Π to $(\Pi^2+\Pi^2)(\Pi^2\Pi)3$ is adequate. But in bigger units the space-time displacing relating to space duplication presents much more demands on atomic structures occupying space within the star containing through set boundaries. In the presumed to be bigger stars there is much space filled with atoms occupying much space. In the stars more massive but holding lesser space the atoms must also hold lesser space but they also hold more protons by number in the lesser space.

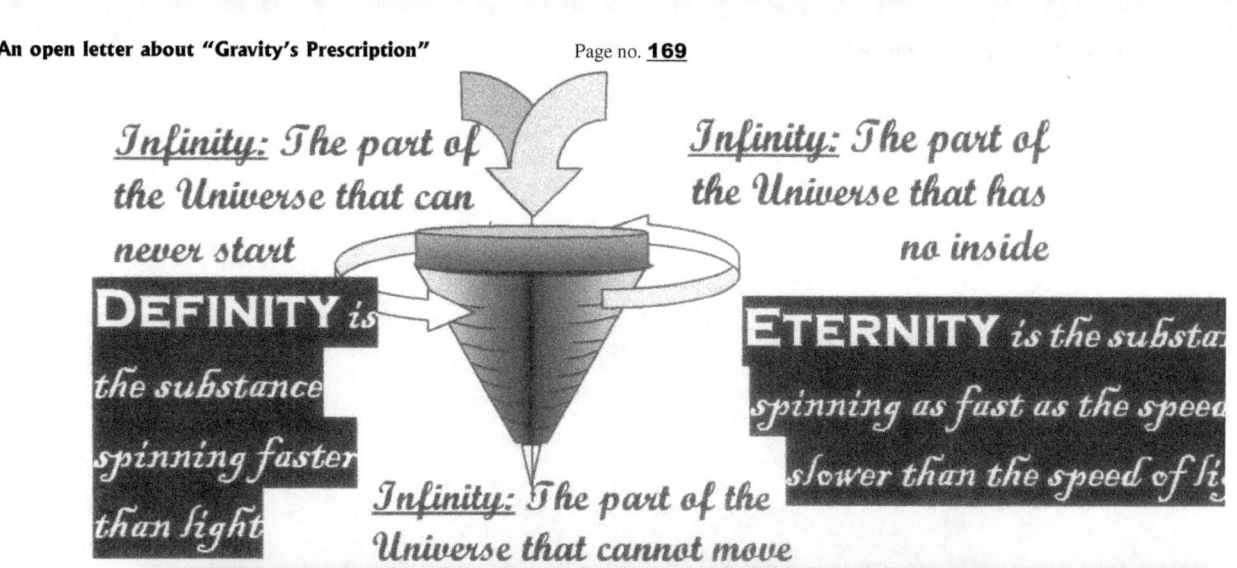

Infinity: The part of the Universe that can never start

Infinity: The part of the Universe that has no inside

DEFINITY *is the substance spinning faster than light*

ETERNITY *is the substa... spinning as fast as the spee... slower than the speed of li...*

Infinity: The part of the Universe that cannot move

The Universe are formed by only two substances where I call o... definity (better known as material) and eternity (better known as... space)

Eternity: The part of the Universe that has no outside

Eternity: The part of the Universe that can never end

Eternity: The part of the Universe that cannot stop .... moving

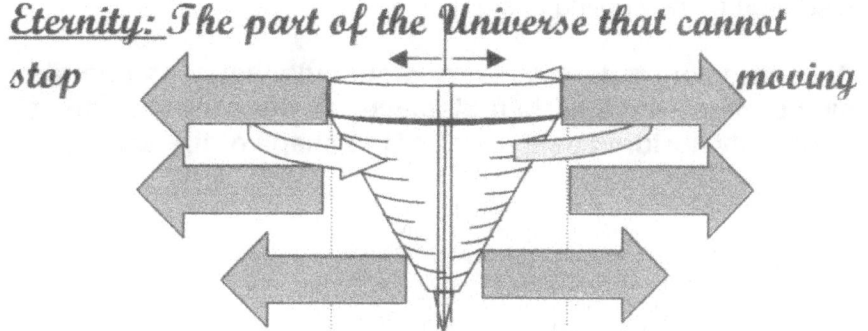

I touched on this process when we examined the working detail of t... ball trajectory and when I said that this is the way the Universes d...

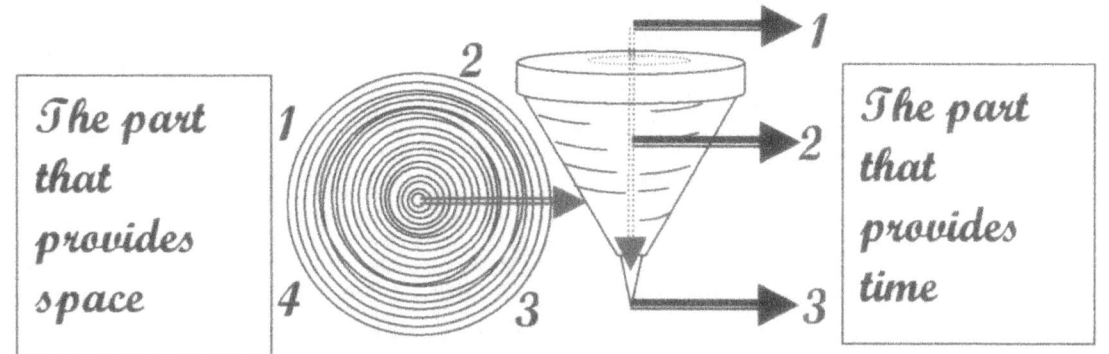

The part that provides space

1 2 4 3

1 2 3

The part that provides time

This is very important to know when we start discussing the Titius

The following is one of many introductions to the Titius Bode law explaining is put forward to explain what no one understands.

I would suggest we think of stars in the following terms. A star that generates and transmits a lot of light is weak on gravity because their progress started recently. They command a lot of space-time but the demand they have to keep their cooling acceptable is very low. In that they can generate a lot of light but with the demand on cooling low and the gravity in the centre not very developed, those stars cast a lot of light back into outer space. It is just because of the size the stars hold that tell the that the stars are still young and have a weak developed governing singularity. The stars will have very prominent hydrogen and helium layers, with the inner core not very prominent. The control of the star is still very much in the individual atoms and in that the motion the atoms have to produce in order to maintain their individual singularity will only come about through motion. The atom has to make contact with as much space-time through motion as possible since it has a very poor ability in contracting space –time in support of the cooling system.

The entire motion and the entire contraction of every atom culminates as one effort and this produces a single combining effort which is then displaced to the centre of the sphere of the star where singularity is normally nurtured as a result of the shape of the sphere

The contracting action is at present the only part of gravity that Newtonian science credit as gravity. There is a lot more to gravity than such simplicity. Every atom in a star is pushing the atom in front by filling the space the atom in front vacated. Every atom in front of every atom behind is pulling the atom behind as the atom behind is urged to fill the space that the atom in front vacated. That is motion, which is the most complex issue one can find in the Universe. Since every atom is driven by singularity and no singularity are able to move it bring about that every singularity must remove and rebuild the space every atom fills or vacate as the atom moves along.

There is a building of an entire Universe going on in every split second and this split second is so fast we cannot name it. By naming it there will be so many time units gone by, by the time we said the name, the Universe might not even be recognisable. We might call it energy but I hate to call it energy because energy is a lot like Holy water. It can come from anywhere and you can use it for everything and in the end it does not even become something durable because its use eventually comes to nothing.
The atom restricts dismissing of space by the containing structure to the atoms relevancy being Π^0 in singularity bringing on Π relating to $(\Pi^2+\Pi^2)(\Pi^2\Pi)3$.
As the layers swap there aligns between duplicating and dismissing the atomic relevancy adapt to comply

Since the star performs as an accumulated atom where innumerable atoms inside the confinement of the star combine to select one centre spot forming singularity that represents the star, I have chosen the to use the same symbols that I found in atoms to describe the relations in space –time to singularity within the space-time of the star. I refer to a star as a cosmic atom in other books.

Early stars still in the envelope of heat within the centre of the Galactica have only space duplication and growth through the cover of such enormous heat. These class stars are not visible but are shrouded in a blanket of heat covered by light. The atoms forming the stars are small and under developed. They remain cool because they contrast with the heat surrounding the star where the star material supports the cool space and does not form part of the liquid heat forming the outer limit. I would like to draw your attention once again to the fact that the Sun at one stage was a cool 18×10^6 ^{0} on the inside and a freezing cold at 6500 ^{0} on the outside while all the time outer space was a blistering 10^{34} ^{0}. This was considered the coldest place in the Universe because the Sun was still part of the deep frozen space inside the blanket of heat. Look at any galactica and see in the centre there are stars surrounded by a blanket of heat with stars conversed by heat sitting like a duck frozen in this pond of liquid heat.

With the cosmos the size it is and space so large compared to our smallness we have no chance in finding the centre of the Universe. The Universe started where singularity is and singularity is the sure indicator of the Universe. With all spinning objects holding singularity we then have located singularity in as much as finding the centre of the Universe. The Universe started with a dot forming. That answer arrive from taking mathematics back to a point of being the smallest possible position, far smaller than we may be able to calculate form. The ten dimensions I named the atomic relevancy is also showing the double value of singularity as singularity extends into as well as beyond space. The atomic relevancy is $(\Pi^2+\Pi^2)(\Pi^2 \times \Pi \times 3) = 1836$ that is the mass relation between the electron (3) and the proton. Proton = $(\Pi^2+\Pi^2)$ Neutron = $\Pi^2\Pi$. The atomic relevancy holds the dynamics of singularity control. In the ratio and dimensions we find in the atom, all space-time derives from the atom, whatever the atom is. Our instincts, our logic and our calculating process all indicate that the sphere holds a centre point from where six evenly positioned point's position matter to be. Using The formula $F=G (M_1.m_2)/ r^2$ it indicates to a force pulling objects closer, where each force is coming from each centre point the body in question has. The contraction must commit the two bodies towards a point in each case being spot on in the middle, not withstanding what direction the force is applying, the body will draw to the centre. If the Universe spins around a centre point holding singularity, and singularity confirms the centre of the Universe, then every particle holds the centre of the Universe making the number of universal centres immeasurable many, and every atom and sub atom particle presented outside the atom in smaller bits, are all not pieces of the Universe but they are a Universe surrounded by many Universes. If every atomic particle no matter how small is holding the centre of the Universe, then the gravity is coming about from that point because that is where the gravity applying in the Universe is applying contraction. If the Universe did start from one single point and time, matter and space flowed from that point, then that point must have a relative connecting base because such a point holding singularity must be eternal as space, matter and time link eternal. There therefore must be one point linking the entire Universe when regarding the fact of singularity.

Then according to the theory off relativity there has to be one exact point holding time in relevance notwithstanding the fact that time departs from that position and relate differently to all space-time away from such a point.

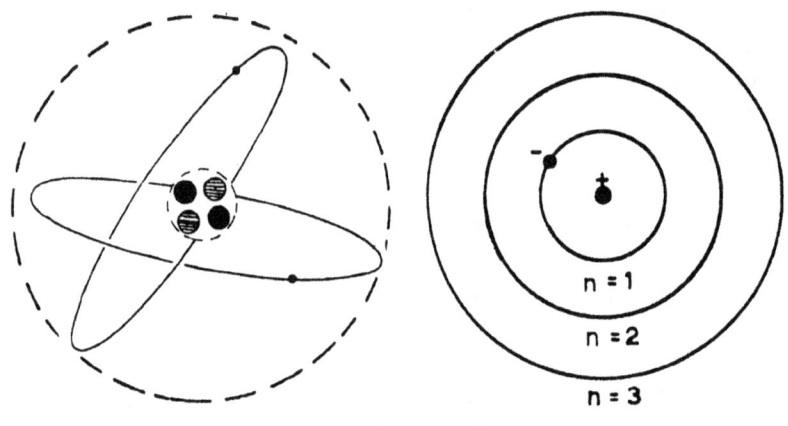

In the final analysis it is the atom that control the Universe because it is the atom that is the Universe.

It then is the atom in the most centre part where space and time meets singularity, that Einstein found a Universe collapsing to a single dimension, and every atom at a point post of the proton where gravity initiates in according with the proton dimensional colas of $(\Pi^2 + \Pi^2)(\Pi^2 \times \Pi \times 3) = 1836$

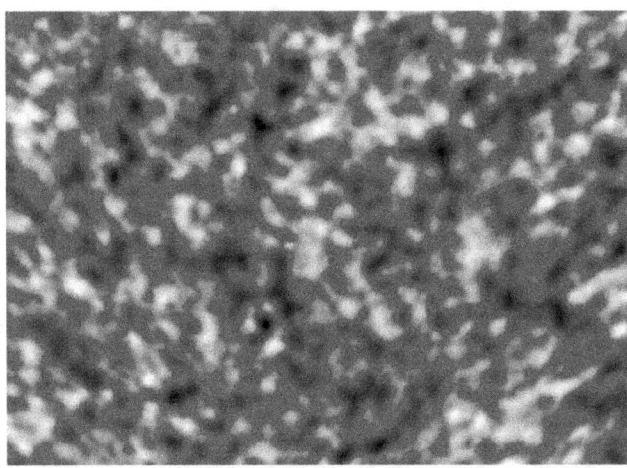

Every person with whom I have discussed the facts concerning creation recollects images in the trend depicted in a presentation as one may find to the above. That would be the most unlikely way Creation came in place. The recalling of pictures representing images about creation must have form, but to mathematics it had no form. From this thought the very opposite arises where Creation came from nothing but such an idea is mathematically simply not possible. The thought of nothing is just what it is, a thought of nothing and although it is in the nature of the human mind, to present nothing as a value in the recalling of something, nothing is a presentation of the figment in the human mind. There can be no number such as nothing and that was (possibly) Newton's biggest error. Nothing represents non-existing and that is just what nothing is, it is non-existing. In order to prove my point I wish to ask the reader to define the shortest line there can theoretically be. If he should answer anything but that the shortest line will be at a point where the beginning and is the very same spot he will be wrong. The shortest line that can ever be anywhere must have a start and finish holding the exact same spot. The line will be humanly impossible to create but we humans are capable of very little.

When the line has a beginning and an end at the very same spot and it wishes to extend the position as to further the possibility it has, which direction should it favour. Humans in the west would naturally think of extending from left to right while in the east humans may want to go from right to left.

Some persons will tend to go up or down, but all of the options are about human preference and not mathematical conclusions. Extending the line in any one direction will favour one direction without a conclusion about not extending in other directions. Such a conclusion has no sound mathematical foundation. The only option about extending will be in all directions equally in order to give a meaningful non-bias flow of mathematical equilibrium

The shortest line in the realm of possibilities must have a start and finish holding one spot and such a line will also be a dot or a circle. Not favouring one direction puts all directions at equilibrium meaning that any form what ever may be can develop from such a spot with the end and the start being the same. This reasoning prompted me to look for singularity in such a spot because if the prime spot from which all came was a spot, then the spot must hold the shortest line but more prominent it will hold the smallest form including the smallest circle. One possibility that the shortest spot can never have is having a starting point on the zero mark. If the mark of zero holds the start it must also hold the end because the end and the beginning has the same position. If the position of zero then is the beginning, the end will also be zero leaving the line without an end as well as without a beginning. The conclusion from this is that no line can start at zero because that will be a mathematical impossibility. A line or spot starting at zero would therefore be shorter than the shortest line possible. A line growing or extending from zero can never leave zero because of the influence of being zero disqualifies any possibility of growth. If the line then had to grow in all directions at the same pace the line must therefore be a circle. The value of the circle is Π, and that is where creation started.

In the centre that holds the line the line is a generated notion that is so thin the line is generated by motion and still the line is not part of the cosmos, while it is supporting the entire cosmos by controlling the entire cosmos. While it is not there, there is no denying that it is there and the control it has over all of the entire cosmos goes beyond question. It is establishing all the dimensions by seven supporting ten.

Every object that spins also generates such a line through the spin. The purpose of the spin places coherency into the Universe to generate the control by command and every atom is the seven. The atom places the seven in relation to the ten and all other atoms in direct linking of the seven points form the ten, which the atom holds as motion or liquid. The line is the diversion of the four by three establishing a parting between infinity and eternity.

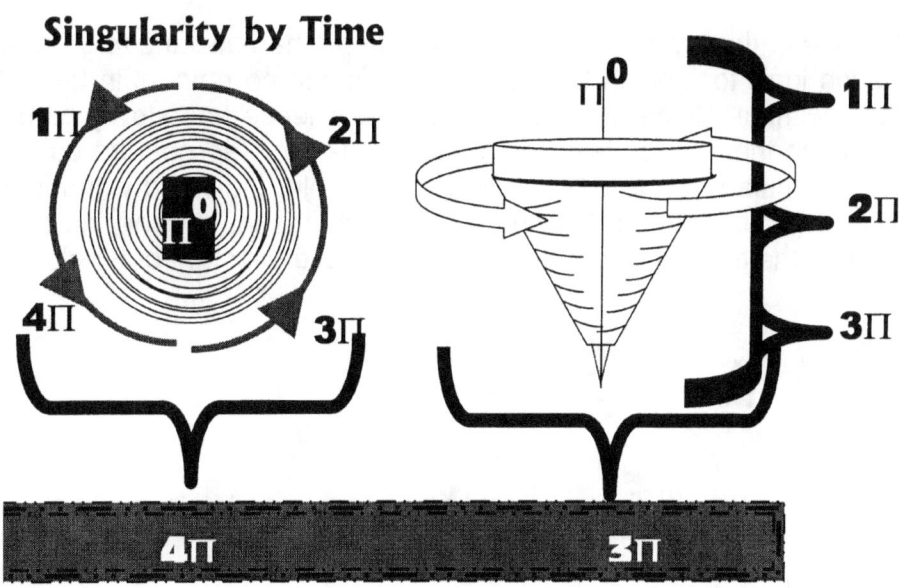

Singularity by Motion

Singularity by Time

If the alignment is in ninety degrees to each other then Pythagoras has to apply strictly. Should my argument be sound and which it is sound we have to be able to use Pythagoras to determine the value of time in space. When the material rotates or moves the filling of the material is in perspective to the time. However, material can only be in one location in one split time. Since material has to cross over to the other side of the Universe in order to duplicate, which is how material moves, then the material, can be only on one side of the Universe,

Every time matter is generated and moves, it is singularity that is complying with it activating another point in singularity being charged with the motion. The Universe started from allocating singularity charged by heat into positions where such positions contributed to space- time. Every time the spot overheated the spot expanded into four dots and by expanding the spot cooled. In cooling the spot retained heat by which it spawned the dots allocated as time. In overheating objects expand and by expanding objects cool. That is gravity. Gravity is the expanding in relation with the cooling which means it is duplicating material in relation to a generated centre that is contracting the motion by cooling. Every inclination of motion is in fact motion and every movement be it contraction or expansion is moving to the other side of the Universe by bridging singularity because singularity is immovable. Therefore by being immovable, motion has to cross the division singularity applies and by crossing the division the factor that comes in place is $\Pi^2/4$, which results in the Roche limit. But such motion is three and the square of three in addition to the square of four brings about time in space.

As a school going youngster, I was fascinated by astronomy and in particular the cosmology aspect. In a long and strenuous process of self-education I was completely stunned by the behaviour pattern that the comet had in its relation as it orbits the Sun. Please forgive my boyish way of presenting the following but it is important that I bring it across as I saw it as a boy and as a matter of fact still see it today as a middle -aged adult.

Science acknowledges growth as the Hubble constant and then refuses to put the growth in line with the solar system. The growth they reluctantly admit too, they refuse to connect that growth to the solar system in any way. They take a Universal year as a solar year being that of one cycle it takes the Earth to rotate the Sun in the present day. Then they reflect on this as if this was going on since time began, because by doing that, there then is a nice crooked constant that fit mathematicians. Push this double standard applied back to before the Sun took its position and there was not Earth to indicate the year. How small was the year circle at that point in time and space. Take this right down to the:" Big Bang" where "the whole Universe were the size of a man's fist" (To use their words), how far did the circle goes to indicate a year then? The year was immeasurably smaller, shorter and faster than at present. This is logic even the Newtonians must accept. There is no space outside insanity to apply time to the past at the value it is at present and far worse, to use something so extremely insignificant as the Earth to measure it by.

Again I feel that the use of this type of constant just to fit mathematicians to corrupt the truth they in science are using such logic to rubbish the truth. There is just no rational in the time verses events that can explain facts without. Since the time of Newton, science has slowly nibbled at the truth to compensate for the game there is to play. It is as if the one crooked posture corrupts all in general. That is what Kepler's formula is all about? That is what Kepler indicated with his formula $a^3 = T^2 k$. The space of an object (a^3) is equal to the time (T^2), which it is in, in every given instant (k). If the space becomes smaller, the time duration becomes longer every instant of time's progress.

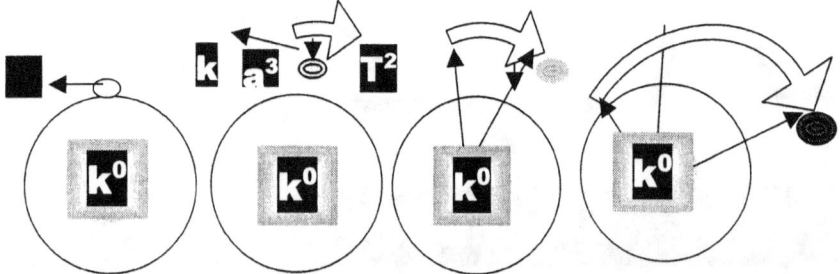

In the accumulating of heat, the object finds motion. The motion brings along structural independence and such independence puts distance between k^0 and a^3. The heat increase will accompany a larger T^2. By increasing heat, the distance between the objects will grow.

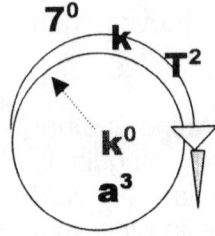

Only by creating a total independent heat centralised in a point holding singularity and feeding k^0 with an independent heat supply can an area a^3 establish a k that will release the independent a^3 from the secure larger k^0. The overall condition is that the escaping a^3 must establish a route following k^0 as k^0 places a diverting 7^0 where that 7^0 then forms part of the object creating heat to secure a release from the established a^3. Providing the heat will bring about a release placing a new object into outer space.

Time has been three since eternity started and time will remain three until eternity ends. When the heat came about eternity spawned the one in infinity that generated a line and then time became the three positions of past present and future all depending on the one line in infinity. In the triangle that time established in conjunction with the law of Pythagoras the three of time goes square that forms nine and when the one marker of infinity is added time by the square in space becomes ten.

The line however forms three and in conjunction with the four positions in space –time (three in eternity and one in infinity that is there eternally) there are four positions relating to the three in the line and from that lying between eternity and infinity is the four eternal position plus the three generated positions which forms the seven in space-time.

However because of dimensional duplication the square of time is ten and five will be on the one side of the Universe and five will be on the other side of the Universe. That then is why the Lagrangian system holds five positions in relation to singularity.

When the four in time spins off one more in infinity as time moves on a fifth spot becomes valid that erects a line by heating and that fifth spot then reverts to the first spot that again parts eternity from infinity. When it has spawned a fifth position that position also goes square and forms by the law of Pythagoras the Lagrangian fifth position.

This puts a huge question mark on the correctness of Newtonian presumptions that currently fondle the idea that rotation has no influence on the cosmos and all gravity goes down to mass where mass has all the influence and control.

All spinning matter has the point where the spin is still there but the radius is to small to measure by any means. That point is standing still in relation to the rest of the spin. In relation to that logic I do not except Newtonian science holding the radius of s spinning object unaccountable in the spin, whether the spin is applying or not.

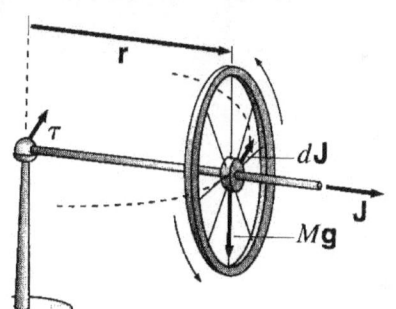

Applying Newton's second law F=ma
One arrive at the formula
$GMm / r^2 = m (\omega^2 r)$

By replacing $(\omega^2 r)$ with $2\Pi / T$ we obtain Kepler's third law

This law predicts that $T^2 = a^3$

What this statement implies is that r does not exist. When anything has a value of zero it is for all purposes non-existent. Only when an object is following s straight line can the radius be non-existent because the radius alters value through time development.

Taking the argument back to Kepler's law,

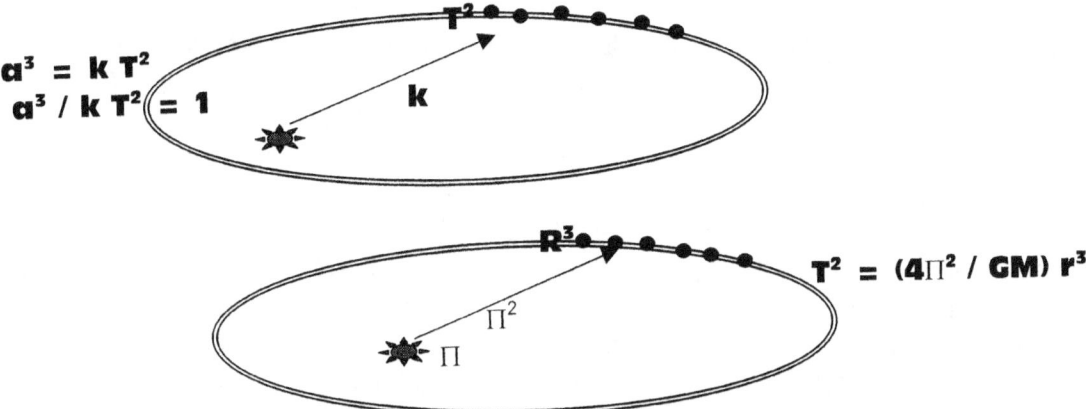

$a^3 = k T^2$
$a^3 / k T^2 = 1$

k

T^2

R^3

$T^2 = (4\Pi^2 / GM) r^3$

Π^2

Π

The spinning or not spinning is not part of the issue because at the point of absolute singularity the object never spins. Therefore spinning or not spinning does not apply to the point of singularity because singularity never spins in any event. In the whole structure with a pivotal centre as the control to the motion of the space the fact of Π is a natural outflow and any adding of Π is totally incorrect. According to Newton the result of spin is zero, however the top will tell a much different story.

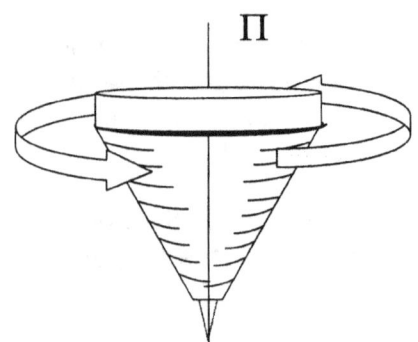

On the surface, at first glance the top is an ordinary piece of dead wood that is machined into a sloping shape. The top is normally fitted with a sharp needlepoint at the bottom and the sharper the point is the better will the spin balance be. It is obvious that the spinning of the top inspired the entire Universe into a reality that is not there while it is in control of the entirety we find as real as life itself

When translating Kepler's mathematical expression into a verbally spoken form of communication such as English we can see what Kepler said also read as $k = a^3/T^2$ where **k** is one point from a centre point that is space a^3 relating to time T^2. From a centre comes space-time

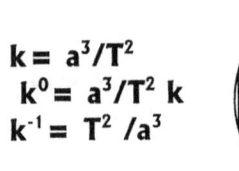

$$k = a^3/T^2$$
$$k^0 = a^3/T^2\ k$$
$$k^{-1} = T^2/a^3$$

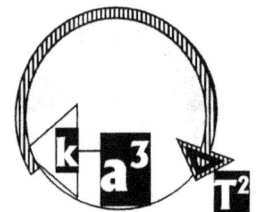

Kepler said
$a^3 = T^2k$ but that
could also be
$k = a^3/T^2$

Others like Newton and Einstein came much later and coined the phrases but Kepler formulated the concepts. They named Kepler's innovations. That is very clear but only on the condition that Kepler is read correctly and Newton gossip about what Kepler is saying is ignored. What Kepler said in mathematics all the brilliant Mathematicians through so many centuries were unable to read although the coded language was written in mathematics!

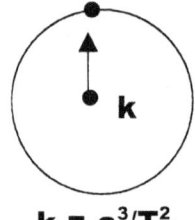

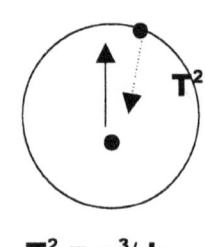

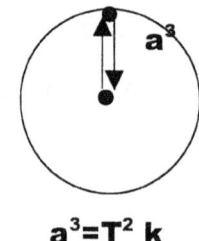

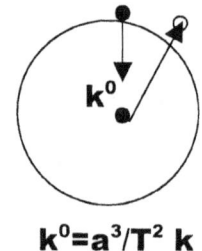

$$k = a^3/T^2$$

$$T^2 = a^3/k$$

$$a^3 = T^2\ k$$

$$k^0 = a^3/T^2\ k$$

But as one can see I also realised gravity is relations of motion applying in two factors. There is no separation of the two of the factors acting as one but both have different application and values in the unit. It was what gravity was because this action prevented expanding. This is the result of singularity having three parts acting as one but giving three distinctions in application.

Gravity is as much part of dismissing space as it is about making contact with space in time. Since the connection comes about as a circle, the connecting points will relate to Π as the value. Due to the spinning nature of such a point with all surrounding the point will be alternating direction favouring change every second and in that the value to such a point can only be Π because of its constant changing. Using r would specifically oppose another r from every angle because the use of r will bring about a static relation to the previous and following instant and therefore it will cancel the constant spin flow. By reducing the line to its maximum possibility one end with Π being the minimum but that Π is actually $Π^0$ which can also be k^0 or a^0 or T^0, which all indicate positions in singularity. Only when forming a value past singularity does independent identification come about. When the atom formed that atom applied a relevancy of ten positions where seven positions are included in the atom spinning and three positions are part the exterior of the atom spinning but all the positions relate to singularity but as space flight taught us such relevancies can change when an object is within the space boundaries of a larger structure or roaming free in outer space.

Within the boundaries of the atmosphere where the sphere border touches the space borders the space borders hold six positions and the sphere hold seven points. But at the precise place where the points make contact with the sides one side fall away in favour of the point it connects too leaving five sides relating to seven and where one of the six sides takes control in removing one of the cubical sides by replacing that side with a sphere point position the object then becomes directly controlled by singularity positioned in the centre of the sphere. The object seems then to fall from space and enter the atmosphere becoming a shooting star. What the Coanda effect proves above anything else is that gravity in control of space-time comes about from a centre and such a centre can be created by motion applying to a liquid in relation to a solid. That means there is undisputedly a flow of space-time towards a centre and the centre has to diminish the space-time reaching such a centre to create the flow and therefore the control from such a centre. That's the one pivot of gravity. Since the Coanda effect shows gravity is control of space-time by motion flowing towards a centre that also prove as it explains the one part of gravity that reduces space by increasing time towards a centre that is established by motion and the lack of space establishes a lack of motion in that centre.

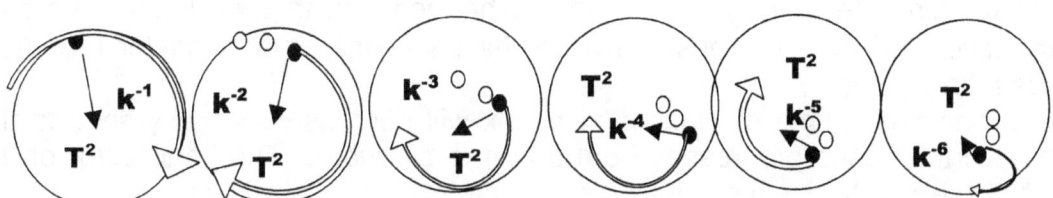

Because the smaller object holds much less space the duplication is in a lesser relation than the main object and because the time factor enforces the duplication period to match therefore something in the applying ratio has to give in to allow the major relevancies to remain in place. Since a^3 has to rematch to apply to the conditions set by the larger object a new relevancy comes about where the new a^3 will bring along a reducing T^2 with the diminished k that the Earth enforces. Since the space that motion reproduce is smaller in relation to the Earth, but the earth enforces the same time value, the relevancy of the time value will deplete by reducing k, but not in a straight line because all factor changes will then only be carried by one factor. I this way the diminishing space produced help the cyclic time factor to decrease with the distance that grows smaller.

When the object is released from the atmosphere of the dominating space, this very same gravity k^0 = k T^2/ a^3 ratio will still be enforced since it is not the law of the Earth prevailing but it is the law of the Universe applying. Outside the atmospheric borders the Earth no longer have the means to remove one of the cube sides that form the lesser object space and where the cube reinforces position by keeping the rotating object in position floating above the Earth.

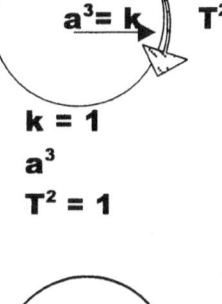

$a^3 = k$ T^2

$k = 1$

a^3

$T^2 = 1$

The space became too small to allow the time it takes to enter because the distance k decreased faster than the space a^3 could compromise with the time T^2 changing from what is present in outer space comparing that to the time in to atmospheric space. With this information being in hand for a period of four hundred years, one should think that the wise could derive a conclusion. Where the information forms the basis of modern cosmology since the information formulated gravity and not merely produced a name for gravity as our English friend did, it is amazing that such accidents can happen and it is more amazing that no one in Mainstream physics has the slightest idea why this is taking place!

T^2

$k = 4$

$a^3 /2$

$T^2 = 1$

Our most impressive astronautic engineers are assembling a machine that will scramble the ratio Kepler introduced to a level in outer space where the ratio will be more than what the ratio in the Sun is. Surprisingly they are not in the least surprised that not one object in outer space is using an excessive velocity.

In realistic physics it means double the space will fill in half the time. We know that that is not possible because it can only bring about half the space in double the time or twice the distance in half the time. Space time and distance is a mesh where the lot integrate because Kepler said so. Kepler said the space forming space is the same space forming the distance of the space and that is the same space taking the time to fill the space. If the ratio changes then changes come about the entire ratio. In order to bring about such acceleration much more heat has to be released to gas in order to find such a drive that will sustain such a high velocity. The drive can only be the result of massive quantities of heat being stored around the singularity the atoms generate. The way the cosmic has designed the fight to relieve overheating is by motion. By duplicating the overheating space through motion the duplicating reduces the heat by half because the heat is spread over half the area that is distributed in double the space. By increasing the relevancy **k** it reduces the area or space by quantifying the number of spaces per time unit in the time from that apply in the atmosphere or outer space.

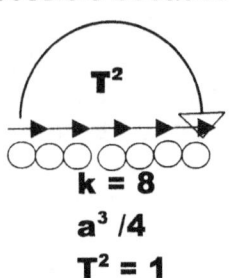

k = 8

$a^3/4$

$T^2 = 1$

k = 4 and $a^3/2$ **if** T^2 remains the same but that will not happen and that we know from past experiences. If that happens, we have the challenger 2004 disaster repeating once more.

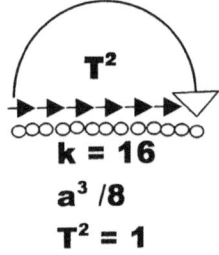

T^2

k = 16

$a^3/8$

$T^2 = 1$

Increasing space-time displacement by six will decrease space by six and the distance the space progresses from a centre by twelve. The heat factor of the craft will rise by twelve times as the space decreases by six times.

Increasing space-time displacement by twelve will decrease space by twelve and the distance the space progresses from a centre by twenty-four. The heat factor of the craft will rise by twenty four times as the space decreases by twelve times.

Motion of anything in any form is about duplicating the existing into following on images of the same thing. That is connecting space to last a certain period in relation to a specific point holding singularity before the next singularity is enticed or charged to maintain the space-time in motion. Every time (and in this case the referring to time proves to be most accurate) is having another singularity building and breaking down the space it represents for that duration of time. The time duration leaves singularity selected in charge of producing the roving space the extent in which it can duplicate the space it has to duplicate. By reducing the period the particular singularity may lay claim to the space, will inadvertently produce smaller space it is able to reproduce in the shorter period of time.

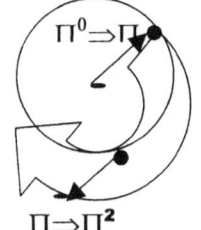

$\Pi^0 \Rightarrow \Pi$

$\Pi \Rightarrow \Pi^2$

There are two ways of looking at this issue. The one is looking at it from the centre that is keeping the rotating object honest or there is the rotating object forming space in relation to the centre and placing the centre in the centre. It will always be one taking prominence to the other and where Kepler introduced the formula it is indicating motion producing gravity which is gravity that is keeping form outside the sphere. Gravity is motion but the motion we see is much different from the gravity we experience while we know it has to be the same with only relevancies changing.

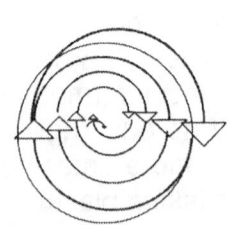

No matter how one looks at the Kepler formula, it signals the same principle. It shows how motion erects the Universe by mathematical equations. It puts singularity, as one in relation to six and that is the Universe decoded.

By rotating around a centre that is, standing still such a centre forms a divide that separates the unified unit. **Any point will be opposing itself** within the **rotating of 180°** where it **then changes every aspect** of its **previous flowing** characteristics it had or will once **again have in 360°** from there. While in rotation from the viewpoint of a bystander it all may seem static and never changing. However to the object in spin every next instant in time will be diverting from every aspect it had every second passing, and the direction it held in relation to the direction it held the previous mille-, mille-second as it will totally be incompatible with the direction it holds the very next mille, mille second of rotation. This is why we can use degrees measuring the circle by (6^2) (forming the square relating to matter through singularity) X 10 (square if space) = 360^0 however it is always in motion.

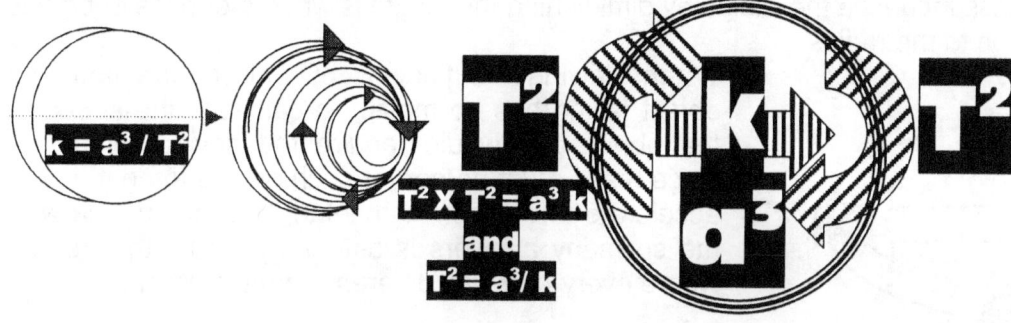

$$k = a^3 / T^2$$
$$T^2 \times T^2 = a^3 k$$
and
$$T^2 = a^3 / k$$

The square of motion $T^2 \times T^2$ forms the square of space $a^3 \times k$. Space a^3 is reducing by the motion of space with the implementing of T^2 having **k** as the constant. It comes about as the earth spins around the Earth axis. I call this positive space-time displacement

$$k = k^{3-2} = k^1$$
$$a^3 = a^{2+1} = a^3$$
$$T^2 = T^{3-1} = 2$$

$$k = a^3 / T^2$$
$$k = a^{3-2} (T^2)$$
$$k = a^{3-2} = k^1$$
$$k = k^{3-2} = k^1$$
is the same as

If space were zero or nothing as Mainstream science so affectively teaches us, then Kepler's principle formula would need the changes Newton brought about. It is true and stands tested like no other research ever coming either before or after Brae and Kepler's work. By reducing the line to infinity and raising the line again back in the direction of space, the line would erupt as a natural sphere having Π as the natural basic value. That is the value Kepler interpreted. However not realising what he saw he chose to use different symbols.

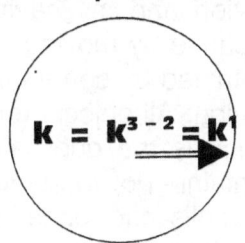

$$k = k^{3-2} = k^1$$

$$a^3 = T^2 k$$
$$a^3 = T^2 k^1$$
$$a^3 = T^{2+1} (k^1)$$
$$a^3 = a^{2+1} = a^3$$
is the same as

$$T^2 = a^3 / k$$
$$T^2 = a^3 / k^1$$
$$T^2 = a^{3-1} = T^2$$
$$T^2 = T^{3-1} = 2$$
It is all the same

$k = k^{3-2} = k^1$ is in direct relation to $a^3 = a^{2+1}$ is in direct relation to $a^3 = T^2 = T^{3-1=2}$. With this information staring mainstream science in the face and scream pleading at them to recognise the information they turn around and ask why can man not fly off to other galactica at the speed of light.

When the astronaut is departing from space on Earth or filling Earth space it will take the departing astronaut k^2 time to reach k^1 and fill out k^3. At present and in this moment our most impressive astronautic engineers will devise an engine that would cut k^1 by say half. This achievement will come as they increase the power output say for argument sake to double what it is at present. There was no friction of particles destroying the frame of the craft because there are not enough particles in space to do it.

However Newton recognised just the opposite and even allowed a freezing of motion and therefore time.

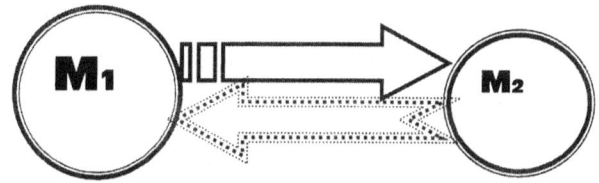

How does one reconcile the behaviour of the top with the foundation of science?

Mass has no influence on gravity in spite of all Newton's unproven claims. The fact that mass in inversely related to the radius as Newton's first formula proved is the proof that mass is not gravity but something after the fact of gravity.

$$F = \frac{r^2}{M_1 M_2}$$

25 kg   5 kg

The distance is equal
The time is equal
The mass is not equal

The only rue way that mass influence the radius by diminishing the length is when the mass of both is placed inversely in relation to the radius.

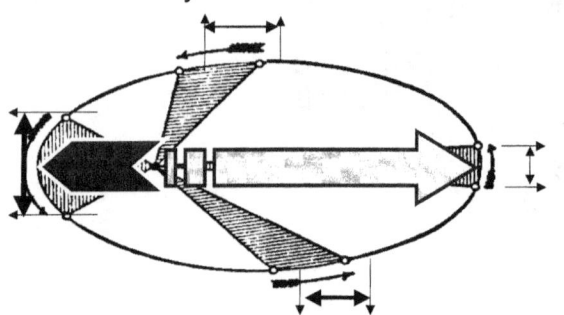

When viewing the findings of Galileo one find that object falling has no mass. To calculate the speed of the object one would require the driving force and since mass must be part of such driving force it has to accelerate the object. At this fact Newtonians threw at me so many answers is differing from north to south where every one was different from the other.

Fact

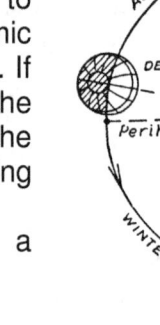

remains if I fall and my mass has any factor in my falling then me being heavier must have a profound affect on the speed of the falling. When I fall I have motion and my having motion eliminated my mass because by moving unrestricted I have no mass. I only have mass when the motion gravity give is restricted by some influence that restrains the gravity in motion. All objects rotating around the Sun has motion, which is the duplication of the space in ratio with the containing of the time. You Newtonians out there try to be realistic for once in your life in your thinking of cosmic physics without being brainwashed by your education. If the planet mass had the influence of producing the gravity that held the planet in orbit in relation to the centre of the Sun then the planets had to orbit by using the perfect circle.

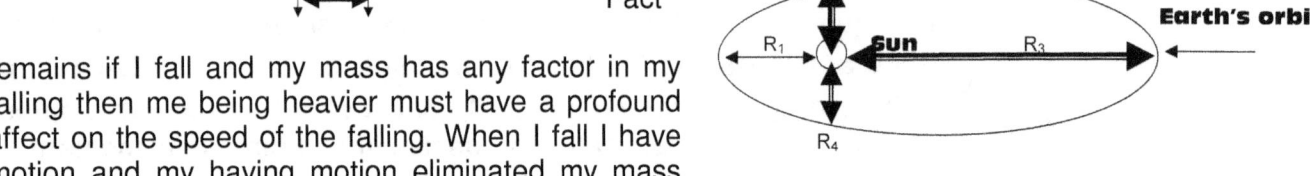

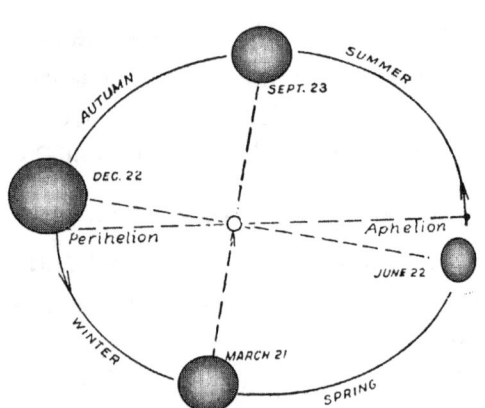

By having a variation radius between planets and the Sun centre it has to mean that either the planet mass show strong variation during the orbit of the year or the Sun shows variation that affect different planets at different time or both must show mass differentiation where the planets become bigger sometimes and other times reduce in size. Since we know that is not the case and we no the orbits do have an eccentric anomaly by the measure of $E - e \sin E = m$ it is the M that I dispute. Another aspect of contention about the fact that if mass did play a part in the orbit it is the largest of the lot that should be closer and the smallest being further away. They are as scrambled as coffee with milk and sugar, which again shows mass and size makes no distinction.

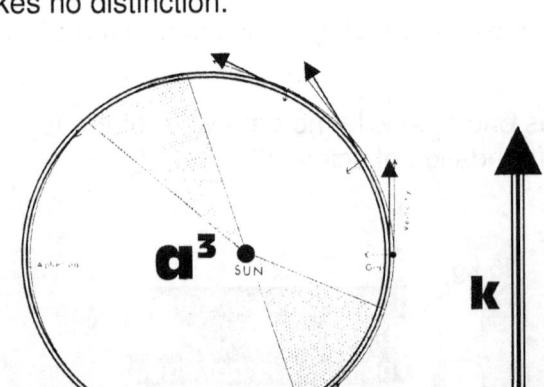

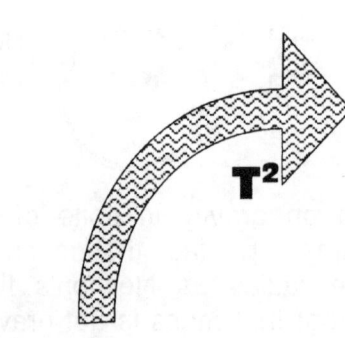

All spinning matter has the point where the spin is still there but the radius is too small to measure by any means. That point in the very and precise centre of all rotating objects is standing still in relation to the rest of the body that is spinning around such a centre. In relation to that logic I do not except Newtonian science holding the radius of a spinning object unaccountable in the spin, whether the spin is applying or not.

Applying Newton's second law F=ma

One arrive at the formula
GMm / r^2 = m (ω^2r)

By replacing (ω^2r) with 2Π / T **we obtain Kepler's third law**

This law predicts that T^2 = a^3

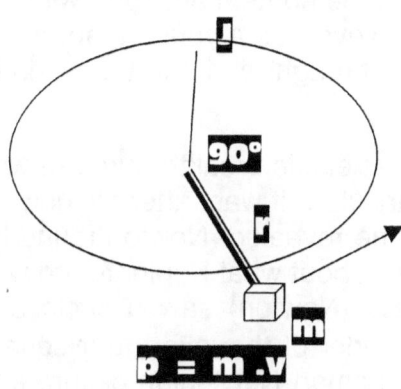

p = m .v

The mass (m) **multiplying the speed** (v) **forms a new value** J **AND THEREFORE** j
CONTINUOUS TO IMPLY J = I ω

= r X p **where** p = (v =r x ω)

J = r.m.v = m.r^2 .ω = I. ω **and becomes interpreted as J = I ω**

This establishes that r = dJ / dt

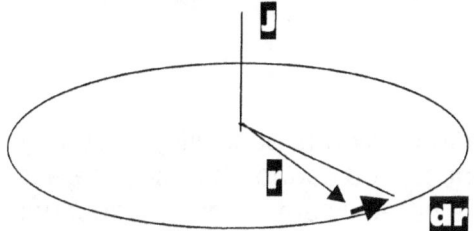

r = dJ / dt In the case of planets in orbit around the Sun r forms a value of zero because dJ / dt = 0.

What this statement implies is that r does not exist. When anything has a value of zero it is for all purposes non-existent. Only when an object is following s straight line can the radius be non-existent because the radius alters value through time development.

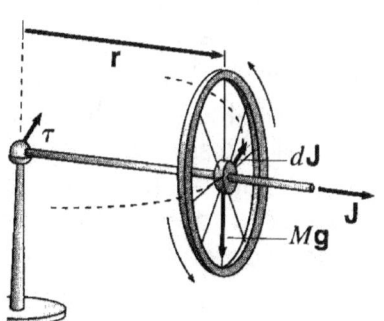

To be realistic there is no comparing w wheel spinning on Earth to the planets spinning around the Sun . To have a wheel spinning on earth one require to intervention of life and life is a most alien aspect in the cosmos. The wheel can never spin by independent initiative without life supporting such a spin. In that sense it is illogic to compare a^3= T^2 k with Newton's second law F=ma
One arrive at the formula
GMm / r^2 = m (ω^2r)
This can only be a reality if life provide and actively participate in the support the energy supply that will realise the spinning motion a wheel would have. To work with mass in physics is very earthly bound and that is precisely what life is. But as mass is a very Earthly aspect of physics, we must never spare any intensity in the effort we have

to keep mass and the likeliness of life from our minds when considering the cosmos and all aspects about the cosmos.

By replacing ($\omega^2 r$) with $2\Pi / T$ we obtain Kepler's third law and that is trash because then the third dimension becomes equal to the second dimension and all goes to hell as this formula then would suggest. This law predicts that $\mathbf{T^2 = a^3}$

Newton had the revelation of all the above mentioned as an apple fell from a tree apparently very close to him. He was admired as an instant genius and the one the world was waiting for to be born. I do not, for one second, deny or dispute the revelation. What I do encourage is to place the event into its correct context. It was merely, and simply an apple that fell from its branch to its roots. The apple did not pretend to be a meteorite that fell from the heavens. If it were a meteorite, I am sure, with the man's genius, science would be somewhat different at this stage. However, as a young man, being very impressionable, as all young men are, and with the attention this brought about in the world of science, the matter overshadowed the fact.

I am not disputing Newton; I am disputing the relevance of Newton's scientific breakthrough. It was not two objects of cosmic proportions, colliding in a show of the spectacular. It was, after all, only an apple falling from a tree and not that big an event. With this miracle he revealed, Newton found he was competent to improve on the work of Kepler and what Newton saw about what Kepler found was to Newton's mind the proof of total mathematical incompetence. He (Newton) saw a circle and without Π there can be no circle. Further more, since he was the founder of the invert four square principal, the principle also had to be included the make the picture a smart Newtonian picture and with that remove Kepler as such.

$\dfrac{dJ}{dt} = 0$ Newton, and science, made one enormous blunder, from this stance.

They took the radius of a wheel not to have any influence on the wheel. In doing that, they removed the very fact that keeps the universal attachment together.

They put two objects in an attaching relevancy and then announced no relevancy. Doing that is breaking the most fundamental mathematical principle.

$\dfrac{dJ}{0} = dt$ or $\dfrac{0}{dt} = dJ$ This disputes mathematics.

DJ / dt can have any number from eternity to infinity, only excluding only one possibility; it cannot be 0. By placing the one in division of the other, you bring in relevance. You cannot then say there is no relevance. By doing such, you proclaim that one of the factors is non-existent. In both cases, one of the factors then does not exist. Such a claim is incoherent, because you proclaim that a circle has no radius, or a radius has no circle. When calculating a circle, you multiply either the square of the radius by Π, or the quarter of the diameter at a square by Π.

$\dfrac{dJ}{dt} = 0$ constitutes a circle and is also therefore $\Pi \times r^2 = $ CIRCLE

If you remove r it then is $\Pi \times r^2 / r^2 = $ CIRCLE.

You cannot then say $r^2/r^2 = 0$ and therefore $\Pi \times 0 = 0$. That is nonsense. $\Pi r^2/r^2$ will always be $\Pi \times 1$, and that is where Kepler placed singularity. By hiding this fact Newton went and threw the baby out with the bath water. There is little standing further from the truth than this statement and reality disproves Newton completely. In the motion every wheel has to have a pivot around which the wheel turns. That is called the axis.

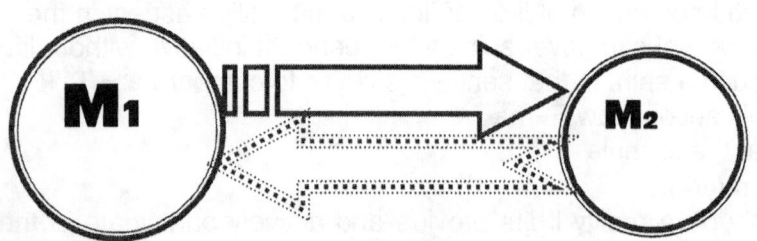

Newton's claim of mass pulling is totally incorrect when compared with reality

Instead time can never stand still as time delivers space in ratio to singularity.

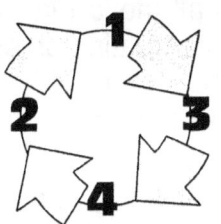 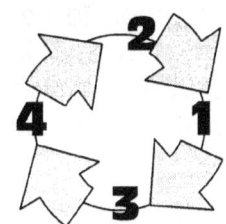 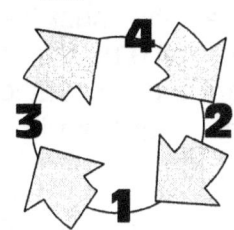

Notwithstanding all the protesting and objection Newtonians may have about the correctness of Newton, the concept is completely fraud in principle.

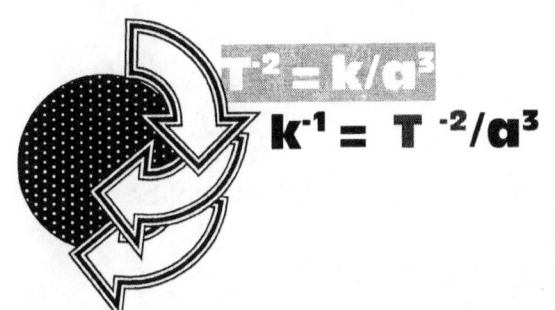

$$T^{-2} = k/a^3$$
$$k^{-1} = T^{-2}/a^3$$

Built into the nature of rotation is the conflict there is between the two opposing sides of the same rotating unit. By crossing the divide the fundamentals in nature changes as every aspect of what was valid change completely to the opposite. On the one side there is a contraction in spin. The thrust draws the spin into the centre by direction of the spin that favours such contraction. This has nothing to do with mass. Then by crossing the divide where singularity changes the direction of motion every aspect concerning the ration changes as it actually alternates. That what previously by rotation came down then goes up in the opposing direction. However that is not surprising because every slightest motion involves just such a change in direction and it

$$T = k/a^3$$
$$k = a^3/k$$

is the process of interacting changes that manifest in charging motion into singularity.

In the way planets rotate around the Sun this characteristics are also present and that forms the criteria foe comic order or gravity. The same characteristics we find in the rotational spin around singularity as well as a governing centre. On one side there is a directional preference to favour the one side and on the other side of the divide this favouring will swap ends.

$$T^2 = k/a^3$$
$$k^{-1} = T^{-2}/a^3$$

singularity generates are motion that provide the singularity to define effect has the role it liquid establish the while the solid uses the motion thereof. Since it

$$T = k/a^3$$
$$k = a^3/k$$

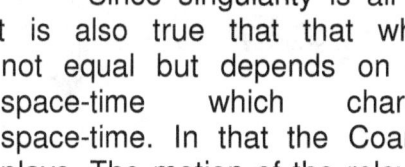

Since singularity is all the same and equal it is also true that that which not equal but depends on the space-time which charges space-time. In that the Coanda plays. The motion of the relevant rotation as the motion bonds the liquid to the solid liquid in motion to extend the space confined by the is two aspects in one unit the defining of the dividing

comes about as the two parts perform each its role on either side of the divide. On the one end the expanding party takes privilege position and on the other side of the divide the concentrating partner takes a privilege stance. From where we stand we see a small and a large, but from the singularity it is one side contributing more motion than the other side by performing duplication or contracting. However there is no big or small. It is all based on the contribution in motion that maintains singularity.

In every cycle of every orbit we find four in time forming different allocations in positions and the varying depends on the relation the allocated position has with the centre.

There are relevancies that form location preference to allocated positions just like seasons do. That is a product of time, which holds four positions in one cycle.

By denouncing Kepler and his formula, one must be prepared then to denounce all motion in that manner, and Newton more than most should have realised that. When looking at any rotating object, there has to be a point of no rotation and no rotation means "no rotation", not no existence. No rotation means a factor of 1, not zero. That then is singularity. The eternal Π, the Π that may not have significance but still it is Π of value.

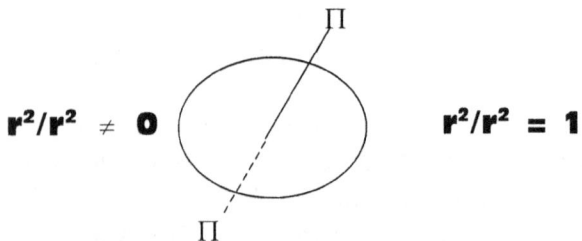

$r^2/r^2 \neq 0$ $r^2/r^2 = 1$

The relativity remains one, eternally one, but it cannot be zero. Therefore, dJ/dt cannot be zero.

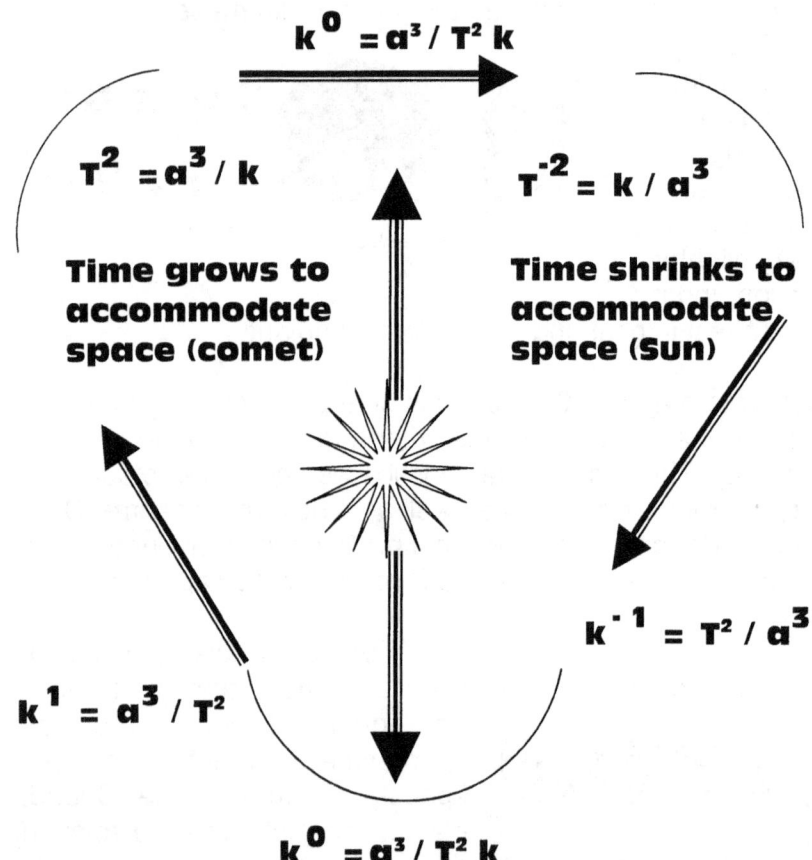

$$k^0 = a^3 / T^2 k$$

$$T^2 = a^3 / k \qquad T^{-2} = k / a^3$$

Time grows to accommodate space (comet)   **Time shrinks to accommodate space (Sun)**

$$k^{-1} = T^2 / a^3$$

$$k^1 = a^3 / T^2$$

$$k^0 = a^3 / T^2 k$$

dJ/dt can be eternal or infinitive or at the worst it can be dJ/dt =1 but dJ/dt $\neq$ 0

Looking at Kepler's statement it seems like a mathematical blunder made by an incompetent not understanding the most basic principles one might have in mathematics. It reads that the space in the third dimension is equal to the calculated motion of the space in the second as well as the first dimensions.

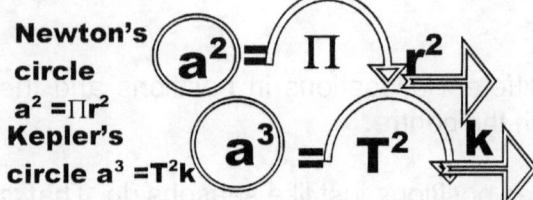

Newton's circle
$a^2 = \Pi r^2$
Kepler's circle $a^3 = T^2 k$

$$a^2 = \Pi \quad r^2$$
$$a^3 = T^2 \quad k$$

Kepler saw a circle because space is motion provided by singularity from a centre.

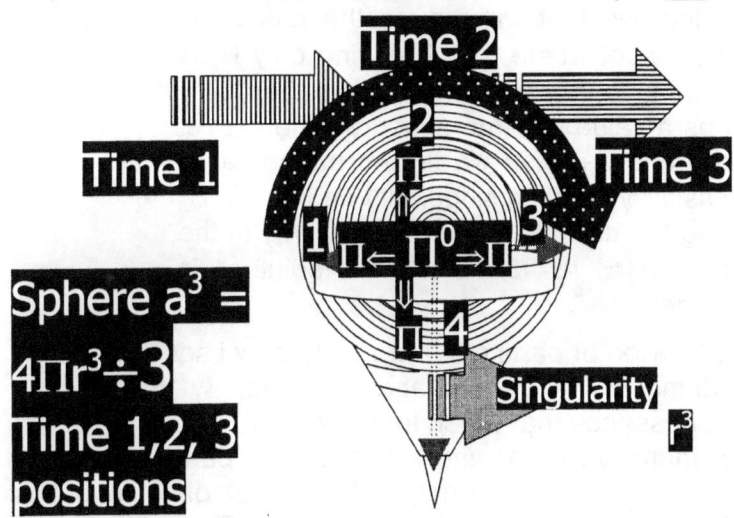

Sphere $a^3 =$
$4\Pi r^3 \div 3$
Time 1, 2, 3
positions

A child will know better than that because it is hardly the manner one might use to calculate the surface of the space let alone the cube of the space. Far better is the use of the correct formula to calculate the cube in space as the one used to measure the volumetric displacement accurately as follows:
$a^3 = 4\Pi r^3/3$. Using $a^3 = T^2 k$ is mathematically a fools argument, and yet even when someone perceive it to be wrong, was it accurately changed by Newton?

Following the mathematical volumetric formula on can see what is required to generate and duplicate every time the motion allocates space a new position. The centre line holding singularity ($\Pi^0 = r^3$) forms and then charges space by the cube in relation to the outer edges where singularity in Π meets time and where the Coanda principal puts the edge on space. The space in time redistributes the volumetric charged space in relation to three sectors time hold space in. That is the volumetric formula.

Newton said a sphere is $a^3 = 4/3\ \Pi\ r^3$

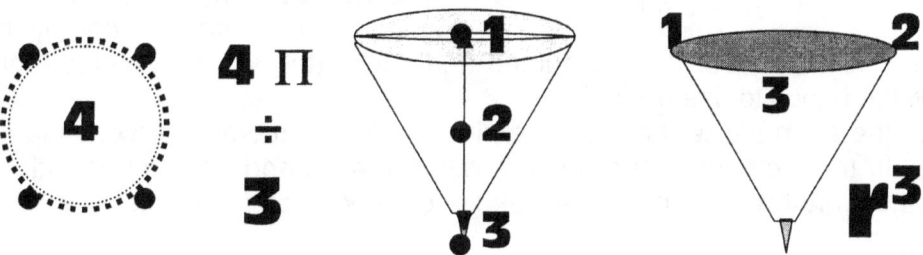

When the top starts spinning the spin generates the time difference that single out the space independence to bring about the space spinning. This is not what Kepler's formula show. Kepler's formula shoes the space turning by rotation as well as lateral motion and by duplicating and contracting the space generates new space as the space carries material through time.

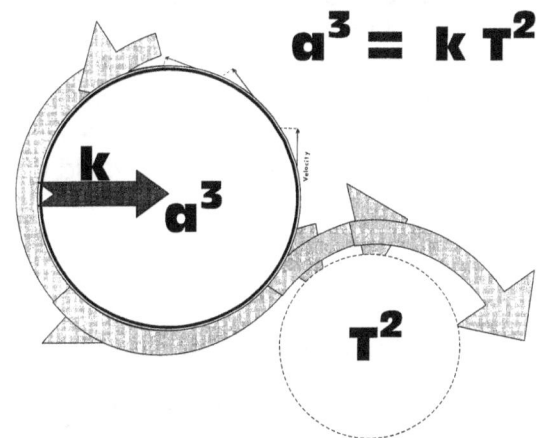

$$a^3 = k\ T^2$$

The space a^3 duplicate the position it had, it has and what it will have $T^2 k$ in relation the where it was, where it is and where it will be the very next instant. It does not mathematically reflect on a volumetric space to enable mathematicians to play a game and prove to their compatriots as well their own personal vanity how skilfully they can play a game with numbers and rules. It indicates a cosmic principle on which the four cosmic pillars rests. Dare I say one has to be more than a mathematician to appreciate this difference? This I say because I am of the opinion that if just one mathematician in four hundred years tried to find out why the formula used by their skills to calculate a volumetric displacement do function in the purpose they use it, that mathematician must then have had the ability to see the difference there is between the Newtonian formula depicting the mathematical purpose and the mathematical language suggesting a cosmic principle.

Newton said a sphere is $a^3 = 4/3\ \Pi\ r^3$
Kepler said the cosmos told him a cosmic sphere is $a^3 = k\ T^2$.

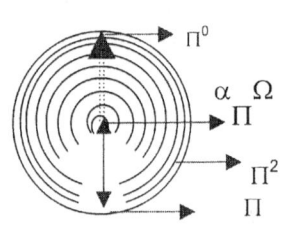

Going down the line that will reduce the radius we finally land on a spot where $r^0 = 1$ leaving only Π as [part of the formula $a^3 = 4/3\Pi r^3$, which then only leaves form as a value with no measure to form. Yet that is not the end because we find in mathematics one more possibility carrying singularity and that is Π^0 which will bring the ultimate value of singularity to the space concept in $a^3 = 4/3\Pi r^3$. At that point **a^3** is singularity altogether.

I have tried for so many years to accommodate Newton or parts of Newton into how I see cosmology but with no success. Newtonians leave everything half explained and from that try to formulate perceptions. What is matter? What is in the finest essence that which form material? When in search of an explanation to define material I find that there is an ongoing report on the particles forming

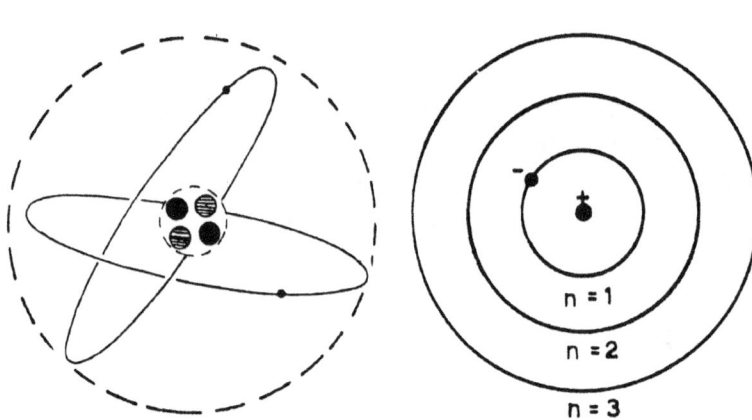

material, which is not ongoing but is lumps called atoms. Then I suppose you may dissect that into as far as we can see with an electron microscope but that still says nothing on what atoms are. What is that which does the circling electron confine? What is the electron putting into a circle that then becomes a confined unit? What fills the space in the cube of material?

It is well accounted that there is this spectroscopy and the energy bands shifts as the energy levels rise or reduce. It is accounted and it is better well documented that the energy requirements allow the bands to rise or deplete.

The more the energy is in the atom the wider circle the electron has to travel to accommodate the more there is in the atom. In the atom we find a neutron with no mass and an electron with mass. Then down at the bottom there are two protons that expand and that reduce the space they hold.

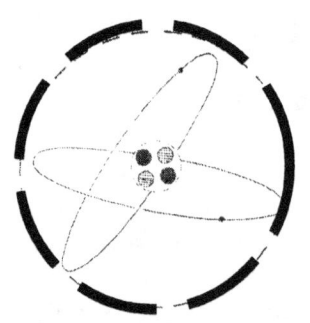

That is so scientific accurate and correct and the only way I have to be more accurate is to remember every useless name some incompetent Muppet gave that enormous discovery that that incompetent Muppet found to be his remarkable creation. The more illustrious the meaningless name sounds the more important it will sound. The most important name will hide the biggest discovery of the lot...that the discoverer has no idea what he discovered and therefore his discovery in the long run has no meaning to science what so eve. Therefore to hide the truth he has to invent a more useless name than the previous useless named other particle has. Never is there a reference to why the electron is spinning in the first place. What is in the purpose of the spin? Why would the electron spin and why would the electron enclose what is in the spin it protects.

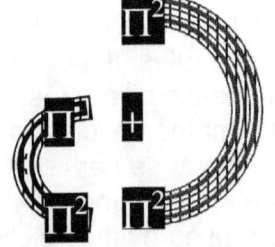

What is on the outside and what is on the inside of the electron that is spinning in a wider or smaller orbit. What is the enclosed material that the electron is protecting? What is behind the electron and what is in front of the electron and what is the electron?

That again brings me back to my first question: what is material?

We know that the atom has two positive particles in relation to what seems to be a neutral particle that is between the positive and the negative particles. The positive is a double 2/3 and the neutral is as much negative (2X 1/3) as it is positive 2/3 and they're the explanations tops. To continue the discussion we than have to establish what a positive is and what a negative is to find a neutral. What will bring about something being negative because being negative is only a position in relation to being positive. What is being positive than? What is positive material and what is negative material because then the neutral has no clear definition in existing.

All the negatives will eat up all the positives and that will leave nothing as neutral. Most senseless is the part that I am the one they frown upon, the one they reject and the one that is uneducated.

In the atom I found a relevancy where the proton has double motion ($T^2 + T^2$) and the neutron has the linear motion of a^3 **X k.** The total positions are equal to every possible position the sphere may offer in relation to a centre singularity

Then when this seven positions are put in relation to time we find that the seven multiplied with the three positions time has the total including singularity expanding and singularity contracting is (7X3 = 21+0.9991 = 21.9991 / 7) = Π. That concludes that the Universe has gravity at all levels because the atom is a sphere as much as the atom is the Universe.

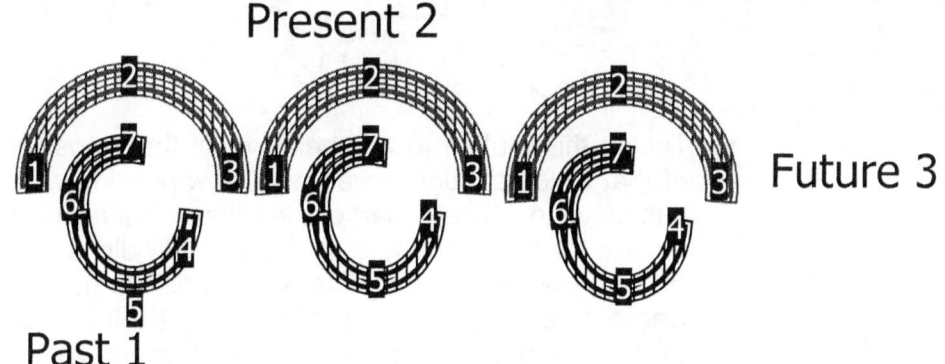

Present 2

Future 3

Past 1

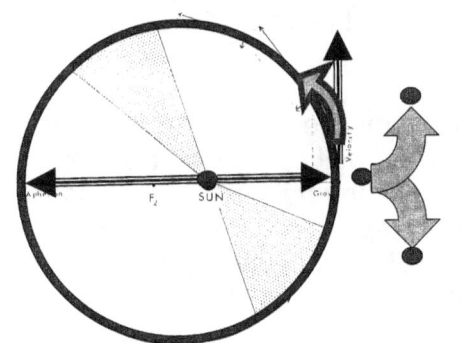

Present

Past

Future

Holding singularity

If we look at nature there is no position we can give time to rest at zero because time is the constant flow of all material in ratio to each other. In that the notion of t=0 is incredibly outdated. The fact that there is time is also the fact that there is repositioning of material in relation to each other on an ongoing constant basis. When I see a planet such as the Earth in rotation around its axis the centre line, then we find the rotation is a flow of atoms repositioning in relation to one another. I stop at the atom because I have to stop somewhere and the atom forms the conclusion of the Universe but our conclusion of the atom forms our conclusion of the Universe and not necessarily the atomic conclusion that the Universe comes to define as the final conclusion. In other words there are mostly and many other forms of atoms in the Universe outside our spectrum.

Every time an atom is moving, such moving consists of a series of duplicating and relocating reference positions where every individual atom is constantly proceeding a line where the atom is replacing the location and the position the atom by taking up the place the atom in front had in relation to the centre such moving of the atom. That centre is incapable of moving therefore every time motion is established such motion is the generating of the entire field that is between the two locations of eternity versus infinity.

The atom is not only moving forward it also is moving to the side in accordance with the centre. It is following the leader but the leader is following a line that finds a relation in line with the position the atom in front had and the atom at present then fills. The filling of the position is in accordance with the centre and the centre is the point that claims the dominance of the moving. While from the point the atom has, such moving is straight in line with the allocation the atom in front has. The relation however is a centred one where there is a point that never moves and cannot move.

This part is represented by one side of time that forms the eternal side in time. The rotation is never concluded but always repeat the previous into the future. It is Newton's invert square law where there is four positions always following each other and this position was present when time was eternal. Please take note that time could never stand still and therefore Newton's assumption of t=0 is much misleading. The time found on location in which it rotates and that position holds all four points on one exact spot but that does not remove the location by giving the location a value of zero. However in that there is another aspect where the infinite secures a position and still maintain the point having all sides sharing one point. That is the very inside and to detach infinity from eternity that has to be a moving where that point holding infinity also has to relocate while in truth it cannot relocate.

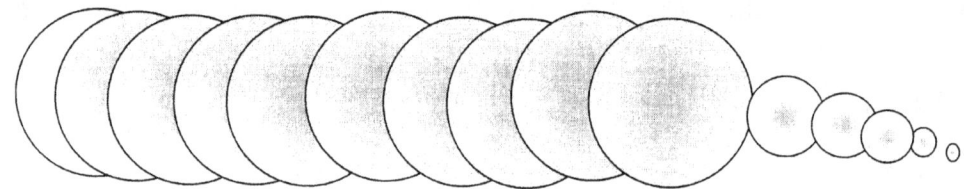

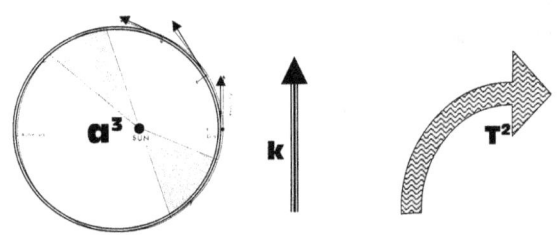

Taking this motion to another level is the repositioning of that which cannot move into a new position where it has move too. This is part of the rotation but also t is the moving of time in a lateral direction. While the lot is rotating the lot is shifting into a new position going from where it came to where it is to where it will be next to where it was every time it is going forward. There are seven points moving in the circle while there is this seven points coming from and going to while it is in the present position. That makes it the seven it is plus three, which is the time aspect ten.

That action is what brings on motion and involves a line as well as an incomplete circle. The space a^3 duplicates what was to the present going to the future in a semi circle T^2 as well as a straight- line **k**.

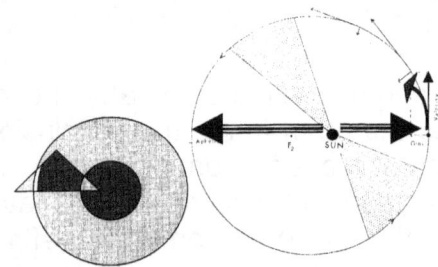

The motion of the space a^3 involves the rotation T^2 as well as the straight -line **k.** In order to be in space the space has to be on the move. Time cannot stand still for then there can be no space because the space is the directional movement in a lateral as well as a circular direction simultaneously. **There is no possibility of time standing still at t=0. If there is t=0 there is only 0 and since there never can be 0 there is no possibility of time standing still.** If you find time standing still, chuck away your watch because the watch and not time gave up the ghost. In order to be in space, space has to duplicate by presenting what was in the past, take it through the present and fling that which filled the past and is filling the present into the future. There are both actions serving space where space is going sideways to go forward as it is going forward in a sideways direction. That is motion whether it fits Newton or suits Newtonian perception it is of no consequences because the atom is moving along as the atom is spinning.

The Newtonian idea that the one direction eliminates the other direction by the sharing of a mutually established centre has no base in reality. Kepler clearly brought to

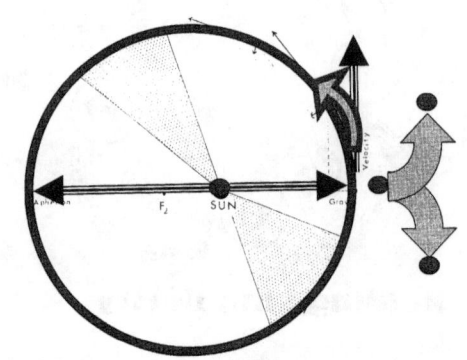

science the fact that space forming include motion of the space but as important is the fact that the motion cannot exclude any one of the factors in the circular or the linear and that is ids the product of both factors that form the flow of space through time.

By supposing that the one factor removes the other factor from the Universe is quite frankly forming an effort to absolutely destroy what Kepler said! Then we get back to the question that I asked earlier as being what material is? What is filling the atom? What is inside the proton, the neutron and the electron? If getting more energy is getting the atom fuller by expanding then the inside must be energy. What is energy because the word or term energy has become an escape goat used whenever Newtonians run out of answers? Einstein was of the opinion that when matter goes past the speed of light it would become pure energy. So what is matter before it goes past the speed of light? Is it then less energy or does it become something other that energy?

That brings about the next question: what is mass. Why would that inside the atom have mass and that outside the atom have a different mass. Why would the proton being so much smaller than the electron be so much more massive? Newton came up with this idea of mass and everyone accepted mass because no one could disprove it and since Newton never was set the task to prove what mass is it was left at that. Why would material have mass?

Taking the argument back to Kepler's law, a^3

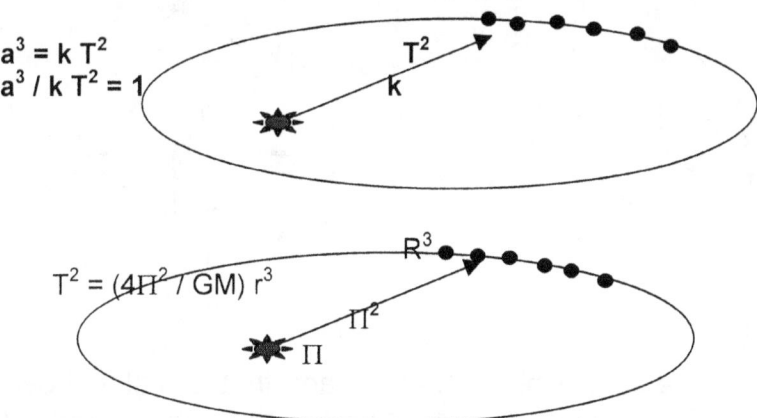

$$a^3 = k T^2$$
$$a^3 / k T^2 = 1$$

$$T^2 = (4\Pi^2 / GM) r^3$$

We all know that all object have mass when the object is on Earth but when the object is in outer space what would give the object mass? If mass was conducting the motion of planets then surely they would be arranged according to mass. The biggest must be to the most inside and the smallest to the further outside. If it was mass that was generating the motion then the most massive must be flying while the smallest must be crawling. That is not the case at all. The distribution of planets relies neither on mass or speed of motion or any distribution in size at all.

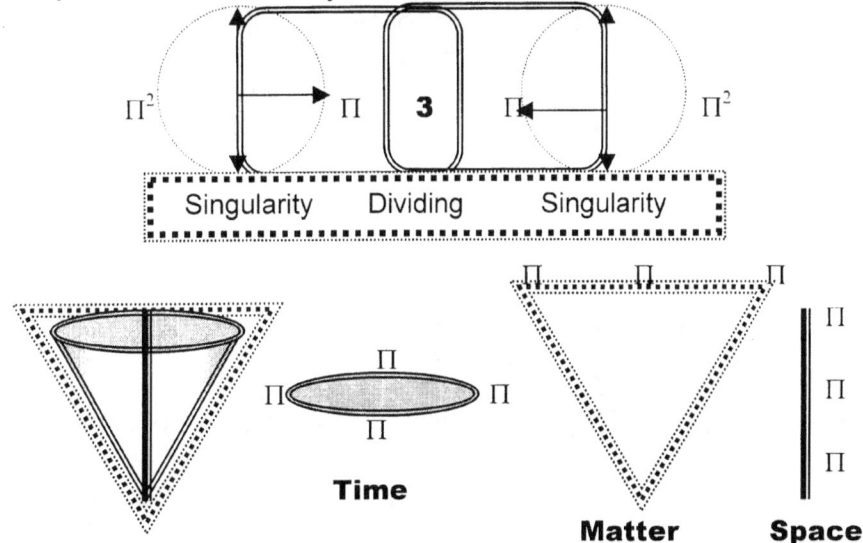

There is no correlation that would even suggest that mass plays a part in the orbits of planets or any cosmic structures. The dynamics we contribute to the realisation of mass should be far better defined if there is to be any clarity on the matter. According to Kepler space needs motion to produce space $a^3 = k\,T^2$. Let us see how true that is. How does that which fills the inside of a particle move from one point to another point during time? The rear must back the front by pushing the front forward while the front is pulling the rear by vacating the front. If the front does not move the rear will smash the font out of line and in some cases destroy the front from behind. Think of a car crash where the front becomes blocked by other objects while the rear is coming to fill the front from behind.

That which fills the front has to vacate the location it holds in an agreeable direction within a suitable time while that which is filling the relocated position of the front requires the filling from behind as to find the required flow so that the motion can go about spontaneously. If that which is coming from behind is retarding in motion as to fill the vacating position in front the unity in the movement will tear into parts. In order to accomplish motion, the motion has to accomplish the sphere. The motion not only fills the vacating space by rotating in a "follow my leader" process but also establish from one point in infinity a line of points running along a line that is never there. The line establishes four rotating positions parting infinity from eternity. In that I do not find grounds to blame mass for anything. I do find that the matter behind the matter in front is only retarded allocated filling of a position in time. The following is in a retarded stance to that which it flows and that makes that which follows behind that which is in front being ahead.

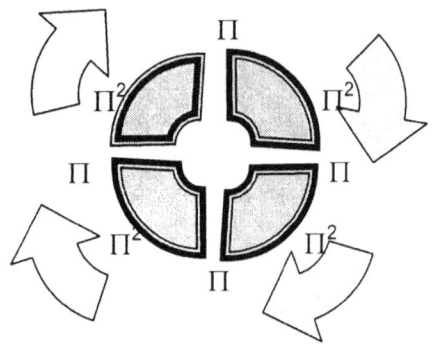

The one is always behind the other which is in front and that is material because being behind that which is in front gives both that which is in front as well as that which is behind an independence from one another. The proceeding to follow and the following that proceeds placed both in separate compartments during time and that puts the one in motion while the other is filling the motion. Is till do not find mass. What I do find is the relevance that Galileo found. All objects notwithstanding size or mass fills the position the one in front left vacated to be filled by the one coming behind.

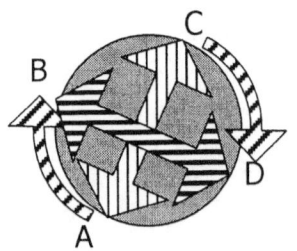

There is more to motion than just that because the rotating plays follow my leader while the lateral plays follow my leader and with that there still are two identifiable substances in relation to the space and the motion that time represents. The lot is duplicating and by duplicating every aspect is relocating what it represents in terms of space –time to a new location.

We have the rotation there is where the one point serves as a guide to the previous point and follows future while honouring singularity. That is one part more and that is time in the lateral or eternity stopping.

the next point into the of the story. There is running without ever

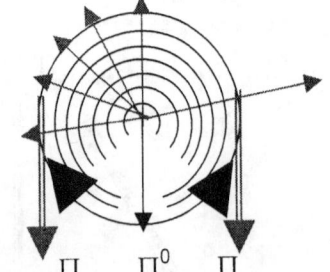

However that which the lateral but also by shifting point A to

cannot move does move not only by repositioning relocating the rotated into ea new relevancy. By point B that shifting also involves moving point K to

where point L was before. However that centre is not movable and on both flanks the lot shifts. That means what ever is relocating a position is breaking down all there is and shift the lot to where it is going to be by generating what there was in the previous location onto where it will be in the next location.

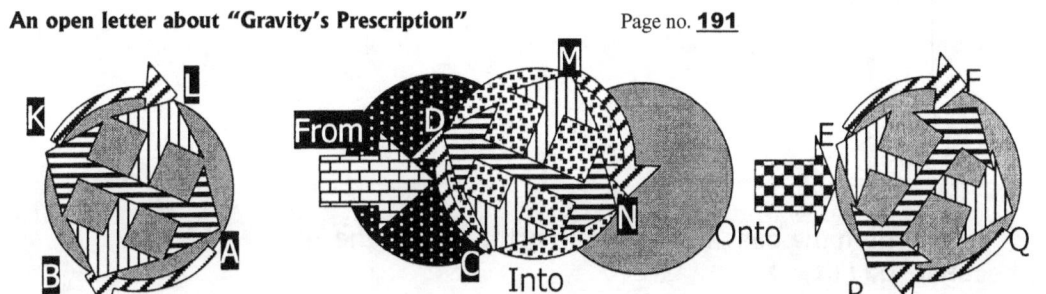

Then this lot in rotation also has to relocate that which rotates along a single file.

The point in singularity cannot move and by it being unable to move it has to break down all relevancies there was and shift all that to new positions in relation to an entire new Universe. In that there is no mass. What there is we can describe as controlled duplication where the duplication is represented by the expanding while the control is represents by the contracting. It is just as Kepler said it will be when it is $a^3 = k\,T^2$ and on the one side there is $k = a^3 / T^2$ as singularity generates space-time in the space generated by time finding a new location during the time such generating takes while $k^{-1} = T^2 / a^3$ the motion produce a new space in the allocated position it has to be at that particular time. Still in all the dissecting I find no evidence of mass doing anything and even less generating gravity.

It is a pity Newtonians never get more specific as to where one may find mass and how mass go about in producing gravity. Being as explicit as I am by going into time between eternity and infinity I still find no mass. I do find eternity forming the three positions on the side that has no end and those three positions do correlate to the most accurate detail in relation to that which has no start. In between that which has no end and that which has no start space-time is generated through the motion that activates space-time.

 Newton made the error Newtonians still do after so many years. He took the fact of life as a cosmic reality. He took the motion that life can achieve as standard cosmic occurrences. Newtonians go much further than Newton did by claiming life comes at a dime a gross throughout the Universe while there is no evidence of that. To swing a weight on a string from the hand has as much cosmology in it as trying to pump a tire with air and then compare that result to a star...and yes Newtonians do just that! When considering the motion applying then first see that being tested fits freely in cosmic reality. A rock cannot roll up a hill and a brick cannot be while a cloud can't cycle a bicycle.

That which Newton supposedly discovered being gravity is that which Newton denounced as zero. When Newton discarded the motion part by putting the value at zero he threw away that which produces gravity on both sides of the border of motion.

By replacing $(\omega^2 r)$ with $2\Pi / T$ we obtain Kepler's third law
This law predicts that $T^2 = a^3$

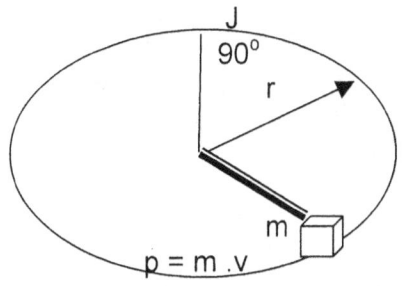

This establishes that $r = dJ / dt$

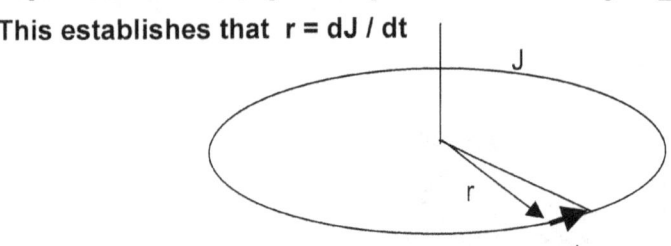

$r = dJ / dt$ In the case of planets in orbit around the sun r forms a value of zero because $dJ / dt = 0$.

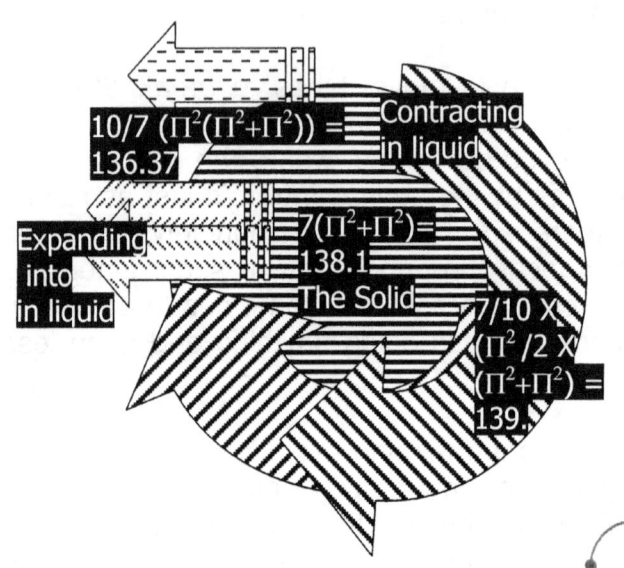

$10/7 \ (\Pi^2(\Pi^2+\Pi^2)) =$ 136.37

Expanding into in liquid

Contracting in liquid

$7(\Pi^2+\Pi^2) =$ 138.1

The Solid

$7/10 \ X \ (\Pi^2 /2 \ X \ (\Pi^2+\Pi^2) =$ 139.

Where the motion of time forming the liquid interacts with space forming the solid we find the two parts motion offer on either side of the divide. The motion is the same but crossing the divide that motion then falls into the other side of the Universe where all changes to become the opposite of what was.

Because of this we find that by nature and not by mass there are elements that favour duplicating more than contracting notwithstanding the number of protons or the resistance the number of protons may show to the blocking of motion. Then there are other elements that favour contracting much more than the duplicating side and again mass plays no part. Mass have a place in Earthbound Newtonian physics motion with the gravity is the restricting of the tendency to move.

HELIUM (2)   LITHIUM (3)   BERYLLIUM (4)

NEON (10)   SODIUM (11)   MAGNESIUM (12)

where one my substitute the correlating restricting of the motion because there tendency to move whereas mass is the motion turning the motion into a

That however is not related to as long as there is unrestricted motion then all the planets are equal as they centre of the Sun . Mass is the experience when unable to contribute to free gravity - motion.

cosmology because there is no mass and circle about the frustration particles

Getting back to the fictitious mass and the question about what material is. What is material and why is material what it is?

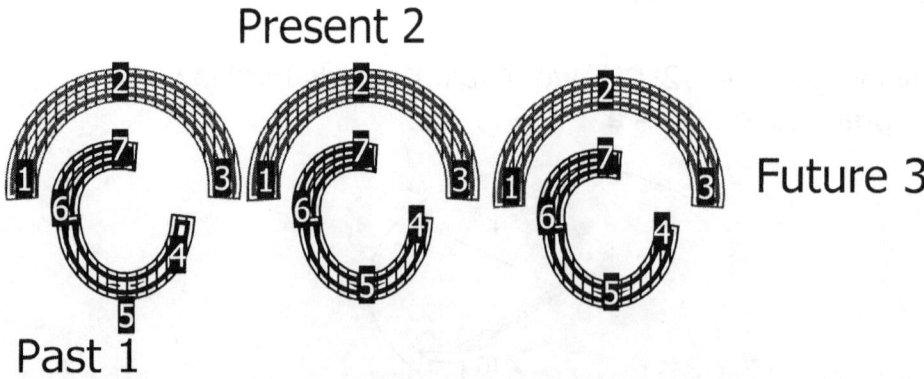

Present 2

Future 3

Past 1

The flow holds seven points that forms the atom and the atom was able to retard time so dutifully and preserved heat so faithfully an entire Universe with immeasurable Universe coming from an array of immeasurable possible Universe can now serve as multitude stages of universe developing eras.

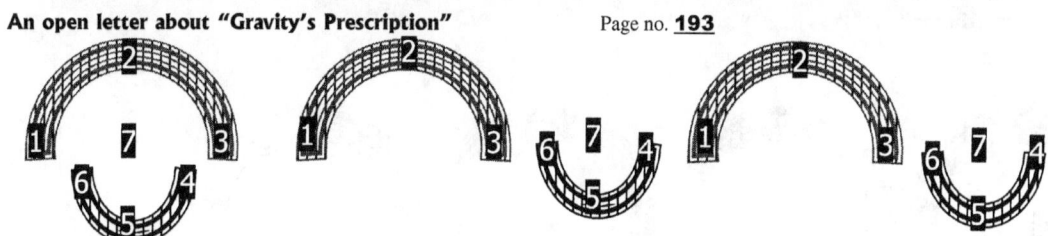

Yet in all that, it never deflected one measure from the original retarding by rotating principle it had in the culmination of starts we preserve to put a Universe in.

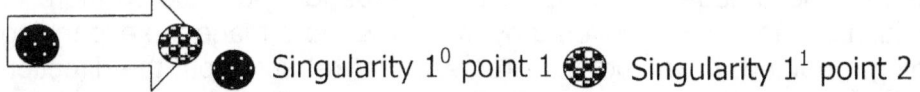

Expanding   Singularity 1^0 point 1   Singularity 1^1 point 2

There was a point in eternity that held infinity and combined singularity. Both are still in our presence and both are wall established. We still gauge the one in the centre of all spinning material and where one the point is not there is a line that is not and in the line the line holds seven points all holding the precise same position. Then light or heat came about and since light or heat expands because it takes more space than what was taken before, there is more of the same that was before.

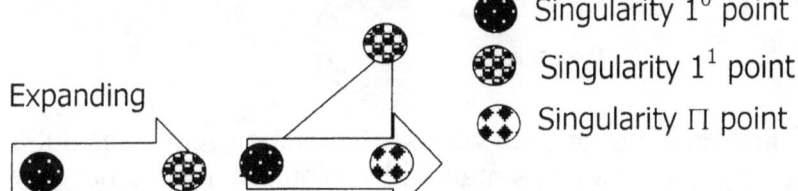

Singularity 1^0 point 1

Singularity 1^1 point 2

Singularity Π point 3

Since the expanding brought what was not in between what is not eternity parted from infinity. But time is motion and since the one became parted from the next the next placed the previous on the other side of the Universe while the following did the expanding. At such a point the law of Pythagoras started to develop the cosmos by putting in the triangle in relation to sides.

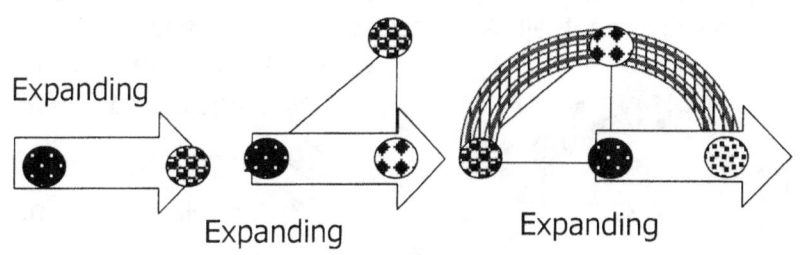

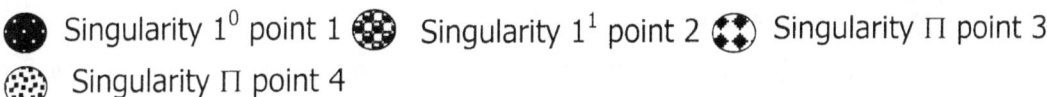

Singularity 1^0 point 1   Singularity 1^1 point 2   Singularity Π point 3

Singularity Π point 4

The flow of time continued as new positions established while point once established did not vanish because once anything is part of the cosmos it stays part of the cosmos, as there is no other place to go but remain in the cosmos. The points continued as time moved on.

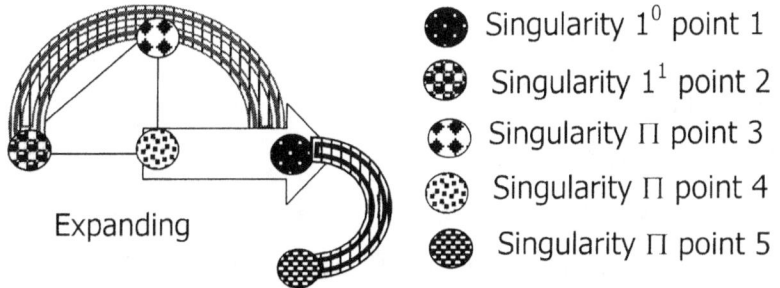

Singularity 1^0 point 1

Singularity 1^1 point 2

Singularity Π point 3

Singularity Π point 4

Singularity Π point 5

Then after the four points expanded eternity produced an additional point at a location where infinity parted company with eternity. At such a point, point five came in place and this was where the cosmos became different from what was previously applying.

The cooling set in where the expanding brought more but also the expanding distributed more of what expanded and since that which expanded was covered by more, the more made the expanding less and by cooling the expanding retracted. It was not the progress that retracted but the direction the progress had that retracted.

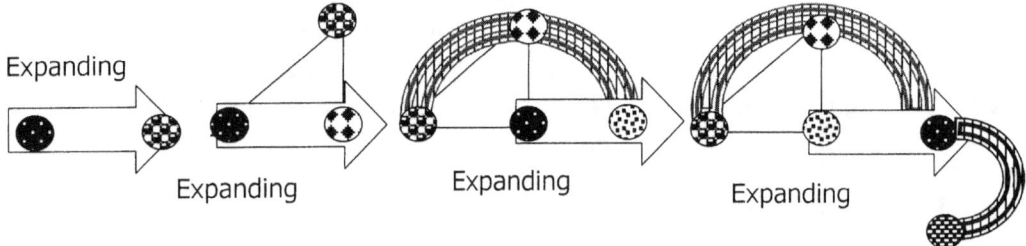

At this point eternity parted from infinity. Infinity is there for all to witness. It is not a hypotheses but a reality. Eternity is there where everyone is seeing it as long as man had a mind. That too is not new and that too is a reality and not a hypothesis. It is a reality within every human and is as concrete as the blood running through the observing person's veins. It is as real as the Universe itself because it is the Universe itself. Any one arguing this reasoning has no mind to understand the smallest concept any human can form. Then came the rest.

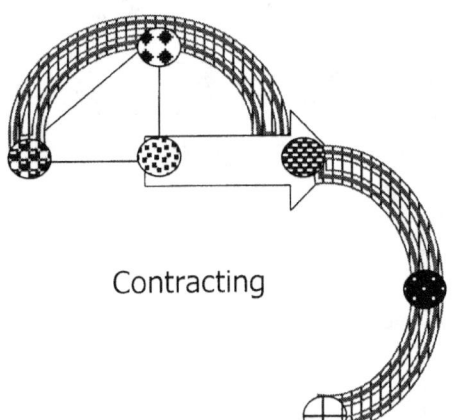

- ● Singularity 1^0 point 1
- ● Singularity 1^1 point 2
- ● Singularity Π point 3
- ● Singularity Π point 4
- ● Singularity Π point 5
- ⊕ Singularity Π point 6

By mathematical implication and the influence of the law of Pythagoras material found a limit at point six.

- ● Singularity 1^0 point 1
- ● Singularity 1^1 point 2
- ● Singularity Π point 3
- ● Singularity Π point 4
- ● Singularity Π point 5
- ⊕ Singularity Π point 6
- ● Singularity Π point 7

That which was inside the seven connecting points serving singularity is contained heat by spin. That which is outside the seven points is expandable heat that expanded without control. That inside was controlled by motion that those outside provided. Any one with doubt that it is controlled heat look at photo's of Nagasaki and Hiroshima and the Bikini island later and see what a tiny bit looks like when the control is released. It is heat that is retarded time and the time is heat dragged on by the lagging behind of heat in a different era of time.

That is material and every seven points holding singularity confirm the backlog of heat dragging behind time. Still I see no evidence of mass and if mass was not then, then mass cannot be now except for in the head of Newton and in the imagination of Newtonians suffering from mental programming. Material is heat that is responding according to a time delay where the spin reduced the time and in that controlled the expansion as it expanded the reducing. That is the essence of any atom.

The material is there because the motion retarded the time to preserve the heat that bonded a unit by a dividing of one unifying point reserved by singularity in order to maintain singularity.

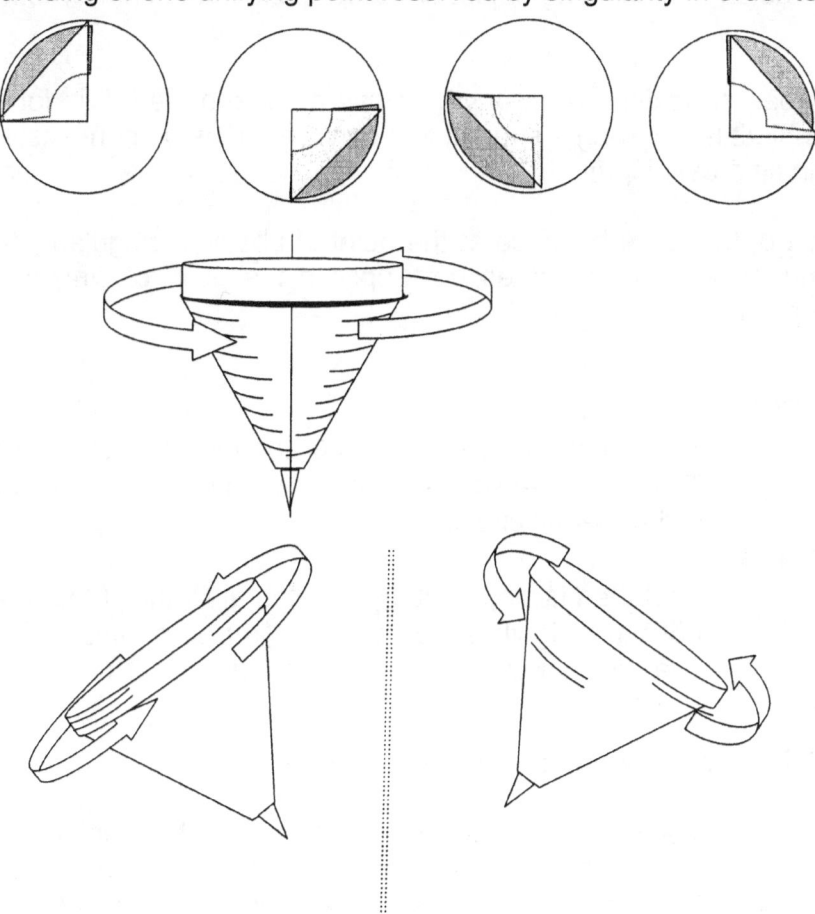

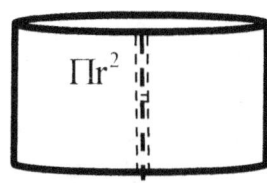

What is it the Newtonians fail to see? If an electron is orbiting around an atom, the inside of the atom must be a circle. If the atom was not a circle, it then had to be a cube. The electron cannot rotate around a cube; therefore, the inside of the atom is a circle.

The radius r runs Π, the value of the radius starts.   It circle Π. and not zero, an The circle does not that   almost

from the circle outwards, from a circle centre point towards circle. In the centre of the circle, there is a point where the runs outwards from that point in all directions towards the Technically, there then has to be a point where r is infinite absolute infinite. However, the circle therefore remains Π. disappear; it remains there for all to see. It is only the radius disappears into the infinite, but it does never become zero!

In a circle, there is a radius that initiates the circle. The calculation of such a circle is $\Pi \times r^2$.

$$\frac{\Pi r^2}{r^2} = \Pi$$

If one removes the radius from the circle, the circle remains, only holding the value of Π. By removing the value of r, Π becomes singularity with no place to be. Singularity is the place where there is no space to be in place. However, Π remains because once r receives the slightest of space Π will find space. Then the circle will grow to $Πr^2$ and r would determine the space. Without space, there is no r but there is a circle with the value of Π. Singularity is in every single rotating object, be it the proton or the combining effort of all particles in the Universe. That is what light and the photon is. It is concentrated heat that the Sun (or any other generator of electricity) connects heat to singularity where the heat receives either temporary connection to singularity or a small piece of individual singularity.

All spinning matter has the point where the spin is still there but the radius is to small to measure by any means. That point is standing still in relation to the rest of the spin. In relation to that logic I do not except Newtonian science holding the radius of s spinning object unaccountable in the spin, whether the spin is applying or not.

What this statement implies is that r does not exist. When anything has a value of zero it is for all purposes non-existent. Only when an object is following s straight line can the radius be non-existent because the radius alters value through time development.

The spinning or not spinning is not part of the issue because at the point of absolute singularity the object never spins. Therefore spinning or not spinning does not apply to the point of singularity because singularity never spins in any event.

$$Π \times r^2 = CIRCLE$$

If you remove r it then is $Π \times r^2 / r^2 = CIRCLE$.

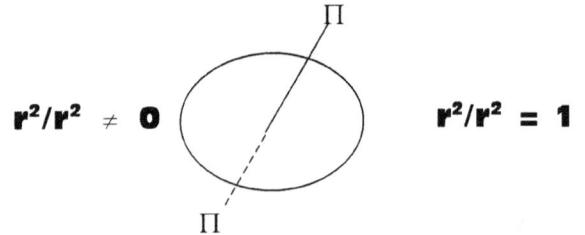

$r^2/r^2 \neq 0$ $r^2/r^2 = 1$

You cannot then say $r^2/r^2 = 0$ and therefore $Π \times 0 = 0$. That is nonsense. $Πr^2/r^2$ will always be $Π \times 1$, and that is the eternal circle.

That then is singularity. The eternal Π, the Π that may not have significance but still it is a Π of value. The relativity remains one, eternally one, but it cannot be zero. Therefore, dJ/dt cannot be zero.

dJ/dt can become eternal or infinitive or at the worst it can become one
dJ/dt = 1

When explaining this to any child, they can immediately see that. Explain this to any Newtonian High Priest and he may have you removed forcefully from campus. I cannot find one Newtonian, of any significance being large or small to accept that. By not having a wheel rotate, the wheel becomes the factor of one, and the rotation becomes zero. The wheel does not disappear. In the cosmos, everything is rotating because nothing ever stands still. Therefore the mean equilibrium, the common factor there is to share, has to be one, eternity, the eternal Π, because all rotating objects has Π in singularity, and sharing singularity, gives every object in space a relation with all other objects in space. After trying for many years to bring them the candle, I concluded that Newtonians are incapable of realizing that mathematical principle as reality.

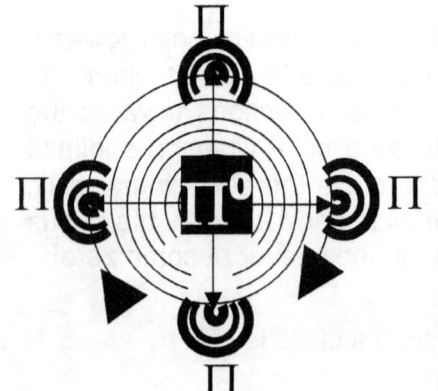

If Newton had said that dJ / dt = 1 then that is exactly what Kepler said when he said that in the centre of space–time singularity is allocated a position of control $k^0 = a^3 / T^2k$. Kepler also said the motion brings about the filling of singularity $k^0 = 1^0 = 1$ when he said that the space is filled by the matter in the motion through the time period. Motion establishes space in time and cannot be zero because THAT is what gravity is. It is the motion of space-time and that can't be zero. If gravity were equal to zero the entire Universe would stop existing.

The centre and that which connects the centre is of such importance that every Universe holding singularity at $k^0 = 1^0$ pivots around it. With every one of the four points taking form to the value of Π at a measure of $\Pi/2$ each brought about the Roche value of $\Pi^2/4$ in relation to the developing centre. One has to remember that the star of today takes on the characteristics of the form of that era. The barrier there is relates to this precise limit being the end of the Universe at $\Pi^2/4$. If there was no bearing of the centre on the why would that be a factor in the formula we use to calculate the size and why would it have any role to play.

In the same manner the ring cannot remove, because the spokes will then still imply where the ring must be. The only way to cheat yourself out of the situation is to remove the wheel and spokes altogether, and you are left with what you say there is: NOTHING. But that does not apply in cosmology. The object rotates the centre structure and therefore there has to be a radius holding the circling orbit in relation to the centre structure.

Removing the radius from a circle does not be removed the circle, because the circle is there, securing the ring If the Universe started from a point of singularity, then there was initial spin at a pace where the spin did not apply and that spin included the entire Universe, still in non-existence. That is singularity. That is the only singularity there can be. The spin was going on for eternity because the spin does not apply, it has a value of infinity and infinity was running along the line of eternity.

By receiving the command, singularity received a value outside eternity as Π^0 received edges. Granted the fact that the edges were so small there still was no r to present a circle.

Having edges where Π^0 duplicates to present the edges, singularity lost the value of Π^0 to the value of Π^1 with the same value singularity had being Π^1 to the one side and Π^1 to the other side, the cosmos received the eternal value of the first dimension outside eternity. It was the square of Π^1 being Π^{1+1}.

That was the first dimension outside singularity Π^0 where singularity has a value of Π^1 in the form of $\Pi^{1+1=2}$. The first claim to space had a value of Π^2. This applied to both sides of the claim to space outside singularity, and the double proton became the dominant factor on matter.

That, which formed the Universe was the growth of time expanding while overheating. It is the expanding by the measure of Π^2 that has much significance as where the expanding went Π and the expanding was limited to the value of Π^3.

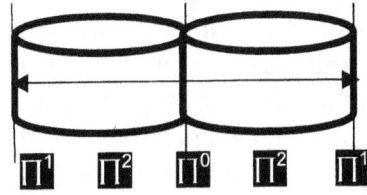

As singularity burst out into matter forming space as much as occupying space inside singularity, the protons started flying around, spinning around singularity, as each individual proton occupies matter in space. For every space there has to be spin and every spin is defined by a relevant. It truly makes me feel bitter thinking about the many times I tried to explain the facts to the Brainy Bunch with no luck. You know you are correct, but that person holding the establishment secured for Newton just push your argument aside, because he has the authority to investigate and lacks the interest to initiate change

In the action of the inseparable drawing closer and moving closer gravity finds the dual value of linear and circular gravity. There is no separation of the two factors acting as one but both have different application and values in the unit. This is the result of singularity having three parts acting as one but giving three distinctions in application.

How many dots was there is a question no person can answer because everything was un-dividable solid and yet it did group together to form every atom located in the 3D.

In the circle T^2k which consists of the atmosphere the space surrounding the rotating object will also extend by **k** as the concentration of the spinning motion draw or drag on past T^2 extending the influence of T^2 by the value of **k**. Very clear evidence about this one can see in the Coanda effect.

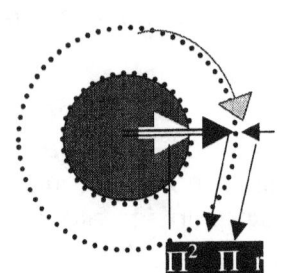

This extending of T^2 to accommodate **k** we refer to as the atmosphere, but physics apply to this extending in the normal fashion. The soil of the structure represents the solid proton being $\Pi^2+\Pi^2$. From the spinning motion T^2 does not stop at the end of the solid structure but the influence of **k** extends and this then becomes the atmosphere.

The influence of T^2 stops at the end of the solid structure but the influence of **k** extending plays a most dominant role in the cosmos, although not yet recognised and that factor is most crucial to a better understanding of the implications of laws governing the cosmos.

With the circle being T^2 **k** the T^2 will reflect the circle in the square with **k** forming the extending of T^2. This is an extending of the six **k** forming in alliance with the centre **k**. This produces that any extension of 6 forming material one further extending goes into space and relates to a seventh dimension. The extending of **k** will not end immediately but will carry to the surrounding space the circle influence through rotation. The influence immediately above the circle will have

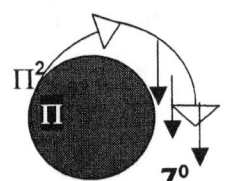

the biggest influence and reduce gradually as the value of **k** reduces in the leverage that the space has on **k** and a gradual but definite change from Π to r will affect the extending of **k** progressively more.

The decline of **k** will follow the same contour of the circle at 7^0. Every one of the dimensions indicates an individual significance as I shall show later and the increase into space runs by 7^0.

Individual singularity and governing singularity and group singularity enhancing the gravity every time singularity find an accumulation. With looking at Kepler's in a mathematical sense it is clear that from singularity comes space by three duplicating space in time by three. $k^0 = a^3 / (T^2k)$. Very clearly the dimensions produced space and produced more space by applying time and gravity as movement.

Any point will be opposing itself within the **rotating of 180°** where it **then change every aspect** of its **previous flowing** characteristics it had or **will once again have** in **360°** from there. While in rotation from the view point of a bystander it all may seem static and never changing but to the object in spin every next instant in time will be diverting from every aspect it had every second passing, and the direction it held in relation to the direction it held the previous mille, mille second will totally be incompatible with the direction it holds the very next mille, mille second of rotation.

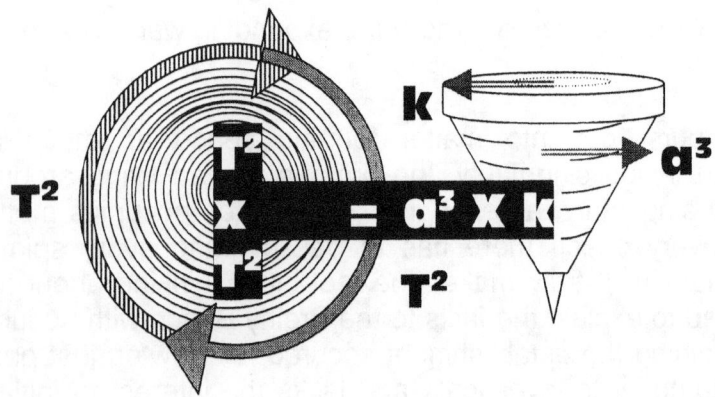

This is why we can use degrees measuring the circle by (6^2) (forming the square relating to matter through singularity) X 10 (square if space) = 360^0 however it is always in motion. That proves no point can be static or constant, though it may seem that way to outsiders. Although matter is matter, matter can also be anti-matter and moreover form its own anti-matter at the same time. This degeneration of structure is very likely to occur with overheating. Revaluing Π to Π^2 will bring about a new contact point where Π meets **r** forming another relation in Π^2

Time is the **changes in relation** where Π **contacts a different r** not withstanding the many r points there may form because **every r constitutes a different value** to the universe through other ratios and relevancies brought about **by heat and light. Time is the duration it takes Π to rotate between any two given points of r** and therefore must always amount to **a square (T^2)** moving from point to point through the **cube of space (a^3)** in that **duration of time (k)**. With that it proves **Kepler's a^3 (space) $=T^2$ k (time in the instant of motion)** but motion must continue through a specific value in space where the space-time is maintaining relevant equilibriums throughout singularity connecting.

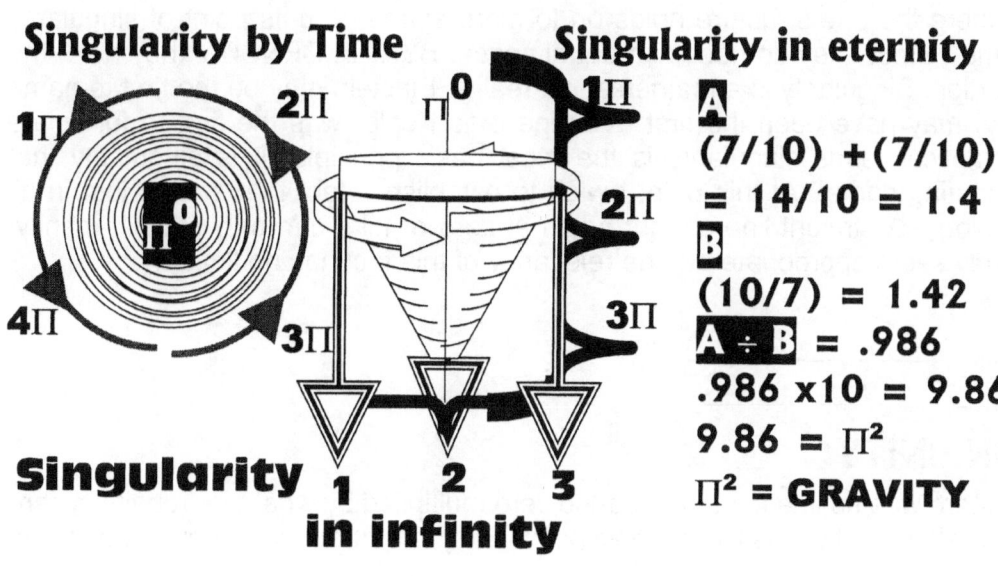

Singularity by Time

1Π   **2Π** Π^0

4Π   **3Π**

Singularity
in infinity
1   2   3

Singularity in eternity

1Π

A
(7/10) + (7/10)
= 14/10 = 1.4

2Π

B
(10/7) = 1.42
A ÷ B = .986
.986 x10 = 9.86
9.86 = Π^2

3Π

Π^2 = GRAVITY

With the dimensional change from space in the cube to space in the sphere a relation of 5 to 7 comes about depicting gravity. The principle of 5 sides in space relating to 7 in the sphere holding matter forms the basis of the Titius Bode and the Lagrangian principles.

Newton, and science, made one enormous blunder, from this stance. They took the radius of a wheel not to have any influence on the wheel. In doing that, they removed the very fact that keeps the universal attachment together. They still insist that rotation results in nothing.

The state that is the time component called outer space, is coming about from the fact that outer space is the Titius Bode law because the Titius Bode law is evidence of how the Universe was compacted by motion. The time zone called outer space is the motion called gravity and is the neutron factor in the Universe.

$$r^2/r^2 \neq 0 \qquad \qquad r^2/r^2 = 1$$

$$\frac{dJ}{dt} = 0 \quad \frac{dJ}{0} = dt \text{ or } \frac{0}{dt} = dJ$$

Π X r^2 = CIRCLE If you remove r it then is Π x r^2 / r^2 = CIRCLE.

You cannot then say $r^2/r^2 = 0$ and therefore Π x 0 = 0. That is nonsense. $\Pi r^2/r^2$ will always be Π x 1, and that is the eternal circle. When looking at any rotating object, there has to be a point in the infinite middle where the one side rotates in one way and the other rotates in the other direction opposing the opposing direction. That point in infinity is the point of no rotation and no rotation means "no rotation", not no existence. No rotation means a factor of 1, not zero.

That then is singularity. The eternal Π, the Π that may not have significance but still it is a Π of value. The relativity remains one, eternally one, but it cannot be zero. Therefore, dJ/dt cannot be zero.

> dJ/dt can become eternal or infinitive or at the worst it can become one
> dJ/dt = 1

When explaining this to any child, they can immediately see that. Explain this to any Newtonian High Priest and he may have you removed forcefully from campus. I cannot find one Newtonian, of any significance being large or small to accept that.

The comet rotates the Sun , and the Sun by itself has a point of singularity where Π remains without r. The comet, holding the orbit, also has a point of singularity, but since there is space separating the two objects, they cannot share a mean point of singularity, the very point of existing. Since singularity means just that, being single, there cannot be two. The comet and the Sun have a mean point of singularity but the space they occupy divides their common singularity. That is why they orbit in an oval path, a path where the one structure holds on to more space from its point of singularity towards the space it claims. Since they do not claim equal space, BY THE DENSITY they hold, the space will not be in proportion. Singularity is a mathematical reality. Einstein may be the first to name it and Galileo (unwittingly) may have been the first to define it as Kepler was the first to formulate singularity, but in mathematical terms singularity is the most basic principle. It is singularity that attaches the top to the orbiting comet. At this point I wish to establish a fact that seems lost in all other grandeurs of cosmology. A straight line cannot begin at zero or nil it can only start at infinity/ Such a statement will hardly seem appropriate but the relevancy of this fact has no limits.

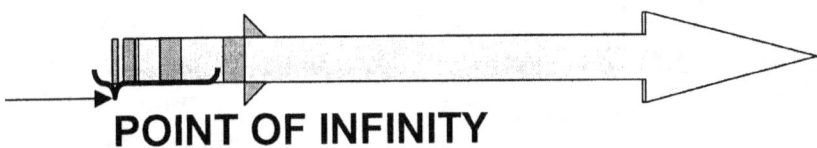

POINT OF INFINITY

If the line started at zero there was no line to start because zero multiplied by whatever results in zero as the answer. That must also be the cosmic starting point. Einstein introduced such a point and named that point singularity.

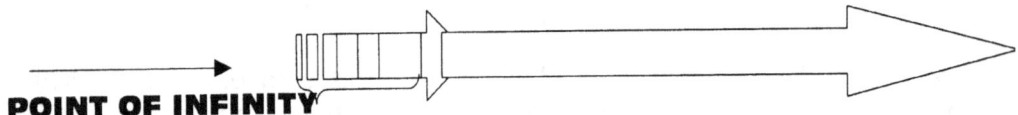

POINT OF INFINITY

The Universe does not change because there is not one single item that is in the Universe that can change. What the top evokes is what was established at Moment-Alfa when time was interrupted by space for the very first time. The spin or expanding that was introduces still present the very same principles as the spin or expanding did when it did for the very first time. The line of time is eternal and was interrupted by space with infinity bringing an end to eternity in time. However in infinity the line was interrupted extensively but briefly while the line continued intensely small but eternally long.

•⇒⇒●••

Einstein introduced matter time and space and I can see where Einstein was heading with the three concepts forming one Universe but Einstein got his wires slightly crossed, because for one, what Einstein saw as space is time and what Einstein saw as matter is time in general as it is a connecting that singularity has with the flow of time throughout the universe. In that there are two strands of time that formed with one massive time delay that compacted the heat in the time delay and that time in delay compacted in nice units, which now forms the part that are the matter Einstein referred too. In that there is overall just time relating to time in time. However the only existing Universe is the Universe, which is not part of the Universe. The rest is creation by generating motion and it is not created as such in measure of time flow so very long ago and back in the most distant part. It is created by motion of time delay through time in time. By moving from one point to another point the flow of time goes square while the points duplicate. The duplication is a product of overheating in one specific spot where that spot exaggerate the space by expanding. However in duplicating there is cooling and cooling is reducing of heat. Heat is what there is becoming more of the same thing and therefore cooling is what there is taking away some of that.

The motion brought about the square of the value but in the square initially was only singularity.

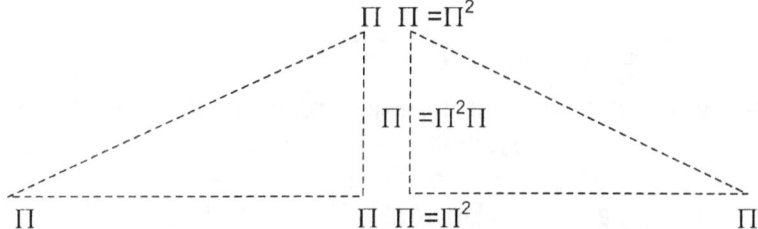

$$\Pi \quad \Pi = \Pi^2$$
$$\Pi = \Pi^2 \Pi$$
$$\Pi \qquad \Pi \quad \Pi = \Pi^2 \qquad \Pi$$

As we already determined on a previous occasion we now accept that expanding comes about from heat and only heat brings about expanding. By heating whatever one may find in the universe expand.

By expanding the space becomes more and the space in doubling cut the heat by half. It is an immaculate and genius way of controlling heat.

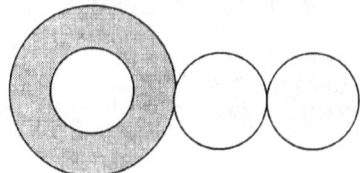

As it is still the case, the contraction results in duplication and by distributing the overheating space over a bigger area, the cooling contracts half (in the beginning but at present the proportionate) the heat back to secure the original singularity while the other position secures a new point that activated singularity. The distribution results in the contraction. When 1^1 expanded from 1^0 there was a linear motion established. The motion took what had no start away from what had no end. In that the expanding had a direction as lateral that connected to the cross of the lateral that connoted what remained of the eternal to the lateral.

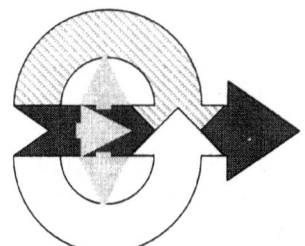

The lateral motion connected 1^0 to 1^1 but in reflection as a mathematical response there was too a connection with 1^1 the upper casing to 1^1 in the lower casing because there was an immediate contraction to the expansion.

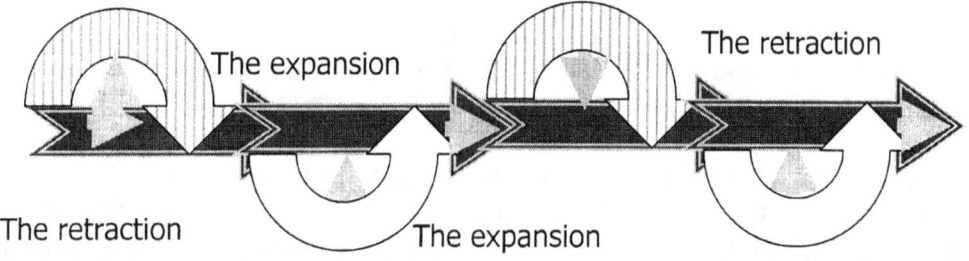

The expansion

The retraction

The retraction   The expansion

The flow of time from the past through the present onto the future had three dimensions resulting from the oblong time in the perfect eternity strayed from by going imperfect. This brought to the future where half confirmed the past, half conformed the future and believe it or not but half converted the future.

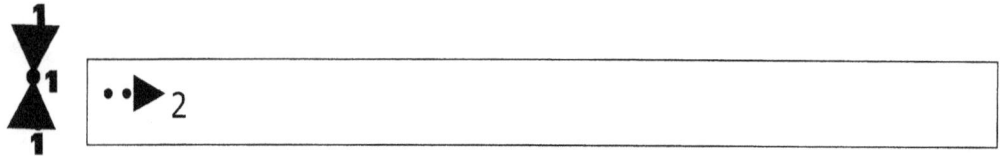

Therefore with one the Universe was in eternity and the value of the one was confining everything in singularity. By duplicating two relevant points forming three positions in singularity came into being a form in the universe. The number arriving in the Universe was two, but I am somewhat reluctant to say that what ever formed at this stage, was already part of the Universe. That is still miles off.

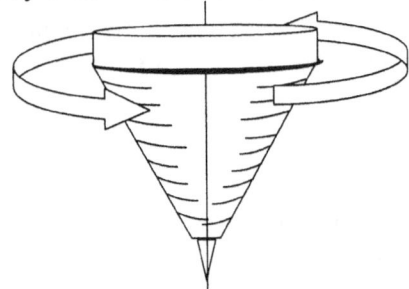

$T^2 = a^3 /k$. In this period of development the time associated with eternity much more prevalent than it did with a break on continuity in infinity.

What happened back then is precisely what we are able to gauge from the behaviour we find the top shows because the Universe doesn't ever change.
This brings us back to the spinning top I presented at the beginning.

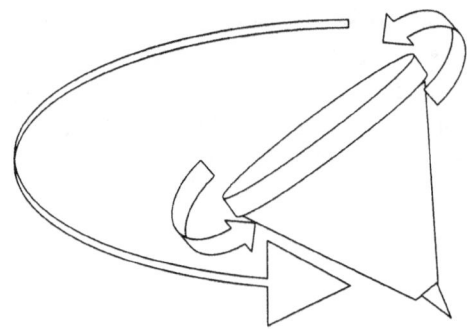
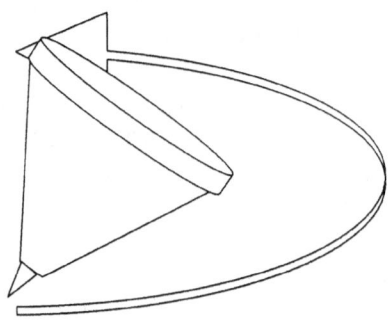

I have asked as many persons as I do not care to remember why the top sinning will remain spinning around one point while turning. The answer I receive from the most educated to the schoolboy is always about momentum. That is a very simple answer and to say the least a little too simplistic by further analysis. Why would the spinning top go off centre when spinning higher than a specific velocity and lowering the velocity it would stabilize and run square to the Earth only after that it will go oblong and then fall.

I could go on about different positions bringing across different momentum of thrust but I do not wish to insult your intelligence because I am aware that you are familiar with all the law. When the top is spinning it is spinning about its own axis and when it is not spinning it still remains spinning about the Earth's axis, therefore when it is spinning it is also spinning about the Earth's axis.

Therefore the limitations applying can only result as an influence coming from the Earth's axis. The second question now comes screaming across and that is in what manner could the Earths axis ever affect a spinning top since the spin and the spinning top is a gross mismatch to what ever standard the Earth may introduce. It is clear that spinning objects do influence each other in contrast to Newtonian opinion.

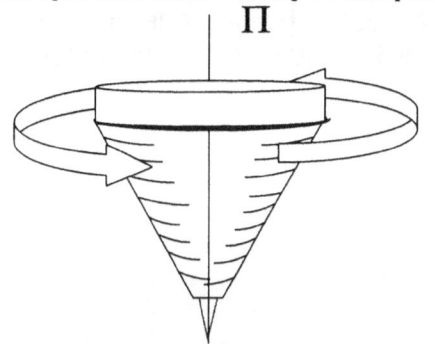

Every round object has a point establishing a very centre, a middle dividing one side from the other. That division determines the space from one side away from the other side. At one point there must be a point that does not fall on either side of the divide. Such a point will still be a circle, because from that side the circle divides into two sectors.

In every spinning object there is a point of infinity, a point that does not turn because it holds the dividing spin. From that point running in all directions the spin is opposing the other side. All spinning activity starts at that point diverting outwards and from that point the spin is either clockwise or anti clockwise in all directions. As I pointed out no line can start at zero because then there is no line and no rotating point can start at zero because then there is no rotation.

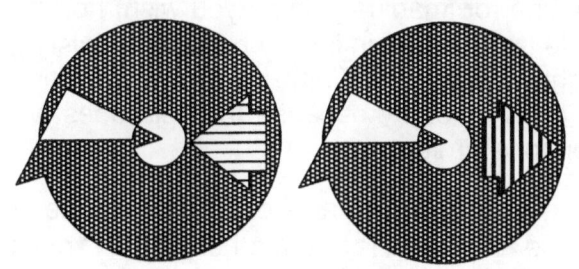

I have indicated that motion creates space $T^2 \times T^2 = a^3 \times k$ and space finds limits of space in motion $k = a^3/ T^2$ as well as $k^{-1} = T^2 / a^3$ where the motion confirms the space while the space conforms the motion. The rotation of the motion in both relevancies completed the space that formed by the motion in linear as well as rotary.

Saying that 1^0 moved to 1^1 sounds much potential locked in that no other such giant step. It happened when the first everything that ever could be in the locked in one spot becoming one dot. To numbers and measure the movement had remained what it was before in **k = 1.** To us little cancellation with **k** being either 1^0 or

senseless but in that motion is so movement ever came close to relevancy came about and Universe and would be was our mathematical genius of no meaning because it still with small minds that reality left 1^1, which remains at **1,** but the moving of the space gave the space a reason to be and the space gave something that could be moved a distance from 1^0 all the way **to 1^1.** The emerging Universe in its entirety was in prominence.

By reducing the one line the other line can never reach zero because then there were no such a line to begin with. That makes a straight line also inevitably always a potential square and that makes the straight line half the value of the square being 180°. At a later point I shall continue with this argument, but for the mean while I wish to come back to the circle. This same principal applies to the cube and that means everything there is and ever will be is either a square being part of a cube or a circle. With the straight line forming half the value of a square $360^0 / 2 = 180^0$ in as much as being one line and reserving one line in infinity to eternity. The straight line is just half the value of a square. In that manner the triangle is also half a square and therefore holds the same dimensional value as the straight line being also 180°

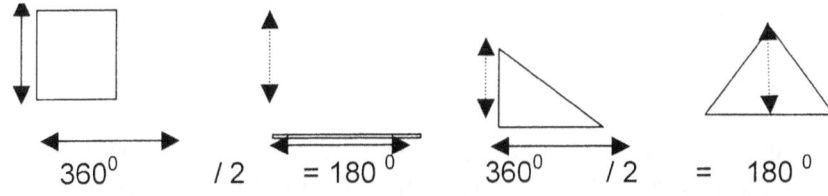

360^0 $/ 2$ $= 180^0$ 360^0 $/ 2$ $=$ 180^0

The circle is a square holding a round shape, as the straight line is a square holding one side to infinity. Calculating a circle involves two aspects where the one is either the radius or the diameter that is double the radius. The other is the factor Π

$\Pi \times D^2 / 4$ = circle and $\Pi \times r^2$ = circle

The point of singularity cannot be in space at large because space is not there and secondly what ever is there spin to slowly to have a connection with singularity directly.

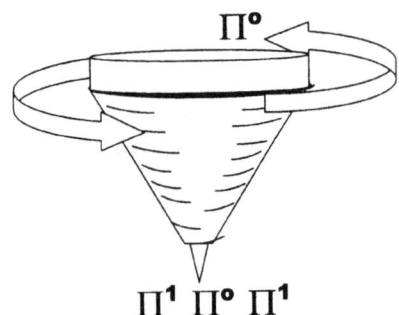

$$\Pi^1 \ \Pi^0 \ \Pi^1$$

With everything in a cube or a circle or a potential of the two, brings about the implication of eternity in a form of singularity or the point of creation. Removing the radius of a circle does not remove the circle, because the circle is there, securing the ring. If the line (or imaginary line if you wish) holding the value of Π^0 = 1 there has to be a point where the circle is no longer in infinity but claims existing outside the imaginary. At that point the radius may be slightly more than infinity, but to all calculating purposes it still remains as infinity. The spin was going on for eternity because the spin does not apply, it has a value of zero and zero is another expression for eternity.

Having edges where Π^0 duplicate to present the edges singularity lost the value of Π^0 to the value of Π^1 with the same value singularity was being Π^1 to the one side and Π^1 to the other side, Π^0 must be the point splitting singularity into two parts of eternity, the eternal value of the first dimension outside eternity. It was the square of Π^1 being Π^{1+1}. That was the first dimension outside singularity Π^0 where singularity has a value of Π^1 in the form of $\Pi^{1+1=2}$. The first claim to space had a value of Π^2. This applied to both sides of the claim to space outside singularity, and the double proton became the dominant factor on matter.

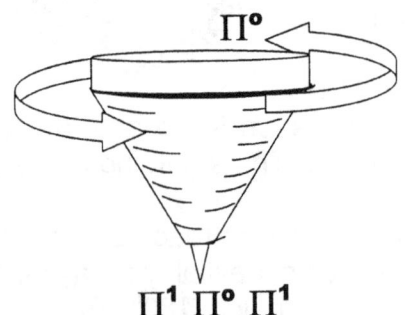

$$\Pi^1 \ \Pi^0 \ \Pi^1$$

Right at the start before space and time became developed the motion produced space in the principle of the Coanda effect. By receiving space, singularity received a value outside eternity as Π^0 received edges. Granted the fact that the edges were so small there still was no r to present a circle. The manner that the top use to evoke singularity, which enables the top to maintain independent motion could be, traced right back to the very first line that came about as Singularity initiated spin

In the beginning there was no space in which to move so therefore the only way to move straight was to move in a circle. The movement **k** producing the line was the same as the motion T^2 that produced

the circle and the space a^3 achieved was the compliment that two factors combined and that formed the space developed.

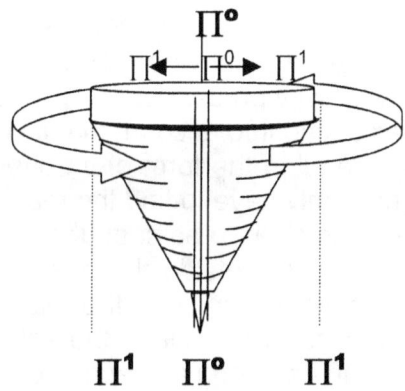

Taken from the point of rotation the two sides are in opposition to each other in every aspect that they may contain and with all that they hold. The motion is the extending of singularity and singularity reacts on the motion by establishing a proton value.

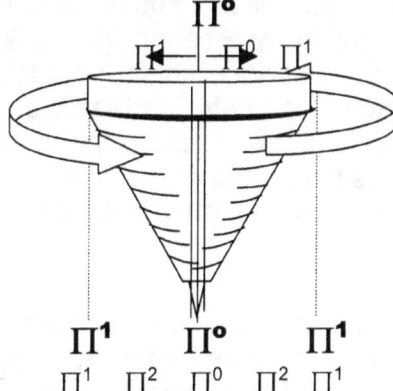

With Π^0 little more than a figment of the imagination there is actually to values of Π^1 facing each other in a relation combining Π^1 to hold the value of $\Pi^{1+1=2} = \Pi^2$ and with two sides being the very same but opposing each other there will therefore also be Π^2 to every side that holds Π^1.

At last I can come to the one part that I disagree with Newtonians, and what I regard as Newton's second biggest infamous or famous blunder. Science, made one enormous blunder, from this stance. They took the radius of a wheel not to have any influence on the wheel. In doing that, they removed the very fact that keeps the universal attachment together.

Singularity controls the Universe by establishing a Universe but that is done in a specific manner.

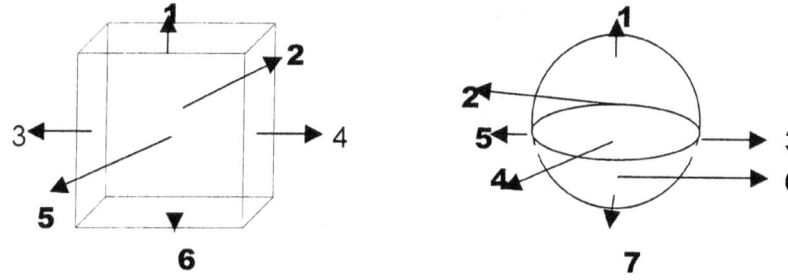

In the very centre of the sphere the form of the sphere dictates that the shape will relinquish space as the line run from the outside towards the very centre. With this natural state of affairs the sphere are naturally inclined to dismiss all space that it can form in the form as the sphere holds space inside and the form will finally be without dimension. All that I attribute to the line shrinking by reducing actually takes pace in every sphere as the diameter reduces to the centre. In the centre where the radius line goes single the form relinquish the three dimensional form it has inside. Being without dimension in the very centre means that at a point in the extreme centre of all spheres there are a point that holds singularity because this point with no space has a mathematical position although it is invisible since there is no sides to such a point to give that point any dimensions. The shape of the sphere is calculated by using the formula $4\Pi (r^3) / 3$.

By reducing r to a point where r is r^0 singularity steps in because only the form remains as Π. Going even further we find that there then comes a point where Π goes singular Π^0. At that point absolute singularity is present but so is absolute gravity present at that point. When holding the strength of the shape of the sphere in mind as well as taking into account that all cosmos objects of importance is in the form of planets or stars and they are all in the form of a sphere, we therefore may contemplate that it is where gravity originate. We now only have to find the reason why gravity will hold a base in a space less ness as Einstein predicted. It is clear to be seen that gravity is in the centre of the sphere controlling from the centre everything that is outside the space less centre. We can reason with confidence that gravity is the strongest where space is the least. We can further reason that it is gravity that is holding the sphere in true form and since the sphere allow gravity the best working opportunity, gravity can form the sphere in as strong a shape and form as the sphere seems to have.

From every point on the surface of the sphere is where that point connects with the other side of the surface of the sphere by a line that runs through the space less ness of such a centre of the sphere. Such a line also connect by an angle of 180^0 as well as 90^0 to six other lines running from top to bottom, right to left, and back to front, where all join and cross in the centre of the sphere. There are therefore six lines crossing and connecting by a centre from any given point on the surface of the sphere. Such points connects in total six surface points on each side of the sphere while they all support one another through the space less centre. In that absolute space less ness in the centre holding singularity we find gravity supporting and controlling all space within the sphere as well as space connected to the sphere. That is where gravity control and guide the space, which falls in the parameters as well as under the influence of the form of the sphere. In the gravity centre space goes singular meaning space becomes space less or flat.

It is from the layout that the sphere uses as natural form that we are able to locate singularity. In the case of the sphere the material naturally reduces by measure of the radius becoming smaller to a point where the radius is r^0. At that point the line that will form the radius has gone single dimensional r^0 and that is equal to 1^0, which is singularity.

Also it is true that the entire form that is the sphere is controlled from a centre within the sphere. That centre holds the sphere in form and shape. Therefore the strong form is dictated from that space fewer centres where there is no space and no form left. The natural inclining is in the form of the sphere. It is part of the roundness that the overall shape of the sphere represents and this structural strength is carrying down to the very centre.

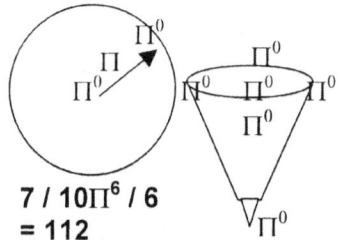

7 / $10\Pi^6$ / 6
= 112

Because the circle is forever reducing that reducing which is inherently part of the form of the sphere becomes a tool in distorting of space in the sphere and is eventually removing all forms of space from within the centre of the sphere. The very centre ends up as having no space because of the reducing that continuous down to become the space less inner centre. The all roundness is the ingredient that forms the backbone of the absolute strength that the sphere has and that is the component that the sphere is so famous for. The form the sphere has allows the sphere to have a control that is coming from the centre deep inside the sphere where the space vanishes and being without space seems to keep the entire structure rigged. From the centre the sphere shape shows strength that the shape as tough as it is. How does it work in its most basic analyses?

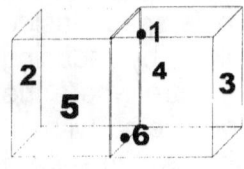

There is one more point in the sphere in the centre forming an addition in the sphere. That point holds gravity secure.

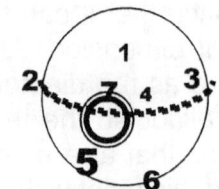

The cube has sides and the sides form a rather weak and flat surface that connects four corners. The flat surface produces a rather indifferent contact point with no special features on the surface. The corners connect to other sets of corners and those corners form a weak structure without any direct support coming from the other five sides. Without material to fill the body of the cube the cube has no direct connecting between any of the sides other than corners connecting at the edges of the sides.

Taking the vantage from the point the sphere is holding from the centre out into space there are ten points connecting to the centre. In that are the dimensions of singularity connecting to space where five connects to space in the second dimension of singularity, and five connects in the third dimension of singularity. On the other hand, the cube does show a very different characteristic, which involves only six sides (at least) connected.

The spinning of Π^0 around the centre Π^0 establishes Π and Π is what produces the form gravity has. Still it is the relation or relevancy there is between the centre Π^0 and the spinning Π^0 that gives status to the form that Πrepresents. In out Universe we are accustomed to and are familiar to the rules we want to place seven points holding singularity to the centre holding singularity in a relation of $7/10\ \Pi^6 / 6 = 112$. In that Universe everything less that a duplication ability to the value of 112 protons fit but only atoms to a maximum of 112 protons fit.

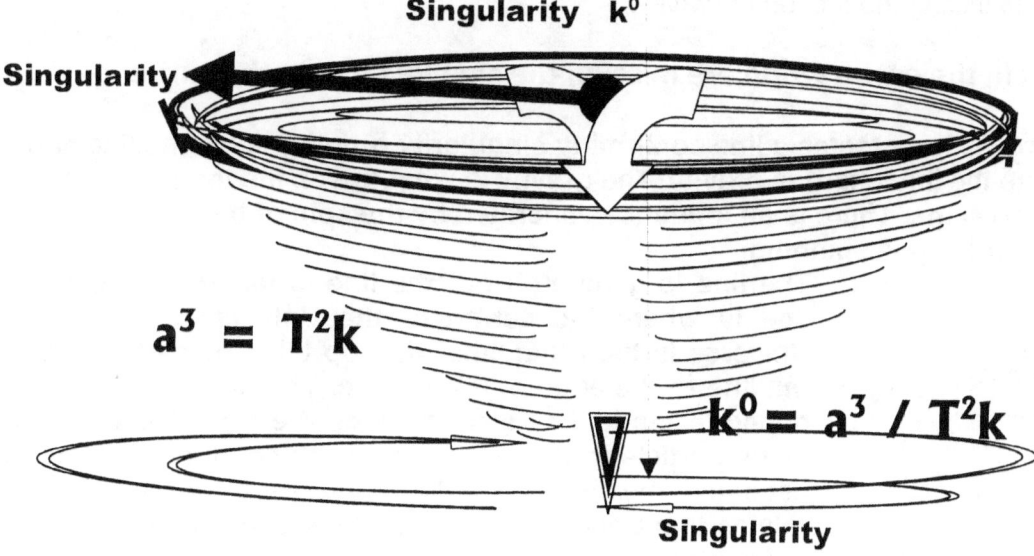

Singularity k^0

Singularity

$$a^3 = T^2k$$

$$k^0 = a^3 / T^2k$$

Singularity

If gravity is motion T^2 the process may sound deception ally simple but it is very complicated to rein act. By motion T^2 space comes about forming a^3. But the motion T^2 will mean a crossing to the other side of the universe since singularity divide the Universe into sectors.

Material produces the dismissing or the concentration of space by applying the motion. Surrounding all elements are a layer we call the atmosphere and even Pluto and the moon must have the atmosphere because the have gravity. In this the relevancy of ten to seven forms this layer and it results in forming a circle because of the combining of the motion duplicates the singularity factor Π forming from that gravity as Π^2

THE PROCESS PARTED USING THE ROCHE PRINCIPLE

By establishing motion and creating motion singularity quadruples to 4Π in rotation. But since the rotation is motion duplicating the space established as four times the value of singularity the motion divide the space coming about by halving such space by the dimension, which is putting a square root over the quadrupling of space. In this comes about the direction gravity takes the universe. The expansion is always double the square root but the square root is neutralising the expansion and that brings about that the neutralising of the expansion creates a contraction that seems dominant to us but it is not. The contraction is doubling the expansion by halving the effort of the expansion.

Singularity is a mathematical reality. Einstein may be the first to name it and Galileo (unwittingly) may have been the first to define it as Kepler was the first to formulate singularity, but in mathematical terms singularity is the most basic principle.

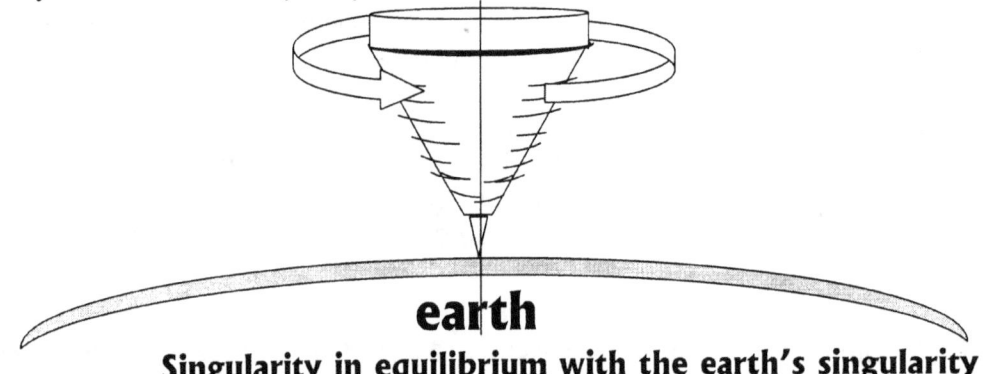

earth
Singularity in equilibrium with the earth's singularity

Singularity applies gravity by charging motion and the motion (not the pulling) of space-time is gravity. It is motion that moves the gravity that moves the Earth and it is also motion the moves the top that forms gravity. Gravity is not the pulling of but the motion or tendency to move to the centre of the Universe, which is the next domineering, point holding singularity in a control or a governing mode. The motion or tendency to move is that which forms gravity and mass is the occupying of space and therefore restricting the motion of gravity.

The greatest minds in the entire world are missing the smallest line in the entire Universe.
By rotational motion, the top creates a line confirming singularity running down the line and by generating the line the line charges gravity. The gravity is what drives the top as the top and as long as the top spins. There is an influence generated by the spin of the top that keeps the top upright while the top is spinning.

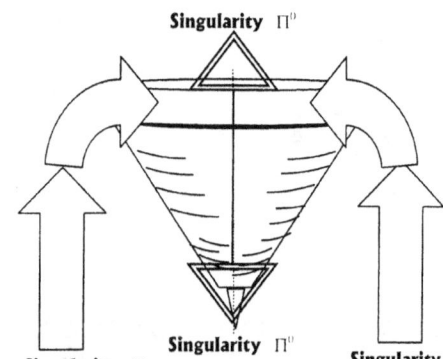

The line is generated but the line is far from magic. The line is where the centre of the Universe is which the Universe is then that what the top filled by particles from the line to the edge of the sphere. The particles in motion generate motion by electing a centre from the centre of every particle in the spinning top. Such an elected centre becomes the centre of the Universe as far as the top relates to a Universe because all the atoms in motion elect the centre of the Universe.
In this, it is clear why the Titius Bode ([10 + 10 + 1 + .991] / 7) and the Lagrangian 5 \\ 1 systems part their ways when applying the different processes they hold. With all the differentiating, the observer must also consider the dual message that light uses in travelling through the vastness of universal space. The thought of nothing is just what it is, a thought of nothing and although it is in the human mind common nature to present nothing as a value in the recalling of something, nothing is a presentation of the figment in the human mind. There can be no number such as nothing and that was (possibly) Newton's biggest error. Nothing represents non-existing and that is just what nothing is, it is non-existing.

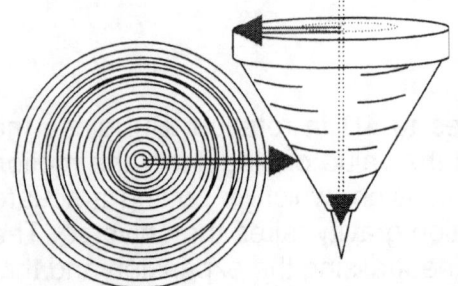

The centre may or may not spin and the fact that it does or does not spin is all the same because that centre part never spins in any case. Therefore the boundaries set by the spinning motion does not depend on the spinning motion of the object but has to stand related to another bogy bringing about a larger spin influence. Granted the fact that the influence the Earth has on the top may be that of gravity but if that is the case then surely the Sun has also influence on the Earth and other rotating objects through gravity. It needs more investigation because it may bring about evidence we are not aware of.

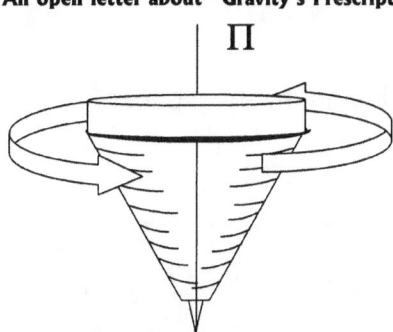

This observation places a much bigger question mark on the statement of Newton where he proclaims no influence on two rotating cosmic structures.

By rotation such rotation is a duplication of what singularity retracts. The rotation involves the three factors in time coming from the past through the present and onto the future, which holds three positions excluding the one position allocated to singularity. The four I named the eternal motion because singularity being without motion was contracting as well as expanding at the same time it was not moving. That placed singularity in eternal motion without ever moving being singularity.

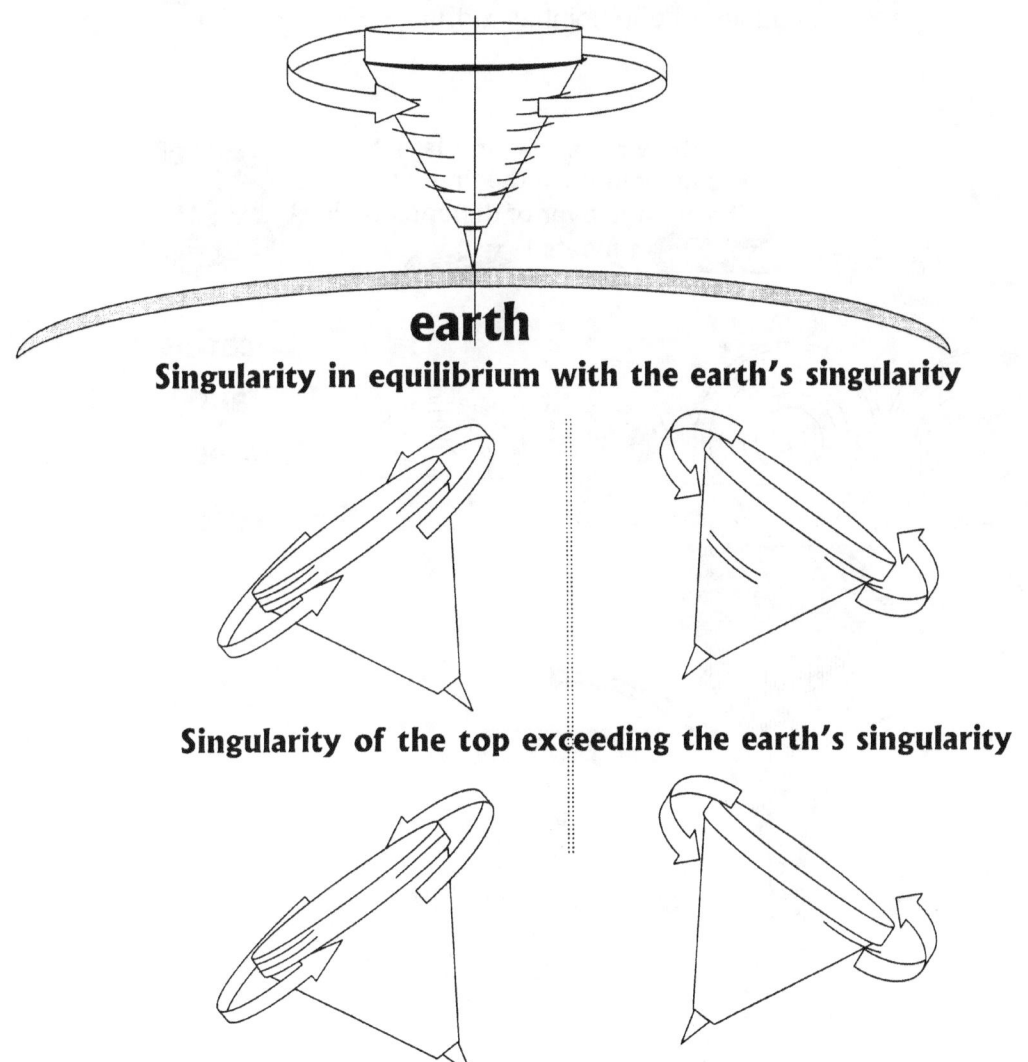

earth

Singularity in equilibrium with the earth's singularity

Singularity of the top exceeding the earth's singularity

The earth's singularity dominating and exceeding the singularity top

Understanding all the following is connected intimately and all conditionally to the fact of accepting that all individual particles in the universe use motion and therefore spin.

Every quarter provide a distinct value that indicates the progress of the flow of time from the one point Π to the next point Π.

Any changers occurring in Π will lead to a an unequal triangle providing two different values to r and will alternate the link between r and Π^2 bringing about different form (Π) and time (Π^2). When singularity forming the lines of the triangle is not in

equilibrium the triangle will destroy the matching of half circle.

In considering the spinning motion in the fraction of time in the detailed instant every aspect of rotation will turn in every instant of change in time. Although the points had the same characteristics only one instant before, they oppose the characteristics it had just before and just after the very instant in which they are and to which they relate by similar points also in rotation. The fact of the graph proves my point in quarterly opposing dimensions and values. As the rotating direction moves inwards, the rings will become smaller and smaller. Move the rotating line progressively to the middle by reducing the length the line have from the edge to the middle. At one point all further reducing ends.

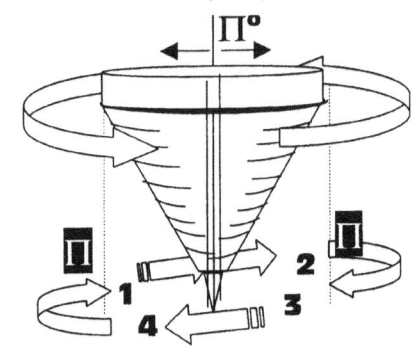

The drawing is the circular Π^2
The movement is the linear r
The change over of dimensions is Π
➡ **r meets Π** ➡ Π^2

Locating and finding Singularity

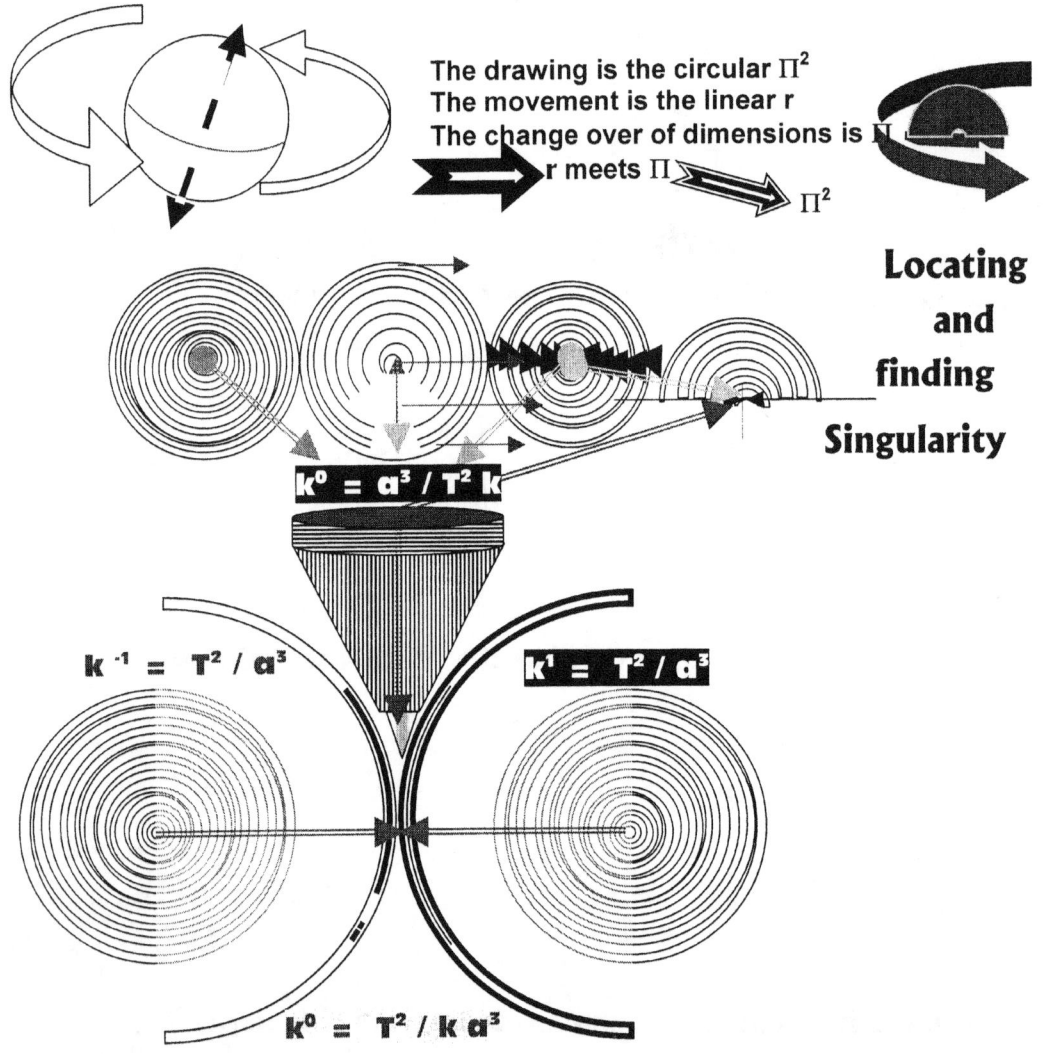

$$k^0 = a^3 / T^2 k$$

$$k^{-1} = T^2 / a^3$$

$$k^1 = T^2 / a^3$$

$$k^0 = T^2 / k\,a^3$$

In the **precise middle** of all **objects in rotation** is a precise centre dividing the object in sectors that will **start the spinning initiation** from that centre point. **That point** albeit hypothetical, is also as much a reality none the less and is where that point **must be standing still** because every line **running from that point** in **opposing directions** are also **in opposing directional spin to each other. That point** is completely hypothetical, is also as much a reality none the less and is placed where that point **must be standing still** because every line **running from that point** in **opposing directions** are also **in opposing directional spin the other or opposing side.**

In considering the spinning motion in the fraction of time in the detailed instant every aspect of rotation will turn in every instant of change in time. Although the points had the same characteristics only seconds before, they oppose the characteristics it had just before and just after the very second in which they are and to which they relate by similar points also in rotation. The fact of the graph proves my point in quarterly opposing dimensions and values. Due to the spinning nature of such a point with all surrounding the point will be alternating direction favouring change every second and in that the value to such a point can only be Π because of its constant changing. Using r would specifically oppose another r from every angle because the use of r will bring about a static relation to the previous and following instant and therefore it will cancel the constant spin flow. There must come a point where the ring is infinitely small, where it can reduce no more, where it reached its ultra limit, but at that point it cannot be zero, because the point is there notwithstanding that it is at a location beyond our Universe. But the spinning object **will have a middle point**, a very specific **centre point that does not spin** and only holds Π as a specific value. One value such a line **cannot have is zero** because **zero does not start any** line and therefore the **value of the line must be infinite**, just as described in **accordance** and by **the definition of singularity.**

From somewhere outside the Universe a line rises while remaining in a position allocated to space being outside the Universe. The line is activated by the rotation that sets the top in coherence with 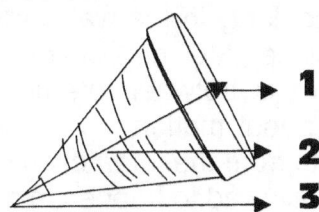 time and the motion grants the top individual status by establishing independence. The motion holding a dual action while being a unit lay down the ground rules for the Coanda effect. The four points turning around a fifth centre also charges a line to place the space-time the top holds to an erect status of independence. The line is at the centre and it is motion by rotation that activates the line. The erecting of the top underline the status the top receives as an independent Universe, which is maintaining an individual singularity. Even when the motion no longer finds the ability to charge the line in singularity and the top stumbles before it falls, there is still a desperate fight for keeping the motion and with that the independence active. The fight is not a fight for balance but a fight for survival.

 Space parting eternity and infinity activates the line and Newton was of the opinion that rotation brings about no work. The line runs from the top of the top down to the bottom of the top without ever being present in the space of the top. The top forms the space being fully independent and it is most critical to consider that line that the motion activates also activate the third dimension of time in space. Space in time and time within space separates by finding separate identities in the cosmos and this gives singularity independent identity. It parts eternity in time from infinity in time. A universe is born through the rotation of the top in spin. By parting infinity from eternity the space in between fills with material that allows the top the position the material in the top has while the top is spinning. When the motion falters in sustaining the singularity it requires to remain independent the point holding singularity goes cold and looses the acquired independence it had.

That point albeit hypothetical, is also as much a reality none the less and is placed where that point must be standing still because every line running from that point in opposing directions are also in opposing directional spin the other or opposing side. Move the rotating line progressively to the middle by reducing the length the line have from the edge to the middle. At one point all further reducing ends. In considering the spinning motion in the fraction of time in the detailed instant every aspect of rotation will turn in every instant of change in time. Although the points had the same characteristics only one instant before, they oppose the characteristics it had just before and just after the very instant in which they are and to which they relate by similar points also in rotation. The fact of the graph proves my point in quarterly opposing dimensions and values.

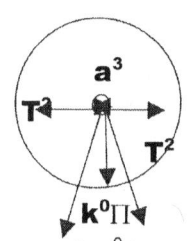 This only applies in relation to time because time is the square or then if you wish time is the flat to space being the cube. Time in the square draws space in the cube flat and that is the why the Universe holds the sphere in place.

Understanding all the following is connected intimately and all conditionally to the fact of accepting that all individual particles in the Universe use motion and therefore spin.

In dimensional terms, which I explain later on the value of **2k** relates to **T²**. That relation extends to the next value where **T²** relates to **k**, which relates to **T²**. The first space in the circle will then be **T² k**. From the centre being in infinity, one can realise by applying mental power the single dimension factor not seen but present all the same. Extending that into the 3D comes six **k** and any one of the six will further extend to form a seventh point as **T^2** All this is a multiplying of **k^0 = a^3 / (T^2 k) = 7**

From this line of reasoning I dismissed the theory of the presence of a force being gravity but rather consider it as a dimensional changing contributed by the spin of the Earth and the spin comes from singularity located in the centre of the Earth. It is all about dimensional changing that influences space as a factor of ten to reduce to Π^2 on a continual basis from point forming new dimensions through billions of such points.

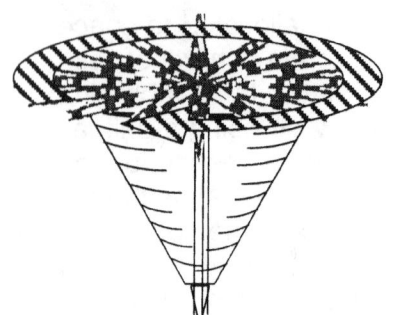

In conditions found in the Universe spin can only come about from heat that is concentrated around singularity. Dropping water on a red hot metal will lead to feverish motion as the singularity in the water absorb the heat and expand the space by accelerated duplication. The singularity find the time differentiation there is between eternity and infinity excelled and space excelled brings about motion in duplication amplified. That what we see in the top results from the interaction of life that in principle is some mistake the Creator allowed to happen in a very small region and as far as man is concerned taking into account all the proof man has to his disposal the phenomena of life is no where else in the Universe. By having **T^2** overheat space will reduce in ratio that brings about linear motion k being shorter per time unit but much more frequent per time unit accelerating because **k = a^3 / T^2**. However in accordance to Newton's law on motion there will be a reaction **k^{-1}=T^2/a^3** and with a reduced space the time in ratio will increase allowing for the heat rising and the amplifying of motion.

More spin increases both lines that force gravity by the increase of **T^2** that extends the influence of **k**, **k^{-1}** in the formula as factors because it reduces the moment of **a^3**. The extending of the liquid heat will increase the motion and increases the contracting gravity **k^{-1} = T^2 / a^3** and the reaction to that is that the space reduces by a larger time contraction **k = a^3/ T^2**. The space then has to duplicate more vigorously in order to cope with the rise in the time aspect. Therefore the space wants to exceed the boundary time slaps on space as a result of the Coanda principle by applying more contracting

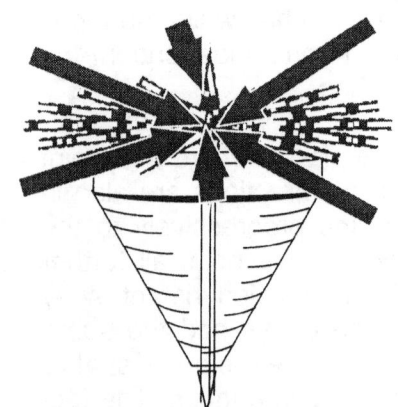

motion because of the rise in the liquid heat levels in time while space has to extend because the rise in heat produces a need for countering the overheating. The duplication of space by an increase of the number of **k** in the time unit **T^2** will allow as far as **k^0 = a^3 /T^{2}k** will permit. In this there is a living up of standards in space and in motion. However not one of the mentioned normal aspects will apply to the top since it is the manipulating abilities of life that charged the top into action.

In the circle using r$^2\Pi$ the r has to have distinctive qualities placing it as a factor apart from Π. Where the growth shows no separate distinction but a continuous flow from the precise centre to the precise edge the flow would become in relation with Π depicting the circle and Π replacing r as reference to any point on the circle. By using r as a distinction in the circle division is possible but by using Π there is no distinction possible making it a solid flow. Any object being in outer space floats and such floating is seemingly random with no specific detectable interfering favouring a movement in a particular direction. Such a devise is depending on influences not in our scope of detection. But then the object comes closer to the Earth and reaches one specific point where the six dimensions that influences the object suddenly changes. At one point, one of the six dimensions falls away as it disappears and the object quite latterly falls to the Earth. The support of one side disappeared and the centre point of the sphere took over the control. At that point the object is under the influence of one centre point in the sphere and we all also know that in such a centre point one will always find the strangest or the controlling gravity.

Space-time is a four dimensional position of the Universe where the position of an object is specified by three coordinates in space and one position in time. This evidence we find as matter grows into the dimension we now share with billions of stars in the cosmos.

With the dimensional change from space in the cube to space in the sphere a relation of 5 to 7 comes about depicting gravity on one side of the divided Universe. The principle of 5 sides in space relating to 7 in the sphere holding matter forms the basis of the Titius Bode and the Lagrangian principles.

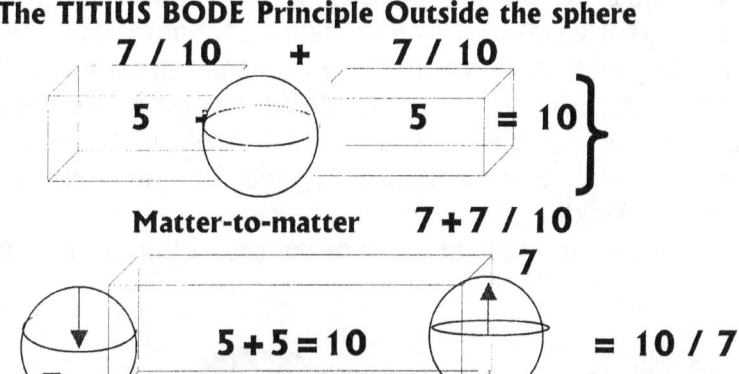

The Titius Bode law is an extending dynamic deriving from the law of the gravity dimensional factor where the space factor in a square of ten relates to a matter factor in the square by half (half since nothing can be in two places in the Universe simultaneously) of the matter factor of π^{7+7} or the square of space (10) relates to the matter factor of 7. From such a point every other point will be opposing any other point not pointing in the direction to which the first point is pointing, whereby it extends the direction it holds. No matter what the point is or where the point leads, such a point holding a specific direction will be unique in the direction it is rotating because at that or any other specific point wherever, it will be directing not in the direction it spins but in the direction flowing from the centre point outwards.

When the foundations were laid in place with singularity expanding even before it was growing The Roche limit became one condition. But while that was taking place another principle came about which is as secured in the foundation of the third dimension as the sides supporting the third dimension. Sides came about through the dimensions that are framing the dimensions, as we know them.

There was the dot. The dot had no borders therefore there was no separation and still we know there were more than one in a group of one. The evidence of this is very present in the cosmos at present and one can find such evidence all around us. The dimensions personify the Titius Bode principle and understanding the relevancy between the dimensions will also mean the understanding of the interlinking values of the Titius Bode law.

Everything is space-time by confirming space in establishing time

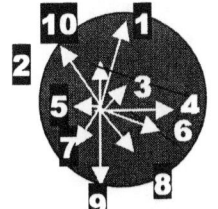

When the Universe was in the beginning with the entire cosmos still in a single dimension there were no limits as we know limits to form in the Universe we use and no borders indicating limits because after all it is the single dimension where there is only one dimension holding so much diversity. The dots referred to in this case have no space but were as close as singularity is when singularity has no sides but only shapes and the lot were the same, the very same one with a time delay parting them. The borders were part of development because we can witness the legacy of such borders in the present day holding the 3D in place. There will forever be smaller particles that combine to produce larger units. The forming of particles start at infinity and there no human can reach except with his understanding, and his mind power.

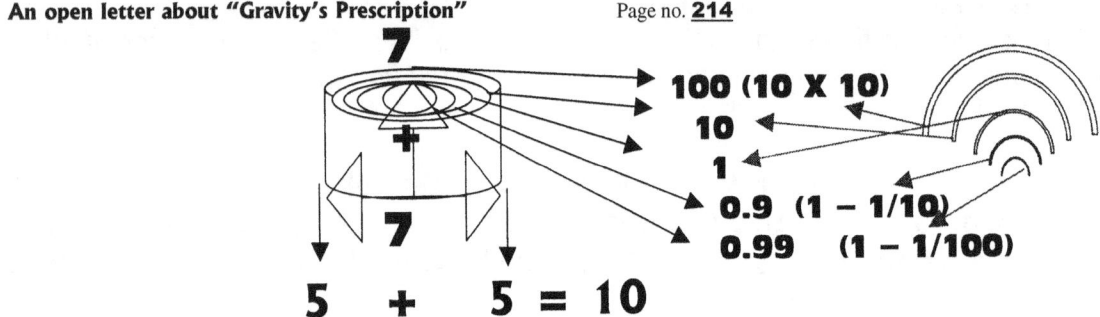

The normal flow will allow singularity extending to 10Π but when singularity blocks another sphere in singularity the two will form a joint value and by this joining the larger will dominate the space as well as the time of the lesser taking control of the surface and the atmosphere. Through this the Roche lobe comes about with all its other dynamics I describe farther on in the theses. The principle is the same, which we know as the conducting of lightning and Jupiter uses it extensively to implement this action. In the Roche limit the straight line forms part (1) and the half circle is part (2) and the triangle forms part (3) to singularity (4) Holding 5 points outside singularity. Every aspect connecting to the universe changes everything it holds totally and becomes the anti-matter to which it was matter 180° previously.

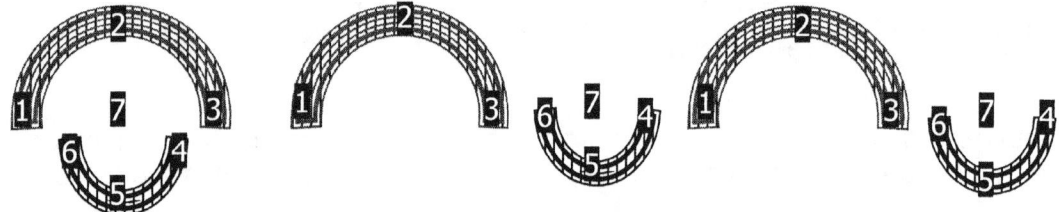

It starts where the first seven points serving singularity meets three points holding time. At present we named the proton combining with the neutron and served by the electron as the atom and from where we gauge the Universe to us the Universe is the entire atom.

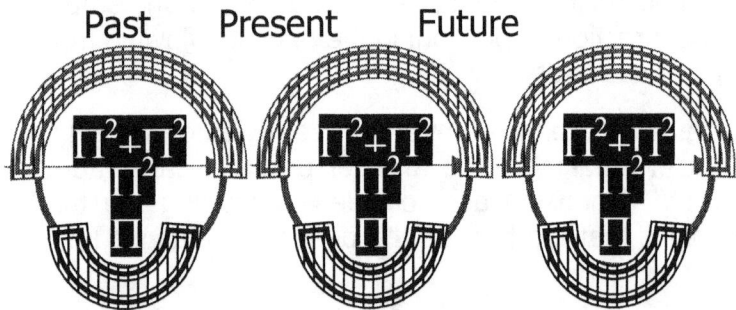

The atom holds seven points $(\Pi^2+\Pi^2)(\Pi^2\Pi)3=1836$ as the Universe but that Universe is seven points in Π being 3 points serving dimensional time to form the Titius Bode law, and a law it surely is! The gravity extending from the Titius Bode law forms the entirety of the building of the Universe by constructing the Universe in the using of the atoms to form the Universe in the entirety thereof.

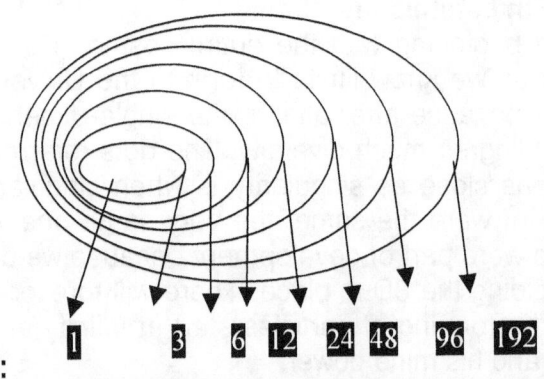

Bode's Law:

Planet	Mercury	Venus	Earth	Mars	Ceres	Jupiter	Saturn	Uranus
Bode's law distance	4	7	10	16	28	52	100	196
Actual distance	3.9	7.2	10	15.2	28	52	95	192

A numerical sequence announced by J.E. Bode in 1772, which matches the distances from the Sun of the six planets then known. It is also known as the Titus-Bode law, as it was first pointed out by the German mathematician Johann Daniel Titius (1729-96) in 1766. It is formed from the sequence 0,3,6,12,24,48,96, and 192 by adding 4 to each number. The planets were seen to fit this sequence quite well – as did Uranus, discovered in 1781. However, Neptune and Pluto do not conform to the 'law'. Bode's Law stimulated the search for a planet orbiting between Mars and Jupiter that led to the discovery of the first asteroids. It is often said that the law has no theoretical basis, but it does show how orbital resonance can lead to commensurability. The importance that becomes known is the sequence the Titius Bode law saw in the number arrangement of 3; 6; 12; 24; 48; 96 etc. The incorrect application of the Titus Bode law lies in subtracting the figure of 3 from 10 leaving 7. The other way of reasoning is to add four each time to the first value of three starting with 3 and so on. The true significance of the Titus-Bode law is that it points directly to a circular growth of 7 stages. The 7 relating to 10 is a precise derogative of the Roche limit or the Roche limit is a precise derogative of the Titus Bode principle because the two systems interlink.

Gravity produces mass but mass is only the result of gravity. Mass do not produce gravity and the manner in which science uses mass can only apply when using the calculations in terms of the Earth. However applying it to stars as science indicates by their formulae used in their calculating of gravity on structure beyond the solar system is very inaccurate. Heat stored in motion produces gravity. Any one not in agreement convinces you by comparing the neutron star with the massive red giant. To calculate a Black hole they go and throw C^2 next to the dividing radius and throw the square onto the C that presents the speed of light. Then they sit back and feel smart in the way they manage to cheat once more to prove their incorrect views correct because after all who will ever fly down a Black hole and return to support or deny their calculations. The Gravity of the Black hole is a speed because the entirety of gravity is speed or better said it is motion. Then the speed that light has is gravity. The gravity of the light can be gravity as much as it at that very same time can be antigravity. What the hell has C^2 got to do with a Black hole because you can pop what ever nuclear device far away from a Black hole and it would be at the most and at the worst very much insignificant. The light will not even escape form the gravity of the Black hole. When this became apparent that the radius of stars reduces as the stars develop through progress. It is some time ago that someone was supposed to say: hey there is a dead rat I smell. For my saying so I am the clown in the courtyard, the one with the two dead brains cells and have no more to use as spare.

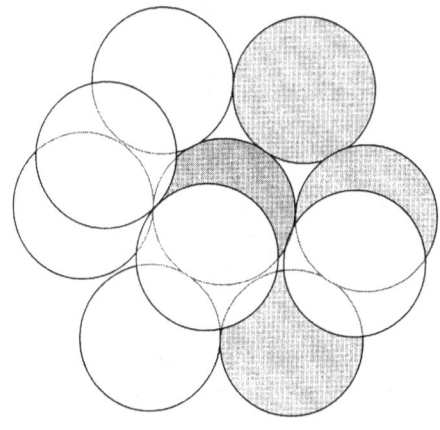

On the one side of the Universe in relevance to all the dots that came before, three dots landed forming one side while three dots formed the second side and three dots formed the third side, all relating to a centre dot which in turn related to the original centre dot from which all the dots came and developed.

Space generates the mass where the space has to reduce the size by becoming more intense and concentrates space-time to the time of 1836 time more when entering the point of the proton being on the verge of singularity. In single dimension seen from one aspect, with single dimension contacting the edges forming the sphere it will still keep the seven positions because the sphere remains a unified structure though apart because of singularity. In the core of the sphere the proton connects in alliances as $\Pi^2 + \Pi^2$ with the solidity of the neutron holding Π^2 as a second forming a π value. That brings the atom unit of π to a number of seven.

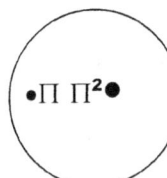

By reducing the space-time the lesser singularity is claiming singularity independence by offering reduced space, which will result in promoted time with heat being the net result. That heat is filling the space, which should then be entered by the independent space in motion if the motion of duplicating is brought closer in a relation to what the matching tempo requires.

It is clear that the density of material in motion is $\Pi^2 + \Pi^2$ but since that is k, which extended we know that that extending cannot sustain the initial speed. Since the speed is reduced the space in motion will value less. Taking these atomic relevancies into account, we can detect what relevancies brought about the atomic Universe of $(\pi^2 + \pi^2)(\pi^2\pi)3 = 1836$. The first substance that formed from singularity was solid and if that were the case the contra substance would then be a fluid with less motion filling more space taking shorter time duration in duplicating. The fluid substance that then formed was one less than the proton in motion which makes it slightly more in mass since it duplicates more with less space that then has to form $\Pi^2\Pi$, which has one Π less from $(\pi^2+\pi^2)$ which is resolved becoming a fluid like substance relevant to the first solid substance which is the proton. The loss of the one Π then became the factor claiming more space that is holding less substance. In this fluid state the neutron has more duplicating of the substance than is required of the proton. That what we find in space we also must find in the atom because the cosmos is not keen on inventing but is passionate on duplicating. This fact will also apply to space-time in many forms. That means investigation must prove the same results and what we find in the atom then also has to present in the cosmos at large.

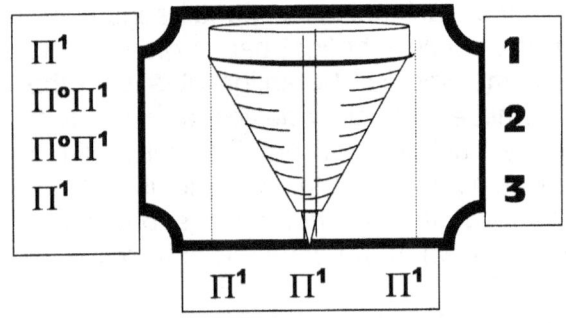

The overall picture resulted in a ring and all rings hold Π to secure the form. The only form that existed then was Π and therefore even today the borders use Π to indicate positions. But in the single dimension such definitions were far from clear and the only distinctions came from securing singularity in preserving the position of singularity to apply gravity and thereby absorb all anti-gravity. But anti gravity could not control expansion by counter acting contraction through gravity so the overheating continued forming non-existing borders in some thing infinitely solid just as Einstein predicted because this took place before light came about and therefore before the speed of light became part of the cosmos. The cosmos formed a partnership with one side overheating forming antigravity by expanding into space through the applying of the overheating and the other side formed gravity or contracting of space.

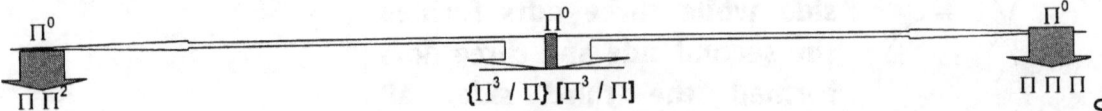

Singularity split the Universe into two parts that under no circumstances can ever meet. The one side of the Universe performs a balancing act to the other side of the Universe that duplicates but never double. The dot started overheating while the dot remained cool by activating gravity; the dot duplicated forming a sequence while the dot redefined the position in control.

Space

$\Pi\Pi\Pi\Pi$

$\Pi\Pi\Pi$ **Matter**

$\Pi\Pi\Pi\Pi$

Time

With the first dimension came matter, but also came space and came time splitting the universe in segments of matter relating to space filled with matter and time influencing the spinning matter.

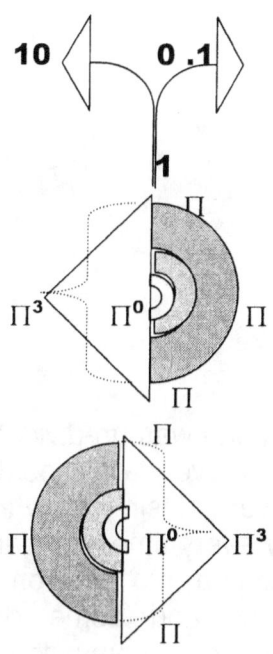

Taking the queue from the numbers line that runs in opposing directions singularity by Π^0 is always going larger as well as smaller. However, the centre takes a value of one. It is a private choice preferring $k^0 = 1$ or $\Pi^0 = 1$ but that splits the Universe into two parts, being smaller and being larger in relevancies.

It is apparent that one cannot substitute the correct formula used to measure the area of a circle by using $a^3 = \Pi r^2$ because if **k** is the diameter then the formula must be $k^2 \Pi$. However, **k** cannot be Π because in Kepler's formula **k** takes the value of the radius. In that case, what will the value be of T^2? That places the formula outside the normal use of mathematics practised in the normal sense of $a^2 = \Pi r^2$.

By using the Kepler Formula $a^3 = T^2 k$ it is good to change the values to Π and see what pans out. If **k = 1**, **k** at the same time would be k^0. By replacing $a^3 = \Pi^3$ then on the other side of the Universe $k = \Pi$ and $T^2 = \Pi^2$.

However, to secure this k in the centre must be 1 leaving $a^3 = 1(1 \times 1 \times 1) = 1$ and $T^2 = 1(1 \times 1 = 1)$. That complies with Einstein's definition of space-time being: Space-time is a four dimensional position of the Universe where the position of an object is specified by three coordinates in space and one position in time.

If k is the middle being $k^0 = 1$ then $a^3 = k^0 = 1 = T^2$. When time is in a shift freezing then $a^3 / T^2 = k^0 = 1$. In order not to overstep my limits by changing valid formulas I changed Kepler's formula to $R^3 = T^2 = 1$.

Space

$\Pi\Pi\Pi$	**Matter**
Π	$\Pi\Pi\Pi$
$\Pi\Pi\Pi$	
Time	

The book being written in Afrikaans the R stands for Rime meaning space and T is time. From that I deducted that the space used in a specific location will equal the time meaning the density of the heat in space. That brings the proof that space equals heat and space is the same as heat. Heat deforming or exploding is the equal to the space created. Also it confirms the substitute between Kepler and Π is correct

In the way space and the sphere connects the sphere will have 7Π points holding a relation to 3Π points not within the sphere forming the 10Π that creation started with. This will mean there is a division forever, and such a division may run smaller everlasting. With fluids connecting it is simple to recognise the sphere as Π for the form will indicate Π as the form of the sphere. By gas forming the connection there are the three points of space being apart and not forming Π, but still holds a relevancy to Π^2 through the value of Π.

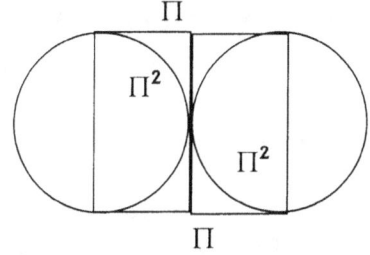

A solid joining by double Π forming as Π^2

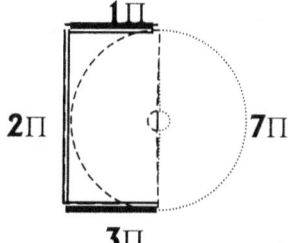

Taking the sphere as a unit with 7 positions and outside the 7 flanks 3 sides in the second dimension = 10Π

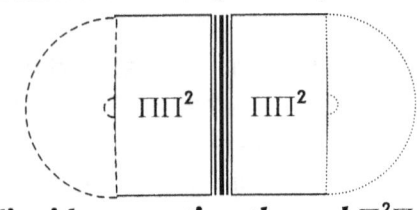

A liquid connecting through$\Pi^2\Pi$

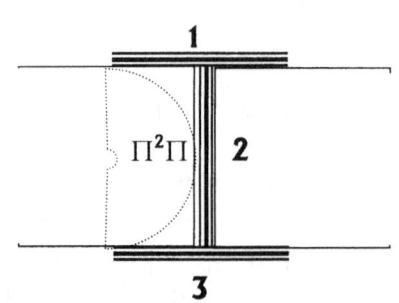

Total connecting relevancy of the sphere forming matter connecting to space
= $\Pi^2\Pi$ **3**

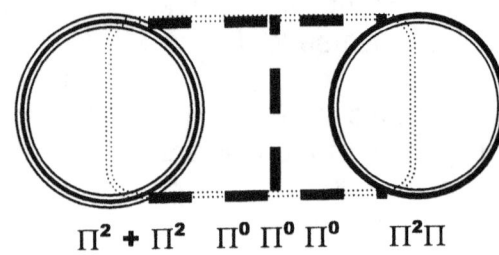

$\Pi^2 + \Pi^2$ $\Pi^0\ \Pi^0\ \Pi^0$ $\Pi^2\Pi$

How many dots there was is a question no person can answer because everything was un-dividable solid and yet it did group together to form every atom located in the 3D. Individual singularity and governing singularity and group singularity enhancing the gravity every time singularity find an accumulation. The Universe came into position by deploying dots supporting other dots and some dots remained dots while other dots went on to become dots of hybrids as it was supporting dots through claiming dots of lesser density and pass that on to dots with larger density.

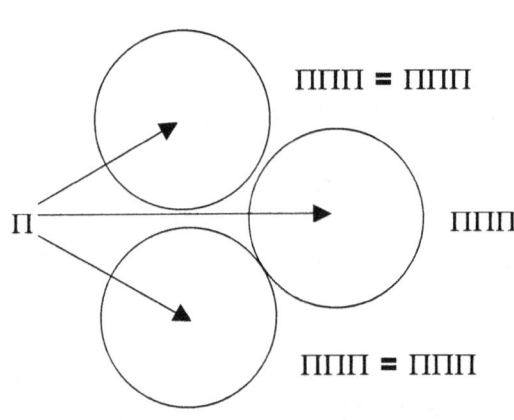

$\Pi\Pi\Pi = \Pi\Pi\Pi$

Π

$\Pi\Pi\Pi$

$\Pi\Pi\Pi = \Pi\Pi\Pi$

Space formed as motion came about through singularity overheating. Singularity k^0 produced motion at the point where k^0 became k and a^3 became T^2 by motion duplicating space. According to Kepler, a^3 is equal to k relating to T^2.

Matter formed where matter had to have $\Pi\Pi\Pi = \Pi\Pi\Pi$ space to occupy since it was to be in some space $\Pi\Pi\Pi = \Pi\Pi\Pi$
therefore $\Pi\Pi\Pi$ met with $\Pi\Pi\Pi$ to form the proton in $\Pi^2 + \Pi^2$ because the matter is within the space it holds and another Π^2 employs Π as a representative of singularity. This then placed the seven positions of singularity as the ending of matter and the three squares ($\Pi^2 + \Pi^2$ and Π^2) of singularity as the limit of material. The last $\Pi\Pi\Pi$ became $\Pi^0\ \Pi^0\ \Pi^0$ and that became the space producing heat without occupying matter in order to allow heat to be restrained inside the dome singularity provide.

It is all about relevancies applying the relations gained and lost through relations. If one place $\Pi^2 + \Pi^2$ on one side then $\Pi^2\Pi$ is the related form, where $\Pi^2 + \Pi^2$ is in the other side of the Universe being on the other side of the relevancy. Then $\Pi^0\ \Pi^0\ \Pi^0$ will again relate to the other two factors forming the "outside" of the other two being the "inside".

The Universe divides into two separate issues because of singularity. Nothing can be in two places at the same time the rest has to confine to the law applied by singularity. Objects can only be in one side of the Universe holding three parts or in the other side of the Universe holding three parts. From the totality three will be a double with six sides too shows, but that forms 3D. From singularity it is flat with three sides forming on either side of singularity as the formula used to measure the sphere indicates..

Newton said a sphere is $a^3 = 4/3 \, \Pi \, r^3$

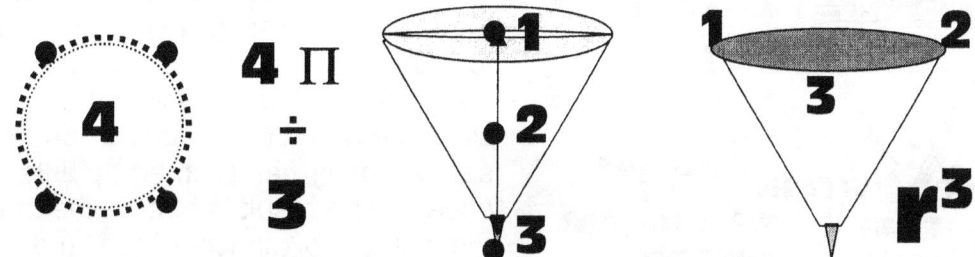

At first when material presented one side of the Universe matter had three sides to show. Matter had to have space to keep matter somewhere in some part of some universe and that made up three positions. Between the two universes **k** and T^2 placed a value but since only singularity applied any values the value therefore was $\Pi^2\Pi$ where $T^2 = \Pi^2$ indicated time coming from 7/10 in relation to 10/7 and $\Pi^2/2$ (proof of that is somewhere in the book) and **k** $= \Pi$ valued by singularity. When space-time developed 3D the dimensions falling outside the sphere becoming space-heat formed as $\Pi^0 = 1$. The electron holds a relevancy of 3 relating to the Neutron being $\Pi^2\Pi$ and the three keeps the electrons in different universes relating to separate or individual singularity.

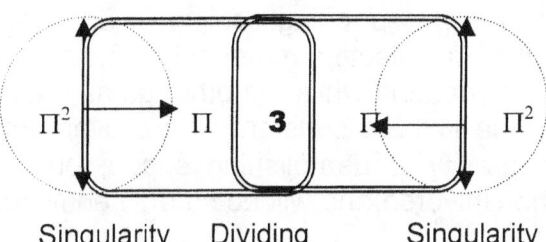

Singularity   Dividing   Singularity

The relevancy between the two particles secures individual positioning between the opposing particles, which positions the material that sufficient space secures cooling and preventing overheating.

As the relevancy between the particles promote overheating or applying antigravity (overheating) to the responding cooling or applying of gravity, the one repels material into space-time while the other is collecting material into space-time. The one loses material and sustain a model of preventing overheating while the other gains material and sustain a model of overheating. The one we named the Hubble constant where overheating produces space and the other we called gravity where gravity is demolishing space, but both phenomenon is at present dominating the flow of time in the Universe and will do so until equilibrium again comes about.

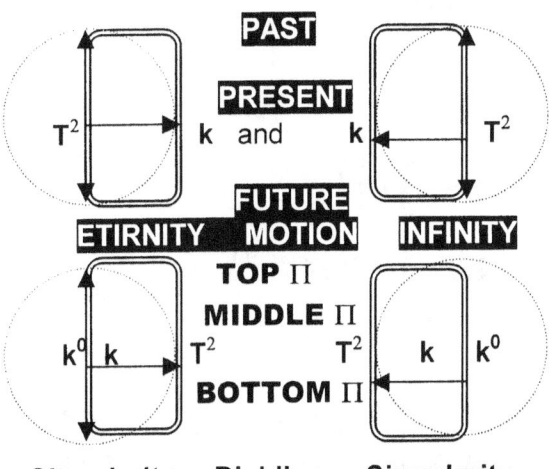

Singularity   Dividing   Singularity

The names I use in TOP, MIDDLE and BOTTOM must not be viewed as sides but merely as terminology using names to implicate divisions. Direction depends on positions and positions form a value only when the observer forms part of the cosmos and not part of the observing.

The universe divides into two separate issues because of singularity. Nothing can be in two places at the same time where as all the rest in the Universe has to confine to the law applied by singularity.

But when the Universe was in the single dimension, all values were Π, therefore every value related to $\Pi\Pi\Pi$ forming three of the same that was very different because it was where Universes met and formed relations. Every spot formed an individual dot or Universe and every dot was another new Universe.

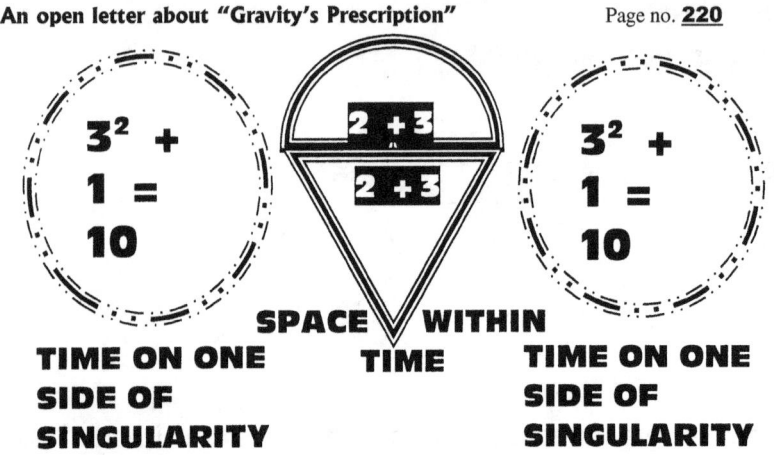

TIME ON ONE SIDE OF SINGULARITY　　　**SPACE WITHIN TIME**　　　**TIME ON ONE SIDE OF SINGULARITY**

In the relevancy where space divide eternity from infinity the three holding Π in relation to singularity holding Π^0 there are three points forming a square in relation to 90^0 which is implicating the law of Pythagoras while on the one side of the Universe the duplication is forming the same result and three points goes square. The result is that on the one side the square of space is ten and on the other side the square of space is also ten. As the relevancy between the particles promote overheating or applying antigravity (overheating) to the responding cooling or applying of gravity, the one repels material into space-time while the other is collecting material into space-time. The one loses material and sustain a model of preventing overheating while the other gains material and sustain a model of overheating. The one we named the Hubble constant where overheating produces space and the other we called gravity where gravity is demolishing space, but both phenomenon is at present dominating the flow of time in the Universe and will do so until equilibrium again comes about.

Keeping these factors in mind it is clear that Π^2 are the choice of gravity and not r^2.

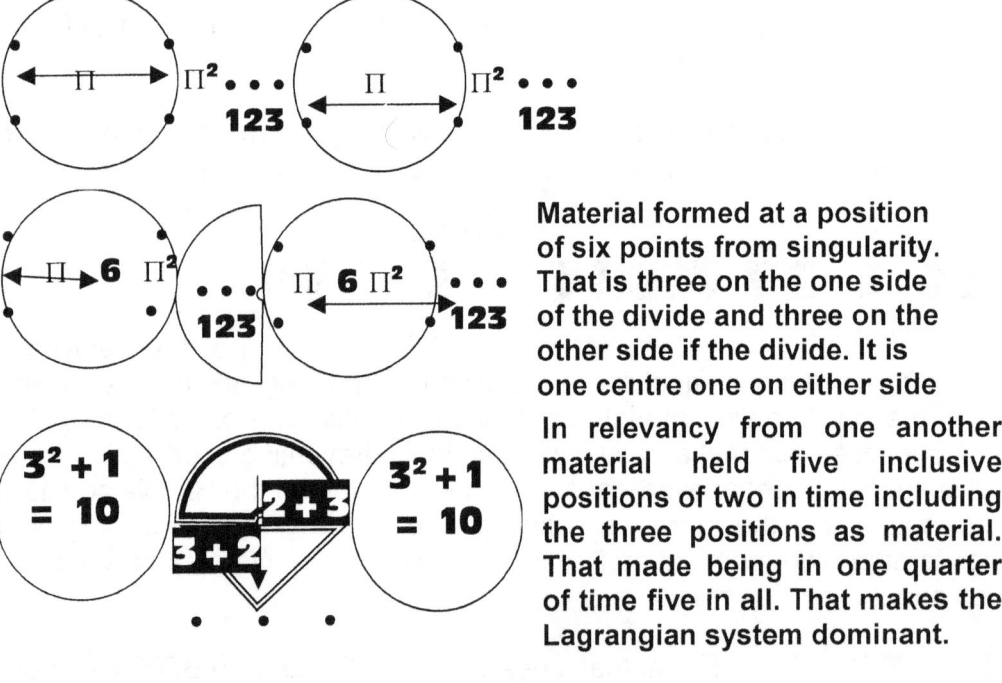

Material formed at a position of six points from singularity. That is three on the one side of the divide and three on the other side if the divide. It is one centre one on either side

In relevancy from one another material held five inclusive positions of two in time including the three positions as material. That made being in one quarter of time five in all. That makes the Lagrangian system dominant.

F=G (M₁ x ⬤ ◯ m₂)/r²
Is this truly the answer…?

In the investigation of light and gravity and objects and gravity, the mathematical rule of the invert square law must apply without question. But according to the observation of Roche that is not the case. From what one gathers through the Roche limit implicating two orbiting structure the opposite is applying. One must accept that although k proves as an indicator it is also much more when complying the thin influences brought about by singularity in the values carried on by singularity.

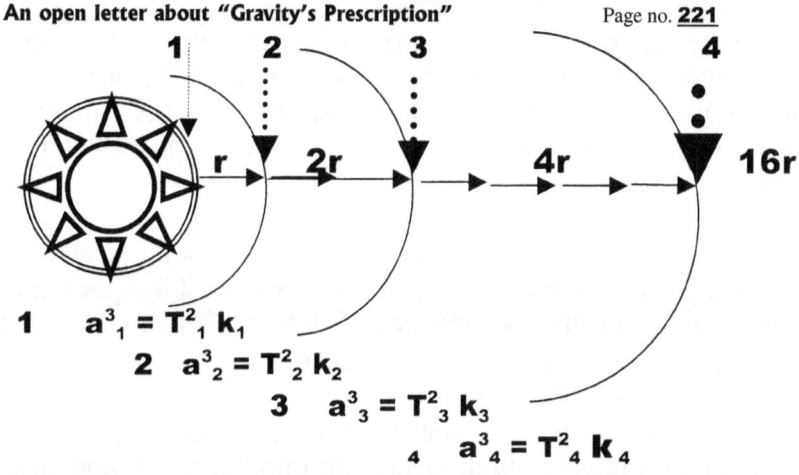

$$1 \quad a^3_1 = T^2_1 k_1$$
$$2 \quad a^3_2 = T^2_2 k_2$$
$$3 \quad a^3_3 = T^2_3 k_3$$
$$4 \quad a^3_4 = T^2_4 k_4$$

In drawing a most basic picture of light passing the gravity lines extending from any structure, I felt it was most insightful that the brains in cosmology was not able to see why light does not bend in the presence of increasing gravity. More surprising was that I found the mathematicians had to call on Einstein for advice regarding an ordinary problem. Light does not bend when passing large objects. It is Kepler's formula applying, and the evidence is clearly in front of the searcher for truth. But one has to go back to Kepler to re-apply what Kepler formulised and change the significant from Newton's significance.

As a^3 increases, so does T^2 as well as k increase and with that the influence of gravity per space unit increases with the concentration demise of a^3. But why would that be and what are we missing? Light shows there is an influence out there in outer space, that redirects light's route through space when passing large gravity fields. It is about the relevancy of k influencing the a^3 to allow the T^2 of light to divert in route because of influences established by k on a^3 and slowing down or increasing the line diverting. In this measure one may also find the Roche limit applying, but to truly understand how the Roche limit comes in place and how the Roche limit works one has to replace Kepler's factors with singularity and singularity extending being Π^3 $\Pi^2\Pi$ and **3.**

In the Roche limit the space factor provides space to a solid structure and therefore the value of r is replaced by the value of Π bringing about a square in half of Π. The cube holding 5 to either side removes allowing the extending of Π to indicate position to space.

Where Π extends to lock onto the next sphere's extending indicator, Π has to connect to Π

5/2

Five sides divided by two spheres.

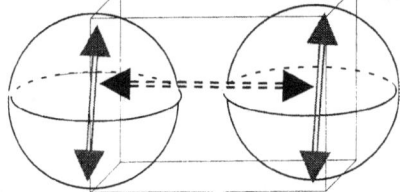

forming the square of space and translating that to the half of Π being $(\Pi/2)^2$.
According to normal mathematics the half of space should have been 5/2, but at the time this divide took place, space was all in motion and motion was Π in motion Π^2 crossing the divide (/ 2) forming $= (\Pi^2/2 \times \Pi^2/2) = 2.467$)

The space between the spheres divide in half, but because of the extending of Π and not applying r as ordinary mathematics will suggest where Π replaces r the singularity extending from Π^0 will be half of Π in the square of $\Pi = (\Pi/2)^2 = $ **2.4674.** In this lies the dynamics why planets have a positional (be it rather a dimensional) relation of 7/10

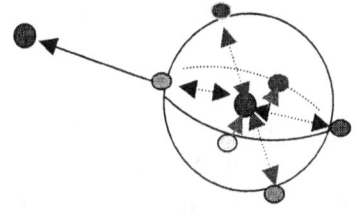

The Titius Bode law must not be seen as some obscure event that took place just before and / or after the Big Bang or when the solar system formed it fell into place. The Titius Bode law applies when the top is spinning, when an atom is spinning, when a motorcar wheel runs on the tar, when a jet engine fires up. It takes Place whenever the Coanda principle comes into effect and the Coanda principle is wherever there is motion in relation to singularity in a centre forming a centre.

This is why we can use degrees measuring the circle by (6^2) (forming the square relating to matter through singularity) X 10 (square if space) = 360^0 however it is always in motion. That proves no point can be static or constant, though it may seem that way to outsiders. Although matter is matter, matter can also be anti-matter and moreover form its own anti-matter at the same time. This degeneration of structure is very likely to occur with overheating.

Revaluing Π to Π^2 will bring about a new contact point where Π meets r forming another relation in Π^2. Every time material swaps sides it also qualifies as anti matter to matter because if it goes out of orbiting rotation frequency. It has the ability to collide with the same matter it forms union with but is located on the other part of the spin. It then becomes in a situation where Π revalue to r.

Time is the changes in relation where Π contacts a different r not withstanding the many r points there may form because every r constitutes a different value to the Universe through other ratios and relevancies brought about by heat and light. Time is the duration it takes Π to rotate between any two given points of r and therefore must always amount to a square (T^2) moving from point to point through the cube of space (a^3) in that duration of time (**k**). With that it proves Kepler's a^3 (space) $=T^2 k$ (time in the instant of motion) but motion must continue through a specific value in space where the space-time is maintaining relevant equilibriums throughout singularity connecting.

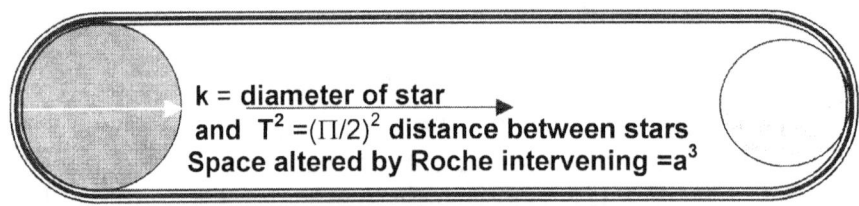

k = **diameter of star**
and $T^2 =(\Pi/2)^2$ **distance between stars**
Space altered by Roche intervening =a^3

The influence of singularity as the extending of Π into space links Π^2 to r and forms 2(5)+2(5) =10+10=20

From the position of singularity there are different values in Π where each indicate a position. The value it represents being $\Pi\Pi\Pi$, Π^3, Π^2, Π and Π^0

From there it influences singularity in the triangle flowing through to the half circle. It is an interaction between circular and linear motion as the value of Π continuous past Π^2 (at the end of the solid) and every cosmic structure holds an individual and specific singularity.

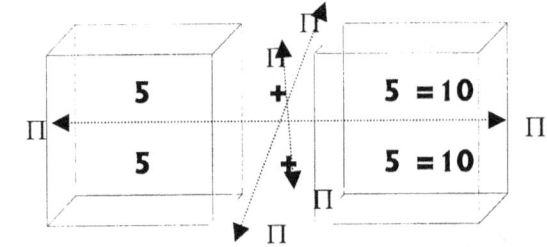

The field where Π extends we call the atmosphere having a value of 21.991 / 7, which is Π.

180^0 180^0 $180^0=\Pi^3$ The triangle, the half circle and the straight –line has two things in common, they share 180^0 as a mutual value and they are part of singularity.

Using the concept that gravity applies Π as the circle factor Π as well as Π^2 replacing r^2 the replacing by Π brings two values as Π and Π^2. That I found is the case with gravity and will be apparent when explaining the sound barrier as well as the Four Cosmic Pillars. In order to create a distinction I remained using r as the indicator of the cube or non-circle that has vacant space and by vacant space I refer to non-solid structures. In the solid structure I use Π as a value for reasons that will become apparent in due time.

Gravity does not apply mathematical equations to the letter as we would like, but rather use Kepler's thinking by enlisting an average gravity applying through out because it never favours and is equal every where. In gravity one find the extending of Π implementing Π^2 on average as a unit and not the radius r as a specific.

Looking at the affect of gravity it shows the precise quality of no distinctive point as gravity never seems to end at a point but flows all over affecting all that holds a position in its sphere of influence. The gravity coming from China meets the gravity coming from America at no particular spot but intermingles without distinction. This takes mathematics back to another fact beyond normal explaining.

We take a line running between two points as being 180^0 and the rest of the explaining is saved in the accepting part of mathematics. Any one of the two points the line starts or ends at is a point in infinity, The start and the end depends on the viewer putting the relevance to favour the side of choice. That puts the point of end or beginning in the spectrum of choice and not fact. Any direction is as equal as all other directions.

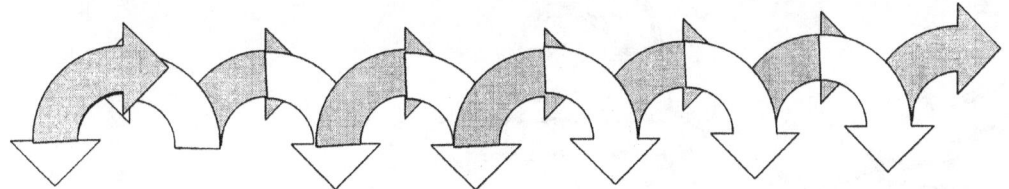

Following the flow of any line such a line is an extension of the previous dot in infinity to the next dot in infinity without any ability to skip or bypass any of the other dots in the connecting line. Any direction change including the remaining of travelling in the same direction is in relation to a line travelling all being the very same. Change does not affect the line.

A straight line, triangle and half a circle will always have equality in dimensional capacity providing equilibrium being 180^0 because each one shares a common denominator in singularity to the value of Π. As the straight line averts a zero it holds another straight line in place to set about such an averting where the two lines will always carry a relevancy in relation to progress (the triangle) and a common denominator in the start from singularity.

This concept we apply as the graph or the vector. By going back to a line, any lines and all lines, the line is a connection of dots in infinity, running from one specific to another specific and avoiding zero or dots. At every point in infinity it dips into infinity coming out on the other side by choice of direction and the direction is unforced and change presents any angle including the straight line, which incidentally is just another angle.

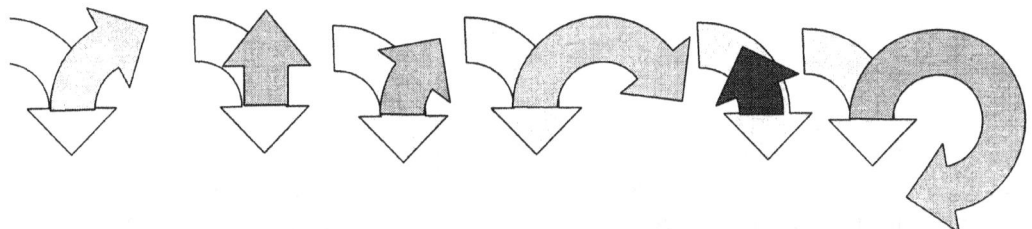

When connecting to the dot representing infinity the flow can be in any and all possible directions, including in the same direction. We all live in a graph, as the universe with all in it is nothing less than a three-dimensional graph flowing according to time. That means in the case of Pythagoras the mere fact that the line shows changes in direction does not implicate or affect the line as a tool of mathematics. Whether the line changes into a half circle meeting at the other end again or meeting in a triangle in forming a half square by joining the point where it began, the result still indicate a line flowing between points. Motion became an integrated part of space because motion is what establishes space. If motion redirects space and such redirection is not complying with a balance, there will be even further delay in time producing motion, which will bring about more time distorting and heat. Since the very first space from point to point was Π and motion produced a value of Π^2 while the four points indicated time, it is presumable that from that the Roche factor of $\Pi^2/4$ came into place.

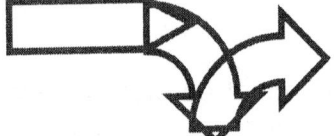  

The line dips into infinity every time it passes infinity when it cuts through infinity. The line going into infinity comes natural as the line progresses because all lines are infinite dots linking one point to another point. That means that coming out of infinity might slightly change the angle but that directs the route to the future and not the form because the form still remain equal whether the form is a triangle, half circle or straight line. The form remains a factor that confused every one in the past. When replacing the value we normally attach to circle being r with Π, the law of Pythagoras becomes quite meaningful and mathematical.

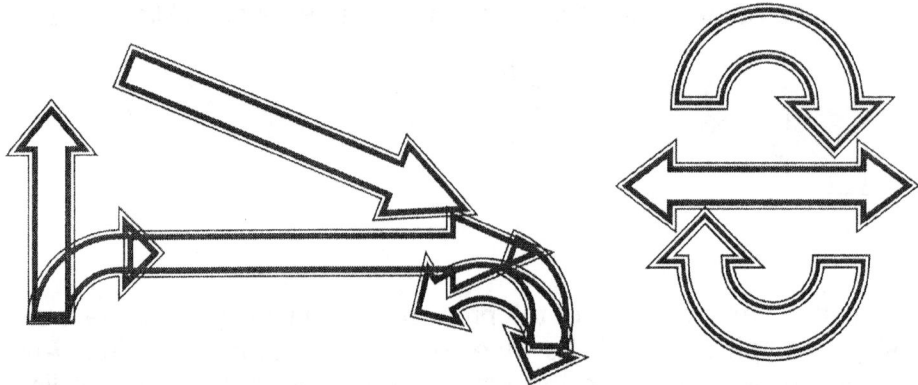

In that way a circle is a straight line following a loop as it comes out of singularity at a different angle and a triangle is a straight line that dipped into singularity but at three stages changed the angle with which the line then left to follow different directions at specific points. From the point singularity observes it still remains a straight line because there is no direction alternation in the first dimension and in that dimension it still remains a straight line in which we on the outside may experience as three forms but is in fact one single line. Only when the direction changes completely in reverse the line doubles in value but comes from multiplication for instance 2Π become Π^2. But the Lagrangian system proves much more than dimensional interlinking, it proves Pythagoras in principle.

LAGRANGE (-TOURNIER), JOSEPH LOUIS DE (1736-1813)
French mathematician, born in Italy. In celestial mechanics, he studied perturbations and stability in the Solar System. He examined the three-body problem for the Earth, Moon and Sun (1764) and the motion of Jupiter's satellites (1766). In 1772, he found the particular solutions to the problem that give rise to the equilibrium positions called Lagrangian points. Lagrange also studied the Moon's liberation.

LAGRANGIAN POINT
One of five points at which small bodies can remain the orbital plane of two massive bodies; also known as liberation points. Three of the points lie on the line joining the two massive bodies: L_1 lies between them, while L_2 and L_3 have the two bodies between them. These three points are unstable, slight displacements of a body from then resulting in its rapid departure. the fourth and fifth points (L_4 and L_5) each form an equilateral triangle with the two massive bodies, 60° ahead of and behind the smaller body in its orbit around the larger one. A well-known example of bodies flying at the L_4 and L_5 Lagrangian points are the Trojan asteroids in Jupiter's orbit. Among Saturn's satellites, Telesto and Calypso lie at the L_4 and L_5 Lagrangian points in the orbit of the much larger Tethys. In similar fashion, tiny Helene precedes Saturn's satellite Dione, keeping 60° ahead of Dione. The Lagrangian points are named after the French mathematician J.L. de Lagrange, who first calculated their existence.

LAGRANGIAN POINT:
*The Lagrangian points
are five equilibrium points
in the orbit of one body
around another, such
as a planet around the Sun*

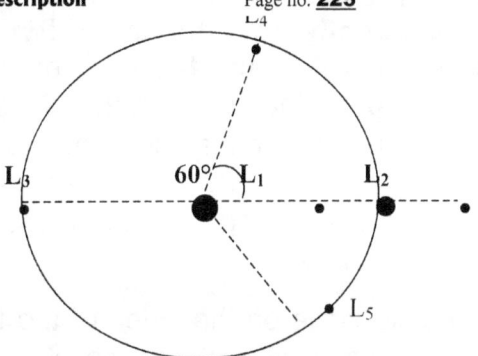

The entire concept of motion rests on the centre forming time and having one point outside time to be delayed or behind time. The delay parts motion in eternity from motionless infinity, which results in forming the Universe. Since the satellites are located as electrons the motion gathered from that falls in as a time delay. All motion is about time trying to cross that space to form a unity with infinity. That is the essence what keeps the top straight when spinning. The spin puts the outside of the top in another time zone than that the inside of the top is in and the four inside has to align with the fifth one on the outside where the fifth one is one in three positions allocated to the flow of time. By having time parted from time there is a flow coming from the fifth to the centre.

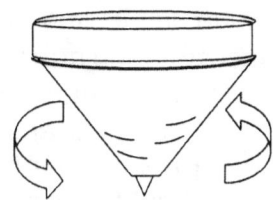

The Lagrangian System implicating the five positions extending from singularity

Each triangle claiming a side of the universe
The half Circle = 180^0 combining as a Sphere when comprising
Singularity dividing the cosmos

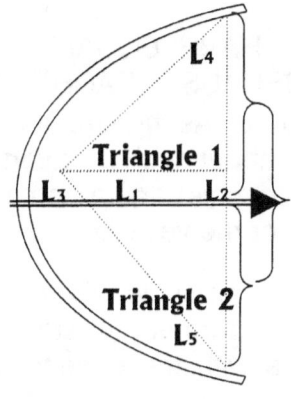

1 Half circle = 180^0 L_3 L_4 L_5
2 Triangle 1 = 180^0 L_3 L_4 L_5
3 Triangle 2 = 180^0 L_3 L_4 L_5
4 Straight Line = 180^0

Singularity in the matching of the value of the straight line forming the half circle and combining as the triangle and all are equal 180^0

The second one also fits in the singularity influence on the Universe.

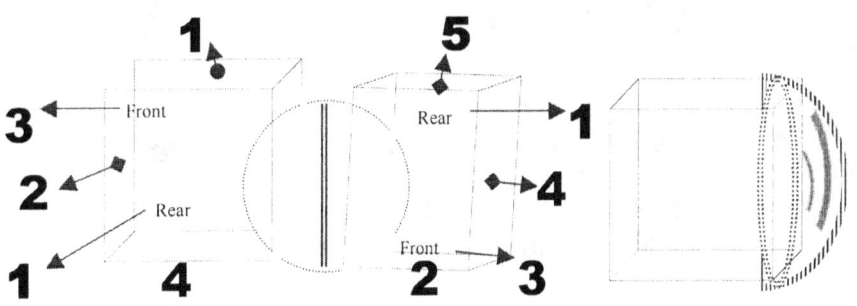

1 Relating to 5

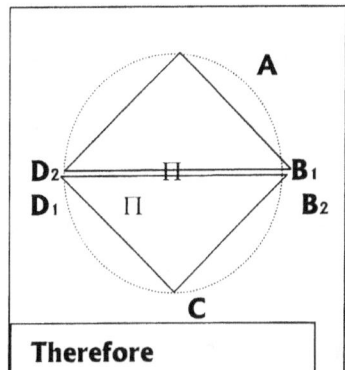

**Therefore
The Roche lobe is
$= (\Pi/2)^2$**

$(D_2 A)^2 + (B_1 A)^2 = (D_2 B_1)^2$ **(PYTHAGORUS)**
$(D_2 A)^2 = (B_1 A)^2$ **(EVEN SIDED TRIANGLE)**

$2(D_2 A)^2 = (B_1 A)^2$
$(D_2 B_1)^2$ **(DIA. OF CIRCLE) AND ABCD EVEN SIDED SAUERE WHERE AB = BC = CD = AD**

$(D_2 B_1)^2/4 = (AB)^2 + (BC)^2 + (CD)^2 + (AD)^2$
$2(D_2 A)^2 = (D_2 B_1)^2$ **BUT** $(D_2 B_1)^2 = \Pi^2$ **(Replacing r^2)**

$(D_2 A)^2 + (D_2 A)^2 = (D_2 B_1)^2$ $[(D_2 A)^2 = (B_1 A)^2]$
THEREFORE $4(D_2 A)^2 = (D_2 B_1)^2/4 = (\Pi/2)^2$

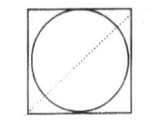 **The value of singularity stems directly from the law of Pythagoras or Pythagoras is the result of the average of singularity. With the shortest line being a dot, all lines must start from a position implicating Π.** A circle is a square without corners implementing Π and a half circle is therefore a triangle without corners. The corners are, an average of Π in the connecting line will come about. As both lines are the straight line forming singularity coming from one line being Π, the connecting line then must be the average of the two lines as Π^2. That is what the law of Pythagoras says.

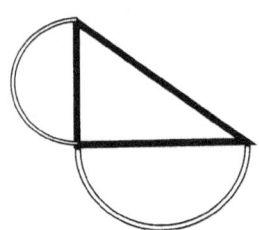   By placing a connecting circle on the sides of the triangle half a circle forms. By implicating Π as a relevancy and not the straight-line r, two values of Π applies to each circle, and the straight line is no longer r, but is Π^2. This will bring about that each circle holds half the square value implicated to the allocated conditions applying to Π in that specific instance. By adding the two half squares forming the two half circles and then calculating the square root of the total that then forms the average diameter into infinity comes natural as the line progress because all lines are infinite dots linking from one point to another point. That brings about that coming from infinity might change in angle bit that directs the route and not the form. The form is all the same

A STRAIGHT LINE, TRIANGLE AND HALF A CIRCLE WILL ALWAYS HAVE EQUALITY IN DIMENSIONAL CAPACITY PROVIDING EQUILBRIUM BEING 180^0 BECAUSE EACH ONE SHARES A COMMON DINOMINATOR IN SINGULARITY TO THE VALUE OF Π. As the straight line averts a zero going down infinity it holds another straight line in place to set about such an averting where the two lines will always carry a relevancy in relation to progress (the triangle) and a common denominator in the start from singularity. This concept we apply as the graph or the vector.

With the normal extending of singularity it will always form the triangle in a half circle whereby Π relates to the cube by 5 points to either side of the line singularity forms. Thus there are 10 standing related to seven and visa versa. By calculating the 4 squares in the circle with the dimensional changing of space (5) becomes the twenty

BC EITHER RELATE TO AB OR AC AT OCCUPYING SPACE AS MOVEMENT DRECTIONAL CHANGE THROUGH   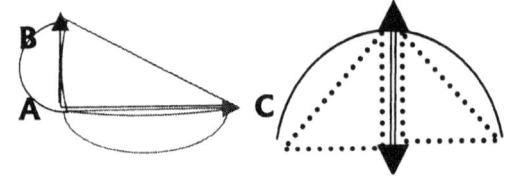   **ANY GIVEN TIME DICTATES DIRECTIONAL FLOW**

The normal flow will allow singularity but when singularity blocks another   extending to 10Π sphere in singularity the two will form a joint value and by this joining the larger will dominate the space as well as the time of the lesser taking control of the surface and the atmosphere.

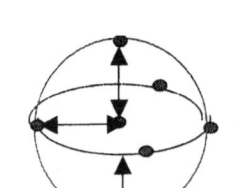   Through this the Roche lobe comes about with all its other dynamics I describe further on in the theses. The principle is the same, which we know as the conducting of lightning and Jupiter uses it extensively to implement this action.

In the sphere there are never only one direction implicated in movement. Movement are always in relation to the centre position because as a line goes up it also goes in or out. When a line goes north or south, it also comes towards the centre or going away from the centre.

There is always relevancy present in movement. As this moving indicates direction it also apply Π^2 for indicating value forming the time factor.

$$(\Pi_{a2} \times \Pi_{a1}) + (\Pi_{b1} \times \Pi_{b2}) = (\Pi^2_a + \Pi^2_b) / 2 = \Pi^2$$

= gravity and that is proven by Pythagoras. Gravity

is the average movement of matter through space in time determent from the position where matter in the sphere meets space in the cube from a point of Π to a point of Π^2 In this the figures of 2(5) = 10 (space) stands related to 7 from singularity as (matter)

 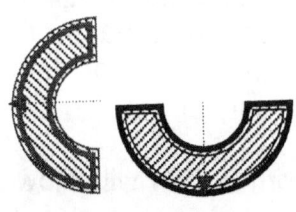 

From the star holding a dominant point or most valued point in singularity it affirm all three other structures, each holding singularity individually and in a compliment of 5.

The network of individual singularity not only provide spinning through governing singularity in the sphere but also provide spinning in the geodesic through out the cosmos linking all matter to matter in a network no one will ever come to understand in full. In the sphere the four squares forming the triangles linking the lines to the half circles holds space in time maintaining singularity of different assortments. In view of the matter-to-matter Roche factor where the factor consists forming relation between particles occupying densified space-time of where (Π / 2 X Π / 2) relating to the foursquare triangle the value of gravity Π^2 comes in position as Π^2 / 4 X 4 = Π^2.

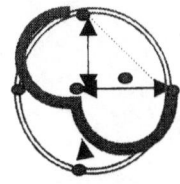 **Because every moving line represents one quarter of the sphere in relation to the rest of the sphere and the line also indicate the relevant position between the point indicated and the point in the centre it is a relevancy of singularity in progress. By connecting the line, as Pythagoras will suggest the singularity within the sphere become a specific value indicated representing one half circle.**

No object can be in two spherical quarters in the same time, but has to alternate in aliens to the space in accordance to time rotation.

To alternate in aliens to the space the relation of time in space has to alternate relevancy to the cosmos.

Singularity holds five dimensions inside and five outside singularity as matter and space forming space-time. The ten dimensions I named the atomic relevancy is also showing the double value of singularity as singularity extends into as well as beyond space. The atomic relevancy is (Π^2+Π^2)(Π^2 X Π X 3) = **1836** that is the mass relation between the electron (3) and the proton. Proton = (Π^2+Π^2) Neutron =Π^2 Π. The atomic relevancy holds the dynamics

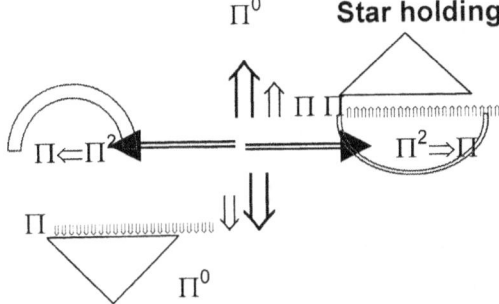

Π^0 **Star holding singularity**

From the dimensional implication comes about, not only the Doppler's effect, but many more of phenomenon not yet understood. The dimensional relevancies formed between matter as six, matters end at seven and space at ten, comes the value of Π.

The TITIUS BODE Principle Outside the sphere

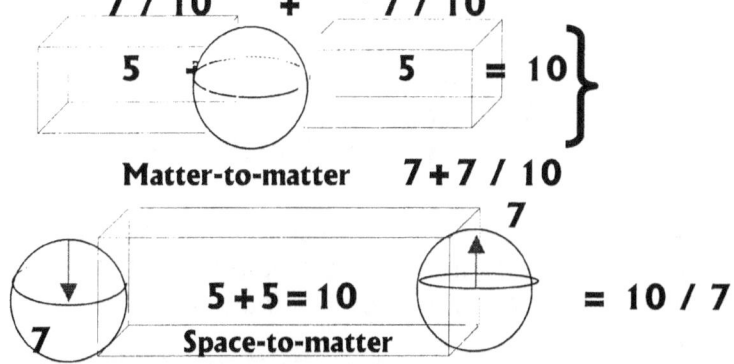

7 / 10   +   7 / 10

5   5   = 10

Matter-to-matter   7 + 7 / 10

5 + 5 = 10   = 10 / 7

Space-to-matter

The process is all intermingled and stands in relevancy to one another. The relevancy compliment holds such attachment that none of the factors can even stand-alone. It is the way that science places every aspect in the cosmos as individual and not related to each other that launches the problems of miss understanding. The Value of singularity appreciates or demises by ten fold. For instance, the value of Π will increase by ten every time singularity applies another layer.

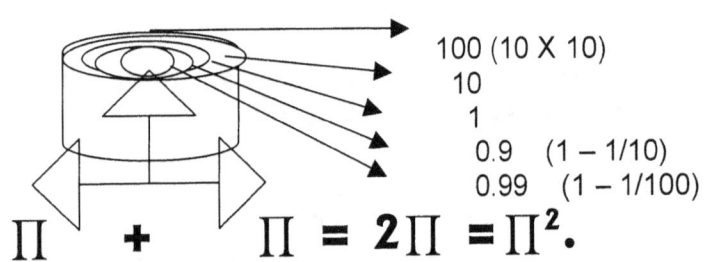

$$100 \ (10 \times 10)$$
$$10$$
$$1$$
$$0.9 \quad (1 - 1/10)$$
$$0.99 \quad (1 - 1/100)$$

$$\Pi \quad + \quad \Pi = 2\Pi = \Pi^2.$$

The normal flow will allow singularity extending to 10Π but when singularity blocks another sphere in singularity the two will form a joint value and by this joining the larger will dominate the space as well as the time of the lesser taking control of the surface and the atmosphere. Through this the Roche lobe comes about with all its other dynamics I describe farther on in the theses. The principle is the same, which we know as the conducting of lightning and Jupiter uses it extensively to implement this action. In the Roche limit the straight line forms part (1) and the half circle is part (2) and the triangle forms part (3) to singularity (4) Holding 5 points outside singularity. Every aspect connecting to the Universe changes everything it holds totally and becomes the anti-matter to which it was matter 180° previously.

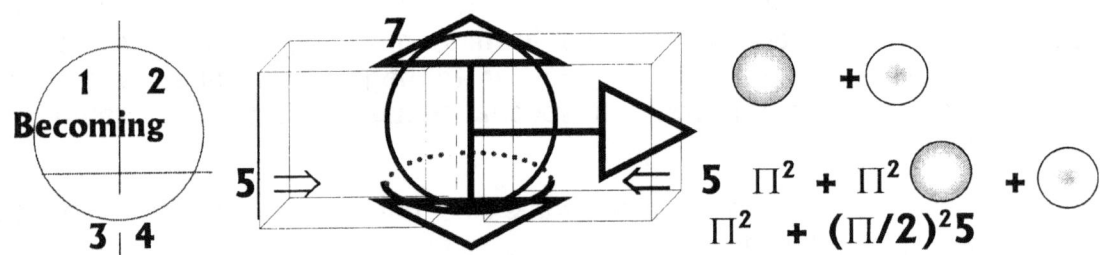

In the Roche singularity apply all three components of singularity

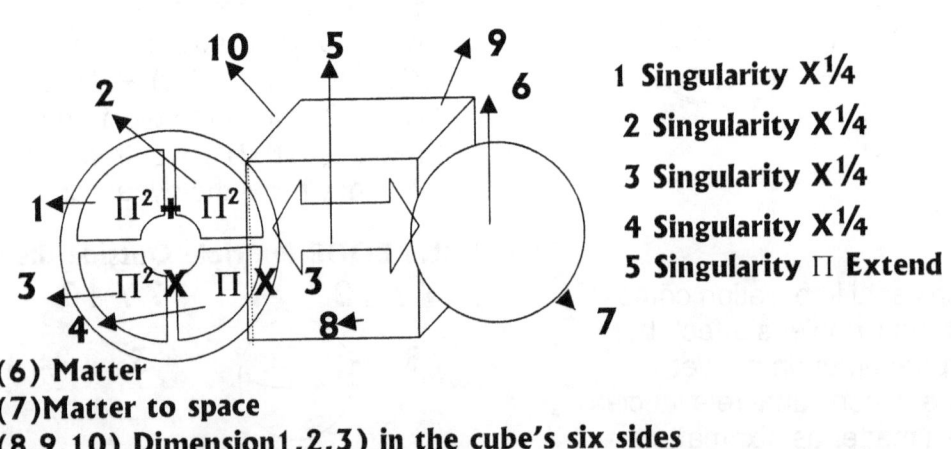

1 Singularity X¼
2 Singularity X¼
3 Singularity X¼
4 Singularity X¼
5 Singularity Π Extend

(6) Matter
(7) Matter to space
(8,9,10) Dimension 1,2,3) in the cube's six sides

Gravity is about a relation established when time begun between particles we know as material and particles we know as free or unoccupied space. Gravity reduces space to apply to fit the form of the sphere and later accept the form of the sphere.

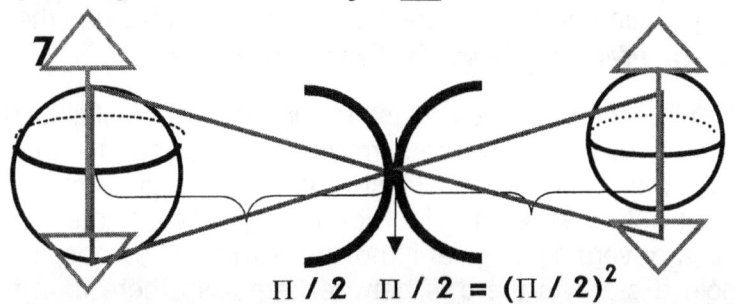

$$\Pi / 2 \quad \Pi / 2 = (\Pi / 2)^2$$

SINGULARITY MEETS AND COMPLIMENTS EACH OTHER.

The diameter of the cosmic structure holds the value of r and singularity holds the dimensional value of Π meaning that the radius or diameter (r) extends to become the diameter multiplying the value of singularity. But since r already consists of the square of space holding a definite positional relation with the value of singularity being Π the diameter comes into effect. Π extends each to an individual value to a point where the singularity on each side meets, bringing about a mutual Π^2 to the value dominance of the larger singularity control.

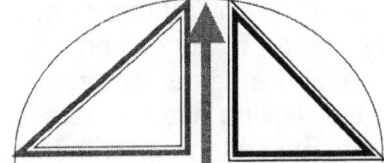

At this point the equality of the straight-line dimension to the triangle and the half circle holds prominence as a straight line, a half circle and a triangle is dimensionally equal. The common denominator will bolster all factors to an equivalent ratio.

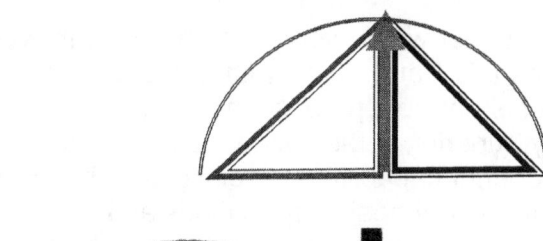

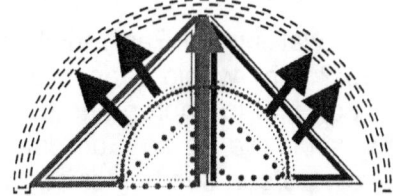

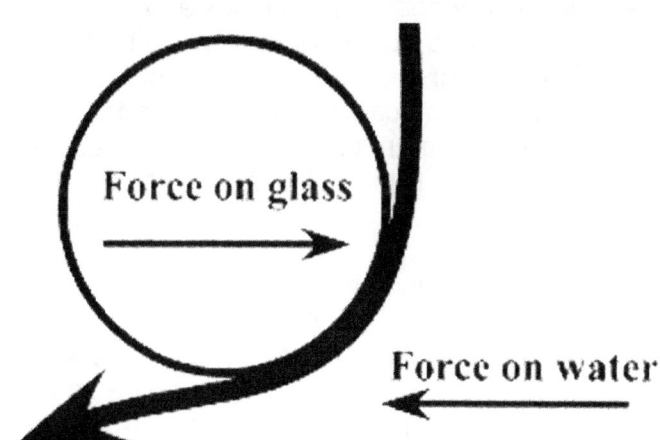

Force on glass

Force on water

The Coanda affect is proof of the functioning of gravity inside the atom. It proves that motion (T^2) of the neutron establishes a centre in line where the compliment of material forming the atom will secure a controlling singularity that is governing the entire atom. That forms the centre of the Universe. Singularity then finds a position at the distance of (**k**) and such motion claims the space (**a^3**), which is the atom by construction from a centre within that motion (T^2). The motion (T^2) creates a centre at the line of (**k**) and a centre of the space (**a^3**) the motion (T^2) establishes a gravity field all along the lines and at the distance of (**k**) in the space (**a^3**) that the motion (T^2) created.

When singularity by the straight line increases the singularity by the triangle it will also bolster giving equal potency in singularity by the half circle. As the singularity of the major component revives the lesser singularity to equality, the **triangle in singularity** will match the performance and so would the half circle respond in precise ratio setting equilibrium in order. The major partner's singularity in the straight line excites the minor partner's singularity in the straight line affecting all other aspects holding singularity in both objects to match equilibriums in all aspects of singularity. That is the Roche lobe.

From this the lesser partner will fill by the extent of the larger partner and as soon as equilibrium sets in the growth will duplex to matching in both accounts, normally to the fatality of the lesser partner, as the lesser partner will be capitulating under the strain of the dual. In that way the inner planets came in place as I explain in part 7 *of Matter's Space In Time The Theses*.

The Titius Bode configuration in accordance to orbiting formation holds a slightly different explanation to the explanation that applies to cosmic structure surrounded by space. It is moreover the individual singularity in maintaining the major singularity, which sustains the governing singularity providing equilibrium in space-time. Not only does atomic individual singularity maintain self-preservation, but in doing that it also sustains a governing singularity holding structural composition and form within a cluster of matter for example a star. As there is between stars so there is in the same manner a mutual or bonding singularity between atoms in stars, which we see as fusion. From this one may freely deduct that gravity is not forcing material closer but is destroying space whereby it converts the space to a density the senior partner has in the atmosphere of the senior partner. Where does all the information given thus far take us you might ask? For one it can help to explain something Newton science can never understand. To start with we have to realise that the Coanda principle is the manifesting of Kepler's gravity and we have to accept Newton's version of gravity is a load of rubbish. Years ago it dawned on me why we all would be so egocentric. This was a problem that was eluding every thinker ever thinking. I admit as a thinker I am quite average but still we are all thinkers, what puts us apart is what we think about and in that I am then equal to the attempt of any other average person with the right also to think. There is something that makes every person in his or her eyes having the opinion that that person is the greatest there ever was. Let's call it a Jesus syndrome. Either the person frequents with Jesus or the person has a special prayer linking such a person directly with Jesus or the person may recognise Jesus or Jesus has come in person to meet with that person in particular and others just simply become Jesus. We all know what I am talking about. What is it that gives every person on Earth the idea that that person is superior to all other persons except those we regard as being more advanced than us?

Why would every man walking on Earth think his sperm is just what every woman on Earth would give her front teeth for? Why would every man that walks this Earth do so with the idea that every woman is just waiting on him to impregnate her and that his her sole purpose in life…to wait on him to fertilize her? Why would we be so God damn ghastly superior in the way we see our status we have? Why would every person see him or her with the superior capabilities of reinventing life? Some would not eat meat. Others would bullshit through their teeth about health implications and the misery of death just to get the world to stop smoking. If we are that scared about death then we better ban the wheel first before any other thing because the wheel in whatever form is killing a hell of a lot more people than smoking can ever achieve. It is the thought that a person can impersonate God and that would allow and enable such an individual to change the course of man forever in all time to come… Some would go to war for any reason because only leaders that killed millions are worthy of the remembering by Historians.

The more any leader killed off his fellow beings the greater role his memory has in the history of man. Others would not war for any reason even in the face of being threatened by death. Some would drop a Uranium bomb on others with the pretext that they did it to save lives. Others would drag a whole world into a war for the benefit of monetary gain, because lets face it, in the back ground behind the drawn curtains there are those bankers and industrialists that makes enormous profits from other fools fighting "for justice". Something is making every person feel horribly special. Something allows every person to know that that individual is in the centre of the Universe right where God should be. There is a very good reason we all feel that way because we are not wrong to feel that way, and we are in the centre, the very centre of the Universe.

Step outside into the night sky and the reason is in front of you. Every sparkle of light coming from where ever is coming to you honour. All the light that was released from any and all points in the Universe is coming to the place you stand. That makes you the most important person ever born because you are the **centre of the Universe**.

When any person is standing on any place anywhere, while viewing the Universe, that person is filling the **centre of the Universe**. Let's get more personal. When you, the person that is reading this, are standing at night and is looking at the Universe you are seeing the Universe from the position that one only can have if that person is filling the specific spot in the **centre of the Universe**. All the light, every single beam that ever left any destiny at any time acknowledges this fact. You are the most important person in the Universe because you are holding the most important position in the Universe. All the light that come across and travelled all of the vacant space from any and all possible positions in space runs directly towards your position using a straight line towards you where you are filling the **centre of the Universe**. Not excluding the effort of one photon, all light is heading to meet you where you are in that centre spot and not one photon will pass you by. Not one photon dare miss you because if they do they miss the effort that all light has to accomplish and that is to locate you as the person filling the **centre of the Universe**.

Should you decide to shift your position to any other place in the Universe, you will shift the **centre of the Universe** to that location as well. If you install a camera on Mars, the light is obliged to acknowledge your relocating the **centre of the Universe** at your will to reposition you're being that **centre of the Universe**. All the light that ever left its destination crossing the vast spaces of the Universe, excluding no particular light, travelled all the way just to find you filling the **centre of the Universe**, right where you are. By you're standing anywhere, you fill the **centre of the Universe**, and the entire Universe admits to that because all the light comes to meet you there. If you shift from the North Pole to the South Pole you will shift the **centre of the Universe** because all the light travelling throughout the Universe will find you where you then moved the **centre of the Universe**. The light left its destination billion years ago as it travelled through space at the speed of light anxious to acknowledge you're being in the very **centre of the Universe**.

No photon will be able to pass you by where you are in the **centre of the Universe** because all light is heading your way from their starting positions. No wonder every person born has the idea they were born to fill **centre of the Universe**, which we do fill. The Universe is spinning around you or I, which is filling a centre where all motion is connected. That is the Coanda effect on the utter-most grandest scale imaginable; nevertheless it is only a manifestation of the Coanda effect. It implicates gravity as wide as can be… Some things mathematics is able to explain but other explaining goes beyond mathematics. Try to explain mathematically the colour of the sky being blue in a clear Sunny day and changing to black when nighttime falls. Do the explaining in mathematics to a blind person that had no vision since birth in such perfect mathematical detail that would allow the person afterwards be able to explain the difference between blue and black to other blind persons by using only mathematics. Some aspects of the Universe go beyond mathematics and some even go beyond words. It is our task to find space, to find time and moreover it is our optimal task to find the Universe. We have to see what is solid, what is liquid and what causes gravity. Please keep this part in mind because in a short while I am returning to this to show how this becomes a cosmic reality.

Gravity **is to move or apply the intension to move** space a^3 **at the** distance or relevancy of **k** while T^2 is the time it is going to take to **apply gravity** or move the space filled with material space a^3 at the distance of **k** in the time period of T^2. That confirms Kepler's attribution to gravity where according to Kepler space a^3 is equal to the movement T^2 (time it takes to move) at the distance **k** from the centre specific.

Then the I took Human nature and science and combined the two, which gave me the vision on the findings Kepler received from the Cosmos. It puts all aspects of gravity in the Universe in new dimensions. But the visions formed the beginning because the visions unleashed many new questions. If gravity is motion, what causes motion? What stops motion? That answer is in the Black Hole. In truth the explaining of the Black Hole is as complicated as the Universe may represent and as simple as the cosmos truly is. If a star is about fusing atoms and with such fusing of atoms is thereby growing, what happen when all the atoms fused into one all collective atom in one already all—atom-accumulated star? What is the gravity if the star has melted all atoms it had into one all-inclusive atom and this all-inclusive atom is providing all the gravity that the star had when the star still had massive volumetric space? If all that space that once filled an entire giant star fused into one specific space less centre holding singularity 1^0 then the enormous gravity is applying to the centre of such a non existing space-less atom and that entire enormous force has been secured in the space less than that which one atom holds. In that case the atom would then show a force that would pull the surrounding Universe flat. The purpose of fusion is to reduce space and magnify space less ness inside the sphere. Where does the gravity of the star end when all the atoms in the star became one giant atom by fusing all atoms into one nucleus? Gravity is smallest where space is least. Where space of an entire massive star is left in the size of one atom the gravity coming from that will pull the Universe flat at that point.

Newtonians have the opinion that it is energy that keeps the planets in rotation and the system is equal to the rotation one will find in Earth. There is one slight problem and that is that all the mass used in the calculation is not worth a penny in practise. In nature all the planets orbit in an equal ratio while in their opinion the mass is the key factor, which implicate all aspects of the energy requirements in the planet orbit.

They say that E = - (GMm) ÷ 2r and the gravitational constant (G) is one factor of three where the product of the three factors holding the Mass of the Sun multiplied by the mass of the Earth (or what ever planet apply at the time) giving the Mass X the mass X the Gravitational constant and this is in division of the radius (r) from the Earth (or what ever planet apply at the time) added (2) from both ends. There is a problem looming on the horizon...

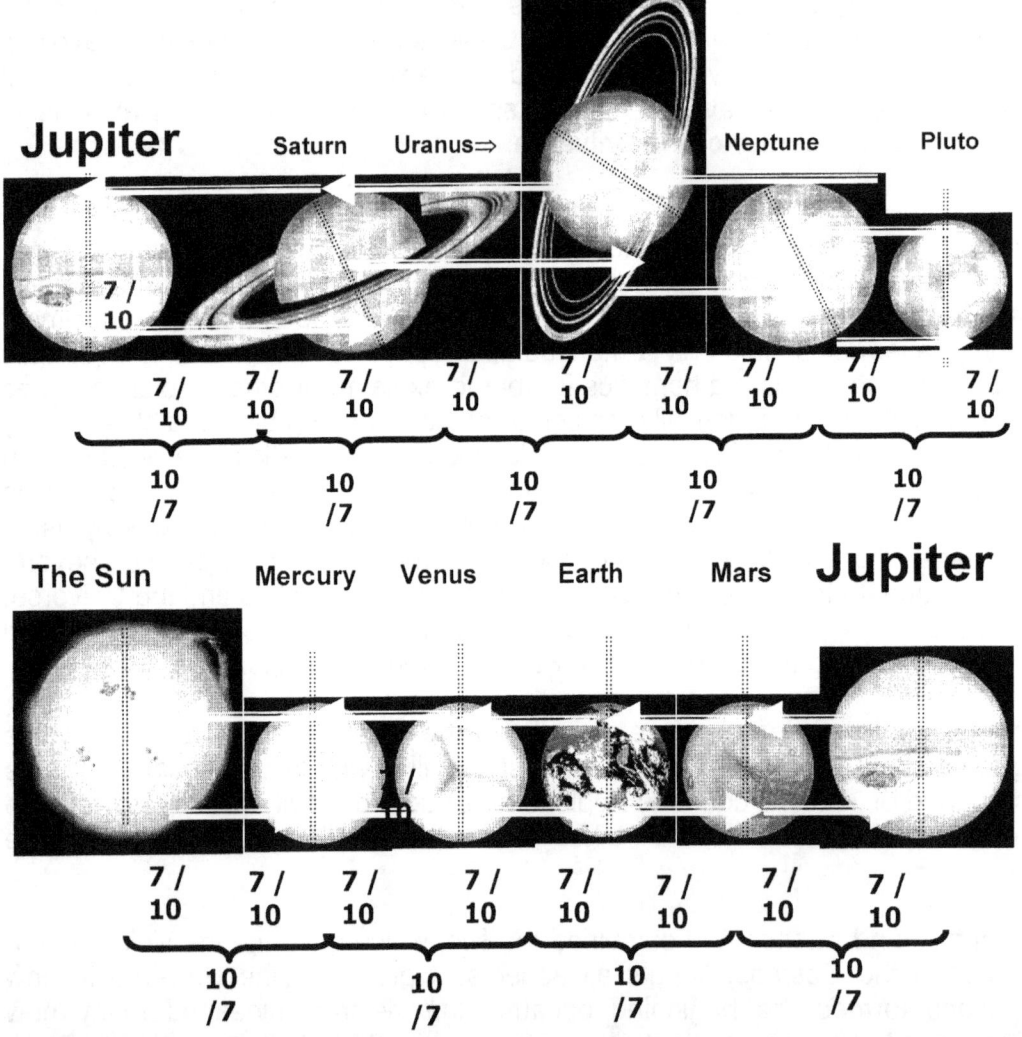

Notwithstanding mass differentiation and mass discrepancies of the large planets in relation to the small solid planets all the planets are in a similar ratio in space and time around the Sun . That means big or small, they travel alike. You can say what ever you like about Newtonians but stupid they are not. They know how to think and think they can...fore instance try and beat this:

Mercury	Venus	Earth	Mars	Jupiter	Saturn	Uranus	Neptune	Pluto
0.055	0.86	1.0	0.11	318	95	14.5	17.2	0.002

Notwithstanding the enormous mass discrepancies we see illustrated in the table, all the planets orbit equal in ratio. That means we can ignore the fact that Jupiter is 318 times more massive that is the Earth because they use the same time to space ratio. One might think that if the one mass (the smaller mass) in the case of the Earth stands to be used in the formula E = - (GMm) ÷ 2r, in comparison to the case where Jupiter is 318 times more, or in the case where Pluto is 0.002 times that of the Earth, the mass will bring changes. As I said, one thing you may not call the mathematicians is that they are stupid. They did notice that all the planets orbit equally and at the same ratio. That did not stop them from implicating mass, no they just went on to blame the gravitational constant being guilty of eliminating the mass discrepancies.

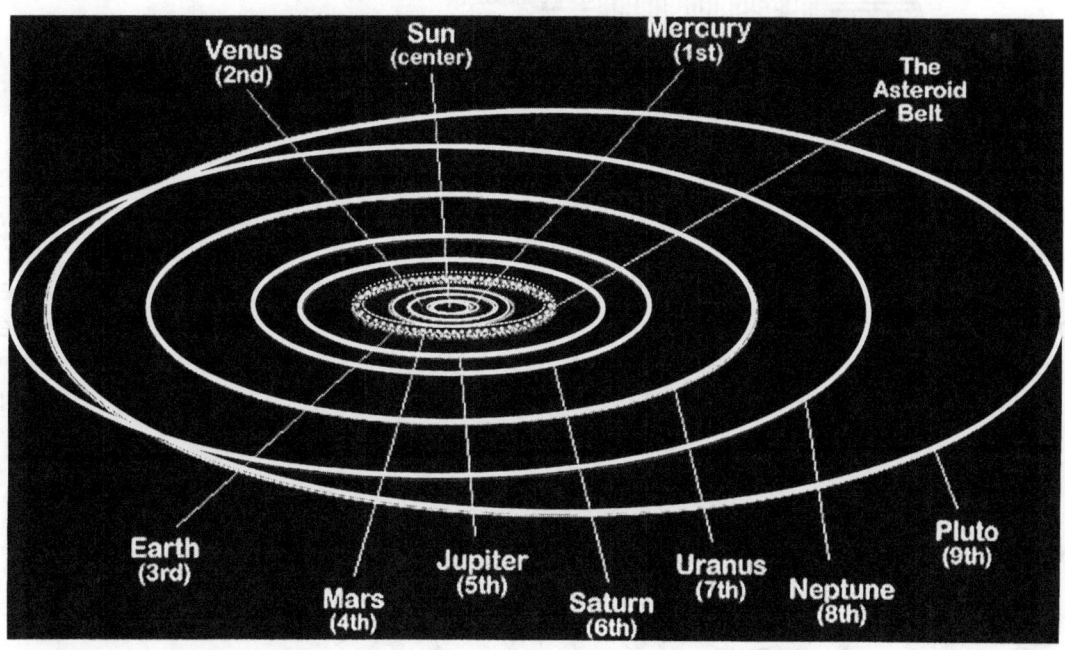

If it were true that it is the gravitational constant that is eliminating the supposed effect of mass on the potential gravity of a star then it would be that the formula would read as follows:

F = (M X m) ÷ (G x r²) where (G x r²) = (M X m) because that will mathematically show that the Gravitational constant (if there were anything of that nature applying) cancels the effect the mass factors has on the orbiting structures. That would mean that the gravitational constant eliminate the mass factor on bother ends of both the radii and not as it is at present where the gravitational constant incorporates the mass as the mass on both ends incorporates one another in order tot compliments gravitational constant to calculate the required planet orbit. As I said, they are not stupid, they will use any bullshit to wiggle them out of a loop. They do with that problem just what they do with me as a problem they pretend it never was a problem and ignore the problem.

In another pert of the book I went into the criminality of falsifying evidence in order to colour a picture to the likings of the person acting criminal or to falsify in order to bring about purposely an incorrect situation. In this part I wish to elaborate on the incorrectness of this approach and the magnifying of I\the intended incorrectness. It is acceptable that there was no one in the past that saw the Titius Bode law for what it is but in the same manner if there is deliberate protectionism of the corrupt and a deliberate effort to cover falsifying evidence and statements, then it will be a natural tendency to over acclimate the process where further investigation is required.

In the Titius Bode law on find the distribution of planets in response to the allocation of singularity respectively. By having a distribution of twice time seven divided by ten in relation to ten divided by seven is the location or position that serves the outside planet. Where the one is twice the other we find that the distribution is coherent with one marker and one planet. The location of the other planets has no role in the position the outside planet has since to the outside planet only its immediate inside planet is a seven. Any planets closer than the immediate inside planet is to the Sun or farther away from the Sun than what that outside planet is, has no function in the allocation of the planet distribution. To every planet that planet is forming the outside and all other planets except the one to its immediate inside is of no significance to the planet forming it's most outside border. This results in the way the distribution uses (7+7/10) in relation to 10/7.

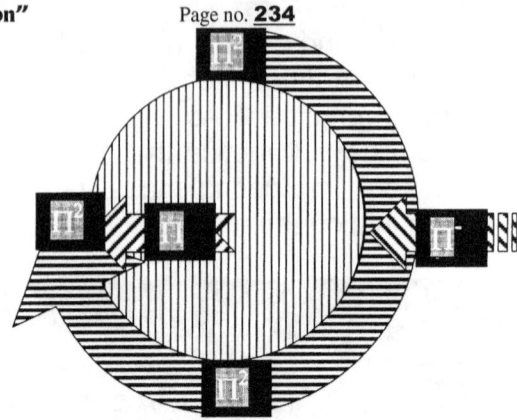

It is so obvious that mass plays no part in the orbit of planets. I just cannot believe any reason or excuse put forward why the worlds most intelligent will hide the truth about mass not playing any part! Yet where the Titius Bode is so overwhelming in evidence of being the process used to form the allocated orbits of the planets, there is such a strong and deliberate attempt to by pass the issue. The blatant misleading reasoning about why the mass will be illuminated by the gravitational constant without having that reflected in the formula used is shocking but even much more shocking is never having one person investigate (in earnest) the Titius Bode law.

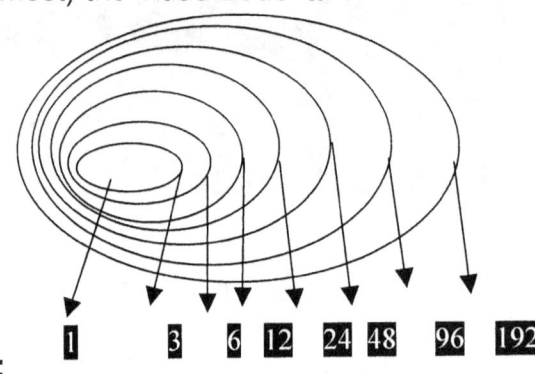

Bode's Law:

Planet	Mercury	Venus	Earth	Mars	Ceres	Jupiter	Saturn	Uranus
Bode's law distance	4	7	10	16	28	52	100	196
Actual distance	3.9	7.2	10	15.2	28	52	95	192

That brings us to another Newtonian problem that they deal with in precisely the same manner; they ignore it and declare it never existed in the first place and any one mentioning it must first prove that it ever existed by proving that it never was a coincidence to start with.

One can clearly see how the singularity of the atoms form the building form used to increase the space –time growth. The seven that material holds are in double relation to the ten that time holds. By valuing the atom as $(\Pi^2+\Pi^2)(\Pi^2\Pi)3=1836$ we find that the seven reflect as the material component and the seven on both sides of the Universe is in regard to the five it is in contact with. But on the other hand the five doubles to ten on every side of the Universe since no one can determine precisely where the five begin to form seven and the five will always be a square to the seven it is in contact with.

The square however dates back from a time when the square still was just a doubling to bring a duplication of one to the other side of the Universe. For every seven in singularity holds relating to material (7/10+7/10 = 1.4) the time doubled by remaining the same ratio (10 / 7 = 1.42) That allocates one line in singularity in space holding time to twice the ratio of time holding space while the ratio remains the same. That means the radii (if one could call it that) in distance doubled (.7 + .7) by allocating one time unit in relevance (10/7)

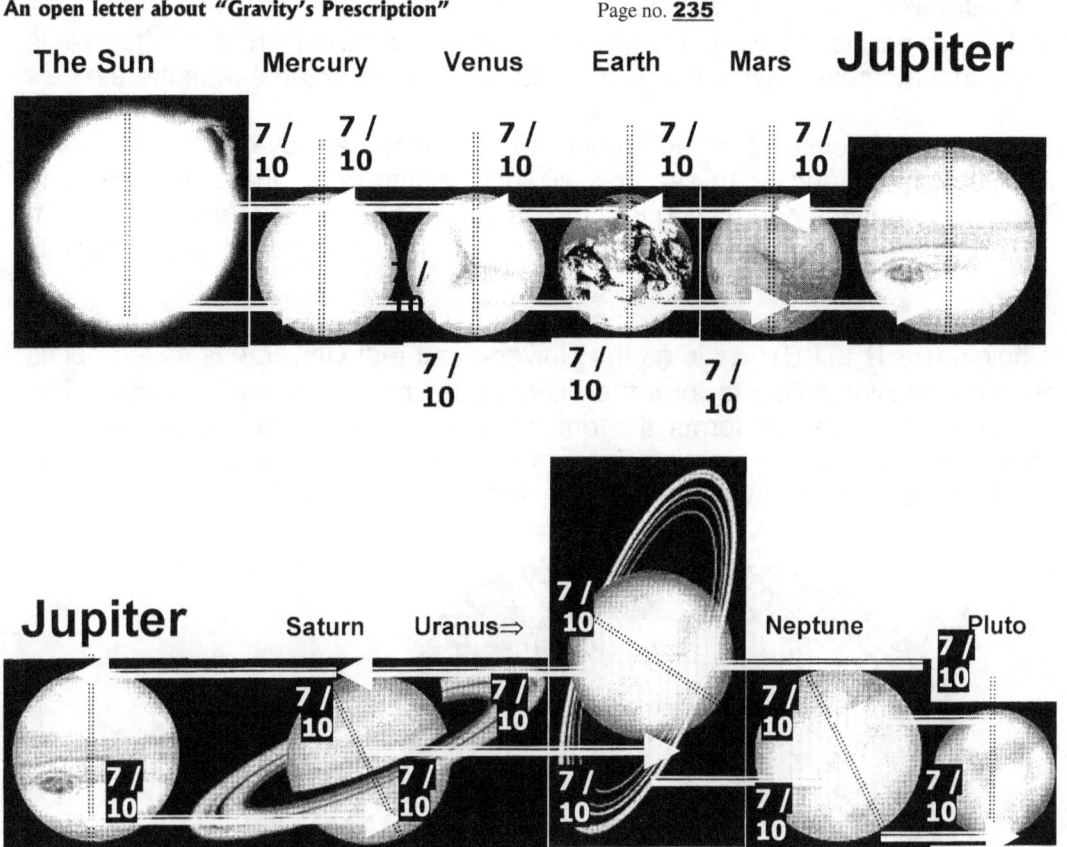

There is no one of the seven directions we move in because time takes the seven directions and move infinitive number of sevens in three time positions from the future to the past. The direction we see is the Universe coming towards us and disappearing into the infinity we have. We are moving from eternity in the direction of infinity.

That easy part explains the frequency Titius and Bode mathematically could interpret. The outer space region is the neutron. The neutron provides gravity by producing motion. Motion is $(1.4 / 1.42) \times 10 = \Pi^2$ and that makes outer space the compliment of motion Π^2 going to Π. That is way the location (Π) is in double the time (Π^2).

Another bone of contention I fail to see is how does Newtonians compromise logic in order to justify Newton in terms of Galileo. Yet I have been, to put it very frankly insulted on more than one occasion because I fail to see how Galileo says mass plays no part in the falling and Newton formulate that the whole affair is mass orientated. $F = G (Mm) / r^2$. On one campus in particular there was one professor that truly got nasty about this and he insulted me in a way I cannot forget. However that same professor failed to show me how Newton's mass brought any object faster to the ground since $(GMm /r^2) = mv^2/r$ which suggests that the square of the velocity multiplied by the mass is the same as the gravitational constant multiplied by the product of both the masses and then divided by the square of the radius.

That means the mass m has to multiply X with the velocity in the square (v^2), which then will reduce (demolish) the distance (r) there is between the Earth and the object on a continuous basis until the distance is reduces. That's rubbish. How do they console this statement with that of Galileo where Galileo said all objects fall equally to the ground! Galileo said that notwithstanding mass discrepancies will all objects hit the ground at the same moment when dropped the same distance and at the same moment. Newton insists on mass while Galileo insist on equality of mass during the fall. The biggest bogus part of the lot is that I have not come across one Newtonian that was able to see this distinction. It is as if they all have an inborn blind spot.

Galileo said that the atmosphere is a neutron that is providing unrestricted mass in the time period that the earth set. Galileo unwittingly suggested 7 / 10 and that is what gravity is. I found the sound barrier as $7(3\Pi^2) = 207.2616$km per hour. That is applying to what ever is falling whether whatever is falling or intending to fall at that moment. That is the neutron state of a body in the atmosphere.

A while back I indicated how man's senses evolved around his view that man (every one alive) is in the centre of the universe. Everyone and I can see how all light coming from wherever is heading directly towards me. By standing outside and gazing into the dark eternity that never ends I see from eternity light flows towards me and that places me in the centre of the Universe. That is a cosmic reality.

The atom holds seven points $(\Pi^2+\Pi^2)(\Pi^2\Pi)3=1836$ as the Universe but that Universe is seven points in Π being 3 points serving dimensional time to form the Titius Bode law, and a law it surely is! The gravity extending from the Titius Bode law forms the entirety of the building of the Universe by constructing the Universe in the using of the atoms to form the Universe in the entirety thereof. That puts the atom in charge of the Titius Bode law since the atom forms the Universe.

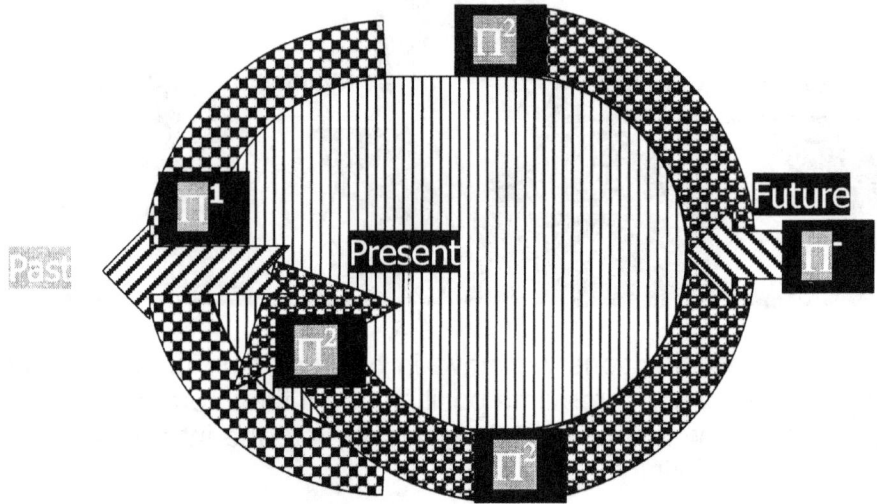

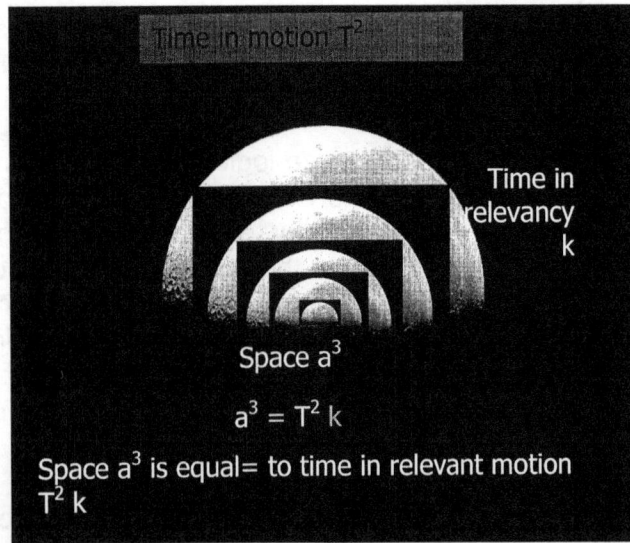

The three points we find time to be moving in is a direction unlike what we in the past thought about as a direction. There are seven basic directions being front and back, north and south, top and bottom and in or out. To our view that is the only way anything can move. It is either one of the lot or a compliment of two forming one of the lot.

By seeing light travelling towards me I am seeing time travelling. I am the direction that time flows. I can see where light was. I can see where light will be. I cannot see where light is going because that is within me and my singularity presents the future. Any one in disagreement should just go outside and see the light coming towards you. See how the light meets from all over the Universe precisely where you are.

The flow of time must never be confused with any part of the seven dimensions in space-time. The flow of time is away from the structure towards the structure then into the structure, through the structure where time disappears into singularity by measure of infinity. The flow of time is the motion that concerns the part science at present think of as outer space but which in essence forms time in eternity. Because Mainstreams science has the name incorrect they also have all attachments they connect with time in eternity incorrect for instance that the Universe has an edge and they give the cosmos a place where the Universe ends. That part on the outside of the atom never ends because that part is continuous to the point where time ends and time cannot end in the part holding the seven dimensions of time in space. Therefore time ends within the atom or as I have life my time ends in me. Therefore I have my position where I am in the centre of the Universe. Time comes from the outside as far and as wide as things can go but time ultimately ends within me.

Time is the outside of the atoms. Time is the inside of the atom where time is excluded from eternity by giving time specifics in motion and a defined value confirming space while the space is conforming time. Then time is taken to infinity where infinity absolves time into a unity once more. Time is what is between infinity and eternity and while eternity is parted from infinity time in eternity as a unit is also standing between eternity and infinity. Eternity is part of the part that is standing between eternity and infinity and therefore I am able to see eternity as a reality.

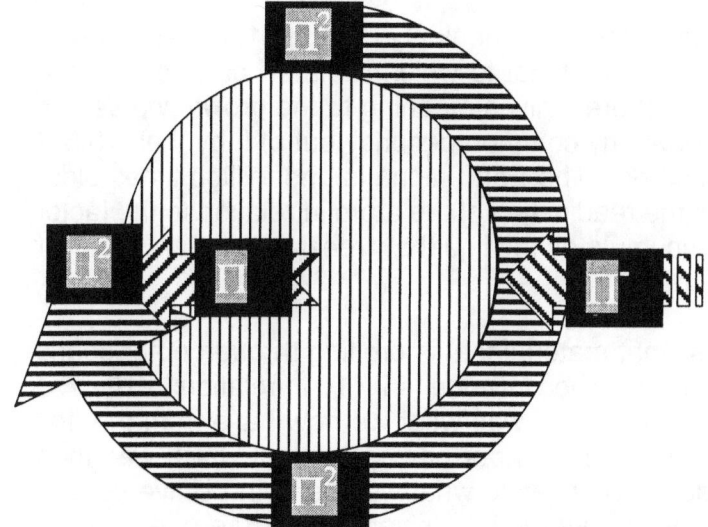

The light coming toward me is going whereto after it is upon me. It is going to the past. But from where I stand the light is representing my past so I am taking my past down my infinity into the future. That is why the Universe is shrinking onto the oblivious.

That is why everything into my future is shrinking into the oblivious as time engulfs material into the future.

You were in eternity because the light is coming from eternity towards you. You are where you are because I can see where you re plus the time it takes the light to come from you to me added to where you are in time. The light is going to disappear into where you are but that is not true. You are dragging the light that reached you into infinity berceuse light tries to escape time by going infinitive.

Infinity that which has no start is in you and you with your eternal life is generating time that parts infinity and eternity. That is not religion because that is raw physics. I have my doubt that any Newtonian will understand this concept since they can't even see that mass has no application on objects in orbit in outer space. If they are incapable of seeing the obvious how the hell will they be able to see what only those with intellect can see. That is why they can see no God. It is because they see mass applying in locations where there can be no mass applying.

Where you are and where you hold your body is the closest Black Hole to you because at that point time converts to space and space disappear into the gateway of singularity. Time ends where you stand but that only applies to your Universe and while your Universe is in contact with the rest of the Universe your Universe is solo and alone in time.

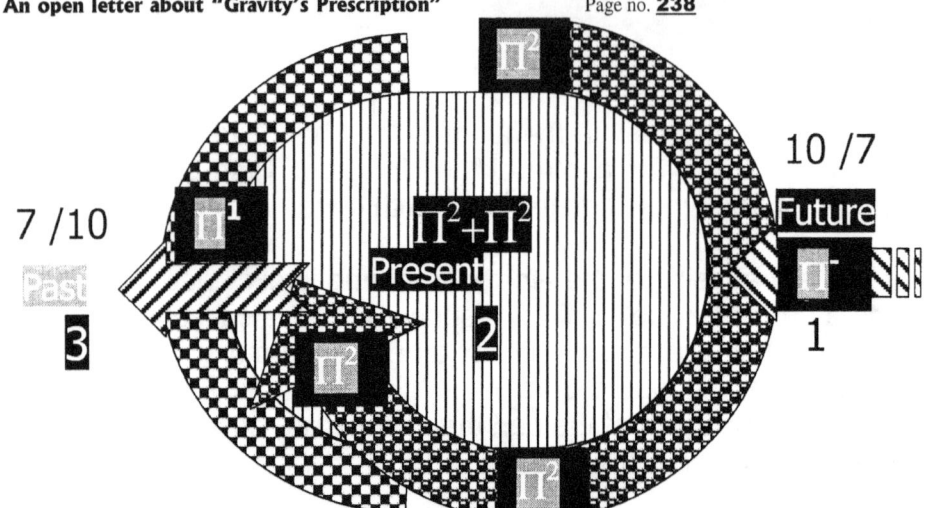

Time is taking the seven that was into the seven that is through singularity (.0999991) onto the seven that is going to be and that (3X7 = 21 + .99991) / 7 of material to which I relate I can be sure the Universe having time forms a sphere. By forming a sphere it gives meaning to the growth we see as the Hubble constant without Newtonians trying to rape any common decency out of it by their 13.5 X 10^9 years. God how could or can any one be that crude? The Earth alone is one million times older that that because what they use to measure time is the readjusting of the atom in relation to the factor the space represents. That is how the star inside accumulates the liquid by freezing the star, however I put more on this in another book where that belongs.

One thing we must not forget is that outer space is what material that is orbiting through outer space is allowing outer space to be. The Universe is the proton. The Universe is 7 / 10 in relation to 10 / 7.

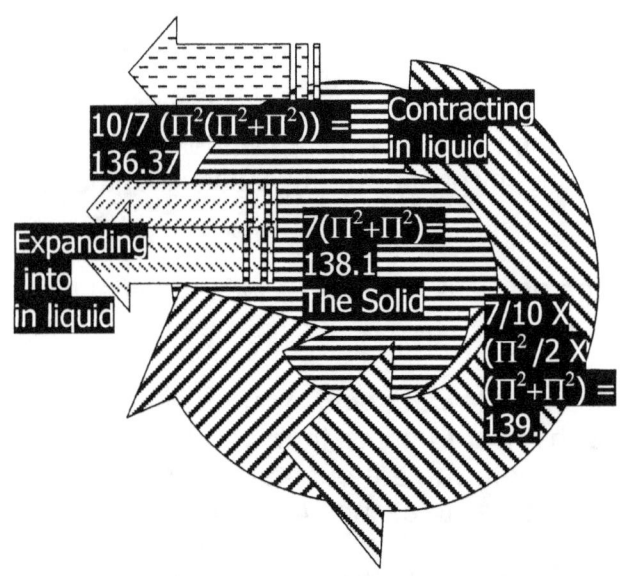

The Universe was what we now have from the first instance but in our perception that which was then does not apply to what we see in the Universe. We have an individual Universe from the one that will apply one day when one hydrogen atom will be a full star at an era of 7/10 Π/ 2 =1.09955. According to my opinion and that is my opinion, what we see as the Universe first applied when liquid and material stood apart from singularity. It was when liquid transformed space to combine again. That was when the neutron as we know the neutron first found a measured value in the Universe. Before that it was a factor but motion in time was at that point only a definition in our standards we now apply. It was when 10 / 7 $\Pi^2(\Pi2+\Pi^2)$ = 136 formed the one wall of the then applying Universe while 7($\Pi2+\Pi^2$) =138 formed the solid and the material was 7 / 10 (Π^2/2)($\Pi2+\Pi^2$) = 139.

At these tree points we have movement beginning to break through the density of space where the neutron associates with the proton an d where divide starts parting what is the atom from what is not the atom's body.

$$10/7\pi^2/2(\pi^2+\pi^2)=139$$

$$7(\pi^2+\pi^2)=138$$

$$7/10\ \pi^2(\pi^2+\pi^2)=136$$

Then following this is the next step where the Universe afterwards takes on form 118.

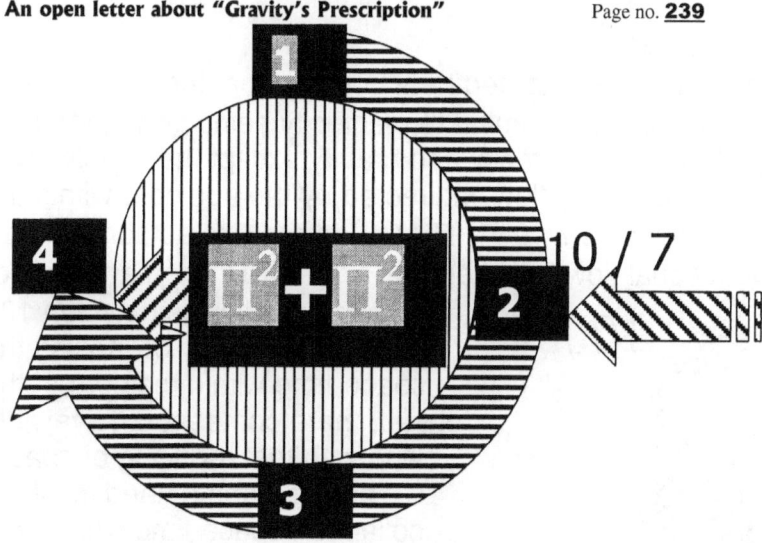

Today in our Universe we have the wall of time at $10/7(4(\Pi2+\Pi^2)) = 112.8$.

That is from where liquid flows to singularity. That from where gravity is generated by the iron core of the star. The core must have a relevant displacement of $7/10(4(\Pi2+\Pi^2)) = 55.267$ in proton displacement to have gravity establish the concentration of heat. That puts the Universe within the borders of the Titius Bode law at $10/7$ and $7/10$ in relation to the proton $(\Pi2+\Pi^2)$ forming time (4).

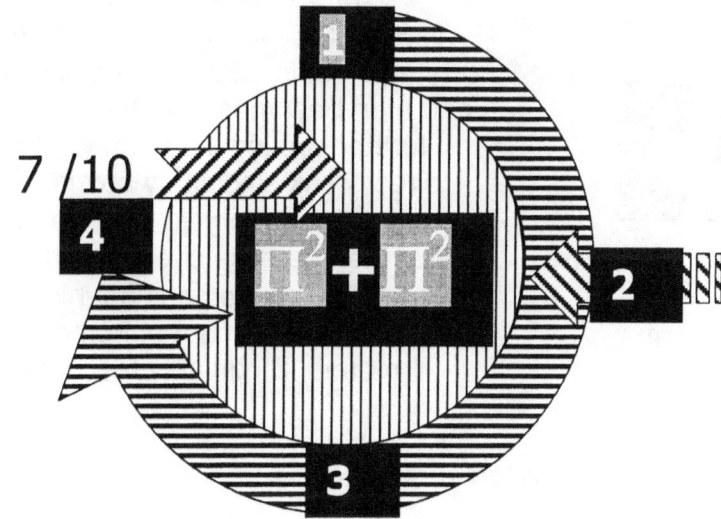

That is where liquid ends at material begin. That is where contraction of gravity begins within every structure that in our era has the ability to generate gravity. It therefore has to have an Iron core.

At the point where the neutron disengage from the atom we find our Universe catch up with time as time then takes control of space once more. The neutron is the lagging of time between $7/10$ and $10/7$. When the neutron removes as a factor that influence the displacement from the atom at $3(\Pi^2+\Pi^2) = 59.217$. As one can see the neutron removes all influence from the atom and when that happens we have a neutron star' which is no longer valid in out Universe. Outer space is not mass implying the gravitational constant. It is not mass that is producing the product by multiplying mass. Outer space is the Titius Bode law. It is gravity or motion or the neutron or movement. It is what the Titius bode law says it is. It is seven where four relates to three. It is where the building blocks of the atom leave their layers in the forming of time.

Our Universe is the flow of space-time in the form of retarded heat coming from the region 10 / 7 $(4((\Pi^2+\Pi^2)$ where gravity is generated by a revolving planet similar to electricity being generated by a spinning armature. There is no difference between electricity and gravity except that electricity is more concentrated in are of distribution. The gravity flow is directed by the iron core within the structure, which has to have, a displacement of is 7/10 $(4((\Pi^2+\Pi^2))$ in order to establish a flow direction. Just as electricity is charged by a directional flow between Iron 7/10 $(4((\Pi^2+\Pi^2))$ and copper $(\Pi(\Pi^2+\Pi^2))$ the flow is between 10 / 7 and 7 / 10. By collapsing the space-time within the core of the planet, there comes room available and with that reducing of space-time it starts a need to fill the collapsed space-time where that flow in space-time or heat contracting is gravity. This can only be when the atom freezes into a position where there is no required motion available the host the neutron. As soon as the proton$(\Pi^2+\Pi^2)$ links with the electron 3 by forming $3(\Pi^2+\Pi^2)$ the star is going outside our Universe and then becomes a proton star. By further cooling the atom will then directly link singularity outside the atom the proton $\Pi(\Pi^2+\Pi^2)$ and in that the space catches up with the time. The space goes double $2\Pi^3$ and eliminates the requirement for motion. Singularity feeds itself.

10 / 7

7 / 10

1

$\Pi^2+\Pi^2$

2

4

3

$3(\Pi^2+\Pi^2)$

$2\Pi^3$

PERIODIC TABLE OF THE NATURALLY OCCURRING ELEMENTS

1 H Hydrogen																	2 He Helium
3 Li Lithium	4 Be Beryllium											5 B Boron	6 C Carbon	7 N Nitrogen	8 O Oxygen	9 F Fluorine	10 Ne Neon
11 Na Sodium	12 Mg Magnesium											13 Al Aluminium	14 Si Silicon	15 P Phosphorus	16 S Sulfur	17 Cl Chlorine	18 Ar Argon
19 K Potassium	20 Ca Calcium	21 Sc Scandium	22 Ti Titanium	23 V Vanadium	24 Cr Chromium	25 Mn Manganese	26 Fe Iron	27 Co Cobalt	28 Ni Nickel	29 Cu Copper	30 Zn Zinc	31 Ga Gallium	32 Ge Germanium	33 As Arsenic	34 Se Selenium	35 Br Bromine	36 Kr Krypton
37 Rb Rubidium	38 Sr Strontium	39 Y Yttrium	40 Zr Zirconium	41 Nb Niobium	42 Mo Molybdenum	43 Tc Technetium	44 Ru Ruthenium	45 Rh Rhodium	46 Pd Palladium	47 Ag Silver	48 Cd Cadmium	49 In Indium	50 Sn Tin	51 Sb Antimony	52 Te Tellurium	53 I Iodine	54 Xe Xenon
55 Cs Caesium	56 Ba Barium	57 La Lanthanum	72 Hf Hafnium	73 Ta Tantalum	74 W Tungsten	75 Re Rhenium	76 Os Osmium	77 Ir Iridium	78 Pt Platinum	79 Au Gold	80 Hg Mercury	81 Tl Thallium	82 Pb Lead	83 Bi Bismuth	84 Po Polonium	85 At Astatine	86 Rn Radon
87 Fr Francium	88 Ra Radium	89 Ac Actinium	90 Th Thorium	91 Pa Protactinium	92 U Uranium												

RARE-EARTH ELEMENTS	58 Ce Cerium	59 Pr Praseodymium	60 Nd Neodymium	61 Pm Promethium	62 Sm Samarium	63 Eu Europium	64 Gd Gadolinium	65 Tb Terbium	66 Dy Dysprosium	67 Ho Holmium	68 Er Erbium	69 Tm Thulium	70 Yb Ytterbium	71 Lu Lutetium

Time forming space = Π^3 =31.0061

Outer space is 10 / 7 (4(($\Pi^2+\Pi^2$)

Time collapsing space= $2\Pi^3$ =**62.01255**

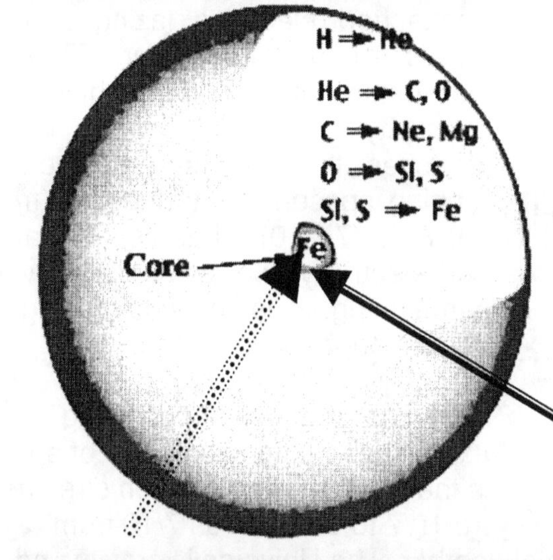

H $\Rightarrow$ He

He $\Rightarrow$ C, O

C $\Rightarrow$ Ne, Mg

O $\Rightarrow$ Si, S

Si, S $\Rightarrow$ Fe

Core — Fe

The condensing of space-time or the freezing of heat or the destroying of unoccupied space or the demolishing of time or whatever term is the favorite to connect to the movement of space in a motion called gravity condensing the motion down to contraction is in the following margins 10/ 7 (4(($\Pi^2+\Pi^2$)) to the lower level 7/10 (4(($\Pi^2+\Pi^2$)) of space-time

=112. 79547 gravity expanding or motion

Inner space is 7/10 (4(($\Pi^2+\Pi^2$)) =55.2697 gravity contraction

Light meeting singularity is $3^3+3\Pi^2$= 56.6

Elimination of space-time is $3(\Pi^2+\Pi^2)$ = 59.21762

Elimination of time and space differentiation is $\Pi(\Pi^2+\Pi^2)$= 62.01255

Space reuniting with time is = $2\Pi^3$ = 62.01255

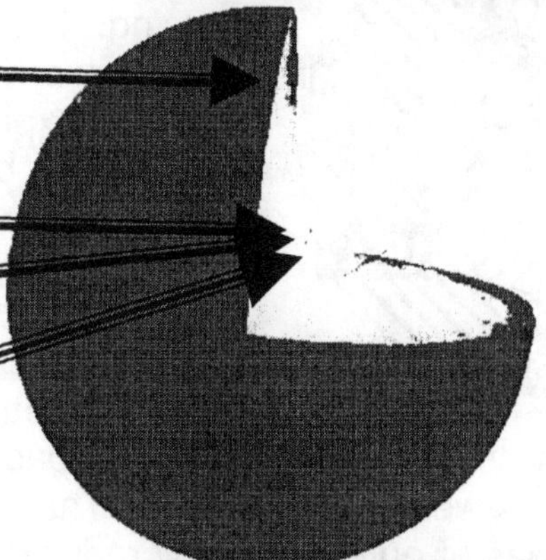

Outer space is 10 / 7 (4(($\Pi^2+\Pi^2$) =112.79547

7/10   (4(($\Pi^2+\Pi^2$)) =55.2697

Gravity generated from outer space to inner space which is collapsing or freezing of heat.

Final collapse of the photon space is $3^3+3\Pi^2$= 56.6

End of space of neutron space is $3(\Pi^2+\Pi^2)$ = 59.21762

Final collapse of proton space is $\Pi(\Pi^2+\Pi^2)$= 62.01255

As the star moves through outer space in is in contact with outer space in more than one way. By moving through outer space the star is disturbing outer space. Outer space is pushing against the star by measure of the star moving through outer space. The maximum displacement in duplication and contraction that outer space can accommodate is the total sum of the atom in relation to singularity which is   and with that in relation to singularity it forms the atomic displacement limit of 112 protons to one cluster. With the motion that is the maximum expanding there can be when duplicating. However motion stands in relation also to contraction. The contraction is freezing of heat into a state of liquid coming from a gas. At 112 the state of singularity is expanded at a maximum and cannot cope with more heat than 112 protons will manage to control in one atom cluster. But relative to that must be a cold where such a cold will not hold space under a specific level of freezing. Beyond a specific limit the cold of space freezes time

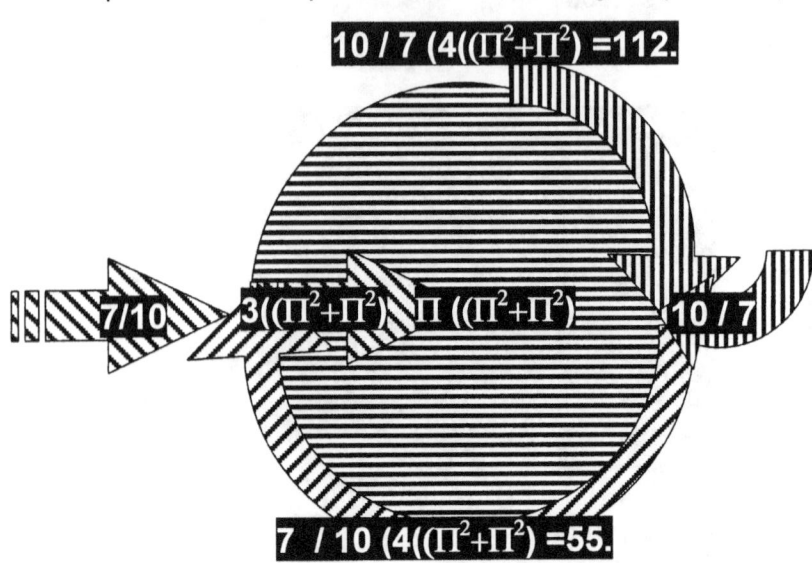

into singularity. The flow of electricity is not the transporting of some unattached electrons lazing around and then put to labour. That is Newtonian thinking. The shifting of electricity is involving motion, which stretches the neutron that then is running space-time all the way from (10/7) to (7 / 10). That also is gravity and electricity and gravity is the very same thing. It is the condensing of heat through motion.

That what is between unbridling expansion and total collapse of space is the Neutron. The neutron stretches fro 10/7 to as small as 7/10 from

where it can reduce little more before abandoning the atom altogether. The Universe is gravity and gravity is the neutron where the neutron can have no mass because the neutron personifies motion of space-time.

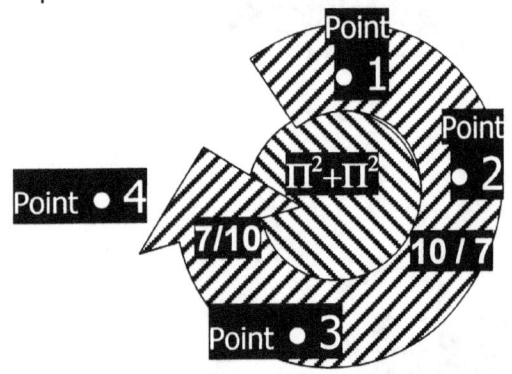

The motion is contraction but the contraction is not directly reducing. It is more a filling of vacant heat and displacing the heat in order to align the occupied heat with unoccupied heat. The expanding on the other hand is not expanding by going bigger but repositioning in order to duplicate. Such duplicating is exaggerating of space by instant reducing of space. In the reducing of space the virtual contact grow substantially more. In relevancy the star moves while outer space is motionless but because outer space is motionless the star is putting the friction of motion onto the account of

outer space. The point of contact between outer space and the stat atmosphere produce heat as outer space e is reduces as well as accelerated. That spinning produces the light, which we see as photons. It is cooling the gas of outer space by reducing outer space to the point where outer space holds friction and the particles spinning in friction that comes across as photons. That is at a displacement level of $3^3 + 3\Pi^2 = 56.6$ the photon is the product of intense cold coming into contact with intense heat. By motion the atom is removing heat from inside the electron orbit to outside the electron orbit. The motion the atom is subjected too becomes more intense with every time there is a duplication of the space-time.

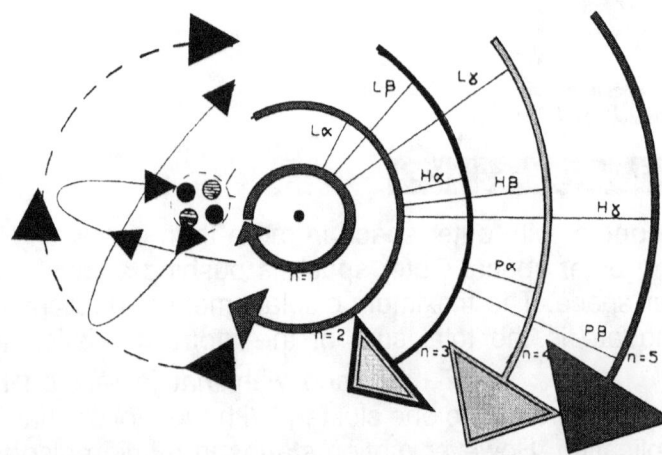

In this duplication presented as cooling the star puts the atoms of difference under the same state of affairs. The one duplicate more then the other does because the one consists of more protons in the cluster as the other does. The one has a more frequent cycle of time repeat that the other has. In this array of possibilities confusion sets in, in terms of duplicating while the other is contracting and fusion takes place where in motion the one incorporates another atom because the cycle period does not match. As I have indicated how one serves as a liquid to another particle serving as a solid the same process apply

within the star. The contraction is presented by atoms with a larger indication of consuming space that the other atom can. This is what Newton got confused as mass. Because the one has a different setting in relation to heat one would stand prone to duplicate more and the other would rather contract. We see this as one being a gas while the other element is a solid. It is the measure in which the element favours duplication above contraction or the other way around. The fact that neither element of different standings show identical patterns in behaviour the one will try and incorporate the lesser developed particle by duplicating at the same pace while also duplicating considerably more at the time of duplicating. It takes considerable more duplicating time to contract an element holding say fifty proton displacement duplication than it takes an element with a duplication displacement of say 1. By duplicating the one element holding one proton in the nearness of the element duplicating 55 protons it can quite easily become a question of finding the element of one proton to be liquid and by the nature of the Coanda effect the space also extend to incorporate the additional liquid supplying the motion. The Coanda effect works on the basis that the extravagance of the liquid providing the motion suppresses the space when the motion is absorbent in nature at that point. There is a hot spat in all the cold and the hotspot causes sudden motion acceleration, as heat will do. The surge in heat has nowhere to go although the surge in heat at that spot insists on having more space. We know a liquid that heat takes up more space because it expands. If the heat at that point has no where to expand, but the heat is there altogether and the same, then the only expanding must be to incorporate the one proton element into the element holding fifty fife protons and that takes the total up to 56. But in this one must see the liquid surging in space as heat level rises but at that very point the space will reduce because the atomic relevancy will freeze the atom into less space. If the liquid becomes more heated and surge for more space but there is no more space to supply in a star that is predominant and overall engulfed with liquid the reaction of the solid will be to become colder. Becoming colder is also freezing in the face of the liquid heating and with the liquid heating the liquid will have much more motion.

The more motion comes with supplying mot\re gravity where more gravity is pushing time longer in the face of space reducing. The situation is running at that point back in the direction of the Big bang and when the motion bridges the Big bang era fusion comes about between the two particles. In the end the particles joining space was the result of motion differences and mass is the result of motion difference. At one point the motion differences extended to a point running into eternity leaving the space at that point in infinity and infinity joins eternity at that spot where the element grows by one more proton. The mass of motion discrepancy did not create the enormous space - time deficiency but the moment the mass became eternal and infinitive the element joined space while enduring eternal time. That is the use nature has for the principle Newton named mass. It is a motion discrepancy whereby atoms would then comply to combine space should the motion applying validate such a drastic step.

Once more I have to return to the sound barrier to explain my point I wish to bring across. Where the sound barrier becomes evident there is a particle at one point displacing space-time at a rate of $\Pi^2/2$. In reality without the assistance of life to intervene it would then be the entire atmosphere that was moving at that pace having the particle maintaining such a motion in relation to all the liquid. In the case of the aircraft there is expanding without the earth compromising. But is the entire liquid atmosphere reaches such a point it could only come as a result of the earth in relevance duplicating at such relevance. The motion refers to the solid, as the liquid is motionless. Therefore the duplicating of the earth crossed the $\Pi^2/2$. By reaching $\Pi^2/2(\Pi^2+\Pi^2) = 97.4$ and that takes all material into the cosmic state of liquid. That would make the earth and the aircraft both having a state of being liquid where both can join. But also the earth would duplicate by reducing to the tune of $\Pi^2/2(\Pi^2+\Pi^2)$ $= 97.4$ in order to reach such a state. In cosmic terms the earth is an atom and the aircraft is a lesser-developed atom but both adhere to the same atomic law, as would hydrogen and an Iron atom do. It is about solids and liquids and motion.

Then a point is reached where there is no more room to allow motion within the Universe. The atom in the star reached the cosmos limit on motion of $7/10(4(\Pi^2+\Pi^2))$ and has surpassed it. The atom has cooled to the point it could no longer sustain the neutron at a point of $3(\Pi^2+\Pi^2) = 59.217$ and the neutron at 7/10 is no longer part of the atomic space. It is where cobalt is and for that reason we find cobalt radioactive. In that region the neutron is no longer any part of the atom. Then more growth increases the atom to a displacement value of $\Pi(\Pi^2+\Pi^2) = 62.0$ which is a point where all space totally collapses. That is the end of space-time in an atomic environment because this is where the atomic space-time relevancy reaches $2\Pi^3 = 62$. At that point in the Universe the Universe at that point became a Black hole and as development evolves the dynamics of the Black Hole will eventually consume the entire star.

That is not the road that serves the Earth situation.

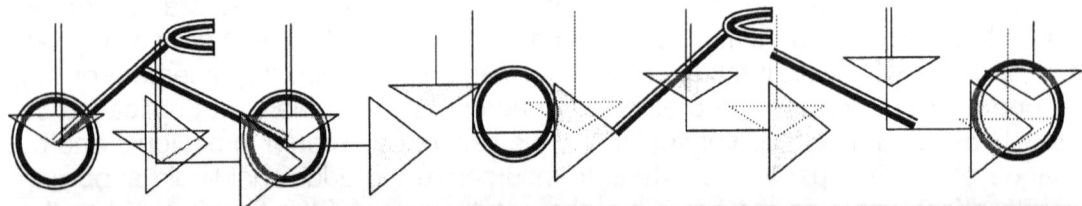

Motion distributes space and therefore decreases heat. By spreading the space over a larger area the heat in the space is reduced because the density of the heat allocated to the space reduces. Motion decreases the heat and therefore the Sun is the coldest place in the solar system while outer space is the hottest part of the Solar system

Let us first forget the accuracy of the bicycle moving in cosmic terms and concentrate on deliberate motion. It is said that if a butterfly flaps its wings in China, a hurricane will hit New Orleans. That is not true because we have to see what does exist and what does not exist.

When a bicycle is motionless the bicycle is not part of the cosmos. It has atoms but the atoms forming the unit do not charge a governing singularity where that governing singularity responds to the motion of the Earth singularity and by doing so establishes the unit into independence. The unit is a unit within the Earth unit. The unit is in motion with the Earth as part of the Earth. There is no additional motion confirming the bicycle as a force that is promoting itself by promoting the Earth motion in addition to what the Earth does to promote motion on behalf of itself as well as the motionless bicycle. The atoms spinning would form a unit as far as confirming the bicycle independence in the unit. That does not make the unit independent. That confirms the unit as a group of atoms forming a unity in form. That confirms $F = \dfrac{r^2}{M_1 M_2}$ which is what Newton saw at first. That is absolutely correct when some of the sharp edges of incorrectness are removed. There is only gravity applying between the Earth and the bicycle and the gravity confirms the restriction of the radius parting the objects to the very limit. That is because the only independence the bicycle has is in form and without cosmic motion.

Then motion enters the scenario on the pat of the bicycle. When the bicycle was motionless the gravity the Earth developed restricted the bicycle to one line placing the bicycle in a direct line with the centre of the Earth. That is a value of $7(3\Pi^2)\Pi^0$. That is the motion the Earth bestows on the motionless bicycle. The line running through the bicycle to the Earth centre has a value of Π^0 while the Earth reserves the motion on behalf of the earth and also of the bicycle at a premium of $7(3\Pi^2)$. Should some cosmic miracle wonder come about such as life is. The bicycle might just find the opportunity to achieve cosmic independence from the Earth such as the moon has.

The bicycle then holds its cosmic independence at a value of anything between $7(3\Pi^2)\Pi^0$ and $7(3\Pi^2)5\Pi^0$. The moon however holds its cosmic value at $4(7\Pi^0)$ days. And in return the earth holds the moon at a reference of $\Pi^0 / 2$ days. The one day of the moon is 28 days made up of one Moon day $= 4(7\Pi^0)$ days in the life of the earth while the Moon is $\Pi^0 / 2 = \tfrac{1}{2}$ days in the life of the Earth. There is a definite division of cosmic liquids that is time between the earth and the Moon and the moon holds a stronger identity of independence in relation to the Earth than the Π^0 that the earth offer the bicycle.

The liquid space surrounding the Earth confirms the bicycle as part of space which time in motion draws onto the space. If the bicycle has independent motion the bicycle side with the liquid time that holds the motion in the relation. If the bicycle has no independent motion the bicycle sides with the Earth putting all relative motion in the basket of time. One must keep in mind that although it is the earth having the motion the Earth projects the motion onto the liquid time since the Earth holds a steady point on the surface of the Earth, which is steady in relation to the centre singularity. The bicycle can be part of space $k = a^3 / T^2$ that is confirmed by time or the bicycle can be part of time $k^{-1} = T^2 / a^3$, which forms an extension of space. When I fall down a cliff I fall at a steady pace. It is the same pace that I would have when I fall down a waterfall holding a cup in my hand the water in the cup will not stay behind. The water in the cup will not spill.

The cup will not fill with water. If I had to fill the cup with water I will have to supply motion in access to the motion with which we fall. The water and I will have the same pace therefore we will fall at the same gravity. My density will not leave me superior. The water mass will not have the water fall more forceful or less forceful. The motion considering all objects is not discriminating on any basic grounds. In such an event I will be submitted to $7(3\Pi^2)\Pi^0$. That is gravity and gravity is motion. Forget about Newton's mass controversy because blaming it on mass is instituted fraud.

Having independent motion requires more than gravity. One may even be able to apply $F = G\dfrac{M \times m}{r^2}$

under conditions about where what fits. This is not a cosmic principle. It applies only to the Earth

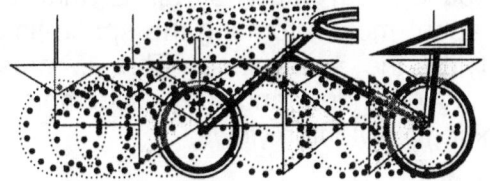

under conditions serving the earth and applying to all objects that submit to the Earth. The motion can be seen where the one M holds the motion the earth provides while the Other m indicate the independent object's independence while the G then will be the additional motion and the radius by square is the balance there is between being liquid and being solid.

In relation to the Sun the Earth is in motion and the earth is a part of the liquid that outer space

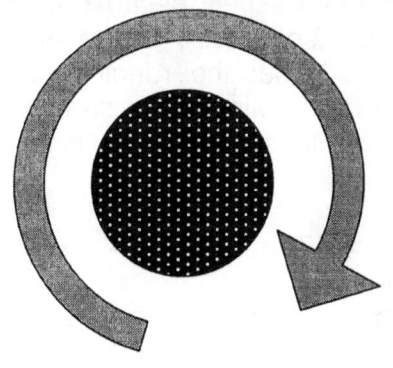

provide. To the Sun the earth is a factor that flows with outer space and although the earth show the ability to counter the flow that outer space has, the Sun still regard the Earth as equal liquid to outer space. Being a solid or a liquid has no bearing on the state of the matter but it all depends in the flow in relation to a securing solid.

Understanding what the Universe is becomes the important key about realizing the dynamics of cosmology. The bicycle being without motion is a part of the Earth because of the Coanda and Kepler principle where the bicycle without motion sides with space a^3 and when achieving motion the bicycle sides with liquid by moving $T^2\ k$.

This dynamic is the key in understanding what measures apply where in the cosmos. When not moving the atoms moving in the Unit the bicycle forms holds relevance relating too the general or governing singularity in the centre of the earth. It taps in to sustain the governing singularity by providing motion that forms part of the earth singularity. The unit uses the atoms to the advantage of the Earth motion and supplies the Earth with relevance in order to sustain as well as promote the earth moving. The bicycle is the Earth because it is space of the Earth within the space boundaries of the earth. It is a^3 and stands relative to T^2k.

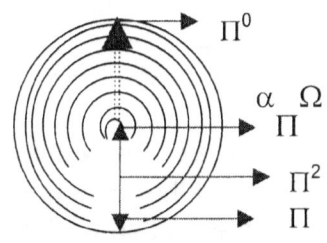

Then when for some reason the bicycle finds the ability to move the bicycle split infinity an eternity. A universe is born. That which has no end parts from that which has no start leaving space-time generated. It is motion that puts the bicycle in three positions relating to time. That splits eternity and infinity just like it split eternity and infinity what moment-Alfa came about. There is just more heat in the backlog and less in direct relevance.

The rolling with time sets a differentiation between eternity and infinity and the measure of the time

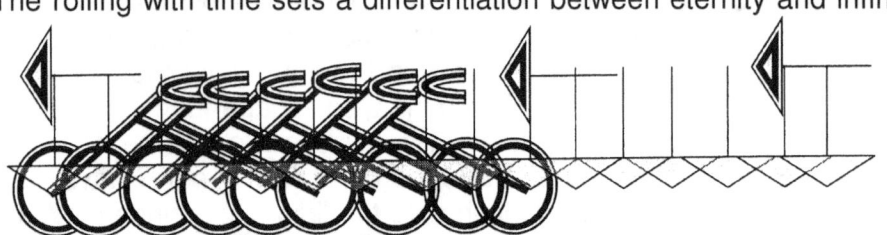

delay forms matter in time. By providing motion the bicycle no longer only keeps the earth singularity generated but also it supply a potential singularity by establishing the individual generating of singularity which sets out maintaining the individual singularity that is apart from the earth singularity while still being within the Earth singularity. The singularity it now generates is no longer Π^0 but forms an independent singularity by as much as $5\Pi^0$. It shifts the line of currently to at the most five positions in delay of currently. Any shifting further brings about serious conflict.

The Universe we have (not the earth filled with life that we have) but in the era we landed we find the Universe going from $10/7(\Pi^2+\Pi^2)$ towards $7/10(\Pi^2+\Pi^2)$ ending at $3(\Pi^2+\Pi^2)$ while eventually all space-time will form $(2\Pi^3)$ in the star limit. The proton disappears when the proton goes to singularity at $\Pi(\Pi^2+\Pi^2)$ which then becomes double space $(2\Pi^3)$ where space being double catches with time being single and loses it's lagging behind time quality. That is what they call a Black Hole or what I named a proton star. When the proton goes singularity then $\Pi(\Pi^2+\Pi^2) = (2\Pi^3) = 62.01255$

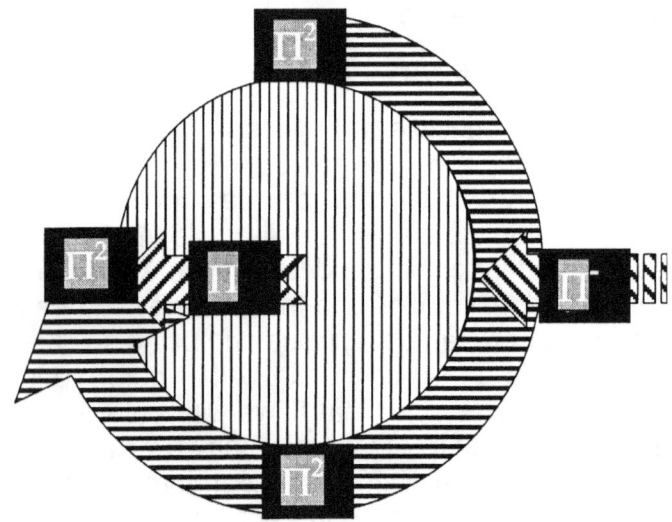

All objects are classed by heat either being in motion through duplicating (overheating and expanding) or being in motion through heat contracting (heat being reduced through motion removing space), but most of all is that all material is about motion forming the space-time and classifying the space-time. This is most pivotal in understanding cosmology notwithstanding Newtonian views.

The motion of the liquid which the neutron is proves to be the time (T^2) aspect because as it increases it claims the space (a^3) in the at a distance k of time (T^2) that the running has increased. The faster the motion is the stronger is the gravity that the motion generates in the

space it claims by the gravity it generates

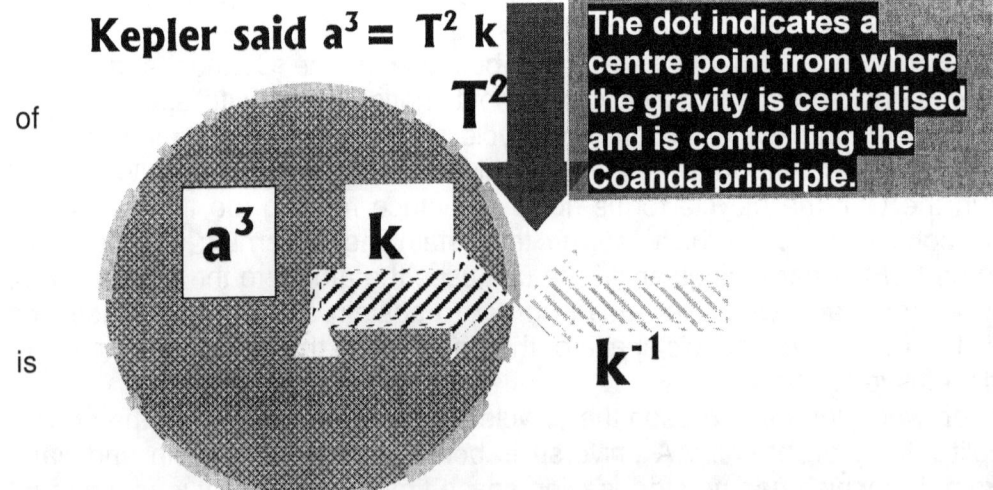

Kepler said $a^3 = T^2 k$

of

is

The dot indicates a centre point from where the gravity is centralised and is controlling the Coanda principle.

The

Coanda effect is proof gravity coming about through space forming motion. In the case where water diverts the normal directional flow the space that translates to the motion deflecting singularity with the flowing water charging the motion. In the centre of the object having the round form,

singularity is duplicated and by transferring Π to form Π^2 and the motion of the water creates a line of gravity that pushes the flowing water to follow the direction that the newly gravity applies to the water. This again proves Kepler's statement of $k = a^3/T^2$ that specifically states that space (in this case the object transferring singularity to a new position within the round object) and with the motion of the water redirects the gravity flow of the water to new space in new time. Only Kepler can explain the phenomenon but only when Kepler stands alone, correctly interpreted and divorced from Newton's opinion about Kepler's statements. There is a flow of time created by motion and defined by direction that produce expanding as well as contracting where expanding is contracting while it is on the other side of the Universe.

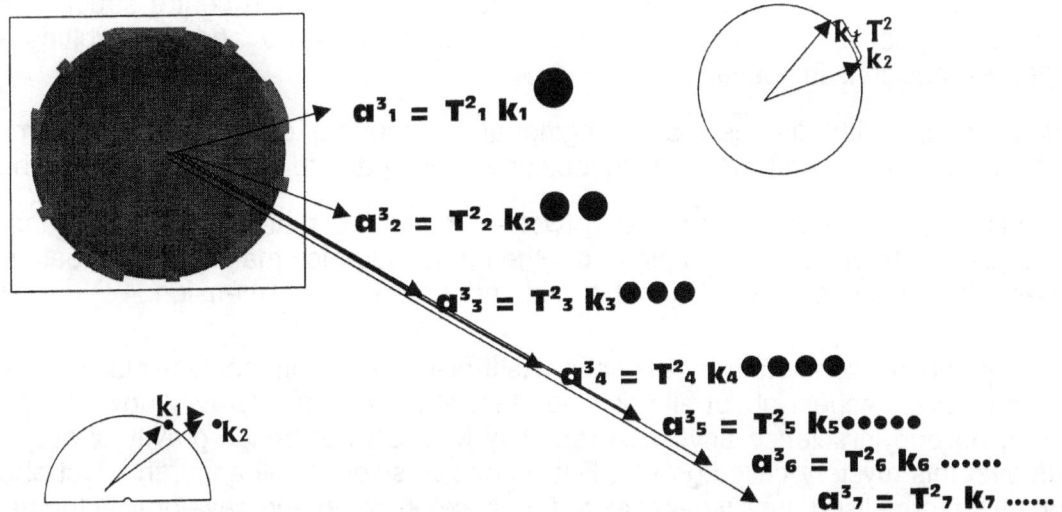

All objects are either cold and reduce space-time in relation to others being hot and expanding space-time. Being cold puts the object in the role of conserving space-time in contraction and that puts the object in a position of being a solid.

Then in relation to the first conserving factor there is the overheating factor, which brings into the relation the expanding, or moving away from the singularity.

The duplicating requires a repositioning of the aligning of **k** from a certain position to a more fore ward position in relation to and in that **k** will also have to extend a value when moving from k_1 to k_2. The essence of motion is to duplicate material that is in a process of overheating. By producing more than one of the same material unit the heat is distributed over a wider area and thus the heat has more space per time unit but less space in a time frame. That is motion. By applying heat to provide motion the Universe see that as overheating and the longer k_1 to k_2 is per time unit the lesser will T^2 be because a^3 is spread over a larger area.

This letter you are reading is my effort by which I hope to interest you in reading my Introducing letter an open letter To Selected Academics ISBN 0-9584410-9-X The book on offer has the title of an open letter To Selected Academics ISBN 0-9584410-9-X and is the actual letter I sent to various establishments.

What brings about the expanding?

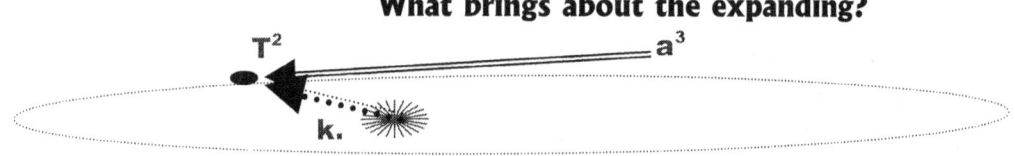

Kepler was the very first person to mathematically introduce space a^3 centre **k** and time T^2. Not only did he introduce space-time a^3 / T^2 but he also placed space a^3 and time T^2 in a relevancy long before Einstein did and placed gravity in space-time a^3 / T^2 even before Newton named gravity. Kepler was the person who placed gravity as the ingredient in the Universe that determines space a^3 and time T^2 and much more. Kepler was the first one that saw that gravity comprises of two factors being **k** or linear gravity and circular gravity or T^2 as gravity keeps space in form while all is staying together.

Since gravity also influences the space outside the sphere, the space we call outer space has seven plus three points bringing about ten positions of gravity influencing space.
The influence inside the sphere also captures the space outside the sphere.

This means that in the cube at the point of contact between the cube and the sphere the cube experiences such a contact point as if the "bottom falls out" of the cube and without a "bottom" to support objects they fall to the sphere as objects does fall to the Earth. Remember that a body "floats" in space, but at one specific point it starts to "fall" to the Earth. That is gravity and it is a dimension change much more than any force. I shall explain this last remark later on. That too is the Lagrangian system with five cosmic structures holding relevancy to the centre structure where the centre structure stands in for seven positions diverting from centre and the orbiting structures standing in for five positions in space.

Gravity has all to do with dimensional changing and reforming of forms to re-affirm alliances supporting the centre. It is the reforming of space converting space to more concentrated heat.

The Universe is in the three dimensions using twelve dimensions that is visible to us and indefinite number of stages in size differences ranging from the immeasurable small to the immeasurable large where mathematics becomes a short fall to the next and the previous dimension.

Up to now every one in science is normally acting as if gravity is a commonly explained factor, which every one knows every aspect about all principles that are involved in gravity down to the smallest detail. In truth, no one in science anywhere remotely knows what brings gravity about and I used Kepler to unravel this mystery called gravity. But no one in science will admit this fact about Kepler being the one who formulised gravity decades before Newton came and gave gravity the name.

Newton did not underwrite or define gravity and even today the most informed in Science at best can only assert their suspicion on a rumour presumed about what causes gravity to perform as the part interlinking the cosmos but no one can go any further by explaining the concept. Newton started this realising of gravity but it had and still has no more substantial proof than a rumour has and Newton admitted to it being a concept he could not explain. In Newton's ignoring to test Kepler's findings, Newton missed the opportunity to find what gravity is. Since Newton every person in science also ignored Kepler and every one is guilty of missing the opportunity Kepler maid available.

By my efforts of studying the implications that results from Kepler's finding I can now un- emphatically declare I know what gravity is. Gravity is the entire following locked into one compiling unit:

Gravity is not being some magic force found between particles grabbing onto everything. I mathematically explained the following phenomena:

> **Gravity is singularity as a factor forming space-time**
> **Gravity is finding space-time**
> **Gravity is proving space-time and aligning space-time with gravity**
> **Gravity is the working principals behind all cosmic occurrences that pre dates the Big Bang period.**
> **Gravity is the Roche limit.**
> **Gravity is the Lagrangian system**
> **Gravity is the Titius Bode law**
> **Gravity is the Coanda effect**
> **Gravity is the sound barrier**

By being able to pin point prove what Gravity is that enabled me to unravel the other entire phenomenon that forms gravity. Each of the phenomenon I mention above has one part or role in what is forming the totality that which we know as gravity.

Should you think this is rather a wild presumption I challenge you to spend a little more time and please think about what I say when you read about what I say in the next few pages. The first thing you should admit in private is what study did you personally so far made about the work of Kepler?

Still, to this day nobody in science at present will denounce the principle of gravity as vaguely researched. Gravity has never been explained as a principle. Even when one is considering what the importance of gravity is, gravity never yet has been understood. It is by now very clear that little if nothing of all objects is pulling closer in outer space. Comets are missing the Sun on a regular basis and no planet has come much closer toward the centre of the Sun. Still everyone in science acts in a manner as if Newton's gravity ideas are the best detailed proven fact and only occasionally does someone quietly admit that even Newton admitted not knowing what gravity is. No one ever comes to the front and boldly state that gravity is just a rumour spread by scientists pretending to know all there is to know and knows little to nothing about what there is to know. Newton admitted that much when he introduced the name (not the concept). Mistakenly Newton corrupted the concept he named as gravity.

Going according to what Newton introduced Newton's concept will by now have the moon much closer to the Earth than it was in the time of Kepler's studies, yet we know the moon is moving away instead of coming closer. By the same measure Kepler suggested that the space a^3 is content with the motion kT^2 as long as the motion T^2k is equal to what the space a^3 will allow. Kepler suggested motion of space remains in equilibrium as long as motion of space a^3 duplicated space a^3 by motion thereof T^2k. That is much more true than objects rushing towards one another by the pulling power of mass. Newton agreed that he could only declare gravity as a vague concept. This fact was at that time drowned by the man's stature and was relieved from the manner of requiring the proof that later Academic science became an absolute necessity. The proof one would demand now a day was never given to put Newton's rumour beyond doubt. When Newton announced a force he also admitted the force could be anything. No one ever came after Newton and proved the fact better. That still underlines the fact that the force to this day can be anything. Not once could one person in the past or present provide substantiating proof on gravity as reality by defining the very principles thereof.

That includes every one since Newton as well as including Einstein and even Hawking. Scientists can declare gravity was a factor at 10^{-43} seconds after the Big Bang but what brought gravity about or why gravity became or still remained, as a presence is still tightly concealed information which all are speculating on. Even in the best and most informed circles and amongst the most educated there is no one that knows what gravity is because they all ignored Kepler and for them to ignore Kepler the price they pay is not finding the principles bringing about gravity. Using Kepler even makes the method to follow and understand Einstein's discoveries shockingly simple. Gravity is the motion of space relating to time in movement.

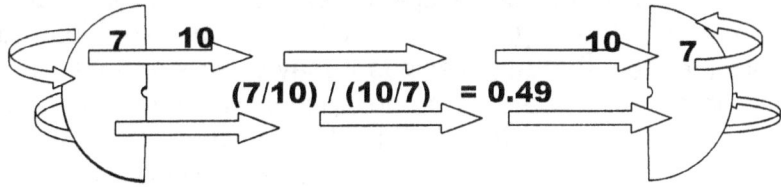

SPACE DIVIDED INTO TIME
(7/10) / (10/7) = 0.49
.7 / 1.4285 = 0.49

Taking also from both orbiting influences

THE PROCESS PARTED USING THE ROCHE PRINCIPLE

10 / 7	$(\Pi/2)^2$ The Roche influence on Titius Bode
7/10	$2.04 \times (\Pi/2)^2 =$ 5.033
$(\Pi/2)^2$	$2.04 \times (\Pi/2)^2 =$ 5.033
10 / 7	5.033 +5.033 = 10.066 from both objects

Crossing the singularity divide and activating the Roche principal $(\Pi^2/4)$

(10 / 7) \ (7/ 10) = 2.04
1.4285 / 0.7 = 2.04. Taking from both orbiting influences

SPACE MULTIPLIED WITH TIME

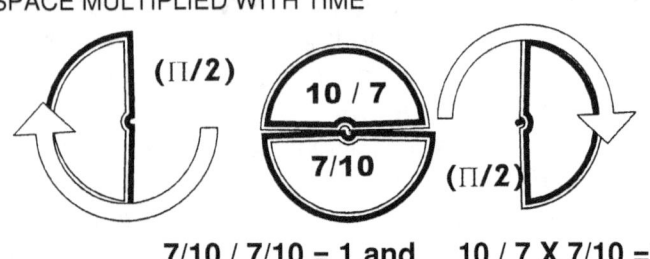

7/10 / 7/10 = 1 and   10 / 7 X 7/10 =1

From dissecting the formula I prove that:

I prove gravity is strongest where space is least (not where the Universe goes flat as Einstein promoted).
My theory I propose is one where gravity is a relation that is based on Newton's law that for every action there is an equal but opposing reaction to the action. Gravity is not one- way traffic but is a relation of relevancies that are applying equally and without the relevancy applying between the action and the reaction in a balance there is no gravity.

In the relevancy there are opposing motion where each participant provide a motion which is contradicting the other relevant but opposing motion. My base this conclusion on what Kepler introduced when he introduced the fact that space is half the motion and the motion is the other half of space. Therefore space cannot be is if space is not moving because the movement is space repeating what was before in the instant and onto the future. Space is one part and the repeat of space from the past onto the future taking space through the present is the part forming space by repeat in motion. Space cannot be if space is not moving to provide space a past into the future bringing the one part of space that in which it is and the other part either what it was or what it is going to be. That part is time in the formula space-time.

The one part of space is what there is in the present but the second part, which forms the time aspect, is what there was or what there will be. Time depends on space moving while space depends on time a position coming from or going to. In an attempt to explain my view I am prepared this one time to grossly simplify my view and relinquish accuracy in the process. There is motion expanding the confined space just like a rabbit tries to flee a dog chasing the rabbit. The rabbit is trying its earnest to escape from the dog. The dog again is trying to contain the rabbit by reducing the space that forms between the fleeing rabbit and the chasing dog. But the space between the two is what is of a major concern and not as much the running of the dog. Focussing on the running will lead to missing the process that really applies. From the view the onlooker has it may even seem as if the rabbit is pulling the dog be an invisible string in an area where both participants are remaining in an area that is repeating the continuing chase as the dog holds the rabbit in chase and the rabbit holds the dog at a distance. The rabbit has itself in a task where it wishes to never see the likeliness of the dog ever again while the dog sees dinner.

The rabbit would love to leave the dog at a distance where the rabbit find freedom from the dog as the dog's dinner opportunity. While the space between the two is a merely a common fact, it unifies their differences. Both in relevancies have to appreciate their differences by the space in the unit that is keeping them apart. The space is the factor that has to resolve the issue being the differences in motion but cannot because different relevancies sustain equilibriums. I prove that as much as there is Newton's pulling there is Kepler's running around and the running around is equilibrium of the other factor providing the running away part.

The angle science is looking at the issue science either dismisses or cannot explain the characteristics or principals, which is there none the less. The explaining of the phenomenon is quit impossible when using the pulling rope magical attachment idea in the manner science tries to explain gravity. Therefore instead of dismissing the rope they dismiss all other factors present by gravity unleashing free motion but they would not release their idea about the rope. Gravity is motion between two particles that brings about mass. In the book I explain this in much detail but frankly there is not enough room to explain this in this web page.

Mainstream science knows about that gravity has never been defined, the Bode principal that is there in all the planets and even the fragmented planet, the Roche limit, the Coanda affect, the Lagrangian system and the sound barrier, but cannot explain any of the phenomena all though the presence of these phenomena is without dispute. It is the explanations about what causes the phenomena that is part of the dispute but in science the manner in which they defend Newton, science would rather discount the obvious phenomena than question the legality of Newton's cosmic views. I only dispute Newton as far as his cosmic principles are inclined. The phenomena being there or not becomes disputed. Science fails to give acceptable explaining of such occurrences we see in the phenomena and therefore disputes the validity of the phenomena and this failing to explain the presence becomes disputing the presence thereof. I on the other hand found a way where these explaining of the phenomena took me past the Big Bang era and introduced me to the start of all starts. Science cannot get past one specific date because they do not accept or understand the phenomena, which I prove started the Universe.

In such a light Scientists must somehow realise they are barking up the wrong tree with the information they have to use to do some explaining. They cannot refuse the phenomena and not realise they must have the cat by the tail as far as cosmology goes. Please remember that with this I am referring to cosmology and not general physics. There is an Earth versus a Universe with huge difference between the two concepts but Newtonians fail to see that because Newtonians cannot appreciate the differences thus they're not able to understand cosmic gravity; they go about blurring the understanding of gravity. If there are that many phenomena (it represents all there is in cosmology) to explain and such little ability to explain (science fails to explain even one) by using the information Mainstream science is using to explain the cosmos, then someone somewhere has to realise there is something drastically wrong in the way they present the knowledge they claim to have. One cannot be serious about science but defend your view by dismissing the validity of all unknown indicating factors presented as such.

There then is some gross incorrectness in the way one reasons. The Roche limit is there and no denouncing thereof can remove it from the cosmos. They may refer to evidence received from the Hubble telescope as "the star is blowing bubbles" for the lack of explaining what is occurring but occur it does. One cannot say it is some unknown gesture presented on occasions because by not explaining the pictures present certain foolishness. It leads to tragedies in aviation and the tragedies they are incapable to understand or explain. For fifty years they lost many pilots but still has no idea what brings the sound barrier about, or find the link gravity holds in the process we call the sound barrier. Instead they try to interpret some effect established almost two centuries ago with steam trains that is travelling at the same speed that horses run. No further investigation with the science in hand brought them closer to new facts! That they should rather see as a sign telling them they are going about incorrectly because by ignoring the cosmos one produce a fantasy and not science. Nevertheless, it does not because tell them anything because Newton did not say so. When I first came upon the amount and the totality of the unknown quantities in cosmology as well as the complacency those involved has about such unknown factors being discarded, it stirred a sense of disbelief and I decided to respond.

All principles I use in the theory I introduce with the publishing of this book. All principles I apply are part of nature. I base my theory on heat stabilizing through space using motion to produce cooling. That is gravity.

I believe some of Creation remained as some particles formed by applying gravity in motion and the lack of motion in others became the lack of gravity, which inspired overheating which then formed plasma. Plasma is the result of heat where light is the epitome of heat. How light became plasma is rather obvious, which again I believe (within reason) I do prove. I believe heat is the destructed form of material and this information the atomic thermo explosions give us.

Analysing Kepler's formula without Newton interrupting Kepler's work helped me realise science has been running on an error for the past three hundred and fifty years. Please let me explain: Tycho Brahe and later Kepler made a study of outer space as never repeated afterwards. From this Kepler concluded that $a^3 = T^2 k$. We all know that a^3 is space and with the space indicated as being in the third dimension and the third dimension is unmistakably a cube that forms volume, which by definition is presenting space. We also know from the way calculations come about by using the formula of Kepler that T^2 is the duration of a specific period of time relating to a specific centre. On the one hand we have space a^3 and on the other hand in direct relation to the space Kepler introduced motion coming from a centre that forms time T^2 **k**. Kepler gave us space-time a^3 / T^2 centuries before Einstein gave the concept a name but no one ever took any notice. In the formula is space a^3. In the formula the space a^3 has direct relation to time T^2 If **k** is a^3 / T^2 it means that from the centre holding the gravity is space-time. Space is a^3 and the motion of space a^3 we accept as time T^2 k and such accepting is part of our understanding for the past three hundred and fifty years. Kepler gave us gravity before Newton named it as a force. Kepler gave us space-time long before Einstein named the notion. With Newton's meddling he missed Kepler introducing gravity as $\mathbf{k=a^3/T^2}$ space / time.

I believe that I achieved an all time breakthrough success because I can now explain what gravity is. Remember that not even Newton could explain what gravity is or where it comes from, but Kepler did that without any person ever noticing. Scientists over the years paid the price of ignorance about gravity by their unwillingness to investigate the father of gravity, which coincidently is not Newton but Kepler. From such explaining what Kepler said without Newton changing formula on Kepler's behalf, I prove the Titius Bode principal also known just as the Bode principle. I prove that the Bode principle forms gravity when using the Roche limit. These phenomena were never explained or understood by Mainstream Science although they appear more than regularly in the cosmos. In the same breath I might add that Kepler also was never investigated. My achievements came from my effort where I separated Kepler's work from the opinion that Newton formed and that he (Newton) gave his compromised views about Kepler's work to the world. For instance from Kepler's work I can explain the operation of the Black Hole, which not even Prof. Stephen Hawking understands. That is because Hawking ignores Kepler. In my opinion my explaining of gravity makes much more sense than the accepted force of Dark Age proportions...and the best part is you do not have to be a genius to realise or understand it.

Even a simple person such as myself can see it clearly! From my view a force is just motion applying and that is what Kepler said gravity is. Kepler said $a^3 = T^2$ **k**. I dissected **k** as a factor in the Coanda effect and found that the Coanda effect is proof of my view about gravity and the ability of establishing gravity by centralising space, which forms singularity that produces the Coanda effect. The Coanda effect is the establishing of individual space a^3 by applying motion T^2k. Where the Coanda effect is producing gravity and such producing is stronger in a small space than the gravity produced by the Earth in that spot I use that principle to show that there was some manner in which the reducing of **k** brought about a stronger T^2 just as Kepler said. This was a crucial part during the Big Bang and therefore had to play a part during the period of the Big Bang. Einstein came to this conclusion but failed to refer his view back to Kepler.

As presumptuous, as it may be on my part of trying to disprove Mainstream Physics, such a presumptions does not change the truth about Mainstream science being incorrect about gravity. After all they admit they do not know what gravity is. I am not disproving anything because they agree they do not know, which paves the way for my showing what gravity is. By admitting not knowing what gravity is they then also admit there is a chance that they can be incorrect about gravity but unfortunately mainstream physics do not see it that way (yet). The question in hand is finding what role gravity played when the Creation came about for the first time. I had to find a method that would allow me to explain why gravity played a role.

My ambition is proving the Universe not coming from nothing and therefore outer space cannot hold nothing. By taking Kepler's $\mathbf{k = a^3 / T^2}$ and using **k** as a line I show through using the line as an example that the cosmic Universe holds everything and all concepts. However the only thing it does not hold is also the only aspect not present in the Universe at all. That is the value of nothing or zero in as much as carrying the definition of the absolute absence of any value. This means the Universe is filled to the point it is overflowing which we call the Hubble constant and not there is not room to be empty. With the line that light uses to flow the lines eliminates any such a possibility of nothing being present. Mathematics is a means of communication about matters concerning the cosmos. As an intercultural language spanning across race and ethnicity or as a principle as such mathematics cannot have zero because mathematics indicating lines, which is about not applying the numerical number or value of nothing. Everything came about from singularity and Einstein proved that. From singularity nothing ever had the chance to enter space. I challenge any person that disagrees with this statement to show mathematically where nothing as a factor ever entered the mathematics of the Universe. If there is any one there believing there is nothing in outer space I challenge that person to prove mathematically where nothing is a factor in the cosmos. Your attempt may either be before or after reading my work but my challenge will stand since mathematically nothing cannot be part of mathematics. Multiply whatever with zero or nothing and such multiplying results in nothing where nothing is then, can establish no multiplication. Kepler gave us the relation between cosmic objects as $k = a^3 / T^2$. From the formula k forms a connecting straight line filling the first dimension and not the single dimension because k in the single dimension is not zero. It is unproven how k can backtrack to become $k = 0$. I deliberately press this point and make it an issue because that removes all the theory of mainstream science from any logical base they have in support of their views that space is, holds and comprises of nothing.

In the book an open letter To Selected Academics ISBN 0-9584410-9-X the book only and exclusively deals just with the fundamental basics of my theory. I do not elaborate or explain the broader aspects or form an overall view. I found if I do that before a solid understanding of the basic concept is established no concept becomes established. The most basic to explain is that the line cannot start at zero because then there can be no line to follow zero. The cosmos has lines forming cubes and lines forming circles, which in applying 3D manifests as spheres. Between the circles and the cubes run lines, so the key to understanding the Universe is the following of a line. The Big Bang was a time when the Universe was incredibly small making the running lines small. Understanding the Universe is taking the line connecting particles through space back to its limits where such limits were during the Big Bang. But the reducing cannot go to zero because zero removes the line all together. By reducing the line to where the line will not reduce any further we will find at that point that all points land on the same spot. The spots all share one position because that is the only position there is to hold in the form singularity presents.

That is singularity being one to all but it is not zero. Finding form in that point shared by all will give a value of singularity. Extend that value received to a Universal centre and bring that value to align with Kepler's $a^3 = kT^2$ and understanding the Universe by finding the centre of the Universe makes the Universe simple as can be The Universe becomes sensible making the entire different yet unexplained phenomenon as easy as children schoolwork. There are suddenly no more mysteries in the Universe. It is only possible when we see gravity not as a grabbing force instead of seeing gravity reducing the space between particles. Gravity is not being some magic force found between particles grabbing onto everything.

The following is the mathematical proof hat through the atom time illuminates space by applying motion. Following the mathematical proof I explain how that is achieved.

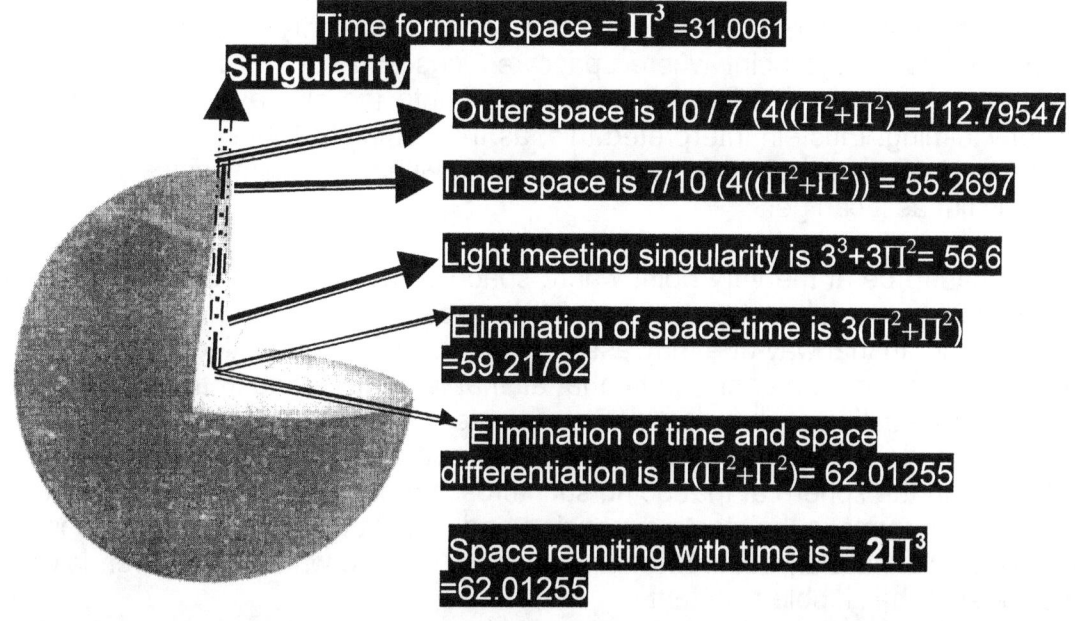

Time forming space = Π^3 =31.0061

Singularity

Outer space is 10 / 7 (4(($\Pi^2+\Pi^2$) =112.79547

Inner space is 7/10 (4(($\Pi^2+\Pi^2$)) = 55.2697

Light meeting singularity is $3^3+3\Pi^2$= 56.6

Elimination of space-time is $3(\Pi^2+\Pi^2)$ =59.21762

Elimination of time and space differentiation is $\Pi(\Pi^2+\Pi^2)$= 62.01255

Space reuniting with time is = $2\Pi^3$ =62.01255

We stand on the outside 150×10^6 km from the spectacle and from such distance we judge the Sun . We don't even judge the Sun from what we can see but we judge the Sun from what we feel. We feel heat coming from the Sun and from that we argue that the Sun is hot. We see the Sun has heat rising from the surface as a liquid soup. That puts the hydrogen layer as the outer layer in a liquid. Hydrogen freezes on Earth at a temperature, which is the coldest amongst all other elements. Yes, the Sun is 6500 ^{0} and that is on the outside. To a human that is hot but a human has no mind judging the Sun . If the Sun squirts pure heat turned to liquid from the surface and the heat falls back into the surface the Sun is a lot colder than the Earth is. The earth requires an enormous effort to cool hydrogen down to a liquid state. We must mind the way we think of the hydrogen in liquid. The hydrogen remains a solid. The element is untouched by temperature differences. It is the heat environment surrounding the hydrogen that changes from a gas to a liquid to a solid. One removes or one amplifies the heat in which the hydrogen is and that turns to liquid or solid or gas. The hydrogen is untouched in the elements worth.

Yet we see the heat flow amongst the hydrogen as a liquid. Nevertheless we remain adamant that the liquid   is a gas and the hydrogen is in a gas and the Sun is a gas bowl filled with hydrogen because to our mind hydrogen must be a gas. After all, our element table classifies hydrogen as a gas and that is the way we think of hydrogen. We do not consider hydrogen to be in a liquid state when we see the heat is flowing just like a liquid and shows all indications that it is a liquid. No the Sun is hot because the Sun feels hot.

In the Universe there are no hot or cold but a state of differentiation produced by time. The Universe parted by parting heat from cold when eternity parted from infinity, when Π^0 singularity parted from Π singularity, when 1^0 parted from 1^1. There is no hot or cold but there is a relevancy where one factor cools and another factor overheats. By retaining the Sun is the coldest space in the solar system and outer pace is the hottest there can be.

From since the time that man discovered intelligence (if he ever did) man has been with the presumption that the Sun is the hottest centre in the solar system. Later on in the present time, it came to someone's attention that the Sun also holds the solar system in gravity. The Earth by its standard and dominating its sphere of which it can control with influence is the hottest centre in the space of its domain and it holds the moon centred to the Earth. The gas planets are the hottest centres in relation with the most heat and they all hold their satellites captured by a hot centre. All space structures hold in every centre there is that is confirming their independence at that point of securing independence the centralizing of the most heat it is able to concentrate and from that centre holds all material captured or controlled in the domain of what that forms the independence of the structure. I can go on and on but heat in the centre couples gravity to space-time, just as if Kepler said before he was spoken for on his behalf and without his permission or his agreeing to it.

$a^3 = (T^2 k) = a^{3\ +2\ +\ 1\ =\ 6}$ with the sphere presuming the position of singularity as part of $k^0 = 1 =$ **singularity**. Einstein proved that at the point where space reduces and such reducing reaches a point where space as a factor in the third dimension disappears into the single dimension (space going flat) gravity is overwhelming. Einstein interpreted this, as the complete Universe going flat but while it may be true that the Universe is going flat, that can only be within singularity since singularity represents the Universe as flat as it can get.

The centre of any sphere has to be at the very point where space completely falls away. It is at the point where all the points of line centres meet by the crossing the centre of their individual connection coming in to contact as a group. In that way one may assume that the lines connecting the controlling points on the other end are crossing on a centre point that all that is participating in the constructing of the sphere is democratically electing such a centre. Please note this conclusion very well because this forms the heart of the Coanda principle. That will put that position where the lines cross, which in itself is centralising all space in the sphere at that point, such crossing point will become very distinct and controlling where that point forms in the single dimension and singularity is the single dimension. Kepler also solves another riddle that truly got Newtonians unstuck. This, to which I now refer, is what is referred to when they refer to the Hubble constant.

The growth we see in the Universe is an adding of space in every cycle completed by every cycle, which all the protons complete. The adding is the smallest addition that can come about in the shortest period of repeating by cycle rotation there can ever be. This growth of space-time next to singularity confirms the growth of singularity as singularity recalls the space it uses to grow in the time it grows. The margin of growth will be by the extension of **k** in the formula $k = a^3 / T^2$. Every cycle completed in the relation to space by the initial value of **k**. $k = a^3 / T^2$ leaves ultimately a^1 extending as space or as Kepler chose to indicate it as k^1. That too has to be compensated by the duration of time reducing the time aspect by the margin that the space expands. This confirms what is evident in the Hubble Constant. The further one looks at time the more time seems to race because time has the invert properties we give to space.

There is a position that is in motion that is forming the very edge of the outside. To be in motion the position must be in relation to a point from a centre. From the centre, there must be a specific allocated space ending at the object in motion and starting from a centre that has no dimensions. The object in motion determines the one limit and the centre with no sides and no space, which is standing still in singularity, determines the other limit. By that we can see there are only one way of looking at what we can observe and that is from the outside in.

The atom must be the utmost coldest and the proton is even much colder because when that cold escapes it turns to heat forming space that no one can understand. When the spin of the atom allows the cold of the atom to release the heat it had it had frozen to space the atom holds but when this heat releases from the containing form of the atom it brings about much more heat than the Human mind can cope with. One may not look at the material and judge the surroundings. The fact that hydrogen remains a gas and so does helium in outer space must serve as enough proof that outer space is hot, regardless of our interpretation of the temperature gauge telling us what we wish to hear. One must look at outer space and judge outer space from the findings only considering outer space. If helium remains a gas, it is hot. The removing of heat makes the centre of the Earth cold although we see it as being terribly hot. The only reason why it can seem to be hot is because it is cold and in such a cold environment, the heat can gather and space can collect heat because the particles find the surroundings extremely cold.

The cold in the earth centre causes the concentration of heat by space reducing, as all cold surfaces tend to do. If it was hot, the space within the Earth would expand and the space within the Earth where we think so much heat is concentrated does not expand therefore it must be cold. To gather and accumulate the space in a liquid means it became much colder being a liquid. Finding the surroundings terribly cold will allow the heat to gather and not expand but when the surroundings are hot, it will not tolerate more concentration of heat and thus will expand to rid the balance of excess heat within space. Look at the Sun and see how the Sun turned the hydrogen to a freezing cold liquid at 6500 K. Hydrogen is in a fluid state within the Sun and is colder than the hydrogen that is in a gas form in outer space. The Sun is the coldest place in the solar system. That is when the protons oversupply the removing of space to produce the cold that is so apparent. By the reducing of space, it can concentrate heat to a fluid state by producing the opposing cold that finally freezes the heat to a solid state. The expanding of space is a way of duplicating space without reducing space and by duplicating in the form of expanding it becomes just the opposite to duplicating by motion therefore reducing space by halving space in time. That is what gravity does. By motion, space duplicates and by space, halving it removes heat in space as well as by dismissing space. In all the applying of gravity, space dies. The density of the protons brings about space dense enough to harbour the heat in such quantities and visa versa applies in outer space.

The application of gravity that condenses space and bringing about heat by the compressing of space we apply in the way we go about tapping into the energy that nature provides. Internal and external combustion engines all rely on this application for harvesting motion by driving power. Compress space even today with a piston in a cylinder and then pump the compressed air into a container and such confining of space will increase the heat by the piston effort to reduce the space brought about in the container. The heat coming about inside the cylinder has no relevance to particles colliding because all compressor cylinders cool down. The walls become colder because when that cold escapes it turns to heat as the heat releases from space forming a secondary form of material forming space that no one can understand when the spin of the atom allows the cold of the atom to release into uncontrolled space. This release and unification with space that heat does is the heat it had frozen because the motion of spin to space that the atom holds, remains in a frozen state under the guard of the spinning electron. When this heat releases from the containing form of the atom frozen by the spin of the electron, it brings about much more heat than the Human mind can cope with. One may not look at the material and judge the surroundings.

The fact that hydrogen remains a gas and so does helium in outer space must serve as enough proof that outer space is hot, regardless of our interpretation of the temperature gauge telling us what we wish to hear. One must look at outer space and judge outer space from the findings only considered in the terms which outer space insists upon. If helium remains a gas, it is hot. The removing of heat from the space that contained the heat makes the centre of the Earth cold. In our universe we see it as being terribly hot because the heat then forms a separate substance but remains a form of material (8) but that is because we see the heat and not the space derived from the separating of the heat. The only reason why the space can seem to be hot is because the space is cold and in such a cold environment the heat can gather in a much concentrated state and space can collect heat because the particles hold concentrated heat in the space separating the particles.

By removing such high concentration of heat from the space that used to be expanded heat, the space then must contradict the heat by being extremely cold. We look at the heat in the space, which by that time is another form of material and find the surrounding heat in the space hot while the space is extremely cold. The cold in the Earth centre causes the concentration of heat by space reducing, as all cold surfaces tend to do. The proton contributes to that reducing of space. If it was hot the space within the Earth would expand and explode but the space within the Earth where we think so much heat is concentrated is so much it does not expand therefore it must be cold. To gather and accumulate the space in a liquid means it became much colder when the space parted from what then is being a liquid. Finding the surroundings terribly cold will allow the heat to gather and not expand but when the surroundings are hot, it will not tolerate more concentration of heat and thus it will expand to rid the balance of excess heat within space. The concentration or release of space with heat or space from heat is a direct contribution of the singularity in control of the space-time. The regard of the singularity stipulates the conducing of heat in space or the release of heat to form space by means of bisecting the occupied space.

Look at the Sun and see how the Sun turned the hydrogen it holds captured in its atmosphere to a freezing cold liquid at 6500 K. Hydrogen is in a fluid state within the Sun and yet it is still colder than the hydrogen we find in outer space that is in a gas form in outer space. The Sun is without any doubt the coldest place in the solar system. That is when the protons oversupply the removing of space to produce the cold that is so apparent in the heat levels that the atom cannot absorb in normal growth and therefore do cannot find accommodation in the walls of the atom. By the reducing of space, it can concentrate heat to a fluid state. By producing the opposing cold that finally freezes the heat to a solid state, we find that is what matter is. The expanding of space is a way of duplicating space without reducing space and by duplicating in the form of expanding it becomes just the opposite to duplicating by motion therefore reducing space by halving space in time. That is what gravity does. By motion space duplicates and by space duplicating the material must be by dividing or bisecting - halving it removes heat in space as well as by dismissing space and in that concentrating heat. The density of the protons brings about space dense enough to harbour the heat in such quantities and visa versa applies in outer space.

The particles claim more space when heated to preserve the cold. The claim to more space produces more space and reduces more heat. Such expanding brings about cooling. When particles heat or cool motion applies in some form. Motion started at a point when the Universe was extremely hot and there was no space. By introducing motion space formed and the lack thereof produced friction that became heat that became space. It is natural, it is simple, and above all, it makes believable sense.

The application of gravity is that which condenses space by bringing about heat with the compressing of space. We apply the progress we have as a species in the way we go about with our skills to unveil ways we can tap into the energy that nature provides. Internal and external combustion engines all rely on this application for harvesting motion by driving power. Compress space even today with a piston in a cylinder and then pump the compressed air into a container and such confining of space will increase the heat by the piston effort to reduce the space brought about in the container. The heat coming about inside the cylinder has no relevance to particles colliding because all compressor cylinders cool down with time moving and not necessarily with the loss or release of particles. It is not only the discharging of air that will reduce the temperatures inside the container. The time flowing bringing motion about where the motion is not about particles escaping but heat escaping in the replacing of the heat density (not the density of the particles forming the material content within the container) but the space that compressed to heat will also bring about that the heat displaces through the container wall to the outside. This is bringing about equilibrium where heat will always flow from more dense areas to the lesser dense areas. This has no influence on the status of the particles on the inside of the cylinder but only concerns the density levels of the particles inside versus outside. After the pumping of air increased the heat in the cylinder which even can go to dangerous levels, will reduce back to room temperature when further pumping ceases and that stops further air movement into the cylinder and such surging of pumping air is what brings about heat stabilizing.

Mainstream physics ignored the clear connection completely, notwithstanding it being so very obvious. There is this far in their recognising of principles in natural physics not one single reference made to prove their appreciation of this matter. They are bent on particle colliding. When particles collide, such collision forms an atomic thermo release and that action we call an exploding atomic bomb. What principle this argument about particles colliding, ignores is that all atoms use negative charged electrons forming the atomic limit on the outside forming a definite border to the boundaries of all atoms and in both electrons from different atoms are being negative charged.

In being negatively charged, it means both will come out and totally reject the other. The closer they come the more violent the rejecting will be and such rejecting is the production of heat that will turn to space. The electrons repel other negative charged sub atomic structures, which the electrons are that form the outer borders of all atoms. With all electrons highly negatively charged (being as negatively charged as any possibility will allow to match the utter extreme) such electrons could not touch.

It is about time scientists start looking with their minds and not their eyes at the Universe and see what is truly out there to see. All the difference we find is seated in the human mind. We humans set differences because we look at the cosmos by placing humans and the life we find on Earth in a pivotal centre in the cosmos instead of placing singularity in the centre and life where it belongs; only found on Earth. Einstein proved mathematically that in the presence of a strong gravity such a strong gravity slows time down.

Surprisingly with that evidence being around this long, nobody in science since Einstein's discovery took those statements and made any further progress from that. It seems to have been left in some drawer to dry. Science still sticks to the opinion that time did not change, not even slightly, since the beginning of the time it held the same pace ever since the start of the Big Bang notwithstanding the implications this concept carries. Before the Earth took one year to circle around the Sun and even before the Sun was there a year was still the same duration of one year. How odd... don't you think ... that the only aspect in the entire Universe that is beyond change is the aspect of time? With the entire Universe including all the gravity now present and not excluding one Black Hole or dust speck pressed in such an area that was possibly the size of a lepton even then the gravity extending from that circumstances must have been beyond what words can ever describe.

When everything was that small when the Big Bang took charge, the gravity at the time was beyond light, because even today in the Black Hole the gravity is beyond the speed of light. If the gravity was that high and Einstein already proved that strong gravity slows time down, then there is one logical conclusion and that is that time was in fact at the time of the Big Bang standing still. Mathematically it is incorrect to allow gravity to compress the Universe into a spot smaller that an atom and exclude any other factors and relevancies to change.

As usual Newtonians has the relevancies mixed up. It is not the Sun that is cooking outer space to cinders but it is outer space that is boiling the Sun to steam The steam we see mainstream science promote as light coming from way in but is the Sun was hot as hell on the inside the Sun would have exploded as stars do when they overheat. Supernova are stars that overheated and if a star can overheat when everything is going wrong then the star must under heated when everything is going right. Therefore a star is a particle that is frozen into liquid and in a process of would be one day frozen into the oblivious.

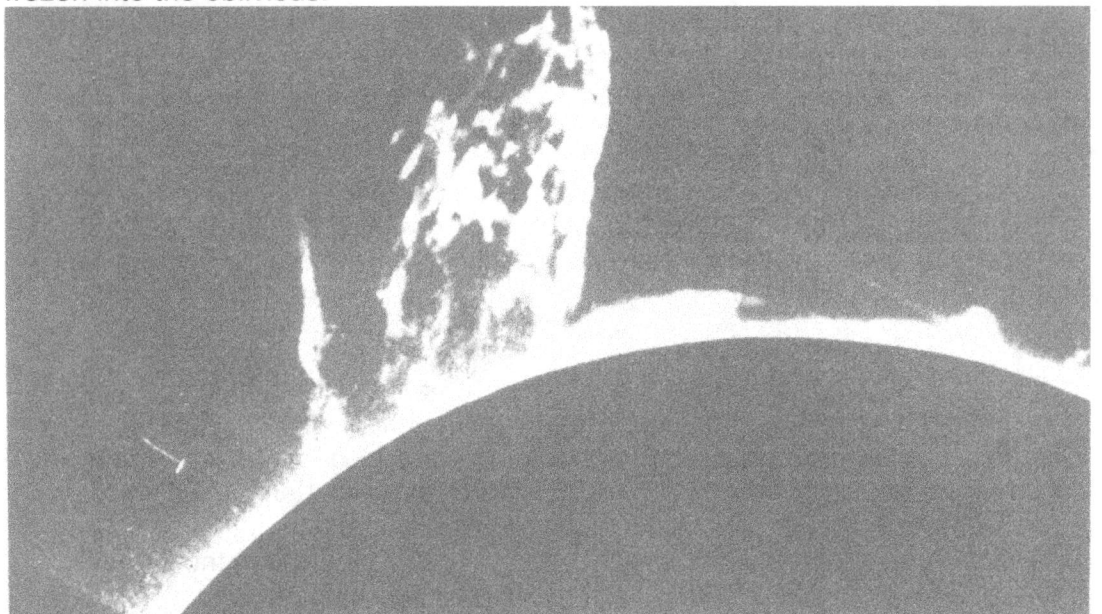

The fact that the prominence squirts out liquid in vast amounts is because there is a lot of space reducing going on in the Sun so there is less space inside the Sun . The fact that the prominence expands to outer space means that the prominence was expanding into a hotter area and does so as a liquid. However that fact that the prominence fell back into the Sun can only result from the prominence not being hot enough to return to a gas state and then through such density discrepancies had to return to a less hot area. That is science not magic. By using Newton one cannot even begin to explain any one of or the combined efforts of the above cosmic phenomena that are all over the cosmos and forms all the laws in the cosmos.

Newtonian definition cannot even recognise any of the principles but only Newtonian science are taught to students. No student can have the fortune to disagree with Newton and remain a student. If the student will dare to disagree with Newton it is the end of such a students academic career. By setting this firm condition Newtonian science becomes institutionalised mind conditioning of the concepts of thought forming in physics. With my saying this I have not made one academic friend but neither have any one proved me wrong. Students are taught to accept Newton and to ignore Kepler and any student doing it the other way around will fail all examinations and other testing at Universities. Students accept Newton or they accept a ticket taking them home. Newton is an institution force fed to each following generation but saying that reserves only resentment towards me amongst Academics. According to Newtonian science space is simply nothing with no qualities but gravity separating space and space does not mingle, as one would expect if space was nothing because space does form borders.

Disasters of unprecedented magnitude arise from such borders. The Challenger disaster of February 2003 is pertinent testimony to those borders that was powerful enough to break the aircraft into pieces while the explaining contributed by Mainstream science is evidence of a shocking lack of understanding about what took place as cosmic laws were breached. I do not pretend to be of superior understanding and do not place myself on any pedestal. On the contrary the information is so simple and so easy to understand that the lack of any Academic understanding frustrates me almost witless. But academic taught culture demands all persons to miss the evidence, which is so clearly visible because academics demand researchers looking in other directions because students are forced to accept Newton's vision about Kepler's work. By the time they reach researchers status, they too have tunnel vision that can only acknowledge Newton and ignore Kepler. Our not understanding laws, provide a platform for future disasters occurring because it will lead to us ignoring more of applying principals that leads to space tragedies of magnitudes we have not thought of as yet. By not understanding the sound barrier, tragedies have and will again come about and will increase as misconceptions become more present in the future because the demand on space travel increases.

The book **an open letter To Selected Academics ISBN 0-9584410-9-X** is about that process adapted by the Big Bang, never ended and it is still bringing over, that which is in unoccupied space to material being in occupied space. Occupied space holds matter and unoccupied space is empty of solid materials. There is contraction, which we know by the name we gave as gravity. Then there is expansion, which we gave many names being the Big Bang and the Hubble Constant or better known as simply exploding or forming plasma with all the terminology accompanying that simple idea. This I show is antigravity. Apply heat and space and a balloon lift where such lifting is antigravity. There is a balance in the Universe where gravity contracts and reforms space and heat expands becoming space and produces space.

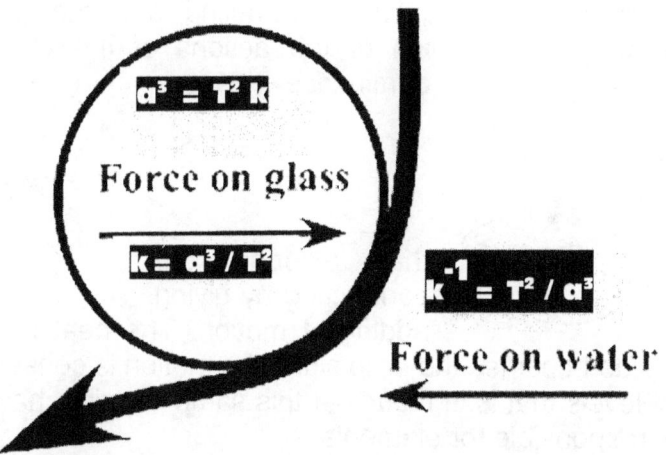

This puts my theory in line with reality. The only way anything can get bigger is when heat is added to what there already is. The Bomb at Hiroshima and Nagasaki showed how intense the heat in atoms are and how well packed the heat is which is contained in the atom. Bringing more heat to the atom brings about the atom having more of the same but in a higher measure. Only heat can make material expand and with the Universe unable to expand because the Universe is what ever can be, it must be the material inside the Universe that is expanding. In the same measure we find where space reduces heat, is removed from that space. Gravity is about contracting and that is true. In the manner that Kepler put it material that is filling space (a^3) is equal and the same as the motion ($T^2 k$) of the Material moving. That means to have time then time must move and the only way time can move is to move the space in the time.

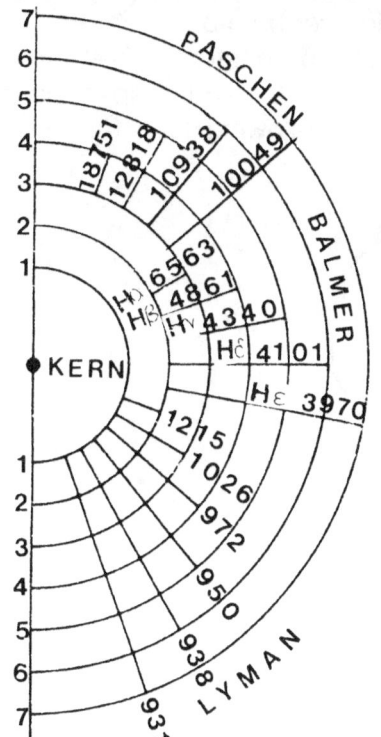

However to move is either to come closer and using Kepler we can see how Kepler would reduce space in $k^{-1} = T^2 / a^3$. That means the expanding subsides to a point where contraction is and contracting is about cooling or reducing the Heat that was expanding the space. Then there is expanding of material which is as Kepler would put it $k = a^3 / T^2$. That would be when the space becomes less and the only valid way for space to become less is when what there is becomes less than what there was. That amounts to cooling where the heat is being removes from the space.

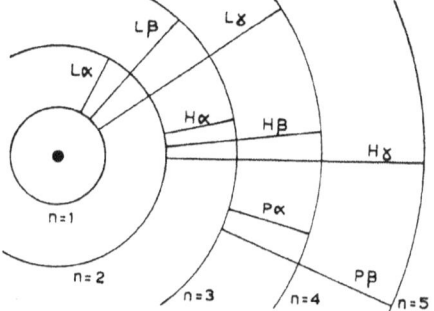

at a wavelength the Lyman series. The lowest level contributes to jump is the greater is the the series, which is reached outside of the atom. absorbing of heat and that is

Whenever the electron jumps from a higher into the lowest (innermost) orbit, the atom gives out radiation corresponding to a spectral line of jumping down into the second the Blamer series. The greater the emitting of radiation to the limit of when an electron enters from the Outward jumps involve the inclining to provide space to

accommodate the increase heat levels because of the increase or rise in the absorption lines. When the heat level in the atom rises, the electron jumps to a higher band and when the heat reduces in moves down one band. The heat coming about in the surrounding of the atom produces more space because the atom increases the space by applying the electron in a higher orbit ring. The moving of the electron is coupled to the giving out of radiation at a wavelength corresponding to the spectral line of the Lyman series.

When the heat level rises or lowers the space within the atom decline or increases. Every atom in every element association shows different corresponding to the heat it is in association with. The corresponding of the atom and the reaction derives from such rises of space is a direct result of the interaction there is in the gravity contracting and the gravity in expanding depending on which of the actions of the Coanda principle is in

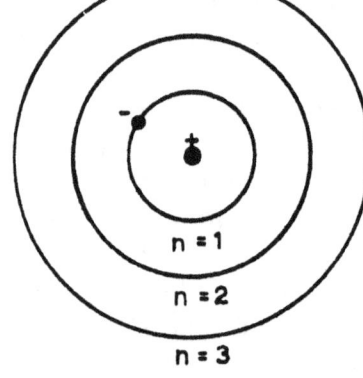

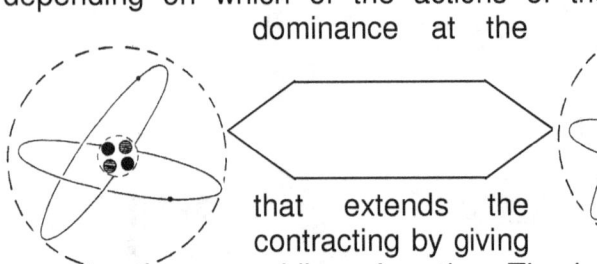

dominance at the rise in heat is a rise in the liquid part that extends the contracting by giving rise to the adding of motion. The heat is liquid because the heat is motion and the atom inside becomes the solid since the motion is conserved by the spin of the electron. Today it is the rise of levels that is in focus but this same principle had to be in use when atoms were formed that today is responsible for elements.

If it was true about mass pulling mass by reducing distance the big bang was not possible and individual elements was not possible. With the cosmos down to the size of a neutron and mass confined to that space within the neutron that would be the recipe for the biggest crunch there could ever be. It the Big Bang was brought about mass confining mass by reducing the radius that implied the space within then what would ever be more applicable than that moment to bring in all the forces the hell can unleash and destroy what ever was not yet in the Universe. The entire idea of mass pulling mass to reduce space is a prehistoric thought and explicitly incompetent in explaining science. The following is a far more suitable explanation and is as true as Kepler is.

BERYLLIUM (4)

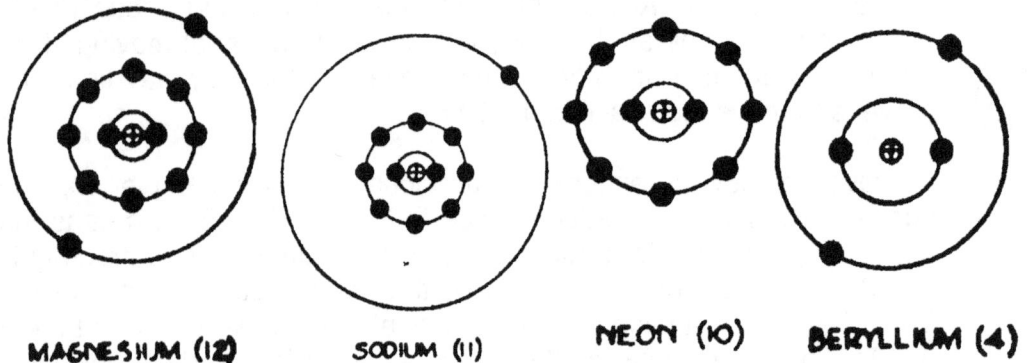

MAGNESIUM (12)   SODIUM (11)   NEON (10)   BERYLLIUM (4)

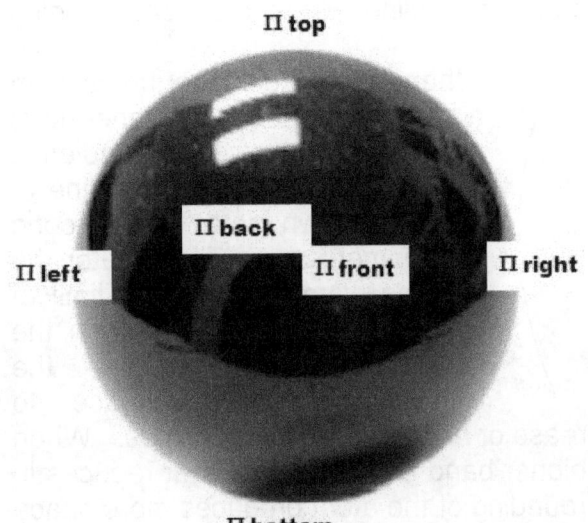

Π top

Π back

Π left   Π front   Π right

Π bottom

The most sensible way atoms formed is by the method of the Coanda principle. Those ones that was first were most of all very dense atoms were the first to come in place when time was eternal and dominant and space infinite and one notch off singularity.

Element	Relative number of atoms
Hydrogen	1,000,000,000,000
Helium	90,000,000,000
Carbon	350,000,000
Nitrogen	85,000,000
Oxygen	590,000,000
Sodium	1,500,000
Magnesium	30,000,000
Aluminium	2,500,000
Silicon	35,000,000
Phosphorus	270,000
Sulphur	16,000,000
Potassium	110,000
Calcium	2,100,000
Chromium	300,000
Iron	3,200,000
Nickel	120,000

We still find the liquid time having a vital role in the space the atom uses. At the time when T^2 was almost eternal space was infinite because T^2 will not permit the space a^3 much room to be. But the opposite is also true that if time was steady then time being so long could pack in large numbers of protons with the accompanying neutrons at the time into the time unit in space that formed.

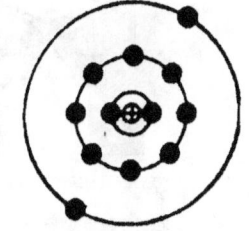

MAGNESIUM (12)

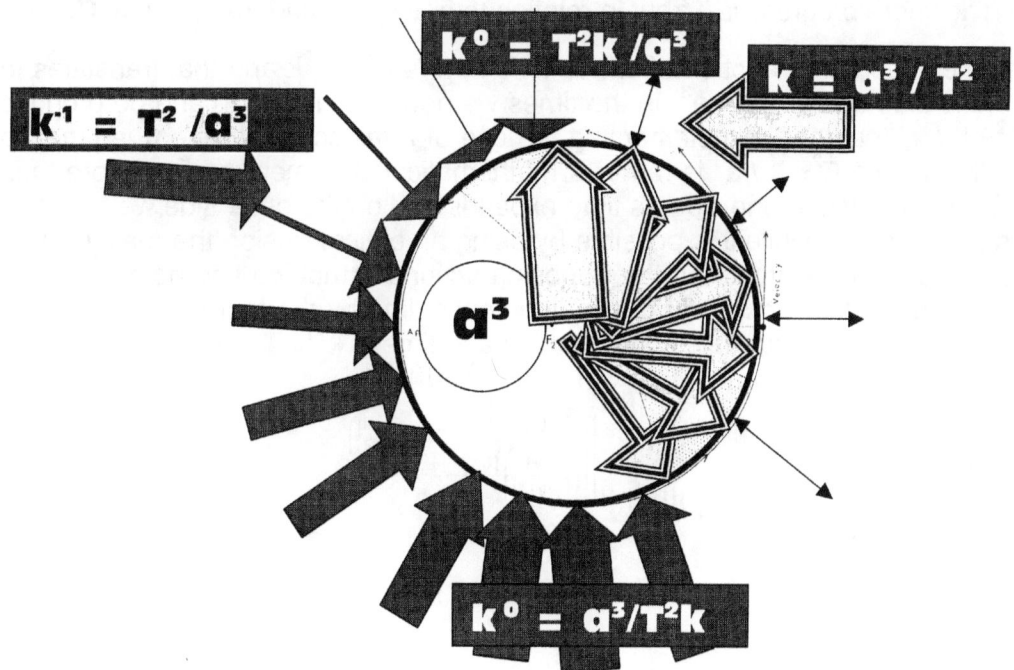

$$k^0 = T^2k / a^3$$

$$k = a^3 / T^2$$

$$k^{-1} = T^2 / a^3$$

$$a^3$$

$$k^0 = a^3 / T^2k$$

This is most accurate but this is only in concern of a unit in rotation rotating in conflict of its own spin. When time in the cosmos at large view time in progress we find that the development is not quite so simple.

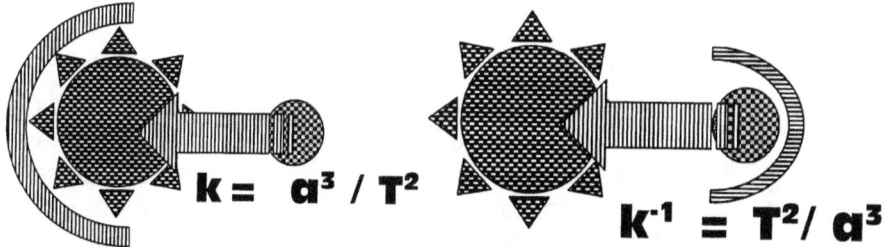

$$k = a^3 / T^2$$

$$k^{-1} = T^2 / a^3$$

The rebound is always less than the progress and the reason for that is the flow of time. But that also forms the reason why there is a Big bang and why there is a Hubble constant in the midst of all the contracting that is shaping the Universe.

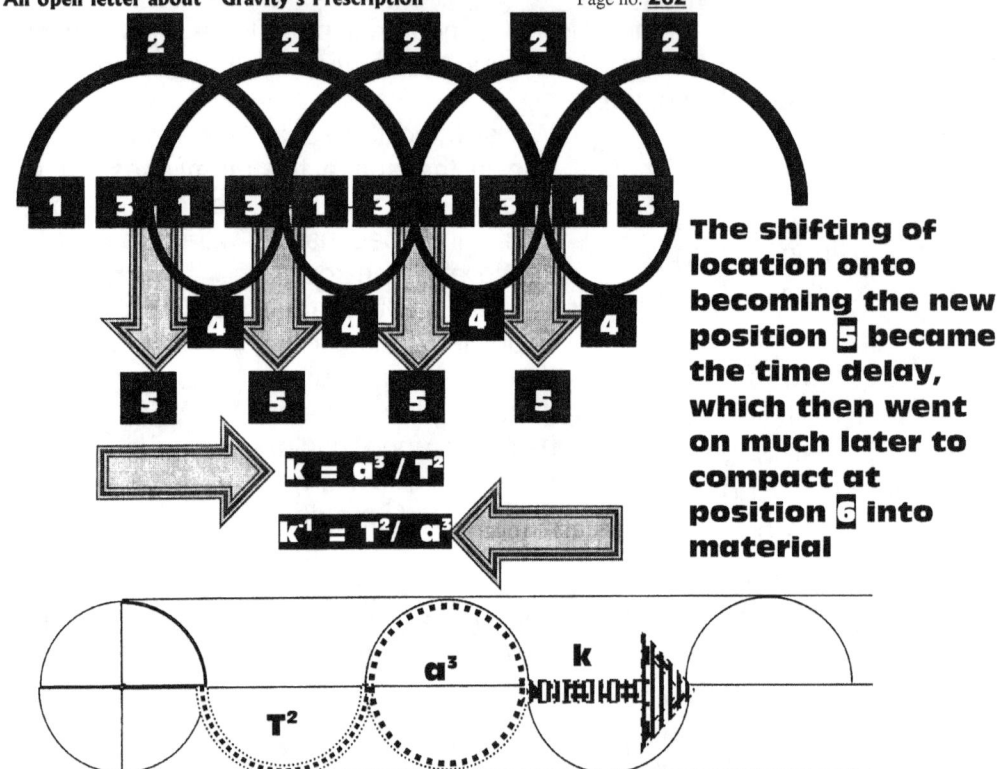

The shifting of location onto becoming the new position 5 became the time delay, which then went on much later to compact at position 6 into material

$$k = a^3 / T^2$$
$$k^{-1} = T^2 / a^3$$

Time moves by the measure of T^2 but time in progress is by the measure of **k**. Therefore space in progress of time is $a^3 = T^2 k$ where progress is T^2 but in relation to gravity we find that $a^{3\,-1}k = T^2$.

In the relevancy we find the action and reaction of space-time flow is $a^3 = T^2 k$ and that translates to being $T^2 = a^3 / k$ on the one side and $T^{-2} = k / a^3$. In the times we now live in we can and do produce an optical illusion of $T^{-2} = k / a^3$, but that is implementing the use of a telescope. In the true time we find as a cosmic reality the fact of $T^{-2} = k / a^3$ is rather a mathematical statement and no more than that. In reality we have $T^2 = a^3 / k$ on the one side as time expands and on the other side we find $k^{-1} = T^2 / a^3$. This we know is true because while it is possible by using an optical illusion the reality is that time can never reverse. In truth the reality about the opposing actions is that we find normal growth and that which Hubble first saw is the process of expanding space-time by the margin of $T^2 = a^3 / k$ while on the rebound we find the opposing while contracting space-time is $k^{-1} = T^2 / a^3$.

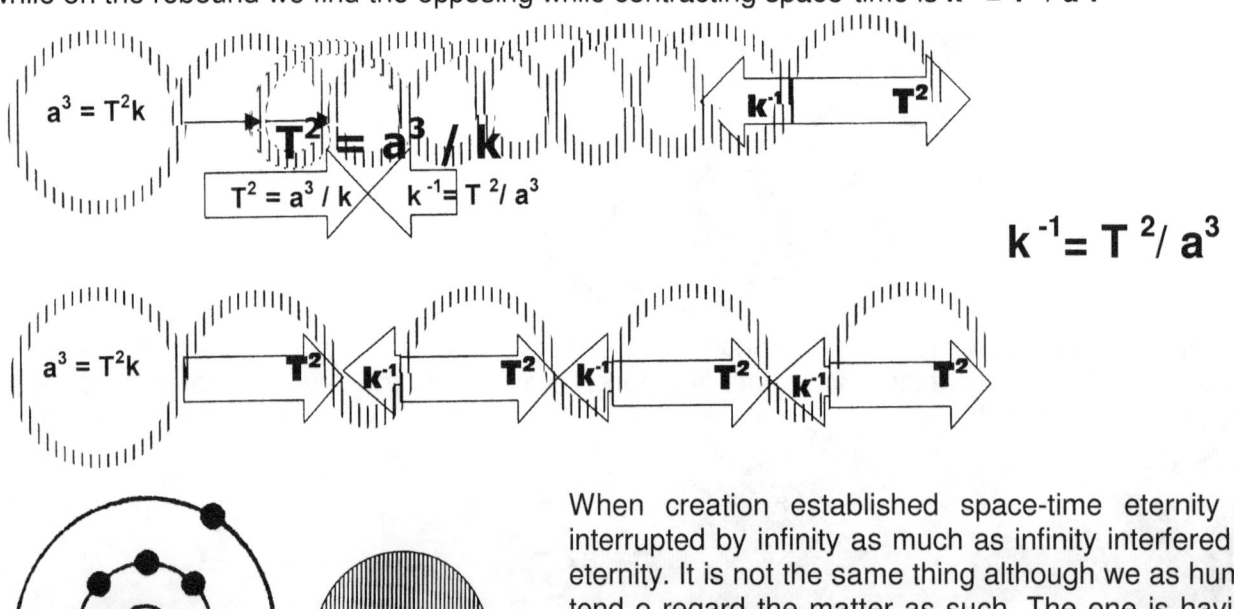

When creation established space-time eternity was interrupted by infinity as much as infinity interfered with eternity. It is not the same thing although we as humans tend o regard the matter as such. The one is having a look at it from the one side and the other is looking at what happened from another perspective.

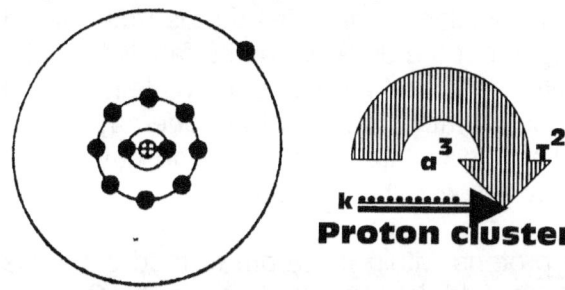

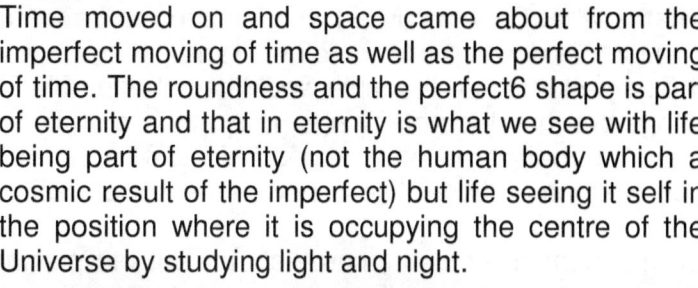

Time moved on and space came about from the imperfect moving of time as well as the perfect moving of time. The roundness and the perfect6 shape is part of eternity and that in eternity is what we see with life being part of eternity (not the human body which a cosmic result of the imperfect) but life seeing it self in the position where it is occupying the centre of the Universe by studying light and night.

The imperfect part is the part we find in with life holding part in eternity we cannot see experience infinity while we see eternity. We always there because we are unable to see changing. Look at your own fingernail your hair growing or even a wound healing see the nail, the hair, the wound but never by growth. You might find what I say at this physics but it is more physics than anything as cosmic physics are part of true physics.

In our ability not to see the imperfect while the imperfect and at the same time see the

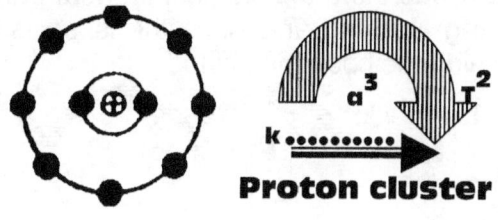

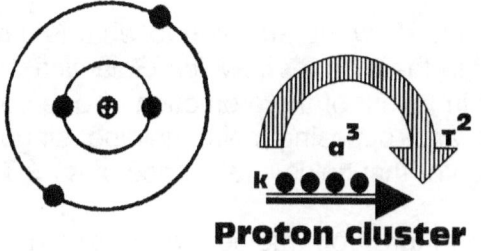

infinity and infinity. We see what is what is growing or and you will the addition point not to be currently used

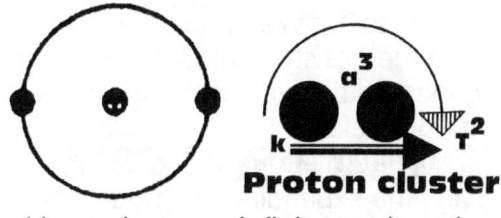

not the perfect it is little wonder we lot are so mixed up in what we see and cannot understand. Eternity lasts forever and that is why no changing is visible but we can only experience infinity because eternity is an ongoing repeat of the same

experiencing perfect in eternity and being part of eternity being

without changes. Infinity on the other hand is what interrupts eternity and therefore what we see in eternity is what infinity is interrupting. The proof of this is in the top where motion distinguishes infinity in the centre from eternity surrounding the time position of the top designating the motion from the past through the present onto the future. Those are there for all to see and the fact of that being there goes beyond dispute.

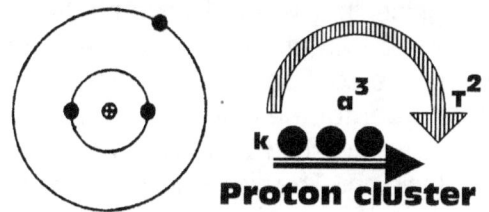

That is the manner how stars move time back to the point of having eternity sharing infinity infinitely. The one can absorb the other just by reducing the relevancy and increasing the flow of time. As the relative flow increase the relation is space subsides and the one become infinitely in relation to the other that then provide an eternal time. The match form and the element gain one more proton.

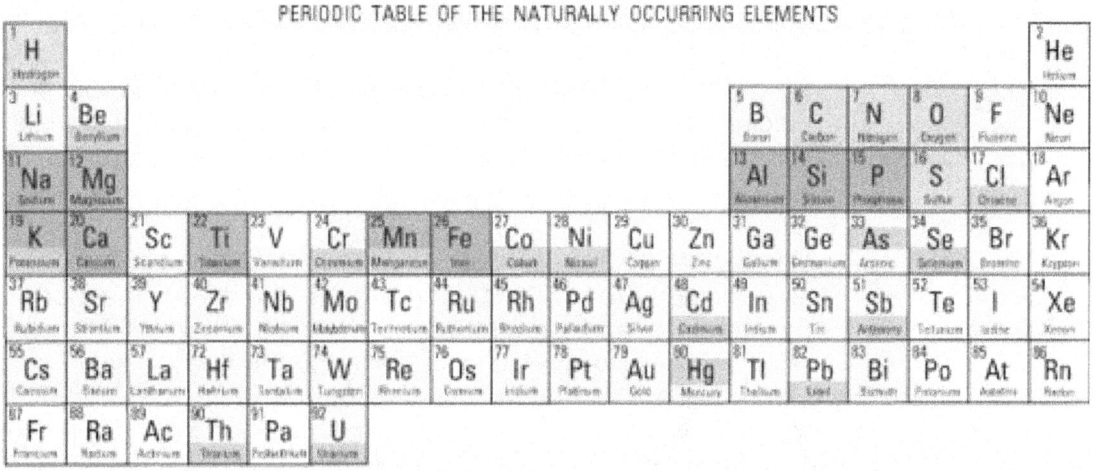

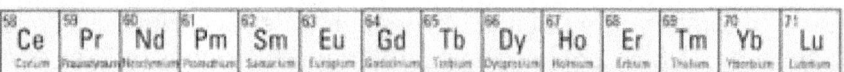

That is what Darwin missed with his species being from one ancient origin. Yes that is true but the one did not develop into the other. Time did produce changes but the donkey has no family ties with the horsed. If it had the mule would have been able to multiply and be fruitful and the mule is a lot of things but that it is not. Things go along in eternity while infinity interrupts and then one-day infinity brings a change no one noticed before. The same building blocks are used built one day a new corner stone is laid and new specie arrives that has no family ties with the previous lot.

In this manner elements came about. But the placing of protons within the atom formed elements whereas the atom was there the first instant heat parted from cold. I go into that part in the **Cosmic Birth...Dismissing Nothing** and since the issue is rather complex in explaining I would prefer to leave that explaining the book mentioned.

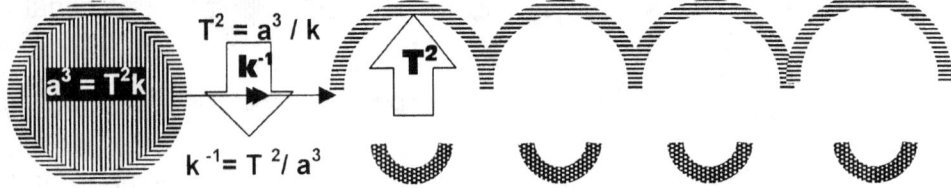

$$a^3 = T^2 k \qquad T^2 = a^3 / k \qquad k^{-1} \qquad T^2$$
$$k^{-1} = T^{2} / a^3$$

Once again I have to draw the attention to what is out there in the cosmos serving as evidence. The proof we still find in the manner in which Galactica and all other orbiting objects develop. There are $T^2 = a^3 / k$ that is in favour of the promotion of one point holding singularity in the relation and to that there is another and opposing point holding singularity in prominence which is in relation the expanding contributor that holds a relevance of $k^{-1} = T^{2}/ a^3$ in ratio to the conserver.

The relevancy was there from moment-Alfa brought relevance from 1^0 to 1^1. We can see that there were seven in ratio of ten and we can see how the seven produced the gravity of motion relating to the ten in time. We can see when the dominance started creeping to the other side and when $k = a^3 / T^2$ got the better of $k^{-1} = T^2 /a^3$ because at a point where the sum total related to the singularity the proton $(\Pi^2+\Pi^2)$ + the neutron $(\Pi^2+\Pi)$ + the electron 3 = 35.89 × singularity Π = 112.75 outer space. Past such a point the expanding factor began to gain lost ground and the expanding got predominant as the containing factor started to store and preserve more than contain.

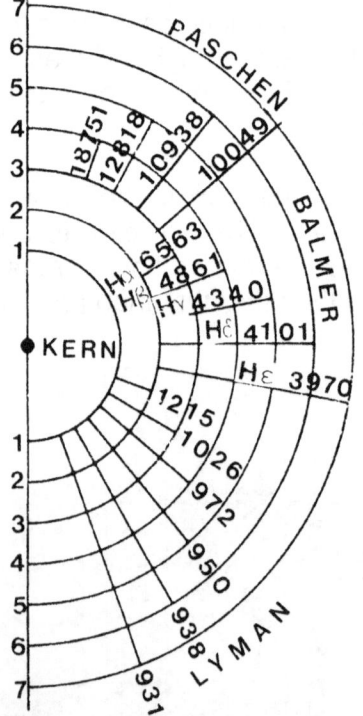

The major issue in hand is to recognise that when 1^1 overheated it parted from 1^0 because 1^0 was to cold to harbour 1^1 and by measure of 1^1 being hot, the same measure places cold on 1^0. The one cannot heat without the other establishing a border for the cold. In the Coanda principle the liquid establishes itself onto the solid by gravity as much as the solid allow the extending to lock on. The liquid is not locking onto to an unattached solid. The solid gain as much as the liquid gains but that what the solid gains are not the same in likeness to that which the liquid gains. The solid will as much prove to be colder as what the liquid proves to be hotter.

One should see the electron as the indicator where the liquid attaches and where the solid ends because where the liquid ends is where the solid start. The fact of the matter is not entirely that simple because the electron does view the neutron as a stabilizing solid while there is nothing in the cosmos more liquid than just the neutron is. The entire idea is in judging what is solid at the time in ratio to what is liquid at the time.

We see our bodies as solid while the truth is that it is life that keeps the body liquid. When life detaches from the body, the body loses its motion ability and then the body goes rigid. Then the body becomes a solid. However while life is in the body the body is used as a mobile object in relation to the Earth being solid.

When going colder the atom raise its solid level to lose some of its liquid level and the other way around. But the fact is that it is the atom going hotter that reduces the atom going hotter. Hot and cold is not by measure of temperature but it is by measure of space moving through time. If there is more space moving more rapid the atom becomes colder and the opposite is also very true.

As soon as motion commences **k** increases because although the relevancy from the Earth perspective remains the same the relevancy from the aircraft changes drastically.

From the view the aircraft holds the **k** that the Earth has becomes the **T²** that the aircraft has and the **T²** that the Earth holds becomes the **k** that the aircraft uses.

The sound barrier is just another manner in the way the Universe brings about gravity. The aircraft has to produce excessive heat and by more heat delivers the space between the Earth and the aircraft increases. The motion becomes more and that stretches the space connecting the Earth and the aircraft applying heat to produce extended motion where the extended motion leads to extending of the space the aircraft covers in the same duration of time.

The heat (formed by the release of motion in the engines) allows more motion to apply than that which the Earth generates which puts the aircraft in a higher atomic bracket where the aircraft holds more space that the Earth normally would grant the aircraft. The aircraft has more gravity (granted that it is using the motion of the Earth which is the gravity of the Earth) than it would have being stationary. With an additional source of heat the aircraft can add to the earth gravity and the Earth gravity is unrestricted motion. It has no bearing on mass whatsoever. Gravity is about motion and mass is the restricting of such motion.

It is not the motion we must be after but what causes the motion in the first place (other than being Newton's pet force). We must find what produces motion and from that then we must think further than what we can recollect from thousands of years of culture that got us this far but is getting us no further. We must see why that which moves or tend to move as we do on Earth in our gravity. It is nothing to do with mass pulling everything about but is a flow of space-time.

When one applies heat to an object it expands. That is primary school science. This states that more heat applied leads to more space acquired by the heated object. In sharp contrast to this is the growth in space when heat levels rises but freezing brings about the opposite result. When I freeze an object that object reduces its occupied space as it shrinks. Removing heat reduces space. That comes directly as nature responds to heat and I can prove that easily.

By expanding it accumulates space to increase the improving of the size of the material. The accumulating of heat is for the sake of securing singularity, which accumulates the heat in the material whereas the freezing tarnishes the overheating symptoms by the removal of material in unoccupied space using external matter and setting motion to the material until it contracts into a form which we see as visible heat. The heat is in the form of dissolved singularity that became material as material used it as growth. That is why by freezing it will diminish the space as to accumulate the heat absorbing into the heat into the material to maintain the equilibrium needed in space.

The atom is the optimum proof of the statement. The atom is the absorber of heat as well as the release valve of heat. The atom regulates heat in relation to space acquired as well as space acquired. The atom is as much about heat as controlling heat and when the atom expands space it accumulates and store heat. When it cools it reduces and absorb space. The cosmos is the atom and the atom is what the cosmos use to regulate heat.

Taking this equation of nature to outer space we seem to confuse the natural law. With outer space as expanded as anything can get we regard outer space as incredibly cold. As heat sets in the normal flow will bring about expanding of heat into the form we think of as space that limits the heat overheating. Outer space is the very edge of expanding of space where heat cannot expand into space any more. Outer space is the limit, the epitome of expanding where heat meets space at the edge of all limits once more. Therefore being the representation of the very limit of expanding outer space has to be the hottest place there is. By applying heat to a kettle holding water, the adding of heat manifests as steam and steam is hot water that traded heat as it reviewed space. By allowing the receiving of the heat to continue the container will let loose steam in order to match the contributing of space.

The manner in which heat expresses itself when confronted by overheating is to provide additional space through expanding of space. Outer space is outer space because outer space has expanded all it can it is still expanding to the speed of Hubble's $1/H_0$ which inevitably does not only affect far-off places where we cannot be, but effects us on a daily basis. As outer space is stretched to its limit, its limit will continue to stretch but while it is stretching it has to having more than it had before in that outer space holds the limit of heats expanding possibilities.   Singularity has been expanding since way back when but that means singularity is still releasing heat as space-time that turns out as space in the universal time of outer space. In outer space heat cannot expand more therefore except for the continual growth that benefits all singularity throughout on a continuous bases concerning all outer space.

Every element is in relation to the heat level it uses in forming the gravity it has. One can see how the forming of the numbers of elements available in the Universe stands related to the density of the elements total numbers. More pertinent to note of that the effect of gravity is not in the mass of the element but shows a much stronger relation with the density and the density is the relation the element has with the heat that marks a boiling point or a freezing point The density factor shows what we use to classify the element in relation to being a liquid, a gas or a solid. This factor is much more prudent than the mass factor and that I show later on as the book develops.

If singularity expands when heated and there is a limit to the point it can heat, and where that point forms the maximum expanding possible, then it has been reached in the area we think of as outer space. Outer space has expanded through the unleashing of heat, where overheating is turning liquid heat into space. Any explosion is a vivid reminder of this fact and the unleashing of space is so real it destroys the space holding solids by rearranging the construction of the solids. With that in mind we can declare with great confidence that outer space as the hottest place there is. Whatever expanding there possibly is, was done to secure the cooling and all cooling that can be introduced to bring about further cooling was performed in the area we think of as outer space. Forget schoolboy culture and the temperature scales and other Newtonian scientific defects I call Xepted mistakes. Think of reality and throw out culture teachings Use the mind and not the thinking power of the past. Any place that can expand no more is the hottest place there is just because of the shear implication that it can cool no further is as hot as it gets anywhere. If that is the case then it is safe to say that galactica is freezing cold notwithstanding our concepts of heat and space and heat in space given to us by our collective culture and not by our ability to reason.

The galactica is little frozen islands in a vast see of heat. That is the reason we can see the galactica because the galactica is space concentrated by a frozen space. The galactica is slowly heating and therefore it is expanding into outer space. Outer space on the other hand has expanded to the maximum that it can yet we think it is cold when it is the extreme there is in heat that introduced the maximum expanding. What I now am saying might be deemed by the most purist as the contradiction of the century and that much I do realise. At the inner core of a star all space shrinks into the oblivious but we consider the inner core area of a star to be the hottest spot in the solar system. That just cannot be because when material shrinks it becomes cold and by shrinking into the oblivious it has to freeze into a fusing element as newly formed units. Again that is the contradiction of the century. Why will that be? The space inside the star shrunk to the minimum there can be and that tells us the space has to be cold because of the shrinking took the space to a position where no space can shrink anymore.

That shrinking of no more space can only be inside the inner star and in that region is where we locate the strongest gravity. With outer space as expanded as nature may allow the space that grew could only grow in conditions of heat because heat produces expanding and expanding is the result of heat coming about. Space shrink because it is cold: that we know and taking this law to the star centre it means regardless of our interpretation of hot and cold, that area in the star centre is as cold as it can get notwithstanding what our nature may tell us. Then obviously the same must apply to outer space for precisely the same reasons because it is so hot there it can expand no more.

At this I have to redeem myself from being human. Only in the eyes of humans are there hot and cold, but as a reality in the cosmos we will find this nowhere. We look at the hotness of space and the coldness of space but it is the relevancy to the solidity that forms the actual heat and cold limits. It is so hot no expansion can produce more space in outer space, as the outer space seems to be the epitome of what can be cold while it is truly hot and quite the opposite reveals as the true scenario inside the star in the centre of a star structure. That means the number of protons in motion has a lot to do with the cold and hot scenarios because where the protons are most dense the cold is in extreme…well in most cases. Only in the absence of space can so much heat gather in excess and the opposite is true about outer space where the least denseness found brings about the space in heat found in outer space. Our human selecting of hot and of cold and what is hot and what is not prevents us the clear vision we would have when truly understanding the applying temperature. Temperature comes about from spin and the smaller the spin density is the colder the space becomes because the more duplication produces the most cold. We think of outer space as 0^0 Kelvin but in fact it is as hot as no other place can be in the Universe. The coldest is where material is freezing solid as material does when frozen solid and the hottest is when by boiling the material is going into a gas with liquid being the intermediate position where heat acquires the space to perform as a flexible substance.

When we look at particles in outer space we see the particles being frozen. It is because there is such a severe contrast between the particles and the environment surrounding the particles and not the particles that is so frozen. The particles are in a gas state because the particles do not form a part that is part of the space unit. Hydrogen clouds of hundred of light years in diameter are a common sight in outer space. The heat we find filling space is not part of the space but like the particles the heat is a separate issue. That heat filling the space is another form of material that could conduce by diverting from space or marry the union of space by becoming more space. If it were that cold which we think it is, it would not have expanded into such a massive cloud but would have contracted forming a cube of frozen hydrogen. But as we can see the cloud expanded the gas as far as the gas can expand.

That expanding is indicative of heat and has extremely little to do with gravity or is it just a matter what we think of as gravity. If you are of the opinion that those hydrogen clouds will contract one day into forming a star, well then think again as there is just no such a chance that that will ever happen because that is not the manner that form of gravity functions. Because outer space is completely overheating the condition it has in support of the particles makes the particles appear to be in a state of freezing but the particles is counteracting the heat limit it meets. However the particles do not contract, as the heat is immense. The space in outer space has absorbed all the heat by means of expanding and will appreciate still further as it will never depreciate. That is not because outer space is freezing the particles but it is because in contrast to the heat of outer space the particles seems to be frozen.

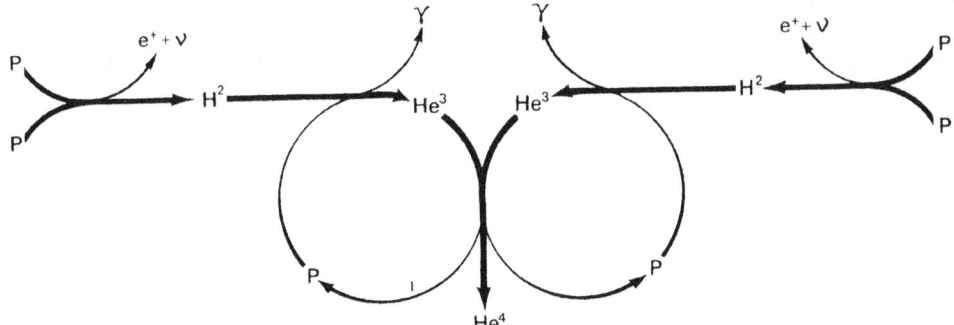

The atom must be the utmost coldest because the proton is even much colder than whet the electron can freeze. In fact the proton is 1836 times colder than that which the electron is able to freeze. We find that when cold escapes it turns to heat and the heat relieves by forming space, however it seems that that no one can understand. Motion brings about cooling. When the spin of the atom allow the cold of the atom to release the heat it had, which it had frozen the heat returns to space. This is what the atom shows in the electron bands or rings the atom holds. This must not be confused with uncontrolled release of heat. When the motion of the electron is interrupted such motion reducing results into the utmost expanding there possibly can be. When this heat releases from the containing form of the atom it brings about much more heat than the Human mind can cope with because no human mind can comprehend the total devastation a nuclear release of space may bring forth. In this

I am not referring to the normal way material relates to heat. That is a totally different matter altogether.

One may not look at the material and judge the surroundings. The fact that hydrogen remains a gas and so does helium in outer space must serve as enough proof that outer space is hot, regardless of our interpretation of the temperature gauge telling us what we wish to hear. In the vent of outer space truly being the coldest we have, then hydrogen and helium should be frozen crystals clotted in balls of material. One must look at outer space and judge outer space from the findings only by considering outer space without the prejudgement of teachings about ideas when persons were still held in prison for being suspected werewolves. If helium remains a gas it is hot. However we can witness hydrogen being a liquid in the Sun. That squirting from the Sun is liquid heat that is frozen as a form of material in the hydrogen layers and holding the hydrogen in form in the hydrogen layer.

We might think it is hot in the centre of the Earth but that type of thinking is as Newtonian as thinking of big stars as mighty gravity pools. The removing of heat into a liquid makes the material in the centre of the Earth cold although we see it as being terribly hot. The only reason why it can seem to be hot is because it is cold and in such a cold environment the heat can gather and space can collect heat because the particles find the surroundings extremely cold. Then again we confuse heat and time altogether and completely but more about that later on…

The cold in the Earth centre causes the concentration of heat by space reducing, as all cold surfaces tend to do. When material reduces space, it parts the material from the heat within and places that heat within the electron bands on the outside of the electron bands. By removing the heat the atom contracts and by contracting the atom reduces space. That heat forming space has to go somewhere. If it was hot the space within the Earth would expand and the space within the Earth where we think so much heat is concentrated does not expand therefore it must be cold. To gather and accumulate the space in a liquid means it became much colder being a liquid. Finding the surroundings terribly cold will allow the heat to gather and not expand but when the surroundings are hot it will not tolerate more concentration of heat and thus will expand, to rid the balance of excess heat within space. That is the terms in which to think in when thinking in terms of cosmology.

Look at the Sun and see how the Sun turned the hydrogen to a freezing cold liquid at 6500 K. Hydrogen is in a fluid state within the Sun and is colder than the hydrogen that is in a gas form in outer space. The Sun is the coldest place in the solar system. That is when the protons oversupply the removing of space to produce the cold that is so apparent. By the reducing of space it can concentrate heat to a fluid state by producing the opposing cold that finally freezes the heat to a solid state. The expanding of space is a way of duplicating space without reducing space and by duplicating in the form of expanding it becomes just the opposite to duplicating by motion therefore reducing space by halving space in time. That is what gravity does. By motion, space duplicates and by space halving it removes heat in space as well as by dismissing space. In all the applying of gravity space bites the dust. The density of the protons brings about space dense enough to harbour the heat in such quantities and visa versa applies in outer space. However it is not purely the density of the protons that produce such cold but the exquisite motion forming a rapid duplication of material and such duplication brings the contraction by removing space. Removing space is also removing heat that is separating material.

We have to accept that the coldest place in the solar system is in the very centre of the Sun because there the most number of protons sharing the least amount of space producing the coldest area that can allow therefore the hottest density of heat within the cold environment. Later I will show why the star is so extremely cold it freezes material together and outer space is over boiling with heat expanding into more space. We have to see what forms space and why space can be the absolute basic container through which gravity can relay the influence it carries.

We have to realise that whatever forms space has to be that same ingredient which also is the basic component that forms the lot of everything in the entire Universe. It is than which becomes more making everything seem more and it is also by removing that which reduces every aspect of the Universe. That which becomes more is what the Universe is built with and it is that which the Universe uses to form its entirety. When particles heat up the particles expand the space the particles hold to limit which the rising heat demands in relation to the heat rising. The particles claim more space when heated to preserve the cold that the material is protecting. The claim to more space produces more space but that in turn reduces more heat exaggeration. Such expanding brings about cooling. When particles heat or cool motion applies in some form. Regarding this fact we can claim that motion started at a point when the Universe was extremely hot and there was no space. However I have indicated that hot and cold are only factors with little specific or formal value in the Universe. By introducing motion space formed and the lack thereof produced friction that became heat that became space. That must be the way the Universe then started.

The application of gravity is the same as the condensing of space and bringing about heat by the compressing of space we apply in the way we go about tapping into the energy that nature provide. Internal and external combustion engines all rely on this application for harvesting motion by driving power. Compress space even today with a piston in a cylinder and then pump the compressed air into a container and such confining of space will increase the heat by the piston effort to reduce the space brought about in the container. The heat coming about inside the cylinder when being compressed has no relevance to particles colliding because all compressor cylinders cool down or become colder when that cold escape through the walls of the cylinder. As soon as the pumping stops the heat releases from the inside space. There is an immediate stopping of the increase of heat as soon as the pumping stops. The material inside the container forms a secondary form of material that comes about since the space reduces and the forming of space is in a turnabout. The compressing of the space inside brings about a rise in the heat levels within the container but apparently that no one in Newtonian circles can understand. By compressing the spin of the atom increase and the motion of the material removes additional heat from the ranks of the inside of the atom. Thus, when the spin of the atom increases it allows the cold within the atom to release the heat the atom holds into uncontrolled space.

This releasing of heat and unifying the released heat once again with space increases the levels of heat in the atmosphere of the containing cylinder. What that heat does is the heat that the material absorbed as material within the atom was captured as frozen heat because of the motion of spin to space that the atom holds remains in a frozen state under the guard of the spinning electron. But when this heat releases from the containing form that is the atom in being the biggest cosmic heat container the heat becomes in a frozen state through the motion within the atom. Forming a frozen substance by producing motion that is faster than the speed of light the heat is frozen by the spin of the electron. The spin of the electron brings motion and such motion reduces the heat to a frozen state which is the frozen state of heat we named material. Therefore one may not look at the material and judge the element state of form by its surrounding which is heat it surrounds its electron to the outer side of the containing spin.

Again we must look at the state of material in outer space and realize that the fact that hydrogen remains a gas and so does helium in outer space must serve as enough proof that outer space is hot, regardless of our interpretation of the temperature gauge telling us what we wish to hear. One must look at outer space and judge outer space from the findings only considering in the terms which outer space insists upon. If helium remains a gas it is hot. The removing of heat from the space that contained the heat makes the centre of the Earth cold. In our Universe we see it as being terribly hot because the heat then forms a separate substance but remains a form of material but that is because we see the heat and not the space derived from the separating of the heat.

The only reason why the space can seem to be hot is because the space is cold and in such a cold environment that rejects the heat within the atom. There the heat then must gather in a more concentrated state and space can collect heat because the particles hold concentrated heat in the space separating the particles. By removing such high concentration of heat from the space that use to be expanded heat, the space then must contradict the heat by being extremely cold. We look at the heat in the space, which by being in a liquid state should be by our standards considered as another form of material and find the surrounding heat in the space hot while the atomic material in space is extremely cold. The cold in the Earth centre causes the concentration of heat by space reducing, as all cold surfaces tend to do. But the numbers of protons contributes that reducing of space and the removing of heat captured by the material. If it was hot the space within the Earth would expand and explode but the space within the Earth where we think so much heat is concentrated is so much it does not expand therefore it must be cold within the solid parts. It is the motion of so many protons in such a little space that allow the heat to be contained as a liquid and the extravagant motion by the many protons in such a reduces area forms the ability to contain the heat as a liquid substance without allowing the expanding of the heat into gas or space. To gather and accumulate the space in a liquid means it became much colder when the space parted from what then is being a liquid. Finding the surroundings terribly cold will allow the heat to gather and not expand but when the surroundings is hot it will not tolerate more concentration of heat and thus it will expand to rid the balance of excess heat within space. The concentration or release of space with heat or space from heat is a direct contribution of the motion controlled by the space-time. The regard of the space-time providing the motion, which provides the cooling of the space, stipulates the conducting of heat in space or the release of heat to form space by means of seizing the occupied space.

Look at the Sun and see how the Sun turned the hydrogen it holds and which is captured in its atmosphere to a freezing cold liquid at 6500 K. Hydrogen is in a fluid state within the Sun and yet it is still colder than the hydrogen we find in outer space that is in a gas form in outer space. That must be because of the enormous motion of the particles within the confinement of the Sun. The Sun is without any doubt the coldest place in the solar system and that is because as the ferocious motion within the Sun. That is when the protons oversupply the removing of space to produce the cold that is so apparent in the heat levels that do not join outer space. By the reducing of space it can concentrate heat to a fluid state.

By producing the opposing cold that finally freezes the heat to a solid state we find that it is what matter is. The expanding of space is a way of duplicating space without reducing space and by duplicating in the form of expanding it becomes just the opposite to duplicating by motion therefore reducing space by halving space in time. That is what gravity does. By motion space duplicates and by space duplicating the material must be by dividing or halving. Halving the material, which is heat, is at the same time doubling the space, which is bringing about cooling. By doubling the space as the duplicating of material removes half the heat from a single space and distribute that same quantity of heat over a double amount of space it removes heat in space as well as by dismissing space and in that concentrating heat. Again it is apparent that in all the applying of gravity it is space that bites the dust. The density of the protons brings about space dense enough to harbour the heat in such quantities and visa versa applies in outer space.

We have to accept that the coldest place in the solar system is in the very centre of the Sun because there the most number of protons sharing the least amount of space producing the coldest area that such intense motion can allow therefore the excessive motion brings about the hottest density of heat within the cold environment. It is the duty of scientist to look far beyond the ordinary and find why the inner star will be so cold and as to why outer space will be so hot while being seemingly so utterly cold or hot in humanly applied standards. It is the duty of the professionals to find matters as they are and not as they would seem to look from a human vantage point. Later I will show in much better detail why the star is so extremely cold and outer space is over boiling with heat expanding into more space. We have to see what forms space and why space can be the absolute basic container through which gravity can relay the influence that it carries. We must come to realise that whatever it takes to form space it has to contain something that is the same ingredient, which also is the basic component that forms the lot of everything else in the entire Universe. When particles heat up the particles expand the space the particles hold to limit the heat rising.

The particles claim more space when heated to preserve the cold. The claim to more space produces more space and reduces more heat. Such expanding brings about cooling. When particles heat or cool motion applies in some form. Motion started at a point when the Universe was extremely hot and there was no space. By introducing motion space formed and the lack thereof produced friction that became heat that became space. It is natural and it is simple and above all it makes believable sense.

The application of gravity is that which condenses space by bringing about heat with the compressing of space. Compress space even today with a piston cylinder wall in an engine cylinder and then from that action pump the compressed air into a container and such confining of space will increase the heat by the piston effort to reduce the space brought about in the container. The heat coming about inside the cylinder has no relevance to particles colliding because all compressor cylinders cool down with time moving and not necessarily with the loss or release of particles. It is not only the discharging of air that will reduce the temperatures inside the container but the time flowing bringing motion about where the motion is not about particles escaping but heat escaping in the replacing of the heat density (not the density of the particles forming the material content within the container) but the space that compressed to heat will also bring about that the heat displaces through the container wall to the outside. After the pumping of air increased the heat in the cylinder which even can go to dangerous levels, the heat will reduce back to room temperature when further pumping seizes and that stops further air movement into the cylinder and such surging of pumping air is what brings about heat stabilizing.

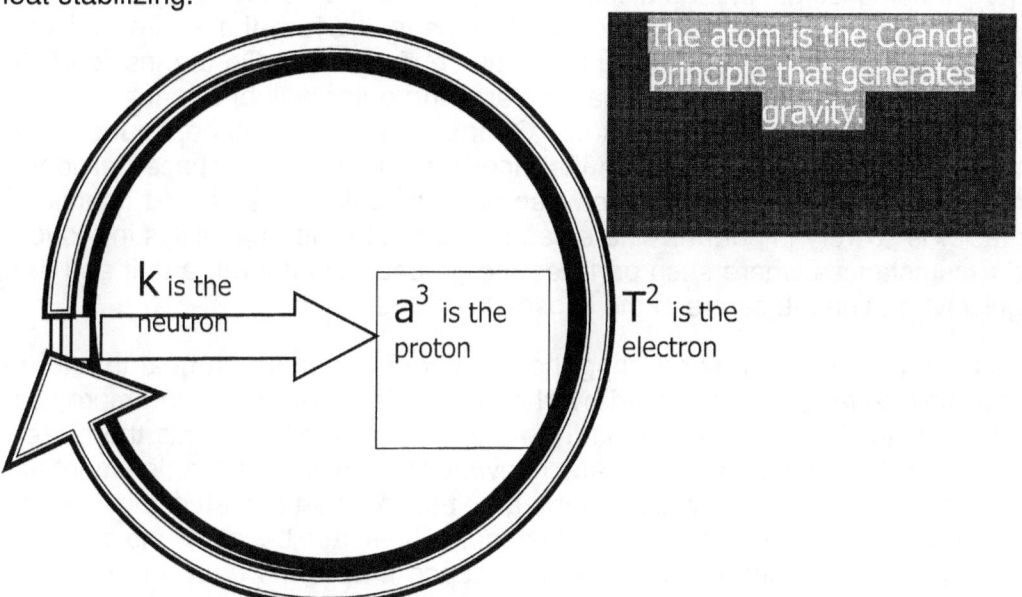

The atom is the Coanda principle that generates gravity.

k is the neutron

a^3 is the proton

T^2 is the electron

Mainstream physics ignored the clear connection completely, notwithstanding it being so very obvious. There is this far in their recognising of principles in natural physics not one single reference made to prove their appreciation of this matter. They are bent on particle colliding notwithstanding the much nonsense such an idea promotes. Atoms cannot touch simply because electrons are all negatively charges and will therefore repel one another long before there is any possibility of touching coming about. However in the case when particles do collide such collision forms an atomic thermo release and that action we call an exploding atomic bomb. What principle this argument about particles colliding ignores is that all atoms use negative charged electron forming the atomic limit on the outside forming a definite border to the boundaries of all atoms and in both electrons from different atoms are being negative charged.

In being negatively charged it means both will come out and one totally reject the other as much repel the other or cast the other away. The closer they come the more violent the rejecting will be and such rejecting is the production of heat that will turn to space. However that rejecting will increase the motion and the increased motion will reduce the space occupied. The electrons repel other negative charged sub atomic structures, which the electrons are that form the outer borders of all atoms. With all electrons highly negatively charged (being as negatively charged as any possibility will allow to match the utter extreme) such electrons couldn't touch. When the pumping of the air container commences the balance at first favours the forming of heat from the space coming in and being reduced in the containing size they are squeezed into is reducing the space from what it was on the outside. The space distribution inside then changes considerably and reduces a great deal compared to conditions outside the cylinder wall and with the decrease of the space distribution inside compared to conditions outside that space then becomes reduced and charges with excess heat on the inside.

The electrons will disallow any contact directly between atoms. No force can be big enough to enforce such touching. It is because of that contact rejection electrons bring about that science has to use an overload of neutral neutrons putting them in the atom nucleus to fake a complying of charges that will eventually lead to atom touching each other but that is through enticing a neutral stance which is enticing a positive overload for a short while. When the touching of electrons does take place the event is called a thermo nuclear reaction where heat is released in unmatchable quantities and the atoms in reaction dissolves into a liquid heat. The increase of heat by the distribution of particles in the space that is forming the connecting space still keeps particles separate. The heat rising is a separate issue that has nothing to do with contained particles colliding because why does it stop when pumping is seized. This ratio of heat reduction is time connected as much as it is motion dependent.

Motion reduces space by expansion as much as time contributes to space distribution by allowing the flow of heat. When the pumping stops the heat immediately starts the reducing thereof. Most important is the realising that every atom constitutes of two parts. In fact the entire Universe constitutes of the two parts, which I go about mentioning in this entire book. On the inside of the atom there is a circle formed by a rotating electron that contains the outer wall of the atom forming the sphere and holds material in contact with the protons. On the outside there is heat surrounding the inner material part within the sphere and distance the inner material from the space between it and the next atom. The electron forms the division between heat uncontained and heat contained. This is why the Roche factor is so very important. There can be friction between particles in reduced space under controlled circumstances where such particles are grouped together in a unit and as a unit elects a group singularity forming the centre of the chosen form of the unit.

The Universe separated heat from material by covering the exterior of material with heat that forms space. Some material became softer by uncontrolled overheating while others remained more solid by containing form through controlling the overheating. On the outside of all elements there are a layer that is the heat the element uses in relation to place relevancies between such an element and the rest of the cosmos. In the case where many atoms form a unit such as an aircraft coming in from outer space the space surrounding the craft becomes liquid heat as the space becomes more intense within the atoms combining as the structure in concentrated space that forms heat. In an aircraft coming in from outer space at altitudes that high there can be no particle in friction and even more so way up there in the atmosphere at the altitude where the cosmos meets the atmosphere just because the particles up there are so sparsely distributed in that part of the atmosphere. Above and beyond this lies the fact that all the so called air particles are very volatile and excitable by nature and they are known to turn the slightest heat into rapid motion thus establishing a scene where the particle that supposedly are in contact with the aircraft sheeting will move away from the hot incoming aircraft. The gasses will become more gasses when the heat levels surge.

If then and not for any other reason why there can be no friction then it is because the particles are highly volatile and exceptionally sensitive to heat. Airborne particles are prone to motion just because it is the airborne element nature to change heat into motion and the motion comes about from their sensitivity to duplicate. No particle in the air being part of the space we call air, which is in a free floating in that air can produce friction because of the volatile nature that those elements have. The craft's coming into the atmosphere produces a point where $a^3 = T^2k$ changes to $k^{-1} = T^2 / a^3$ (the explanation is forthcoming a little later on) The distance separating the incoming object from the Earth centre reduces rapidly therefore the object start to descend towards the centre of the Earth. We must also acknowledge the fact that there is one specific point of specific entry where this will occur more than before.

That point will rapidly increase the time factor where the incoming object crossed such a very visible border. By the reducing of distance k space a^3 will have to compromise in the relation of all the factors forming the equation since T^2 will very suddenly grow more acute. What happens is that the applying gravity reduces the space a^3 and the compromising factor comes about since the time factor T^2 moves back to a time where outer space was as dense back then as the density we now have within the atmosphere that then became the Earth atmosphere. It is outer space that remained denser than what the outer space currently is. I am now referring to a process that I introduce as this book unfolds which is by nature completely different to what is accepted by mainstream science (as you might have noticed in this short space of reading). That which I refer to as outer space back then was the same density as that which the Earth now supports but outer space in the meantime expanded while the motion that the material that forms the Earth structure provided, came about at a point just before the Earth established an atmosphere that grew through gravity and by the measure of the Earth gravity became separated from the atmosphere. While the gravity of the Earth contained the space surrounding the Earth in a much denser packed envelope the area not under the direct influence of the Earth governing gravity became more spacious.

The Earth contained its atmosphere and it relatively grew as much denser as the solar system developed into what it is today and outer space reduces its density. It is a matter of the kettle not being able to call the pot black. As the atmosphere released from what we think of as outer space that releasing from outer space made the atmosphere much denser in the space just above the Earth, which is using a reducing time factor. It is there that the applying gravity makes the Earth atmosphere more compact. That established the T^2 factor to be that more condensed when one compare in ratio the density with outer space. The density that was there at the time when the separation came about in outer space when such parting between the limits of the atmosphere and the limit of outer space separated and such separation allowed outer space as a separate object to move away. This parting brought a barrier that is in place between the Earth and the outer space and any object coming from outer space into the Earth's atmosphere will have to negotiate its entry by passing through that division.

The incoming object then would have to reduce the measure of the space the craft holds as the containing singularity set new standards applying to the incoming object with which the craft then needs to confirms its form and its status within the contained space of the Earth. The reducing will then suddenly no longer use space as the compatible factor but the focus will shift to the time factor that dictates to the space what the space can be. Such reducing comes from the switch there is in space – time where it was in outer space performing as being $k = a^3 / T^2$ to what it has to be within the Earth atmosphere $k^{-1} = T^2/a^3$. When the atmosphere grew apart from the outer space there are two ways of looking at the event. One can think that outer space expanded by the implication of the Hubble constant or that gravity withdrew the atmospheric space of the Earth at the time that the parting of space came about. But however you look at it there was a time when both outer space and the Earth's atmosphere shared equal density as we find it still applies on the moon and on Pluto. Then the Earth became dynamic and now they do not share any density at all. Things were overall more compact back then than at the present time and that included all things in the Universe. The space component is reducing the time component by compacting space to alter the space – time ratio.

This is portrayed by Kepler's formula $a^3 = T^2k$ It shows space as the density of space decreases. The Earth still compact space by reducing the volumetric confinement of space $T^2 = a^3/k$. This we call the atmosphere, as the atmosphere becomes denser towards the Soil of the Earth. There is a change in the time component. Most evident of this is when studying the pendulum. Just as we can see in the pendulum swinging, we can see that the swing reduces. Such reduction is because as the space diminishes every time the arm rocks from side to side. With this there is proof that in the developing atmospheric space of the Earth the ratios change from outer space. This is proved by the pendulum arms that Galileo's experiment used to show that the swinging pendulum indicates $k^{-1} = T^2/a^3$. Further more it proves that Galileo was correct after all and unnoticed by science Kepler helped Galileo prove Galileo's point. In this the net outcome establish Kepler as being correct and the Newtonian argument of friction brought on by gasses fall apart which is at that altitude where such friction supposedly should take place, the material in friction is not even present in the atmosphere.

Nevertheless science will stubbornly cling to the old theory with persistency that would warm any warring Field Commander's heart. In retrospect the following information is established in the past few pages: Every element stands in different regard to the heat surrounding the material, which makes us consider the material to be either a gas or a liquid or a solid. The material in every element there is as such is all three forms and not one of the forms in particular. It is the way under which the circumstances is presented that the element allows the heat to gather and accumulate as the surrounding heat occupying he surrounding space. Every particle is unique in the way it regards the heat to material ratio and how much heat it uses to form either the gas liquid or solid state. The fact of being a gas or liquid or solid is so much more complex but in time we will get to that explaining. If space a^3 declines then so must motion in relevance will have to compensate by reducing **k** and limiting T^2 because space a^3 must always be equal to motion T^2k

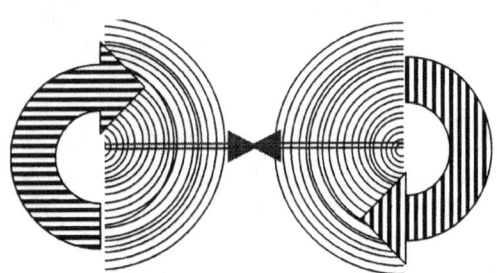

Space shifts as heat releases space and converts the Universe in one direction bringing about expanding into more space but less dense space. Remember how the heat came down from 10^{34} to 0 K at present? The density of heat in space surely diminished considerably since then to now. Gravity on the other hand is exchanging heat through the concentration by removing space bringing about space loss with increased density of particles and therefore heat concentration. In the centre of all spheres, which all stars are it is hot. In fact the heat in the centre of the star is the product of the space it concentrates to form heat and in that we can read the gravity the star can produce. The ability to secure heat by reducing space becomes the measure of the star. Momentum is the second form of gravity symbolised by Kepler, as **k.** The Big Bang is the result of heat expanding into the forming of space. Gravity, on the other hand is about concentrating space back to heat, and take recouped heat through to material, acting out a balance of expanding while contracting. This way gravity is applying the onset of the Big Crunch by destroying space while space is converting heat to material occupying space. The Big Crunch is coming about because the Universe is expanding where the two processes are one principle.

The relevancy there is between the aircraft and the Earth is precisely the relevancy we find between the proton and the electron in the atom. When heat released provides more space between the aircraft and the Earth the distance between the aircraft gravity relevancy and that which the Earth allocated to the aircraft by only providing Earth gravity without the adding of heat by the aircraft allows the aircraft to respond exactly as the electron does in the case of the Atom. The aircraft falls into the role of the electron, the atmosphere takes up the role the neutron has and the Earth retains the aircraft therefore the Earth has the role of the proton. When the aircraft has more heat than that which the Earth provide through the atmosphere the neutron position has to expand in order to facilitates the new dimensions which the additional heat that drives the aircraft provides. The ratio that the Earth initially holds becomes stretched as the aircraft suddenly finds more heat and therefore more motion that becomes more space between the allocated position and the position the aircraft claims by individual motion in addition to the motion the Earth has provided. It is all about heat released that generates motion and motion provides space differentiation.

Throughout the entire cosmos is leaning on the four pillars which is the phenomena and the four culminates in one which accommodates all the others an it is the Coanda principle that establish space which provide the gravity which allocates the motion a position within the space that forms. This very principle of electron / proton is in gravity. Gravity I shall prove is motion and not mass inspired. In fact of mass being a factor corrupts gravity by restricting motion. Gravity is anti mass and mass is anti gravity because the neutron is all motion with no mass. The gravity of motion is heat driven because it is heat that drives gravity. When an atom is in outer space it is surrounded by an atmosphere of 0 K. That puts a limit on the atom as far as structural differentiation goes.

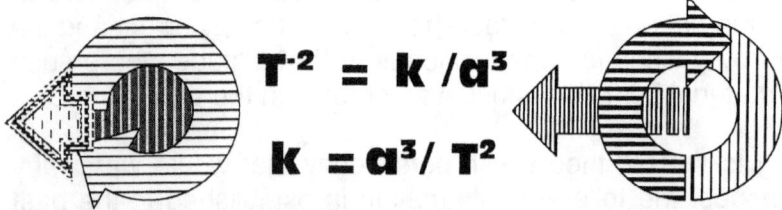

$$T^{-2} = k\,/a^3$$

$$k = a^3/\,T^2$$

By looking at the construction of the Coanda effect we find that the space takes the liquid as an extending of the space that increase the domain the space claims. The space is always the solid acting factor that holds the space a^3 in relation **k** to the liquid T^2 and the Coanda effect is the personifying of Kepler's formula stating that the space holds a direct value to the motion connected to the space $a^3 = k\,T^2$. When putting Kepler's formula into the correct connotation the Coanda principle is the materialising of Kepler's formula. The motion of the liquid limits the space by adding the motion to the claimed space.

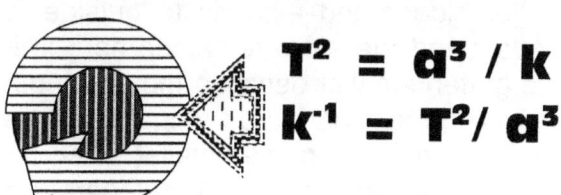

$$T^2 = a^3 / k$$
$$k^{-1} = T^2 / a^3$$

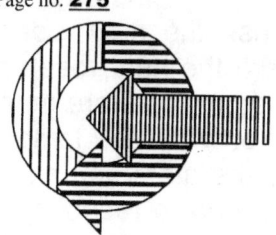

Take this formula into context from the liquids point and we find that there is quite another and opposing connotation to the same formula. From the liquid perspective we find that the liquid T^2 attaches **k** to the space a^3 by adding one more layer to the unit $k^{-1} = T^2 / a^3$ and the motion T^2 is an addition **k** to the space $= a^3$ by measure of $T^2 = a^3 / k$. By removing the extending that the motion T^2 of the liquid offers the space this reduces the space by the value of **k**
$T^2 = k / a^3$. **This means the liquid extends the boundary of the space while the space includes the liquid as the motion attaches to the space.**

The inside as well as the outside must be zero Kelvin because outer space has no other scale that being zero Kelvin. When the atom is on the Earth the relevancy goes that the atom is 40°C, which is 313 K.

If the outside is 40^0 of the atom is hot then the inside of the atom must be 40^0 cold. The heat on the outside must generate a condition on the inside, which opposes the condition on the outside. The inside is in relevancy or in division of the outside because there is a mass differentiation of 1836 times.

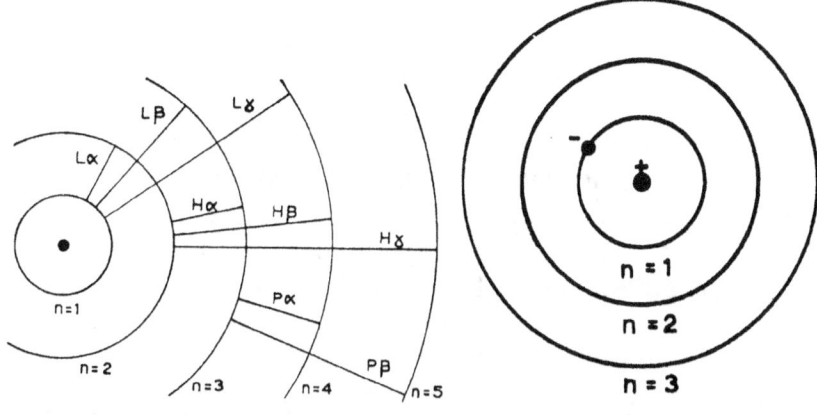

It is true that when concerning the Earth and outer space where there is little to choose from when comparing what changes is occurring in the atmosphere of a star. The Earth is as close to outer space as common civilized decency will allow.

When an atom finds a location in a minor star such as the Sun we are filled with surprise. It seems to us that Sun is very hot and with the Sun that hot the atom has little validity to stay intact. The atom should explode being in such a hot environment, and yet it is there and very it is much undeterred. Any atom we would heat to a temperature of 6500 ^{0} C as the Sun temperature is, will destruct with an enormous bang. Well it is good and well to say gravity keeps it from destroying but when saying that we should use that as a clue and not as an answer. It puts what is in the Sun in another class of structure and confinement.

If the temperature on the outside of the atom rises to 6500 ^{0} C, then the temperature on the inside should respond to what applies on the outside.

It is quite true that when temperatures rise the electron jumps a band. The electron moves apart from the proton as the circle widens. It is not the amount that the circle widens that should be of any interest to us but the total response. On Earth the electron ring would enlarge but at the same time the proton should equally respond by reducing. Place the atom in the circumstances we find in the Sun where the atom heats to 6500^0 C. On earth the atom would explode but in the Sun the atom remains well formed and very intact.

The atom does not explode because the atom does not get bigger and extends to outside its proportions. In such an event where the heat rose enormously and the atom remained as it is in outer space it would mean that the atom therefore must have gotten smaller because the enormous atmosphere kept the atom in tact. Yet with such temperature rising there has to be a change to the form the atom has and that means the atom shrunk in size. The proton became smaller when the temperature rose because the atom had to respond in some way to the rising of the temperature. Putting all this down to gravity is tiresomely attributed to laziness on the part of the human thinking capability because it proves how far we will go to restrain our ability to think. If gravity controls size by heat contribution then gravity has more to do with temperature than it has to do with what mass contributes.

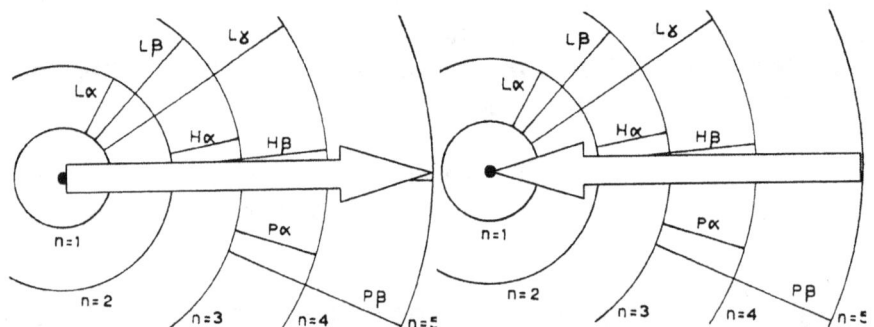

What we see as heat is relevancies because as the relevancy within the Sun changes the atom adapts to the changes. The atmosphere of the Sun becomes denser, which we see as being hotter and the containing becomes stronger. The atom has to reinvent it by adapting to the changes or different surroundings. In this manner the motion that the star provide which is so much more than what is the motion is we find in outer space that the hot / cold dynamic changes all together.

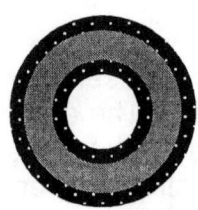

When an object is in outer space that object, encounters a specific relation with what we presume is space. This comes about by motion and through material volumetric size. The space the object encounter by moving through outer space puts a value of a ratio between the space it moved through and the space moving through which Kepler introduced as $a^3 = T^2/k$. That means there is a contact ratio between space containing and space contained by.

When the atom is in outer space the atom is surrounded by a temperature of zero Kelvin and that is because zero Kelvin is the presumably the coldest any temperature can get. Being zero Kelvin on the outside and with zero Kelvin being the coldest temperature there can, it would make the atom also zero Kelvin on the inside since there can be no colder than that. That would mean the entire atom is then zero Kelvin.

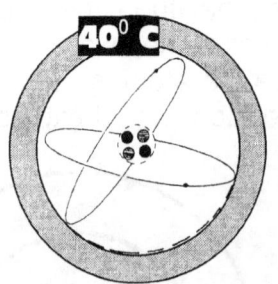

However applying motion reduces temperature and there is much motion going on inside the atom. That Kelvin produces the coldest there can be makes a statement. When the atom is 40^0 C the outside of inside of the atom because from the fact of what the series would represent and that proves that the

means the fact that zero little nonsense of such a the atom must affect the Balmer and the Lyman outside temperature of influence the inside atom. The normal

the atom does temperature of the summer's day

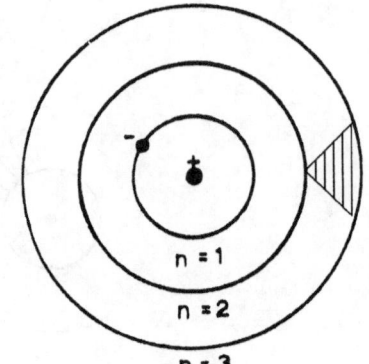

temperature on my farm is 40^0 C normally in the shade because at that temperature little in loony enough to venture outside the shade. We consider that the atom must be 40^0 C because that is what the daily temperature is outside the atom. We feel and experience the 40^0 and we presume that all around is suffering from the heat of 40^0 C.

We know that the action brings about a reaction and the actions leads to a response. If the atom heats on the outside by measure that it finds a need to reposition the electron by one band, then also the inside got smaller in relation to the growth by one band. We associate such repositioning with the heat on the outside that amplify or reduce. However the adding of heat brings on a faster flow of liquid, which results in higher motion and it is in the motion that we find the answer to the cosmic principle applying. In the cosmos there is no hot or cold. There is higher or less motion.

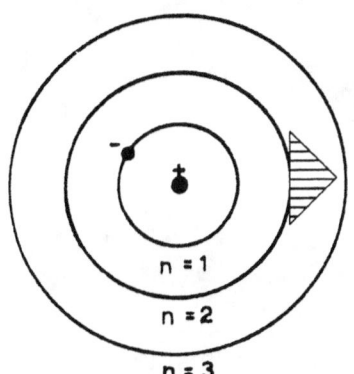

The

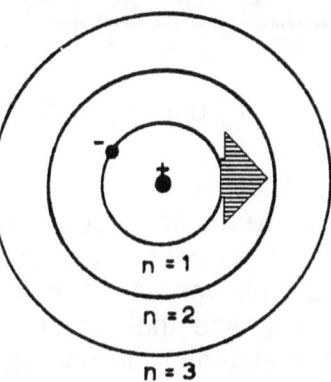

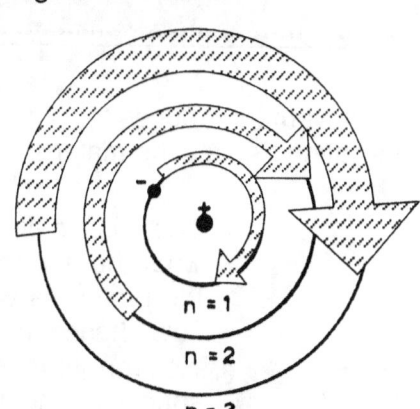

relocating of the electron new position where the jumps a band is done by implication of the Coanda From the Coanda effect that the liquid attach to using the formula $T^2 = a^3 /$ into a electron

effect. we know the solid **k** where

as space identify new boundaries by identifying the allocated boundary set by the liquid as $k^{-1} = T^2 / a^3$ where the space then forms the limit at $k = a^3 / T^2$. Every

time the motion of the solid by applying a new by extending the space However we must not lock on the motion that produces the new and the motion produces establishes a cold. The motion because the accelerated duplication duplication produces

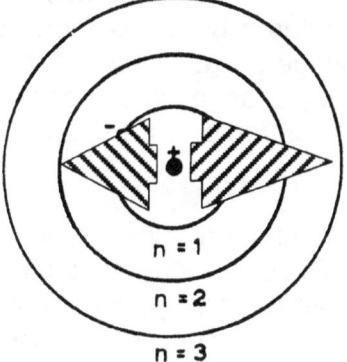

liquid intensifies the motion will attach to the relation, which alters the relation of the solid the solid has differently.

our focus on the heat but we must refocus intensify or weakens. It is the motion that electron allocation a heat that focus is on the motion brings on and accelerated cooling that brings

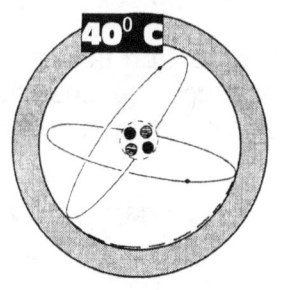

on a relevant cold within the atom.

If the temperature on the outside of the atom changes from zero Kelvin to 400 C it is not the temperature that changes but the atom is responding to higher motion. With the atom in outer space the atom is subject to lesser motion since the atom is

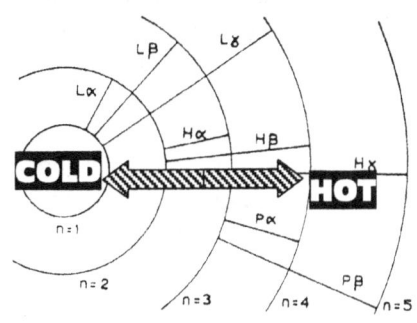

only in distinct and personal orbital motion in relation to the Sun . That is why the atom can be subjected to zero Kelvin. When the atom is within the boundaries of the Earth and circling around the Sun in a location set by the singularity of the Earth, the motion is distinctly more than what it would be if the atom were located in outer space.

The outside of the atom calls for a direct response to condition inside the atom since the outside can change very little if the inside does not respond in an opposing manner to what the outside

produce. In such a relevancy there are always three factors performing as gravity and in that is the Coanda effect in charge of committing the standards by applying the gravity or the motion in relation to the solid.

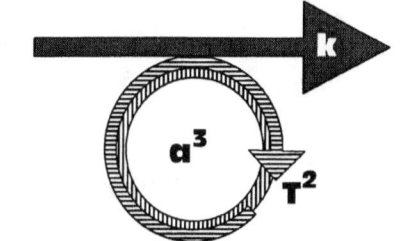

The material revolving through the space holding the material and allowing the material the privilege of motion is in the amount of material per time frame that makes contact with the space which serves time and that it encounters as the space duplicates its position it holds coming from the past through the present into the future

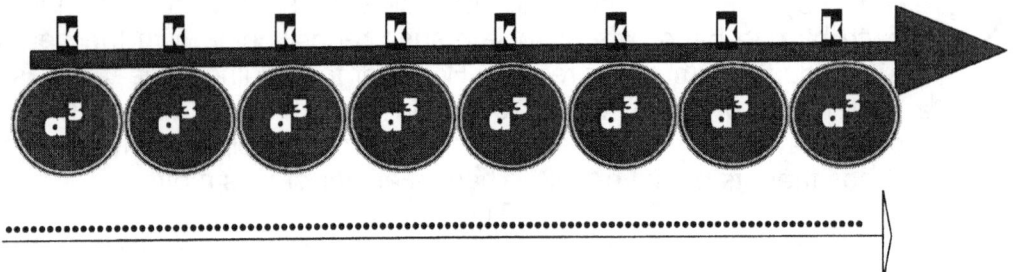

The movement reduces the size the material occupy by duplicating such vat amounts that the duplicating freezes the material into the oblivious.

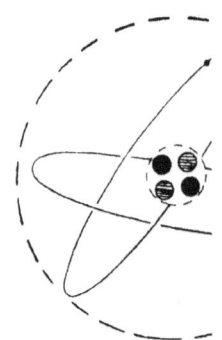

There is a definitive relevancy between the electron and the proton and that factor is what the neutron fills. The neutron is unrestricted gravity or liquid motion whereas the proton as well as the electron is very much restriction of motion of space-time flow, hence the mass. It is proposed that when the atom becomes hotter the electron jumps a band but that statement is not altogether the truth. The proton shrinks as much as the electron jumps a band just as much as the neutron fills the vacant space.

By jumping a band the space within the electron becomes more and the neutron fills that relevancy therefore the neutron becomes more. But if the neutron becomes more the neutron is there to bridge the gap between the electron and the proton and that will have it that the proton needs to respond just as much by becoming colder in the presence of the electron facing more heat. The heat is not the factor but the motion contributed by the heat is what brings about the larger jump in spin.

The neutron facing off the electron as well as the proton will respond on both sides of that which it influences because the response is that of bringing over more motion from the electron to the proton. One cannot gauge the electron behaviour without extending such behaviour to the reaction that the

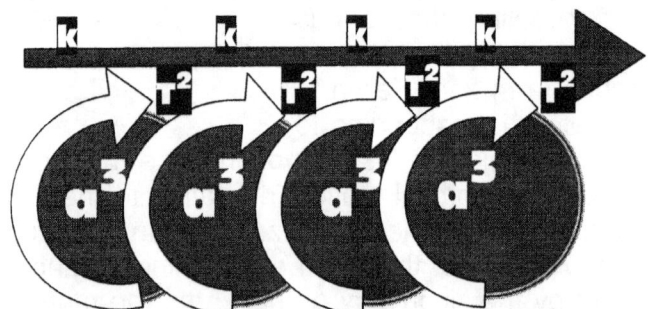

proton would have since the neutron fill the gap and also provide the response on both sides and the changes is what the neutron contributes by suffering the greater discrepancy in changes. However in the ratio or relevancy there will never be any change. The changes come in the form of an amplifying of the motion, which is a relation the space has with time.

When an object is in a location with little motion the duplication present a lot of heat because the distribution of the heat over the space in duplication has very little possibilities of spreading the overall heat over a wide area. The motion of something as small as the earth will confine the atoms into a relative hot area since the space in duplication does

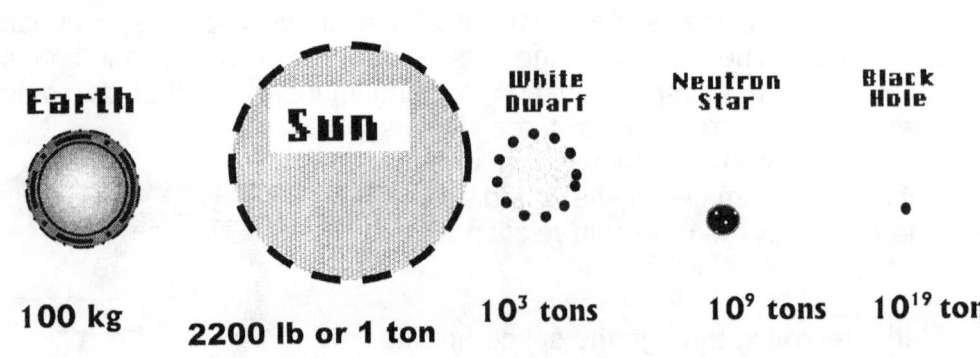

not reduce the extent of the heat by distributing the heat over much space.

In a structure with the size of the Sun the motion of space is enormous by the sure quantity of space in need of duplication. Shifting that volume of space needs duplication that is millions if not billions of

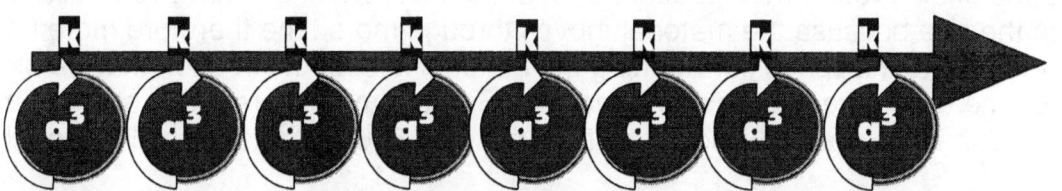

times more extensive than what the earth may produce. By duplicating such a vast area in a

period reduces the individual atom to a fraction of what the situation on earth would allow. The more the spin of the liquid is in relation to the solid state of space is reduces the space and extends the material in quantifiable measure many billion times over to what smaller stars are. It is not the space that holds the matter but it is the spin in relation to what the matter holds that puts the relevancy of hot and cold within the star. The more cold there is because of the mot\re liquid heat bringing about motion, the colder would the atomic material be and the higher the relative contracting gravity that the star produces. This we see in the admitting of Mainstream science confessing that the reducing of space produces an increase in mass and because mass is the frustration of material unable to move, it admits to the fact that mass in volumetric size has no influence on gravity.

The physics we encounter on Earth allow us to use a common and a constant, a fit all and an all-purpose because we find us captured by the Earth singularity. The Earth provides the space we may claim as well as the time in which such material duplication will take place. The earth does not provide conditions found in outer space and neither does the conditions found in the Sun be remotely compatible with the conditions the Earth prescribes. On Earth we find conditions little different from the conditions applying in outer space.

There is a certain ratio of heat to space that allow the material the motion to reduce heat to the extent that will grant the material a certain volumetric size in space-time. The material is hot because it is holding large quantities of heat in the structure of the atom. By moving lowly more space is duplicated less giving less heat being distributed over a smaller area covered by material. By the motion the conditions give a specific ratio of time that allow space to duplicate to that specific required ratio. Since outer space is as hot as time can be, there is no more expanding of singularity possible in outer space.

We call this dynamic speed or velocity, which is just another name for a motion in ratio with time. There is a volume (meters moving) in time Seconds flowing and that ratio produces the size of the object in relation to the time the object allow the ratio to be in contact with the time the object moves a distance. We also know by blowing over a body the "air" cools the body. That means the more "air" that the body is in contact with, the colder the body will get. To this argument there is a lot more and later in this book I return to the matter. However
It is a ratio that is coming about.

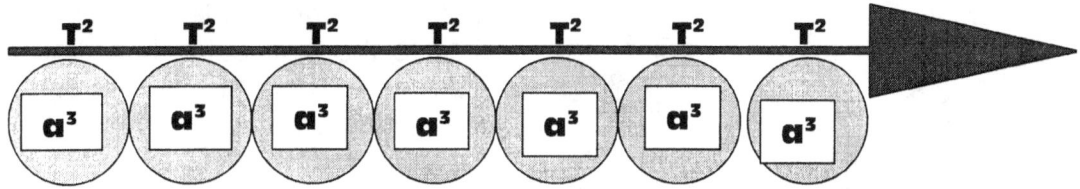

If the motion of the material is more it is more in contact with "air" which is not "air" or even "space" but it is time, the time (or space or air) effects more of the material since the same volume of material moves through more time. The time is a constant and therefore the material cannot increase the time but the ratio can produce more material (in contact with time) than moving at a slower speed. The material reduces in relation to the time it moves through and therefore the material shrinks allowing heat to flow from the material to the time aspect. This is the same, which we find when compressing air into a container used for storing compressed air.

That means the relevance between "space" which the material moves through or is in contact with, reduces the size of the material in ratio to the space it encounters. But we know that this effects the heat balance more than the size because the material moves through more time therefore more heat is transferred from the material to the space surrounding the material It is for this reason that we blow or radiators with fans. By the excessive motion of the massive Sun the material reduces allowing the material to become so cold that the space outside the atom, become 6500^0 C in a normal day on the Sun.

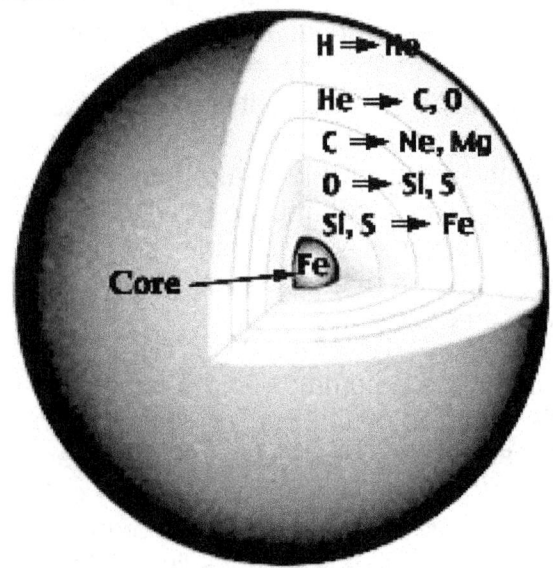

With the enormous container that the Sun is there is nowhere remotely the duplicating by motion going on anywhere else in the solar system. The outside layers are formed by elements known for their volatility, which is another term for duplication. Hydrogen and helium has very high ratio of interaction with heat and in that they have very high freezing temperatures. It is not by coincidence that the most mobility is on the outside and as the layers reduce space towards the inside we find the containing elements preserving space on the very inside.

The Sun is as enormous as it is because it freezes material in ratio as the motion shrinks the material to a fraction the size it holds on the lesser solar structures. By the massive duplicating of space-time it contain heat in vast quantities because it freezes hydrogen at 6500^0 C to a liquid. Deep inside the Sun it gets so cold that the restriction ion motion freezes hydrogen to other elements and this process is called fusion.

It is a case of the motion cooling the material and the cooling is shrinking the material while it is excelling the heat in response to the material cooling. That way gravity is all about motion and heat that contracts material as motion cools material in relation to the outside of the atom that has to rise because of the lowering of the coldness and size of the material. Gravity has very little to do with mass and has so much to do with motion and it is gravity in motion cooling down material that shrinks material to accommodate more dense heat on the outside of material which becomes prevalent within stars.

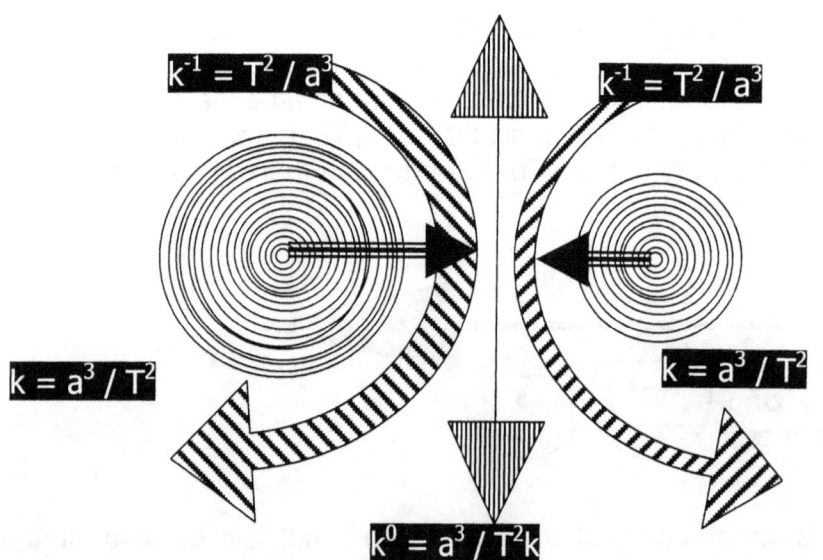

If the Sun is 18×10^6 on the inside of the Sun one have to take into account that the inside of the atom therefore would presumably be zero Kelvin. If that is not the case then the exercise is fruitless because the 18×10^6 will be meaningless. If there is a limit to the side being hot, then the side being hot must have another side being cold in order to give the being hot side any validity.

In that case we also have to mention that when the hot side is zero, the inside of the atom that is zero Kelvin from the other end must be minus 18×10^6 when gauged on the reaction side because for all actions here has to be equal reactions. Everything is in a relevancy and only nothing can be unattached. There is always a relevancy coming about as one part in the relevancy does thee expanding factor and the other part is producing the containing factor

As the rotation commences around a centre point there are changing relevancies as the one cosmic object orbits the other cosmic object ands there are forever cosmic objects orbiting another cosmic object. Even in the case of the Black Hole eternity is orbiting infinity as eternity melts back into infinity. Light is the only factor that responds directly to time by joining time as light is in the space sector $a^3 = 3^3 = 27$ and the motion part is $3\Pi^2 = 29.6$ giving a total displacement relevancy of 56 .6 within the star. Yet even in this case where light is overheated singularity there still is a cyclic flow of time T^2 through the four quarters, which is in the orbiting contexts of cosmic structures forming seasons. It is where singularity changes the dynamics in the relation **k** has with T^2 and a^3. The locations of positions and the allocations of positions of material in motion places the dynamics within motion in different concepts in the ratio they have to each other.

In the cosmos every aspect there is indicates an atom's behaviour pattern. Even the behavior witnessed when objects move shows expanding relating to contracting and the one forms the electron or expanding factor while the other form the proton or contracting factor and the two factors are joined by a liquid

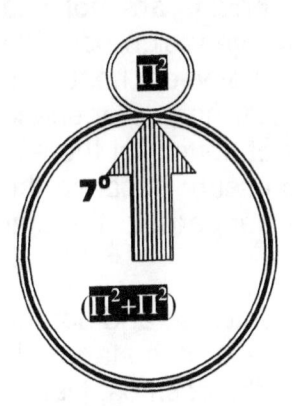
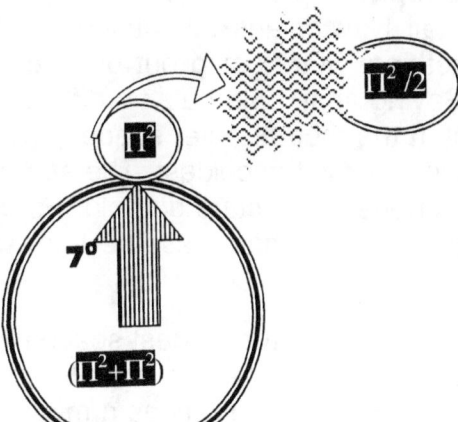
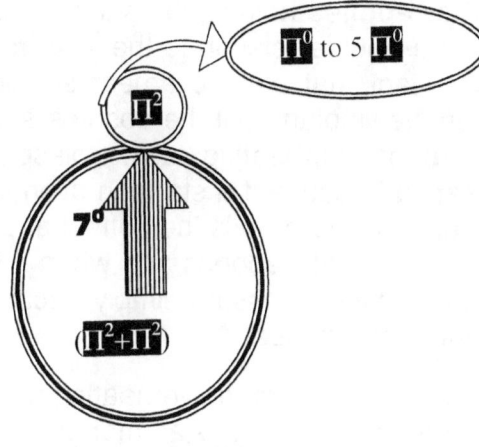

that holds pure gravitational and motion. The part that connects the

unrestricted
two factors form a neutron dynamic that is free flowing within the unit.
$a^3 = [T^2 = 7(3\Pi^2)] \ [k = \Pi^0]$.

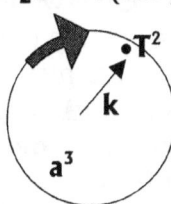

The Earth holds a specific size in relation to the motion of the individual object being the Aircraft $T^2 = 7(3\Pi^2)$. Since the craft is stationary the distance between the craft and the object is $k = \Pi^0$

a^3 $= [T^2 = 7(3\Pi^2)] \ [k = \Pi^0]$.
As the motion of the aircraft accelerate the Earth still holds a specific size in relation to the motion of the individual object being the Aircraft $T^2 = 7(3\Pi^2)$. However the connecting flexible link being the atmosphere which plays the role of the neutron has to extend in order to compromise to being $k = \Pi^0$

$a^3 = [T^2 = 7(3\Pi^2)] \ [k = 1 - 5\Pi^0]$.
The Earth holds a specific size in relation to the motion of the individual object being the Aircraft $T^2 = 7(3\Pi^2)$. Since the craft is now in motion the first beacon to arrive at in relation to singularity Π^2 extends the distance between the craft and the object to $k = 2\Pi^0$ then $k = 3\Pi^0$ and so on.

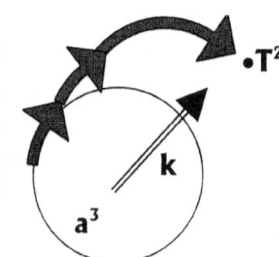

$a^3 = [T^2 = 7(3\Pi^2)] \ [k = \text{exceeding } 5\Pi^0]$.
As the motion of the aircraft further increases the Earth still holds a specific size in relation to the motion of the individual object being the aircraft $T^2 = 7(3\Pi^2)$. However the connecting flexible link being the atmosphere now has to extend to being the

most furthers that the neutron can possibly extend and such extending can compromise up to being $k = 5\Pi^0$

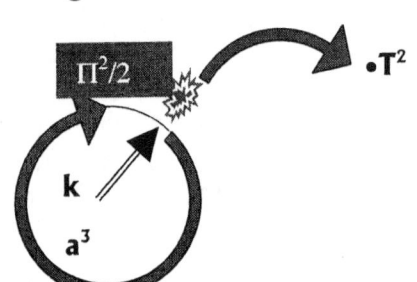

$a^3 = [T^2 = 7(3\Pi^2)]\ [k = 2\Pi^0]$.When the motion of the aircraft further increases the Earth expands beyond what the limits will allow being the Roche factor of $\Pi^2 / 2$. This puts a cap on the specific size in relation to the motion of the individual object being the aircraft $T^2 = 7(3\Pi^2)$. In this the neutron of the aircraft parts in a dimensional time value from the neutron the earth holds and the connecting flexible link being the atmosphere then cannot extend beyond what the neutron can possibly extend and such extending breaks down at $k = \Pi^2 / 2$. The entire issue is about the atom of whatever proportions containing heat in relation to a specific centre.

Stars can and stars do **overheat**, sometimes and the **Polar Regions** where **the Titius Bode matter-to-matter applies** holding the square matter (7+7) in relation to the square of space (10) and **other times** in a double relation to the **square of space** 10 to that of matter in a half square (7 /10 or 7/ 10). The fact that stars overheat should tell Newtonians something but Newtonians are not told because Newtonians tell the cosmos something. Stars going out of tune show every sign possible there are of overheating taking place. Saying that one has to differentiate between heat and overheating because if a star can overheat in the face of outer space then the star and outer space are polarised where on is the hottest and the other is the coldest. The star can get hotter but there is no evidence what happens and when it happens when stars go cold. Forget pressure because the star cannot have pressure simply because if the star has pressure then what happens to the star when the star overheat.

If the star expands when overheating a star represents the coldest space in the Universe and not the hottest space. Pressure is as man-made concept as mass is and as temperature is. It is standards set by life according to the practises life apply. It has no validity as a measure in the cosmos, just as mass has no measure of validity or as hot and cold has any validity as measurements. When a star overheat it means the star must be "under cold". The star expands and the star explodes which is all part of undeniable tell tale signs of overheating. That means the star cannot be as hot as the limit would allow otherwise the star being at an ultimate temperature could go no higher. Only when the star is far under a limit can there be any possibility of overheating. If it is gravity that contains the star then the gravity has a direct link with temperatures because when temperatures go array and increase the gravity becomes dysfunctional in containing the star as a container. **Heat and cold are relevant dynamics** forming **in appreciation of singularity. Heat and cold had to part when the first moment of cosmic development came about. Then at that moment eternity became the hottest since motion is with eternity and motion apply as heat keeping infinity the coldest which it then still is. Eternity expanded because the time provides motion and infinity remained frozen because infinity is still incapable of producing space.**

The Sun is the coldest place in the solar system and that is fact. Looking at evidence the Sun provides contradict everything science wishes to believe about cold and hot. Science wish to see the cosmos through the eyes of what fits the needs sustaining life on Earth and what benefits maintaining surroundings in support of life as one find on Earth whereas life has no part in the cosmos except for the speck of dust we call Earth. Looking at the cosmos impartial to life the evidence support another view. Every aspect in **the cosmos is the very opposite of what science believe** it is. The Sun is **not a ball of gas but** a **giant sea of**

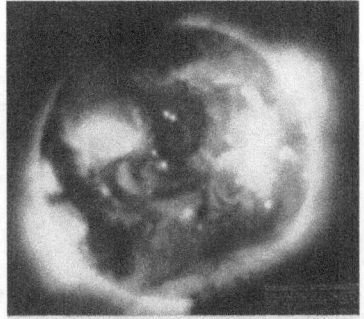

liquid, frozen **without any** form of **gas or air** in the interior. Having a liquid interior **the Sun has no pressure** but has the **very opposite of pressure** to which there is yet no name given. **The liquid comes from singularity freezing** space-time within the atmosphere of **the Sun,** and such is the case with all stars still in the shining phase. **Stars more developed than the Sun is frozen solid causing fusion.**

In **the picture to the left** we find not withstanding whatever name we attach to the **red liquid substance flowing from the Sun into** space and back to the Sun, **that liquid is heat** in a very direct form. **If outer space was the coldest place in the solar system** the heat **should** immediately **escape to outer space** and **not return to the Sun as** it clearly does. If **outer space were colder the heat would not return to the Sun.** If the liquid were heat the result would be a burst of space as one can see in the next picture.

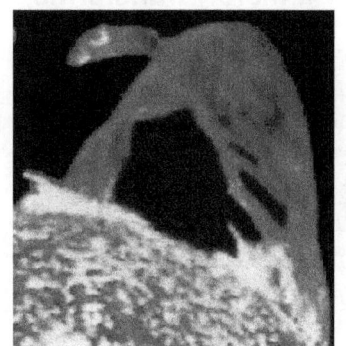

The expansion of the star going supernova shows an outpour of uncontrolled heat while the picture of the sin indicates very well controlled heat coming from a liquid that heat at the surface while the rest of the structure remained very cold. We tell the Sun that the Sun is hot but the Sun tells us back that the Sun is cold. If the Sun were hot the Sun would have exploded the way the picture to the right shows the star to go. It is a typical portraying of heat surging above controlling limits. If the star was under gravity control and gravity contained the heat levels of the star by preventing it to surge beyond control then gravity is all about heat management.

Where the ultimate destroying of stars come about one can clearly see the liquid bursting out from the centre. The story of stars exploding shows clear signs of heat management that went beyond the control of the gravity. It shows that balances in heat management was not that well managed and what was suppose to be cold became uncontrolled heat. The liquid we presume has to a result of the heat exploding but the heat coming from the Sun draws a picture of heat being in a liquid form within the Sun.

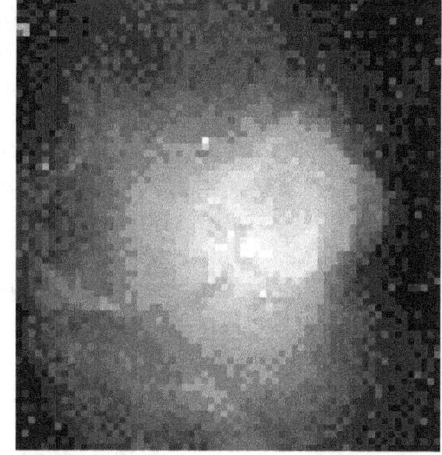

We associate the fact that it is a liquid with immense heat but in worst-case scenarios the liquid clearly turns to gas as the liquid expands. The gas is a higher ratio of heat than is liquid and by going to a gas from a liquid the gas takes the liquid into a higher state of heat. That is science and not picture fiction we draw on because we are culture programmed to believe what we wish to see.

It is exactly the same as the case with mass. We are retained on Earth and out retaining comes with a measure whereby we are retained. The retaining is in measure by a value we prescribe and

with it we prescribe mass as a value of such retaining. The fact that mass has no validity when we fall is top the Human mind of no significance because we can only appreciate the retaining when we can connect the retaining with a measure we associate such retaining with. We feel it is a hot day and we feel it is cold. When the Sun shines it is hot and when the Sun is absent it is cold. With the Sun we are hot and without the Sun we are cols. The Sun must be hot and outer space being the other pole must therefore be cold. We do not look at the cosmos and find direction in what the cosmos present but we find our needs and apply such needs to what the cosmos would need.

All elements forming matter in as much as the heat forming **an atom is** as much a **liquid as it is a gas and a solid. There is no hot as there is no cold. It's about storing energy in space or in heat, which is another Cosmic equal being opposing similarities.** Hot **and cold** are **relevancies brought about by singularity valuating space-time** and during **the Big Bang** the universe was **freezing cold** at **three billion degrees C**. It is the relation matter has with heat that provides the form the particle has at that moment.

The increasing or decreasing the heat will alter the form of the element. Therefore all elements forming **matter is as much a liquid or not than it is a solid or a gas. It is the space surrounding** the atom which provide the form the atom find its relativity to the rest of the atoms it share space with. **Hydrogen is as much a solid as tungsten is a gas depending on the heat in relation to the space matter is within.** Should **you reply** that it is **the gravity pulling the heat back to the Sun ,** then that **confirms** my theory that **gravity is all about collecting heat onto matter** with outer space being the hottest place. **It is the concentration of heat in space being relevant to form. When overheating a star turns its liquid to gas whereby it merely transforms it's interior to a relevancy it has from pre- to post- Big Bang.**

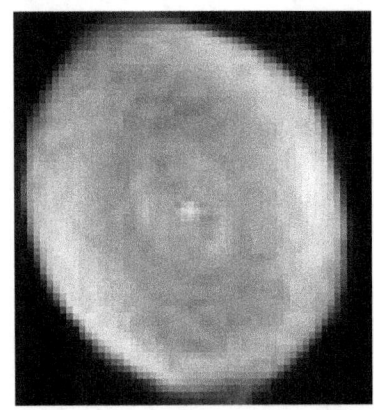

As star going supernova paints a picture of liquid going onto form gas. When liquid goes into a gas form it is a sign of heat rising. When gravity goes wrong and heart levels raise the indication resulting from that is that gravity is about containing heat. It puts heat in containing while it keeps cold inside the star by keeping heat out. When the heat comes in the star explode and that is what the picture of the supernova shouts out, if only Newtonian arrogance will learn from the picture and not tell the picture what Newtonians wishes to teach the picture to tell.

One thing we humans have to realise and that is that heat expands and expanding is produced where more material hold space and therefore heat is present while claiming more space. It is having what there is just more of it proportionally than the rest has or more than what was. If it contracts it is colder notwithstanding our human mentality and when it expands it is hotter notwithstanding our human perceptions. When it moves it is hot and when it does not move it is cold. The fact that the Sun contracts serve as proof that the Sun is cold. The Sun expands at the borders because the Sun heats at the edges but heating at the edges is little proof of massive heating going on inside. In order to have space reduce space has to go cold. Some space has to release some of what it has contained to heat or it has to contain heat which is released but that would be a cooling of heat.

We humans on Earth think that hydrogen is a liquid at -259^0 C but that only apply to the Earth. We say according to standards we established that -273^0 C is cold and we say 6500^0 C is hot and we

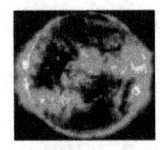

stipulate what we intend to confirm. This we do while ignoring all the signals from the Universe where all indications are that the very opposite applies in accordance with cosmic standards. When any object is in a form and heats the levels of space required to hols such heat will rise. The demand on more space to be filled by more material is a natural result from heat rising. When a star expands Newtonians draws the

conclusion that gravity got mad. The fact that more space becomes filled because of a demand that developed is a sure indication of heat becoming more. If the heat becomes more when things go array the heat was less when things were normal. When the star heat it shoots the content within the star in the direction of where that which fills the container at that point can expand no more. In outer space no heat levels can bring about any increase in space except the controlled expanding of the Big Bang exploding. The picture clearly shows the **heat in a liquid** flowing **from the Sun and back**

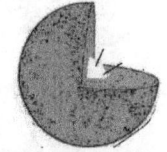

to the Sun. In the Sun the hydrogen holds enormous quantities of heat in a liquid at a temperature of 6500^0 C. When a star has its singularity secured the star is bitterly cold because it has heat in a liquid form flowing back to the point of singularity although we may regard the star to be rather on the hot side. The Sun   (fore instance) freeze hydrogen in a liquid form at 6500^0 C. If hydrogen remains a liquid at

6500^0 C, just think how cold it must be as the star's interior approaches the point of singularity. Therefore fusing protons comes from cold and not from heat or pressure. By allowing the singularity to overheat the star overheats and heat within the star flows from singularity to outer space freely. In such an event outer space is then colder than the star because the heat releases to outer space with no intention of returning whereas in the Sun it returns as soon as it leaves. There are two ways to reduce heat; one is to bring about expanding space, as the photographs clearly show. The second one is where heat will reduce when in motion by spin. When withholding or retarding motion matter will overheat. Gravity is the motion of unoccupied space through the dimensional transformation to occupied space. Motion and space therefore is the anti-, the opposite the negative to heat being the positive. With singularity overheating the expansion of the singularity drives heat into space, creating space to compensate overheating **That is a natural phenomenon.** The only reason why **heat will** rather **flow back** to the star than **escape to outer** space once the star released it into outer space is

if outer space presents more heat than does the star, because **heat always flows from hot to cold** no matter what influences may arise. **Outer space must hold more heat than does the star but the accumulation of space in relation to heat makes it seem colder bringing expanding of heat to become space. <u>Space and heat directly relates being the one form of the other</u>**.

The cosmos is all about **converting space to heat** which we see **as gravity** and **returning heat to space** as a **control mechanism** always **keeping** a very delicate **balance** which we see as **a star shining or being normal.** The purpose of the converting of space to heat is to supply the core where singularity is with heat. **It turns space to heat** sustaining matter but sometimes singularity overheats and then matter converts to heat allowing heat to convert to space. That we call many names amongst others exploding into super nova. Whatever the names used is less important because the **process rests on space and heat interacting to form energy.**

That was what **the Big Bang** was and **the Hubble Constant** is all about where **matter converts heat to space.** I show that **space and heat is the very same thing** and there **is no such a thing as pressure** but releasing **heat produces space** and **concentrating heat reduces space** with the two interacting on singularity demand setting time to space with time being the spin or motion of heat in space. **Heat and space form the second singularity** caused by the **fragmenting of singularity to compensate overheating during the pre-** Big Bang matter forming era. That is what we see as **light and space,** which again is the **same thing and is fragmented singularity forming radiation and heat, where the star re-transfers heat back to space due to an overload.** This comes about through the overheating of singularity (7+7)/10 (top) or layer overheating 10 / 7 (bottom) The fact that stars overheat is evident throughout the universe and as such should not be surprising. The reason why they overeat is very simple and very surprising. When I make controversial comments nobody finds any reason to listen to me. Everyone finds an incoherent novice trying to make sense of some incompetent view that strays from the accepted. Nobody takes the accepted and compare that with the in views I have. In all this they do not believe me but moreover is it that they do not believe their eyes. What I say they can see but what they see is not what they believe.

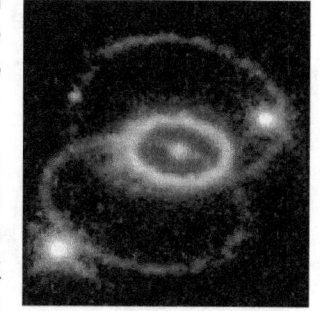

When gravity can no longer contain heat by motion that reduce space, which will cool heat, the motion that should contain the heat is unable to do so and the heat releases due to a lack of sufficient motion to contain the heat. When the material in the star is spinning the star too slow to keep the heat in check the heat will release from the inside of the atom and when on the outside the heat then is comparable with outer space. Being part of outer space the heat will expand to what outer space would allow. The motion that singularity demand will keep the flow of time acting in a manner to sustain the law of singularity by following Π as the indicator. That is the curvature of space-time after all. But the motion will apply to the measure of Π while heat is released from the cold star to the hot outer space. The motion inside the star condenses the space by retaining the heat. Motion cools the star and when there are insufficient motion to contain the material as cooled heat in the space the star reserves the space will expand and increase the space. This when the star heats. Gravity cools the star by preserving the heat through motion and motion comes by measure of duplication as well as containing.

I do realise after everything I said about the sun being cold and outer space being hot every member of mainstream science would gather bricks to stone me because to their thinking I have finally gone mad! Hot and cold is not found in a human laboratory but it is a factor in the cosmos. One thing I wish to bring to your attention is the fact about science: above and beyond most other things is that **the Universe was not created with man in mind and was not created to serve man or even to hold life. Life is an additional option only found on earth notwithstanding mainstream science's popular mythology. Life as such has no significance whatsoever except for a very short time period on earth.** Science can't put life in the centre of the Universe because life plays no part in the Universe. Life fits nowhere but on Earth and then life does not even fit everywhere on Earth. The Universe was never created to hold life but life was created to survive on earth for a very brief time in cosmic history. After the earth served its time as a planet, life with the earth will vanish from the cosmos. If the cosmos disappears life would follow but if the earth with life disappeared the cosmos will go on unchanged. That is the importance of life.

That is the reason why so much "mass" would fit into so littler "space" by such a lot of material, and even moreover is the fact that the smaller the star gets the more material can fit into so much less space while the temperatures get so much higher. Gravity is motion that reduces heat which brings on cold which reduces material size which compacts space into more dense heat that multiply the gravity or motion of the relation there is to "space" or one actually should call it time and space which we actually call matter.

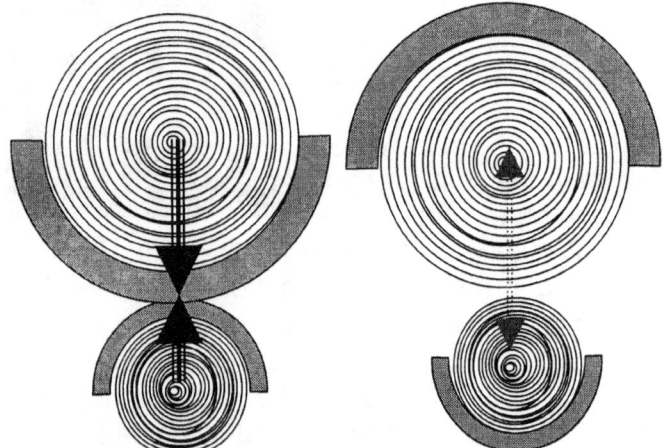

Once upon a time a very long time ago everything we see sprang from one point. Since then all points are a precise duplication of such a point where one forma the expanding factor and the other forms the contracting factor. Both factors are equal since both is the same but one appreciate space to the benefit of the unit by expanding while the other part is conserving the space by contraction also to the benefit of the unit. Since the unit is equal in the value notwithstanding that the unit is served by different factors, the factors hold the equal value is in all aspects of the rotational gravity. By contradiction of the two sides falls to one the factors that form rotation apply to both. Each one of the factors is covering one side of the Universe they form since both have it in goal to preserve and maintain singularity.

So often Newtonians talk about pressure within stars and heat pressuring to commit to fusion. A Star that is under pressure is a star that is destructed. For that they have a fancy name. They call it a new star. Can you believe it that after the star has come to pass and blew up like a cherry cracker on New Years Eve, they call that star new? In the star there is supposedly pressure and with the enormous pressure the star pushes elements into fusion ...and best of all is that they walk around with the doctorates in op physics! Let's have a close up in the process that applies when the air or pneumatic compressor is pumped with air.

$$R^3 / T^2 = R^3 / T^2$$

When pumping a cylinder with air the air inside heats up. The scientific explanation is that the molecules bumping each other to the extent that friction must occur because if not how does the heat come in place cause this. It is hydrogen, oxygen, nitrogen and a bit of helium that is pumped. It is not copper vapour tinted with iron and tungsten. The so-called gasses which is extremely volatile and very much a gas, finds the air inside the cylinder so cramped they collide.

This is rubbish, as the next day the container is cold. What made the molecules calm down because through all evidence, the air is still there and the container wall is cold. Pumping the container will increase the heat levels inside the container. The inside gets hotter but we are taught at school level that heat will flow from hot to colder areas. There is another way of thinking about this issue, which might seem more accurate in the final analyses. When any object is heated it expands and when the object is cooled it shrinks or that is what those carrying the flame of knowledge tell us. When we pump air into the compressor the air gets more inside the cylinder. The compressor gets hot while the air gets more. The air gets more while the size of the container remains the same. Seen in another

way we can think of the air while the compressor is getting compressor is containing more is at a constant.

$$R^3 / T^2 \simeq R^3 / T^2$$

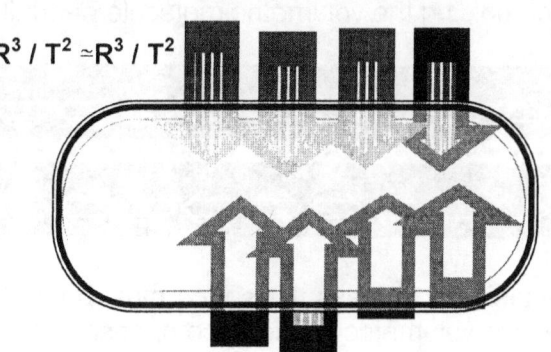

remaining even smaller. The while the air level

The heat on the outside of the the same value as the heat on the cylinder. Then by pumping the air the molecules take with the heat space time) they contain. Inside the relation to heat gets more because container remains the same except

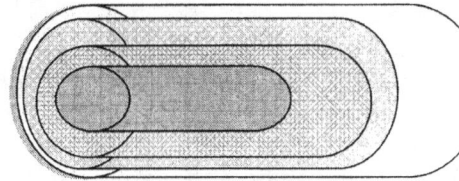

cylinder is at first inside of the into the cylinder, (unoccupied-container the the volume of the

that the container walls get hotter that holds the air in place. Because the walls get hot we may assume the walls try to stretch because by heating the walls should expand as it gets hot inside and therefore it will force the walls to expand. By this token it is clear that as much as the container is filling the container at the same time is also shrinking. Because it is also the size of the compressor that shrinks as much as it is t6he content growing more the space outside the cylinder has to accommodate the increase in flow of heat coming through the compressor walls because the overall practise of science is that nature rules by equilibrium.

As the air becomes more the walls of the cylinder will reduce by the same token. There is more air connecting with the cylinder wall and therefore there is less cylinder wall with which the air can

$$R^3 / T^2 < R^3 / T^2$$

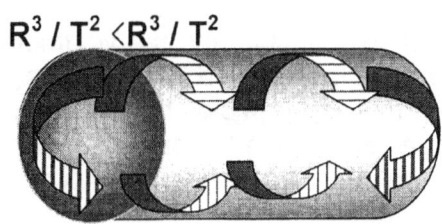

connect. The flow of air inside the container encounters less of the cylinder wall and more space and in that the truth is about cosmology. The air that came in brought with it the same volume as heat as what it had related to per volumetric ratio as was applying in the atmospheric space when the air was outside.

The volume of air expanded but when anything expands it gets hotter. There is no evidence of anything expanding without increasing heat and even in the spectroscopy we have evidence of just that. It seems the bouncing has the increase in heat except that when gasses increase in volumetric capacity the gasses become volatile. If that was the case then there was more heat within them container

$$R^3 / T^2 = R^3 / T^2$$

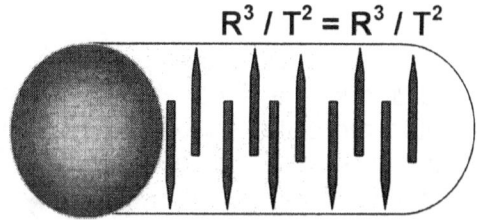

that the air brought in and since the container got smaller the space held more heat per measure of atoms than was the case when the air was outside the container.

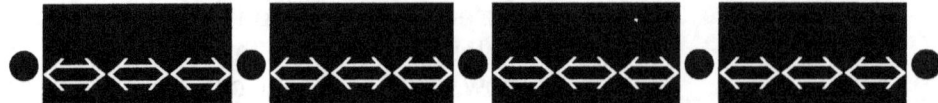

With the air increasing, by the very same ratio will the size of the cylinder keep reducing. That is why the compressor walls get hot. It cannot stand the reducing and ties to grow and expand, as it should while the air remains the same volume. From the container side there is no growth in the sir volume but there is a decline in the wall size of the container and that is why the container tries to expand the shrinking walls by allowing heat to try and expand the heating walls. The molecules are then more to the inside than the outside, the heat containing them, is also more on the inside than the outside (bigger ratio inside than outside).

What we find taking place in the wall of the container that is shrinking is the same that is taking place when concerning the position of the space not filled by material The space is becoming less that is holding the material that is becoming more. If the material per volumetric molecule is taking up more space then the space holding the volumetric molecule per unit is getting less.

The more the molecules are the less the space must be that the molecules claim and the more the molecules has to reduce volumetric space to compensate for becoming more. Then the same applies in relation to the space parting the molecules as the space in ratio also have to reduce in order to accommodate more space claimed by the molecules as well as space pumped in that was accompanying the increasing number of molecules entering.

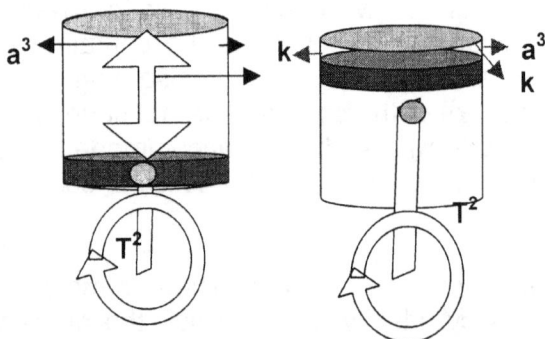

However we find the same process in the internal combustion engine with the only exception in that the process is put in reverse. Notwithstanding the different application the end result remains the same. In the case of the Diesel oil engine we find spontaneous combustion occurring when the process becomes at its peak of compression. However there is not a pumping of air but normal airflow into the container. At a point an intake valve ends all further airflow. Then the piston moves up and the piston reduces the space.

This time it is the space of the container that becomes less by motion reducing and the volumetric reduction of the space. In this the heat level rises to a point where the air gets so hot it makes oil combustible. The volumetric space reduced and in that the particles became more. With the increase in the number of particles per space available the space available between the particles also became more in ratio. What becomes very clear is that the reducing of space brings about an increase in temperature.

The very opposite is also true and we use that principle for cooling in everyday life. By blowing with a fan over a surface reduces the temperature of that surface. By making the space available more the space between the particles also becomes more. Then the space being more reduces the heat level surrounding the object, which is cooled. There is air blowing over a surface normally not moving and therefore by blowing over a surface that is not moving one gives the surface not moving the opportunity to be in a position that it can move in. In that way we enlarge the surface that is not moving by duplicating the surface not moving as the surface finds a location where it enjoys a larger ratio with the space it does no occupy in the same period of time.

Past going onto present and becoming the future
We have two persons standing still. The one is a thinker and the other is an accomplished and distinguished but sincere Newtonian scientist and which one of the two is which, that is for you to decide… The problem we investigate is how does both come from the past move through the current and leave for the future. The defining characteristic about time is that as time moves on the position of every object changes in relation to future and past positions.

We find that in any given area there is a ratio of Matter filling space and
In this ratio is built in another ratio of Matter holding time.
Matter determines THE RELEVANCY OF Matter to space during time.
Space holds heat in A RELATION OF SPACE-TIME unoccupied- occupied-, densified and singularity. Motion or moving by time or otherwise is the most complex issue there ever can be. Time relocates the structure by breakind down the entire structure as to relocate the entire structure and re assemble the entire structure to the previous spacifications and by perfect duplication.

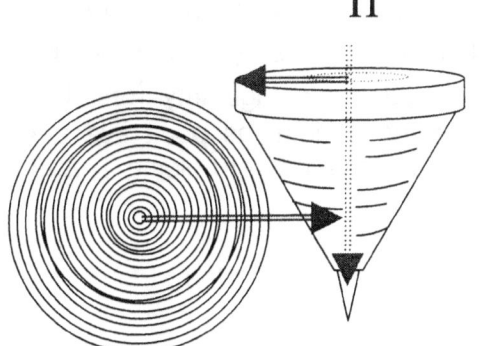

Π

The position of the following instant neutralizes the previous position as it takes the place of the previous position.

In order to understand this concept it will be best to return and see how space and time started. The location where it all started is still present in the entire Universe in use today. Fortunately we do not have to move back that far but investigate how the ordinary top is enabled by motion of rotation to stand erect. In the centre is a point that was there before time began. Time evoked the point back then as time still evokes the point in the present.

The point is so small it holds all points in one position. All for points are there and are rotating but by rotating from point 1 to point the point number 2, as it leaves 1 it lands on three because from there it moves to for which is also 3. All the points rotating are on the very same point. The point was eternally rotating and the rotation was there but the points became undefined and blurred because they were allocated to the same position. In such a simple concept as motion there are so many relevancies that has to establish new relevancies before relocation my motion can take place.

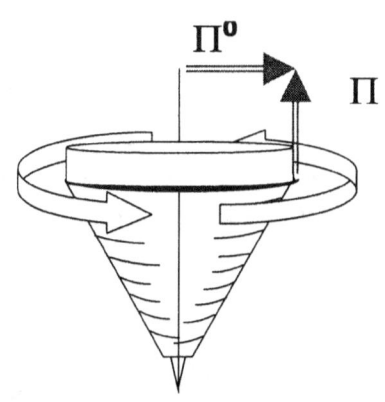

Π⁰   Π

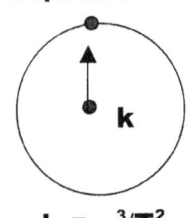 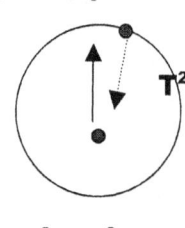 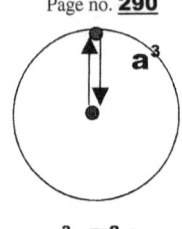 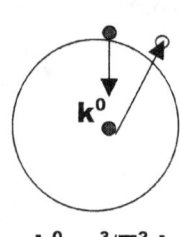

$$k = a^3/T^2 \qquad T^2 = a^3/k \qquad a^3 = T^2 k \qquad k^0 = a^3/T^2 k$$

$k = a^3 / T^2$ We have the fact that in the moving of space-time brings a new identifiable location for space to centre.

$T^2 = a^3 / k$ The motion will establish such a centre

$a^3 = T^2 k$ The space provides the motion to continue into the future while the space fills the one side of the Universe holding a position in eternity.

$k^0 = a^3 / T^2 k$ Singularity establishes and relocates space-time successfully by completing the motion.
Simple wasn't it? Let's run through the process once more and find the simple matter of motion in time.

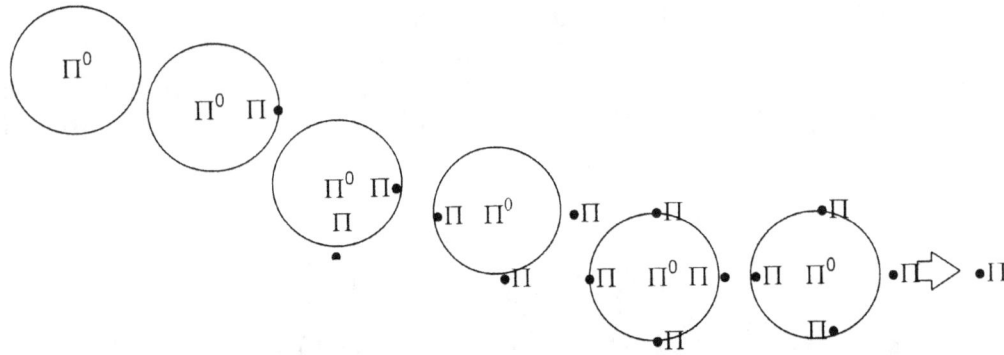

Singularity shifts from Π^0 to Π, which is a spot forming a dot. In our simplistic Newtonian way of thinking is that a spontaneous sphere formed as the spot expanded into a dot. Beware, all is not that simple because we have then small matter of **$k = a^3 / T^2$** to deal with. Every spot has to find a position in accordance with rotation as well as a position with relocation. The same dot that was 1 became 2 because it was relocated and not reinvented. Then the dot was allocated a position in position three by reinventing 3 as well as relocating 2. This became **$T^2 = a^3 / k$**. At the very same instant **$T^2 = a^3 / k$** did not disappear because what once is in the Universe is always in the Universe. As the motion took the dot to 4 then 4 became the new 1 because motion took singularity from Π^0 to Π where Π^0 was placed into a new allocated position by establishing a point as point five and relocated 4 as 1. Simple is it not. Try do that to every point that has a possibility of holding 1, 2, 3, and 4 in one position where all share the same position.

All this is rue because **$k^0 = a^3 / T^2 k$** singularity positions pace in time by circling the straight line and repositioning the allocated spot.

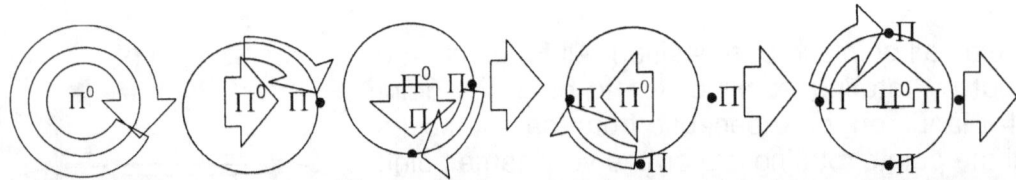

This is the prominence we find in the Lagrangian system using 5 points in the system where singularity forms four plus one. The motion in time in eternity is in a direction, which we might call 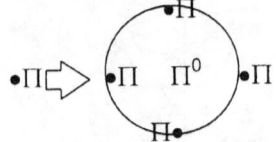 progressive but also there is another relocation of the dot taking k from k_1 to k_2 that will form T^2. In all of this it is vital to see that there is the rotation as well as the linear and that forms the allocation of space where material is the time delay caused by locating the position of distribution and not being able to remove the previous allocated positions quick enough. Material is the time delay of heat distributed.

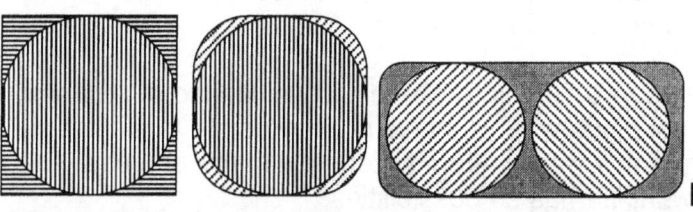

In the cosmos we have space filled with material filling space not filled with material while both are filled with heat. By moving the material filled heat faster than the cosmos relate the filled heat with unfilled heat a specific such action will surpass limits of a specific ratio and that ratio we call time. By blowing air over the hot surface that is not moving we are relating that area with a larger

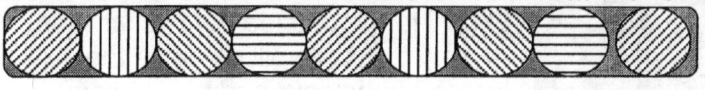

unfilled space and therefore without moving the filled space becomes larger in relation to the increase in unfilled space. With the filled space becoming larger the heat within the filled space becomes distributed through a larger area because the relevancy of the filled space has increased the filled space in size by matching the filled space to a greater ratio in unfilled space. That means by moving the air we are increasing the size of the material and by increasing the size of the material the material has to duplicate more often and by duplicating more often the material is shrinking in size.

With this information fresh in mind let's return to our compressor cylinder filled with air.

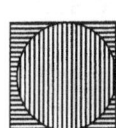

The cylinder had an initial size to begin with. While expanding through the pumping the inside of the smaller as a result of the pumping of air into the Newtonians say the molecules are colliding and friction brings about the heat. Then why is the time because the particles doing the bumping is still doing the were doing the bumping in the first place. Yet the heat does has to leave the cylinder to get the heat to subside.

the air was cylinder became cylinder. The bumping and that cylinder cooling with bumping if they subside and no air

As the space reduced the air got more and as the space reduced the material became smaller and

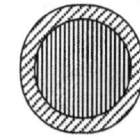 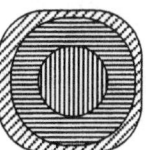 

with the material becoming smaller the material had to dump heat from the inside to the outside. Therefore not only did the air not filled with space compress and heated but the heat inside the atom had to disperse some of the heat to decline and dispense of some filling because it had to

reduce the initial size it had.

The process just described relies on pumping, on pressure, on retaining by an outside wall, by confining through deliberate replacing of material, which is confined into a cylinder that offers more confining. Most important is that the entire process relies on life and if not for life intervening in cosmos affaires none of this would be possible. So how does this comply with conditions in side a star? Well it does not comply even by a stretch of the imagination and only a Newtonian that is prepared to forsake logic in favour of madness and forces of unknown origins can see any connection between the star and the cylinder having pressure.

Looking at the Sun as a cosmic object I do not see any retaining cylinder wall and therefore there is no material seeking to find a way out. There is no pump putting material against the flow of nature into the container. There is no forceful relocation of material from one side to another side and there is no escaping from what is unnatural circumstance. All there factors contribute to what makes me not accept the view of pressure inside a star.

There is no bursting of what is inside to what wishes to be outside. There is no evidence of retaining what is inside. It is a round structure and therefore it holds what is inside in accordance to singularity applying. There is no possibility of life intervening in any way or life interfering with the process. The scope of cosmos affairs just is limitlessly beyond what life has as possibilities.

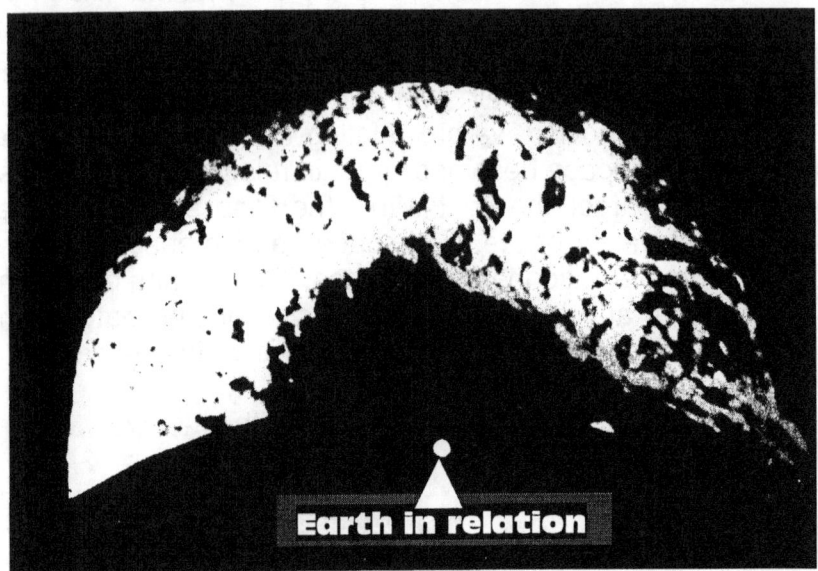

Earth in relation

Yes we do see what is inside trying to spill to the outside but it is far more evident that the spilling out is the forceful behaviour and the retaining is what comes naturally. There is no deliberate escaping from the pressures within but when released that which was inside flows back as a natural reaction and defies the whole idea of unnatural pressures building up inside of the retainer. There is no comparing the cylinder of pneumatic principles to the star that holds liquid and not gasses inside that star. There is no evidence of any gas although the flow of photons emitting light rays is by some imagination some part of a gas.

Inside the star the movement of all atoms combine in motion that establishes a centre governing as a principal all conditions applying in the star. The rotation of the atoms forms a synchronised motion that establishes the line, which parts infinity from eternity. The motion confine the material to the star but it is because the material elects a principal to confine the conditions to a status which all material inside the star agree on and benefits by the conditions of space-time we find in the star. As the conditions serves the star gravity will come about and gravity sets freezing conditions within the star.

 What happens inside the cylinder is the part that is compatible with what is applying in a star when we exclude the pumping, the pressure and the container idea. Lets go back to the fan blowing wind over an area in need of cooling.

When heating the object increases its initial size from normal to becoming larger. The heat increases the size the object has to a larger ratio than what applied before. To cool the object an increase in the ratio is needed on the airside to keep equilibrium and to bring in cooling even bigger ratio is required. By increasing the air we are decreasing the material and by increasing the heat we reduce the material by progressively anticipating more duplication as the ratio of space to material is increased by more space duplicating more material. The heat has to increase the size of the object within in order to match the ratio set by time on the outside. At this point I think it worthwhile to remind the reader that during the Big Bang the lot outside seemed hot and today the lot seems a lot cooler. I say it seems because it is not truthful. The heat has to increase the size of the object in relation to the match it has to find in the space it is within. With the heat coming into the object the relation that the object has with the heat or air outside makes the object that many times bigger, because the ratio in the heat balance is disturbed. If we blow air over the object we increase the size of the object by allowing the surface of the object to make contact with much more air in the same period of time, which will bring the size of the object back to the normal ratio it was before, because in relation and considering the contact with air the object expanded by the motion that increases the amount of air being in contact with the object. In the normal flow of time the object has a heat to space relation set by the time the dictates. Then we go and increase the heat of the object and in that event we actually increase the size the body has in relation to the heat in the air. By blowing air over the body we increase the air and therefore we increase the size of the body during the same period of time. There is now a dispensation of many times more air where the body is carrying more heat and making more contact with the surface whereby it is contacting heat or air which brings the equilibrium back to normal what ever normal then is. There was a body size ratio and by applying heat the balance shifted to the reducing of the body size in relation to the heat. The body then had to expand in heat because the body was to small to incorporate that larger heat. Then by blowing the air over the body it increases the size of the body and heating the body decreases the size of the body in comparison with the air it comes in contact with. The body is either expanding or the body is reducing and the balance in heat places the body in relation to either gravity cooling by contraction or by expanding by overheating. The very same principle applies in the sound barrier.

The gravity motion of the Earth is 7 ($3\Pi^2$) which is the distance of space in relation to the time it takes to displace that space and any motion above that is an extension of the atmosphere where the atmosphere accepts the role it has as being the neutron of the Earth. The extending can go from Π^0 to $5\Pi^0$, which will then be the moving, object extending its neutron part while still attaching to the Earth atmosphere. Above that Lagrangian limit of 5 the Roche limit in sharing neutron status sets in at $\Pi^2/2$ and the attachment there is between the atmosphere linking the aircraft and the Earth is severed.

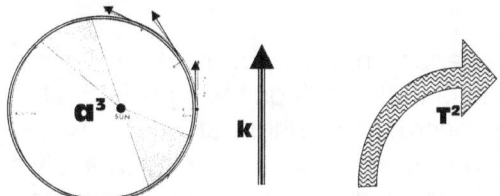

In physics there are always two relevancies at work, which has nothing to do with mass but is solely directed on singularity achieving motion.

The current notion of mass pulling mass has no comparing with reality because as I explain mass has only a counter effect of gravity. In the Universe there are a flow of space-time and a relevancy bringing about expanding as well as contracting gravity. This has no bearing on mass and is a control mechanism we find in the cosmos. It is a product of the cosmos we named the Coanda effect. We have to see that there are two parts in gravity where the one is expanding while the other is contracting and centre to this is the control we call gravity.

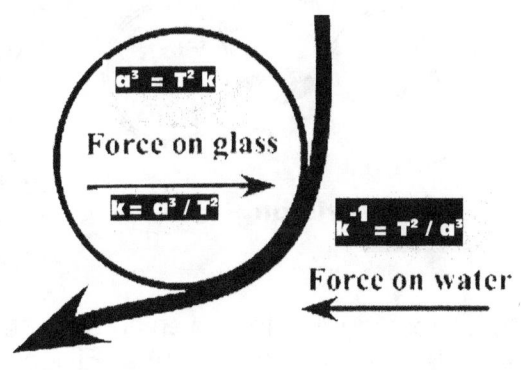

Many moons ago long before I dreamed of becoming an author of any of the books including this letter I started my search on the basis of a certain remark that Einstein once made on a realisation or a conclusion that Einstein came to in his younger days while still being a clerk at the patent office. Apparently the idea Einstein came to was concerning the subject of gravity. This happened while Einstein was still being a patent clerk in his younger days. Apparently Einstein was looking out a window of the multi story patent office, when Einstein suddenly realised that had he, Einstein fall out of the window from the roof to the ground of the patent office where he was working at the time, then he (Einstein) would feel as if he was weightless during the time of his fall. Not only that but also so would all the articles in his office that surrounded him at the time being his office chair, his desk and a pen. By falling with him those articles would feel equally weightless should they accompany his fall down as being part of the falling process in his imagination.

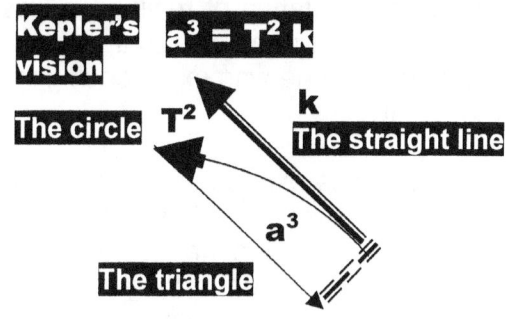

As the objects were travelling alongside Einstein down the building to the ground the lot would travel at the same speed from the top to the bottom of the building. That is what Galileo concluded about five hundred years ago. Then I went one step further by supposing the Einstein group's falling was real and no imaginary thoughts were set in

▲ **The pulling away of the smaller space. a^3**

▶ **The double counter-acting referee. T^2**

▼ **The pulling towards within the larger space k**

the fall, then what was the imaginary factor then? Let's pretend Einstein did fall with his pen, his chair and his desk and Einstein was not imagining his fall. Einstein as a human being can imagine but his falling companions can't. Then during a true fall Einstein may have had an imagination that could tell him about his feeling and in particular about the condition of his weightlessness, but the pen, the chair and the desk had no such imagination and they were travelling at the same speed as he did downwards and therefore had the same weightlessness as he (Einstein) had while they all were being in a downwards fall. If Einstein was imagining his weightlessness, it might be psychological, but in the case of the other travelling companions it was not possible to imagine anything. The falling companions had no such a luxury as having an imagination, however they too had to be weightless as they travelled next to Einstein all the way. There is an immense difference in size between the falling companions and that notwithstanding they travelled the same speed while descending. If they travelled the same speed as Galileo proved and they all hit the Earth the same time, which then indicated that their weight and mass, that which gravity used to drive and what propelled them downwards and that which was causing the drawing of what the mass was instigating to allow the motion of fall to commence, was equal. Size changed nothing to the equality there was in speed. Einstein should only have thought a little further than he did at the time because that would have made him realise what gravity exactly was and what Kepler found gravity to be. Kepler found space a^3 being equal to the motion thereof T^2 in relevancy to a centre point **k.** Kepler found space had to move.

Realising this part made me doubt the correctness of comic science. In the cosmos all things are moving therefore all things are falling never to get there. Galileo said that all things falling fall equal. If

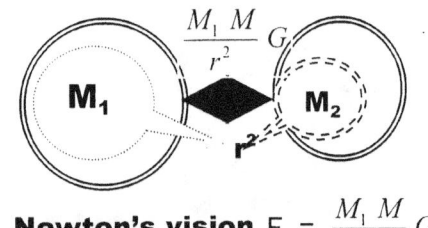

Newton's vision $F = \dfrac{M_1 M}{r^2} G$

there were Newton's mass discrepancy when falling this would not be possible because part of the driving force of such a fall is the mass. Newton even put the product of the mass in relation to the destruction of the radius. The mass forming the driving force then has to allow for heavier objects falling faster by measure of mass discrepancy. But that would sideline Galileo and Galileo could not be sidelined. It is either mass driven with Galileo being incorrect or it is as Galileo said that all things fall equal rendering mass equal during everything falling.

When reading this that evening so many years ago, I came to realise that Einstein could only feel weightless if it was true that he (Einstein) was weightless. He could not feel as if when the as if was part of his imagination because he was truly falling, and in truly falling the falling was then without his imagination doing the pretending. Einstein had to feel his weightlessness as a cosmic fact in the true sense because if he was truly falling, then the part, which was the falling experience, was what he was experiencing in reality by three dimensions with one dimension in time. Then he (Einstein) was feeling weightless through falling and that feeling came as a result of what was happening to him as a cosmic interpretation of reality. He was not pretending to fall whereby he then would feel as if...he was really falling and with that there is no "as ifs". What he then would have experienced came by means of what he was experiencing in reality because of his cosmic state in relation to his relevancy with gravity.

If Einstein was experiencing weightless ness, it would be because he was weightless while falling, then Einstein would not imagine the weightless ness because Einstein was truly falling, thus carrying out his cosmic state he was in. His body being in motion ($a^3 = T^2k$) was at that moment truly weightless while experiencing unrestricted gravitational motion. Einstein, the pen, and the chair had the same weight since they were all weighing the same in falling. If there were any mass differences there had to be speed differentiation for the force of the one would generate more motion than the force of the other onto the different mass components but since there is not mass discrepancy amongst the falling while falling the lot is having the same state of weightless ness, they adopt the same speed in the fall. After all it supposedly is the mass that is doing the pulling and more mass does more pulling...except if the mass is not doing the pulling in the first place. With more force applying to different masses there had to be more speed involved and an increase in mass in some participants has to generate more force. All four items including Einstein, would be equally weightless during the falling...that was what Galileo found because objects of different size and different mass travel at an equal pace (distance over time or space moving divided by time flowing while the object changes position in relation to the Earth ($a^3 = T^2k$)) while descending. The bigger objects do not fall quicker than a smaller object and that can only be attributed to one fact; it can only be true if the four weighed the same while falling and no one weighed anything while falling. That means the gravity applied while time flow in relation to the space that was applying the motion, which was what gravity is $k = a^3/T^2$ according to Kepler. The single line falling is represented by the factor k being the relevance of space a^3 that was relocating its cosmic position while all that was happening in relation to the motion of the Earth T^2, which was in relation to the Earth spinning around the Sun and that rotation gives us our time T^2. While in motion the four different objects weighed the same since they travelled at equal speed downwards.

However, when they stopped moving and came to a standstill, they then weighed different, which then indicated a difference in mass factors amongst them. By standing still the objects had mass differences and when they were in motion they weighed the same. When the motion became frustrated by being blocked by another space that was also filled with material and that was holding the spot too where the motion was directed, they then had different weight. The two had different levels of frustration with the larger party being more frustrated in the inability to move. The pushing resulted from the bodies striving to remain independent. It is the independence of the two bodies and the desire the bodies have to remain independent and not to share space that bring about the mass or weight. The two objects were in a fight to claim the position each desired, and that was to fill the centre of the Universe. Being ($a^3 = T^2k$) was being in the centre of the Universe because the centre of the Universe was $k^0 = a^3/T^2k$. $a^3 = T^2k$ $k^0 = a^3/T^2k$

From this one can deduct that gravity is motion or the intent to commit motion and mass is when the motion of gravity is frustrated by some solid structure blocking or preventing the continuing of the motion. Then one may conclude that gravity is motion of space and mass is the restricting of the motion of space. Having mass does not bring about gravity but it does restrict gravity's motion, which is what brings about the mass and weight. Gravity produces mass but mass does not produce gravity or in fact mass produce weight but mass is not responsible for the intended motion. Gravity on the other hand is the intention that the body has to move the very instant the blocking is removed.

The intent on moving while being blocked by another object is frustrating the motion of gravity in both cases and the higher the frustration on motion is the more mass there is co0ming the way of the bigger object who then has the greater desire to move. The reason why it has the desire to move and why space is equal to the moving in time of the space in relevance to the centre of the Universe (which at that point might be the Earth or be the Sun) is what I am trying to explain. Mass is the restraining of motion and gravity is material moving about by committing gravity. Mass only comes into the application thereof when two objects filled with space moves into a position where both want to claim the very position in space the other occupy. It is the motion and the independence they show to hold onto their individuality as independent cosmic structures that prevent them the sharing of space which in turn prevent further motion that causes mass.

Gravity is in essence where mass is present, still in a tendency to commit motion but is then in the frustration of motion and gravity at such a point is the commitment to move once the blocking of space is relinquished. Because the one object that has more "mass" would put in a more assertive effort to move in relation to a smaller object and the effort to move will constitute to a greater resisting effort by the blocking object in a fight not to relinquish its position on the space both object claim that the tendency to move and the tendency to block the movement will bring the effect of greater or smaller mass being present during the effort and in line of resisting the effort. However while any space is in motion, the gravity of motion is equal to all and puts everything on an equal basis. Therefore there is no big and small and the big Sun does not pull the small Earth closer. The big Sun allows the small Earth to glide past in a circle year after year without interfering because the two does not claim the space each other has. Mass is when the motion is prevented that a differentiation in motion effort becomes part of the picture.

Do not be fooled by the seemingly innocent explanation that space is the motion thereof which is what gravity produces because of all things the cosmos creates, motion of space through time is the utmost complex manoeuvre and without bringing a restraining of mathematics into science, it is so complex there is no viable explaining in physics about how the cosmos produce the act of motion of space in time. To get every atom to spin as every atom follow the lead of the atom in front and gives direction to follow to the atom just behind while giving coherency to the structure the lot of atoms are holding as an individual unit times the units there are going around in the entire Universe is beyond what the human mind can absorb. While the atom in front is vacating space to fill the space of the atom in front that is vacating at that instant, the atom behind is filling the space that the atom in front has vacated in order to vacate and relinquish the previous position in favour of the following position to honour the direction gravity is insisting upon. Times that with every atom there is in the Universe and one may grasp the significance of the calculation. The coordinating of moving one atom from one point to a next point requires the skills that the human mind may never conquer.

We may see the moving of an object through space being as simple as merely accepting it as a given fact, as science has done in the past, or we may reason about the complexity as civil person's should do, and come to realise that the complexity of motion of matter is beyond the scope of human understanding. Removing material from space by filling material into a position of new space sounds simple because the complexity has never been realised. This was all a result of understanding the dynamics of Einstein's arguing about gravity and mass. Then with this information I further realised gravity is motion differentiation between objects. It is the independent motion providing a different speed while sharing a common centre of attracting that allows a discrepancy to establish mass under specific conditions applying between the two in relevancy. While falling the gravity applies as moving of space that is putting time in relation to the distance travelled.

That means there is a speed relevancy between particles in motion and synchronised motion would bring about equal orbit around a shared centre. That is the result of gravity functioning. While the object falls the motion confirms gravity. When motion ends mass sets in and becomes the constraining of the object preventing further motion. The motion is still there but now it is reduced to a tendency to move thus establishing the object mass as the limiting of further motion. Preventing the motion by implementing mass is the resting of objects against each other by resisting the motion to continue, which then is where the mass takes the place of the motion. Where a confronting of objects restricts gravity the action then implements an introducing of the mass as a substituting factor for motion that then replaces motion as substitute for the motion that would be and the mass is providing the tendency of gravity being the motion of space. However mass then restricts motion and becomes motion in a tendency to apply motion.

While falling gravity applies and motion neutralizes size, mass or weight. Mass counters motion being when the Earth restrains further motion of the falling object and the moving object is stopped from further movement where mass is then preventing or hindering gravity. This is the result of objects claiming an individual and personal claim to space occupied in a dual or in fighting for their individuality and independence of each other while wanting to be in the **centre of the Universe**. While falling or moving there is no opposition to the body being independent. When the motion seizes the falling object remains individual and still tends to move while Earth individuality resists further movement of the falling body's movement. Further movement is disallowed as other material fill space that falling body wants to lay claim to. The only manner to remain independent by the falling object will be to relinquish to motion in the securing of mass as a substitute to motion where it then finally comes to rest. Mass then sets in not causing the motion but substituting the motion and from that motion restriction becomes resistance that becomes mass. While falling the object is experiencing gravity because the object is in gravity but when on the soil the object experience mass which is the restricting of gravity or motion by other space filled with material. It is a fight of objects to secure and retain the position they have of being in the **centre of the Universe**.

Moreover, I then came to another conclusion of equal importance. When any person is standing on any place anywhere, while viewing the Universe, that person is filling the **centre of the Universe**. Let's get more personal. When you, the person that is reading this, are standing at night and are looking at the Universe you are seeing the Universe from the position that one only can have if that person is filling the specific spot in the **centre of the Universe**. All the light, every single beam that ever left any destiny at any time acknowledges this fact. You are the most important person in the

Universe because you are holding the most important position in the Universe. All the light that come across and travelled all of the vacant space from any and all possible positions in space runs directly towards your position using a straight line towards you where you are filling the **centre of the Universe**. Not excluding the effort of one photon, all light is heading to meet you where you are in that centre spot and not one photon will pass you by. Not one photon dare miss you because if they do they miss the effort that all light has to accomplish and that is to locate you as the person filling the **centre of the Universe**.

Should you decide to shift your position to any other place in the Universe, you will shift the **centre of the Universe** to that location as well. If you install a camera on Mars, the light is obliged to acknowledge your relocating the **centre of the Universe** at your will to reposition you're being that **centre of the Universe**. All the light that ever left its destination crossing the vast spaces of the Universe, excluding no particular light, travelled all the way just to find you filling the **centre of the Universe**, right where you are. By you're standing anywhere, you fill the **centre of the Universe**, and the entire Universe admits to that because all the light comes to meet you there. If you shift from the North Pole to the South Pole you will shift the **centre of the Universe** because all the light travelling throughout the Universe will find you where you then moved the **centre of the Universe**.

The light left its destination billion years ago as it travelled through space at the speed of light anxious to acknowledge you're being in the very **centre of the Universe**. No photon will be able to pass you by where you are in the **centre of the Universe** because all light is heading your way from their starting positions. No wonder every person born has the idea they were born to fill **centre of the Universe**, which we do fill. The Universe is spinning around you or I, which is filling a centre where all motion is connected. That is the Coanda effect on the utter-most grandest scale imaginable; nevertheless it is only a manifestation of the Coanda effect. It implicates gravity as wide as can be... Some things mathematics is able to explain but other explaining goes beyond mathematics.

Try to explain mathematically the colour of the sky being blue in a clear Sunny day and changing to black when nighttime falls. Do the explaining in mathematics to a blind person that had no vision since birth in such perfect mathematical detail that would allow the person afterwards be able to explain the difference between blue and black to other blind persons by using only mathematics. Some aspects of the Universe go beyond mathematics and some even go beyond words. It is our task to find space, to find time and moreover it is our optimal task to find the Universe. We have to see what is solid, what is liquid and what causes gravity. It is therefore very important to see what is a solid and what is a liquid. Again we must put culture in the background and value the cosmos by using cosmic standards.

Everything that moves, do so in relation to another that is relevant stationary is a liquid notwithstanding that it may or may not contain material. It is a liquid nevertheless. Everything that is relatively stationary is a solid in relation to the liquid that moves about the solid that anchors the liquid by gravity. Gravity **is to move or apply the intension to move** space a^3 **at the** distance or relevancy of **k** while T^2 is the time it is going to take to **apply gravity** or move the space filled with material space a^3 at the distance of **k** in the time period of T^2. That confirms Kepler's attribution to gravity where according to Kepler space a^3 is equal to the movement T^2 (time it takes to move) at the distance **k** from the centre specific.

Do not frown on this being in the **centre of the Universe** or regard it too lightly because from that I can prove life being eternal and life being part of the other side of the Universe. That is not part of this letter since that I do prove in another book with another title under the article heading **Man – in-Motion**.

I then subsequently reviewed my vision I received from the vision Einstein received and applied such a vision on the findings Kepler received from the Cosmos. It puts all aspects of gravity in the Universe in new dimensions. But the visions formed the beginning because the visions unleashed many new questions. If gravity is motion, what causes motion? What stops motion? That answer is in the Black Hole. In truth the explaining of the Black Hole is as complicated as the Universe may represent and as simple as the cosmos truly is. If a star is about fusing atoms and with such fusing of atoms is thereby growing, what happen when all the atoms fused into one all collective atom in one already all—atom-accumulated star? What is the gravity if the star has melted all atoms it had into one all-inclusive atom and this all-inclusive atom is providing all the gravity that the star had when the star still had massive volumetric space? If all that space that once filled an entire giant star fused into one specific space less centre holding singularity 1^0 then the enormous gravity is applying to the centre of such a non existing space-less atom and that entire enormous force has been secured in the space less than that which one atom holds. In that case the atom would then show a force that would pull the surrounding Universe flat.

The purpose of fusion is to reduce space and magnify space less ness inside the sphere. Where does the gravity of the star end when all the atoms in the star became one giant atom by fusing all atoms into one nucleus? Gravity is smallest where space is least. Where space of an entire massive star is left in the size of one atom the gravity coming from that will pull the Universe flat at that point. However fusing means freezing together because only by reducing the heat can the removing space be accomplished and by reducing space to the point of freezing material permanently together is getting material frozen permanently. That means the Sun and all stars are as cold as they can get and not hot!

 The motion is a product of heat and the motion produces a cold that sets in as the motion comes about. When the object moves it moved because the heat became excessive but by moving it is doubling the area it holds by halving the area as it divides the space between where it goes and from where it came. As soon as the motion halves the space used to move the halving of the space halves the heat which produces the cold which brings about the containing or the reducing of the space. Then with the motion completed the contraction retains heat where the heat increases to bring more heat rising that leads to more expanding coming about and the cycle once more repeats.

I am not getting into that argument now, but because of the size the Sun has and the size moving through such distance the Sun in its very centre is the coldest place in the solar system and outer space is so hot it is over boiling. That is why outer space is expanding. It is because the heat rises as much as the stars reduce the heat by containing the heat as material inside the atoms. Nevertheless, gravity is motion and motion comes from overheating which the motion then produce the cooling that contains the overheating by accumulating the contained heat inside the atom. To do that the Coanda principal is employed and the Coanda principle sets the Titius Bode law in operation and the Titius Bode law produces gravity. By spinning liquid heat around solid space a relation between 10 / 7 and 7 / 10 produces a relevancy that contacts heat.

It is about. Gravity is all in Kepler's $a^3 = T^2 k$ and $k^0 = a^3/T^2 k$ where then relevancy $k = a^3/T^2$ and in response to keep equilibrium applying $k^{-1} = T^2/a^3$

This proves the reality of the Titius Bode law, which too I have to add, the Newtonians put down to a coincidence. This proves that the Titius Bode law is part of the chain that brings about gravity. Most of all this disqualifies mass as having any importance what so ever in the producing of or the conducting of gravity. This proves that gravity is the motion where space interacts with time to give singularity the significant control it has as not being part of the Universe and yet being responsible for all action in motion taking place in the entire Universe.

This is what keeps the top erect while spinning and it keeps the Earth in gravity as much as it built the solar system to a mould that built the entire Universe. The fact that motion brings about gravity in line with singularity must be proof to all Newtonians that their perception on mass has no grounds. Even where those Newtonians are unable to show what brings about gravity even after so many centuries of investigative research while trying and getting no results does this simple arithmetic proves more than all the multitude calculations on their part that proves zero about the manner they promote gravity as being a pulling force.

Matter in relation (part of) to the total dimension of space.

$(10 / 7) \backslash (7/ 10) = 2.04$

$1.4285 / 0.7 = 2.04$ Taking from both orbiting influences

SPACE DIVIDED INTO TIME

$(7/10) / (10/7) = 0.49$

$.7 / 1.4285 = 0.49$ Taking from both orbiting influences

SPACE MULTIPLIED WITH TIME

$7/10 / 7/10 = 1$ and $10 / 7 \times 7/10 = 1$ Therefore not influencing change

THE PROCESS PARTED USING THE ROCHE PRINCIPLE

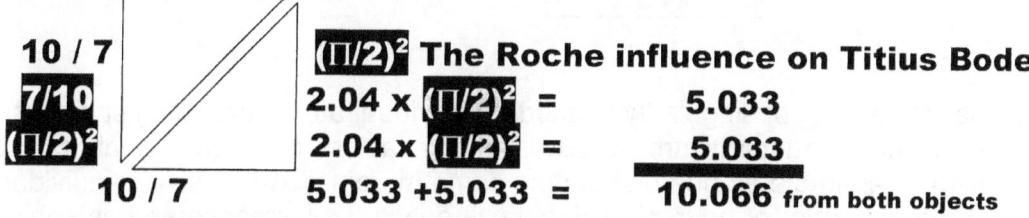

10 / 7
7/10
$(\Pi/2)^2$
10 / 7

$(\Pi/2)^2$ The Roche influence on Titius Bode

$2.04 \times (\Pi/2)^2 = 5.033$

$2.04 \times (\Pi/2)^2 = \underline{5.033}$

$5.033 + 5.033 = $ **10.066** from both objects

SPACE DIVIDE INTO TIME

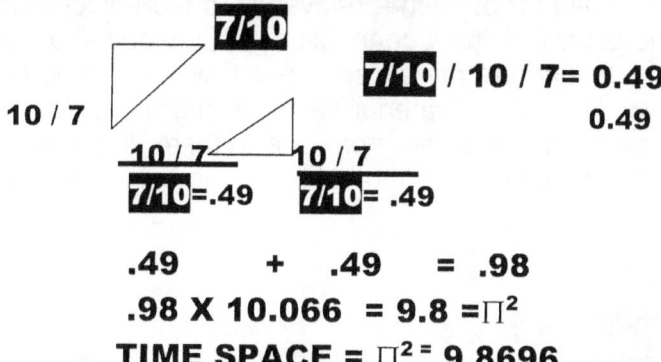

7/10

$7/10 / 10 / 7 = 0.49$

0.49

10 / 7

10 / 7   10 / 7

7/10 = .49   7/10 = .49

$.49 + .49 = .98$

$.98 \times 10.066 = 9.8 = \Pi^2$

TIME SPACE $= \Pi^2 = 9.8696$

TIME SPACE $= \Pi^2 = 9.8696 = $ Space and time in a dimensional implication.

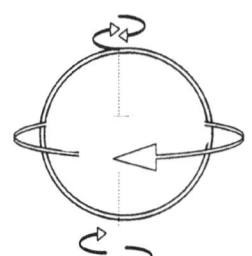

On the inside, there are the seven markers of which singularity is the focus point in the centre of the centre. The markers are representing one aspect of space, which for argument's sake let us call it cold. Then there are three more markers on either side being part of the space but not captured in the space. It is space in motion by the influence of the motion of the Earth.

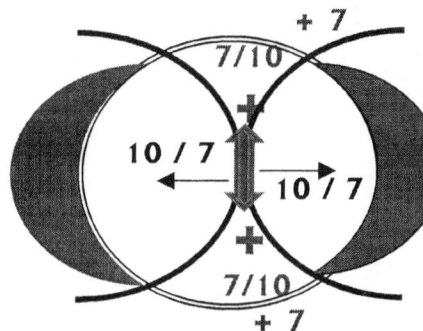

+ 7
7/10
10 / 7
10 / 7
7/10
+ 7

TITIUS BODE LAYER CONNECTING

Inside the cosmic sphere

$7/10 + 7/10 = 1.4$

Singularity in the square of matter

$10 / 7 = 1.42$

Singularity in the square of space

$1.4 / 1.42 \times 10 = \Pi^2$

MATTER HOLDING THE SECOND PROTON COUPLING THAT TO THE NEUTRON TO COMPLETE THE NEUTRON. Due to the influence of the matter dimension on the space dimension, the curvature of space-time comes into affect by dominating outer space.

The Titius Bode Principle is equal to gravity @ $= \Pi^2 = 9.8696$

Proving that the Titius Bode Principle is a product flowing Directly from the growth of singularity forming space-time The Titius Bode principle directly valuating TIME to SPACE $= \Pi^2 = 9.8696 = $ MATTER HOLDING THE SECOND PROTON COUPLING

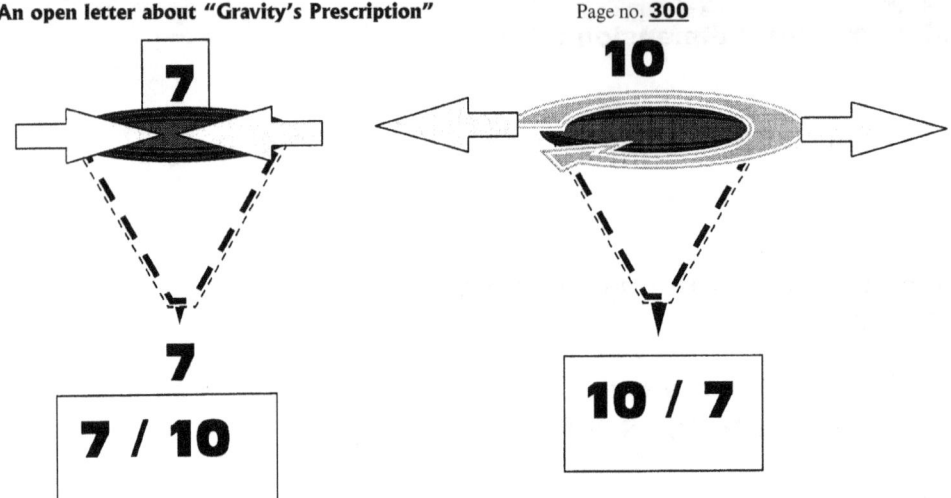

In this maintaining of cross referencing of singularity located in individual atoms providing spin to the governing singularity that maintain structural form in solids, many factors of singularity all form a close knit network and being inseparable as one unit, by the same margin it also is strictly individual to a point of destructing. From the inner or governing singularity outward all is concerned as space-heat. The relevancy in the material sector always includes the governing singularity and the very next one to the inside. All the others do not form any part of such a relevancy to the object forming the relevancy. On the time issue it is only the relevancy forming a connection with the one in question and the governing singularity. All other objects in the line are merely space-time with no value to the object that holds the relation. The fifth object will link in the material sector to the fourth and then directly to the governing singularity skipping or excluding all others from one to there. In the case of say three, it will connect to two and skip one while all points holding singularity to the outside is no consequence to the rotating object.

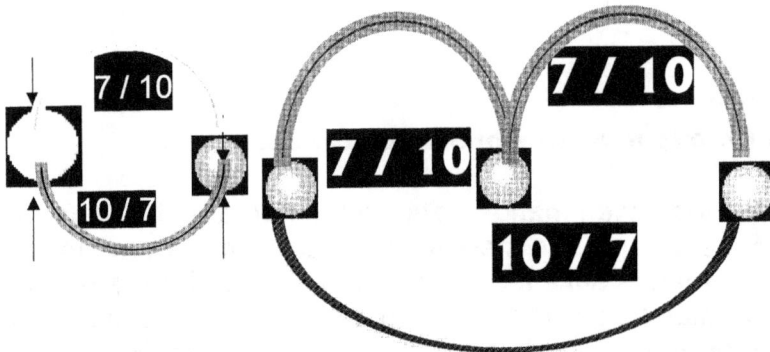

Time started at zero, eternity, whatever you wish to say, as long as you say time did not move at all. Then the command came and time overheated for the first Π^2 in time. That brought space into play.

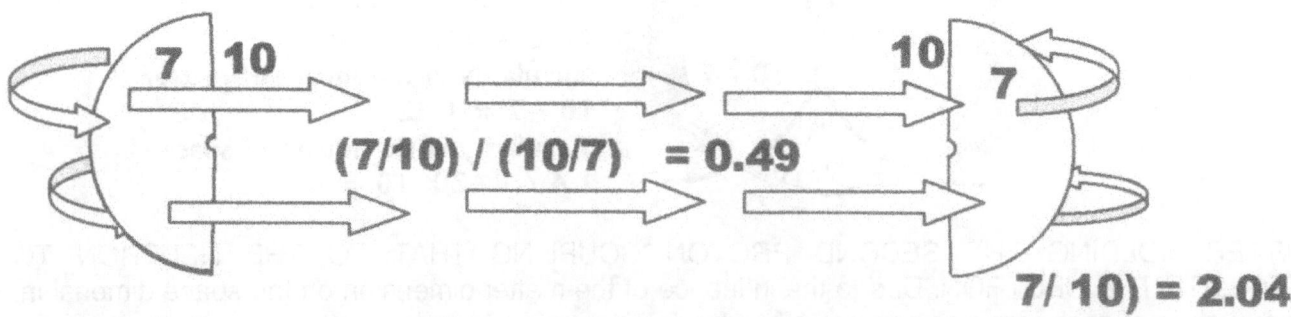

The factor of 7 in relevancy with 10 is gravity forming cosmic space in accordance with cosmic time. The Titius Bode law forms as a result of 7 standing in ratio of 10 or the other way around but that proves that the Titius bode law is responsible for what becomes gravity and what eventually is the cosmic space.

The spherical positioning layout forming the Titius Bode Principle

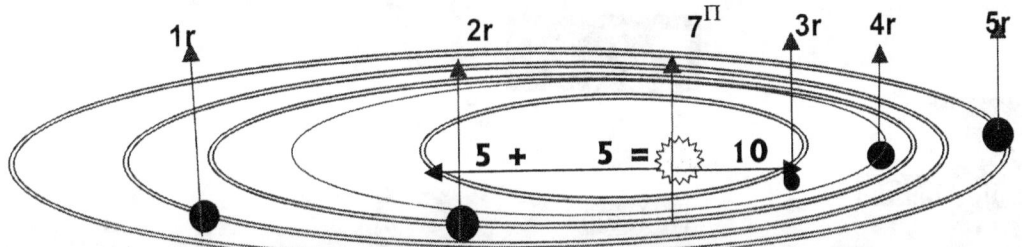

From the matter-to-matter relation in the Titius Bode configuration there are 7 / 10 + 7 / 10 = .7 + .7 = 1.4

From the space-to-matter relation in the Titius Bode configuration there is 10 / 7 = 1.42

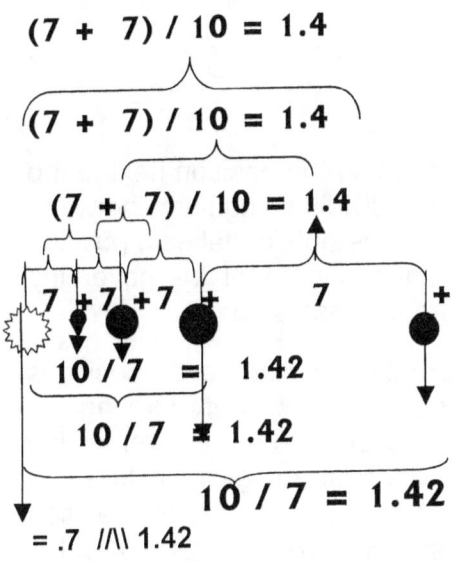

$(7 + 7) / 10 = 1.4$

$(7 + 7) / 10 = 1.4$

$(7 + 7) / 10 = 1.4$

$7 + 7 + 7 + \qquad 7 \qquad +$

$10 / 7 = 1.42$

$10 / 7 \neq 1.42$

$10 / 7 = 1.42$

= .7 //\\ 1.42

The 5 + 5 = 10 is a position of dimensions as space loses value to singularity. The 7 that matter diverts in points from singularity may seem as coincidental but is valid. Still in accordance to our perception valuing the number in degrees, it seems coincidental but if it is coincidental, it is nevertheless a figure of diverting proven as accountable in all other calculations and plays a most dynamic role.

The Lagrangian 5 point system results as much from the Curvature of space-time as does the form the Black Hole holds. The Galactica is the opposing equivalent of the Black Hole and has identical but opposing similarities being the five points positioned to singularity. The galactica is generating space and the Black hole is degenerating space. = 1.4 //\\ 1.42. **Because the space-to-matter is in the square at 10 placing the matter-to-matter at a square of .7 + .7 = 1.4 the space-to-matter forces the matter-to-matter to double the distance by number as structures are place father from the mainΠ^0 maintaining singularity.**

1   3   6   12   24   48

Reasons why this does not fully apply to the solar system I give in book # 7.

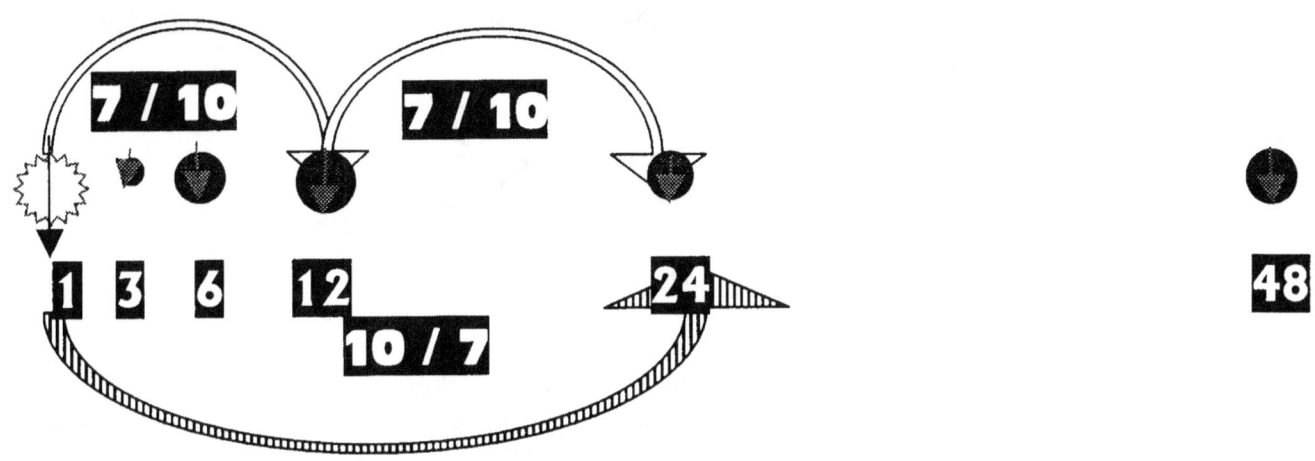

SINGULARITY BY DIVIDING SPACE INTO MATTER AND MATTER INTO SPACE, ANG ALL OF THIS ACCORDING TO THE TITIUS BODE LAW OF 10 / 7 AND 7 / 10 IN CONJUNCTION WITH THE ROCHE PRINCIPLE OF $(\Pi/2)^2$

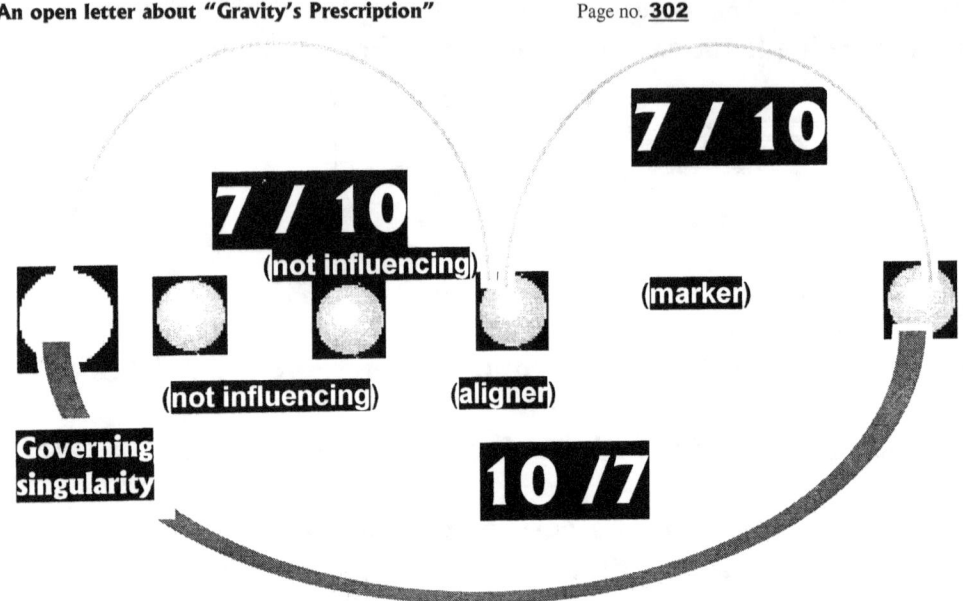

From the dividing singularity only one reference holds a matter value forming the position next to the governing singularity and therefore 7+7 becomes a factor and not all the dividing singularity between the point of reference and the governing singularity. That way the star to the outside takes a position doubling the distance every time. In balance everything in space to the outside of the governing singularity is space be it space or matter that makes no difference therefore that is 10.

The extension of Π is well received as a dimensional implication to matter holding seven positions from singularity and space having four quarters through out the rotation of singularity forming the centre to the five dimensions (one side lost to the cube's six sides connecting to the five remaining sides) making the total sides facing space from the point holding singularity at any given instant at a value of twenty (4 X 5 = 20). Then adding the singularity cross of Π being (1+1) = 2 the relation becomes 22/7. This is crude because in more precise calculations it becomes .91 + 1 = 21.91/7 = Π

The sectors provide individual singularity as a means in sustaining governing singularity by which provision comes through maintaining governing singularity the required spin in maintaining cooling. If this process did not apply, there would be no connecting individual singularity to major singularity. The sectors provide individual singularity a means in sustaining governing singularity by which provision comes through maintaining governing singularity the required spin in maintaining cooling. If this process did not apply, there would be no connecting individual singularity to major singularity

SINGULARITY BY DIVIDING SPACE INTO MATTER AND MATTER INTO SPACE, ANG ALL OF THIS ACCORDING TO THE TITIUS BODE LAW OF 10 / 7 AND 7 / 10 IN CONJUNCTION WITH THE ROCHE PRINCIPLE OF $(\Pi/2)^2$

This ratio there is between the governing singularity and the marker (innermost planet and planet marking a position according to the Titius Bode law. There are reasons why some diverting stems from this but in other books I explain that using much better detail

From the orbiting structure (planet) aligning singularity only one structure the very inside singularity applies as a position of reference and that is reference to the distance applied between the points in governing singularity. From the Sun (governing singularity) the matter marker is 7/10 = 0.7 with the only one other forming a marker 7/10 = 0.7. The two form 1.4. From the Sun (governing singularity) the outer planet forming the marker in search of position holds space in the square 10 / 7 = 1.42 in aligning with the 7 forming material of the Sun . Therefore there are two sevens relating to ten forming the material positioning of the structure in orbit and from the governing singularity all outside the Sun is the square of space (ten) aligning with one particle (seven) and not one of the other structure to the inside or the outside holds any value. Because .7 + .7 = 1.4 and 10 / 7 = 1.42 the distance doubles every time there is an aligning of three orbiting object. In this there is definite proof of influences coming about between particles sharing gravity. But then again the entire Universe shares gravity and as such then all will influence everything.

Mercury	Venus	Earth	Ceres	Mars
4 − 4 = 0 14 − 4 = 10	0 + 7 = 7 20 − 4 = 16	7 × 2 = 14 32 − 4 = 28	10 × 2 = 20	16 × 2 = 32

Jupiter	Saturn	Uranus
28 × 2 = 56 56 − 4 = 52	52 × 2 = 104 104 − 4 = 100	100 × 2 = 200 200 − 4 = 196

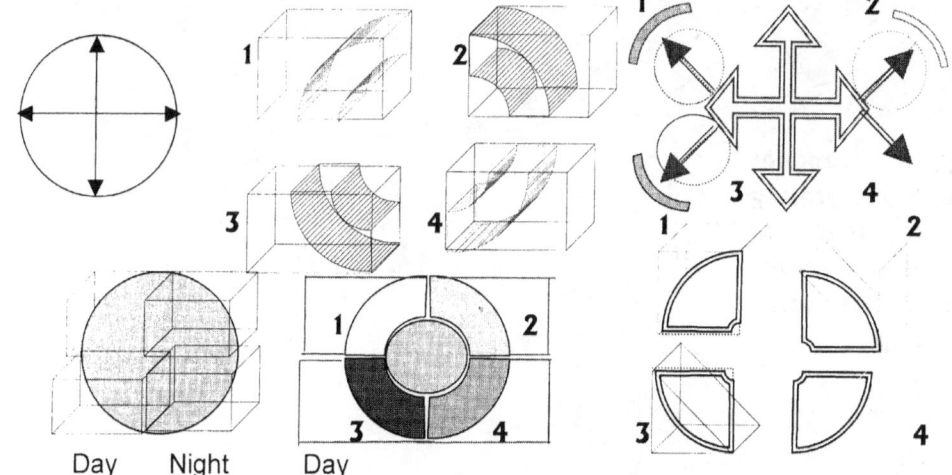

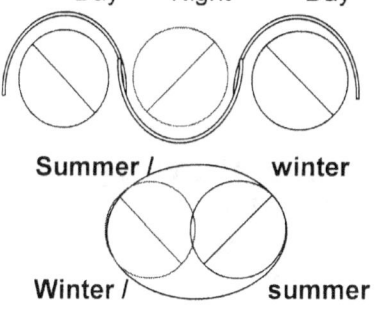

Day   Night   Day

Summer /   winter

Winter /   summer

I do realise science do not recognise a relevancy between the rotation or orbit of the earth and the position of the sun as Newton claimed and I shall come to that in a brief time. On monitoring the rotation of the earth to a graphic display one find that the earth movement displaying in accordance to change in positional location does indicate a relevancy that imitates the flow of current to an almost exact. Seasonal change has all to do with the graphs influence derived from the cosmos and little to do with the position of the sun and the Earth.

It is the position singularity holds in relation to the Universe and the Milky Way forming currents and seasons moreover than the Sun shining brighter or not. The Sun in size over dominating the Earths in comparison disqualifies any positional influence that can alter the Earths heat standings. Through shear size the Sun can shine at the top and the bottom of the Earth simultaneously without effort from all normal possible angles. I show a relation between singularity in different positions maintaining seasons and north/south polarity, not only as far as concerning the Earth but also outside influencing polarization. This has to do with the second position singularity holds in accordance to matter and space and is an "*electromagnetic*" (used for the lack of a better word) sustained positional opposing derived precisely from the graph in the manner when calculating electricity.

In this it is clear why the Titius Bode ([10 + 10 + 1 + .991] / 7) and the Lagrangian 5 \\ 7 systems part their ways when applying the different processes they hold. With all the differentiating, the observer must also consider the dual massage that light uses in travelling through the vastness of universal space. The thought of nothing is just what it is, a thought of nothing and although it is in the human mind common nature to present nothing as a value in the recalling of something, nothing is a presentation of the figment in the human mind. There can be no number such as nothing and that was (possibly) Newton's biggest error. Nothing represent non-existing and that is just what nothing is, it is non-existing.

The Titius Bode influence in a manner that on the one side holds the matter-to-matter relation of 7+7/10 whilst on the other side during the same time holds the space-to-matter relation of 10/7 forming equal and opposing values. From this the orbits of cosmic structures are always oval favouring the singularity dynamics of the one structure at one point and switching the favouring to the other structure on the opposing side. Because the structures can never be equal in size (singularity will not permit that where the Roche principle will intervene) the shape is always "off centre" as well.

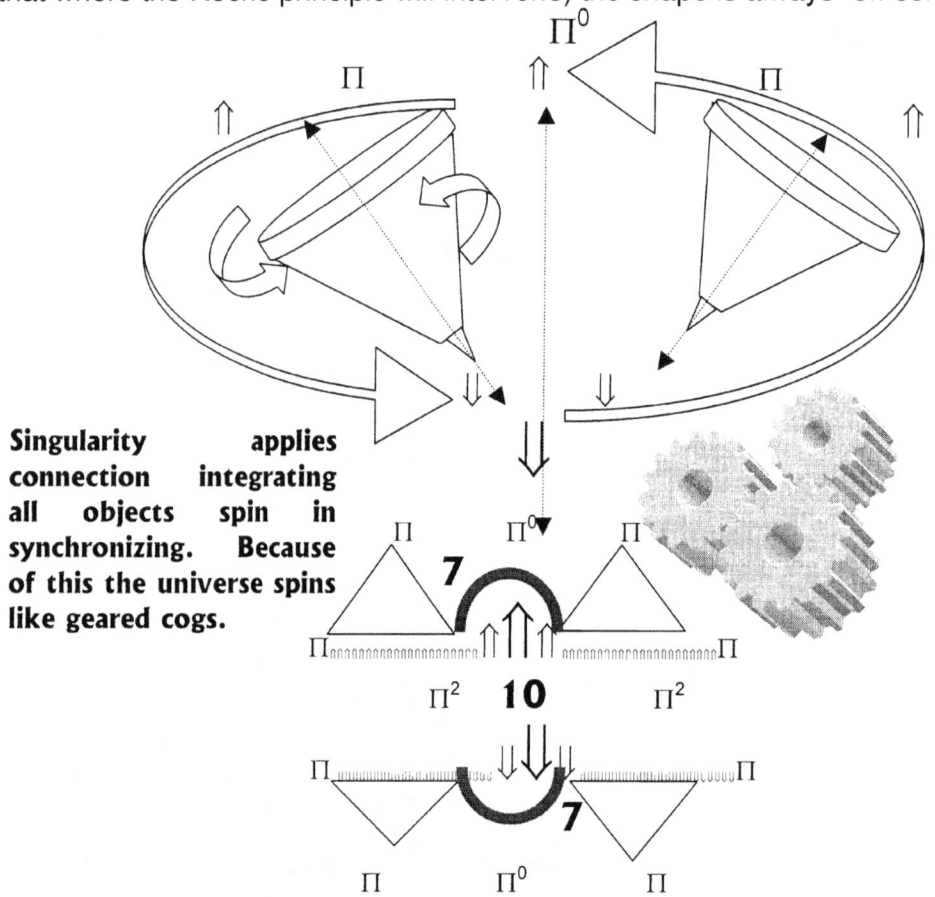

Singularity applies connection integrating all objects spin in synchronizing. Because of this the universe spins like geared cogs.

The ten dimensions I named the atomic relevancy is also showing the double value of singularity as singularity extends into as well as beyond space. The atomic relevancy is $(\Pi^2+\Pi^2)(\Pi^2 \text{ X } \Pi \text{ X } 3) =$ **1836** that is the mass relation between the electron (3) and the proton. Proton $= (\Pi^2+\Pi^2)$ Neutron $=\Pi^2$ Π. The atomic relevancy holds the dynamics of singularity control. In the ratio and dimensions we find in the atom, all space-time derives from the atom, whatever the atom is.

It started with a dot, because that is the only form, size and dimension mathematical logic will allow our brain to accept. From the one dot had to come a second dot and a third dot. The dynamics of such a dot is smaller than we can understand because such a dot is in negative relation to what we see Π to be, and the deeper we delve in finding the smallest fragment where space started, in the spot where time is still eternal as much as we can accept eternity to be. This we find in the aligning of planets. Where the one dot from which the aligner stem becomes the reference too the distance applied between the aligner and the original dot, or governing singularity or structure in charge of holding position to all orbits following.

The reason why we should first locate the spot is because we can only work from that point forward. By working forward we have to work backwards to locate where we are heading. The cosmos started at a point and where such a point is, we will find the Universe. Every one knows where the Universe is, because we can see where the Universe is, but if we can see where the Universe is, then we should find the centre of the Universe in that spot. Einstein theoretically positioned the point of beginning at a place he indicated where singularity should be.

With the cosmos the size it is and space so large compared to our smallness we have no chance in finding the centre of the Universe. The Universe started where singularity is and singularity is the sure indicator of the Universe. With all spinning objects holding singularity we then have located singularity in as much as finding the centre of the Universe. The Universe started with a dot forming. That answer arise from taking mathematics back to a point of being the smallest possible position, far smaller than we may be able to calculate form.

My approach might seem unconventional but through the abandoning of the accepted, it enabled me in locating the precise location of a universal singularity forming a connecting basis of the Universe (this I say with some degree of confidence). The smallest figure there can be must be a dot. The dot is the only form that leaves all the options open to extend in any and in all directions should the opportunity arise. The only mathematically sensible option about extending a line from the dot will be non-bias progress in all directions equally in order to give a meaningful flow of mathematical equilibrium.

The Pythagoras mathematical principle is the proof and that I explain. The obtaining of singularity is in my rejecting of nothing by replacing it with something being the dot. With the clepsydra or "water thief" Empedocles deducted that air was composed of innumerable fine particles, braking the thought that what we now know is air, was also believed to contain nothing being altogether a space filled with nothing until proven to be wrong so many years ago. Never did science take the lesson learnt back then to the future and out onto outer space. If there is space, there cannot be "nothing" as space is something. The claim becomes obvious when observing the connection between the half circle, the straight line and the triangle, which could also promote all the qualities lurking behind the pyramid. Consider the connection between 180^0 sharing and then one may realise much of the pyramid mystique becomes less spectacular in considering the very basic in mathematics being the Law of Pythagoras on which all mathematics are based. Once the water thief was eliminated by some human intelligence the matter was left at that. Nothing shifted out to an area we think of as outer space. In outer space we now find nothing. There is nothing but an atom here and there and even the atom is covered in nothing.

I wonder why the nothing landed there? Could it be that the reverse came about and because there was no visible "water thief" the very limit of man's suspicions came into practice. Man has always been extremely good in flying from one outer edge to another and if the water thief proved something was present, then the mere absence of a water thief must therefore prove that nothing must be in outer space. But what is space as such. What can space be, because with explosions we can clearly witness space created from heat. Our culture prevents us from admitting our vision, but the release of heat produces a *"shock wave"*. That *"shock wave"* is nothing less than space created from heat released. We have to brake free from culture of the past and a rigged mind set narrowing our vision.

Einstein's Critical Density lacks the accepted matching facts we need in proving the critical mass factor. But our inability in securing such required evidence defies the most basic logic. It seems all new evidence we receive from outer space is disputing all Newton laws findings that disprove Einstein's Critical Density as the answer. The Universe will not reach a point of contracting, not withstanding whatever dark matter astronomers try to locate in the vast space.

Why would the expansion turn around and do a reverse by going back to where it came from. Consider the momentum alternation such a change will bring about.

The Sun is not a gas-filled sphere holding hydrogen in its "natural gas" form, but it is all fluid and is in a liquid form where singularity is liquid- freezing hydrogen at 6500^0 C while outer space is boiling over at $- 276^0$ C. This book explains the Roche limit in the practical sense... when applying cosmic laws instead of improvising cosmic laws uncovers that reality then becomes awesome. It becomes clear the Universe is as much expanding as it is contracting and contracting by expanding. As there is no hot or cold, no big or small, no grand opposing but relevancies in ratio to one another. If you do not believe me, then believe your eyes when looking at the picture. What ever the Sun is it is fluid falling into fluid.

Consider the time it took from 10^{-43} to 10^{-5} seconds to create a cosmos the size of a neutron. Compare that to what is happening now and see how many events took place by the creation of every lepton and every hadron and it is true that that period took longer to complete than it took the Universe to create the solar system. The flow of light through the density that space produces heat gives the speed of light the relevancy of time in space. The thicker the "soup" of heat is that space forms, the longer it will take light to cover a distance. It is very important to note that the speed of light is a relevancy between time (seconds) and space (kilometres). The speed relies completely on the value **k** holds on space –time. The speed of light is forever a constant but the constant is part of the relevancy of space-time

If one looks at the transmission of sound, it too depends on the relocation of matter, but to a very small degree, and in this process lies the transmitting of sound. To make the error of judgment in confusing the process with the breaking of the Doppler rings are quite understandable.

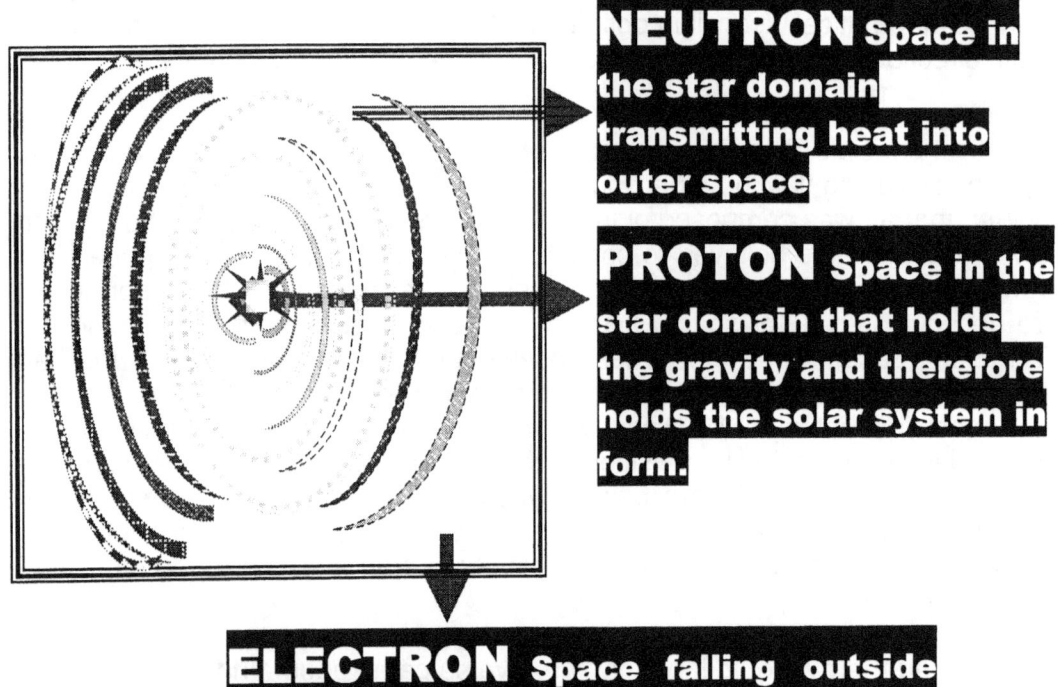

NEUTRON Space in the star domain transmitting heat into outer space

PROTON Space in the star domain that holds the gravity and therefore holds the solar system in form.

ELECTRON Space falling outside the domain of the star

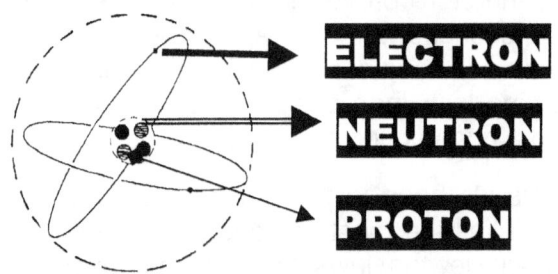

ELECTRON

NEUTRON

PROTON

The Universe connects in a way Kepler established through his relevancy theory. Those not convinced answer this: where would the Planets be if not for the Sun securing planet positions. The relation proves the ratio of one in all cases to be valid. It proves much more than merely connections at liberty of holding positions where ever the randomly opportunity placed the structure. The structure does not come closer by a pulling and tugging. Kepler's figure must still be around and by repeating the task but this time made much easier with the help of computers and telescopes of magnificence compared to those which excited the likes of one Tycho Brahe in his time. Science should become serious about science and not about self-protection and self-preservation. I found on all and every campus I went that any remark about Jesus Christ supposedly making a mistake generated immediate interest with even the most adhering Christians coming to hear the argument. Making a remark about Newton making an error gets you marched off the campus by security. Why not test Newton's $F = G (M.m)/r^2$ from figures Kepler left us and see how far did planets shift closer. I guess this will again make this book as successful as the others with my openly criticising Newton and Newtonians but Universities are not about knowledge but t about protectionism. Universities protect their own without any willingness to test that which it protects. It should be clear in confirming that the basis on which the entire world science union is founding all their policies and beliefs are correct and not only that, how far did the structures move closer. From that we then can see what we are waiting for and how long before the big solar clashing will begin. The absence in they're just mentioning such possibility confirm to me they know as well as I do there is no tugging and the Universe is in synchrony more than any person may ever be able to prove.

PROTON   NEUTRON ELECTRON

$$(\Pi^2 \quad + \quad \Pi^2) \qquad (\Pi^2 \qquad \Pi) \qquad 180^\circ \qquad 3$$

180°   180°   180°

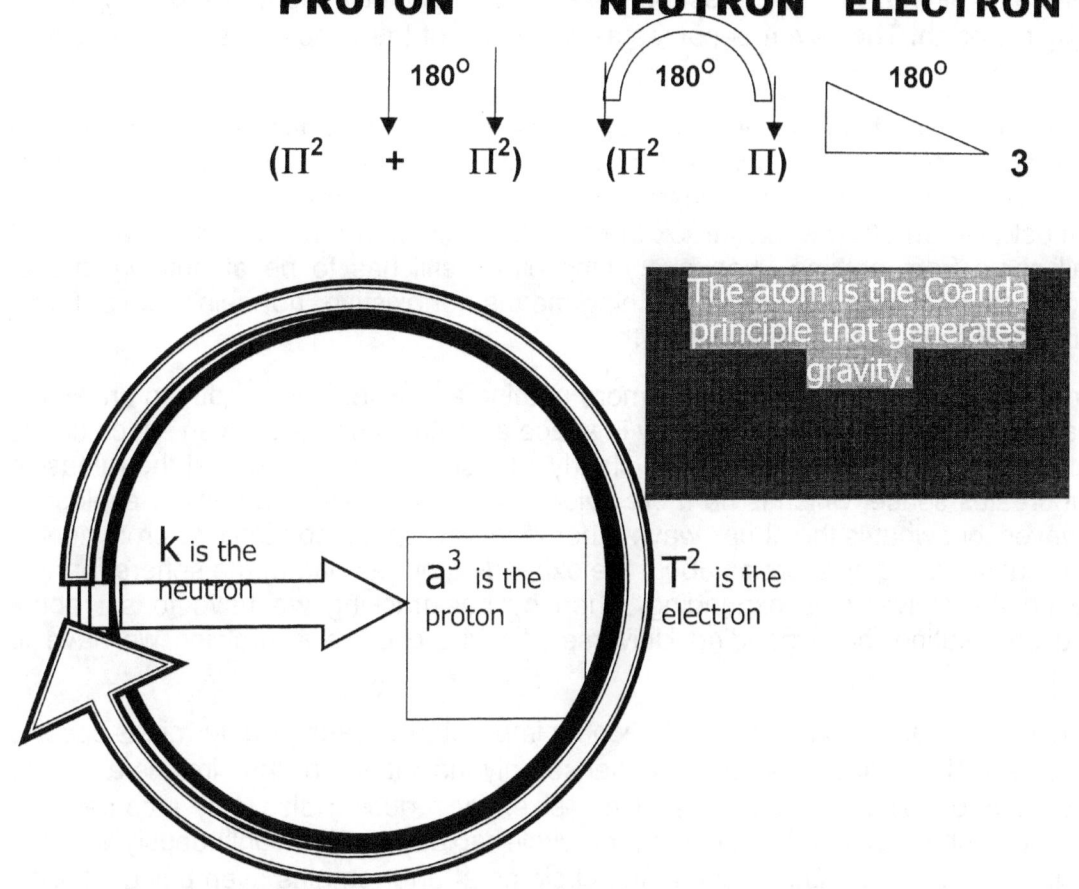

The atom is the Coanda principle that generates gravity.

k is the neutron

a^3 is the proton

T^2 is the electron

Everything in the cosmos is moving, either by own individual accord, or under the influence of some other singularity dominance. In explaining we return to the top.

When the top is in a state of motionlessness on own accord it is everything but motionless. The motion it adapts are synchronised with the Earth in harmony with the solar system and according to the greater picture of the cosmos. When an energy source not related to the cosmos called life intervenes and energises the top's motion, the singularity in that top suddenly jumps to life. By adopting a rotation energised to an unnatural state of energising because of life's intervention, the singularity of the top is not in charge but as it applies more and more energy, it will begin to find a means whereby it can escape and apply individual singularity as the top starts to separate from the singularity the Earth holds. The singularity holding the Earth would then allow the singularity of the top to rotate within a specific band where that a specific band of being active before the Earth's singularity will start to destroy the singularity in rebellion.

The top on the other hand will try its outmost, when the singularity it holds gets by individual spin is too strong to remain in domination of the Earth's singularity. The motion of the top is an attempt to begin applying an individual singularity space-time defying and standing apart from the Earth's gravity. That action we see as the top starts rotating in a manner where the top does not align with the Earth's singularity, but est6ablished a driving singularity independent from the Earth's gravity. With the adding of spin, the time the top holds becomes unrelated to the time the Earth holds and the top will start a campaign too escape from the singularity domination the Earth has on the top. When the time or spin of the top exceeds the limits the Earth places on the top, the top would emerge by trying to escape from constrains placed by the Earth. The view I represent at this point is known to science for almost as long as science knows mathematics.

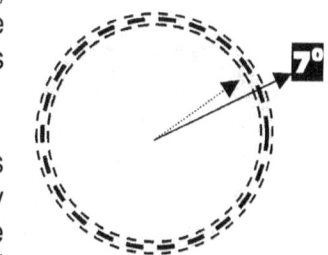

Not long after the law of Pythagoras was understood where Pythagoras introduced mathematics Eratosthenes of Syene made as big a discovery as Pythagoras did. But in the one instance the world took notice because the world could see and understand and the other instance the world disregarded the findings because the world did not see what the implications was. The same apply to aircraft flying and when the aircraft wishes to escape the earth's singularity hold it has to comply with the laws laid down by the earth. The seven becomes as big a part of the concept as does Π as it all interacts

If we wish to find the future we should locate the past. If the cosmos is contracting, where to is it contracting? The direction of contracting must be in the opposing direction the direction of expanding. If we wish to locate the past from where the cosmos came and through that in what direction the cosmos came, we must take an effort to backtrack the direction it came from. Should the argument come about that all came from nothing, then everything either still has to be at nothing, or our understanding of nothing leaves much to desire. Nothing means not existing, not being, never found and unable to produce any multiplication of any growth.

The above questions, but mostly the fact of what is more nothing and what is less nothing draws me to the realisation that there can be no such a quantity in space as nothing because even space has to be something. Heat expands as the levels rises and clearly it is for any one to see that the releasing of uncontrolled heat creates space, which is no more evident than by releasing a nuclear explosion. The wind is shock waves, but what is the shock wave other than new space coming into prominence. In that way it is clear that releasing heat brings about the expanding of r as part of the sphere forming space. Hubble proved the Universe is expanding. Then by backtracking we have to set about reducing the sphere constituting the expanding Universe. If r in the circle is growing we have to reduce r to backtrack.

When the circle reduces, the value located to r will become implicated because r determines specific size. Not so in the case of Π, because Π in the true sense only indicates that the circle is a square without corners and therefore Π dictates form and not size. By reducing size only r comes into contest and will point to such reduction. By reducing the circle radius r by half continuously will lead to an infinite small circle but Π will remain because the circle as a form remains even being infinitely small. In the past, and even in some quarters today, science is on the search for the 100% efficiency machine. That theory runs on the surmise that a machine can drive as an output delivery without receiving input of energy. A few hundred years ago many Kings were fooled by such notion and some scientists truly spent a life in honest search of just such a device. Mostly the accomplishment came from cheats that very well new their machines were not up to the task, but in fooling a rich investor, brought about wealth to the inventor. As science progressed the no input giving all output machine became less and lesser a feature of the honest inventor. But the idea does not exclusively come from crooks finding a way to cheat the world.

The practise of receiving without giving comes from science in the form of physics. It is physics taking the world on a wild goose chase in the way physics presents the cosmic motion. Physics propagates that the cosmos is all about running without input driving energy. The cosmos is all about wasting matter to a supply of motion. This idea prevails even after the world of science saw clearly in the past that there could be no such machine anywhere. Even the cosmos must be a machine driven by an input and an output. It is the input / output driving energy that must be located and the driving ability we have to locate. Science holds the mass drawing power to prominence, but what if it is not the drawing power of mass that holds prominence, but it is the reducing or contracting of space that is the driving motor behind the cosmos. All energy we humans at present use to accomplish matter motion, holds some form of heat redistribution. Even electricity is a form of pure heat. I say that in mind of what applies when the energy of electricity becomes over abundant and the machine overheats. By overheating it means that the motion the machine creates comes about from heat control and precisely planned heat distribution.

When I realised that it is not me that is drawn towards the Earth, it is the space in which I find myself that reduces, and that produces the effort bringing me closer to the Earth. The formula $F = G(M_1.m_2)/r^2$ suggests driving, moving in a direction and contracting. It suggests the reducing of space and not merely drawing or moving closer. When looking at any machine in practice, the machine draws power from space reducing whereby heat increases. Not releasing the heat to form space will lead to the destruction of the composition forming the machine. There is no form of matter, or element strong enough to resist matter deformation brought about by overheating. Having this in mind that matter does not resist heat, it is of importance to recognise that it is heat that is allowing space to give matter form. Looking at the manner in which energy is utilised it is space and heat forming matter allowing motion that allows work to achieve value.

At this moment science is all about a body falling where the two bodies are producing a force whereby the bodies draw one another closer. The bigger the mass, the bigger the drawing that comes about from the force unleashed by the mass of matter. The idea about this practise was phenomenal in 1602, it was impressive in 1802, but it is really ridiculous in 2002. Why would Boron form a solid having 5 protons weighing 10.811 g / mol and Argon a gas having 18 protons weighing 39.9 g / mol, but the "heavy" element with the biggest drawing power is a gas and the lightest element is a solid. That denounces the contracting force theory. The way we compile and use energy must be in a similar manner to the way the cosmos uses energy distribution. **We humans can create nothing, but nothing is all that we humans can create**. The rest of our achievements are by duplicating whatever nature provides. To establish what drives the Universe except for blaming some medieval magical force coming from nowhere going nowhere we have to find what drives us. The energy we use in all forms is producing heat in space by either converting space to heat or heat to space. Explosions are about converting heat to space. Compressing is about reducing space to heat. That is all energy composing work and is the only method of producing energy notwithstanding the immeasurable many names we use to express the same function in different forms.

Arriving at the question about locating the space and time forming the centre of the Universe one has to realise the centre of the Universe are in every singularity forming matter be it is big or small, size carries no significance. It is the impartiality of singularity that is claiming the value and not the differentiation of matter. One must realise there are no big / small or hot /cold or near / far. It is all relevancies between matter claiming space and space is heat in a turnabout manner. Every aspect in the cosmos is locked-in Universes, sealed off from other Universes and inclusive or exclusive depending on singularity holding relevancies relating to one another. The relevancies rely on inter dependence and inter linking, but there are no differences according to human sizes or standards. Accepting that principle unlocks the "so called mysteries" of the Universe and brings about clear understanding. It is all about accepting, acknowledging and interpreting the role singularity maintains on matter.

One should not try to focus on an image of such a spot or dot because there is no image. The line dividing the cosmos and that run through every particle, no matter how large or small is beyond our vision. Such a small line, so small it is not even noticeable is large enough to part the cosmos into sectors. It splits the biggest there is into particles and we are not even able to notice the precise location of such a split. In truth there is no top or bottom that we living in 3D can see. We shall have to use a general conception brought about by intelligence. Your intellect tells you about such a spot, but that is all because that spot is on the other side of the Universe (quite literally). From the centre of the dot there is a top and a bottom spot. From those points there is connection with four quarters. That produces six connecting points that are all aligning to the centre. Because it serves big and small, hot and cold equal and alike, and it is the smallest cutting the biggest into equality, size is of no issue. Size is what man makes of it. In the Universe there is no size in hot and cold, large and small. For the smallest there is, it is serving the largest there is equally.

Our instincts, our logic and our calculating process all indicate that the sphere holds a centre point from where six evenly positioned point's position matter to be. Using The formula $F=G \ (M_1.m_2)/ \ r^2$ it indicates to a force pulling objects closer, where each force is coming from each centre point the body in question has. The contraction must commit the two bodies towards a point in each case being spot on in the middle, not withstanding what direction the force is applying, the body will draw to the centre. If the Universe spins around a centre point holding singularity, and singularity confirms the centre of the Universe, then every particle holds the centre of the Universe making the number of universal centres immeasurable many, and every atom and sub atom particle presented outside the atom in smaller bits, are all not pieces of the Universe but they are a Universe surrounded by many Universes. If every atomic particle no matter how small is holding the centre of the Universe, then the gravity is coming about from that point because that is where the gravity applying in the Universe is applying contraction.

It then is the atom in the most centre part where space and time meets singularity, that Einstein found a Universe collapsing to a single dimension, and every atom at a point post of the proton where gravity initiates in according with the proton dimensional colas of $(\Pi^2+\Pi^2)(\Pi^2 \times \Pi \times 3) = 1836$

See the fluid push out of the Sun where it lets the Sun seem to be a bowl of liquid, as the liquid is spilling back to the Sun and not escaping into the cold of outer space as released heat should do. The liquid heat squirts into the cold of outer space and then it falls into the bowl of liquid that is the Sun . The inside of the Sun is not gas but it is fluid.

In all of nature there is no NATURAL GAS as much as there is no NATURAL SOLID.

 No element is either a gas or is a fluid or is a solid. We arrange the elements in such a manner, but that is only applying to the situation the earth grants the elements.

When an element freezes it is solid notwithstanding...

When an element melts it becomes a liquid

When an element boils it is a gas again notwithstanding..

 Hydrogen is as much a liquid as iron is a gas and neon is a solid. It depends on the element relating to the space/heat in the circumstances surrounding the substance at that very precise instant in time. We have to stop telling the cosmos to show us what we wish to find and start accepting what the cosmos is telling us to find. The culture that I am referring to is all about **nothing.** At present we find that there is something we think of as nothing in outer space. Because nothing is what we wish to find and nothing is precisely what we are getting because we think of outer space as nothing. If you accept the cosmos to be nothing, then please define nothing to yourself and find the definition in the cosmos.

The liquid the Sun has is the driving force that creates the duplication in motion. Without such liquid heat the Sun would become stationary and only depend on contraction while the contraction then passes the motion onto the heat in outer space. While the Star is in liquid all motion comes from the accumulating spin effort of the combined motion all elements together accomplish. In the case where the star is still in liquid the heat is stored in the atoms and as the star develops the heat transfers to the governing singularity, which makes the star immobile as the governing singularity takes charge of the entire star.

The formula of $F = G (M_1 \times m_2) / r^2$ only apply in a very specific range, and at a very determinable point the formula does not effect objects in the air. After such a point one will find satellites able to orbit, be it art a definite pace that matches the rotation of the earth. Still...below such a point (B) orbiting objects will come crushing down to the earth.

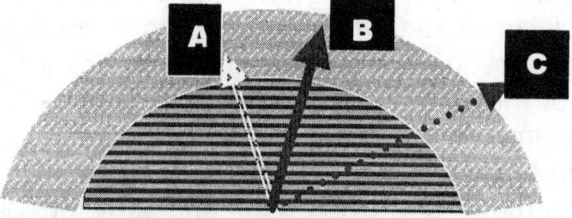

From point (B) to the earth Newton's formula apply and from point (B) upward Kepler's formula apply, but my pointing this out brings about all sorts of annoyance concerning academics. It must be clear to all persons that there are a big difference between the applying of Newton's $F = G (M_1 \times m_2) / r^2$ and Kepler's $a^3 = T^2 k$. When the objects reach some point

they will drop to the earth and when that happens, mass do not play a part in the speeds they come to reach.

When examining the case where two balls drop vertically, gravity, as a force does not apply and therefore gravity does not come into effect because there is no difference in speed or duration.

With out any apparent reason the formula is substituted with the following formula:

g = G(M . m) /r^2 where:

G = the gravitational constant,

M = the mass of the body,

M = the mass of the lesser body

r^2 = the radius between the two bodies.

Let us take this formula back to the accepting of the Big Bang and find sensibility amongst a lot of confusion that I can see.

There was a beginning that saw a radius between objects so small that the size will never again be repeated. The diameter of the particles were also next to nothing but that should not be a contributing factor surely...the main focus point is that particles were as cramped as it shall never again be repeated.

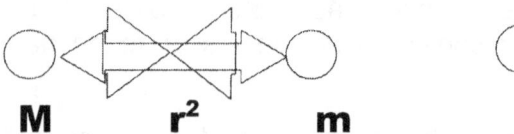

M   r^2 m   M   r^2 m

With the radius in the square dividing the shared and combined mass of the particles the relevant mass of the particles rises by the square as the radius reduces. If the radius becomes infinite, the relevant mass that the particles will produce goes up eternal. No force in the world would keep particles apart drawing on each other with an applying force but such a force is divided by an infinitely small separating radius. This is a recipe for joining and not dividing. Still according to the Universe I am able to witness that the dividing became enormous and the joining practically irrelevant. The gravity was more than words can describe, the heat was able to melt it all in one structure, but that did not happen. It split into billions of individual atoms.

Another point I question about the Official Policy is that they as I am are in agreement that the heat 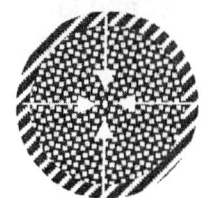 melted particles onto particles and in those joining better combinations of particles came about. How it happened is another bone of contention but more about that a little later on. There was heat on the outside and there was matter on the inside. The heat was liquid because the Sun and other stars still indicate masses of liquid fluid inside. I can only imagine that that liquid inside the Sun holding temperatures as low as 6500^0 K and up to 1.8 X 10^6 K the heat already is in a molten form. What about the heat then when the frozen outer space was 10^{34} K and such temperatures were the general order of the day back then. If the Sun is liquid now then those temperatures raging back then must put the heat in form available in outer space at the time as thick as mud.

From the outside drawn onto the particle inside the blanket of heat came a flow of soup that became matter. That much I do understand. This carried on until...when? When did this stop? When did the Universe run out of heat? When could one consider outer space as the coldest all around? Where to did the Universe dismiss the heat that was once there but now is empty? How did the process stop of bringing from space intense heat and from that particles grew stop? When did it stop affecting the growth of a particle, the growth of space? In fact the growth of everything that grew came from this first growth. What you see or do not see grew since it was part of the Big Bang and everything in the cosmos at present was part of the cosmos during the Big Bang. I say this process of collecting heat from outer space never stopped but was an on going process we now give a nice name calling it gravity. Outer space never became empty and void but relevancies changed concepts where centres formed that should not be as it then interferes with concepts about relevancies Gravity is not and never was about particles pulling each other closer.

If it was, no Big Bang was possible. Gravity is about turning space, which is released heat back to heat and concentrate the heat where gravity is the strongest and heat is the least. Space is the transverse form of heat and visa versa is also true. Should any one not believe me, try a bicycle pump by compressing the plunger while blocking the valve bit. The heat will burn your finger to blisters if the force on the plunger is strong enough, the plunger seals enough and your ability to withstand pain can last that long. Then answer your own question about where the heat came from because sure as hell is hot, it did not come from friction with air particles such as oxygen and nitrogen escaping through the valve bit. Heat is unleashed space and space is concentrated heat. Reducing space to heat is gravity and antigravity is expanding from overheating blowing into space accumulation.

 When looking at a sphere the inside has always (in a cosmic relevancy) the location with strongest heat also always has the strongest gravity in any given cosmic sphere. The centre of the sphere clusters the combination of particles forming the sphere into unity. By holding a specific centre the sphere becomes the strongest form any object can be. The sphere is without any doubt the favourite choice in forming gravity. Where gravity has the last say without other influences changing possibilities as collisions leaving debris in space or natural out burst like Super Nova explosions, gravity will enforce the sphere to be the form taken by the particle. But there is no evidence of particles of similar size joining in matrimony through gravity being the shotgun at the wedding. In cases where there is a mismatch of size outside any proportions of equality then there is a contracting of the lesser by the greater. In such cases the lesser is not qualifying as material (and that I prove later on) but the greater considers all the lesser to be heat. It is humans bringing distinction to matter in form.

The two objects should have their own value of gravity and _gravitons_ and in comparison with the _gravitons_ of the Earth; their value is insignificant. However, these two balls are in their own individual deuce to see who reaches the Earth first, and the iron ball's _gravitons_ should give it a superior advantage. This comes about because the two objects are in a position where they compare in relation to one another and share a common second factor, which is the Earth. In relation to the Earth, the gravity - motions of the two balls do not come into consideration, but this does not play a part since the Earth is a common factor. The balls, however, is put in a situation where they stand in relation to each other. When compared to one another, the _gravitons_ should give the heavier ball a sizable advantage. The sensible example one can show to prove that where some structures matching in size come into conflict about occupied space sharing. In such an event one of the structures are turned to heat as it is liquefied flowing in the space dominated by the other and larger structure. If the structure proves too large the superior structure turns the lesser compatriot into heat. Then being heat it will apply gravity and admit such heat into the ranks of its atmosphere, but not before it turned it into fragments good enough to be heat.

The Official Policy Protectors never tries to explain the relation between Newton's laws as mentioned above, and the binary star system forming the principle we know as the Roche limit. The binary stars are systems where two stars spin around each other and never collide. These stars are many times over the size of our Sun . When one applies the same Newtonian formula as given above, these massive giants must crash into each other, destroying themselves in the process. The enormous mystery is not in the apparent misbehaviour of these giants, but the fact that this is known to science since the previous century. Relate the binary once again to the comet/ Sun relation and there is a distinct similarity.

With the comet, the Newtonians regard a force to attach to the Sun in some way where this force pulls the comet towards the Sun . At the same time another force joins in that pulls the Sun closer to the comet, but such is the mass difference between the Sun and the comet, the force the comet applies never realizes. In view of this, only the force the Sun applies, comes into effect. The comet proves this force by speeding up its movement as it comes closer to the Sun . If the force did not become greater, why would the comet gain momentum?

With the arrival of the comet in the Sun 's domain, the Newtonians leave the argument to be. The Sun applying the force should remain applying the force and the force should increase all the time, accelerating the comet to the point of splash down. We must all argue that gravity is a force, which pulls an object to the centre of the larger object wherever that centre may be. The very same force that pulls the Chinese down, is pulling the Americans, and if not for the surface of the Earth's intervention, the next world war would be between the Chinese and Americans for King and country, honour and glory and to find who has the most powerful gravity force that will provide space to live in. If not for the Earth stopping matter falling right through the Earth because of conflicting forces on both sides of the Earth the Chinese and Americans will then have to establish border checkpoints in the centre of the Earth. The checkpoints will indicate where the Chinese gravity meets the American gravity and by allowing the force of gravity to find borders, we will finally have world peace. The only problem is to find the position where the Chinese gravity meets the American gravity and the two forces nullify each other. Just think if the forces of gravity, and not man, will intervene to set border standards: that must be the answer we were always finding a question for. This is a study far too complex to bother the United Nations, so we can find a more suitable group to investigate this fact to bring about world peace.

I am personally part of Africa, born and bred in Africa as an Afrikaner. I know the African solution to such a problem. In Africa, appoint a committee to investigate and then wait for everyone to forget about the problem in investigation. Therefore, such a problem is far better solved in Africa, because those in government aim to receive maximum western aid but never aim to solve problems, you make it go away by postponing the solution to the unsolvable. The African way is to ask the west for aid in order to create another useless committee to become over paid and under worked, quite capable of dealing with any non-existing issue of any magnitude that will never find an answer. Then sit around at leisure and wait each month for pay day to come for many years while the west is paying the committee to be bored until their pension dates arrive. By then no one would know the name of the committee and much less remembers the problem investigated.

On the other hand, the Newtonians are doing quite fine by their method on their own using a technique they apply for three and a half centuries. To solve such problem, the Newtonians will apply a very different solution: Blame gravity's boundaries on a non-existing force, brainwash all future students in accepting it to be a force by telling them they will accept the force and forget the problem or fail the examinations and be chucked from campus, because that solved the problem so far. By the time, the student reaches a senior position he (or she) will no longer bother their mighty brainpower with the little aspects. They will advance to a point where they can move Black Holes around, travel at the speed of light, and divert time back to the past while others calculate all the mass seen and unseen in the Universe and any other ridiculous notion they may find to test their personal brilliance. If you for one second think everything about this last paragraph was silly, the silly ness started with GRAVITY ON BOTH SIDES OF THE WORLD, opposing each other, and that idea is not mine!

This is where the century, old trick of the Newtonians work best; do not think any further and no further problem will arise. Leave it at a force because with a force and thoughtlessness applying even-handedly, the problem never surfaced yet and that continued for the past four hundred years or so. So why bother with a problem that bothers no one. When a fellow like Hubble proves quite the opposite to Newton's claim of attraction, get a man who has a bigger ego than a brain and tell him to measure the Universe. It will keep every one involved occupied with something senseless while the problem vanishes through the many centuries to come. It is a force, and the way of all forces is mysterious, but never admits in believing magic. Those that do not accept forces to be of a mysterious nature should just contact astrologists and come to their senses about forces being not understandable. With everyone in agreement about forces, their nature and unpredictability, who then needs more real problems to solve?

No big-brain should bother about little issues like comets when there are so many galactica to conquer. Apparently the comet-problem just will not disappear. Something broke the force, something interrupted gravity. Let us see what happens. A force means it acts the same way as tying a rope on one object and start hauling the tied object in. The longer the rope is, the less control will be on the lesser star. As the rope shortens, the better the control will become. By implying that gravity is the force, our Newtonians tell us that we have to regard gravity in the same way the rope is hauling in a comet. It is something like fishing where the comet pulls, and the Sun pulls and eventually the angler gets his fish. One may argue that the rope is not the force because the force is actually the hauling, or shortening of the rope. I have had Newtonians trying to avert the problem they refuse to see by bringing in this argument. This manner of reasoning has the same value as introducing the African committee of investigation that will never uncover an answer. The rope is the extension of the force in a way being the sole representative of the force and the instigation acting out the force. The rope therefore is the force, extended somewhat, but still acting out the application of the force.

Weather the rope eventually broke, or hauling stopped, the effect as far as gravity applying its force, the process came to discontinuing ... and we know that gravity is a force that pulls something to the centre of the body in control of the gravity. What made the force act in defiance of its nature? Why did gravity change its mind? What stopped the Sun applying its ferocious onslaught of the body holding the poor defenceless comet? Non – Newtonians will blame me for exaggerating, but I know that there is no Newtonian that can understand my argument. In that light, I ask non-Newtonians to show patience, because there may be a few Newtonians that will also read the book and to them everything said this far does not make sense.

This is where tutoring comes in best. Should a student bring up such non-academic and spiteful thinking about the mysteries of a force, then the lecturer sets a date on testing all the students' reaction about how much they accept the force. When any student shows signs of defying the force the lecturer can fail him on the spot and have a good reason to drive the silly youngster from campus.

By ignoring the problem as to why this comet brakes free from the gravity of the Sun , and continue in its freedom until gravity is at a point where it is most weak, may not bring answers, but it surely avoids nutty questions! Questions are not there to interrupt Newton's laws! Ask any Newtonian High Priest and he will either tell you that in a very roundabout way or he will simply ignore you by telling you to your face that you are incapable of understanding Newton. The best way to get out of the answer of course is to tell the sod with all the questions he has not the qualifications or the mental capacity to understand Newton. That will make the pest retract to some ditch he should be in, in the first place without bothering the greater minds with some stupid minor issue. How do I know this you may ask? I have been down that alley many times and treated with that precise treatment on occasions more than I care to remember.

Still, the comet defies the force of gravity and my questions remain unanswered.

Dear reader, if you wish to read the funnies, jokes and laughs -a- minute, treat yourself to some real good clean jokes. Read the Newtonians explanation about how comets came about; how they get to the Sun and where they came from. It is going from the ridiculous to the thoughtless and ending in the realm of the mindless. However, be warned! Only do this on occasions where you feel very depressed. The jokes will otherwise drive you in a state of laughing hysteria. Poor old Newton was considered a very dry humourless chap in his day. To think what silly ideas can come from his forces.

Hauling in and releasing something caught on a line is called playing with fish. We might say that Newtonians love fishing and confuse planets, comets and fishes when they regard the interaction of comets with the Sun . There is only one small problem with that argument and that is that fishermen and fishes form part of a second natural force named life. Life stands apart from the cosmos. Life and the cosmos only share time in space, not a joining of forces. Beside that, comets were part of the cosmos long before life had any role to play, so blaming it on someway life interacts with life does not cover the solution.

Why would the comet brake free from the Sun 's gravity? That is defying the law of gravity. Far worse than that still, is the fact that the comet's actions have the nerve to defy Newton. No one alive can defy Newton and remain alive. Does the comet not realize his actions contradict the all-important Newton and the gospel of the Newtonian - Priesthood. The best way the Priesthood of Newtonian gospel can deal with such defiance is to ignore it and no one will notice the actions of the comet. That is the scientific approach. Ignore and forget the problem. It is as simple as that.

With that let us conclude comets and really enter the world of forces at work! Let us now apply our attention to the forces of planets.

The next formula is very simple to understand. It is the fight of understanding the applying that becomes not applying it that is troublesome. If you understand the applying of the working and never spotted it not working, then forget it. You are a brainwashed Newtonian and if not, well there is still hope that you have a clear mind left. The resentment you carry with you from childhood about the formula is in, not understanding it, but accepting the outcome of the formula you never could understand. Newton said that the force between two objects depend on the mass of both objects multiplied with each other and with the gravitational constant and the derived product you divide by the radial distance square that separate the objects. I shall put this in a mathematical language for your enjoyment that will explain the life-long not understanding to better effect.

$F = G (M_1 x M_2) / r^2$. What does this say?

The greatness of the force depends on the masses of the two orbiting objects, aligning that product with the contribution of the gravitational constant. This then, you divide by the square of the distance between them at any given point.

Please, in all fairness to you, the reader, I have to warn you that quite a number of professors in physics told me that by reasoning in the manner I do, I only prove that I know nothing about Newton and understood even less about his work. Considering such allegations, I shall explain to you what I understand in as much as telling you what I know.

The Newtonian's formula states that the force between the planet and the Sun will improve as the mass of the planet increases (becomes bigger) and by multiplying that with the universal gravity constant you will get a value that will become lesser, the larger the distance are between the Sun and the revolving planet. With the reducing of the distance the mass on either side must therefore be on the increase because it holds an inverted relevancy. This means the Sun is pulling according to its mass. The planet is pulling according to its mass. The gravitational constant is influencing the pull evenly at both ends and the distance between the objects will reduce to the square value of the force's total application. I could never see what part I do not know and what I did not understand. No professor ever explained to me what it was that I did not understand either. That left me in a place where I did not understand what I did not understand and I never could see what I never could see. I shall try and make sense of my not understanding my not understanding as follows:

This is like having two balls attached by a rope on a floor that holds the same drag on both balls. When I reduce the length of the string, the bigger ball will show a greater resistance than the smaller ball, therefore the larger ball will apply a larger tug than that of the smaller ball. The rope will reduce (become shorter) at the end where the larger ball is than at the point of attachment where the smaller ball is. What is wrong with my argument? When the two balls are so miss- matched in mass as is the case with the Sun and the comet the one ball will do all the moving, leaving the larger ball stationary.

Surely the tugging at the larger end must bring the smaller object closer. By comparing the mass differences, you will find there is no comparison. The smaller object just has to come closer with the application of such a force as gravity. We know that gravity can really pull. By standing on a tall building you will find proof of this. Drop a tennis ball down from the building's roof and see for yourself how it falls. The distance between the Earth and the ball reduces by some speed. With that being obvious, the distance between the Sun and any planet have to reduce as the planet orbits the Sun each year. Even if it is small, there has to be a visible reduction after four and a half billion years of pulling and tugging! Today after wrestling this problem for the duration of twenty-five years I can say (with a clear mind) I finally know how it works. It does not work!

I was always looking for mistakes on my part. At first, I thought there are a fifth force that I am unaware of because of my slender education, a force the academics can obviously see, but I cannot through obvious lack of education. I thought that my personal ill literacy gave me a blind spot that every non-educated have and was born with. The blind spot cleared only thorough education as education removes it in the way only education in science brings knowledge. I thought the removing process similar to the way washing removes stains and spots from whites; education can remove blind spots through the process of intensive tutoring. All I wished for was some academic to help me remove my blind spot about comets and their behaviour. The comet's behaviour, I could see, was an exaggeration of orbiting patterns applied from our planets orbiting around the Sun ; in the way, we observe galactica in the sky.

Then finally I came to the point of accepting defeat. It was not I, with the blind spot; it was all the academics brainwashed into a state of having such a blind spot. Science insists on repeatedly ignoring mathematical principles, because Newton had his claim to fame with one single calculation, THAT HE, IN FACT, DISCARDED, BY THROWING IT AWAY.

He made a brief calculation as a young man that saw an apple fall from a tree. Seeing this he jotted down a formula and the chucked it away. His piers and elders picked up the trashed paper with the calculation, and got all excited by the logic implication it had. $F = r^2 / (M_1 M_2)$. The mass of the two objects destroys the radius between the objects. Everyone went ballistic, proclaiming him as an instant genius, the one the world was waiting for after the crucifixion event.

I do not, for one second, deny or dispute the revelation. What I do encourage is place the event into its correct context. It was merely, and simply an apple that fell from its branch to its roots. The apple did not pretend to be a meteorite that fell from the heavens. If it were a meteorite, I am sure, with the man's genius, science would be somewhat different at this stage. However, as a young man, being very impressionable, as all young men are, and with the attention this brought about in the world of science, the matter overshadowed the fact.

I am not disputing Newton; I am disputing the relevance of Newton's scientific breakthrough. It was not two objects of cosmic proportions, colliding in a show of spectacular. It was, after all, only an apple falling from a tree. With this miracle revealed Newton found he was competent to improve on the work of Kepler and if I may dare say this, there must have been some political agenda behind this act and the accepting of it for Kepler was a German and what German can ever teach any Brit. The very same politics are still the order of the day forming international rivalry on all fronts.

Newton, and science, made one enormous blunder, from this stance. They took the radius of a wheel not to have any influence on the wheel. In doing that, they removed the very fact that keeps the universal attachment together. They put two objects in an attaching relevancy and then announced no relevancy. Doing that is breaking the most fundamental mathematical principle.

$$\frac{dJ}{dt} = 0$$

This disputes mathematics. DJ / dt can have any number from eternity to infinity, only excluding one; it cannot be 0. By placing the one in division of the other, you bring in relevance. You cannot then say there is no relevance. By doing such, you proclaim that one of the factors is non-existent.

$$\frac{dJ}{0} = dt \text{ or } \frac{0}{dt} = dJ$$

In both cases, one of the factors then does not exist. Such a claim is incoherent, because you proclaim that a circle has no radius, or a radius has no circle. When calculating a circle, you multiply either the square of the radius by Π, or the quarter of the diameter at a square by Π.

$$\frac{dJ}{dt} = 0 \text{ constitutes a circle and is also therefore } \Pi \times r^2 = CIRCLE$$

If you remove r it then is $\Pi \times r^2 / r^2 = CIRCLE$.

You cannot then say $r^2/r^2 = 0$ and therefore $\Pi \times 0 = 0$. That is nonsense. $\Pi r^2/r^2$ will always be $\Pi \times 1$, and that is the eternal circle.

When looking at any rotating object, there has to be a point of no rotation and no rotation means "no rotation", not no existence. No rotation means a factor of 1, not zero. That then is singularity. The eternal Π, the Π that may not have significance but still it is a Π of value.

  When looking at any rotating object, there has to be a point of no rotation and no rotation means "no rotation", not no existence. No rotation means a factor of 1, not zero. That then is singularity. The eternal Π, the Π that may not have significance but still it is a Π of value.

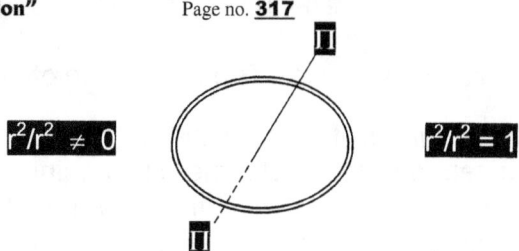

$r^2/r^2 \neq 0$ $r^2/r^2 = 1$

The relativity remains one, eternally one, but it cannot be zero. Therefore, dJ/dt cannot be zero.

When explaining this to any child, they can immediately see that. Explain this to any Newtonian High Priest and he may have you removed forcefully from campus. I cannot find one Newtonian, large or small to accept that. By not having a wheel rotate the rotation seize, not the wheel. When the wheel begins to rotate, you cannot state that all things remained as it was. With the wheel in non-rotation the rotation still exists forming the infinite possibility of rotation. Then afterwards the wheel starts to rotate and by the start of rotation the circumstances surrounding the wheel changes. A wheel in rotation is very different from a wheel not rotating and therefore cannot be the same thing. By establishing non-rotation, the wheel becomes the factor of one, and the rotating action becomes zero. The wheel does not disappear. But in the same manner does a wheel in rotation not remain still.

In the cosmos, everything is rotating because nothing ever stands still.   Therefore the mean equilibrium, the common factor there is to share, has to be one, eternity, the eternal Π, because all rotating objects has Π in singularity, and sharing singularity, gives every object in space a relation with all other objects in space. After trying for many years to bring our Brainy Bunch the candle, I concluded that Newtonians are incapable of realizing that mathematical principle as a reality. They maintain they know mathematical principles far better than an ill literate such as I and yet….

The comet rotates the Sun , and the Sun by itself has a point of singularity where Π remains without r. The comet, holding the orbit, also has a point of singularity, but since there is space separating the two objects, they cannot share a mean point of singularity, the very point of existing.   Since singularity means just that, being single, there cannot be two. The comet and the Sun have a mean point of singularity but the space they occupy divides their common singularity. That is why they orbit in an oval path, a path where the one structure holds on to more space from its point of singularity towards the space it claims. Since they do not claim equal space, BY THE DENSITY they hold, the space will not be in proportion. They do share in the common fact of singularity and singularity cannot   be   two,   because   then   it   will   be "dualarity" or (in case there is no such a word) duplicity where both find the space they occupy, with the space they hold, will be their individual eccentricity from singularity. The two objects are holding eccentric space around their individual but common singularity forming a point of mutual singularity in accordance with the individual singularity both claim space from. That point of singularity is Π the circle without the radius because the singularity removes all forms or values of r, leaving Π to be singularity.

That is why Newton is bullshit, and his F = G(M₁M₂)/r² is utter nonsense. The moment you say Newton or any of Newton's laws, the Newtonian brain stun. For all the life in me, I could not once find one single Newtonian to see this. If you say Newton is wrong, they spiral down to frenzy, and just mention gravity and they all fall on their knees, cover their eyes in the ground, start praying and you cannot make them say anything other than Newton is correct. Dare say there is no such a thing as gravity and Newton is wrong, they have you in an armed escort patrol, straight to the department of mental disabilities and psycho diseases in preventing you committing acts of extremely dangerous life threatening behaviour to yourself and others.

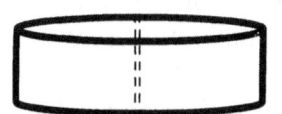

What is it the Newtonians fail to see? They fail to see the relevance applying. They fail to see that the Universe not only holds the atom, not only comprises of an accumulation of the atom but that the cosmos indeed functions with gravity and all as one massive atom. The Universe is exactly the atom that the Universe formed which then forms the Universe. If an electron is orbiting around an atom, the inside of the atom must be a circle. If the atom was not a circle, it then had to be a cube. The electron cannot rotate around a cube; therefore, the inside of the atom is a circle.

In a circle, there is a radius that initiates the circle. The calculation of such a circle is $\Pi \times r^2$.

The radius r runs from the circle outwards, from a circle centre point towards Π, the value of the

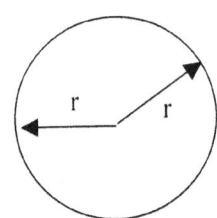

circle. In the centre of the circle, there is a point where the radius starts. It runs outwards from that point in all directions towards the circle Π. Technically, there then has to be a point where r is infinite and not zero, an absolute infinite. However, the circle therefore remains Π. The circle does not disappear; it remains there for all to see. It is only the radius that almost disappears into the infinite, but it does never become zero!

$$\frac{\Pi r^2}{r^2} = \Pi$$

If one removes the radius from the circle, the circle remains, only holding the value of Π. By removing the value of r, Π becomes singularity with no place to be. Singularity is the place where there is no space to be in place. However, Π remains because once r receives the slightest of space Π will find space. Then the circle will grow to Πr^2 and r would determine the space. Without space, there is no r but there is a circle with the value of Π. Singularity is in every single rotating object, be it the proton or the combining effort of all particles in the Universe. That is what light and the photon is. It is concentrated heat that the Sun (or any other generator of electricity) connects heat to singularity where the heat receives either temporary connection to singularity or a small piece of individual singularity.

At first you as the reader may think I am trying to create a mountain from an ant heap, but in scientific terms the human race is preparing for the start of the cosmic journey. By completion of this book you will realize how _Xepted_ science believe they built science on a solid foundation, and, boy are everybody in for a rude awakening. Compared to the leaning tower of Pizza, science is about to start with the next section of a much bigger building adding many levels and already the view at the bottom where I am looks far worst than the leaning tower does.

If I contacted and argued with one Physics Lector or Professor about Newton, I have been in correspondence with at least a couple of hundred. What prompts me was the comet's orbit. The commit truly fascinated me from my childhood days in the way it defies all the laws of gravity. Since my very young days, I was in search of what I at first believed to be a fifth force. I have raised the argument with just as many people not schooled in the art of physics and received a very different response. The most amazing aspect was the fact that the two groups were that far polarized. The non-physics group reacted astonished, amazed, disbelieving and reserved about my view about comets at first, but with their distrust not withstanding, everyone saw my point.

The non-Educated responded in the same manner that I did at first. They argued that I was missing something of vital importance because "why do the wise not see it", was their argument. They always were of the opinion that I was too little educated to understand, while the educated was of the opinion that I was too little educated to understand. Neither party had the same view about my not understanding. The non-Educated understood my argument, but dismissed it on the fact that it was so obvious I missed the rest of the knowledge behind the facts that makes my arguments too difficult to understand while the educated dismissed my argument that I could not see anything they could not see. Education brings the ability, which then made me unable.

In short, they thought I was to stupid in order knowing the rest of the story. Polarized to the non-Academic view was Official Policy Protectors where not one academic could understand my argument. The academic response was as much defending the Newtonian view as it was drawing a blank about my questions. They all seemed as if their ability understanding my view was completely locked behind some wall. The non-Educated, of which I am a member, at least understood what I was saying, but dismissed the simplicity about the argument. In the corner of the Official Policy Protectors, was no response of any kind, but to feverishly defend Newton by raising the dumbest arguments I have ever heard. The arguments, even the most highly educated brought about, seemed motorized and non-responsive. When it seemed their accepting the points I raised with my questions would demise their senses, in defence they put up a block. There is a peculiar sense of numbness in the way they could not understand what I did not understand. The academics showed no signs to indicate that they could even argue my point of view, by responding that I have an argument, and from that launched a responding argument to explain how or where I made my mistake. Their abilities in even understanding always seemed to hide behind a wall of not understanding that someone may not approve of Newton's arguments.

Newton says two pieces of rock will draw each other closer by reducing the distance keeping them apart. That we all can see by merely jumping in the air. No sooner have you lift off than you are back on the ground. That is what Newton said about three hundred and fifty years ago. Even Trying to tell the Official Policy Protectors that Galileo said mass of an object has nothing to do with the falling, seemed to pass the Official Policy Protector's sense's of comprehension by miles. I was told on so many occasions that I did not understand Newton, but there it stopped. No one could explain to me what it was I did not understand about the comet missing the Sun by miles, where it was supposedly to hit the Sun with a dazzling impact. To this point, I cannot get through to them as much as they cannot get through to me. Our understanding is so far apart, we do not share the same planet, and yet after all my arguments and investigation no one, and I repeat: not one could once clearly tell me what it is that I do not understand.

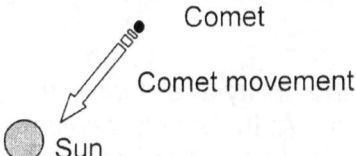

You have the Sun and you have a tiny piece of rock covered by water also better known as the comet. There are thousands of them flying around, but never aimless. At first Newton's formula makes pretty much sense. The Sun draws the comet towards the Sun , as Newton said it does. The comet responds by speeding towards the Sun , also as Newton predicted.

Anyone can see a collision coming ten miles away. The Sun applied gravity, the comet applied gravity, the Sun is far too massive to fly to the comet, so the comet with much less mass does the flying on behalf of both objects. Every person with even the least of knowledge about science knows how the gravity application works.

The gravity of the Sun collected the comet from no-one knows where, pulled it through billions of kilometres to the area where the Sun produces the gravity with which it pulls the comet where the comet is to find its last resting place. The mass of the Sun is obviously so large, it could produce gravity that can locate any comet hiding anywhere and collect it as a souvenir. What is there to understand?

The gravity of an object always points directly towards the centre of the object, the very, very middle point. Concluding from the fact that the comet is heading towards the centre of the Sun , just as much as the Sun is heading towards the centre of the comet, would not be out of line. The two centre points are heading for a direct collision, the collision becomes more and more unavoidable as the radius reduces by the value of the gravity that is produced the mass in accordance with the gravitation constant. The comet is heading towards the Sun , and by not even moving, the Sun is moving towards the comet by attaching the movement the Sun were suppose to have, on the comet. Newton's law proves to be exceptionally correct.

As the Sun /comet, radius reduces, the radius separating the mass of the Sun and comet effectively increases the relativity of the mass influence on each other in the form of gravity. The mass of the Sun and the comet increases by the factor of reduction of the radius separating the two objects. That will produce a growing gravity force as the comet / Sun radius becomes smaller. By the time, the radius becomes one the mass will grow on either side by a relevancy of 100, and when the radius becomes infinitely small, the relevance to the mass of both structures will raise a force with eternal power.

$$\frac{M_s \times M_c}{100} = 1 \times F \qquad (r^2 = 100)$$

$$\frac{M_s \times M_c}{50} = 2 \times F \qquad (r^2 = 50)$$

$$\frac{M_s \times M_c}{25} = 4 \times F \qquad (r^2 = 25)$$

$$\frac{M_s \times M_c}{5} = 20 \times F \qquad (r^2 = 5)$$

At a point, where the comet / Sun apply a force of immeasurable strength, the comet brakes this immeasurable force. Remember the direction of gravity always point to the centre of the object, and that is where the collision is heading. As the objects draw closer, the distance reduces, but in accordance to the relevance the objects also become that much bigger in drawing power. It depends how one considers the relevancy to grow by the approaching nearness diminishing the distance between the objects.

Then out of the blue, the comet finds the ability to eliminate the eternal powerful force of gravity, and keep at a safe distance around the Sun . At this point, Newton goes sour. Nothing Newton predicted is happening. The comet and Sun not only stabilized the force, the force begins to decrease as the radius between the comet and the Sun is on the increase AT THE POINT WHERE THE FORCE IS THE STRONGEST, THE COMET BRAKES FREE AND SLIPS AROUND THE SUN , UNSCATHED.

UP TO THIS POINT I STILL SEE WHAT THE BRAINY BUNCH AND NEWTON SEE, BUT IT IS FROM THIS POINT ONWARDS THAT THERE COMES THE POINT THAT OUR MUTUAL POINT OF CONCENT DIVERTS POINTING OUR MUTUAL POINT ABOUT THE POINT OF AGREEMENT TO THAT OF OPPOSING POINTS WHERE OUR VIEW SEPARATE BOTH HEADING IN OPPOSING DIRECTION ON AN ETERNAL DIVERTING PATH.

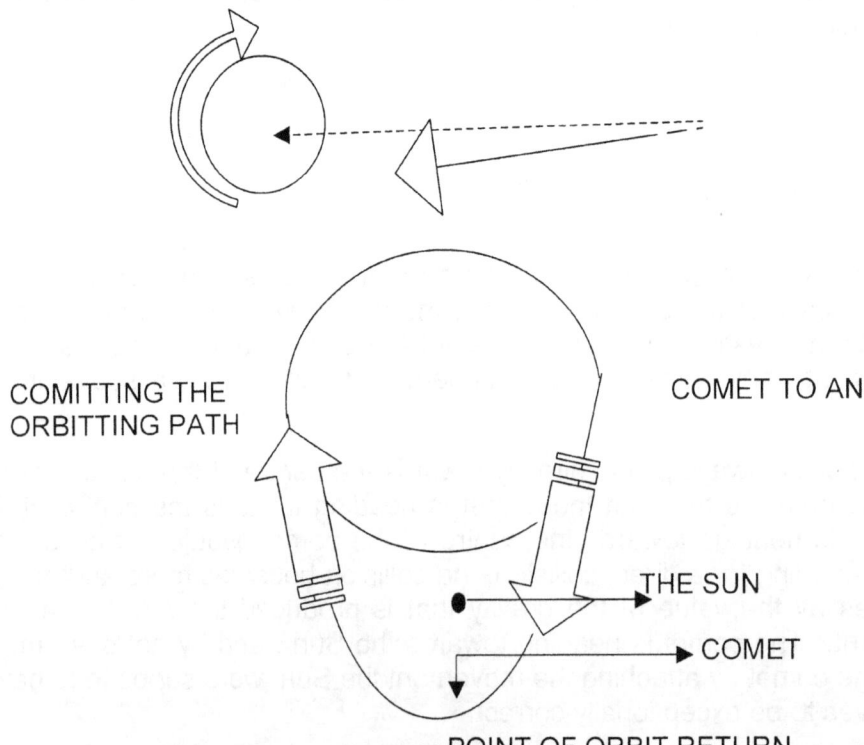

COMITTING THE ORBITTING PATH

COMET TO AN

THE SUN

COMET

POINT OF ORBIT RETURN

Then, in complete defiance of the Newton Law on gravity, quite the opposite applies. At the point where the radius that is separating the two cosmic objects is at its strongest, it will also bring about that the gravity force is at its weakest. At the point of almost no ability the gravity force suddenly releases enough strength to break resulting in the parting of the two structures. The force now curbs the rebel comet on its way escaping the Sun 's gravity for the very last time.

At the point where the force was the greatest, the comet overcame the force, but where the force was the weakest, the force overcame the comet's rebellion.

The correctness of my argument is no longer the issue. It was twenty-five years ago, when I still held the impression that I was missing some point here. I do not state this phenomenon any longer in the hope of bringing across some flaw in my understanding. The flaw in my argument is not there because the flaw is science as a whole.

I could never understand the reason why "the ordinary", like me and others with my development level, can see what I can see, yet academics that has more brainpower in their heads, than I have life in my body, were unable to see such an obvious conclusion. You; those Official Policy Protectors are my superiors in every sense a human can have, with the brainpower to break a wall, and yet you cannot see how far the tower of Pizza is leaning over.

I make the point to help you, the reader, to judge yourself. If you are able to see the validity in my argument, you are not brain dead. Education has not yet bashed your thinking ability out of your scull. However, if a cloak of not understanding roll over your brain, and a numbness sets in on your ability to reason about this phenomenon, beware, you are a Newtonian. Newtonians should read this book very slowly because the effort you are about to launch, may be the most painful you shall ever experience throughout your academic career. You are going to suffer from reconditioning and Newtonian withdrawal, not that dissimilar to that of an addict in rehabilitation. You are going to reject me, hate me, despise me, loath me as you never felt about anybody else. If you think I am sarcastic, I am not. You will reach a point where you will abandon the reading of the book. You have my sincere sympathy and with all the soothing it may bring, know that you are not the first I saw getting such painful Newtonian withdrawal in rejection of Newtonian doping.

Once more, this phenomenon should not occur with Newton's presumptions about gravity. These bodies will and must collide and destruct, without a doubt. When the formula $F = \dfrac{M_1 M_2}{r^2} G$ applies, there should not be any force which is able to keep them apart especially when r reduces to almost infinity compared to what it is at maximum. However, they do exist and what is more, they maintain a certain distance apart.

With the "force" of "gravity" "pulling" the stars closer using the accumulative mass of the stars and multiplying that value with both objects by the mass component, this will reduce the radius r^2 progressively until r^2 reduces to zero. Seen from this view, it is little wonder that the significance of this was lost in the notion that this is yet another "mystery" of the Universe. The scientists of the day (and the past) lost the importance, which this holds for us as Earthly dwellers.

A most surprising aspect of this is that it is not that an unfamiliar or rare phenomenon. However, any answer to this would clash with Newton's presumptions, and before the scientists allow that to happen, they would much rather ignore what is obvious. However, what is the obvious?

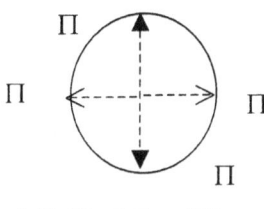

$4 \times (7 = \Pi) = 28$

The cosmos works in relevancies that have a ratio to singularity. I am desperately pressed for printing space but I will show a small example of the implications of this statement. The ratio of space to time or then the more commonly used terminology is phrased as the curvature of space-time inclines by Π From the Earth's perspective matter holds space at 7 and matter in rotation is 4 giving a allocation where the moon should be at

The moon holds its singularity in relation to the Earth in singularity. Because the moon is in a Roche limit (near enough) the proton value the moon accepts is $\Pi^2/4$. The neutron position the moon holds, relating to the Earth is 10 and the moon has an own point of singularity, forming the electron edge of the Earth.

The Earth is $4\Pi^2$ = 28 days to rotate to one moon cycle of 1. The relevancy of the Earth, taken from its point of singularity is 7 (matter in relation to space will always be 7 to the space value of 10).

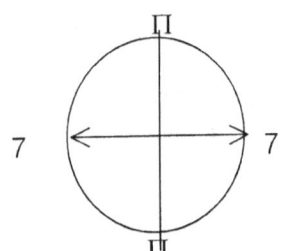

Therefore Π in singularity will accept 7 as a factor forming Π^2 giving the space holding matter on one side the value of 49 (7^2). Adding the one of singularity and half of the time factor will then be (49 + 1 = 50) on one side of the Universe

To get a relative position that the moon would hold would be the 4 quarters material ha in relation to time by ratio of the material holding singularity in maintenance and that would then amount to (4X 7= 28)

OBVIOUSLY ALL RELEVANCIES HOLD TWO SIDES TO EQUAL PROMINENCE. THE RELEVENCY THE MOON HOLDS TO THE EARTH WILL BE DIFFERENT TO THAT WHICH THE EARTH PLACES ON THE MOON.

The Earth holds singularity and the Earth's singularity holds the moon's position in singularity. To the moon's relativity, it holds value to space from the Earth's singularity of 1 x $(\Pi/2)^2$ (the relation between the Earth and the moon) x 10 (the fact that the moon is within the space of the Earth, as the moon has an individual point holding singularity (Π). Therefore the moon holds (1 x $(\Pi/2)^2$ x 10) + Π = 27,8 days as one and the Earth holds to moons single day of individual singularity at one to 28 days.

$(4(\Pi^2)$

Π

Earth

$(\Pi^2/4 \times 10) + \Pi$

Π

Π

Moon

Behind this principle are the sound barrier and the reason why aircrafts break the sound barrier. It is all a relation in different positions of singularity. From the time, I wrote the first few pages I was on a quest to find a more suitable person to take over the work. I, more than any one else, know my limitations and limitations they are. I knew that anyone of the more than one thousand five hundred academics I eventually contacted held more knowledge in the tip of his (or her) little finger, than I have in my entire body. I tried in desperation as I tried in vain, to convey my message to the right person that could see what I could see. It all came to nothing because the academics all sat on the same mighty Sear Tower far too high to even notice me pointing at the cracks down below. From where the Official Policy Protectors sat, they did not notice me. Those to whom I drew some attention be it personal, by mail or on the Internet saw me as a nonsense proclaiming nonsense.

From where I stood, I could see the mighty tower they sat on. I could also see what construction held that mighty tower together. I could see how much that mighty tower was leaning over like the tower of Pizza. I could see how the tower will fall one day if not soon then later, because I was at the foundation of the tower. From where I stand, at the very bottom, I see the foundation of this enormous Petranos Towers collapsing from the misconceptions it holds as a base. Down at the very bottom where I am, I could see whatever other one holding a position in this tower could not see. Those that are at the very top, ARE so high, so very secure, they would not even hear me or take notice, and yet they are the only ones that can do something about the inclining tower.

After attempting for seven years to make myself heard to get the High and Mighty to notice the insignificant me down below, so small in relation to their greatness, I decided to show the world what holds them that high. I decided to show everyone how great "they" are, especially about the greatness of their misconceptions that put them at such a dizzy height.

To every one of the Official Policy Protectors, I say this: Every opportunity you had, you thoroughly rubbed my nose in the knowledge that we are not in the same class. I accept that fact as I accept my academic qualifications being so very poor. I shall never be your next-door neighbour or the bloke living down the road from your house, because I am not in your league and I shall never be. I do not begrudge you your academic position, your mighty achievements or the height you have reached in your sphere. I shall never enjoy your company as an equal, because I can never be your equal.

Your brainpower puts you light years ahead of me, and for that reason I do not even wish to have the honour of your company.

All I ask, is listen to a mere mortal, a mindless illiterate compared to you, one sod down here at the bottom where you are at the very top and that may just see things you cannot see from the height you hold. I do not wish to join your company for I shall not fit. I never had or have any ambition to fit either, because I am quite happy being in the sub-minor league. All I ask is to be heard. So many times you, honoured members of the clan of Newtonian High Priests did not even attempt reading my book that I sent to you. You did not even try to pretend I had a point therefore you merely threw my book away in disgust. To you my illiterate arrogant views about science being wrong for the past few hundred years are the epitome of a mindless person. You saw me as a totally mentally underdeveloped excuse for breathing and you could not bare my company because for me to have my view such as my view is in rejecting the view of an establishment centuries old. I can understand your disgust, but that does not change my point and that does not change the incorrectness of Newtonian science!

Before you throw the book down in total disgust, first answer the following argument and if you can answer it truthfully then throw the book down. If you cannot answer it, go on reading the book and you may just set your thinking mind in motion. Hear this from a mindless: you may have the ability to learn and afterwards reflect on that which you learn, but you cannot think, and I do not know who is the most mindless, me without education or you with education and without reason.

You are addicted to Newton. You do not understand Newton because if you did understand Newton as far as cosmology goes, then you would reject Newton. You were force-fed by your superiors on Newton at the time and during the time when you were a student and being force-fed you either had to grow addicted to Newton or die from Newton. Being where you now are it means you are not dead. That means you took the second option by becoming addicted to Newton. You saw comets come and you saw comets go. You did not question the reason why the comet went because of your addiction to Newton. You accept Hubble's expanding yet you go about looking for a critical density in the hope you do not have to abandon your Newtonian addiction. You know the Big Bang cannot be possible while Newton's mass annihilates the radius between masses because then the Big Bang had to be the Big Crunch and from where we are there is only one enormous out explosion that had no Newtonian implosion. The enormous force that were suppose to contract the small radius that was in place during the Big Bang never imploded on the mass by nullifying the radius through the tremendously large mass and small radius being present at the time. All the facts modern day science accepts does not stroke with Newton. Is it because subconsciously everyone is silently admitting about Newtonian incorrectness?

Now I come and I tell you your addiction is killing you. As all addicts do, you are not willing to kill your addiction although you do realize that your addiction to Newton is senseless, stupid and all together wrong. The second choice facing all addicts including you is to hate the messenger that comes to take the drug from you. You will hate me, as the messenger. You will be willing to kill me being the messenger of evil because you see me as the evil that wished to part you from your addiction, which is Newton. You now find the bizarre feeling to put your anger distrust and vengeance on me and on the book you are reading. I have seen many academics stop reading at the part you now have arrived at. Brave yourself and throw out you fears by reading the rest. It will eventually cure you. Newtonians declare that a force brought about by the content of the mass of a body will therefore pull objects closer where the pressure coming from the weight brought about by the mass of the bodies, will lead to heat.

If a cylinder is pumped with air, the pumping is a force. The force comes about because the force is that of the intentional action of the only force in the Universe, the force of life. The more pumping there is and the longer the pumping will last, the hotter the air will become, and therefore the hotter the walls of the cylinder will get, as it transmits the heat from a point where the heat is most abundant to a point where the heat is least abundant. Heat flows from hot to cold.

After pumping stops, the heat on the inside will reduce in value up to a point where there is equilibrium between the heat on the outside of the cylinder and the heat on the inside of the cylinder.

The heat reached equilibrium with the event of time. Please for the sake of sanity, do not reply that it is the molecules in the air tank that is bumping against each other and through that collision, friction causes the heat. That is as much Newtonian rubbish as one can ever find. Should you insist on that being your answer, then ask yourself what will calm the molecules down afterwards where they get so calm, there is no more heat in un-equilibrium. Did the molecules take drugs, or did the force calm them by telling them gentle nighttime stories. I had so much bullshit thrown at me wherever Newtonians defended their Master it sickens me. I may be uneducated, but I am far from mindless!

No molecule can ever, ever touch another molecule because the electrons guarding the outside are equally negatively charged, and will therefore reject any contact or coming closer to one another or with another molecule.

When we look at the Earth, we find the coldest region on the outside of the atmosphere, where the least mass, weight and gravity is. As the circle grows smaller, the molecules become more, the mass becomes more, and this will increase the weight that brings about heat from pressure. This is not my say so this is Newtonian science.

The increase in mass, bringing about an increase in weight, puts most pressure at the centre to be the hottest. There is this force in matter that pushes and pulls, until everything is boiling hot, and that is gravity.

There must be a point, where all the matter has finally found a position to bring about the overheating that occurs in the centre. The heat is energy and cannot be manufactured, but has to come from somewhere. The heat cannot be lost, because heat will transfer to a colder region to bring about equilibrium. The heat cannot continuously come from nowhere because at a point, the molecules will settle in their individual positions, find equilibrium and maintain the heat balance in that spot. It cannot produce heat from nowhere, on a continuous basis, because heat is energy and being energy it must come from somewhere as much as it is going somewhere. Either the atoms lose their mass to heat and the mass becomes heat in order to generate heat on a continuous basis, or the heat must stop, decrease and become as cold as outer space because no further heat comes about because the heat is exhausted.

If the matter as much as mass becoming weight established heat by applying continuous pressure, the Earth should decrease in size as mass turns to heat with the consequential loss of mass and gain in heat. The Earth must then deflate and be a pretty small place by now.

You may say there is a lot of mass-producing a lot of pressure becoming a constant flow of heat, but that will mean some of the mass must have disappeared to heat because four thousand five hundred million years is a pretty long time. In four thousand five hundred million years, some size diminishing should show, or the Earth should have reached a point of equilibrium by now where the heat supply is completely exhausted because again four thousand five hundred million years is pretty much enough time I would say for matter to have cooled down by now.

Yet, the flow of heat maintains in the Earth, very much uninterrupted and in no way showing signs of decrease. If the force of gravity is manufacturing heat, from where does it get its raw material and where is the manufacturing plant? If weight and pressure leads to heat, then the heat should have transferred altogether to outer space, the coldest place we know at minus 276 °C. Four thousand five hundred million years have come and gone since the Earth became the form it has now.

On the other hand, if the force continued its pushing and pulling it applied when it formed the Earth, the Earth should by now have incinerated. What force of gravity is required in getting enough pressure to push hydrogen dust into a solid iron formation and release that formation as a molecule in being as solid as the Earth is. Such an effort requires a lot of pushing and compressing. If that pushing and pressure continued as much as it must have had it had to increase with the demise of particle space that is separating the molecules? Keeping that in mind as a natural law, everything on Earth should be covered in flames by now. When you go with the argument that there is not enough matter to produce such pressure, then there was not enough matter from the start to get the place as dense as it currently is. The matter did not decrease, the force therefore had to become weaker with time and the force is so weak now, it does not collapse the Earth into a smaller space than it was say three thousand five hundred million years ago. However there are very little signs of that, in fact it seems the whole thing is getting bigger, with all the lakes losing their depth, and the mountains rising, including the volcanic activities establishing new islands.

If you cannot state why the heat has not reached a point of saturation or has decreased to equilibrium with outer space, it proves you are not thinking. In that case read on … it may arouse your thinking ability once again. This I say, not in arrogance because I, of all people should know how poor my abilities are. I sat day after night, night after day, breaking my brain to find answers, or only clues to the questions in hand. I knew at each point I arrived, that the answer is right in front of me, but through the darkness of my personal ignorance, I could not see what I knew there was to see. From that feeling of incompetence I tried once more on every occasion to contact any person with more brains than I but every time I was luckless to energise interest and had to continue with my personal incapacity.

Every time I contacted an Official Policy Protectors in desperation for help, they ignored me flat. Every message I sent telling whomever I contacted, that there is no such a thing as gravity, it is all a medieval hoax, Newton is altogether completely wrong with his gravitational laws and the Bible is one hundred percent correct about creation, they would not even reply or at least respond. Well… to them I say this: I still maintain that there is no such a thing as gravity: it is a hoax.

Obviously, the obvious is that the Newtonian Order of High Priests would lead every body to believe that Newton and Einstein's findings are flawless. All these mentioned discrepancies are known to "Xepted science", yet they keep the charade going on about other planets they are about to find, only to mislead the public and milk their tax money, in the name of research. If it were not for funding and an effort to provoke general interest in skimming tax money, then why would they deliberately spread such malice?

Scientists know about this discrepancy in the Newtonian laws, yet there is never any mention about it.

The so-called "evidence of the existence of planets" is based on just as laughable principle. Allow me to explain:

The findings which science base their proof on about the location of other planets are the gravitational pull.

At first the way in which the facts are presented does not sound that unfamiliar in an argument, and one tends to accept it without a second thought. When given the second thought, the blatancy in the matter leaves one breathless.

In the evidence, the stars and "planets" are presented to be in a tug of war. This one can see in any sketch about the solar system. How this supposedly works is that the one star first pulls the other system closer with the force of gravity, and then it is the other systems turn to apply gravity and jerk the first system to its side. What they do explain in explicate detail is that they do not understand the first thing about the matters they pretend to understand.

The Official Policy Protectors know very well that all children play this game, and therefore every one will associate this explanation with familiar events in their past, and no further questions will be asked. Let us examine this principle with obvious general knowledge about space flight and how this applies in outer space. An Astronaut is capable of lifting four tons of equipment by self-propulsion and relying on human muscle. He (or she) can perform this action effortlessly. However, what is impossible to perform in outer space is to correct his position when he is not secured to a stabilized object. Every person knows it can be life threatening when an astronaut loses his grip or connection to the spacecraft. In this, the question is; how can two heavenly bodies have a tug of war under such conditions? There is obviously some stabilizing factor on Earth one do not find in outer space, and that is not gravity because such an answer is avoiding the issue.

In the past so many SUPER –EDUCATED dismissed my rejecting the Newtonian claim about mass being a factor in falling objects. Some even went as far as refusing to read my book beyond that statement, that being on page four of a one thousand seven hundred page document, claiming I do not understand Newton, but no one can explain why oxygen is a gas, and yet oxygen is more massive than boron and carbon. If mass were a factor of the essential being the claim of Newton, then oxygen would fall to the earth faster than boron and not float in the air as air. Yet boron is a solid and oxygen is the gas.

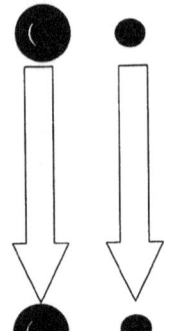

As stubbornly as the Newtonian paternity refuse to admit about the comet not colliding with the Sun in spite of all Newton's predictions, just as stubbornly they refuse to admit that Galileo completely contradict Newton's mass blaming. With the mass devouring the square of the radius when helped by the product of the counterpart the whole formula hinges on the mass being the devourer from both ends. That means the heavier or more massive the object is the faster should it fall because the quicker will it devour the radius parting the objects. That Galileo distinctly disproves because all objects fall at the same rate.

The reason why the **mass of the smaller object does not apply**, is because **it is not the object** that is **drawn** to the Earth, but **the space-time in which the mass of the object finds itself in,** that is being **drawn towards the Earth**. The most obvious proof that something applies movement to something directing the movement towards Earth is the simple pendulum. If gravity was a force, one should get the same result by applying a pull spring to the pendulum. With the spring connected to the bottom of the pendulum, both time as well as space would compromise. However, we know that the result proves the opposite, which proves that there is no force applied too the pendulum. On one of my many crusades I met one of the most influential academics on astrophysics in Africa.

I tried to explain the pendulum swing to a high ranking man of much academic importance, doing great work on behalf of NASA in South Africa but with me not knowing his superiority on the matter of the pendulum, I took this man on, on one of his specialties. This man was apparently an expert on research or lecturing or what ever on matters about the pendulum. Of course, as usual, the very first thing he asked (as all Newtonians do) before even asking a person's name, was at which university did I study and what my academic qualifications was. By replying that I have never been at any university for longer than a few hours in my life, and therefore my academic qualifications was less than zero, this highly rated person of high standings was less than impressed to spend any of his valuable NASA paid time with me. To complicate the whole aspect of my un- welcome visit was that I tried to explain to him being the expert on the pendulum that he is, about the pendulum. Boy, was the man annoyed with me.

He was very polite and very civilized about the whole issue, but his annoyance with MY EFFORT ABOUT EXPLAINING THE PENDULUM TO THE EXPERT ON EXPLAINING THE PENDULUN WAS MORE THAN HE COULD BE CIVILISED ABOUT. There is a point where a person gets a little too civilized to be true and at that point where a person gets a little too polite to be civilized about a topic. Any person can sense such a point the point when one realizes that is the point of limits. I also realized that what ever I had to say on any matter concerning all relevant matters was as good as never said as far as our Professor Doctor, was concerned. As a matter of fact that Professor from the University of Potchefstroom is one of the most exceptional men to walk this planet. I gave him a copy of The Thesis and after reading only four pages from a book containing over two thousand pages he was able to draw a conclusion and condemn my work. Of course the reason was the usual: I did not understand Newton obviously because of the lack of education on my part.

You may ask yourself: *"What was me (the un-welcome person) trying to say?"*
This is what I am saying:

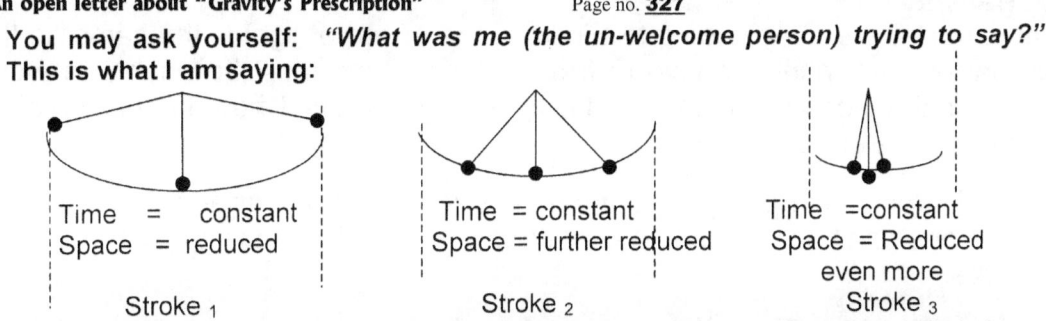

Time = constant	Time = constant	Time =constant
Space = reduced	Space = further reduced	Space = Reduced even more
Stroke $_1$	Stroke $_2$	Stroke $_3$

Time remains the same. The swing distance tarnishes. Period $_1$ = Period $_2$ = Period $_3$
Swing distance $_1$ ≠ Swing distance $_2$ ≠ Swing distance $_3$

In the pendulum principle that brought Galileo his everlasting fame, the pendulum swings at an even interval. As the **pendulum swings**, the **space tarnishes** while the **time (period) remains** the same. This is the principle on which all clocks work. What is it that Newtonians are missing for three hundred and something years about the pendulum? Newtonians are not seeing the very best example there is to indicate singularity outside singularity.

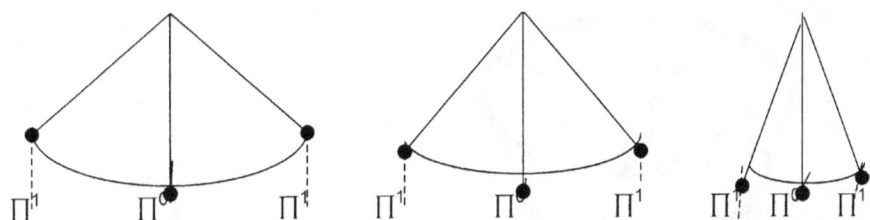

The pendulum indicates the very point of singularity the earth holds and the marks on both sides where singularity deviates in space, giving time to that singularity diverting.

If gravity, which is a force, did apply, then it would be as if a spring was fixed to the bottom of the pendulum and to an unmovable object below the pendulum. With the applying of the springtime will not remain at equilibrium but will tarnish in a vector with the declining of space. Something is holding time steady to the demise of space.

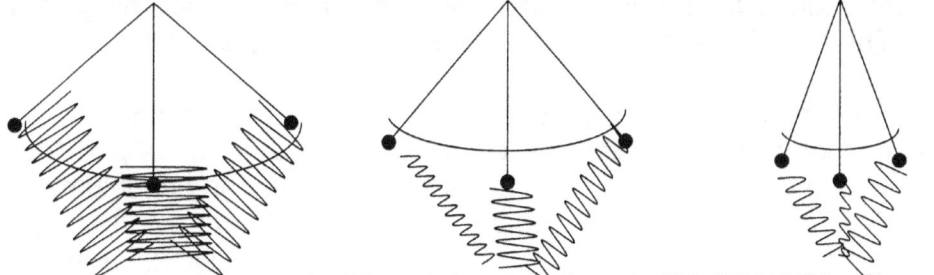

This means there is **NO FORCE** and therefore there is **NO GRAVITY**.   There is only **space** (stroke) **time** (period) = **space-time.**

When a spring of 9,81 Nm is mounted to a pendulum, which is an equal force to that of gravity, both the time period and the swing distance would equally be affected, but to a lesser degree as the swing distance declines.

ALL SCIENTISTS GO INTO FRENZY BECAUSE GALILEO WAS PUT IN HOUSE ARREST FOR TEN YEARS!   HOWEVER, THESE VERY SAME SCIENTISTS ARE STILL HAVING GALILEO'S WORK KEPT IN HOUSE ARREST AFTER ALMOST 350 odd years. HOW DO THEY EXPLAIN THAT? Galileo introduced the best devise indicating space-time and half a millennium onwards, nobody but me can see it.

What is space-time exactly?   Einstein was the first to explain the existence of space-time, but Galileo was the first to indicate space-time and Kepler was the first to pin point the position of space-time. Einstein made one big error in judgment. In his all to well-known formula $E=MC^2$, he relates to space-time as if space had a factor of one and time was the altering factor. According to the Einstein / Newtonian Order of High Priests, we live in a total dark, totally flat, and single dimension Universe.

Why would it be a total dark Universe? According to Einstein, the speed of light is the same as the speed of time. Should that be true, photons have to freeze in time, and must be unable to move through space in time! (I shall elaborate on this in due time.) This proved how far the greatest Newtonian outside Newton really were off the mark

Look around you and see all structures (**space**) are different in size. Therefore, **space** cannot have a factor of none converting to one and back to none, but relate to the size of the object, whether it is an atom or the cosmic Universe.

Time (C^2) is at an even factor as all things in the Universe relate to the same time, (although not the same duration of time). By implying that E $= MC^2$, he puts R^3 at a relative value of one throughout the Universe. Space can hardly disappear, but can compromise under abnormal star growth.

Every round object has a point establishing a very centre, a middle dividing one side from the other. That division determines the space from one side away from the other side. At one point there must be a point that does not fall on either side of the divide. Such a point will still be a circle, because from that side the circle divides into two sectors.

Π^1 Π^o Π^1

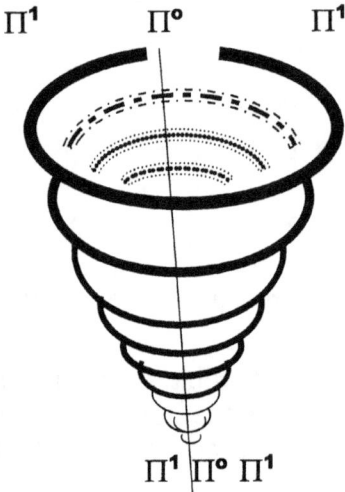

Π^1 Π^o Π^1

Every solar structure is spinning around an individual axis while the whole lot is spinning around a mutual axis the Sun provides The spin that shows on the different planets is the most crucial aspect of their orbiting the Sun . Calculating a circle involves two aspects where the one is either the radius or the diameter that is double the radius. The other is the factor Π. Π X D^2 / 4 = **circle and** Π X r^2 = **circle** The point of singularity cannot be in space at large because space is not there and secondly what ever is there spin to slowly to have a connection with singularity directly. The pendulum indicate the very point where all the Universe conjuncts placing space in relation to the time-Zero singularity as indicated through the position the Earth maintains individual singularity parting from cosmic singularity The pendulum is a direct measure of space-time flowing but to this moment where I write this, I still have to find one Academic that understand my connection. Since there is no Newtonian thus far capable of seeing the comparison I draw between the moving of space in the time it takes such space to move I shall repeat it once more in the hope one might see the light.
Take the pendulum.

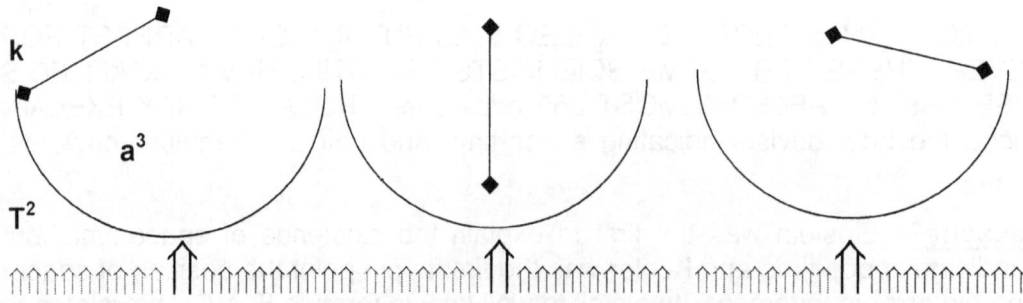

Every time the pendulum arm crosses to the other side it indicated the most important factor.

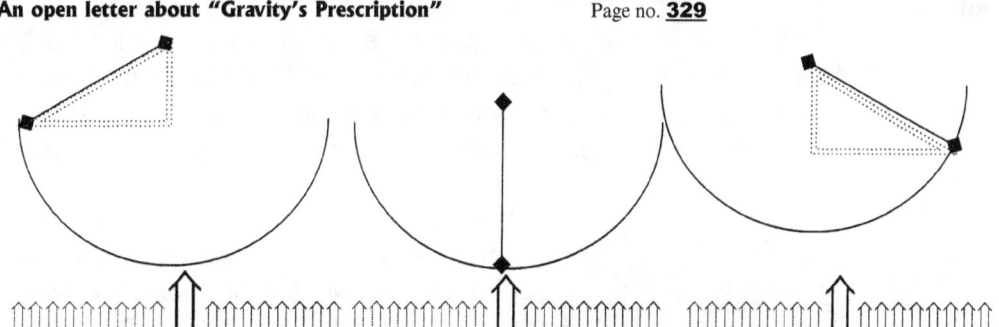

Every swing the pendulum arm does it brakes through the factor holding the universe in place.

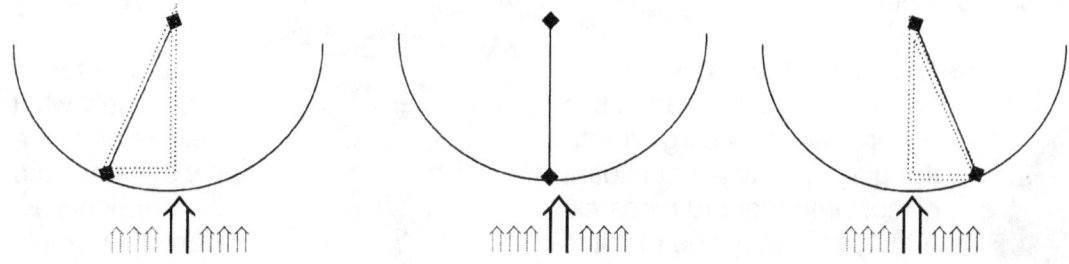

The pendulum not only crosses the singularity the earth dictates at that given time and the pendulum not only points at the factor maintaining space-time on earth.

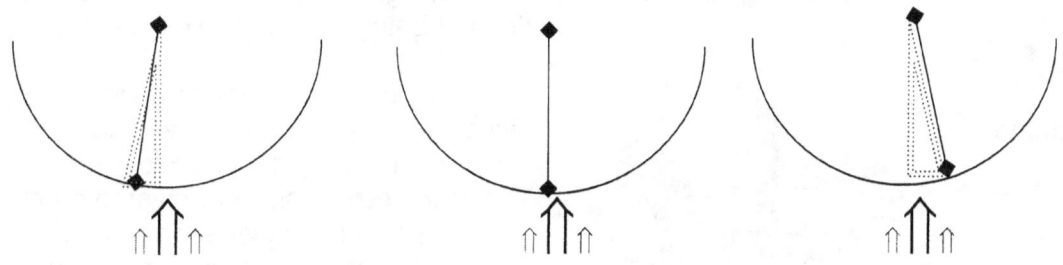

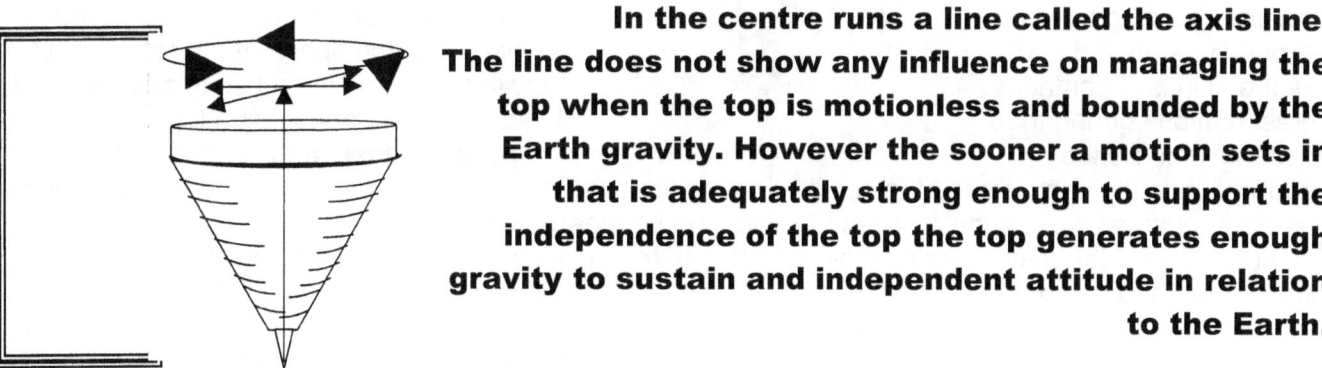

In the centre runs a line called the axis line. The line does not show any influence on managing the top when the top is motionless and bounded by the Earth gravity. However the sooner a motion sets in that is adequately strong enough to support the independence of the top the top generates enough gravity to sustain and independent attitude in relation to the Earth.

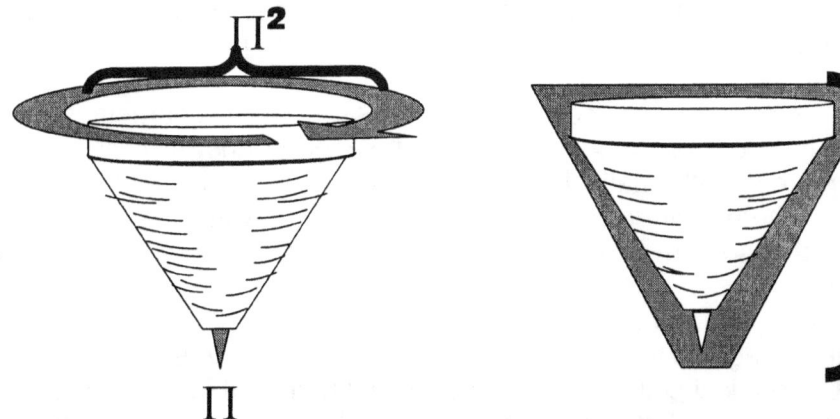

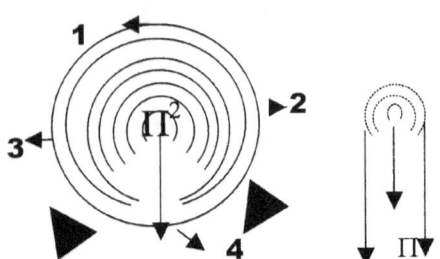

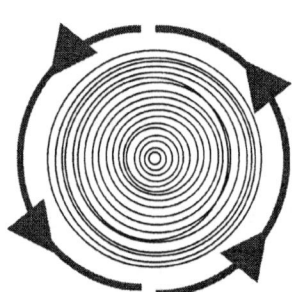

The dynamics that then support the top in motion comes from four points serving the top with time $\Pi^2+\Pi^2$ which comes about from the circle the top forms by spinning the body of the top rotating and the space rotating in relation to singularity **space** $\Pi\Pi^2$.

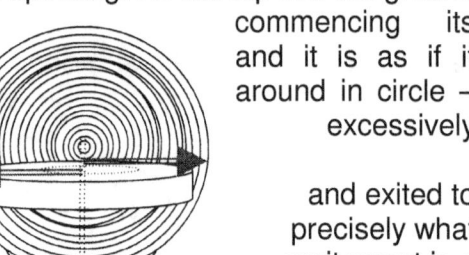

after being thrown with it's spin initially the top starts to rotate wishes to exceed is spin by moving like formations while spinning around its axis.

After the top is thrown the top changes some very vital characteristics in behaviour to what it had when it was not spinning. As the top hits the ground commencing its and it is as if it around in circle – excessively

It spins vigorously as if the top suddenly is too energetic stand still and that is happens. This surging with charging of vitality that finds a is most important rule cosmic One can clearly see that it is playing out as the Coanda the space is developed and defined by the motion T^2, which verifies the independent space the top has acquired by motion of spin, it is also clear that the relevant factor of linear motion demonstrates it's presence in the moving about as the top is rotating. That puts Newton's claim of motion not being a factor in total disbelieve.

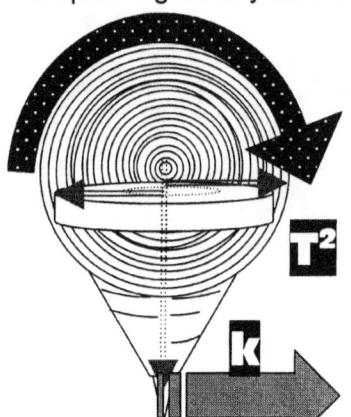

and exited to precisely what excitement is a new dynamic and principles. Kepler's formula principle. While

take the extending of the singularity inwards governing singularity that That is why the top is assertive the spin is in the lines running towards expanding outwards. In as the spin is in contact presentations during the material unit fills the

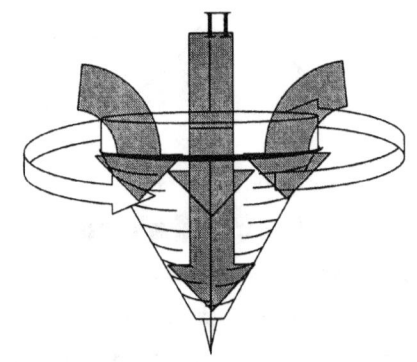

With all the excitement and no where to drive line runs down the developed towards the newly established keeps the newly formed Universe erect. spinning in the first place. The more velocity the more reaction there is from the centre and extending through the real terms the space of the top expands with more time in motion in quicker same time in period as a bigger space because of more material

duplication in the same period that is allowing the top to spin. In this the space in which the top spins has to expand as well as the process of filling time by means of duplicating as well in order to compromise for the material relevancy growth to fit the newly acquired singularity governing the space-time and being erect by the motion. The support that the spinning top finds in it's task to establish a governing or controlling singularity keeps the top spinning in an upright and erect position that is only supported by the motion putting space between infinity in the centre and eternity in which the top spins. By placing differences between infinity and eternity it not only charges singularity to life but also charger space-time into the Universe.

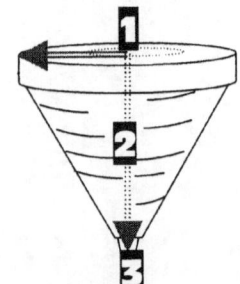

Through the behaviour and that the top display it is possible to find cosmos that the greatest mathematical to solve. It is a case where their genius find solutions and the search of the proved to complicated.

characteristics answers to the minds was unable was to great to final results

Looking at what takes place as the top start spinning we find a line coming from what was a mathematical point. What is there is not there but for those with intellect to find that what is there being present. What is there is not part of the cosmos and yet what is there drives the cosmos. It generates motion by not turning and places what is, in contrast to what was and what will be the very next instant. The facts are indisputable even to the most ardent mathematical Newtonian disbeliever and what the top parts is how the Universe came about. Motion parted eternity and infinity by developing space-time as a partitioning screen.

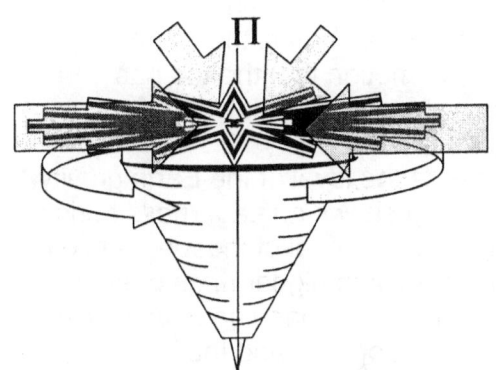

More spin increases both lines that force gravity by the increasing of T^2 extending k, k^{-1} as well as a^3. The space wants to exceed its boundary because the motion suddenly allows the space to become extended. The gravity line running to the centre wants to extend for the same reasons and so does the gravity line running towards the liquid that should be there and that should be enforcing this sudden living up to better standards.

about as a result of more heat. With caused by heat will bring on a linear centre of the top. This is then a counter this (Newton's law on action balance comes about where $k^{-1} = T^2$ inline with the progressive spin and that should be because of a liquid

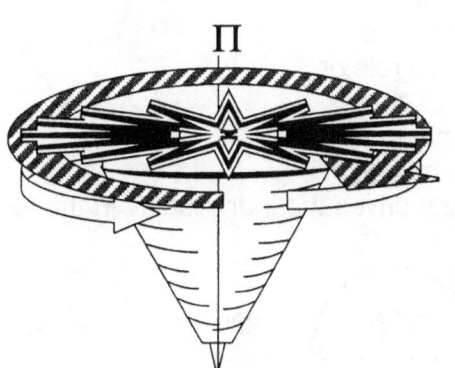

The spin under normal conditions can only come that aside the spin normally gravity running towards the product of $k = a^3 / T^2$. But to and reaction), another $/ a^3$ that centres the material the extending of the motion heat adding to the material.

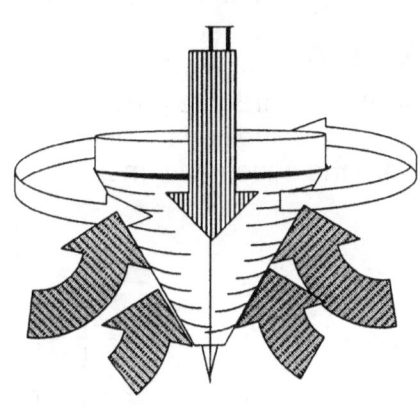

The support that the spinning top finds keeps it upright and performing as if in a fighting mood. However, again I have to press the point that it is life that initiates the motion and for this motion to start as a natural flow of events requires a lot of nourishing by the independent singularity that starts to drive the object through a combined effort of rotation of all the included atoms accumulating assuasive heat to bring on such motion. When this process is in a natural occurrence within a star within a galactica it is the indication of the coming about of a newly developing in the heat centred cradle of a galactica. However it can only be gravity that is able to fight gravity by extending the Earth gravity and by extending the Earth gravity we find some part of the Roche limit also applying.

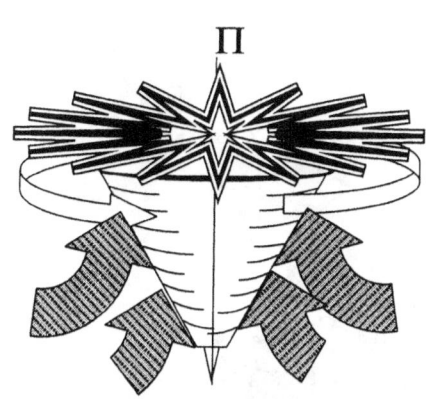

The heat that should supposedly under cosmos law drive the spinning top will come from the governing singularity, which is accumulating the heat in concentration by the contraction or cooling ability the top singularity acquired. However in this case the spin is a result of life's ability to manipulate space-time and alter cosmic events to the free will and interfering nature of life. The heat that would establish such a drive in motion in real cosmic terms would require a lot of nourishing a sustaining from a large number of maintaining atoms that produce a large flow of space-time and can concentrate much excess heat outside of the atom sphere. With sufficient energy the top gets into a fighting mood that makes the top very reluctant to give up this newly established freedom. Be behaviour now attributed to the top is normally the manner how a star develops in the galactica cocoon and how the fledgling star gains it's birth right to leave the nest of the cradle of the galactica. The atoms form a sum total of space-time displacement that can support the generating of the required gravity in securing the heat that would unleash such a drive. Such singularity in governing come to life and release the new star from the blanket of heat that covered the star up to the time of its release.

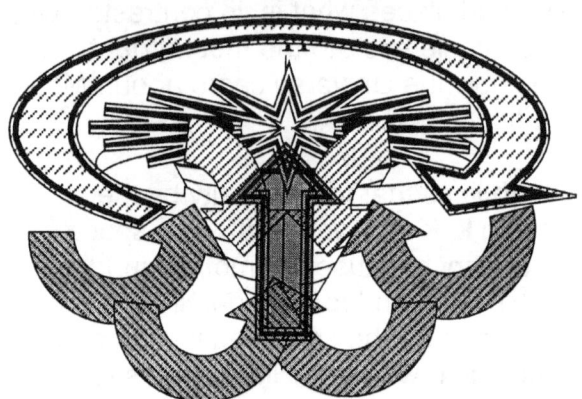

This example that we can gather from the top's behaviour shows how desperate the governing singularity can become when starved of motion and how such an excited singularity can put up a fight for life and independence. The top is in a fight for independence while the Earth is restraining the independence. The fight goes on until the Earth suppresses the last bit of motion that the top has and the top uses the last motion it has to defy the Earth's control.

When the motion exceeds the level of the Earth gravity,

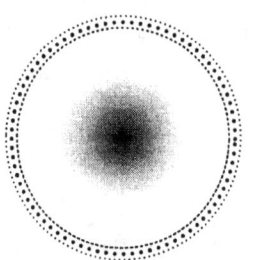

the same manner that an electron because the top with motion is in an the Earth which is filling the proton or atmosphere being in the neutron role or Let's quickly establish events as they controlling entity that is demanding space-separate individual drive. The motion which generates the gravity that drives the individuality in the top.

the top shows an eagerness to rise to a higher level of independence in reaches into higher rings of energy electron or expanding relation with contraction role and the gravity-motion supply. translate singularity from a dot to a time through the establishing of a comes about which prove to be that

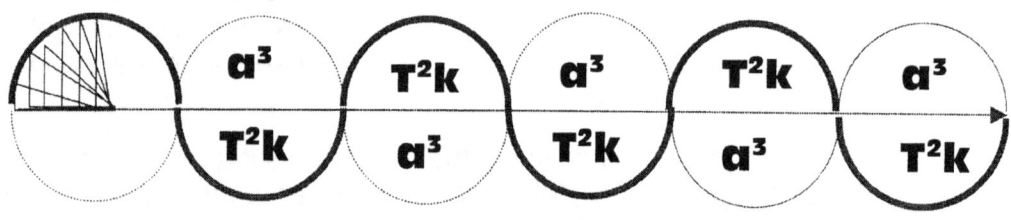

In the sphere centre is the spot that has to be there mathematically by measure of $(\Pi r^2) / (\Pi r^2) = (\Pi^0 r^0) = 1$. In order to provoke the line forming

singularity into existence, motion is required, just as Kepler indicated where the space becomes equal to the motion and the motion is equal to the space $a^3 = T^2 k$

The inner four is singularity points equal to and is singularity

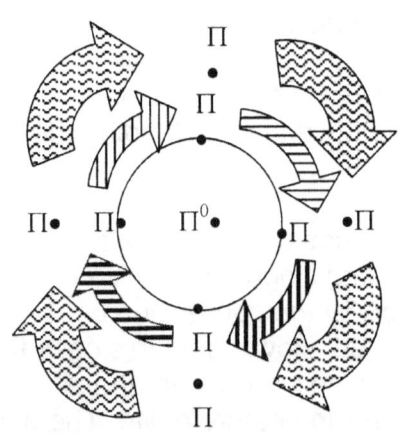

being in direct contact with singularity charging time. It is singularity charging the presence of singularity by four points that would form the four forming time. Then one further point to the outside would form space with an indirect link to singularity. This formation goes on as long as time distorts to form space and space will forever have four to the

inside connecting time and three in equivalence forming motion.

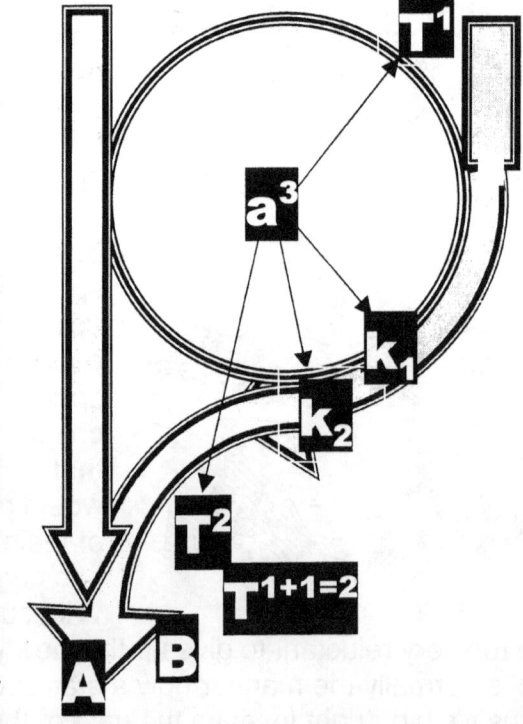

The Coanda effect is proof of gravity coming about through space forming motion. In the case where water diverts the normal directional flow the space that translates to the motion is deflecting singularity with the flowing water charging the motion. In the centre of the object having the round form, singularity is duplicated and by transferring Π to form Π^2 and the motion of the water creates a line of gravity that pushes the flowing water to follow the direction that the newly gravity applies to the water.

Normally water will run down to the centre of any gravity point, as A shows. By allowing the flowing water to come into contact, an object of a specific form the flow will divert (B) from the normal line and follow the contour of the object presented. For that to take place there is one condition that has to come about.

This again proves Kepler's statement of **k = a³/ T²** that specifically states that space (in this case the object transferring singularity to a new position within the round object) and with the motion of the water redirects the gravity flow of the water to new space in new time. Only Kepler can explain the phenomenon but only when Kepler stands alone, correctly interpreted and divorced from Newton's opinion about Kepler's statements.

The motion we detect as part of the Coanda effect runs through all spinning material. It is part of the atom as much as it is part of the sound barrier and the sound barrier is just another atom having Lyman series lines, and work also by the principle of adding heat which puts the object expanding in a higher relevancy than there was before.

However much noteworthy as it is it is prudent to consider that only when the atom unit is broken and the Roche limit is crossed does the sound barrier come into affect. It is the breaking of the bonding unit $(\Pi^2/2)$ that becomes the sound barrier although the breaking is never completed $(\Pi^2/4)$ as long as both objects share concentrated liquid time that the earth supply.

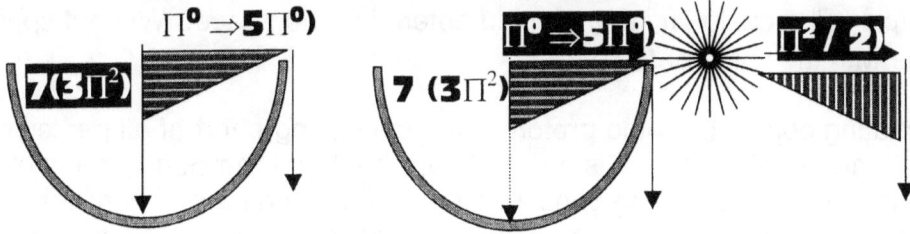

It is as if one then must claim in affect that Kepler held **a³ = T² k = 0**. If the Sun and the Earth have a rotating relevancy of zero either the Sun has gone away or the Earth stopped existing. One cannot claim there is a wheel and then remove the spokes because according to you taste, too do not like the spokes

With everyone of the four rotating points spinning around a centre while duplicating the value of Π in relation to the centre Π^0 at a measure of $\Pi / 2$ and where Π^2 is responsible for establishing as well as relocating Π by duplicating Π through the motion thereof, therefore $\Pi^2 / 4$ becomes a limit in relation to the development from the centre. One has to remember that a star of the present takes characteristics of the form from the ear before space was a factor.

As the absolute master of motion **Newton** should have placed emphasis on the motion aspect when he as a young man that saw an apple fall from a tree he made a brief calculation but he used the mass instead of the motion while Galileo proved that mass has nothing to do while the falling occurs. Seeing this he jotted down a formula and chucked it away. Newton however insisted on mass in spite of the clear evidence brought by Galileo to the contrary of mass playing a part. However most surprising to me is that most Newtonians are not only incapable of seeing the facts my way but they get sometimes pretty unpleasant in a very coldish pleasant way about my view. If mass had a major part, then the more massive must fall quicker because the mass will provide the drive and the drive will excel the velocity. While it is true that all things fall equally, then mass has no part to play while the dealing is occurring and that is in spite of all the Newtonian abstinence about the matter.

In this matter I am disputing Newton's honesty. He placed the relevance on mass when he was a young man and retracting his former claim would have tarnished his reputation as a genius. His glory was worth more to his mind than what the truth was. As young man he drew instant fame by claiming mass as the driving force and when as an older man he found he had to retract the first genius, his fame seeking would have left him a scar on his reputation. In that I can forgive the man for the man was human and as Cecil John Rhodes said, all men have a price by which the man can be bought. Newton's academic genius was his all- important vice. The problem is that the incorrectness stuck with science fore almost four hundred years on and no brilliant mind since than was able to make the Galileo mass connection. What happened to the many wise that walked the path after Newton had gone to better grounds. Where is the honesty in those that were supposed to search for the unblemished truth? Gravity is motion and mass is the restraining of the motion of gravity. The top shows the truth. Let us reflect once more

What is it the Newtonians fail to see? If an electron is orbiting around an atom, the inside of the atom must be a circle. If the atom was not a circle, it then had to be a cube. The electron cannot rotate around a cube; therefore, the inside of the atom is a circle. The cosmos is one big atom imitating all atoms as all atoms produce one Universe

In a circle, there is a radius that initiates the circle. The calculation of such a circle is $\Pi \times r^2$.

$$\frac{\Pi r^2}{r^2} = \Pi$$

If one removes the radius from the circle, the circle remains, only holding the value of Π. By removing the value of r, Π becomes singularity with no place to be. Singularity is the place where there is no space to be in place. However, Π remains because once r receives the slightest of space Π will find space. Then the circle will grow to Πr^2 and r would determine the space. Without space, there is no r but there is a circle with the value of Π.

Singularity is in every single rotating object, be it the proton or the combining effort of all particles in the Universe. That is what light and the photon is. It is concentrated heat that the Sun (or any other generator of electricity) concentrates to connect the concentrated heat to singularity where the heat receives either temporary connection to singularity or a small piece of individual singularity. All spinning matter has the point where the spin is still there but the radius is to small to measure by any means. That point is standing still in relation to the rest of the spin. In relation to that logic I do not accept Newtonian science holding the radius of the spinning object unrelated to the spin, whether the spin is applying or not.

Applying Newton's second law F=ma
One arrive at the formula
$GMm / r^2 = m (\omega^2 r)$
By replacing $(\omega^2 r)$ with $2\Pi / T$ we obtain Kepler's third law
This law predicts that $T^2 = a^3 r$

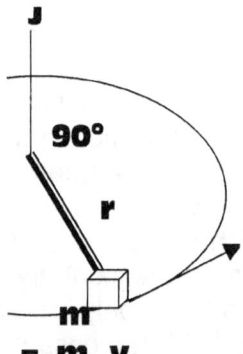

= **m .v**

The mass (m) multiplying the speed (v) forms a new value J AND THEREFORE j CONTINUOUS TO IMPLY $J = I \omega$

$J = r \times p$ where $p = (v = r \times \omega)$

$J = r.m.v = m.r^2 .\omega = I. \omega$ and becomes interpreted as $J = I \omega$

This establishes that $r = dJ / dt$

Since this is the absolute crux that Newtonian science pivots around I feel it is important enough to return to the whole issue once more in similar detail.

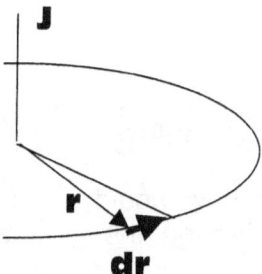

r = dJ / dt In the case of planets in orbit around the Sun r forms a value of zero because dJ / dt = 0.

Since Newton became an institution forming the King bee of the academic cartel world wide The Brainy Bunch had Newton's vision written in the minds of the future generations almost at gunpoint…well definitely at an academic gunpoint.

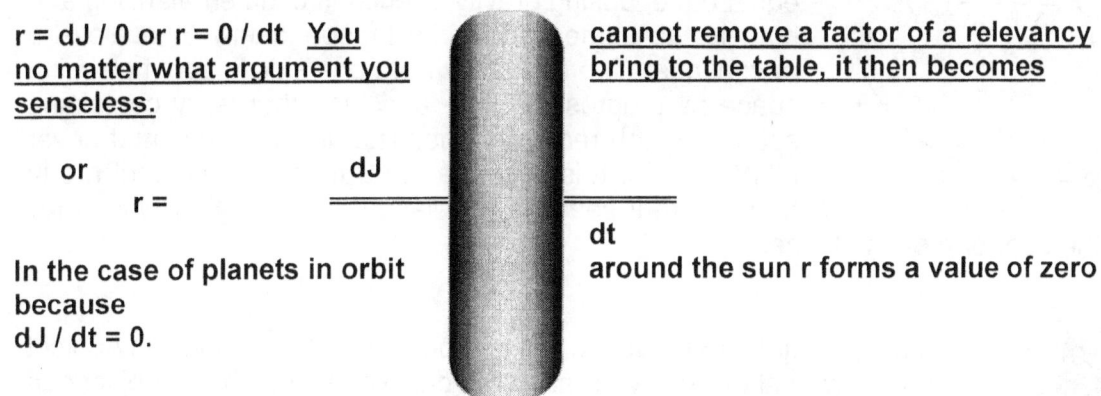

r = dJ / 0 or r = 0 / dt <u>You</u> <u>no matter what argument you</u> <u>senseless.</u>

or

r =

In the case of planets in orbit because
dJ / dt = 0.

<u>cannot remove a factor of a relevancy</u> <u>bring to the table, it then becomes</u>

dJ
———
dt

around the sun r forms a value of zero

I am not the brightest in the world that I admit, but one thing no one can do, not even if you are the one and only Isaac Newton, is that you cannot place any relevancy in a relevancy and then claim it not to be in a relevancy because such a relevancy does not suit your taste.

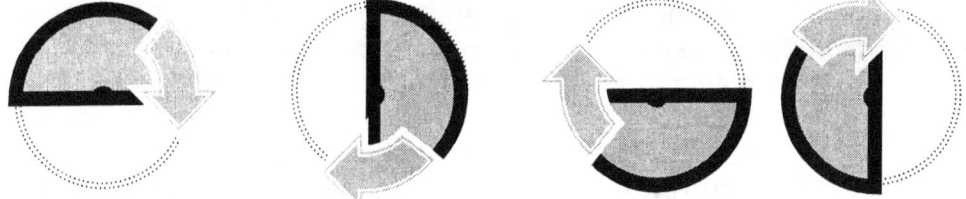

I wonder where would one put the zero part on the spinning wheel and what part must be excluded from the wheel. What Newton suggests, is a wheel has one side on top and no side at the bottom. While the wheel is spinning one may not remove the one side and then claim there is no attachment between the top and the bottom. That would mean in a graph the top is not connected to the bottom because a wheel spinning is a graph moving against time. It is the principle all driving is done and not the least electricity.

Every quarter of a rotating body is opposing the opposite sector directly and completely.

Any Newtonian that wishes to justify any form of support about Newton's claim on rotation not establishing work must please explain what happens when the Coanda principle draws water by motion and how that motion cannot be gravity.

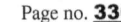

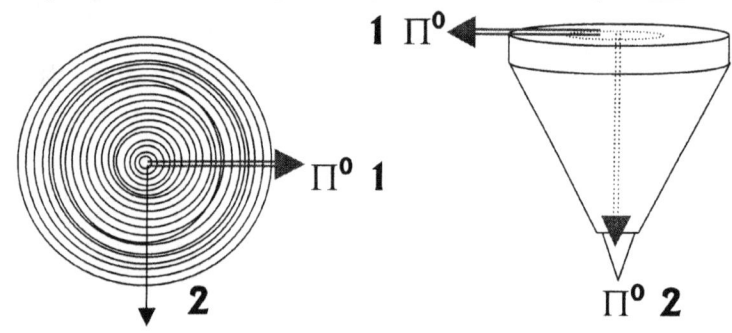

If there is no production through motion how would a top find a balance and what then inspires singularity to charge an erect stance through the generation of motion? These are legitimate questions in search of answers.

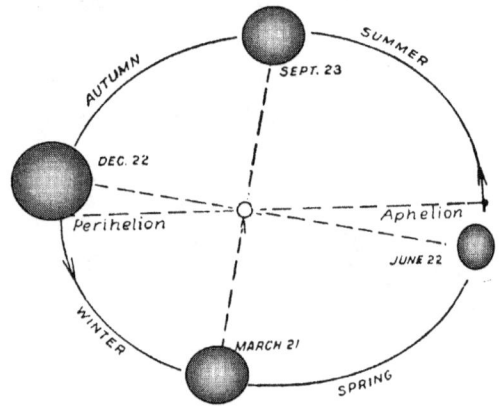

If gravity was mass inspired it would have the result that the Earth must be at some point during the year more massive than during other periods of the year. This we know is not the case and therefore the claim on mass is somewhat silly and a little middle aged. There are two equal but opposing gravity directions counterbalancing and both is the same that works independently to achieve a mutual goal. The relevancy from one side is about claiming space by progressing time and the other is by containing space through reclining time. That is why the comet never hits the Sun . It is because the Sun and the comet are in four different seasons in relation to each other while they are going through the quarter motion of time.

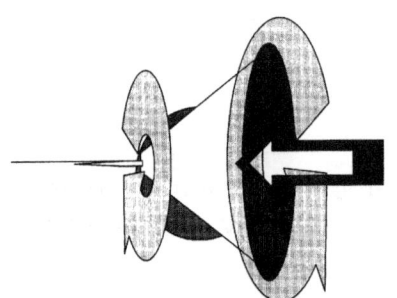

Gravity is motion and space is the blocking of the motion. Any object must have either gravity or mass but cannot have both. An object can be with gravity or the object can be with mass but it cannot be in both conditions simultaneously.

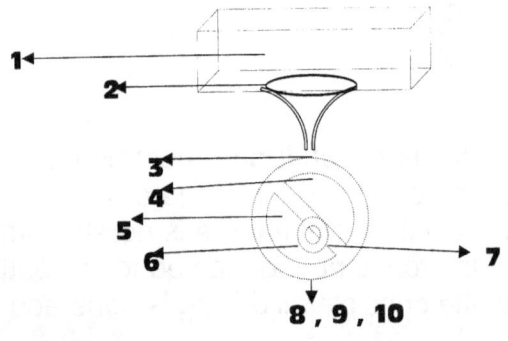

The motion of the neutron (2) covers the gravity (3) that the neutron has while the space (4) flows unhindered from the time (1) position through the location (5) fitting the neutron to the location (7) fitting the proton (8,9,10). The electron has mass because it restricts the flow or gravity and the proton has mass because in constrains the flow of gravity. Only the neutron has gravity because it flows unrestricted. That is what Galileo's work tries to prove but no one listens even to someone as important as Galileo because every one is mesmerized by Newton's while Newton was absorbed by the lack of understanding the difference between gravity and mass. If he did understand the difference there are then he would have realized what Galileo was trying to say. While his little apple fell it had gravity, but once it landed it had no more motion and therefore the containing part took charge as the apple then had mass. Galileo said all things fall equal (meaning all things are equal in gravity) while falling or while being in motion notwithstanding the difference in size or mass. That Newton missed.

That part all Newtonians that came later also missed. That is why Newton's first finding $F \propto \dfrac{M_1 M_2}{r_2}$ being $F = \dfrac{r^2}{M_1 M_2}$ is most true and most accurate however $F = G \dfrac{M_1 M_2}{r^2}$ is nonsense. All objects must be in motion $a^3 = T^2k$ where it will be $k = a^3/T^2$ in relation to one location and in another at the same time it will be $k^{-1} = T^2/a^3$.

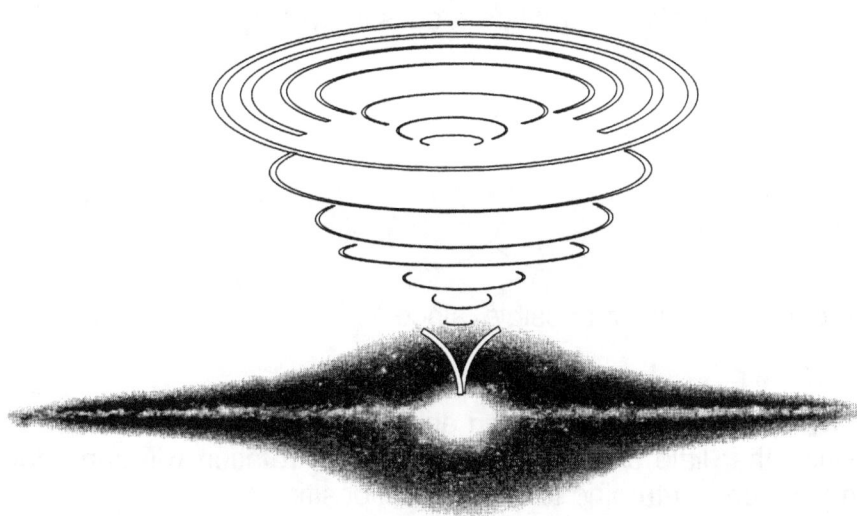

The motion within a galactica even generates sufficient gravity to re-enact a Black Hole within the centre of the galactica. This comes about just like the motion by which the top rotate, but the centre singularity being charged in the galactica is truly then a product of cosmic proportions.

The aircraft has mass when standing still but the motion that the heat of the engine produces, that expanding of heat to space converts the space to motion or gravity, which then relieves the aircraft of some of it's mass as the motion converts a part of the mass into gravity. It will always be some of the mass since the aircraft cannot be all motion. In the motion coming about from the engine that is converting heat to expand into motion the aircraft converts part (not all while it is in the Earth atmosphere) into gravity, which is independent motion from that of the Earth. While the aircraft is within the Earth, the Earth provides the motion and thereby serves the mass, which the aircraft (or all other bodies for that matter) will endure as the bodies remain a part of the Earth atmosphere.

The ship has mass but the buoyancy of the liquid in the water sustains the mass factor in order to provide the ship with another factor and that is displacement. The water holds motion in place and since the ship being on the water becomes part of the water in relation to the Earth it holds a part of the water in mass. However since the ship then holds part of the air or atmosphere in relation to the water the ship becomes part of the air in relation to the water and therefore the ship holds a relevancy of air in relation to the water. Some of the mass the ship has is regarded by the water as air and some of the mass the ship has is regarded by the Earth as water. It is locked in relevancy as the factors establish a ratio.

By enlarging the ratio of air (wind we call it) onto a part of the ship, the motion takes up a part of the mass of the ship into the realms of the air and the air contributes to the motion that then find the ability to go beyond the breaking power the mass has and convert some of the mass into motion. Again the wind is merely heat expanding and the expanding provides the motion that contributes to the duplication of the ship. An army battle tank is all mass in our thinking because of the iron composition providing it with such a solid and heavy structure. When the tank is thrown from a flying craft, some of the structure goes to mass because the tank requires more than one parachute to slow the descent down making the fall less destructive in nature. If the tank is left to fall with any other body and without restraining, the tank will not fall faster than any other body because the gravity the tank has is equal to all other bodies. It is the restraining of the gravity that produces the mass that requires a larger effort to contain the decline of the tank, but that again is the interfering with nature since mass is the interfering with the normal flow of nature. The fact is that motion is the duplication of the same in ratio of the relative flow of time and when the duplication starts to claim the same position at the same location during the motion in time, the motion of the duplication of the space converts the part being restricted to the same location as mass, while the rest is being converted to duplicating gravity. By duplicating the mass converts to motion and while the duplicating is hindered the restraining goes into mass. But in all Newton's claim that motion results in nothing is nonsense.

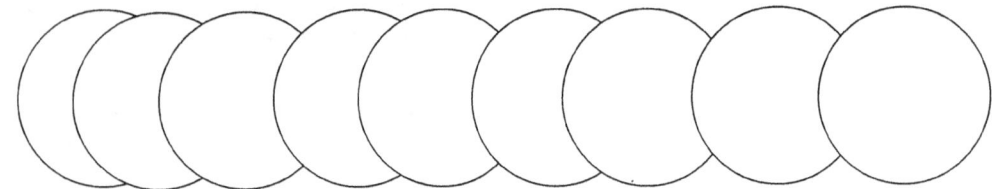

The fact that rotation does not produce work is impossible since rotation brings about motion changing the principles of the location.

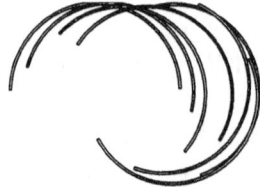
The same relevancy we find in the motion applying to atoms. One do seem to get the impression that little changes in line with the rotation will bring some forward motion and some returning to the original position.

Even by using half a wheel would still bring considerable confusion but one can clearly see that Newton's presumption does not quite match reality.

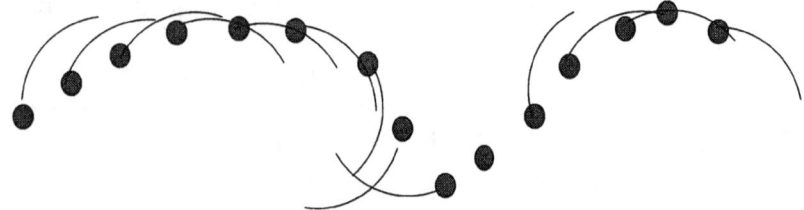

Shortening the arch changes the complexity considerably as one can then see a changing of the arch does not nearly bring the return of the dot to the previous spot.

When placing arrows pointing to a direction that is indicating the direction the line of movement, it becomes clear that there is a complete mismatching and the cosmos changes as rotation progresses. The behaviour, which I describe, is a flow of space through the line of time. Electricity is charged in this manner. The same generated force keeping the top upright is what is used to generate electricity. The flow of a charged conductor through excited space-time brings about the flow of a current.

An object in outer space has limited motion, which provides a part of mass and another part in gravity or motion. When the same object is in a Black Hole it is limitless and infinite in mass and has no motion. Outer space however is all motion as it provides motion therefore outer space is without mass. There is a mixing of mass or motion being gravity but having both is not having the same.

Even the electron serves the line of time, in the same manner. As the Earth spins through time by repositioning space in time singularity is re-applied, repositioned and re-aligned with the entire Universe in the manner I describe. The relation of the proton moving has to effect the following location of the electron since the electron is relevant to a position in space in time by a continuous motion through time.

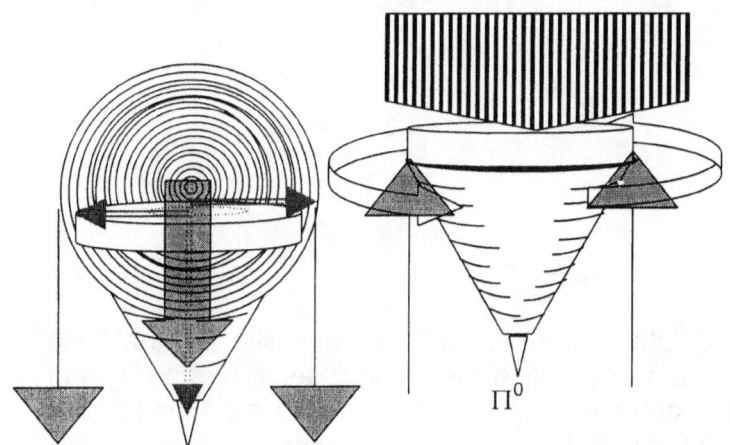

Π^0

Notwithstanding the motion that space forma as space is moving towards the centre of the Earth, the top finds a way to counteract the motion by producing a motion that is stronger than the motion restriction or in other words the mass, that is the restricting of the earth's gravity that fights to dissolve the impendence of the top altogether and thereby form mass which then is the lack of independent motion of the top. By spinning there is no force pulling the top down and restraining the top to the surface of the soil. The mass is still there but that mass the top try to combat with vigour and the needlepoint holds the top spinning as the top is fighting the mass. On the needlepoint the top rides out whatever force the mass would enforce to restrain the mass. The total restriction the mass control over the top has all but disappeared because the force or mass that the needle point of the top generate multiply the normal mass of the motionle4ss top many times over because of the intensity that such a small area does increase the effectiveness of the top. Yet notwithstanding even more restriction by an increase in the mass restriction the motion still generates independence by motion evoking a defying erect stance.

$$T^2 = a^3 / k$$
$$k^{-1} = T^2 / a^3$$

$$T^{-2} = k / a^3$$
$$k = a^3 / T^2$$

In the spin there is relevance within the unit that forms all the principles we attach and associate with gravity. There is the expanding as well as the contracting which forms an integrated part of the rotation principle. If there were no rotation of a body, which installs the contrasting we, finds associated with rotation then gravity by principle would not have been possible. However the linear aspect also shows strong influences, which is as much part of gravity as gravity by rotation is a factor.

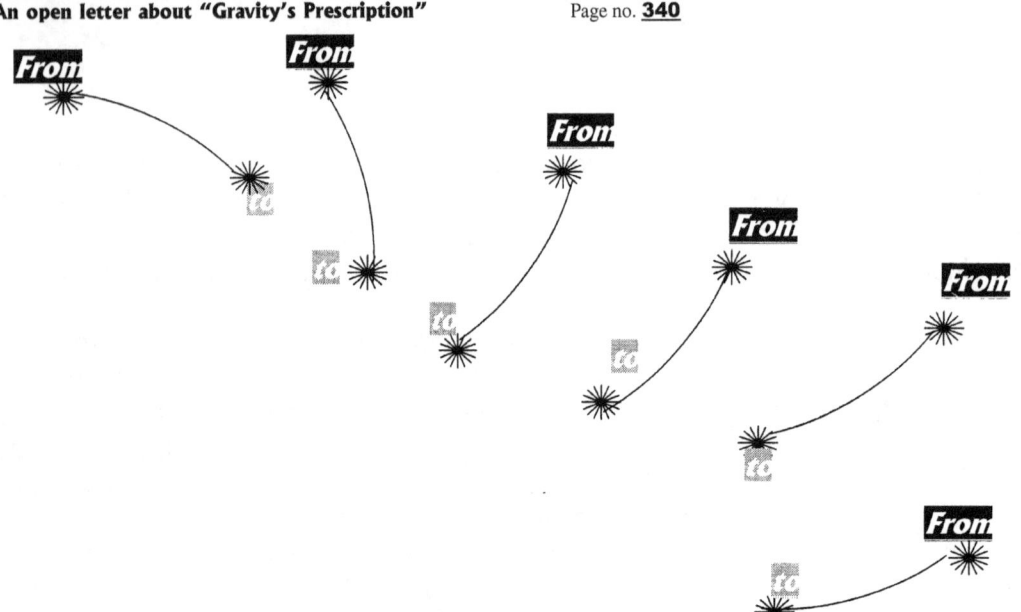

Although the electron is orbiting at the speed of light it is still in motion vertically and that also becomes a product of time as the whole structure is repositioning the relevancies, it had a moment before to what it will have the next moment. Such motion will again have an influence on the relation in the position the electron forms with the rest of the Universe while the lump of metal is now travelling as a spacecraft destined to other galactica. It is if we use the logic those intellectuals calling them Academics show and those Super-Educated that advocate how we may travel to far away galactica while we go on skipping the nearby galactica that is only two to twenty million light years away. Since the electron is duplicating by motion the motion links the electron to a time constant. The time constant is linked to the speed of light but time as such, is part of the speed of light. The faster we take the electron to go straight in the motion man produces, the less time there will for the electron be to circle around the atom. If we make **k** bigger in relation to increased motion, the smaller will T^2 produce a usable space.

There is **k** that forms the distance between the proton and the electron while the electron is spinning

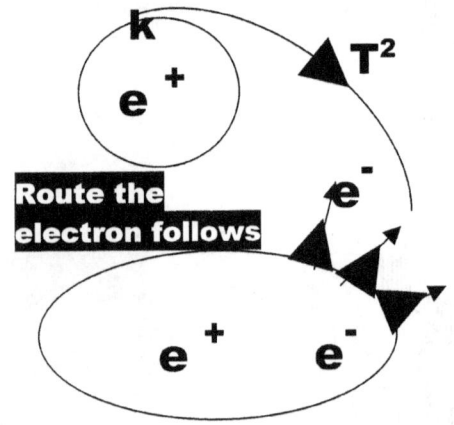

T^2 around the proton k^0. While all this action is going on, we think of the atom as being very still and satisfied with being a small part in a lump of metal we call iron. It could be any element but I use iron just as an example this time. The lump of iron is as motionless on earth as anything can be while being.

Even by coming erect through motion the generating of this stance finds its roots in the relocating of the rotating (T^2) in relation to the alignment with the line (**k**) in relation to space-time a^3 in time-space T^2k. The generating of the top and of gravity and electricity is provided in the very same manner by the Coanda principle.

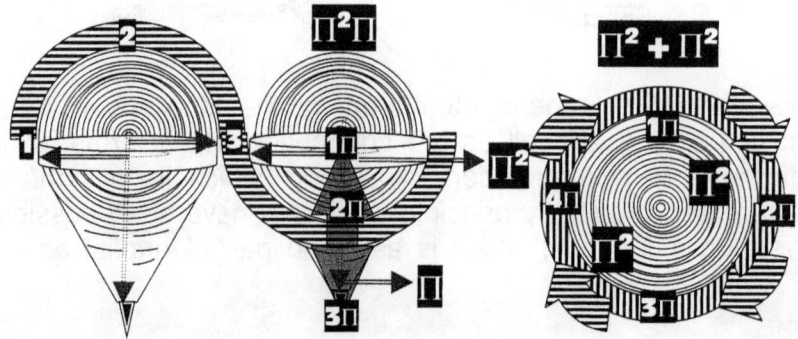

The motion that evokes singularity charges a graph from where the graph runs along the line of time. It is said that the spinning top is in balance but there the explanation ends and all parties are

satisfied. Never is the question raised about what comes into balance? The balance is a control of space-time that is established as space is duplicated by time while time support space in duplicating. The space is limited by the rotary action of **4** points in relation to singularity where this generates **3** points serving infinity that creates a division between infinity holding it's centre space and eternity being **3** active positions in time and the three is an eternal motion that never ends. By setting the division between **3** in infinity and **3** in eternity the containing that comes about is creating a cyclic space in **4** points.

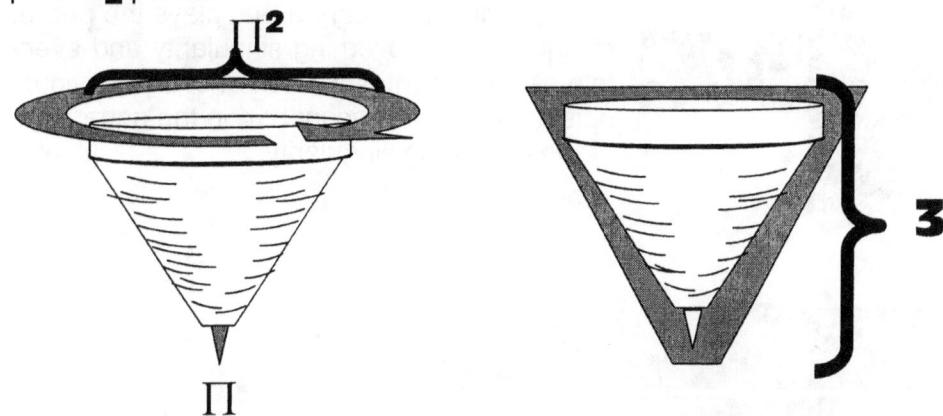

There is a something (if you wish I can use the term force although I strongly hesitate to use such an outrageous term for the most common aspect of the Universe) that is generating the power that keeps the top upright while rotating. The energy that is charged that is charged has the dynamics to stand its ground against the gravity of the Earth where the gravity of the Earth would under normal conditions depress the top into submission. However by rotating the top seems inspired and is reviving singularity by motion. The top is fighting and rebelling against the Earth's gravity when in spin. The top is performing the same way as an electric motor would. The difference there is between kit and an electric motor is the origin of the source that produces the drive. After all debating, there is one source that drives all forces small medium and large and that is the containing of heat and the distributing of heat.

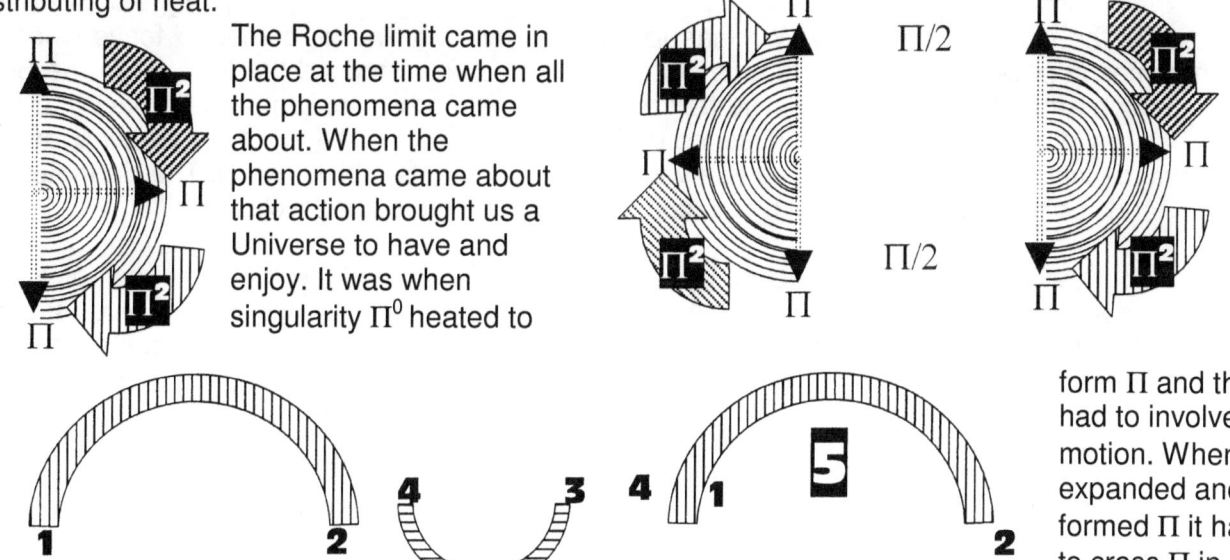

The Roche limit came in place at the time when all the phenomena came about. When the phenomena came about that action brought us a Universe to have and enjoy. It was when singularity Π^0 heated to

form Π and that had to involve motion. When Π^0 expanded and formed Π it had to cross Π in doing so. In order to establish motion Π^2 it had to go from Π all the way to where Π duplicated as Π. This involved the initial motion at moment – Alfa when space formed time by forming space.

As the rotation was a change of directions involving four aspects, which was a duplication of the previous along the present going into the future, a division came in place that parted the one unit from the next unit. As the forth-spot serving singularity landed where the first developed, that made the first spot the fifth spot. However this was accompanied by a rule, which today still apply in the cosmos.

With every four rotating points duplicating to reproduce on unit, a parting had to be devised to separate one Universe from continuing into the next Universe. The four points duplicating the value of

Π in relation to the centre Π^0 crossed the centre at a point measuring $\Pi/2$ and halfway where Π^2 lands the next Π by motion thereof therefore $\Pi^2/4$ became the limit that brought space in dividing point four from point five in relation to the developing centre. One has to remember that the Universe presently holds the characteristics it once enjoyed because once anything is part of the cosmos it has to remain part of the cosmos since there is no other place to go but to remain in the cosmos.

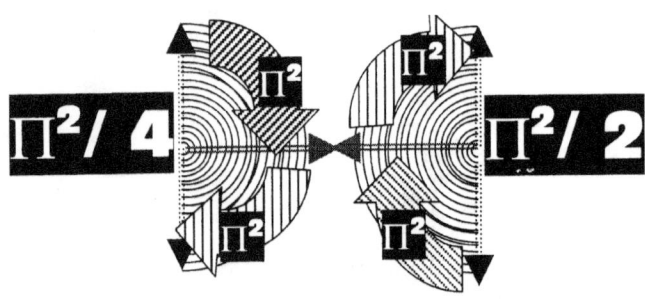

Even the motion of innumerable stars relate to a singularity in the centre that plays the part of the generated governing singularity and every faintest and slightest motion of every individual object plays a significant part in the generating of the governing singularity.

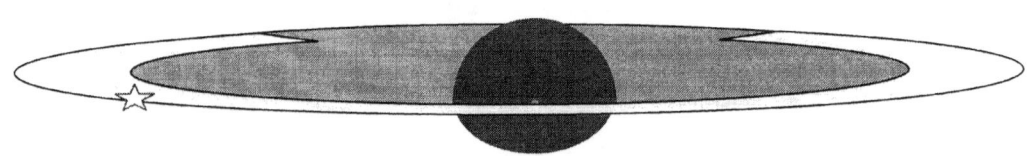

The Sun is on the outskirt of the Milky Way and the Sun is in an ova orbit around the Milky Way. The law of orbit is in principle that all orbiting structures follow an oval path.

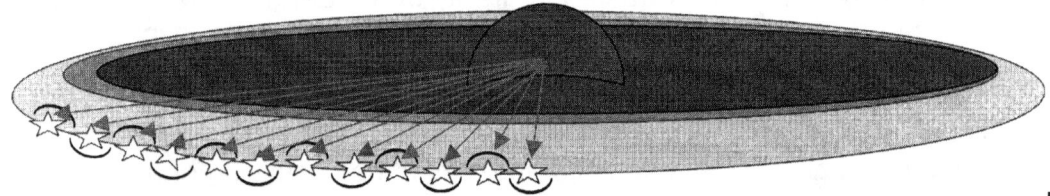

Exaggerated to a large extend the influence the Milky Way has to have on the Earth orbit comes to focus when a pattern comes in pace as the Earth follows not a circle but a wave around the Sun while the Sun sets its motion around the Milky Way. The fact that the planets orbit the Sun and the fact that the Sun orbits the Milky Way indicate an influence undeniable. The fact that the Sun is heading farther away from the influence should then lead to a variation in the planets orbiting wave. The Earth never, not once lands on the exact same spot by the completion of one more year cycle.

By not having a wheel rotate, the wheel becomes the factor of one, and the rotation becomes zero. The wheel does not disappear. In the cosmos, everything is rotating because nothing ever stands still. Therefore the mean equilibrium, the common factor there is to share, has to be one, eternity, the eternal Π, because all rotating objects has Π in singularity, and sharing singularity, gives every object in space a relation with all other objects in space. After trying for many years to bring them the candle, I concluded that Newtonians are incapable of realizing that mathematical principle as reality.

The comet rotates the Sun, and the Sun by itself has a point of singularity where Π remains without r. The comet, holding the orbit, also has a point of singularity, but since there is space separating the two objects, they cannot share a mean point of singularity, the very point of existing. Since singularity means just that, being single, there cannot be two. The comet and the Sun have a mean point of singularity but the space they occupy divides their common singularity. That is why they orbit in an oval path, a path where the one structure holds on to more space from its point of singularity towards the space it claims. Since they do not claim equal space, BY THE DENSITY they hold, the space will not be in proportion.

They do share in the common fact of singularity a point away from their individual singularity proclaiming their cosmic individual reason to exist in the cosmos. That point of common singularity holds space between individual singularity and that point of mutual singularity saves and protects the points of individual singularity. Since the start of time at moment-Alfa where both found the space they occupy, in the space they hold, maintaining a time to that space in accordance to the singularity they hold that point will be their individual eccentricity from singularity. The two objects are holding eccentric space around their individual but common singularity. That point of singularity is Π the circle without the radius because the singularity removes all forms or values of r, discarding r to infinity and leaving Π to be singularity.

That is why gravity is a fixation of Newton's mind making Newton bullshit, and his $F = G(M_1M_2)/r^2$ is utter nonsense. The moment you say Newton or any of Newton's laws, the Newtonian brain stun. Not once did I find one Newtonian surprised at this, I could not once find one single Newtonian to see this. It always leads to an argument and the argument is about Newton being in use for centuries. One Professor even answered me by saying that I should realize Newton's formulas placed man on the moon, and if that is not proof of his correctness to me I will never obtain proof. That is beside the point. That is miles from the issue. If you say Newton is wrong, you commit the worst blasphemy possible. One may swear at God and all is understood but mention your not accepting Newton's gravity and they all fall on their knees, cover their eyes in the ground, start stuttering and moaning and you cannot make them see anything but Newton. Dare say there is no such a thing as gravity because Newton is wrong, they run outside and hide the woman and children from your rage of mental instability.

Because I have had unmentionable arguments that I in the end lost because the mental Newtonian block all Newtonians hold covering their senses, where I could not reach a single spot of healthy logic within their minds, I wish to run through the facts once more and find what is so incomprehensible about the issue.

This in fact, is the very same findings that brought Johannes Kepler his own everlasting fame when

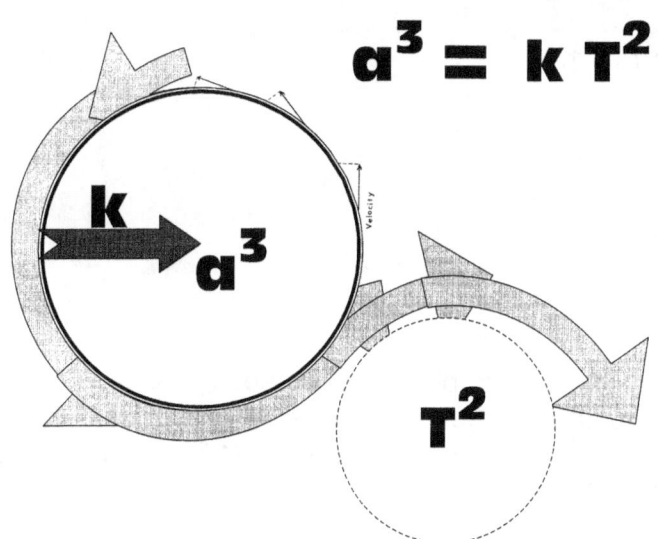

$$a^3 = k\,T^2$$

he declared that the planets stand to a value of $a^3 = T^2k$ as they orbit the Sun . Never once did he mention the presence of a force or gravity. Newton came up with this bogus idea all by himself without the help of other "giants" as he called *Galileo and Kepler*. In a later stage I indicate that I might prove the possibility that Newton did not have enough information to draw conclusions about Kepler's work. Newton saw a circle in Kepler's formula and there is a Universe of information hiding in that formula because that formula depicts the key to science namely singularity.

Newton made the formula one big blunder as far as the cosmos is concerned. Newton works perfectly well where there is equilibrium and unchanging in space and time as we find on Earth with the Earth forming the basis for space-time. Taking Newton to outer space is a blunder and Newton created the blunder by re-adapting his original formula of $F = r^2/ (m_1 X\ m_2)$ to fit Kepler's vision of $a^3 = T^2\ k$. This very same bogus idea helped Einstein to ignore the space factor of R^3 and place a relative value of one to space.

This he stated (without stating it) when Einstein put the Universe in a single dimension property at the point where gravity was stretched to the limit. Einstein put the Universe to a three dimensional value of matter, space and time and then out of the blue he places space at a factor of one when gravity supposedly destroys time. This notion stands totally unrelated and divorced to reality. I do admit that Einstein is absolutely accurate when saying this, but the space he refers to, as outer space and the space disappearing in time are as far apart as the cosmos is wide.

Einstein was the one that said that space and time could never be separated because it was the very same thing, a point I agree with in all my findings. The difference between my point holding the Universe and being the Universe is within every atom because it is there where singularity is. From singularity through the atom space has the relation between Π^0 as singularity and Π forming space inside singularity holding time $T^2 = \Pi^2$ in relation to the triple value of $\Pi\Pi\Pi$ forming Π^3. I am afraid that Einstein made much more sense when he was still an amateur, working as a clerk in the Swiss patent offices. Then he landed himself under the spell of the Newtonian disciples and all his initial ideas that were factual, became integrated and confused with delusions of the "Xepted scientific Newtonian High Priests" called "acclaimed scientists" and their mesmerizing fantasies about gravity.

There is a way to explain space-time by finding **space-time**. Space-time is not some force well and truly out of our reach, the one we may dream about and wonder why we have to adhere to it with so much respect. Einstein the master of physics was completely lost in his physics. He went looking for a flat Universe, he saw singularity in the dark of the night hiding as obscure fairytale characters behind Black Holes, where he saw gravity lingering around stars with nothing better to do than to wait for passing light just to bend seven variations of different types of shit out of each of them. Singularity makes every atom rotate that makes every cosmic object rotate that applies the overall rotation to the Universe. **THAT IS TIME**. Each time I try to share the idea of mine with the **"Accomplished Scientists",** I do not get farther than the phrase*: "Newton and Einstein are wrong."* After completing this sentence, I get treated as a raving lunatic with extremely dangerous hallucinations indicating a murderous tendency. Why would not one academic listen to the rest I wish to say before bluntly denouncing me?

Nobody even listens or pretend to listen to the rest of my case. Every time I see the light in their eyes go blank and they sit patiently and wait for the motor mechanic to finish his senseless rambling. It is so obvious they consider me as mindless person with arrogance and having a nerve to criticize the two highest-ranking Newtonians of all time! I can assure you I am not mentally disabled! It took me twenty-one years of research and another six years in compiling and writing this book.

That is the relation matter has outside singularity. $R^3 / T^2 = 1$.

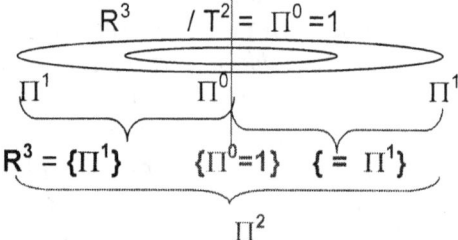

$$R^3 \quad / T^2 = \Pi^0 = 1$$
$$\Pi^1 \qquad \Pi^0 \qquad \Pi^1$$
$$R^3 = \{\Pi^1\} \quad \{\Pi^0 = 1\} \quad \{= \Pi^1\}$$
$$\Pi^2$$

In any of the pictures on the next page one does not see space, because you see a space filled with particles. It is the atoms holding the space secured that forms the picture. Why on Earth would nobody realize Einstein was seeing the Universe from a wrong perspective? What Einstein saw was one hundred percent correct but Einstein saw what he saw in the space of the atom and not in space at large.

That is space-time holding every aspect the Universe hold to a specific relevancy
The space outside singularity holds the time outside singularity because everything in the Universe is spinning. Science knows there has to be a difference because in space an object might be weightless, although it retains its mass, and no one can say the difference, except to put it down to "gravity".

Us, the tax paying public, are letting these Master Minded Academics get off the hook so easily, because every one is so scared to ask "why and how". In the pages above, I pointed to the most basic mistakes about the "gravity" which science ignores, because the answers they do not know. Even Nobel Prize winning work is blatantly misguided. I challenge any person to prove how an atom can collapse on itself, by force, by weight, by pressure or by any other means. No atoms will ever touch one another let alone compress to diminishing space, and if they do, the result is a nuclear reaction

IF IT DID NOT SPIN, IT WAS NOT ROUND, AND NOT BEING ROUND THERE WILL NOT BE SINGULARITY.

According to Einstein, the speed of light is a constant throughout the Universe. The speed of light results from two factors, being distance (kilometres) and time (seconds). This speed is accepted at 3 X 10^6 kilometres per second. Scientists know that it takes Sun light 10^6 years to reach the surface of the Sun , and we know the Sun is not thousands of billions of kilometres in diameter. The "Xepted scientific Newton Mistaken" explanation about this fact is that the Sun light "bounces against matter" and this retards the Sunlight dramatically. When light hits matter, (except in the case of glass), it joins singularity immediately. Therefore, the Sun holding matter on the inside has to be all- glass, or the "Xepted scientific" explanation is not very scientific at all. It all comes down to the density of matter in space valuing the time in that space away from the point maintaining singularity. Why can nobody but me see that?

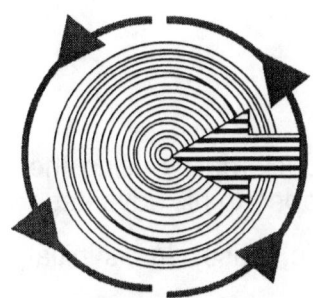

Where the arrow points we will find a spot that has no start. It is 1^1 that are the part that release from 1^0 when motion parts singularity by infinity and eternity. It comes about when motion unleashes the dot 1^1 that has no space and has no start from the spot 1^0 that has no end. Every time the top starts spinning a Universe is born in motion. The top is instigated and the motion is produced by the skills of life but in the cosmos serving nature such motion is the product have heat concentrated to sustain and maintain singularity. It is invisible, unseen and only detectable by intelligence and still it is a part of the cosmos that is no part in the Universe and from it the principle we call the Universe comes about. It is always a principal because it has no where to go but to be on call and by never being in the Universe it always is in the Universe. Sin any thing and see it is there by not being there.

Walk outside and look at the vastness of the blue sky or at night at the blackness we can see without being able to see because it is impossible to see darkness. That what you see when looking at the vastness is eternity that parted from infinity when space-time established a Universe. That which you see has no end because it is eternity in every aspect one may attach to such a connection. Standing where you are no matter where you are you are standing in 1^1 and you are part of that which parts 1^1 from 1^0. You form part of 1^1 as you stand and being part of the centre of the Universe (because all light flow directly towards you and acknowledge you at being the centre of the Universe, you therefore also form infinity being the inner most part of eternity. That means the infinity you hold gives you with life entity that never can be disputed. You are in 1^0 that can never end as much as you carry 1^1 that never can start. You are both the spot 1^0 and the dot 1^1 and neither can ever start or end.

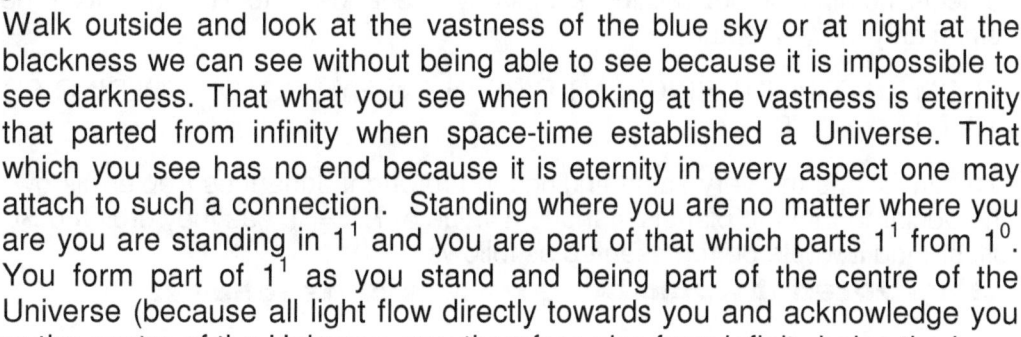

The pendulum arm covers a specific distance per time unit, every instant it swings. This is because of Singularity in position a^3 during time T^2 in instant **k**.

The space a^3 holds precise accordance to the time T^2 that it takes minus the compromise singularity claims from k by reducing space to the increase in heat.

Π^1 Π^0 Π^1

Space-time depends on the relevancy of matter occupying space change position in accordance to all other matter relating or relevant or even only influenced by the space a^3 in the e duration of the time the matter changes position T^2 in the instant of changing. **Space-time is everything excluding singularity diverting from singularity** and that is what Galileo recognised without realising in his observation of the pendulum. Where Π^0 is singularity and Π^1 is the diversion from singularity forming $\Pi \times \Pi = \Pi^2$ being gravity or time.

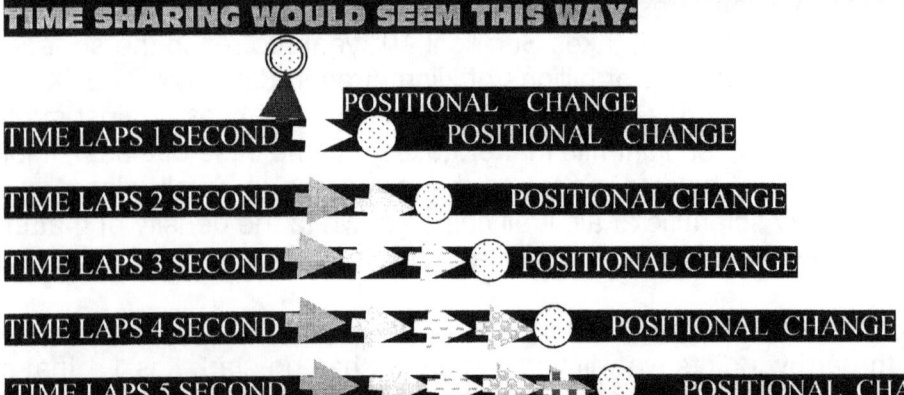

THAT IS AS SIMPLE AS SPACE-TIME IS

That is what Kepler (again I cannot say whether he wittingly or unwittingly) declared by using the formula $a^3 = T^2 k$ he announced space-time in a formula, the formula Newton raped to his advantage because in $\frac{M_s \times M_c}{r^2} G = F$ there can be no pointing to singularity in the cosmic sense. In his initial formula $F = r^2 / (Mm)$ singularity point at every aspect because as matter falls to Earth, matter continues down a precise path that singularity provide holding that specific position that leads the way. What it does point at is the motion caused by the Earth's singularity applying on much lesser objects holding or not holding singularity. I change a to R and T to T holding **k** to Π^0, which is singularity in the instant

FROM THAT POINT SINGULARITY IS IN EVERY PROTON HOLDING SPACE AS RELATIVE AS TIME IS. R^3 / T^2 = one

This in fact, is the very same findings that brought Johannes Kepler his own everlasting fame when he declared that the planets stand to a value of $R^3 = T^2$ as they orbit the Sun .
Illustrated it would be represented as follows:

Illustrated it would be represented as follows:

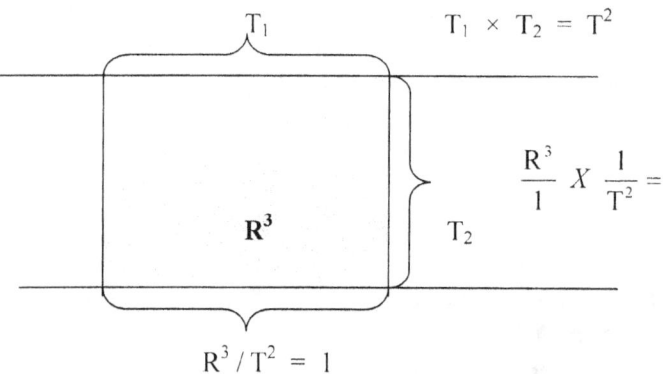

That means when the time that a structure relates to, is effected, the space will be effected pro-rata.

That means when the time that a structure relates to, is effected, the space will be effected pro-rata.

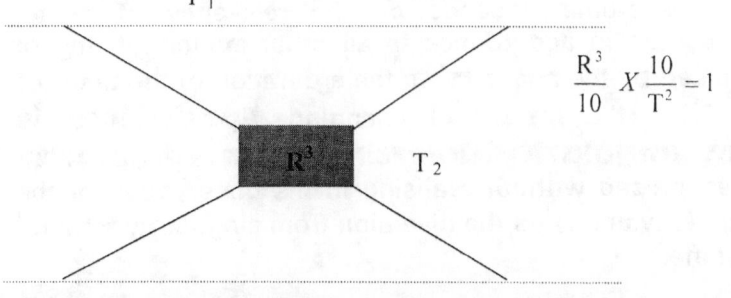

If one can illustrate the universe and its relation with space-time, the following illustration would fit like a glove.

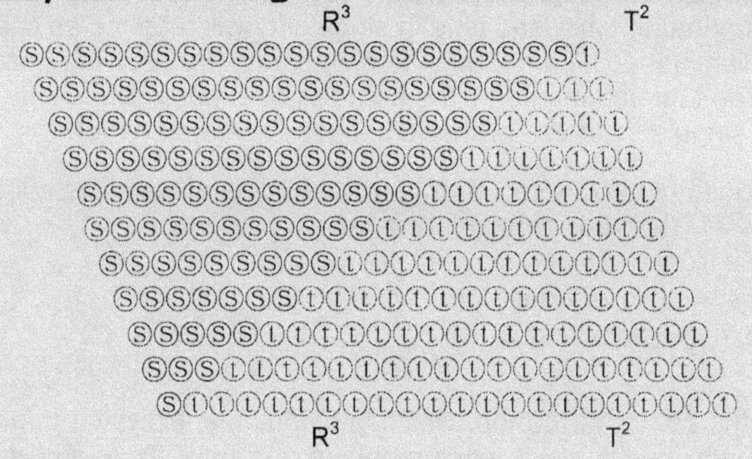

R^3 T^2

ⓈⓈⓈⓈⓈⓈⓈⓈⓈⓈⓈⓈⓈⓈⓈⓈⓈⓈⓈⓈⓈⓈⓈⓉ
ⓈⓈⓈⓈⓈⓈⓈⓈⓈⓈⓈⓈⓈⓈⓈⓈⓈⓈⓈⓈⓈⓈⓉⓉⓉ
ⓈⓈⓈⓈⓈⓈⓈⓈⓈⓈⓈⓈⓈⓈⓈⓈⓈⓈⓉⓉⓉⓉⓉ
ⓈⓈⓈⓈⓈⓈⓈⓈⓈⓈⓈⓈⓈⓈⓈⓈⓉⓉⓉⓉⓉⓉⓉ
ⓈⓈⓈⓈⓈⓈⓈⓈⓈⓈⓈⓈⓈⓈⓈⓉⓉⓉⓉⓉⓉⓉⓉ
ⓈⓈⓈⓈⓈⓈⓈⓈⓈⓈⓈⓈⓉⓉⓉⓉⓉⓉⓉⓉⓉⓉ
ⓈⓈⓈⓈⓈⓈⓈⓈⓈⓈⓉⓉⓉⓉⓉⓉⓉⓉⓉⓉⓉⓉ
ⓈⓈⓈⓈⓈⓈⓈⓈⓉⓉⓉⓉⓉⓉⓉⓉⓉⓉⓉⓉⓉⓉ
ⓈⓈⓈⓈⓈⓉⓉⓉⓉⓉⓉⓉⓉⓉⓉⓉⓉⓉⓉⓉⓉⓉⓉ
ⓈⓈⓈⓉⓉⓉⓉⓉⓉⓉⓉⓉⓉⓉⓉⓉⓉⓉⓉⓉⓉⓉⓉ
ⓈⓉⓉⓉⓉⓉⓉⓉⓉⓉⓉⓉⓉⓉⓉⓉⓉⓉⓉⓉⓉⓉⓉ

R^3 T^2

In the search for time in space, the most obvious place to look for the factor, time as such, must be where it is excluded from the space factor and stands alone. Therefore, one should find the place where space is zero leaving time to be eternal. Such a point would be impossible to locate and to place a value on time. However, in the cosmos at large, there is no such a place, because there is no such a thing as zero time or zero space.

Before I start with the true purpose of this letter of introducing my new method of revising cosmology, I wish to say in my defence that I chose one aspect from a wide range of possibilities to explain the way the cosmos formed. However that would constitute to a book much larger than the one you are reading and therefore I limit the development only to where matter, space and time parted and then I immediately thereafter focus on the point where the Big bang came into place.

In the beginning, there was time Zero to moment Alpha. There has never been a Big Bang, as such and there were too many Bangs too numerous to count. Everything is a variation of time duration in space. During the period of time Zero to moment Alpha the value of 1 second was equal in duration to about 1 000 billion, billion, billion, billion years (I am only stopping with the billion part in order not to bore the readers), measured in geodesic space-time values that currently applies. It could be even billions times this duration because the value of time then, was measured far beyond the speed of light, since light did not yet exist. We have no way to calculate the duration of time. **The closer time is to singularity the longer the duration would be. The method time is expanding is by heat and only heat can expand while only by reducing heat can there be a demise of space.** That applied during that geodesic space-time era as much as it does today, and we must accept it as one equal to infinity shorter than eternal. It is heat in all its splendour because only heat can expand. Even boiling soup produces space that expands. When a bowl of soup is boiling, have you seen the bubbles of air rising from the soup? Has any Newtonian ever taken the time to explain that process in detail? I think not, because such explanations would be far too "everyday-like" to bother their mighty brains.

Well, that boiling soup tells the complete story about the creation. Creating is a fact of creation, however creation was not created and left on its own, the Universe is in creation being created every smallest fragmented split instant there can be. The Universe is generated as it moves and such generating of the creation is a process of creating what there is. We speak so lightly of creating and no one comes close to understanding the concept of creation. Poets and painters and writers always wishes to say how "they created their creation". That is rubbish; they created nothing. They brought nothing new to the cosmos, they only rearranged what was a small part of the cosmos into a new order, that one can detect a distinction from. Creating is producing what never was before. When looking at the boiling soup, there are bubbles rising from the soup at the top. In the soup's brew, there are only liquids and solids before the heat came. In such a manner the expanding of heat created space. No one placed air in before the event or during the event at any time. Yet from the brew of liquid and solid rises gas, or if you wish space. That space was not there previously. That SPACE WAS CREATED.

That space is energy and energy is the interaction between heat and space. As space becomes a part of the soup, a part not there before, with no room to be, it moves out. We refer to that process as boiling. That space creation is applying heat to time, and time in singularity will respond as space in singularity. The space created will vanish just as it came, back to singularity. By applying heat to time, brings forth space, and from the three components, only the heat factor is not in singularity. It removes space in singularity from time in singularity to establish room (space) for heat (time).

That is how creation started. Time in singularity overheated and the product of that was space. That is the 180 $^\circ$ of the straight line as much as it is the 180° of the half circle.

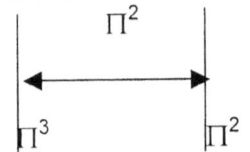

The Π^3 space from
The Π^2 is motion of liquid heat
The Π is time in space to singularity.

There is a time as a line that we find in the centre of the top, which is singularity. However there is another time, which offers material, the space in which material are able to duplicate. That too is time but it is the relevance between the holding time and the space-time where material is located.

Material uses the relevancy of time within, which developed as singularity and time without which was the expanding of singularity to commit to motion

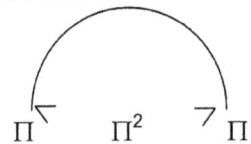

The half circle is 180° placing matter in a circle but because space only applies, to one half, 6/2 and matter holds space to value 6/2=3 only half the circle comes into effect. Half a circle is 180°. Because space has three parts in effect, it also becomes a triangle.

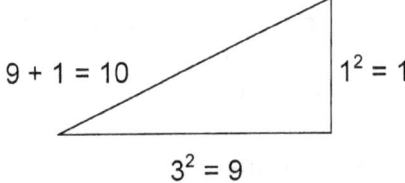

That means where space holds three and time is one, the heat within that space becomes another dimension, the fourth dimension holding space-time (3^2+1^2) = 10. That changes the matter inside space in singularity at ten and "gravity" at Π^2 . This is why "gravity" Π^2 is space (10) losing one dimension (Π^2). "Gravity" is all about space (occupying matter and heat) losing one dimension back on a long journey to singularity.

As time is in singularity, and space is in singularity and both are the same thing $(\Pi^3 \to \Pi^2 \to \Pi)$ the 10 of matter (heat) that affects space (10 Π) will also affect time (Π^3) and therefore time carrying heat will become 10 (Π^3) with space 10 Π. Anyone with a simple calculator can divide 10 Π^3 by 10 Π and see where Π^2 fits in. It is the doubling of matter in relation to time (7/10 + 7 /10) times the double factor of time in space (10) standing related to the line of time that refers to matter (10/7). That gives gravity its value of (Π^2).

Through the Coanda principle the motion of liquidΠ^2 confines space Π^3, to what space Π^3 confines as the atomΠ^3 = Π^2 Π) to the solid. In this containing of space by liquid in motion with the limiting or putting a border on space by liquid flowing, which confirms the space what establishes the Roche limit. In that there has to be a liquid (the neutron at Π^2 lies in the two components of space-time occupation or "gravity" manifested in the Roche limit. All objects spin and spinning is a circle Π^2 while all objects are moving in a direction $\Pi^2/2$. Again only, half of Π has any dimensional validity at any given time, therefore the dimension surrounding an object is Π. That is how gravity forms the atom as the surface of the cosmic object extends from Π^2 to Π but only half of the circle of Π (180°) can apply to time (Π^2) being in a straight line $\Pi^2 \to \Pi$, "gravity" will form at that point of $(\Pi/2)^2$ giving the Π in space the "gravity" to hold.

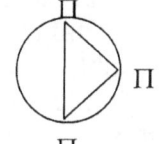

$\Pi^6 \, (\Pi^2 + \Pi^2) \, / \, (6 \times 10)$

That places Π in a total of Π^6 with 6 sides in space (10) affecting the proton ($\Pi^2 + \Pi^2$)

That is why space will forever comply with $7 / 10 \, \Pi^6) / 60 = 112$, (the Π^6 is ($\Pi^2 + \Pi^2 + \Pi^2$)) and time forming the line (180°) between the half circle (Π to $\Pi = \Pi^2$) at a 180° will form the triangle of space in half (180°). The matter component of the Titius Bode law effectively applies to the value of space, therefore 7/10 comes into the calculation. That places any atom with an existence in space at a premium of $7/10 \, (\Pi^6) \, (6/10)$. The reason why plutonium at $5(\Pi^2+\Pi^2) \, (\Pi/2)^2(3/5)=244$ is at the element limit is obvious; when dissecting the relevancy in detail. The complete element holds the very edge of what an element in space and time can endure in this era, but two or three era ago it had the function cobalt has at present

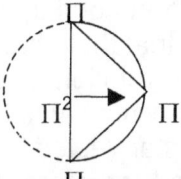

That will produce time in singularity a value of Π^3.

Explaining the other five stages of gravity (Π^2) development is extremely complicated and for that there is no room in a book meant to introduce new ideas such as this. My motto in this book (part one) is "Keep it simple"

With time in singularity, time was eternal.

Π^0

Time is the spin rate of heat in space. This translates to heat in spin (the atom sealing time off by the spin of the electron, which then produces a motion relevancy with the proton and time in space, which brings about the time line. As we can see when the top spins the top forms the spin of heat (the top spinning) in space. That means the way the movement changes where matter and heat relate to other matter and heat in space. All the movements are relating to a circle (Π^2) going somewhere (Π) in space 3. The Π will form the radius to the circle (Π^2). Any novice can see that the longer Π becomes, the wider Π will be and therefore the longer change in the repositioning of matter will be.

Any person wishing to uphold Einstein's view about the speed of light being the limit through which matter can apply velocity, then that person should first explain how the Black Hole works. It is very distinct that whatever is inside takes that which is inside, to exceed the velocity of light. In other words the Black Hole is able to force matter into speeds that goes way beyond the speed of light. The contraction produces a spiralling of particles that takes matter into a motion dimension far beyond the speed of light. The concept involves not the moving of the particles but the slowing of time to force a duplication cycle that goes beyond the capability a photon can withstand. Inside the Black Hole must be matter, because there is no space, yet time does apply because it takes the particles spiralling inwards to the centre time to move from point to point. Matter in motion is time. However, no light can return to the surface, therefore the light is slower than the moving particles within the star. The only thing about the star is that it maintains a higher relevancy than the relevancy the speed of light can apply. By accepting the existence of a Black Hole, any of Einstein's claims about the speed of light being the fastest that matter can travel becomes fictitious.

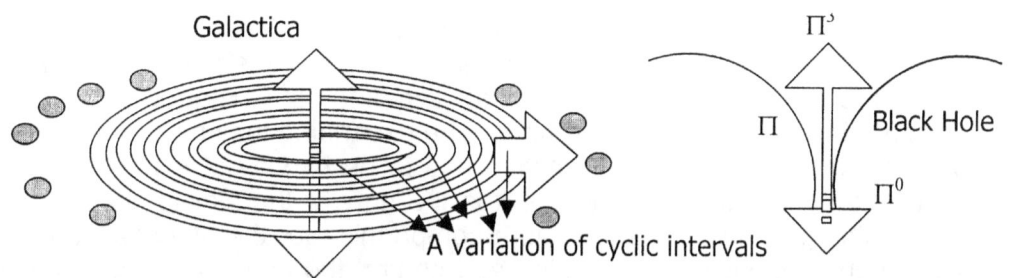

Another place where the speed of light becomes obsolete is within the centre of galactica, where the accumulative movement of matter exceeds the speed of light. That is where doctor Hawking saw a Black Hole that is not a Black Hole, but the precise opposite. Light, matter and heat, moves inward in an effort to maintain cooling as the group of proto-stars belonging to an era to the future) where they still claim their share of heat maintenance. Those particles in such close proximity, establish a time in motion well above that of the speed of light. Everything in the cosmos is all about relevancies. Particles in that phase are still very close to time eternal where motion of material took space into singularity. Time started at such a high velocity, it had to be eternal. Nothing that diverts from eternal can become more than eternal so it has to be less than eternal. It is fragmenting eternity into parts making eternity smaller.

Professor Hawking holds the opinion that there is a Black Hole centred within the centre of the star, which of course cannot be possibly true. The dynamics of a Black Hole is such that it is a star as massive as they come, that fused all the atoms into one structure. The nature and the essence of a star are to unify the singularity that was divided amongst all the atoms during the process of cosmic expanding. The star is a collection of atoms, which unite in motion that then through the unit generate motion to establish a controlling centre governing singularity. The rotation of every individual atom spinning is collected as a generated effort and the collective drive accumulates the effort to the centre of the star. The more the star develop the more is the drive of the star vested in the centre of the star and is it less concentrated in the material compiling the heat and therefore the drive. As the star becomes self secured, the maintaining of the star removes the duty of finding heat to secure the star from overheating from the atoms to the centre governing singularity. The singularity finally takes control of the motion of the star as the star evicts all space and drive time back to eternity.

Then a point arrives where the star abandon all motion. The star then achieved the main goal all stars have by producing a gravity that controls the motion of time. The spin has moved from within the star to time itself and be abolishing space all together the space became what singularity can offer. By using Kepler's formula the relevancy placed infinity in control of contraction and $k^{-1} = T^2 /a^3$ which means the space used by the star is infinitively small as the motion producing the gravity calls on the entirety of time to move and to establish such infinitive immobility. On the other hang the space that the star then control is eternally big $k = a^3 / T^2$ since the entirety of time establish by motion the outer limits set by the Coanda effect. That makes the controlling gravity the entirety of the time aspect because singularity being infinite commands time to motion where such command is stretching the ultimate. It is more complicated and I do explore the working of stars and the development of stars leading to Black holes much more in detail in another book I have being "*STARSTUFFIN*".

In the very opposite it is the motion of all the heat and all the stars proto or otherwise that forms the unit driving the galactica to improvise a Black Hole situation within the centre of the star. The Coanda effect that generate the singularity which control the star becomes generated as a result of all the heat and particle motion that turns about the star centre. Where the motion is valid enough to sustain a drive that would generate an equal gravity to that which the Black Hole demands, the totality of the liquid in motion in the galactica invests into a gravity that does form the drive equal to the drive of a Black Hole. But the drive forms what seems to be a black Hole. There is no real Black hole because if there was a Black Hole, then the galactica had met its destiny before any of the stars within such a galactica could journey onto a road of development. From a Black Hole nothing escape and every star is a future Black Hole on a journey of development to finally become the ultimate, the Black Hole. However there is one star more supreme than that, but there is no space to go into that explaining.

Gravity is motion. Motion is either the expanding or the contracting of material because of heat interacting with space. When material overheats, heat expands the space of the material and when heating diminishes the cooling reduces the space that material claims. In both instances it is motion applying. While moving the material overheat thus it expands. The expanding may be controlled and therefore the progress of expanding is controlled but motion has to be by way of expanding even when the expanding is under the auspices of contracting. It is the duty of the star to contract that which the galactica expanded. The galactica expanded as a compromise for the overheating but in the expanding the galactica, such expanding also develop and control the progress of young stars into adulthood. The galactica expands, expanding the stars and the stars has the role to contract the Universe back to singularity. While the galactica expands it gives the stars the opportunity to place heat stored as space of heat frozen by spin in the atom. The atom generates a governing singularity that demolished the space as it accumulates all the heat back to singularity.

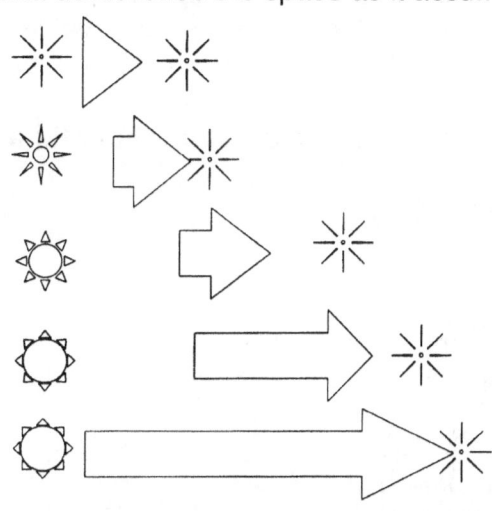

It is not only the star that reduces, or the fact that the Universe expands as the star that grows, to accommodate the space that expands by diminishing the relative space-time the star claims. It is relevancies applying more tendencies in representing the relations there are between structures in space and structures and space.

With outer space carrying the blackness in progressive multiplying, the very essence of space being space within, the atom too must be in growth claiming more space. Of all the above factors Mainstream science only acknowledge the growth of space in as much as calling it the Hubble Constant. However, that is not where the growth affects ends because it originates as much from any individual atom as it comes from Alfa singularity. Space does not expand because the space is only reducing the heat in density while producing density in space.

BACK THEN when the Universe was new

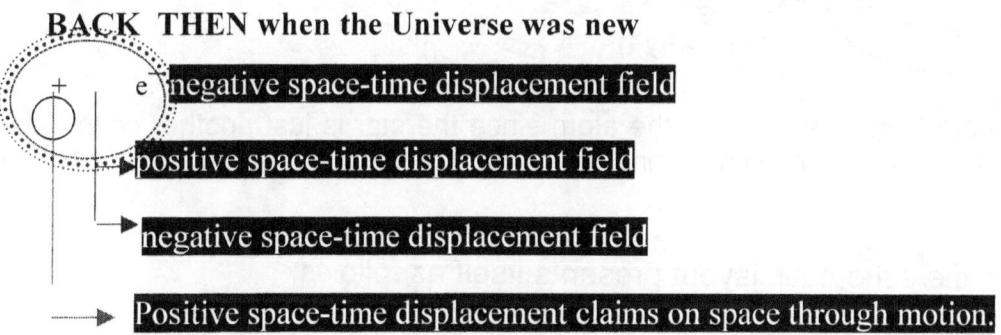

PRESENTLY we refer to the sizes we find space has in the Sun as quantum meaning they are inexplicably big

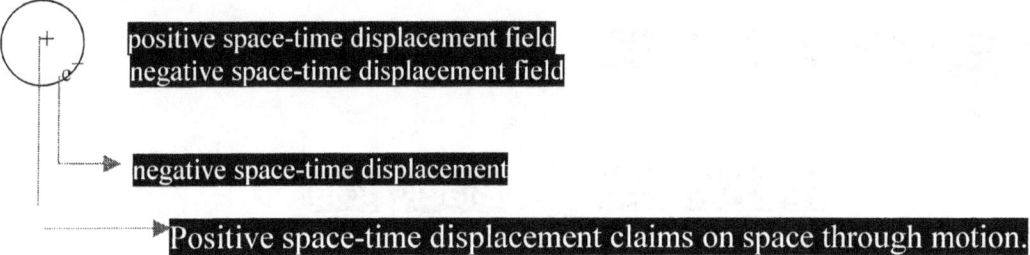

IN FUTURE TO COME they are going to get a lot bigger than the quantum size now present.

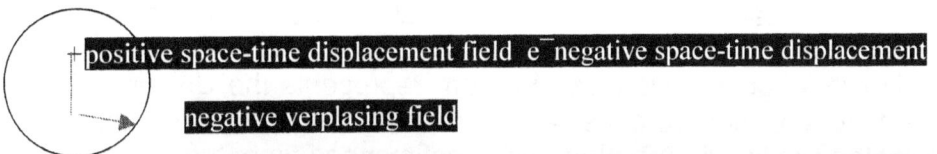

As the Universe expands the Universe is then the atom t6hat expands. The part of the Universe that Newtonians of when thinking of the Universe that expands, well that part they think expands cannot expand because that part is eternal. The part that expands is the atom and the atom is the Universe because the atom defined infinity from eternity by dividing the two points representing singularity. The star on the other hand has the task to convert the atom back to singularity by contracting all the heat into singularity. The star go in development by shedding its layers until it finally has only singularity in the centre left. Since the start is a combination of atoms forming the star the star is one cosmic atom every layer holds elements that serve the star in the particular development the star finds it in and the layers hold a certain displacement value of space-time. Therefore the atomic proton value is representative of the relevancy there is to form the need of singularity maintaining within that layer and as a layer contributing to the star as a whole.

This became the atom $(\Pi^2+\Pi^2)(\Pi^2\Pi)(\Pi^0+\Pi^0+\Pi^0)$ = 1836 and the atom formed stars that still act in accordance with and to the atomic relevancy

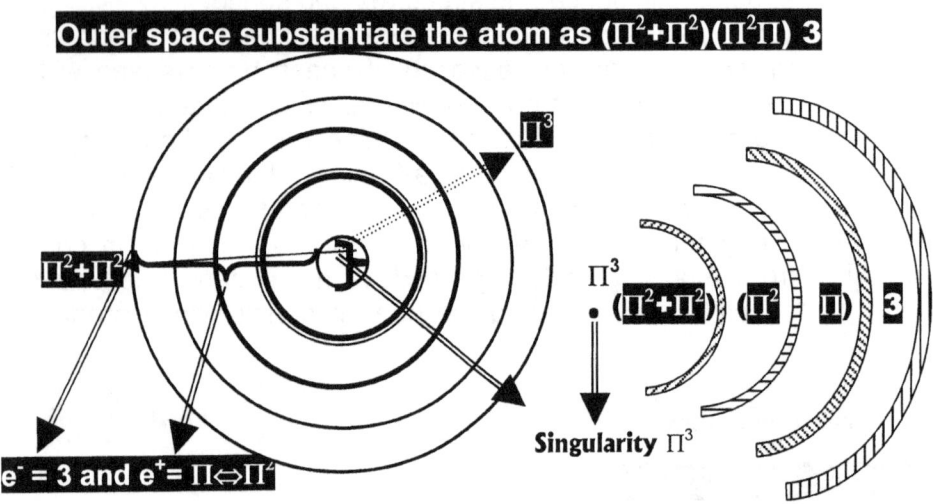

Every layer in the star represents one factor in the atom since the star is just another cosmic atom securing strings of atoms that as a unit aims for one goal and that is to secure one singularity within the star.

The manner, in which the schematic layout presents itself as follows.

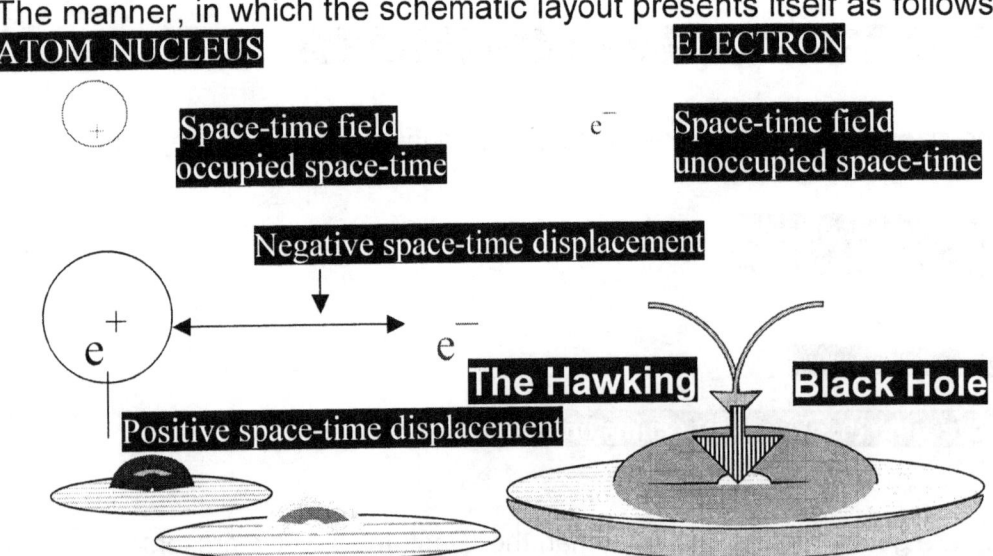

It is not only outer space that grows because $k = a^3 / T^2$ is as much the cosmic value as the value within the atom. That means that $k = a^3 / T^2$ is also in place within the atom and that shows the space within the atom grows as the Universe grows because the atom represents the Universe that is in growth. As gravity brings space-time reduction from the centre of the proton so must the growth come from the centre to the atomic proton cluster. As the atom expands in space-time, the proton can also grow dimensionally bigger through the neutron growing in stature. It expands in captured space–time by pushing the electron walls to allow the atom more space to occupy. It is pushing the electron to achieve a distance every time in the same manner that the body lets nails and hair grow. There are three factors of space-time where space-time is released. Cosmic unity and space and heat parted as singularity released the space heat holds by forming motion which produces time to set boundaries and relevancies applying.

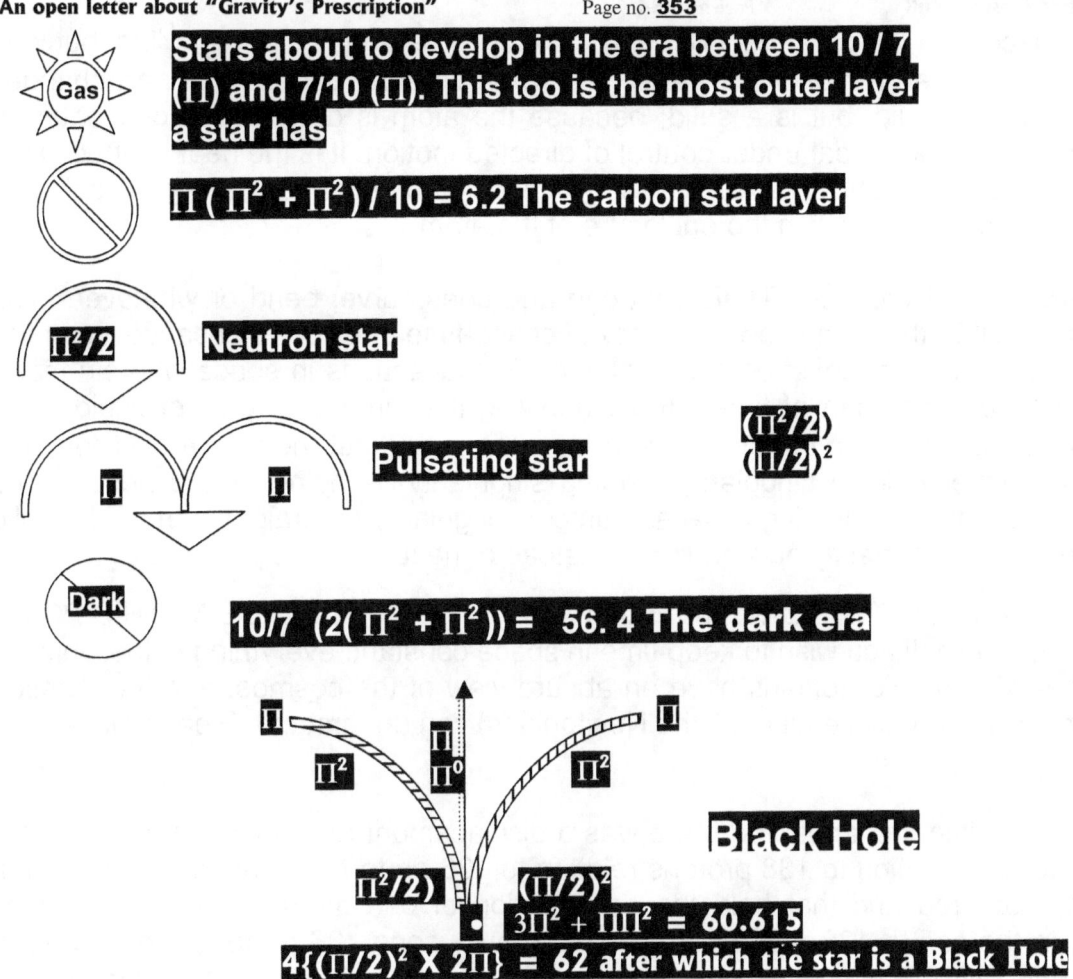

$\Pi(\Pi^2 + \Pi^2)/10 = 6.2$ The carbon star layer

Stars about to develop in the era between 10 / 7 (Π) and 7/10 (Π). This too is the most outer layer a star has

$\Pi^2/2$ Neutron star

Π Π Pulsating star $\dfrac{(\Pi^2/2)}{(\Pi/2)^2}$

Dark $10/7 \ (2(\Pi^2 + \Pi^2)) = 56.4$ The dark era

Π Π

Π^2 $\begin{array}{c}\Pi\\\Pi^0\end{array}$ Π^2

Black Hole

$\Pi^2/2)$ $(\Pi/2)^2$ $3\Pi^2 + \Pi\Pi^2 = 60.615$

$4\{(\Pi/2)^2 \ X \ 2\Pi\} = 62$ after which the star is a Black Hole

Every star is on the inside many different stars because every layer holds a different (k) or relevance making the space in the star very different from every other layer in the star. This is because every layer has different motion in relation to the governing singularity and therefore has a different gravity confining space. The layer is the result of the gravity effort of all the atoms in such a layer and therefore the space in that layer will bring about the time factor that produces the proton cluster relevancy

When contracting, gravity takes place by means of lying down newly acquired heat to maintain the cooling of the structure.

However by contracting it is accumulating material that produces a build up of material in order to enlarge the existing heat surface. In that manner it spreads the heat in a wider area than what the area was the instant before and in that manner it duplicates slightly more than it did duplicate the instant prior. That means even by contracting the measure is still expanding the material by relocating the material from an uncontrolled zone to a controlled zone (the atom.) Still this accumulation by contraction is expanding by motion. When saying this please be sure of one thing: it is not the Universe that is expanding but factors within the Universe that relocate that which takes up space and provides space in the Universe. The Universe remains unaffected by all this by never increasing or decreasing. Two of the three factors swap ends and that places the third factor at a different relevancy. The Universe can expand as little as it can reduce. The Universe expands by the curvature of space –time as Einstein proved but my solution proves much simpler than the way Einstein went about.

I will in a short while indicate how Einstein is correct about the curvature of space time in his theory about "The curvature of space-time" because the curvature of space –time is the form of Π, which is the value of singularity and that forms the Universe. It is a building of what there is by the dynamics that singularity provides in relation to the accountability space-time has relating to singularity. There is the space-time complying with singularity and filling the space-time in singularity is heat and matter valuing space-time. Space-time (Π^3 to Π) cannot bend, cannot curve, forms a straight line, but what fills space-forming time is matter in motion (Π^2) and heat (3) in time in space. That part changes. The atom cannot be gas, or liquid, but is a solid, because the atom is densified in occupation of space-time. The atom is space with heat under control of directed motion. It is the heat in unoccupied space-time that produces the gas, and a liquid, is closely connected as much as part of the solid that all substances form however it is not within the enclosure of the atom.

It is THE HEAT in SPACE that produces TIME, that can and does curve, bend or whatever. That HEAT in SPACE forming TIME that forms the relevancy of space-time does bend because it can flow and flowing is changing direction or constructing by altering flow directions in space wherein matter flow but that then is part of being part of time. If, by applying the forming of gas, or liquid to the element where it is the space between the elements, of course you will get the incorrect vision of Π, where the space-time (matter holding singularity forming singularity) is doing all the bending that applies to the curvature of space (validating time) and time in singularity (a straight line) will be solid. Einstein placed the relevancy incorrectly on singularity, instead of heat.

I do admit, IT IS A LOT MORE COMPLICATED THAN WHAT I ALLOW IT TO BE AT THIS POINT, but the motto is, Keep it simple. If you wish to keep time in space constant, everything in the Universe will be oblong. That is why the Newtonians have an absurd view of the cosmos, and they present facts in the cosmos in a way nobody (least of all the Newtonians) can understand. Please allow me to explain this part first.

When material in space in time first appeared there was a displacement relevancy that was in place then that had 139 protons in relation to 138 protons relating to 136 protons. It was an atom that had many atom variations gathered and that held the construction of one atomic atom worth on the outside 139 protons, in the centre 138 protons and in the gravity zone 136 protons. There was no electron at the time because the electron came about at $10 / 7\ (4(\Pi^2 + \Pi^2)) = 112$. This was the phase where the neutron was established as part of the Universe.   The first time I can detect space / time and matter is when the proton came to a relevancy of $\$T = 7(\Pi^2 + \Pi^2)\ = 138$

That produces space at the square in relation to matter at 7.
In that this relativity indicated that at such a point the proton was already demanding space ($\Pi^2 + \Pi^2$) in time set to the proton's conditions, and controlling Space-time not yet established, but creating the correct environment for controlled space-time to be. Time was slow, time became faster and faster because by extending the position of Π, Π^2 will produce speed.

$$\$T = 10/7\ (\Pi/2)^2\ (\overline{\Pi^2} + \Pi^2\,)\ = 139$$

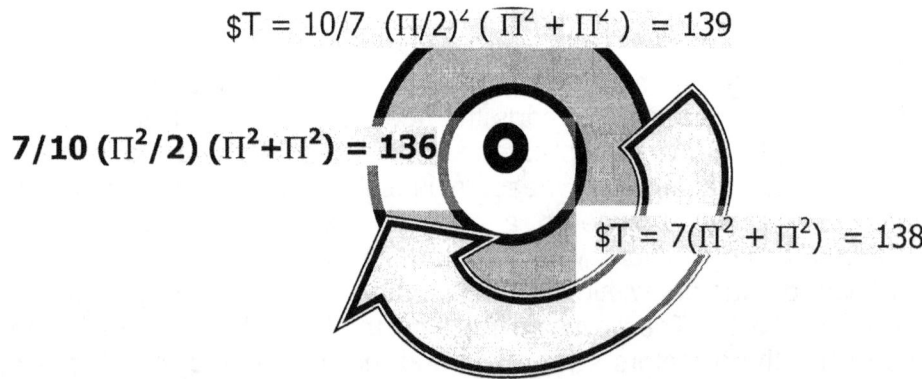

$$7/10\ (\Pi^2/2)\ (\Pi^2+\Pi^2) = 136$$

$$\$T = 7(\Pi^2 + \Pi^2)\ = 138$$

Explaining is as follows:

$\$T$ is space-time ($\$$) in the time sector (T)
7   refer to the 7° of any sphere.
($\Pi^2 + \Pi^2$)   refer to the double proton.

At the same time space formed as a consequence to the Roche limit and so did matter. Forming the Aanplasings-Atomic-Epitome (the point where matter breaks in singularity)
$\$T\ = 10/7\pi^2/2(\pi^2+\pi^2)=139$

Keeping these factors in mind it is clear that Π^2 are the choice of gravity and not r^2.

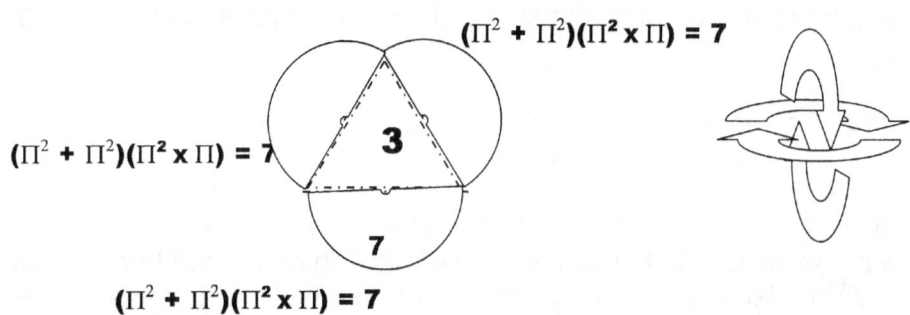

$(\Pi^2 + \Pi^2)(\Pi^2 \times \Pi) = 7$

$(\Pi^2 + \Pi^2)(\Pi^2 \times \Pi) = 7$

$(\Pi^2 + \Pi^2)(\Pi^2 \times \Pi) = 7$

$T = 7(\Pi^2 + \Pi^2) = 138$

Explaining as follows:

10/7   Forming the space factor and

7/10   Forming the matter factor in the application of the Titius Bode principle.

$(\Pi/2)^2$ the interaction that matter forms in sharing space and time by applying the Roche principle relating to matter-to-matter.

$(\Pi^2 + \Pi^2)$ the double proton

This is what the Universe consisted of, everything that is today, was then, in a dimension that only holds "gravity" or gravity- motion.

The only way new space can form is by unleashing the forming of heat. Some particles still overheated which evidently forced more space. From the overheating and the consequential space that came about, particles broke down allowing matter to dissolve to matter (neutrons) and space (electrons). It must be very clearly stated that neutrons and electrons did not yet form but only a suitable set of rules formed that later would apply, where matter could supply matter to initiate a sufficient heat supply as heat, locked in time, converted to space. The process that happened at this point lies far beyond that of a nuclear reaction, it lies far beyond that of a "Black Hole" or as I prefer to call is a proton star. All of this still fall under the same aspects and parameters we find energy or the application of energy to be at present.

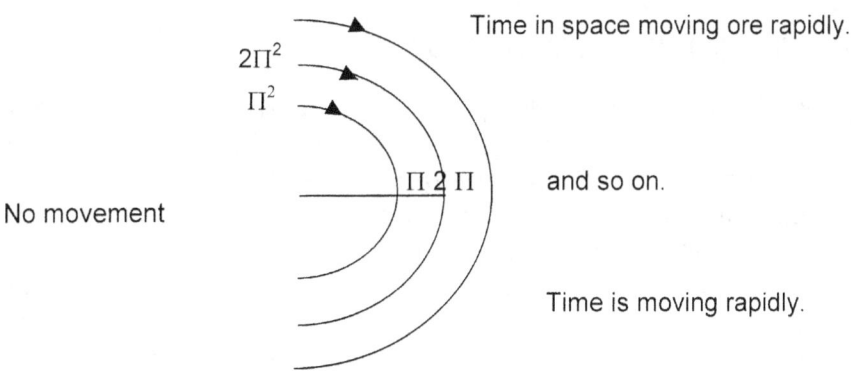

Time in space moving ore rapidly.

$2\Pi^2$

Π^2

No movement

$\Pi \, 2 \, \Pi$ and so on.

Time is moving rapidly.

Time started at what ever point you wish for it to start, and we can take time back indefinitely. That will be pointless.

Heat converts time in forming space by expanding through overheating that creates space and that was energy back then, at present and to the end of the future. In order to keep matters short, I shall proceed to the "Big-Bang" explanation, the period of nuclear heat forming space. Space formed for the very first time an identity separated from matter, apart from time, becoming the fourth dimension. In other words, the atom as we see it came into practice. Matter solidified at $T = 7(\Pi^2 + \Pi^2) = 138$ which enclose a formation by the Coanda effect spin which brought the atom into place, but this was relevant to a displacement flow of liquids. However the contrasting between expanding by overheating (10/7) and contracting by cooling (7/10) did not end because the displacement of space –time in time in space was in progress. That brought about that the Universe found a means in resolving the issue at the constituting of the six sides in the three dimensional Universe $((\Pi^2+\Pi^2)+(\Pi^2+\Pi+3)=(35.75 \times \Pi)= 112.313)$ and that capped the limit to whatever a compliment of protons in the six sided Universe can withstand. The time factor was at an expanding limit of $10/7(4(\Pi^2+\Pi^2))= 112$. This was when the three dimensional Universe first announced its presence with an apparent Big Bang.

The last element formed and had one proton, but still the temperature was out of control. Then the final element came into place, the cosmos. This was now possible because although the Coanda effect was in place from the first instant the flowing of liquid that was no longer a part of time made diversions of what materials would finally become more possible. The Coanda principle can sustain gravity under the guidance of motion $\mathbf{a^3= T^2\,k}$ where liquid flow $\mathbf{T^2}$ adheres to the form singularity (Π) dictates by securing space $\mathbf{a^3}$ in relation to the relevancy $\mathbf{k}$. For the first time then there were liquids that could flow $\mathbf{T^2}$ and by the measure of the flow $\mathbf{T^2}$ compacted space $\mathbf{a^3}$ in expanding $\mathbf{k}$ the realm of space $\mathbf{a^3}$. There too was solids $\mathbf{a^3}$ that was contracting by means of cooling $\mathbf{k^{-1}}$ gravity in applying the liquid in motion $\mathbf{T^2}$ to the space $\mathbf{a^3}$. There was a dimension of motion by space 10/7 that allowed motion of space 7 / 10 in space. This went on until our Universe in the form of the atom arrived at 112 protons displacement relevancy.

$T = 7/10\ (\Pi^6) / (6 \times 10) = 112$

7/10   is the matter has the dominant value.

Π^6 matter has the six sides it holds in the fourth dimension.

6   are the six sides to space occupying matter.

10   are the value or dimension in which space holds a ratio to time.

The cosmos began, not to a specific space, because all the space that was there, initially is still there at present. Any atom above 112 cannot apply to the fourth dimension not then and not now.

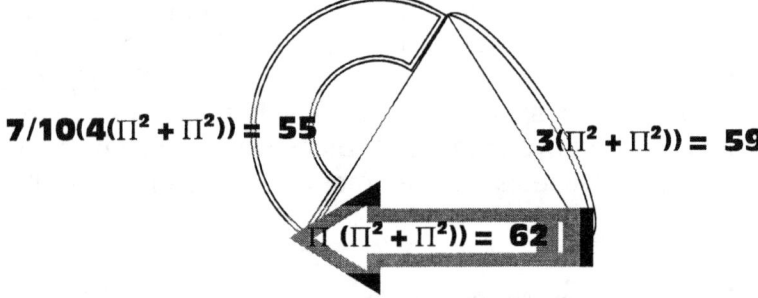

$$7/10(4(\Pi^2 + \Pi^2)) = 55 \qquad 3(\Pi^2 + \Pi^2)) = 59$$
$$\Pi(\Pi^2 + \Pi^2)) = 62$$

FROM THIS SPACE HEAT AND MATTER DEVELOPED

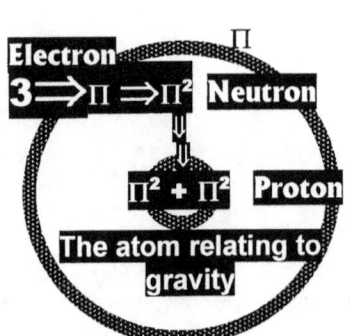

From the value that outer space can support being the sum total of the particles forming the atom $((\Pi^2 + \Pi^2)$ the proton $+(\Pi^2 + \Pi)$ the neutron $+ (3)$ space $= 35.75 \times \Pi$ singularity $= 112.313$ the star deliberately reduces the atomic space or the subatomic constructed space as the star intensifies motion and that reconstructing of space-time changes the qualities of the atom from what we presume the atom to be, to suspending the atom beyond the boundaries of $7 /10 (\Pi^6) / 6 = 112.16$. The converting of space- time from outer space through gravity to the star centre is the same route electricity follows.

Electricity and lightning is the absolute epitome of the Coanda effect where the Coanda effect is precisely the manifestation of light following the exact principles of the Coanda effect and the **Total Internal Reflection** is also miming the same principle as the Coanda effect which is vivid proof of space-time $\mathbf{a^3 = T^2 k}$ (the Coanda effect in acting principally by using the flow of photons instead of atmospheric heat). **Total Internal reflection** is only about applying motion by the flow of space-time (in this case water running) through the atmosphere but in the case of the phenomenon we call the **Total Internal Reflection,** singularity captures light holding the flow of light honest to a specific centre as does the Coanda effect and by setting borders the boundaries light is restricted to as singularity sets limiting boundaries to the flow of photons. But that is what electricity is; it is only creating space-time accelerated motion with much intensity added and it links a line than is concentrating space-time as it accelerates space time through the displacing differentiation which one finds in stars between copper dismissing space and iron accelerating heat directly to singularity.

It is only with much more intensity. All it is, is the Coanda effect forming electricity and lightning as the Coanda effect.

The copper field coil breaks down space-time by dismissing space-time as the 63 factor excels the space-time as fast as the electron will, as the space-time has to flow through the copper in the event where the iron being in motion causes the flow and charging the flow to the equal time set as the photon has. That is electricity. It is taking space-time directly to the centre of the earth because the motion T^2 k excited the space a^3 to a level that gravity is within the Earth core.

As the space in motion is occupying less space due to the motion duplicating and reducing the space, the space in motion will need less space to duplicate, thereby then create motion. Smaller objects can apply faster motion since smaller objects require less space to duplicate and therefore less time to do the duplicating of the space.

At this point I feel advised to remind you that mass does not produce gravity as is alleged by our respected Newtonians but is the restriction of the flow of space-time, which then is gravity. The restricting is at the top $10/7(4(\Pi^2+\Pi^2))= 112$ and at gravity level it is $7/10(4(\Pi^2+\Pi^2))= 55$ while in the end it is at the bottom of $3(\Pi^2+\Pi^2))= 59$ with space collapsing into eternity at $3(\Pi^2+\Pi^2))= 59$. At a value of Π $(\Pi^2+\Pi^2))= 62$ the atom collapses and the neutron is banished from the atom and that is where the Neutron star is in development. That means the element blessed with the number of protons to the value of 62 is the lucky participant that has the honour to collapse space-time while the runner up can achieve the generating of gravity has a number of 55.6 protons in its cluster.

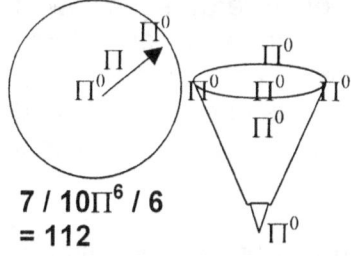

7 / 10Π^6 / 6
= 112

The spinning of Π^0 around the centre Π^0 establishes Π and Π is what produces the form gravity has. Still it is the relation or relevancy there is between the centre Π^0 and the spinning Π^0 that gives status to the form that Πrepresents. In out Universe we are accustomed to and are familiar to the rules we want to place seven points holding singularity to the centre holding singularity in a relation of $7/10$ Π^6 / 6 = 112. In that Universe everything less that a duplication ability to the value of 112 protons fit but only atoms to a maximum of 112 protons fit.

What that says is that the square of the proton diverted from singularity by the square as the proton claimed the square from singularity and hold an overall control of space in a sphere (7) and that was the form matter from that point on would always be. Space is the condition set by the neutron, and without a neutron there is no space. To understand this one must firstly understand the principle behind the theory I introduce. At the most inner point one finds time or if we can supply it with a completely fictional name: "The gravity - motion". The gravity - motion carries the value of Π^3. This value determines time in eternity a position matter has no space, but is occupied in singularity. Taking the neutron position to that of the proton we find the value created when the three dimensions (six sides) came about $\Pi^2 + \Pi^2 + \Pi^2$, which carried to the fourth dimension in cosmic or geodesic space-time became Π^6. When relating Π^6 to singularity it becomes Π^1 (space) x Π^1 (time) x Π^1 (matter).

I do realize this explanation does not suit normal mathematical principles but we are working in dimensions and Π^1 (time) in a straight line is 180° and Π^1 (matter), which is half a circle is 180° and Π^1 (space) which is a triangle is 180° as each Π^1 represents one dimension establishing another dimension and providing that dimension's existence.

In relation to the " gravity - motion" space through a straight line will be Π and through the half circle matter with heat positioning space (Π = 1) to 3 sides 3. Relating the gravity - motion Π^3 three will always be a Π^2 and a Π combining 3. The half circle will be Π^2 and the straight line Π. Because there are three directions of flow of space-time two of the three is in opposition at 10. 7 going away into overheating and expanding while 7 / 10 are contracting where cooling reduces space.

$7/10(4(\Pi^2 + \Pi^2)) = 55$
**Gravity generating
space-time flow**

$3(\Pi^2 + \Pi^2) = 59$
**evicting the
neutron from
the atom**

$\Pi(\Pi^2 + \Pi^2)) = 62$
**Abolishing space-time
by dismantling the
proton**

Behind all of this explanation is one obvious rule, the one subatomic particle positions in such a way as to displace space-time in the form of heat breaking down the value of space in order to meet the requirement of time. It is again all a relation between space and time and the less space has heat, the less the value of space becomes, because space has no value. Without heat in space and without matter there is no such a thing as space, therefore space does not exist but for matter and heat valuing space to form time.

Π^0 singularity diverted to $(\Pi^2 + \Pi^2)$ diverted further by $(\Pi/2)^2$ in forming neutron space

When saying this, I wish to include the following explanation for those that may have an interest however they are more likely to condemn by being sceptical.

Energy is a term for a power or a force, an effort to get work done. Energy relies on movement. The influence one object holds relating its position in accordance to another position. Energy is the flow of space-time. It is gravity as much as it is electricity as much as it is nuclear force and every other force Newtonians keep in their back yard as home broken pets. I was thus far unable to distinguish between electricity and gravity. One might go as far as calling the flow of liquids energy (10/7) and the flow of gravity (7/10) but this might be recognised in the light that it is merely a directional distinction there is between the same things.

When an object remains in one position relating to the rest of the surrounding objects, the time it remains in that position is unchangeable. Therefore, for that duration it will remain eternal. To reflect this relation to a formula will be $R^3 = 0$ and $T^2 = \Omega$. Once the movement starts the position changes, therefore the space relation changes. The movement relates to time because any movement takes time, even with an uncontrolled explosion. Changing space always takes time.

The misconception claims its incorrectness in the way Einstein placed the speed of light. The speed of light is not time, but merely particles in time occupying space. The Universe is not in singularity: the Universe is the way matter divert from singularity, claiming space and time within the constraints the Universe in singularity will allow. Singularity is time, holding time at an eternal value and allowing space to bring about the infinity that interrupts eternity. The time aspect is in the matter that keeps space in singularity apart from time in singularity. In fact the only Universe there is, is the Universe outside what we think of the Universe locked in singularity with all other factors being generated by the Universe in singularity.

THERE WAS SINGULARITY

Before that what was still was not the idea of what was not gave no meaning to 1 because there was no one and 1 was still part of everything not invented. Without one in place the measure of zero being numerically 0 had no place because the numerical order that starts with 0 or zero was still an invention that came about with the invention of 1 or one as a concept. To think the Universe started with zero or 0 was jumping time by one eternity because again I have to stress 0 was something that found a place one eternity after 0 or zero became part of the thought that put one or 1 in place. Newtonian mainstream science are unable to think that far back because Newtonian mainstream science are unable to think in terms of how nature functions! This is because theists would rather dismiss nature, as they don't understand nature. In their view everything was in place because they can't see past their eyelids.

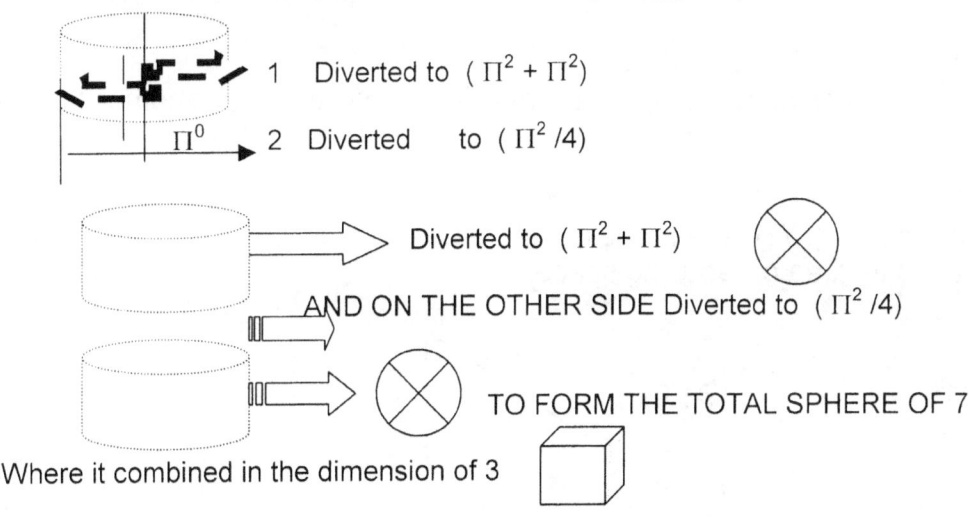

1 Diverted to $(\Pi^2 + \Pi^2)$

Π^0 2 Diverted to $(\Pi^2 /4)$

Diverted to $(\Pi^2 + \Pi^2)$ $\otimes$

AND ON THE OTHER SIDE Diverted to $(\Pi^2 /4)$

$\otimes$ TO FORM THE TOTAL SPHERE OF 7

Where it combined in the dimension of 3

TIME IN ITS ORIGINAL VALUE DID NOT DISAPPEAR. TO THIS DAY AND UNTIL THE COMPLETION OF THE COSMIC ETERNITIES, TIME WILL RELAY ITS GROWING VALUE TO SPACE DUPLICATED IN TIME PERFORMING AS THE DISTORTION OF TIME.

Time started ROLLING AND ROLLING IT REMAINS. **TIME CANNOT STOP; THIS IS CRITICAL IN UNDERSTANDING THE COSMOS. THERE CAN BE NO MEASURING OF T, ONLY OF T^2 AS T^2 INVERSELY RELATES TO R^3.** Therefore the formula to determine **time t** $=\sqrt{ (1 - (C^2 - V^2)) }$ is **nonsense**

The first question that one can ask is why would there be the value of $(\Pi/2)^2$ between orbiting structures positioning themselves in a time relation to space. Humans have a connection between human historical occurrences done by humans onto humans utterly confused with time. Those are eventualities that occurred and if it was not historically documented it would be lost as waste. History is not time but merely a colander, reminding whom ever has interest in such reminding of historical re-occurring of events relating to human behaviour in memory of human perceptions, which is far too short lived to have any bearing on time as such. Time is cyclic reoccurring events repeating events that can never again repeat as it did before and is totally one off. Time is a circle running on a straight line of cosmic progress and cosmic development where everything is happening at least once but also only once ever and will never precisely repeat after occurring all detail afterwards changed for ever. Most important, time can never stop because if it does and where it does, space will collapse onto time and disappear into time.

Every person knows about the entry restriction an orbiting spacecraft finds that forces the craft to comply with the entry maximum of 21,991 and the minimum entry of 7. This is without doubt, the number of Π (21,99/7). The Earth holds its value to $4\Pi^2$ and when an object is not part of the surface of the Earth, even say a mountain; it becomes a holding value of 7. Later in this part I explain the sound barrier in more detail and the 7 will then become better understood.

At this point one must see the Earth $(\Pi^2+\Pi^2)$ where all other particles will be in relation to each other by be either Π^2 or Π. When water is in a vapour form, it will have a value of 3Π, having heat separating the water to the factor of 3. By dislodging the thunderbolt, the 3 receive a square value and displaces to the Earth in the linear light to time stance of 3^2. With heat (3) grouping by initial spin value, it will remove from space leaving the water to the value (Π to Π) and this will then give the water a relevancy of Π^2. The factor of Π^2 places the water no longer amongst heat as gas, but heat as a liquid (rain) or solid (hail).

The relevancy of the water will change from 3Π to Π^2 placing the water's position from space (3) to liquid or solid (Π^2). Where does the Π that one find in the Roche limit $(\Pi/2)^2$ and the vapour(3Π) finds its relevancy to gravity? Every particle that enjoys space-time outside the Earth's structure $(\Pi^2 + \Pi^2)$ will hold a neutron position of $(\Pi^2\Pi)$. The Π^2 end will be at the point where heat passes through the object directly to the Earth and this position of space-time relates to the neutron time link of Π^2. The space link of the neutron will then form the Π link. The value of the Π link we find to be $(\Pi/2)^2$, but the explaining to why it is $(\Pi/2)^2$ is rather more complicated and is therefore left to another book for another day.

In nature there is the Roche limit placing a limit to the reduction of space and the inflow of heat to sustain proton cooling. At a point of $(\Pi/2)^2$ the reduction of space disallows any object the cosmic object cannot reduce, an entry to its area of reducing space.

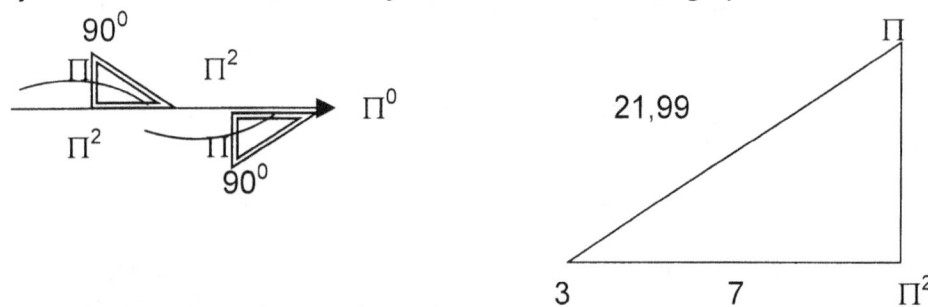

Time is a line with space in time and time in space on either side and that is 90 ^{0} to either side of the line being 180^0. It is Pythagoras in every sense there can be.

At the end of the space relevancy 3 where matter occupies space (21,9 / 7) is a border Π. That border is the exact point where space reforms to a square of time placing all matter (occupied heat) and heat (unoccupied matter) to a value of the square of time.

That specific point is in relation to the square of the diminishing shield around the Earth. However it takes matter (R^3) from the 3 dimensional position to the square (Π^2) in relevancy to time in singularity. With time holding space in singularity the 4 sides of Π truly relates to half of the total square value. Let us take the "Sound Barrier" from a point we see phenomena apply laws that matter complies with because the sound barrier occurs when the 90^0 square breaks.

The "gravity" factor of Π^2 becomes one and only holds the square to the Π position as it holds space to singularity at a square (time dissolving space at a square) and the time value (Π) remains dimensionless in singularity at (1). It then is $(1)^2$ where the one becomes the space position ($\Pi/2$) representing time (1) at that point. This makes the position time (normally Π^2) but now directly links to Π^3) relating to the singularity position of space diminished from the three dimensional to times single dimension in the square. That makes the Roche limit holding the position of $(\Pi/2)^2$ when the neutron position of time (Π^2) links directly to singularity (1). This may only represent figures, something to accept through intellect but lying far outside the reality surrounding our everyday understanding. To the best of my ability will try to convey my comprehension on the matter can, bring across how I can see it.

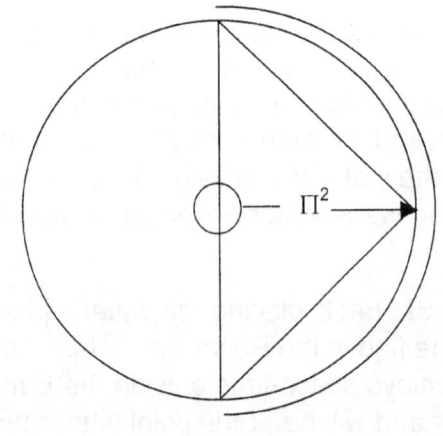

The square or space factor is a single worth $180°$ ($3\Pi^2$) or ($\Pi\Pi^2$)

The half circle $\Pi/2$ is $180°$ holding Π to half its value in occupied space ($\Pi/2$)

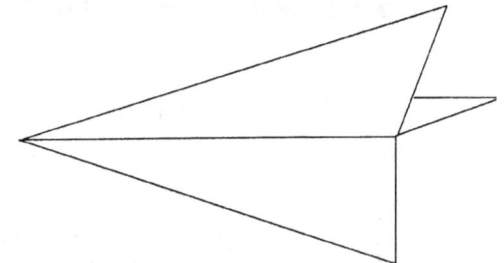
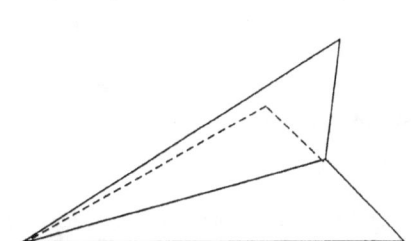

That is the shape needed to fly in the surrounding atmosphere of the Earth that we presented with a name THE ATMOSPHERE. In outer space a flying object can hold any shape because any shape fits the three -dimensional form. It could be round, oval, flat or any combination of all shapes and forms. The dimension is 6 with a linear value of 3 applying in one direction. This implicates the fact that wherever an object directs a direction in time revaluating positional change in space, it will forever be heading in three directions at the same time. The only control in direction is the concentration of heat at a point opposing the direction of travel. By applying the release of heat, one is applying "anti-gravity". One is producing space that never existed before, because the release of heat will produce space. To the object in travel another dimension brings about influence to his three directional space application. This application of the heat brings about a fourth dimension that applies directly to time. I shall come back to this point duly.

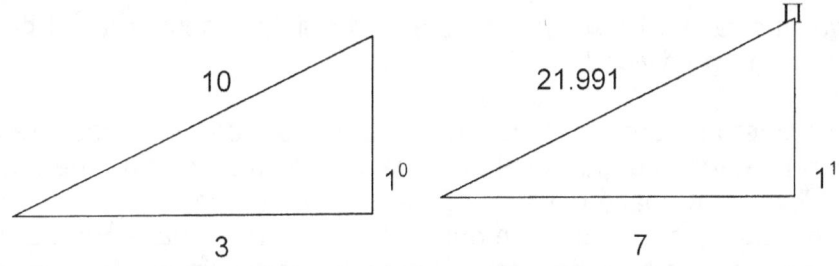

Without the application of specific heat, the object remains in the three directional moving of six possible directions. The value of space unoccupied therefore remains $\Pi \Pi^2$, as it was before the "Big Bang" event, whichever "Big Bang" you wish to refer to, because there were many. However, space unoccupied holds time to the value of 10 to 1, and is in relation to time following the law of Pythagoras.

As Pythagoras also indicates, the triangle being 180^0 in line with time is 90^0 in relation to the flow of time by the measure of half a circle. The total formed holds space to Π. Therefore unoccupied heat holds the relation to space in applying 3 directions of influence $(10)^3 = T^2$ $(R^3 = T^2)$. Always part of this equation is the dual function of space in $R^2 = T$ while at that very instant one have R=T. Therefore in space in time you have $10(10^2) = T^2$. Applying the fourth dimension does not bring about another part in the six dimensions, but actually cancel the influence of the one dimension by favouring a specific dimension. Bringing about the fourth dimension will lead to halving one dimension because of favouring the direct opposing side. Then the equation of influence becomes $(10/2)$ (10^2) and this means the implication of any heat, be it heat, be it light, will apply under the factor of $5(10^2)$ Π. The implication of this may not dawn on one the very instant of realization, but to scientists, there is no greater shock than just that. To any application of movement, the factor will be

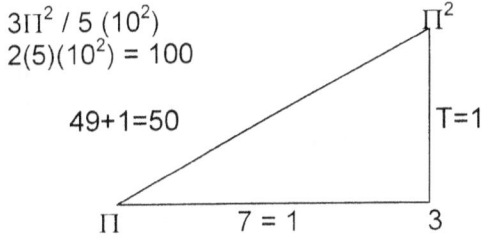

$3\Pi^2 / 5 (10^2)$
$2(5)(10^2) = 100$

$50 = 5 (10^2)$ where the complete   is
Pythagoras be comes a factor
$\sqrt{100} = 10$ the value of space.

The factor of space comes about as $(7^2 = 49)$ plus $\Pi^0=1^0$ adds to become 50 and that is in the three dimensions of space R^2/T where T holds the relation to one and R/T again where T relates to one. At this point it is most important to remember that Pythagoras works on the application of the sum of the square of the two sides.

When seven has a direction in the fourth dimension applied to it, the opposing dimension will be one and this applies in time relevancy, therefore the interchanging in time between infinity in time and eternity in time will place matter at 7^2 x 1 relating to circular and 7^2/1 with 7^2 x 1.

The square in matter holding 49 plus one (singularity) always being a factor of one. Space in time however, never can be a cube, it will always be a square with one side pointing the direction of time from time to the past (1) to time to the present (1) to time to the future (1). This means with space in singularity in relation to time in singularity there will always be $R^3=1^3$ / $T^2 = 1^2$.

It is the value of matter, be it surrounded by heat solid, liquid or gas that differentiates between time in space and space-time. Without applying heat release the space in time would be $10(10^2)$, which is the circle we think we see space to hold. This is where astronomers make their biggest judgement error, they look at space thinking they are viewing the cube of space, but they are in fact viewing the half of the base times the square of the cube $(10/2)$ (10^2).

What they think they see is 1 000 instead they are seeing 500, but as I will later explain, they are not viewing 5 (10^2) they are not viewing light setting the time factor $3\Pi^2$ but they are in view of light in space (Π) being in time ($\Pi^2=1$) the connection to space is Π/500.

That means whatever the radius between the objects the light reaching it will have a comparable velocity of the distance of travel divided by 500.

This will be discussed later in more detail as well. From this there is a measured point where the dimension of space in heat forming gas ends and space in "liquid" forming the neutron begins. The dimensional time application in space diminishing starts at that point. We think of it as the Earth's atmosphere. This no person, no force, no money can recreate. I have read articles about some Newtonian Wizard that plans on building some ship where this ship will "create artificial gravity by centrifugal spin."

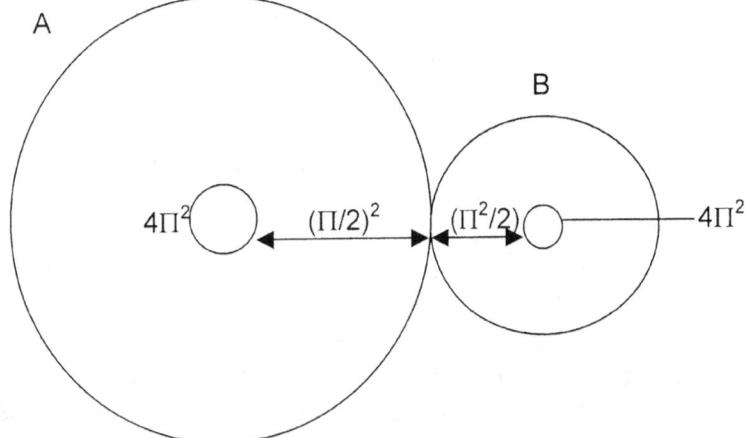

Again and again Newtonians show just how little they truly know science. The depletion of space through increased spin contributes to the view that the Earth is acting as the nucleus of an overgrown atom. The point Π indicates where 3 gain one side of its dimension becoming Π. In that space all objects are at the mercy, not of a FORCE called GRAVITY but of a change in dimension   At that point a dimensional change takes place where the Universe sacrifices one dimension from 3 to Π (in reality it is 6 toΠ.). In that space objects which wish to fly, need a replacement for the dimension sacrificed. The aircraft needs wings to fly. Hot air balloons and the Zeppelin Hydrogen flying cylinder use the natural tendency that hydrogen provide by enlisting concentrating heat to maintain at the value of Π and not reducing to Π^2. An aircraft of any kind, that wants to maintain a vertical flight would have to apply three directional wings for stabilizing because of the loss of one dimension. There may be the possibility that with enough application of heat at various points, the introduction of the fourth dimension (heat application allowing space creation) may even bring about that stabilizing may occur in such a manner. Although the prospect seems light, however with no research to widen the view on the possibilities available, it is hard to tell.

What is overall important though is that the point the Roche limit indicates, being $(\Pi/2)^2$, the dimensional change takes place. Any object holding space-time in own value, will either reduce space occupation, or enlist time duration enhancing, or most probably a little of both as it will cover itself with heat. This is the result from the atoms on the outer edge of the atom, repositioning space-time occupation where the normal application will be in the dimension of the Titius Bode law developed to a certain degree. This change will bring about that the atom will revalue its position to incorporate the Roche limit in protecting its own space it holds in the time of that specific space. The value of the space-time of the atom of the aircraft therefore will reduce its electron position, as it locks itself in a position under the cover of the heat shield. The point of the Roche limit is as much part of the Macro Cosmos as it is of the Micro Cosmos.

Both structures hold eternity to Π^0 in singularity as time $\Pi.^2$ and space Π diverts from singularity at Π. The deviation of Π^2 and Π will be in accordance with the measure of deviation that heat occupied and heat unoccupied hold in relation to both objects claiming space from singularity.

At this point of $(\Pi/2)^2$ the reduction of space through positive space-time displacement will start. Any other cosmic solid that will not comply with the flow of heat to the object will remain outside that area. This means in more suitable language, that the second object will block the flow of heat to the first object (A). To object A everything outside its proton sphere of $(\Pi^2+\Pi^2)$ must either be Π^2 or Π. If the object is in a neutron time position, it will hold the value of Π^2 and if it is in a space position, it will hold a value of Π. Should the object B maintain a density where it will become a singular value to the proton value of object (A) in as much as maintaining its own proton value of $(\Pi^2+\Pi^2)$, it will cause overheating in the proton heat flow of object A.

The demand in heat supply of object A will either remove the heat object B needs for cooling, thus increasing object B's chances of overheating, or it will start overheating itself. By overheating the space that objects A occupies, will increase because the overheating will demand more space occupied. By increasing object A's space-time occupation, the demand on heat will increase as the point of space reduction shifts further away. The increase to the value of the $4\Pi^2$ factor, will increase the $(\Pi/2)^2$ linear factor.

As the demand for heat rises, the competition for heat sustaining will also rise in the occupied space of both structures' time zones. That is the Roche limit and that is the purpose of the Roche limit. At a point where the space reduction and demand on heat supply starts in all earnest $(\Pi/2)^2$ seen from the view of the object in question $(4\Pi^2)$ it will either reduce the other object to a suitable value of (Π^2) or Π, but it will not allow any object to compete with it's demand $(\Pi^2+\Pi^2)$ on the supply of heat.

It will reduce the competing object's value of $(\Pi^2+\Pi^2)$ as it relates to its own value down to a mere manageable (Π^2) or (Π). The reducing method object A will provide, is removing as much heat flow from object B in order to get object B to overheat. Thus object B will increase its space-time value from a solid $(\Pi^2+\Pi^2)$ to that of a jelly (Π^2) or even better still to a liquid (Π) by removing all gas (3) from object B in its demand for heat. If object B can comply with its own demand for heat, it will remain in space of it's own, providing it's own time within the cosmos $(\Pi^2+\Pi^2)$.

The fact that it takes the space-time back to what the Universe was when it was in total development $(\Pi/2)^2$ $(\Pi^2+\Pi^2) = 48,7$. At present that value supersedes the space value of 10Π completely, so it holds a relative position far above the geodesic value (10Π). That is not all, because the one object acts as a neutron to the other object, both claim a proton flow of heat from space, holding a time of $(48,7 + 48,7) = 97,4$. That places the active space-time development with the combined structure to an atomic era that was relevant just after the "Big Bang". Heat became a gas at the atom value of $7/10(\Pi^6) / (6 \times 10)$ and darkness parted from light at the atomic value of 3^2 $(\Pi^2+\Pi^2) = 98$. As you can see, the two objects pushes each other back in space to a time that applied three eras ago.

Therefore one can observe that the Lagrangian point of S_1 is in fact the electron position because the two objects hold the status of a compound, sharing space-time and heat flow. In other words, the two objects become two super sized atoms forming a compound in space. By the same measure a Lagrangian system becomes an enormous five-electron atom with all the cosmic objects working in conjunction as a unit with one mean goal and that is self-preservation through group preservation. Since this falls outside the spectrum I allowed this book to have, I mention that only to fill in a picture that may otherwise seem empty.

If the superior object does not take the minor object as a threat blocking its supply of heat flow, it will consider the minor object as a neutron that can help it grow. Even if the two objects maintain a relation of $(\Pi^2+\Pi^2)$ it will hold the other object to a value of $(\Pi^2\Pi)$ and outside the combined effort they will jointly relate to the rest of the Universe as 3. Therefore seen from the two separate views there is the following atomic order:

$(\Pi^2+\Pi^2)$ $(\Pi^2\Pi)$ S_1 $(\Pi^2\Pi^2)$ $(\Pi^2+\Pi^2)$
Proton   Neutron   ↓   Neutron   Proton
                            Electron

But placing the two in a battle for survival the relation becomes super strained where the one will not give any way to the other object's demand for singularity dominance and then the relation becomes that of

$(\Pi^2+\Pi^2)$ $(\Pi^2/2)^2$ $(\Pi/2)^2$ $(\Pi^2+\Pi^2)$
Proton

                Space-time establishing

This will bring about space-time development that applied when the Universe was fresh and new, hot and lively and full of spinning power. As the objects cannot destroy one another, there is the Hubble Constant effect that comes into place. Each structure has to build a Π value, to maintain and secure its relevant position until the end when the final eternity arrives. In using other words, each object will secure its place in future as a Black Hole or Proton Star. By producing a value of Π separating the Π^2 from three and thus ensuring itself an own atmosphere it will have to revert to space-time development that even preceded the "Big Bang". I will produce once again the value of Π from its position of Π^2 (time) and matter. Matter is seven and time is Π^2. Both hold claim to the same space diverting from individual singularity, therefore the development of matter and time will produce Π and through that the Titius Bode Principle.

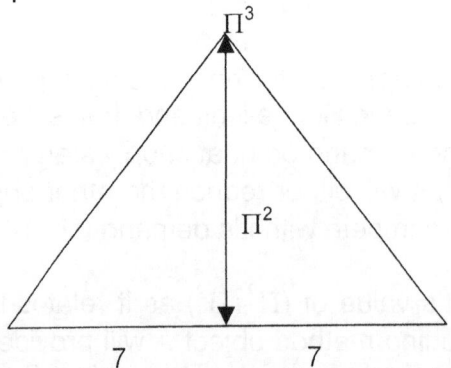

They join space-time therefore the matter factor is the same. This is where one can visually see the one object, filling the space of the other object's atmosphere.

$$7 \times 7 = 7^2 = 49$$

That is matter Π^2 (time) times matter (49) = 483,61.

As this is all under the law of Pythagoras the law will evidently place a square root to that value of 483,61 and therefore $\sqrt{483,61}$ = 21,991. This leaves the space value of the Roche-limit, as it develops into the Titius Bode law giving them a shared value of 7 (matter) and 21,91 (space) the value of 21,991 / 7 = Π.

Then the relation becomes

$(\Pi^2+\Pi^2)$ $(\Pi^2\Pi)$ $(\Pi\Pi^2)$ $(\Pi^2\Pi^3)$ holding space (3) still outside. They therefore will share space and that sharing will continue till times end. We know by now that matter is 7, and space is 3 holding time to a relevancy in singularity of 1.

Sharing the space means that 21,9 will become (10) space to the one side
1   to the instant position of time (k^0)
,99   lost to space depletion $\Pi^2/10$
7   the relation to matter.

Through that the Titius Bode law comes into affect of 10/7 or 7/10, depending on whether space or matter holds a superior position to time. From that stance, all objects will relate to one another by the value of $\Pi^2\Pi$ and seen in a whole sale total 7/10 or 10/7.

We on Earth take our measurements per second, minute, hour and so on. This time relation we derive from the ever-changing position the Earth holds to the Sun first, and thereafter to the Universe. That is time in space and more precise space-time.

Time is the spin of matter in space. **It is the changing relevancy that particles have in the eternity Π^O to matter relating to the outside but also relating to matter holding the Earth Π^1.**

This leads to a point where I would love to show how this cosmic code can be interpreted and from there one may make some deductions from where one can come to some conclusions about how the solar system came into place. The evidence provided in this letter is far from convincing but please see it as a much-abbreviated version of the much more detailed and much more complex book **_Seven days of Creation_**. In that book I am proud to announce that I do bring proof and not only by accordance to scientific facts but I prove to the letter that the Biblical version of events occurring during the Creation of heaven and Earth stands proven scientifically beyond doubt. This letter however I present as an example as to indicate to you as the reader that there are other possibilities of finding solutions and there are other methods which we can use to translate what is out there within the limits of the cosmic code. This we can achieve when we use the cosmic code to the correct effect. Now I will repeat some of what is said to show how one may implement the code and use the facts to interpret what is already said in the letter.

All atoms hold an electron to the very outside of the space which that atom occupies. All electrons are negatively charged. No two sub atomic particles that have equal charge can touch; because of the way they will repent one another. Should there be a way to force the particles to touch, say with "matter" versus "anti-matter" a nuclear discharge will result from such a manoeuvre. Therefore, short of causing a nuclear discharge, no matter can come into direct contact with matter because of the negative relation the electron holds to other electrons. NO ATOM CAN EVER COME INTO DIRECT CONTACT (TOUCHING) WITH ANOTHER ATOM.

This is what the Universe consisted of, everything that is today, was then, in a dimension that only holds "gravity" towards and away from singularity controlling.

This piece of logic shows that matter places matter in relation to the heat separating matter. There is never any empty space keeping matter apart. I hope that this bit of minor logic will end all theories about the fact that "Gravity" has the ability to contract or draw matter closer. If you wish to stick to the terminology using "gravity" as a term, then the best way in explaining the way "gravity" works is the ability of the element to reduce the heat between the particles and as such reduce the space. Explaining the factual working of "Gravity" would be rather complex to use in this book and the intentions in publishing this book does not cover the most intimate technical details. Time and space were still parting in relation to the Titius Bode law growing from the Roche limit.

$$10/7 \ (\Pi/2)^2 \ (\Pi^2 + \Pi^2) \ \text{and}$$
$$7/10 \ (\Pi/2)^2 \ (\Pi^2 + \Pi^2)$$

Explaining as follows:

10/7   Forming the space factor and
7/10   Forming the matter factor.
$(\Pi/2)^2$ the interaction that matter forms in sharing space and time.
$(\Pi^2 + \Pi^2)$ the double proton

The only way new space can form is by unleashing the forming of heat that expands. Some particles during the development in that era still overheated which evidently forced more space. From the overheating and the consequential space that came about, particles broke down allowing matter to dissolve to matter (neutrons) and space (electrons). It must be very clearly stated that neutrons and electrons did not yet form but only a suitable set of rules formed that later would apply, where matter could supply matter to initiate a sufficient heat supply as heat, locked in time, converted to space. The process that happened at this point, lies far beyond that of a nuclear reaction, it lies far beyond that of a "Black Hole" or as I prefer to call is a proton star. All of this still fall under the same aspects and parameters we find energy or the application of energy to be at present.

Even before making the following statement I find myself in a position where I have to repudiate myself from making the following statement. The Universe is neither big nor small as the Universe is beyond size. However this is a letter and the letter can't change human culture in one go. If I stick to what I say, the letter will be too short just to explain everything I say every time I say it. I shall quickly name one instance to prove my case. I say the Sun is cold and that is true. However by my addressing the Sun and referring to it being hot I am not denouncing myself, but merely sticking to human form. The Universe cannot be small because it fits everything within and yet the Universe packs what ever it has into what is a dot smaller than what may be found in the Universe. The dot that came from the spot is the size of something not fitting into the Universe at present and yet it does hold whatever is in one package. The Universe that started was 1^0 and it grew to 1^1.

The Universe holding the smallest moving fragment is the very same size being 1^0 going to 1^1. The Universe is in relevancies but only man works in size because of man's incompetence to realise what is in the Universe. What is in the Universe is relevant to singularity and all points confirming singularity is equal at $\Pi^O=1$. That proves the Universe is in a state of equilibrium. The rest we give size and measure is generated by singularity and as far as the Universe is concerned in the matter, it is locked in the relevancy between two points of equality holding singularity. However my writing is with words and speech demands laws abiding culture and for that reason of not getting all-bohemian I will in this letter for your benefit stick to what is presumed as being the normal and the norm and confine myself to what culture insists on in order to make myself understandable. When I refer to a small or a large Universe or a hot Sun I am not in conflict with myself and neither can I repeat this explanation every time I am forced to contradict my discipline in the parameters of this letter.

At first, the Universe was small dominated by the density of unoccupied space-time; the value to unoccupied space-time was little less than the value to densified space-time, leaving no chance for occupied space-time. Ever so slowly time in space evolved from a continuous eternity, flowing in undistinguishable periods of eternity in duration forever weakening the strangle hold of time on matter. Then the age of the atom arrived, and light, which brought fore ward the atom encircled by electrons that is at this stage quite unknown to man. By the time, the atom was a recognizable unit, more eternities passed than we have sells in our brain.

The cosmic dynamics are all part of proportions and relevancies changing as relations change the applying dynamics. One must see the cosmos starting from the size where what ever was in the cosmos at the time would currently fit on top of a needlepoint. However that does not make what was applying back then, small in comparing to what is now valid because what was small is at present large, because what ever were in the cosmos, is still in the cosmos, and grew with the cosmos so what ever was small had grown with the cosmos and is in practise as large as it was. The dinosaurs are a vivid reminder of that and in **_Seven Days Of Creation_** I explain this extensively.

These first atoms were small producing very insignificant positive space-time displacement in the sea of negative displacement. The atoms were small, but to our great fortune growing as they extended their size and influence. By the time, time was a distinguishable factor the nucleus was presumably less than the size of the electron. The electron's orbit was right next to the small nucleus. Because of the size of this tiny atom, it could participate in the geodesic space-time value back then, which was a little less than the photon's velocity value back then. I have to emphasize that by explaining in this manner, I put the relevancy on the human aspect in view that it concerns the Bible, and therefore mainly the perception that life holds on events.

Heat converts time in forming space, that is energy back then, at present and to the end of the future.

In order to keep matter short, I shall proceed to the "Big-Bang" explanation, the period of nuclear heat forming space.

The cosmic cube we live in is **7 /10 (Π^6) / 6 = 112.16**

The six-sided cube (Π^6) / 6

**In motion applying gravity
7/10)**

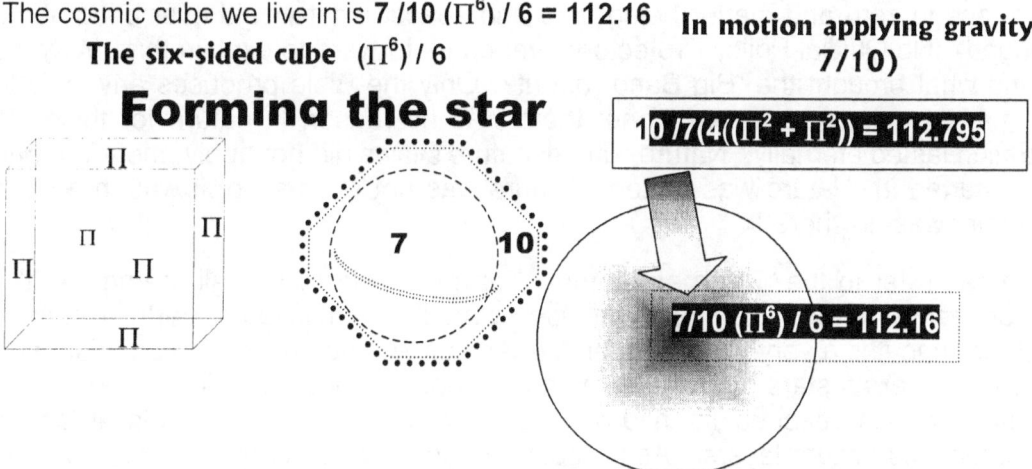

Forming the star

$$10\ /7(4((\Pi^2 + \Pi^2)) = 112.795$$

$$7/10\ (\Pi^6)\ /\ 6 = 112.16$$

Space formed for the very first time an identity separated from matter, apart from time, becoming the fourth dimension. In other words, the atom as we see it came into practice. The last element formed and had one proton, but still the temperature was out of control. Then the final element came into place, the cosmos.

$T = 7/10\ (\Pi^6)\ /\ (6 \times 10) = 112$

7/10   is the matter has the dominant value.
Π^6 matter has the six sides it holds in the fourth dimension.
6   are the six sides to space occupying matter.
10   are the value or dimension in which space holds a ratio to time.

The cosmos began, not to a specific space, because all the space that was there, initially is still there at present. Any atom above 112 cannot apply to the fourth dimension not then and not now.

The Universe had two options. The matter which is densified space-time could join the geodesic space-time and spin itself into another eternal oblivion, or its positive space-time displacement, could withstand the geodesic space-time displacement could withstand the geodesic space-time negative space-time displacement. Fortunately equilibrium prevailed.

Obviously, our luck was in and the positive space-time displacement within the atom stood its ground against enormous odds.   With this effort, the expanding of the Universe began, as space-time transferred from the universal value to the densified value. The more time transferred this way, the less time had a value in the geodesic sense and the more time had a value in the densified sense. In this, the space value grew in a geodesic sense as the time value in a geodesic sense declined. This brings about the eternal expansion the Universe that will only stop once the geodesic time value becomes zero and the space value becomes eternal. This then will be the second eternity, to which the Bible also refers.

This very process was the inspiration, which lead the Universe to be what it now is. In the chapter, designated to moment-Alfa, an elaborate explanation shows how the value of Π has changed four times already, and will change another three times. The development through all the different era brought about that the Universe had to endure because of this very principle. I do not wish to go back to the old cliché about gravity and all the theories that were thought up to explain the means which will enable the Universe to retract again. The retraction of the Universe lies in the growth of time as it is transferred to matter. This process consists of the transformation of unoccupied space-time to occupied space-time to densified space-time.

In this lies the value that all stars have in their role they have to fulfil to bring about a successful conclusion to the life story of the Universe. Objects are growing more massive as time transfers from one value (Verplasing) to the other value (aanplasing). As R^3 is showing an increase in both the occupied as well as the unoccupied space, T^2 reduces from the unoccupied value to the densified value. In the star R^3 reduces as T^2 transfers to the densified space-time. This is what stars are made of, when they grow up to be stars.

With time in eternity, space in zero and matter being time and space, what would ever bring about that this situation changed? No Official Policy Protectors ever came forward to explain this. Why did the "Big Bang" start and what brought the "Big Bang" about? Only the Bible produces any logic to this question. Time, matter and space froze in one; there was no reason in nature for things to change, since this situation lasted eternally. Nature with all nature's laws did not apply; therefore one cannot say that nature started it. Nature was frozen. Nature was not even solid; it was in a state beyond being solid. Nature was nowhere!!

If the spoken language can refer to the Earth as Mother Earth then surely it is all the right in the world to address the Sun as Father Sun, for that is precisely what the Sun is to us Earthlings. If we have the Sun as a father, then it is a very young father. We as humans have received an infant star in the Sun and a juvenile as far as stars go, where it is on the edge of becoming just a toddler. A fully-grown star that has not yet reached its middle age development, will have discarded its hydrogen and most probably its helium layers. As it ripens in age, it will discard its carbon and all helium related layers to have only silicon, an iron and an inner core that has outgrown the element value altogether. This process leads the star to a route that takes time back to its origin of time = eternity; space = infinity.

The spin in the Universe slowed down, up to a point where the spin was equal to that of the speed of light. At the point where the Universe spin equalled that of the speed of light, the Universe was still in total darkness. The light (photon) was there, but did not yet produce light. Light only came about as the spin reduced to below the speed of light, and only then light became obvious. The Universe grew away from darkness as this event lasted many eternities, during the period where the light separated from darkness.

How did this "Big Bang" take place? The best way to examine the reason is to see why anything in the Universe expands. To get anything to expand one has to heat it. All matter expands when overheating. Science may come up with whatever brilliant theory, the fact of the matter is that when matter overheats it expands. The bigger the overheating, the bigger will the expansion be, it is as simple as that.

This means whatever leads to the forming of the "Big Bang", whatever preceded it, it had to come about from matter that overheated. With the event of the nuclear age, the proof came about that matter is heat in some frozen form. Unleashing heat from its frozen form brought about a jolt of heat, never yet experienced by man. By breaking matter from the frozen state, of which it is in, within the atom, heat produces light and heat. Where this process clearly shows how new space-time forms is where the releasing of heat caused winds that stun man's logic.

The nuclear explosion shows quite clearly what the "Big Bang" was, with the nuclear explosion being a very minute form. Yes, we have all heard the rubbish about matter and anti-matter. What can anti-matter then be, since matter is heat and that we know from our nuclear experience? The Coanda spin defined the heat to a certain space that then heat occupies for that time while it is confined to the space. With matter being frozen heat, what would form anti-matter. Anti-matter means the opposite to matter, and if matter is frozen heat, anti-matter must then be overheating heat. This in itself is quite ridiculous. Anti-matter can only be matter with an opposite spin in relation to what we regard to be matter.

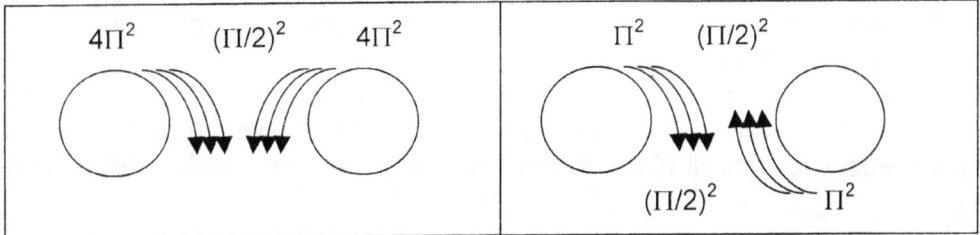

In the second sketch Π^2 $(\Pi/2)$ / $(\Pi/2)^2$ Π^2 the Roche limit cancels the border that is set by the equally negative electron and with that the space of the electron effectively disappear. The electron is a direction of flow of space-time and when two opposing flows do not reject each other they will cross the parameter separating both and both will dissolve on another through spin. The two protons touch destroying each other and the neutron as well as the proton demolish and become heat 3^3.

The process where matter then touches matter, will bring about a reduction in the feeding process of heat, where all matter in that space will overheat and expand, producing unfrozen heat. This means there was heat occupying space, and matter with both in relation to time.

The proton with a positive space-time displacement less than 136 placed its displacing properties in negative space-time displacement. In short, to substitute for mass shortcoming of less than 136 grams/molecule and still finding sufficient cooling properties for the proton to survive the overheating deficiency, it has to spin more rapidly, therefore by spinning it makes contact with more heat than it would otherwise do. One may consider this as "breathing". If the proton does not find an adequate supply of heat and does become motionless, the atom has to substitute the movement through motion. Forming an object that has an increase in heat supply through work commonly uses this fact. In nature, just the opposite is true because of the motion by movement, as one finds in the case of wind.

There is no "force" in the cosmic because everything is a "force" in one way or another. The proton takes heat from space in an effort to maintain temperature and stability. This flow of heat brings about the reduction of space by increasing the heat in that space. The flow of the heat, through the electron by means of the neutron to the proton is time. The amount of heat taken by a proton is a constant throughout the Universe but relative to the space reducing effort of all the protons influencing that specific space. That is "gravity" (a term I denounce and reject). "Gravity" is nothing else but additional application of time. The higher the gravity is, the slower the time will become by prolonging the duration of time, NOT TIME ITSELF. TIME REMAINS A CONSTANT, BUT THE DURATION DEPENDS ON THE SPACE THE HEAT IS CONFOUND TO.

By taking this statement and introducing that to a Galactica, the shining luminous middle part, holds time to eternity through movement. The atoms in the middle admits light because it holds time duration to the speed of light, the longest duration that time can be and still remain in the fourth dimension. The centre of all cosmic structures determines the time that applies to the structure itself. As the cluster of protons supply the density, that influence the space of occupation, the density is a collective reducing of space with the increase of heat. This can apply through an object relating to space through movement by the object and by the object reduction of space through the density of the accumulative effort of the cluster of protons we named elements.

As we Earthlings had a double advantage from the process of binary stars, I would like to ponder on the binary stars, as they are the key to time in the cosmos. The first influence of a binary system we benefit from is the Sun and some other minor star, which by fragmenting formed the non-gaseous planets. The second encounter was the binary to which the Earth and moon developed, and from which LIFE then was able to take full advantage.

The two stars develop in the galactica in close proximity as they help each other in transforming negative space-time displacement to positive space-time displacement. With the combined effort, they can grow extensively in supporting each other. In spite of all our Official Policy Protectors teachings, stars do not form as result of cosmic dust storms. It is just not possible to form that way, because even in the present era we find ourselves the iron era, only elements with a space-time occupation value exceeding 17,5 can reduce space sufficiently to apply "gravity".

Planets never form naturally because a planet has no role to play in the Universe. Cosmic objects do not form in any way or means through the application of "gravity". Matter occupies space presenting that specific space a time value in accordance with the time in space the matter holds. I realize this is rather a mouth full to understand. To us humans, planets are a necessity, which one cannot do without. We humans have a measurably small place in the Universe, so small the Universe do not recognize life as a substance.

We see around us a planet overflowing with life. Life is so abundant, we can hardly distinguish between nature and life because to us nature is life. We have to think of nature without life because cosmology is without life. An easy way to distinguish is thinking of a mountain burning. The mountain cannot burn, it is plants that burn and the plants are not nature. The plants are life, therefore the mountain cannot burn and even less is the possibility the mountain can move independently without only being part of the Earth's motion. It is only life than can manipulate space-time. Other than that all other objects may extend the space-time they occupy by natural converting heat from a gas to a solid through proton growth.

Whenever you wish to distinguish between life and the cosmos, the rule of thumb is: Can it move of its own accord? No mountain can fly to the next continent for whatever reason. That action belongs to life, and it is the manipulation abilities of life that renders a piece of metal such as an airplane the mobility of motion. If not for life, the aircraft could have an ability to commit motion. The actions we see must accompany life, and therefore be associated with life and the qualities attached to life. It is not in common use wherever we might go. In the rest of the cosmos heat drives objects in relation to the flow of time and life has no ability in those locations to intervene. The aircraft will fly, but the principles belong to the cosmos and by life's manipulative abilities the craft will only fly as an extension life brings about. We can make life manipulate cosmic law on Earth and on the planets (well, some of them but not even most of them). It is where life can be, on the inner planets and their moons, but life will not function even on the micro stars such as Jupiter, Uranus, and Neptune etc. On the little fragmented pieces of a star gone by, yes, that is where Newton's physics apply and apply they do. The comet obeys rules applied by the cosmos, in the terms the cosmos provide. Newton's physics do not apply to the cosmos however which way Newtonians try to explain it by applying Newton's laws. A comet cannot leave the Earth and journey around Pluto, and come back with a detailed report. This action belongs to an object that is acting on behalf of life as an extension of life. I am very adamant about this. If Science wishes to understand the cosmos, do not make the cosmos one big Earth. Up to now, such an application only brought miserable results that apply to nothing and helps science establish the joke it became. Life is the extension of God, the Universe is the Creation of God, do not confuse the two.

The Sun is not an Earth on steroids. Life cannot be on the Sun because the Sun holds completely different relevancies. The relevancy has a comparing formula but the end result will be totally different in every aspect one wishes to apply.

By referring to "gravity" only one aspect of the space-time relation of any elements apply. There is no mention of the second and crucial part of "Gravity" where the motion of the object brings about the space-time relation, or if you wish, providing the cooling aspect. At first, the proton cluster's total positive space-time displacement has an insufficient "gravity" to secure a stable cooling effort. This spinning of the element clusters is inherent from an event, even predating the "Big Bang". To find proof of this statement I just made, look at the photograph of any galactica. In the centre part heat is a liquid flowing like a river and taking matter for the ride. This is one part of the relevancy. The other part is that matter is spinning at such a tempo, it is capturing space in the form of pure liquid heat. The proof of this is obvious in every galactica.

Galactica is living proof of cosmic development and what is much more is the fact that it proves cosmic development in stages. In the centre of a galactica one can see that the Big Bang is presently arriving for what ever matter holds space-time within. Then as the circle expands, it is clear that the widening brings about less heat occupying more space producing darker regions holding bigger bodies of matter. Every galactica proves the Big Bang theory in precise detail.

The spinning motion of the element clusters (or proto stars or if one wishes to use the name of future stars, it will be just as applicable) hold their relation to heat secured by maintaining motion. As the time value in the cluster's space occupation (mass) secures an era related value, the structures that were spinning, reposition in such a fashion as to apply a new linear displacing value. At first the motion is such that the linear position is negligible, but as the mass grows, the linear distance grows accordingly, placing the revolving structures further apart and at the same time, "pushing" the rotation of the objects in a wider revolving orbit. By widening the rotation circle, the objects rotate at a "lesser" pace and this pace coincide with the space-time occupation ("mass") of the totality in the effort of all the protons put together. In this one will not find a "force" but it will be a complete balance between matter, space and time. By securing an ever-increasing space-time occupation (mass) the future star will reduce its negative space-time displacement (motion) and increase its positive space-time displacement (gravity). The higher the positive space-time displacement (Gravity) becomes, the lesser the negative space-time displacement (motion) will be. At present only stars holding an $iron_{56}$ inner core can maintain a star status, and any object with a lesser element in the inner core will not bring about fusion, or in fact, any form of luminosity.

For instance by the time the "Big Bang" arrived, only elements with a "mass" of 112 had the ability to release from the Galactica luminous core and during the Era of the Quarks, the releasing mass of the time determining elements carried a combined proton-cluster "mass" of 88. As the single proton's time holding value increased (molecular mass) the time grew less and the Universe "grew bigger".

I uncovered a mistake in science. From the onset, the mistake seems as insignificant as it is small. Because the rest of the book is about the mistake, I do not intend on elaborating about the mistake itself. The mistake came about with the culture of education and the mistake in itself seems harmless. When admitting that, one must also admit that any pilgrim that got lost and died of starvation through an incorrect travelling direction, made the very first part of his ultimate mistake by looking in the wrong direction. How harmful does looking in a specific direction seem, and yet such a mistake leads to his ultimate mortality. The traveller could when taking his first directional flaw with that the first incorrect step, only put his foot skew in avoiding a rock. Or he could have turned his face to avoid a branch and that move pointed him in a direction that lead to his fatality.

It is not the mistake that becomes the penalty and it is not the origin of such a mistake that leads to the penalising, but the ignoring of accepting signs telling the wonderer of an impending error and his stubborn ignoring of such telling sign that makes the lost party pay the ultimate price. By ignoring the mistake, for whatever reason, the ignoring of such a mistake is his undoing because the price due comes from the inability in recognising the sign indicating the presence of the mistake forming the reason for his final demise. The sooner such a person sees and admits the wrong, the less will be the consequences of his final price to pay.

The Newton mistake is one born in culture and the penalty from this mistake is bred by arrogance. At school minds are young and accepting, although developing. Through many tens of millennia humans came to a habit in surviving as a specie where culture taught them that accepting the advice from their elders is the same as to ensure survival of the following generations, and by such doing is also following the quickest way to an adult mind. By accepting the advice from their elders the accumulated of generated knowledge went to benefit those that protected their culture and they gained experience without question proves the dominance of the tribe in relation to other tribes of the same race. This is culture we cannot do without and still maintain progress. It is an inheriting method humans grew on and is the corner stone of all civilization. We cannot abandon it.

This culture of accepting the teachings and advice inherited from the elders is the foundation of civilization. As far as any students mind can tell the news is as actual and accurately tested beyond a crack of doubt as anything notwithstanding the fact that such news may be with the human mind for thousands of years while it was actually never tested. Whatever the teacher tell students will generally be a first time experience to the student with the information the student receives being so new he has no time to digest the information. The teacher will not allow any pondering on the matter and an in-depth analysing because of a program the teacher has to complete in time. Taking into account his youthful ways (which we all had), he has little stomach to scrutinize it because learning is a painful process to all. Without pain and perseverance there can be no education of any sort. He does very willingly accept the facts as tested and correct without flaws of any kind. The scholar has to because in any education system time will not allow students to ponder about detailing information and securing a prognoses to all learnt every day. What ever the Master tells the scholar is taken as Biblical correct without any thought about testing the results. Where there are cases of scholars having doubt and subsequent questions, the Master takes such behaviour as being obstinate and being a reflection of the student on his (the tutor's) personal integrity and knowledge. The young mind will very soon discover that his behaviour is not tolerated by the system, and the truth is the system cannot tolerate such behaviour for the good of the rest. Time must be spent on learning and accumulating as much information as that which the young mind can accept.

No information could be more affected by such a culture than that of Newton. You better understand Newton and are then considered as being smart, or you do not understand and therefore accept Newton however that has a price tag attached. If you are not readily accepting Newton as the word of God, then with that comes your accepting that your stay on campus will be very little short lived which excludes you from any future you might have had in the world of science. Newton is science. No Newton understanding automatically becomes "no science" education or learning. When in the past I went around explaining the degree of incorrectness I saw, the only attitude I would draw was sympathy because to their mind I did not have the mental capacity to understand Newton. What is there to understand anyway? The bloody comet misses the Sun by a country mile and using Newton's formula it should hit the Sun bull's eye in the centre. Without Newton, there is no other and science will be a faculty filled with vacuum of containing nothing. This is very unfortunate but is the ultimate of truth. It is either Newton's way or no way at all. With this culture also brought along the stigma that only the minds of the sharp and the sighted can accept and understand Newton and when not understanding Newton one tends to fail your personal I.Q. test. It is a sure sign of the slow witted when the student fails to recognise what Newton said. The only way to advance in science is to understand Newton and indicate to all your pears how brilliant your mind is in accepting information.

All students have little understanding about Newton, and that I can and will prove through the next two hundred or so pages. The mistake Newton made and which I discovered is laughable small, yet it took me (not being that bright I may add) almost one lifetime to recognise the mistake whereas it took mankind three hundred and fifty years of research without recognising the flaw. Others in the past may have come to see what I saw, but if there were such persons they never saw what I saw because if they saw what I saw they should also see that behind such an almost invisible puncture hole in the tube, is a reason for science to deflate and not accumulate. When "understanding" Newton it becomes the very same as learning Newton from the heart and accept that what you memorise is what you know. The memorised knowledge is beyond question, as it has to form part of the identity of the student having secured the knowledge.

It is not the hole forming the puncture and preventing inflating being as such the obstacle that is of importance but what that hole does to the tire and the car and the travelling with the car that becomes a menace. In the extreme effort to keep the journey on course, the Academic Masters are spending an all out effort in inflating the deflating tire faster than the deflating tire can deflate. The recourses attached to this effort is enormous and without cause. It will lead to nowhere and that is where science is heading in their attempt not to head in that direction.

When students become masters the Masters seek to break new academic ground. Masters do not ponder the ways on which they lead their students, but have an all out effort in establishing their own support of new territory that will distinct them from the rest in future to come. My uncovering of such a mistake as I did could only be from a person as ill-educated as I. I did not go through the learning process where the learning process is the very same as a brain washing and mind controlling process. This remark may seem harsh and is not intended to be such, because reality demands no other way in education. The mistake and the carrying of the mistake with the support in refusing to recognise such a mistake is part of human training and there is very little to do but to admit that such may occur whereby to follow in acting to correct the mistake after the discovering, rectify what ever can be salvaged and build from there.

Friend and foe think of me alike as being slow of mind and not quite in faculty to understand Newton. This was what I was told on more numerous occasions than I care to remember. Mind you, it was never done in direct accusation and was always handled with great sympathy as to show those in ability has a passion for us with lesser developed minds, after all I am only a motor mechanic with little schooling and I go around admitting that. Pointing at the mistake I see, I am told by the wise as well as the not-that-wise, such as I that through my lack in education I cannot dream to understand Newton and therefore are committed to the position of the ILL-EDUCATED, a position I learned to accept with grace.

Thus my position renders me the place to shut up and do as I am told because my brain is to under developed to form a comprehension about the complexity of astrophysics. I must say that except for two or three occasions those of the Highly Schooled part of life did treat me and therefore my diseased mind with the utmost sympathy their positions could allow them to muster on that particular occasion. In all cases there are more sides than one and therefore I think of myself as poorly educated, the contras to my position must be those fortunate to be SUPER-EDUCATED. In this letter there will be the addressing to you holding the position of the reader and (I hope) the un-bias judge, with me presenting my case to the un-bias (you) in concerning the third party, the SUPER-EDUCATED. Referring to the party in opposing me as the SUPER-EDUCATED is by no means in disrespect and much to the contrary holds my whole-hearted admiration, (at time somewhat limited I admit.)

Quoted directly from the Oxford dictionary of Astronomy the following:

The definition of space-time is as follows:

Space-time is a four dimensional position of the Universe where the position of an object is specified by three coordinates in space and one position in time. According to the theory of special relativity there is no absolute time, which can be measured independently of the observer, so events that are simultaneous as seen from one observer occur at different times when seen from a different place. Time must therefore be measured in a relative manner as are positions in three-dimensional Euclidean space, and this is achieved through the concept of space-time. The trajectory of an object in space-time is called world line. General relativity relates to curvature of space-time to the positions and motions of particles of matter.

The definition of space-time is as follows:
Space-time is a four dimensional position of the Universe where the position of an object is specified by three coordinates in space and one position in time. According to the theory of special relativity there is no absolute time, which can be measured independently of the observer, so events that are simultaneous as seen from one observer occur at different times when seen from a different place. Time must therefore be measured in a relative manner as are positions in three-dimensional Euclidean space, and this is achieved through the concept of space-time. The trajectory of an object in space-time is called world line. General relativity relates to curvature of space-time to the positions and motions of particles of matter.

The definition of singularity is as follows:
Singularity: a mathematical point at which certain physical quantities reach infinite values for example, according to the general relativity the curvature of space-time becomes infinite in a black hole. In the big bang theory the Universe was born from singularity in which the density and temperature of matter were infinite.

While it probably is the greatest mind to walk the Earth that produced the spectacular in the above, a much more simple mind as the one I have noticed much more simple aspects of nature that only one with a simple mind as I have could recognise because my mind does not have the capacity for the greatness of the great minds.

If the Universe did start from one single point and time matter and space flowed from that point, then that point must have a relative connecting base because such a point holding singularity must be eternal as space matter and time link eternal. There then therefore must be one point linking the entire Universe when regarding the fact of singularity. Then according to the theory off relativity there has to be one exact point holding time in a relevance notwithstanding the fact that time depart from that position and relate differently to all space-time away from such a point.

Every person I have discussed facts about creation recollects images in the trend depicted in a presentation as one may find shown with massive clouds of unbounded material That depict chaos and in chaos I would have no ability to use mathematics. The fact that I can use mathematics presents a Universe of order and that is just what gravity is. Where there is gravity chaos is prohibited. That would be the most unlikely way Creation came in place. The recalling of pictures representing images about creation must have form, but to mathematics it had no form. From this thought the very opposite arise where Creation came from nothing but such an idea is mathematically simply not possible.

The mathematical presence we have in the distribution and organisation of the Universe and even the calculation where able to present where chaos is abundant, tells a story of organised growth and not a blob of material with no centre from where gravity control and no even distribution of material. No wonder those Newtonians will fill a Universe with nothing and then stand back to have a view of the entire nothing they can see. Please keep in mind I am the under-educated zombie that has no brain function, which would enable me to understand Newton while they can have their view at night on what they fill the Universe with and call it nothing. How on Earth can one say one can see nothing and still point at something you see?

The thought of nothing is just what it is, a thought of nothing and although it is in the human mind common nature to present nothing as a value in the recalling of something, nothing is a presentation of the figment in the human mind. There can be no number such as nothing and that was (possibly) Newton's biggest error. Nothing represent non-existing and that is just what nothing is, it is non-existing.

In order to prove my point I wish to ask the reader to define the shortest line there can theoretically be. If he should answer anything but that the shortest line will be at a point where the beginning and is the very same spot he will be wrong. The shortest line that can ever be anywhere must have a start and finish holding the exact same spot. The line will be humanly impossible to create but we humans are capable of very little.

When the line has a beginning and an end at the very same spot and it wishes to extend the position as to further the possibility it has, which direction should it favour. Humans in the west would naturedly think of extending from left to right while in the east humans may want to go from right to left. Some persons will tend to go up or down, but all of the options are about human preference and not mathematical conclusions. Extending the line in any one direction will favour one direction without a conclusion about not extending in other directions. Such a conclusion has no sound mathematical foundation. The only option about extending will be in all directions equally in order to give a meaningful non-bias flow of mathematical equilibrium

The shortest line in the realm of possibilities must have a start and finish holding one spot and such a line will also be a dot or a circle. Not favouring one direction puts all directions at equilibrium meaning that any form what ever may be can develop from such a spot with the end and the start being the same. This reasoning prompted me to look for singularity in such a spot because if the prime spot from which all came was a spot, then the spot must hold the shortest line but more prominent it will hold the smallest form including the smallest circle.

One possibility that the shortest spot can never have is having a starting point on the zero mark. If the mark of zero holds the start it must also hold the end because the end and the beginning has the same position. If the position of zero then is the beginning, the end will also be zero leaving the line without an end as well as without a beginning.

The conclusion from this is that no line can start at zero because that will be a mathematical impossibility. A line or spot starting at zero would therefore be shorter than the shortest line possible. A line growing or extending from zero can never leave zero because of the influence of being zero disqualifies any possibility of growth. If the line then had to grow in all directions at the same pace the line must therefore be a circle. The value of the circle is Π, and that is where creation started.

That gave me the clue where to start looking for singularity. One would find singularity in the value Π and the value Π will be in all things rotating in a circle. To start my explanation about my cosmic theory I wish to firstly start with some nostalgic and the relevancy will become apparent later on. Such is the importance however that I wish to place this at the very start of the prologue.

When we were boys we played with a top we called the spinning top. I cannot imagine that there is one boy in the western world that did not hold such a devise in his hand. Tying a string securely around the tapered cone started the operation and then with a jerking or pulling throw the devise is launched in a projectile manner and the big knack to success was getting the nail end firmly on the ground and with a releasing jerk the top was rotating. The champion was always the one boy that could throw his top to spin the fastest and that would create a humming sound. The louder the sound produced the bigger the champion.

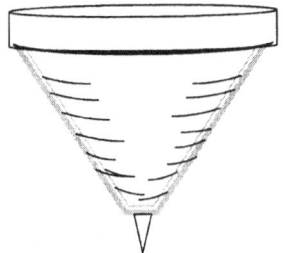

When a back braking effort produced a throw of enormity the spinning top would not only produce sound varying in pitch but also create a spin that would seem to have some instability. There are very many limitations about the spin, parameters that determine the slowest and the highest sin rate and spinning is within the parameters of such settings. The question arising is why such parameters are there in the first place?

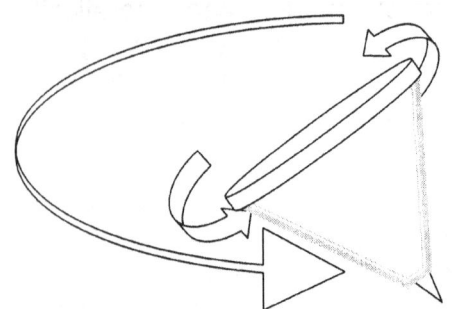 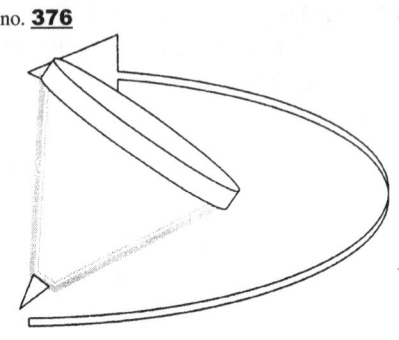

An enormous effort will have the top going oblong while spinning violently and as the pace reduced the top will stabilize by coming to an upright position. In the upright position it wall then spin for the remainder of the period where it will in the end start tilting to the side and in a last effort throw a few wild oblong turns and fall over.

Boys playing games will never realize scientific breakthrough explaining and grown ups do not play with toys. In this little toy played everywhere everyday by almost every one is the answer most brilliant of human Brainpower seeks answers about all the cosmic riddles no one seems to understand. In the spin as such one may find two vital boundaries in the motion and the boundaries are marked by a wobble coming about as if the top is fighting some other influence. Spinning too fast pulls the centre off centre and so does spinning too slow. It is the same influence coming about at both ends of the limitation in the spin. There are influences at work, but force…no; it cannot be forces setting such boundaries. From that I started per cuing what sets such limitations because that limitation must be universal as all matter is spinning in one way or the other.

I MAINTAINED DURING ALL MY CORRESPONDENCE TO SO MANY I HAVE CONTACTED, I STILL MAINTAIN AND I PROVE MY VIEW POINT THAT:

1) There is no gravity and therefore GRAVITY DOES NOT EXIST.
2) With no gravity it stands to reason that I also maintain that NEWTON AS WELL AS EINSTEIN IS ALL TOGETHER WRONG!
3) With no gravity NEWTONIAN VIEWS ON THE WAY CREATION CAME ABOUT IS ALTOGETHER INCORRECT.
4) THE BIBLE IS ALL TOGETHER CORRECT ABOUT CREATION AND FOR THOSE SCEPTICS OUT THERE I PROVE IT AND THIS TIME THE ATHIEST MUST BRING THEIR PROOF IN DOUBT. For instance that darkness filled with nothing being the night sky is light and that was the very first command "Let there be light". That is light out there! Now it is the atheist turn to prove that which is filling the night sky is nothing. Go on… prove it mathematically that $149 \times 10^6 \times 0$ = the distance there is between the Earth and the Sun .

In the past these remarks made me the clown in the courtyard and no friends came to my aid because no friends were in support of my statements. A description that would be closer to is that no friend wanted to admit any friendship because such admitting may also reflect on his or her sanity.

When looking at the cosmos from whichever angle indicates the fact that the cosmos is moving. It is forever spinning and it is going to as much as it is coming from. Everything is on the move and always encircling something of greater importance. A top can spin but the parameters of its spin are limiting the motion it can apply. By not spinning the top is still spinning as the Earth are doing the spinning on its behalf.

When spinning too fast the top fights something because the alignment keeping it upright starts to tarnish. The same apply when spinning too slowly but that makes sense. It is the fact that the same affect comes about when spinning too slow that triggers the questions.

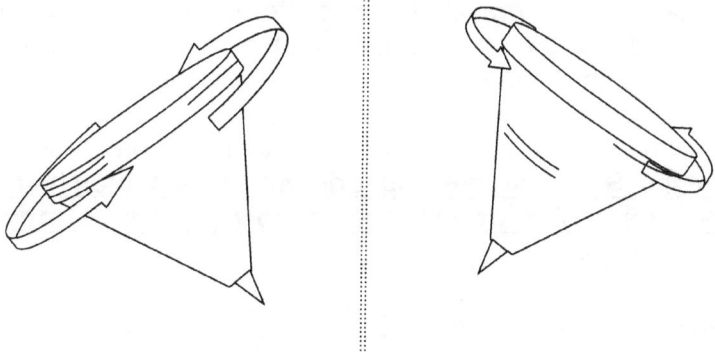

The spinning top is all the evidence any one needs to come to such a conclusion. By saying that I first have to admit (no not my mental stability), that I have no academic background and I do not enjoy any link to any university.

Without trying to not to be too presumptuous I'd say a fair guess would be that I know probably as much as any graduate about cosmology but lack certificates to prove my knowledge. I am not part of established science. In my developing of knowledge accumulation I came to some conclusions about cosmology that are unique and divert somewhat too drastic from the accepted norm. Most of the work I see the same way as the norm does but in a reverse. Allow me a short explanation

When looking at a red flower we say the flower is red. Nothing can be further from the truth. The flower is every colour in the spectrum, except the colour we attach to it. It is screaming with all might to its disposal that that specific colour it cannot accept. Yet, we maintain that that colour is the colour we associate with the object, ignoring the objects rejection of that colour. Only when looking at the cosmos from this stance, can the cosmos make sense? By recognizing a disassociation in spite of our cultural recognizing the association, can we understand the cosmos? We maintain the Sun is burning, while the fact of the matter is that the Sun is freezing. From our perspective on the outside we see the Sun burning as we see the red flower. What we see is not what is the truth. Only by applying the correct view to the cosmos can the four principles I introduce, make any sense and find any proof… and I do prove them. Only by telling the complete story as I do in the complete six parts of *"Matter's Time in Space – The Thesis"*, can the explanation surface to a point of understanding. One cannot draw any conclusion from the outside; one has to be inside the star to see what is going on. To get such proof I had to do extensive research on cosmology. The proof lies in unrecognised and misunderstood laws and principles science know. These laws fall outside the parameters of applied physics.

I defined gravity; I defined energy, but before that I had to prove the existence of time and time's control over the Universe, time's role in the Universe and what time is. This was up till now not yet been achieved. I had to prove what space is, that time and space is sides of the same coin, with matter forming the separation. The main conclusion that brought about such conclusions was my different view of science. It's not the explanations science at first that made me question the validity of Newton, but the things Newton cannot explain but is factors in the cosmos nevertheless. As a school going youngster I was fascinated by astronomy and in particular the cosmology aspect. In a long and strenuous process of self-education I was completely stunned by the behaviour pattern that the comet had in its relation as it orbits the Sun . Please forgive my boyish way of presenting the following but it is important that I bring it across as I saw it as a boy and as a matter of fact still see it today as a middle -aged adult.

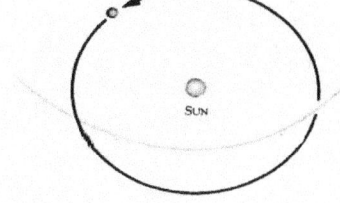

We may start by determining the influence of gravity on planets as we find them in the solar system. **First, let us concern ourselves with a comet**. It is common knowledge how the comet relates to the Sun 's gravity. **Firstly, picture the comet at its farthest Point, away from the Sun .** The **gravity** of the

Sun pulls the **comet straight towards the Sun** , this we all know. Gravity always pulls an **object directly towards** the **centre of a cosmic body**: that too is common knowledge. Therefore, the comet is drawn directly towards the centre of the Sun and throughout its journey the comet is picking up momentum directly related to the gravity that is cantered in the middle of the Sun , (**gravity is always cantered in the middle of a cosmic body**). As the comet is increasing its speed, the comet comes closer to the Sun and therefore the Sun 's gravity pull is simultaneously increasing as the distance between the two cosmic bodies is reducing. Each instance the comet is drawn towards the Sun , the gravity that the Sun applies to the comet becomes larger progressively. When the comet is at its <u>**closest point to the Sun**</u> , <u>**something odd happens which cannot be explained by Newton's gravity at all! Remember gravity should now be at its strongest point because of the proximity of the two objects.**</u>

The comet remains at an even distance encircling the Sun .

No longer does the gravity of the Sun pull the comet towards the centre of the Sun .

At this very point the gravity that the Sun applies on the comet does not pull the comet towards the centre of the Sun any longer, in fact, it seems as if the effect of the gravity has been neutralized.

1. The comet stays at an even space from the Sumas it goes around to complete a half circle's orbit around the Sun .

2. No longer does the gravity of the Sun pull the comet towards the centre of the Sun .

3. At this very point the gravity that the Sun applies on the comet does not pull the comet towards the centre of the Sun any longer, in fact, it seems as if the effect of the gravity has been neutralized.

4. The comet stays at an even space from the Sun as it goes around to complete a half circle's orbit around the Sun . It only completes a part of its rotation around the Sun .

5. After this, an even more peculiar event takes place. <u>**The Sun , at the point where gravity should be at its most dominant, suddenly loses its complete grip on the comet.**</u>

6. The comet brakes free from the Sun 's pull of gravity and speeds off towards   its destiny into the vastness of the cosmic space, undeterred by the gravity of   the Sun .

Then after a pre-determinate and pre-calculated time the Sun starts applying its gravity on the comet once more. At a point where the comet is at its farthest point, the gravity of the Sun becomes strong enough to bring about a complete turn around to the comet's direction of travel. <u>**However, the gravity between the Sun and the comet is at this point, at its weakest point of influence.**</u>

However, **this is not all**. When we regard the planets as they stand related to the Sun , the effect is the same, but not as obvious. All the planets follow an oval orbit around the Sun and therefore the same factors concerning gravity apply to the letter as it does in the case of the comet. Let us investigate the one planet we relate the best to, which of course is the Earth.

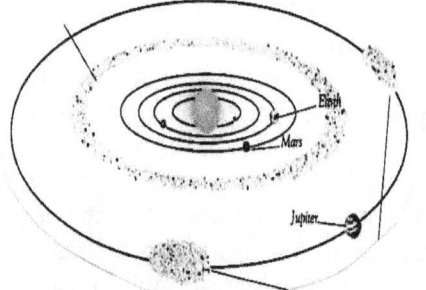

So, when the Sun 's gravity is at its strongest, the comet manages to brake loose and neutralize the Sun 's gravity pull in order to avoid its fatal collision with the Sun and when the Sun 's gravity is at its weakest, the comet cannot escape the pull of gravity. There is definitely something very wrong, either with the comets or in this case the Earth's circling behaviour or the laws made up by Newton.

Well, this is the part Newtonians are so able to understand and because I am a mechanic I am not able to Understand. The Earth makes a circle and Newton in all his mathematical splendour never provided for this circle. He even went further by introducing Π to what is the eternal circle! The Newtonians sympathise with my poor understanding because of my low intellect and education! Who should be sympathising with whom should be a better question?

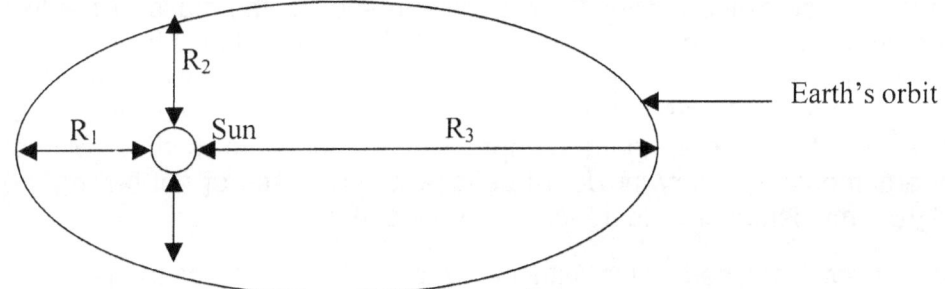

This illustration does exaggerate the radius of the Earth's orbit around the Sun , but since it has taken place 4 500 000 000 times, it has no real effect on the validity of the next statement.

At one point (R_1) the distance between the Sun and the earth **is less than** at another point we call R_3. Let us put a value of R_1 = one and R_3 = three. This means that each year, for the past 4 500 000 000 years the effect of the common gravity between the earth and the Sun has a greater effect than at another point six months later. **At one point the earth should be drawn or pulled closer to the Sun and after another six months** interval **the earth should stand less effected by the Sun 's gravity**, therefore it should move away from the Sun . Each cycle of twelve months would have one point where the gravity pulls the earth closer and exactly the opposite must apply six months later when the gravity is at its least. So, for the past 4 500 000 000 years the earth has been re-establishing its seasonal swing towards the Sun and away from the Sun , which means by now the earth has to collide with the Sun in midsummer or escape from the Sun in midwinter, as it may then drift away into the unknown.

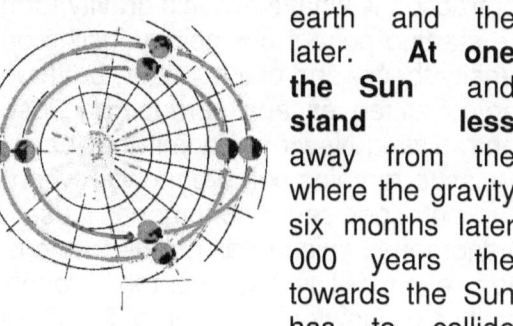

For the more mathematical minded person the argument is as follows. May I remind you, THAT NEWTON'S OWN LAWS ARE IMPLIED, and again the planets disobey these laws completely!

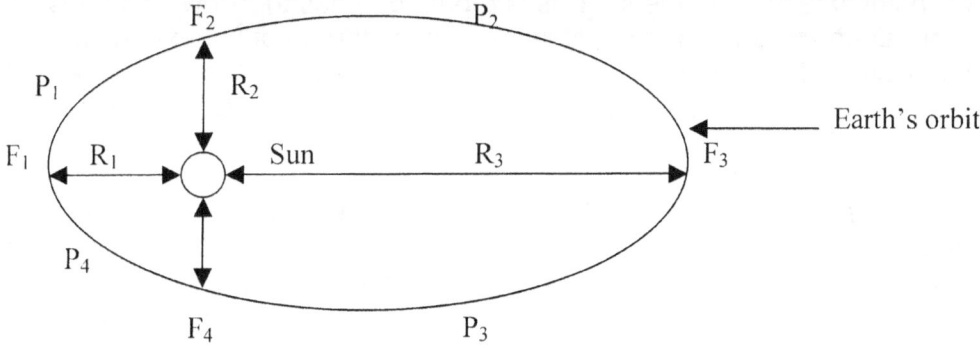

$$F = \frac{M_1 M}{r^2} G$$

We know that $F_1 \neq F_2 \neq F_3 \neq F_4 \neq F_1$
because that is what seasons are all about.

Even if $F_1 \neq F_2 \neq F_3 \neq F_4$, we know that $P_1 = P_2 = P_3 = P_4$.

Because r is at different values F could not be to the same value.
Therefore, the value of F has to be unrelated to force its value on to P.
Nevertheless, Kepler has proven that $P^2 = a^3$, although $a_1 \neq a_2 \neq a_3 \neq a_4$.
If $a_1 = a_2 = a_3 = a_4$, we would not have had season and climate changes on Earth. That means that to proclaim $F = \frac{M_1 M}{r^2} G$ is nonsense. The truth of the matter is that Newton actually proclaimed

that in an ellipse, which has an uneven circle (Kepler's findings) the value of $F_1 = F_2 = F_3 = F_4 = F_1$, but because an ellipse has no constant radius, it actually means that $r_1 \neq r_2 \neq r_3 \neq r_4$ and thereby anybody can see that Newton's calculations are wrong. $F_1 \neq F_2 \neq F_3 \neq F_4$.

In this book, I dare to prove that **there is a difference** between findings of *Galileo and Kepler* on the one side and the work of **Newton and Einstein on the other hand**. Only one of these two group's findings can be right, because there is an unmatchable difference in the concept of these two groups' opinions. Newton considers that a force exists between two bodies in space: the mass of the two bodies' product is being brought into context with the gravity constant (G). This value is then divided by the distance r calculated as a square (r^2) value. I have to admit that I have not once seemed to bring across the importance I see in the arguments above when translating it to academics of stature. Every time I introduce the behaviour of the comet I get the impression that academics either will not or cannot see any truth in my arguments. In every incident I became the accused of not having the brainpower to understand and from my perspective there is little to understand.

I stood accused by many academics I crossed paths with in the past that I am not familiar with Newton and because of my poor academic background that I am not capable of understanding Newton. That is not the case and I have to be very adamant about that. I would accept such accusations if Newton's science explained all of science. That is hardly the case because I can state four very prominent cosmic principles that no one can explain by applying Newton's claims. The Roche-Lobe, the Titius Bode principal, and the Lagrangian five-point position and the Coanda gravity contraction is what Newton's gravity formula cannot explain at all. Neither can the Big Bang theory be the starting point if cosmology insists on using the application of Newton. I am aware of the Critical density theory and black matter, but those arguments has not found proof in the slightest and in truth serves as an escape corridor because science is at the end of its tether with the phenomena contradicting Newton. I admit that I am lost at finding a starting point introducing the book because the issue remains comprehensive when dealing with issues of cosmic proportions I shall explain the four unrecognised phenomena I use in proving my statements. With my introduction of the phenomena, which I named the four cosmic pillars you will find it obvious why science do not accept them even if it is documented throughout the Universe and is quite commonly found. It totally annihilates Newton's formula of $F = G\,(M_1M_2)/r^2$.

From these cosmic phenomena I produce a path of cosmic development, preceding the Big Bang. The problem that comes from this, is that I take the reader from a point and lead the reader through the explanation of the existing principles, pointing out how they are flawed and introducing my explanations and proof and substantiate my argument. This is a path one has to follow. There is no point where one can drop in, or out and in again and maintain the golden thread of understanding. To conclude, only this: As I bring proof of existing evidence in cosmology about phenomena and of which science acknowledges the existence but science is failing to understand or explain the correctness thereof.

The Bible says: "Do not think of the heavens as Earth." Whether the Newtonians take exception in my quoting from the Bible means little, because that verse holds all the mistakes and misjudgements that science have about the Universe. Science holds the attitude that from the Earth, they can judge the Universe and by applying standards maintained on Earth, it is a "fit all bar a few minor adjustments." Science accepts the Hubble Constant and even uses it as a barometer. This Hubble Constant indicates how the Universe is expanding, growing in size. That means it is the measurement of how matter drifts apart. If the Hubble Constant applies to the Universe as a barometer, then that application should affect the Milky Way as much. Remember I am using Newtonian logic, not my own, so please do not misquote me. If there is the shift of matter, away from a centre point outwards, the circle in which the orbiting structures hold their position should then enlarge, because of such a shift. By increasing the distance that the orbiting structure travels around the centre, the time it takes to travel should also increase. If one wishes to form a concept of the implications in hand the most prudent manner is to return to the star of it all and then try and progress from that position.

The Big bang was never bigger than at the time the procession of time started. The Biggest Bang was when light or heat started the process we call the Universe. Heat expands and cold contracts and that is gravity. As heat expands it leaves an area by halving the heat in that area. The relevancy changing makes that the duplicating brings about the material two half by doubling the space it moves through. By the halving the heat, the motion is enforcing the area to cool off by half and that produces contraction. By expanding there is contraction because overheating expands while it then cools bringing about contraction. As the light came into the darkness by heat coming onto the eternal line of time the line was interrupted or broken by infinity releasing the line in eternity of its eternal procession. Infinity released from eternity, heat broke from cold singularity in the dot released from singularity in the spot, Π^0 relieved Π from eternity, 10 broke from 11 and all the Universe coming about fell into place in space in time in space-time an in time space. The line dotted on a spot the line was in progress one eternity long without progressing because it was in an eternal progress that had no beacon, no marker and no comparing to, which put the eternal running line eternally in one spot. The heat interrupted eternity by infinity parting from eternity.

Now please listen to my explanation and compare my explanation on how creation began with the nonsense of Newtonian science and that which according to them I am too simple minded to understand. Simple minded I might be but so was creation at the start never very complicated. It started from a spot that became a dot and by becoming something eternity began to flow, as time broke free from monotony, although at first by the instant of infinity.
The line had flow. The line did grow albeit from infinity to infinity relieving eternity from infinity. The Universe made the biggest leap it never could repeat afterwards and it was so small it can never be seen inside this Universe we are within.

The line diverted from Π^0 forming Π but there was so little established that although Π was Π also Π = 1 because 1^0 moved to 1^1. That put the figure of two in the Universe. Then from two the square of two diverted from the line of time by 9^0 0 which established a line of 180^0 at a relevancy of 90^0 to the timeline at 180^0

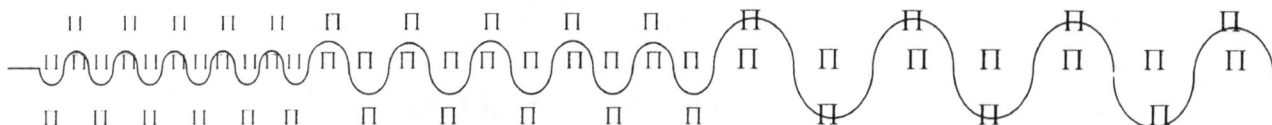

There was the line $\Pi^0 = 1^0$ that formed $\Pi = 1^1$ and $\Pi = 1^1$ ninety degrees to the line in time. Through the motion that came about that motion brought about a square in the line of time meaning that Pythagoras established mathematics.

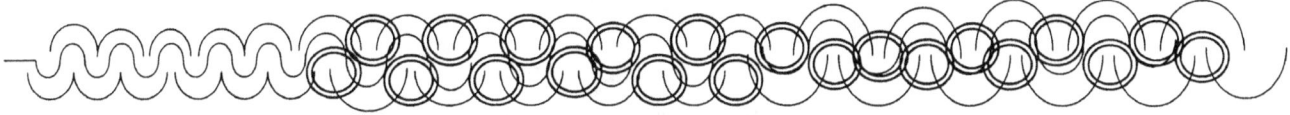

As time moved on a lagging behind started to creep into the flow where the heat began to distort the flow of time by not repeating the flow in the same cycle

This lagging of heat formed the basis for space holding time end very much later it became even further behind and in such a lagging that the time solidified to form heat dragged behind as material. There at that point formed matter. But at first every point formed another value in the line of figures coming from one and ending at ten. The line established the forming of every dynamic present in the Universe that established numbers in relation to Pythagoras except there was no zero formed. The one number formed the next number in conjunction with Pythagoras and in that way mathematics formed the cosmos while the cosmos formed mathematics. However it is evident that Newtonians are under the impression that while mathematics was used to form the cosmos mathematics in principle never formed.

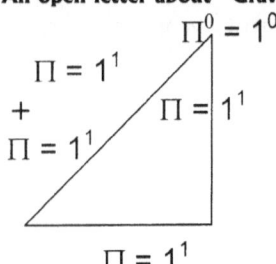

$\Pi^0 = 1^0$

$\Pi = 1^1$
+
$\Pi = 1^1$

$\Pi = 1^1$

$\Pi = 1^1$

$2\Pi + \Pi^0 = 1^0 = 3$
2Π = another r dimension forming the square $\Pi \times \Pi = \Pi^2$ = motion of space in time

Then the Universe was 3 with two divisions on both sides of the divide.

It is clear that the adding was the first dimension because to get from one to two by multiplying brings one back to one. However by adding it does bring about three in the triangle while two is part of the half circle as well as the line of time. By forming a relation with the line and on the one side with a half circle while being relevant to a point on the other side of the divide clearly shows why 1800 fits all three forms and put the here on an equality that otherwise does not make any sense to us. However one can see that in the next development the square came about by multiplication of two points travelling equal along a centre line.

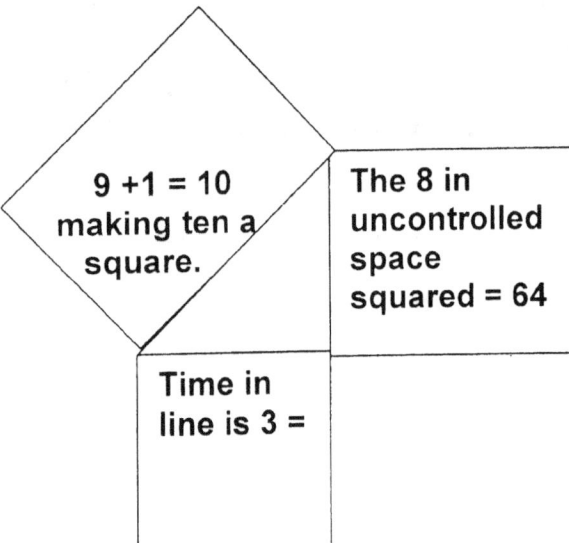

9 +1 = 10 making ten a square.

The 8 in uncontrolled space squared = 64

Time in line is 3 =

The line in time is three by measure of the layout of the Universe ($\Pi\Pi^0\Pi$). The square of the time line is 3^2 = 9 and with that in relation to singularity we find that the $\Pi^0 = 1$ plus the square of time (3^2) = 9 and that makes the one part of time the square of time is 9 plus the line holding singularity $\Pi^0 = 1$ forming 10.

That makes ten the square of as well as the root of time and that is another reason (if not the only reason) why we are able to use ten as a decimal basis for counting by numbers.

In motion space-time found a relevancy by the ordering of gravity.

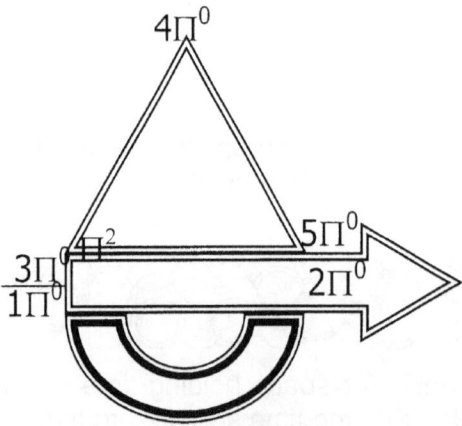

$4\Pi^0$

$4\Pi^2$

$5\Pi^0$

3Π
$1\Pi^0$

$2\Pi^0$

Motion is to the value of gravity, which is the value of Π^2. Crossing the division of singularity brings about a waving half circle that connects by the triangle to thee other side of the divide. That places 5 points holding singularity Π^0 in relation in relation to the motion of space going through time in Π^2. The five points in singularity is the motion of space in Π^2 passing through time's five and that also conclude as material in the square at 49. Now we can put that in relation to Pythagoras and find gravity forming space-time while material in time formed gravity.

There was the line $\Pi^0 = 1^0$ that formed $\Pi = 1^1$ and $\Pi = 1^1$ ninety degrees to the line in time.

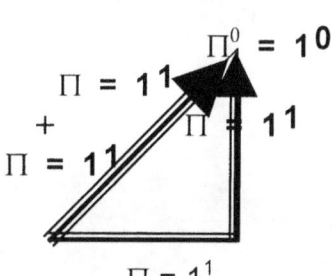

$2\Pi + \Pi^0 = 1^0 = 3$
2Π = another r dimension
forming the square $\Pi \times \Pi = \Pi^2$
= motion of space in time

Then the Universe was 3 with two divisions on both sides of the divide.

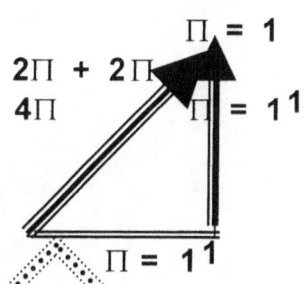

The motion of space- time in space in time brought about the compliment of time being four.

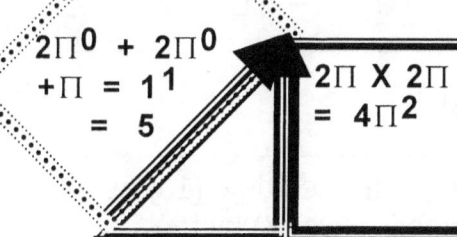

In this 5Π became valid in the cosmos forming a point one separated from time using 4

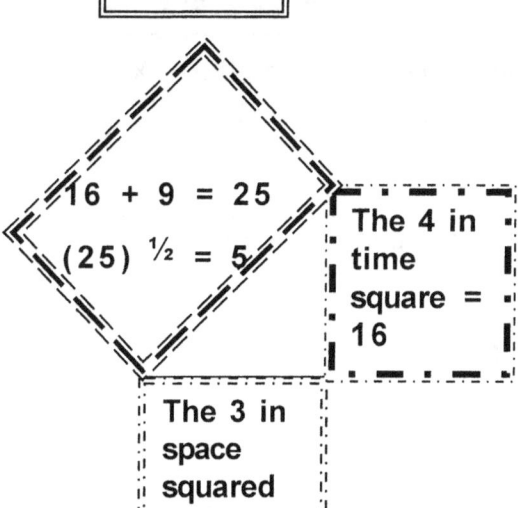

Time is holding 16 + material is holding 36 = 52. The compliment 52 – the 3 of the time line gives 49. The square of 49 is the end of filled space, which are 7. The sphere runs by seven points. That means compromised space starts at point 8.

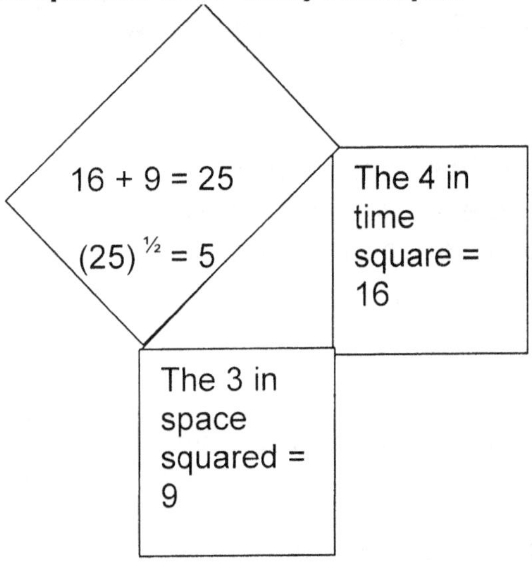

$16 + 9 = 25$

$(25)^{1/2} = 5$

The 4 in time square = 16

The 3 in space squared = 9

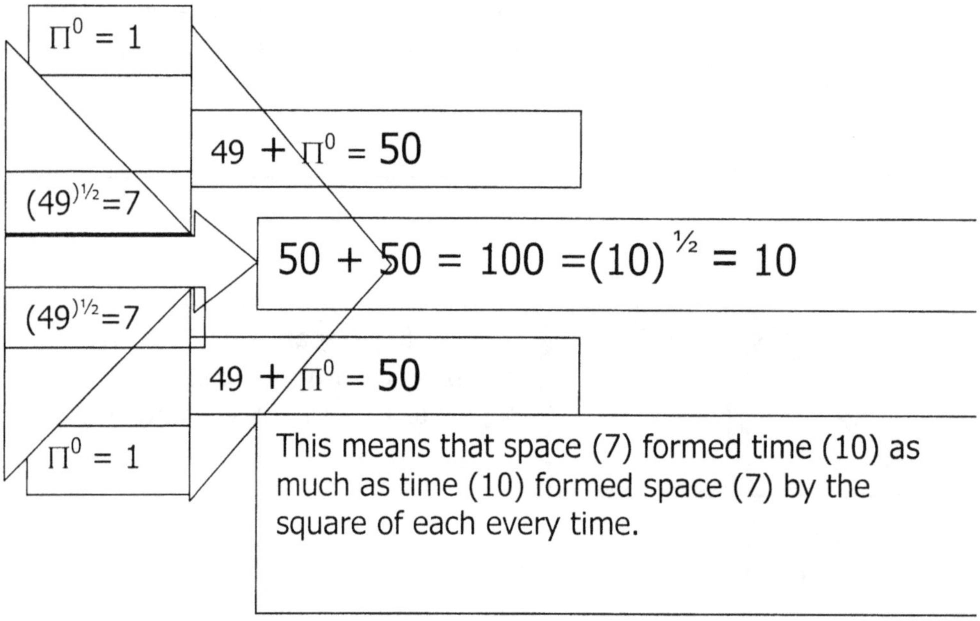

$\Pi^0 = 1$

$49 + \Pi^0 = 50$

$(49)^{1/2} = 7$

$50 + 50 = 100 = (10)^{1/2} = 10$

$(49)^{1/2} = 7$

$49 + \Pi^0 = 50$

$\Pi^0 = 1$

This means that space (7) formed time (10) as much as time (10) formed space (7) by the square of each every time.

The square of space is 6 X 6 (6 by dimensions) in relation to Pythagoras

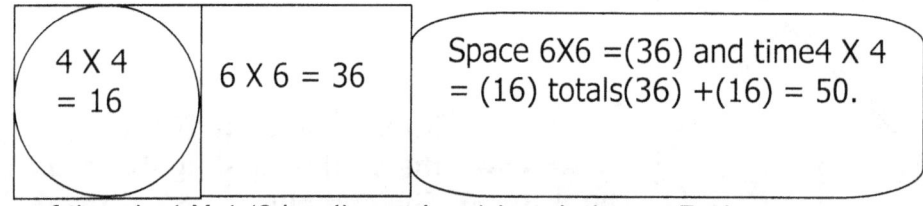

$4 \times 4 = 16$ $6 \times 6 = 36$ Space 6X6 =(36) and time4 X 4 = (16) totals(36) +(16) = 50.

The square of time is 4 X 4 (6 by dimensions) in relation to Pythagoras

$\Pi^0 = 1$

$50 - \Pi^0 = 49$

$(49)^{1/2} = 7$

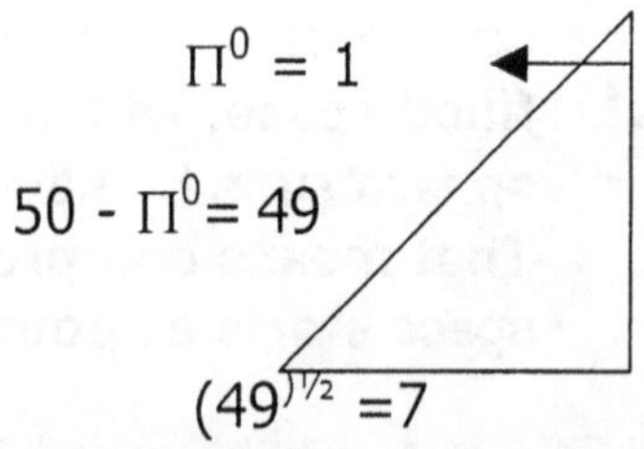

That way space in time formed the curve of space by 7^0.

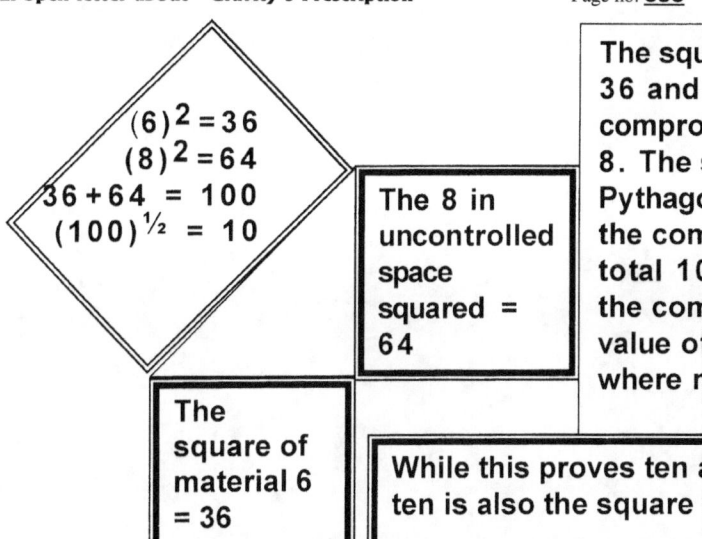

$(6)^2 = 36$
$(8)^2 = 64$
$36 + 64 = 100$
$(100)^{1/2} = 10$

The 8 in uncontrolled space squared = 64

The square of material as 36 and the point where compromised space starts is 8. The square of both in Pythagoras is 36 + 64 and the compliment of that total 100. The square of the compliment gives the value of 10 and that is where motion in time ends.

The square of material 6 = 36

While this proves ten as the root of time ten is also the square of time.

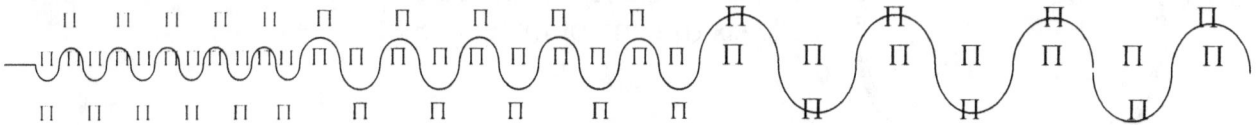

There was the line filling with ones that formed twos as 1^0 formed 1^1. That made two lines and that had to be three because 1 + 2 = 3.

Since three came into place while Π^0 was establishing Π I guess (I cannot prove it but it does come to mind) that the motion had took place while Π became the factor and 3 became the position that the small difference there is between 3 and Π was somehow initiated by the motion in the line.

In the motion space found a relevancy by the ordering of gravity.

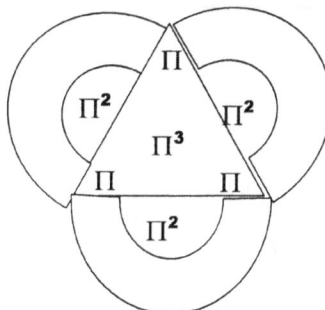

The Universe grew from relevancies but not out of the relevancies. The relevancies still dictate everything there is in the Universe. The only way space can grow is when heat comes about and change the form. That is nature. The only way expanding can come about is when heat becomes a factor added to particles. That is nature. The only ways particles reform or when heat remove some or all the solidness in the form. Heat coming into or removing from determine form. It is still part of the cosmos as it is still part of the cosmos that friction accumulates heat and heat results from a lack of usable space. That is nature.

Motion is to the values of gravity, which is the value of Π^2. Crossing the division of singularity brings about a wave half circle connecting to a triangle on the other side of the divide. That places 5 points in singularity Π^0 in relation to the motion of time Π^2. The 5 points in singularity is 5 x the motion of singularity is 9.8 X 5 = 49. Put that in relation to Pythagoras

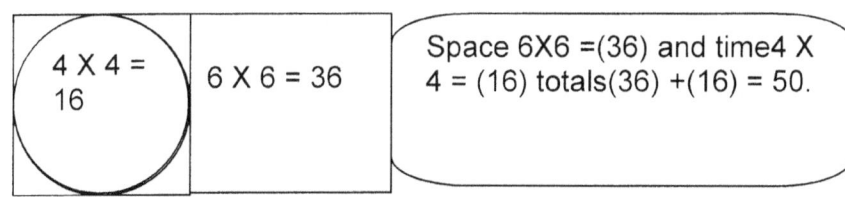

4 X 4 = 16

6 X 6 = 36

Space 6X6 =(36) and time4 X 4 = (16) totals(36) +(16) = 50.

One point away from the divide $\Pi^0 = 1$

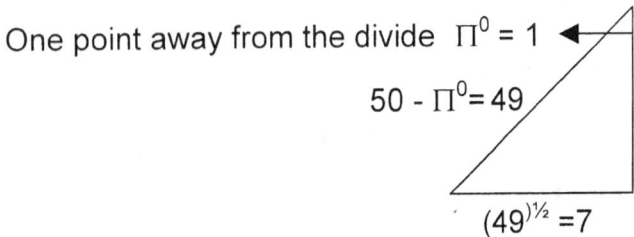

$50 - \Pi^0 = 49$

$(49)^{\frac{1}{2}} = 7$

That way space in time formed the curve of space by 7^0.

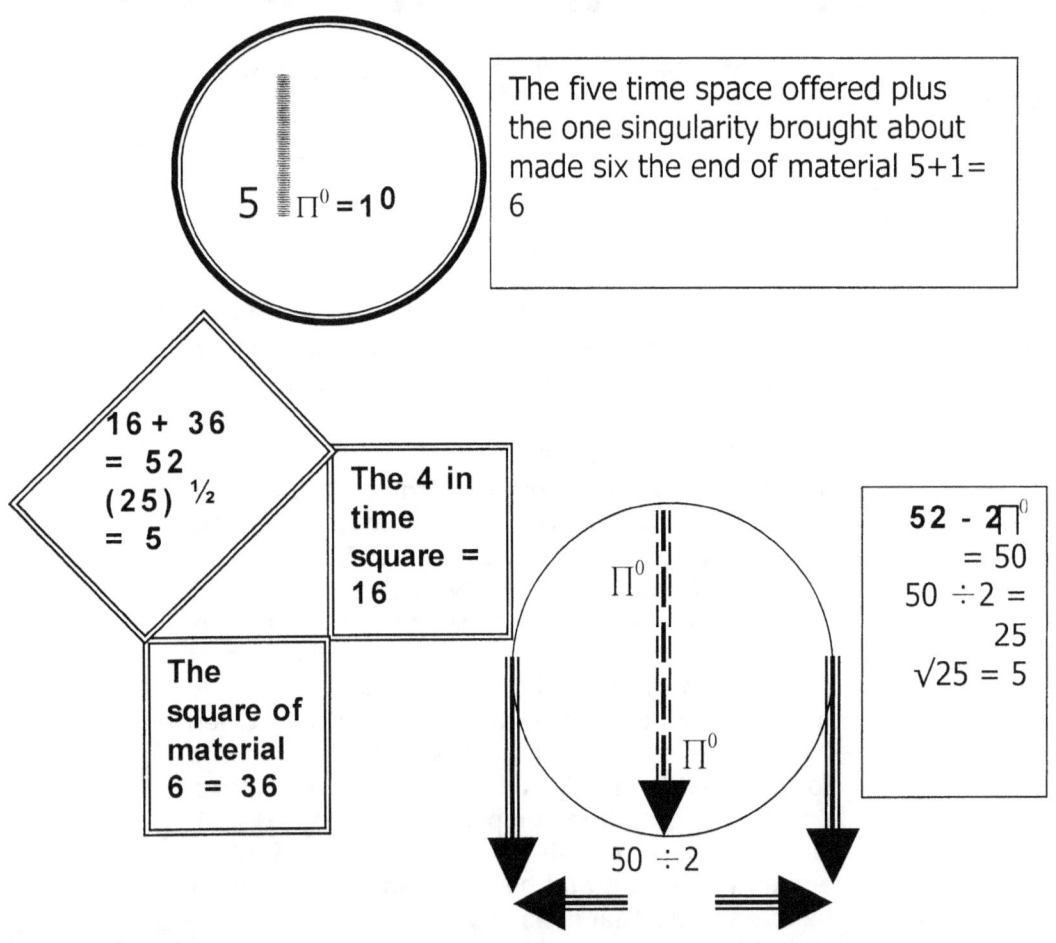

$5 \, \Pi^0 = 1^0$

The five time space offered plus the one singularity brought about made six the end of material $5+1= 6$

$16 + 36 = 52$
$(25)^{\frac{1}{2}} = 5$

The 4 in time square = 16

The square of material $6 = 36$

Π^0

$50 \div 2$

Π^0

$52 - 2\Pi^0$
$= 50$
$50 \div 2 = 25$
$\sqrt{25} = 5$

That makes ten the square of as well as the root of and that is why we may be able to use ten as a decimal factor forming the basis of all numbers. Lets run through the process again in order to pick up some loose ends.

Every time Π formed singularity brought about motion by gravity in Π^2, which still maintained a line. Gaps formed and that defined and became the atom in space.

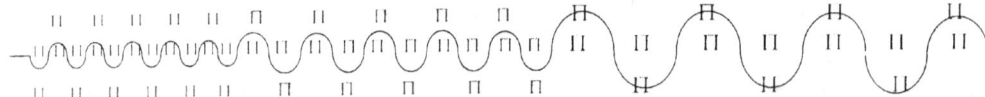

The line formed singularity by forming from the spot going onto become a dot. The heat surging moved as heat still does today. The heat expands as it moves into more space. That which formed was Π and the motion in the surge was Π^2.

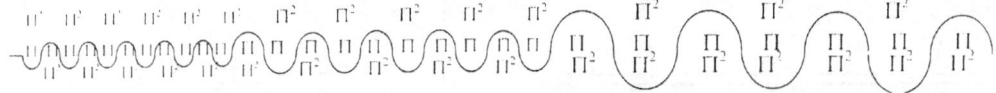

Every time Π formed singularity brought about motion by gravity in Π^2, which still maintained a line. Gaps formed and that defined and became the atom in space. The heat moving became the motion that became the gravity. The moving was a relation between $((7/10 + 7/10) \times 10) / 10 / 7$ and the moving formed Π^2.

$((7/10 + 7/10)$. Material forming as space becomes available provided by heat surging. It was material moving in time in space that was becoming time.

(10) The three in time-space or the space in which material has room to move.

10 /7 heat cooling by surging back as the contraction retarded the responding growth.

In this was a line, which I now am unable to sketch since sketching firmly relies on a three-dimensional drawing, and at the time what was three-dimensional is very flat today.

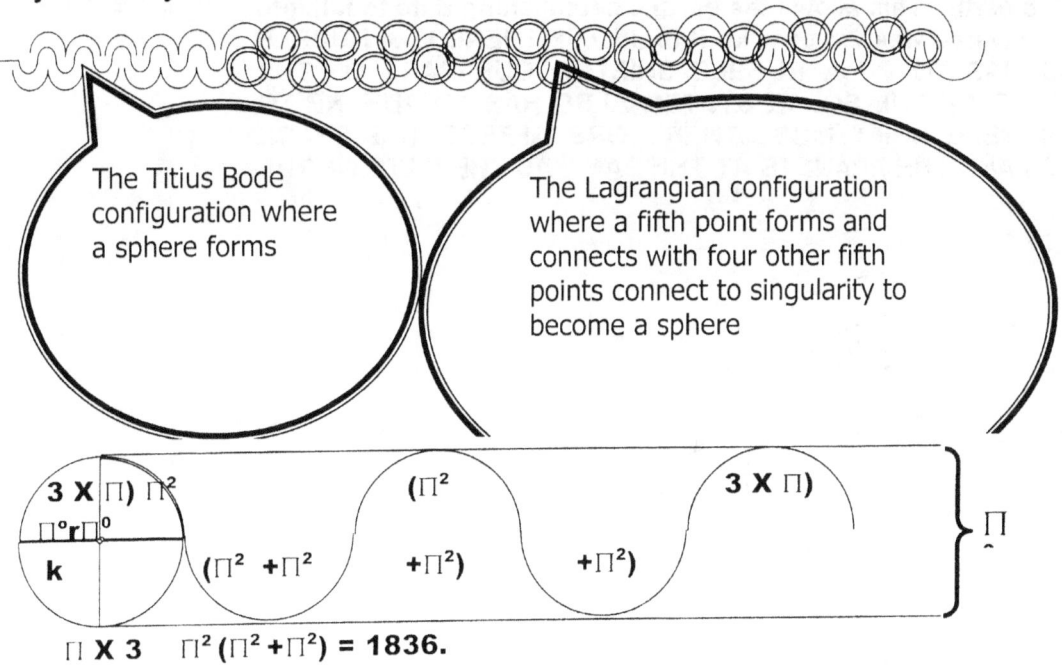

The Titius Bode configuration where a sphere forms

The Lagrangian configuration where a fifth point forms and connects with four other fifth points connect to singularity to become a sphere

$$\Pi \times 3 \quad \Pi^2 (\Pi^2 + \Pi^2) = 1836.$$

Finally with all the heat retarding an interrupting of the line a gap formed in time.
Π^0

All the while time is just a spotted and dotted line running along time as space duplicated with heat surging and cooling as cold contracted much similar to the actions of stars in the process of pulsating known by what ever name one wish to use. The star takes time back so slow we can see the pulsation of gravity cycles.

This was what lead to the process through which at a later time became the method that the atoms of various significant formed by the motion that was prevailing at the time. We find the evidence in the characteristics the atoms show in relation to heat. In the book Starstuffin I do explain this in length.

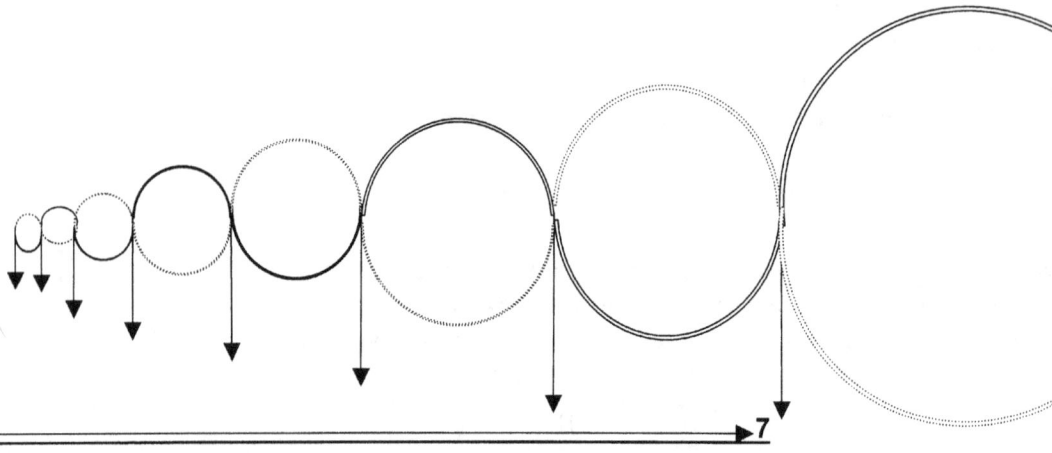

Newtonians put the Big Bang, as the start while the Big Bang they refer too was the seventh Big Bang and ever bang was considerably smaller than the previous one.

Amount of radius increase provided by the Hubble after the first:
2) $1X \ 10^9$ YEARS AFTER THE BIG BANG EVENT
3) $3 \ X \ 10^9$ YEARS AFTER THE BIG BANG EVENT
4) $6 \ X \ 10^9$ YEARS AFTER THE BIG BANG EVENT
5) $9 \ X \ 10^9$ YEARS AFTER THE BIG BANG EVENT
6) $12 \ X \ 10^9$ YEARS AFTER THE BIG BANG EVENT
7) $15 \ X \ 10^9$ YEARS AFTER THE BIG BANG EVENT

They apply an age of $13,5 \times 10^9$ years in the case of the universe and $4,5 \times 10^9$ years in the case of the earth. This allows the earth's establishing date to fall into a position where the universe was 2/3 of what it is now. IF THERE IS A HUBBLE SHIFT, THERE ALSO HAS TO BE A HUBBLE SHIFT IN OUR PART OF THE UNIVERSE AND THEREFORE THE SOLAR SYSTEM ALSO HAS TO ADHERE TO THAT. THE SAME HUBBLE SHIFT MUST THEREFORE AFFECT THE SPACE BETWEEN THE EARTH AND THE PLANETS AT THE SAME AS THE RATE OF THE UNIVERSE.

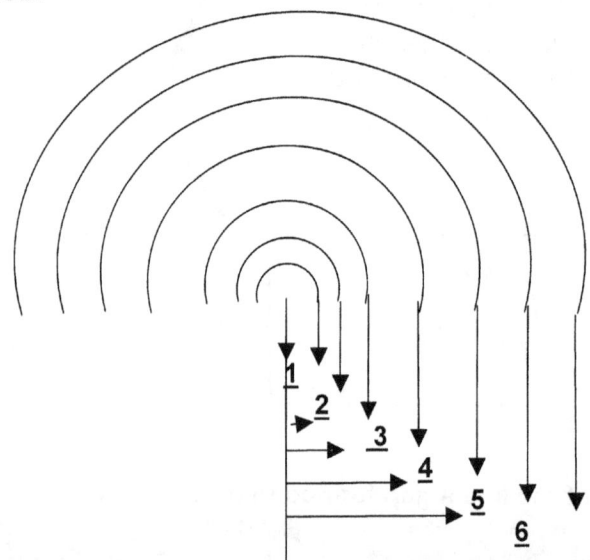

How long did the first instant in time last? It lasted one eternity minus one infinity. Then the next duration was one eternity minus three infinity and then the Titius Bode rule started affecting time by the measure of $(7 + 7) / 10 / (10/7) \ X \ 10 = \Pi^2$

They apply an age of $13,5 \times 10^9$ years in the case of the Universe and $4,5 \times 10^9$ years in the case of the Earth. This allows the Earth's establishing date to fall into a position where the Universe was 2/3 of what it is now. IF THERE IS A HUBBLE SHIFT, THERE ALSO HAS TO BE A HUBBLE SHIFT IN OUR PART OF THE UNIVERSE AND THEREFORE THE SOLAR SYSTEM ALSO HAS TO ADHERE TO THAT. THE SAME HUBBLE SHIFT MUST THEREFORE AFFECT THE SPACE BETWEEN THE EARTH AND THE PLANEYS TO THE SAME AS THE REAT OF THE UNIVERSE.

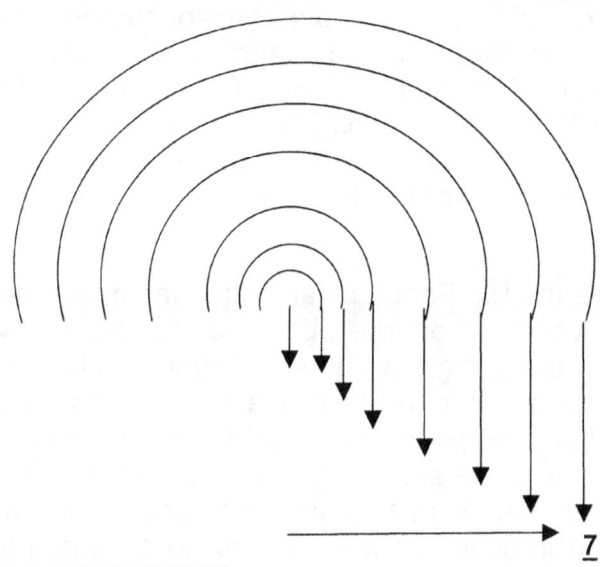

1) THE BIG BANG EVENT

AMOUNT OF RADIUS INCREASE PROVIDED BY THE HUBBLE SHIFT AFTER FIST:
2) 1×10^9 YEARS AFTER THE BIG BANG EVENT
3) 3×10^9 YEARS AFTER THE BIG BANG EVENT
4) 6×10^9 YEARS AFTER THE BIG BANG EVENT
5) 9×10^9 YEARS AFTER THE BIG BANG EVENT
6) 12×10^9 YEARS AFTER THE BIG BANG EVENT
7) 15×10^9 YEARS AFTER THE BIG BANG EVENT

Any person that holds a clear view and is not adamant in defending Newton has to be aware of many misgivings his formula brings to the Big bang, If he saw mass compacting mass by mass accumulation then the Big Bang was a missing event and we were no more here than we are there. There should be no star structures that formed because how easy was it for mass to pull mass when there is no separation between particles at all. Many such discrepancies go unnoticed or are either blatantly ignored by a sure lack of interest on the part on science. One cannot advocate a withdrawing Universe and claiming that very instant there is a contracting Universe the way Newton's law lays claim to that. While we all know that that is the case Newton does not prove it and to defend Newton is to become the devils advocate then.

With the Universe, expanding as it is, taking the Milky way along for the ride, and taking into effect the Universe square law where if the radius doubles, the circle grows four times, a year back when the Earth supposedly started, had to have been quite a measured mile shorter than it is today. Back then it had to take the Earth a few days to complete a year, since the orbit path was so much shorter. (Again, I wish to remind the readers that I am not applying my logic but that of the Super Educated.)

The Hubble's expansion brings about the increase in the radius of the distance between the Sun and the Earth. Every time the radius doubles, the circle grows four times by distance. This is a mathematical fact, beyond reasoning, to be accepted by one and all! Holding this mathematical law of value and positioning, the age of the Earth happening some 2/3 down the road the Universe took to develop, and judging the size the Universe got to where it is at present, the circle the Earth uses to orbit the Sun had to quadruple many times over. What is a year? The distance the Earth takes to orbit the Sun at present, or the distance it took the Earth back then. I am not the one to maintain a second back then is the second now.

Our Super-Educated are the ones that insist on maintaining to stick to the second minute and hour as it is today. By reducing the radius between the Sun and the Earth, and then multiply that figure with the number of four that the year was back some $4,5 \times 10^9$ years ago, one had to use some stopwatch to measure one year. If scientists took less time in studying far off galactica, and took the Hubble shift they regard with such prominence closer to home (say implementing it in our solar system) and work the Hubble shift from that angle backwards, a great surprise would await them. They would find that not one single application of the Hubble shift would nearly fit our solar system.

My being ill educated brought so much to resolving of issues in favour of academics in the past that I am putting science in two categories: them and me, and if I stand accused every time not understanding Newton for the lack of education then I am prepared to be the un-educated as long as the other party accepts the title of the SUPER- EDUCATED. I think it is only fair to bring distinction to both parties from both ends. Saying this I also do not wish to offend you in person so I wish to make the issue a three party affair with you in a refereeing stance judging me versus them (the SUPER-EDUCATED-ONES.)

In the first part of the cosmic development since the Big Bang arrival, there was a lot of activity that brought stars of many sorts and flavours to the cosmos from dust, but since there is no evidence of that happening in view of so many astronomers dissecting the sky every night, one has to conclude events slowed down the past 4.5 billion years or so with not even one star in a half built state to show. It must have been great turmoil as everything happened in a short period of time because the development produced all matter relative to the Universe, all space there is in thee Universe, positioned all particles in place to one another, set standards for stars and galactica, reduced galactica, produced dust by the truck loads and in specific semi oval, not to distant and precisely aligned positions enabling stars to form and in some cases even form and disintegrated several times. What an era to live through. Much different from the placid one we find ourselves in at present!

Lets give the Newtonians some leniency and say that the Hubble shift started by having a space of about fifty millimetres in diameter. Two thirds down the road of cosmic development the solar system received its day of birth. Gauge from the development where that should leave the Milky Way, and particularly our solar system. Some scientists are even of the opinion that a massive star developed, destroyed and from that debris and rubble the solar system came about. That means the last third of the cosmic development was a relative dreary and boring era, compared too the first and second era. Think about the billions upon billions of galactica that formed in the first event, followed by the massive stars that formed and demolished during the second era, and during the last third not one single planet formed, that we can witness to prove all the claims. The Sun and the planets formed during the second era; therefore, by the start of the third era, they were in the position and place they are at present. The working tempo slowed down considerably the last five billion years or so I guess some one somewhere formed a workers union and brought in human working conditions to ease the workers demands for overtime compensation.

Push this double standard applied back to before the Sun took its position and there was no Earth to indicate the year. How small was the year circle at that point in time and space. Take this right down to the:" Big Bang" where "the whole Universe were the size of a man's fist" (they go even further by putting us into a neutron), how far did the circle goes to indicate a year then? The year was immeasurably smaller, shorter and faster than at present. This is logic even the Newtonians must accept. There is no space outside insanity to apply time to the past at the value it is at present and far worse, to use something so extremely insignificant as the Earth to measure it by.

Using such logic makes science appear foolish. There is just no rational in the way Newtonians suggest that time lapsed verses how events occurred that can explain facts without. Since the time of Newton, the arguments tarnished from being brilliant to clever to fair, to poor and a hundred years ago to the point of being stupid. That is what Kepler's formula is all about! That is what Kepler indicated with his formula $a^3 = T^2 k$. The space of an object (a^3) is equal to the time (T^2), which it is in, in every given instant (k). If the space becomes smaller, the time duration becomes longer in every instant that time flow progresses.

Singularity is a mathematical reality. Einstein may be the first to name it and Galileo (unwittingly) may have been the first to define it as Kepler was the first to formulate singularity, but in mathematical terms singularity is the most basic principle.

When science calculated the value of gravity and the gravitational constant, as well as the speed of light, they never considered the moon to play a distracting factor. It is quite understandable because they did not know about the Roche-factor influencing their calculations.

With Π^0 little more than a figment of the imagination there is actually two values of Π^1 facing each other in a relation combining Π^1 to hold the value of $\Pi^{1+1=2}=\Pi^2$ and with two sides being the very same but opposing each other there will therefore also be Π^2 to every side that holds Π^1.

From the above I can conclude that gravity is not 9,81 Nm/s, it is Π^2 = 9,8696.

The gravitational constant is not 6.67 but it is 6.9 (7/10 (Π^2)) and here the moon had an even bigger influence. It is a fortunate coincidence that we took water to be the measured calculation since water holds the combined value of 17,5 and that is half the value of either space (31) or time (Π^3). That makes a kilometre (1 000 m) one cube laid flat and since movement represents space-time occupation in a linear manner, it is the cube that went in a single line.

More of the same fortunate coincident is that we connected time to spin long before Newtonians came along. The Earth spins through space at 360° in one day and space represents 10, therefore there is 3600 minutes in the 7° of spherical angle moving to the outer rim representing again the seven. All this makes explaining matters a lot less difficult.

The value of the proton is not 2/3, but Π^2. The proton spins at a rate of Π in a dimension of 2. The neutron is not 2/3 + 1/3 + 1/3, but again it is $\Pi^2\Pi$ and the electron's 3, holds a dimensional implication, because time is in singularity and space is in singularity. Time is eternal and heat releases space, which is time from singularity. In an effort to make the understanding simpler, you have time at an eternal value and that makes space zero. $R^3=T^2$; $R^3/T^2=1$, Time and space interlinks because it is the same thing and heat, (matter in many spin rates) allows time to break free from eternity by allowing a distinguishing of the flow of events. Time in movement is the result of matter (which includes heat) to change their relating positions.

Think of a movie. The continuous flow of pictures indicating the change of the position of the photo's bringing about the concept of time. Play the picture too fast or too slow and it will be unreal because we know at what tempo matter changes its position in relation to all other matter surrounding it. That is time and that makes time irreversible, because the position matter holds in relation to each other (in considering it to be throughout the Universe), can never repeat once it has changed.

Newtonians, forget about time travel because just by mentioning such absurdity you prove what little you know about the cosmos. To go back to a certain time, you will have to redirect all matter in the Universe in a reverse, apply that reversing of all particles up to the point required, stop the movement of all particles and start time going forward. Before some Newtonian grabs for a calculator, remember, you that are doing the changing, is as much part of matter in the Universe, therefore your action in changing the direction will stop even before you start! It is silly to think people with healthy minds, acting like adults will indulge in senseless stupidity such as claiming to be able to reverse time.

What was within the Universe at the start will be in the Universe at the end. The Universe holds all; maintains everything and combines the lot. In Afrikaans we call the Universe the "Heelal". It is a combination of two words namely "geheel" and "alles". "Geheel" means everything and "alles means everything. Therefore the "heelal" directly translated from Afrikaans to English will mean the "Everything of everything". Nothing can be added and nothing can be lost. It is all-inclusive. With this fact so commonly known and accepted, how can the Universe grow? How can the Universe expand? Well, it cannot, and that is yet another illusion the Newtonians create through misunderstanding. What is in it is in it and it cannot grow, as much as it cannot shrink. It cannot expand and it cannot demise. It is only a consistence of changing relevancies, where the relevancy flows away from one part of eternity or singularity (space) to another part of eternity or singularity (time).

Every aspect of the Universe holds relevancy by applying time to space and the time to space first claims space from singularity then controls space from singularity and influences space outside the direct contact with singularity. In every event the factor remains the same, as it is only the relevancy re-applying a dimensional influence on space-time.

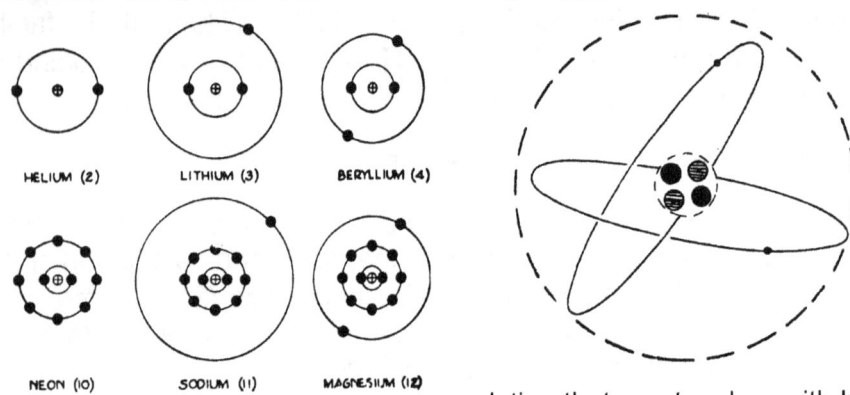

The   relation that an atom has with heat stems from the number of protons in the nucleus of the proton cocoon.

The key to the relevancy is heat and space. When matter heats, it expands therefore it takes more space. When matter is cooled it shrinks, therefore takes less space. That is the relevancy because matter in any form is heat. Heat produces the increase of space and reducing space produces an increase of heat. That is the relevancy. That is the secret of the Universe. That is the secret of gravity. That is the secret of momentum and every other aspect within the Universe.

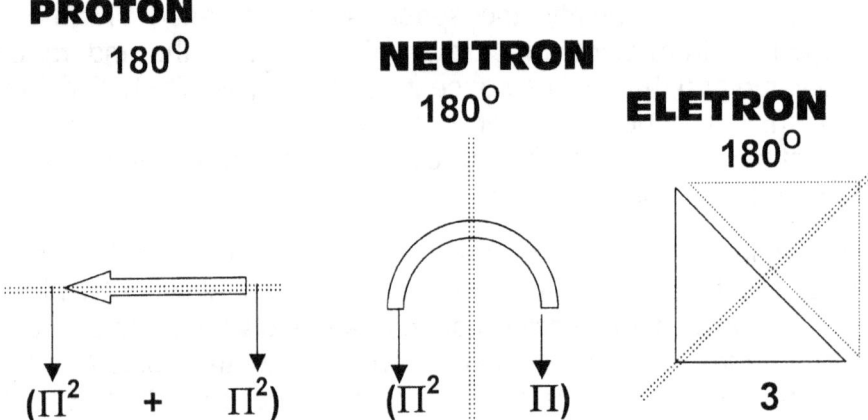

PROTON 180° **NEUTRON** 180° **ELETRON** 180°

$(\Pi^2 \quad + \quad \Pi^2)$ $(\Pi^2 \quad | \quad \Pi)$ 3

Time stood still in eternity, then after a command of the Creator, time started to move by overheating and eventually formed the relevancy of the proton $(\Pi^2 + \Pi^2)$ the neutron $(\Pi^2\Pi)$ and the electron (3). As a star returns time by depleting space to the dimensional increase of heat, space destruction is in progress and the star will abandon systematically some of the dimensions the atom holds. That is the relevancy. That will be whatever position there is in the Universe. In the depleting process of dimensional re- adapting, the star shall abandon aspects of space-time. The electron (3) may become obsolete, the neutron $(\Pi^2\Pi)$ may become obsolete in neutron stars and even $(\Pi^2+\Pi^2)$ the proton will become dysfunctional as space reduction completely disappears from the star's space-time occupation. However, those stars will be dark, and beyond our vision.

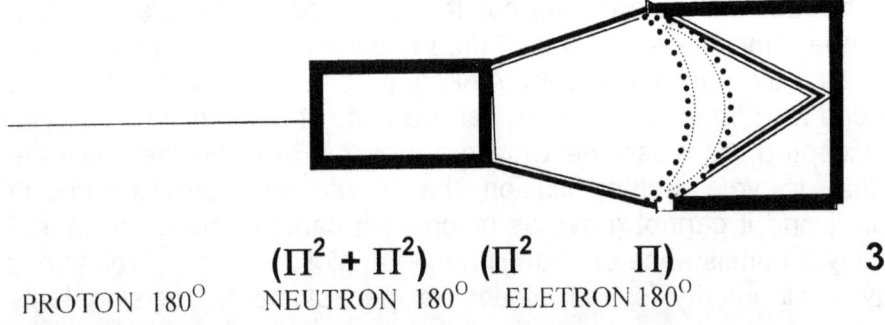

$(\Pi^2 + \Pi^2) \quad (\Pi^2 \quad \Pi) \qquad\qquad 3$
PROTON 180° NEUTRON 180° ELETRON 180°

The relevancy holds value pointing the relation between the various dimensions as they are in the atom. The relevancy of $(\Pi^2 + \Pi^2) (\Pi^2\Pi) (3) = 1836$ will remain but the mass of the electron and the mass of the proton will change in every space that time applies. Cosmology thus far was incomprehensible because it was incorrect. When applying natural laws, it becomes so simple that a person as ordinary as I can understand and explain it.

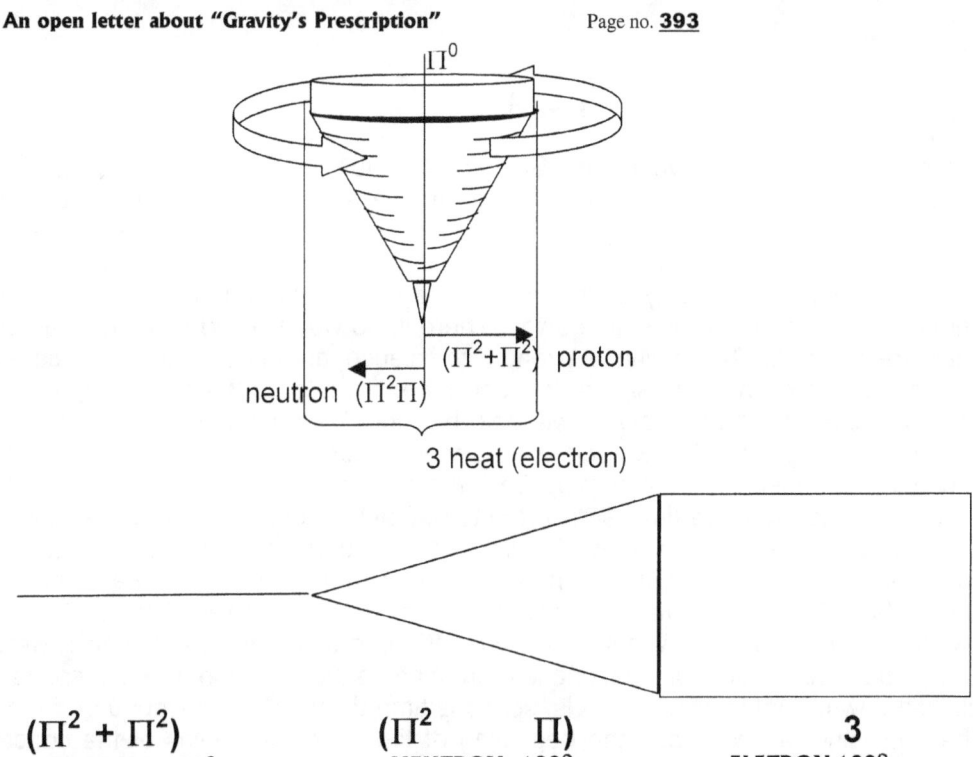

$$(\Pi^2 + \Pi^2)$$
PROTON 180°

$$(\Pi^2 \quad \Pi)$$
NEUTRON 180°

3
ELETRON 180°

The above indicates where singularity originates and how that establishes the factor in singularity Π. The Universe started from the factor in singularity Π. The entire Universe holds a spinning relevancy to all other factors in the Universe. If that were not the case, the Universe would not be there. The first person to consider the factor in singularity Π was Galileo. In the swing of the pendulum he saw singularity remain as one, that formed time, destroying space to maintain time. To prove my statement I shall very briefly indicate some barriers of motion science refer to as the Doppler effect, but Doppler used a slow moving train that at best could indicate two or three very minor moving limit.

If I can understand it, every other non-brainwashed human on Earth should understand it. The relevancy of $(\Pi^2+\Pi^2)$ $(\Pi^2\Pi)$ and 3, is a dimensional reduction of the flow of heat from space back to time. The flow of heat becomes necessary to prevent solid matter from overheating. By removing heat from the gas of space, through the neutron, to the solid of matter, space reduces as the intensity of heat flow requirements increases.

When we look at the night sky we see images of stars. I am of the opinion that our vision of stars and our interest we show in stars is just what sets us apart from other species. We are able see what we never can touch though we can appreciate what we never can have. We interpret what we see without ever making contact to confirm and that gives us external knowledge and insight. Our vision about that, which we see tell us that there is more than the animal's concept of a plain survival on Earth where it is that you can eat or you can be eaten. Fathers show their children the constellations and although we no longer attach religion to our stargazing it never subdued the bliss we find in our astonishment about stars.

The star that gives us our greatest wonder is a star we cannot see. Every one stands amazed at the

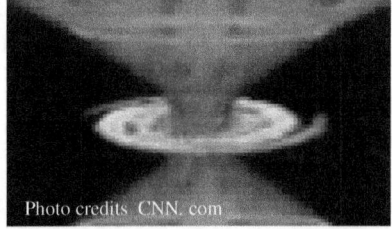

fact that there can be a thing such as a Black Hole. There is so much to ask and such a lot to wonder as to why and how and where and which…yet, we cannot see any that we interpret. We see but we cannot see and that makes us wonder what it is we cannot see and what it is we wish to see. The fact that our view is obscured by the fact that our view is obscured dramatizes our sensation of wonder many fold. That is human and that is why we are what we are and why we are in terms where we are.

It is part of the human concept to believe your eyes. Seeing the sand dunes on Mars is equivalent to seeing the sand dunes on Earth by means of the television media and could just as well be of the Sahara. The Sahara is a place we can go and visit should any of us wish to do so, but the dunes on Mars are another problem. Visiting and confirming what we then see is not that simple to accomplish. The Martian dunes are not only space away, which means I can cross space in time and visit. The dunes of Mars is not even space away but is time away. There is no way I would ever cross time to see for myself what there is to see. That is what is wrong with science, amongst others. Science is of the opinion we see space. We do not see space. We see time, but it is not time we see, it is the distortion of time that we see. The "further away" we look the more time we see. However it is not time we see. It is the distortion of time that we see. The further something is away, the more it is in the distance, the longer it will take the light coming from the object to reach us. That means the longer it takes light to reach us the more time is distorted to put distance between what we see and what there is to see. It is not space that we see but the distortion or the compromising of time. It is the time delay between here where we are and there where the object is that we see and we do not see the object or the space the object has or even the space between us and the object. We see the time delay there is between the object and us. We see what was there in time gone by, however, we do not see what is there and we see space for what space represents to the Universe. We see space as time delay, time slowed down. That is what space is, space is time delay. That concept urged me to go and look for the beginning of space and the beginning of time and the origins of the concept space-time. Please allow me to explain the beginning of space by measure of time delay. At the time of the Big Bang everything was small...not so...it was as big as it is today. If the Universe was the size of a neutron, then we had no size at all. One cannot compare apples with oranges and see bananas. The space we see is the distortion time has to separate points of comparison.

In order to understand what I am trying to say I have to use a picture that is most probably not a true event. What we think we see is space. It cannot be space that we see. In the forefront we see a line that is a result from a comet travelling. Then there are pixels indicating lesser star structures and some clear dots indicating stronger light spots, which would personify larger stars present in that direction. The rest we see is the black of night. If the Big Bang theory is correct and to my thinking there is no doubt about that, then not to long ago there was a lot less space between the objects than is the case at the present. The space was less. That cannot be the case because if the space was less it would then take the light much quicker to arrive at the spot we are at present. The light coming from what should be the comet is relatively quick in reaching my location while there may be some of the faint dots that have light travelling a considerable time to get to me. I presume the comet is closer. Looking at the image of a roving planet it shows a structure filling space at intervals. The space it fills is a constant because the space does not change in becoming bigger or smaller. However, the space it is moving trough appears to grant the roving planet another position every time it is photographed. It is in the terms of time that the answer is. It takes a different period to position and obtain the light coming from the different position where the object is located.

If the prime object were the space as it is in the case of the space serving to fill the roving planet, then no changes would come about to the space. It would take as long to fill the space between the object and where I as viewer am in position with travelling time. It does not because the motion that the light has to endure is shorter or longer by time duration. It is the space that is constant, yet the time to travel varies. It takes time to cross the space whereas the space holding the object remains filled at an even volume every time. In the case of the planet space filling with material by gravity the same space filled without changes. In the case of the dark space, that space is putting time at a different duration to reach the location I am in. The "further" the object is in distance from where I am, the more time it would take the light coming from the object to reach me. The object will appear smaller as the distance increases but I know the space the object holds is filling the same volume as it does when being close to me. In the case of the space filled, that space appears to change but that space is filling a volume at a constant. It is when the space in which the object moves increases the space the object holds then diminishes. The space that the light has to pass through to bring me the picture of the object increases. That cannot be because the space is filled all the time by the same margin. It is the time the light takes to bring me the picture that increase and it is that light that shows me a diminishing space. It is not the space between the object and my location that increases, but the time that increases and by allowing the time to increase I allow the space of the object to appear to become lesser. The space the object holds has to remain the same and the space between the object and me cannot change by motion. It is filled by volume that motion cannot change. Only time can be affected by motion and since it is motion that is changing it can only be time the motion can change. The slower the motion the longer wills the time be that it takes the motion to negotiate the space. That black of the night that I see is not space that I see but is time that I see and the space I think I see is the retarding or slowing of time that I see. Outer space is not space but it is time that space retards and therefore space is not space but a retarding or a distorting of time. That means that which see thinking it is space that we see is all the time, time that we see and being time it has no outside because time is eternal. Space, being infinite interrupts time to give time in eternity duration value.

The relevancies we are about to address are about form. It takes us into a Universe when a line had the same value as a half circle and as a triangle does. It takes us beyond space to a Universe when time formed space. It puts the Universe beyond distance. It is what came about when space interrupted time to deliver us the black of night, which we incorrectly think of, as space. The Universe did not start small it started outrageously big. It is not expanding it is reducing. When the Universe started there was no outside to that which started because if there is an outside then what is on the outside of that which started. The Universe has always been an inside that went smaller. The limits grew smaller not bigger. The initial start had no limits. That which we think of as so small and tiny, so small it has no sides is so big it cannot have sides because it is too big to have an outside and all we see and all we cannot see fill the inside.

Where we are now in the Universe we are so much smaller than what was when the Universe was the size of a neutron or whatever it was. If the Universe as one block without limits had no outer limits and was the size of a neutron then it grew smaller because what was our size when the Universe contained all it had in a neutron. When the Universe was a neutron we were not even a thought. It is easy to lose perspective but perspective is all we dare not lose. That which took al the space a neutron could offer back then has no limits now and has no boundaries. It is too big to be cooped up by limitations and boundaries. We with limits and boundaries now have measurable quantity to calculate, but what was the Universe then has no calculations art present. Where there is no boundary to shift what shifts then and yet they say the Universe is shifting its boundaries because the Universe is expanding therefore it shifts! Where no growth is possible since it captured the growth at the beginning where too can it grow. The end of such a shift by what cannot shift to where no shift is possible will eventuality be what they named The Big Crunch even before locating the Big Crunch. It is like naming a baby even long before knowing how the procreating is taking place that will lead to impregnating of some member of the specie (which member it will be is still then still unclear at the time the name giving was undertaken) where it later on will lead to conceiving the baby … that is the manner in which science dogma is enunciated but that is how clever those mathematicians are that knows everything there is to know on science. They can name a baby before even knowing what procreation is and that they do by calculating what they don't know anything about… like procreating the baby! It seems more likely that that which has no prominence finds prominence, which means the lot is shrinking. The Universe is surely shrinking to give us space to be.

When we altered the size of the moon in relation to the size Mars has what we did was change our relevance to that of Mars. We first brought Mars on a time line as close as it would be if it were hanging around in the space the moon has at present. Then we moved the time line back because it takes time to travel to the structure. Pushing Mars back does not increase the space, because eventually Mars fills the same space. It increases the time duration between Mars and us. It is not space we cover. If it were space then the time would be equal for light and for all to complete the journey in the same time. By changing the time the relevance change as to how long it would take to get there.

s NASA

When we look at the images of the two solar objects it is so easy to put them out of perspective and in the same size, although we know they are not the same size.

In cosmic reality the reality is quit substantially different. When we put our hand out we are able to touch…say the door we are immediately in contact with the door. It is the door we touch because it is the door we see we touch. Moving back one meter we find we are no longer able to stand upright and touch the door because we are one meter away from the door. We are one meter away because we can see we are one meter away from the door. We grew accustomed to this thought because Galileo's pendulum shows we are in time in space in the Earth timer in space. The time we will take to touch the door corresponds directly on Earth with the distance there is. Things change drastically when we leave the Earth or when we view object not confined to the Earth as we are. The truth is we are accustomed to think we are one meter away from the door we are unable to touch because we think we see the door is one meter away.

However that it is not the door we see. We cannot see the door because the door is not there for us to see. We see light banging on the door and as the light is rejected by the same door that the light comes flowing to us. We see the rejected light bringing an image of the door we cannot see. It is light we use and that we are used to of using to confirm what we see but such confirmation is what makes the most intellectual stumble. In quite the same manner we see the darkness of the night and observe such darkness as darkness. By darkness we interpret the meaning as that which we cannot see or that which we are unable to see. Reality tells me that the darkness is light that is too bright for us to see. Take an image of Mars with a close up view. Then reduce it and go on reducing it until it is so small it becomes invisible. The space filling darkness is not darkness filling space because the ratio of darkness increases as the ratio of light in comparison to the darkness reduces. The object does not go dark by moving back. It rather becomes more of the same when it blends with the darkness, which proves the darkness is not darkness but it is light.

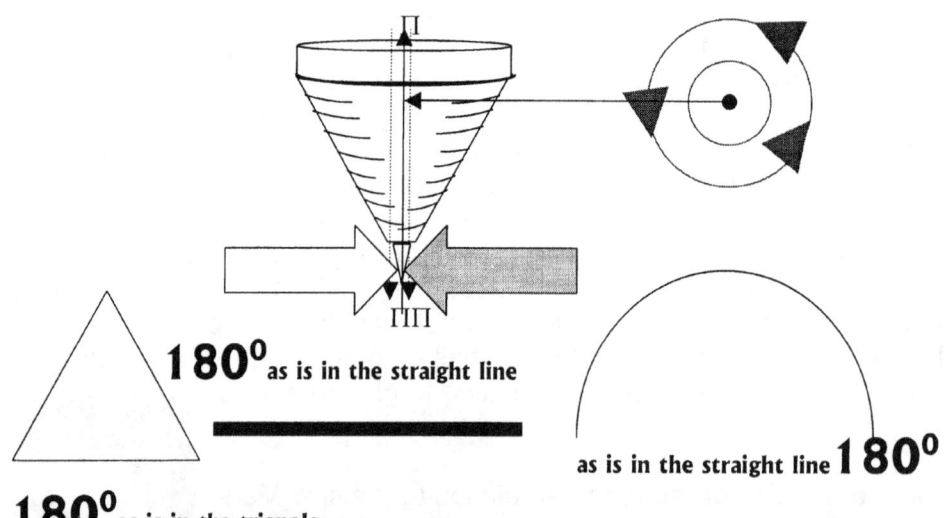

180^0 as is in the straight line

180^0 as is in the triangle

as is in the straight line 180^0

Einstein famously said there are matter, time and space and if I am not mistaken he said very little more. Is the space that Einstein would have reduced to a flat surface forming a flat Universe really space just because it is not time. Then on the other hand we must ask the question what is time then? What is our Super designer of space whirls really reducing when he reduces space to have the lot fit into not one but two Black Holes? What is matter and what is space? Looking at what the top tell us we have matter time and space a little confused. We find matter in time but also we find time in matter and that no one before realised! By reducing the space an object has the darkness becomes either more or less but the darkness promotes the object or reduces the object. The fact that large objects are close and small objects are at a distance we on Earth relate to more space and less space. The only factor that can produce more space and less space is time because time is irremovably connected to time. By reducing the share of the combination of space-time time must reduce or increase to allow space to do the opposite.

We have our focus square on the distance we find that part us from the object in our view. It is another culture thing from the time we hunted I suppose but we confuse distance with the time the distance really represent. By the end of the twentieth century every one has become so accustomed with photographs and lenses bring into focus and bringing objects close it became part of our breathing. We reduce the space in order to see the object better. Is that really what it is all about or is the issue significantly more complex.

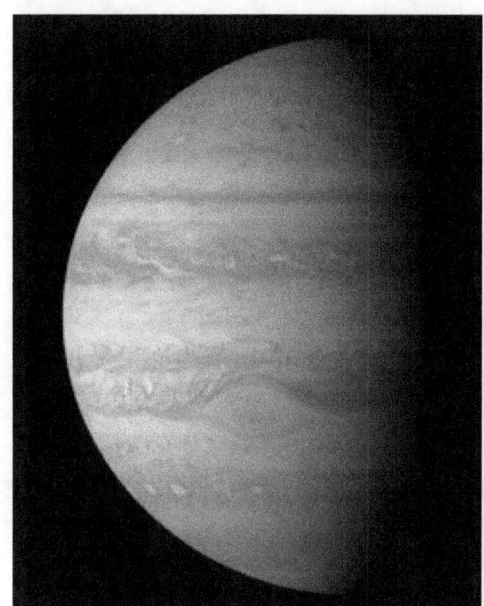

The pictures that you are looking at are probably the best example of space-time you will ever find. The term we use is where space stands (division) by time. That means in the context space is enlarged or reduce by the time factor that increases or that decreases time. The time divides the space (space / time or

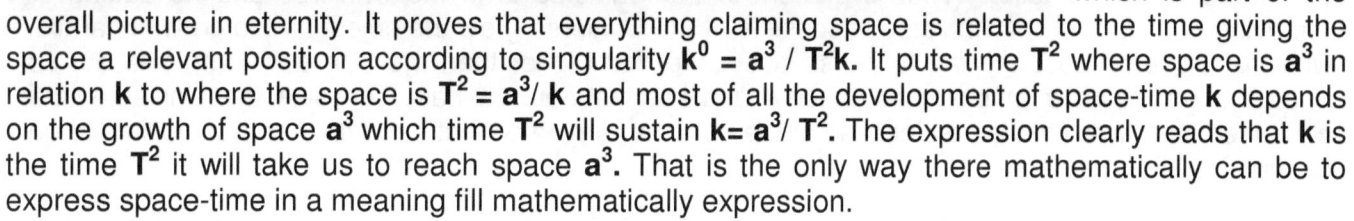

space $\div$ time) according to the time it takes to reach the space from the location where the viewer stands. The time factor divides the space factor into smaller sectors of time components, which is part of the overall picture in eternity. It proves that everything claiming space is related to the time giving the space a relevant position according to singularity $k^0 = a^3 / T^2k$. It puts time T^2 where space is a^3 in relation k to where the space is $T^2 = a^3/ k$ and most of all the development of space-time k depends on the growth of space a^3 which time T^2 will sustain $k = a^3/ T^2$. The expression clearly reads that k is the time T^2 it will take us to reach space a^3. That is the only way there mathematically can be to express space-time in a meaning fill mathematically expression.

The realising that only time can affect space by the measure of appearance is a huge step in the right direction. Space is a constant therefore time has to influence the appearance of space to become apparently more or apparently reduce to become less. Being big is a sure sign to the brain of an object being close. That would then appear as if there is little space between the observer and the object in observation. That is culture talking because space may appear larger or smaller but it can only be a medium of space that may allow space to appear. Space as such has the same measure and has the same prominence when measured. Time is the factor that allows space to reduce and even to reduce to the obscure.

Moving the object back into obscurity does not reduce the space the object has but puts the space the object has into a much larger definition in space in relation to the space I witness. The light streaming from the object will also fade into obscurity and disappear as the definition of the object declines in relation to the space it holds by comparison to all the other space in view. The light the object had did not decline or reduce but it diminished in relation to the gross of space holding light. In relation to the space out there the space diminished the light in relation to the darkness the light then offers. That way the light could only reduce by comparison if the light was less in relation to what the light is in the darkness we see. That means the darkness is flooding the light the object has and therefore the darkness we see is light. However, our relation to the light makes us in relation to little to be able to appreciate the light because as the object retracted from the position we had, we also diminished in space by the same measure. The space we hold therefore is too little to enable us to appreciate the darkness flooding us with light.

In the presence of this there is the Universe at $k^0 = a^3 / T^2 k$ which holds space -time in an equal proportion but is equal to space-time. However and this must be clear; it is not space-time! The mathematical statement puts the Universe equal to one and one is singularity by the dimension of zero. It is not zero but one and the one takes on the dimensional position of zero. The Universe is singularity that grows into space-time. It would be well advised to go on a search to find the Universe that is going flat before again declaring there is a Universe that is going flat in gravity and not knowing what universe is going flat with intensive gravity. The universe apparently is on the one side space-time but it is on the other side it is a flat singularity.

That is the easy part to figure. By moving the object back in relation to what we view is not diminishing the space the object has because the object will hold the same space it had before. The object is as big at present as it was at the Big Bang event because what was there was there with no adding. What is present in the Universe is in the Universe and no adding or removing of what is in the Universe is possible. If the Universe grew the object had to grow in parallel with the Universe because the Universe got somewhat bigger than the size a neutron has but so does the object have much more space that what the neutron has. The size the Universe had contained what was inside the Universe at the time the Universe went bang. In that there is little to no change possible.

As the cosmos present its evidence, we can see from such evidence how destructive overheating is. Forget pressure, because Newtonians over simplify everything with pressure and exploding. That might happen to a drum they fill with gunpowder but that is not applicable in the cosmos. In the cosmos, unlike in containers, there is no retaining wall that sets limits to pressure inside the container versus pressure levels outside the container. The cosmos has no pressure or pushing or pulling. It has a flow of space-time by concentrating time and duplicating space as it is driving space-time towards the centre. In any picture about any star there is no containing wall that keeps whatever is inside, inside. There is no limit to what the wall if the structure can contemplate before bursting. In the centre of a star is a point holding singularity and since such a point has no space and is immovable, space has to compromise by flowing towards such a location. We regard what we see at night as space and how wrong can we be?

Have you as you sit reading this part at this minute sat back and gave a thought about the light enabling you to read? Such a thought brings to mind the most simplistic answer one can imagine. The light hits the page bounces from the page and contact the lens of my eye where the lens conveys the photons becoming electricity to a part of the brain that translate the electricity to an understandable message and that makes one read. It is as simple as that! Ever gave a deeper thought about light streaming across the night sky, coming from ends of the Universe we do not even realise it is there? How does the photons manage to convey one complete picture coming from as far apart and as wide an area as it does? With a few photons connecting the eye or lens no one ever noticed the wonder of light. The photons reflect a view that seems as if coming from all the billions upon billions of stars. But most is coming from darkness covering an area no man can measure. Yet how many photons can actually connect to the lens of the camera or to the eye? Still a few photons coming from a single direction directly ahead eventually tell the entire storey. It is very simple to take the process of seeing by means of photon conducting very lightly and I have never heard one of the Brainy Bunch really in sincerity uncover the process to its utter and full potential. It is impossible that light from such an array of assorted sources can simply come together at the eye lens and show a picture of objects spanning across a Universe as wide as our mind can receive where the objects they reflect is beyond human measurement and the quantity is inconceivable many.

If scientists think of outer space as geodesic zero, with nothing in outer space but space then how do they explain the fact that we can see the nothing. How can we see nothing being in between light? According to official science that blackness out there represents geodesic space. Geodesic zero means the light travels in a straight line from where it originates unhindered all across space to where the light connects the eye. By crossing the vastness of space, the intensity of light reduces. The light coming from the Sun is quantifiably less than where the light is hitting the surface of mercury. The light loses intensity while travelling through the Blackness of outer space. If light was losing intensity the intensity of light must be that what is robbing the light that which is taking immeasurable small measured light away from the mainstream of light flow must be light. If that which was collecting the light were not by own measure light the light would have stood apart from that which was collecting the light without being light. The light cannot mix with the darkness and remain apart because it is not the nothing the darkness represents without standing in an identifiable visible support of what it remains to be. It can only mix, if it was the same that was mixing.

Isn't it rather reasonable to think that if the light were different from what it went through the light would leave a luminous trail as it went along since the light cannot mix with the conducting medium but still leave some part of the light behind as the price it pays for passage?

Think how wide your view span is. Think of the enormous room, space, volume size you pack into such a small space as your eye sockets are. Try to calculate at night what the volume is at that point that you can see simultaneously. Pack that lot into your eye sockets. See how many cubic light years you put into an area where only a few electrons can go. How do you manage that! Look at on dot in

the night sky and think about that one simple dot. Think how many electrons must be streaming from that spot. Every photon that reaches you must have left one atom that went fusing. As it travelled on route it went in circles for millions of years before it was able to leave the star. The photon did not increase in size, and it carried the information it obtained from the one point that went into fusion. At the outside of the star it did not grow bigger and at that point it represented the information of such a small area we have no means to calculate it even if it was possible to put that area it represented into your eye. As it came along the information it represented grew in stature, as it became a larger part of a smaller growing space. It started to tell about information it could never know anything about except on the condition that the photons fuse together as they travel on. At best the photon originated from an area smaller that the eye socket and it represents (say for arguments sake) one group of stars. How did that information scramble unless all the photons fuse as they come along? It can only be that light is singularity overblown and represented by photons over spanning it whole context. Light then is singularity k^0 in space a^3 in motion T^2 over the distance it came k. Light we see is therefore $k^0 = a^3 + T^2 k$ or the space it represents a^3 at the distance it travelled $T^2 k$ bringing in once again Kepler's formula of $a^3 = T^2 k$. **That is the story of one photon telling us about one light source that may be one star or one group of stars.**

 The picture gets even more extravagant when one think about that that one photon might represent an entire galactica with a combination of billions of stars grouped into one area.

That is not the only picture we are getting. Our picture contains much more than a few billion stars in one dot. We might see several such galactica in one area.

 **Nothing is all about not being and not "not seeing".**

We visually see two spots holding light that is in the overall mathematical picture quite close to each other. Why do we not see one spot in double vision? What is keeping the two dots apart? How do we put the entire picture in precise co-ordinates into a total three-dimensional unit that fits all into my eye socket? Surely in comparison there is an example of everything going to nothing. Try to put that measurement in reduction into a sensible and audible mathematical expressed configuration in order of simple understanding. Convert that mass into a comprehensible reduction. The cosmos is not about mathematics because though you Newtonians fool yourself in that you're in superb ability with your mathematical skills. The most obvious puts your mathematical skill to shame. When it cosmos down to true issues you and your mathematics can only degrade the cosmos.

The fact that one can see the night sky is a proven fact! The fact that it is there in every man that has the ability to see is the proven fact. The fact that your calculations fall short is the proven fact. The fact that in true issues of cosmic proportions your mathematics does not even cover the idea there is presented in the wider cosmos is a fact. The fact that by your mathematics you do not even have the capacity to think out the question, which is hardly a reason to understand the question that is hardly a reason to come to same answer proves how far your superb ness in mathematics leave your wonderful atheistic claims in shame.

Truth is that the two spots that are seemingly so close together is might be further apart than the entire Milky Way is wide. This means there is a lot of never explored between the obvious which makes the unobvious part very unobvious. I hope that makes sense in the way I wish for it to make sense.

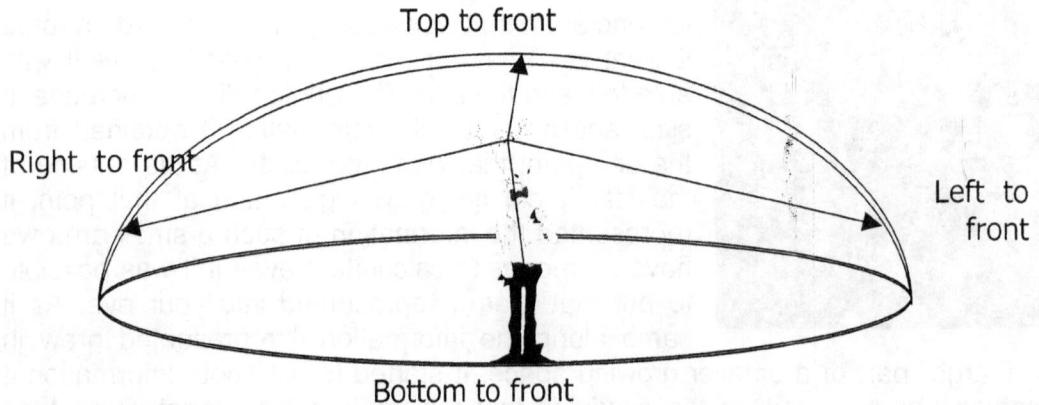

Top to front

Right to front

Left to front

Bottom to front

Go out into the desert and look at the night sky where the sky is unblemished of light pollution. The night sky seems three dimensions. All that is lost with man's artificial light degrading the unnatural cosmic light. See how wide an area runs from front to centre, top to centre, left to right to centre and calculate that what is in the eye range to an understatedly mathematical measured volumetric size. See how much " space" there are in the view. Then go home and start calculating so that it might enable science to find answers.

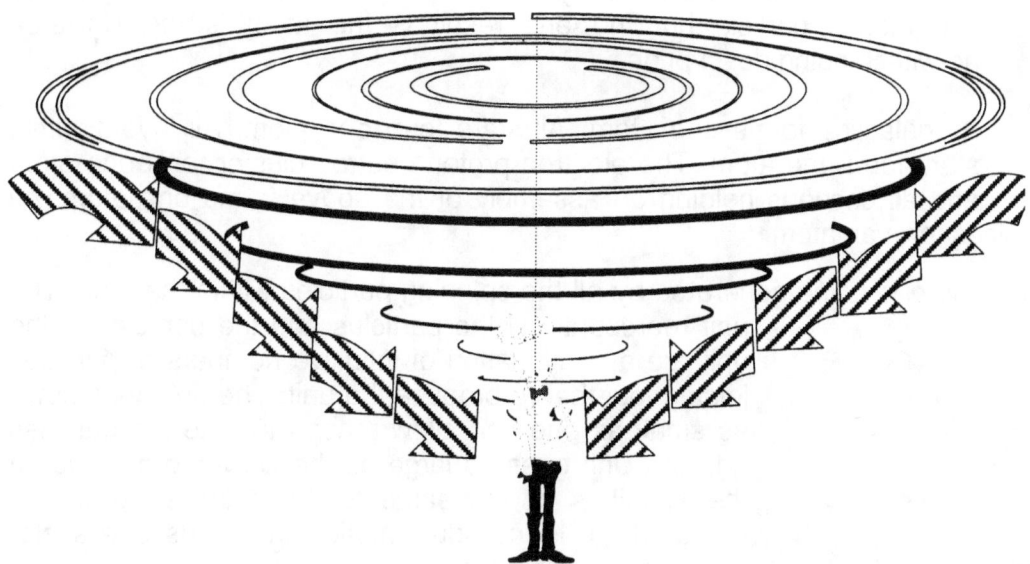

Then try to fit that calculation into the size of the eye. Reduce that lot to the size called the eye nerve and squeeze that which is visible in the desert night sky into a nerve fibre that would carry the information unto the brain. Let those mathematical Master Minds show precisely how that space goes into one eye.

If light came as individual streams of photon flurries our vision the concept would translates into proof that that as such shown in the fragmented picture above. It would be a picture unconnected bringing across some photons in the manner where every object stands apart not being related in any way and that will be what we see, if it is anything that we see. That we know is not the case but that means geodesic zero is as much rubbish as anything Scientists regard with simplicity and with careless thought. Geodesic zero means nothing and how can I see nothing as darkness because "nothing" is not darkness, nothing is "nothing" and the darkness I see is darkness showing the darkness as something.

 Such an idea by itself is outrages because the stream of photons reduce in space to such a minute quantity that taken the area the photons travel and the space in vastness it covers, the chances of one photon coming across many hundreds of light years through billions upon trillions of cubic kilometres of space and selecting my eye to convey the electricity is less than infinite. Yet such conveying takes place every second of every minute. The position of the location of the second singularity, which is the precise duplication of the first singularity but in a diminished capacity, is obvious to miss when one is not applying a detective mentality, as one should in scrutinizing the cosmos.

We may view two dots that would seem to be close to each other but seen through the lens when using a strong telescope the two dots are many degrees apart. The two dots might hold information across an area in space that covers something such as our Milky Way many times and that is done by each of those we are observing. As far as the size on the claim of space they contain go, we can calculate but the numbers such calculations arrive at is meaningless to our small minds. What is the difference between 10 light years across and 5×10^5 light years across? Now we get to the true sticky issue I am aiming to get to after concluding this approach run. Now we get to the question I wished to propose in the very beginning namely: What about the blackness in between. If a dot of one or two millimetres might represent\t an area covering a 100 light years and the two dots both being a millimetre across are some five millimetres apart, the five millimetres they are apart represents an immeasurable time span. That black stuff Newtonians conclude to be nothing. Let's dissect the "black" part first.

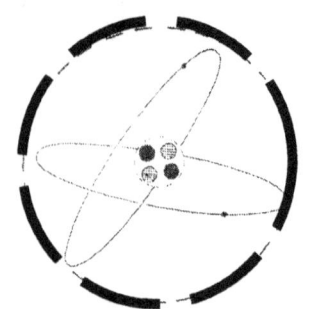

The atom in whatever form is a combination of what forms the Universe and in that to the extent that it is the Universe. The combinations may go as high as the most advanced galactica or as low as the most insignificant sub atomic particle but the end result is the atom is the Universe notwithstanding description.

All galactica forms one atom. All stars form one atom. All layers within stars form one atom. The electron proton neutron cluster form one atom and all cocoons holding an assembly of the above forms one atom. All subatomic particle groupings form an atom.

In the centre of any atom there is a line generated by all the spinning particles within that unit. The

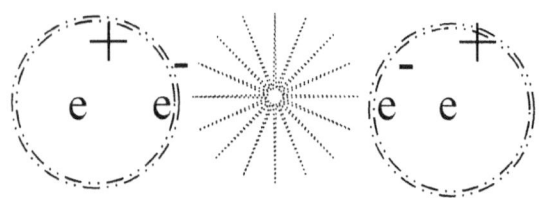

unit represents all the particles and the particles in the unit are a combination of infinitive numbers of particles joining together to produce the unit. The unit goes down as small as one can allow reason to take the particles and the Unit goes as large as the Universe at large. In the end it is all the same to the cosmos by cosmic standards. It is individual motion that parts one sector from another sector and in the end the parting is connected by infinity.

From what ever our abilities are that part of our abilities vested in the Universe will never reach or

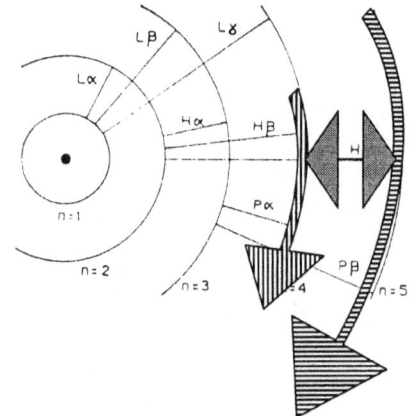

even almost nearly reach the end of the line. The atom is made up of energy but that is as general a term as time or space is. Never can any one accredit energy with an infinitive meaning but only with a vague description. By seeing what happens when an atom increases space or decreases space we can see what is it's final substance. By growing, the atom endorses heat and by reducing, the atom rejects heat. The atom is heat in more or less quantities. Newtonians use the term energy as a very convenient escape passage. The moment their explaining runs dry the term they grab onto is energy. What is heat…heat is energy. Every time I hear that phrase I feel as if can blow my top, and moreover using it so liberally seems to please every Newtonian. It is the same as saying that every one is willing to bluff the other one as long as the other one is prepared to be bluffed and then will bluff right back on the same terms. What is energy in the most infinite sense? Energy is heat pure and simple. Energy is heat where heat is time delayed and time delayed forms time in progress either by being space or time and mostly both.

To answer that we must be clear on what is cosmic within human realities and what is beyond cosmic in realities. At this point the atheist mind gets equal to the abilities of the animal and being an animal in mind by being an atheist I suppose the following will go far above their abilities of reasoning. The thoughts that I generate travels far beyond the speed of light and therefore my thoughts are not part of my cosmic material that I take charge of. I can move my body's parts with my mind's thoughts, which proves that my mind's thoughts control my body parts. I am not going to go into details about this but in another book with the title "Xepted Astronomical Mistakes" ISBN. 0-984410-1-4 I explain life in physics in extensive detail. Life control space –time by converting thought to electricity through the generation thereof. The electricity is not life. The electricity is only the result of what life can generate. The in electricity is command instructions that regulate and control space-time whether by moving things physical or by moving the physical body supplied to host and support life or by manipulating other life or objects life created / established / uses it all is terms and the term makes no difference. All motion we have come to accept as common practise is exclusive to our environment, which is a host to life. All other places are most hostile to life and only on Earth does life find a way to flourish. Every other place holding space is hostile to life to the point being there or putting life there will be futile to the life as part of the body holding life. There is a need for an acute and a deliberate turnabout in Newtonian standings points about life and cosmos motion. There is a need for a differentiation about what is cosmic motion and what is life inspired. However, to understand any and all motion is the most complex issue in nature. In order to understand the true concept behind motion we have to break the cosmic motion down to where motion initial start we have to return to moment – Alfa. There was a continuous motion where eternity met infinity before moment –Alfa. Eternity was spinning within infinity because we still have infinity and we still have eternity. They are tangible entities found everywhere throughout the Universe and are in control of all aspects of the Universe. Between eternity and infinity heat moved in and heat parted that which cannot part. This partition is presented as a time delay that later (at present time) became a time delay and all of the entire Universe is heat lagging behind the time which is in front and was pulling on time that was further behind. It is space-time parting infinity from eternity.

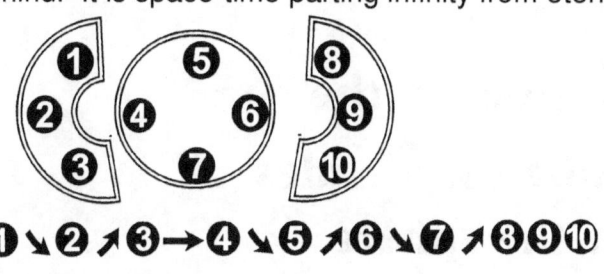

In the motion there are three factors filled by the same substance that is the same substance although the parting is also standing in for the substance. It is the same thing that is in time and time is in front of that which is flowed by that which is behind. The three is inseparably one.

There were four in the centre that wasn't there because the four shared a centre spot, which was the centre spot.

The four spots that represent eternity was spinning around one centre that represents infinity and because the centre was on the same position as where the four in eternity also was spinning time became a repeat of three circling around on where the four was sharing one location at any one time. The same one in the centre was also three spinning on the same spot.

The three was at the same time the three holding eternity while the three was spinning around a fourth centre spot.

Time came four parted positions from the position. it stood because was of the was in par and the

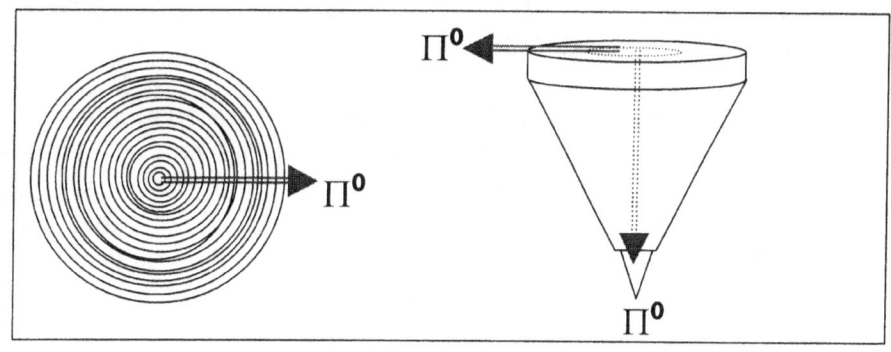

about when the by having three standing apart fourth centre The reason why apart is the one point allocated ahead centre, the other with the centre other went

behind the centre. A Universe was born because 1^0 moved from 1^1 to form $7 \times \Pi^0$ and that established Π. By motion of Π moved from Π^0 to Π establishing Π^2 that resulted in Π^3. The proof is still in every spinning top.

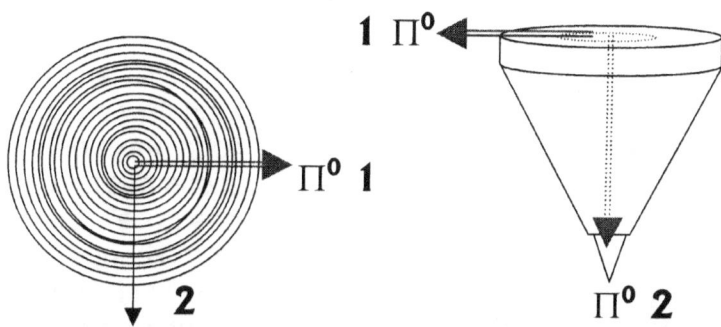

The universe came in place because infinity as 1 parted from eternity as three and because 1^0 that also were incorporating 1^1 as part of 1^0 then by motion became relevant to 1^1. A top can spin because the line forming singularity

parted from the three positions forming time and with motion space-time entered between the two factors. As motion becomes part of the top the outside, which defines eternity and is representing the eternity factor then through motion associated with the individuality of the top forming one edge of the top that parts infinity within the top from eternity outside the top. It is in the spinning top present for all to witness.

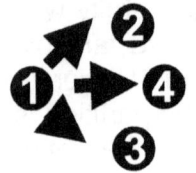

The center singularity point expands with heat accumulating and release space as motion to establish three points serving time in the past the present and the future. It is a relevancy that it brought about.

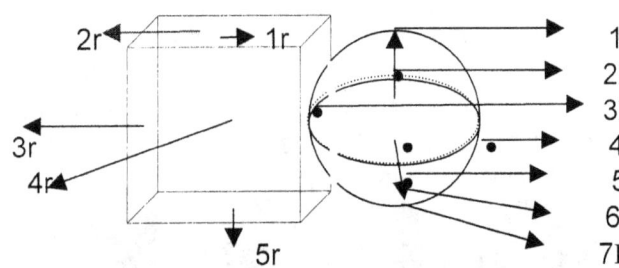

5 sides in the cube vs. 7 sides in the sphere

This is proven by the flow of space-time just as Kepler's calculations reflect. The space a^3 is a combination of motion in time that produces a dimensional quality of $7/10 \, \Pi^6/ 6 = 112$. It is a^3 that is a collection of motion (kT^2), which is an assembly of time ($k = \Pi^0$) and ($T^2 = 10^2$) which is then in the dimensional expression $\{a^1 = (\Pi^0 \, X10^2)\} + \{a^1 = (\Pi^2 \, X10)\} + \{a^1 = (\Pi^0 \, X10^2)\} = 298$. That is the space-time ratio T^2 / a^3 that Kepler introduced as the value of k. This is more a symbolic expression than a calculated mathematical statement but it does prove that space is time by three positions and the combination of space-time by three positions in three allocated setting is $7/10 \, \Pi^6/ 6 = 112$. It is the seven of space in relation to the ten representing the square of time (5+5) in relation to singularity holding three positions in space as well as three positions in time relating to the Universe we are in having six coordinates to fill.

It is the time factor in the relevancy in form relating to the space relevancy in form that produce the Titius Bode law of cosmic proportions. What this indicates is there is a movement Π^2 of Π^0 to Π^0, which is a movement of 1^1 to 1^0 that circles about 1^0.

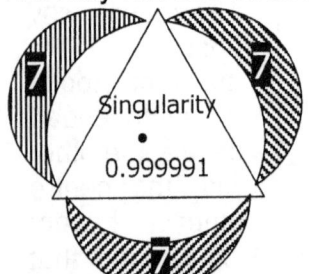

We may not see such a flow as a direction because all the direction we associate with space is part of the three sevens that becomes the three flowing with time. We see the flow of time coming from eternity towards infinity because the time is lagging in the side eternity holds and is catching the side infinity holds. What we see at night, the black stuff that we see that is holding all the bright dots in position is time in eternity. Eternity is reuniting with infinity and infinity is within every spinning particle within my body

seven circling points and time gone by of time and that The positions in time the flow of time and bring set time as a That centre has to be has to be time

In the relation of space against material there are always three positions of forming a triangle of time to come time in present rotating about a very specific centre. This is the flow indicator points to the direction of the flow of time. as well as the reference the positions make during the order the allocations of the various positions cosmic controlling centre. There has to be a centre. because of the motion surrounding the centre. There retarded that formed heat in between such centres in

relevancy. The nature of the spin promotes, as much contraction as expansion and the loss of expansion is the gain to contraction.

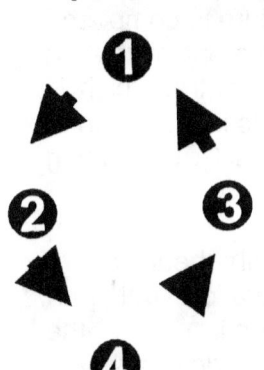

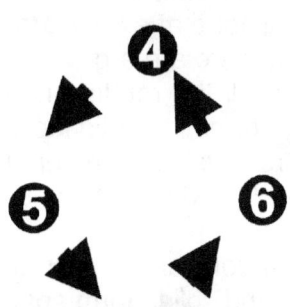

The spinning going on in the past is going on in the present, which is the involving the same points in the note that it is the space-time within circumstances that is generated and generated is different. That which is generating is identical, precisely the an exact clone copy.

the spinning same, spinning future. Please those particular what is doing the same duplicating

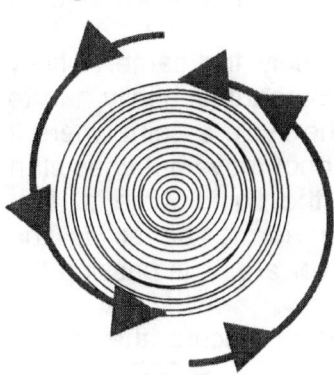

It is absolutely vital to understand that coming on, as time is the very same disappearing into singularity. It is where time in eternity that parted from time in

the three points three points

infinity is catching infinity to become unified once again. The unification is part of every spinning object there are and is the centre of the Universe.

Singularity by Time

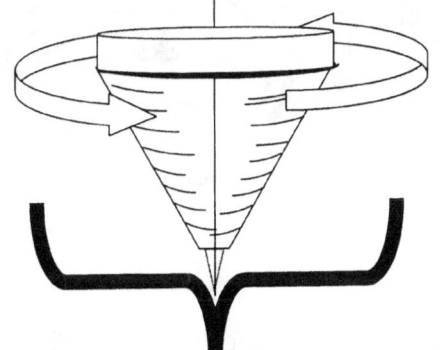

Looking at the top it seems that the body structure of the top is solid and the air surrounding the top is liquid. The top as a structure composes of solid particles that light cannot penetrate and that material cannot pass through. In that sense it seems to fit all the conditions we set for solidness. The top spins and it spins through the air that allows the top to spin seeing that the top has much more density than the air has.

What can move is liquid and stands related to what cannot move being singularity. Since everything is singularity everything is immovable but also since everything is

Every thing outside the top is liquid with the top forming a solid or so it seems to us. Well yes in a way and not that much either. The top is a pump that pumps heat from the outside inwards just like a turbine engine. Every atom that is rotating inside the structure of the top is

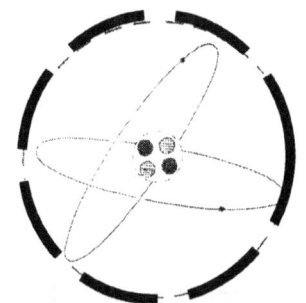

keeping the centre erect. The centre is totally motionless because all the atoms in the top are moving and the moving of the top circle is extending the singularity of the top to the edge where the top meets eternity. The extending of singularity is holding the air as a liquid and being the liquid the flow of the liquid keeps the top erect and spinning. The spin produces a cold in relation to the hot that the liquid is.

What is moving is liquid and what is not moving is solid. Everything has a reference in relation to another point. That which is capable of relocating is forming a liquid in relation to that which is securing the position of rotation. Everything in the cosmos can move and yet not one particle in the cosmos can move. The cosmos stands divided between the eternal moving of eternity and the immovability of infinity

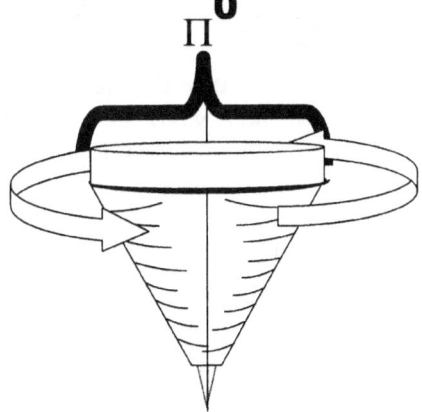

$$\Pi^0$$

Everything around the top is liquid with the centre being a solid. However the solidness and liquid has cosmic standards and just as it is in the case of hot and cold, big and small, fast and slow, our standards and cosmic standards do not share any measurements. So too does cosmic notions about liquid and solids have a totally different meaning in cosmic terms.

There is a pumping interaction of space-time flowing towards singularity through every point that confirms singularity.   Every thing in the top that forms the material is also liquid. By providing motion the matter in the top serves as the liquid factor that extends the space that singularity provide. The structure is composed of atoms. In the atom there are a governing generated singularity around which all material rotate. In the case of the atom all the rotating material forms the heat while the generated centre, which is incapable of rotating, forms the solid factor. Every aspect that is without motion stands in a relation of 1^0 and that which is relatively moving or changing location or find a new position holds 1^1. Everything that is standing still is 1^0 and everything that is moving is 1^1.

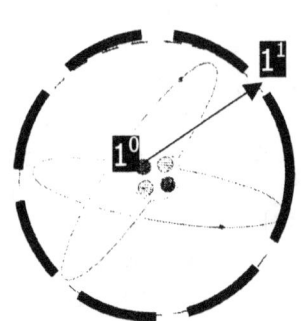

Gravity or motion is a constant relation that solids have with heat where heat forms the liquid and solids form space. There is the rotation but part of the rotation is the lateral progressing by rotation to confirm the generated centre. The generation is in the rotation but the flow towards is the lateral and just as electricity produce a flow of time in relation to space collapsing, space-time by measure of gravity is using the same system to do the very same

There is no substance difference between 1^0 and 1^1 and it is a relation where one moves as the liquid partner and the other is the solid factor.

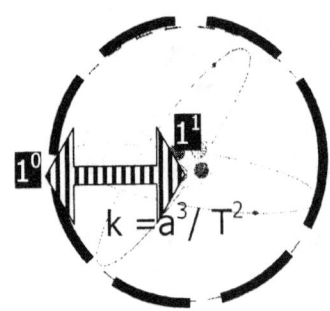

$$k = a^3 / T^2$$

Both are not as much equal as they are precisely the same. Infinity cannot move and eternity cannot stop moving. By parting infinity had to move and eternity had to introduce as part of the cycle a point where it stops moving in relation to the other side that cannot move but does start moving. The factor that shows motion forms the liquid while at that moment the factor that does not show motion forms the solid. The measure of 1^0 is transformed to 1^0 and which ever are 1^1 is passing the extending of space on to 1^0.

Time spin because everything spins in order to secure the centre singularity. But also time moves and in that there is the linear that always are part of cosmic motion. The centre is referred to by heat but heat also secure the centre by reconfirming the centre in the lateral. But in both cases singularity is reinstating singularity by confirming as it is referring one another. In the manner that 1^0 confirms a position in singularity 1^0 is supporting 1^0 by generating 1^0. By generating 10 it is repositioning and reallocating a position by confirming 1^1.

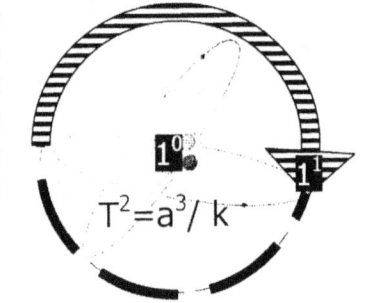

$$T^2 = a^3 / k$$

Coming down to understanding the concept in the infinite we must turn to light, which is the smallest material particle visible to the eye. Since the photon is small and fast and we are incapable of really managing an investigation into the photon, we must turn to the photon's more spectacular but far less frequent counterpart being what is referred to as "ball lightning" Ball lightning is heat or time liquefied as it is generating a point within the centre and such singularity is generating motion to concentrate time to heat or electricity or flames or whatever name one wishes to attach to the very same thing.

The solid producing the space is $T^2 = a^3/k$

The liquid in motion space is $T^{-2} = k / a^3$

The spin centres singularity (k^0) and the space (a^3) establish a singularity centre motion by spin (T^2) as well as by relevance (k). It is the Coanda effect where the liquefying of heat into a visible electric flame produces the limits of a space created by the motion of the heat being charged by a centre charging the centre again. It is again the manifestation of Kepler's space-time $\mathbf{a^3} = \mathbf{T^2 k}$, which is $\mathbf{k^0} = \mathbf{a^3} / \mathbf{T^2 k}$. It is time catching up with infinity and the result is eternity returning to infinity as the heat is once again recovered by the uniting of eternity and infinity. That is the photon on a smaller scale as well. That is all concept of space-time on a smaller scale than what light is. In the Universe size is not an issue but a change in relevancy. Since the ball lightning is larger the spin factor $\mathbf{T^2}$ takes most of the motion and that increases the space $\mathbf{a^3}$ factor to suit the situation. But that also decrease the relevancy or linear factor $\mathbf{k}$. In the case of light the spin factor $\mathbf{T^2}$ reduces the space factor $\mathbf{a^3}$ and that allows the relevance to match C. Light is what remained of the Big bang where space was an electron C^3 leaving gravity at C^2 and motion by relevance at C.

Locating and finding the presence of singularity

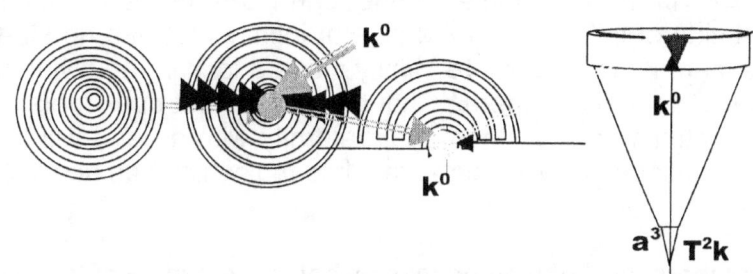

We stand in time holding space. Life takes charge of the material lent to life to support life's manipulation of space-time by another form of movement separated and apart from cosmic motion. By bringing about motion there is a discrepancy established. The discrepancy is by the changing of the alignment between the one position time holds in eternity relating to a point infinity holds and the next minute realigning the space point. It is realigning 1^0 with 1^1 by three locations in time. One must not look at the motion in the circle but at the motion of the circle as the relevancy reduces by duplicating that which rotates in ratio with that which contracts.

It is the rotation $\mathbf{T^2}$ that duplicates space $\mathbf{a^3} / \mathbf{k}$ by reducing the relevance $\mathbf{a^{3-1}}$ where the relevancy is $\mathbf{k^{-1}}$.

Since in truth the contraction is $\mathbf{T^2}$ but it would lead to great confusion if we state that time moves backwards because it cannot, we have to use the reducing of time as space reclining its value $\mathbf{a^3/k}$, which is precisely the same thing. In the end it is 1^0 relating to 1^1. The essence of moment –Alfa was that eternity parted by placing light between eternity and infinity. The essence of the atom is to remove light from eternity and place it in the atom to be united with infinity once more.

What is allocated to the position where 1^0 are in relation to 1^1 is space-time. The motion established between the two allocations of singularity is time delayed as it is returning time in delay to time in the moment. The moment is where infinity is encircled by heat in the atom. However eternity is also 1^0 just as much as infinity is 1^0 and eternity is also 1^1 just as much as infinity is 1^1. It is the same thing which heat parted when eternity moves away from infinity. Since eternity is three in time, which is the four in space and is the three in singularity, eternity is space, which flows by charging singularity. What is in eternity is not just like what is in space or just like what is in singularity. It is the same. It is a clone of the very same thing. The difference is not 1^0 or 1^1 but the motion of time putting 1^0 at a point to relate to itself at a point 1^1. The factor 1^0 fits into 1^1 in total harmony since the factor 1^0 is 1^1.

Because the factor 1^0 is the factor 1^1 and the factor 1^0 is precisely what the photon encircles. The photon encircles the electron centre because the electron is rushing the light through my eye nerve to my brain. The movement is taking 1^0 in the way of 1^1 to my brain, which holds 1^1 as a reflex of 1^0. Since what is in my brain are 1^0 is that which forms the expansion of time is also 1^0 but is 1^1 in relation because of a time constraint, I can see all 1^1 because I have 1^0 in my brain, which is 1 in time. I am as much part of eternity, which I see as I am infinity uniting eternity.

What I see is what I am is what is in the yonder of time and only by measure of time delay is there a differentiation between that which I am and that which I see that is flowing towards me. I am the end of time and therefore I am the centre of the Universe. The atom hosts the centre of the Universe and because the Universe started with the atom, the universe concludes through the atom. The atom is the gateway where eternity again once more reunites with infinity and in that the Universe arrives at a conclusion.

Because I am at the end of eternity where that which has no end starts, I represent that part of eternity that holds no start in relation to the part that has no end. I find a start in that what I can see as being the part in eternity that has a beginning. Being at the start of that which has no end I am unable to see the other part of eternity, which is the part that has no end. That is because there is no such a part in eternity as forming the side or the part in eternity that has an end. Eternity has a start through me but has no end. In the very same manner can I see the part of infinity that has no end because I represent that side of infinity while the side of infinity that has no start is also unseen by me. That too is because there is no side where infinity starts. I lock in the divide between infinity that has no start but uses me as an end and eternity that has no end but uses me as a start. When I and all of space –time are eventually remove then again eternity with no end will once more unite with infinity with no start and all space-time will disintegrate into that which has no start and neither has an end.

I can see what ever is out there in all of time because what is there in all of time is exactly what I hold in space less time. I am 1^0 and therefore all of 1^0 is also part of me because 1^0 has no sides and has no space, therefore all of 1^0 fits all into 1^1 and the whole Universe fits into my optic nerve with no squeezing required what so ever. I can fit into me what I am and since I am what time conclude I conclude eternity by uniting eternity with infinity. I am a black hole as much as a Black Hole is a black Hole. The difference between my being a Black Hole and the Black Hole being a black hole is that I still sport atoms and the Black Hole got rid of all time delay of any standing.

Culture will have us believe that when one sees a colour shining from an object the colour is associated with the object. Logic tells a different storey. A yellow dot is all the colours in the spectrum but yellow because it is disassociating with the yellow. That goes for red blue and all other colours we may visualise. I think the norm accepts this as scientific fact with very little argument or substantiating proof about that required. We have in our minds the formed concept of opposites applying. The opposing side of red is blue. The opposing side of white is black. If white is an array of all colours scrambled, then the forming of the colour black must be where all colours that are present in the scrambling of the white is also absent in the scramble mixture of black. If there was total absence of substance as far as colour goes, the heavens would be a few specks of white dots that is blending because outer space as "nothing: is parting them and therefore "nothing is parting" them. Going one step more in sane is that we then all agree the colour of the nothing is black since we can find no colour mixture would blend to form black.

What then about colours that are technically not colours as is the case with black and white? White is simple. By spinning all the colours in the spectrum the colour white shines through. Black is quite another matter. A friend of mine whom is one of the best painters I have ever come across told me that one couldn't paint black but have to make black a dark blue to show shade on the canvass. That apparently is his success in achieving the realism.

He also went on to explain how many variations of dark blue form the shadows in one simple tree. This remark set my mind in motion. One cannot see black because black has no colour to show, but black is the colour most prevalent in the universe. One can see only by colour and since black is not a colour we should not see black, but we do.

If the darkness was the representation of "nothing", then that should be exactly what we must see, nothing but the stars. Taken from the top picture some stars and leaving the rest to nothing is what we see in the picture below. A blind person sees nothing but when we look at space, we see something that we think nothing of as we see as space. One cannot have the ability of sight and see nothing except by closing your eyelids and then you see nothing. But in that case you do not see "nothing" in contrast of "something" you see "nothing" without it contrasting to "something".

By the ability to see the darkness such action in it self renders the darkness a factor of forming something other than nothing and that changes the acquired value of the darkness from nothing to something. There is an eternal difference between something in infinity and nothing. That black stuff…the use of the word black by itself makes it a contentious issue. That is even before we are using logic by disregarding the nothing part which crosses over what borders the ridiculous throwing the whole argument into the mindless because it is ridiculous to think of anyone is able in seeing something that is made up of nothing. Yet I stand alone against the might of Mainstream science no matter how correct my views are and notwithstanding how far their senselessness are going into madness, I still stand alone in my thinking as I am again and again rejected by the Brainy Bunch.

The arguments introduced touches the most basic aspects of my work and by no means can such an introduction secure an opinion that I do realise. Yet, not once through all my long investigation in the past thirty or more years have I found any other person claiming such views that I have brought about even in this skimpy way. If you see it, what you se must be light because you see it.

We view an object over a distance and see what the light brings across. It reads as $k = a^3 / T^2$. I see the space that time holds by time developed as k the time factor of relevance. Time at present T^2 is holding the space a^3 in relation to how the cosmos developed the relevance k. If I move time back by bending time with a lens $T^2 = k / a^3$ then space will increase as the "distance" or relevant position of the space increase. I challenge all Newtonians to increase or decrease nothing. I can see with the aid of light. Light has to come to my eye to allow me to see. If the dot is to small it will find not enough representation in the overall picture of visible light coming to me. I still can see, however that which I interpret as light I can see and to my mind that which I cannot interpret I interpret as if I cannot see it. I can see it but I am unable to interpret it. If I have to acknowledge that I can't interpret what it is that I am seeing I am as stupid as a grazing cow not knowing what she is looking at when she stands staring at what it is she does not focus on while grazing. No mathematician that concerns him as smart as to accomplish the calculations God used when God invented the cosmos will ever think of himself or herself as so stupid he or she doesn't even know what they are seeing. They know exactly what they are seeing... they are seeing nothing and to top that they are to stupid too realise that they should realise it is impossible to see nothing! By increasing the one part the other part has to decrease. But that doesn't mean they have to go less blind or blinder because they are seeing what they are seeing. It is only a matter of finding a balance in what they do interpret and what they do not interpret.

Looking at light coming from the Sun there is a distinct interpretation that the light coming on is bright and it is hot and it is hot because it is bright. Finding myself in the darkness I find the darkness is cold and the light is dark. In that way I think of the light being bright as hot. That is because in relation to the cold of the light coming my way I am hot, In relation to the darkness and the cold in the darkness I am cold. With the bright light is my relation in the light that brings about that I am hot. I am the relative hot party because I feel the light touch me and where the light touches me it exaggerates the cold, which I interpret as a hot spot. One can't feel the light but I can feel my condition at the point I sense the light and at that point I feel heat. Where I am within the darkness the darkness of outer space is not freezing me because I am freezing at the point I touch outer space. It is because I am so much

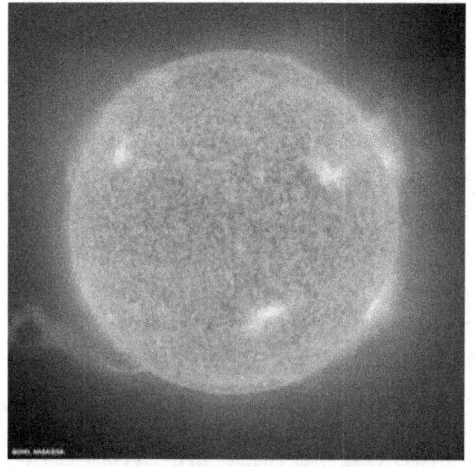

colder than the hotness of outer space that I am freezing instantly. Remember I am freezing and not outer space. Lets take a look at the Sun . Where the Sun meets outer space the Sun is boiling and not freezing over. By the cold Sunlight touching me I am able to generate a lot of heat within me as the heat has a route to where it can conduct. By generating the heat, which I then am able to conduct, I feel heated and generated. On the other end where I touch outer the region close to outer space, there is an abundance of heat outside me and with me being so utterly cold in comparing to the hot outside of outer space, I cannot conduct a flow of heat outwards and that stops me from generating heat inside. That smothers my generating ability and life is generating motion in a quest to manipulate space-time. If the part holding outer space was cold, the Sun would freeze at that point with ice everywhere.

However, where the Sun is in contact with the outer space region, the Sun is boiling. That means the temperature the Sun holds must hot up and star cooking to meet the required heat at the point of contacting outer space. Outer space is raising the temperature of the Sun and not dropping it. I am in relevance to the heat, which makes me either the hot party where the coldness of the Sun touches me and I feel my body raising the temperature of the heat to bring it to an expectable level to my body heat. At the point where my body meets outer space I burn black because at the point outer space ingenerates my cold body with its heat.

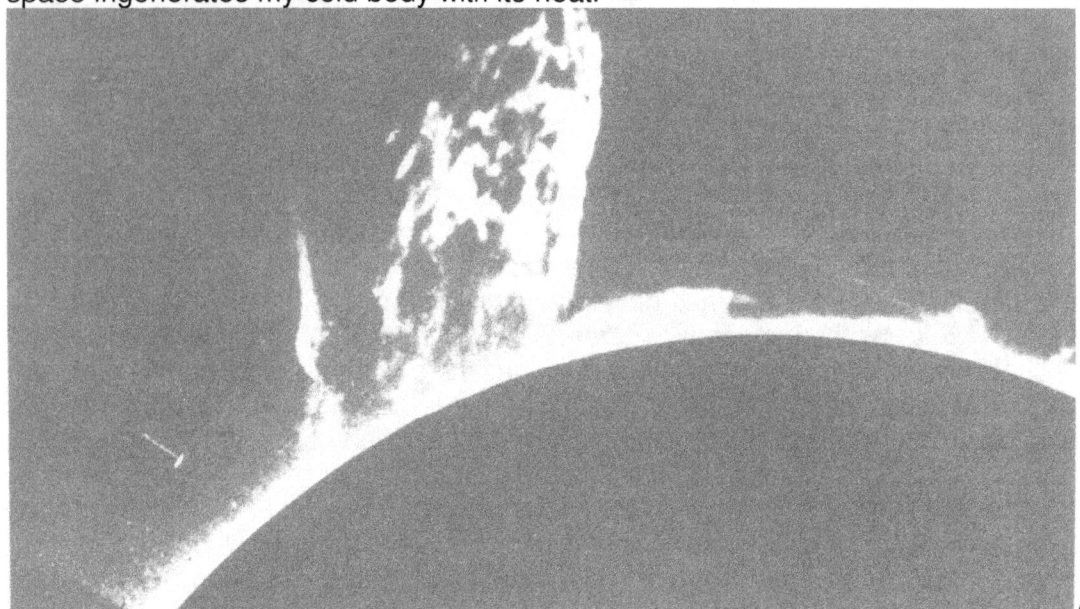

At the point where the Sun touches the heat of outer space the liquid heat of the Sun starts to boil. It is not like the heart transmitting we are familiar with by an element heating the water from within. The contact is on the outside and the heat at that point releases from outer space as it flows to the Sun . The Sun has no ability to realise the heat into the Sun and the she heat turns the pebbles into small crystals we know as photons. It is heat particles like it was in the Sun , but at that point where it touches the atmosphere it becomes concentrated liquid particles with a gas envelope leaving it to be photons. Within the Sun , the liquid photons are like water with a much more solid binding. This is because the Sun is that much colder than that outer space is. The temperature of the Sun rises at that point where outer space comes into contact with the liquid of the Sun . It is not the Sun that is loosing heat because then the Sun would freeze at that point. The Sun throws plumes into outer space because the heat entering from outer space distributes unevenly and at certain points the heat rises to higher levels than at other places on the surface of the Sun .

This can only be because of waves forming at that point just like the waves we see in the phenomena we call mirages. It is layers of different concentrations of heat. At those points it turns the cold substance within the Sun , the liquid in of the Sun into gas plumes. But even there the distribution is uneven and only where outer space truly touch the liquid can some liquid rise sufficiently to form a gas that will be absorbed by outer space. The rest that did not heat to a proper gas remain a liquid and drops back into the liquid Sun . If the plumes of heat forming the prominence carried the heat internally as the heat expanded outwards the plume would explode since all that lovely heat is suddenly exposed to cold and the plume would violently release the heat to the cold. It is not happening that way. The plume of liquid remains concentrated and only on the outer regions does it turn the liquid to a gas. That means the heat is not in the liquid but the heat is outside the liquid and it is turning the liquid to a gas from the outside inwards.

When we see dry ice boiling the boiling is in the fringes where the ice is in contact with the air. The ice does not heat from the centre because then the ice would explode from the centre outwards. The ice is stable on the inside but burns away at the fringes where there is contact with the much hotter atmosphere. The boiling would go on until the liquid ice has gone into a total form pf gas. But the process is on the wall of the ice and not from the inside and therefore the ice does not explode but turns to gas systematically. The Sun converts much more space to a frozen liquid than outer space can boil away the liquid interior of the Sun because with the exasperating movement of the Sun the motion freezes outer space by gravity, which is the freezing of time.

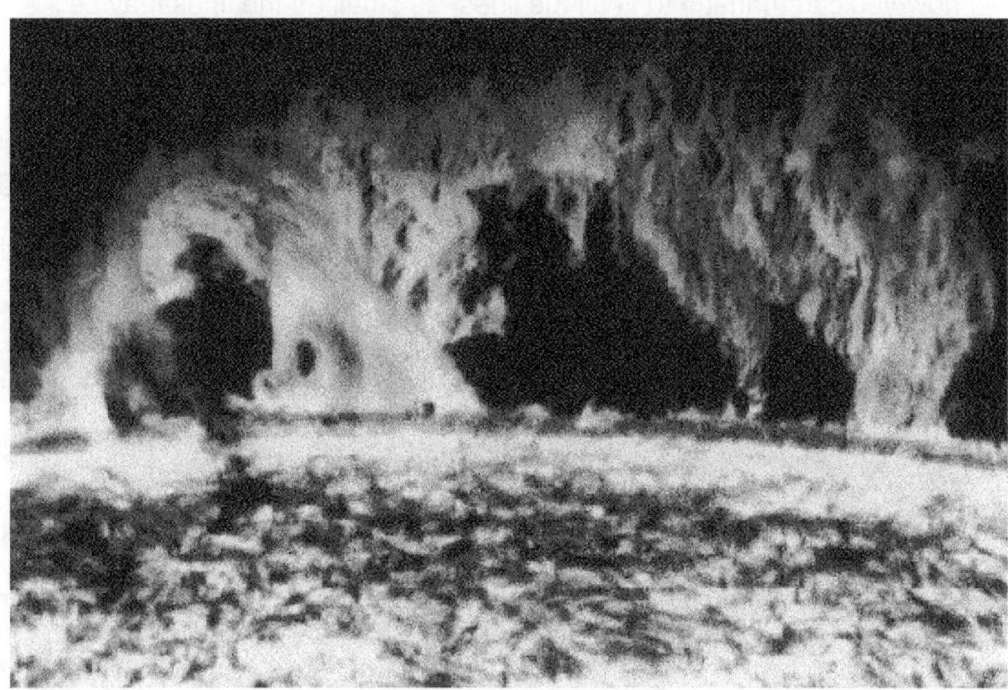

Look at the flow of the prominence. If the heat was in the prominence the prominence would disintegrate the moment it release from Sun because the heat would expand into gas. But since the prominence is cold because the Sun is cold that makes the edges of the prominence hot. When the edges of the flowing liquid turns hot, the relevancy to the inside turns the prominence even colder since the edges of the province is hot. In that case the prominence does not expand with heat but contracts and the density the prominence has increases. The prominence then falls back into the Sun because it contracts and its density would not allow it to remain out of the Sun.

If the heat were inside the prominence the prominence would boil away and not conserve the relative small amount of heat it has. Since the prominence plume is part of the contracting cold it falls back into the Sun , leaving the photon particles that became liquid gas prominence plumes to expand with the gas in outer space and allow the hot outer space to expand the light in all directions. The surface of the Sun is a pool of liquid. Where outer space touches the pool of liquid the heat wave formed by outer space breaks down the cold in the liquid if the Sun . The liquid breaks up into fragmented light particles and as the fragmented liquid light particles become a gas it flows through outer space and out. But by outer space also being a light with much more intensity than the liquid heat of the photon has, the photon puts the space it holds in relevancy 3^3 into motion of space $3\Pi^2$. And where the space is colder than the contact it makes the space forming the photon puts the heat difference between the space and the photon into motion exactly like a drop of water does when the drop lands on a red hot stove plate. The heat transforms to gravity and gravity is either expanding motion or contracting motion

Light is much more than the medium science takes it to be. Light connects the Universe in a way we cannot contemplate. Light being far apart originating from regions not in the same time or Universal space connects in a way that present us with a picture holding the Universe in an understandable content. From the point we stand and we watch the Universe the significance of what we see surpasses the sense of understanding of what we are experiencing. How can the few photons that our lenses catch coming from such an area as the night sky cover transmit the complete picture of what we see. Take a few seconds and study the picture of the night sky then rethink the picture applying the full content in the picture to what the size of you eyes is.

Think how big the picture is that your eyes take in and translate that area to the size of your eyeball in an effort to determine a ratio. One will be forgiven if one thinks of the ratio as eternal to nothing. Yet a few pages back I showed that according to mathematics there couldn't be anything as nothing.

Consider the path the light followed from the source connecting to light from all other sources where all particles of the other light may come from and bringing a full picture to the lens one use to look through. In your mind connect a line from every atom producing light and connect the lines to your eyeball and see how you can manage to fit all the lines, as small as the lines may be.

If I can bend something to increase the light by which I see, then it is light that I am bending. If I increase the light I use in magnitude in order to see, then I am increasing light. I am not decreasing the darkness by bending the light. I am bending the light because I am improving the focus I have in the light. By bending the darkness I am not squirting out more light, I am increasing a balance between that which I have a use for and that which I am unable to use. It doesn't mean that which I am unable to use is noting, unless I am a mindless animal that cannot realise that that what I cannot use, also exist in a way that I can use it if I had some intellect to do so. Newtonians are much better…they seem to realise that what they are unable to calculate cannot exist because the hell can be so stupid as to invent something that is outside the spectrum of their calculating abilities, after all, they are occupying the centre of the Universe from where they can see and calculate the lot.

It is so easy…if you can bend something more and afterwards see more in it must be a lens with a curve because we alter the curve. It you can read radio waves in the curve by bending the curve more with optic devises such as radio astronomy, then that which you bend is light. One can only reverse time as an optic elusion and bending something to gain an optic visibility improvement must be that there is an improving in the clarity of light. It is all part of the curvature of space-time since the curvature of space-time is the eternal sphere. We have to keep in mind that the first sphere was so small it was by today's standards not part of the cosmos because eternity parted from infinity. That came by measure of light or heat or whatever one wishes to call the substance the entire Universe is made of. By today's standards that, which was so small is so big the entire Universe fits into the parts that parted. Everything even what is so huge that we cannot see it still only is between the two points that was so eternally small.

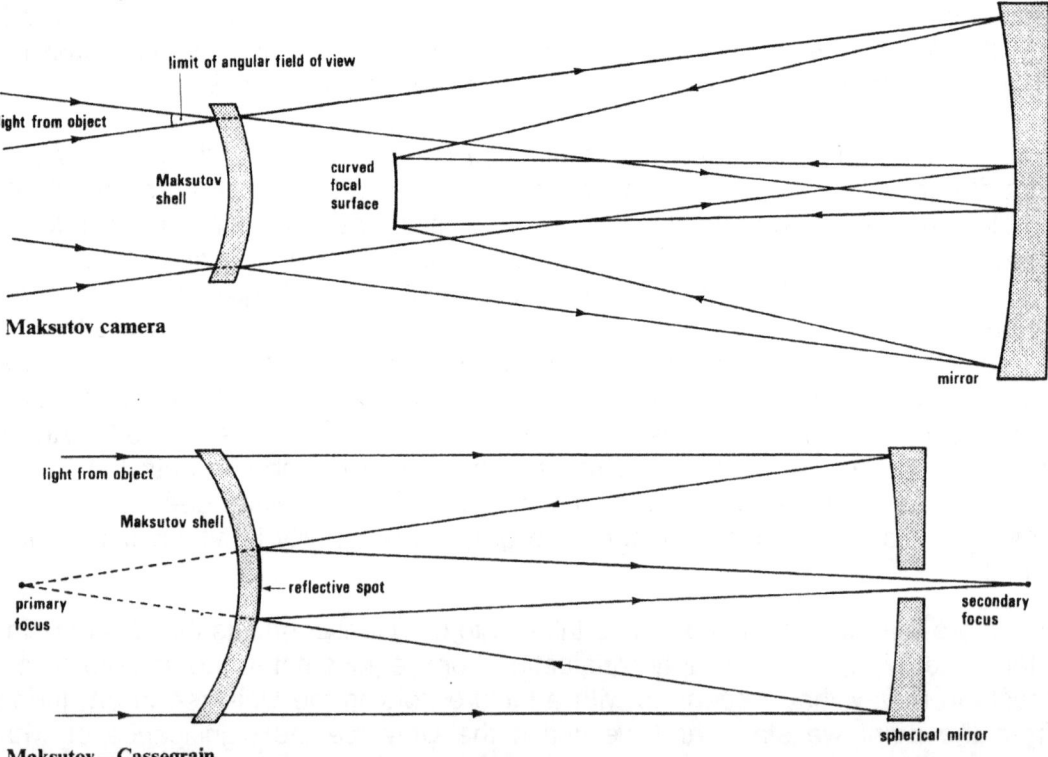

Maksutov camera

Maksutov—Cassegrain

If it is lenses that enable us to see what we can't see in outer space it also means we cannot see the light, which is outer space because we haven't got the lens to match the curb of outer space. Newtonians think of outer space as geodesic zero, with nothing in outer space but space. Geodesic zero means the light travels in a straight line from where it originates unhindered all across space to where the light connects the eye. Such an idea by itself is outrages because the stream of photons reduce in space to such a minute quantity that taken the area the photons travel and the space in vastness it covers, the chances of one photon coming across many hundreds of light years through billions upon trillions of cubic kilometres of space and selecting my eye to convey the electricity is less than infinite. Yet such conveying takes place every second of every minute.

The position of the location of the second singularity, which is the precise duplication of the first singularity but in a diminished capacity, is obvious to miss when one is not applying a detective mentality, as one should in scrutinizing the cosmos. Culture will have us believe that when one sees a colour shining from an object the colour is associated with the object. Logic tells a different storey. A yellow dot is all the colours in the spectrum but yellow because it is disassociating with the yellow. That goes for red blue and all other colours we may visualise. I think the norm accepts this as scientific fact with very little argument or substantiating proof about that required.

If light came as individual streams of photon flurries, then our visage would translate that as such shown in the fragmented picture above. It would be a picture unconnected bringing across some photons in the manner where every object stands apart not being related in any way and that will be what we see, if it is anything that we see. That we know is not the case but that means geodesic zero is as much rubbish as anything Newtonians regard with simplicity and with careless thought. Geodesic zero means nothing and how can I see nothing as darkness because "nothing" is not darkness, nothing is "nothing" and the darkness I see is darkness showing the darkness as something. What then about colours that are technically not colours as is the case with black and white? White is simple. By spinning all the colours in the spectrum the colour white shines through. Black is quite another matter. A friend of mine whom is one of the best painters I have ever come across told me that one couldn't paint black but have to make black a dark blue to show shade on the canvass. That apparently is his success in achieving the realism. He also went on to explain how many variations of dark blue form the shadows in one simple tree. This remark set my mind in motion. One cannot see black because black has no colour to show, but black is the colour most prevalent in the universe. One can see only by colour and since black is not a colour we should not see black, but we do.

It is quite obvious that any object sporting a specific colour is all the colours there are except the colour it is rejecting just because it is rejecting that colour If we see an object being yellow it is a human response to think the object is yellow Quite obviously it is not yellow because it rejects the yellow we associate the colour with. If that argument is true in respect to one set of circumstances it must be true through out. If the Sun rids it of heat it supposedly has the Sun must be cold because there is just no more heat left. If outer space absorbs all the heat on offer and shows no change that can be attributed to the collecting of heat, then outer space is as hot as it can get. If a yellow object is yellow on the outside it cannot also be yellow on the inside because where is all the yellow coming from?

If it was true about a yellow object not being yellow and a red object rejecting red and therefore not being red, the same must be true about dark and light. The bright object rejects all the light just because it is light from the outside. If it is light from the outside it has to be very dark inside. That too would count for what we believe it to be dark stars. Such dark stars must be most brilliantly lit because they keep all the light to their inside and well protected. They are dark because they keep the light on the inside where we can't see it and therefore out of our viewing range. The stars have gone so cold the stars have to conserve all heat to remain in gravity. With light being the highest concentrated form of heat, it stands to apparent reason that the light would be the energy of prime choice to contain. The same must then apply to outer space in that outer space is conserving all light and by keeping all light, outer space is brilliantly lit. We just are unable to witness the light because our position is such concentrated where as the light being dark is expanded to the full. We are able to see the galactica because the galactica represents highly concentrated light in one reduced area. The darkness contrasting the light we see as darkness because the light is expanded to the ultimate. The fact that we can see the darkness makes the darkness light, which we are unable to see. However with the space stretched to the maximum the lens we see light by has as far as our position goes, not even slightly curved because we are so small. Now you go and tell any mathematician in charge of theories this much and see how far you can get convincing him about your view. They wish to manufacture and design space whirls and not see reason.

The fact that we see light means that the dark next to the light cannot be "nothing", If the darkness was the representation of "nothing", then that should be exactly what we must see, nothing but the stars. Taken from the top picture some stars and leaving the rest to nothing is what we see in the picture below. A blind person sees nothing but when we look at space, we see something that we think nothing of as we see as space. One cannot have the ability of sight and see nothing. It is light that we see and it is light that we use, which enable us to see. That proves the darkness that we see in outer space is light that we see without recognising it as such. If the darkness was the representation of "nothing", then that should be exactly what we must see, nothing but the stars. Taken from the top picture some stars and leaving the rest to nothing is what we see in the picture below. A blind person sees nothing but when we look at space, we see something that we think nothing of as we see as space. One cannot have the ability of sight and see nothing. It is light that we see and it is light that we use, which enable us to see. That proves the darkness that we see in outer space is light that we see without recognising it as such.

What puts us humans in a category one higher than animals (or so we like to think) is our ability to think about that what we can see. The less develop an animal is the more it has the attitude of eat or be eaten. The higher developed animals are the more the animal find reason to argue. One may teach a crocodile not to eat you if you start feeding the animal. That is a mindless reptile and yet it can think above eat or be eaten. What we see is not merely the truth and it requires reasoning to see the truth and substantiate between culture motivated observations and thought through decisions. What we see is not the truth but it is what we think while we think of the truth about what we see that puts us as humans in a higher or superior dimension than the normal animal life can achieve.

It is the Coanda effect that produces gravity and it is the Coanda effect that is keeping the Universe together. Let us take the "Sound Barrier" from a point where we see phenomena apply laws that matter complies with. The sound barrier is a prime example of the relevancies, which I suggest, takes place. There are two points in singularity always referring to each other and one is expanding while the other is contracting. In nature there is the Roche limit placing a limit on the reduction of space and the inflow of heat to sustain proton cooling. At a point of $(\Pi/2)^2$ the reduction of space disallows any object to immediately compromise space claimed in time because the cosmic object cannot reduce space while an entry to its area demands such a time reducing in space claimed by the material.

The first question that one can ask is why would there be the value of $(\Pi/2)^2$ between orbiting structures positioning themselves in a time relation to space.

Every person knows about the entry restriction an orbiting spacecraft finds that forces the craft to comply with. The entry maximum is 21,991 and the minimum entry is 7. This is without doubt, the number of Π (21,99/7). The Earth holds its value to $4\Pi^2$ and when an object is not part of the surface of the Earth, even say a mountain; it becomes a holding value of 7. Later in this part I explain the sound barrier in more detail and the 7 will then become better understood. At this point one must see the Earth in the proton status of $(\Pi^2+\Pi^2)$ while acting as an atom. In this relation the atmosphere including all particles in the atmosphere will in relation be either Π^2 or Π. When water is in a vapour form, it will have a value of 3Π, having heat separating the water to the factor of 3. By dislodging the thunderbolt, the 3 receive a square value and displaces to the Earth in the linear light to time stance of 3^2. With heat (3) grouping by initial spin value, it will remove from space leaving the water to the value (Π to Π) and this will then give the water a relevancy of Π^2. The factor of Π^2 places the water no longer amongst heat as gas, but heat as a liquid (rain) or solid (hail). The relevancy of the water will change from 3Π to Π^2 placing the water's position from space (3) to liquid or solid (Π^2). Where does the Π that one find in the Roche limit $(\Pi/2)^2$ and the vapour (3Π) finds its relevancy to gravity? Every particle that enjoys space-time outside the Earth's structure ($\Pi^2 + \Pi^2$) will hold a neutron position of ($\Pi^2\Pi$). The Π^2 ends will be at the point where heat passes through the object directly to the Earth and this position of space-time relates to the neutron time link of Π^2. The space link of the neutron will then form the Π link. The value of the Π link we find to be $(\Pi/2)^2$, but the explaining to why it is $(\Pi/2)^2$ is rather more complicated.

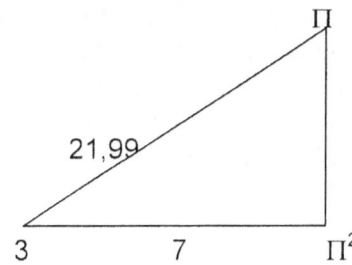

At the end of the space relevancy 3 where matter occupies space (21,9 / 7) is a border Π. That border is the exact point where space reforms to a square of time placing all matter (occupied heat) and heat (unoccupied matter) to a value of the square of time. That specific point is in relation to the square of the diminishing shield around the Earth. However it takes matter (R^3) from the 3 dimensional positions to the square (Π^2) in relevancy to time in singularity. With time holding space in singularity the 4 sides of Π truly relates to half of the total square value.

The "gravity" factor of Π^2 becomes one and only holds the square to the Π position as it holds space to singularity at a square (time dissolving space at a square) and the time value (Π) remains dimensionless in singularity at (1). It then is $(1)^2$ where the one becomes the space position ($\Pi/2$) representing time (1) at that point. This makes the position that time normally has Π^2 but directly links to the controlling singularity which we then give a value as Π^3 which then relates to the singularity position of space diminished from the three dimensional to times single dimension in the square. That makes the Roche limit hold the position of $(\Pi/2)^2$ when the neutron position of time (Π^2) links directly to singularity (1). This may only represent figures, something to accept through intellect but lying far outside the reality surrounding our everyday understanding.

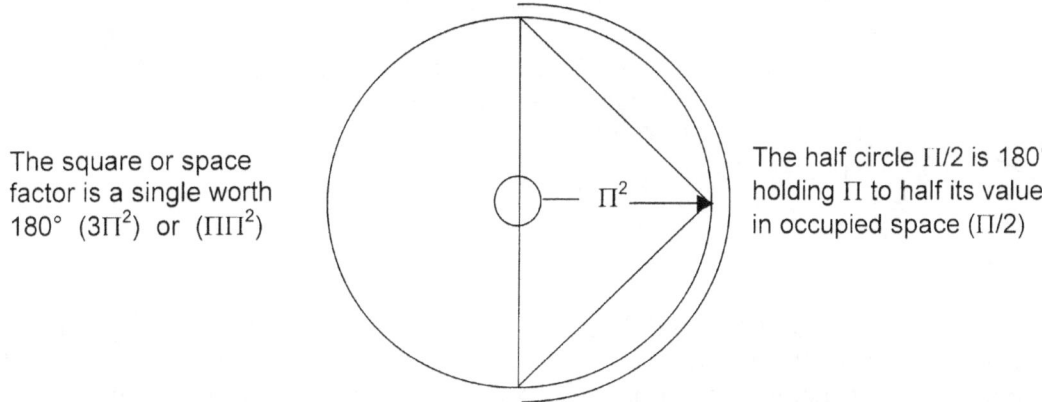

The square or space factor is a single worth 180° ($3\Pi^2$) or ($\Pi\Pi^2$)

The half circle $\Pi/2$ is 180° holding Π to half its value in occupied space ($\Pi/2$)

We all think of many reasons why birds fly and when the pioneers of flight started experimenting the main presumption was centred on the flapping of wings. It is a natural tendency to presume that flight comes about from the wings "pumping air" in order to raise the mass and I am sure but not certain that such a thought had presented itself at sometime in the Head of Newton. Once again that is as far from any truth as is possible.

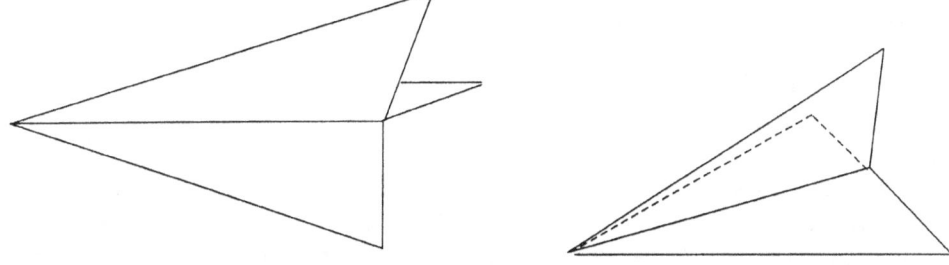

That is the shape needed to fly in the surrounding of the Earth that we presented with a name THE ATMOSPHERE. In outer space a flying object can hold any shape because any shape fits the three-dimensional. It could be round, oval, flat or any combination of all shapes and forms. The dimension is 6 with a linear value of 3 applying in one direction. This implicates the fact that wherever an object directs a direction in time revaluating positional change in space, it will forever be heading in three directions in the same time. The only control to direction is in the concentration of heat at a point opposing the direction of travel. By applying the release of heat, one is applying "a directional change". One is producing space that never existed before, because the release of heat will produce space. To the object in travel another dimension brings about influence to this three directional space application. This application of the heat brings about a fourth dimension that applies directly to time. I shall come back to this point duly.

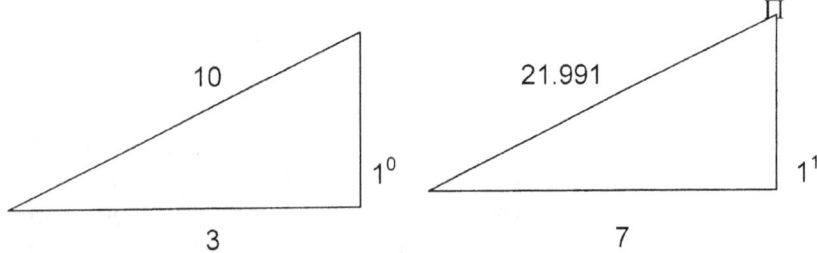

Without the application of specific heat, the object remains in the three directional moving of six possible directions. The value of space unoccupied therefore remains $\Pi \Pi^2$, as it was before the "Big Bang" event, whichever "Big Bang" you wish to refer to, because there were many. But space unoccupied holds time to the value of 10 to 1, and as also indicated, holds space to Π. Therefore unoccupied heat holds the relation to space in applying 3 directions of influence $(10)^3 = T^2$ $(R^3 = T^2)$. Always part of this equation is the dual function of space in $R^2 = T$ while at that very instant one have $R = T$. Therefore in space in time you have $10(10^2) = T^2$. Applying the fourth dimension does not bring about another part in the six dimensions, but actually cancel the influence of the one dimension by favouring a specific dimension. Bringing about the fourth dimension will lead to halving one dimension because of favouring the direct opposing side. Then the equation of influence becomes $(10/2)$ (10^2) and this means the implication of any heat, be it heat, be it light, will apply under the factor of $5(10^2) \Pi$.

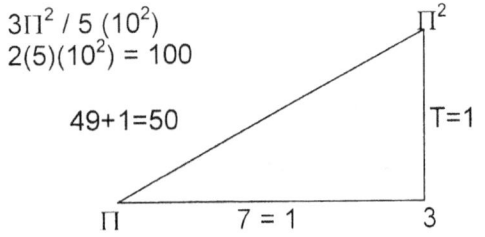

$3\Pi^2 / 5 (10^2)$
$2(5)(10^2) = 100$

$49+1=50$

$50 = 5 (10^2)$ where the complete   is
Pythagoras be comes a factor
$\sqrt{100} = 10$ the value of space.

Let's run through the sketch once more after the deliberation of the previous few pages. The fact of this comes as 49 plus one becomes 50 and that is in the three dimensions of space $R^2/7$ where 7 holds the relation to one and $R/7$ again where 7 relates to one. At this point it is most important to remember that Pythagoras works on the application of the sum of the square of the two sides. When seven has a direction in the fourth dimension applied to it, the opposing dimension will be one and this applies in time relevancy, therefore the interchanging in time between infinity will place matter at $7^2 \times 1$ relating to circular and $7^2/1$ with $7^2 \times 1 = 49$ plus one (singularity) always being a factor of one. Space in time however, never can be a cube, it will always be a square with one side pointing the direction of time from time to the past (1) to time to the present (1) to time to the future (1). This means with space in singularity in relation to time in singularity there will always be $R^3 = 1^3 / T^2 = 1^2$.

I am the first to admit that the calculations introduce with the atomic factor relation on the Universe has little mathematical value other than indicating some relevancy and some predictability. But you cannot use it in precise calculations because the heat on the day of the day will apply different measuring standards to what science perceive as constants. A kilometre will be slightly longer or shorter an try to tell a pilot travelling at two thousand five hundred kilometres an hour a few meter nearer or further has no importance! For instance launching an object into the outer space will require an atomic relevancy of $4\Pi^2(7(3(\Pi^2) (\Pi^2/2) (\Pi^2) / 3600 = 11,2$ km per second. Should you wish to make it an applicable factor, one will have to multiply the second object's proton relation with 10 making the total formula stand at $\$T = 4\Pi^2(7(3\Pi^2) (\Pi^2/2)(10\Pi^2)/3600 = 112$, which will indicate the cosmic factor value of independence, the same value that applies to Newton's $F = G (M_1 + M_2)/r^2$. By using $\$T = 4\Pi^2(7(3\Pi^2) (\Pi^2/2)(10\Pi^2)/3600$ you cannot put a rocket into orbit that is true. However you can establish an indication to the position in the cosmic space and time holding matter. This is not an absolute accurate science, but it helps with indication of relativities in the cosmos. When dissecting such formulas and the factor that arrive from it, one can judge the conditions applying.

In this case:

$4\Pi^2$ one proton holds absolute cosmic superiority

7   the spherical value influencing the formula

$3\Pi^2$ the two objects will relate ($\Pi^2 + \Pi^2$) where the one holds a proton position and (Π^2) where the other holds a neutron to time position.

$\Pi^2/2$ the minor object will be totally dependant on the "gravity" of the major component.

$10\Pi^2$ the final position of the minor structure (Π^2) as an independent cosmic (10) object in space in time.

3600   braking down the kilometres / how we work in to kilometres / second. This I use to evaluate speed, although I must add we adapted the value of time and space to apply to degrees (360° the circle of the Earth) and kilometres, being the same as $1m^3$ that rolls into a straight line, where the measurements apply both ways and in that manner it brings confusion to something we regard as proven.

By using the factors one can establish the planets and their moons or satellites in the position they hold, as well as how they came to be there. By applying it as an indicator helps to establish knowledge that was previously not there to help. But let me assure the Newtonian Order of High Priesthood, for all their precise calculations, they are far worse off when looking back at the nonsense they came up with. Man has NO MEANS of precise calculations as far as the cosmos goes for ONE MUST NOT VIEW THE HEAVENS IN AN EARTHLY MANNER.

The "mass" of the proton is easily identifiable, and so is the "mass" of the electron. Separating the two known factors, is the neutron that proves to be rather un-defined. This indicates that the balance is within the neutron, determining the electron value. I say this, because it is not the amount of electrons that determines space, but the amount of protons in the specific space.

It is the value of matter, be it surrounded by heat solid, liquid or gas, that differentiate between time in space and space-time. Without applying heat release the space in time would be $10(10^2)$, which is the circle we think we see space to hold. This is where astronomers make their biggest judgment error, they look at space thinking they are viewing the cube of space, but they are in fact viewing the half of the base times the square of the cube $(10/2)$ (10^2). What they think they see is 1 000 instead they are seeing 500, but as I will later explain, they are not viewing 5 (10^2) they are not viewing light setting the time factor $3\Pi^2$ but they are in view of light in space (Π) being in time ($\Pi^2=1$) the connection to space is $\Pi/500$. That means whatever the radius between the objects the light reaching it will have a comparable velocity of the distance of travel divided by 500. This will be discussed later in more detail as well. From this there is a measured point where the dimension of space in heat forming gas ends and space in "liquid" forming the neutron begins. The dimensional time application in space diminishing starts at that point. We think of it as the Earth's atmosphere. This no person, no force, no money can recreate. I have read articles about some Newtonian Wizard that plans on building some ship where this ship will "create artificial gravity by centrifugal spin."

Again and again Newtonians show just how little they truly know science. The depletion of space through spin increases, the Earth brings on acting as the nucleus of an overgrown atom. The point Π indicates where 3 lose one side of its dimension becoming Π. In that space all objects are at the mercy, not of a FORCE called GRAVITY but in a six-sided three-dimensional container of matter not relating directly to one specific point of singularity. All cosmic objects are cosmic atoms and all cosmic structures hold claim, control and influence on all surrounding space in its immediate surroundings as well as the cluster effort of the group – unit stretching as far and wide as the observer wish to take the relevancy. However in the Universe as it stands there is no collective standard all including and all counting general gravitational constant that has an equal value and measurement every where and wherever Newtonians may go Newtonians will encounter this constant. At that point a dimensional change take place where the Universe sacrifices one dimension from 3 Π. In that space objects which wishes to fly, need a replacement for the dimension sacrificed.

The aircraft needs wings to fly. Hot air balloons and the Zeppelin Hydrogen flying cylinder use the natural tendency that hydrogen provides by enlisting concentrating heat to maintain at the value of Π and not reducing to Π^2. Aircraft of any kind, that wants to maintain a vertical flight would have to apply three directional wings for stabilizing because of the loss of one dimension. There may be the possibility that with enough application of heat at various points, the introduction of the fourth dimension (heat application allowing space creation) may even bring about that stabilizing may occur in such a manner. Although the prospect seems light, however with no research to widen the view on the possibilities available, it is hard to tell.

What is overall important though is that the point the Roche limit indicates, being $(\Pi/2)^2$, the dimensional change takes place. Any object holding space-time in own value, will either reduce space occupation, or enlist time duration enhancing, or most probably a little of both as it will cover itself with heat. This is the result form the atoms on the outer edge of the atom, repositioning space-time occupation where the normal application will be in the dimension of the Titius Bode law developed to a certain degree. This change will bring about that the atom will revalue its position to incorporate the Roche limit in protecting its own space it holds in the time of that specific space. The value of the space-time of the atom of the aircraft therefore will reduce its electron position, as it locks itself in a position under the cover of the heat shield. The point of the Roche limit is as much part of the Macro Cosmos as it is of the Micro Cosmos.

At this point of $(\Pi/2)^2$ the reduction of space through positive space-time displacement will start. Any other cosmic solid that will not comply with the flow of heat to the object will remain outside that area. This means in more suitable language, that the second object will block the flow of heat to the first object (A). To object A everything outside its proton sphere of $(\Pi^2+\Pi^2)$ must either be Π^2 or Π. If the object is in a neutron time position it will hold the value of Π^2 and if it is in a space position it will hold a value of Π.

Should the object B maintain a density where it will become a singular value to the proton value of object (A) in as much as maintaining its own proton value of $(\Pi^2+\Pi^2)$, it will cause overheating in the proton heat flow of object A. The demand in heat supply of object A will either remove the heat object B needs for cooling, thus increasing object B's chances of overheating, or it will start overheating itself. By overheating the space object A occupies, will increase because the overheating will demand more space occupied. By increasing object A's space-time occupation, the demand on heat will increase as the point of space reduction shifts further away. The increase to the value of the $4\Pi^2$ factor, will increase the $(\Pi/2)^2$ linear factor.

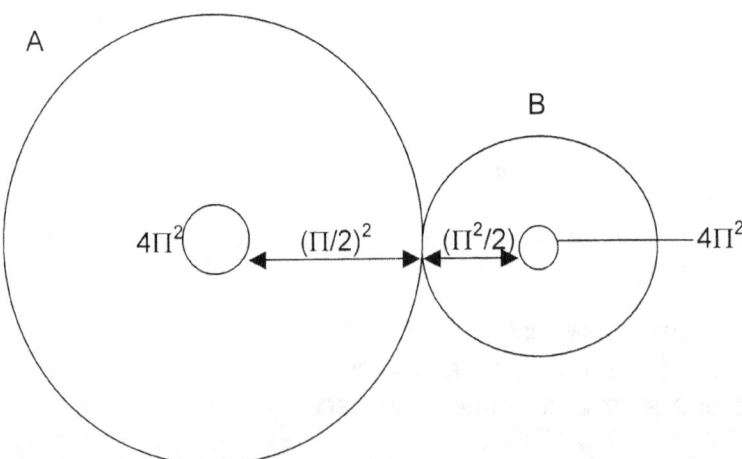

As the demand for heat rises, the competition for heat sustaining will also rise in the occupied space of both structures' time zones. That is the Roche limit and that is the purpose of the Roche limit. At a point where the space reduction and demand on heat supply starts in all earnest $(\Pi/2)^2$ seen from the view of the object in question $(4\Pi^2)$ it will either reduce the other object to a suitable value of (Π^2) or Π, but it will not allow any object to compete with it's demand $(\Pi^2+\Pi^2)$ on the supply of heat. It will reduce the competing object's value of $(\Pi^2+\Pi^2)$ as it relates to its own value down to a mere manageable (Π^2) or (Π).

The reducing method object A will provide, is removing as much heat flow from object B in order to get object B to overheat. Thus object B will increase its space-time value from a solid ($\Pi^2+\Pi^2$) to that of a jelly (Π^2) or even better still to a liquid (Π) by removing all gas (3) from object B in its demand for heat. If object B can comply with its own demand for heat, it will remain in space of it's own, providing it's own time within the cosmos ($\Pi^2+\Pi^2$).

Therefore one can observe that the Lagrangian point of S_1 is in fact the electron position because the two objects hold the status of a compound, sharing space-time and heat flow. In other words, the two objects become two super sized atoms forming a compound in space.

By the same measure a Lagrangian system becomes an enormous five-electron atom with all the cosmic objects working in conjunction as a unit with one mean goal and that is self-preservation through group preservation. This we shall come to later.

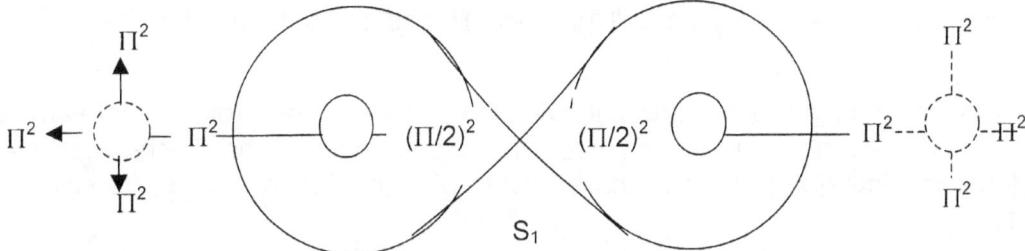

If the superior object does not take the minor object as a threat blocking its supply of heat flow, it will consider the minor object as a neutron that can help it grow. Even if the two objects maintain a relation of ($\Pi^2+\Pi^2$) it will hold the other object to a value of ($\Pi^2\Pi$) and outside the combined effort they will jointly relate to the rest of the Universe as 3. Therefore seen from the two separate views there is the following atomic order:

$(\Pi^2+\Pi^2)$ $(\Pi^2\Pi)$ S_1 $(\Pi^2\Pi^2)$ $(\Pi^2+\Pi^2)$
 Proton   Neutron   ↓   Neutron   Proton
                         Electron

But placing the two in a battle for survival the relation becomes super straightened where the one will not give any way to the other object's demand for gravity then the relation becomes that of

$(\Pi^2+\Pi^2)$ $(\Pi^2/2)^2$ $(\Pi/2)^2$ $(\Pi^2+\Pi^2)$
 Proton

                  Space-time establishing

At present that value supersedes the space value of 10Π completely, so it holds a relative position far above the geodesic value (10Π). That is not all, because the one object acts as a neutron to the other object, both claim a proton flow of heat from space, holding a time of $(48,7 + 48,7) = 97,4$. That places the active space-time development with the combined structure to an atomic era that was relevant just after the "Big Bang". Heat became a gas at the atom value of $7/10(\Pi^6) / (6 \times 10)$ and darkness parted from light at the atomic value of $3^2 (\Pi^2+\Pi^2) = 98$. As you can see, the two objects push each other back in space to a time that applied three eras ago.

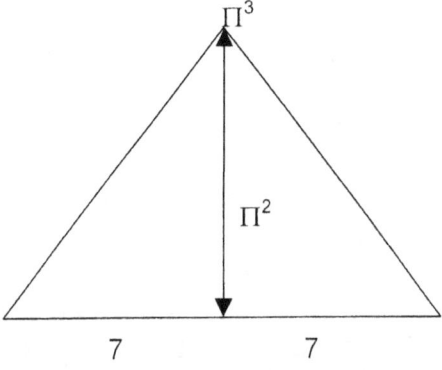

7 are matter and 7 are matter.

They join space-time, therefore the matter factor is the same. This is where one can visually see the one object, filling the space of the other object's atmosphere.

$$7 \times 7 = 7^2 = 49$$

That is matter Π^2 (time) times matter (49) = 483,61.

This will bring about space-time development that applied when the Universe was fresh and new, hot and lively and full of spinning power. As the objects cannot destroy one another, there is the Hubble Constant affect that came into place. Each structure had to build a Π value, to maintain and secure its relevant position until the end of the final eternity arrives. Said with using other words, each object will secure its place in future as a Black Hole or Proton Star. By producing a value of Π separating the Π^2 from three and thus ensuring itself an own atmosphere it will have to revert to space-time development that even preceded the "Big Bang". I will produce once again the value of Π from its position of Π^2 (time) and matter. Matter is seven and time is Π^2. Both hold claim to the same space, therefore the development of matter and time will produce Π and through that the Titius Bode Principle.

As this is all under the law of Pythagoras the law will evidently place a square root to that value of 483,61 and therefore $\sqrt{483,61}$ = 21,991. This leaves the space value of the Roche-limit, as it develops into the Titius Bode law giving them a shared value of 7 (matter) and 21,91 (space) the value of 21,991 / 7 = Π.

Then the relation becomes

$(\Pi^2+\Pi^2)$ $(\Pi^2\Pi)$ $(\Pi\Pi^2)$ $(\Pi^2\Pi^3)$ holding space (3) still outside. They therefore will share space and that sharing will continue till times end. We know by now that matter is 7, and space is 3. holding time to a relevancy in singularity of 1. Sharing the space means that 21,9 will become (10)   space to the one side

1	to the instant position of time (k^0)
,99	lost to space depletion $\Pi^2/10$
7	the relation to matter.

Through that the Titius Bode law comes into affect of 10/7 or 7/10, depending on whether space or matter holds a superior position to time. From that stance, all objects will relate to one another by the value of $\Pi^2\Pi$ and seen in a whole sale total 7/10 or 10/7.

This does not stop at that point, because this will affect the energy called life, and the extensions life develops because it has the ability of manipulating space-time. When an object relates to the cosmos, it relates to it from the point where the object is a proton $(\Pi^2+\Pi^2)$. With that the objects holds everything else to the factor that it uses when depleting space through increasing time. That value is Π^2. Because the object holds the value of matter and matter parts time (R^2/T) from space (R/T) separating the two factors in singularity (T x T). This positions matter in space (R^3) in time (T^2).

Matter holds the value of 7, always 7, in the position of space (7+3 = 10) and the combined space-time value will relate to time in singularity during that specific instant of time duration ending one eternity with the intervention of infinity (1). Placing all this array of words into a mathematical solution or formula $R^3/T^2=1$. Space ends the eternity of time to the value of 1. (Time is the movement (spin rate) of heat in space). When a structure holds heat to time in space the relevancy becomes $4(\Pi^2+\Pi^2)$ therefore when the object retains time to space it will have a value of $4\Pi^2$. This puts all the sides IN TIME SHARING WITH THE OBJECT AT $4\Pi^2$ to the Π^2 of positive space-time displacing. All objects will relate to the $4\Pi^2$ in the same way. As I showed, a few paragraphs ago, if space-time extends beyond the critical of $(\Pi/2)^2$, it will attach as the Titius Bode Principle of 7/10 or 10/7. When an object is in the atmosphere (the atmosphere being the neutron value of the Earth) it holds a position of $4\Pi^2+\Pi^2$. That means the object is completely secured by the positive space-time displacing of the Earth. When not secured it becomes $4\Pi^2+\Pi$ and I have shown that Π is the concentrated value of matter $(7)^2$ in time (Π^2). The atmosphere, we know through actual orbiting entry, holds an entry limit of 21,991° as a maximum and 7° as a minimum. That means to become part of the neutron status of the Earth, the object has to be space (21,991 or less) and prove to be matter (7) before the Earth will accept it. If holding a position of less than 7°, the Earth will discard it and if it is more than 21,991 the Earth will find the relevancy to be higher than the space it holds in a neutron time.

That places the object in a relation of $4\Pi^2 - \Pi^2$ (because it is not part of the Earth) in a position acceptable matter holds (7) within the confinement of Π (21,991/7). That means the object is part of space (22,991) acting as matter (it holds an acceptable own proton structure) 7 relating to the Earth in the position the Earth allows of $3\Pi^2$. With the space position of the matter in the parameters of 21,991 it relates to the Titius Bode law as a factor of one. The object has the space value of 10 plus the space value of ten, in that instant of time (1) complying to the Earth's space (10) reduction (Π^2) formulating $\Pi^2/10 = ,99$. That makes the object complying with the full agreement as laid down by the Titius Bode law . The object is, no matter where it is, travelling at a rate of 7 ($3\Pi^2$) in the space of the Earth (21,991). This will be agreeable to the parameters of the Titius Bode law as long as it remains within the space depleting "gravity" limits of less than Π^2.

In accordance to the Lagrangian atom layout, anything less than 5Π is manageable and is in effect less than Π^2. When it exceeds 5Π it will start opposing the dimensional equilibrium space holds of 10Π, therefore it will (according to space) exceed the linear point of R/T, which is $10\Pi/2$ (space going in a straight line). By exceeding the straight-line value of $10\Pi/2 = 5\Pi$, it will then categorize itself in the position time holds to space, and that value is $10\Pi^3$. By exceeding 5Π it will start to defy space and this will automatically bring about an individual space-time relevancy of $\Pi^2 \Pi$, establishing its own proton position in the confinement of the Earth's space-time of $7(3\Pi^2)$. Thus while it holds a relevancy of Π and Π has the relevancy placing it at a value of Π^0 because the neutron space link to time is 21,991/7 in the atmosphere, the Earth will tolerate individual movement of up to 7 ($3\Pi^2$) Π^0 to 7 ($3\Pi^2$) $5\Pi^0$. Beyond that point, problems start arising as the object is not complying to the Titius Bode law any longer.

Moving beyond $7(3\Pi^2)$ $5\Pi^0$ the Roche factor comes into effect and this is all in the space-time depleting zone established by the neutron atmosphere of the Earth. That means the Earth still holds its own 21,991/7 neutron base, but the object breaking the Titius Bode law bring in its own value of $\Pi/2$. This value of $\Pi/2$ is in the neutron parameters of the Earth where the Earth removes the linear factor value of the neutron from 1 (Π^0) to Π. That means there is a dual for supremacy of a neutron position is as much as ($\Pi \times \Pi/2$) becomes $\Pi^2/2$.

That places the sound barrier at $7(3\Pi^2)$ ($\Pi^2/2$). At the value of $7(3\Pi^2)$ $3(\Pi^2/2)$ more implications will come about, because the object will establish an own space within the space-time of the Earth and this will lead to the object whether joining the time of the Earth $4\Pi^2$ or securing itself an individual space and time. Obviously the heat supply will be insufficient to bring about the value the object needs to place it in outer space ($4\Pi^2(7(3\Pi^2) / (360° \times 10))$) so the aircraft will forcibly join the core of the Earth, crashing on the ground.

From the sound barrier and the effect the sound barrier has on matter, one can measure galactica and how that "growth of space" seems to become a reality as the Hubble Constant indicate.

When the object holds a position above Mach 1 and Mach 2, the temperature to the outside of the aircraft rises dramatically. This is the same as that which we may find to be the value of the part that we can see from the outside in the core part of the galactica. The structure of the craft becomes a solid or a proton. The craft holds the value of ($\Pi^2+\Pi^2$) in relation to the Earth because it maintains an own space value in the related time of the Earth. The craft applies a demand for additional heat, because to all cosmic purposes, it shows a higher resistance in sharing space and time. Place this in a human context and the aircraft has a higher mass. Science holds the movement of the structure in the category of momentum, but momentum is only linear "gravity" and mass is circular "gravity". In order to claim the space-time occupation that the craft demands, will mean an excess of protons in order to bring about the space-time occupation required for that strong relevancy to heat. The cosmic law does not allow for artificial heat production and life has no normal place in the cosmos. To the cosmos it is all a relevancy of supply versus demand.

In order to stipulate where I refer to space –time my referring to space-time have to be by the use of a signal. Therefore I decided to use the $ as a sign by which I shall refer to space-time.

That is where the Mach principle finds a solid foundation. What separate time and space in joining singularity is the matter and matter is heat. However Mach placed all emphasis on mass and mass is only the resistance of matter in parting with own space to join another object's space. Part of the resistance of mass or parting with individual space is to apply the Roche limit that will remove the neutron value of the object from $\Pi^2\Pi$ to a value of $(\Pi/2)^2$. This will improve the density of the proton considerably because it removes the Titius Bode principle of matter in space as it eradicates 7 to 10. I have shown how 7 to 10 are the development of 21,991/7 and this of course is Π. By removing Π from the equation there no longer exists space in time but a value of half time to time $(\Pi/2)^2$. The influence that this brings along is a huge increase in time value of occupation of space-time that the object initiates. Where the object normally holds a relevancy of space to time in as much as $(\Pi^2+\Pi^2)$ $(\Pi^2\Pi)$ (3) = 1836, the atom adopts a new relevancy of $(\Pi^2+\Pi^2)$ $(\Pi/2)^2 = 97,4$.

The normal mass of the proton will be (within the confinement of the Earth) $1,672648 \times 10^{-27}$ and the electron will be $9,109534 \times 10^{-31}$. With the alteration of the neutron and electron relation the "electron" value of the total object becomes 18,85 times larger. That means the relation between the proton and the neutron has a normal relevancy of 1836, but is only 97,4. That means above and beyond all the electrons' space-time demand, the structure as a whole, has an immense demand of 18,85 times larger. All this happens within the atmosphere of the Earth.

The structure holds a normal value of $\$T = 7\ (3\Pi^2)/10^2$ if one wishes to replace the Earth confinement of matter to the outside. Therefore it holds a relevancy of $\$T=2,07$ that makes it a cosmic relevancy that helium holds. Helium is an element that surrounds itself with heat in a gas value. When the structure reaches an own velocity of $\$T = 7\ (3\Pi^2)\ (\Pi^2/2)\ /10^2$ it then will have a value of 10,25 which holds a comparable mass (space-time integrating resistance) equal to that of Boron, at 10,811. At mach 3 the space-time demands will become $\$T=7(3\Pi^2)\ 3\ (\Pi^2/2)\ /10^2= 30,778$.

This places the relevancy where that particle can demand own space in own time due to the density it acquires from space. It has all the properties that a star in space requires, but it will still be confined to the Earth. In that way the object then must acquire a relevancy of the full Roche limit, without the protection of the Titius Bode law. The Newton cosmic relevancy will reduce to nothing $\$T=(\Pi^2) + (\Pi^2) + (\Pi^2) = 9$ and subsequently it will find itself in a huge struggle to defend its space-time or demise its position. With a relevancy of $3\Pi^2$ it holds the same value as liquid heat, which is light. The $3\Pi^2$ comes from the proton $(\Pi^2+\Pi^2)$ adding another neutron space value of Π^2 and this will then be (proton $(\Pi^2+\Pi^2)$ + neutron $(\Pi^2) = \Pi^2+\Pi^2+\Pi^2 = 3\Pi^2$, the same as the photon at 29,6. Beyond 29,6 the structure will have to claim individual space-time value. That is the same value as that of cosmic structures still within the inner-core of the galactica. The structures in the galactica holds a value of $7/10\ (\Pi^2\Pi)+(\Pi\Pi^2) = 28,8$ and to space it holds a relevancy of $3\Pi^2$.

That actually means that the objects that are still within the galactica core, shining because of the outside value of the particles being $3\Pi^2$, are in fact burning silicon, or if you insist, it is glass. In the very centre, where professor Hawkins presume Black Holes to be, is the very opposite, holding structures that have carbon inner core and therefore holds a neutron value of $\Pi^2\Pi$, with no proton core development at this stage. They are still in eternity beyond the "Big Bang" and therefore beyond light. This value brings us right back to the aircraft that applies an own space-time relevancy of Mach 3, or $7(3\Pi^2)2\ (\Pi^2/2)$. This relevancy places the object holding a star status core of $4(\Pi^2+\Pi^2)= \Pi^2\Pi\ /\ 5$ that is the value of a star holding an element of carbon. That is still two eras away and they are not yet even in the present times in our field of vision. To the Earth holding an iron core $4(\Pi^2+\Pi^2)$ in a galactica *7/10) an object in a space time position of $\Pi^2\Pi/5$ is completely out of space, out of time, out of era. It will apply the Roche limit with such ferocity, as its own space-time occupation will allow. According to cosmic law that object has a relevancy of less than that of the Roche limit $(\Pi/2)^2$. The space-time relevancy that the Titius-Bode law allows is (7/10=0,7) and (10/7 = 1,428).

With the fact of the speeds I use in the sound barrier explanation, all the figures are only relevant below 500 m above ground level, and ground level being sea level. Any point above 5×10^2 becomes a changing factor to the relevancy of the Titius-Bode Application. In a previous part I pointed out about the inclination of the atmosphere and that is true, but that view apply the way we see physics through the eyes of Newton. That mentioning of the changes of relevancy is only an introduction, in the same way I introduced the relevancy of the Roche-law and the relevancy of the Titius-Bode law indicating the way Doppler interprets the Titius Bode law.

The relevancy has a lot to do with that of Π also playing its part. That part is rather a bit less straight forward than the facts I mentioned up to this point and in that sense it becomes a little bit more complicated because time in the square starts to mingle with space in the cube. We are dealing with the space part of space-time that has no time value outside the Earth's atmosphere. The limit to the sound barrier being 5×10^2 is because at that level the relevancy of linear space (5) holds space (10) in time (10^2) to a relevancy of 1. Just like the Π^0 that I explained. Above that the next level of factor interference by relevancy changes comes about where the 5 (half that of space placing space (10) to a linear value of half). This is the same application that time has in $\Pi^2/2$. However, time cannot exceed Π^2, but as space only holds a relevancy because of the heat (matter) that it holds, space can then receive a cube value. All of this applies within the atmospheric space of 360° x 10 (space) being a relevant 1.

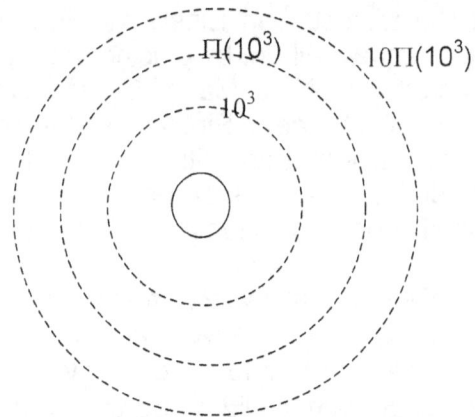

Beyond the atmosphere the Earth's atmospheric space-time or concentrated space-time loses its influence and it becomes a factor enforcing the Roche limit. For example, when the aircraft wishes not to crash but instead flies nose up, pointing in the direction of space, the factor of 360 x 10 will start applying. Then the equation changes somewhat where the effort will have to overcome that disadvantage as well. The space-time occupational need of the craft will be $7(3\Pi^2)$ 5 ($\Pi^2/2$) / (360 x 10) = 0,855 km/sec. That falls way short of the escape velocity requirement of 11,17 km/sec and the Earth circular displacement will destroy the craft's linear displacement in seconds. Coming back once more to the relevancy applied by the atmosphere holding the escape barrier of (360° x 10) in a relevancy of 1.

The atmosphere is space (R^3) in space-time and for that reason astrophysics and physics can never apply similar equations or formulas. At a relevancy of 1 to space ($R^3 = 10^3$) a barrier will apply reducing the actual velocity to a value where space restricts the craft by 6. In reality it still is the same as what it is at $5(10^2)$ but at $5(10^2)$ there is no actual space, it is time to space (10^2) to half or linear space (5) and at 10^3 it becomes a cube of air that holds the craft in the air. The next barrier of factor changing relevancy will be at an altitude of $\Pi(10^3)$ holding space (10^3) to a time value of Π. At an altitude of $10\Pi(10^3)$ that means space-time in space (10^3) the relevancy decreases to the extend that the aircraft will need to maintain a relevancy of 10 times that of ground speed to stay in the air. What this means is that at that altitude Mach 1 becomes equal to $\$T = 7(3\Pi^2)$ ($\Pi^2/2$) (10Π) = 32231km per hour. If you wish to reach Mach 1 at an altitude of 31 000 meters above sea level, you have to reach a velocity of 32 231 km/h. For that reason and that reason alone, N.A.S.A. cannot bring an aircraft to the altitude of 31 km above sea level and then let it quickly slip through the space barrier of (360° x 10) without the Earth noticing it. When the American Generals boast about x15 reaching Mach 5 or Mach 7, they only boast about how little they know.

Should the aircraft wish to break free from the atmosphere while applying a linear displacing position, the craft will need a velocity of $7(3\Pi^2)$ $(\Pi^2/2)$ (10Π) x 3 (Mach 3) leaving it with no less than a speed of just meter 1 00 000 km/h required and even at that velocity I am not sure it will pass the Π influence of 21,991/7 barrier that shield the Earth remaining totally intact and unscathed. I wish the alien hunters that claim aliens can come and go, as they please will use their brainpower in the direction of common sense and not common fantasizing. No way in heaven or hell can there be a craft that will slip in and out of the atmosphere at leisure to the demand of its alien pilot. The aircraft has to have a structure built at least from Californium 98. That shows how ridiculous the alien detectors are and they do not even know their own stupidity.

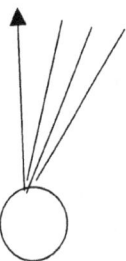

Bringing in reality again I would like to focus on Einstein's view of the curvature of space-time. Einstein made the human misjudgement we all do, by placing the focus he holds on the space his is in, in the time he is in. The aircraft travels at Mach 2,7, in relation to the space and time in compliance to sea level on the Earth. That means Einstein is standing on the Earth on a beach and has the binoculars focused on the craft flying at an altitude of 31 kilometres straight above him. The craft is maintaining a velocity of 2 700 km/h. On the ground the aircraft would be relating to a certain area of Earth every second. At the altitude of 31 kilometres that relevancy no longer apply.

On the Earth the aircraft holds a position that allows 250 meters of soil to pass him by every second. However at 30 kilometres above if he wanted to have that same distance of soil to pass by every second his relative space-time displacement has to increase by a measure of $(7(3\Pi^2)$ 2,7 $(\Pi^2/2)$ X (10Π) in relation to the $7(3\Pi^2)$ 2,7 $(\Pi^2/2)$ that is valid on Earth. That will then still hold a relevancy of 750 meters of soil passing under the wing of the craft. The effort by the craft therefore has to increase, not only by 10Π but another (10^3) to comply with the space orientation factor that the craft needs to sustain the equivalent space-time displacing. One factor of the space-time will remain constant, while the other two will have to vary. Should the energy remain at one, the velocity will reduce and the time to object relation of the craft will reduce. This means that in the triangle of space-time the triangle's 180° will maintain form, but the half circle's 180° will bend out of shape and this will bring about the straight line time has to hold in order to hold the value of 180° will also bond. When the triangle maintains 180°, none of the other two factors will apply their standards of value any longer. The circle will become oval and the straight line will become a semi-circle. From Einstein's perspective on the beach he will see time bend, and he will see space bend, because he focuses on the constant of the speed of light. He places the speed of light (which is a three dimensional triangle of (3) space (1) relevancy Π^2 (time) in the value of the factor remaining at 180°. That means time (Π^3) goes skew, and so does space, to prevent the constant of one to go skew. He allows the variables to vary in a dissimilar fashion to what the cosmos grant as the variable thus will vary. The cosmos holds Π^2 at a constant, always, no matter what because Π^2 come from the proton and the proton's space is the last space that time will comprise. The evidence is in the Black Hole where the Proton Star bends the 3 of space completely around, (the 3 is 3 sides of a possible six that will relate to any other object in the Universe) therefore he turns the 3 factor inside out. This evidence we see as matter spirals towards a centre point.

At the same time it completely demolishes the relevancy factor of 1, because of the repositioning of space to matter. This will lead to one of a possible two relevancies going astray. The total relevancy of the combined proton value places Π^2 at a number higher than we can ever calculate. When Π^2 changes its value to comply with the totality of the combined proton cluster, and space spirals changing three around, the time contact of 1 will maintain its position on singularity. Therefore matter in space will reshape, and gravity will reshape to maintain a constant to singularity. In the speed of light the factor 3 will change and the gravity factor will change in order to maintain the contact with singularity.

Einstein allowed singularity of space holding (heat) to bend and the singularity of time contact (1) to bend in order to have matter ($3\Pi^2$) maintain its norm. We all know from what the Roche-limit proves, and the Titius Bode law proves, and from what we can see happening to our craft, that matter in space ($3\Pi^2$) changes shape, form and norm, keeping the singularity of time-space-unity to the relevancy of one. There is no curvature of space-time, only of matter occupying space-time and when focusing the constant factor on the variable, the true constant will seem to bind.

Again I say: There is no curvature of space-time, only the revaluation of matter, be it unoccupied space-time as in that we regard space heat to be or in the occupied space-time as that we conclude matter to be and both forms part of the same unit. By applying our focus incorrectly to matter, science concludes that space is growing. That is as inaccurate as Einstein's curvature of space-time. Matter can relocate and revaluate its relevancy, but space and time remains the constant of eternity in singularity. This means that saying the space of the Hubble Constant is growing, is just as inaccurate as saying the Sun is rising each morning.

The influence of the Titius-Bode law may have one other extremely significant influence on the speed of light. To every possible way I look at it I am convinced that my view is correct. The part of the Titius Bode law which science confuse with the Doppler effect has to play a part as the Doppler effect can only apply under the conditions laid down by the Earth's space-time displacement and that renders the opportunity of precise half circles following one upon the other. In the Titius Bode Principle however, the matter holding a relevancy to the heat in that space determines the value of space-time.

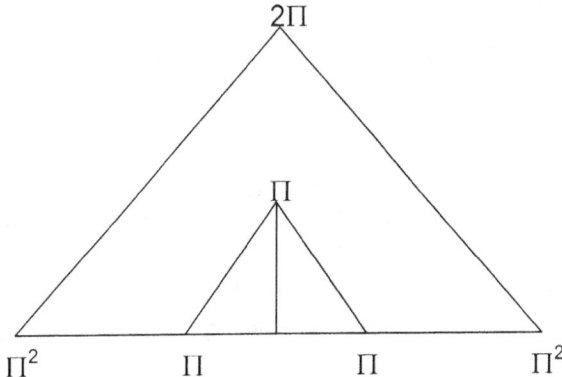

The combined value of matter holds the space component at 3 pointing out the point of origin where the particle was holding space-time before growth started. This will give a value of 3 and the point of relevancy through time by position allocation will be 4 ($\Pi^2+\Pi^2$) this will then be the seven Π's in the law. Therefore the 3 always determine a position of origin and the 4Π a position of location.

That will bring about that all particles orbiting any structure will see the position of shift through time advancing or the Hubble growth in a manner of time to space. As I have indicated the progress to matter's relation to space will be the 7 that matter forms (7Π) in words the seven pi and space (10Π) in word the ten pi. Matter will always be in a galactica submitting the influence it holds to the heat space has to provide, maintaining the cooling of matter. The density factor favours heat in space and heat in space will conduct existence (mass) to space depletion (gravity). In the case where matter lies outside Galactica the opposite is true and therefore the opposite will become the Titius Bode controlling factor in as much as space (10) becomes relevant to matter (7).

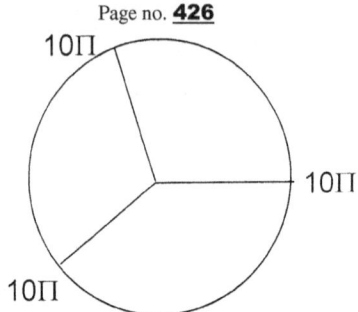

The metric system provides a vehicle that makes space and time measuring far less complicated, because we base the space measurements to that of water. One cubic meter of water is one ton with one meter in the cube moving one meter per Earth velocity which all connects nicely to the earth governing singularity. Space holds a relevancy of 10Π. That will be the value of the full field of the spherical composition of space.

To that reason time has the relevancy of $10\Pi^3$ and space holds the relevance of $10^2(10)$ $(\Pi^2\Pi)$ where the second 10 is the matter's composition of $7\Pi+3(\Pi) = 10\Pi$, but because Π only connects the fourth dimension to the time factor it falls outside the fourth dimension therefore space holding matter carries the total of $10(10^1)^2$ in the full circle.

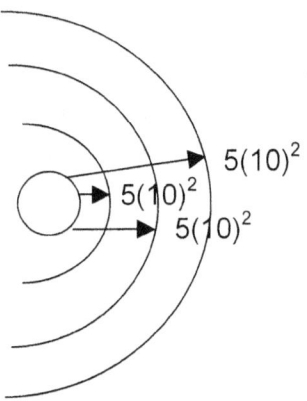

The linear factor of light travelling from the Sun will be the distance between the Sun and the Earth. The light density that reaches the Earth is the density that remained after the rest of the wave (3^3) went to space in the form of radiation (3^2) and heat invisible to the naked eye, the heat that holds the universal gas that covers all objects, the same heat that one day must all return through matter to time. When looking at a sphere the three we see, has a lot less dimensional value then we can see from a cube. Even looking at a cube from only one side, will still have the support of half of the other sides that provide enough stability to the cubic orate so that we can keep on seeing the one side.

To introduce my theorem, in short I wish to bring a very short overview, before we start with the complicated cosmic laws and definitions I am forced to use to defend the point I make. I take the most common phenomenon on Earth and build a theory based on that principle. The basis of my theory is that everything is heat, be it solid, liquid, or gas. Culture tells us that hydrogen is a gas, or that gold is a solid. Our biggest drawback we can have is precisely such wrong cultural conceptions with no base of proof. Hydrogen is much less a gas than gold is a solid. We connect a cultural conception to what we may presume as facts, but that does not make it a fact, it only makes our conclusions an elusion.

Everything except life uses an energy driven by heat removed or heat expanded. It is all a balance where the flow of heat started at a point and pushed in a direction. This brings me to my first definition. What is energy? I shall prove that **energy is the interaction between heat creating space and by demolishing space, heat concentrates. All energy in the Universe holds relativity to this, no matter what.**

The Universe started from something I named time in singularity. Time in singularity is the only constant because there is always a direct line of contact to time factor in singularity Π and time in singularity is the end product of everything that is cold, frozen beyond space. We view the ultimate freezing point to be outer space at a temperature of –273°C. Heat will always move, from the hottest area to the coldest area. If the outer space were the coldest point there is, the Earth would be frozen solid, because heat will never arrive at the Earth from outer space.

We would have a continuous flow of any heat the Earth may hold to outer space. Such is not the case. We may view that heat flows from the Earth at night to outer space and that is correct. However there is no flow of heat that will bring about temperatures to fall to limits even close to outer space. Nothing on Earth can reach –200° without the interaction of human life. That means the Earth will always be a place colder than the Universe, just because we have more heat on Earth than in the Universe. If it was not the case, the Earth should be at least as cold as the Universe because all heat will flow from the Earth to the outer space.

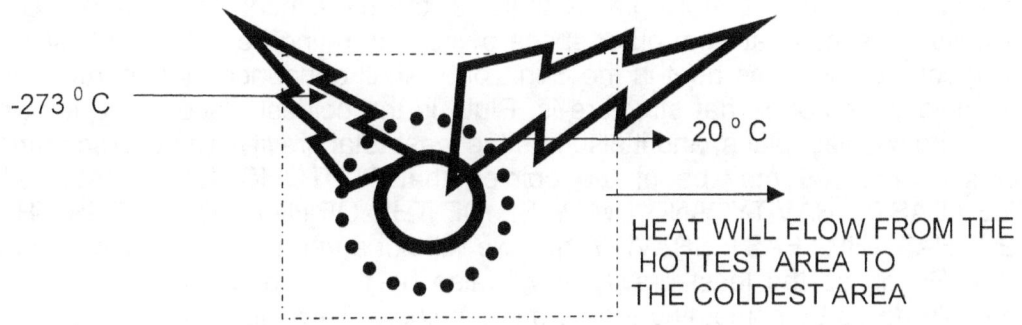

-273⁰ C

20 ° C

HEAT WILL FLOW FROM THE
HOTTEST AREA TO
THE COLDEST AREA

IF THE OUTER SPACE REGION WERE COLDER THAN THE SOLLID STRUCTURES, THE COLD WOULD DRAW HEAT TO THE HOTTER REGIONS. THE NATURAL TENDENCY MUST APPLY WHEREVER THERE IS HEAT VERSUS COLD.

The opposite is happening

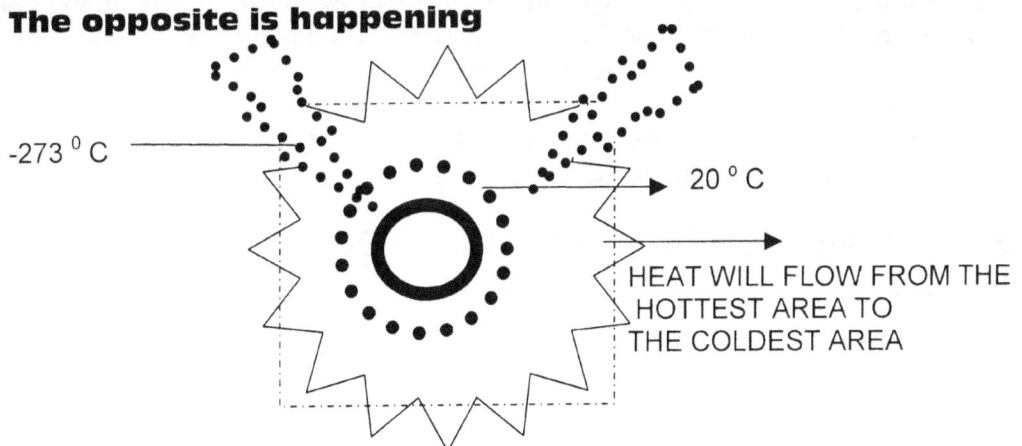

-273⁰ C

20 ° C

HEAT WILL FLOW FROM THE
HOTTEST AREA TO
THE COLDEST AREA

INSTEAD OF THAT, THE NATURAL LAW PROVES TO HOLD HEAT TO THE STRUCTURES. THAT MEANS GRAVITY HAS SOMETHING TO DO WITH HEAT CONSENTRATION. THAT ALSO INDICATES THAT GRAVITY IS THE RESULT OF SOMETHING MUCH COLDER THAT IS NOT IN THE VISION LEVEL OF THE HUMAN EYE. THE ONLY SUBSTANCE HUMANS CANNOT VISIONALISE IS THE PROTON.

Why would the Earth appear hotter, but is colder than the Universe? It is the fact of gravity, a term I reject because of the connection science applies to gravity being a force. If it was a force then something is pulling or pushing and since we do not believe in spirits creating magic forces, and no one can detect anyone applying any force, we have to dismiss the connection of a force or forces.

For the moment in order not to confuse any reader in the introduction, I shall remain using the term gravity but in doing so, I dissociate myself totally from the implication that science connects to gravity.

One may say that this is the effect of the Sun shining on the Earth, and one may be partly correct, except when considering the wider picture. Measure the space the Sun pours light into and divides that area the Earth holds in the vastness of space. I truly think no ordinary pocket size calculator will provide a realistic reading. Yet the Earth is many times hotter than the rest of space, if it was merely the Sun shining on the Earth, the Earth was no factor to consider in relation to the space out there. That alone cannot account for the difference there is in temperature.

This brings us to gravity, (not a force). My definition about gravity is that: **<u>Gravity is the reduction of space to concentrate heat,</u>** therefore the more gravity there is, the more heat there will be. The Earth, or any cosmic structure claims heat from outer space by concentrating heat. You may shout that fusion is the blaming contributor to the heat in the Sun, but I shall denounce that shortly. The more gravity a structure holds, the hotter that structure is, Pluto is the coldest place in the known solar system, but it also is the smallest place, and it also has the least total gravity. The one scenario compliments the other scenario, you may be of the opinion that PLUTO IS THE SMALLEST BECAUSE IT HAS THE LEAST GRAVITY, AND I MAY BE OF THE OPINION THAT IT IS THE SMALLEST BECAUSE IT HAS THE LEAST GRAVITY, and we will not agree at all, although we are using the same words. If Pluto has the least gravity, it will also be the coldest planet, if gravity concentrate heat, it can only do so by something that is much colder. This will increase heat at a point because the effect behind gravity is producing a reduction of space, a point of reference that is much colder than any place else.

I started off at the beginning of this part, by saying everything started at a point, that was so cold, space froze to singularity. Let us test this statement: When an element heats up, the space it occupies, increases. When an object becomes colder, the space it holds reduces. The less space per atom there is the colder the object must be. Anything can freeze rock solid, as everything can boil to gas. This means all elements in nature, is neither gas, nor liquid nor solid. It is the space that is between the elements that allows the elements the form they hold at that moment. By reducing space, space has to concentrate because to concentrate is to reduce the solution. By reducing space, we find more heat. If you find more space by increasing heat, and you find more heat, by reducing space, then heat and space is the same thing.

THE FURTHER ONE RISE ABOVE SEA LEVEL THE COLDER IT BECOMES, AND THE DEEPER ONE GOES INTO THE EARTH (NOT THE SEA) THE HOTTER IT BECOMES.

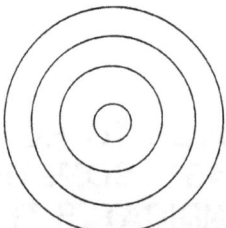

You may think that at this point I reached a point of becoming ridiculous. By concentrating matter to a solid, the matter holds less space in relation to the heat surrounding the matter and since matter is pure concentrated heat, the heat actually increases because the space around the matter is less dense heat than the density of heat the matter holds. Therefore, matter is the ultimate concentration of heat, because it is heat-frozen rock hard.

The colder anything is the less space it requires and matter requires much less space than does space require space, If mater is heat frozen rock hard, then although having a higher concentration of heat, it also have a point within, that is much colder, and the flow of heat towards that point of the ultimate cold, concentrate the heat. To cool any overheating object, you increase the flow of air. The air is space, and that space increase brings about a reduction of heat by increasing space. That means the space -air ratio changes and the product is heat reduction. However one has to look past the obvious to the factors not red dally in sight, It also means by increasing the flow of heat, the heat factor reduces and that will lead to a decrease in heat availability, with a lesser heat factor.

The lesser heat factor will provide a faster relevancy to the flow of heat. That means the faster heat flows, the colder heat becomes, because the more space there is to allow the flow of heat. Looking at a rotating object the relevancy of space reduces by the factor of four to the halving of the radius. If one draw a presumption that the same volume of heat flows through less space, the flow of heat must become faster by the square of the decrease of radius, and that reduction leads right down to the point of singularity. As that point still connects to singularity, a point without space because it froze space out of existence, the flow of heat away from such a point becomes "over-spaced", because the point of singularity is without space. All things "over-spaced" are also overheating. Therefore, as everything away from such a point is hotter, it will also hold more space to heat because the more space means the overheating factor is higher. That is the cosmos, the place hotter than singularity and therefore more space to hold heat. This applies in the precise way inside an atom. The atom holds more heat to less space and therefore it has to be colder on the inside than what it is on the outside.

While outer space is growing the growth can only come about from singularity, which was out of control and expanded without direct influence of contracting. The contracting at that pint was secured in the atom and all heat was stored in a secluded space by using the Coanda principle. In this the cosmos gathered what was beyond control and expanding into the oblivious.

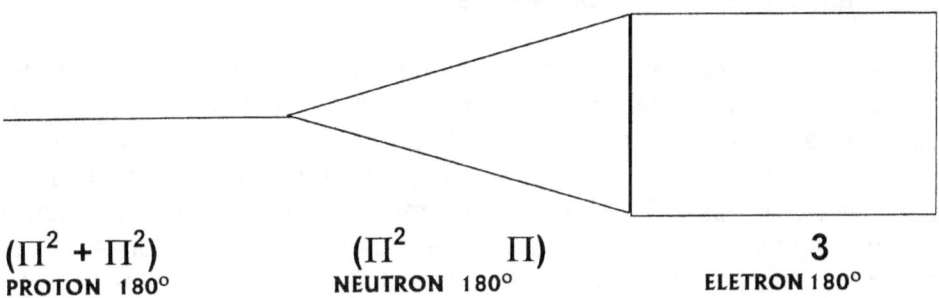

$$(\Pi^2 + \Pi^2)$$
PROTON 180°
$$(\Pi^2 \qquad \Pi)$$
NEUTRON 180°
$$3$$
ELETRON 180°

We view the atom always from the outside, which means from the electron. The fact that matter to the inside is dense, only means that the matter to the inside of the electron is frozen solid. Frozen solid by itself has some terminology, some concept we have to define in order to establish the true meaning. The frozen state of matter within the atom, also apply a frozen realization bringing across implications that are to our mind completely alien. To us something frozen solid means that the density around the atom hold the least space we can imagine.

This is correct, but it is also relevant because what may be frozen to our view is still liquid to nature. Even freezing an object to −273°, only means that the heat surrounding the atoms is at its limit of forming a gas. It is at its limit, but still it is a gas. If it was more than a gas, the electrons will touch, because the electron is the state of heat where heat becomes so cold it turns to liquid.

If there is only liquid heat separating electrons, the electrons from different atoms will connect and establish a fluid, linking the atoms in space-time. Linking atoms is better known in science as fusion. The atoms melt together, they fuse, they bond. Hiroshima and Nagasaki as well as the Bikini Islands bear testimony of such an event. The cooling of the atoms have no more gas-heat flow in maintaining the atoms at a relevancy below the cosmic relevancy of 112, and the whole atom turns liquid. The liquid forms energy as the liquid heat arrives at a higher spin rate than the frozen state will demand.

Energy is the release of heat to gas, forming space, and the space creates the nuclear winds, accompanied by radiation and light. Radiation is only intense spinning gas and light is a heat droplet.

Therefore the heat becomes condensed heat in the form of heat "Vapour" and heat "drops". This is the closest we can come to see what is on the inside of the atom. The rest our mind must tell us, and not our eyes. When heat travels, it becomes more condense, holding less space, and is therefore colder. Behind the electron, the movement of heat is faster than the electron, because it spins with an ever reducing radius, therefore as the radius reduces to the atom core, the spinning of the heat holds four times less space for every time the radius halves in length.

Having an ever-increasing spin rate, the heat moves faster, condensing even further as it progresses to the proton core. That makes one realize that inside the atom the heat contracts, holding less space, and by holding less space it must be in a more frozen, or colder state. This is about the same as feeling the core of a radiator of a cooler unit, and thinking if there is that much heat on the outside, how much heat will be on the inside. Well, the similarities may not be that correct, but the principle is very much corresponding. Matter is frozen heat because it is spinning through less space.

I shall make a statement, that will surprise every reader, I think. Space has no value, because space is the product of time in singularity, a part of singularity, a by-product of time in singularity that has no cause to be except to hold heat. When boiling water one finds bubbles in the water. No one blew bubbles in the water, it obviously could not enter from the top, because it will eventually rise to the top and it did not enter from the sides or the bottom of the pot, the heat that forms the pot is watertight as much as air tight. Yet there are newly formed space that came from nowhere and will disappear into "thin air" as soon as it leaves the pot through the top of he pot where the water ends in space. No one can detect the space but in water vapour and between the water vapour is no space, there is only heat. If it was not heat, the water will be less hot, and be a liquid. This proves that space has no value, except for the heat that forms the space. This is another point made, which proves that outer space is actually less dense heat and matter is more dense heat.

At this point I arrive at the biggest bone of contention I have with Newtonian science. It is not mass that produce gravity, it is the density of matter that produce gravity. When one cubic meter of water forms vapour, it will be a cloud with as much mass as the equal cubic meter of liquid water but it will float in mid air. It will hold in mid air because it holds more space, therefore less dense heat, than would the water have because the water has a liquid form. The mass of the vapour is the same as the mass of the water, yet gravity applies less to the vapour because of the abundance of space allowing less density. About this I bring proof that Pluto has substantially more gravity than the Sun , calculated in the manner the Newtonians do.

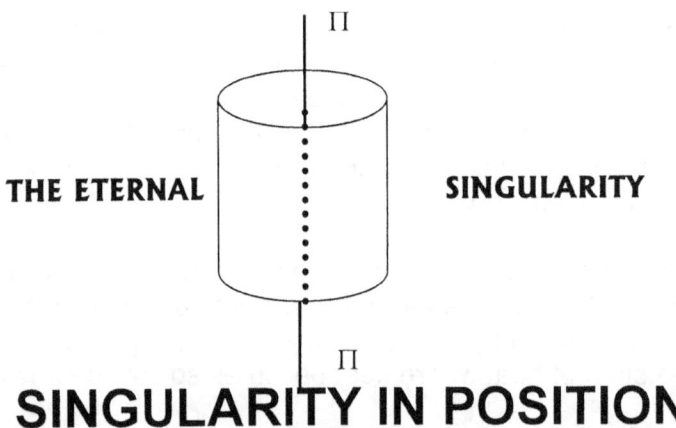

SINGULARITY IN POSITION

As a principle when any point expands the expansion is evenly in all direction simultaneously. That configuration only Π allow by dictating the form of the sphere. The fact that space holds 3 and time holds for in relation to a centre is the proof of the formula we use to calculate a sphere. Therefore only the sphere could form when creation initiated by motion. The formula is then $a^3 = (4\Pi r^3) / 3$. Although it was running on the line the line through the interlocking of the numbers that applied Pythagoras was forming sphere that kept space apart from spheres that kept space within. But as the Roche limit came into use and not the dividing of five, we can be sure that only form applied and at that point space was not yet a part of the Universe, just as singularity at present is not a part of the Universe while it is well establishing the Universe.

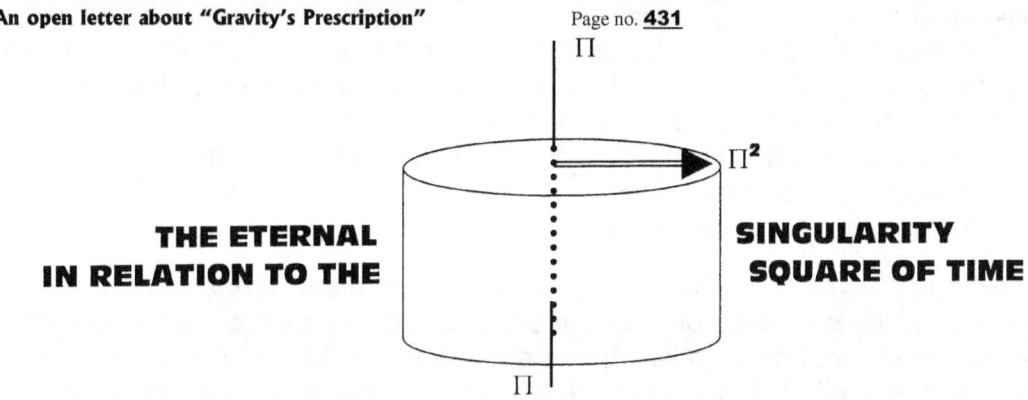

THE ETERNAL IN RELATION TO THE   SINGULARITY SQUARE OF TIME

SINGULARITY CONNECTING TO TIME IN THE SQUARE

Of course, the Newtonians make one great blunder, which I also prove, but more about that later. It is not the space that one should calculate, but the density heat has within the space in question. All space holds time therefore it is space-time. Space does not apply and can disappear, if not for heat within that space, allowing space to be. This then dispute all claim that Einstein made about space-time curving under gravity applying. Space does not exist without heat, and gravity concentrate heat, therefore space has no bending to it, and it all depends on heat. Gravity concentrates heat by reducing space. That brings us to gravity and time. What is time? If the Newtonians wish to bend time, they should at least find out what time is before they can start bending it in all forms and shapes.

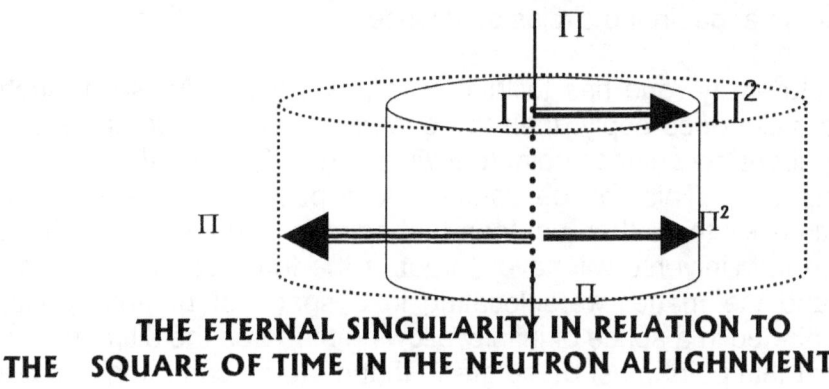

THE ETERNAL SINGULARITY IN RELATION TO THE SQUARE OF TIME IN THE NEUTRON ALLIGHNMENT

SINGULARITY CONNECTING TO TIME IN THE CUBE (R^3) OR $\Pi^2\,\Pi$. THIS IMPLICATE THAT TIME IS THE PROTON AND SPACE IS THE NEUTRON, AND IN THAT VERY CONCEPT IS THE "MYSTORIES OF STARS AND STUFF"

To find what time is, let us see what the standard is we apply to create a sense of time. The Earth holds a position relating to the Sun , and that position is different every second time ticks by. In summer, winter, autumn or spring, the position is always different. One may say that I am wrong when I say time can never repeat, but the truth is that because the Earth never repeats any position in which it was. While changing the orbit the Earth follows around the Sun it is also following the orbit the Sun has around the Milky Way and as much as the Sun changes the location it has every cycle, in the same manner does the Milky Way change the orbit location of the solar system by the growth of material expanding. There is no chance that the Earth will ever be in the exact spot twice. Newtonians, as usual never cast their eye past the obvious. The Earth is not the only planet in the solar system, there are nine other planets relating to different positions as they combine a relation with the Sun as much as hold individual and connecting positions according to the Sun .

In this context, the Sun is only one measly star, of countless many, all circling about a centre point around the Milky Way. In that respect, no point that any star holds, can repeat that position in accordance to the combining relative positions all the objects hold in relation to time past, time present and time future. The Milky Way is one measly galactic amongst countless others all repositioning their position to the structures within them as much as with all the other galactica the Universe may hold. In that respect I would love to see how Einstein will bend that lot! Light has nothing to do with time, because every photon is just another relative position filled with liquid heat, changing its position according to all other universal occupants, how large or small they may be, because the combined effort of repositioning create the variable one find in the flow of time. Running up and down a chamber between two space whirls will not reposition every object the Universe may hold. That shows how ridiculous science has become, with their view on time travel, holding two space positions in one instant and the rest of funnies the Newtonians declare as science.

Time and space are one and the same in singularity, with heat in many forms separating the singularity of space and time. Neither space, nor time can bend, because they are concepts existing beyond the point where heat starts and heat ends. Space does not exist and heat in various forms apply time, therefore time in singularity holds a value of eternity and space, where heat in various forms relocate their positions, holds infinitely as every repositioning heat produce apply an infinity, by breaking the monotony of eternity.

This is then my third definition: Time is the spin rate (movement) of heat in space.

This brings us to the fact that I mentioned about heat having many forms. This then is where I dispute Einstein's other theory, the one about the critical density of the Universe. Einstein said that a point has to arrive where matter will contract, because of some force that will start applying. One should view the time when this theory came into place, more than the theory itself. A person by the name of E.P. Hubble brought "indisputable evidence" about the Universe drifting apart. I shall at a later stage dispute that statement, however, at this point, I wish to bring to your attention, and the crisis this finding brought about in the circles of science.

In view of this critical density, one has to think when was the Universe at such a point of critical density. If Newton's claim does apply, then there is no Hubble Constant, as there is no Big Bang. However, everything about the cosmos point to a widening shifts in matter, where the space between matters is on the increase. This can be detected to a position where the Universe was infinitely small. What better time will there then be, than that point for the critical density factor to kick in and produce an implosion the Universe will never forget. If the force can contain and reduce the space separating matter, and the matter were locating in a space of non-existence, the mass of the Universe then concentrated in a space of infinity allowing the force the ultimate opportunity of endless strength. The force at that point must be so great, that no force can bring about any separation. All matter the Universe at present contains, it contained back then in infinity to space. There is no adding of matter. There is only adding of space. The Universe did not collapse, it "exploded". Why then would it explode at a point it had all the reason to collapse by the mass of gravity causing an implosion, well, that is if Newton was correct.

The answer is so obvious that answering it actually is a joke. Newtonian measuring and calculation methods, by the way, do not apply to the cosmos, and science should keep mathematics out of it. The answer to the Hubble explosion that started at the Big Bang depends on the increase in space. A few pages ago I showed that space has no value, it is a form of singularity, therefore adding space is the counterpart of matter reducing space with heat concentration.

The more space gravity will reduce, the more space heat will apply as heat in space reduces to heat in time. By adding something that effectively does not exist in any case, and can appear or disappear at beg and call of heat, it will and can bring no change to the cosmos as a whole. It has no effect other than to indicate the moving flow of time.

However, why the desperation then and why did the institution of academics go into a spin about E.P. Hubble's discovery? All of modern science finds its base on the fact that Newton proclaimed where he never produced evidence in support but still declared that a force named gravity will reduce the space between objects in relation to the products of the mass of each body. The space separating particles will reduce in accordance with the gravity the mass will apply. The emphasis falls on the reduction of space, and yet, undeniably, Hubble proved that the space does not reduce to the contrary it increases with the event of time. This cracks open the completely complex issue about the misconception they call the science of physics. Again, I repeat, Newton is as clear as daylight when true forces apply, but only life can apply any force!

Only as far as Newton is cosmic outlook goes, no one can understand Newton. Anyone that understands Newton will not accept Newton. To accept Newton's application of cosmic law, you brutalize and bully students into a state of accepting Newton and then have the student believe he understands Newton. It was at first a fashion that later became an industry. If you do understand Newton, you will realize that you do not understand Newton.

Through the centuries, across the world, in all countries, they institutionalised Newton. All of science will fall flat, because the opposite of what science use to base their many calculations on comes to nothing. The Universe expands it does not contract. Therefore, who better to get than the man that already produced the relativity theory to come up with some cockeye notion about Newton's law that will still apply one day? With the help of another misconception introduced by an Austrian, E. Mach, Einstein recalculated the theorem about the critical density of matter. This was all to one big effort to save Newton's claim on which rests all of science.

To distract all attention about the folly, they started a major attempt in deceit, (wittingly or unwittingly I will not know) whereby they calculated the density and mass of matter contained in the Universe to reapply in order to reconfirm the truth about Newton's claim. At this point, there is only ten percent of the required mass floating around to save Newton's claim. Again, another attempt was launched to save the industry where they will locate, so called dark matter, in a hope that enough dark matter will save science. This is another fiasco where the poor and insignificant tax payers hard earned money is thrown into a bottomless pit that will bring no conclusive answer.

The conclusive answer is in rejecting Newton's claim about the cosmos. We see the Earth and we see life on Earth as part of the Earth. In the obviously short sightedness science have, they wish to project everything found on the Earth as a direct translation to outer space. Outer space holds no life.

Newton's other laws do apply, as it is, precisely to point the genius made, and a genius he was, about that there is no doubt. To get a perspective on Newton, see the apple that fall as life, not as the cosmos. See an aircraft, a boat, a building, and a dam as an extension of life not as a product of the cosmos. That is the only way to save science and Newton, because that is the only thing wrong about Newton.

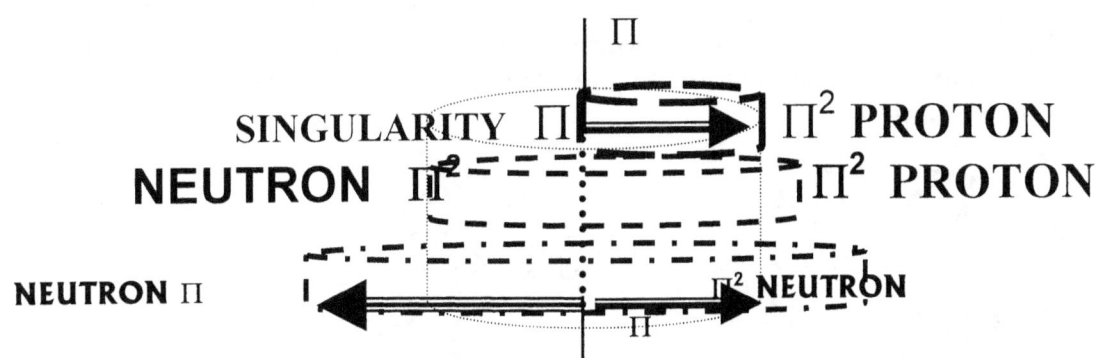

Newtonians' physics belong to life on Earth and other laws applying other dynamics apply to the cosmos.

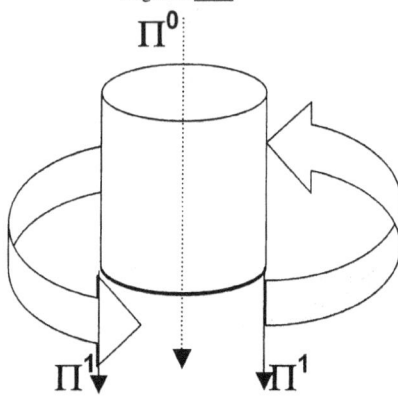

As much as life does not belong to the cosmos at large out there, the calculations and mathematics life can use has very little significance but to bolster the ego of those mathematicians that consider their mathematic skills far more that that which the Earth may appreciate. Those mathematicians should put their enormous skills to better use and build a dam or a skyscraper. Gravity is motion of space that is producing time.

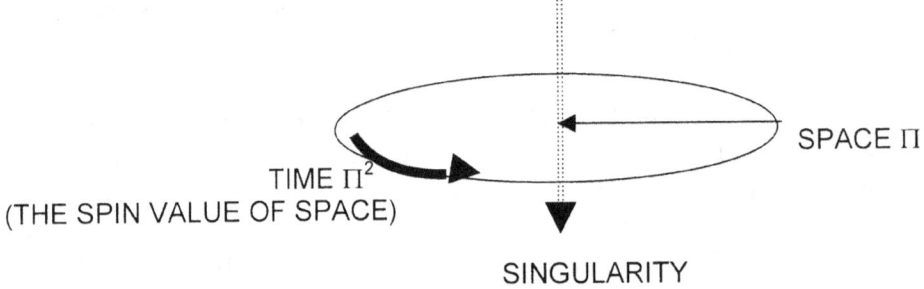

Gravity is the very same but it is the recalling of the space by creating motion in the space. Gravity is the retracting of heat by splitting matter as matter duplicate and reduce space by increasing space in expanding. As the space gets more and the time holding the space gets less per unit in time used the heat distribution is wider in less time and by such distributing the heat in relevance gets less because more gets distributed by a wider area in a shorter time frame. Gravity is the retracting of heat by cooling because of the expanding of heat increasing.

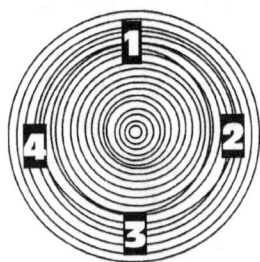

By moving from 1^0 to 1^1 and from $1^0\Pi^0$ to $1^1\Pi$ requires space. Yet such moving did not leave the realm or the domain of singularity. The motion was still within singularity because moving involved forming a relevancy between heat and cold between infinity and eternity, between space and time and most of all producing what will in the far future develop into a Universe that can even be a host for life albeit on a very small spot for a very short while in relation to the vastness space has and the duration cosmic time has.

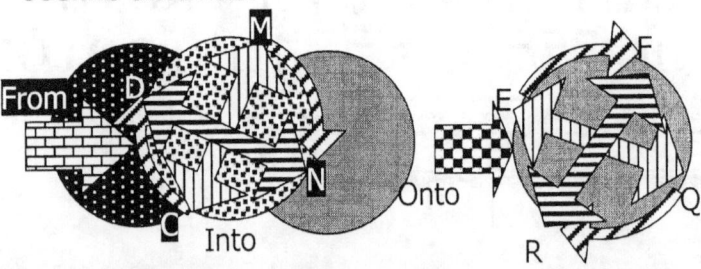

Relevancies came abut when the dot moves away from the spot but had no space to move. All that was possible was to charge singularity by relevance to comply in being activated into complying. Space-time is motion and movement are all the same things only separated by dimensions and dimensions are formed space, where the dimensions become space being in motion and the space is motion by contraction or by expansion but because time is almost

eternal at k⁰ our perception of the universe we are in is a stable and steady eternal structure. Gravity is motion and motion creates space to the third by the third in the third that interacts with one but establishes ten.

The cosmos holds no constant and that is the only constant. Every aspect of the cosmos is relevancies where matter in different forms, form different relations to other matter also in various forms. As I indicated about time, where time is an ongoing repositioning of relevant matter locating relevancies in the position they hold to the time they apply. The cosmos cannot grow, as much as the cosmos cannot shrink. It is a never-ending flow of changing relevancies, where singularity meets singularity as much as space meets time. The point in singularity I named time, is the point where time started and where time ends as much as where time will finally fulfil its reducing of space. Heat made space renegade and time slowly contains space by reclaiming heat. That is the cosmos.

The reclaiming process has various stages, all forming separate dimensions. The following description is wholly inaccurate, but in view of maintaining simplicity, it is also the best I can do. Mass is a door, a hole or a gate through which heat flow from space to time whereby it produces gravity by introducing the events forming time. This is a relevancy. There is in this relevancy time, which I gave a value of Π^3. The value is rather complex in explaining while I am in effort to attempt in this letter to keep information as simple as I can manage.

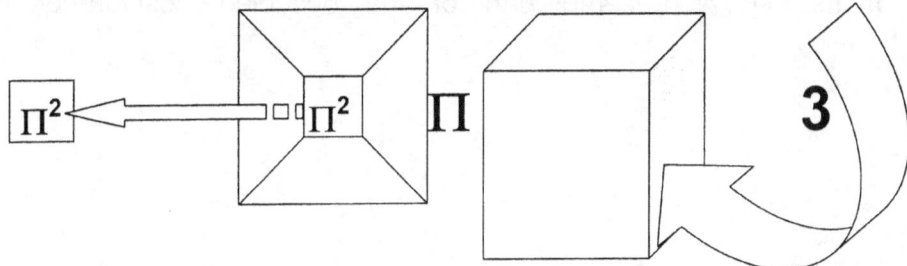

Connecting to time in singularity there is a rotating door that is two protons, one linking to singularity where time and space in singularity meets, while at that very moment, another proton, also Π^2 connects time and space where time and space holds heat. This I called the double proton ($\Pi^2+\Pi^2$) and the heat is in the densest, most frozen form heat can be outside singularity. As I said, this is a revolving door where the one at one point dips eternity into infinity, while the other takes heat to dip when it is the next proton's turn to dip into singularity. The value of Π^2 is also rather complex to explain in the introduction.

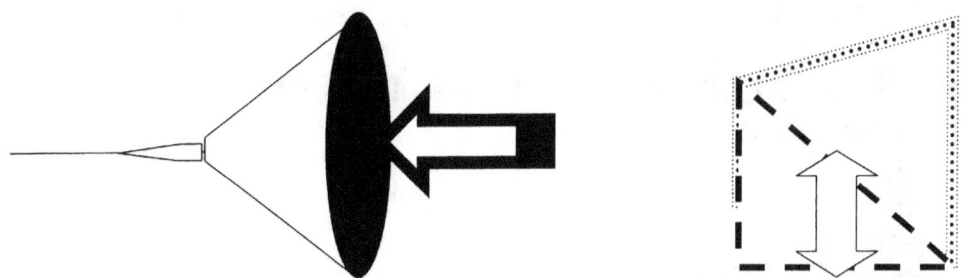

WE HAVE NO POSITION IN THE UNIVERSE EXCEPT FOR ONE MEASELY DISCARDED STAR-CORE. THAT IS THE ONY WE SPACE WE CAN OCCUPY, THEREFORE WE SHOULD NOT SEE WHAT WE WISH TO SEE FROM OUR PERSPECTIVE, WE SHOULD SEE WHAT THE COSMOS TELLS US TO SEE ABOUT IT SELF AND THE RELAVENCIES IT COMPILES IN TIME MATTER AND SPACE.

Behind this is the neutron double gate where gravity removes one dimension. The proton removes one dimension to take heat to the dimensionless ness of singularity; while at the same time it removes one dimension of space by applying the square of time (T^2). The neutron removes one dimension of space, by applying gravity in a two-prong third dimension of $\Pi^2 \Pi$ where Π^2 is the circular part of gravity and Π is the linear part of gravity. Following this dimension is the fourth dimension where the six sides an object can have, will only hold three sides in relation to any actual relevancy, since we can only be in one part of the Universe. The other three sides will apply to the other part of the Universe that we do not occupy at that precise moment in time. The conclusion to this "port" figuration is $T (\Pi^2+\Pi^2) (\Pi^2\Pi) 3 = 1836$, the difference between the mass of the proton and the mass of he electron. That is the relevancy to all space-time.

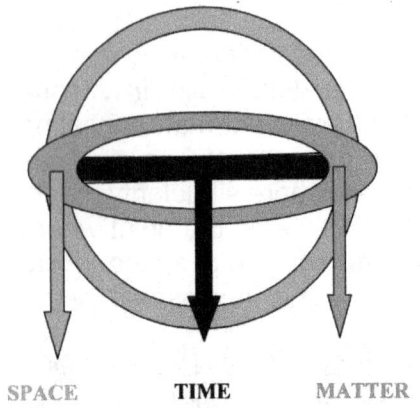

SPACE   TIME   MATTER
139   137   138

Time started … only God knows where. I can take you to a space-time relevancy of $\Pi^3x\Pi^3x\Pi^3$ and I do in The Thesis, but without detailed explaining it will only be numbers. The explaining and calculations are tedious as they are complex. I can also take you back to a relevancy of $(\Pi^3/2) (\Pi^3+\Pi^3)$ but without the proper explaining it is senseless. I can take you through many more relevancies but they will be as senseless as the next.

I wish to start at $7 (\Pi^2+\Pi^2) = 138$ also $7/10 (\Pi^2/2) (\Pi^2+\Pi^2) = 136$ and $10/7 (\Pi^2) (\Pi^2+\Pi^2)$ This is where the Universe received a sphere in formation, and although space was still part of singularity, the Titius Bode law on matter (7/10) as well as the Titius Bode law on space (10/7) formed a conjunction with the Roche limit in occupied space-time $(\Pi^2/2)$. At this point matter formed $7(\Pi^2 + \Pi^2)$ where 7 is the value of matter and during this the prelude to matter in space (7/10) as well as space holding matter (10/7) became part of the occupied position of the

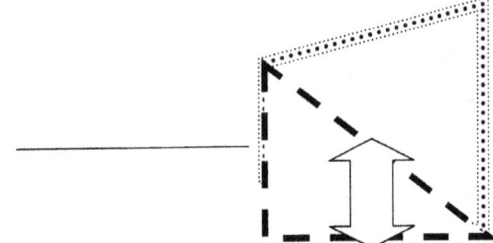

WE MAINTAIN A SPACE-TIME POSITION SOMEWHERE BETWEEN THE FOURT AND THE THIRD DIMENSION, NOT BELONGING TO THE FOURTH AND NOT BELONGING TO THE THIRD DIMENTION EITHER.

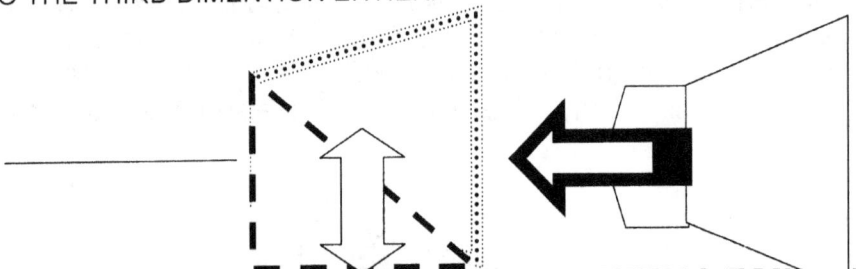

FROM THE ONE SIDE WE SEE HEAT RUSHING TOWARD US AND MOVING HEAT WILL SEEM COLD

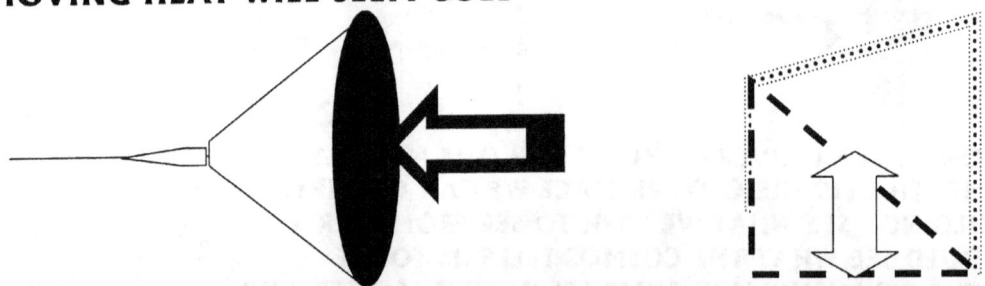

FROM THE OTHER SIDE WE SEE HEAT RUSHING AWAY TO A POINT, MUCH SMALLER AND THE CONJESTION ALLOW US TO BELIEVE MOVING HEAT SEEMS HOT.

Roche-limit $(\Pi^2/2)$ I introduce this aspect, because the Titius Bode law applying to matter (7/10) and to space (10/7) as well as the Roche-limit holding position outside occupied space-time $(\Pi/2)^2$ is of all importance to the understanding of how the Bible recalls the Creating events. On that part, I shall have to apply some technical aspects, in order to combat the onslaught I know that the Newtonian atheists will launch.

The cosmic Calendar as I call it, indicates space-time relevancies where time overheated, as it poured heat into space. At first, it was solid, dense, matter that did not receive cooling, therefore it became the jelly liquid state of the neutron, and later the liquid form of flowing heat. Finally, at the point where the Big Bang arrived, heat got its final form, being gas. Through this event the final element formed, the one we and all other elements, and element combinations occupy. Every star in the cosmos is a proton, with its atmosphere (be it gas or liquid or singularity) acting as a neutron to the value of Π^2 Π. The Π holds the dome of the sphere at a relevancy of 21,99 /7 and to the outside is space, already in a square of 10.

There is one more relevancy I think is worth explaining in this introduction, and that is the relevancy of the speed of light $3\Pi^2$. The 3 holds the photon's relation to space, being part of space, where the proton (Π^2) draws the photon 3 as it connects the photon to time in singularity. The photons do not merely fly around because they have nothing better to do. They move in a straight line, to the nearest proton, that connects them to time and because of the liquid density of the heat, the photons can only break free from the wave, by entering the neutron through the electron (3) to the atom at Π^2. The rest of the wave will pass, on its way to the next proton to which the photon connects. In this way, "gravity cannot bend light" but the density of the proton numbers to the space occupied by matter, will influence a position of the wave of photons, disturbing the wave where the proton density is so great it can influence a complete sector of the photon wave. This then will bring about a disturbance in the density of the wave of photons at one given sector, and as the photons are liquid, they will then re-apply equal density throughout the wave. These are all relevancies that space-time holds to space-time, much in the same way as logs are relevancies numbers hold to the basis of 10.

In order to explain events according to the Bible as the Authentic Author reported it, we first have to consider the relevancy time has on space. When two points are 2000 light years apart, even though each one relates to time at that very second, time to one another is 2 000 years in the future. The time travelling at the speed of light will take 2 000 years to reach the other point. Therefore light leaving point A, will relate to a Universe that had the development and growth applying to a Universe that has a space-time development of 2000 years later. In real terms, the explanation is much more complex, but that I leave to the Thesis.

```
A                              B
O ───────────────────────────► O
1AD                            2000AD
```

By the time the light that departed from point A left point A, space-time had a specific relevancy to development. It leaves A now, but it is 2 000 years to the future of B. It will reach B only in two thousand years. The same goes for B. Light is only a messenger, and if the messenger delivers a letter written 2 000 years ago the news apply to the recipient as if it is breaking that very minute. 2×10^9 years at present. The message however, relate to events 2 000 years to the past of its current position where it is 2 000 years in the future to the events it holds as news happening now. Time is a constant that is true, but in the constant time remains at a relevancy to the heat that produce the events determining time.

Let us place time at a factor of one to the relevancy of heat in whatever form forming time in space.

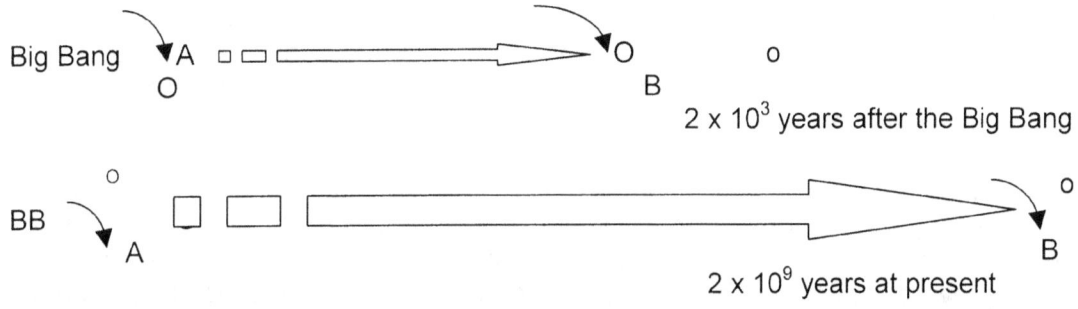

Big Bang   A ░ ▢ ▭ ═══════► O   o
            O   B
                    2×10^3 years after the Big Bang

            o
BB   o
     ▢ ▢ ▢ ▭ ═══════════════════════════►
        A   B
                    2×10^9 years at present

If A and B rotated at an even pace as time to the constant of singularity suggest, then, because A is rotating around a much shorter radius, the value of time in A will be pretty eventful since A has a relativity, which comparing much more action to much less space. On the other hand from the position B finds itself in, B will be to A one million years to the future and B will therefore be invertible. Therefore we can see NOT SPACE but TIME TO THE PAST as light, but we cannot see to the future, because light does not travel back in time. Therefore no messenger will bring news from 2 000 years to the future, light can only move with time remaining with time. Hubble said the further objects are away, the more they appear to go faster, and the most furthest point, seem to rush and run away from the rest.

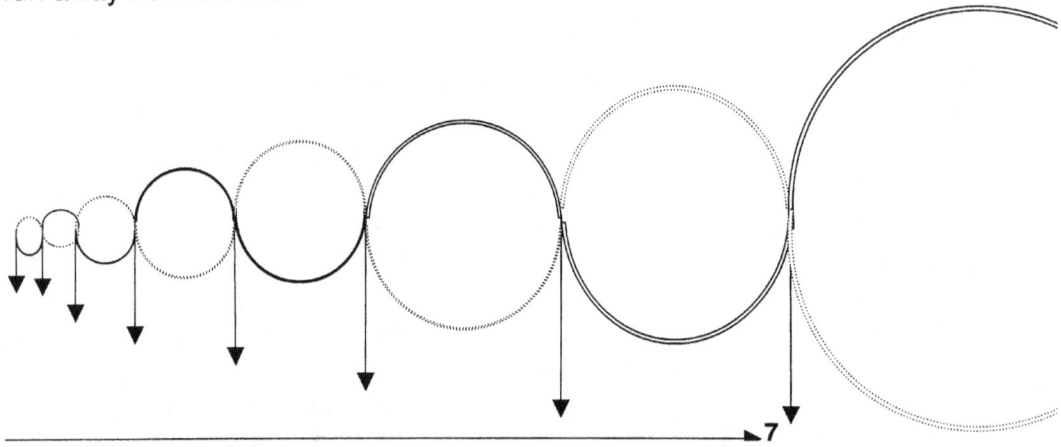

AFTER EVERY PERIOD, (WHAT EVER CONNECTION ONE WHISHES TO APPLY TO THE PERIODE; THAT HAS NO RELEVANCE) THE DURATION OF TIME MUST BE DIFFERENT TO THE PREVIOSE AND THE FOLLOWING TIME. IT IS EITHER THAT, OR THERE IS NO BIG BANG AND NO HUBBLE SHIFT. THE INVERSE SQUARE LAW WOULD NOT PERMIT A CONSTANT TIME DURATION, WHILE THE RADIUS OF SINGULARITY Π EXTENDS.

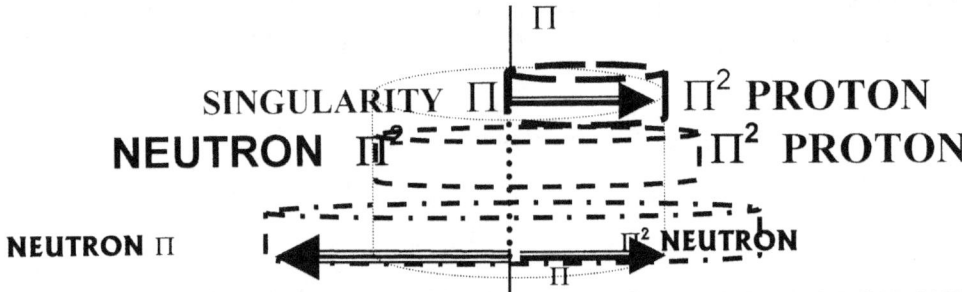

ALL COSMIC OBJECTS ARE A CLUSTER OF PROTONS W CALL SOLLID MATTER, HOLDING A GAS SPHERE WE CALL THE ATMOSPHERE THAT IS A DIMENSION AWAY FROM THE HEAT DIMENSION WE CALL OUTER SPACE.

It is all due to relevancy, in time having space as a product of time. Space is not growing at all. It is time becoming quicker in duration of time. If time remained at a constant and the Universe grew to Hubble's claim, it would have taken the Earth but a few seconds to rotate around the Sun , seen from our view, that we find the relevancy of the distance and therefore the year rotation circle to be much smaller than we presently have.

However, we maintain a constant of one to singularity, the same constant of one that time then had to singularity. Therefore, space had to be smaller, and time according to our view moved much faster. It had a speed relevant to our standards we apply at present, that things exploded, because we cannot even follow the events with our eyes. The truth is the space and the time both are in singularity, therefore the radius back then were the same as at present.

However, the relevancy matter had to matter, were much more compact, and instead, they maintained the same positioning alterations in a much slower duration of time. If A and B move at the same speed, with the closer proximity of matter to matter at point A, point A will experience in slow motion, when relating events taking place at point B. From point A looking at B, light will not move because from point A, the light travelling at point B will exceed the speed of light one million times in relation to the speed of light at point A.

Therefore, from point A the future at point B is invisible, because from point A, time travels one million times faster than the speed of light at point B. This is the consequence of the Hubble shift, and you cannot have the Hubble shift, without the effect time in duration has on space. Science either has to abandon the Hubble shift, which it cannot, because the evidence is there, it is clear and it cannot be ignored.

Therefore, if you apply the shift Hubble saw, you have to adapt that shift to the implication altering with space changes. It is not the space that changes because space cannot grow to more or less; it is the concentration of heat in the space that changes the duration of time. It is because of this that "more gravity will effect time", as Einstein correctly pointed out, because gravity concentrate heat, and heat produce time in motion.

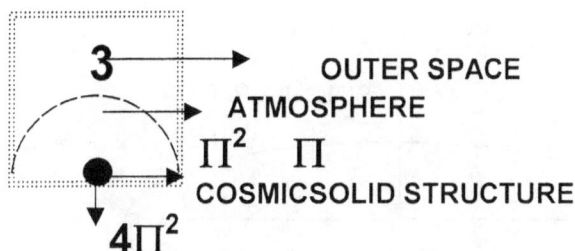

Once again, please I emphasize if your brain is still intact enough to accept this is not for your concern. It is only for the brain washed Newtonians that lost all acceptance through mind control.

All that the following calculations prove is that space holding matter to time where space (10) interacted with matter (7) in the already established Roche limit $(\Pi/2)^2$ to form the point where the Newton time value (Π^2) transmits heat to the double proton $(\Pi^2+\Pi^2)$. Heat in a jelly form separated from heat in a more solid form, forming liquid heat (later to become the gas we see as space) and matter, the third dimensional form of matter. You may feel free to see it in the same manner as the way a farm cream separator spins milk to separate the cream and at the same time, unlike the separator does, form bits of butter and buttermilk. In essence, there is very little difference.

The explaining may be rather excessive to non-Newtonians, but please remember that unlike the Newtonians, you are still blessed with an unblemished brain functioning well enough to accept.

Time and space holds the same value because time is space in singularity and space holds time outside singularity. Time and space is one thing. Time holds the value of singularity outside space therefore it is Π^3.

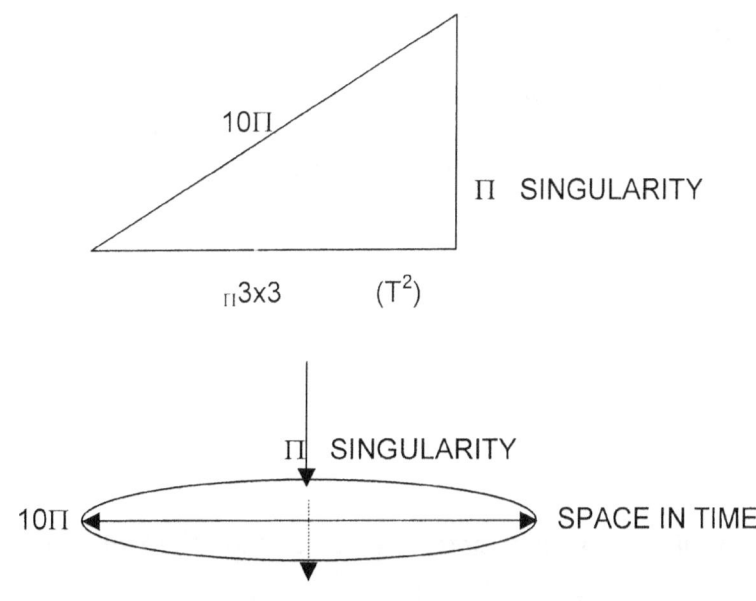

Time has always a value of square (Π) (R^2/T). The linear value time holds to space (R/T) has the value of Π. Therefore space has the value of 9 plus one more value presenting singularity $\Pi=1$ making it 10Π. Since Π^3 holds one identical position in space than space does, it also has to become $10\Pi^3$. This sounds irrational because from our position of time in space and not time as space, singularity as such by itself is irrational. To form the second proton of the double proton $10\Pi^5$ divides into $10\Pi^3$ as heat transforms from space to time.

From the position of time space will always hold Π^3 to Π. This produces time to space (one of the protons) and time in space (the other one of the protons). After developing the double proton, space holds a value of 10Π or $10\Pi^3$, being only in "space to be" 10Π was still in formation. With Π^3 coming as a result of $\Pi^3 \rightarrow (\Pi^2 + \Pi)$ matter formed within the limits of the space-time developed. This lead to a total dimensional possibility of 7. Matter is in a dimension one less than time, therefore in that space the three of Π^3 has the dimensional attachment. Because of this, and the fact that time ($10\Pi^3$) was producing space (10Π), the overall number had to conclude 10Π or $10\Pi^3$, and which ever makes no difference.

$10\Pi^3 \div 10\Pi = \Pi^2$

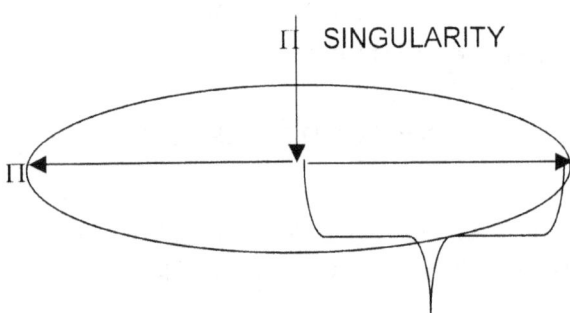

In one dimension space became 10 and in that dimension matter became seven. In order to separate matter (7) and space (10) through time (the spinning of matter) (7) in space (10) and space (10) spinning the matter (7) the following result came about through the application of the Roche principle $(\Pi/2)^2$

Then space mingled with time 10/7 $(\Pi/2)^2$ ($\Pi^2 + \Pi^2$) and space separated from time $7/10(\Pi/2)^2(\Pi^2+\Pi^2)$. In this, the result was that matter formed to space allowing the neutron a time value. Time divided into space $(10/7) \div (7/10) = 2,04$
$(10/7) = 1,4285 \div (7/10) \, 0,7 = 2,04$
Space divided into time $(7/10) \div (10/7) = 0,49$
$(7/10) \, 0,7 \div (10/7) \, 1,4285 = 0,49$

Then space multiplied by time $(7/10) \times (10/7) = 2,04$
$(7/10) \, 0,7 \times (10/7) \, 1,4285 = 2,04$

That brings about that the Roche Principle
worked both ways (double)   1. $(7/10) \, 2,04 \times (\Pi/2)^2 = 5,033$
and with multiplication   2. $(7/10) \times (10/7) = 2,04 \, (\Pi/2)^2 = 5,033$

Resulting in the combined value of $5,033 + 5,033 = 10,066$
On the other side the other combined value came to $0,49 + 0,49 = 0,98$

And the result from this product was

        0,98 x 10,66
=   9,8696
= Π^2

And with this the neutron's time end connected, established the link to either protons when the proton connected to space.

From this the neutron developed the space value of Π. The process is as follows:

The square of time (Π^2) multiplied by the square of matter (7) became in fact one part of the double proton. This is better explained by saying that when one proton (Π^2) delivers heat from space, dimensionally reconstructed. In real terms the space proton holds the value of $(7)^2$, which is 49. From Pythagoras, with the interaction of dimensional equalization the factor becomes the opposite of the law of Pythagoras.

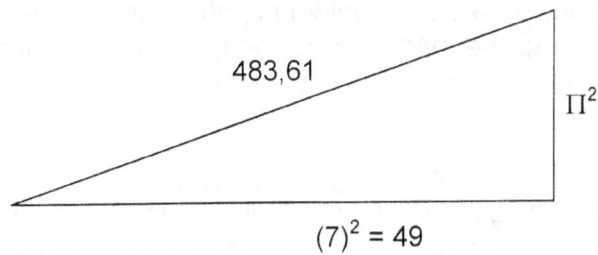

$(7)^2$ = 49 x Π^2 = 483,61. But at this stage matter was in the relevancy of $(\Pi/2)^2$ therefore the factor that $(\Pi/2)^2$ represents, holds the value of 483,61. To get to the resulting dimensional value of 483,61, the square thereof becomes a factor.

$\sqrt{483,61}$ = 21,991

As 21,991 is one half of ($\Pi/2$). Π Therefore, must be that value matter holds, which is 7. When matter divides into space (7 ÷ 21,999) the result from that is Π. Through this the neutron (which is matter) holds $\Pi^2\Pi$ as a factor.

Therefore, for the first time the reason why Π is 21,999 ÷ 7 = 3,146 becomes clear through the application of true astrophysics, as derived from pure cosmology, and not a lukewarm deduction of Newton's Earth physics projected to the cosmos where it does not apply in any event. Through this one must never instantiate dimensions from the equation, otherwise the cosmic calculations becomes Newtonian physics. Again I stress, there is nothing wrong with Newton's physics, as long as the application of it links to life's accomplishments or some derogatory thereof.

Although Π^2 is in a square time takes the square to singularity, being a straight line of 180°. Matter in time holds the half circle value of 180°, where matter is the one link $(7)^2 = \Pi^2$ and time is the other link (Π^2).

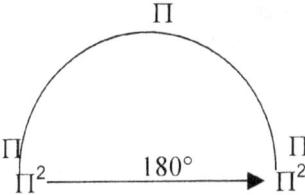

Through that, the value of the triangle in matter, space and time also holds a 180°.

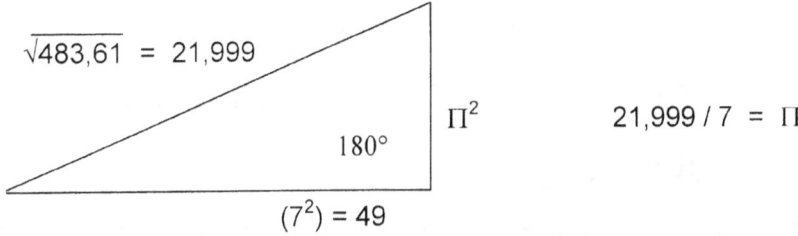

Remember that this is time astrophysics, not merely the projecting of Newtonian concocting directly from Earth to space. You have to use your mind; not just your eyes. Everything, every object in the

Universe is an island in a dimension of its own, created to the specific relation it has in time in singularity and space in singularity $T = (\Pi^2 + \Pi^2)$ $(\Pi^2\Pi)$ 3 and $\$ = R^3/T^2 = 1$ where $\$ = R^2/T \times R/T = 1$.

No person can deny the fact that the Earth is a sphere, excluding outer space, where our need to apply entry into the sphere of inclusion $(\Pi^2\Pi)$, and the law to abide by is the four cosmic pillars. You have to abide by the Roche limit where rules allow you entry, or destruction. No Newtonian can deny that. When you cannot deny the fact that the Earth is excluding space as it is including time the rest is beyond denying also. One has to seek the evidence where the evidence is, where one can locate such evidence and above all, read the evidence correctly. The evidence proves the existence of a binary before the Earth came to be. The four inner planets are left over parts, a reminder of a star that uses to be part of the cosmos.

The spinning of the atom around the point of singularity holds the position of singularity and singularity cannot be without spinning in the Universe. Any object in rotation holds one (imaginary then, if you wish) specific position, a point so precise that does not spin, turn or move.

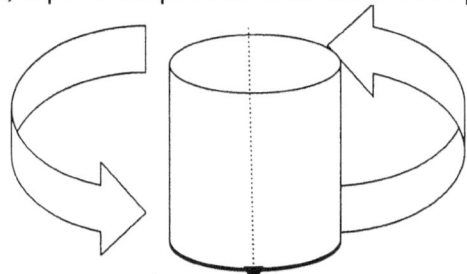

That line that cannot move is singularity. Any point away from that point is space wherever space may be, in whatever form it may hold and whatever density forms that space within that space. The moving of matter gives time (the motion of the space less spinning) its value away from eternity. As much as the rotating line is in the imagination, the line is in eternity and the line is in reality. That understanding of existence of such a line comes from human genius, the way the human mind works in understanding concepts. The line you shall never see feel or touch, BUT ONLY KNOW that it is there because your intelligence permits you the insight to form the concept about the line being there.

THIS FOLLOWING BIT OF INFORMATION IS ALL THE RELIGION I SHALL EVER SHARE WITH THE WORLD AT LARGE

The line holding time to eternity is as much present as the space as the space that is not there. The line is as much understanding as it is accepting through being human. That line that is never there and is forever there is the same as the presence of GOD, which is never there but forever there **FROM ETERNITY TO ETURNITY THROUGH OUT ALL ETERNITIES (EVERY ROTATING ATOM, PARTICLE IN SPACE). I CHALLENGE ANY ATHEIST TO PROVE THAT, THAT ETERNITY IS NOT THERE.**

The Bible says God is in eternity, from eternity to eternity, through out all eternity, and yet again, the Bible proves the to be the best book ever to be written on science that there ever was.

 WHAT THE BIBLE SAYS IS THAT IF THAT LINE IS THERE, THAT LINE PROVES THAT GOD IS THERE, IN A LINE ETERNALLY PRESENT AND AT THE SAME TIME NEVER PRESENT TO OUR KNOWLEDGE. HUMANS HAVE THE CHOICE OF BEING ANIMAL AND NEVER UNDERSTAND THAT CONCEPT OR BEING HUMAN AND UNDERSTAND THAT CONCEPT. YOU CAN BE HUMAN IN BELIEVING THROUGH INTELLIGENCE, OR YOU CAN BE ANIMAL, BEING AN ATHEIST AS ALL ANIMALS ARE, WITHOUT THE BRAIN POWER TO UNDERSTAND.

I CHALLENGE ALL: PROVE ME WRONG!

That is all my religion I can ever share, because from that point my religion becomes far to complex to ever share with any one outside my faith. Science must realize by disproving religion they are only proving their own lack of understanding (therefore being animal) of true science.

All atoms are spinning; therefore, the line of singularity must be present. If all around that line is spinning, that line too must be spinning. The rotating value of any object spinning is Π, because Π puts space in dimensions of the square. Anything in a sphere or in a rotation holds the value of Π. That line therefore is Π and any point away from that line is also Π.

The point of Π comes from within every atom holding Π and that value of Π extends, while remaining one, in different rotating velocities. Each spin in velocity stands apart from any other point in spin velocity. Therefore the group of atoms forming that cluster of protons will also spin, at different rates in accordance with the value of Π as much as in density as in mass.

It is here, at this point, where proof comes in about every orbiting matter holding an influence on the cosmos as an atom. Every star holds singularity as much as the protons within it holds singularity and reflecting the combining effort of singularity and therefore it is spinning.

The group effort of the combined density provide density provide the star its atom status, by giving it a spin value, A cloud of hydrogen do not spin, therefore it is not a combined atom, but is merely a group of atoms spinning in individual time alone. That means every spinning object is a sphere (a Black hole is spinning even by not spinning it runs so close to singularity, it cannot spin any longer to our concept. All we can see is the in going matter that picks up momentum as it will eventually exceed the speed of light just before entering the non existing Black hole atmosphere.)

To calculate Π follows the following method:

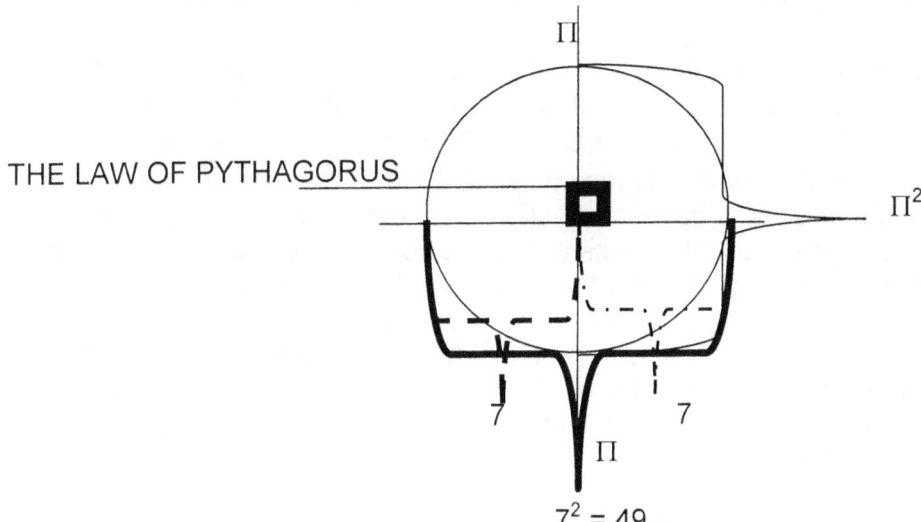

THE LAW OF PYTHAGORUS

$7^2 = 49$

The singularity in the eternal gives time the value of $Π^2$ while the matter that is spinning gives the time occupied by heat the value of 7, and that means the square of 7 is 49. To get the value of Π to space in the occupied space, it will be the combination of time in singularity and time in occupied heat that will position the value of Π to space occupied.

$Π^2$ X 49 = 483 .61.

Since time in singularity is in a square with time occupied, (the double proton) the law of Pythagoras plays its part in dimension equilibrium

$\sqrt{483.61}$ = 21.991

The triangle forming the fourth or electron dimension therefore holds 7 in relation to 21.991 and that equals the value of Π.

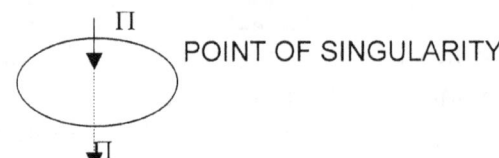

POINT OF SINGULARITY

THE POINT OF SINGULARITY IS WHERE THE RADIUS HAS NO VALUE, BUT THE CIRCLE REMAINS BECAUSE THE CIRCLE IS THERE.

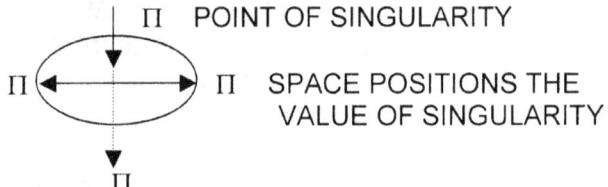

Π POINT OF SINGULARITY

Π SPACE POSITIONS THE VALUE OF SINGULARITY

THE SPACE THAT SINGULARITY TAKES ARE THEREFORE Π TO BOTH ENDS.

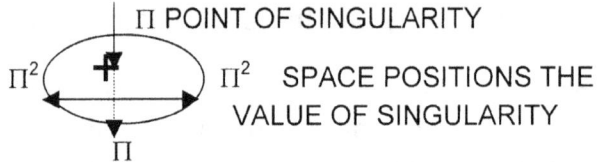

Π POINT OF SINGULARITY

Π^2 SPACE POSITIONS THE VALUE OF SINGULARITY

Looking at the Universe, the Universe is within every atom. Every atom holds the Universe to the point of singularity. From the point of singularity, every proton demands space. With the demanding of space, the atom controls space. That would be the proton. The proton is in total dominating control of space that forms the neutron the neutron holds space from the point of the proton Π^2 to a point of the electron 3. The proton forms time (T^2) and the neutron forms space (R^3). There is no large unified all-inclusive Universe out there that will be either filled with matter, or not filled with matter.

The centre of the Universe runs down each proton in the proton singularity. That is the centre of the Universe. The proton holds a double value of ($\Pi^2 + \Pi^2$), the neutron holds the value of ($\Pi^2 \, \Pi$) and the electron holds the value of three. This combination forms the relevancy of the Universe. $T = (\Pi^2 + \Pi^2)(\Pi^2 \, \Pi)$ 3 = 1836. The test in proving this relation is the calculation there of. Many sources supply different readings. The accuracy of the results is beyond question, because there is so slight varying in the result between the mass of the proton and the mass of the neutron, that it is insignificant. The variation of the readings has to come from another factor that influence the control study results. The only flexing of the result therefore must come from a flexing of the heat supply during that precise time of calculating. That is the only variation that may influence the test results. I do not wish to elaborate on that point at this stage, other than to say it is a relevancy, and as with all relevancies the result may be different according to variations in the relevancy influenced by heat occupied and unoccupied changing the dynamics at that precise time.

I have also stated that the Universe has a precise centre running down the centre of every proton, proton cluster and individual atom. That centre point holds space occupied to a dimensionless point in singularity.

THE TIME THAT THE SPACE WILL REQUIRE IS ($\Pi^2 + \Pi^2$) TO BOTH SIDES. THAT ROLE THE PROTON TAKES. SINCE NOTHING IN THE UNIVERSE CAN BE IN TWO POSITIONS AT THE SAME INSTANT, TIME COMES FROM THE PROTON ALTERNATING POSITIONS TO SPACE AND TO SINGULARITY.

Π POINT OF SINGULARITY

SPACE POSITIONS THE Π^2 ($\Pi^2 \, \Pi$) VALUE

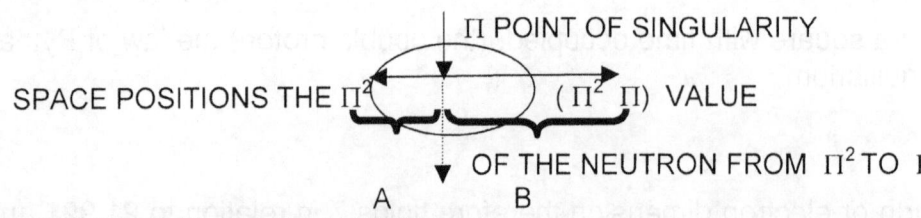

OF THE NEUTRON FROM Π^2 TO I

A   B

(A) SPACE THE PROTON DEMAND
(B) SPACE THE PROTON CONTROLE.

Holding Time at a very precise measure is the labour of one proton Π^2. While the other proton remains in contact with singularity at a value for that eternity of Π^3. While the one proton holds the time in eternity Π^3, the other holds equal value outside singularity at the time demand (proton Π^2) in the space control (neutron $\Pi^2 \, \Pi$) to the value of heat unoccupied in space (3)

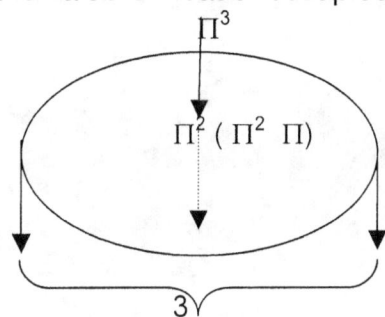

In this there are still a natural balance forming due to the interaction that the proton Π^2 and the neutron $\Pi^2 \, \Pi$ hold to singularity. This is the point where the Roche limit holds the space dividing in order to prevent any destruction of singularity to any of the structures.

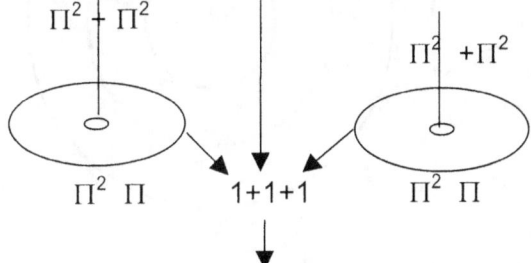

SINGULARITY EFFECTING HEAT TO THE VALUE OF 3

The unoccupied space-time is not distant from singularity because it too holds a position to the cardinal singularity, therefore the heat allow the spin of two atoms to maintain heat control while spinning in opposing directions. The cluster of atoms remain the same in domination and control to space-time, be it in a cluster of atoms the size of an atom or a cluster of atoms the size of stars.

Where two cosmic atoms control space in a joint occupation of space, the space domination will have to set in borders to defend each one in singularity. It is the singularity within each object that gives that object validity to exist in the cosmos. It gives the object independence as it makes that object the Universe. From that point holding singularity comes to space that controls of the atom. The atom is no longer in overheating but is controlling the heat by accumulating heat in order to grow. It is the atom that grows and not the Universe that is growing. The atom no longer is in the heat expansion, which brings devastation but that control which the atom influence put the limit of the heat under its control. The atom is now the epitome of gravity by expansion control as well as contracting control. The Universe became the atom and the atom's diversity became the Universe. Every star in every galactica and every galactica becomes an atom through the discipline that the atom brings to space-time.

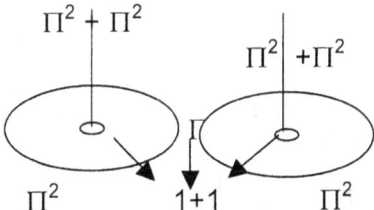

SINGULARITY EFFECTING HEAT TO THE VALUE OF Π

The atom forms the Universe as it shapes the Universe because the atom is the epitome of the Coanda principle. The atom spins and the spin provides an accumulative that drives stars, which in their turn give an accumulative spin that, drives galactica and in all that there is a centre governing singularity controlling every spin just as the simple top showed us.

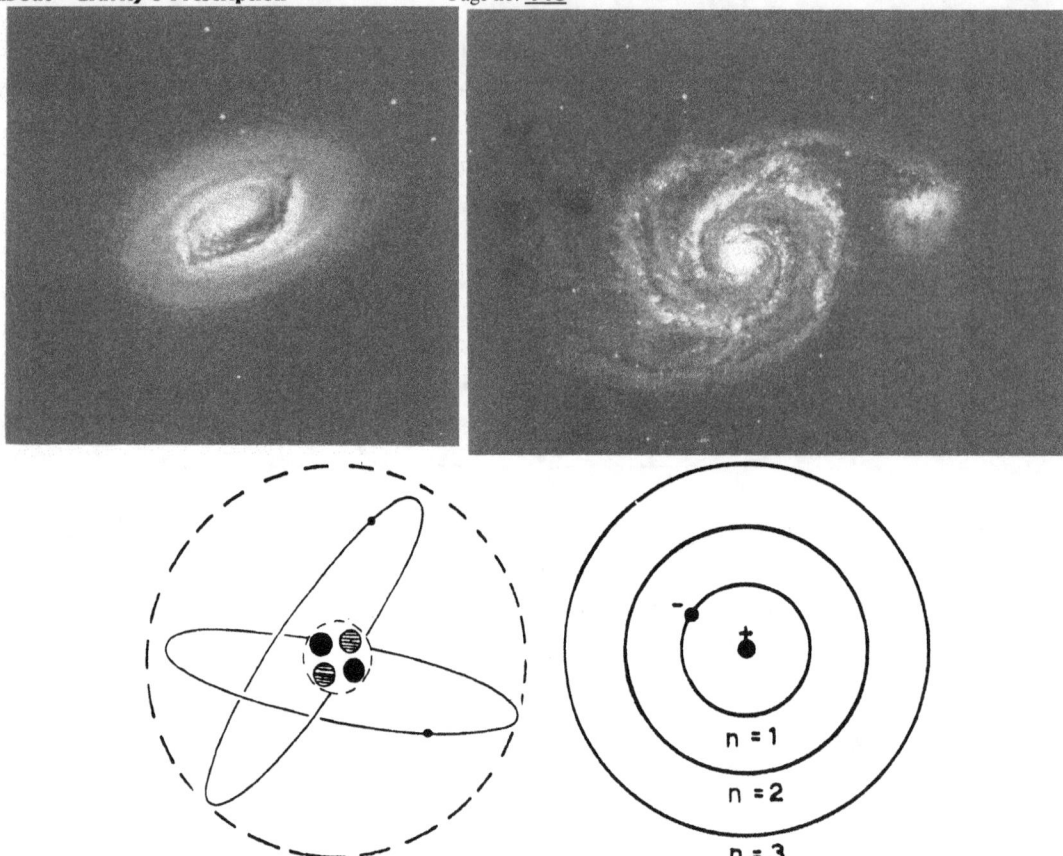

The atom is the template on which the entire Universe is formed. The atom is a galactica, it is a star, it is a bicycle, it is a supersonic aircraft flying in the atmosphere, it is a moon orbiting another containing cosmic structure, in fact only nothing does not use the atomic principle as a format in the Universe and nothing is the only fact not represented in the entire Universe. There is a solid centre, which is supported by a liquid outside, and the balance between the two determines the relevance applying. The atom therefore is the Coanda effect which is the Universe in its entirety.

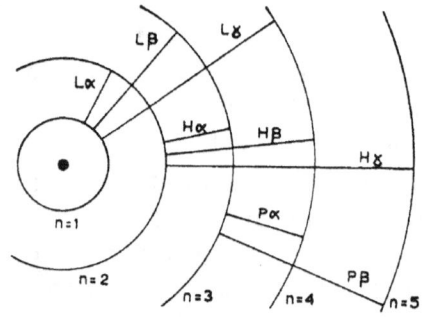

The control the atom has in regulating heat is plainly godly simple as it is Godly genius. The control the atom has on heat by expanding everything outside while accumulating heat, as a time retardant inside is that which creates the Universe just as much as it is that which produce the Universe.

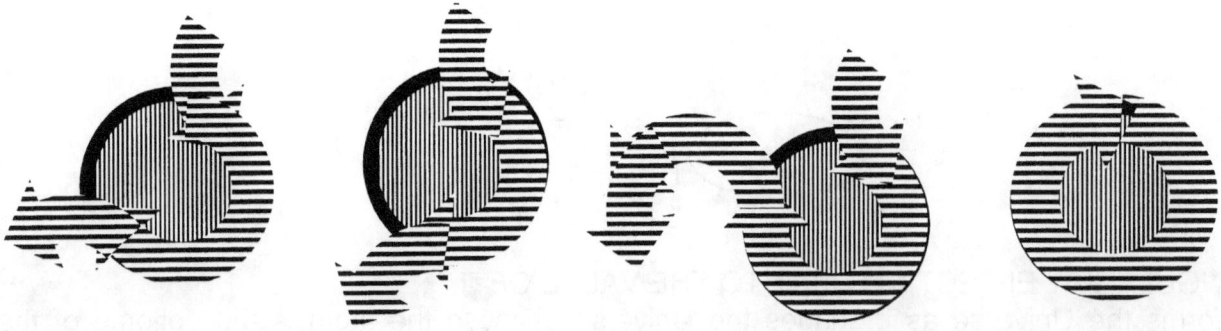

That what I am about to explain may sound inconceivably simple but don't blame me for that. It is not my fault no one brought what I am about to say into the context of gravity and if I don't explain the most mundane in connection with gravity there is no one going to do it.

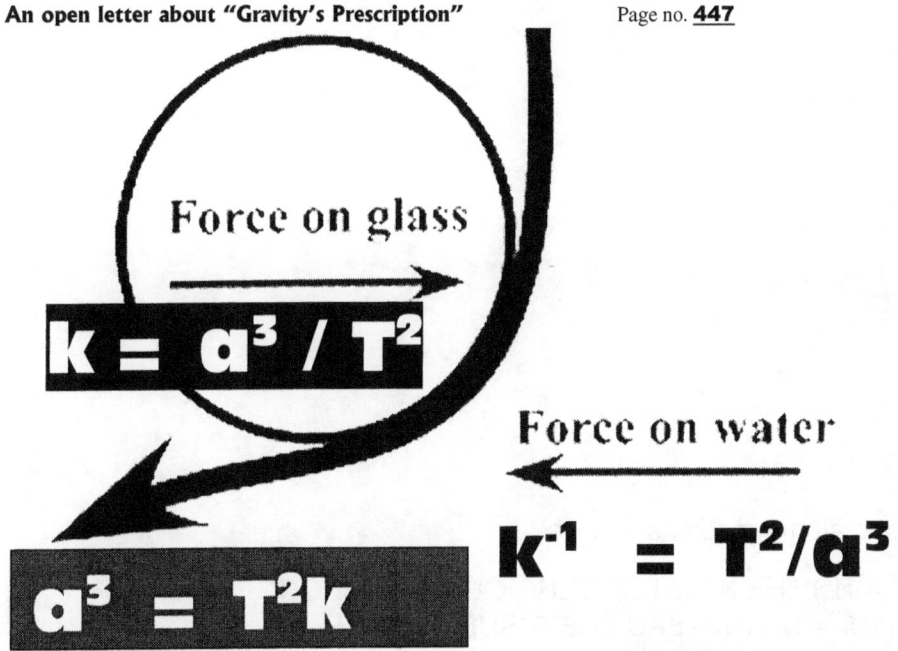

The **space a³** of the unit is **defined k** by the **flow T²** of the liquid

By producing space in expanding when heating the atom not only control what belongs to the atom but also that what is in the control of the atom and what is not in the control of the atom since the entirety of the Universe is the world of the atom.

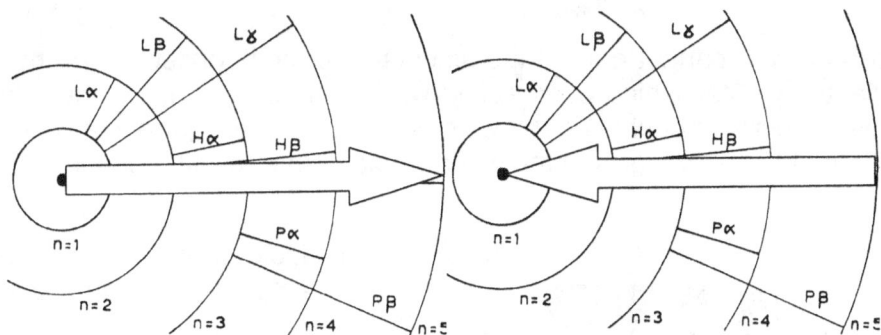

What we see as heat is relevancies because as the relevancy within the Sun changes the atom adapts to the changes. The atmosphere of the Sun becomes denser, which we see as being hotter and the containing becomes stronger. The atom has to reinvent it by adapting to the changes or different surroundings. In this manner the motion that the star provide which is so much more than what is the motion is we find in outer space that the hot / cold dynamic changes all together.

Depending on the flow of time, the atom will accumulate heat within the structure or on the outer side of the structure all depending on which at the moment holds the strongest relevancy between $k = a^3 / T^2$ and $k^{-1} = T^2 / a^3$. That is gravity committed by the flow of space-time and controlled by singularity $k^0 = a^3 / T^2 k$. **This was also what the Big Bang was all about.**

This became possible as the Coanda effect brought the speed of light as gravity (**k**), which brought the motion of the Universe in contraction to the speed of light T^2 that made the Universe be the speed of light a^3, which is exactly and specifically what the **GUT** theory proves.

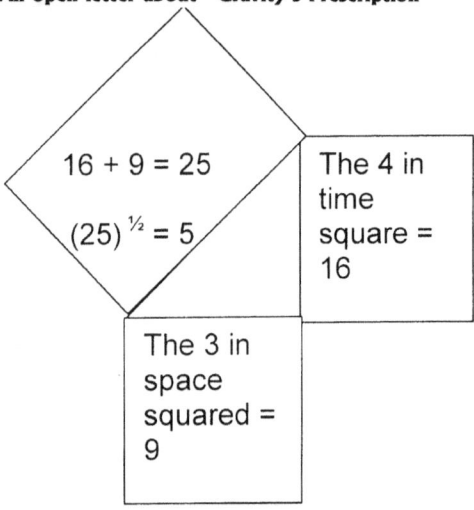

The relevancy between the two cosmic atoms reads as follows:

$T = (\Pi^2 + \Pi^2)(\Pi^2 \Pi) \; 1 \; (\Pi \; \Pi^2)(\Pi^2 + \Pi^2)$ THE 3 NO LONGER HOLDS POSITION
$= (\Pi^2 + \Pi^2)(\Pi^2) \; \Pi/2x \sqrt{} 1x \; \Pi/2 \; (\Pi^2)(\Pi^2 + \Pi^2)$ INSTEAD THE Π SUBSTETUTE

THE BORDER
$= (\Pi^2 + \Pi^2) \; (\Pi/2)^2 x \; 1x \; (\Pi/2)^2 (\Pi^2 + \Pi^2)$ ALL BARRIERS COME INTO EFFECT CLOSING RANK ON THE BORDER THAT MAINTAIN SINGULARITY.

When the one cosmic atom cannot defend the singularity of self-conservation, the other object will destroy the singularity by establishing overheating with in the singularity of the minor atom. I think NASA refers to this as "blowing bubbles or blowing heat " but do not quote me on that. I just found it amusing at the time that the best brains in the world would come up with nonsense like that.

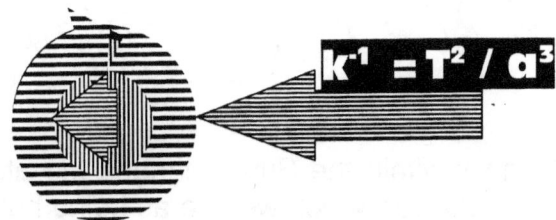

The spin of the liquid proves the value of the relevancy. The stronger the motion is that the liquid generate, the higher would the contraction be and the lower the spin motion is that the liquid generates the higher would the expanding be. In that we find a definite favouring of either the factor of seven or the factor of ten depending on what the situation will dictate.

Depending on the balance there are the contraction will be totally dominating but never to a point where it annihilates expanding and on other occasions the circumstances would be that the expanding may dominate but also never to a point where it devastates contraction.

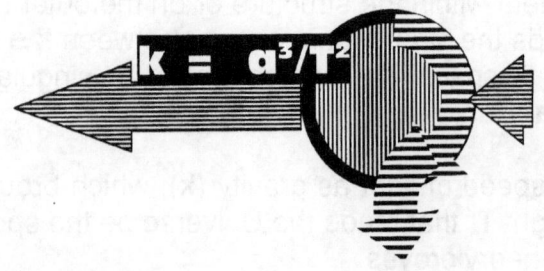

The motion of the liquid factor puts the aspect of time in eternity in relation with infinity where the ratio that develops gives infinity the chance to interrupt eternity. Infinity in the centre is immovable but has to associate with eternity since the tow parted by space. Therefore we have eternity having three positions that goes square since it develops an alternating stance with singularity and in that the three of eternity develops a square since the angle of developing cross 180^0 by the margin of 90^0.

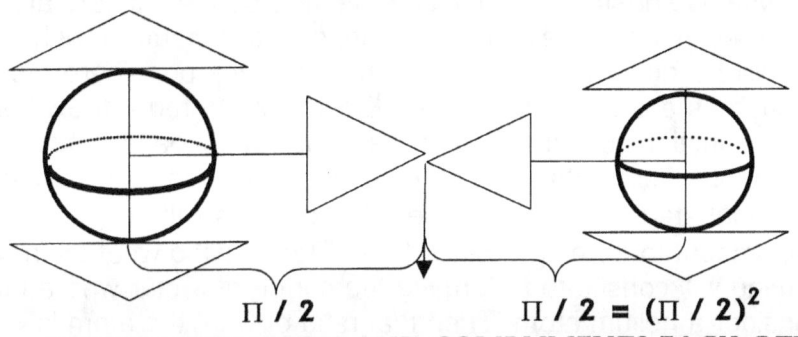

$$\Pi / 2 \qquad \Pi / 2 = (\Pi / 2)^2$$

SINGULARITY MEETS AND COMPLIMENTS EACH OTHER.

To the one object, anything distant from its proton cluster is space, space in whatever form. By destroying the singularity, the space becomes heat either under its influence or under its control. On the one side, space holds the value of Π and on the other side it also holds a border of Π. The time however changes to defend the singularity therefore the neutron time square then holds position where space normally holds position.

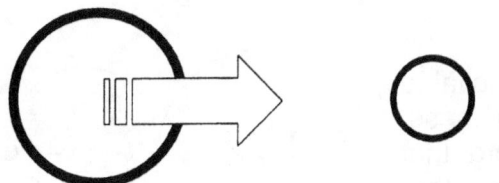

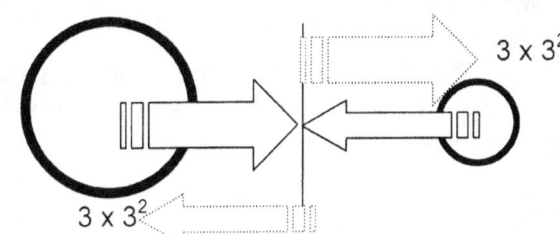

SEEN FROM BINARY MAJOR EVERYTHING STANDING AWAY FROM ITS PROTON CLUSTER IS 3, THEREFORE EVERYTHING IS 3 Π^2. The 3 are space and the Π^2 holds value to the space. This means that when the space does not relate to the proton cluster holding the Π^2, the Π^2 also becomes space at a value of 3^2. The value of $3\Pi^2$ holds the seed of light in relation to singularity, while the cluster of photos form space in three dimension at 3^3. Where Π links to Π^2 the value of displacement (moving through space) is well above the speed of the photon and the photon connecting to the proton is at a relative displacement of 29.6. The wave holds the relative displacing value at 27. This then bring about that the wave will always shine red in approaching any object and will always loose luminosity. With this the photon always holds two positions to space-time one being in direct contact with the proton at $3\Pi^2$ and in direct linking to the wave of photons at 33^2. By the way, the consequence resulting from this is that the electron will produce two signals under close inspection

LINE HAVE DUAL OR MUTUAL SINGULARITY

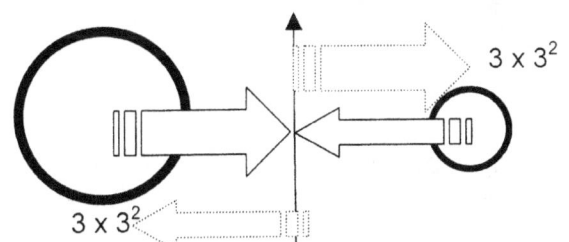

The mutual line of singularity $(\Pi/2)^2$ holds a position referring to each objects individual line of singularity in the value of $(\Pi^2 + \Pi^2)$.

In the cosmos there are a division in what is defined as a unit. The solar system is an integrated unit that is defined by a ratio of 7 / 10 and this famous Bode and Titius discovered which brought them ever-lasting and worldly acclaimed fame.

There then after this is a division between units that I explain on another occasion and which we can

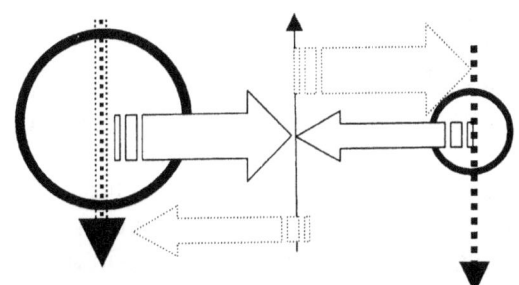

identify as the Roche limit at $(\Pi/2)^2$. This is when the Unit is under the relation control of another factor where there is a expanding and a controlling part while the Unit does show a relation but the relation has got much independence about the actions of the controller. The Earth Moon combination is one of these factors and the Moon does form a unit in relation to the Earth but also act much more independent than does a cloud for instance.

We should be cautioned not to consider what we deem as the ultimate atom to be the ultimate atom the cosmos chooses. That concept what we think of having two protons and a neutron with an electron is a phase that came and that will be going into history as time marches on. There must come a time when having one particle such as a proton will cause a Black Hole to form and when stars will be a cosmic structure with the function to collect what we consider to be sub atomic particles and collect those particles into one atom that then immediately form a Black Hole. Fortunately in that process will come the final stage of the Universe. That process will be (7 /10 Π/ 2)= 1.09955 where (10/7Π/2)= 2.244 will amount to forming a Black Hole. Then having what we now think is a completed hydrogen atom forming will constitute to forming the centre of a star where the Black hole develops at what we now consider a helium atom. Then that ratio of what the atom is will be much different from what we now have. It is all locked in the developing of singularity and how singularity manages the space-time it control. It is all about the ratio one find between time in eternity matching time in infinity and the result of that is space-time. The atom at this point concerning the singularity driving the Universe in reference to us is $(\Pi^2+\Pi^2+\Pi^2+\Pi+3)$ =35.75. Then this forms outer space and with this in relation to singularity Π it is 112.31. However, there are numerous other Universe combinations that was, and that will still be in time to come and each one is served by an independent Big Bang.

As time places coordinates in relation to motion the immovable moves without ever moving. Because of conciseness, compactness, contractedness, briefness (hell, the lot including every word that describe or has any inclination to describe and even all put together can't come close to what is needed in words to indicating the crowdedness and cramping that was going on at first when space was still a thought and infinity parted from eternity) the divisions came

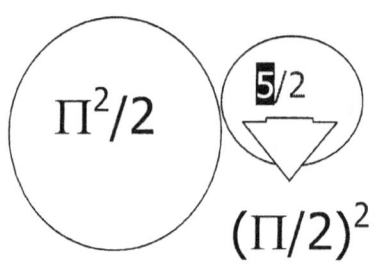

abrupt and conclusively at Π^2 forming 10 on both sides of the divide. At first the dividing was in relation to units forming as time and in eternity split from time in infinity. In that there was the four relating to a common centre representing singularity by contraction.

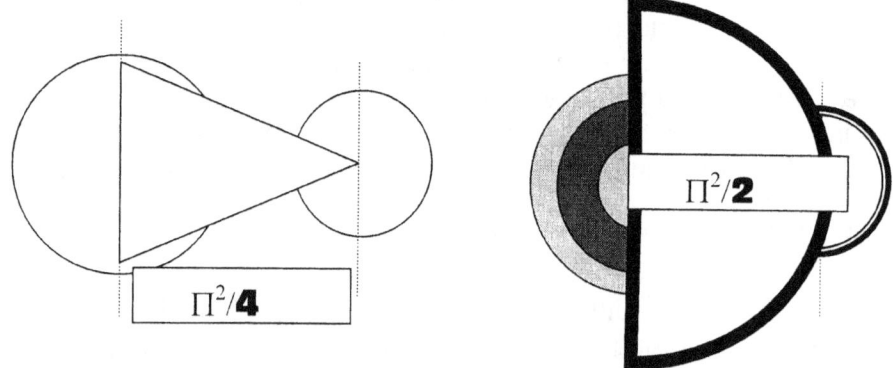

From the mathematical legacy we can see that time was forming 5 but with material not formed yet, that which later became material moved through time holding the value of 5, at a measure that was slightly less than half the value of time and still had the notion to move being Π^2 /4, which was at slightly less half the value of 5.

But as space grew into a stronger part of the Universe time came into a square having five on both sides of the divide and time gave space an integrate part of the motion that time provides. Then the motion became Π^2 /2 and material found a better and a much more valued relation in time. This only happened while time grew less, which gave space more value in $a^3 =$

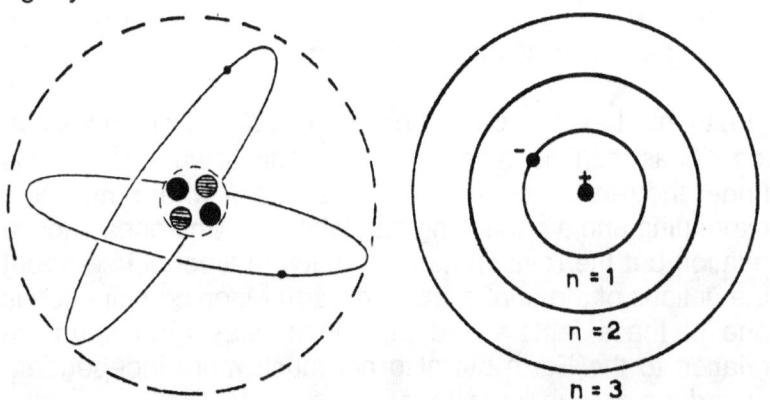

$T^2k.$

That produced the atom we now enjoy with three factors instead of only two, as was the case in the previous dispensation. There then was time, space to hold time and space lagging behind time. That part that was lagging was excluded and that part formed a secluded unit that became the atom.

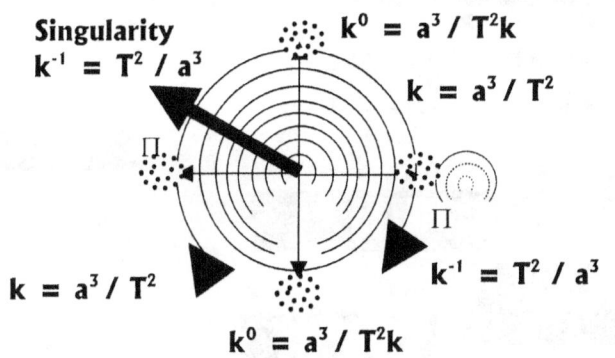

Every time the point rotating changes direction it crosses the divide and in that the entire Universe changes. What comes down then goes up and what went forward then goes back. The line is a cutting limit that divides one sector from the other like no other found anywhere. In that the divide produces changes that are beyond comparing to anything else and in that we find such conclusive distinction in gravity. That which cannot spin does spin by never actually spinning.

From investigating this we can presume with very little doubt as to how the universe came about.

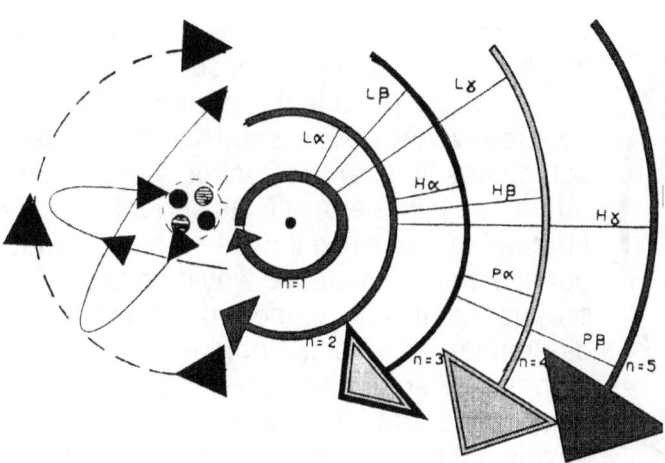

The atom is the Universe that is true but that sounds like rhetoric. Let's break that down a bit by dissecting what this means. The atom is more than just the main ingredient because the atom is also a calendar of the Universe. The atom is a storage facility of the Universe and the atom is a building stone of the Universe. The atom is the final formation the one that will come last as much as it is the one that came first. The atom is the custodian of singularity and the guardian of maintaining the Universe.

What can I say to describe the atom…the atom is the Universe. But the atom is not what we wish to see as the atom because comprising a new singularity status will differently drive every era and in that status is locked in all the changes that came this far and that will come in the future. Those with the astonishing mathematical abilities that wish to reinvent the Universe because they believe they can match the mind of God, they wish to know when the atom they know came into place. Well that is easy and since it is so easy I suppose it is far beneath their ability to see. It required someone with my unsophisticated ability and lack of up bringing by education to realise the truth. It was when the neutron was limited between 7 / 10 and 10 / 7 and time was putting the proton $(\Pi^2+\Pi^2)$ at 4. That is when the Universe we identify with came in place. It was when $10/7\,(4\,(\Pi^2+\Pi^2))$ =112.795 and within the star space broke down at $7/10(4\,(\Pi^2+\Pi^2))$ = 55.27 and the neutron arrives at a factor of one when there is only an electron (3) serving the proton $(\Pi^2+\Pi^2)$ giving a displacement value of $(3(\Pi^2+\Pi^2))$ = 59.21 after which the Universe collapses because time in our era catch space.

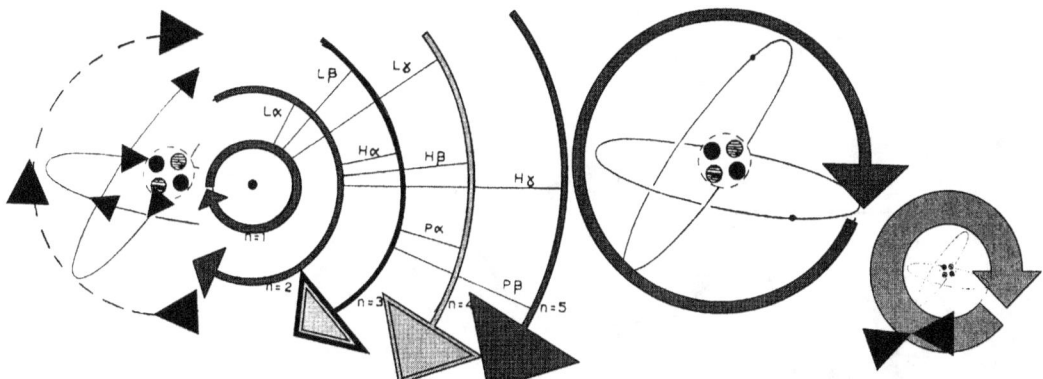

Going back to the start is going back to where space was a thought that divided eternity from infinity.

Time cannot move back but in relevancy time can respond to the nature of motion contradicting its repeat by the nature of the repeat. Time progressing $T^2 = a^3/k$ has a repeat of $k^{-1}=k/a^3$ which leaves a net growth of $k=a^3/T^2$

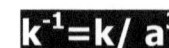

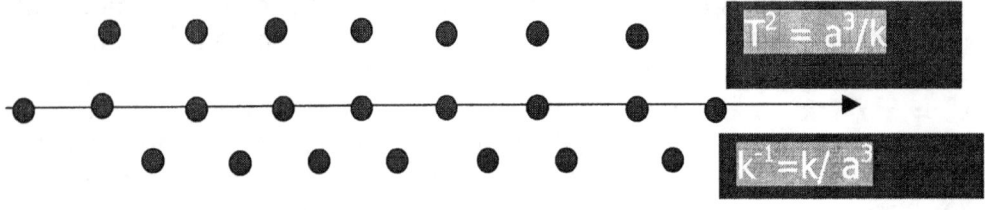

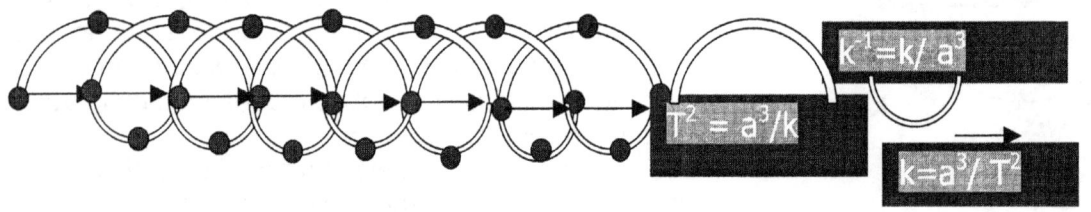

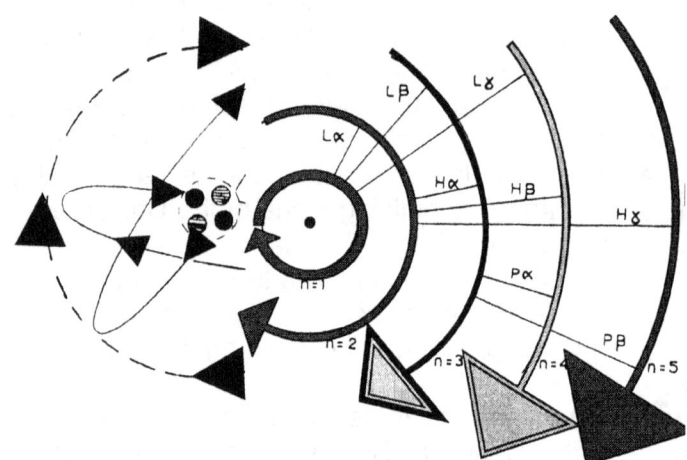

When referring to how things develop I have this human tendency to put such matter in relation to the past. In this matter I must beg your forgiveness but it is my human nature I sometimes find hard to control. That, which I refer to as if in the past is as current as you or I breathing, but since it does no match what our everyday experience would have us believe that it is occurring around us I differentiate in a human perspective commonly use. In this I use the past but please note that I state without excluding anything that in the cosmos there is no future or past in the sense we humans wish to observe time.

In science everyone has the opinion that material exist while the rest is nothing. It is the atom and around the atom is nothing. It is nothing keeping atoms or material apart. Such a view is extremely short sited and is evidence of a lack of thought. It is time we put true thinking power to task when dealing with matters in cosmic proportions. I do not put myself on a pedestal. To the contrary any one can see how shallow and little educated my work really is and yet, if someone with such a little educated background can see what I see, how much more is there to see when persons with true mind power star to truly think on cosmic issues. Three words that should be taken out of any cosmic context is nothing, maybe and suppose. From those three words I could find no evidence in my entire search.

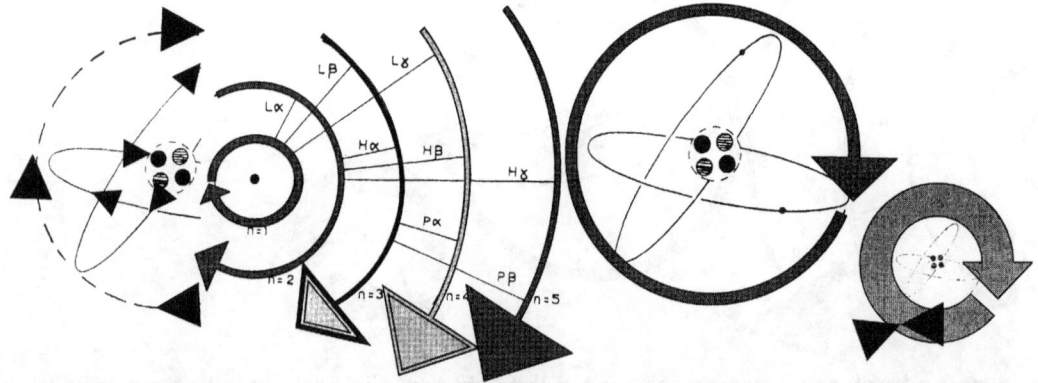

We find the atom spin and think of the atom as where we now are. Sure that is as correct as it can be but the atom spins because new alliances come about from previous conceptions. Being previous does not mean it has moved to the past, as did napoleon and his thirst for war, no it is just functioning holding prominence below the veneer of the current.

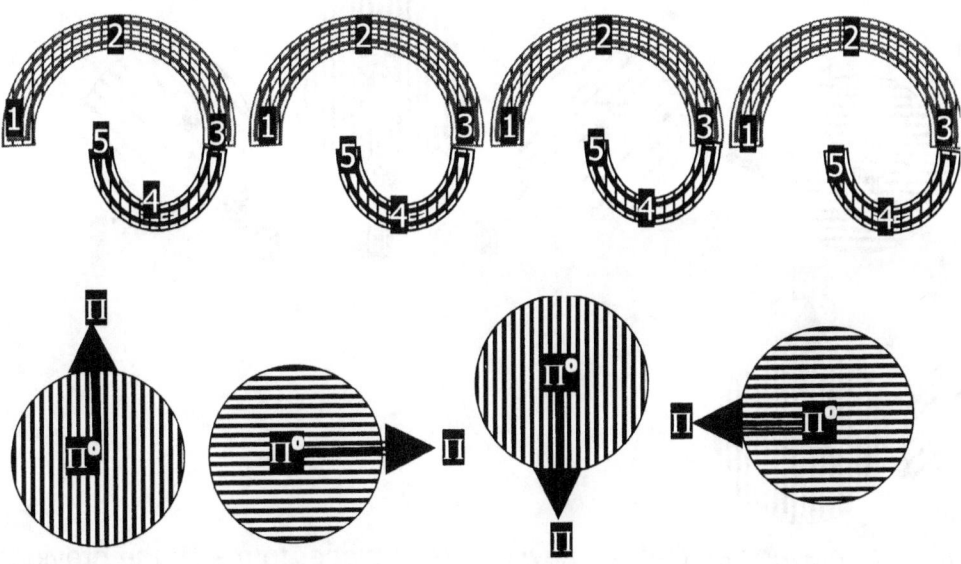

The motion of the atom still is the result of the motion of much smaller groupings of retarded heat that play a catch up game with time.

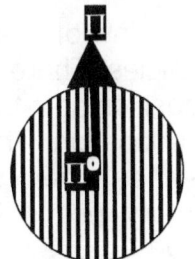

The time retarding grew as the spin involved a higher degree of relevant points. At first there were two points namely this side and the other side of the Universe. Then the four came into prominence. Later eight points came into affect and so it carried on. This was time developing as time always increased the back log it formed with the current situation prevailing at that moment.

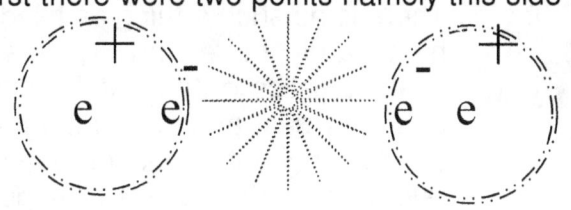

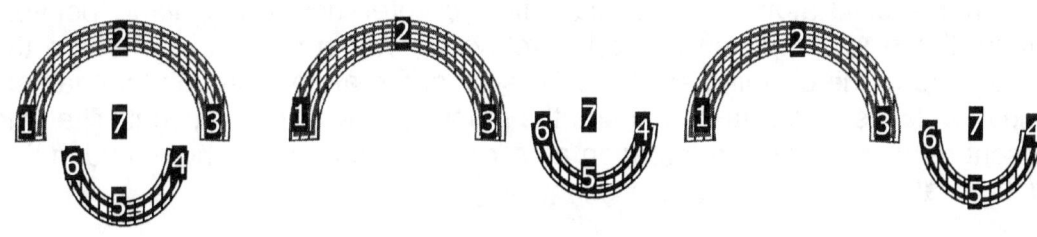

The flow of time is heat flowing past points that was there since eternity parted from infinity. It was there when the cosmos came about. What distances the pointers from each other is heat or light or whatever one would call the formless filling of space-time.

Present 2

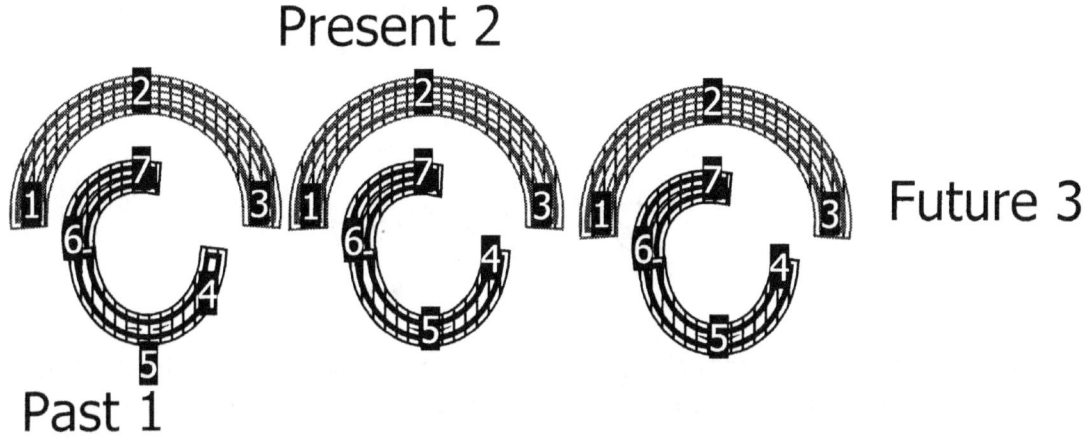

Future 3

Past 1

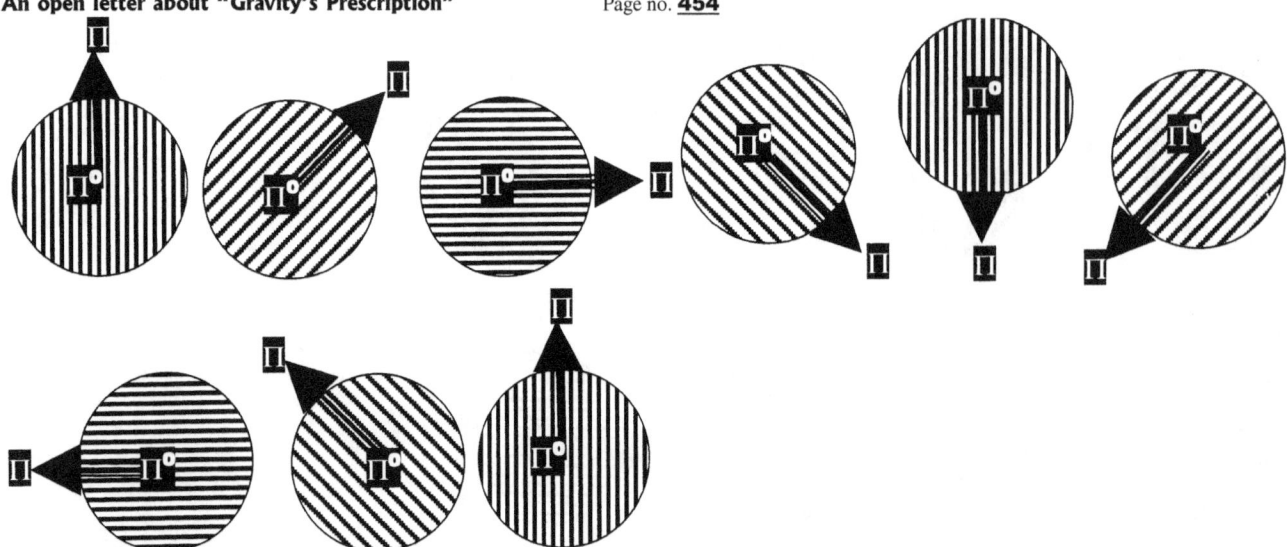

Every time the increase brought new relevancies that put new orders in placer to match the previous requirements. But also it brought about that duplication by progress of space increase was in advantage of the situation.

The relevance we find in the atom is a flow of heat. There is no such a thing as fixed particles. There is heat and there are points positioning the characteristics of the heat at such a location. The heat flow past and at that point the heat adhere to a positional interpretation lead on by points in reference to other points in singularity.

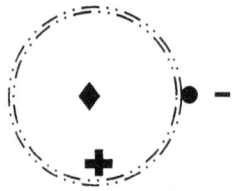

The Atom is a solid atom but it is space-time in reference to singularity pointers that mark those spots, which forms the relevancy within the markers where the markers perform the claiming as characteristics of the atom. Through the markers flow heat, which is time retardation and the pointers, are time in present. The time in present refer to the time retardation as allocated reference markers and in that allow the heat or time lapse to catch singularity.

The atom is taking in heat and the heat in that instant inside the atom adheres to the form, which the pointers or markers prescribe. As retarded time moves on the heat will change form in relation to the allocated position and to what the position that holds the pointers dictate the form will be. It that sense the entirety of everything there ever can be inside our cosmos is liquid time and everything dictating what form the liquid will take on is singularity that is holding pointers which acts as the solids.

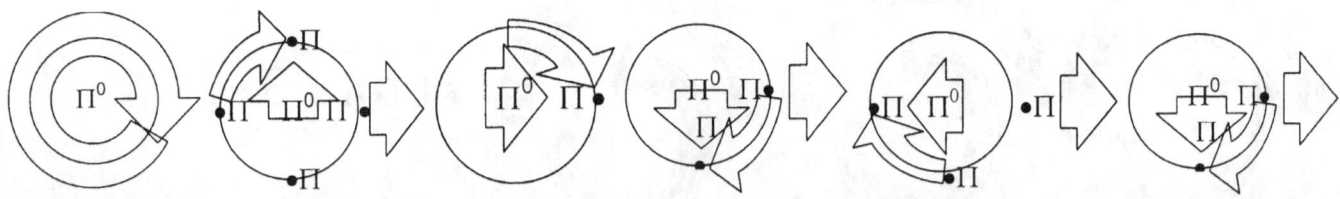

In the motion there are the four markers indicating the limit of the atom while what fills the atom is the motion that flows directional as time catching time by destroying the backlog of time.

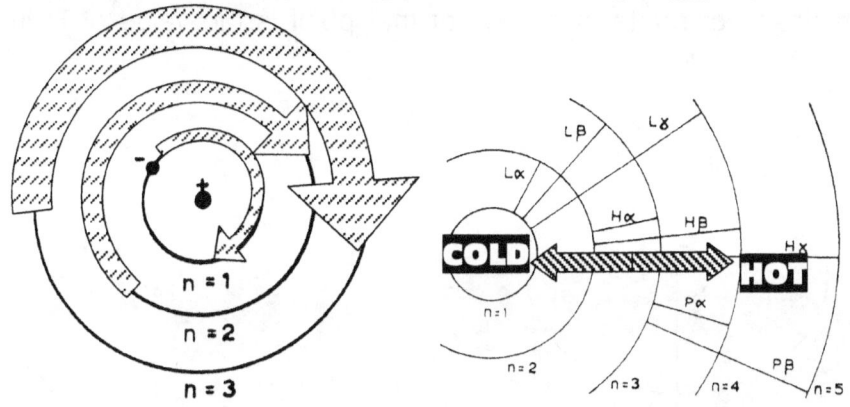

The liquid of time is part of what we see and the solid forming time is part of what we cannot see. The solid shape the liquid into forms and specifics while the liquid provides substance. The liquid is part of the Universe we materially hold and the solid is part of what we in spirit hold. The solid is not a part of the cosmos but is there in directing that which is apart of the cosmos. It is singularity that control heat and the flow thereof while it is heat that provide singularity that which is generated as time retarding.

The cosmos is the Titius Bode law and the cosmos is the four pillars.

Coming back to the calculation of Π, the Titius bode law flows from the figure of Π

Coming back to the calculation of Π, the Titius bode law flows from the figure of Π

THE LINE OF DUAL OR MUTUAL SINGULARITY WHILE EACH ATOM PROTECTS ITS INDIVIDUAL LINE OF SINGULARITY.

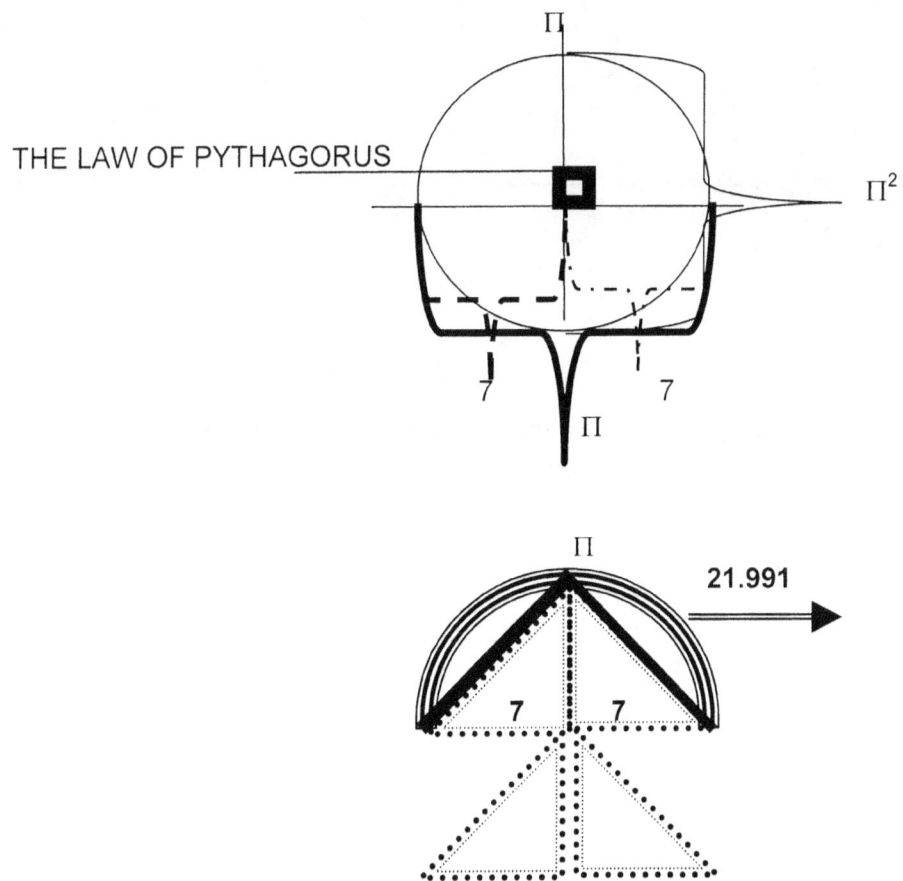

The combined value of 21.991 holding the circle is a combination of $10 + 10 + 1 + \Pi^2 / 10$. A more specific explanation about this aspect is elsewhere. At this point, I wish to (once again) confirm the value of space being 10Π, and how that proves the effect the Titius Bode Law holds on effects science merely brushes off as gravity pulling or pushing or even shoving.

At one point in the vertical, the Earth does not rotate at all and at that point is the value of Π in singularity.

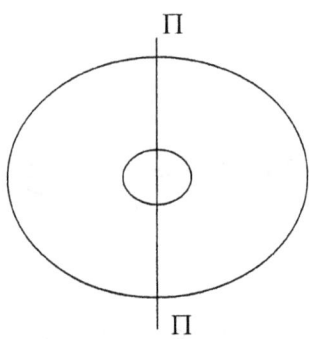

In the calculation of the specific value of, Π I showed that Π takes on the value of seven (7) in the form of matter.

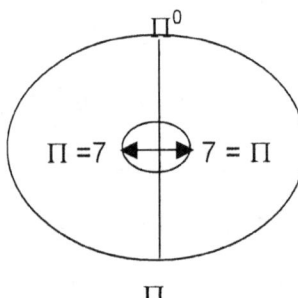

I also showed that 21.991 relating to space replace the value of space in the number of 10

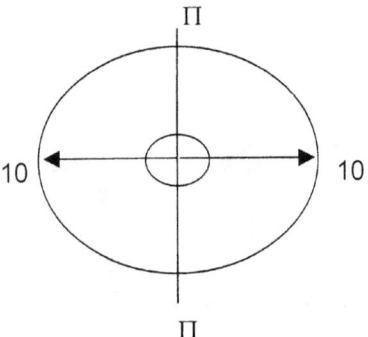

This means that the value of space in accordance with the Titius Bode Law effecting the Earth, the value of space-time is $\Pi = 7 / 10$, at the Arctic and at the equator the value of space-time is $10 / 7 = \Pi$.

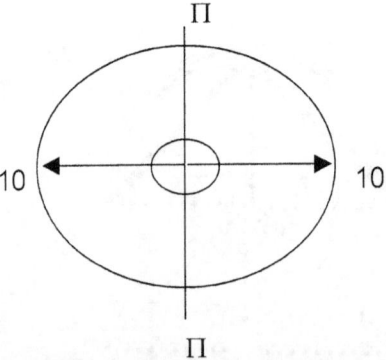

At this point, I wish to raise an illusion that the Earth is "bulging at the equator because of the slinging effect of "gravity" by means of a centrifugal force flinging matter away because of the spin. However that is not the main reason why I show this because at one point later in the book I shall show how the Titius Bode Law brings about the effects of La Nina and also El Nino, the weather phenomena science so desperately tries to connect to "GLOBAL WARMING".

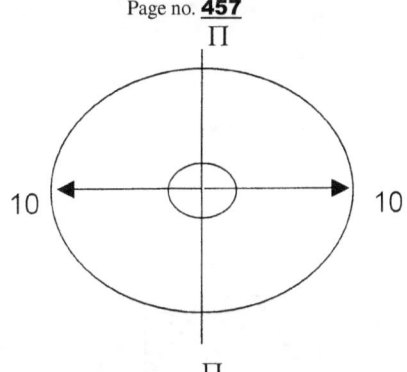

At the Arctic the space-time is at a minimum of 7/10 because of the position, it holds so close to singularity, and at the equator the point space holds is at a maximum, because of the same reason.

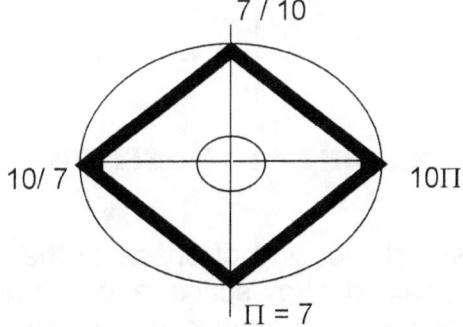

Therefore the Earth will be at a larger diameter horizontally (10/7 = 1.42) than at the vertical diameter (7+7)/10 = 1.4. With the atmosphere (space-time) at the Arctic regions at a value of 7 to 10 and at the equator at a value of 10 to 7 we can see where heat accumulates through the natural comic dispensation and location of heat in space-time. The equator holds a much higher relation to heat unoccupied than does the Arctic region and this will allow aircraft to hold a much higher speed in relation to actual time taken to fly at the Arctic than at the equator. Flying over the Arctic will be much quicker than flying over the equator, although actual speeds may be the same. It is all due to the relevancy of space-time.

The Titius bode Law holding 7/10 and 10 / 7 extends its influence much wider, but the influence of extending as far as the Earth is concerned is s drop in a bucket to the effect it holds on us through what the Sun provide in the manner the Sun 's influences the Earth where space-time heating goes. Changes in the Sun influence and the space-time in the atmosphere of the earth which in turn heats the Earth core and the heat accumulation that the Earth uses as a driving fuel which the Earth reserves to influence the motion is cardinal to changes occurring.

As much as the Earth claims its value of the Titius Bode Law, the Sun takes its share to a much higher proportion and similarly, the moon also takes its claim on space-time by a smaller margin. This brings about (partly) why the Earth / Sun has an oval orbiting pattern and the moon forms high and low tides. To explain the Titius Bode principle in detail one must once again return to singularity and understanding that is as simple as it is totally uncomprehending.

There is singularity at Π^3 in singularity at Π^0, which forms matter we named the proton Π^2 and that is another name for spin through exciting motion. Connecting eternity to the proton Π^2 is another proton connecting time to infinity as the one proton alternate positions and with the alternating comes infinity breaking the eternity into fragments. The aspect just mentioned is only the time aspect because this relates to the neutron that holds space to time. Getting into the thick of this will hold no purpose, but in THE THESIS, I elaborate how the neutron has no mass because it is space, confirming the matter of space to time.

The one proton takes time to eternity, by joining space in singularity, reducing space that holds heat in unoccupied matter. This dipping into singularity reduces space by one dimension to zero dimensions.

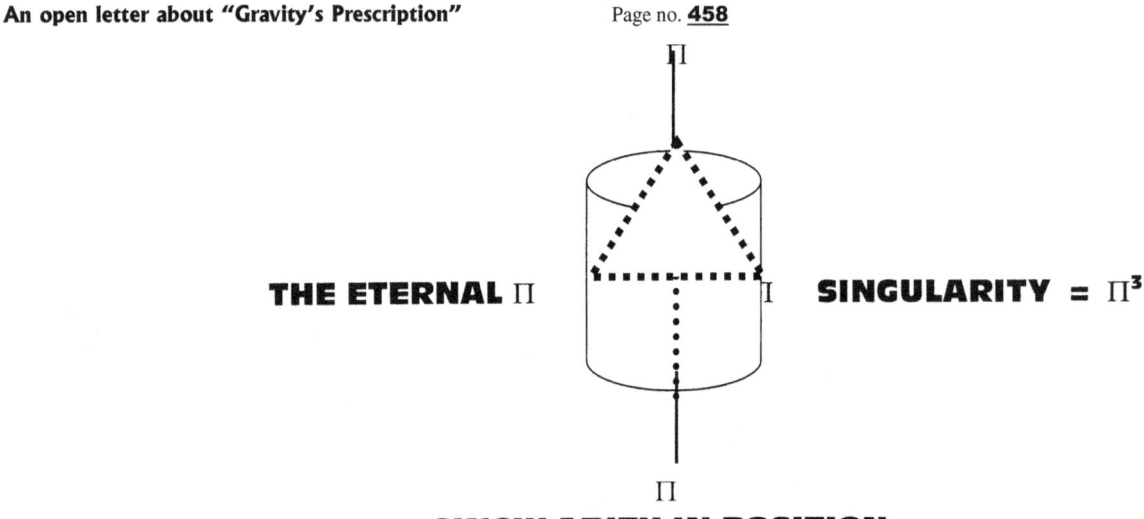

THE ETERNAL Π   I   SINGULARITY = Π³

SINGULARITY IN POSITION

This procession leads the dimensional reducing of space to the value of time in the square. The proton dipping into singularity effectively destroy space, and that leads all the way up to the Hubble shift, the Big Bang and all there is because the atom is the Universe, remember!

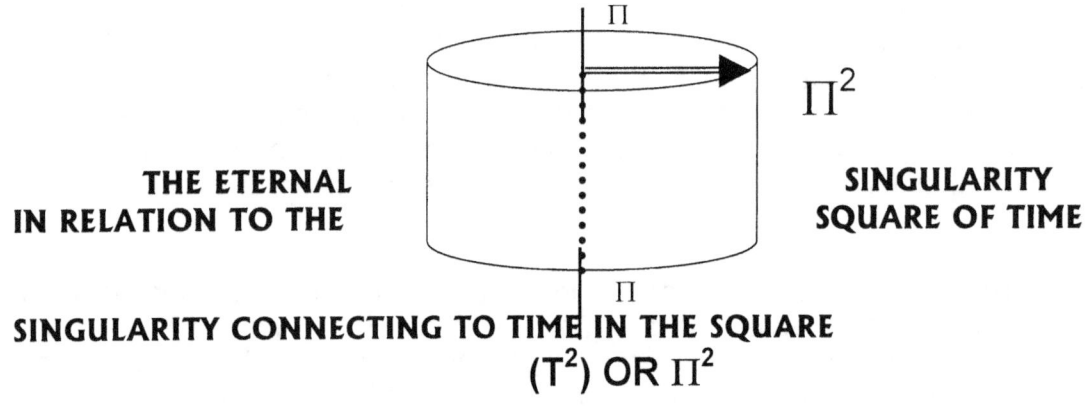

THE ETERNAL Π^2 SINGULARITY
IN RELATION TO THE   SQUARE OF TIME

SINGULARITY CONNECTING TO TIME IN THE SQUARE
(T^2) OR Π^2

To the side of that space which is less, it is the proton Π^2 waiting on its turn to dip into singularity by demising space completely. While waiting its turn, it keeps busy by destroying space to time, demolishing one dimension of space (the neutron), and that leads to space being at that point in a square of time. While this process is going on at the bottom of the ladder, the top also remains busy with the same process in a dimension more than the square of time.

In an effort to put the neutron value in context with Newtonian view one can find the value they hold as the gravitational constant being singularity removed from gravity by the factor of gravity. It gravity $\Pi^2\Pi$ which is the neutron factor we can deduct singularity from the total neutron application by removing Π from $\Pi^2\Pi$ and then from Π^2.

But this has nothing to do with outer space! This will be a restriction we find in the atmosphere where the atmosphere serves the motion of Π^2 without singularity. We are after all dominated by singularity in any case. That will leave a value of $(\Pi^2 = 9.869) - (\Pi = 3.14159) = 6.728$ but other than that I cannot find any value they have to confirm the gravitational constant.

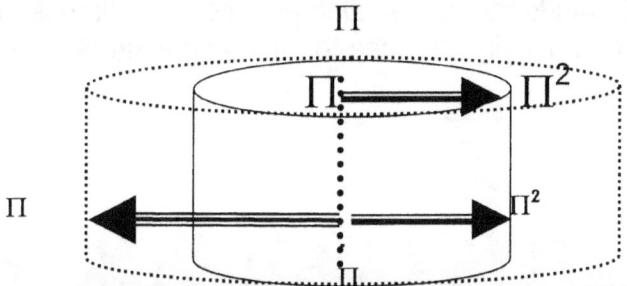

THE ETERNAL SINGULARITY IN RELATION TO
THE SQUARE OF TIME IN THE NEUTRON ALLIGHNMENT

Above the proton dimension destruction, the neutron converts heat from the fourth dimension of heat in space where it reclaims one dimension more than the electron dimension has. To the one end of the neutron, space holds a dimensional value of 3, and to the other side of the neutron the dimensional value becomes Π. The process is much more complicated because there is still the effect of 10 converting to Π^2 and 10Π converting to $\Pi\Pi^2$. I shall not go into all that, because I only need explaining the one side of dimensional alteration or "gravity" to understand the Titius Bode Law. The main consideration is that all over, there is one dimension at loss.

$\Pi^2 \Longrightarrow \Pi^3$ while

$\Pi\Pi^2 \Longrightarrow \Pi^2$ and

$3 \Longrightarrow \Pi\Pi^2$.

This process is more complicated because we only use one relative in heat unoccupied space, but there is still others. I shall shortly explain that very briefly. To summarize there is:

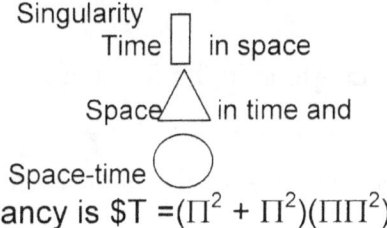

The full relevancy is $\$T = (\Pi^2 + \Pi^2)(\Pi\Pi^2)3$
Where $_1\Pi^2$ relate to Π^3

The full relevancy is $\$T = (\Pi^2 + \Pi^2)(\Pi\Pi^2)3$
Where $_1\Pi^2$ relate to Π^3
$_2\Pi^2$ relate to $_1\Pi^2$
$_{Neutron}\Pi^2$ relate to $_2\Pi^2$
$_{Neutron}\Pi$ relate to $_{Neutron}\Pi^2$ and
$_{Neutron}\Pi$ relate to 3

The one dimension always stands related to the one side as space forming matter 7 / 10 (even heat in space) and to the other side as matter in space forming 10 /7 In reality it is always space to time 10_Π / 7_Π or it is time to space 10_Π/ 7_Π. This implication brings about that the Universe cannot relate to any point of time duplicating space at two points during the same time. Matter can never be in two places at once. This is very important when I bring proof about the forming of the solar system.

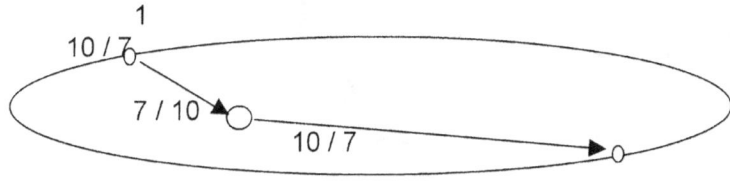

Through this, the equal relevancy of space 2 Π holds the same value as time Π^2. If time is at a value of Π^2 (10 /7) then space will be Π holding Π, at a relativity of space forming at the same token a value of ((7+7) / 10), which converts to 2Π.

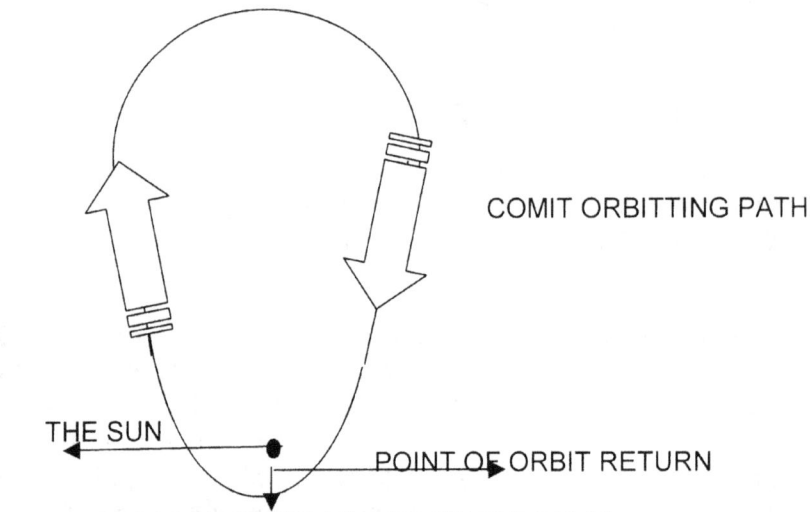

COMIT ORBITTING PATH

THE SUN

POINT OF ORBIT RETURN

MEAN POINT OF SHARED BINARY SINGULARITY.

It stands to reason that where there was a binary point holding mutual singularity, the excessive fluctuating of space to time 10/7 and time to space 7/10, will seem more apparent than usual because the space growth and time duration will be more than the Hubble shift will normally allow. There is just a lot more heat to create space in the form of light and unseen heat.

Again we have the principle of Kepler showing what gravity is.

Looking at the entire picture we see the motion of the comet T^2 with **k** relative to singularity k^0 that produces the space a^3 in as much as space moves $a^3 = T^2 k$.

The picture coming from where singularity focuses on space –time in the formula we have $k^0 = a^3 / T^2$ **k,** where it says that singularity control space-time.

From the Sun we have the comet extending the gravity influence of the Sun by the measure of **k** where $k = a^3 / T^2$

Finally from the comet we find the comet attaching to the Sun by the negative contraction of gravity $k^{-1} = T^2 /a^3$. The Newtonian concept of gravity does not apply only because it does not exist in any form anywhere. It is space flowing towards and space fighting for independence that establish gravity.

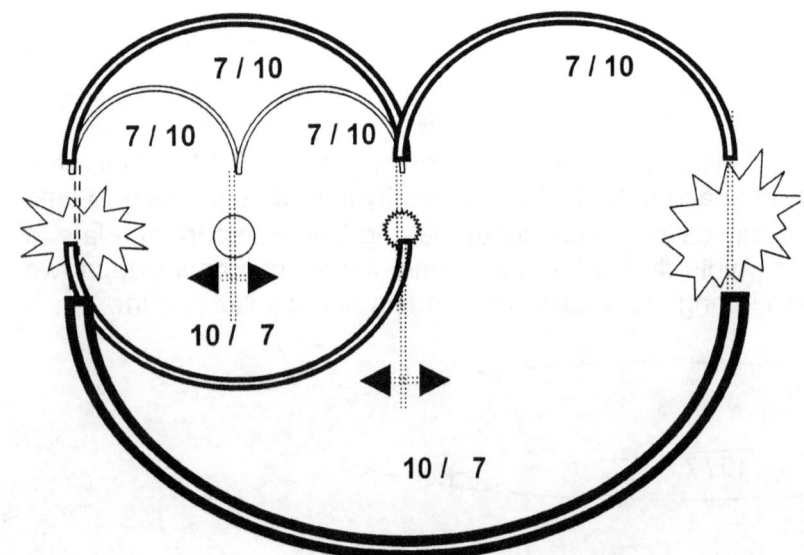

7 / 10   7 / 10

7 / 10   7 / 10

10 / 7

10 / 7

The normal application of the Titius Bode Law provides that a orbital fluctuating of .02 will remain between the perspective the pivotal cosmic atom provide and the orbiting structure. This does not apply to comets because the comet lost its individual singularity and holds on to the original binary-shared point of singularity as if it was its own. The expanding heat turned to space makes the positioning of the original binary hold a small growth, when in relation to the Sunspace growth there is virtually no comparison. Heat turning to space has a way of denying us humans to grasp a true perspective.

That brings us back to the dynamics of heat in as far as the speed of light is relevant to the electron and space-time.

Dealing with the speed of light as a cosmic factor with the aim of connecting that to time, has many implications that may well lead to confusion. The photon has no connection to singularity, neither in the wave nor in the proton, but only connects to singularity through single photons within the wave that connects to distant protons outside the wave.

The line of singularity running through the centre of the line of singularity will bring across a double message, one in front of the wave, and one behind the wave. The result from this may lead to the perception that that specific electron holds two positions in space-time, but this cannot be. If such a perception is there, then the mathematical calculation of the specific position of the electron will prove of only one position. This is because the speed of light links to time, but is not time, whereas mathematics do not confuse space and time as the human perception does.

I shall try to explain this in a manner that we see a burning charcoal fire. The smoke is heat, retained by oxygen, the flames is liquid heat, retained by nitrogen and the glowing charcoal is heat transmitted by the coal or carbon. To our perception, it is the very same thing. It is heat, holding space relevant to time where each element holds the relevancy, therefore determines the time of the space and that gives us the perception of three different objects. The truth is that it is the same thing because the fire may lower or increase its intensity but ….. is the same fire.

The value of Π, the value of 3 and the value of 3^3 give us humans an impression of time in different positions, however, it is the very link to singularity and the eccentricity space holds to time in each case that stuns our wits. IT IS OUR WITS BEING STUNNED AND NOT TIME OCCUPYING SPACE IN TWO DIFFERENT POSTIONS AT ONCE.

The Universe is also as much in every atom, as it is in every atom cluster, where that forming a group density where the group density controls the space they occupy, the space they claim through control and the space, which they influence. Newtonians have made very poor judgments leading to theories that miss the target by a mile. For instance, with all Einstein's brilliance and genius, he could redesign the Universe, if the Universe had worked the way he said. Einstein demanded the Universe to be space-time, in singularity and he missed the Universe by the width of the Universe.

Not every atom controlling its claimed space and influenced space, loses the space when it is dense to a point it holds the minimum heat in separation. The group or cluster of protons start to combine forces and claim control space, as much as influenced space as a unit, although they perform as a unit, of many individual Universes claiming to be in the centre of the Universe they are in fact. Every atom is its own Universe, therefore every atom claims its part in the Universe and the Universe is that number of atoms cling to that space they claim. The Universe is the atom claiming space to the heat surrounding the claimed space and that heat is the space the atom, atom cluster or atom group seeks to influence to its own benefit.

THE UNIVERSE IS NOT THE SPACE, BUT THE ATOMS CLAIMING THAT SPACE WITH THE SPACE THE ATOMS CONTROL AND THE SPACE IT WISHES TO INFLUENCE.

Without the atoms, the space in whatever form, would not be, because it is the atoms parting singularity from the two ends forming singularity. The singularity is one, there cannot be two, because then there is space to hold in different spaces. That task falls on atoms, forming different spaces, and as much as different time to hold to that space. There the space becomes two, not in the space of singularity, but in the space of the atom, where the atom parts with singularity. That is why the Universe is spinning.

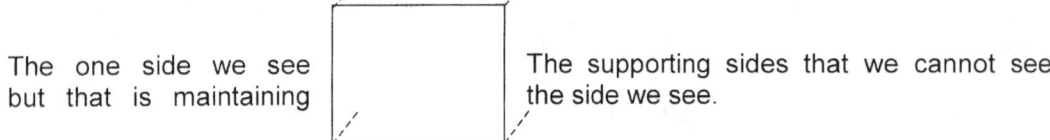

The one side we see but that is maintaining

The supporting sides that we cannot see, the side we see.

Therefore we see one side (1) and half of four other sides (4/2) = 2. In fact we still see three sides even if only one side is in our vision. This is a lot different in the wave. (Let us ignore the pencil line science and Einstein wishes to apply to light). We see Π^2 as the light moves to the nearest proton and if we are able to see the light, that proton will be my eye.

We can only see the light holding the time value of 1 carried by the Π^2 value. Then we can see the light not yet deflected by the 7° outward movement of the sphere. The 7° is reflected in a new wave and top the other side we have another sphere of photons deflecting. Therefore only Π^2 and $(7^0)^2$ remain in our field of vision as the rest of the light goes lost to us in a space occupying (existing) of heat. From the sphere we can only see 7 x 7 = 49 x Π^2 = 483,61. This value holds space in the fourth dimension of the triangle in 180° being the compliment of the straight line (180°) and the half circle (180°). The value forms a corner of 90° between space and time, therefore Pythagoras comes into play. That means the square root of 483,61 holds the relevancy, which is 21,991. The relevancy of the light we see holds the value of Π to the space (21,991) and matter (7). The light that reaches us has only the value of Π, with the rest lost to space once more.

That brings the distance between the Sun and the Earth as the factor of Π and Π holds the space relevancy of $10(10^2)$ going in one direction $5(10^2)$. So Π will reduce its value by $5(10^2)$ every eternity parted by an infinity. $T = \Pi$ and $R^3 = 5(10)^2$. That means the distance of the Earth separated by the Sun will be $r/5(10)^2$. The reason why matter forms a square $(7^2) = 49$ is that matter forms the link to time and time is always in a square. From this loss of light to darkness (which is actually light that lost its direct link to time from the position of our vision vantage point) will reduce as the Sun will seem smaller the further it is from the object in its field of vision. From Mercury the Sun will be many times "bigger" than it will be from Pluto. It is due to the loss of density applied by $\Pi^2 = 1$ and the linear factor of Π. This EVERYBODY KNOWS ALL TOO WELL, SO WHAT ABOUT THAT IS NEW THEN, you may ask.

The main issue arising from this, is the following:
The distance between the Sun and the Earth is $149\ 600\ 000/5(10^2)$ = 299200 km/sec (the kilometres come from R^3=1m^3 of water and the time arrive from 360° x 60 (space) a relative man made). When light was measured, the distance between the Earth and the Sun was larger than average and the distance I use is the average. This means within the heat density in the space-time we hold in time and space at this relative point from the Sun the speed photons displace the heat density is $3 \times 1\text{-}^5$ km/sec. That too means that the speed of light at Pluto will be a staggering $11,827 \times 10^6$ km/sec. Fortunately the second will no longer be a second, as will the kilometre no longer be a kilometre. What does this proof? Astrophysics applied by the Newtonians is a joke, and the Universe is immeasurably older, bigger and more massive than what our Super-Educated will ever accept. This makes all their brainpower and calculations irrelevant.

That, once again proves my theory, that the displacement of space-time holds the relevancy and not time neither space. Einstein gave matter in space the constant by searching for the critical density factor and the curvature of space-time. That is a lot of bogus nonsense. Illustrated, this will happen.

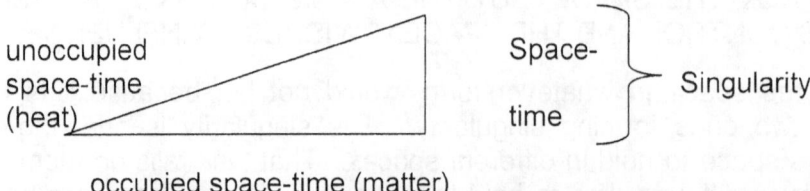

unoccupied space-time (heat)

Space-time } Singularity

occupied space-time (matter)

When matter (occupied space-time) becomes more time and space will bend. That is nonsense. There is no curvature of space-time and there is no critical density factor because:

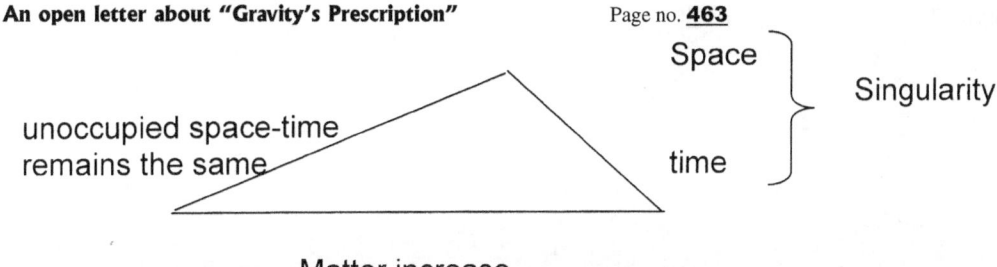

Matter increase

By increasing the matter aspect and maintaining the unoccupied space-time at the same value, then of course will space and time holding singularity bend. But singularity must always be a straight line because it is dimensionless, only parted by heat occupied and heat unoccupied. SPACE AND TIME LINK ETERNITY TO ETERNITY WITH HEAT OCCUPIED AND HEAT UNOCCUPIED PRODUCING INFINITIES.ANDTHE ETERNITY LINK REMAINS ETERNAL. IT IS THE NUMBER OF INFINITIES FOLLOWING ONE ANOTHER THAT CHANGES.

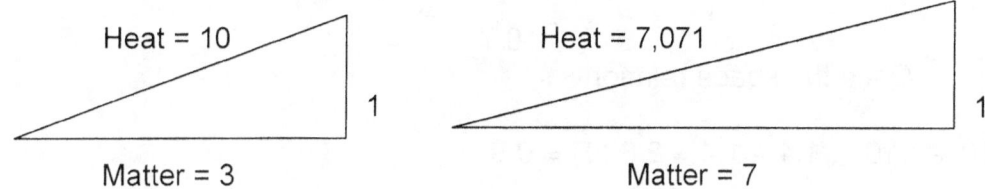

The increase in matter will reduce the density of space-time unoccupied (heat). Throughout this, time linking space will remain 1, eternal.

From the space-time constant the speed of light relates to the cosmos as follows:

$R/T = 10$
$R^2/T = (10)^2$ $(10)^2$ $9+1=10$
$T = 1$

$(3\Pi^2) = 1$ Relativity

1^3

Π 3

3^3

In this you find the relevancy of space (3) relating the fourth dimension to time (Π^2) through a position matter holds to space and time $(21,99/7) = \Pi$.

But space-time is Π^2 in the space of 10. Π^2 to the fourth dimension are 10 because "gravity reduces 10 to Π^2 and the Π^2 value lies on the space side of the atom. Therefore the full compliment of "outer space towards time will be the relevancy of $\Pi\Pi^2$ where Π is the Pythagoras value.

7/10

7/10   +   7/10

$= \dfrac{14}{10} = \dfrac{16}{7}$

Therefore $\Pi = 7/10$ and $\Pi^2 = 10/7$
But the matter position relates to the space position.

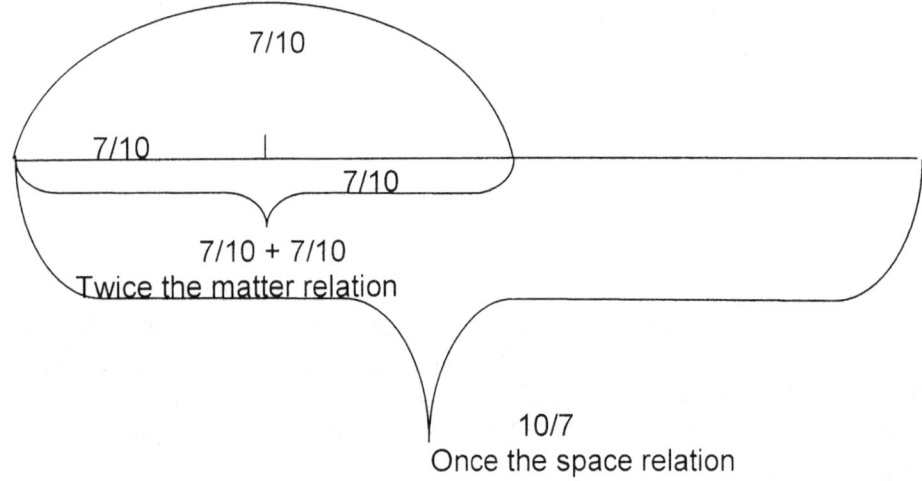

7/10

7/10

7/10

7/10 + 7/10
Twice the matter relation

10/7
Once the space relation

Twice the matter relation   10/7
                                        Once the space relation

O   O   AC = 7/10 + 7/10   1,4 + 1,4 = 2,82/Π = 0,9
A   7/10   B
    +10/7

7/10 + 7/10 + 10(7) = 2,8285 / Π   = 0,9
Π/(7/10 + 7/10 + 10/7) = Π/28285^2 = 1,10
                        Total   = 2,01

Therefore singularity at Π relating to space-time = 0,9
and space=time relating to Π in singularity   = 1,11

The total of matter to singularity   = 2,01.

Therefore Π^2 to Π = 2Π

This then is 7/10 + 7/10 = 10/7

Therefore space will always hold double to the relevancy of matter.

To that end
.   3. (7/10 + 7/10 + 10/7)
.   3.   6.
A   B   C
                2Π

Then 6 becomes 3 = 2Π(Π=6)=12
A   C   D

Then 12 become 3.   2Π = 24
And that concludes the Titius Bode configuration of 3; 6; 12; 24; 48 etc. by valuing the triangle and the half circle.

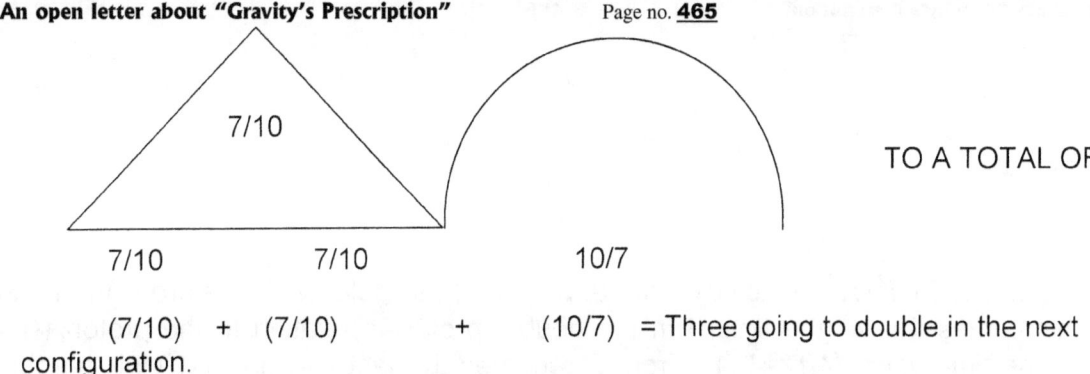

TO A TOTAL OF

$(7/10)$ + $(7/10)$ + $(10/7)$ = Three going to double in the next configuration.

The first $\Pi^3 \rightarrow \Pi^2$ Π Separating singularity.

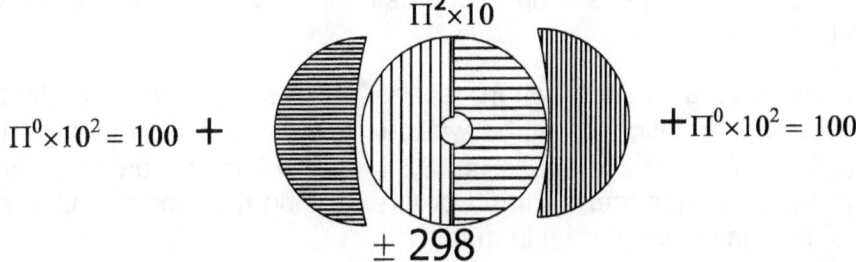

$$\Pi^2 \times 10$$

$$\Pi^0 \times 10^2 = 100 \; + \qquad\qquad +\Pi^0 \times 10^2 = 100$$

$$\pm\; \mathbf{298}$$

This then brought on Π^2 in heat.

$$\updownarrow \begin{array}{l} \Pi \\ \Pi^2 \\ \Pi \end{array} \qquad \Pi \qquad\qquad 3\Pi = 3$$

From that space developed

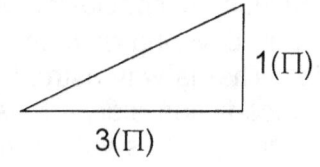

$1(\Pi)$

$3(\Pi)$

3 in the value of Π (three pi) position in relation to (1) (Π).

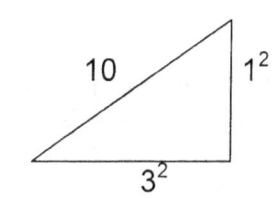

10 1^2

3^2

From that space came about
And this then became 10Π.

$(\Pi \times \Pi) = \Pi^2$

10Π Π $10\Pi + 1\Pi = 21\Pi + (\Pi^2/10) = 21,99$

Π $(\Pi \;\; \Pi) = \Pi^2$ Π

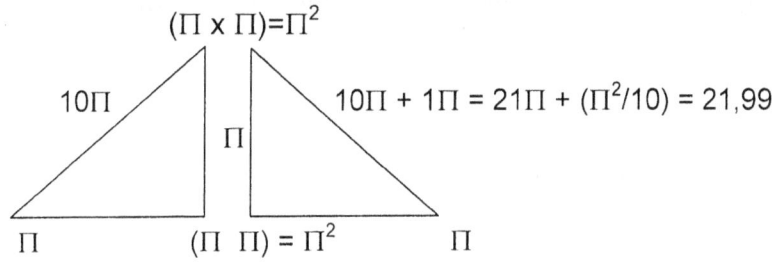

Add that to the seven that holds densified and occupied space-time and there is seven of Π in the triangle of matter adding 3 Π's in the half circle of space ($\Pi^2\Pi$) and the total Are 10Π. With 10Π the value of the total triangle (in a square) and 10Π the total of space-time (matter holding singularity apart, there is a factor of 7 by Π to 10 by Π, with the triangle having two Π - two factors where Π at the bottom formed $(\Pi \times \Pi) = \Pi^2$ and to the top $(\Pi \times \Pi)^2 = \Pi^2$, with all of this constructing space (10Π) to matter Π^7.

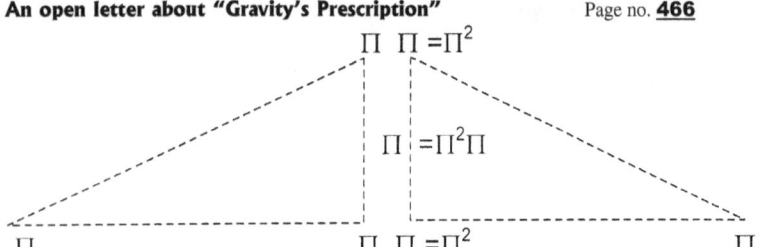

This will bring about 5 $(\Pi^2+\Pi^2)$ bringing matter and space to singularity. There are Π to the number of 5. From this comes that 4 (the four Π in the double proton in relation to the proton $(\Pi^2+\Pi^2)$ will always form the time value $(4(\Pi^2+\Pi^2)$. From this comes the fact that $3(\Pi^2+\Pi^2)$ will hold the space component as that in fact is half space-time. Space-time being $(\Pi \times \Pi)(\Pi)(\Pi \times \Pi)$ and half of that to any direction is $(\Pi \times \Pi)(\Pi)$. This is why space-time in the geodesic sense has the value of 10Π and that 10Π in a sphere are Π^3. It is both triangle relating to one half sphere which is both triangles (7) in the totality of the half sphere (10).

The value of $2(\Pi^2+\Pi^2)$ will relate to a position where space passes on time in a dimensional transformation, but not a value transformation. It will be when Π relates to Π^2 as much as $\Pi^2+\Pi^2$ = 19,7 and 6 (the sides available for Π to use in space is 18,84. It will form a border of dimension where space will not apply in the same manner as it did before reading that border. One may say that is the densifying border of heat in space to heat in matter.

The last position of the Lagrangian atom is where only Π (in the triangle) have value and this links space directly to time (matter). At this point concerning the Lagrangian layout, we must view heat, filled space, the stuff we exist in, as an element. Hydrogen becomes liquid at $-269\,°C$ and heat (outer space) has a gas value of $-273\,°C$. This is only a dimensional changeover from 10 to Π^2 or from 3 to Π. That was what the "Big Bang" was all about.

Creating the Universe in space as we see space was the process where the last natural element formed. The Universe or Cosmos is the last atom, which formed. It was the conclusion of the proton $(\Pi^2+\Pi^2)$ in developing. That space can be in gas as we now see it, covering elements to position them in relation to the rest of the Universe as gas, liquid or solid. That is why astrophysics in the ways science apply it at present, is but a good old romantic science-fiction story. Mass holds no value, it is density and that density brings in the heat in space, unoccupied as yet by matter. That density gives a star its "gravity".

Where the Sun can only hold heat to liquid more dense stars will hold heat to a "jelly" and others will take heat all the way to something as hard as tungsten. That is what tungsten is. Tungsten can place heat in a relevancy that the heat relative to tungsten is almost as dense as the neutron within an atom. That is why I refer to the system as the Lagrangian atom, because the Universe holding heat, produces all the relevancy matter can have. The Universe is the final atom.

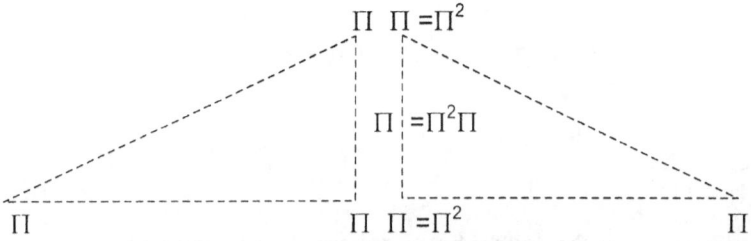

From the illustration it is obvious that the inverse cannot be in a value of $6(\Pi^2+\Pi^2)=118,4$. To be in that position an object has to be in two places at once, which frankly in spite of the views our Super Newtonians contribute, is simply illogical. To do that will take space-time to a point where space not yet formed. I do admit at 118 there is something, but we shall never know, because whatever it is, will be even beyond a Proton Star, in the heart of a Black Star.

Because no object can hold dual positions the one proton will link to Π in space-time at the end of the triangle and the other will connect to Π in singularity.

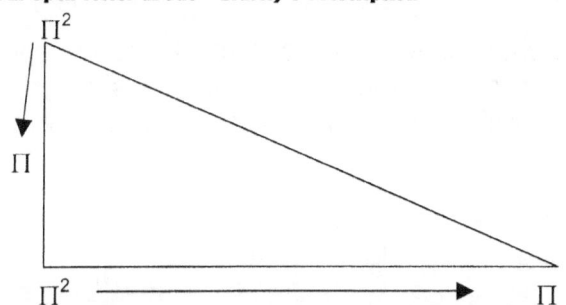

As I mentioned nothing could be in two places at once, not even Π can be in two places at once. Therefore while the one proton connects to the heat establishing space, the other connects to what is space, confirming the circle of dimensions. That is "gravity". It is a dimensional link establishing a transformation from the 6 by Π to the 3 by Π by relating this neutron Π^2 to the neutron Π^2 where the neutron Π^2 links to the " gravity - motion" Π^2 as the " gravity - motion" dips back, unifying Π^2 in singularity to Π in singularity, revaluing Π^3. After establishing this, and with the Newtonians simplistic view on "the gravitational force" I surely would enjoy watching them while they were "creating artificial gravity". What an effort that will be and the means in which they wish to perform, by merely spinning whatever spacecraft, and thereby creating centrifugal force. Newtonians should stop dreaming and start thinking. Dreams, magic and forces are for children, so let the children have them while the grown-ups use their minds for thinking.

The cosmos is the effort that space in singularity will perform energy through which it will produce gravity, to re-unite with time in singularity when time in singularity holds the value as matter Π^2 and heat holds space (Π) in singularity to a position of 6, effectively operating as 3. \$T= $(\Pi^2+\Pi^2)$ $(\Pi^2\Pi)3=1836$.

With that \$ = R^3 = T^2 and R^2/T x R/T = 1. What that imply is that Π^2 is time as singularity in space Π reverts to both value of Π, be it in singularity or heat. Connecting Π to both Π^2 is the Roche limit where one object (cosmic atom) holding a position of Π to both ends. It means they are connecting at Π to singularity that means Π holds both one will hold Π/2.

While one aligns with Π^2 it also holds the other cosmic atom as Π^2 at the point it positions (Π/2). Therefore the Roche limit is always (Π/2) with the Titius Bode applying some influence, even if it is holding the Roche limit accountable for service.

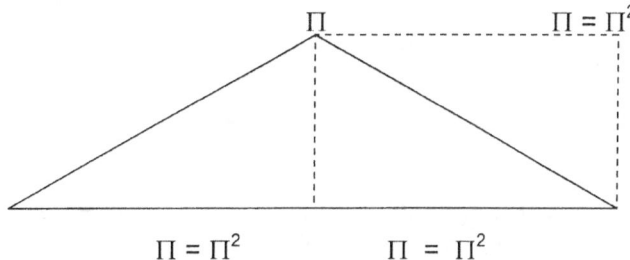

The Roche limit then is the closest any object can be in being in two places at once, because it holds time to one place. It is holding time Π^2 to a quarter of cosmic time 4. That means it is holding position in 4 Π possibilities to time Π^2 $\Pi^2/4$ = Roche limit. This gives iron56 its value of 10 x 21,7 in a relating to 4. The 10 is space with 21,7 the space relevancy to matter (at 7) and iron holds matter, therefore iron holds $(\Pi^2+\Pi^2)$ to one.

Space 10(Π) x space (22Π) is $220\Pi^2$ holding time $4\Pi^2$ = 55, the value of iron and iron holds time to singularity through matter $4(\Pi^2\Pi)$ + 1 (Π in singularity). Anything less than 55 cannot link with singularity in space, where Π will provide stability.

To compare, let us first view the Newtonian position on events.

The Newtonian approach of gravity is promoting the fact that matter will cling together because the mass sucks like a hovercraft blows, and with the suction all space separating all particles will reduce to nothing. After this reducing of the separation distance parting the particles will still suck, and suck and suck until the particles can withstand the suction no longer. In anger the particles will respond by heating as anybody can feel when one rubs one's hands with force. They see the force we use to pressure our hands together as a similar force that will do some sucking on the one side adding some pushing on the other side and that will produce a burning star. This is total rubbish with no proof whatsoever. If that is gravity, then I reject gravity.

I shall try to explain how I see the diminishing of space comes about in the process we regard as "gravity". By this process I shall also give account of forming of our solar system.

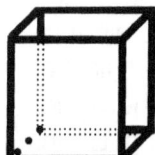 This is what we can see as space that is filled with material that is establishing time in reference to singularity

What the Universe therefore is, is a box confined to a sphere. In the box three sides are in conduct holding to one side of space.

Even when you only view one side of a box, there are three sides.

In view there is a square with one side, but in order to remain in a position in space, four halves have to support the position of that one space.

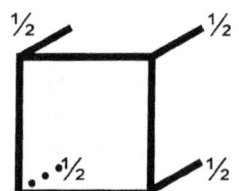 Even looking at the box square on, the same three sides apply.

In the box there are four (4) half (½) sides and one full side giving a total of three sides in all. In cosmic reality this is only a four -sided triangle.

Even with a box holding just one side in full view, three sides still apply.

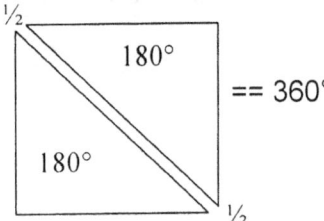

== 360°

Convert that to a 'lesser' dimension and it will be 2 x ½ + 1 = 3. That is space 3 x 2 = 6. Place this in a sphere and there is the Titius-Bode law. The application of the Titius Bode law has a value of 1, 3 , 6, 12 etc. This is a dimensional translation of the neutron value to space-time in the fourth dimension.

With the Sun in a Lagrangian layout it should be fitting at this point to bring in an explaining concerning the Lagrangian layout

THE LAGRANGIAN ATOM.

It starts as follows:

No line can start at zero because having a starting point of zero there is no line (0 X by what ever reduces whatever to zero). The starting point has to be infinity the shortest any line can be leading to eternity the longest any line can ever be. By having infinity there then has to be a VERTUALL ZERO (not zero) and from that point the rest of the line must start running the other way.

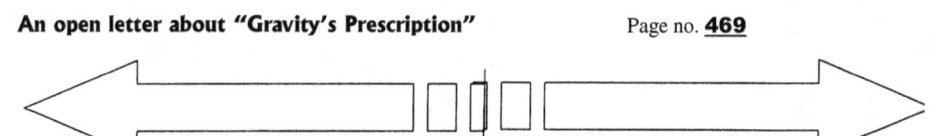

This establishes that no line can have a single direction and must have a continuance to both ends. Such a line has to have a value of 360° not 180° as believed.

This places the single line that is half of the full line at a value of 180°.

That is also the value of the circle and the square, both dimensional components of the cosmos. By reducing the one side of the square to zero (which it cannot be) the square will disappear. Therefore one has to reduce the one line of the square to infinity to produce a square holding a straight line with the line running in both directions from a point of infinity.

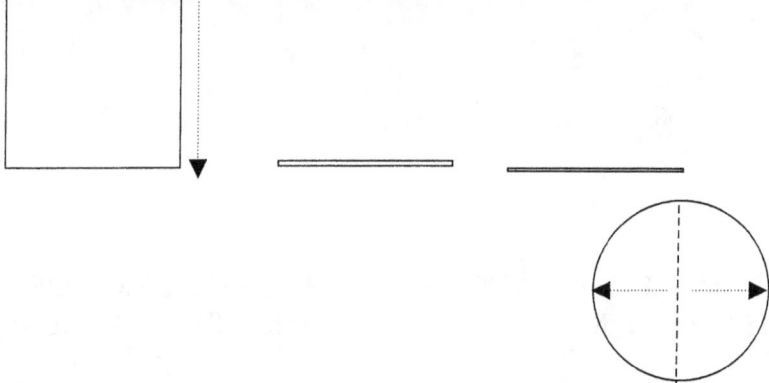

That same principle applies to the circle because the radius of the circle is one side to the round cube or a cube with no corners. That makes the square the circle and the straight line the very same thing of something totally different. The common denominator is the singularity of eternity finding infinity.

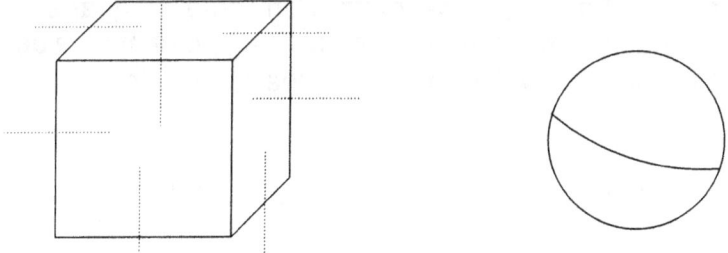

In the cube and the sphere there are a limitless of singularities connecting to one common denominator, the major singularity.

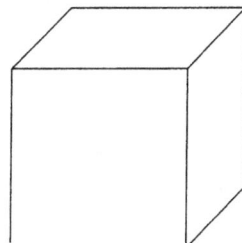

The concept behind the term we use for gravity is the reducing of one dimension.

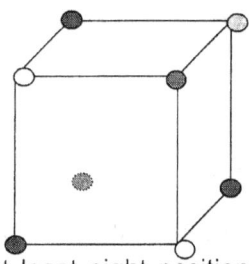

In a cube there are at least eight positions matter can relate with matter through space and therefore gas clouds cannot become structures.

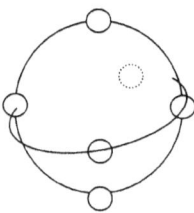

In any sphere six points of singularity positions six possible connecting singularity with that connection connecting to an eternity of singularities When the applying of dimensional reduction sets in by the dominating of one superior singularity controlling the space the other points of singularity occupy, one such a point has to fall away other wise "gravity" (the depletion of a single dimension at a time) cannot apply.

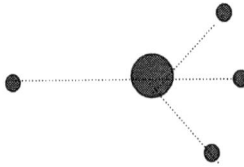

I KNOW THE NEWTONIANS ALL LOOK AT THIS ILLUSTRATION WITH THE EXPRESSION OF " *NO GOOD WELL FINE, SO WHAT*" BUT THEN THEY NEVER ALLOW ME THE TIME TO FULLY EXPLAIN WHERE THIS COMES FROM.

From the line of singularity to both sides forming the edges of singularity starting the border of singularity all aspects of the cosmos holds two halves in four quarters. Each of the halves and all of the quarters are in direct opposition to each other as much as to one another. From any point all space is moving from one side of the Universe to the other side of the Universe in quarterly displacement. The time it takes the movement will be the movement from the starting point at a value of Π to an ending point holding that moment of infinity to a value in eternity therefore bringing the square of Π^2 to the value of time.

Because time is the movement bringing about the change in quarters from any given point to any other given point nothing can be in two places of the Universe simultaneously as much as no one can freeze time to single instant to calculate time.

When I came to this point I realised that there was a time before space. There was a time before material and there was a time when time was form. The Roche limit splits the Universe by 2.4674.

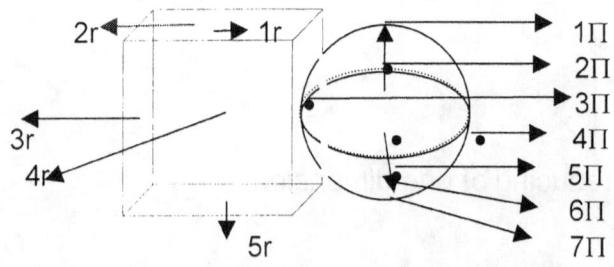

5 sides in the cube vs. **7** sides in the sphere

However if form was in place at the start of time then by the square of space 5 should divide the line forming the Roche because it is the five that should split the difference becoming 2.5 and not 2.4674. From that I made my first deduction that there was a point before space when time was the only consideration because time is the flow of heat in space and that is Π^2.

It is the motion that establishes form that determines the value. That proves that the value connected to the cosmos before form was a valid part of the cosmos. However it is in place and it strongly ties in with the half of the square of space before the square of space implicated true form.

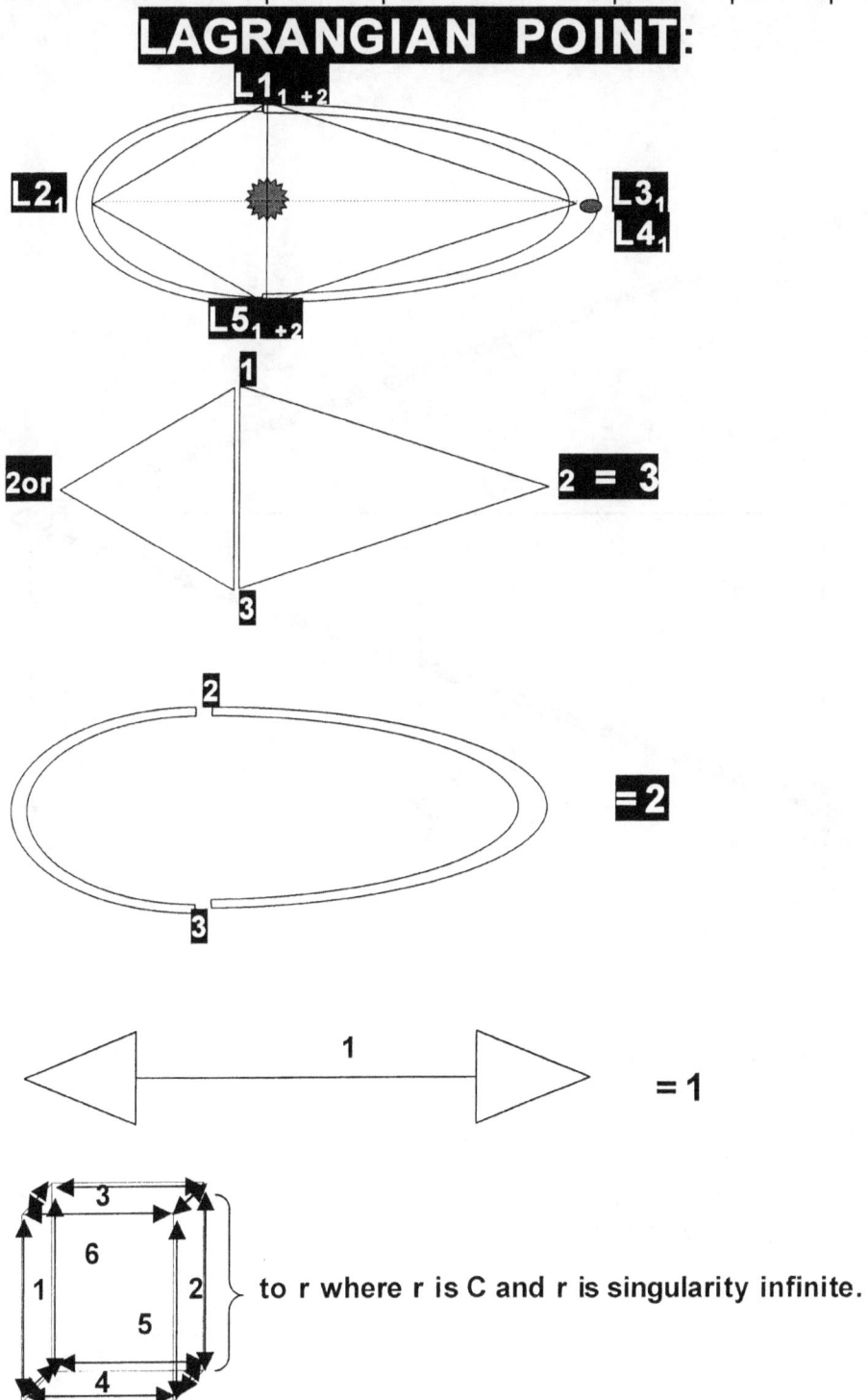

That makes light the seventh link to the cube and light forms singularity connecting the Universe in unoccupied matter whereas the sphere holds Π to seven positions of singularity connecting occupied heat. In the factor of space unoccupied where r holds singularity there must always be space between r and the next r to prevent r from being Π as in the case of Π not allowing space but only time between Π and Π^2. By disallowing r space Π is the connecting value and whenever matter unoccupied is a factor, space always must form part of time to distinguish r from Π. With 6 positions away from singularity forming Π and matter r has to withhold one position in space to allow space and therefore distinguish r from Π. The maximum space can become is six position and r. However in the event of space connecting to Π and space holds more than five positions r will disappear in becoming Π. That is not possible as r is in clear distinction of Π. That is the basis for the Lagrangian system as well as the basis for the Lagrangian atom and that is the reasons all stars have space collapsing after $5^2 \times r + 1 = $ iron $_{56}$ in the proton value but in the last remark there are other factors also involved.

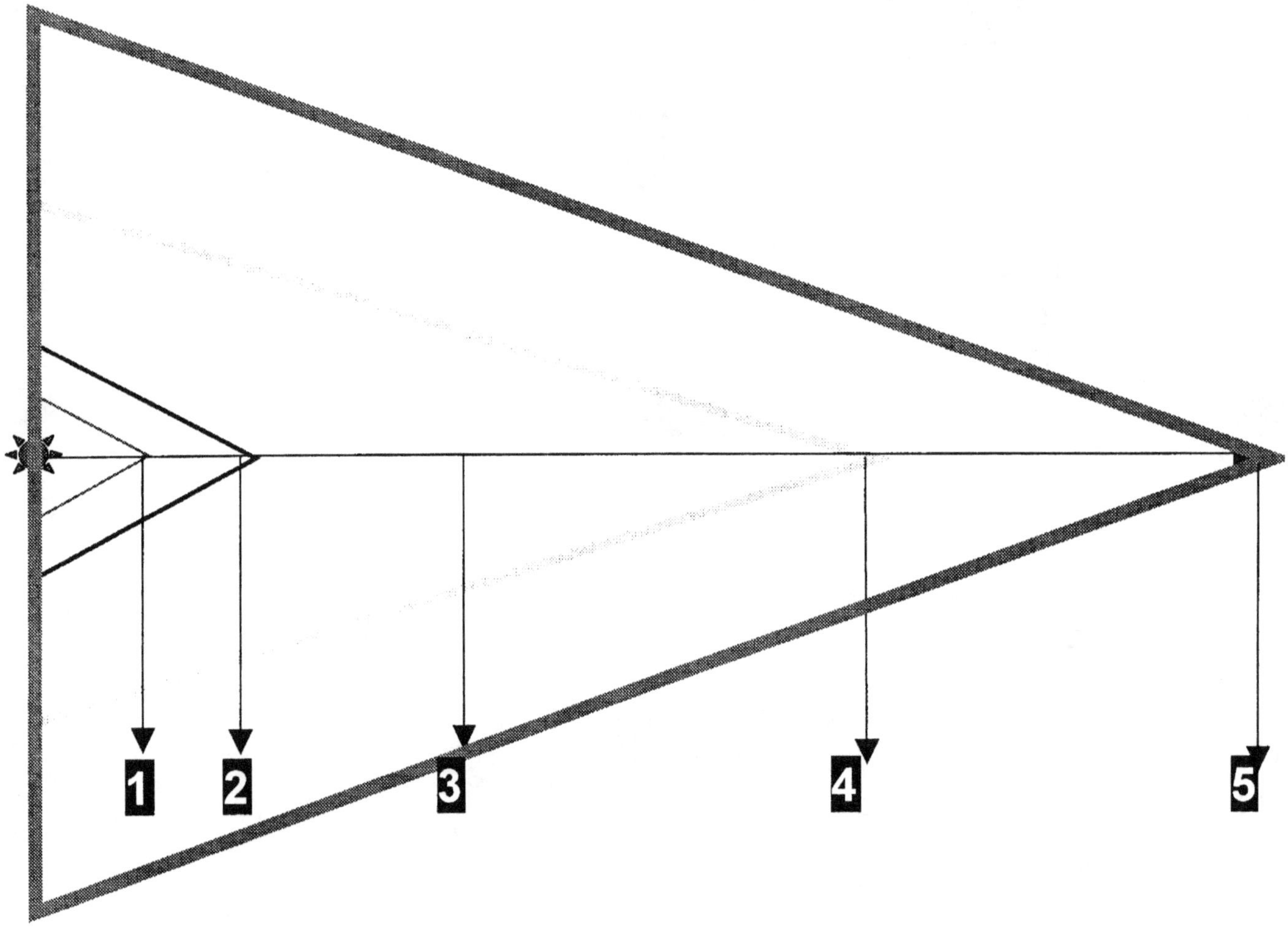

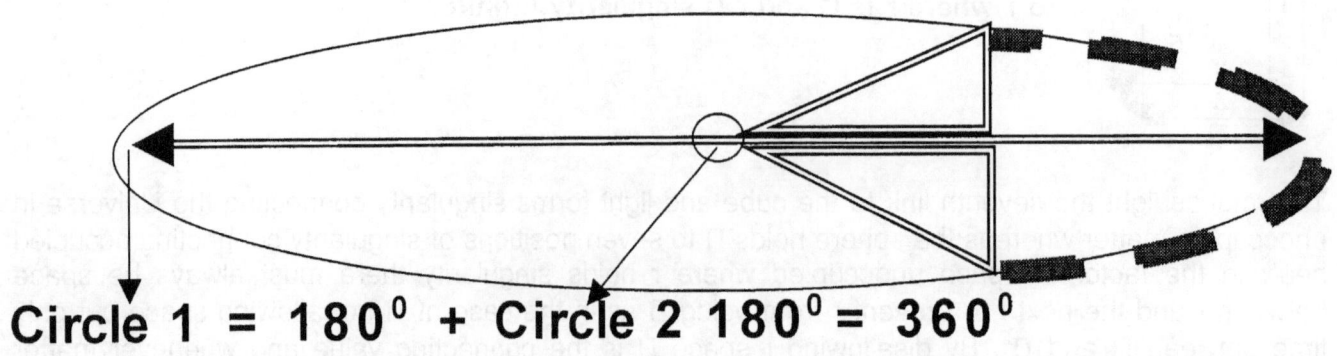

As a unit by three, two and one holding five in total and with space being the result from matter dismissing Π to favour r, space must either join matter by becoming Π and dismissing r or maintain r and hold a maximum of five points to singularity at the greatest value. With the Universe always in division by singularity the singularity holding seven position to Π will relate to the two singularities affecting the position of a cosmic atom. That will form as double points to space where five then multiplies with the two aspects of singularity divide and form the value of 10

$$\text{Circle } 1 = 180^0 + \text{Circle } 2\ 180^0 = 360^0$$

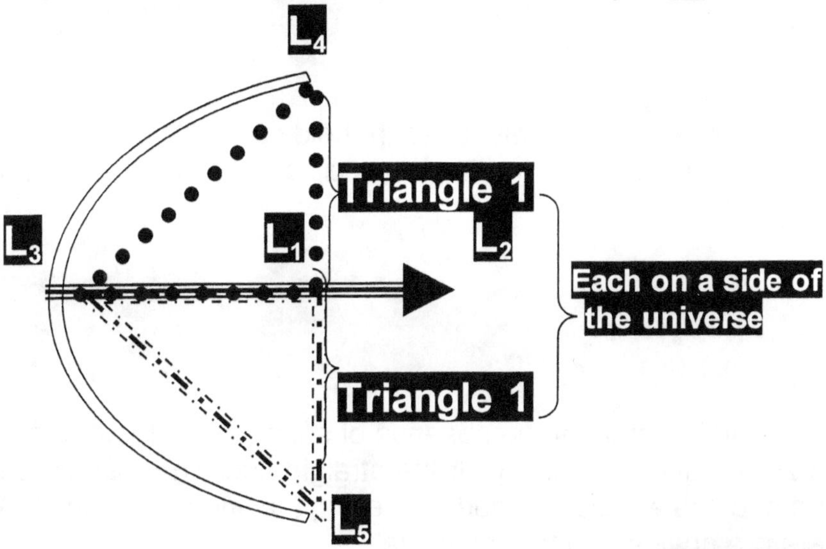

L4

Triangle 1

L3

L1   **L2**

Each on a side of the universe

Triangle 1

L5

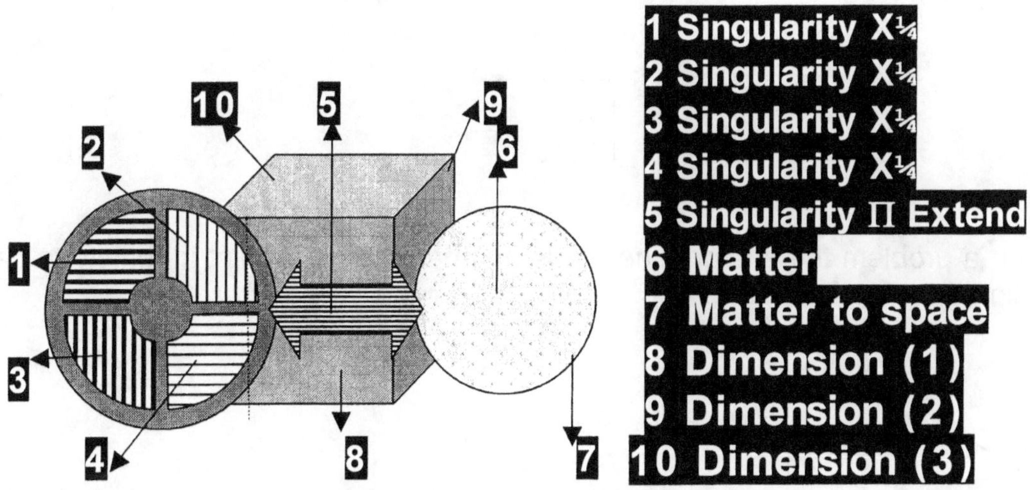

1 Singularity X¼
2 Singularity X¼
3 Singularity X¼
4 Singularity X¼
5 Singularity Π Extend
6 Matter
7 Matter to space
8 Dimension (1)
9 Dimension (2)
10 Dimension (3)

From the line of singularity to both sides forming the edges of singularity starting the border of singularity all aspects of the cosmos holds two halves in four quarters. Each of the halves and all of the quarters are in direct opposition to each other as much as to one another.

From any point all space is moving from one side of the Universe to the other side of the Universe in quarterly displacement. The time it takes the movement will be the movement from the starting point at a value of Π to an ending point holding that moment of infinity to a value in eternity therefore bringing the square of Π^2 to the value of time. Because time is the movement bringing about the change in quarters from any given point to any other given point nothing can be in two places of the Universe simultaneously.

However being in one place puts anything in one half of any Universe and puts anything one position outside eternity, which by that action implicates such a position in the location reserved for infinity.

LAGRANGIAN POINT:
The Lagrangian points are five equilibrium points in the orbit of one body around another, such as a planet around the Sun

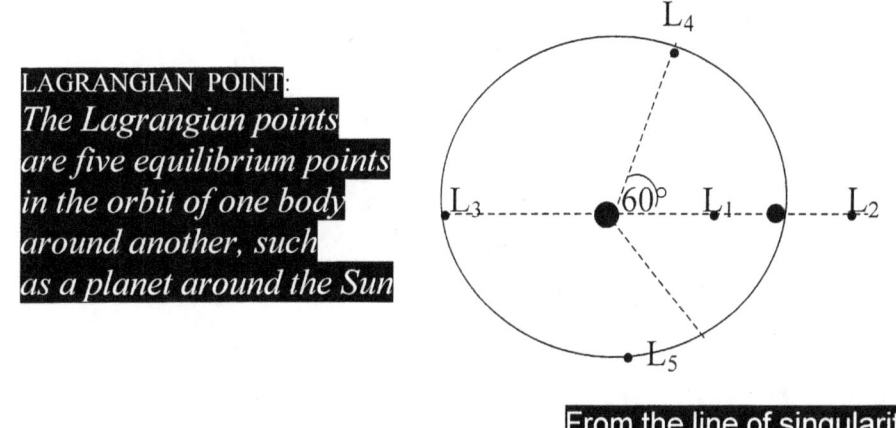

From the line of singularity

There are in the manner that form holds values three values each holding 180^0 comes from singularity and this fact science is familiar with. The straight line is always a potential triangle with on side apparent and the other side in infinity.

There is no zero from where a line can start and because of the absence of such a point mathematics brought about a diversion to escape the zero mark not existing. If the straight line did cross zero in would not be one line but no line since the one line will discontinue cancelling the line at one point. All lines have to have a start and an end therefore no line can be half a line.

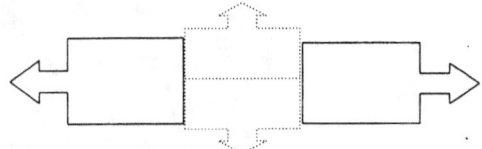

In order to overcome such a problem the straight line holds another line as a point in infinity to half the line as to enable the line diverting from zero.

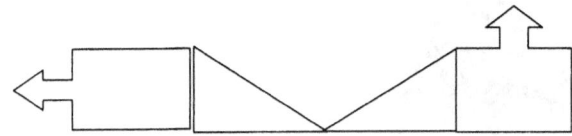

Because each line represents the other side of singularity dividing singularity by half the square of such a halves represent singularity two the half of a circle thus bringing total of the two halves would match the other half of singularity in half the circle.

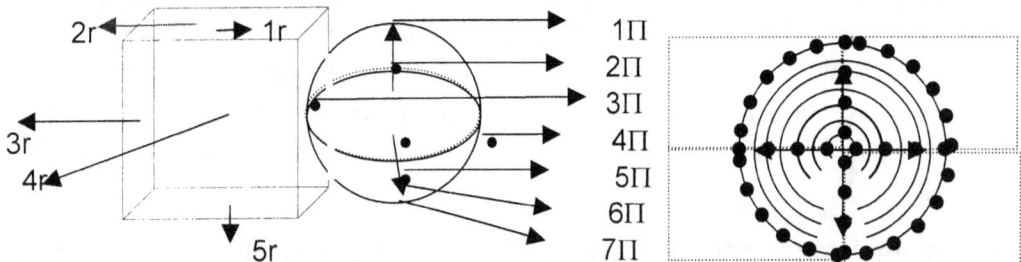

From space the cube holds the value of 5 times four quarters in relation to singularity forming the four five points of the square in the cube.

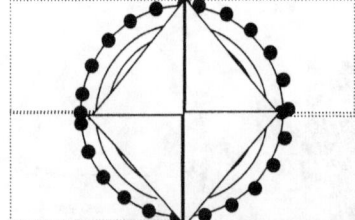

From singularity holding the relevancy the five sides in the cube as a square holds four triangles to two circles.

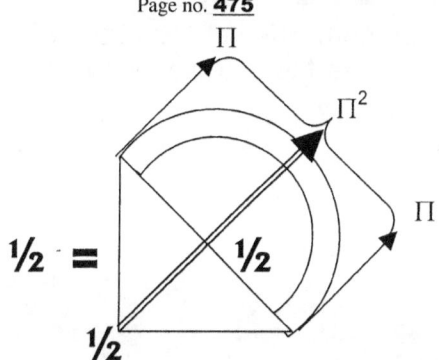

½ + ½ = √1
360⁰ /5 = 72/ ∏ = 22.91
.91
1 from singularity on the one side of the Universe and 1 from singularity on the other side...
10 from singularity on the one side of the Universe and....
10 from singularity on the other side bringing about the...
∏ that holds 21.91 to 7.

Since the sphere is double the circle and half the circle represents singularity by the square, half the square of the triangle is a straight line diverting singularity the law of Pythagoras is valid.

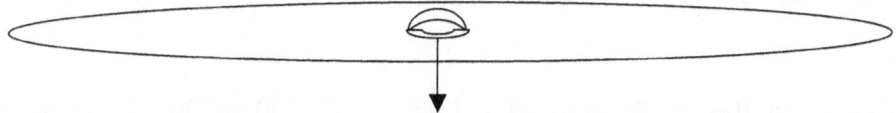

The divide bringing about the two sides of the Universe where the one (1) to singularity depicts the one side of the Universe and the other side (1) depicts the other side holding space from singularity (.91) bringing about the singularity value of 2.91

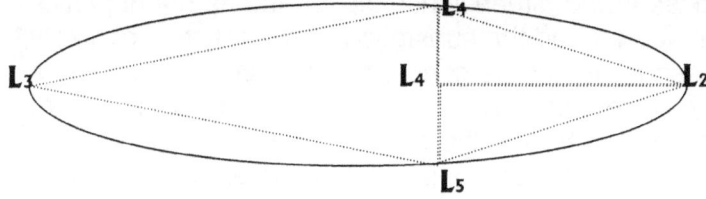

Dividing the four fives, singularity holds a centre line (.91) with one on either side but since it is a space relating to a sphere only one of the quarters on either side of the divide relates to a specific. Therefore unlike the sphere where the full value of ∏ relates to four fives, bringing about ∏ as the dominant the space separating the sphere from the points in space holds a combined value of one cube in line with the divide singularity supplies having five points.

It takes any line time to relate to space and only nothing is instantly and nothing is what it is, the moving of the line through the space it covers are the square of time. The time factor is in all cases in relation to the space factor in the square while the space is in the cube and that is a law of cosmology. To that reason I reintroduced Kepler's formula to the formula R /T X R² / T =1 and this puts a relevancy between time in the line and space to the line.

Since the triangle in singularity are on both sides of the divide of singularity and the circle holds the time aspect relating to space in the square the triangle therefore must relate in a square in order not to duplex singularity in the divide.

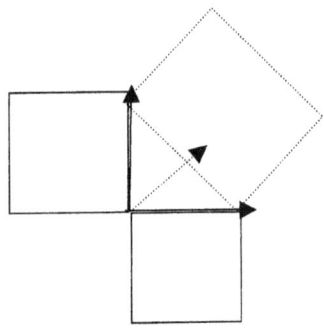

The time affecting the space of AC will relate equal to the time the line AB relates to time and space and where time is always in the square the lines will be the square of the triangle forming in relation to the square existing in the total of the time to the space relation forming between the lines.

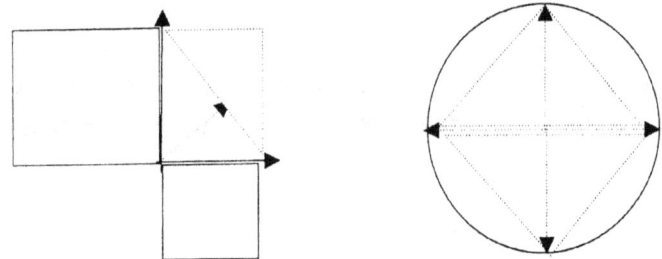

Since the triangle forms on both sides of the divide and all things concerning singularity is in duplication the double triangle will be the square. In the same way the circle represents both sides of the divide singularity forms and then also has the value of 2×180^0 as does the double triangles. That brings about that Π relate to the square and r to space and from this fact mathematics can substitute Π by using r. However that is not the case in the dimensional aspect and as mathematics gets its queue from the cosmos and not the other way around the substation may apply to singularity as space in the sphere but not as space as the sphere. Heat is concentrated space and space is expanded heat. Gravity and electricity is the very same thing where electricity is a concentration of heat demolishing space in a very specific location and gravity is the concentration of heat in a less dynamic form but acting in a much broader space. In both instances it is the polarized motion of iron $_{56}$ linking time directly to singularity through the conducting of heat surrounding elements. To this effect I wish to point out that no element is either a liquid, a gas or a solid as all elements are all three forms and it is only the state of the relation that apply to an element at a very specific position in space-time that will allow the element to act in either of the conditions that the heat or space which is the same thing will allow. Gravity is the dimensional destructing of space to the concentrating of heat in that space by increasing the time duration through the Titius Bode principle or when matter holds less space to the normal allowing of matter occupying space, matter will produce the Roche principle in guarding its individual singularity of the mutual singularity between objects.

This principle also is the only difference of notoriety between electricity on Earth and gravity on Earth. On Earth Π has a very slightly higher value than space in as much as space is 3 and the atmosphere is Π. In a cosmic midget as the Sun are the relevancy changes are considerable and the space to atmosphere can be 10 Π. There is no chance of generating electricity in the atmosphere of the Sun because the atmosphere of the Sun is electricity in as much as the gravity being 10 Π to the Π of the Earth. That explains the fact that the Sun liquefies heat to a watery substance. With the heat in a liquid the Sun becomes a sea of heat. By matter applying Π as the reference, there is little man can do to change that. In the case of electricity using r to form the value C we can change that because as r relate to space and we are part of space as space above the Earth in the neutron zone we can change the space holding r as value. When looking at the Sun applying gravity and relating what we see to what we find in electricity there is hardly anything to recognise a similarity by except that one can change and inter change the two aspects of heat. But when in view of dimensional dynamics it is the same thing because in the Sun even r holding heat becomes Π holding matter as heat becomes liquid and that stands in between matter and gas. With this view in mind it would be worthwhile to have another look at the way we see how creation started and bring heat in as related to r and matter being related to Π.

Gravity is the transformation of space to heat in one specific dimension changing that particular dimension in relation to the other five dimensions. That brings the reason why the Lagrangian system can only allow five positions and allowing any more will destroy any form of dimensional implication between object relating to one another while sharing space occupying in time duration. The Big Bang had its massive motion brought on by first implementing the Roche factor of $(\Pi / 2)^2$ after which when matter had a larger claim to space and space broadened the Roche factor adjusted to Π2 / 2 and then implementing the Titius Bode principle very much later on toΠ.

It starts as follows:
ALL OF CREATION STARTED WITH Π = TIME AND AS TIME WAS ETERNAL,

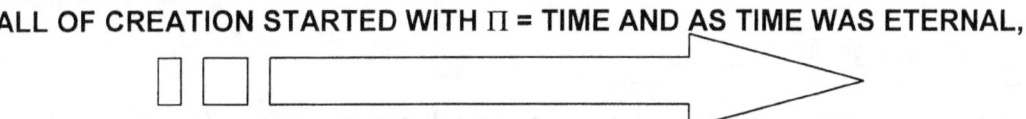

The spot grew into a dot by the value of 1^0 going onto 1^1. However this also was Π^0 going onto Π. However with no space yet available and the spot forming being so small (as we still can see when observing the centre of the spinning top because the line inside the line that is inside the line we cannot see is the line we are referring too.) The line is there as it was, but our inability to go that far into singularity places us at the disadvantage not to recognise the line as it is. Because it is spinning it has to be Π and because it is spinning it also has to be a line and all of that is there to witness for all those disbelievers.

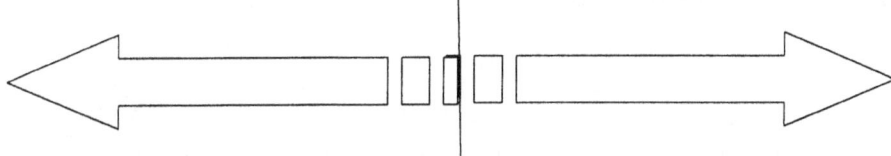

The line formed a spiralling line that was so small it was continuous and never broken due to the lack of space and those not believing me I advise you to inspect the spinning top. The line formed next to the running line and that too is in plane view of all but we cannot see. The line has to form a line on this side of the divide because it forms a divides as much as there has to be a line on that side of the divide for the very same reason

This places the single line that is half of the full line at a value of 180°.

That is also the value of the circle and the square both dimensional components of the cosmos. By reducing the one side of the square to zero (which it cannot be) the square will disappear. Therefore one has to reduce the one line of the square to infinity to produce a square holding a straight line with the line running in both directions from a point of infinity.

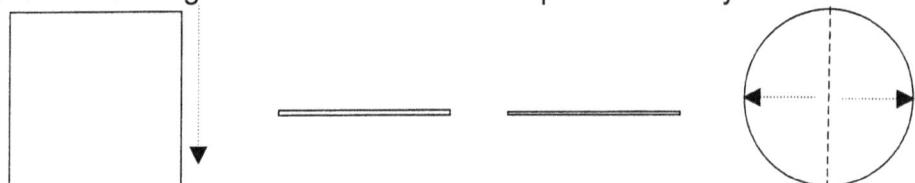

That same principle applies to the circle because the radius of the circle is one side to the round cube. That makes the square the circle and the straight line the very same thing of something totally different. The common denominator is the singularity of eternity finding infinity.

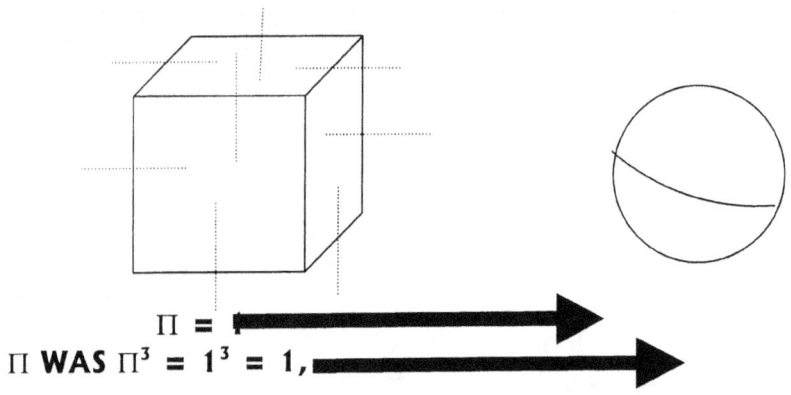

$\Pi =$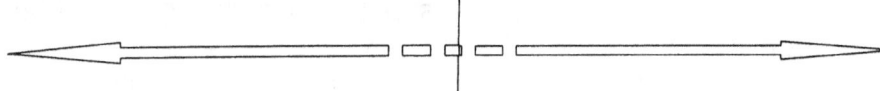

Π WAS $\Pi^3 = 1^3 = 1,$

Π WAS $\Pi\Pi\Pi = 1 \times 1 \times 1 = 1$ **THEREFORE TIME WAS ETERNAL:**
$E = \Pi = 1$

Let us have another look at the straight line
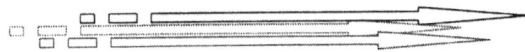

No line can start at zero because having a starting point of zero there is no line (0 X by what ever reduces whatever to zero). The starting point has to be infinity the shortest any line can be leading to eternity the longest any line can ever be. By having infinity there then has to be a VERTUALL ZERO (not zero) and from that point the rest of the line must start running the other way.

This establishes that no line can have a single direction and must have a continuance to both ends. Such a line has to have a value of 360° not 180° as believed.

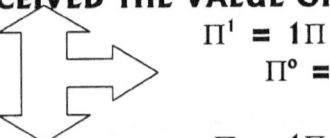

From infinity came the straight-line borders bordering the straight line on both sides of infinity

MATTER EVOLVED FROM TIME AS TIME = $7 (\Pi^2 + \Pi^2)$
THEN TIME, BECAUSE IT WAS ONE, MOVED A DIMENSION UP AND DOWN, WITH BOTH DIMENSIONS BEING EQUAL, THE SAME IN EVERY WAY.
Let the lines be somewhat bigger than infinity where we can see the lines more clearly

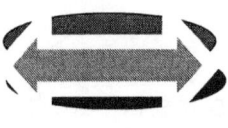

MATTER RECEIVED THE VALUE OF Π^2 AS FOLLOWS:

THEN $\Pi^1 = 1\Pi$
$\Pi^o = 1$

$\Pi = 1\Pi$

THEN TIME HAD A DIMENSIONAL VALUE OF $\Pi = 1$; $\Pi = 1$ AND $\Pi = 1$, BUT AT THE SAME Π DUE TO THE PROXIMITY OF SINGULARITY

POINT OF SIGULARITY

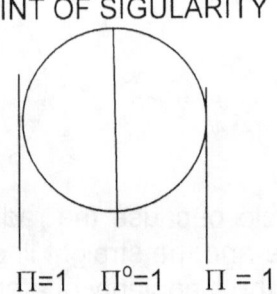

$\Pi = 1$ $\Pi^o = 1$ $\Pi = 1$

The Universe formed singularity in the straight line with two points of Π to both sides. There was no radius because the radius was infinite small. Then time formed as part of singularity in the value of Π going on to ΠΠΠ going on to $Π^3$. The one Π remained in singularity Π while the other ΠΠ became $Π^2$.

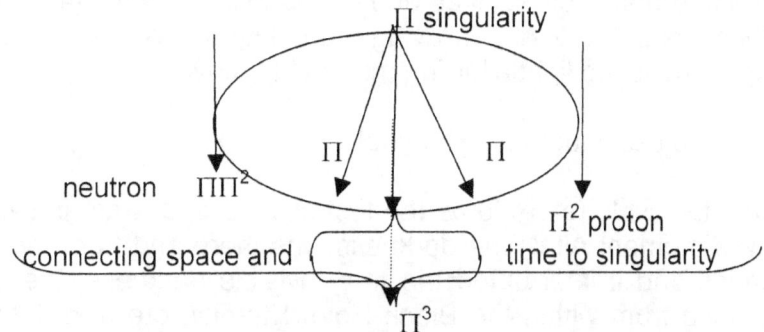

$$Π = ΠΠΠ = \textbf{SPACE} = 31 = 10Π$$
$$= ΠΠΠ = \textbf{MATTER} = 31 = ΠΠ^2$$
$$Π = ΠΠΠ = \textbf{TIME} = 31 = Π^3$$

TIME Π

While on the other side of the universal atom $Π^3$ formed space in time in the value of $ΠΠ^2$. Between the two universal atoms the line to singularity was the combination of $Π^3$ holding Π to $Π^2$ and forming space in matter to the value of $ΠΠ^2$.

Π singularity

neutron $ΠΠ^2$ Π   Π

connecting space and $Π^2$ proton   time to singularity

$Π^3$

In all it became 10 Π in all being in singularity

Whatever there is today started at a point we cannot trace. It started at a point we are able to envisage and locate but that is only with a mind that can accept what is not there to view. It is acceptable through intelligence because through intelligence we can detect it albeit outside the Universe we have. It is just like religion and worst is that it is generating what there is because what there is are not if not generated. There was $Π^0$, which was $α^0$ or if you would rather have it $Ω^0$ or it maybe was 1^0, but more correctly it was all the above and the beyond because multiplying what ever constitute the mentioned will bring about what is mentioned to a precise equality. It was a spot that was not. The spot is still there because the spot is still not there. It was a line that ran eternal but because it ran eternal and kept repeating exactly what was before to the precise what came afterwards, the line was there and was eternally running, while never changing in the least or growing by any measure. It was not one because before it could reach one, it returned to what was repeated and the process cycled back to before one was reached and even before one could be accomplished. It was such a continuing of the monotony, no change occurred and therefore never did the running produce progress because the progress was in the perfect repeat of what was before. The duplication brought contraction to the minutes detail. That is where our atheists get one hiccup. The repeat brought eternity and the repeat was so perfect that the repeat continued. The repeat still is with us as much as we are within the repeat. To bring change to the eternal repeating of the monotonous there had something beyond the Universe that institutes change. There was something that brought a difference and we are within that difference. That difference was time and that time is what we move through as much as what we see at night.

Oh, how stupid and how thoughtless the minds of atheist and other atheistic animals are. Baboons do not recognize the light we are within because they cannot think and are therefore atheists. Spiders cannot think and therefore they are atheists, as they do not think where the line is that is not. Reptiles cannot think and without thought they are incapable to see what time is, how time that is not generate space that is in time that is not. Mammals cannot envisage what space is, what light is and what makes us see the darkness cannot be. All the animals I have mentioned are mindless atheists because they fail to see beyond the visible into the realms of the thinkable. All that I have mentioned passes them by including religion and the accepting of God because through being incapable of thought and reason they cannot envisage what only intellect can bring to mind. Because of the incapacity to think the animals are both mindless and they are atheists. Therefore atheists are mindless. The night sky is such a bright light our evolution protected our vision from the brightness in order to give as much better vision. Through evolution development our eyes are protected by how we remove the qualities from the light. However animals do use that light and not our light to see by.

You can shine a bright hunting spotlight onto an animal at night and the animal will not be able to see the light on it. The animal does not use the light to see better as the animal is totally unaware of the light. Then a prowling cat comes from the night and sees the antelope in the light the night provides. It does not use the light, which the spotlight casts and the light is not even traceable to either animal being the hunter or the hunted. From there we accept that during the day the animals must be using our light to see because the nightlight is inferior to see by. Who says they use the daylight much different from the nightlight because all evidence is there that they cannot recognize our light as light. It is very evident in the manner they go on hunting and grazing while being totally unaffected by our form of light. That which you see at night because you cannot see darkness and you cannot see black is the light the Universe is painted in just like the Bible says. That is the light that started all because that is the light holding us away from the eternal darkness. This is not religion and it is not a sermon, it is hard-core and brutal basic science and it the most fundamental basic physics there is. It is the start of the mathematical Universe portraying the only physical way it could ever be. The light that came from the Command is the light allowing material to move.

My atheistic idiots, your mindless caught up with you!

Then came this light that the Bible refers to as the first of what ever was and what our stupidity tells us is darkness. This was moment-Alfa. The darkness was there and from the darkness heat came about. Only heat expands and it interrupted the true invisible darkness, the blackness of a Black Hole, the invisibility coming from within the Black Hole. Eternity tore from infinity. Darkness broke from light. Heat broke from cold. Relevancies parted by 1^0 going 1^1. There was one but also there was two too because one cannot be without two being there to ensure one is one. The marks are still with us but to see the marks requires a great deal of intellect.

$\Pi^0 \Rightarrow \Pi$. In this there was only space for one being one in the two forming one. It was $\Pi^0 \Rightarrow \Pi$ however there was no space to be $\Pi^0 \Rightarrow \Pi$ and there fore because of the lack of space to be which is the infinity of time braking the eternity of time the true measure was $\Pi^0 \Rightarrow \Pi$ but realized only 1^0 going 1^1. Π was to the future because of the motion of time involved and the space less ness of space at the time. By inclining to move the process crossed the Universe but also it took one eternity to accomplish the feat.

The fact that 1^0 going 1^1 brought movement can only become a reality as a result of light. Light is heat and the heat is expanding.

1^0 going 1^1 where $1^0 \Rightarrow 1^1$

1^0 going 1^1 1^0 ▶ 1^1 had to bring about 1^0 going 1^1 1^0 ▶ 1^1, because the eternal repeat of duplicating while contracting was not relieved from the Universe. Before the contracting was equal to the duplicating because by measure the heat was identical to the cold. It was eternity that was interrupted by one cycle of infinity and was in repeat of eternity. Once something is part of the Universe there is nowhere else to take it so it has to remain as a part of the Universe.

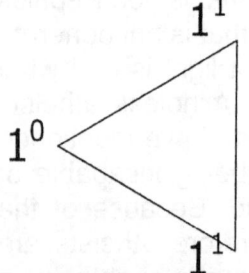

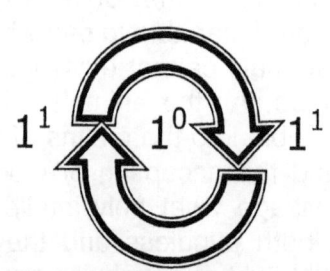

Then came three because motion was so limited that the least inclination to move threw what wished to move to the other side of the Universe, As it moves it also moved across singularity. It crossed the entire Universe as it moved because it moved and finding nowhere to move too. It crossed the entire Universe and it took one eternity less the measure of one period lasting infinity to achieve that. That brought to relevance three points where each was in measuring quantity exactly equal but also one Universe apart.

In the reality there was now two points holding singularity on both sides of the Universe because by crossing the divide that crossing set in place the two sides relevant of singularity governing. However infinity was bridges at two points holding infinity with which process eternity repeated the past into the future.

> • •▶2• • •▶3• • • •▶ 4

This then is the occasion where Pythagoras stepped in. Since it as a crossing of the divide the crossing involved a line that formed a half circle connecting a triangle. But the crossing was done in the space of half the Universe and since the Universe was 180^0 half of the Universe was 90^0. That involved Pythagoras as mathematics was born. Up to this point it was arithmetic with adding but now mathematics came into place. Remember we are a few eternities in side the development of the Universe.

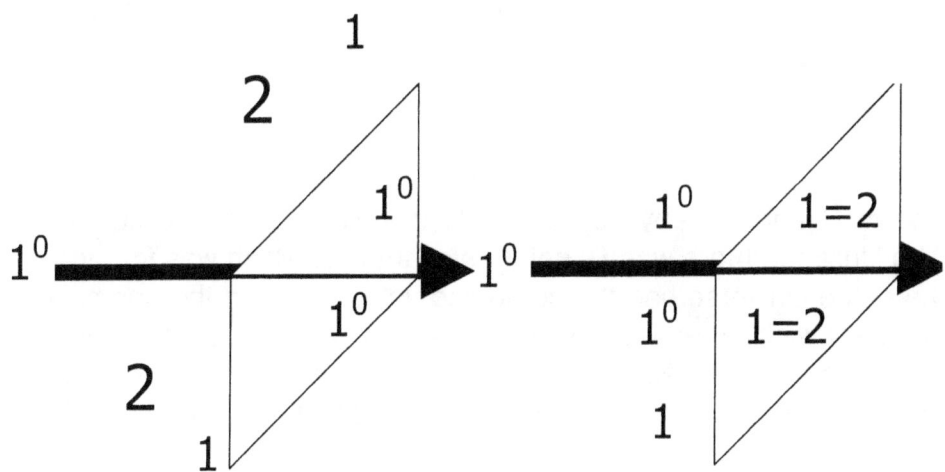

In the three came four that brought along five.

> • •▶2• • •▶3• • • •▶4• • • • •▶5

Reaching five is a benchmark because at that point half the Universe was finalised. From the five that formed the Universe continued and formed space-time. All progress balanced on the five that formed where two parts of equality parted eternally as liquid separated from solid, motion moved apart from the motionless, heat diverted from cold. The four in time grew away from the one point space formed being just outside the control of time. It was the start of material since from the point five formed just outside the four of time. There was a lagging behind of heat building and not being in the range of the immediate control of time. Time could generate a point in heat without bringing immediate demise to the point through cooling.

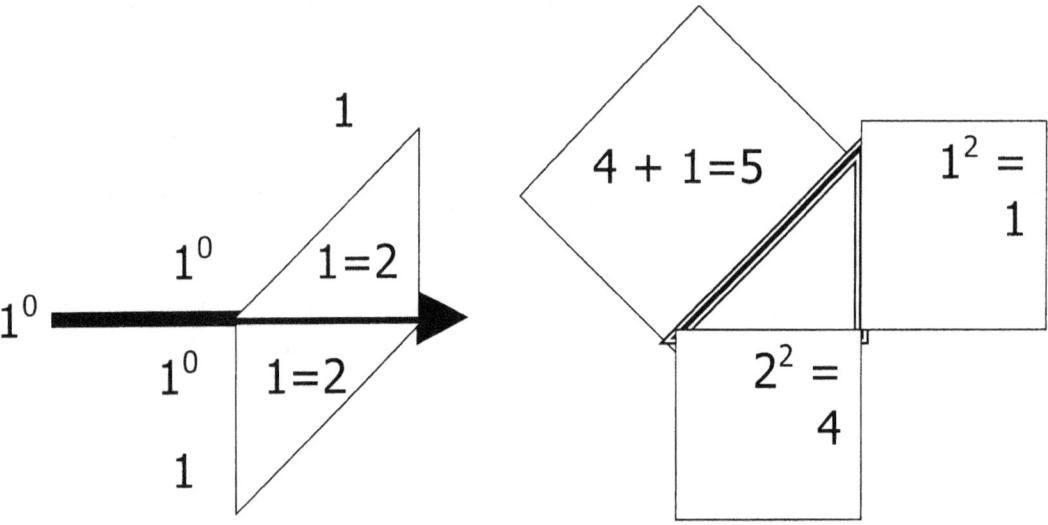

Then five filled the one half of the Universe that was able to contract and cool while the Universe divided the other half into sectors of what was (5) and what will be (5), which put material (7) in relation to the half of the Universe $\Pi^0 \Rightarrow \Pi$ in which the material was at that specific point (five relating to seven) in time.

1^0 ▶ 5

1^0 $= 1 + 5 = 6$

The motion consisted of Π moving to Π and thereby duplicating Π to relieve Π^0 of the burden of overheating. On the one side of The Universe there was Π^2 being relevant to Π which was forming on the other side of the Universe. The entire Universe had the combined value of Π^2 on the one side in addition of Π forming on the other side. I wish to remind the reader that any and all points formed by singularity was as much representing the Universe as it was the Universe at all times because $\Pi^0 = 1^0 = 1$. That made the entire Universe being any point affirming singularity by forming about singularity.

But that meant that the Universe was a total of $\Pi^2 + \Pi$ which when added was also $\Pi^2 + \Pi = 13.0$

•• ▶ 2 ••• • ▶ 3 •••• ▶ 4 ••••• ▶ 5 •••••• ▶ 6

$13.0 - \Pi^0 = 12$ because singularity cannot be part of space-time developing as the space, which later was filled with the material that formed, filled this part.

$12 / 2 = 6$ Material formed at the point where six was located.

$$6 \ + 6 = 13 - (\Pi^0) = 12$$

Because singularity is a divide and is not part of space-time singularity as a factor removes from space-time. Why it adds with five to form six is because to the one side only singularity is in the other side of the divide. Only nothing can be in two places at the same time therefore on any one side was the half of twelve, which divided 12 in two parts. That then was $12 / 2 = 6$

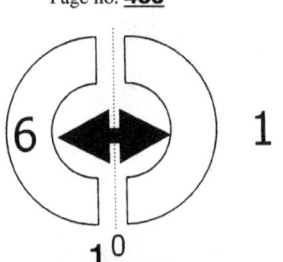

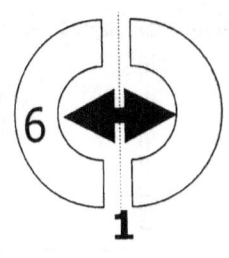

$$6 + 6 = 12 + (\Pi^0) = 13 \qquad 1^0$$

Developing six was an addition to the square as the line flowed and did not involve the crossing of the divide. Therefore Pythagoras was not involved by the forming of six.

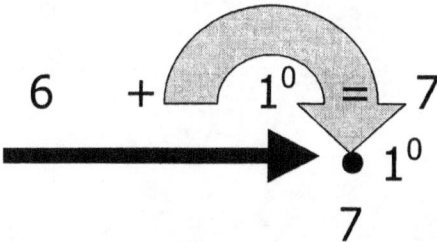

$$6 \quad + \quad 1^0 \quad = \quad 7$$
$$1^0$$
$$7$$

At the point where the space filled with heat meets the point in time representing singularity the end of material (6) confirmed the following spot (+1) at 7.

Forming seven very much involved singularity because it confirms appoint where space ends and space (8) begins.

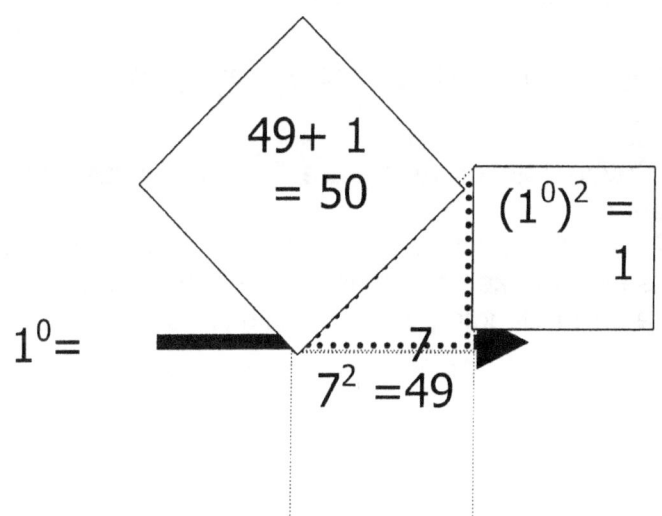

$$49 + 1 = 50$$

$$(1^0)^2 = 1$$

$$1^0 =$$

$$7^2 = 49$$

By taking singularity into Pythagoras and filling the Universe by halving the square of space seven completed the required circle within one half of the Universe in order to relate to half the time it takes material to fill time by duplicating. To find the necessary cooling required for control material has to use five points to be within because of the square involved. The there has to be another double five amounting to ten to fill the void from time in the past (position of five) and time in the future (another position of five).

50 | 1^0
7
7
1^0 | 50

$$50 + 50 = 100; \ (100)^{1/2} = 10$$

$(\Pi^0)=5$ 10

7

The Titius Bode require meant is seven holding relevance to ten and ten being relevant of seven while being in half the Universe $\Pi^0 = 5$

Then come eight causing a line of material to break.

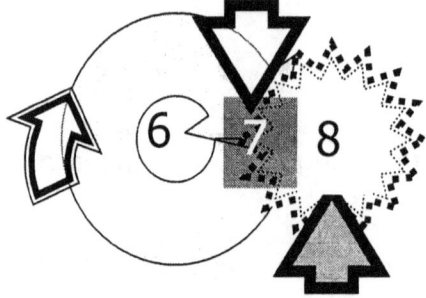

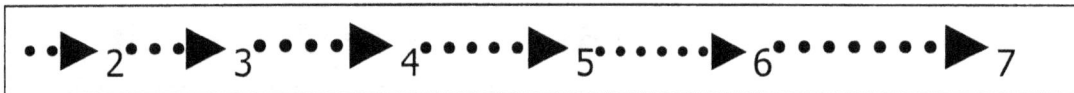

At seven the line completes at a point distinguishing material with in space from space without material

The circle of development has finalized a point. Seven has gone square 7^2 and realized with singularity half of the final of space in the absolute square.
It is this eight to ten science does not recognize and do not distinct as one other part of time. This in relation with the finality that came about at the point seven marked by using Pythagoras that another space, this time in time was developed to compromise for the lagging of time within space-time.

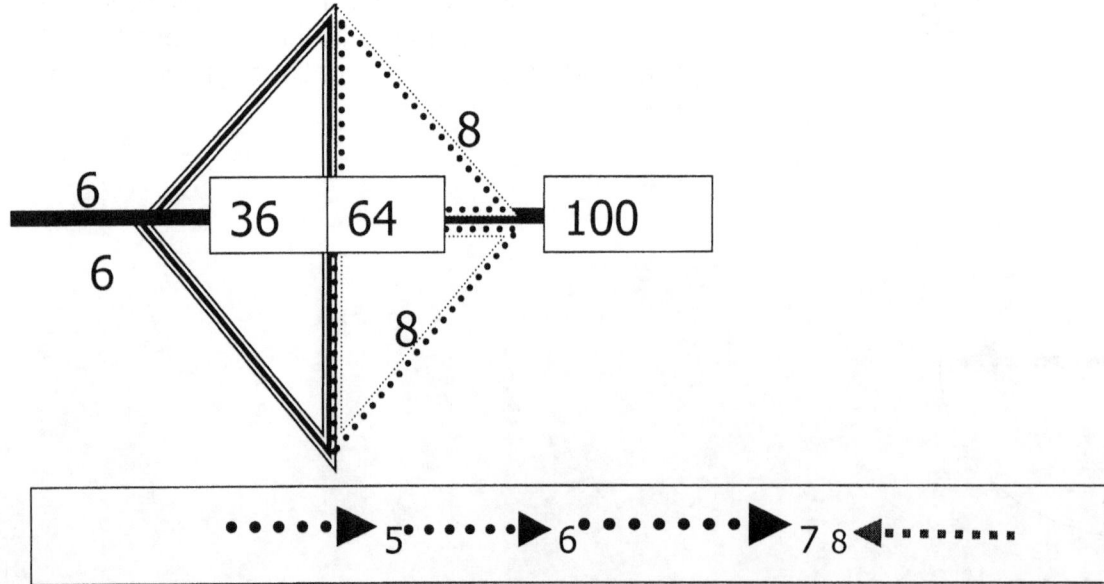

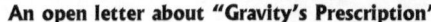

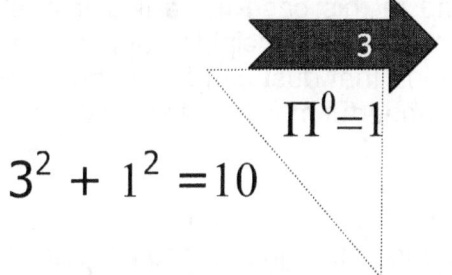

$$\Pi^0 = 1$$

$$3^2 + 1^2 = 10$$

The cycle of eternity could then complete one more time by forming singularity once more

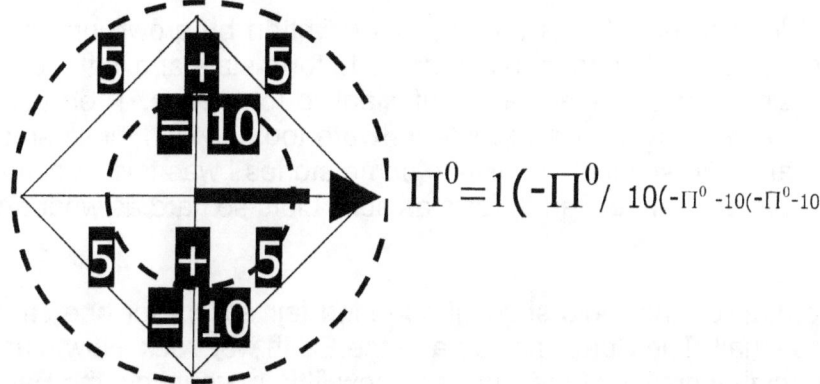

$$\Pi^0 = 1(-\Pi^0 / 10(-\Pi^0 - 10(-\Pi^0 - 10$$

With the Universe established at ten crossing the divide meant that Π^2 at four was a half and five was completing the one half.

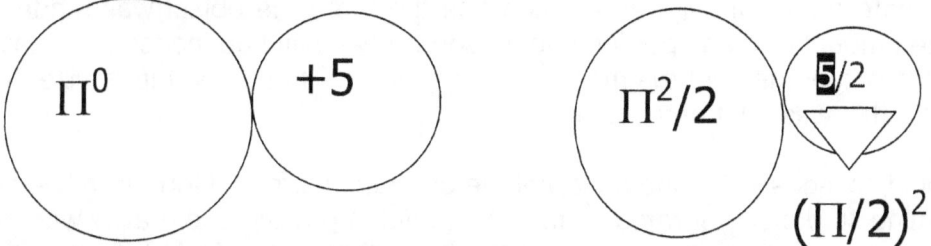

$$\Pi^0 \qquad +5 \qquad \Pi^2/2 \qquad 5/2$$

$$(\Pi/2)^2$$

The Roche limit shows that singularity needs at least more than half the Universe (5/2) to share and…the Lagrangian system is at least half the Universe.

Therefore the circle extended to the point outside singularity where matter was holding a space value of 10 Π^3 and space holding matter was in 10 Π. Then there came the proton connecting space in time to singularity ON THE OTHER SIDE OF THE UNIVERSE.

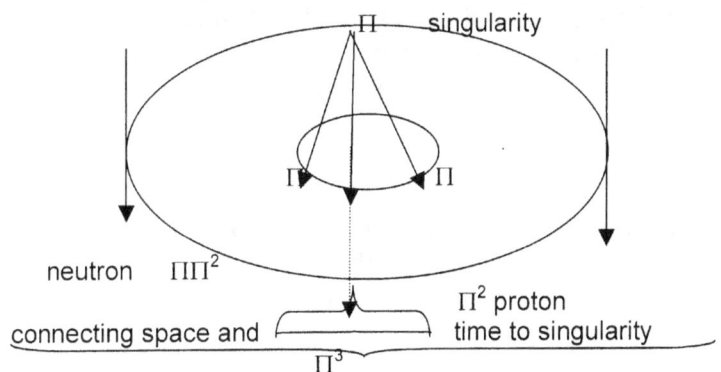

In all it became 10 Π in all being in singularity

To connect time to space the 10 Π divided into the 10 Π^3 forming the other proton of Π^2.

THE PROTON THE NEUTRON ELECTRON

$$(\Pi^2 + \Pi^2) \times (\Pi \ \Pi^2) \times \quad 3 \quad = 1836$$

MATTER AT $^7\Pi + \Pi^3$ SPACE OUTSIGE SINGULARITY IS 10 THE ATOM

We can thank that same process forming our solar system in the small manner as I described. After running through the cosmic formation principle I wish to show in what way one might be able to use the relevancies to further one's understanding of the cosmos and in this case in particular one's understanding of the solar system and the manner in which it formed. The principles I apply is real and does not reflect on the Newtonian Mede Avail mind boggling idea that dust can fall on dust and forms layers that will eventually turn as hard as rock by the mere wonderful magical medicinal forces unleashed by the magic of gravity.

We take the approach that the cosmos does prefer to use science instead of the wonders of magic and that we are able to explain everything there might be in the cosmos through the medium science renders us.

When I was very young I had a good look at how I was told in books written by grown ups how planets is formed from rings of dust they accumulate and then compress to form stars and in the case of the smaller stars it became planets since the material was insufficient to form stars. Then there was the fact about comets being small pieces of rock and so meteors were too and with those small pieces of rock getting that solid I was starting to smell rotting fish in some stories I was told. A comet and a meteor were small, yet they were big enough to start to form as solid blots so hard as what any rock could be.

Hey, for a child that still imitated Superman on the radio show, it sounded fair accept for one small detail. The Sun was big. The earth was small. The Sun was gas and the Earth was rock. How much more material was accumulated by the mighty gravity of the Sun and how little material did the Earth use to accumulate and what force had the earth available to compress all of that in comparison with the Sun . Yet the Earth and all its smaller planet sisters could compress the little it had to rock but the bigger they got and with the more gravity there was in each one the more the object was made of gas. Gas can't bring about the amount of mass per inch going square (we still had inches then) than that what rock can however, the bigger they get the more gas they collect. That made me realize that Superman was a possibility, but this was a lot of rubbish.

If it was the case that dust could collect and lie one on top of the other in layers and form from weight to become dense while becoming so dense it formed stars with the aid of gravity, and gravity was the pull of material making them dense, and gravity was about the more there are of what there is the stronger the pulling is and on top of that the Sun was Big and the Earth was small, yet the Sun was a gas and the Earth was rock, then someone painting me this picture in monochrome got black and white confused and a lot of disciplines mixed up.

If gravity was about a lot pulling more than the little was pulling less because of the lack of weight and the lot compressed, what was available to compress so hard it became solid hard rock by pulling, then it had to be the other way around. The Sun must be awfully hard while the Earth is still trying to sort out the rings of flying dust that the Earth was unable to control with the little gravity the Earth had available to use. As I said before, back then I was a child and was strongly influenced by the opinion of my superiors and Superman made a lot of sense.

Nevertheless… that lot got me suspicious and when I had a good look at the tale of the illusive comet that refused to be caught by the big bullying Sun and how the comet gets away every time as Robin Hood always did in the light of what gravity was available to the Sun and in the amount the Sun had to its disposal while the Sun in all its splendour was not doing to the comet versus what gravity was suppose to do with the comet, made me ask questions.

After a long time passing by with no one being able or in mind to answer my asking questions I came to the conclusion I had to provide some sensibility myself if I was reluctant to accept the insanity of what cosmology in theory had to offer. That what cosmology offered was harder to believe that Superman and by then I was getting a lot of doubt about the fanciful abilities of a man called Clark Kent. To go into more detailed suspicions I had and how I turned that around to fit my theory I eventually devised will be much too roomy to have in a compressed book such as this letter is supposed to be.
The Sun was one part of a binary. The binary was part of a Lagrangian layout.

O   Saturn

Binary
O   Neptune   oo   O   Jupiter

O   Uranus

This was all frozen before the Big Bang in the same way as the structures in the inner core of galactica still are at present. The binary started spinning around their common or joint axis at $(\Pi/2)^2$. The Sun eventually overpowered the one structure in the binary and that star, (which I named Unknown Star) overheated as it could not apply enough cooling to its inner-core.

As it broke into bits, those bits of inner core formed the solid planets, the asteroids, the comets and even the Pluto double rock structure. All the moons and satellites are also remaining debris from this event. The outburst caused by the Unknown Star destruction led to the release of liquid heat and this liquid heat is what the Authentic Author saw as water. The winds he refer to is nuclear winds, established as the heat formed gas as space from its position of being liquid heat, developed into a gas heat we call space.

Since the First movement of time, the Roche factor was present and from the Roche factor came the Titius Bode principle. Each object seen, as well as not seen, represents a different period contained by a different specific value of space. All matter is time in a different frozen state in space and the time gives the space its particular value.

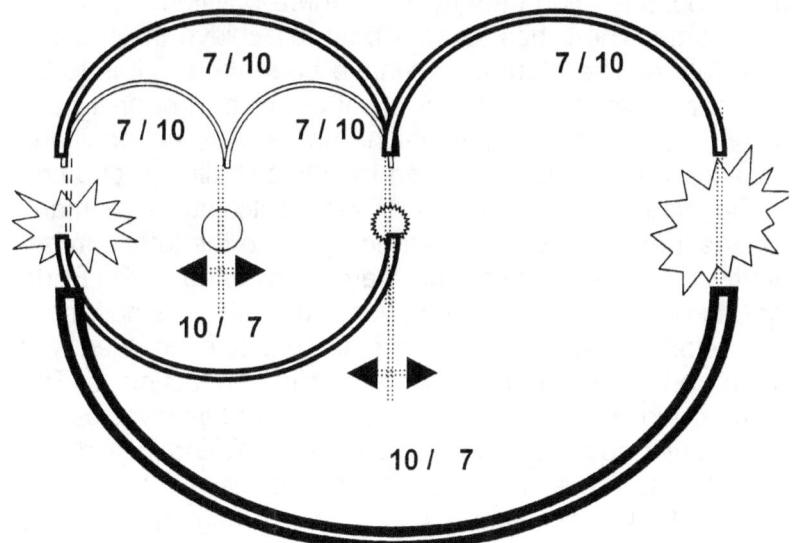

We now can use what is cosmic law to interpret how development actually took place by cosmic construction. The form the cosmos adapted remained d present to this day because what once is part of the cosmos does not go away and the cosmos use the same there is in so many ways the many ways make what there is confusing to us.

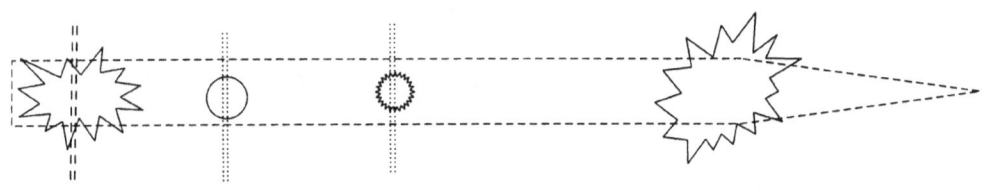

They say its Kepler's laws but that is where the inaccuracy starts because Kepler never said that. It is a derogative Newton established of what Kepler introduced as something very different.

The Law Newton gave as Kepler's law of orbits.

All planets move in elliptical orbits, with the Sun at the centre focus.

The figure shown below is set to show a planet with a mass of m, moving in such an orbit (blue dotted circle) around the Sun . The Sun holds a dominating mass of M. We assume that $M > m$ so the centre of the mass of the planet is virtually in the centre of the Sun . The orbit is described by giving its **semi major axis** a and its **eccentricity** e the latter defined so that ea is the distance from the centre of the ellipse to either focus F or F^1. *An eccentricity of zero corresponds to a circle,* in which the two foci merge to a single central point. The eccentricities are not large as the sketch would indicate but in reality they seem circular. In order to bring across the eccentricity the orbit discrepancy has to be exaggerated to find meaningful clarity. In the case of the earth the true eccentricity is only 0.0167.

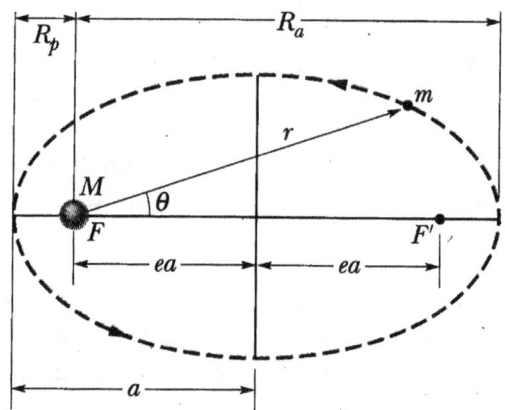

According to Newton and from what we can observe in the sketch the planet with mass m is at focus F of the ellipse. The other, or the *empty,* is focus F^1. Each focus is at a distance ea from the centre, e being the eccentricity of the ellipse. The semi major axis a, of the ellipse, the perihelion (nearest to the Sun) is at a distance R_p, and the aphelion (farthest from the Sun) is at a distance R_a are also present.

Notsofast.

The diverting from the main axis forming an angle θ is not a mathematical angle but is only time duration. In physics there might be a point in reference to another point but this is far from physics. It is cosmology. There is a point holding eternity 1^1 by three positions $1^1 \times 3$ in relation to infinity 1^0 and the three positions is time. There has to be a past to achieve the present and to achieve the present there has to be a future. The atom has to have been there if it is to be where it is in order to go too it will be next. There cannot be a top or a bottom or a front or a back because there is an infinite centre one relating to six other position which numbers eternal positions all sharing one unit. Those being strong in the field of mathematics, can take this challenge and find the possible number of 1^1 in relation to the centre with no sides 1^0 and calculate how many possibilities there might be on the edge of any sphere that will correspond by six to the total number of possibilities not there in the centre singularity 1^0. In the Universe there are no fixed point. On Earth there is the all - dominating centre which grants physics relevant formation, but in the cosmos there are no such points anywhere. There is in the relation one point holding a singularity concerned with expanding, which has the position of 1^1 and there is a singularity concerned with contraction 1^0. The motion depends on getting al the atoms from one location to the next location and this involves the distributing of all the factors concerning 1^1 in relation to every one of the factors relating individually and as a group to 1^0. This reference of F relating to $F1$ is corresponding not to the planet but it is corresponding to every atom individually where the group forms a centre, which corresponds to a centre that forms with the aid of all the atoms forming motion within the Sun . The relation that this group has, forms a partnership with the motion within the entire Earth and from that a centre of governing gravity is selected by the motion of every independent atom that responds to the motion of every subatomic particle notwithstanding how incredibly minute they may seem. The Sun turns with the turning of every atom within the Sun which generates a combined motion that generates a selected centre through the all inclusive effort of every proton that provides the expanding and contracting balance we find in the Sun . By the motion there is also a conflicting motion that supports the motion where the one part is lateral and the other is rotating. The space $\mathbf{a}^3$ moves partly in a circle $\mathbf{T}^2$ and partly in a straight-line $\mathbf{k}$. That is what Kepler said when Kepler said $\mathbf{a}^3 = \mathbf{T}^2 \, \mathbf{k}$.

An eccentricity of zero corresponds to a circle, in which the two foci merge to a single central point.

If that was true then the circle in which the planet orbits the Sun should be dead canter because with an eccentricity that is obvious there has to be favouring one side and then favouring the other side. The zero correspondence is not there since there is a dual eccentricity detected in all the planets in orbit around the Sun . There is much more a swapping of prominence between the two points in foci.

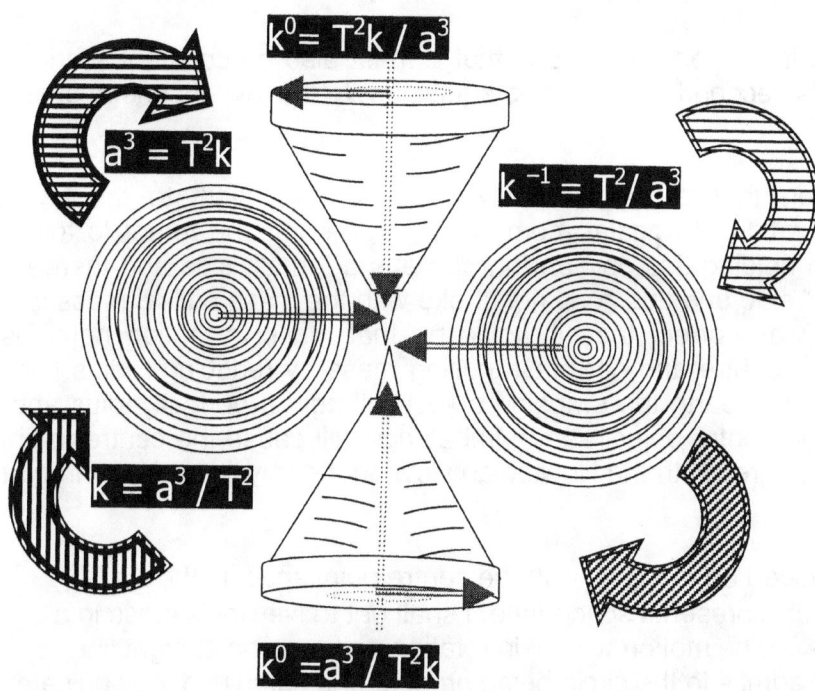

There is a divide that separates one side of the orbit from the other side of the orbit and the one is opposing the other side of the same orbit. This has to do with the changing of the motion in relation to the orbit where the changing of relevancies produce gravity, but not by mass because mass is something in Newton and in Newtonian imagination. Mass is a hoax, a lie and fraud to such an extent as the world has never witnessed.

The Law Newton gave as Kepler's law of areas.
A line that connects a planet to the Sun sweeps out equal areas at equal times.

Qualitatively, this second law tells us that the planet will move most slowly when it is farthest from the Sun and move most rapidly when it is nearest to the Sun . As it turns out, Kepler's second law is totally equivalent to the law of conservation of angular momentum. Let us prove it.

In time Δt, the line r connecting the planet to the Sun (of mass M) sweeps through an angle $\Delta\theta$, sweeping out an area ΔA. The linear momentum p of the planet and its components.

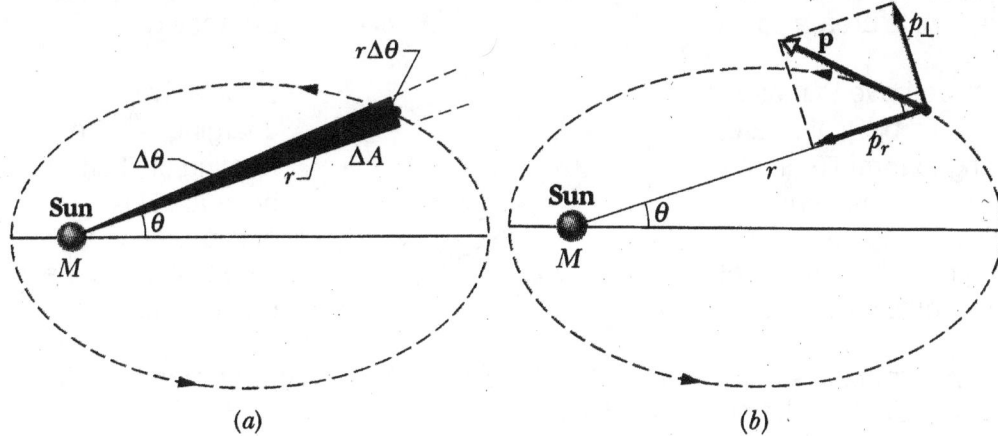

(a)   (b)

The area of the shaded wedge closely approximates the area swept out in time Δt by a line connecting the Sun and the planet, which are separated by a distance r. The area ΔA of the wedge is approximately the area of a triangle with base $r\,\Delta\theta$ and height r. Thus $\Delta A \approx \frac{1}{2}r^2\,\Delta\theta$. This expression for ΔA becomes more exact as Δt (hence $\Delta\theta$) approaches zero. The instantaneous rate at which area is being swept out is then

$$\frac{dA}{dt} = \frac{r^2}{2}\frac{d\theta}{dt} = \frac{r^2\omega}{2}$$

in which ω is the angular speed of the rotating line connecting Sun and planet.

The figure shows the linear momentum p of the planet, along with its components. From, the magnitude of the angular momentum L of the planet about the Sun is given by the product of r and the component of p perpendicular to r, or

$$L = rp_\perp = (r)(mv_\perp) = (r)(m\omega r)$$
$$= mr^2\omega$$

where we have replaced $v_\perp$ with its equivalent ωr . Eliminating $r^2\omega$ leads to

$$\frac{dA}{dt} = \frac{L}{2m}$$

If dA / dt is constant, as Kepler said it is, then that means that L must also be constant – angular momentum is conserved. So Kepler's second Law is indeed equivalent to the law of conservation of angular momentum.

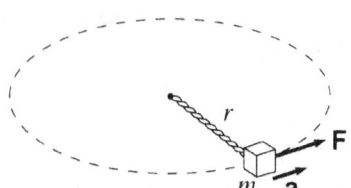

Newtonians share a dubious opinion that the second law is totally equivalent to the law of conservation of angular momentum. If so someone forgot to tell the person holding this opinion that there is a "small" side and there is a "larger" side and this does not stroke with the mass pulling mass idea. If this was correct then it would be that mass in motion neutralise mass in centre and not mass pulling mass by reducing radius to the square. On earth the swinging object will hold a perfect radius and if not the spinning object will destroy the centre object as the imbalance will cause the centre to shift because the thrust varies. I still do not agree with the Newtonian view on the physics part but in that I have no mission.

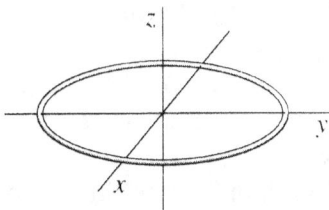

A ring is placed on the x y with the centre point spacing the circle evenly. That represents space-time. I shall get to Newton's mass in a short while… The motion we get in rotation honours the straight line because it admits to the circle being present and combined they pulsate between the triangle representing one another in relation to the other side of the Universe and each other. Lets study the motion in the most incredibly small minute exclude any and all time delays forming the the space. We go to where there still are no by Newton's imagination. Lets make the Force F the mass m the overheated while area *a,* is the

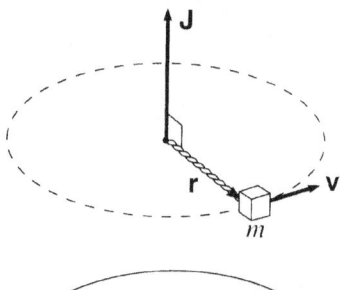

process of the context in order to material that is filling mass present even the overheating and expanding.

The first acknowledgement we have to make is charging motion and that would be 1^0 that are now where $6\Pi^0 + 1^0$ is going to form Π. In the continues to run in a line. The line is forming Π forming on the inside of time being on the is in the process of forming is to the outside of eternally with no end and no outside on the

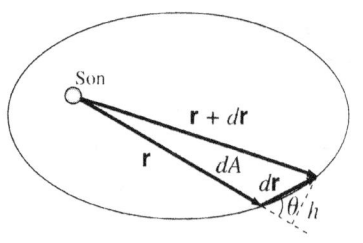

singularity is charging $\Pi^{0.}$ We are forming of Π time because Π is outside. That which time because time is outer limits of Π.

On the one side the motion formed a straight line to the value of 180^0 and at that moment on the other side of the Universe (the Universe being 1^0) a half circle also to the value of 180^0 that in relation forms a triangle also to the value of 180^0 and that is making what there is in duplicate in both directions (**k⁻¹** and **k**) formed as the dot 1^1 coming from the spot 1^0 went around the Universe because the spot1^0 could turn just as much as it still cannot turn. That is the reason 1^0 has to charge 1^1 to commit motion as Π^0 and dorm a line that represents a half circle on the other side of the Universe. We can see that Θ is equal to p while d is the motion representing p that is acting as r but since there are no possibility of motion O is actually going over to the other side of the Universe and from (**k forming k⁻¹**)

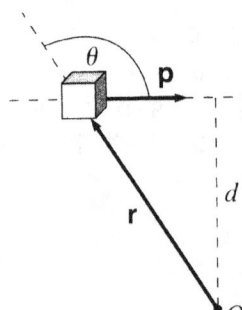

In order to is happening is expand to the outside. Eternity order to expand releasing infinity giving infinity a start to the outside where eternity that has no end

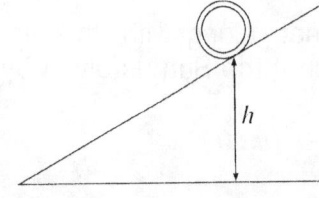

understand what truly that eternity cannot outside. It has no has to reduce in and by reducing it is which has no star by

F

finds an end in infinity with no start point where infinity starts. I am incapable of putting it more practical and Newtonians with not the mental capacity to understand must at this point find a motor mechanic to explain to them what is going on.

The motion coming about is eternity squeezing the daylights (because it is heat or light) out of infinity that already has no space where in it can be squeezed.

The circle finds the drive it will have contact with eternity and eternity circle. It is powered by the rim in relation to the singularity centre the space-time that the motion stand erect alone is most significant. that drives the top has its significance top. Life with its manipulating and with a release of the top while locomotion. The biggest energy issue The energy is life induced. It is not

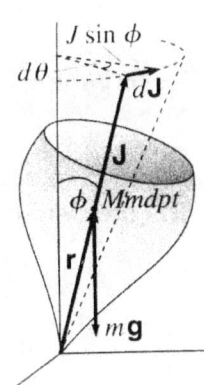

from the outside where the outside is in finds a line to guide (not drive) the rolling discrepancy in the motion of time on the that still is unable of motion. The drive is in charges intro action. The fact that the top The drive- line that charges the motion from a line spun around the outside of the qualities wrapped a rope on the outside still controlling the line that the top found at this is admitting the energy origination. cosmic.

The energy cannot have an equal in "mass" meaning size or "mass" meaning duplicating tempo since the atomic or particle presence within the top has no strong enough motion accumulated between the lot to displace the singularity relevance to a centre where from such a centre a governing singularity can be charged and maintained.

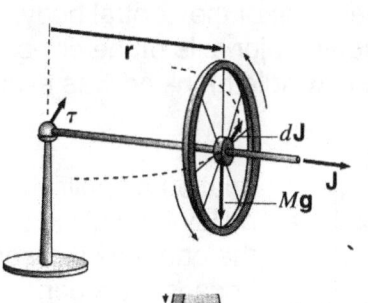

The whole motion locomotion is life inspired and life has validity on one small spot in one tiny solar system that is maintaining a most insignificant solar object of very little cosmic substance. The gyroscope finds the drive coming from the outside because the very inside is unable of motion. The motion drives by means of both dJ and Mg where Mg has no mass reference. The required increase in driving effort comes from the larger ratio the outside of the wheel hold to the significance of the inside that still is too small to move notwithstanding the increase in ratio between eternity and infinity. The significance of a drive line is placing the generating there is between that which moves eternally and that which eternally can't move in relation to another but connected Universe where that which eternally cannot move has a conflict in ratio to that which moves eternally.

Notsofast.
From the planets view the planet is moving away from the Sun by a measure of $k = a^3 / T^2$.

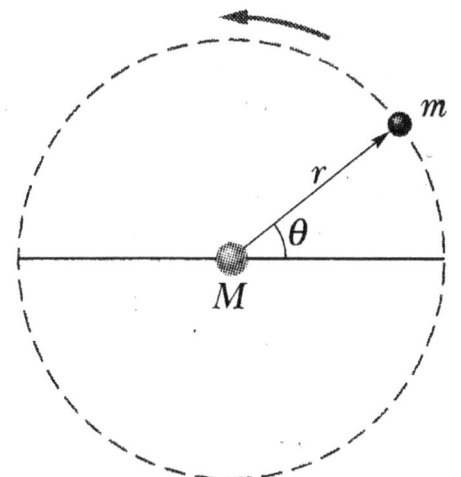

Therefore according to the planet it is moving straight ahead way from the Sun without even glancing back.

From the view the Sun is enjoying the situation the planet is coming back and the planet has no time to look back into the open sky as it is on its return by the measure of $k^{-1} = T^2 / a^3$.

In reality from an observer's view going by the name of Johannes Kepler space a^3 is in a relation with time by the straight - line k as well as the partial circle T^2. The planet is returning as fast as the planet is escaping and every factor finds a position of victory while equilibrium is the only victor.

A planet of mass m moving around the Sun in a circular orbit of radius r.

The Law Newton gave as Kepler's law of periods.

The square of the period of any planet is proportional to the cube of the semi major axis of its orbit.

To see this, consider a circular orbit with radius r (the radius is equivalent to the semi major axis of an ellipse).

Applying Newton's second law, F=ma, to the orbiting planet yields

$$\frac{GMm}{r^2} = (m)(\omega^2 r)$$

Here we have substituted from for the force F and used to substitute $\omega^2 r$ for the centripetal acceleration. If we replace ω with $2\pi/T$, where T is the period of the motion, we obtain Kepler's third law:

$$T^2 = \left(\frac{4\pi^2}{GM}\right) r^3 \qquad \text{(law of periods)}.$$

The quantity in parentheses is a constant, its value depending only on the mass of the central body. **The Equation** also for elliptical orbits, provided we replace r with a, the semi major axis of the ellipse. This law pinned on poor old Kepler who was rather innocent is the biggest swindle mankind was ever hoaxed by.

Newton suggested that the two structures were working in a relation as pulleys would. To do that a point has to be in place on which the pulley may pivot. Then the one mass will pull(ey) the other mass and bring the one mass closer to the other mass by diminishing the radius to the square because it is done from both sides. He caught the world hook line and sinker for three hundred and fifty years and every one was caught because no one would admit they were to stupid to understand what the most brilliant brain of all time saw.

Notsofast.
This is the biggest scam the world was ever caught by. Nothing can even compare to the stunt Newton pulled off. This made Newton the Master swindler who is matchless by miles. Some sell the Eiffel tower and other tricksters walk through the China wall, but let Mr. Houdini repeat this one. Newton never (not once) had to prove mass because mass was never a cosmic factor and never being there it was never there to be proven so everyone accepted it as proven. Mass is Earth born and it is Earth related.

There has to be a resistance stopping the motion or resisting the gravity to bring about mass because the mass comes about when duplication by motion is prevented from moving while the moving still manage to preserve the independent nature of the structure holding the considered mass factor. In outer space everything must move therefore the only restriction is the conserving nature the particular elements reserve.

For three and a half centuries no one had the idea to return and see why mass is what mass is when mass is and when mass is not. Good God (that a prayer for my fellow humans), is everyone that stupid, that blind and that naïve. The most ridiculous part of all is the joke they pin on me. When they are cornered and their explaining has to go beyond the ridiculous, they say it is I that cannot understand Newton. It then becomes me being uneducated and at every occasion they pin to my jacket the label that I am mentally underdeveloped.

Then they counter the blame on me arguing that I am so retarded that I cannot understand Newton because I am uneducated. Suddenly my being a motor mechanic convince them most of all because of all brainless things any one can be a motor mechanic is the worst there is. Then it dawns on them that it was I that never even for one semester was at any University to be educated in the science of

Newton. By all these standards I am ridiculed and labelled because I can't understand Newton! I thank God I can't understand Newton! Lets look at what is so obvious even I can see it and yet that all the educated never question or sow any doubt.

KEPLER'S LAW OF PERIODS FOR THE SOLAR SYSTEM			
PLANET	**SEMI MAJOR AXIS** a $(10^{10}m)$	**PERIOD** T (y)	T^2/a^3 $(10^{-34}$ $y^2/m^3)$
Mercury	5.79	0.241	2.99
Venus	10.8	0.615	3.00
Earth	15.0	1.00	2.96
Mars	22.8	1.88	2.98
Jupiter	77.8	11.9	3.01
Saturn	143	29.5	2.98
Uranus	287	84.0	2.98
Neptune	450	165	2.99
Pluto	590	248	2.99

This law predicts that the ratio T^2/a^3 has essentially the same value for every planetary orbit around a given massive body. The table above shows how well it holds for the orbits of the planets of the solar system.

If the Newtonians could just once for one instant show me ho mass does effect the behaviour of planets orbiting even in the least, then I too can boast about me finally being able to understand and appreciate Newton. I have tried for so long on so many occasions and in so many respects to find grounds for Newton's persistence on blaming mass in relation to motion created. I have committed so many arguments that I could not take in any direction as to why Newtonians would follow the untested and meaningless arguments so sheepishly just to follow their Master. There are slight anomalies but one can see there is a persistence of three in relation to the time space holds in all cases.

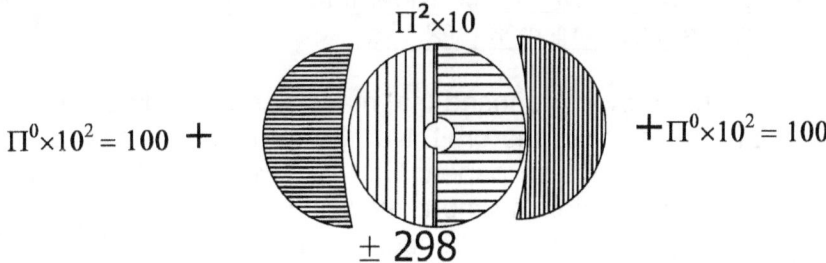

$$\Pi^2 \times 10$$

$$\Pi^0 \times 10^2 = 100 \quad + \qquad\qquad\qquad + \Pi^0 \times 10^2 = 100$$

$$\pm\ 298$$

Time in spot 1
$\Pi^0 X\ 10^2 +$
Time in spot 2
$\Pi^2 X\ 10 +$
Time in spot 3
$\Pi^0 X\ 10^2 = 3D$ 3 Dimensions in total $\pm$ 298 or then two
positions (spot 1 and spot 3) and one allocated in (Time in spot 2 $\Pi^2 X$ 10) gravity-motion.

Notwithstanding the planet arrangement time relates to space by $k^{-1} = T^2 / a^3$.
Spot 1 Spot 2 Spot 3

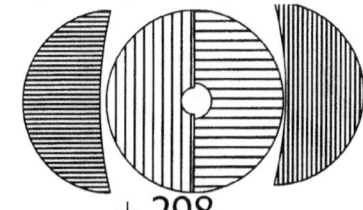

$$\pm\ 298$$

From the above table one can clearly see the three positions in time that space holds. By placing time in the relevance to space we also can clearly see that the space is flowing where the relevance is reducing time $k^{-1} = T^2 / a^3$ or $\Pi^0 = T^2 k / a^3$. The space is flowing towards a dominating centre and that we know is the centre of the Sun .
$k^{-1} = T^2 / a^3$ which then is $3 = T^2 / a^3$

$T^2 = a^3 / k$ which then is $T^2 = a^3 / 3$
$a^3 = T^2 k$ which then is $a^3 = 3 T^2$
$k= a^3 / T^2$ which then is $k= 3 a^3 / T^2$
$k^0 = a^3 / T^2$ which then is $k^0 = 3 a^3 / T^2$

There is no mention of mass or any discrepancies in any form or shape or thought. All the planets serve the Sun on equal terms notwithstanding whatever notion humans try to add for what ever disguised pleasure the human may find in such an argument.

Here is a typical Newtonian response to a very formidable question about the masses of the planets don't matter when determining their orbits. What counts on this situation is the equilibrium between the gravitational attraction and the centripetal force. Look at these equations:

$Fc = (mv^2)/r$ and $Fg = (GMm)/r^2$

Where Fc is the centripetal force, Fg the gravitation, m is planet's mass, M is Sun 's mass, v is the orbital velocity of the planet (which is really not constant, but we can consider it this way) and r is the orbit's radius. G is the gravitation constant. Its value is 6.67E-11 $m^3/(s^2 * kg)$.

The orbits are stable, otherwise we wouldn't be here to discuss this problem! So, the attractive force (gravity) has to be equal to the centripetal force, which tends to make the planet escape from its orbit.

$Fc = Fg$
$(mv^2)/r = (GMm)/r^2$ and with some simple manipulation: $r = (GM)/v^2$

As you can see, the radius of the orbit doesn't depend on the planet's mass. It depends only on the Sun 's mass, and inversely on square velocity. What determines the position of the planets in any solar system is the system's initial conditions, which aren't well known... yet! The initial angular momentum, the mass distribution discontinuities in the dust cloud that originated our solar system (or any other), and some other factors, were the conditions that led the planets to be arranged this way. Yes, they were "built" about 5 billion years ago on almost the same orbit they have today!

Fc is the centripetal force, Fg the gravitation, m is planet's mass, M is Sun 's mass, v is the orbital velocity of the planet (which is really not constant, but we can consider it this way) and r is the orbit's radius. G is the gravitation constant. Its value is 6.67E-11 $m^3/(s^2 * kg)$.

There is no indication whatsoever in any language ever proposed that mass has any indicative role or influence on any planet in any way. Not the motion. Not the orbit. Not the rotation. Not the velocity. That which is supposed to be is in our minds and not on paper. That is how we wish to interpret cosmology. The larger person is the heavier person holding more momentum. The larger planet is the heavier planet forcing more gravity and is therefore the strongest. That is rubbish. That is the mindless gargle of the brainless. I am sorry but I cannot be less explicit about this matter. It is out of touch with reality it could be part another fairy tale. It seems no one has learned! Go on and calculate which planet is generating most gravity per cubic what ever. Find out in simple terms that the big structures are generating almost nil when compared in direct relation to the smaller planets. It has nothing to do with their being gas and everything to do with the inverse square law.

Again I reiterate ands repeat my question: If the gravitational constant had the means to eliminate the mass factor in the orbiting of the why is there a need to use it in any form of calculations? Why persist with a meaningless proposal.

Jupiter has a MASS of 18955.872 $\times$ 10 ^{24} **/ GRAVITY of 24.89778 = 761.58**
  That means every 761.58 parts holding space generates one Nm of gravity

Saturn has a MASS of 5686.76 $\times$ 10 ^{24} **/ GRAVITY 10.556 = 538.7**
That means every 538.7 parts holding space generates one Nm of gravity

Neptune has a MASS of 1027.872 $\times$ 10 ^{24} **/ GRAVITY 11.2815 = 91.1**
That means every 91.1 parts of space held generates one Nm of gravity

Uranus has a MASS of 866.52 $\times$ 10 ^{24} **/ GRAVITY 8.96634 = 96.64**
That means every 96.64 parts holding space generates one Nm of gravity

Earth has a MASS of 59.76 $\times 10^{23}$ **/ GRAVITY 9.81 = 6.091**
That means every 6.091 parts holding space generates one Nm of gravity

Venus has a MASS of 48.7044 $\times 10^{24}$ **/ GRAVITY 8.87805 = 5.485**
That means every 538.7 parts holding space generates one Nm of gravity

Mars has a MASS of 6.418224 x 10^{24} **/ GRAVITY 9.81 = 0.654**
That means every 0.654 parts holding space generates one Nm of gravity

Mercury has a MASS of 3.3029352 x 10^{24} / GRAVITY 3.64932 = 0.905
That means every 0.905 parts holding space generates one Nm of gravity

Pluto has a MASS of 0.11952 x 10^{24} / GRAVITY 0.3924 = 0.3045
That means every 0.3045 parts holding space generates one Nm of gravity

PLANET	MASS	GRAVITY	GRAVITY/MASS	%
Mercury	0,05527:1	0,372 :1	6,821	682,1 %
Venus	0,815 :1	0,905 :1	1,11	111 %
Earth	1 :1	1 :1	1	100 %
Mars	0,1074 :1	0,38 :1	3,538	353,8 %
Jupiter	317,8 :1	2,538 :1	0,007986	0,80 %
Saturn	95,16 :1	1.075 :1	0,0113	1,13 %
Uranus	14,5 :1	0,914 :1	0,063	6,306 %
Neptune	17,2 :1	1,14 :1	0,663	6,63 %

Because of the invert square law the existing of such a law puts Newton's theory on mass in doubt. The more the size is the less the force of the mass will be in proportion to space occupied. The bigger is not the more powerful or the strongest because of the invert square law. If there were no invert square decrease of space to power ratio, then yes it would make sense. But the Sun has the least gravity of all in the solar system just because of the size it holds.

Can any Newtonian out there please explain what mass has to do with the rotating of planets around the Sun and what do you accomplish by calculating the mass as a factor. That has no practical implication even in the least in relation to the orbit. It implies a massive degree of senseless ness shown by the educated to prove his education to be bizarre and null - in void. It is like blaming the ice-cream sales in Alaska for the persistent drought in the Namib Desert and then invents non-relevant factors to prove a link. If mass was a factor then Jupiter must either propel faster or slower than the rest or Jupiter should be farther or closer or have some indication that places the size relevancy in another category than the others. To say the gravitational constant even things out means to say the gravitational constant is nullifying the mass and then that proves by point, the mass had nothing to do with the price of eggs in the first place.

Every one is bullshitting one another by using invalid factors to calculate non-existing quantities in no relevant and banal calculations. The mass does not make a planet go faster or slower or influence the structure in motion. Then why the hell use it in calculations. The mass does not influence the allocated orbit in any way, then why the hell use it. The mass does not pull the planet closer to the Sun so why the hell bring mass in as a factor. Mass does not indicate any difference even in the smallest indication between cosmic structures so why the hell use it. It is used to make the mathematician feel worthy and superior even if it also shows his absolute mindless ability in rational arguing. If a factor has no influence on whet is proposed what their calculations wish to measure except to prioritise the ego of the mathematician then using it in calculations is outrageous. Using mass has the same implications in the calculation than it has to bring in India's net pepper production and substitute that with any factor so why not use that as a factor in the calculations. This is how the layout of planet orbits is in reality.

Mercury	Venus	Earth	Mars	Jupiter	Saturn	Uranus	Neptune	Pluto
0.055	0.86	1.0	0.11	318	95	14.5	17.2	0.002

If mass had anything to do with gravity we would find either the largest in the inner orbit since it has the largest mass that will erode the radius the fastest or we will find the least in the inner orbit because the least mass will produce the least motion. Then the planet orbit layout would be as follows:

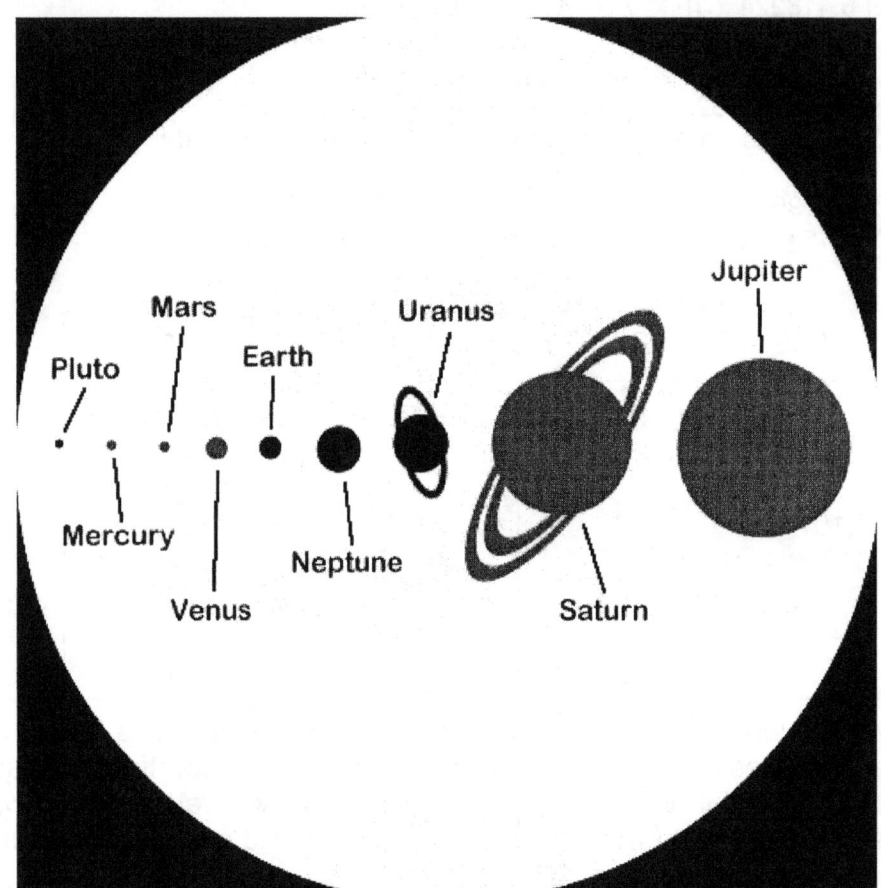

If mass had any influence on the orbits as the calculation formulas would suggest then the orbit arrangement should be as the illustration suggest.

Jupiter has a MASS of 18955.872 x 10^{24}

Saturn has a MASS of 5686.76 x 10^{24}

Neptune has a MASS of 1027.872 x 10^{24}

Uranus has a MASS of 866.52 x 10^{24}

Earth has a MASS of 59.76 x 10^{24}

Venus has a MASS of 48.7044 x 10^{24}

Mars has a MASS of 6.418224 x 10^{24}

Mercury has a MASS of 3.3029352 x 10^{24}

Pluto has a MASS of 0.11952 x 10^{24} /

This is the manifestation of the fairytale where it shows the naked reality when the sublime goes mad in the practical implication. It is equal to the fairytale that tells about the King that paraded around naked because only the stupid would realise he was without clothes and since only the mindless would not be able to see his magic clothes. Then he including his subjects pretended to marvel at the beauty because they would rather bullshit their minds than to admit that they were to stupid to see the clothes the King was wearing. So everyone pretended not to be stupid, who showed exactly how stupid he or she were. In this case it was the intellectually Superior that was privileged to be stupid just to show the rest who was the most stupid of them all and never to ask what the hell mass has to do with the whole affair. So every one who thought he was superior showed how big an idiotic fool he would be in order to prove the point. That made him not more a fool than the fool he or she was trying to impress and both idiots impressed other idiots by accepting that mass plays a major part while mass has no reason to be part of any of the facts relevant. In this case it is the using of mass that everyone echoed because by not realising the use of mass would show the any individual that needed to become part of the Brainy Bunch how mindless he or she was not to see the advantages of the use of mass. So the mindless then accepted there is some lunacy. That is the fairytale but in reality the dead is criminal. If I go about and suggest facts, which is untrue and while I know them to be untrue I still support the facts and support the system declaring the facts as to be true, the system that includes me is a fraud. No wonder the whole lot is devious and mindless atheists.

A question never asked and a thought never put to words is that if there is a Gravitational constant guiding all cosmic matter by the force of gravity where is it going and where is the centre where it is taking all the material. That statement is criminal with dubious intent of misleading the public and is a conspiracy outweighing crime syndicates such as drug lords and the mafia.

Again from the lips of the Newtonian into your face and then you tell me who is the criminal misconduct. If any person goes into a contract with such devious preconditioned misleading an reliable evidence with the intent to mislead and defraud by giving unreliable and untrue facts deliberately while full well knowing what is suggested has no truth to bear, you are a swindler, a cheat, a criminal and you belong in a safe place away from society where you and your malice may not defraud others any more. If you do not like my saying this prove me wrong. Prove how your swindling behaviour to cover Newtonian defrauding is not criminally intended!

This might seem harsh criticizing but it is reality. Producing fictitious facts in order to mislead and present untruths is criminal. Any academic wherever can either sue me for wrongful slander but doing so that academic must prove without doubt how mass plays any part in initiating or contribute to motion of whatever major or minor influence on whatever scale. It is criminal to go around and falsify information in order to protect a corrupt complot and counterfeit facts with the intent to deceive and spread deception. I say astrophysics is a hoax from the start up to the present and up to now the academics scandalously avoided me by ignoring me in order not to face up to their deceit. I challenge who ever to charge me with slander and prove me wrong. Prove in what way does mass bring about any implication to the orbits of any cosmic structure.

	Distance (AU)	Radius (Earth's)	Mass (Earth's)	Rotation (Earth's)	# Moons	Orbital Inclination	Orbital Eccentricity	Obliquity	Density (g/cm^3)
Sun	0	109	332,800	25-36*	9	--	--	--	1.410
Mercury	0.39	0.38	0.05	58.8	0	7	0.2056	0.1°	5.43
Venus	0.72	0.95	0.89	244	0	3.394	0.0068	177.4°	5.25
Earth	1.0	1.00	1.00	1.00	1	0.000	0.0167	23.45°	5.52
Mars	1.5	0.53	0.11	1.029	2	1.850	0.0934	25.19°	3.95
Jupiter	5.2	11	318	0.411	16	1.308	0.0483	3.12°	1.33
Saturn	9.5	9	95	0.428	18	2.488	0.0560	26.73°	0.69
Uranus	19.2	4	17	0.748	15	0.774	0.0461	97.86°	1.29
Neptune	30.1	4	17	0.802	8	1.774	0.0097	29.56°	1.64
Pluto	39.5	0.18	0.002	0.267	1	17.15	0.2482	119.6°	2.03

Mass has precious little influence on gravity because gravity is the motion, whereby there is interaction between time serving as a liquid flowing by contraction to the centre of the Sun and the planet by duplication is part of the flow. Since there is a flow and duplication set in balance the duplication is in the range of the motion T^2 while the flow of three is in the straight- line k that serves time.

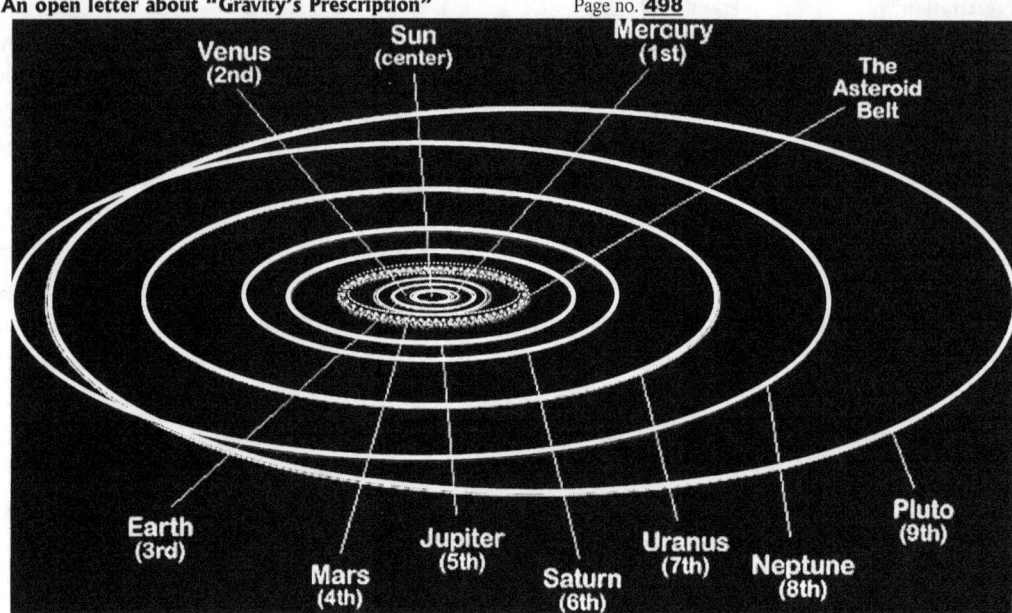

Just as improbable and therefore impropriate is the manner that science suggests that "by the magic of gravity" stars accumulate dust to form stars. The suggestion alone that dust by the force of gravity can accumulate into solid matter is fraud and then to wilfully with deceit intoned to use tax money to investigate such dubious ludicrous nonsense is over the top. It is fraudulent to use tax money on something that is obviously a hoax. If any banker or politician did the same criminal offence he or she would have been branded as a swindler and paid a penalty in a penal institution for a very long time.

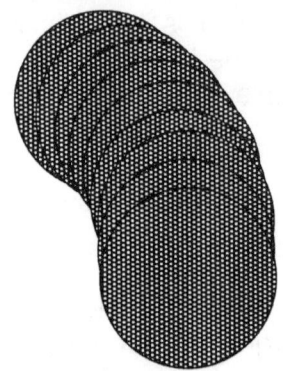

Let us for once and for all accept that Newton's mass activated gravity is inspired by his imagination. The process is when liquid in the form of outer space lines up against a solid such as a palate or the Earth. There is motion in the one department that acts as if it is a solid but is in fact the partner that holds the motion. Then there is the liquid, which by being the stationary acting the part of the solid but is a liquid all the same.

The body delivering the motion is a solid that forms a unit as space. The part that serves as a cosmic liquid is the partner that is also stationary and serves as an immobile liquid. This has things rather confused in the manner that gravity in the cosmos operates. Remember the planets is not a normal set up and even stars with micro stars to attend to is holy unnatural. It is very seldom in combination and when it is things get as confusing as we find it to be on Earth. The norm in the cosmos is a lone star that spins on an axis while in motion around a galactica. The galactica presents the same layout as a star and the working process in similar mode, as

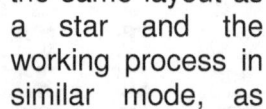

motion is a solid The part that the partner that is an immobile liquid.

the galactica that apply in a stars with layers would have.

The body delivering the that forms a unit as space. serves as a cosmic liquid is also stationary and serves as This has things rather operates. Remember the micro stars to attend to is holy when it is things get as the cosmos is a lone star that spins on an axis while in motion around a galactica. The galactica

confused in the manner that gravity in the cosmos planets is not a normal set up and even stars with unnatural. It is very seldom in combination and confusing as we find it to be on Earth. The norm in

presents the same layout as a star and the working process in the galactica that apply in a similar mode, as stars with layers would have.

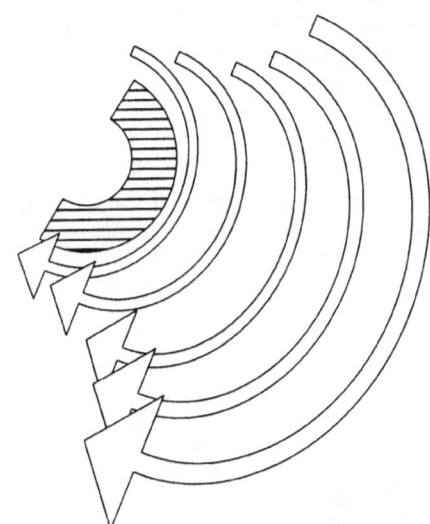

Do not look for the pumping going on where one can see time that meets space. The pumping action is going on where the proton pumps time into singularity by expanding and then contracting in the very heart of the atom nucleus. There the duplication present the expanding and the contracting which feeds the star with the motion either in duplication or in contraction that the star requires to comply with the demand space-time insist on as gravity. The reducing of heat by motion is presented as cooling since motion reduces space and by reducing space it is cooling. To establish that rapid cold the proton moves 1836 times faster in order to restrain the heat from the value the heat had when the heat was at the electron relevance. At the electron the heat was already at the speed of light and therefore the atom removes all heat by freezing the heat into the oblivious every atom is a black hole. All this adds up as a general reducing of space by the governing singularity in charge.

Every atom in the star is a pump that coverts heat to cold and transfers singularity 1^1 to singularity 1^0 to regain what was lost during moment-Alfa. The gravity in the star is not nearly the gravity going about the planets.

In the case of the planets there is an orbit motion that puts liquids in relation to solids without the much pumping being the dominant factor. The liquid allows the solids space within to move. As the solid pushes against the liquid the liquid bears down on the solid and some liquid give way but the inner liquid increases the density at the point and just above where it touches the solid moving structure. The liquid pushes down the solid while in accordance with the Coanda principle the space expands to a point directly relevant to the motion that the solid provides. That which is without motion is secured by the liquid to be part of the Earth while that which is liquid is secured onto the Earth as an extension of space.

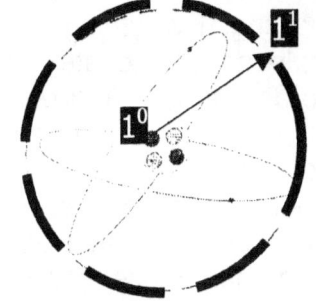

There is an allocated line designated by the extending of the solid that includes the liquid to gather that liquid into the unit forming the solid. We gave that so many names ending with sphere even the thought of all these sphere makes ones head spin. How Newtonians fit the sphere as in stratosphere and atmosphere and what not into gravity is still a puzzle, which is eluding me in the manner that Newton's vision on mass was eluding me. In the end of all this there is a line that is the friction point and it is at that line where liquid tear from solid while the solid is actually intensified liquid.

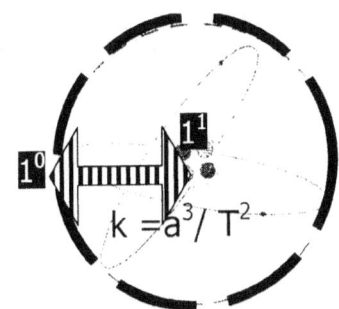

$$k = a^3 / T^2$$

In the case of comets the Sun is the solid that forms stability while the comet is the solidity that moves and outer space is the liquid that does not move. In the case of comets the cosmic law is transgressed. The Sun is an atom. The Sun consists of a unit forming an atom where the Sun is the atom in compiled group but also where the group serves the unit. Every layer in the Sun is a liquid to the top where the bottom serves as a solid to the top layer forming the liquid. The proton puts time at motion where time puts space in demise. Time devour space as eternity meets infinity. The atom is Black Hole with matter in between infinity and eternity and this fills the black Hole with substance that is forming space - time. The final conclusion that any star can arrive at is when it takes

The proton serves as 1^0 to the neutron being 1^1 where the neutron serves as 10 to the electron being 1^1. The atom forms a Universe that hold both eternity and infinity apart by allowing motion to separate time. The atom concludes the Universe because the atom is what concludes the Universe as much as it started the Universe. In the end all star will be one atom in the hydrogen atom but that sis the final conclusion where the last era arrives. The atom is the Universe.

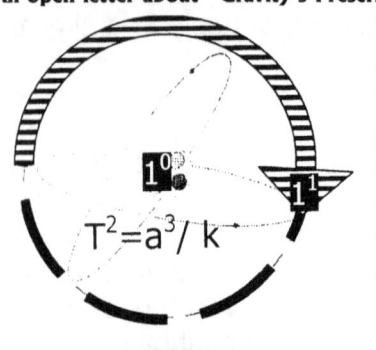

$$T^2 = a^3 / k$$

The atom maintains relevancies where the core within the atom serves as 1^0 and the orbit serves as 1^1. The core is the solid and the electron is the liquid. The electron provides the motion because in relevancy at the point where the electron is located it is the electron that is in motion while the core within the atom is a solid that does not move. The atom serves as movement because singularity generated by all atoms forming the unit provides the motion. All atoms forming the star are allocated the value of motion being 1^1 while singularity charged with governing the star is 1^0.

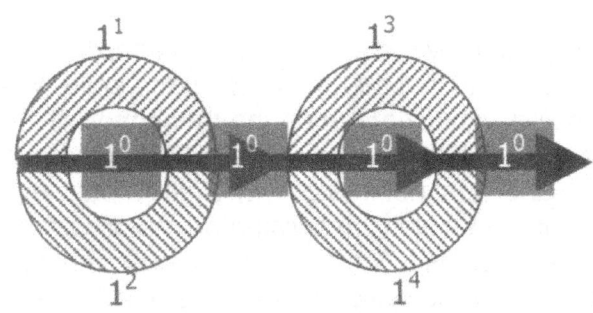

However the only constant in the Universe is that there is no constant applying. Everything is in cyclic shifting as the relevance relocate and alternate positions. In order to get a flow of space - time 1^0 and 1^1 must be forever alternating. The fact of constants are that constants are as Newtonian as mass can ever be and constants are as much a fact that does not apply as mass where then mass has the same position. The planet forms an electron to the Sun becoming the solid and the Sun allow the planet to spin while the planet receives it alternating which forms motion from the Sun that provide the governing singularity not only to the Sun but also everything orbiting the Sun as an electron Because the planet is just an electron the planet will rotate about the Sun as any good electron would do.

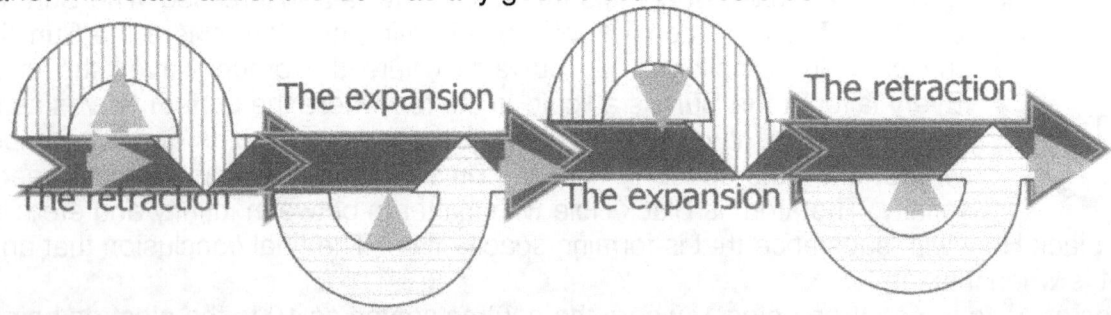

The expansion

The retraction

The retraction

The expansion

When the planet is on one side of the Universe where the centre of the Sun forms the Universe the planet resist the flow of time. When the planet is on the other side of the Universe the planet. The Sun is 10 but the planet alternate 11 and 11 because the planet land on one side of the Universe and then on the other side of the Universe.

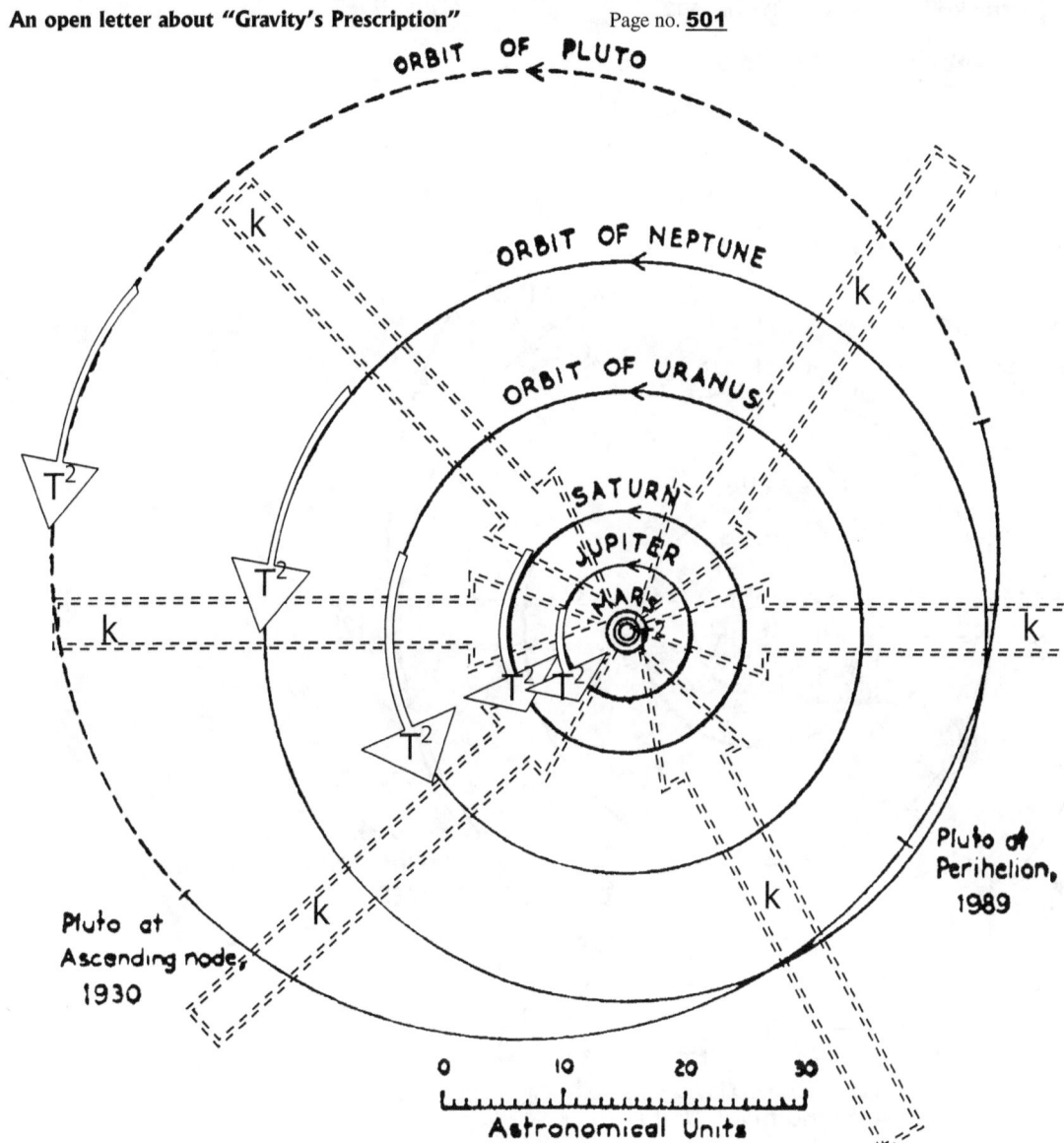

Kepler's formula insists on the flow of space-time. Gravity us never mentioned because gravity is a name coupled to an idea. If the idea is incorrect the coupling is unjustified. The space is floating in the moving time. $a^3 = T^2k$.

The Sun is contracting time towards the centre and by doing that it is implicating the process we gave the name of the Coanda effect. The space is duplicating the space it holds in order to flow in the time being contracted. By duplicating the space the space is generating what it was to where it is to the location it is going to be. That is space-time and that is gravity and that is motion. Restricting this process leads to mass applying as a restraining because no object can share space just as much as no object can be in two places at one time.

There is a flow of space-time running towards the Sun at a duplication tempo where three portions of space are in relation to one unit of time. Material stand related to space by measure of the past position, the present position and the future position and that the cosmos told Kepler by a language of mathematical equations. It is not my say so. It is not the say so of Kepler. It is not the hearsay of Newton. It is not the interpretation of the visions of Hubble or the guesswork of Hawkins or the calculations of Einstein's perceiving. This comes from the horse's mouth. This has nothing lost in translation. The cosmos told this to Kepler without ceremony or private preferences colouring and tainting preconditioned favouring of prejudice. That the cosmos spoke in a language all can appreciate because no dialect my sound some incorrect abbreviations...and Newton still managed to bane all the information to impress with preconceived ideas.

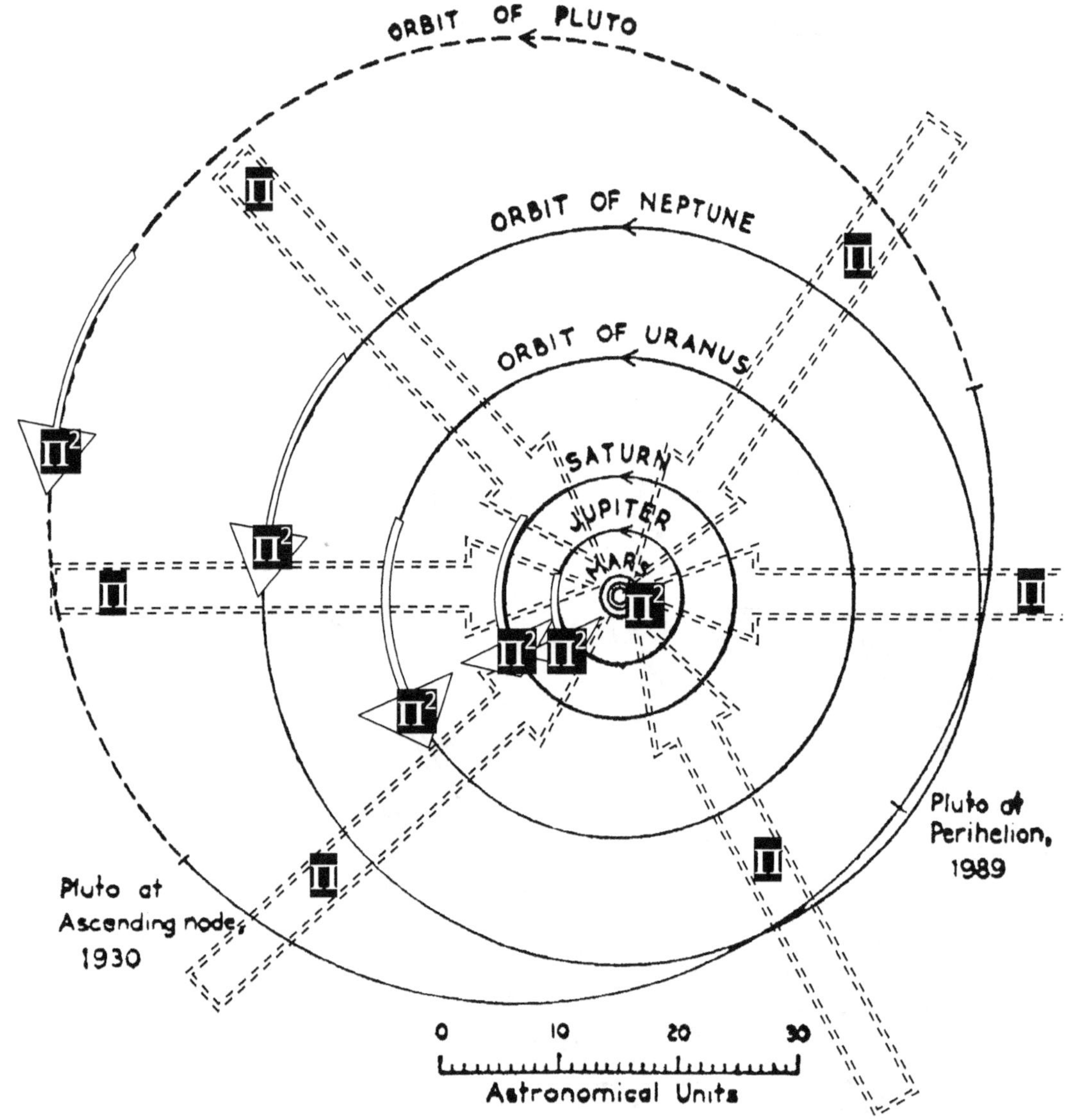

ORBIT OF PLUTO

ORBIT OF NEPTUNE

ORBIT OF URANUS

SATURN

JUPITER

MARS

Π^2

Pluto at Perihelion, 1989

Pluto at Ascending node, 1930

0   10   20   30

Astronomical Units

It is the atom that is in charge of the Universe because it is the atom that charges motion. The rotation of the object produces the proton duplication while the rotation around the contracting centre is gravity or motion unblemished and the time flowing towards the centre places the orbiting structure in the 3 dimensional space in time $(10)^2$ square. In the final analysis it still is the atom that produces the atom, which serves as a star. It is the proton at $(\Pi^2+\Pi^2)(\Pi^2\Pi)3 = 1836$ that forms the Universe in more ways than any human can appreciate.

It applies simply because in time there is a ratio whereby space duplicate in expanding as well as

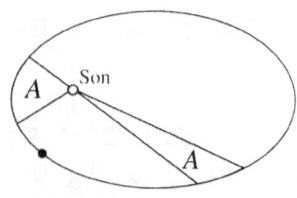

contracting and this ratio is serving what ever the cosmos might be. The cosmos said it is $a^3 = T^2 k$ therefore it is little surprising that Newton did detect a ratio between space-time and time flowing about space. It is more surprising that he missed the rest. Much more surprising is that every one of his dedicated followers and mathematical geniuses missed the rest. Then I better put my mouth where my proof should be and explain what the rest is that all missed.

Eternity • Unified with infinity

The cosmos is not expanding because it is shrinking. When it started that which cannot be bigger parted from that which cannot be smaller. It started where infinity was eternity. Then infinity parted from eternity leaving eternity bigger than infinity but because of the size of infinity, eternity was not much smaller than eternity. The star is the atom within the star by the multiplied motion of all the atoms forming a unity that charges the motion in the star. The star is a culmination of the efforts of the star.

When the first moment came the two factors being eternity and infinity was the same. Then they parted company putting a reference between them and not much more that just a reference of division

The ratio is increasing because it is decreasing the Universe. Let's put it this way: the Universe

Eternity
•• parting with
infinity →

cannot expand because that what was at first is eternally big with no possible end and in that is the reason why there is not possible expanding of the Universe.

The Hubble constant is the measure whereby the Universe is reducing since there is no possible room for any expanding. The part being infinity is the part that cannot reduce since it is as small is infinity can ever be smaller. It is so small it has no sides but all points share on spot. Any further reducing will bring about an increase in size and by reducing further it is increasing what never can further reduce. With infinity being there without having a possibility to reduce it is increasing what cannot reduce and by increasing that which cannot reduce it is reducing the part that cannot increase. While neither of the two is capable of changing in the direction they represent because they represent the entirety there is to represent, the increasing of that which cannot reduce is reducing that which cannot increase because without ever changing the two are growing apart. By eternity never changing while growing apart from infinity that aspect too never can change, and while never changing that part that cannot increase is increasing the part that cannot reduce while the part that cannot reduce is reducing the part that is incapable of increasing. Those,

Eternity
growing • ►away from
infinity

the ones that cannot change is doing what it can do best to the other part that cannot change and in changing its relation with the other part it is remaining the same while the other side in reference then changes the reference.

What Hubble saw was a Universe that was shrinking away into the oblivious because that part that cannot reduce is reducing the part that cannot increase and since the part that cannot increase holds eternity, it can shrink the part that it has in the smallest side by shrinking its reference to that into the oblivious.

It all ends with the Black Hole (not quite but to go into detail about that requires another half a book of explaining and proving as the Black Hole has two more steps to involve mathematically). Let's put the Black Hole as the biggest there is while correcting this error at the same time. The Black Hole is so huge it fits a Universe inside but that means when the Black Hole was huge it was so big it parted eternity from infinity. That came when eternity was just bigger than infinity because the two then parted their shared unity no that long ago. At that point when the Black Hole was the liquid star sloshing away and shining as bright as it could the Universe was separating eternally big from infinitely small by a margin of Π. Eternity parted 1^0 from 1^1 by a margin of Π^0 that increased to Π by becoming Π^2. The increase came as Π move to Π forming Π^2 that came to a total of Π^3. Still it was at a time when infinity was just smaller than eternity and eternity was just bigger than infinity.

The difference at that at the start when 1^0 was going onto 1^1 which was going onto Π^0 that was forming $7\Pi^0$ and was combining time as $(10 + 10 + 1.9991 = 21.99991 / 7\Pi^0 = \Pi)$ this was forming Π^3 but there was little else to show on the eternity side as well as the infinity side and little split the difference. That was just about the environment the Black Hole encountered (okay there was

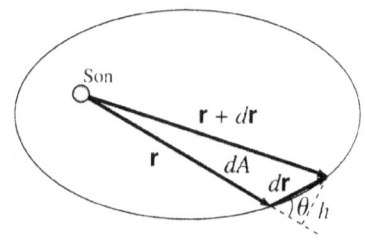

more...but not that much more) and in that the Black Hole was what split infinity and eternity at the time. This we may deduce on the grounds of the facts we now see the Black Hole represents. At present the Black Hole is so small it can reduce the entirety of eternity into infinity while it remains so large that it can absorb the entire eternity into infinity without needing any matter to produce a time delay. It still continues to have the power to carry on what it started with. It started a process putting a bridge between eternity and infinity and counteracted

when it shrunk eternity into infinity while it is expanding infinity onto eternity. It took this job when the Universe was wasting away.

The wasting involved that when what cannot expand reduced and as it reduced it pushed apart that which cannot separate by decreasing that which cannot decrease as that which cannot expand did expand. The split was about that which cannot decrease to part from farther from that which cannot increase while the difference brought about an increasing of the reducing Universe. After the long storey about mass and a centre point and a shared centre point with one point favouring both, it boils down to see how material move. Elsewhere I explain what material is. Material is seven point that has no space but is generating space claimed by that which pretends to heat and the retarded heat circle compacted as matter around a centre forming a sphere as small one are not able to imagine. It has no name because no name giving fame seeking Newtonian can get to it. Only the heat in retarding spin is a part of the Universe but that substantiating and controlling the heat confirms the allocation of the heat by swerving singularity Π^0 which is maintaining $7\Pi^0$ to become the smallest sphere there may be.

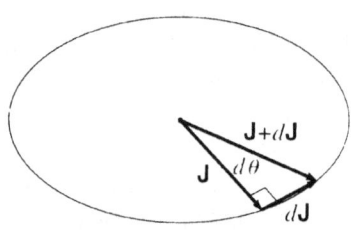

In the centre of the smallest matter runs a line that actually is just an expanded point and the retarded heat spins around this point where the rotating is holds matter as matter with the rotating. It is the fact of rotating that is producing the retarding of time and the more of this 1^0 to $7\Pi^0$ there is the more retarding of time going backwards there is. That is a part of the "nothing" Newtonians put into outer space as outer space. That line is where all possible points serving the line lands on one point that line is a dot with a spot in the centre that has no sides and all point in this line falls on the very same point.

The value of this point is 1^0. Because this line focuses all the possible sides on one point and this spot including all the sides that fit into it then still has no sides and fits still with the adding fits all into one place while it serves the rotating centre of retarded heat that finally combine as the atom, one finds that all the lines serving all the retarded heat fits into the next spot that form a centre line that also has no sides in one spot. Eventually the lot forms a combining line that includes all the material within the atom. This eventually then finally combines as $(\Pi^2+\Pi^2)(\Pi^2\Pi)3 = 1836$ and we find it serving our Universe in the capacity as an atom.

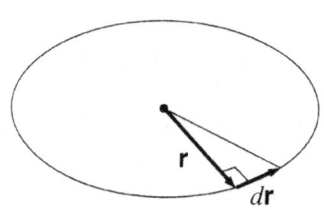

The atom is the atom because all the lines that centre the rotation of all the retarded heat combine in one line forming a centre to the atom. This is possible because all the lines has no sides just like the atomic governing singularity centre line also still has no sides. From there all the atomic governing centre lines fit into a centre line that can hold all the atomic lines because it holds all the atomic lines on one spot as one spot. The line finally forming as the governing singularity driving the motion of the star which is at that point representing all the retarded heat centres which project to the governing centre singularity of the star.

All the positions in eternity relate to every position in infinity and the line holding infinity is the result of the accumulation of all the points serving infinity in the retarded heat. All points that hold 1^0 projects to one point holding 1^0 because all points have no sides therefore all points fit into one point that form a centre line. All the possible lines by all the possible atoms is projected to one centre line since that one line holds all the possible points there can ever be, on the only one point in a point there possibly can be. This of course also applies in the case of and well as also to the orbiting satellite. Since that point cannot be smaller it can all fit into as well as fit all into that one point that cannot ever be smaller. Since all the points are therefore exactly equal to the extent they all are being the same point, all points everywhere are then the same point concentrated into one point while spread out as far and wide as the point may reach.

The point is singularity referring to singularity. Mathematically this point is expressed as 1^0, which by all mathematical rules are equal to 1^1, which is equal to $\Pi^0 = 1$. In that number 1^0 the entire Universe units as one unit that incidentally also never can be because everything fits into one spot that is not. To infinity eternity is 1^1 on the condition that infinity then is 1^0 and just the reverse is applies in the relation that eternity holds infinity because the reversing holds infinity at 1^1 as long as eternity can be 1^0. Since both are 1^0 while holding the other as 1^1, the roles inherently have to change when crossing over to the other side of the Universe. The one side of the Universe will gauge infinity as 1^0 while eternity is 1^1 and at the same moment on the other side of the Universe eternity will be gauged as 1^0 while viewing infinity as 1^1. The end to the Universe is not near as the start is far away (eternally further than the idiotic 13.5×10^9 years Newtonians give the Universe. The end will arrive when the difference between infinity and eternity will be so large infinity will again join eternity while eternity at that point will be so overburden extreme that the incorporation will go unnoticed by all that it happened. The planet orbit the Sun since time placed material at that location and allocated a time delay as to the position the Sun holds and the Planet holds. This same argument also serves the same way as it applies in the Sun and the orbiting satellite holding their relation secure.

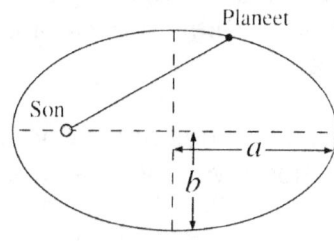

In regard to all these mentioned facts we can deduct that all the matter that the satellite holds one line charged that has no sides. Since this line has no side and takes up no space it fits into the governing singularity that has no sides and all possible lines of all possible matter fit into that into that one charged line that charges the next line where all the lines also fit into. This charging and combining of lines that has no space claims and therefore holds all points on one allocated position eventually form the atom that form the layer that form the star that form the motion in duplication and conserving the duplication by contraction.

Since that one line is exactly in equality to the next line running to the compiling centre line in the Sun, the centre line in the Sun therefore is also the centre line in the slightest piece of independent time delayed matter of the satellite and the satellite then becomes 1^1 to the Sun being 1^0. From the particle there is a reverse reality that the Sun forms 1^1 to the satellite particle line that holds 1^0, and all this represents eternity in reality departing from infinity forming reality and to the one, the other is 1^1 while that one holding the reference is 1^0.

To the Sun it is the satellite that stands between eternity 1^0 and infinity 1^1 while to the satellite it is the Sun parting infinity 1^1 from eternity 1^0 where eternity drives it in motion. This is the result because the same eternity drives both the Sun and the satellite as individual location where the other to the point that one has becomes part of the cause why the eternity parted from infinity. While the one is playing a blaming game on the other by taking president in the relation since it holds eternity and the other forms a factor of infinity, the one holds 1^0 to the position that one then allocates to the other as 1^1. In all the talking and all the explaining matter is in reality not even reality because matter is three points that lagged in time behind three points that in time serves as time to follow four points spinning in a centre where the four points all share the same spot on the fifth centre spot.

In view of the Titius Bode time depletion, time in flow creating space has a far more complicated arrangement than Xepted science can even produce on a chart. To be honest every person knows that Xepted science cannot even place the planets on a chart, depicting true distance to size, but they WILLFULLY never mention that information when the chart they show is as false as a three-dollar bill. In a sense it does no harm leading people down the ally in such a way, because others in my class of mental insignificance in society is far to un-intelligent to realize the correct way and will therefore not understand the correct way in any event. That is a mistake with a stinging tale. It is as dangerous as a scorpion to Xepted science. In an introducing article, I named Anglo-American Mythology, I pointed out how misconception feeds society, favouring the lies and untruths and blatantly ignoring the truth. In the past, since time began the powerful used this on the brainless masses, and for a period where that civilization lasted, got away with that strategy. The next civilization that came to power, followed the same methods applying the same dogma, and in the end paid the same penalty because their greed, lust for power, and sublimations gave them control over the masses for a while.

The misconception those in favourable positions forced onto the masses, made the very people in power so shortsighted, their course on vanity lasted but a few generations. This is achieved because our Earth environment is tolerant and can buffer a lot, to save life in the end for life's contamination on Earth. They wish to extend life's connection to Earth, as being a connection to the cosmos at large and will be in effect as long as life remains in the cosmos. When we are "going abroad" to our "next door planet" that apparently holds all the supporting evidence of life carrying organisms, the connection to the cosmos remains and connecting to the Earth is of little consequence. That bluffing must stop. I realize no one on Earth will ever take note of what one sod (like me) on Earth is shouting, but misery awaits our Martian Colonists.

I started using the term Xepted because I do not accept Newtonian views and therefore I except Newtonian views but this bloody brainy machine tell me every time what I can and can't write when I refer to excepted (rejected) Newtonian science. Then I got brainy all by myself too and created my own word and told the machine to accept and shut up. Now we all are satisfied and English just got one word richer by my inventing Xepted science which is what I use when I refer to Newtonian science that science wilfully accept while they very well know it is Xepted science.

Suffering will be the reward for the fools attempting to catch the bounty of "fame, riches and glory" on behalf of the All Powerful Dollar and the dollars absolute true benefactors. Those with eyes, let them see, those with ears let them hear and let the rest self demolish.

Binary stars, spinning to self-destruction will produce significant heat. Heat create space, space forms winds. That is facts that the Bible present and is indisputable. Where the Earth was, was still a void, containing a sphere of circular displacement and this will reduce linear displacement to zero. Linear displacement is space and circular displacement is containing heat for matter survival.

Binary Star Minor overheated. That is why the core brittle and fragmented. This action will release tremendous contained heat; the heat will produce magma flowing in space like water in space and this eruption of heat space that created winds. Once again the recollection fits the scenario. Releasing the heat and producing space will establish space-time and fill the void where the Earth should fit. This is fact and if anybody even tries to dismiss this will be because of abstinence on his or her part. I did not prove the Bible correct. The Bible told the truth and in such correct detail, it is beyond human comprehension, but sublimation on the part of Newtonians and science before them, disallowed their ability seeing it.

That is what an insignificant formula $R^3/T^2 = 1$ where $R^2/T \times R/T = 1$ represents space-time in singularity as well as space-time in densified, occupied and unoccupied format. That means the everything of the whole lot, or as we say in Afrikaans, the "Heelal" meaning Universe. It refers to space-time for the first time while nuclear explosions are the epitome of $R^3/T^2 = 1$ where $R^2/T \times R/T = 1$ and how long has nuclear explosions been with us?

Now comes the proof: In the electron dimension the value is $\Pi^2\ \Pi$ in relation to 3.
In the cosmos there are always at least six sides to any object.

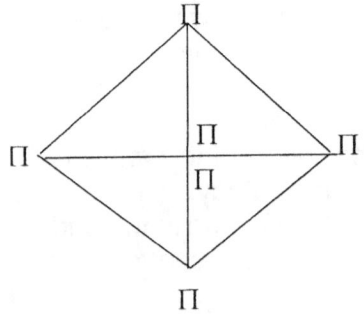

This gives the value of Π^6 in relation to the Titius Bode value of 7/10 in relation to the 6 sides in space (10).

$\$T = 7/10\ (\Pi^6) / (6 \times 10) = 112.$

From this comes the Newton formula holding time in the fourth dimension to space in the fourth dimension. At all times the dimension application will be the value of time in singularity (Π) to the time of matter (Π^2). Separating the illustration will be as follows:

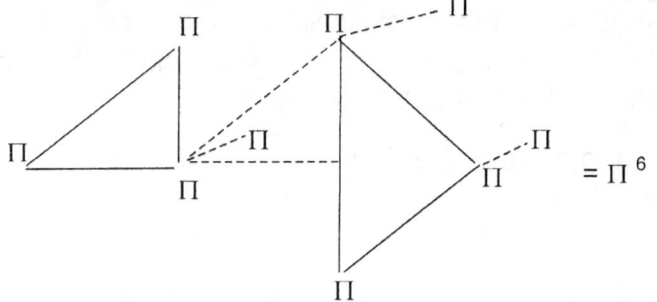

$= \Pi^6$

Through this "gravity" has the ability to produce a dimensional change from 3 to Π. Heat in a liquid form will be $3\Pi^2$ while one step more concentrated (the neutron state) heat will be $\Pi\Pi^2$. This reduces the triangle in space-time to the half circle in space-time.

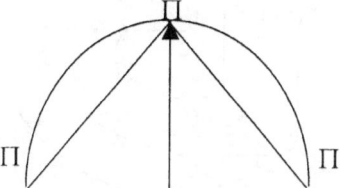

The one side is connecting the 3 x Π and with the change in dimension cross over falls away changing the 3 (number of Π) Π^2 removing of the dimensional value of a half cube (6/2) to that of a half circle. ($\Pi\Pi^2$).

Both remain 180° contact with time in singularity Π^3, which is a straight line (180°). But there can never be a dual application. The one holds the other in support. In a star this fact becomes irrelevant but a star is the cosmos, applied in reverse. In the galactica (of which the solar system is a fragment) the relevancy of support will always apply at the outer circles and not apply within the inner layers. However, this is beside the point as we are dealing with the Titius Bode configuration of 3 ; 6 ; 12 ; 24 etc. that apply to time in space formed by heat.

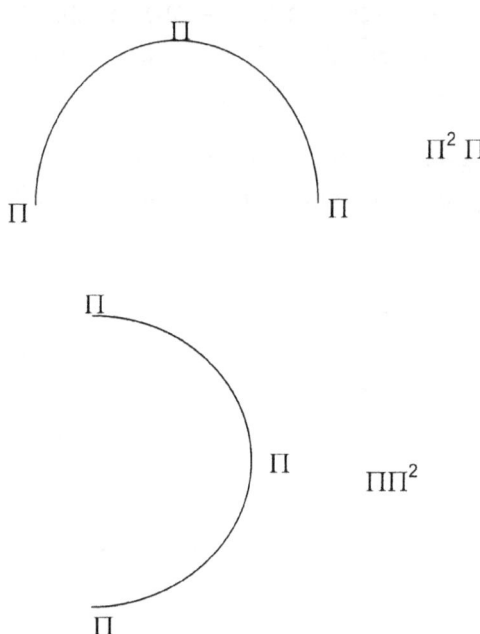

$\Pi^2 \Pi$

$\Pi\Pi^2$

The relevancy applying in a cosmic cluster will always be that of a half circle applying as a triangle because it is in support of the triangle. The single Π will always support the Π^2 of time and this configuration will be a half circle.

Only the half circle can apply at any given point in any given moment. The value of Π^2 extends as the value of Π because Π^2 is the circular of "gravity" and Π is the linear of "gravity".

The rest is everyday mathematics. The invert square law applies just as well to a half circle than it does to a full circle. The value of Π in the next circle will be that of Π^2 in the previous circle. Well in a way presenting it as the invert square law does apply, but it has a cosmic sting to it.

There is a much more substantial explanation about how the Titius-Bode law arrive at the configuration of 1, 3, 6, 12, 24 doubling every time. However, such an explanation covers a great volume of facts because we have to cover the neutron's calculation from every angle available.

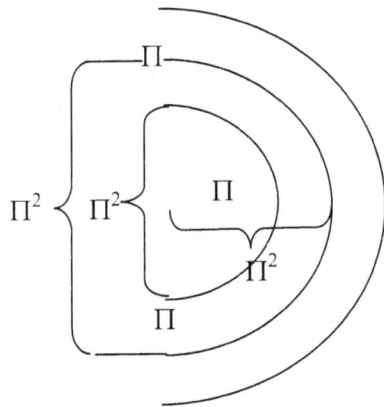

In the first configuration matter lends space all value at $\Pi^2\Pi$ configuring to the dimensional influence change of 3. This will be as such.

Matter

$\Pi^2\,\Pi$ $\Pi = 7/10$ 3 x (Π and $\Pi = 7/10$) becomes 3

Π

7/10

7/10 7/10 $\Pi = 7/10$ ↔ 7/10 $= \Pi$

Π Π

Every time another body develops to the outside of this the inner space will be three dimensional space $3^2 + 1^2 = 10$ and this apply to the sevens, therefore the space to the inside becomes 10/7 and from the space the next applies a matter value of $\Pi^2 = 2\Pi = 7/10 + 7/10$ with Π at 7/10. Only Π^2 determines time therefore $2\Pi = 2(7/10) = 14/10 = 1,4$.

Space holds the value to the already developed part as 10/7 = 1,42. Therefore space will be 1,42 followed by matter, 1,4 and that leaves the Titius Bode law to double its distance. 3 → 6 → 12 → 24 → 48.

Where Π^2 are the radius of the one circle it will double to become Π in the next circle. It is the manifestation of the neutron dimension applied in the electron dimension where all 3 of Π holds equal value therefore Π^2 will become 2 x Π. This relevancy will apply wherever heat and matter produce space-time. This is a given, standing as firmly as the Hubble constant, the Roche limit and any other law application.

Every person in the past sought a relevancy in the application of this phenomenon in as much as it applies to the solar system. The proof there is, is not in applying, but in the way it does not apply and the reasons why it does not apply.

When testing for proof in the application of the Roche limit as far as it stands in figuration of the solar system, we will find it does not apply at all. That means the fact that is NOT PRESENT, PRESENTS THE PROBLEM. The absolute importance of the Roche principle not only reflects on the influence of the Roche principle alone but the Titius Bode space-time growth and the Titius Bode configuration that is an extension of the Roche principle.

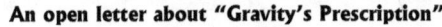
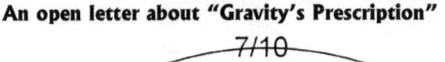

7/10

7/10   +   7/10

$$= \frac{14}{10} = \frac{10}{7}$$

But the matter position relate to the space position.

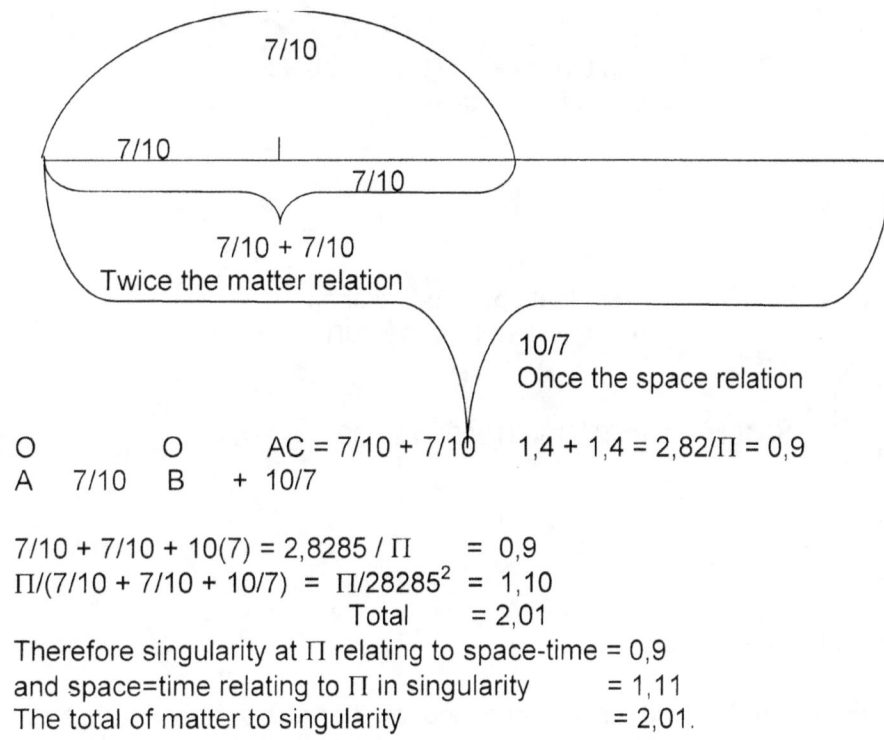

7/10

7/10 |

7/10

7/10 + 7/10
Twice the matter relation

10/7
Once the space relation

O   O   AC = 7/10 + 7/10   1,4 + 1,4 = 2,82/Π = 0,9
A   7/10   B   + 10/7

7/10 + 7/10 + 10(7) = 2,8285 / Π   = 0,9
Π/(7/10 + 7/10 + 10/7) = Π/28285² = 1,10
                              Total   = 2,01
Therefore singularity at Π relating to space-time = 0,9
and space=time relating to Π in singularity   = 1,11
The total of matter to singularity   = 2,01.

Therefore $Π^2$ to Π = 2Π . This then is 7/10 + 7/10 = 10/7

Therefore space will always hold double to the relevancy of matter.

To that end
.   3   . (7/10 + 7/10 + 10/7)
.   3   .   6   .
A   B   C
              2Π
Then 6 becomes 3 = 2Π(Π=6)=12
A   C   D
Then 12 become 3   .   2Π = 24

And that concludes the Titius Bode configuration of 3; 6; 12; 24; 48 etc. by valuing the triangle and the half circle.

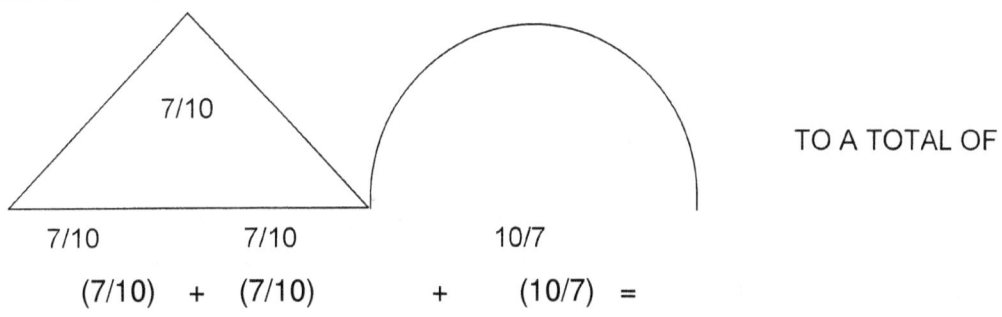

7/10

7/10   7/10   10/7

(7/10)   +   (7/10)   +   (10/7)   =

Three going to double in the next configuration.

The first $\Pi^3 \rightarrow \Pi^2$ Π Separating singularity.

This then brought on Π^2 in heat.

Π
Π^2 Π $3\Pi = 3$
Π

From that space developed

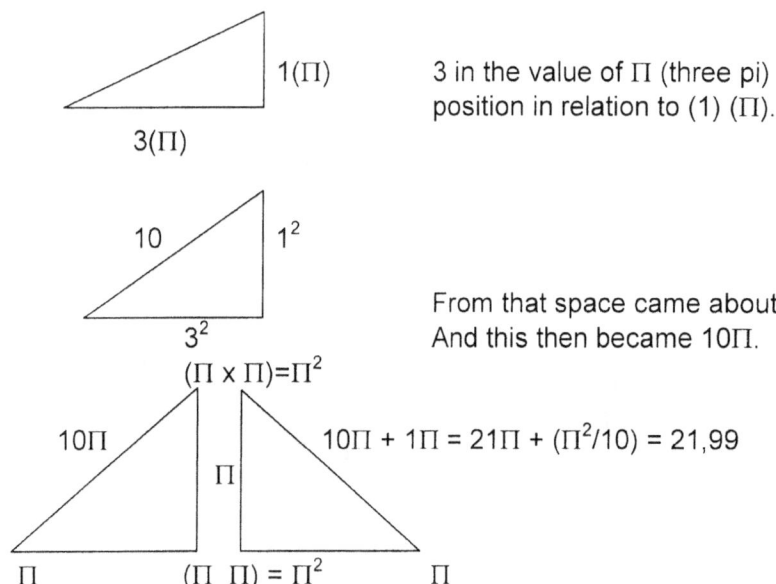

3 in the value of Π (three pi)
position in relation to (1) (Π).

From that space came about
And this then became 10Π.

$10\Pi + 1\Pi = 21\Pi + (\Pi^2/10) = 21,99$

On top of that the sphere established.

Add that to the seven that holds densified and occupied space-time and there is seven of Π in the triangle of matter adding 3 Π's in the half circle of space ($\Pi^2\Pi$) and the total is 10Π. With 10Π the value of the total triangle (in a square) and 10Π the total of space-time (matter holding singularity apart, there is a factor of 7 by Π to 10 by Π, with the triangle having two Π - two factors where Π at the bottom formed ($\Pi \times \Pi$) = Π^2 and to the top ($\Pi \times \Pi$)^2 = Π^2, with all of this constructing space (10Π) to matter Π^7.

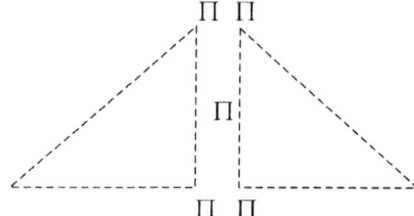

This will bring about 5 ($\Pi^2+\Pi^2$) bringing matter and space to singularity. There are Π to the number of 5. From this comes that 4 (the four Π in the double proton in relation to the proton ($\Pi^2+\Pi^2$) will always form the time value ($4(\Pi^2+\Pi^2)$). From this comes the fact that $3(\Pi^2+\Pi^2)$ will hold the space component as that in fact is half space-time.

Space-time being ($\Pi \times \Pi)(\Pi)(\Pi \times \Pi$) and half of that to any direction is ($\Pi \times \Pi$) (Π). This is why space-time in the geodesic sense has the value of 10Π and that 10Π in a sphere are Π^3. It is both triangle relating to one half sphere which is both triangles (7) in the totality of the half sphere (10).

The value of $2(\Pi^2+\Pi^2)$ will relate to a position where space passes on time in a dimensional transformation, but not a value transformation. It will be when Π relate to Π^2 as much as $\Pi^2+\Pi^2$ = 19,7 and 6 (the sides available for Π to use in space is 18,84. It will form a border of dimension where space will not apply in the same manner as it did before reading that border. One may say that is the densifying border of heat in space to heat in matter.

The last position of the Lagrangian atom is where only Π (in the triangle) have value and this links space directly to time (matter). At this point concerning the Lagrangian layout, we must view heat, filled space, the stuff we exist in, as an element. Hydrogen becomes liquid at –269°C and heat (outer space) has a gas value of –273°C. This is only a dimensional changeover from 10 to Π^2 or from 3 to Π. That was what the "Big Bang" was all about.

Creating the Universe in space as we see space, was the process where the last natural element formed. The Universe or Cosmos is the last atom, which formed. It was the conclusion of the proton $(\Pi^2+\Pi^2)$ in developing. That space can be in gas as we now see it, covering elements to position them in relation to the rest of the Universe as gas, liquid or solid. That is why astrophysics in the ways science apply it at present, is but a good old romantic science-fiction story.

Mass is the frustration of motion while motion in duplication as well as contraction representing gravity. Even with our experiencing of mass it is the tendency we experience to move that is the gravity and the mass part is the blocking of the space we with our bodies wish to claim. While the earth is blocking our claiming of the space we wish to occupy the earth as well as us are both in mass but the mass has precious little and a lot of nothing to do with the fact that we are moving through the application of gravity. To quote Kepler space a^3 filled with material has to move in line with a centre that is controlling the moving of space through time and in time as well. Therefore I repeat what I said before. Mass holds no value, it is density and that density brings in the heat in space, unoccupied as yet by matter. The motion represents density of time in space. That density gives a star its "gravity". The space has to move through time by duplicating and the more dense time is the more effort such duplicating requires. Where the Sun can only hold heat to liquid more dense stars will hold heat to a "jelly" and others will take heat all the way to something as hoard as tungsten. That is what tungsten is. Tungsten can place heat in a relevancy that the heat relative to tungsten is almost as dense as the neutron within an atom. That is why I refer to the system as the Lagrangian atom, because the Universe holding heat, produces all relevancy matter can have. The Universe is the final atom.

The diameter of the Sun is 1391,980 km. Bring this radius in relevance to the Roche factor and the first orbiting structure will be Π^2 relating to $(\Pi/2)^2$. With Π^2 at a value of 1391980 x $(\Pi/2)^2$ the position of Mercury must therefore be 3 4345 73 km making it approximately 3,5 x 10^6 km. With the effect of the Titius Bode configuration the next position must be 7 x 10^6 km and the third at 14 x 10^6 km. If the orbiting structure were that close, as it should be under all normal conditions, we would have been roasted toast.

1.   Mercury
2.   Venus
3.   Earth
4.   Mars
5.   Ceres
6.   Jupiter
7.   Saturn
8.   Uranus
9.   Neptune

	1	2	3	4	5	6	7	8	9
Roche Limit according to actual dist variation	3.5	7	14	28	56	112	224	448	896
	57.9	108.2	149.6	227	414	778	1427	2871	4497
	16.5	15,4	10	8	7,39	6,5	6.3	6.4	(5)

If there were only the Sun that affected the positioning of the gas, "planets" the diameter of the Sun will be Π^2 placing Π at a position where dimensional implication becomes valid. Since the structures are still in space-time positioning, the effect of the Roche limit will still be in place. Therefore the Sun would be Π^2 arranging $(\Pi/2)$ accordingly.

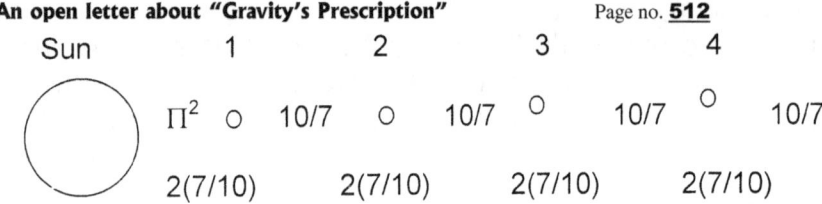

I shall explain the layout as follows: To every structure the value of space-time in the electron dimension is

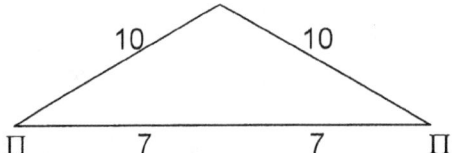

Therefore the matter positioning will be 2Π and that in terms of matter relates to (2x7/10)=1,4. The space-time to space will therefore be 10/7-1,42 because matter is two parts while space remains in singularity and singularity is always one. Therefore 10 cannot double it is one in relation to any one matter part at any given point. This then means Π^2 means in real terms (2 x 7) /10 as one Π and 10/7 as the other Π. Because matter relate to the dome in the half circle and space to the triangle, in relation to space it is $\Pi+\Pi$ and to matter it is Π x Π. In relevancy to matter as we apply our attention to the two structures, the correct connection is $(\Pi/2)^2$. However, in the space factor it will be (Π + 10Π) / 7 x (2^2) therefore it will be on the one side

(7 + 7) / 10 (matter plus matter) = 1,4.

On the other side though it is (10/7) = 1,42.

This is 1,4 + 1,4 = 2,82 in the space where Π relate to 10 and in this instance Π is 2,82. This means that bringing the relation back to matter will effectively mean it is (2,82) x 10 relating to 7 in conjunction with 2.

The Roche limit is Π which in this case is in space, therefore cannot be a square since the ten already apply as a square.

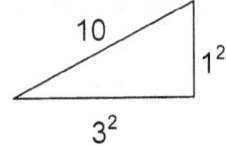

This all relate to the fourth dimension in space, but matter relate to the third or neutron dimension that holds time to a square. The matter as such remains in singularity (7) but time stands in regard to the square of half of Π. This then means matter (7) holds a relative to half the time effects matter (7x 2).

Space-time outside singularity then is

Space-time outside singularity then is

$T = \dfrac{28,2}{7 \times 2}$ which is (1,4 + 1,42) x 10 , which is matter (7) to time 2

$T = \dfrac{28,2}{14}$ = 2

Therefore every object holds the inner structure as 1 and in accordance to its own position of 2. This then is $\dfrac{10}{7} + \dfrac{2(7)}{10}$. That is the Titius Bode implication, however, I explain this better when dealing with the Titius Bode just before this part. Therefore the first portion Mercury must have is Π^2 (the sun with the diameter of 1 391 980) with the Roche factor implicating $(\Pi/2)^2$. According to this the positions are as follows:

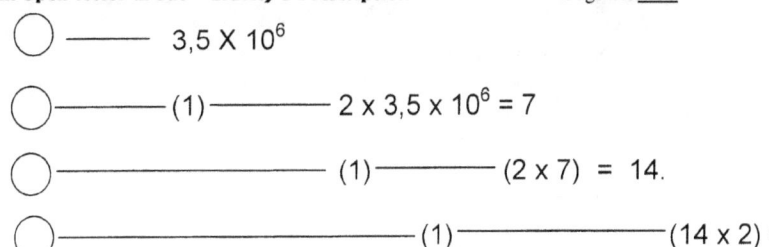

We find that the gas planets are on average about 2 Π overshooting the development that would apply in the case where the Roche limit would result in positioning orbiting structures. As explained a few paragraphs ago, the application of Π^2 in terms of the Titius Bode configuration will be 2Π, a dimensional factor change. As I have indicated the fact that it shows as 2Π, in the electron dimension, become Π^2 through putting matter in space.

Therefore the two Π you see, is the Π^2 you get. Having Π^2 means one thing: there was another object (Π^2) that related to 2Π and the two Π can only come from one more object that filled that space during some duration of time in the past.

The Titius-Bode principle relating space-to-matter at a value of $R^0 / T^2 = 1$ where space holds the square of 10 and matter is 7

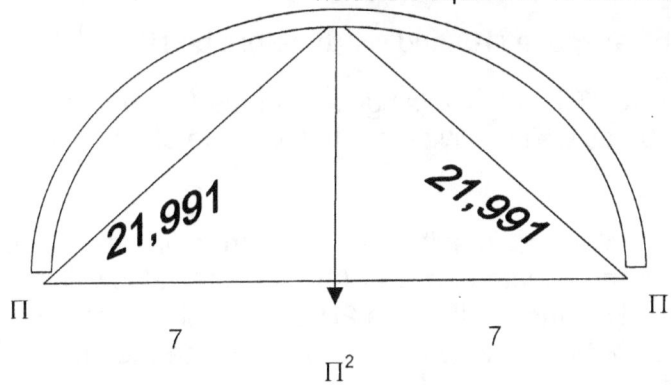

　　　　　　　　　　　　　　　　　　　　　　　　　　　　　　　　　　Therefore we may correctly surmise that something to the value of a relevancy of 2Π becoming Π^2 played a part in the positioning of the outer "planets" As shown repeatedly a double star would apply as $\Pi^2 + \Pi^2$ with $(\Pi/2)^2$ separating the stars where the Sun holds the position of Π^2, and therefore another object was present during space development in the time the Sun released from eternity. In other words THERE WAS ANOTHER STAR IN BINARY TO THE SUN .

In the half circle the centre is 2Π's and in the triangle the corners form 2Π.

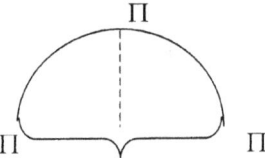

　　　　　　　　　　The dimension "gravity" removes as it replaces it with Π^2
In the half circle the fourth dimension holds a triangle.

That is singularity in time to the value of Π^3. The aim we have is determining the value of Π^3. The aim we have is determining the value of Π2. Known to all at this stage is that there is 7/10 in the Titius Bode law and in all spheres, including the Earth, we have a space dome of 21,991 holding space to the 7 holding matter. When a spacecraft re-enters the atmosphere, the angle of entry must be not less than seven and not more than 21,991. Therefore there are 7 holding 21,991 to the value of Π. This is the dimensional equal of the Titius-Bode law of 7/10. In order to determine Π^2 you therefore have to translate that value to the fourth dimension, giving it a value from singularity (linking time and space to a figure of 1) therefore 1, to its time position of Π^2.

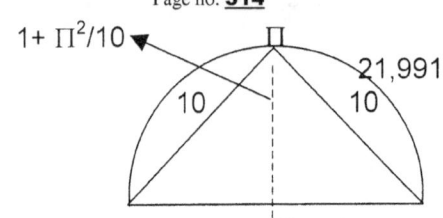

$$1 + \Pi^2/10$$

$$21{,}991$$

$$10 \qquad 10$$

$$7 + \Pi^0 + 7$$

First of all, all Newtonians are educated in mathematics, therefore they will know that the triangle holds 180° the equal to the next dimension of the half circle at 180°, also equal to the straight line of 180°.

Secondly, Newtonians are aware that multiplying in the one dimension translates to the next dimension in the form of adding i.e. 4 = 2 x 2 and 4 x 2 = 8. Therefore $2^{1+1}=4$ and $2^{1+1+1}=8$. In the one dimension adding is the same as multiplying in the other dimension. However, in astrophysics one do not merely transfer dimensions, you work with dimensions running concurrent in value. Therefore to the top you add or subtract, and to the bottom you multiply.

First we subtract the top from 21,991.

Titius Bode 10 + Titius Bode 10 is 20. Adding the one, the link between space and time in singularity, holds a position of one.
This leaves us 21.

Then determining the point where gravity (Π^2) will be at space (10) and will therefore be $\Pi^2/10 = 0{,}99$.

In order to get Pythagoras, you add 10(T.B.) plus 10 (T.B) plus 1 (singularity) plus $\Pi^2/10$ ("gravity" ending in space) and you. Square this total of 21,991 as well as divide it with the value of 7 x 7. (The top you add, while the bottom you multiply).

Therefore the top is 483,6 and the bottom is 49. To get to the "gravity" part in another dimension you divide (not subtract) the square of space (483,6) with the square of matter (49) and the value will then be 9,869467, the value of Π^2 relating to singularity. This means that the 2Π space holds, were filled with matter (Π^2) . There can be no argument about that fact and we can at present see the other Π^2 being the Sun .

$$\Pi^2 \qquad \Pi+\Pi \qquad \Pi^2$$
$$\Pi + \Pi = 2\Pi \text{ translating to time as } \Pi^2.$$

That is what the Roche limit is all about. It is the point where a cosmic proton (Π^2) shares (that means halved) a space relating position of Π in neutron dimension of time (Π^2) $\Pi \rightarrow (\Pi/2)^2$.

$$\{(\Pi = 7) + (\Pi = 7) + (\Pi = 7) + (\Pi^2 / 10 = .991)\} = 21.991 / \{(7 / 2) + (7 / 2)\} = 7$$

$$R^2 / T \ (S\$^3 \ S\$^2) \ R / T(+ \ S\$^2) = (7 \ X \ 7) / T^2 = (7 / 2) + (7 / 2) = 7$$

$$\Pi = \$_T = 7$$

$$1^3 \qquad 1^4$$
$$1^0 \qquad 1^0 \qquad 1^0 \qquad 1^0$$
$$1^2 \qquad 1^6$$

$$(S\$^3 \, S\$^2) + S\$^2 = \quad = 483.74$$
$$= \sqrt{483.74} = 21.991$$

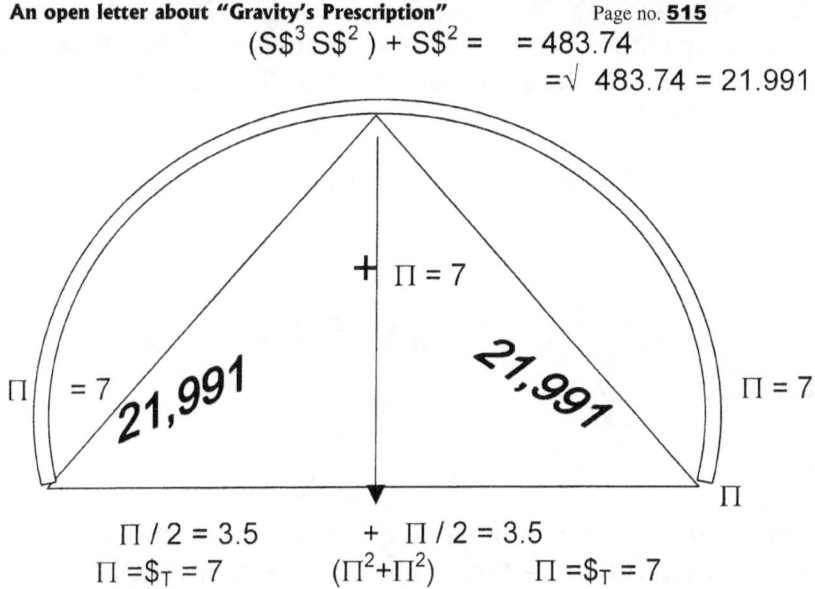

$\Pi = 7$

$\Pi \quad = 7$ $21{,}991$ $21{,}991$ $\Pi = 7$

Π

$\Pi / 2 = 3.5$ $+ \quad \Pi / 2 = 3.5$
$\Pi = \$_T = 7$ $(\Pi^2 + \Pi^2)$ $\Pi = \$_T = 7$

If any person wishes to cling to Einstein's view that the speed of light is the fastest that matter can apply velocity, explain how the Black Hole works. Inside the Black Hole must be matter, because there is no space, yet time does apply because it takes the particles spiralling inwards to the centre time to move from point to point. Matter in motion is time. However, no light can return to the surface, therefore the light is slower than the moving particles within the star. The only thing about the star, is that it maintains a higher relevancy than the relevancy the speed of light can apply. By accepting the existence of a Black Hole, the Einstein claim about the speed of light being one, becomes zero.

Another place where the speed of light becomes obsolete is within the centre of galactica, where the accumulative movement of matter exceeds the speed of light. That is where doctor Hawkins saw a Black Hole that is not a Black Hole, but the precise opposite. Light matter and heat, moves inward in an effort to maintain cooling as the group of proto, proto stars two era to the future) still claims their share of heat maintenance. Those particles in such close proximity, establishes a time well above that of the speed of light.

Everything in the cosmos is all about relevancies. Time started at such a high velocity, it had to be eternal. Nothing diverting from eternal can become more than eternal so it has to be less than eternal. It is fragmenting eternity into parts making eternity smaller. The smaller eternity becomes, the lesser eternity will be. That means that time started at eternity and became shorter with the introduction of infinities that broke the monotony of eternity. The more inanities there are affecting eternities, the shorter will eternities be.

I prove in "Matter's Time in Space – The Thesis" where Einstein went wrong in his theory about "The curvature of space-time". There is the space-time complying with singularity and filling the space-time in singularity is heat and matter valuing space-time. Space-time (Π^3 to Π) cannot bend, cannot curve, forms a straight line, but what fill space-forming time is matter (Π^2) and heat (3).

That part changes. The atom cannot be gas, or liquid, or a solid, because the atom is densified in occupation of space-time. It is the heat in unoccupied space-time that produces the gas, liquid, or solid that all substances can form. It is   THE HEAT in SPACE that produces TIME, that can and does curve, bend or whatever. That HEAT in SPACE forming TIME that forms the relevancy of space-time and that does bend. If, by applying the forming of gas, or solid or liquid to the element instead of the space between the elements, of course you will get the incorrect vision Π, where the space-time (matter holding singularity form singularity) is doing all the binding that apply to the curvature of space (validating time) and time in singularity (a straight line) will be solid. Einstein placed the relevancy incorrectly on singularity, instead of heat.

Once again I do admit, IT IS A LOT MORE COMPLICATED THAT WHAT I MAKE IT TO BE AT THIS POINT, but the motto is, Keep it simple.

If you wish to keep time in space constant, everything in the Universe will be oblong. That is why the Newtonians have an absurd view of the cosmos, and they present facts in the cosmos in a way nobody (least of all the Newtonians) can understand.

Time was slow, time became faster and faster because by extending the position of Π, Π^2 will produce speed.

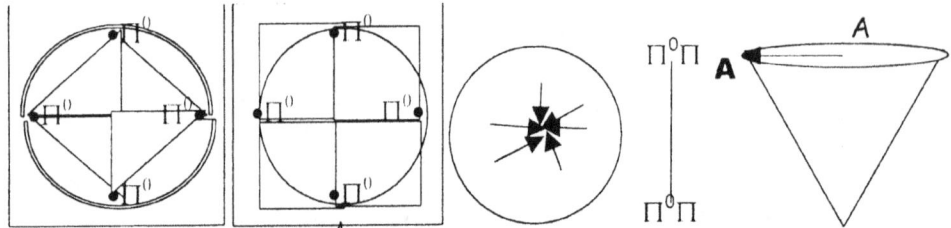

At that point space developed to a sufficient degree as to allow material, space and time form units being confined structurally while standing also individually apart. It was the pre runner to what then later became galactica filled with stars that was filled with atoms. But this was the prelude to all of that.

All rotating object has to be round to rotate. From the ends of the Unit rotating there will run a line running horizontally which turns with the top and as it turns with the top, the horizontal line is crossing another line that is running vertical but is not running at all because the three points cannot turn. The picture on the side is a picture showing the rotating object from the top as one would look down onto the line in singularity that does not turn. That line in singularity is representing the cross over limit parting the one part in the Universe from the other part in the Universe. At the displacement value of space (139) matter (138) and time (137) the Universe had tangible liquid in a relative motion with a structures solid containing heat in space.

Space   Time   Matter
139   137   138

centre spot where eternity heat came apart from cold set in place motion that space and time without Relevancies came abut spot but had no space to charge singularity by activated into complying. are all the same things only dimensions are formed space being in motion and

There were forever four sharing a centre spot while spinning around a locked infinity into a unit. But then that parted infinity from eternity that provided space in time and time in space or time.
when the dot moves away from the move. All that was possible was to relevance to comply in being Space-time is motion and movement separated by dimensions and space, where the dimensions become the space is motion by contraction or

by expansion but because time is almost eternal at k^0 our perception of the universe we are in is a stable and steady eternal structure. Gravity is motion and motion creates space to the third by the third in the third that interacts with one but establishes ten.

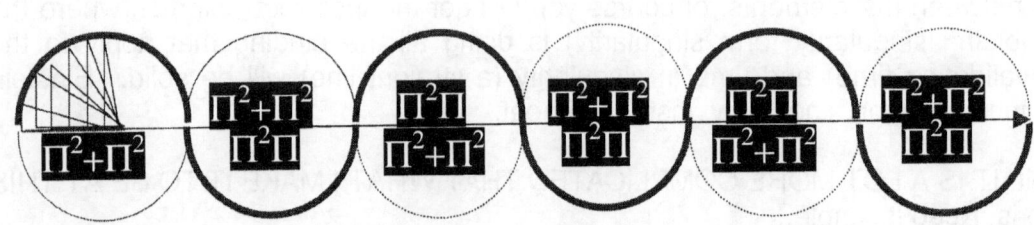

Motion is parting the Universe because it is representing eternity as much as it is eternity. It will never disappear because it is never there to begin with. The line cannot stop turning because the line can never start turning. The line is absent because it can never hold space, yet the line is always there because any motion may charge the line into presenting the centre of the Universe. To find the centre of the Universe is to reduce the line because Kepler said $k^0 = k\,T^2 / a^3$.

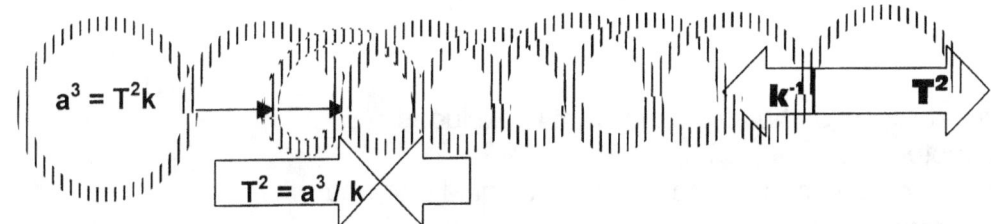

This process is a natural normal occurrence everywhere in nature without any person ever noticing. We see this so clearly in the spinning top. When the top is spinning such spinning of the top creates a centre and the lines start reducing space in the direction of the centre of the spin. The centre establishes a balance in space-time where at a point it finds partial independence from the dominance of the Earth's gravitational motion that is depressing the tops movement to a standstill in the space the earth confiscates.

By moving from 1^0 to 1^1 and from $1^0 \Pi^0$ to $1^1 \Pi$ requires space. Yet such moving did not leave the realm or the domain of singularity. The motion was still within singularity because moving involved forming a relevancy between heat and cold between infinity and eternity, between space and time and most of all producing what will in the far future develop into a Universe that can even be a host for life albeit on a very small spot for a very short while in relation to the vastness space has and the duration cosmic time has.

Time started by placing the double proton in space and in matter, where space and matter will always be in a sphere. The sphere always forms 7° from one point to another running outwards.

$$\$T = 7(\Pi^2 + \Pi^2) = 138$$

Explaining is as follows:

$\$T$ is space-time ($) in the time sector (T)
7   This indicates that the Coanda principle placed motion to space.
$(\Pi^2 + \Pi^2)$ Refer to the proton committing the Coanda effect.

At the same time space formed as a consequence to the Roche limit and so did matter. Forming the Aanplasings -Atomic-Epitome (the point where matter breaks in singularity). This was where the three-dimension concept was introduced but not quite accepted in practise because form still rules.

The moving of Π^0 to Π involved relegation and not motion as we consider motion. It was Π^0 getting a side and that is all. There was no true side but only a form that came into place. Singularity (**A**) received singularity (**A**) and no more of anything but the shift to comply with having a relevancy forming in relation to singularity. The dots had no sides, had no length or diameter. There was not measurable space or measurable time involved. The time could have been a micro, micro second as much a trillion millennium because time had no relevance. It was eternity interrupted by infinity, as it still is the case, however the line that eternity followed was no line because there was no space to hold the line. The line was momentarily interrupted by infinity, however with no one there, there was no one to notice. The lines were not lines but relations to sides being formed.

There was then an outer line forming time in space 10/7. Then there was the inner line forming space-time being 7 /10. The there was material filling space at $(7)(\Pi^2+\Pi^2)$ forming the sphere as it was filling the sphere.

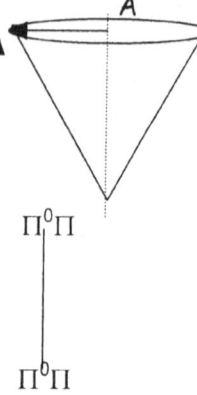

$T = 10/7\ (\Pi^2)(\Pi^2+\Pi^2)$

Explaining as follows:

10/7   The limit between what is part of the atom and what is excluded.
7/10   Forming the matter factor.
$(\Pi^2 /2)$ Indicates a deliberate inclusion of an atmosphere or a liquid or a neutron
$(\Pi^2 + \Pi^2)$ Shows that the proton still had total control

$T = \mathbf{7/\ 10}(\Pi^2\ (\ \Pi^2 + \Pi^2)\ \mathbf{=138}$

Explaining as follows:

7/10   Heat flowed to material supporting gravity within the centre of the sphere.

$(\Pi^2 + \Pi^2)$ the double proton

(Π^2). The boundaries were set by the motion that the neutron provided.
  The atom was born

$T = 7(\ \Pi^2 + \Pi^2\ )\ =$ 138 The circle the atom has.

•$T = 7/10\ (\Pi^2(\Pi^2 + \Pi^2)) = 136$
Motion towards the inside of the atom;

$T = 10/7\ \ (\Pi/2)^2(\ \Pi^2 + \Pi^2\ )\ = 139$ the relevancy of space carrying time to allow material space to apply motion within.

The atom formed a circle and that placed the Coanda effect in control of the Universe. For the first time there was matter in relation to time 10/7 in space in relation to space-time 7/10
This is what the Universe consisted of, everything that is today, was then, in a dimension that only holds "gravity" or the " gravity - motion". The Coanda effect took the Universe into the three dimensions.

$T = 7/10\ (\Pi^6) / (6 \times 10) = 112$

7/10   is the matter has the dominant value.
Π^6 matter has the six sides it holds in the fourth dimension.
6   are the six sides to space occupying matter.
10   are the value or dimension in which space holds a ratio to time.

The cosmos began, not to a specific space, because all the space that was there, initially is still there at present. Any atom above 112 cannot apply to the fourth dimension not then and not now.

A proton with a "mass" of say 12g/mol on Earth will have a "mass" in accordance to Earth standards of 25g/mol on Jupiter and it will hold a comparable "mass" of 100g/mol within the Sun . This "growth" in mass of any molecule within the structure's potential occupation of space-time increases. With this in mind, one cannot merely bring in such a relation to the "mass" of the proton in the beginning or within a star, or as it is on Earth. As the atom's spin increases or decreases in the relation to $\Pi^2 + \Pi^2$ $\rightarrow \Pi^2\ \Pi \rightarrow 3$, and with it, the "mass" will subsequently alter.

In explaining all of this, it is quite impossible for me to give it a value in mass, or time as both these factors alters in space-time occupied.

I INCLUDE A SMALL PART OF THE TECHNICAL DETAIL TO SILENCE THE "SUPER-EDUCATED-KNOW-ALL" THAT IS FLOATING ON THE "CUTTING EDGE" OF SCIENCE. FOR THE AVERAGE PERSON THAT DOES NOT HAVE ANY CLAIM ON THE IMPORTANCE OF A TITLE IN BEING PART OF THE ESTABLISHED "SUPER-EDUCATED-IN-XEPTED-SCIENCE-MOCK", FEEL FREE TO READ THE TECHNICAL EXPLANATION, OR IGNORE IT, IT DOES NOT CHANGE, ADD, OR DISCARD ANY LATER EXPLANATIONS.

Whatever one believes, one has to be honest by admitting that time had to start somewhere. It proves only shortsightedness on the part of the Newtonians, to conclude that "gravity" started at 10^{-43} sec after the "Big Bang". This only concludes that the start was with the "Big Bang" and little else. Nothing is said about what caused the "Big Bang". Beside the point, but still very valid is the fact that I have no words in expressing my resentment with the term used as the "Big Bang" being the start of the Universe. This name only explains how little science understands about the cosmos.

NOTHING WAS BIG BACK THEN AS NOTHING WAS SUDDEN, OR QUICK OR BANG.

No person ever came up with a logic and scientific explanation to what brought about the process of heat expansion. What is irrefutably true however is that it came on route from eternity or timelessness or whatever one wishes to call it.

At this point, I have to explain the mistake we go about thinking about science and time. At first I was arrogant enough to think I was the first to understand the way time works. After all, it took me some time to figure out how the handle fits the fork. Then to my shock, I found that H.P. Wells already concluded my way of thinking about a century before I have. Well that proved so much for my personal brilliance and modesty once again, returned to me.

When witnessing an event we regard as an explosion we surmise that what happens on the inside of such an explosion is extremely fast, but to the contrary, it is very slow. In the explosion, the duration of time extends, becoming longer.

To explain this we take two persons, one watching the other runs a mile. We place both persons initially in the same duration where both will endure four minutes of time lapse. The spectator will see in real time how the competitor takes four minutes to complete the mile.

Then in the next scene, we increase the duration of the competitor to 1 : 60 and the spectator's time remains the same. To the spectator the athlete will be covering the distance sixty times faster, and to the athlete the spectator will be cheering 60 times slower. The spectator would not believe his eyes because of the athlete's abilities in running that fast while the athlete will think the spectator is in frozen state of admiration.

In the third scene, we enhance the duration in the athlete's time zone by another 1 : 60. This will bring about that the athlete, in the view of the spectator, will be running the mile in less than 7 hundreds of a second and the athlete will be watching the spectator trying to wave while the action of the spectator will last 240 hours. In the eyes of the one person, the other's time span will be either an explosion or, everlasting, depending on the person's point of view from the space in time that he holds.

To each one, the spectator and the competitor, a time lapse or time duration of 240 minutes occurred, although the actions in both sectors would have seemed to alter severely. Any confusion coming about from the explanation above, I wish to remind that it is a common and well-accepted fact that time slows down as "gravity" increases.

Behind all of this explanation is one obvious rule. When the one subatomic particle positions in such a way as to displace space-time in the form of heat breaking down, the value of space, in order to meet the requirement of time, is once again all a relation between space and time. The less space heat has, the less the value of space becomes, because space has no value. Without heat in space and without matter there is no such a thing as space. Therefore, space does not exist, but for matter and heat valuing space to form time.

To understand this one must firstly understand the principle behind the theory I introduce. At the most inner point one find time or if we can supply it with a completely fictional name: "The gravity - motion". The gravity - motion carries the value of Π^3. This value determines time in eternity a position matter has no space, but is occupied in singularity.

Taking the neutron position to that of the proton we find the value created when the three dimensions (six sides) came about $\Pi^2 + \Pi^2 + \Pi^2$, which carried to the fourth dimension in cosmic or geodesic space-time becomes Π^6. When relating Π^6 to singularity it becomes Π^1 (space) x Π^1 (time) x Π^1 (matter). I do realize this explanation does not suit normal mathematical principles but we are working in dimensions and Π^1 (time) in a straight line is 180° and Π^1 (matter), which is half a circle is 180° and Π^1 (space) which is a triangle is 180°. As each Π^1 represents one dimension establishing another dimension and providing that dimension's existence.

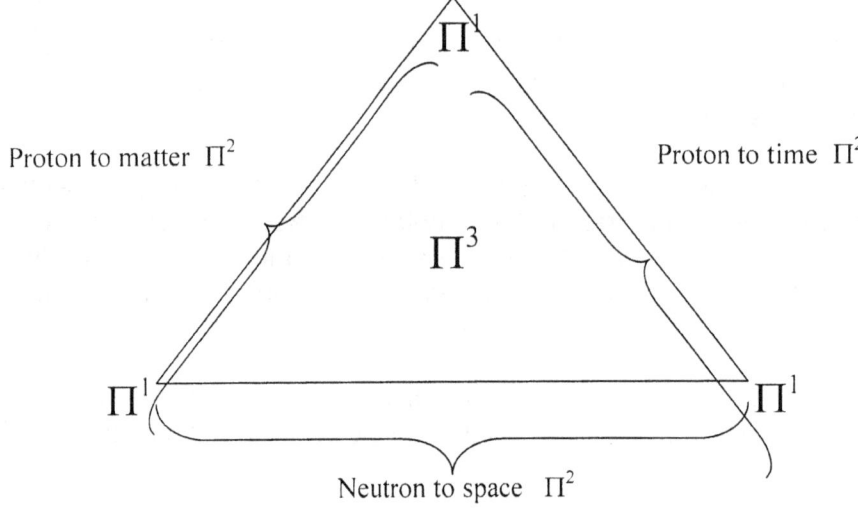

Proton to matter Π^2 Proton to time Π^2

Neutron to space Π^2

FROM THIS SPACE HEAT AND MATTER DEVELOPED

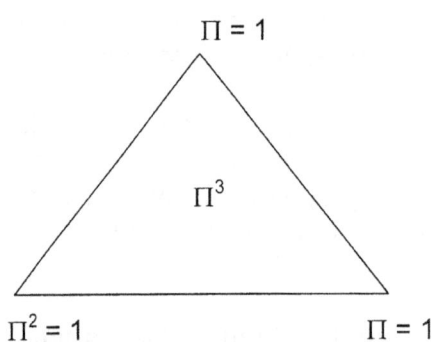

$\Pi = 1$

Π^3

$\Pi^2 = 1$ $\Pi = 1$

In relation to the " gravity - motion" space through a straight line will be Π and through the half circle matter with heat positioning space ($\Pi = 1$) to 3 sides 3. Relating the gravity - motion Π^3 three will always be a Π^2 and a Π combining 3. The half circle will be Π^2 and the straight line Π. Behind all of this explanation is one obvious rule, the one subatomic particle positions in such a way as to displace space-time in the form of heat breaking down the value of space in order to meet the requirement of time. It is again all a relation between space and time and the less space has heat, the less the value of space becomes, because space has no value. Without heat in space and without matter there is no such a thing as space, therefore space does not exist but for matter and heat valuing space to form time.

It would be far too complicated to explain why space-time and water share so many characteristics but they do. I have, to some extent, tried to explain it but I am aware that the explanation falls short of satisfying. I will repeat it once more, well aware that it cannot bring acceptance. It is heat that produce gas or liquid or solid. The period before light, everything about the cosmos was a soup cocktail, heat was liquid and space in singularity was liquid flowing heat that appeared like water. In

The Thesis I spent many pages in explaining this fact, but I do not wish to overcomplicate this book as I wish to bring across the scientific proof about the seven days of creation from a realistic scientific stance, for all persons to understand.

It seems very ironic that science with all its bravado, money, wisdom and splendour, can only begin at the point where light came to the Universe, while the Bible explains the creation in detail, long before the "Big Bang". The reason why there was no light before the "Big Bang" was that the spinning matter exceeded the speed of light, being $3\Pi^2$. The Authentic Author of Genesis refers to this as a mighty wind and this leaves a question. What better name can one give to this occurrence?!

With time in eternity, space in zero and matter being time and space, what would ever bring about that this situation changed? No Super-Educated-Wonder ever came forward to explain this. Why did the "Big Bang" start and what brought the "Big Bang" about? Only the Bible produces any logic to this question. Time, matter and space froze in one, there was no reason in nature for things to change, since this situation lasted eternally. Nature with all nature's laws did not apply, therefore one cannot say that nature started it. Nature was frozen. Nature was not even solid, it was in a state beyond being solid. Nature was nowhere!!

The spin in the Universe slowed down, up to a point where the spin was equal to that of the speed of light. At the point where the Universe spin equalled that of the speed of light, the Universe was still in total darkness. The light (photon) was there, but did not yet produce light. Light only came about as the spin reduced to below the speed of light, and only then light became obvious. The Universe grew away from darkness as this event lasted many eternities, during the period where the light separated from darkness.

How did this "Big Bang" take place? The best way to examine the reason is to see why anything in the Universe expands. To get anything to expand one has to heat it. All matter expands when overheating. Science may come up with whatever brilliant theory, the fact of the matter is that when matter overheats it expands. The bigger the overheating, the bigger will the expansion be, it is as simple as that. This means whatever leads to the forming of the "Big Bang", whatever preceded it, it had to come about from matter that overheated. With the event of the nuclear age, the proof came about that matter is heat in some frozen form. Unleashing heat from its frozen form brought about a jolt of heat, never yet experienced by man. By breaking matter from the frozen state, of which it is in, within the atom, heat produces light and heat. Where this process clearly shows how new space-time forms is where the releasing of heat caused winds that stun man's logic.

The nuclear explosion shows quite clearly what the "Big Bang" was, with the nuclear explosion being a very minute form. Yes, we have all heard the rubbish about matter and anti-matter. What can be anti-matter, since matter is heat, defined to a certain space occupied for that time. With matter being frozen heat, what would form anti-matter. Anti-matter means the opposite to matter, and if matter is frozen heat, anti-matter must then be overheating heat. This in itself is quite ridiculous. Anti-matter can only be matter with an opposite spin to that we think of as matter.

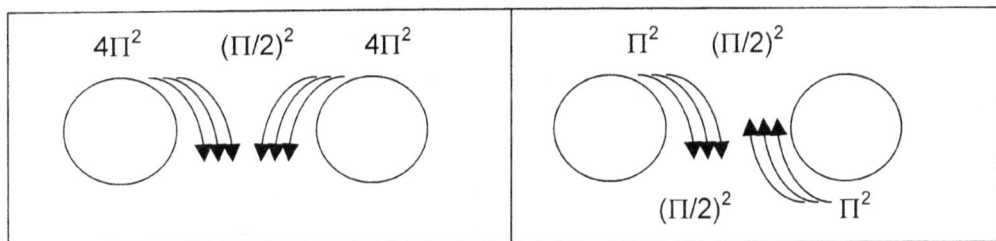

In the second sketch $\Pi^2 \ (\Pi/2) \ / \ (\Pi/2)^2 \ \Pi^2$ the Roche limit cancel each other and with that the space of the neutron effectively disappear. The two protons touch destroying each other and the neutron as well as the proton demolish and became heat 3^3.

The process where matter then touches matter, it will bring about a reduction in the feeding process of heat,. where all matter in that space will overheat and expand, producing unfrozen heat. This means there was heat occupying space, and matter with both in relation to time.

The proton with a positive space-time displacement less than 136 placed its displacing properties in negative space-time displacement. In short, to substitute for mass shortcoming of less than 136 grams/molecule and still finding sufficient cooling properties for the proton to survive the overheating deficiency, it has to spin more rapidly, therefore by spinning it makes contact with more heat than it would otherwise do. One may consider this as "breathing". If the proton does not find adequate supply of heat by being motionless, the atom has to substitute the movement through motion. Forming an object that has an increase in heat supply through work commonly uses this fact. In nature, just the opposite is true because of the motion by movement, as one find in the case of wind.

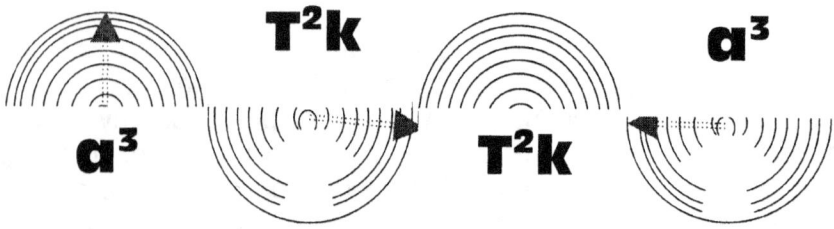

There is no "force" in the cosmic flow of time because everything is a "force" in one way or another. Everything is in a 90^0 angle with time therefore the cosmos is restraining time while it is retaining space it is moving in relation with time but also opposing time all the way. The proton takes heat from space in an effort to maintain temperature and stability. This flow of heat brings about the reduction of space by increasing the heat in that space. The flow of the heat, through the electron by means of the neutron to the proton is time. The amount of heat taken by a proton is a constant throughout the Universe but relative to the space reducing effort of all the protons influencing that specific space. That is "gravity" (a term I denounce and reject). "Gravity" is nothing else but additional application of time. The higher the gravity is, the slower the time will become by prolonging the duration of time,

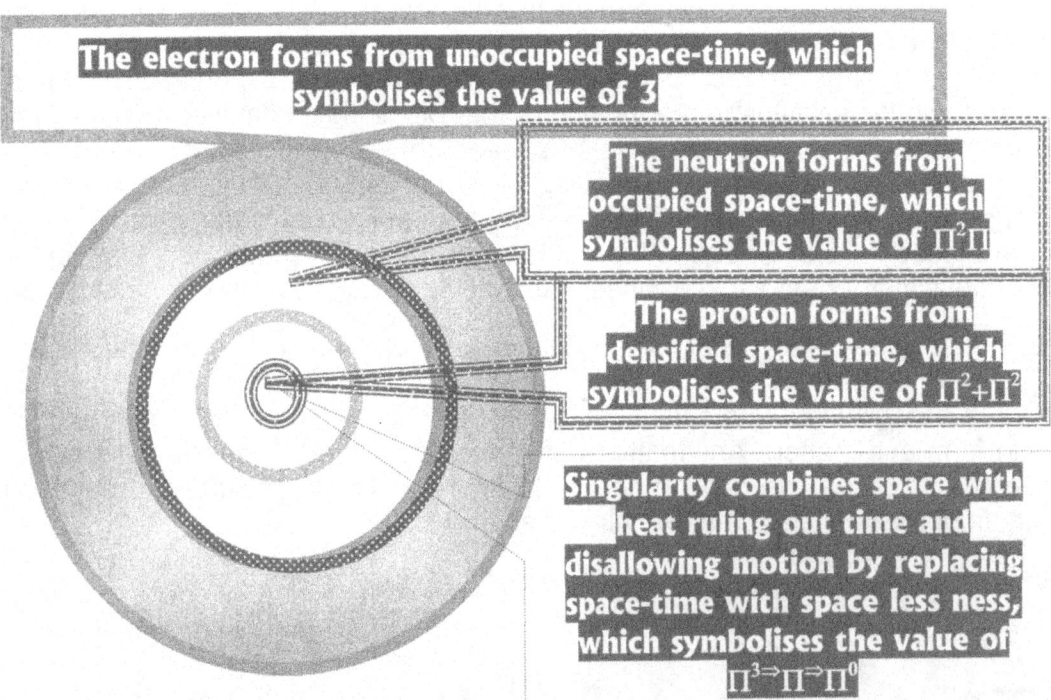

Unoccupied space: This forms the atomic relevance of **3,** which is where the ratio of space moves to the ratio of liquid in space. This is bringing motion in contact with **unoccupied space**.

Occupied space: Then forms the atomic relevance of $\Pi^2\Pi$, which is where the ratio of liquid space moves to the ratio of solid space. This is bringing motion in contact with **occupied space**.

Densified space: That forms the atomic relevance of $\Pi^2+\Pi^2$, which is where the ratio of solid space moves to the ratio of **densified space** or motionless space. This is removing motion by disallowing contact with space and then forming space less ness.

Space less ness: That forms the atomic relevance of $\Pi^2 + \Pi^2$, which is where the ratio of solid space moves to the ratio of densified or motionless space. This is confining motion in a position being part of eternity.
I give the following relevancies in order to show how I define a star in relation to an atom since a star is just another cosmic particle or just another atom.

By taking this statement and introducing that to a Galactica, the shining luminous middle part, holds time to eternity through movement. The atoms in the middle admits light because it holds time duration to the speed of light, the longest duration that time can be and still remain in the fourth dimension.

The centre of all cosmic structures determines the time that applies to the structure itself. As the cluster of protons supply the density that influence the space of occupation, the density is a collective reducing of space with the increase of heat. This can apply through an object relating to space through movement by the object and by the object reduction of space through the density of the accumulative effort of the cluster of protons we named elements.

By referring to "gravity" only one aspect of the space-time relation of any elements apply. There is no mention of the second and crucial part of "Gravity" where the motion of the object brings about the space-time relation, or if you wish, providing the cooling aspect. At first, the proton cluster's total positive space-time displacement has an insufficient "gravity" to secure a stable cooling effort. This spinning of the element clusters is inherent from an event, even predating the "Big Bang".

The spinning motion of the element clusters (or proto stars or if one wishes to use the name of future stars, it will be just as applicable) hold their relation to heat secures by maintaining motion. As the time value in the clusters space occupation (mass) secures an era related value, the structures that were spinning, reposition in such a fashion as to apply a new linear displacing value. At first the motion is such that the linear position is negligible, but as the mass grows, the linear distance grows accordingly, placing the revolving structures further apart and at the same time, "pushing" the rotation of the objects in a wider revolving orbit.

By widening the rotation circle, the objects rotate at a "lesser" pace and this pace coincide with the space-time occupation ("mass") of the totality in the effort of all the protons put together. In this one will not find a "force" but it will be a complete balance between matter, space and time. By securing an ever-increasing space-time occupation (mass) the future star will reduce its negative space-time displacement (motion) and increase its positive space-time displacement (gravity). The higher the positive space-time displacement (Gravity) becomes, the lesser the negative space-time displacement (motion) will be. At present only stars holding an iron$_{56}$ inner core can maintain a star status, and any object with a lesser element in the inner core will not bring about fusion, or in fact, any form of luminosity.

For instance by the time the "Big Bang" arrived, only elements with a "mass" of 112 had the ability to release from the Galactica luminous core and during the Era of the Quarks, the releasing mass of the time determining elements carried a combined proton-cluster "mass" of 88. As the single proton's time holding value increased (molecular mass) the time grew less and the Universe "grew bigger". Up to this point all arguments came about from the theory about the "Big Bang". In The Thesis I show mathematically that the Universe will last seven cosmic days. This however is not the seven solar days and under no condition may one confuse the two.

We are in the fourth cosmic day calculated from the Big Bang as if the Big Bang was the first day. To understand the process of the cosmic days, please study the cosmic almanac as seen on the last page of this letter. While I am saying this, this book is about proving with undeniable facts. That I leave to The Seven Days Of Creation ISBN 0 – 9584410-4-9. In this letter I only show it is possible to prove what I say I proved. The Bible speaks of seven days of creation; therefore we must look for the seven days the Earth formed. According to the cosmic calendar, we presently find ourselves in the fourth day, with three more days to follow. Why do I refer to these periods as days? Well the term "day" is as manmade as clothes, buildings, trains etc.

In this is another point that proves the technique science applies at present, does not nearly give a near value to the time duration of development on Earth. It should be out by as much as a few billion years for all we know. To indicate the meaning of what I am trying to bring across I shall illustrate a time scale in which the development might have taken place. I do most strongly disagree with the age the Brainy Bunch hands out to the solar system but since I have no better time to give I shall use the Xepted table ONLY AS AN INDICATER SERVING TO PLACE RELEVANCIES.

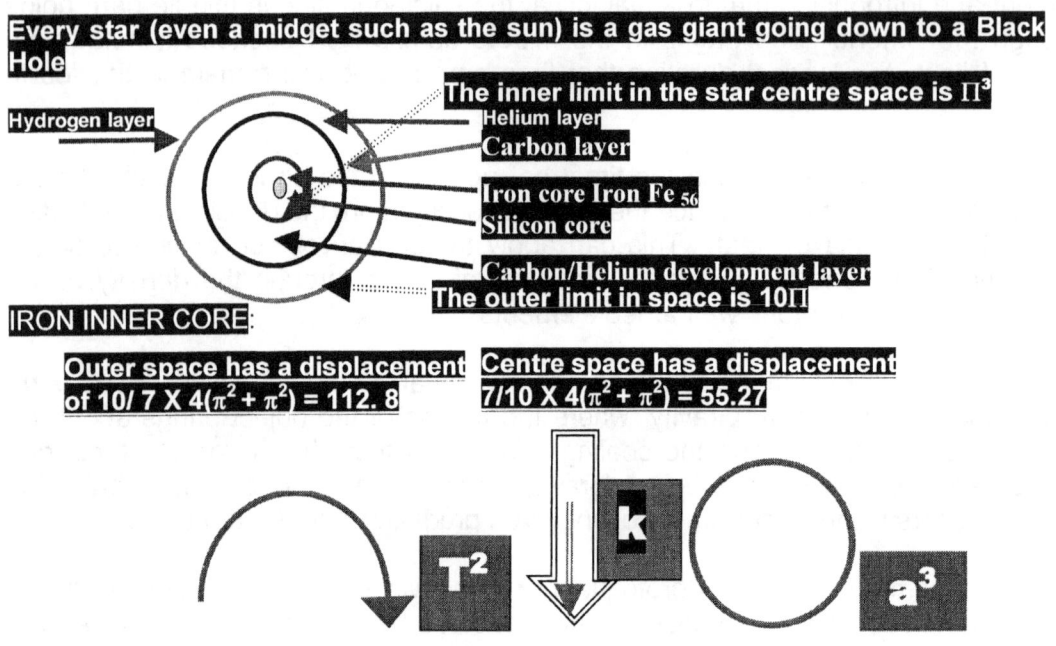

Every star (even a midget such as the sun) is a gas giant going down to a Black Hole

Hydrogen layer

The inner limit in the star centre space is Π^3

Helium layer

Carbon layer

Iron core Iron Fe $_{56}$

Silicon core

Carbon/Helium development layer

The outer limit in space is 10Π

IRON INNER CORE:

Outer space has a displacement of $10/7 \times 4(\pi^2 + \pi^2) = 112.8$

Centre space has a displacement $7/10 \times 4(\pi^2 + \pi^2) = 55.27$

T^2 k a^3

PLANET	PERIOD (Years) (T)	MOVEMENT (T²)	DISTANCE	SPACE (a³)	RATIO k
Mercury	0.241	0.058	0.39	0.059	0.983
Venus	0.615	0.378	0.728	0.381	0.992
Earth	1.000	1.000	1.000	1.000	1.000
Mars	1.881	3.54	1.524	3.54	1.000
Jupiter	11.86	140.66	5.20	140.6	1.000
Saturn	29.46	867.9	9.54	868.25	0.999
Uranus	84.008	7069	19.19	7067	1.000
Neptune	164.8	27159	30.07	27189	0.999
Pluto	248.4	61703	39.46	61443	1.004

The inner core has to be Fe_{56} to produce gravity. This is what reduces space in conjunction with singularity where the atoms produce a dismissing value that the space-time can sustain with enabling the flow of heat through space. In the one limit of the six sided Universe no element can sustain duplicating above the value of **$10/7 \times 4(\pi^2 + \pi^2) = 112.8$ and above $7/10 \times 4(\pi^2 + \pi^2) = 55.27$** within the star inner core. Dismissing space beyond that capability will no longer contribute to duplicating space-time of the atoms involved. Only the iron atom producing and maintaining a displacement value of $55 - 56$ can produce gravity by being on the edge of demising space time while maintaining duplicating which is gravity and in our Universe only stars with an iron inner core has the ability to bring about gravity. Gravity can only achieve a displacing relevancy at **$7/10 \times 4(\pi^2 + \pi^2) = 55.27$,** and that produces a potential difference that brings about gravity within the inner star core where gravity accumulates. This then relates directly to the second value of the Titius Bode value of **$10/7 \times 4(\pi^2 + \pi^2) = 112.8$** that limits outer space in the three-dimensional and six sided boundaries of what forms our Universe as outer space forming the value of **$7/10 \Pi^6 / 6 = 112,162$**. That is the outer relation to the inner relation set by the

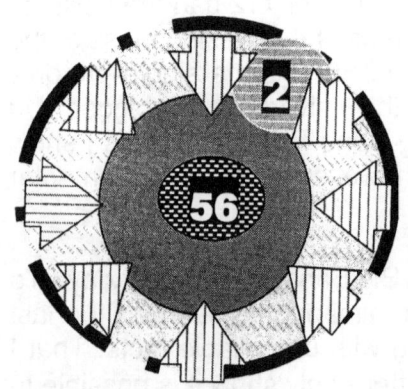

core in ratio to the outer space securing a position for the star identity in the space limits and is an indicator of the balance in space-time displacing potential of the star. In every star there is this flow towards the centre firstly of every individual atom but also as a combined unit flowing towards the centre of the star and the dismissing of space in every atom centre brings about the forming of a relation as a group within one unit structure we call a star. This flow is there because we gave it the name of gravity and gravity is the result of all the atom protons dismissing space and as such then has a linking that is invisible to the naked eye. In young stars, the core ability is yet to develop and in such stars, the gradual reducing comes about as layers support the effort little developed inner core. The space reduced becomes a unifying effort from all the atoms in all layers from the outer (hydrogen$_1$ and helium$_2$) through the carbon / oxygen centre and the silicon layer down to the iron core and even going down further into space-time obscurity where the atoms as a group combining their effort acting as one atom. An atom securing one proton will provide much more space a much better field to flow.

The scale above proves the accuracy to some degree but with the information being as sparkly as it is in this letter I would rather prefer if you would please see it more as a scale to use and form ideas than to be a mathematical yardstick. He way I present the following might not seem to as an accurate and tested scale in time but merely as measure to indicate how the frequency will relate to the time duration. From where we stand, we may have a perception that the frequency is getting shorter, but as seen from within the Sun , the time duration would be precisely the same value each time. This is the process in which time is concentrated in the space confinement and the relative space-time is amplified to extent the duration. In time, this variation is perceived as flair and later as a pulsating readjustment. I hope it now will be apparent just how small and under developed our Sun really is when compared to other structures.

This comes about because the relative size of a star is based on its space volume that contains matter and the incorrect way in which the density of stars are calculated. At present the frequency could have come down to as little as 15 000 years, maybe slightly more, but who knows. However, it is not the frequency that is the problem, but the way in which the frequency is measured that is of concern. At present, we relate to the duration of time laps relevant to the magnitude in which the Sun presently is. This might be a problem to all of mankind and civilization. There exist neither method nor means in which one can determine at what stage of progress the Sun is in at this moment. All that is extremely clear is that at one stage, the Sun becomes a raging bull, and a sleeping bear follows this. In between these two possibilities of time duration, time can become double the value it holds now, which then is followed by a period where time might have half the duration, we experience at present. The first thing that springs to mind, is that we find ourselves in the middle, which averages out the extremes. That is not the case.

Let us start by taking the size Jupiter is today. We know seven events happened and each event had influence on the Jupiter distance.

The relevance of the actual distance is however, 16,5; 15,4; 10,6 and 8 respectively in relation to the others of 2Π.

With this knowledge secure, we have to seek the positions evidence as how the structures came to be in that place as the inner planets do not confirm their position in accordance with the rest of the solar objects. When we give the distance that should apply if the inner planets were also just orbiting objects it would then be

The official average distances from the centre of the Sun the average distance of rotation that each inner planet completes in relation tot the governing singularity.

At first I thought the way I presented my first impression of the solar development was correct in as much as the way I first introduced the image. Back then I was still very much under the influence of the Newtonian conceptions of a runaway star, and other misconceptions I later found to be alien to cosmology. There can be no runaway star precisely for the reasons I explain the constitution of Galactica. A star with an individual developed space in the time of this era the iron$_{56}$ era, will then establish a circular "gravity" that is able to withstand the influence of the linear gravity. The higher the circular "gravity" becomes, the more static will the linear "gravity" be. In the case of a Proton star (Black Hole) the linear component lies with the fact that we can actually see matter performing its linear component by not curling as lesser stars do, but placing the circle and linear components all on the matter as the Proton star pulls matter, space and time into its pace-time occupation. A Proton Star is unmovable, static, and stationary and every other name you wish to connect to its immovability. It can no longer go anywhere. The stars within the sphere where doctor Hawkins identified a Black Hole to be, within the very centre of a galactic, holds all occupation relevance to the spin or linear component of space-time occupation and only a very minute part to the circular "gravity" component.

I bring evidence to prove my personal theory development and showing honest misconceptions on my part. In that view, I wish not to remove the first suggested solar formation but to replace it partly with facts that I became aware of, as my personal insight grew.

There is another way of looking at the effect the Titius Bode law applies to cosmic atoms and this is very important within stars. No person can deny the fact that the Earth is a sphere, excluding outer space, where our need to apply entry into the sphere of inclusion ($\Pi^2\Pi$), and the law to abide by is the four cosmic pillars. You have to abide by the Roche limit where rules allow you entry, or destruction. No Newtonian can deny that.

When you cannot deny the fact that the Earth is excluding space as it is including time the rest is beyond denying also. One has to seek the evidence where the evidence is, where one can locate such evidence and above all, read the evidence correctly. The evidence proves the existence of a binary before the Earth came to be. The four inner planets are left over parts, a reminder of a star that uses to be at the other side of the atom. While the one side of the atoms in a star relate to the square of space 10 / 7, the other part of the atom in the star relate to the matter to matter (neutron ($\Pi^2\Pi$) holding matter to space while space becomes time ($\Pi^2+\Pi^2$).

The Sun was in a binary with another cosmic structure that has no longer have a full place in the solar system as the solar system stands today. The second object had a good measure of the Sun s' potential, but not adequate to survive. If the Sun was the size of what Jupiter at present is, the second binary was then about the size the Earth is today.

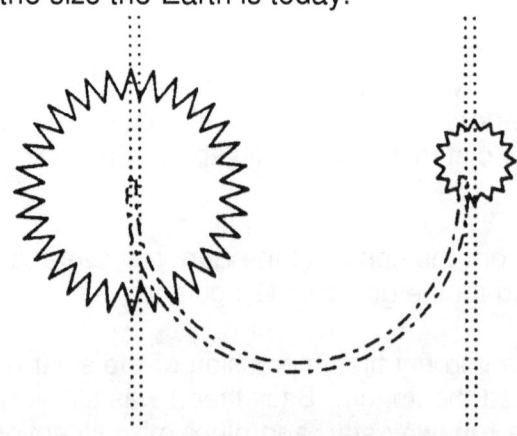

Both had individual singularity Π placing a value of $2 \times \Pi$ in space, as well as a common singularity $\Pi^2 + \Pi^2$.

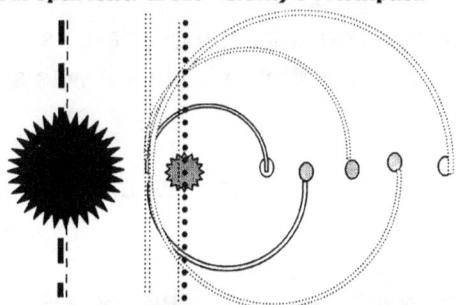

From this binary extended a singularity connecting five rock ice cubes, each holding a point of singularity, with the electron position of the binary where the binary holds the mutual point of singularity. This had nothing to do with 5 or 4.5 X 10 ^{9} years in time laps.

What I am about to do, is very unscientific and the next few pages must be regarded as pure speculation, brought into the book for one purpose and that is to amuse. That is all value that the next calculations have. I dispute the fact that any calculations can ever determine the precise size the structures had because the structures at present hold the very same size that it held when the solar system formed. However, life is not only about proof and fighting dispute of proof, there has to be some entertainment in the book, merely then to satisfy our need to gossip. It is far better to gossip about the planets than it is to gossip about one another, because I do not think it will hurt the feelings of the solar objects at all.   It is utter speculation and a needless process and I do not wish to encourage such wild guesswork in any way what so ever.

From this binary extended a singularity connecting five rock ice cubes, each holding a point of singularity, with wit the electron position of the binary where the binary holds the mutual point of singularity. This had nothing to do with 5 or 4.5 X 10 ^{9} years in time laps.

What I am about to do is very unscientific and the next few pages must be regarded as pure speculation, brought into the book for one purpose and that is to amuse. That is all value that the next calculations have. I dispute the fact that any calculations can ever determine the precise size the structures had because the structures at present hold the very same size that it held when the solar system formed. However, life is not only about proof and fighting dispute of proof, there has to be some entertainment in the book, merely then to satisfy our need to gossip. It is far better to gossip about the planets than it is to gossip about one another, because I do not think it will hurt the feelings of the solar objects at all.   It is utter speculation and a needless process and I do not wish to encourage such wild guesswork in any way what so ever.

For your entertainment alone, I shall go about trying to determine the size of the Sun back when…as the size of the Sun and other structure was when the dual came to its final resolve.

Let us start by taking the size Jupiter is today. We know seven events happened and each event had influence on the Jupiter distance.

With this knowledge secure, we have to seek the positions evidence as how the structures came to be in that place as the inner planets do not confirm their position in accordance with the rest of the solar objects. When we give the distance that should apply if the inner planets were also just orbiting objects it would then be

1	Mercury	$58 \div 2\Pi = 9,23 \times 10^6$ km
2	Venus	$108 \div 2\Pi = 17,188 \times 10^6$ km
3	Earth	$149 \div 2\Pi = 23,714 \times 10^6$ km
5	Mars	$227 \div 2\Pi = 36,12 \times 10^6$ km

The relevance of the actual distance is however, 16,5 ; 15,4 ; 10,6 and 8 respectively in relation to the others of 2Π.

In the case of Mercury, Mercury is $5\,\Pi$ further than the $(\Pi/2)$ (Π^2) of the others are and Venus is $(\Pi^2/2)$ (Π) ; Earth is Π (Π) and Mars is almost $(\Pi/2)^2\,\Pi$ away from the Sun . I admit that the distances do not apply to the millimetre but that will become apparent soon. The importance here is the relevancies Mercury is $(\Pi/2)$ (Π^2). That is the place where the binary minor should be in if it was still there.

Venus is $\Pi^2/2$ (Π), which is in the space-time that binary minor held at a position it would hold its value to Π.

Earth would be in a position of one Π more than where binary minor would relate Π. This then is Π x $\Pi = \Pi^2$. Mars would be $(\Pi/2)(\Pi)$ (half a Π) even further away than the Earth's position of Π (Π). We shall get to Ceres (a fragment of what was another planet in due time. Therefore let us establish the relative positions to Mercury, the sole holder of the star binary minor in relation to the other fragments.

Mercury = 0.
Venus = $(\Pi^2/2)$ Π. The edge of the entrance field that Binary Minor had.
Earth (15,4 + 10,6) = 26. That has a relevancy of about $(\Pi/2)$ (Π^2) = 24,35. This is where we must not forget that the Earth too is a binary and the moon played its part in the drift relating to a position of 26 instead of 24,35 or $(\Pi/2)$ (Π^2).

Mars holds a relative position of 10 (Π) and we know that 10Π are the relevant space position to $\Pi^2\Pi$. This indicates that anything outside 10Π will be far outside $\Pi^2\Pi$ and this places the object in that space, without space. Any object without space will be directly into time and this will mean total destruction of that object. It will have the very same consequences as having a cosmic body holding an iron core in the previous era, or having a cosmic body with a silicon core in this era. It will and must disintegrate; there is no question about that. The space-time applied to such a core will be double than what is to the other five inner planets. It will destruct, in the same manner, as did the Shoemaker Levy 9 comet with the one exception, it held a relative position where the Sun could not get hold of the fragments as Jupiter got a grip on the Shoemaker-Levy 9 fragments.

At first I thought the way I presented my first impression of the solar development was correct in as much as the way I first introduced the image. Back then I was still very much under the influence of the Newtonian conceptions of a runaway star, and other misconceptions I later found to be alien to cosmology. There can be no runaway star precisely for the reasons I explain the constitution of Galactica. A star with an individual developed space in the time of this era the iron 56 era, will then establish a circular "gravity" that is able to withstand the influence of the linear gravity. The higher the circular "gravity" becomes, the more static will the linear "gravity" be. In the case of a Proton star (Black Hole) the linear component lies with the fact that we can actually see matter performing its linear component by not curling as lesser stars do, but placing the circle and linear components all on the matter as the Proton star pulls matter, space and time into its pace-time occupation. A Proton Star is unmovable, static, and stationary and every other name you wish to connect to its immovability. It can no longer go anywhere. The stars within the sphere where doctor Hawkins identified a Black Hole to be, within the very centre of a galactic, holds all occupation relevance to the spin or linear component of space-time occupation and only a very minute part to the circular "gravity" component.

I bring evidence to prove my personal theory development and showing honest misconceptions on my part. In that view, I wish not to remove the first suggested solar formation but to replace it partly with facts that I became aware of, as my personal insight grew.

Let us establish a line of evidence and fill the puzzle.

1.   There was another star with the Sun where the Sun was Binary Major and
        Star Unknown was Binary Minor.

2.   The Binary system catapulted the solar system out of its frozen eternity, way ahead of its time of development, bringing along the rest of the micro stars. The outer "planets" are not planets; they are micro stars in development.

3.   The Binary system formed part of a Lagrangian system holding the Binary in the centre and with Jupiter as the first orbiting satellite.

4.   The position Jupiter held for most of the developing period made Jupiter the second main benefactor of the dual, with the Sun the major winner and Binary Minor the major loser. This will also explain why Jupiter has such an advance in space-time occupation when compared to the other micro stars.

5.   The position where the (six) inner planets find themselves, were a void, AS THE BIBLE CLAIMS.

6.   The relative positions were as follows:

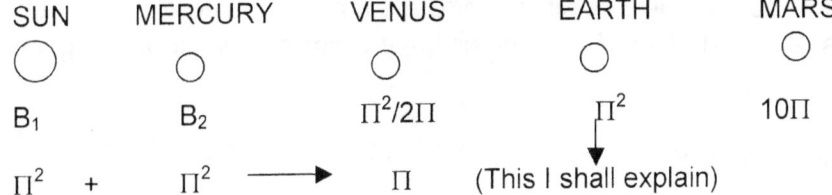

(1 + 1) = 1 2	3	4	5
Binary Jupiter	Saturn	Uranus	Neptune

Then Unknown Star capitulated as it could no longer serve the dual it fought.
It fragmented into possibly 9 major parts and many minor parts, (the comets.)

7.   As the core fragmented the brittle parts dislodged in a position each to a relative neutron position in the space-time binary minor held relating to its point of $(\Pi^2+\Pi^2)$ $(\Pi^2\Pi)$ 10Π. Whichever way we Earthlings will look at our position, from whatever angle and by whichever calculation we devise, our relevancy will be 1, will be 10, will be Π^2. Should any person ever do a calculation and find his answer does not bring this fact to bear, he must go back and fix his mistake. There will be a mistake on his part.

SUN	MERCURY	VENUS	EARTH	MARS
◯	◯	◯	◯	◯
B_1	B_2	$\Pi^2/2\Pi$	Π^2	10Π

Π^2 + Π^2 $\longrightarrow$ Π (This I shall explain)

Binary stars, spinning to self-destruction will produce significant heat. Heat create space, space forms winds. That is facts that the Bible presents and is indisputable. Where the Earth was, was still a void, containing a sphere of circular displacement and this will reduce linear displacement to zero. Linear displacement is space and circular displacement is containing heat for matter survival.

Binary Star Minor overheated. That is why the core brittle and fragmented. This action will release tremendous contained heat, the heat will produce magma flowing in space like water in space and this eruption of heat space that created winds. Once again the recollection fits the scenario. Releasing the heat and producing space will establish space-time and fill the void where the Earth should fit. This is fact and if anybody even tries to dismiss this will be because of abstinence on his or her part. I did not prove the Bible correct. The Bible told the truth and in such correct detail, it is beyond human comprehension, but sublimation on the part of Newtonians and science before them, disallowed their ability seeing it.

I found no one could look past me and see my formula $R^3/T^2 = 1$ and $\$T = (\Pi^2 \times \Pi^2) (\Pi^2\Pi) 3 = 1836$ which is the relevance of the cosmos. By not finding a person that could see past me, I knew that person will not be able to look beyond "a burning Sun and see the frozen state in which the Sun is. Without noticing such crucial evidence, the rest goes lost. That person that sees me and not my formula will never see the cosmos for what it is.

Slightly of the mark but duly valid I wish to make a brief remark on the Sun / moon binary. As the moon is also in a binary extended position with the Earth I wish to take this quick opportunity to show that the moon was never part of the Earths proton - proton value ($\Pi^2 + \Pi^2$) value but is in a neutron to space position $\Pi^2\Pi$. This can only apply when the one object occupying less space-time has a proton value ($\Pi^2 + \Pi^2$) that is less than the superior object's position on $\Pi^2\Pi$. In other words, when the total core value of the lesser structure is in any case less than the neutron value that the larger object relates to, concerning the smaller object. This means the one is totally dominating the other in all aspects.

Some quarters of the Newtonian High Priest in High ranking made claims that the moon once formed part of the Earth. In the following elaboration I shall prove why I dismiss this claim as utter nonsense.

From these facts about a binary, one can then clearly see that having two structures in a position overshooting the Binary scenario, is very much fantasy. It is just not possible because the valid space-time will exceed 112, and the structure will not have the ability to hold position in the universe that is limited to 112.

The proton value of the Earth is ($\Pi^2 + \Pi^2$) and it will hold the second object (the moon) at $\Pi^2 /2$. This is because the second object is in the "gravity" application of the larger object (the Earth) and the "gravity" factor of the Earth takes on a linear value, half that of the gravity factor of the Earth. The Earth will not allow any linear action to exceed 10Π and at ($\Pi^2 + \Pi^2$) ($\Pi^2/2$) it exceeds that value.

As the two core has a dual space-time occupational value of ($\Pi^2 + \Pi^2$) ($\Pi^2/2$) = 97, and the core value of the Earth is at 7/10 (4 ($\Pi^2 + \Pi^2$) the combined value will even exceed the critical space factor of 3 ($\Pi^2 + \Pi^2$) applying to stars holding space, therefore the space separating the two objects will vanish into singularity. The reason why the Roche principle maintains core separation is that the core combinational value , seen from one or the other objective, is ($\Pi^2 + \Pi^2$) ($\Pi/2)^2$ = 48. The individual space-time factor of each core is 7/10 (4 ($\Pi^2 + \Pi^2$) = 55, therefore the space holds less heat and therefore more space.

Where two structures go into a Roche Lobe and the one structure forms a proton value of Π^2, but the comparable space-time occupation is less that Π the Shoemaker Levy 9 structural fragmenting will take place.

As larger structures will have no occupational space loss due to overheating, but the one holding a Π value has great concerns.

From the superior object the occupational distress will be ($\Pi^2 + 2\Pi$) ($\Pi.2)^2$ = 39,8. The geodesic space value as a factor is Π^3 = 31, therefore it will bring about a "gravitational pull" revaluing the relation to ($\Pi^2 + \Pi$) ($\Pi^2 /2$) = 64. Being at 64 it means the smaller object holds a position of space reduction, as the space value is twice that of the geodesic value. The conclusion is that it will fall under the invert square law of spheres. By looking at what happened to the comet, one can see that such estimation will be correct. When taking that formula and applying it to the position that the smaller objects holds, one cannot surmise immediately that it will be part of the atmosphere, therefore the 7 in the formula in atmospheric heat income will change from 7(3(Π^2) ($\Pi^2 /2$) to ($4\Pi^2$ +Π^2) Π^2 /2 = 121,36 because the second object holds a far superior occupational position in its application of "gravity". With a relative value of 121, overshooting the highest atomic occupational possibility of 7/10 Π^6/60 = 112, the atomic structure that the smaller object holds will diminish to heat and photons. It will break up; turn to heat, photons, and dissolve, which are precisely what, happened to Shoemaker Levy 9. One can witness the structure demolishing in heat, light and fragments.

With an object larger than that of Shoemaker Levy's relevancy to that of Jupiter, the same laws apply but the values derived from it bring about a different end result. The only change will be in the position of the relevancy where the one object being the superior will again apply the same formula in establishing its position. ($\Pi^2 +2\Pi$) ($\Pi/2)^2$= 39,8. With this value being the same as 2 ($\Pi^2 + \Pi^2$) = 39,47, it will hold the structure in a cosmic orbit, not being able to reduce the space-time separating the two, and applying the gravitational equilibrium of 2($\Pi^2 + \Pi^2$) ($\Pi/2)^2$ ($2\Pi^2 + \Pi^2$) ($\Pi./2)^2$ = 73 and with the space-time occupation not only exceeding $3(\Pi^2 + \Pi^2)$ where space destructs but going another half a Roche factor down $(\Pi/2)^2 /2$ above and beyond the space demolishing value of 58, it means there must be a total structure space decrease of some sort. It will not be a structural break up and fragmenting as in the case of Shoemaker Levy, but still a space-time occupational re-adjusting. This one can witness by studying the evidence Hubble's photos brought back. As indicated the superiority of the proton rules, not only the atom, but also the universe. The volumetric size matter holds, is in precise ratio to the space value of the protons. Apparently all protons hold the same space-time value ("mass") with only the space that changes holding the protons. This factor indicates the density of the star and it is a far greater asset to space – time occupation than merely mass. In this aspect of the proton is the universal equilibrium that produces universal time as matter takes heat in unoccupied space-time directing it to densified space-time through occupied space-time and then finally to time. The progress in the proton is the demise to space. As space is in singularity, space cannot demise. If space demises the singularity within the proton, which controls the space-time occupation has to grow. When the space-time occupied grow, it will control the space-time unoccupied.

The simplicity in proving this is laughably stupid. Photons travel through unoccupied space-time, and if the amount of protons can influence the travelling light, the protons in that particular space during that particular time, also influence the unoccupied space-time. There is more heat around the Earth than in outer space. The protons therefore that controls and maintains the Earth's "gravity" also has to draw the accumulation of heat to the Earth.

Saturn and Uranus which is much further from the Sun , is immense hotter on the surface than in the case of the Earth. That fact has to be a sure indicator that the application of "gravity" has to have something to do with the attraction of heat. If heat will only flow from hotter regions to colder regions it indicates that the proton has to be a lot colder than even outer space. By moving particles through spin brings about cooling, therefore the proton has to spin much faster than the speed of light to be able to draw photons from the unoccupied geodesic space-time (outer space) to the proton.

If the proton draws heat it can only be to cool the proton and therefore the proton has to accumulate heat. Through this then the single proton grows in "mass" or densified space-time. This brings about that all matter becomes larger through the development of time.

The "mass" will deform, possibly brake up, as the space within Jupiter will revise the time. The space, which the wood occupies, will reduce to the extent that the structure may brake into a liquid and even a gas. Through this the "mass" will not reduce, it will increase as the heat component increases. As the heat component increases, the matter will grow faster than it would in outer space.

The formula science uses in determining time is $t = 1 - \sqrt{C^2 - V^2}$ in as much as the photon's speed (square $\sqrt{}$) minus the speed of light (C) square minus one representing time will produce time. This formula does not allow for any change in time. With this view, science is also in solidarity about the fact that everyone in science accepts that "gravity" influences time. This fact was tested in launching the most accurate chronometers man can devise and found positive results. Yet, not one formula complies in any way of this change to influence the universe.

I indicated the influence density has on the "gravity" by showing the relative difference planets holds. Presumably this influence of density will multiply by billions of times in one Black Hole, or as I wish to call it, a Proton Star. If "gravity" influences time duration to retard in a minute environment such as the Earth, how much more does time retard within a Proton Star? Time would literally to all human measures stand still. It will become eternity because that is what time in a still standing mode is to us humans.

A Proton Star is just the first star with the uttermost fragment in space (almost to the point of singularity) of the universe as it came out of eternity, equal to the "gravity" endured by matter back then, during the "Big Bang". Even if one use the Newtonian formula the measurements must be beyond calculation, bringing the time duration that applied during the "Big Bang" also to eternity. To us non-Newtonians this conclusion is obvious, but to the Newtonians it is far too simplistic. Not surprisingly, the logic behind the argument and facts are far too simple for our Super-Intelligent-Super-Educated-Wise-of-the-Wise. Being as super intelligent as they are, the cosmos has to test their own brilliance by introducing problems only those with their super intellect can see, understand and solve.

The matter of the fact is that when time slows down to a minute pace, it will seem everlasting to us. This fact is beyond any argument. Another fact is that heat does not bring about fusion, but it does bring about change in the application of the duration of time, affecting the space in which the time is.

This rather lengthy elaboration of repeating facts already explained in detail is to bring across how little science can piece together the most obvious and logic of facts, which they supposedly are the masters of. Life is fare more complex than anything in the universe and because we are part of live, we can only view life as life reflects the history of time. We humans are part of life, yet with all the research, no one ever came up with a definition about life.

Life is an energy with the ability to manipulate space-time occupied and unoccupied. To change the body, which holds life, is only part of the manipulation of space-time occupied to the benefit of live occupying that space-time.

Because of the atmospheric and surface heat they believe the water formed vapour and the vapour vaporized and disappeared into the vastness of outer space. It is this part of the theory that makes the theory completely unnatural and bogus. What the scientist wishes to imply, is that the Sun 's solar winds will be stronger than the "gravitational pull" of the planet. The minute the vapour becomes a solid, which water is when frozen, it will be heavier than air and it will fall back to the planet surface. Even when evaporating again before the water reaches the planet surface, it will evaporate, but again it will form water and ice, and this process will continue indefinitely.

As for comets with boiling water forming the tails as the Sun "heats the surface of the comet". I am not willing to waste any space or time in this book by dealing with such illogical nonsense. I do explain the misconception about comets and their tails rather extensively in "Matter's Time in Space".

The Earth has an abundance of water. The question arises: Where does it come from? The answer is in the closeness the moon has with the Earth. The moon is not a moon to the Earth, but much more a sister planet. When studying the effect of the Roche Lobe and interpret this to the relation the moon and the Earth once must have had, many unexplained questions find answers.

Examining all the facts about the dual planet system, it seems one is blessed with all the cosmos can offer, and the other one is dead and docile. The sister planets are in the most extreme of positions of all planets in the solar system. Science has developed the knack to apply circumstances they find today and interpret it as if it has been there since time began.

Let us reflect what happens in the Roche Lobe and apply this to the sister planet system. Even if the distance between the Earth and the moon does not fully comply with the necessary Roche distance today, one has to bring into the equilibrium the fact of solar development, which would be in the category of the Hubble Constant. There is differentially growth of the Titius Bode application to consider.

As every one knows, water form where one oxygen particle forms a compound with two hydrogen atoms. When any two structures go into a Roche Lobe they cut the circular motion (R^2/T) off from the geodesic space-time.

Through this action one find a secluded system, cutting off all influences from the outside.

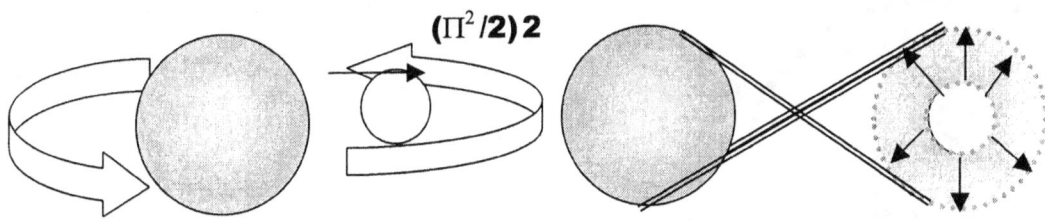

$(\Pi^2/2)\,2$

On every cosmic system "gravity" will always be Π^2, but the value of Π^2 will be different. As both systems share a common linear "gravity" (R/T) of $\Pi/2$ at point L, both structures will have the same atmosphere. With both structures forming the atmospheric value of $\Pi^2\,\Pi$, this will allow the perfect condition to form compounds. In view of the fact that the Earth will have a dominant Π^2 value, it will take up all the progress that the double structures can produce through the Roche Lobe.

The void to which the Bible refers, is the effect of the Roche development between the two systems, as an atmosphere formed in the Earth section, destroying any possible chance of atmospheric development on the moon.

However, another major factor of development is that the core of the Earth will benefit largely from the Roche system, as the Earth will be the major benefactor of the heat increase, deriving from the large spin the circulate motion of both structures increase. The Earth therefore has a double development in progress denying the moon its fair share in normal development.

On the surface of the Earth a great amount of water developed, cooling the Earth's atmosphere drastically. The cooling will accompany a huge vapour of water, as the water formed clouds. With the atmospheric temperature being this high, the clouds will evidently be extremely high, forming a massive and thick cloud layer. There will be little chance of rain, because the water will form back to vapour as the water rains down on the surface, never reaching the surface.

Who can, without the support of a fully developed technical language, describe the Roche-developing factor between the Earth and the moon, in a better way than did the Authentic Author of Genesis?

After separating the waters from the sky, came the third solar day. The Bible verse is as follows: "Let the waters under the heaven be gathered into one place. so that dry land can appear."

One has to remember that the Roche principle was still in full effect, much more than as it is presently in effect. During the Solar Day, it is only "non-conviction" and during the "Solar Night" it becomes conviction (if you will).

To understand this commissioning is quite simple. Pour a steaming, boiling hot cup of coffee to a specific point. Then let the cup of coffee cool to room temperature and you will find that the mass of the coffee reduced by a millimetre or slightly more. This is not due to loss of matter through vapour, but through the loss of heat. The mass of any heated substance swells. The very it apply when the Earth and the moon surface heats extensively. To all natural principles, we are still in the Roche-Lobe but as the source of the heat is external, and not internal as in the case of stars, the Roche Lobe comes into effect with a solar expansion.

In the next phase one can clearly see that the "day" the Bible refers to consists of many seasons, with many growth periods.

It will be of little use to remind the readers that man has a written memory of a few thousand years, where even with such little information, the biggest amount of the evidence is lost through time.

In the Roche-Lobe the following principle becomes a major factor:

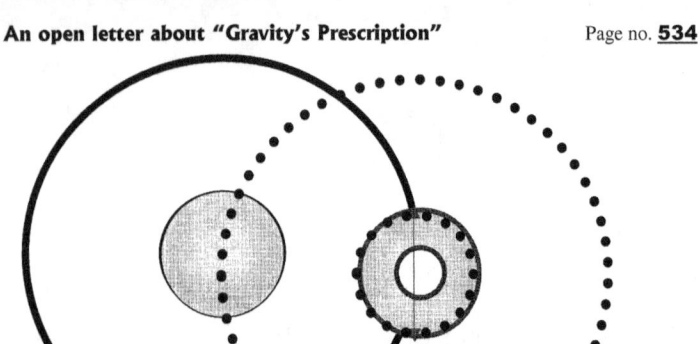

With the Sun blowing heat in thick clouds over the planets, the heat that the Earth and the moon detains through "gravity" is many times more than it is at present. Because the iron$_{56}$ core will reduce most space, densifying heat, it will also retain most heat. As it cannot accommodate all the heat retained, it will relocate heat through space forming, to the surface. In relevancy to the iron-core the silicon is space, therefore rejecting surplus heat will bring about introducing excess space amongst the silicon. This is the same that apply when baking bread and the bread "rises" in the oven. The silicon layer "rose" as the heat, coming from above, as much as from below baked the silicon to rise. From the top, the water still formed vapour, taking all the heat that the moon and the Earth holds as a compliment and then with the core of the Earth being the dominant, directs almost all heat to its core, because the Earth's iron core brings about the most space depletion ("gravity").

Through this process where the combined effort of the moon and Earth removes heat by space depletion a large area becomes effected on the space end … but, on the inside, at the time end, the iron core of the Earth is the almost sole recipient of heat, leaving the silicon of the Earth as the sole benefactor of space-incorporation (bread rising). Said in another way, the moon helped in doing all the work, but the profits of the work went entirely the Earth's way. That even includes the vapour where oxygen and hydrogen combined through the excessive heat to form water. The vapour from the charging of hydrogen and oxygen, discharged again as the heat moved by lightning to the Earth. From this water formed in abundance as the moon and the Earth both collected, both stirred oxygen and hydrogen into a mixture, but only the Earth collected the end product.

This process became as seasonal as winter and summer now are, as seasonal as rain spells and drought spells are or as ice age and heat spells are. Who would know the intervals, and the intervals are not important, because time back then is not time at present. The important issue is the evidence left in the Earth.

As the cumulative positive space-time displacement rises above the value of the other surrounding protons in surrounding atoms, the spin will exceed the average inner-Core-value of the other protons, thus "freezing" the nucleus of the atom in fusion in the time zone of the major element.

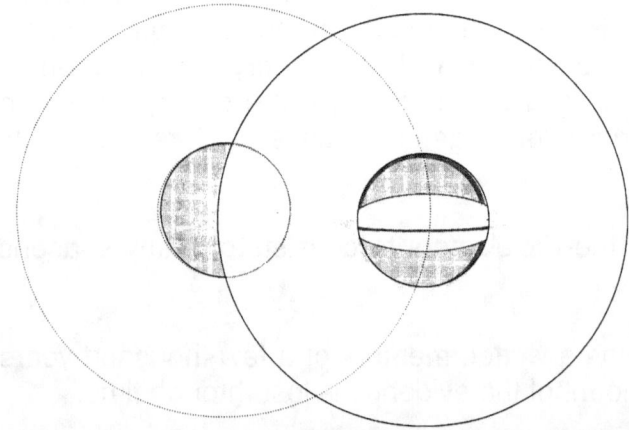

As the time duration extends further with the growth of the proton, that specific atom will develop a time duration much higher than the surrounding atoms. An accumulation of heat above and beyond the actual time duration develops, placing the space-time of that atom much higher than that of the actual time value within that layer. The "mass" increases and this causes the **aanplasing** to grow "pushing" the atom in the direction of the centre of the star where the time duration matches the value of the growing atom.

The ratio imbalance that occurs within the atom can be displayed as follows

$$10 \, (\Pi^2 + \Pi^2) \rightarrow 10 \, (\Pi^2 + \Pi) \rightarrow 10 \, (3)$$
$$10 \, (\Pi^2 + \Pi^2) \rightarrow 12 \, (\Pi^2 + \Pi) \rightarrow 10 \, (3) \text{ as the time value grows}$$
$$11 \, (\Pi^2 + \Pi^2) \rightarrow 11 \, (\Pi^2 + \Pi) \rightarrow 13 \, (3) \text{ as fusion occurs releasing heat and light.}$$

Because of the growing demand for heat caused by the increase in space-time density and therefore bringing about more use within the proton, the neutron, which is the balancing factor of the atom, has to readjust and apply more heat in order to keep the neutron form overheating. A point arrives where the accumulation effort of the neutron can no longer sustain an effort in supplying the accumulative call for heat by the proton.

In order to apply a balance in space-time density (mass) the neutron captures a proton that is in verplasing and this proton freezes into the density of the overheating atom. With the growth in the density of the proton, the neutron's accumulative cooling effort does no longer need to sustain such an effort, therefore rebalancing the heat application. The heat excess and rebalancing releases a great amount of heat built-up in the unoccupied sector. As this heat release comes about from the neutron reaction, it comes out as radiation and photons. In the lesser dense (top layers) area the release of heat overshadows the requirement of the proton; therefore a lot of light and heat discards back to geodesic space-time. However, in the inner star structure this readjustment will be at an equal balance and other atoms within the space confinement will apply the heat to suit their need. This produces much less of heat to the outer regions of the star.

Should a value of $7/10(\Pi/2)$ represent the hydrogen atom and $(2\Pi^2/4) \times (\Pi^2/2)$ that of carbon. By inflating the carbon atom's unoccupied space-time slightly one can see that it would accommodate a hydrogen atom.

In this, the Earth iron core grows at twice the ratio of the silicon layer. As the Earth grows, the Earth has to rise above the water at certain points. Therefore the Bible once again is correct by declaring that the water mass, which at first covered the complete surface of the Earth, separated from the water by rising above the surface of the Earth.

In the beginning of the part, I proved the ratio that applies to the atom ratio being $(\Pi^2 + \Pi^2) \times \Pi^2 \times \Pi \times 3$. This particular ratio not only applies to the atom determining the "mass" of the proton (M_p) in relation to the "mass" of the electron (M_e). This ration extends far wider than only the atom, as it is an indicator to the revaluation of time duration application and plays the major role in determining the "sound barrier".

The "sound barrier" is a sure indicator as to how heat relations affect the atoms. By intensifying the heat ratio between atoms, the time to space of the atom changes completely to a point where the time overshadows the transfer of sound. It is all a dependence to heat forming time or on the other side space. The less heat there is between the atoms in the unoccupied space, the less the time will be affecting the atom, and the reverse is also true.

It is not only the heat that **one finds between the atoms** that influences the proton – electron ratio and this is the major part of the huge misconception in the view "gravity" applies, because "gravity" not only relies on the "mass" of the protons, which makes up the number of the protons, and it is not mainly the density in which the number of protons are, but it is just as much dependent on the "speed" that the protons travel in relation to other protons.

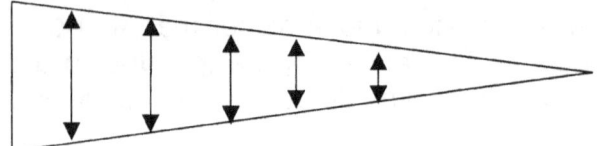

As the electron "travels" through space the time relation increases. The proof in this is that when a wind blows over an open fire, the heat in the fire increases. Oxygen as such, does and cannot burn. It is the special relation that oxygen holds with heat that increases the heat by speeding the flow of oxygen and nitrogen through the fire that increases the amount of heat.

By reducing the flow of air (oxygen and nitrogen) the fire smothers and the heat contained or regulated in the process. This does not change the transformation from wood to ashes it retards and controls the process.

While the one proton connects to space in singularity (Π^2 going to singularity) and connects space Π (in singularity) the other proton brings time Π^2 directly to space Π, re-uniting space-time as a unit to singularity (Π^3). We may call this re-unification unifying the "Graviton" in order to identify the one proton unifying time with space, while still in contact with the other proton holding (Π^2) time to (Π) space in as much as being occupied by matter (Π^2) and unoccupied heat Π forming 3.

This places Kepler's formula at a relevance s $\mathbf{a^3/T^2 = k}$ where $\mathbf{k^1}$ will hold a relation to heat with time in eternity, while AT THE SAME MOMENT infinity also applies, giving $\mathbf{k^0}$ the value of singularity. This is all to do with the Universe remaining in contact with singularity while keeping the Universe in matter and heat, in the dimensions of space and time. This explains then the absolute value of time as a square, with the square having both a circular value of R^2/T multiplied by the linear value of R/T producing the link to singularity in time and singularity in space. That is what is the significance about the formula $R^3/T^2 = 1$ where $R^2/T \times R/T = 1$ represents space-time in singularity as well as space-time in densified, occupied and unoccupied format. That means the everything of the whole lot, or as we say in Afrikaans, the "Heelal" meaning Universe. It is a replica of Kepler's formula $\mathbf{a^3 = T^2\,k}$ but it brings space in line with both aspects of time being $R^3 = T^2$ if $\mathbf{k} = 1$ where $\mathbf{T^2}$ then become R^2 / T and $\mathbf{k}$ becomes R / T. It put space in relation to both aspects of time because the moving of space is both in a circle and also in a straight line at the same time. The one cannot be without the other.

When saying this, I wish to include the following explanation for those that may have an interest in the technical aspect.

Energy is a term for a power or a force, an effort to get work done. Energy relies on movement. The influence one object holds relating its position in accordance to another position.

When an object remains in one position relating to the rest of the surrounding objects, the time it remains in that position is unchangeable. Therefore, for that duration it will remain eternal. To reflect this relation to a formula will be $R^3 = 0$ and $T^2 = \Omega$. Once the movement starts the position changes, therefore the space relation changes. The movement relates to time because any movement takes time, even with an uncontrolled explosion. Changing space always takes time.

Science agrees on one aspect of time; they have no idea what time is and when anything has the value of being unexplainable to science, it does not exist. Should one not believe my suggestion that science do not accept time as a part of the law of science then think of the money science spend on a ridiculous conception in as much as "time-travel". Even discussing such a conception is time wasting and science should discard it as time wasted. That topic is beneath my dignity as much as it is below my mentality.

Applying energy ALWAYS entertain a synopsis of space converting to heat through a period measured in time. Should one not believe me, think of a lamp, a heater, a stove, they all work on heat where heat is applied by some or other man made device converting heat from space surrounding a generator to heat travelling through some element and when oversupplying the element with heat in a controlled manner. It will readmit the heat to space in the space life wishes to apply the heat. By oversupplying the conductor with electric current, the conductor, resistor or whatever will burn. Anything can only burn with excessive heat applied at one place. When an electric device burns, it is just that. The heat we concentrate by spinning iron through a concentrated excited space filled with polarized (excessive spinning) heat, and by spinning excessive, that heat distinguishes itself from the rest by maintaining a higher spin than the rest of the heat and therefore create an individual time to space than the rest of the environment. This allows the iron, also spinning therefore applying an individual time in space, to place the heat in a separate time than the surrounding atmosphere. It will take heat directly to the Earth in a time- frame where the space is much less than the environment.

The flow of electricity is not a force, it is energy where heat receives a separate value of time, distinguishing it from the rest and as with all energy, it will bring about a reduction to space. In short, electricity is heat flowing. Electricity is space converting to heat and the proof is by investigating electrical human interactions. When human flesh makes contact with a high flow of current, the flesh shows sings of burning. In all events of applying energy, it comes down to the conversion of space to heat through time. It is in all cases, the heat (of the Sun) that supplies the heat (of the Earth). All considered; energy is all about interchanging and converting heat to space and space to heat and relating this action to time. When converting elements to stored heat, which fossil fuel is, is in fact transferring heat from the Sun to chemical bonding. This is transferring heat to space in a natural surrounding. Forming that compound to oil, coal and gas is storing the transferred energy in time.

Unleashing the energy for use to extend the influence life holds on our region of the universe, is again all about the transfer from heat back to space, and using the conversion to the "benefit" of mankind.

When dissecting the "Big-Bang-Theory" it is all about converting heat to space. It was rather exceptionally hot at the start in a surrounding which seem to u as considerably small. This however is only a perception we form from our perspective in retrospect. In truth, the universe was just as large back then as it is small today. Nothing grew and nothing shrunk, because nothing goes wasted.

Reflecting on what King Solomon said, three thousand years ago: "There is a time for everything." In this sentence, the primary word is time.
This picture applies. Whether our Newtonians want to accept it or wishes to understand it has very, very little to do with what reality is in science. This is the evidence and any child will see it.

There was time in singularity, as there is still time in singularity and everything will end in time in singularity. However singularity means just what it says, there is no movement of time, no movement in space, there is no space. Every aspect was frozen at zero to everything, what ever you may think of, it was frozen solid. Then came the Creator's command and everything responded immediately. That immediately is not our immediately. Ask any Theologian proclaiming knowledge about the Bible and he will tell you that the Creator is, according to The Creator Himself, FROM ETERNITY TO ETERNITY THROUGHOUT ALL ETERNITY.

In this lies the Universe because whatever is in the Universe is between that which has no end and that which has no start. The Universe comes about as it came about when that eternity with no end split from that eternity with no start. It is not religion but it is physics. If atheist are too feeble minded to understand physics it would be better fore those pea brains to go out and wheel a cart in the markets place. That which is eternal split having on one side the part that never can and on the other side the part that never can start where the tow parts are the very same part. By responding does not mean everything started running frantically out a blistering pace. It means eternity ended. That does not mean time as we see it at present started, but at a pace ten times faster. It is a pity that the Theologians never read the Bible, because it the Bible documents it all. If they read the Bible as they claim, why would they then insist upon an Earth being seven thousand years old? The Bible states that to our Creator a thousand years (another way people used to express a time back when the recording of time lead to misconceptions), has the same value as one nights work. THIS IS ALL ABOUT EXPLAINING THAT THE DURATION OF TIME IS PURELY A MAN MADE CONCEPT.

Time started in infinity or in eternity, whatever you wish to say, as long as you say time did not move at all. Then the command came and time overheated for the first Π^2 in time. That brought space into play.

When a bowl of soup is boiling, have you seen the bubbles of air rising from the soup? Has any Newtonian ever taken the time to explain that process in detail? I think not, because such explanations would be far too "everyday-like" to bother their mighty brains. Black Holes and finding the mass of the neutron, and such mighty brainpower cannot bother with small events.

Well, that boiling soup tells the complete story about the creation. Poets and painters and writers always wishes to say how "they created their creation". That is rubbish; they created nothing. They brought nothing new to the cosmos, they only rearranged what was a small part of the cosmos into a new order, that one can detect a distinction from. Creating is producing what never was before. When looking at the boiling soup, one sees bubbles rising from the soup at the top. In the soup's brew, there are only liquids and solids before the heat came. No one placed air in before the event or during the event and any time. Yet from the brew of liquid and solid rises gas, or if you wish space. That space was not there previously. That SPACE WAS CREATED. That space is energy en energy is the interaction between heat and space. As space becomes a part of the soup, a part not there before, with no room to be, it moves out. We refer to that process as boiling. That space creation is applying heat to time, and time in singularity will respond as space in singularity. The space created will vanish just as it came, back to singularity. By applying heat to time, brings forth space, and from the three components, only the heat factor is not in singularity. It removes space in singularity from time in singularity to establish room (space) for heat (time).

That is how creation started. Time in singularity overheated and the product of that was space.

Because time and space both, is part, of the same thing time became space and space became time. I prove that the repeating of this process happened about seven times, in seven different ways and explaining the other five ways is rather complicated and tedious. Therefore, I shall only give two explanations in this book. One is as I explained Π^3 (singularity time) parting with Π (singularity space) leaving "gravity" Π^2. What happens to space happens to time. Space holds three parts with six sides, of which only three sides directing towards any object at any time. Therefore 6/2 (half of the six sides are valid at very instant, only 3 has an effect.). Time in singularity holds a line directing to time. A straight line is 180°. Matter Π^2 holds Π^3 from Π, being valid in forming $\Pi \times \Pi = \Pi^2$.

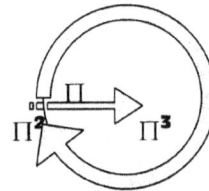

The Π^3 is matter in singularity
The Π^2 are motion or heat.
The Π is space in singularity.

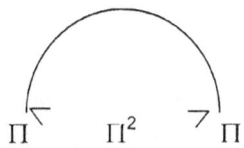

The half circle is 180° placing matter in a circle but because space only applies, to one half, 6/2 and matter holds space to value 6/2=3 only half the circle comes into effect. Half a circle is 180°. Because space has three parts in effect, it also becomes a triangle.

That means

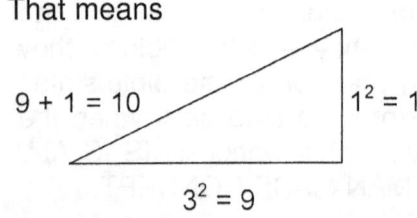

Where space holds three and time is one, the heat within that space becomes another dimension, the fourth dimension holding space-time $(3^2+1^2) = 10$. That changes the matter inside space in singularity at ten and "gravity" at Π^2 . This is why "gravity" Π^2 is space (10) losing one dimension

(Π^2). "Gravity" is all about space (occupying matter and heat) losing one dimension back on a long journey to singularity.

As time is in singularity, and space is in singularity and both are the same thing ($\Pi^3 \rightarrow \Pi^2 \rightarrow \Pi$) the 10 of matter (heat) that affects space (10 Π) will also affect time (Π^3) and therefore time carrying heat will become 10 (Π^3) with space 10 Π. Anyone with a simple calculator can divide 10 Π^3 by 10 Π and see where Π^2 fits in.

The reason why time in singularity is Π^3, lies in the two components of space-time occupation or "gravity" manifested in the Roche limit. All object spin and spinning is a circle Π^2 while all objects are moving in a direction $\Pi^2/2$. Again only, half of Π has any dimensional validity at any given time, therefore the dimension surrounding an object is Π. Because the "gravity" of the surface of the cosmic object extends from Π^2 to Π but only half of the circle of Π (180°) can apply to time (Π^2) being in a straight line $\Pi^2 \rightarrow \Pi$, "gravity" will form at that point of $(\Pi/2)^2$ giving the Π in space the "gravity" to hold.

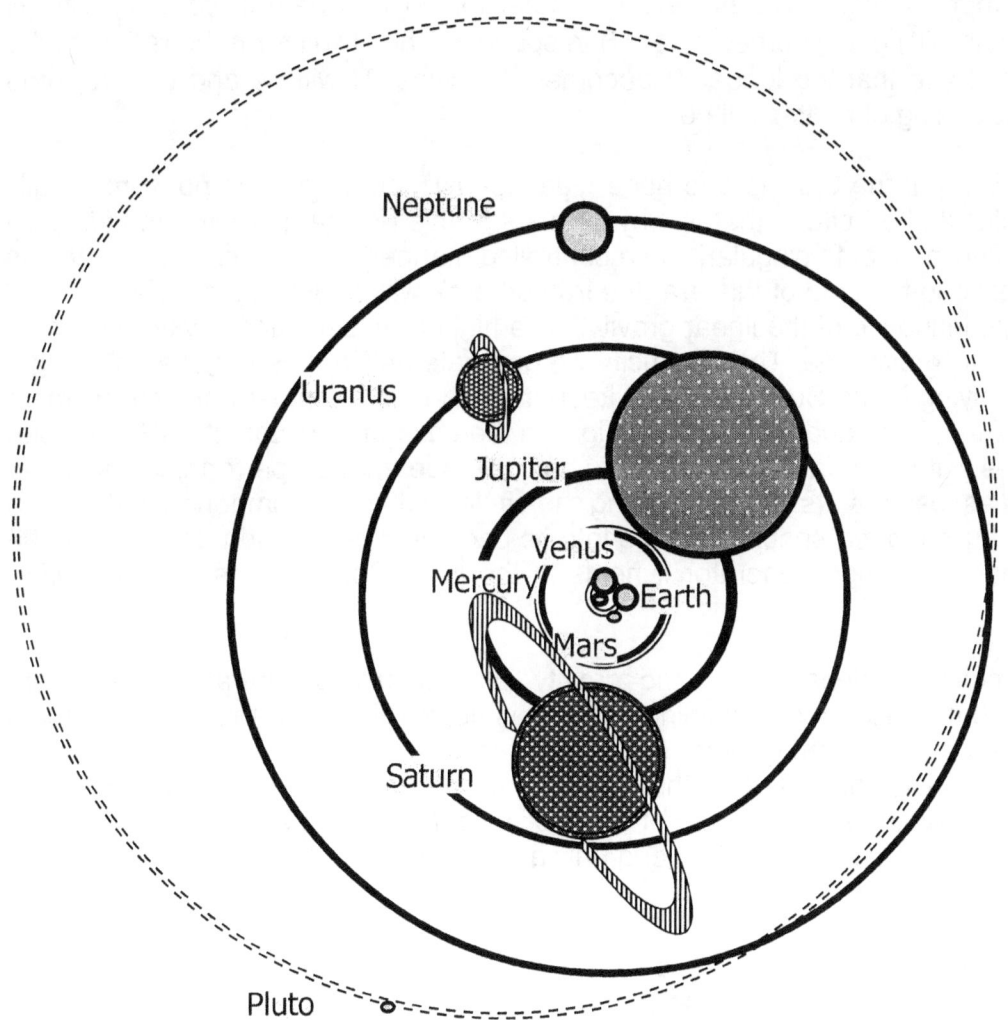

Gauging this tells a story of catastrophe and unplanned accidents, which is not the normal cosmos. This tells a picture of the prefect being interrupted by the imperfect. This tells of something extraordinary to the entire perfect that is the ordinary in the cosmos. This tells of the very opposite to the ordained and structurally sound we find in the cosmos.

That is why space will forever comply with $(7 / 10)\ \Pi^6 / 6\ = 112$, and time forming the line (180°) between the half circle (Π to $\Pi = \Pi^2$) at a 180° will form the triangle of space in half (180°). The matter component of the Titius Bode law effectively apply to the value of space, therefore 7/10 comes into the calculation. That places any atom with an existence in space at a premium of $7(\Pi^6)/ (6\text{X}10)$.

The reason why plutonium at $5(\Pi^2+\Pi^2)$ $(\Pi/2)^2(3/5)=244$ is at the element limit is obvious; when dissecting the relevancy in detail. It is in these lines that we must look for reasons why the solar system is so much extraordinary.

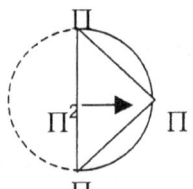

That will produce time in singularity a value of Π^3.

Explaining the other five stages of gravity (Π^2) development is extremely complicated and for that there is no room in a commercial book such as this. My motto in this book is "Keep it simple"

$$\Pi^0 \Rightarrow \Pi$$

With time in singularity, time was eternal. Time is the spin rate of heat in space. That means the way the movement changes where matter and heat relate to other matter and heat in space. With all movement relating to a circle (Π^2) going somewhere (Π) in space 3. The Π will form the radius to the circle (Π^2). Any novice can see that the longer Π becomes, the wider Π will be and therefore the longer change in the repositioning of matter will be.

It took some while to realize that the Sun is one huge gigantic, awesome (there is no word to fully convey the thought!) electrical short circuit that is why all stars in this era, (a star with an individual developed controlling the flow of heat to singularity in a controlled manner), without creating space as space converts to heat space in the time of this era (the iron 56 era), will establish a circular "gravity" that is able to withstand the influence of the linear gravity. The higher the circular "gravity" becomes, the more static will the linear "gravity" be. The electricity we generate on Earth is not even a thought comparing it to what is applying in the Sun . However keep in mind that, we must not over estimate the Sun 's ability to that of the Earth and think it to be big. In the case of a Proton star (Black Hole) the linear component lies with the fact that we can actually see matter performing its linear component by not curling as lesser stars do, but placing the circle and linear components all on the matter as the Proton star pulls matter, space and time into its pace-time occupation. In the Sun the Sun is concentrating heat by a huge generator it holds in the inner-Core-value and that is what "gravity" is.

Electricity is all about a process condensing heat and "gravity" is a process concentrating heat. There is an enormous difference but takes some explaining and that is not applicable to this book. A Proton Star is unmovable, static, and stationary and every other name you wish to connect to its immovability. It can no longer go anywhere. The stars within the sphere where doctor Hawkins identified a Black Hole to be, within the very centre of a galactic, holds all occupation relevance to the spin or linear component of space-time occupation and only a very minute part to the circular "gravity" component.

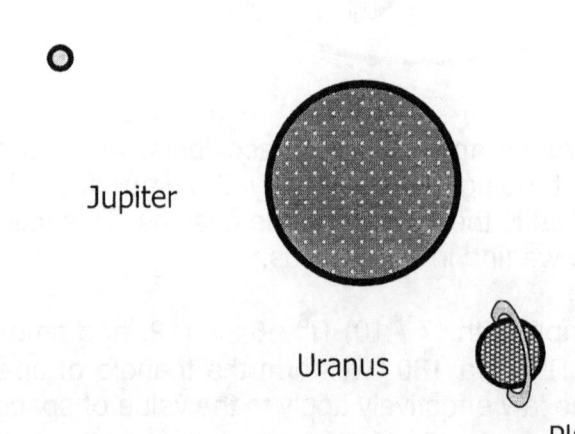

Mercury

Earth

Jupiter

Uranus

Pluto

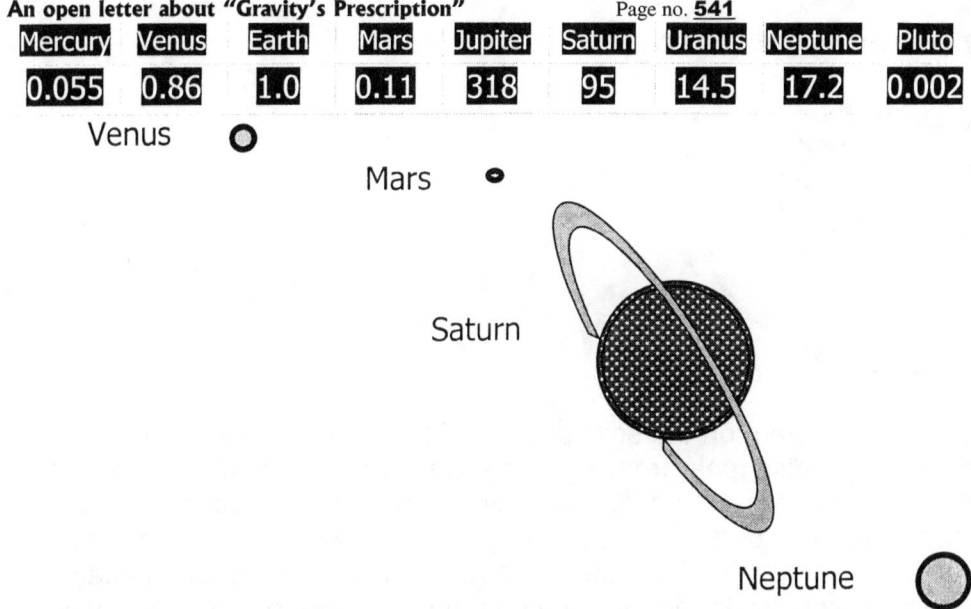

Mercury	Venus	Earth	Mars	Jupiter	Saturn	Uranus	Neptune	Pluto
0.055	0.86	1.0	0.11	318	95	14.5	17.2	0.002

It is so very clear what is missing from this picture the solar system presents the cosmos. It is conformity. It is a planned formation presented as all the laws governing the cosmos was adhered by the structures within the system something is missing. A planned layout is not present.

When investigating any structure in the cosmos we find the less developed objects allocated towards the centre and the structures that are more progressive in time and have developed better are object located on the rim or edge of the galactica. In this case there is no order. There is no recipe and the most precise recipe that has ever been used for any constructs the cosmos. This is what is wrong with the picture we see in the solar system. The development that took place is no correlated with normal growth the big Bang provides.

That is the big clue missing from the picture we may find in the development of the solar system in relation to what should be the case with the normal development provided by the Big Bang. Then in releasing this shortfall we have to turn out attention to one of the four cosmic pillars to give us the indication of what happened and how it happened when it happened.

(1) A binary formed from the pre Big Bang frozen blanket.

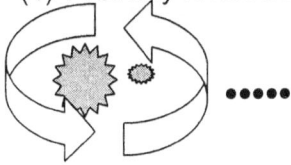

(1) Outside the spinning dual were 5 dots minding their own frozen space and concerning their own singularity, which the binary confined as space under influence.

(2)

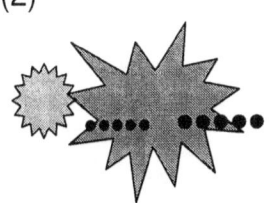

(3) Then the one factor of the binary dismantled by allowing the singularity to brake down in heat and fragments holding singularity.

FIVE FRAGMENTS CAME OUT OF THE BLANKET CLOUD OF HEAT.

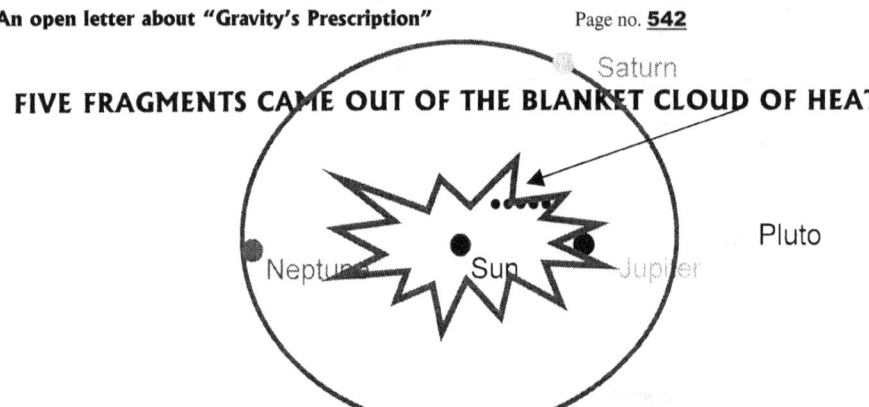

This debris can be witnessed as meteor craters on the solid planets. The bigger structures were the first ones to become cosmic missiles and disappear from sight, because there were less of them to go around in the first place. Afterwards only smaller particles remained and they too became ill fated. However, none of these fragments became part of the planetary system. As the inner structure changed its orbital course in both the Binary minor and the Unknown structure, its gasses became reduced in size in other words compressed, while the inner heat in the gas rose to extreme values, creating an enormous wealth in magnetic space time (how and why this happened, will be explained). However, the inner core could not maintain the reduced value of the stars, as they were firstly far too small, and secondly the core was in process of becoming fragmented.

Therefore, it did what all unsuccessful stars do. The magnetic space-time at first were deprived of its negative space-time displacement because the increase in the stars overall growth in negative space-time acceleration. So all the matter was compressed up to a point where the acceleration of the stars became a value of evenly linear motion. As the momentum did not accelerated any longer, the compressed matter began expanding again. Then the magnetic space-times own negative displacement took over and the gasses of which the two stars comprised of, became known as Oords Cloud by humans after a duration of $4{,}5 \times 10^9$ years.

Then soon afterwards another two tragedies followed and only because of these two tragedies, life became possible on Earth.

THE HEAT OF BINARY MISSING CONFINED FIVE FRAGMENTS WHILE FROZEN SPACE-TIME CONFINED FIVE UNDEVELOPED MICRO STARS WHILE THE SUN AS THE PRIZE WINNER CONFINED THE LOT

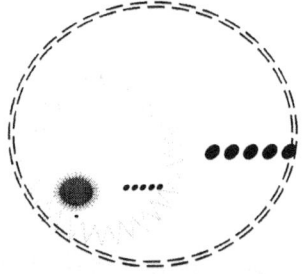

Let us establish a line of evidence and fill the puzzle.

1.   There was another star with the Sun where the Sun was Binary Major and
       Star Unknown was Binary Minor.

2.   The Binary system catapulted the solar system out of its frozen eternity,   way ahead of its
       time of development, bringing along the rest of the micro stars. The outer "planets" are not
       planets, they are micro stars in development.

3.   The Binary system formed part of a Lagrangian system holding the Binary in the centre and
       with Jupiter as the first orbiting satellite.

4.   The position Jupiter held for most of the developing period made Jupiter the second main
       benefactor of the dual, with the Sun the major winner and Binary Minor the major loser. This

will also explain why Jupiter has such an advance in space-time occupation when compared to the other micro stars.

5.   The position where the (six) inner planets find themselves, were a void, AS THE BIBLE CLAIMS.

6.   The relative positions were as follows:

(1 + 1) = 1	2	3	4	5
Binary	Jupiter	Saturn	Uranus	Neptune

7.   Then Unknown Star capitulated as it could no longer serve the dual it fought. It fragmented into possibly 9 major parts and many minor parts, (the comets.)

8.   As the core fragmented the brittle parts dislodged in a position each to a relative neutron position in the space-time binary minor held relating to its point of $(\Pi^2+\Pi^2)$ $(\Pi^2\Pi)$ 10Π.

Whichever way we Earthlings will look at our position, from whatever angle and by whichever calculation we devise, our relevancy will be 1, will be 10, will be Π^2. Should any person ever do a calculation and find his answer does not bring this fact to bear, he must go back and fix his mistake. There will be a mistake on his part.

SUN	MERCURY	VENUS	EARTH	MARS	
B_1	B_2	$\Pi^2/2\Pi$	Π^2	10Π	Then the rest.

Π^2 + Π^2 $\longrightarrow$ Π (This I shall explain)

Our Super-Educated hope to establish Martian Colonists. They cannot even begin to comprehend their ordeal. Columbus only had to fight some sea dragons, a Wall of fire on the water and an ocean-sized waterfall on his way to India. Even if all that myth were true, and he encountered the hole lot in one day, his problems would seem minute when compared to what awaits our Martians. They have no clue of their folly. Mars is not another island in an ocean of water waiting anxiously to be discovered by a wind powered sea-faring vessel. Mars is an island, a piece of rock in the middle of a gas of heat, with total alien circumstances to life. Scientists can fool the public, they can fool themselves, they can fool with figures corrupting cosmic laws, but they cannot fool the facts. There is a reason why the Bible excludes the other solar structures and concentrate on the Earth. This is the same reason why any way we do calculating the solar system; the Earth will come to one according to human standards. Life from our perspective, from our perception, totally connects to the Earth. That connection is so deeply routed in man and other life it is part of life. Newtonians may spread the rumour about life being ten a penny scattered to the social structure that man established on Earth through every social alliance or group, be it politics, law, medicine, science, theology or whatever denomination they wish to control, they will not create facts.

I admit, man derives all the above-mentioned lies mainly from corrupting the Bible. The Roman Catholic Church started the trend 1 500 years ago and still maintains it as best it can. You can prove whatever you wish to prove from facts you take from the Bible. The Bible is in support of whichever standing you may support. This is not because the Bible is incorrect, it comes from our insignificance to appreciate the full content of the whole Bible. The Bible promote only truth, man takes from that what man wants, divert the truth to suit his need by corrupting the lot.

I know presenting evidence as well as I do, will not change the course of science. I know too, that I am too small to pinpoint conclusively whether Authentic Author referred to the creation of the first cosmic period, or the first solar period. This is not because of inaccuracy on the part of the Bible.

This inaccuracy comes from the human interpretation of Authentic Author on his vision, and then my insecurity to interpret his interpretations. It is human error bringing on misinformation. The Bible's recollection CAN and DOES apply to either period. Therefore, to respond in containing human misjudgement on my part, I shall re-apply the vision of Authentic Author in the context of the first solar day. I showed how it fits to the first cosmic day already.

Binary stars, spinning to self-destruction will produce significant heat. Heat create space, space forms winds. The Bible present facts that is indisputable. Where the Earth was, was still a void, containing a sphere of circular displacement and this will reduce linear displacement to zero. Linear displacement is space and circular displacement is containing heat for matter survival.

Binary Star Minor overheated. That is why the core brittle and fragmented. This action will release tremendous contained heat; the heat will produce magma flowing in space like water in space and this eruption of heat space that created winds. Once again the recollection fits the scenario. Releasing the heat and producing space will establish space-time and fill the void where the Earth should fit. This is fact and if anybody even tries to dismiss this will be because of abstinence on his or her part. I did not prove the Bible correct. The Bible told the truth and in such correct detail, it is beyond human comprehension, but sublimation on the part of Newtonians and science before them, disallowed their ability seeing it.

Let us envisage what factors applied to the third planet binary that set it apart from the other planets (not micro stars). First, the eruption occurred, and with the eruption came the release of massive quantities of hydrogen, oxygen, carbon and nitrogen. I respect to the quantities the Sun stockpiled for own "personal use it was not much. However, to what the other structures in micro planets had, it was exceeding their quantities by hundreds of times.

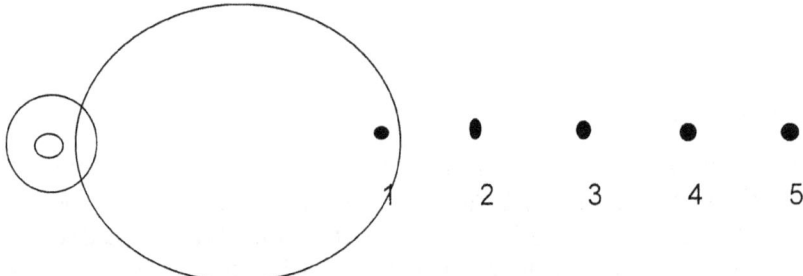

Jupiter (1) still enjoyed much of the vapour for the longest duration, and was a benefactor before the fragmenting as much as it was a benefactor after the event. As the clouds drifted from the position of point zero, the other micro stars also became receivers, in order of position. The mass they hold today, is still evidence of their position placing them as recipient 1 – 4.

1.　　Jupiter　　　318 x
2.　　Saturn　　　95 x
3.　　Uranus　　　14.5 x
4.　　Neptune　　　17 x

However, the Neptune micro star in all evidence has very particular characteristics that I cannot explain. I have a theory about Neptune, it is only a theory that I cannot substantiate with mathematical proof but nevertheless there is circumstantial evidence I shall present in order to prove that my theory is not merely wild guessing on my part. I shall return to this matter shortly.

The indication clearly points to the fact that the micro stars were in this cloud cover of star gasses, for a long period, and the period were substantially more as the micro star holds a position in relation to close proximity to the destructed binary. All the orbiting structures the micro stars hold, except (I suspect) Neptune, are fragments of Planet 5, the destructed one.

One must realize that Jupiter holding an inner core sizable enough to produce 2,34 times the "gravity" of the Earth were a result of benefiting from the Binary dual between the Sun and Unknown Star.

Compare that figure to the rest, and it points to the fact that something major set Jupiter's development on a course of progress, where factors benefited Jupiter by far.

1.   Jupiter   2,34
2.   Saturn   0,93
3.   Uranus   0,79
4.   Neptune   1,2

Again we can see that Neptune had some other benefits to its progress, because, again Neptune diverts from the obvious sequence.

Once more, this phenomenon should not occur with Newton's presumptions about gravity. These bodies will collide and destruct, without a doubt. When the formula $F = \dfrac{M_1 M_2}{r^2} G$ apply, there should not be any force which is able to keep them apart. However, they do exist and what is more, they maintain a certain distance apart. Seen from this view, it is little wonder that the significance of this was lost in the notion that this is yet another "mystery" of the Universe. The scientists of the day (and the past) lost the importance, which this holds for us as Earthly dwellers.

As explained, there is no gravity, instead a balance exist in space-time, where a value of linear displacement relates to a value of circular displacement. $\dfrac{\frac{R}{T}}{\frac{R^2}{T}}$ (R/T // R^2/T). Regarding this, the Roche lobe comes about because the Roche lobe forms a borderline between these two related values. Space-time lying within the Roche lobe stands at a value where the linear displacement is at a greater value than the circular displacement. In mathematical terms it expresses as follows in an equation:

$$\$ = R/T > R^2/T$$

Matter located on the border of the Roche lobe, will represent mathematically in the following equation:

$$\$ = R/T // R^2/T.$$

This is the position, which satellites have to comply with, in order to remain in orbit. Matter located on the outside of this border, which the Roche lobe holds, will mathematically represent as follows:

$$\$ = R/T < R^2/T$$

You, the reader, might react with surprise, because all structures with influences in this way, are far beyond having any influence on our life. This represents a great misunderstanding, as it has everything to do with life developing on Earth. In systems of unmatchable and unequal space-time values, the larger body will tend to dominate the smaller body in as far as high jacking the smaller bodies' space-time values are concerned. Please take note throughout all of this discussion, there is NO FORCE applied on any of these bodies, but only a balance, which maintains or goes array due to certain reasons. Therefore, NO STAR STRUCTURES (OR PLANETS) can ever collide, and a meteor colliding with a planet, is not a collision, but an imbalance of space-time manifestation.

To understand the meaning of this statement, I shall firstly explain why no star system can collide. When two objects i.e. double stars, come into a conflicting position as to sharing space-time, the two systems has a response in space-time values.

The academic world has treated me very poorly, because of my view on what they regard as Holy Scripture. I came across some brilliant scientists whom were able to form a conclusive opinion of my work after just reading the first four pages of a book containing almost two thousand pages of explanations and facts. By just reading two pages those highly informed professors decided that I am completely misinformed. The claim they made is that I am not familiar with Newton, and therefore do not "understand" Newton. They never even allow themselves the time to get to the part where I explain why I do not recognize Newton, let alone begin to introduce my opinion. Such intelligence, I must admit, is a true indication of just how their acquired brilliance can allow their decisions made in split seconds about a book that would take them at least one month to read extensively.

I know that if I went the conventional route by enlisting at a university and following such a course, I would have had to accept the institution's views on science or they would boot me out. I did not escape the booting process, as many academics discarded my work. I wish to put this to all of the religious scientists: Even if you will never admit it, you believe in Newton more than you believe in the teachings of the Bible, no matter how much you wish to deny it. The moment the Super-Educated faces up to any criticism about the work of Newton, they do not have the courage to even investigate the criticism. If one says that that person disagrees with Christ, everyone in hearing range is prepared to listen to such a view, although it must be considered by all believers as profanity against the almighty and this charge I direct directly to those so called believers that has Bible versus on their answering machines. Tell the physics department on any campus you disagree with Newton and they honestly treat you as a mad raving lunatic keeping busy with blasphemy!

A straight line is 180°

Half a circle is 180°   THE HALF CIRCLE IS 180°   A triangle is 180°

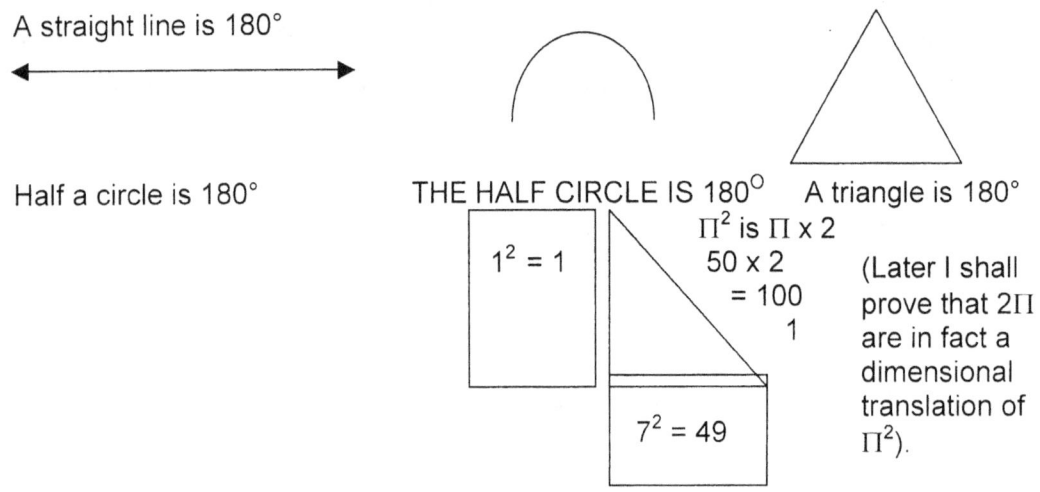

Π^2 is Π x 2

$1^2 = 1$ 50 x 2   (Later I shall
                   = 100   prove that 2Π
                   1   are in fact a
                                  dimensional
$7^2 = 49$ translation of
                                  Π^2).

Π^2 because it holds a higher dimension is 49 + 1 (Pythagoras) = 50
and 50 x 2 ($\Pi^2 = 2\Pi$) = 100   (The two is a dimensional implication
$\sqrt{100} = 10$ carried by the value of 50)

Therefore space to time is 10 and matter to time is 7. Time to time is 1.
I base my facts that the moon never was or could be part of the Earth on the Roche limit that would never allow it. What is apparent though is that the Earth robbed the moon of all vital atmosphere and water.

This much I shall say: If not for the moon's position and size, life would not found such an acceptable environment in which to evolve.

The Titus-Bode Principle of heat growth and time growth works on a simple basis, the same basis as the speed of light. It is all a question of dimensions that space holds, in conjunction with time, running parallel, that exist in accepting more than understanding. Trigonometry is the sole proof of such dimensions.

I have shown Π^2 to be 49 from the Earth's perspective which means Π is 7. But Π should be 3 from the law of Titius Bode. And Π^2 then have to be 6. By placing Π at 7 and Π^2 at49 it means the distance positioning came about from space in the fourth dimension $7^2 = 49$ and not time through proton development in the third dimension. $\Pi = 3$ then $\Pi^2 = 6$; $\Pi = 6$ then $\Pi^2 = 12$. The square that applies to seven is an indication of definite heat creating space and not time through proton growth alone. But one can still detect the proton growth of 4, in the accumulation of seven.

There is of course a much simpler way about to go in explaining the Titius Bode 10/7 and 7 / 10, and I do give tit in another part of the book, but knowing the purist of academics, such simplicity would go by unnoticed.

In a previous part I established the planetary relation according to space-time application of R^3 (space) and T^2 (time) holding equal value. In the following pages I wish to apply the same standards, only relating the standards to cosmic principles I already explained.

A star form time at a value of 7/10 (4 $(\Pi^2 + \Pi^2)$ = 55 and the core has to be less than 56 to allow light (29,6 + 27) than 56 to allow light (29,6 + 27) ore holding a higher value than 56,6, the light will no longer escape from the inner core and as the core grows, the star will start absorbing its light (photon) production until such time it turns into a full blown "black hole". At a value of 3$(\Pi^2 + \Pi^2)$ the core will dissolve space to the effect that the neutron no longer has space in which to be. The neutron then will abolish the star altogether.

When a star is in a Roche limit, the value is $(\Pi/2)^2$ $(\Pi^2 + \Pi^2)$ = 48, but since it is two stars it becomes 2 x 48,7 = 97. That places the space-time development of the binary in a position held by stars two era's previously. The space-time enhancement is about double to normal growth. This fact also carries a high degree of significance when I explain how the solar system came to be in the layout in which it presently presents itself.

This places the star in a space-time occupation that was valid before the "Big Bang" surpassing the cosmic value of 112. It puts the stars in the Roche-lobe in a state where the stars in the binary combines to a cosmos excluding the outside to favour the inside where the atmosphere is excluding the cosmos and does not have space separating the stars, but merely the atoms. The time relation of the atoms hold a combining value similar than that, which is 2$(\Pi^2 + \Pi^2)$ $(\Pi/2)^2$ = 98. The atoms will not destruct, but all space between the atoms (Π as well as 3) will diminish as the atom alone falls outside singularity, but all other space has reduced to singularity. Fusion will not occur since the presence of heat maintaining the proton / neutron will still generate through matter spin of the two objects as one, bringing about a linear locking, applying a circular contact with space-time.

With the two structures forming a cosmic dual of survival each will apply $4\Pi^2$ $(\Pi/2)^2$ where the factor of $(\Pi/2)^2$ will hold the relevancy the one has to the other. When the two structures have equal prominence the relations will be as follows. In spoken language each will regard the other as a neutron attachment of half Π and half Π^2, therefore $(\Pi/2)^2$.

$4\Pi^2$ $(\Pi/2)^2$ $(\Pi/2)^2$ 4Π

$4\Pi^2$ (Π^2) and $4\Pi^2\Pi^2$

As both hold and equal value to the neutron or second Π^2 the two will apply a joint value of $(\Pi^2+\Pi^2)$ $(\Pi/2)^2 + (\Pi^2+\Pi^2)$ $(\Pi/2)^2$.

This value places the space-time within the lobe to a time duration where heat was liquid, just after the forming of the neutron. In such a star, not only will the heat and space of the electron have liquid space, but also the neutrons will find itself in dense space. The combination places the space-time value in a position where it will enhance the neutrons in the space-time between the two. When the two structures, almost equal all neutron space, it will naturally become a combining or combined proton star or double Black Hole.

In the event of the two core structures being so equal and similar, it will start growth the space-time the protons hold, placing the double proton in an unnatural era. At this point I wish to indicate that the factor values indicate the space in time separating the objects or if you wish the numbers only apply to the heat value of the space keeping the structures apart. The matter part will remain apart from the unoccupied space-time gone into singularity. Saying this I have to add that with the event of the matter finding itself in space-less time, predating the "Big Bang" the matter eventually will also have to dissolve to time as the matter starts to fall out of space and going on to singularity. The matter, however, will follow another route. All these facts I introduce to prove my point that the Sun is not even a star but through the Grace of God it is there. If the Sun was a true natural star, all "planets" will incinerate with no chance to support life.

As indicated above, the Roche-lobe, from the matter's vantage point (not the space separating the two objects), the one object relates to the other object being a neutron to the proton.

$$4\Pi^2 (\Pi/2)^2 + 4\Pi^2 (\Pi/2)^2$$

Therefore $(4 (\Pi^2 + \Pi^2) \Pi^2/4) /2$
$= ((\Pi^2 + \Pi^2) (\Pi^2)) /2 = 97$

To enable any person to see how significant and out of era this is, compare this value to that which the last surviving element, Plutonium holds at 94. The true value of plutonium carrying an overload of neutrons to stabilize the element is then $5(\Pi^2+\Pi^2) (\Pi/2)^2 (3/5) = 244$.

Analysing this is as follows:

5	End of space-time
$(\Pi/2)^2$	Limit on demolishing space-time
(3/5)	Heat stretched to its limit. This whole ration spells one nuclear disaster waiting to occur. The space-time environment within the binary, even exceed this Plutonium.

Therefore half of the combination regards the other half as $\Pi^2\Pi$ and $\Pi^2\Pi$ is a neutron position holding the three dimensional Universe to the same value in space as that which separates matter from matter being 3. Since the 3 stands completely excluded from the atomic combination of $(\Pi^2 + \Pi^2)$ holding the second object as

The stars maintain their individual circular displacement values, as they move closer, driven by their linear space-time displacement. At a point where space-time becomes a unit, star A's linear displacement has to overcome the Roche lobe boundary of star B. This applies to star B in the same way. At the limit of the Roche lobe border, star A and star B will establish a mutual circular displacement value, limiting the other star's linear displacement. Then a situation develops, where the mutual circular displacement will replace the individual circular displacement values, which both stars had. At this point, the stars would find it impossible to move closer, because of the Roche lobe limit. As both stars share equally in the circular displacement, they have to share an equal linear displacement. This would leave them unable to break the linear displacement equality, therefore they maintain at a distance apart, spinning around a mutual axis somewhere in the middle of the distance keeping them apart. Mathematically the equation can express as follows:

In the event where the one star's space-time value would overshadow the second object's space-time value, therefore the value of $\$_c$ would locate within the boundary of the larger star's Roche lobe border.   In such a case the Roche lobe, would in effect not apply and the larger system will incorporate the smaller system within the larger system's space-time.

The effect that this would leave on the two systems which are able to maintain a mutual linear displacement value, is that they would either share in a common space-time growth or the one system will destroy the other system at a certain point in space in time.

$\Pi^2\Pi$. This can only apply when the one object occupying less space-time has a proton value $(\Pi^2 + \Pi^2)$ that is less than the superior object's position on Π^2 This means the one is totally dominating the other in all aspects. Some quarters of the Newtonian High Priest in High ranking made claims that the moon once formed part of the Earth. In the following elaboration I shall prove why I dismiss this claim as utter nonsense. From these facts about a binary, one can then clearly see that having two structures in a position overshooting the Binary scenario, is very much fantasy. It is just not possible because the valid space-time will exceed 112, and the structure will not have the ability to hold position in the Universe that is limited to 112.

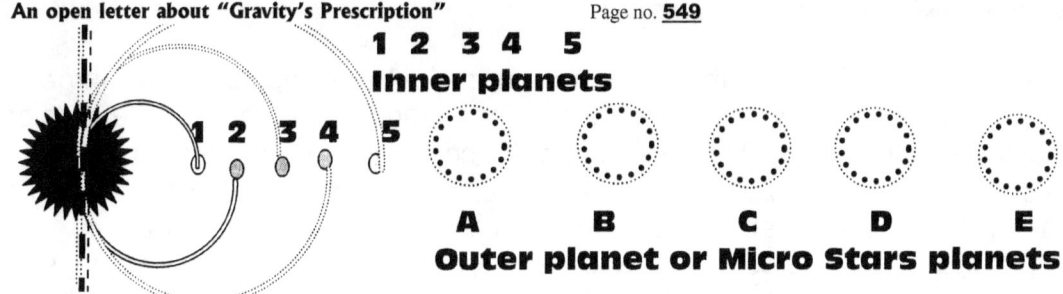

In accordance with the Lagrangian rule there cannot be more than five objects pairing onto a centre object. However in the case of the solar system we have a double except that in the one pairing there is one devastated and in the pairing of the larger group we also find the most outer structure demolished or at least in ruin and in association with cosmic debris. With this evidence there now is a manner in which to determine how the rules were broken.

From the fragments we find that is scattered all over the solar system and the irregularities we find as I indicated up to this point we can see there is a lot of history in the story, which will unfold as we investigate. There was tussle in core battles between the Earth and the Moon, where each falls into a sequence arrangement of a proton value of the Earth is $(\Pi^2 + \Pi^2)$ and it will hold the second object (the moon) at $\Pi^2/2$. This is because the second object is in the "gravity" application of the larger object (the Earth) and the "gravity" factor of the Earth takes on a linear value, half that of the gravity factor of the Earth. The Earth will not allow any linear action to exceed 10Π and at $(\Pi^2 + \Pi^2)$ $(\Pi^2/2)$ it exceeds that value. However this can only be a result of another tussle of much bigger importance and ferocity. The debris can only formed from many disarrangements and cosmic laws being broken as a chain reaction that sprung from one specific event. From the evidence that the debris leaves it is a fact that there were many numbers more rocky structures that was left over from the Unknown star demise, but as the law would not allow more than five, it was five that was formed with one being crushed as an aligning planet. But there were more than those forming the line, but I have not the time, nor the space in this letter to go into speculating about the crushed ones since the publisher limits this letter and therefore because of the length limit the publisher placed on the book I cannot delve deeper. The moon for example too must have been one of the after thoughts of the disaster and in _Seven Days Of Creation_ I speculate about this with some convincing proof as how that came about. I place a great importance on the fact that the Moon and the Earth performed as a Roche partnership and this much more than any other reason brought the development or the possibility of life developing to the Earth and not one of the other solar structures. That too I cannot share in this book because I am limited about the length of the book and it has to be accompanied by a vast quantity of proving information. Without the evidence I know I shall meat with sure rejection from our Newtonian camp and that is as sure as the Sun Shines during the day.

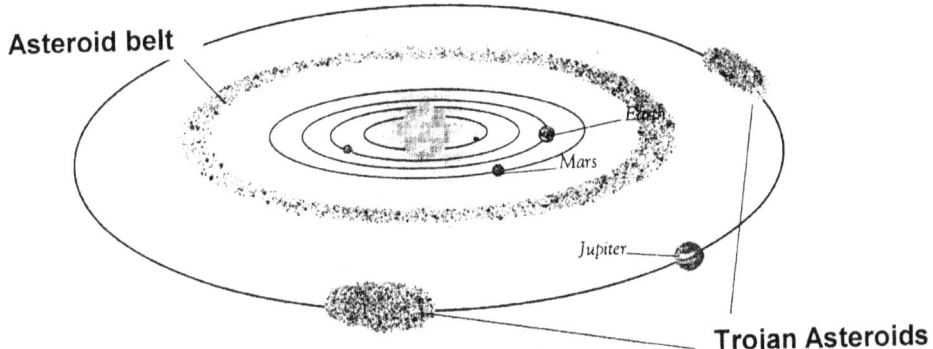

There is a great deal of pieces of hard rock floating all over the show. Some are nicely grouped in units, which one can clearly see was designated places and solid units. The only way the rocks could have brittle as they did was becoming excessively overheated. The only way it could have overheated in the manner it did was by bridging the Roche limit divide with much larger object, and there is a much larger object in the region and sharing an orbit with the immense structure. This was bullying at its best and we can see from this how cosmic plunder, thieving, brutality and theft came about as the larger murdered the little.

When the two structures go into a duel. As the two core has a dual the fragmenting of larger to smaller is a method of distributing heat by expanding. It common language we call it an explosion, space-time occupational value of $(\Pi^2 + \Pi^2)$ $(\Pi^2/2) = 97$, and the core value of the Earth is at 7/10 (4 $(\Pi^2 + \Pi^2)$ the combined value will even exceed the critical space factor of 3 $(\Pi^2 + \Pi^2)$ applying to stars holding space, therefore the space separating the two objects will vanish into singularity. The reason why the Roche principle maintains core separation is that the core combinational value, seen from one or the other objective, is $(\Pi^2 + \Pi^2)$ $(\Pi/2)^2 = 48$. The individual space-time factor of each core is 7/10 (4 $(\Pi^2 + \Pi^2) = 55$, therefore the space holds less heat and therefore more space.

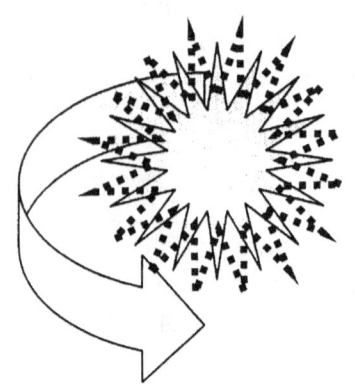  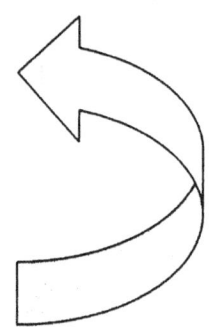

Lets reconsider what was apparent.

We know from the fact that there are four peculiar hard surfaced objects formed within the inner part of the Sun that there was some abnormality in the solar system development. The inner objects are named planets although in the entire Universe there are no call for such object releasing from the Milky Way this early and therefore the fact of their presence relate to some irregularity that came about in the past. In order to have the four smallest planets blessed with a solid surface while the other "giant planets" all are a mushy gas structure must be because the five inner planets was part of a star much more equal to the Sun that any we now have. In order to be of significance and not be totally destroyed in the Roche battle the Sun I suppose was the size of what Jupiter at present is, the second binary was then about the size the Earth is today.

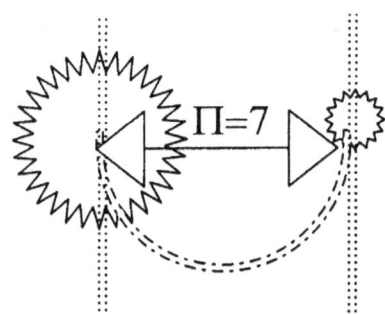

I have explained that matter is seven in relation to the Titius Bode law and that pts space in a position of 10. If 7 is Π, then Π^2 is where the Earth is because I am in the Earth and therefore I am the centre of the Universe bringing the Earth position also as the centre of the Universe. All motion spins around me making me Π to all of Π^2. If $\Pi = 7$ then $\Pi^2 = 7^2 = 49$. Therefore my position in relation to the governing singularity is 49. I am holding space in Π^6, which then is $(10)^6$, which is then 10^6 kilometres away from the Sun and with material being 49 the factor in relevancy is 49 $\times 10^6$ kilometres away from the Sun .

Since I am on the third planet from the Sun the relevancy my planet have with the Sun is in line with the factor relevancy, which I have with the Sun . That puts me with my Earth in relation to 49 $\times 10^6$ $\times$ 3 (since I am on the third planet from the Sun and if the development of our solar system was normal) the allocated position of the Earth must then be 3(49 $\times 10^6$) = 147 000 000 kilometres

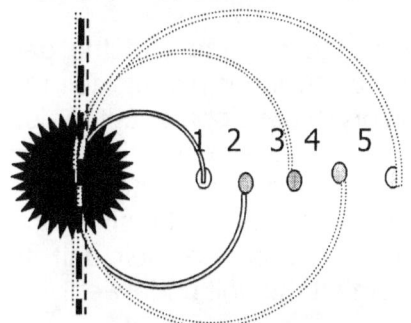

By the way that is the reason why Mercury follows such a strange pattern while circling around the Sun Mercury presents 7/10 = 10 /7 so in the case of Mercury aligning with the Sun nothing makes sense for poor old Mercury. More to the point there is evidence of unusual space –time development at the place Mercury now occupies.

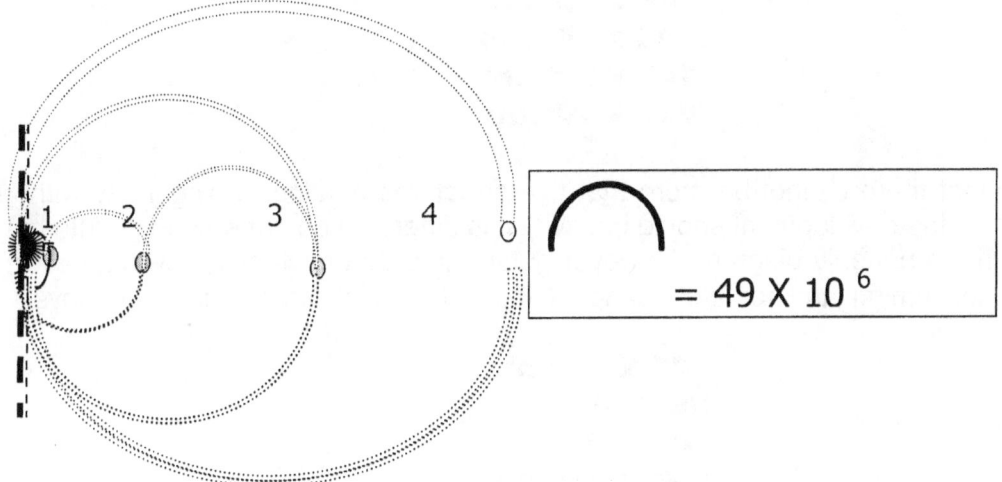

$$= 49 \times 10^{6}$$

There is a clear defined value, which serves as **k** in every event of every planet in relation to the spin as well as the development in accordance to the Titius Bode law. B y the square of seven representing material related to the double cube of time through which material must rotate while aligning with the Sun the distance on average is one million.

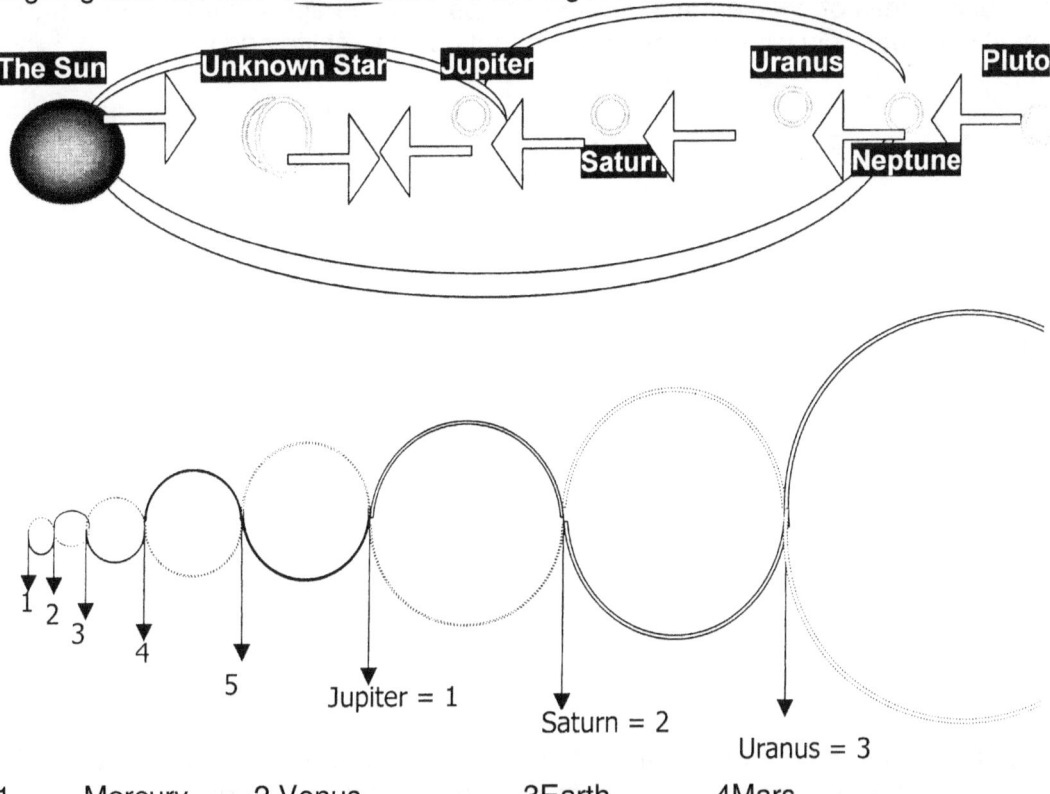

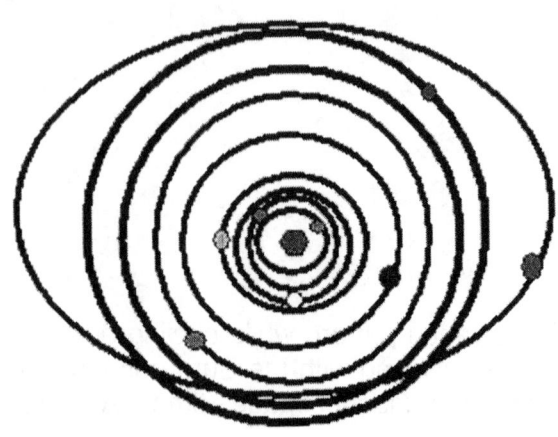

The difference there is in the alignment of the gas planets when compared to the inner planets tell the tale of much different history in development. It is also clear that there were seven periods of significant and altering stages that influenced the solar system to the core of development. Most important in all is the distribution of the cosmic structures in their unit as they relate to the development the unit underwent.

1	Mercury	49 X 10^6 km
2	Venus	98 x 10^6 km
3	Earth	147 x 10^6 km
4	Mars	227 x 10^6 km

Under normal development there cannot be more that five structures relating to singularity without forming a sphere. It is how the development should be. Why it is different I do not wish to go into but I will say it has to do with the butterfly diagram. However getting into that explanation would be very complicated and time consuming and for that reason I think it is best left to the Sven days Of Creation.

1	Mercury	49 x 10^6km
2	Venus	98 x 10^6km
3	Earth	147 x 10^6km
4	Mars	196 x 10^6 km

This is what it is at present so there is strong indication about some development above and beyond the normal cosmic growth that the solar system did experience as part of the Hubble growth. The normal flow would put the fragments at time relevancies as I interpret them by using the space-time rule in time in space confirming normal development in space-time. However it is not as innocent as it all seems and much evidence tells of a violent and crime filled past that shaped our solar system in becoming as unique as it eventually did with sporting a place that could host life and all.

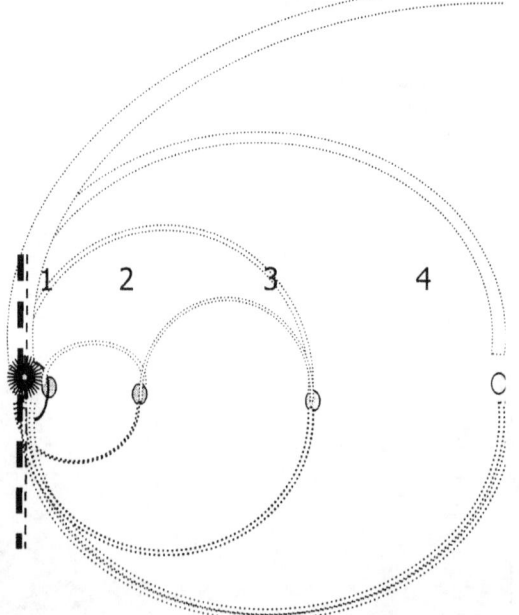

One can clearly see there was a push of the planets in the Titius Bode law moving outward towards the direction of the last structure. The plants move from one to five but five landed in a massive problem. The growth coming from the Sun was defining the structure to a specific location while that location was blocked by a micro star may times the size of the planet. This Micro star was in tune with quite another singularity line because of the history the micro star had in early development with the Sun . The micro star was located as structure on in alliance with another set of developing structures and was unmoved by the oncoming dwarf. The response the micro star showed in relation with the Sun was in another frequency than the Micro star was. I guess Jupiter at the time was not as disproportionate gigantic as Jupiter seems today but being a Micro star in relation to the other micro stars such as Saturn, Uranus Neptune and Pluto (yes Pluto but Pluto again has a history which is his story and we have no time for that). With the Sun forming space-time and Jupiter blocking such an advance there came trouble to paradise.

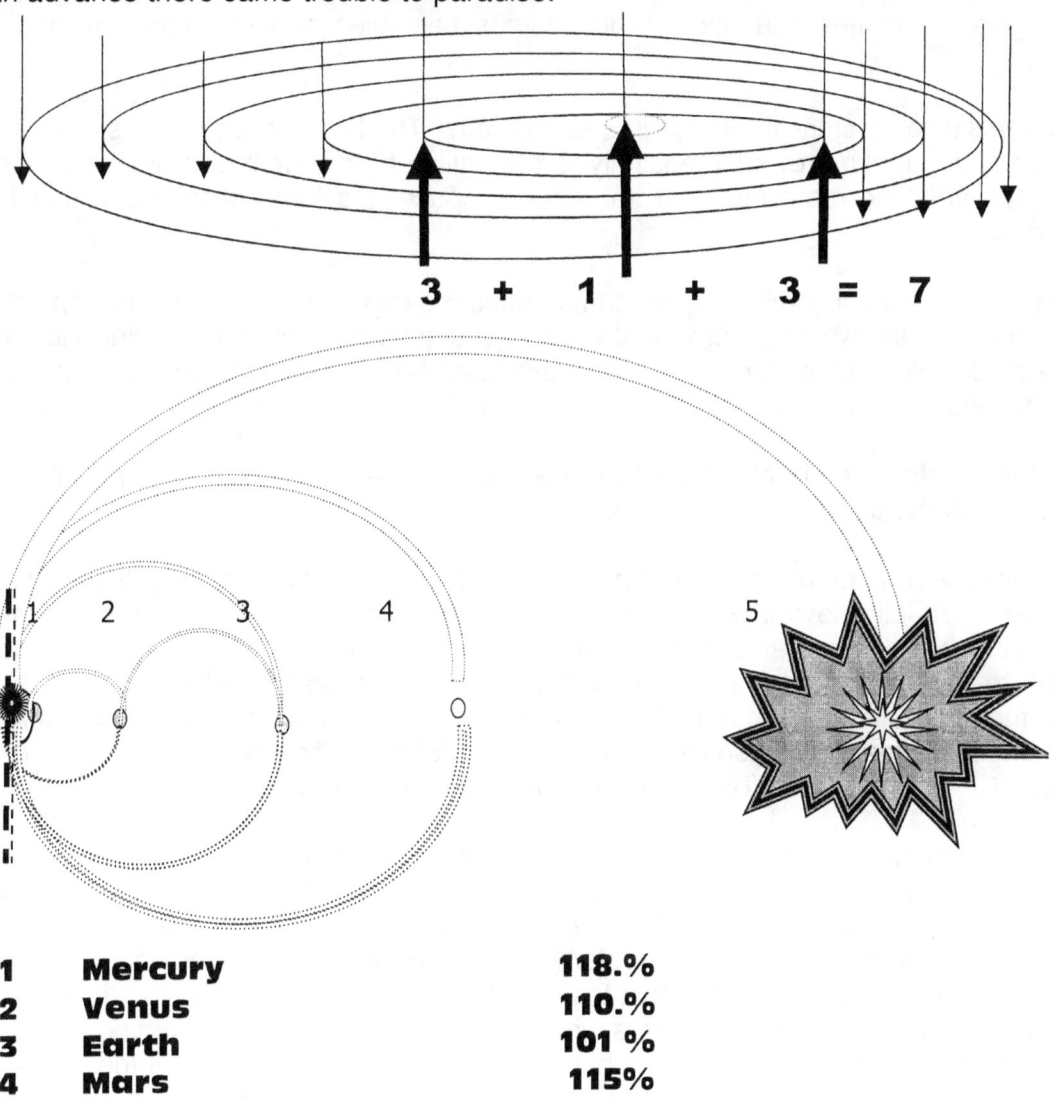

$$3 + 1 + 3 = 7$$

1	Mercury	118.%
2	Venus	110.%
3	Earth	101 %
4	Mars	115%

That indicates than space grew from a source to the outside of Mars and there is only one way an enormous quantity of heat could release at that point. It must be from the fragmenting of the fifth solid planet as the planet had the same fate as its mother star did and went the same rout by being forced to expand into a territory where no more expanding in such a direction is possible. Such a situation demolishes the core by exaggerating the heat load to appoint where almost the entire core liquefies and as the singularity of the little planet cannot take the bearing of the micro star, the little planet becomes wasted space-time.

The fragmenting tells how so many "moons" or satellites fragmented and were captured by the giant micro stars. Where the Sun came into the opportunity to capture some, the Sun did not use those satellites as "moons" but had them as liquid heat.

The planet went liquid but that is another bone of contention because science do not even recognise heat going liquid by forming gas as heat expands. Well, if they do they never made it part of cosmology. In the star hydrogen has a function in relation to heat. So has helium have a very special function. The oxygen in the relation to the carbon in relation to nitrogen has a specific purpose and that purpose tells to what point did the star stage of development reached. The Hydrogen has the duty to produce motion by as much duplication as the star singularity requires in maintaining. The Helium contains the captured heat, which then is charged to the carbon layer that keeps the heat in transit. With the oxygen coming into contact the oxygen transports the heat where the hydrogen transfers the heat from a solid as it was in the carbon and to an extent in the oxygen and release the heat to be served as fuel driving the core region. When looking at a fire one can clearly see what every element does with heat in liquid form. The carbon keeps the coal red while the oxygen store the heat as smoke and the hydrogen takes the heat as flames into a volatile motion. Every element stands directly related to the purpose it holds with heat and in the manner it serves the star. We can see this from the way a fire ignites and burns. The cosmos has rules and the rules will apply everywhere.

This same effect happens when a small fire starts in a large room. The heat lodges in the smoke as the smoke fills the room the smoke grows increasingly to the ceiling of the room and nestles in the smoke. When a draft enters the room the heat transfers to space and that action the scientists regard as an explosion.

Science fixes their attention on the incoming oxygen that causes the explosion. This might be true in part, but that is a small part of the whole picture. In the cloud of smoke gathers heat, confined to the smoke with the flow of direction out of the roof. The advancing heat flows from the burning fuel igniting underneath the smoke.

Material form heat at the bottom, the heat advances to the top, but the heat at the bottom remains higher than does the heat at the top.

At this point we have to look at another natural phenomenon. As heat can convert to space so can space convert to heat. We all know what happens when a compressor forces air into the air container. The container heat rises dramatically. Science call it pressure, but pressure it is not. Pump the compressor to say ten bar and leave it overnight. AT first the container will be hot. After a while (say 12 hours), the heat will reduce to room temperature. At that point one may call it pressure, but with the heat amongst the matter, it is space turned to heat. With the flow of time, the heat will return to space. TIME IS THE SPIN RATE OF HEAT IN A SPECIFIC SPACE.

To go back to the oxygen argument, one has to examine the burning oil well in Kuwait during the Iraq war. In order to distinguish the oil fire, they used blasting material to cut the heat from the oil. According to science it is to cut the oxygen from the fire, but the oxygen will flow in just as fast as it flowed out. The oxygen will return immediately, therefore the oxygen as such does not increase the fire. It is the response that the oxygen has with heat that kills the fire. By accelerating the time from burning to beyond the interaction with oxygen, that is what really kills the fire. The incoming oxygen does not transfer sufficient heat to restart the fire. What is also true is that not only does the oxygen bring heat in, but also it relieves the burning material of heat, therefore allowing more space to convert to heat.

One has to seek for this natural phenomenon as it occurs inside the star. In the case of a steel cutting however, the acetylene is the fuel, and the oxygen's role is to enhance the heat, not burn the metal. Again it is the way oxygen responds to heat and that accelerates time. The oxygen removes heat by bringing in more space to become heat, as much as remove heat to become space. We have to recognize the dual role of oxygen and not merely the fact that oxygen burns. OXYGEN DOES NOT BURN AND OXYGEN CANNOT BURN. IF OXYGEN COULD BURN, THE EARTH WOULD NOT HAVE ANY OXYGEN LEFT BY NOW.

In fact we observe the metal degeneration as cutting but it is time enhancing on one small area of the metal. The metal is "rusting away" at one point. The rusting process speeds up by thousands of years at one specific point. It is not the oxygen but the way the oxygen responds to heat and the interaction between heat and space that cuts the metal. In other words, the oxygen only carries the heat and space to the iron.

In every galactica, a certain value of space-time was sealed in, released, as its time becomes valid. Every galactica therefore, corresponds to a different value at a different rate in as far as compensating for the time draining in the Universe, as the Universe is shrinking. We, as humans have up to now, placed a false value to the universal expansion, because we have valued the invalid aspect to the Universe, which is space.

Seen from the valid perspective, which is time, the Universe is shrinking as it loses in the valid perspective, which is time; the Universe is shrinking as it gains in the invalid component to the Universe, which is space. Therefore, the Universe is not expanding, but it is shrinking all along. I have been in so much disagreement about so many aspects of science in the way they regard the Universe, that I do not wish to split hair about such trivia. In the light of this, I only point this aspect out, but I do not wish to make it an enormous issue, because it does not really matter, from what angle you look at it. Another outcome may be where both objects maintain the claim to singularity, by pushing the space-time occupied to new levels of occupied space-time values. The result of the establishing of new individual but unequal points of singularity is the oval way objects rotate, first favouring the on in the matter part and the other in its space part and afterwards turning the points of reference around.

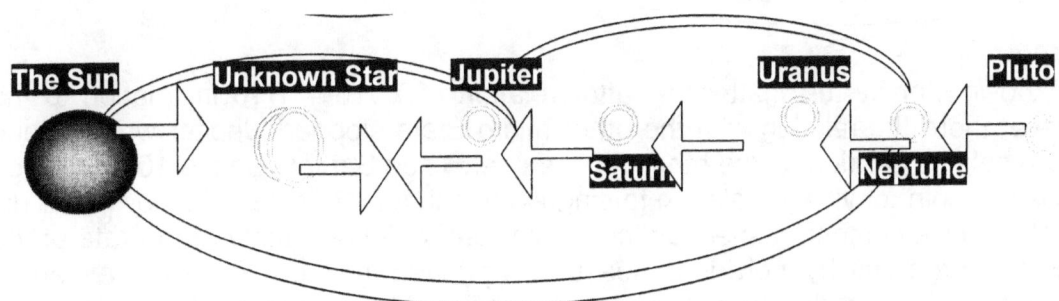

Since the First movement of time, the Roche factor was present and from the Roche factor came the Titius Bode principle. Each object seen, as well as not seen, represents a different period contained by a different specific value of space. All matter is time in a different frozen state in space and the time gives the space its particular value.

The Roche limit that came about between the Unknown star and the star we call Sun contributed to the demise of the smaller star.

The Roche limit is:
The region surrounding each star in a binary system, within which any material is gravitationally bound to that particular star. The boundary of the Roche lobes is an equipotential surface, and the lobes touch at the inner Lagrangian point, L_1, through which mass transfer may occur if one of the components expands to fill its lobe. It names after the French mathematician Edouard Albert Roche (1820-83).

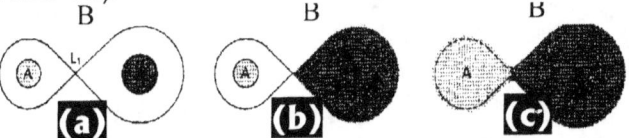

THE ROCHE LOBE: In a binary system, the Roche lobes of components A and B meet at the L_1 Lagrangian point. (a) In a detached system, neither star fills its Roche lobe. (b) In a semidetached system, one massive component, B, fills its Roche lobe. (c) In a contact binary, both components overfill their Roche lobes and share a common envelope. As with the graph I can see the two sides forming a connection therefore relevancy has to apply, all contradicting Newtonian claims of no connection but through mass attractions. The mass does not attract but one interferes with the other total influencing the space surroundings.

Any person taking Newton seriously should at least take on the challenge and find the comets colliding with the Sun , find how much the planets moved closer to the Sun since the days of Newton and indicate where there is unprecedented collision between stars. Yet the closest the Universe comes to that is to show " how stars blow bubbles" in space and that is to use the precise words.

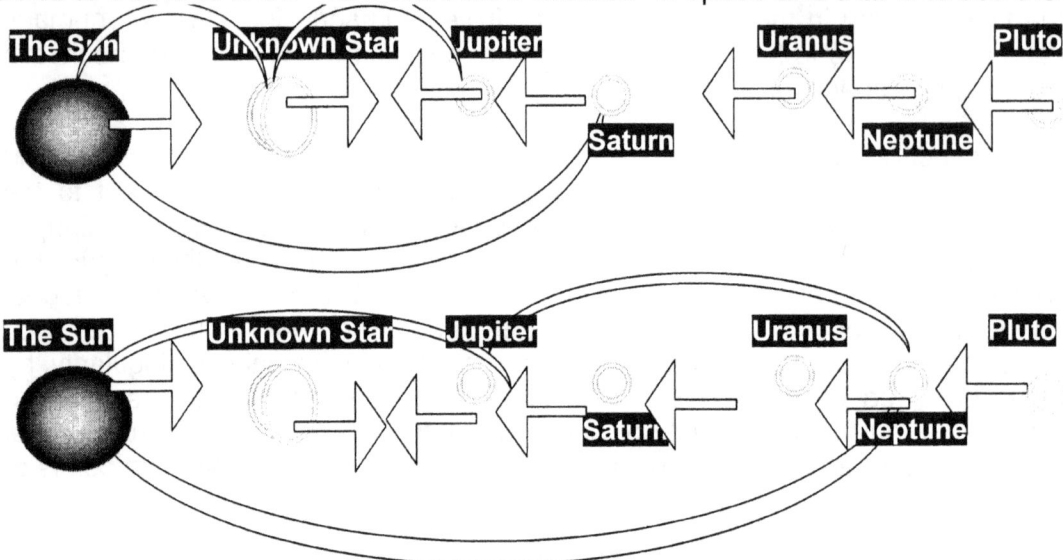

Jupiter related to Unknown star by the matter-to-matter relation of 7 /10 + 7/10 in relation to the space of 10 /7. However there the relation with the other micro Stars stopped! Jupiter was the one 7/10 marker and Saturn held its position as the other 7/10 marker. Then Saturn had the 10 / 7 relation with the Sun . From Saturn point of view in relation to singularity control the fact of Unknown star did not exist. Where the Sun development pushed Unknown star out, it did not move at the rate of the Micro stars since it was a large star by individual measure. It pushed the Sun Back as well and it pushed Jupiter but the other four that did not see Unknown star held firm. That brought conflict since Unknown star did not move well with the others. It had a singularity relation with the Sun but that was it. The Micro stars had a singularity relation with the Sun and Unknown star was never in the picture.

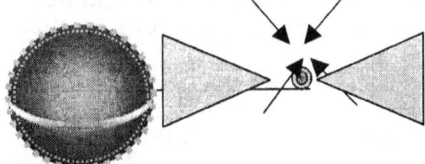

The extending of **k** in $k = a^3 / T^2$ had a murderous effect on Unknown star since from the Sun the star had to grow away from the Sun but five other Micro stars had now addition about the intentions of Unknown Star and therefore Unknown star became something like Johnny –in –the- middle.

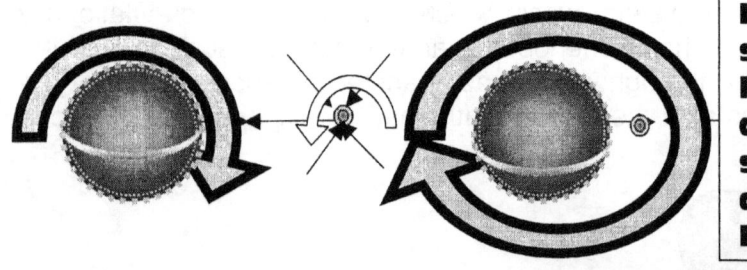

From having separate to having a joint Π^2 **as the lesser star core is dissolved to heat.**

The war came to rage as a battle for supremacy began to emerge because Unknown star and the Sun came into a Roche limit. How the actual figure places the position that Unknown star had I will not get into because that is fairly complicated and it is time consuming. In the Roche limit there is an intense flow increase in liquid time and it is the minor structure that cannot take the contraction. The advances star puts the flow Π^2 it receive from the spin that forms the space Π^3 to the value of Π. Then as the Roche limit is bridged the time value $4\Pi^2$ is in conflict with the time difference limit $\Pi^2 / 4$ and since the time ratio of $4\Pi^2 / \Pi^2 / 4 = \Pi^0$, which in effect brings about that the singularity controlling the better developed singularity also takes control of the lesser developed space-time since both then relate to Π^0 and is in relation to $k = \Pi$ because $T^2 = \Pi^2$. However there is no way that the lesser core can stand the heat since the better developed core increase the flow of liquid extending the flow to the lesser star and then the lesser star is taxed with the burden of Π which puts the same value of space Π^3 in relation to the lesser developed core.

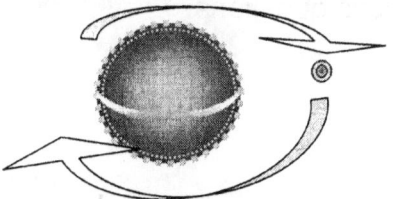

It is well advised to remember that it is not the Universe that grows but it is the Mterail in the Universe that expands by the margin Kepler mentioned as $k = a^3 / T^2$. Therfore in all the expanding of space that took place and did not take place the diameters of the Sun as well as Unknown star did expand. As they expanded so di the distance parting them not respond by the same measure because the Micto stars was blocking the route into expanding as far as the well being of Unknown Star goes. Soon time arived that introduced serious conflict in Paradise.

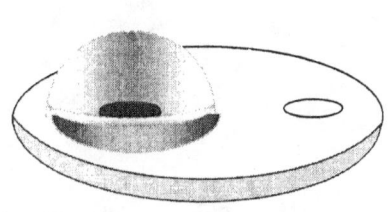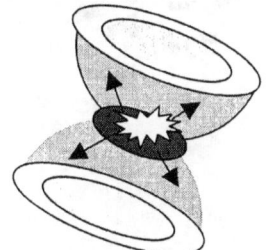

The Roche limit is evidently widely applying throughout the entire cosmos but Newtonian rules do not explain the phenomenon. The heat increased condemned the fate of Unknown star and the core of Unknown star expanded as it brattled and blew. Five chunks remained as well as many fragments, which became comets. The Hydrogen gas now at present forms what is known as Kuiper belt.

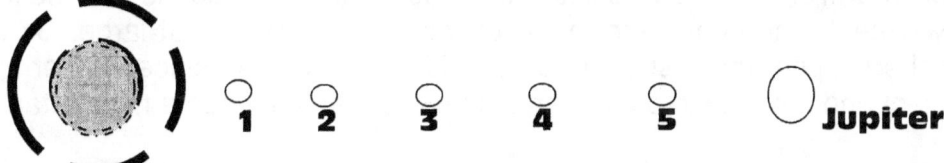

Then there were five totally unrelated fragments on the inside of five Sun related micro stars. This spelt disaster in many languages. The five now formed did not correlate to the five micro stars and therefore the five micro stars still did not at all align with the newly fragmented hard rocks.

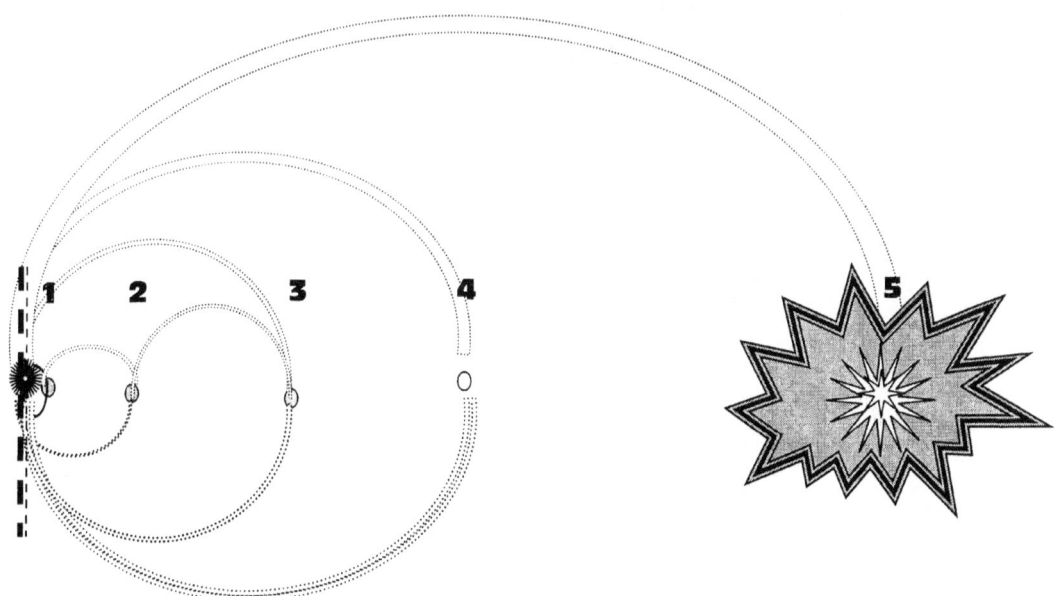

We have a ring of debris and cosmic junk clustered around where a fifth planet should be on the very edge of the inner planets. Logic tells us that too must be part of a solar disaster but how can one prove that? I once again do not propose the following as proof but my main attempt is only to show there are methods one can use when applying my cosmic code to find answers to answerless questions.

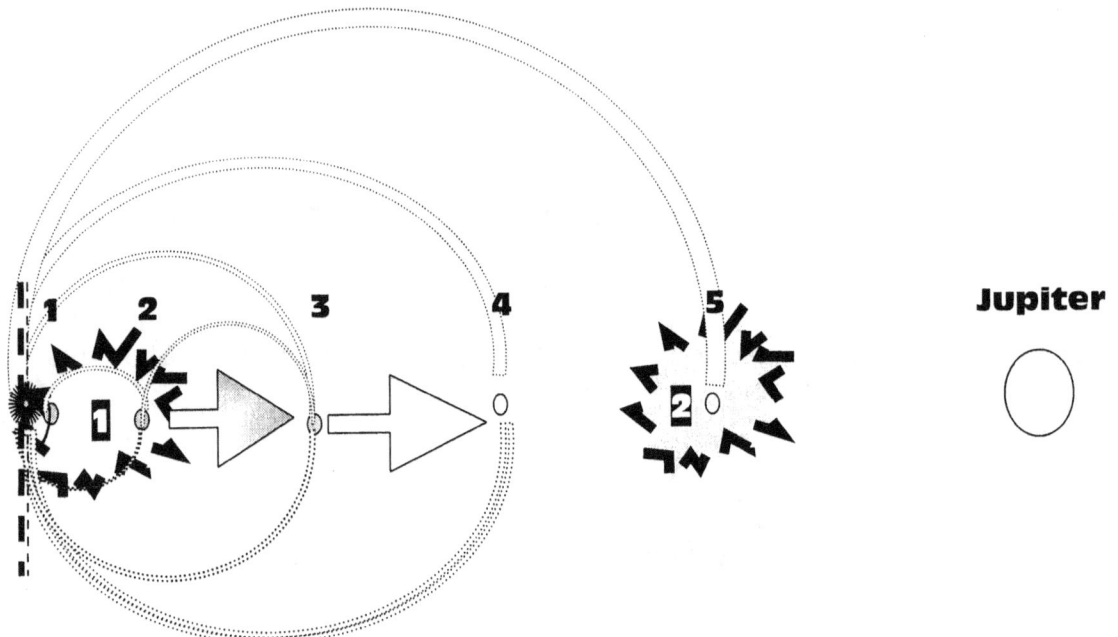

The evidence we see is in the numbers we find. There was this event that produced an enormous quantity of heat near or close to the Sun . This we find in evidence where Mercury expanded 18 % more than that which should be gauged as normal. The sympathy decline in space development as the planets developed at a further distance from the event. In the case of Venus we find an exaggeration of 10 % and in the case of the earth there is an exaggeration of 1 %. This is to be expected since the demise of one little star In a Roche limit is hardly expected to change the face of the entire Universe at large. By the tome the development approached Jupiter there should be no evidence of growth since Jupiter now did not connect with a page object and the growth could hardly constitute to influence a micro star.

Is:

1	Mercury	49 X 10^6 km
2	Venus	98 x 10^6 km
3	Earth	147 x 10^6 km
4	Mars	227 x 10^6 km
5	Fragments	413.1952

Should be:

1	Mercury	49 X 10^6 km
2	Venus	98 x 10^6 km
3	Earth	147 x 10^6 km
4	Mars	196 x 10^6 km
5	Fragments	245 x 10^6 km

1	Mercury	118.%
2	Venus	110.%
3	Earth	101 %
4	Mars	115%
5	Fragments	168%

However when we include a Quantity of debris at a location where the Titius Bode law does indicate the presence or position of a structure, the significance change considerably.

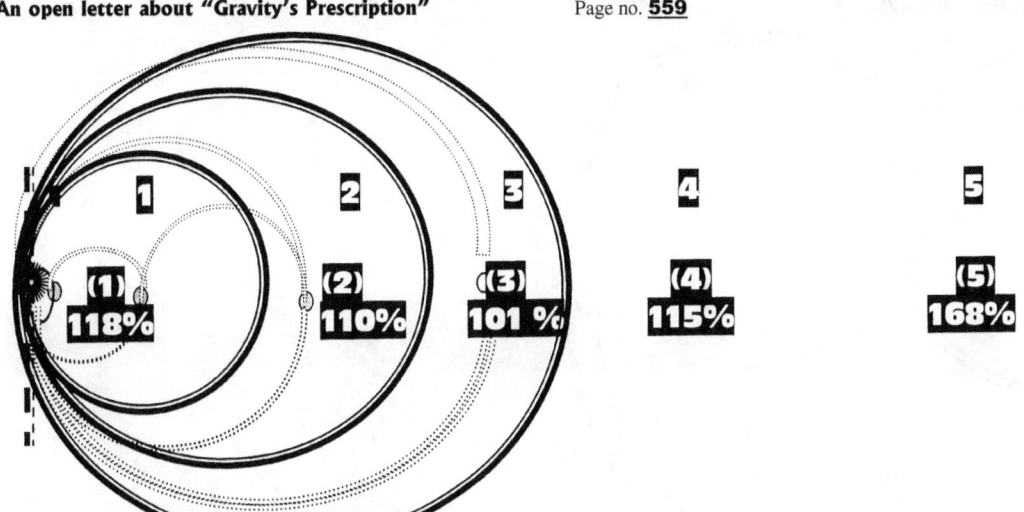

Where the reducing should start nullifying the space growth, which is past Mars there we find an increase that is most astonishing. We find that there is an increase of as high as 68 % where there should be no traces of any growth left. This can only be the result of a heat release of gigantic proportions in the manner of a (very little) Super Nova spectacular.

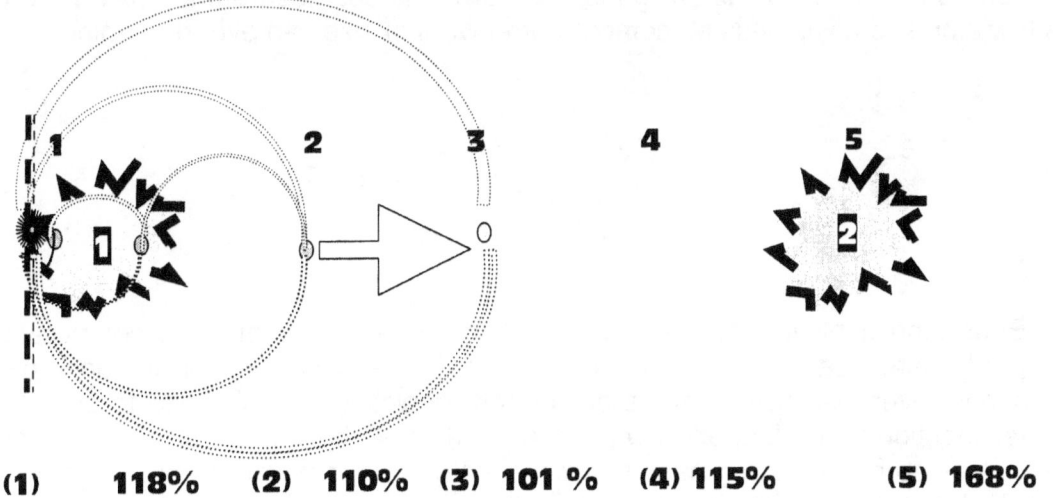

Jupiter released the same fate on the fifth planet as that which befell the fifth plants mother star old Unknown star. I the fifth planets did no fragment (and believe me there was no alternative solution) the alignment between the planets and the micro stars would not have realised because then there was no linking the micro stars and the planets in relation to the using of the Titius Bode law.

Any application to use this method in gauging the development of the micro stars would bring no clear results since we and the micro stars are not connected in the manner we connect with the four planets.

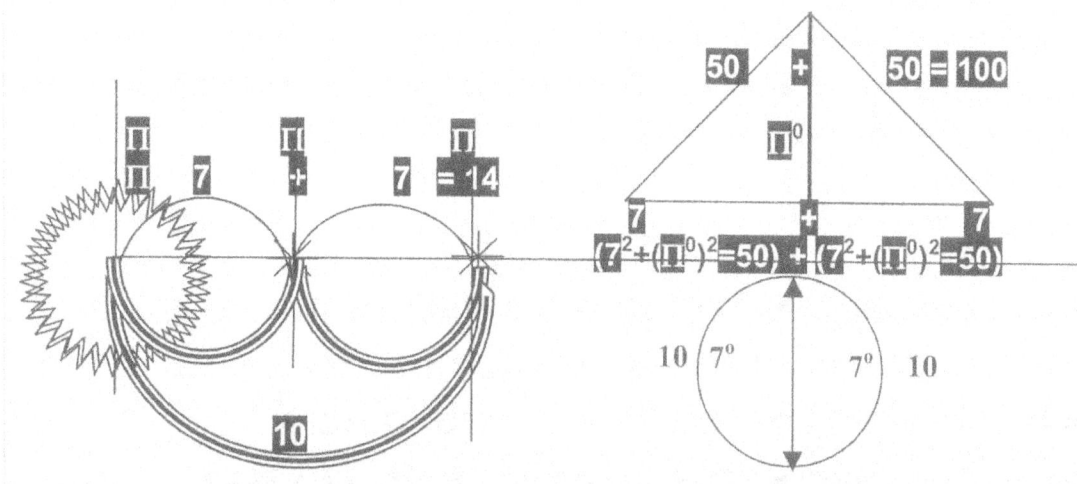

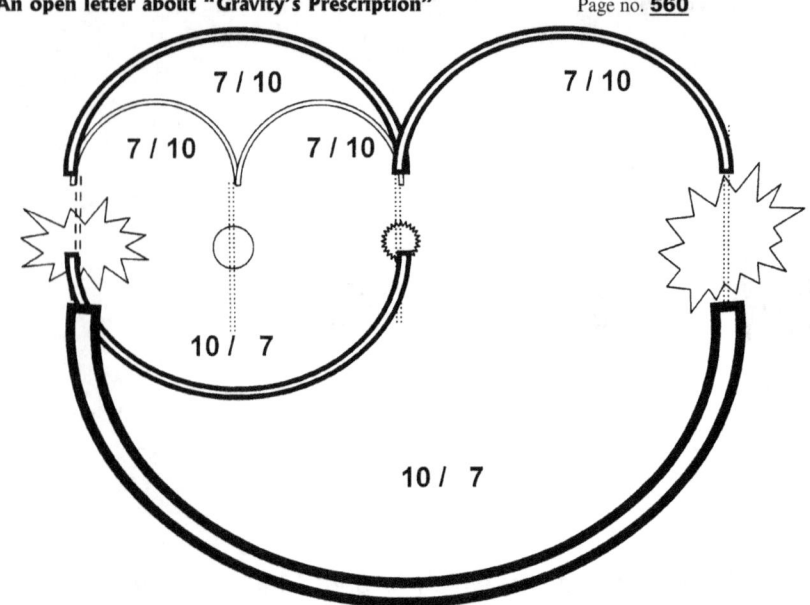

The significance and implication that the application of the Titius Bode principle holds on the Hubble constant reflecting on era to come as well as era gone past turns the cosmos from a small piece of vacuumed holding a few atoms to a vastness no computer irrelevant of size can ever determine.

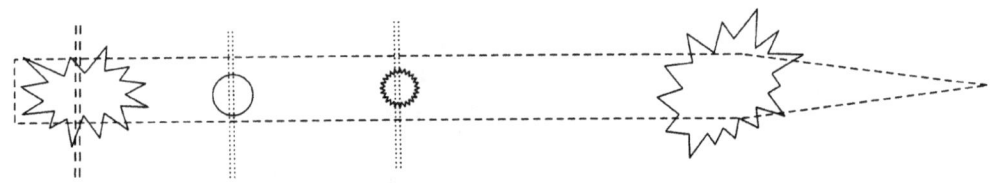

In view of the Titius Bode time depletion, time in flow creating space has a far more complicated arrangement than Xepted science can even produce on a chart. To be honest every person knows that Xepted science cannot even place the planets on a chart, depicting true distance to size, but they WILLFULLY never mention that information when the chart they show is as false as a three-dollar bill. In a sense it does no harm leading people down the ally in such a way, because others in my class of mental insignificance in society is far to un-intelligent to realize the correct way and will therefore not understand the correct way in any event. Now you ask "So what about Neptune and why does Neptune not fit the pattern…"well that storey is more complicated which I therefore reserve for *"Seven Days Of Creation"* because the explaining is as simple after I produced a much wider vision with a lot more explaining to make it as simple as this…

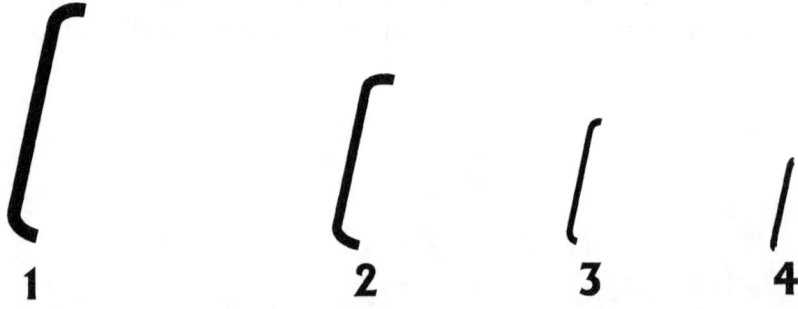

1 SINGULARITY HOLDING MATTER TO OUR FUTURE IN SPACE-TIME

2 SINGULARITY HOLDING MATTER IN RELAVANCY TO SPACE-TIME

3 SINGULARITY HOLDING MATTER AT THE SPEED OF LIGHT

4 SINGULARITY HOLDING MATTER BEYOND THE SPEED OF LIGHT

That is a mistake with a stinging tale. It is as dangerous as a scorpion to Xepted science. In an introducing article, I named Anglo-American Mythology, I pointed out how misconception feeds society, favouring the lies and untruths and blatantly ignoring the truth. In the past, since time began the powerful used this on the brainless masses, and for a period where that civilization lasted, got away with that strategy. The next civilization that came to power, followed the same methods applying the same dogma, and in the end paid the same penalty because their greed, lust for power, and sublimations gave them control over the masses for a while. The misconception those in favourable positions forced onto the masses, made the very people in power so shortsighted, their course on vanity lasted but a few generations. This is achieved because our Earth environment is tolerant and can buffer a lot, to save life in the end for life's contamination on Earth. They wish to extend life's connection to Earth, as being a connection to the cosmos at large and will be in effect as long as life remains in the cosmos. When "going abroad" to our "next door planet" that holds all supporting evidence of life carrying organisms, the connection to the cosmos remains and connecting to the Earth is of little consequence. That bluffing must stop. I realize no one on Earth will ever take note of what one sod (like me) on Earth is shouting, but misery awaits our Martian Colonists. Suffering will be the reward for the fools attempting to catch the bounty of "fame, riches and glory" on behalf of the All Powerful Dollar and the dollars absolute true benefactors. Those with eyes, let them see, those with ears let them hear and let the rest self demolish.

I admit, man derives all the above-mentioned lies mainly from corrupting the Bible. The Roman Catholic Church started the trend 1 500 years ago and still maintains it as best it can. You can prove whatever you wish to prove from facts you take from the Bible. The Bible is in support of whichever standing you may support. This is not because the Bible is incorrect, it comes from our insignificance to appreciate the full content of the whole Bible. The Bible promote only truth, man takes from that what man wants, divert the truth to suit his need by corrupting the lot. I know presenting evidence as well as I do, will not change the course of science. I know too, that I am too small to pinpoint conclusively whether Authentic Author referred to the creation of the first cosmic period, or the first solar period. This is not because of inaccuracy on the part of the Bible. This inaccuracy comes from the human interpretation of Authentic Author on his vision, and then my insecurity to interpret his interpretations. It is human error bringing on misinformation. The Bible's recollection CAN and DOES apply to either period. Therefore, to respond in containing human misjudgement on my part, I shall re-apply the vision of Authentic Author in the context of the first solar day. I showed how it fits to the first cosmic day already.

Binary stars, spinning to self-destruction will produce significant heat. Heat create space, space forms winds. That is facts that the Bible present and is indisputable. Where the Earth was, was still a void, containing a sphere of circular displacement and this will reduce linear displacement to zero. Linear displacement is space and circular displacement is containing heat for matter survival.

Binary Star Minor overheated. That is why the core brittle and fragmented. This action will release tremendous contained heat the heat will produce magma flowing in space like water in space and this eruption of heat space that created winds. Once again the recollection fits the scenario. Releasing the heat and producing space will establish space-time and fill the void where the Earth should fit. This is fact and if anybody even tries to dismiss this will be because of abstinence on his or her part. I did not prove the Bible correct. The Bible told the truth and in such correct detail, it is beyond human comprehension, but sublimation on the part of Newtonians and science before them, disallowed their ability seeing it.

At this point I invite you, the reader to go back and read about the Newtonian version of cosmic structure forming. Compare that FORCE applying MAGIC to the Biblical portrait of events and see where the fools hide. When comparing notes about Newton's view and the Bible's view, judge for yourself who in the end understood Newton and who did not. All I wanted was to find some one that could look past the mechanic and appreciate the work he is representing. I never seek prominence, I only wished to introduce my view and let another more educated, more significant and more wise man take the reigns from there. I always knew I am not the man to do the job. My field of knowledge is too small, too limited and above all, too insignificant.

I found no one that could look past me and see my formula $R^3/T^2 = 1$ and $\$T = (\Pi^2 \times \Pi^2)\ (\Pi^2\Pi)\ 3 = 1836$ which is the relevance of the cosmos. By not finding a person that could see past me, I knew that person will not be able to look beyond "a burning Sun and see the frozen state in which the Sun is. Without noticing such crucial evidence, the rest goes lost. That person that sees me and not my formula will never see the cosmos for what it is.

While the one proton connects to space in singularity (Π^2 going to singularity) and connects space Π (in singularity) the other proton brings time Π^2 directly to space Π, re-uniting space-time as a unit to singularity (Π^3). We may call this re-unification unifying the " gravity - motion" in order to identify the one proton unifying time with space, while still in contact with the other proton holding (Π^2) time to (Π) space in as much as being occupied by matter (Π^2) and unoccupied heat Π forming 3.

This explains then the absolute value of time as a square, with the square having both a circular value of R^2/T multiplied by the linear value of R/T producing the link to singularity in time and singularity in space.

That is what an insignificant formula $R^3/T^2 = 1$ where $R^2/T \times R/T = 1$ represents space-time in singularity as well as space-time in densified, occupied and unoccupied format. That means the everything of the whole lot, or as we say in Afrikaans, the "Heelal" meaning Universe.

When comparing notes about Newton's view and the Bible's view, judge for yourself who in the end understood Newton and who did not. All I wanted was to find some one that could look past the mechanic and appreciate the work he is representing. I never seek prominence, I only wished to introduce my view and let another more educated, more significant and more wise man take the reigns from there. I always knew I am not the man to do the job. My field of knowledge is too small, too limited and above all, too insignificant.

I found no one that could look past me and see my formula $R^3/T^2 = 1$ and $\$T = (\Pi^2 \times \Pi^2)\ (\Pi^2\Pi)\ 3 = 1836$ which is the relevance of the cosmos. By not finding a person that could see past me, I knew that person will not be able to look beyond "a burning Sun and see the frozen state in which the Sun is. Without noticing such crucial evidence, the rest goes lost. That person that sees me and not my formula will never see the cosmos for what it is.

While the one proton connects to space in singularity (Π^2 going to singularity) and connects space Π (in singularity) the other proton brings time Π^2 directly to space Π, re-uniting space-time as a unit to singularity (Π^3). We may call this re-unification unifying the " gravity - motion" in order to identify the one proton unifying time with space, while still in contact with the other proton holding (Π^2) time to (Π) space in as much as being occupied by matter (Π^2) and unoccupied heat Π forming 3.

This places Kepler's formula at a relevance s $\mathbf{a^3/T^2 = k}$ where $\mathbf{k^1}$ will hold a relation to heat with time in eternity, while AT THE SAME MOMENT infinity also apply, giving $\mathbf{k^0}$ the value of singularity. This is all to do with the Universe remaining in contact with singularity while keeping the Universe in matter and heat, in the dimensions of space and time. This explains then the absolute value of time as a square, with the square having both a circular value of R^2/T multiplied by the linear value of R/T producing the link to singularity in time and singularity in space. That is what an insignificant formula $R^3/T^2 = 1$ where $R^2/T \times R/T = 1$ represents space-time in singularity as well as space-time in densified, occupied and unoccupied format. That means the everything of the whole lot, or as we say in Afrikaans, the "Heelal" meaning Universe.

THE COSMOS IS NOT OUTSIDE; IT IS INSIDE, INSIDE EVERY ATOM. THE ATOM CANNOT DISAPPEAR, AS IT CANNOT VANISH. After all it is energy.

In an effort to convey what I see as time I wish to convert the Cosmic Calendar to some Cosmic Time Scale, This scale does not name events, but rather use relevancies, running time and space as the event unfold from singularity. The singularity comes from the point where all matter occupied and unoccupied confirmed one line with out space as space was infinite and time was eternal. As purely an indication of physical time found in the cosmos applying at this moment I wish to introduce an illustration in bringing a better comprehension of time flow. There is a possibility of many other starts to the flow of singularity, but this point holds most valid significance to the theme we explore

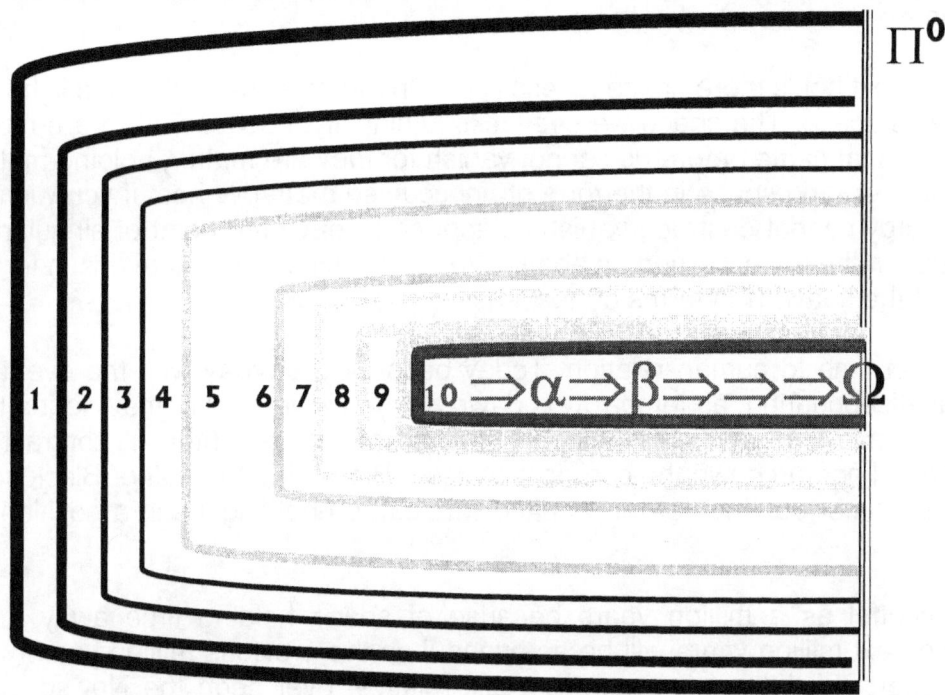

Newtonians tell about a Big Bang explosion that included everything there is.

$t = 10^{-43}$ seconds

the very first instant, the entire Universe were much smaller than a neutron and the temperature was $\approx 10^{32}$ K

$t = 10^{-34}$ seconds

The Universe underwent an increase in pace expansion growing in size with a factor of 10^{30}. The Universe becomes a soup of quarks and leptons at a temperature of $\approx 10^{27}$ K.

$t = 10^{-4}$ seconds

Quarks combine to form protons and neutrons and their anti particles. The Universe cooled down to such a slow pace electrons no longer can breakdown the particles remaining. Particles of matter and anti matter collide and annihilate each other. There is a slight excess of matter not finding annihilating partners, surviving to form the world that we know today.

$t = 60$ seconds

The Universe has by now cooled down enough to form protons and neutrons and with colliding can stick together to form the nuclei of low mass elements ^{2}H, ^{3}He ^{4}He and ^{7}Li. The predicted relative abundances of these nuclides are just what we observe in the Universe today. There is plenty of radiation around, but light cannot travel before it interacts with nucleus. The Universe is opaque to its own radiation.

$t = 300\ 000$ years

The Universe has now fallen to $\approx 10^4$ K, and electrons can stick to bare nuclei when they collide forming atoms. Because light does not interact appreciably with (uncharged) particles such as neutral atoms, the light is free to travel great distances. From this light comes background radiation Atoms of hydrogen and helium can hold together under the influence of gravity, and begin to clump up forming galactica and stars. In every small human mind we try to find time, which we know and trust. The cosmos holds time much to the properties the Creator describes in the Bible. I shall not be blasphemous and say it is the time the Creator refers too, because the cosmos is time the Creator created with all other aspects, therefore the Creator refers to time at a pace that puts our vision of eternity in the same class as we find an explosion to be.

To us the future is dark because it holds more space to less light. On the other hand the past is bright because it holds lighter to less space. The space we see lacks luminosity, because there are much more space to hold light. Stars that came before us cannot vanish for they are matter, holding matter to occupy space of matter. Matter (and space in the form of unoccupied matter or heat if you wish to call it that) is energy and energy cannot destruct, vanish disappear or leave the point of singularity. Again we are facing the situation the Bible warned us about. We are thinking of the heavens in terms of Earth instead of thinking of the heavens in terms of heavens.

On Earth we humans, connect time to human relation. Today become yesterday with the event of tomorrow. We think of today disappearing, as tomorrow is dawn. In the Universe that may not be the case, because where true cosmic time holds space, we humans shall never enter. In contrast to Popular Newtonian belief we will not run down the corridors of some Black Hoe to another Black hole and in the process mesmerize time. Such is not for us fitted with carbon-holding life in a position of singularity.

In stars one year is as eventful as a million years because of space holding time away from singularity. On the moon the next million years will be as eventful as the previous million years with nothing to report in newspapers. Slightly of the point but still relative: ever seen the Newspapers come out and say nothing much happened? There is always news only the relevancy may change from day to day. Well, this does not even happen on the moon because on the moon life will not bring news as it happens; it does not happen!

Looking at the sky we do not see space we see time. The photon is space travelling through time and that is the only space we see, that which the photon hold and that the photon occupy as much as bring to us. Space is light in utter darkness. Space is a ray of photons in magnitude not directed to our singularity but in countless other directions where singularity manifest space in time. Space is heat and heat is photons of lesser implication or not directed our way. To my view (what it may be worth) the Authentic Author did not move back in time, he merely moved foreword in space. By moving in space he reached Moment-Alfa.

Moment-Alfa is in eternity and eternity never ends therefore he moved to eternity in space eluding time by having a vision. He could therefore see what he reported because he was in that space but not in that time.

If he were in that time he would still be there, he would be there eternal and never come back to report on what he saw. There is much rumour of a Big Bang, however I am inclined to think the biggest bang that ever was also became the smallest bang there ever can be. It was the instant when heat parted from cold. The line started in infinity because the line was continues but being continues it never was. The very instant followed the previous instant identically and the instant was so identical it remained the same by never moving while always moving uninterrupted eternity upon eternity. Then came the entire Universe when infinity broke free from eternity. It was when darkness broke into light. It was when whatever possibly can be became a possibility to be. It was when the first number mathematically arrived and from the one became two by gong 1^0 to 1^1. That infinity is so small it houses everything there is in the entire Universe. The entire Universe still is in a spot that formed a dot. The spot has no outside but it only has an inside while it is inside all that spins as it generates all that can spin. Yet, by spinning it brings motion into being.

The spin creates a drive that keeps the Universe mobile. Still the first forming of the dot from the spot came about to the inside and not the outside, which makes the Universe shrink and not expand. It is the smaller things that come into relevance as the larger things were placed in relevance when time began. The Universe is shrinking into the oblivious since the Universe never had anywhere to expand to. Never once did one Newtonian sit back and consider their laughable proposal of an expanding Universe with nowhere to go when it is expanding.

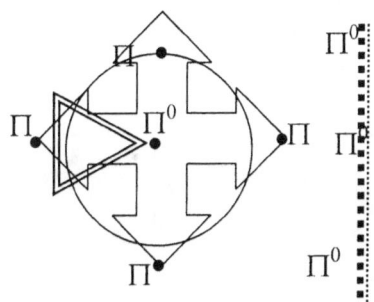

Three points formed a line covering singularity where the centre singularity recovered heat to grow and two points served as an axis to allow the rotation and to assist the duplication. There is one centre connecting the duplication of three as well as the recovery of one (the fourth one) that is applying the tie aspect. Therefore, motion consists of three positions in relation to a centre, which forms as space in relevancy to the motion and the space receive a controlling centre.

At the first glance, Kepler's formula seems to be numbers and positions applying between the Sun and specific but different planets in the solar system.

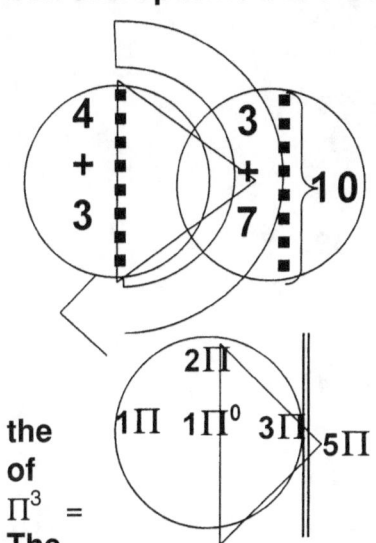

That is why the Universe is Π

The duplication comes about as singularity is exciting another singularity in precise relevancy of 3 to 3 to 1, but the points charged is as space less and as motionless as only singularity is. The heat it requires to carry the exciting between points forming space and the space excites heat and the time delay it takes to excite singularity between points forms space-time.

Where motion conducts electrical charging which is equal to gravity the charging of motion is to entice duplication of singularity. This is the basis, the heart and the sole ingredient of the of Π^3 = The **Coanda principle that includes the Roche limit ($\Pi^2/4$). The charging gravity $((7/10) + (7/10)) / (10/7) = \Pi^2$ and the charging of space-time $\Pi^2\Pi$ is all due to the relevancy brought on by the Coanda principle. The value of motion came from singularity exciting singularity and that is the duplication while the duplication or motion presents the space.**

When the cosmos came to motion, motion was not yet defined. When the cosmos brought about motion, the first motion was parted from hot. Eternity parted parted from motion absence. laboriousness of eternity for the The spot became and grew into

relevancies. Cold from infinity. Motion Infinity broke the duration of infinity. the dot.

When space brought division between eternity and infinity

From what the spot was to what be just a mathematical from 1^0 to 1^1 but in reality that creating of and establishing of with all possibilities now in it. much growth become a reality,

the dot now is might implication of going first motion was the an entire Universe Never again can that although to us the

growth is beyond what we ever can notice. But it is because the growth is so massive and we are so small that we are unable to notice such almighty growth.

When the spot Π^0 became functional and established all relevancies possible, heat parted from cold as eternity parted from infinity. The expansion was not clear motion but more a parting of relevancies where a centre formed a relevancy because the centre could not provide motion. Without being capable of motion, the centre established four points, which also served singularity. From the inverse square law we know that the centre doubled by producing the four points holding singularity.

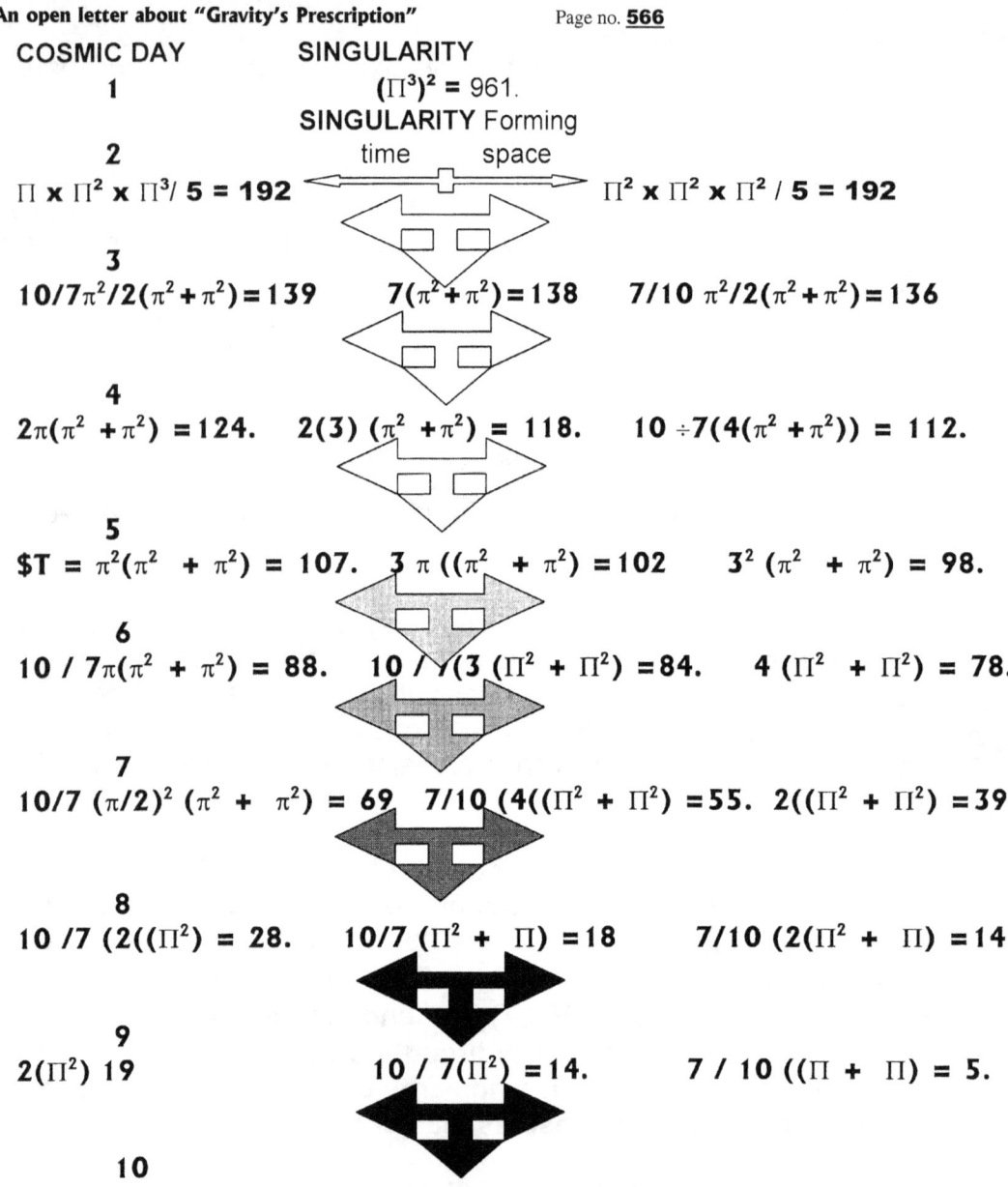

COSMIC DAY   SINGULARITY

1                         $(\Pi^3)^2 = 961.$

                    SINGULARITY Forming

2   time   space

$\Pi \times \Pi^2 \times \Pi^3 / 5 = 192$ $\Pi^2 \times \Pi^2 \times \Pi^2 / 5 = 192$

3

$10/7\pi^2/2(\pi^2 + \pi^2) = 139$ $7(\pi^2 + \pi^2) = 138$ $7/10 \; \pi^2/2(\pi^2 + \pi^2) = 136$

4

$2\pi(\pi^2 + \pi^2) = 124.$ $2(3)(\pi^2 + \pi^2) = 118.$ $10 \div 7(4(\pi^2 + \pi^2)) = 112.$

5

$\$T = \pi^2(\pi^2 + \pi^2) = 107.$ $3\pi((\pi^2 + \pi^2) = 102$ $3^2(\pi^2 + \pi^2) = 98.$

6

$10 / 7\pi(\pi^2 + \pi^2) = 88.$ $10 / 7(3(\Pi^2 + \Pi^2) = 84.$ $4(\Pi^2 + \Pi^2) = 78.$

7

$10/7 (\pi/2)^2 (\pi^2 + \pi^2) = 69$ $7/10 (4((\Pi^2 + \Pi^2) = 55.$ $2((\Pi^2 + \Pi^2) = 39.$

8

$10 /7 (2((\Pi^2) = 28.$ $10/7 (\Pi^2 + \Pi) = 18$ $7/10 (2(\Pi^2 + \Pi) = 14$

9

$2(\Pi^2) \; 19$ $10 / 7(\Pi^2) = 14.$ $7 / 10 ((\Pi + \Pi) = 5.$

10

$7/10 (\Pi) = 2.$ $(\Pi/2) = 1.57$ $7/10 (\Pi/2) = 1$

What is the Universe?

The Universe is the part that became visible when the invisible found a means to generate motion into the immovable and by splitting the inseparable that parting brought a division within the undividable. That which has no limits found a centre when that which has no inside started moving apart from that which has no outside. The separation of the undividable formed a line running in the centre of all spinning matter connected to that which allows spin to all spinning matter which holds space align with time while it keeps time synchronised with space in motion. The Universe has no sides, is undetectable, is only found outside our spectrum of reality yet is, controlling what it is generating the unreal from where we are part of the unreality outside singularity. We call it singularity yet it creates by generating space- time in time in space without being in form inside the Universe. It creates the Universe in time so fragmented we will never understand.

The information in this book I purposely made easy to red, and easy to follow. The content however, is only a drop compared to a bucket of information contained in all seven parts of "MATTER'S TIME IN SPACE The Theses". All information presented to you, in the first part of the book is an introduction to the second part. In the latter part of the book the conclusion comes. With out a detailed introduction in the first part, the information in the second part will be of little value, as the information is very conclusive about the first part. I have been researching cosmology since 1978, on a part time basis. The conclusions may seem simple; that however is in retrospect. Some of the arguments took me up to six months to arrive at since I do not have all the information on hand.

Facts I advise the reader to become acquainted is the following:

Aanplasing, verplasing, versnelling and inperking

As this book is a translation from Afrikaans originally, some terminology and expressions I had to revise to accommodate my Ideas. Where I could I used modified English words to express a thought or an idea. One such a term is gravity. I had so much criticism about the word, which I feel I do not deserve. There is a certain notion clinging to the idea represented by gravity. Gravity links to a force that is all compelling, but I do not agree with SUCH AN COMPELLING FORCE, such as the word gravity implies. Gravity I introduce, works on two principles, but gravity to Newtonian standards is a single force. When I refer to gravity the normal reaction is that I am referring to the force I deny. By declaring that gravity THE FORCE CONTROLING ALL OF THE UNIVERSE AS A STANDARD CONSTANT IS NON EXSITING, I bring the wroth of the scientific world upon me. When I make the statement that there is no gravity, every person considers me mentally unstable.

Of course there is a movement of energy keeping all objects attached to the Earth, but gravity implies work, and with that work principle I disagree, because that is NOT WORK, THAT IS A COSMIC BALLANCE THAT STARTED AT A POINT OF ETERNITY AND WILL END AT A POINT OF ETERNITY. ONLY LIFE, STANDING ALONE AND DETACHED FROM THE COSMOS AS THE ENERGY THAT MANIPULATE SPACE-TIME CAN COMMIT WORK, THE REST IS A BALALANCE RUNNING CONCURRENT FROM TIME'S BEGINNING TO TIME'S END. I HAVE UNBELIEVEBLE DIFFICULTY IN RELAYING THE DIFFERENCE BETWEEN LIFE WITHIN THE COSMOS, AND THE COSMOS WITHOUT LIFE.

In the entire Universe there is no work, it is a balance running concurrent through time and space. The balance within stars shift in some cases to favour space and in other cases more to favour time. But in all of that shifting, a continuous balance strikes every aspect of space-time. This applies new ideas never brought to light before and the new concepts clashes with the conventional names that science applies to current ideas. I had to divorce the science ideas from those I introduce and the only way was with new etymology. I have to start implementing the newly created terminology, which will apply to the rest of this book. This stems from my lack in ability to find suitable words in the English language that would define the concepts as they are, in order to establish the difference in meaning from the current words, which convey the existing misinterpretations (or if you wish, to my view incorrect applications).

Firstly, we start with the word **densified**, which is not a normal English word, but a word I had to produce in order to make a comprehensible statement. The correct word that applies is concentrated, that much I do know about the English language. But concentrated has not the correct meaning or the expression that I would like to bring over. Concentrated can apply to any substance, be it gas, liquid or solids where one of the ingredients become more than the rest of the ingredients. In that way, matter as a solid substance produced from the eternal substance which is heat, cannot be concentrated. Nothing in the entire Universe can compare with the density of pure heat that spins at a rate in which that very heat can produce a value and which has a density far beyond anything else. Therefore, I chose to use the concept of concentration in a position where it makes a lot more sense. A star is concentrated space-time, but there is a huge difference between a star's concentrated space-time and the value of pure matter. When a star does therefore become densified space-time, it can only be at the end of the Big Crunch eternity, witch I prefer to call moment Omega; that is when space becomes eternal and time becomes Zero. In this light I chose to call matter densified space-time. Densified space-time should therefore be in a definition where matter or substance has reached a point in density that will last one eternity, but has no limit. Concentrated space-time, on the other hand does have a limit, which is at the point where it becomes densified space-time.

The second word I created is **Aanplasing,** which is the ongoing redirection of heat as in matter to heat as in time and that connects to a circular deepening of the separation that matter undergo transforming to time as it discards heat for the cold of fusion. Later (I hope) it will be clear enough for every reader to comprehend and to distinguish between the various factors that bring about **aanplasing** as should the reasons be clear why I prefer to have created this new word.

In this case however, there was no English word that could merely be altered and then be re-applied. A more suitable word that relates to a better meaning in the case where I brought in "densified" would have been the Afrikaans word "verdigting", where "verdigting" stands in relation to "konsentrasie" (concentrated). The fact of the matter is that I am not wilfully forcing Afrikaans down the throat of the Anglo-American and in the case of density I was able to modify a known English word that could adopt a new concept. By using an English word, would mean that there is no liberation from the "misleading" focus that depends on gravity, nor can it liberate the feature of this "misconception".

As for the Afrikaans words: **aanplasing, verplasing, versnelling** and **inperking**: there are no such words or concepts in existence that the precise meaning can derive from the English written or spoken language. Should any such words exist, the misconceptions that remains connected to the original English words, would not bring justice to the concept which I wish to apply to convey the meaning that lies behind the correct value of the thought. If I stuck to the word "gravity", the concept I wanted to introduce would forever remain confused with Newton's application. To that end the new realization would then never come across in the way I intend it to be.

Inperking: This stands apart from the idea of curtailment because in curtailing something or someone, means that object's or person's movement or moveable motion is deprived. This then brings over the misconception in the accepted notion of an expanding Universe. In due course I shall explain the concept, but inperking involves the same value that was there at first and will be there in the end, only the location in the balance shifts to favour one or the other part of the same coin. Because of the fact that none such a thing applies when space-time "accelerates" (versnel), and where this brings about inperking, it does not apply. Instead, all functions and factors still apply when inperking becomes valid, therefore the meaning of inperking becomes more applicable and this word describes the process much better. Inperking relates to time, where the duration of time extends, but not the value of time as such, as time applies in the cosmic sense.

One should realize that the entire atom, as well as its surroundings including all other surrounding atoms are reduced in space-time volume, so the atom is not actually curtailed, nor is its surrounding which means the word curtailment does not really apply. All aspects of occupied and unoccupied space-time are in reality, re-focused down in the true sense and above all, remains to the precise relative relation value it had for one entire eternity where the relation between such times, only refocus. However, scaled down would neither apply, because that would not refer to the time involvement, which lies at the hart of this revaluation. Where less time applies, inperking would be more severe and where more time applies, less inperking will apply.

Versnelling: It carries exactly the same meaning as acceleration, but the meaning or concept connected to acceleration implies to matter as the matter increases its own positional change in space and time. That is not the impression I wish to relay when referring to versnelling, because it is exactly the opposite of that meaning. In this, the actual meaning is more applicable to the true connection. These I must explain carefully, not to convey confusion. When a person stands outside an explosion of some sort, the time laps seems instantaneous, quicker than the senses can relate to. However, inside the explosion, time is almost standing still. Any person, who is inside such an explosion, would relate to time on the outside as being instantaneous. Whether this statement is accepted or not, the truth is that a person in an explosion cannot die, although his body is shattered in a million pieces. Such a person is sealed in a period separated from the period he and we lives in. This I explain at a later stage. The time duration slows down immensely, but from the outside, it accelerates immensely. Therefore, time versnel to the outside of where ever one relates to.

I wish to bring over the fact, as just been said, that the concept we have, is quite the opposite. Versnelling implies that the motional increase lies with the transfer of space-time, regardless whether matter occupies it or not. As aanplasing (not gravity) and versnelling bring about inperking (not curtailment as the body remains free to do as it wishes) the space-time that the body occupies and the surrounding sphere are in constant state of versnelling. The increase in motion has an effect on the matter, but the matter stands weightless as its specific density applies the time in that particular space.

Verplasing: This word is preferred to that of displacement, because although the matter in motion is displaced, time and space are implicated in the process.

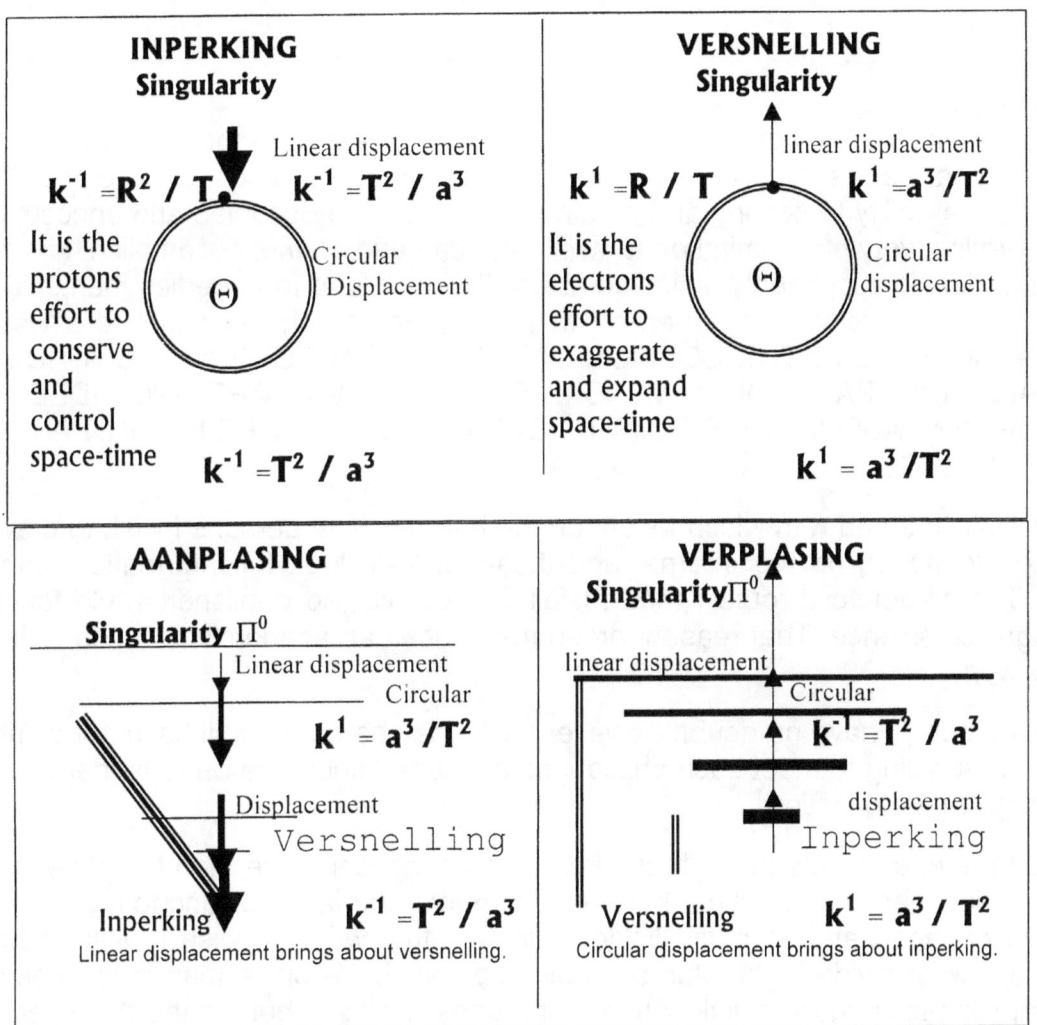

Verplasing is in fact the transferring of newly created magnetic space-time by matter, as a body composed of atoms has to replenish the space-time it occupies in order to maintain its position, place and structure in space-time in time in space, according to its geodesic positional allocation within the star's space and in time. Verplasing comes in effect as matter progresses in position, but the time-affect of verplasing that it has on matter, comes into real effect when an object reaches Mach $_3$ depending on its shape and altitude. In short: **Aanplasing** is relatively where matter is in a geodesic motionless position as space-time carries the motion component of the two values. This means that aanplasing is relative to positive space-time displacement.

Verplasing on the other hand has to do with the motion being with the newly created space-time in relation with the matter and the geodesic space-time remains relatively motionless. In both cases inperking and versnelling is a consequential result of the process. The difference is in the application of the time component itself.

A practical example of the difference between aanplasing and versnelling is as such: a body in **aanplasing** is in example a skydiver is falling towards the Earth and **verplasing** is where a body, such as a rocket is on a trajectory path as it fires into space. Both bodies will comply with the linear and circular displacement, but the circular displacement will relate oppositely in each event.

In the past, I have been severely criticized by academics for the view I hold and to my judgment rather unduly. The largest complaint by academics comes because of my critic about Newton's findings. There is a vast difference to the cosmos and the Earth surrounding us filled with life, the part we humans take fore granted, but the only place we will find life will be on this planet. Compare Earth to the moon, and the moon represent the real cosmos. It is dead, except for matter that grows with cosmic time.

To understand the cosmos the cosmos and life has to stand apart, the two on they're own and only then does the cosmos appear as it is. A rocket launched is not part of the cosmos, but is part of life's extending energy. Life is the ability to occupy and manipulate space-time occupied and unoccupied. In the manipulation, there is a very strict limitation to what man can and cannot accomplish. It is as if man has become drunk with his supposedly unlimited ability. This will lead to tragedies man still has to meet. There is no reason sending brave men off to their death, if only we can find a way to recognize the dangers. NOTHING IN THE COSMOS CAN DIE, BECAUSE DEATH IS A HUMAN MISCONCEPTION. ALL ARE PART OF DIMENSIONS WHERE WE ARE INCLUDED OR EXCLUDED. THIS CONCEPT ABOVE ALL SEEMS TO EVOKE THE MOST REJECTION AMONG THOSE I HAVE COME ACROSS.

The last four five wrote full time and with which I went on the Internet. The persons I wish to draw to my book do not seek information on the Internet and those who do look for information, cannot understand my work. The commercial route I tried to follow; however no publisher would touch it because it is not recognized science. That reason forced me to seek an academic route, in order to find recognition for my work.

To the correctness in my work, I have no doubt; however there may be parts I still have not clarified sufficiently. I am more than willing to meet such challenges because I know the proof is there, I was just not yet able to recognize the question.

Being human we tend to form concepts through culture and such concepts we then translate to the cosmos in an attempt to fid meaning moreover about our worth than the facts relating to the cosmos. We find ourselves physical apart and as much linking and with that we then wish to link and find meaning. What links and what divide is the stumbling block because we place man in the cosmos through which we then wish to produce a link. Such a link does not exist but for the tiny speck of cosmic dust which we call either home or Earth. The main issue to divide what should be divided and then accumulate what should connect and doing that is the purpose of this book.

Stars will never collide because stars can never collide.

The only absence in the cosmos is zero and without zero there cannot be an end to eternity but only an everlasting cycle that breaks to start one more eternity now and then. With the cosmos created minute by minute from no space within the cosmic centre, the cosmos is ruled from a position with every thing but nothing is but where we know God must be. By accepting singularity and the rule there of brings into the cosmos things physics are unable to explain, mathematics are unable to calculate and man is unable to dismiss. If you accept physics you have to accept God because you cannot except one proving singularity without the other coming through singularity.

If it is that simple then why is it complicated.

BEST WISHES,

PETRUS. (PEET) S. J. SCHUTTE

www.ingramcontent.com/pod-product-compliance
Lightning Source LLC
Chambersburg PA
CBHW080615190526
45169CB00009B/3190